BRAHMAND

WORLD DEFENCE UPDATE 2017

Last Updated till 31st December, 2016

DISCLAIMER:

- The data has been obtained from the open domain. Photos and illustrations for which no specific credit is given are understood to be in the public domain.
- The boundaries of the countries shown in the maps and illustrations in this book are neither authentic nor correct and are only representational.
- The authors do not take responsibility for actions arising from reliance on it.
- Assessment of strategic issues are of the editorial staff and not of the Organisation.

ACKNOWLEDGEMENTS:
The data and images for the Country Reports have been obtained from the Ministry of Defence/ official websites of various countries, CIA World Fact Book and US Library of Congress website, besides other sources.

BRAHMAND
WORLD DEFENCE UPDATE 2017

Foreword

Dr. Subhash Bhamre
Minister of State for Defence

PENTAGON
PRESS

Brahmand World Defence Update 2017

ISBN 978-81-8274-944-3

First Published in 2017

Published by

PENTAGON PRESS

206, Peacock Lane, Shahpur Jat,
New Delhi-110049
Phones: 011-64706243, 26491568
Telefax: 011-26490600
email: rajan@pentagonpress.in
website: www.pentagonpress.in

Design and layout: Heena Sodhi

Printed at Aegean Offset Printer

डॉ॰ सुभाष भामरे
Dr. SUBHASH BHAMRE

रक्षा राज्य मंत्री, भारत सरकार
नई दिल्ली-110011
Minister of State for Defence
Government of India,
New Delhi - 110011

FOREWORD

India has undergone a massive transformation and established itself worldwide in each and every spectrum. Today, India is a force to reckon with in the global industrial scenario making great advancements in Defence & Aerospace arena. The country's defence industry, with key focus on R&D, has taken rapid strides in realising many state-of-the-art systems and technologies indigenously. The massive growth of India in terms of economy, defence and its foreign policy towards other countries have led India to play strategic role in many of the International forums.

India having the third largest armed forces in the world is committed and determined to modernise and bolster its military capabilities. The Government of India's **"Make In India"** initiative was launched in the year 2014 to make India a global manufacturing hub. In defence sector, with a strong focus on indigenisation, both multinational as well as domestic companies from the defence industry are being encouraged to manufacture their products within the country, in order to achieve enhanced self-reliance in defence manufacturing.

India is steadfastly marching towards complete self-reliance in missile systems and technologies. A unique example in this regard is **BRAHMOS** supersonic cruise missile system which has added tremendous firepower capability to Indian Armed Forces. The Indian-Russian Joint Venture comprising of DRDO and NPOM technology have successfully developed various versions of BRAHMOS missile that can be launched from a ship, submarine and land. The missile is now getting geared up to be test launched from the Indian Air Force's Su-30MKI strike fighter. BRAHMOS has become a role-model for the country to become self-reliant in defence technology.

The **Brahmand World Defence Update 2017** is an earnest attempt to bring out a comprehensive assessment of the present day global military order with a focus on 113 countries' military capabilities, new arms acquisitions and defence spending along with detailed strategic information on 33 powerful nations. In this latest issue, the yearbook has included a new chapter which emphases on **"Naval Warfare Systems** of **China, France, India, Russia and the USA"**.

I am confident that Brahmand World Defence Update 2017, like all its previous editions will be a useful reference handbook for the Armed Forces, Defence Analysts, Official Defence Establishments, Procurement Officers, Defence Industry, R&D and Production Units all over the world.

Jan, 2017 New Delhi

(Dr. Subhash Bhamre)

114A, South Block, New Delhi - 110011
Tel. : 011-23792128, 23794621 Fax : 011-23016255

Residence : 25 Meena Bagh, Opp. Nirman Bhawan
New Delhi - 110011 Tel. : 011-23063814

PREFACE

Brahmand World Defence Update 2017 is a comprehensive assessment of the present day global military order. The book presents a fully updated Country–by–Country representation of the Armed Forces of the World, including: Geopolitical Situation, Economy, Statistical Data, and Defence Budget; Strategic Analysis; Structure of the Air, Land, and Naval Forces; Defence Equipment in service and in order; Defence Production and R&D, and Procurement Programmes.

The yearbook covers 113 countries and also provides detailed strategic analysis of 33 important countries selected from each continent/region with key focus on their geopolitical/geostrategic importance, defence capabilities, imports & exports of armaments, defence budget trends etc.

The 2017 defence yearbook covers all aspects of modern defence technology, requirements, procurements and programmes and gives useful military data with up-to-date figures and information on defence budgets and equipment holdings.

The sixth edition features a new chapter on Naval Warfare Systems of five strategically important countries namely, China, France, India, Russia and USA.

The arms and equipments are presented in tables with the names of countries arranged in an alphabetical order for easy reference. The book also features images of important political persons and defence persons and equipments and detail analyses on:

Defence & Security – Internal, External conflicts, threat perspective of the subject country, past wars and present military conflicts, its strategic alliance with other countries and with any major international associations.

Defence Capabilities - Military history, structure, total number of Army/Navy/Air Force personnel, latest RFPs/Contracts as well as future procurement plans, modernisation programmes and equipments.

Defence Industry – Its defence production, resource & development, past procurements & future procurement plans and procurement agencies/organisations.

Overall, the book provides the potential market entry strategies for business opportunities and acquisition of military hardware and identifies top companies in the defence industry.

- Editor

NAVAL WARFARE SYSTEMS

The influence of sea power in determining the military might and strategic expansion of a country is quite evident from the annals of history. No nation has been truly powerful until it has aspired to become a power at sea.

Naval warfare strategies have evolved over many centuries from now. Few countries around the globe, owing to their proximity to the sea, have attained global power status in the past by building up a strong maritime posture. Development of advanced battle ships have helped in expanding their maritime reach across the globe.

In the 21st century world too, sea dominance remains the most important aspect in attaining great power status. With swift advancements in military technology worldwide, the naval warfare domain has also undergone a major transformation and now includes a wide range of modern maritime assets for littoral, surface and sub-surface missions.

Development of advanced platforms such as aircraft carriers, attack submarines, missile destroyers, frigates, amphibious vessels, patrol craft, replenishment ships along with state-of-the-art weapons and systems have greatly widened the realm of naval warfare strategies and tactics in modern times.

With these advancements, many nations are now aspiring to attain "blue water" naval capability to undertake both near-sea missions as well as far-off ocean operations effectively with an intent to expand their economic and strategic outreach beyond their immediate maritime periphery.

To highlight these developments in the naval warfare sphere, the latest edition of **Brahmand World Defence Update 2017** has attempted to provide a detailed coverage of Naval Warfare Systems of five powerful countries – China, France, India, Russia and the USA.

Each country's naval & naval aviation platforms along with the major weapon systems are detailed comprehensively with technical specifications provided wherever available in this issue of the yearbook.

CONTENTS

AFGHANISTAN
(Capital: Kabul)

INTRODUCTION

Area: 652,230 sq km
Population: 33,332,025 (July 2016 est.)
Coastline: 0 km (Landlocked)
Maritime claims: None

KEY POLITICAL PERSONS

PRESIDENT & COMMANDER-IN-CHIEF: Ashraf Ghani Ahmadzai

CHIEF EXECUTIVE OFFICER: Dr. Abdullah Abdullah

KEY DEFENCE PERSONS

DEFENCE MINISTER: Gen. Abdullah Khan Habib

FIRST DEPUTY MINISTER OF DEFENCE: Gen. Abdul Khaliq

ECONOMY

Decades of conflict in Afghanistan has affected the economy. According to the World Bank report, economic growth slowed to 1.3 percent in 2014. While agricultural production continued to be high it did not supersede the strong agriculture output in 2013. Growth in 2014 was mainly driven by expansion in industries (2.4 percent) and services (2.2 percent), it said. Afghanistan is endowed with a wealth of natural resources, including extensive deposits of natural gas, petroleum, coal, marble, gold, copper, chromites, talc, barites, sulphur, lead, zinc, iron ore, salt, precious and semi-precious stones, and many rare-earth elements. Despite progress of the past few years, Afghanistan is extremely poor, landlocked, and highly dependent on foreign aid. The reforms initiated by the Government and international community will take time to implement and Afghanistan will remain dependent on international donor support over the next several years.

GDP: $62.32 billion (2015 est.); $61.53 billion (2014 est.)

Real Growth Rate: 1.5% (2015 est.); 1.3% (2014 est.)

Industries: Small-scale production of bricks, textiles, soap, furniture, shoes, fertilizer, apparel, food products, non-alcoholic beverages, mineral water, cement; hand-woven carpets; natural gas, coal, copper

Total Exports: $658 million (2014 est.); $2.679 billion (2013 est.)

Export Commodities: Opium, fruits and nuts, hand-woven carpets, wool, cotton, hides and pelts, precious and semi-precious gems

Major Markets: India 42.2%, Pakistan 28.9%, Tajikistan 7.6% (2015)

Total Imports: $7.004 billion (2014 est.); $12.19 billion (2013 est.)

Import Commodities: Machinery and other capital goods, food, textiles, petroleum products

Major Suppliers: Pakistan 38.6%, India 8.9%, US 8.3%, Turkmenistan 6.2%, China 6%, Kazakhstan 5.9%, Azerbaijan 4.9% (2015)

INTERNATIONAL DISPUTES

- ✦ Afghan and Pakistan military meet periodically to clarify the alignment of the boundary on the ground and on maps
- ✦ Afghan and Iranian commissioners have discussed boundary monument densification and resurvey
- ✦ Iran protests Afghanistan's restricting flow of dammed Helmand River tributaries during drought
- ✦ Pakistan has sent troops across and built fences along some remote tribal areas of its treaty-defined Durand Line border with Afghanistan
- ✦ Russia remains concerned about the smuggling of poppy derivatives from Afghanistan through Central Asian countries

DEFENCE

AFGHAN NATIONAL ARMY (ANA)

The Afghan military was gradually rebuilt by NATO after the removal of the Taliban and formation of the Karzai administration in late 2001. The military is now able to operate independently from the US-led International Security Assistance Force. A new Afghan National Army (ANA) was created along with an Army Air Corps which later became the Afghan Air Force (AAF). Being a landlocked country, Afghanistan has no navy. The ANA is divided into 31 battalions, 28 of which are considered combat ready. The National Military Academy of Afghanistan was built to provide future officers. The President of Afghanistan is the Commander-in-Chief of the military acting through the Ministry of Defence, with the authority to declare war or a state of emergency. The ANA and the Afghan National Police under the Ministry of Interior are referred to as Afghan National Security Force.

Personnel: 195,000

Equipment

Category	Name	In Service
MBT	T55/T62	
AIFV/APC	BMP-1/2	82
	M113 A2	30+
	M1117	18
	Humvee	2000+
Artillery	2A18	
	M114 155mm howitzer	
	M1954	
	ZPU4	
	ZSU-23-4	
	ZU-23-2	
	BM-21	

New Procurements/Upgrades

✦ US-based defence contractor Textron Marine & Land Systems has been awarded a contract for the delivery of 55 additional Commando Select Armoured Security Vehicles.

✦ AM General has been awarded a US$ 109 million contract in 2016 to provide 433 more High Mobility Multipurpose Wheeled Vehicles (HMMWVs) to the US government for delivery to the Afghan security forces.

✦ ScanEagle UAVs received from the US are being operated by the Army.

ANA AIR FORCE

The international Combined Air Power Transition Force (CAPTF) and the NATO Air Training Command-Afghanistan (NATC-A) is responsible for rebuilding the Afghan armed forces. The air force is divided into three wings, with the 1st Wing at Kabul, the 2nd Wing at Kandahar and the 3rd Wing south at Shindand in western Afghanistan. There has been continuing expansion in infrastructure, training and maintenance facilities.

Personnel: 3,600

Equipment

Category	Name	In Service
Aircraft	MD 530F	27
	Mi-24/35	9
	Mi-17-5	86
	Cessna 208B	25
	PC-12 NG	18
	C-130 Hercules	4

New Procurements/Upgrades

✦ Afghanistan has taken delivery of the fourth and final C-130 Hercules military transport aircraft.

✦ India has delivered three multi-role Cheetal helicopters along with allied equipment and spares to Afghanistan. Cheetals are the upgraded version of Cheetah light utility, multi-role helicopters built by state-run Hindustan Aeronautics Limited.

✦ Delivery of the MD 530F Cayuse Warrior scout attack helicopters has been completed in August 2016. The Air Force had ordered a total of 27 such rotorcraft.

✦ Four more A-29 Super Tucano light attack aircraft have been ordered from the Embraer-Sierra Nevada Corporation in addition to the 20 already on order. The first set of the aircraft is expected to be delivered in 2016-17.

Super Tucano light attack aircraft *An EmbraerPhoto*

Defence Expenditure

Total defence spending: $16 billion (2015 - mostly foreign donations)

Defence spending in terms of GDP: 10% of GDP

Defence Production and R&D

The country is rebuilding its defence capabilities after decades of conflict with the support of the NATO. A number of ANA officers are sent to be trained in India. The Indian Military Academy provides a 4-year degree to army officers, while the National Defence Academy provides a 3-year degree.

Defence Procurement

All the major defence procurements are based on the donation to the nation from across the world including US, UK, France and India. Afghanistan does not have any procurement policy as of now. The country's budget is spent in rebuilding of infrastructure and other facilities. India has signed an agreement with Russia to pay for arms and equipment to be supplied to the Afghan military. Under the deal, smaller arms such as light artillery and mortars will be sourced from Russia and moved to Afghanistan. It could eventually involve the transfer of heavy artillery, tanks and even combat helicopters that the Afghans have been asking India.

CONTACT DETAILS

Ministry of Defence

Zia Watt, Kabul, Afghanistan

Website: www.mod.gov.af

Armed Forces HQ

The National Military Command Centre

Kabul, Afghanistan

ALBANIA

(Capital: Tirana)

INTRODUCTION

Area: 28,748 sq km

Population: 3,038,594 (July 2016 est.)

Coastline: 362 km

Maritime claims: Territorial sea: 12nm
Continental shelf: 200m depth or to the depth of exploitation

KEY POLITICAL PERSONS

PRESIDENT: Bujar Nishani

PRIME MINISTER: Edi Rama

KEY DEFENCE PERSONS

MINISTER OF DEFENCE: Mimi Kodheli

CHIEF OF GENERAL STAFF: Maj. Gen. Jeronim Bazo

COMMANDER OF LAND FORCE: Brig. Gen. Nazmi Cahani

COMMANDER OF NAVAL FORCE: Brig. Gen. Ylber Dogjani

COMMANDER OF AIR FORCE: Brig. Vladimir Avdiaj

ECONOMY

Once a closed economy during the country's Communist regime, Albania has transitioned into a modern open-market economy. GDP growth in 2011, as per International Monetary Fund data, was 3.4% with the service sector making highest contribution followed by agriculture. Remittances, mostly from Albanians residing in Greece and Italy, a significant catalyst for economic growth, declined from 12-15% of GDP before the 2008 financial crisis to 5.7% of GDP in 2014, mostly from Albanians residing in Greece and Italy. Albania exports a number of goods and services though lack of adequate infrastructure and poor business environment have failed to attract foreign investments to expand its export base. The country is still one of the poorest in Europe. It continues to face challenges from increasing public debt, having exceeded its former statutory limit of 60% of GDP in 2013 and reaching 72% in 2014 and 73% in 2015. Strong trade, remittance, and banking sector ties with Greece and Italy make Albania vulnerable to spill-over effects of debt crises and weak growth in the euro zone.

GDP: $32.65 billion (2015 est.); $31.81 billion (2014 est.)

Real Growth Rate: 2.6% (2015 est.); 2% (2014 est.)

Industries: Food and tobacco products; textiles and clothing; lumber, oil, cement, chemicals, mining, basic metals, hydropower

Total Exports: $854.7 million (2015 est.); $1.241 billion (2014 est.)

Export Commodities: Textiles and footwear; asphalt, metals and metallic ores, crude oil; vegetables, fruits, tobacco

Major Markets: Italy 42.8%, Kosovo 9.7%, US 7.6%, China 6.1%, Greece 5.3%, Spain 4.8% (2015)

Total Imports: $3.402 billion (2015 est.); $4.17 billion (2014 est.)

Import Commodities: Machinery and equipment, foodstuff, textiles, chemicals

Major Suppliers: Italy 33.4%, China 10%, Greece 9%, Turkey 6.7%, Germany 5.2% (2015)

INTERNATIONAL DISPUTES

Albania has no known international disputes.

DEFENCE

ARMY

Personnel: 8000+

Equipment

Category	Name	In Service
APC	M113	130
ATGW	HN-5 Manpad	
	HJ-8	

New Procurements/ Upgrades

✦ The NATO-standard M4 carbine assault rifles are being procured to replace the AK-47 rifles.

NAVY

Personnel: 1,500

Equipment

Category	Name	In Service
Patrol vessel	Damen Stan 4207	4
Patrol boat	Shanghai II class	2
	Small patrol boat	17
Mine sweeper		2
Auxiliary ship	Diving vessel	2
	Floating dock	1
	Torpedo ship	7

New Procurements/ Upgrades

No major naval procurement plans are under consideration at present.

AIR FORCE

Personnel: 1,400

Equipment

Category	Name	In Service
Helicopter	A-109	1
	AB-205	7
	AB-206C	7
	Bo-105E4	6
	AS-532 Cougar	4
	EC145	1 (VIP transport)+ 2 (utility)

New Procurements/ Upgrades

✦ Two EC145 light utility helicopters have been delivered in 2015.

Defence Expenditure

Estimated defence spending in terms of GDP: 0.83% (2015)

Major suppliers of defence equipment to Albania in the past have been China, USSR, Turkey, Germany and Italy.

Defence Production and R&D

There are no military research and development facilities in Albania. The country has no immediate plans to take up defence production.

Defence Procurement

The European country has procured naval patrol craft, helicopters and other weapons and equipment to strengthen its military which is undergoing a large-scale modernisation phase including a revision in structures and procedures to meet the future challenges as a NATO member. Under the reforms drive, Albania is disposing its obsolete weapons and equipment while refurbishing the existing inventory. The Government has set a plan to reduce the size of its armed forces by 3%. All military reforms are being carried out as per NATO standards.

CONTACT DETAILS

Ministry of Defence

Martyrs Boulevard, Tirana, Albania

Tel: +00 355 4 226601

Email: kontakt@mod.gov.al

Website: www.mod.gov.al

STRATEGIC INFORMATION

ALGERIA

(Capital: Algiers)

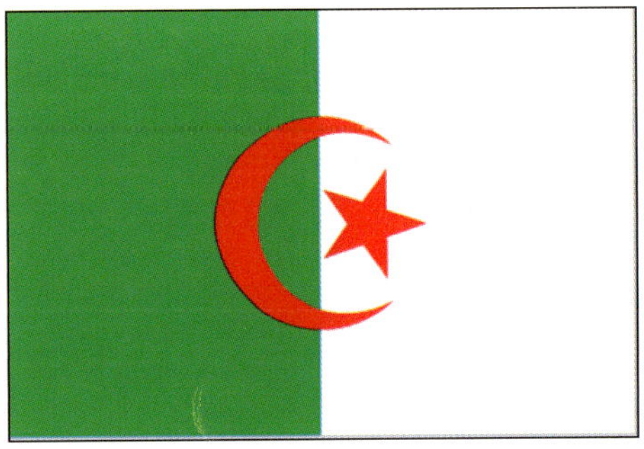

INTRODUCTION

Area: 2,381,741 sq km
Population: 40,263,711 (July 2016 est.)
Coastline: 998 km
Maritime claims: Territorial sea: 12 nm

GEOPOLITICAL IMPORTANCE

The largest country in the African continent, Algeria acts as a gateway between Africa and Europe. The North African country is bordered by Libya and Tunisia in the East, Morocco and Western Sahara in the West, and Mauritania, Mali, and Niger in the South. The Mediterranean Sea is in its north. The Sahara Desert covers around 85% of the country's landmass and is mostly uninhabited. From a geostrategic perspective, Algeria has gained prominence with the discovery of huge amounts of oil and natural gas mostly in the Saharan desert in the 1950s. The African country today is a leading energy supplier, mainly to the European nations. It is the third biggest supplier of natural gas to the European Union.

POLITICAL OVERVIEW

After being ruled by France for over a century, Algeria gained independence in 1962 which was achieved following a bitter fight for freedom. The freedom struggle was led by the National Liberation Front (FLN) which came to power post independence and ruled Algeria as a one-party socialist state until the late 1980s. The FLN had military affiliations which led to military interventions in the political dispensation of the country. In 1989, the country's Constitution was amended to introduce a multi-party system. The election of current President Abdelaziz Bouteflika in 1999 marked a new beginning as he contained civil unrest and also ended the dominance of the military in the country's political system. Constitutional amendments further increased Presidential powers in the political dispensation, thereby ending the Army's direct intervention in the system.

Presently, Algeria functions as a presidential republic where the President is both head of the state and head of the government even though internal unrest persists and is dealt with strictly. President Bouteflika secured his fourth term in office after general elections were held in April 2014.

KEY POLITICAL PERSONS

PRESIDENT & MINISTER OF NATIONAL DEFENCE: Abdelaziz Bouteflika

PRIME MINISTER: Abdelmalek Sellal

ECONOMY

Algeria's moderately-growing economy, the fourth largest in Africa, is highly dependent on oil and natural gas exports. The largest share of its revenue comes from exporting hydrocarbons which accounts for 98% of the country's total exports and 30-40% of its GDP. Algeria has the 10th-largest reserves of natural gas in the world and is the sixth-largest gas exporter. It ranks 16th in oil reserves. After liberalising its economy partially in the 1990s, Algeria witnessed a spurt in economic growth compared to its post-independence socialist phase. Hydrocarbon exports have enabled Algeria to maintain macroeconomic stability and earn large foreign currency reserves and a large budget stabilization fund available for tapping. In addition, Algeria's external debt is extremely low at about 2% of GDP. However, Algeria has struggled to develop non-hydrocarbon industries because of heavy regulation and an emphasis on state-driven growth. Investment has not been diversified into other sectors of the State-controlled economy except energy. The country is gradually introducing economic reforms as it seeks to be a member of the World Trade Organisation (WTO). Since late 2014, declining oil prices forced the government to spend down its reserves at a high rate in order to sustain social spending on salaries and subsidies, particularly since the government has been unable to boost exports of hydrocarbons or significantly grow its non-oil sector. In 2015, the Government imposed further restrictions on imports in an effort to reduce withdrawals from its foreign exchange reserves.

GDP: $580.7 billion (2015 est.); $558.9 billion (2014 est.)

Real Growth Rate: 3.9% (2015 est.); 3.8% (2014 est.)

Industries: Petroleum, natural gas, light industries, mining, electrical, petrochemical, food processing

Natural Gas - Proved Reserves: 4.505 trillion cu m (1 January 2014 est.)

Crude Oil - Proved Reserves: 12.2 billion bbl (1 January 2015 est.)

Total Exports: $36 billion (2015 est.); $60 billion (2014 est.)

Export Commodities: Petroleum, natural gas, and petroleum products 97%

Major Markets: Spain 18.8%, France 11.2%, US 8.8%, Italy 8.7%, UK 7.1%, Brazil 5.2%, Tunisia 4.9%, Germany 4.5% (2015)

Total Imports: $50.7 billion (2015 est.); $59.67 billion (2014 est.)

Import Commodities: Capital goods, foodstuff, consumer goods

Major Suppliers: China 15.6%, France 14.4%, Italy 9.4%, Spain 7.4%, Germany 5.6%, Russia 4.1% (2015)

DEFENCE & SECURITY

The African country does not face direct military threat from any country, barring few of its neighbouring states with whom it is locked in territorial disputes. Nevertheless, Algeria maintains quite a large and well-equipped military force to counter any domestic or external threat. The country's armed forces, called the People's National Army (PNA), constitute the land, naval and air forces along with an air defence command. Algeria's unusual geographic features, with about 80% of its area being desert, steppes, wasteland and mountains, pose a challenge for its armed forces to secure its borders and defend it from external aggression.

The armed forces of Algeria are equipped mostly with Soviet-era military hardware and weapons. They have received training in the then Soviet Union during the PNA's initial phase of modernisation. The PNA has also relied mostly on Soviet military doctrine which emphasised on large-scale, combined-arms warfare and operational warfare.

The Algerian military is primarily engaged in containing domestic social unrest and Islamist armed conflict that have marred the nation since the early 1990s. The North African country is presently undergoing a phase of political uncertainty and military insecurity. In a recent development, it has deployed military troops across its border regions to quell the unrest emanating from terrorism and organised crime as major terror groups are becoming active either within Algeria or on its borders with Morocco, Tunisia, Mauritania or Mali.

KEY DEFENCE PERSONS

DEPUTY MINISTER OF DEFENCE & CHIEF OF GENERAL STAFF: Lt. Gen. Ahmed Gaid Salah

COMMANDER OF ARMY: Maj. Gen. Ahcene Tafer

COMMANDER OF NAVY: Maj. Gen. Mohammed Larbi Haouli

COMMANDER OF AIR FORCE: Maj. Gen. Abdelkader Lounes

Internal Conflict

Internal problems and unrests have plagued Algeria more than its external threats and have often taken dangerous proportions. For over last two decades, Algeria has been affected by an Islamist armed conflict which has taken the form of terrorism by non-state armed groups. The armed conflict that erupted between radical Islamist militants and the country's military following a scrapped general election won by an Islamist party in 1992, has resulted in the death of over 100,000 civilians. The violence, mostly carried out as a guerrilla warfare movement, has given way to organised crime like robbery, abduction, extortion and killing in the later years. While the Algerian Government has contained the decades-long civil war to a large extent, nevertheless sporadic incidents of violence and internal unrest still continue to affect the country.

External Conflict

Algeria has no major external conflict areas barring border dispute with its neighbour, Morocco. Its longstanding border dispute with another neighbour, Tunisia, has been resolved in 1993 and the two countries have also signed an agreement in 2011 to demarcate more exactly their maritime border.

Morocco: The two African neighbours are entangled in dispute over a sparsely-populated Western Sahara Desert region. Algeria, like several other countries, is opposed to Morocco's control over Western Sahara. The 1,601km border between Morocco and Algeria, that includes a 42-km stretch alongside the disputed Western Sahara region, has remained shut since 1994 over the issue. The stalemate over the disputed territory has continued between the Polisario movement, which demands independence and has got strong support from Algeria, and the Moroccan government's offer to grant autonomy to the region.

THREAT PERSPECTIVE

Algeria, like the rest of the North African countries, does not face any direct external military threat, although its relations with neighbouring state Morocco often remains tense. The country is primarily engaged in resolving its internal problems, terrorism being a major threat in recent times, and has never been keen to get involved in a military conflict with any of its neighbours.

STRATEGIC RELATIONS

Russia: As the most important strategic ally, Russia has provided military, technical as well as economic assistance to Algeria since the time of its freedom movement in the 1960s. The African country has received majority of its military assets from the then Soviet Union. Algerian armed forces have also been imparted training by Russian military institutions. To further strengthen this cooperation, the two countries signed a strategic partnership agreement in 2001 with a thrust on strengthening bilateral defence cooperation, particularly in expanding long-term arms trade and exchanges between their armed forces.

EU: The proximity of oil and energy-rich Algeria to Europe makes the African country a crucial strategic ally of the entire European Union which largely relies on Algerian hydrocarbons supplies. From the economic aspect, Algeria has offered the European countries to explore investment opportunities especially in its manufacturing sector and research & development area. The EU has also supported Algeria's bid to become a member of the World Trade Organisation. Some European countries, including France, Italy and Spain, have also supplied military equipment to the North African nation in the past.

China: In recent years, China and Algeria have explored ways to develop bilateral strategic partnership alongside their growing political, economic and trade ties. China is also believed to be one of the major military suppliers to Algeria in recent times. As Algeria's defence expenditure is growing, it has acquired Chinese equipment and systems, including artillery pieces, corvettes and UAVs, as part of its military modernization drive.

Multilateral Alliances

Maghreb: The Maghreb is a region of Northwest Africa constituting of Algeria, Morocco, Tunisia, Libya and Mauritania. To strengthen regional cooperation, the Maghreb nations in 1989 formed the Arab Maghreb Union. The Union was formed to enhance regional cooperation in the economic, political and defence spheres while safeguarding national interests of each country. However, Algeria's internal troubles and strained Algeria-Morocco relations have hindered the growth of the Arab Maghreb Union as a regional powerhouse.

Arab League: Algeria, though geographically separated from the Middle East, yet shares a cultural identity with the Arab-Islamic nations. To affirm this affinity, the African country joined the League of Arab States immediately after its independence in 1962. The Arab League or League of Arab States is a regional organisation of Arab countries in North and Northeast Africa, and Southwest Asia (Middle East). The 22-member Arab League functions with a goal to strengthen relations among member nations and enhance cooperation to safeguard their freedom and sovereignty, and to stand united for Arab interests.

DEFENCE CAPABILITIES

ARMY: The land force of Algeria, with an active strength of over one lakh personnel, is the second largest Army in North Africa after Egypt. It operates with the mission to defend and protect national sovereignty in coordination with other forces of the nation. The Algerian Army possesses a large number of Soviet-era weapons most of which have been obsolete now and are in the process of being phased out.

Personnel: 107,000 (Reserve: 150,000)

Equipment

Category	Name	In Service
MBT	T-90	372
	T-72	280
	T-62	280+
	T-54/ 55	275
IFV	BMP-1	700+
	BMP-2	300
	BMP-3	100
	BTR-50, BTR-60	30 + 400
APC/ Armoured Car	Panhard AML-60	50
	BRDM-2	115
	Fuchs	14
Artillery	122mm SP and 152mm towed	500+
	SU-85	5
	2S1 122mm SP	145
	2S3 152mm SP	40+
	ISU 152mm SP	10+
	Norinco 155mm towed	
	Norinco PLZ 52 155mm SP	
Mortar	120mm + 160mm	400
MLRS	BM-21 (120mm)	48
	BM-14/ 16 (140mm)	50
	BM-24 (240mm)	25+
SSM	FROG-4	18

Army continued...

Category	Name	In Service
ATGW	Sagger	
	Milan	
	Spigot	
AA Gun	14.5mm ZPU-2/ 4	100
	23mm Zu-23	60
	23mm ZSU-23-4 Shilka SP	200
	M1939 37mm	100
	57mm S60	65+
	85mm KS12	30+
SAM	SA-7/14/16 (MANPAD)	150+
	SA-6	40
	SA-8	24
	SA-9	20
	SA-13	30
	2S6M Tunguska	24
	PANTSIR S1-E, SA-22	40+

New Procurements/Upgrades

✦ Russia has delivered 372 T-90 MBTs to Algeria since 2006 onwards.

✦ Local production and assembly of Fuchs armoured personnel carriers in partnership with Germany's Rheinmetall Defence is going on. 54 Fuchs armoured personnel carriers were ordered from Germany's Rheinmetall Defence in 2011. Rheinmetall later announced plans to assemble over 1,000 Fuchs APCs in Algeria under a deal finalised in 2014.

✦ Delivery of upgraded BMP-1 IFVs by Russia's KBP Instrument Design Bureau is going on.

NAVY: The Algerian naval force is tasked with safeguarding the maritime territory of the country. The missions of the Navy include: surveillance and safety of maritime approaches; national maritime space defence; protection of national interests at sea; coastal defence and littoral protection; different kinds of public service missions at sea. The Algerian Navy is equipped with a sizable number of frigates, corvettes, submarines and other support vessels to defend the country's nearly 1,000-km long coastline.

Major naval bases: Algiers, Annaba, Mers al-Kebir, Skikda, Jijel

Personnel: 6,600

Equipment

Category	Name	In Service
Submarine	El Hadj Slimane class (Russian Kilo class)	4
Frigate	Murat Reis (Koni Type)	3
	Meko class	1 (+ 1 on order)
Corvette	Ras Hamidou class (Nanuchaka II type)	3
	Djebel Chenoua class	3
	C28A	3

Navy continued...

Category	Name	In Service
Missile Boat	Osa Class I and II	6
Gun Boat	Kebir class	15
Inshore patrol boat	El Yadek class	6
Training vessel	Shanghai class	6
Amphibious Ship	Kalaat class logistic landing ship	2
LPD	Soviet Pilnochny B class	1
	Kalat Beni-Abbes	1
Mine Hunter	T-43 class	1
Survey vessel		1
ASM	SA-N-4 Gecko	
	SS-N-2C Styx	
	RBU-6000 rocket	
	KH-35 batteries (coastal defence)	4

Naval Aviation

Category	Name	In Service	On Order
Helicopter	AW101	6	
	Super Lynx 300	4	6

New Procurements/Upgrades

✦ Two Project 636M Varshavyanka-class (improved Kilo-class) diesel-electric submarines have been ordered from Russia's Admiralty Shipyards with the contract signed in 2014. The first submarine is expected to be delivered in 2018. The new acquisition will take the total number of submarines in the Algerian Navy to six.

✦ Italy has delivered "Kalaat Beni-Abbes", the new multi-role landing platform dock (LPD) ship, to the Navy in September 2014. Italian shipyard Fincanteri had received the order in 2011 to build the LPD. It was launched in January 2014. The vessel has been built as an Italian San Giorgio-class 8,000t LPD capable of accommodating around 600 personnel.

✦ The first of two Meko-200 class frigates, ordered from Germany's ThyssenKrupp Marine Systems in March 2012, has been commissioned into service in April 2016. Delivery of the second frigate is expected in 2017.

Illustration of Meko-200 class of frigate *An internet image*

✦ The Navy has contracted Russia's United Shipbuilding Corporation to build two Project 20382 Tiger-class corvettes.

✦ Russia's Severnaya Verf shipyard has been awarded a contract to upgrade two Algerian warships – the Nanuchka II class corvette and a Koni II class frigate – both procured from Russia and in service with Algerian Navy since the 1980s.

✦ The third and final C28A guided missile corvette built by China has been delivered to the Navy in July 2016. The first and second ships of the class were commissioned into service in 2015 and early 2016 respectively.

✦ A contract for a new mine countermeasures vessel (MCMV), with an option for a second, has been signed in 2014.

✦ The second upgraded logistics support and landing ship (LST) "Kalaat Beni Hammad" has been delivered to the Navy by Spanish shipbuilder Navantia in Sept. 2015. The first ship of the class, "Kalaat Beni Rached", was delivered in June 2014 after upgradation.

AIR FORCE: The Algerian Air Force is entrusted with the tasks of defending the country's air space, carrying out surveillance and monitoring, coordinating with the other two wings of the military and supplying military transport and cargo airlift. The Air Force is replete mostly with Soviet-era fighter and surveillance aircraft. It also possesses US, Russian and European transport planes. The country has started modernising the air force by acquiring new combat aircraft such as MiG-29 and Sukhoi-30, trainer jets, and is also planning to induct advanced attack helicopters. The old inventory of aircraft and helicopters are being upgraded and some, like the MiG-21 fighters, have been phased out from service.

Personnel: 12,000

Major air bases: Ain-Beida, Bousfer, Umm Bawaki

Equipment

Category	Name	In Service
Fighter/Trainer	MiG-29S	30
	MiG-29UB	5
	Su-30 MKA	44
	MiG25PDS	10
	MiG25U	2
	L-39 Albatros	24
	Yak-130	10+
Ground attack	Su-24MK	30+
Reconnaissance	MiG-25RB	3
	Su-24 MR	4
Transport	C-130H/H-30	12
	C-295	5
	SE-210	3
	An-12	4
	PC-6	2
	IL-76	4
Refueling	IL -78	4

Air Force continued...

Category	Name	In Service
VIP transport	F27-400	2
	F27-600	1
	Falcon 20	1
	Gulfstream III	3
Helicopter (attack)	Mi-24 D/V & Mk3	50
	Mi-171	39
	Mi-28 NE (On order)	42
Helicopter (transport, utility)	Mi-6	2
	Mi-4	4
	Mi-2	28
	Mi-8	45
	Mi-26T2 (On order)	6
	Bell 206 L3	3
	AS-355N Ecureuil	8
	PLZ-W3A Sokol	7
	Ka-32	5
Utility aircraft	Super King Air	2
	Zlin 142/ 143/ 242	50
Maritime patrol	Commander 650 AR	1
UAV	Seeker	
AA gun	85 MM KS 12	38
	100 mm KS19	54
SAM	SA-2	28
	SA-3, SA-6	20
	S-300 PMU-2	8 Battalions (250+ missiles)

New Procurements/Upgrades

✦ 12 Sukhoi Su-34 fighter bombers have reportedly been ordered from Russia in 2016.

✦ A deal has been signed with Russia in 2015 for the purchase of 14 Sukhoi-30MKA multi-role fighter jets. Irkut Corporation, which has received the contract, expects to complete the delivery of the fighters by 2017.

Su-30MK fighter *An Irkut Corporation photo*

✦ Delivery of Mi-28N attack helicopter is going on. A total of 42 Mi-28N attack helicopters and six Mi-26T2 heavy transport helicopters have been ordered from Russia under a deal worth $2.7 billion. The deal also entails upgrading of the existing fleet of Mi-171 helicopters to the Mi-171Sh standard.

✦ 16 YAK-130 advance trainer jets have been ordered from Russia with deliveries going on.

✦ Delivery of S-300 PMU air defence systems is going on.

Russia is delivering the S-300 PMU air defence system. *Internet photo*

✦ Negotiations are on to finalise a deal to procure three A330 multi-role tanker transport aircraft from Europe's Airbus Military.

✦ Plans are on to buy C-17 Globemaster heavy-lift transport aircraft from the US.

DEFENCE PRODUCTION

The defence sector in Algeria is at a nascent stage and the country is looking for thc development of a firm domestic defence industrial base. The Government has set up few defence manufacturing companies in recent years to produce armoured vehicles and electronics equipment. The T-90 MBTs contract Algeria has inked with Russia also entails local production of the main battle tanks in the country. German firm Rheinmetall Defence's announcement of locally assembling Fuchs APCs in Algeria is also set to give a fillip to the domestic defence manufacturing industry.

Defence Procurement

Algeria has imported majority of its defence equipment from the erstwhile Soviet Union, including tanks, aircraft and missiles. In recent years, China and some European countries, including France, Italy and Spain, have also supplied military equipment to it. Germany has emerged as a leading exporter of military platforms and equipment to the North African nation. Algeria is also making efforts to indigenously produce some defence systems. Russia still remains the leading defence supplier of Algeria.

The country's major defence procurement decisions are taken by the President in consultation with the Defence Minister and the Chiefs of Services. The chiefs of the armed forces are individually capable of making small procurement decisions. They further require the consultation of Chief of Defence Staff.

DEFENCE BUDGET

Defence expenditure in Algeria has increased manifold in the last two decades and according to some estimates, it is the top-most military spender in the entire African continent. Its share of defence budget is estimated to be over 4 percent of the GDP and it ranks 29th in the world in military spending. The country became the first in Africa having a defence budget of over US$10 billion in 2013-14. Modernisation of its armed forces and acquisition of new military platforms and equipment have led to a significant rise in Algeria's annual defence expenditure.

Estimated defence spending: $13 billion (2015); $12.3 billion (2014)

CONCLUSION

The resource-rich Algeria has faced frequent internal unrests since the 1990s and is also fighting armed extremism and terrorism. The country has not been engaged in any external military conflict, though the recent turmoil in neighbouring Libya has affected Algeria militarily and economically. The main thrust for Algeria has been to contain domestic unrest and secure internal peace and stability while safeguarding its territorial integrity. With its oil-rich economy growing, Algeria in recent times has emerged as one of the lucrative markets in the African continent for military hardware exports.

CONTACT DETAILS

Ministry of National Defence
B.P 184, Alger Gare, Algiers
Tel: 021711515
Web: http://www.mdn.dz/

STRATEGIC INFORMATION

ARGENTINA
(Capital: Buenos Aires)

INTRODUCTION

Area: 2,780,400 sq km

Population: 43,886,748 (July 2016 est.)

Coastline: 4,989 km

Maritime claims: Territorial sea: 12nm

Contiguous zone: 24 nm

Exclusive Economic Zone: 200 nm

Continental shelf: 200 nm or to the edge of the continental margin

GEOPOLITICAL IMPORTANCE

The second largest country in South America in terms of size and population, Argentina shares its land border with Bolivia and Paraguay in the north, Brazil and Uruguay in the east and Chile in the west. The Atlantic Ocean lies to its east. As a leading economic and military power in the Latin American region, Argentina is replete with vast natural resources. It yielded great influence during the early twentieth century by remaining one of the richest countries in the world and the richest in the Southern hemisphere. However, years of oppressive military rule coupled with corruption and a policy of economic protectionism has resulted in Argentina's burgeoning financial crisis, rising level of unemployment and poverty in present times.

POLITICAL OVERVIEW

The Constitution defines Argentina as a Federal Republic where the President is both the head of the State and head of the Government. The Executive branch of the Government is headed by the President elected for a period of four years. The Legislative power is vested with a bicameral legislature, Argentine National Congress, consisting of the Senate (upper house) and Chamber of Deputies (lower house). The judiciary functions independently with the highest federal court being the Supreme Court. Argentina, however, has witnessed military coups several times and alternated between civilian and military dispensations in the past. The last military rule in the country spanned from 1976 to 1983. The country embraced democracy in 1983 following restoration of the constitutional system.

KEY POLITICAL PERSONS

PRESIDENT & COMMANDER-IN-CHIEF: Mauricio Macri

VICE PRESIDENT: Gabriela Michetti

ECONOMY

The economy of Argentina consisting of agriculture and a diverse industry with rich, abundant natural resources is the third largest in Latin America. It is a leading food exporter in the international market and also gains from a brimming tourism industry. The economic graph of the country has been quite volatile wherein high economic growth has been alternated by severe recessions. Once a brimming economy, Argentina witnessed financial downfall and grappled with recession, fiscal and current account deficits, high inflation, capital flight and mounting external debt in the most part of 20th century. The financial crisis continued till 2002 when the Latin American country defaulted on its debt, its GDP declined by nearly 20% in four years, unemployment reached 25%, and the peso (national currency) sharply depreciated. It was a period of complete economic meltdown for the country. It was only in 2003 onwards that the economy bounced back after new financial measures were implemented, including commodity exports, nationalisation of a number of entities and infusion of money into social development programmes. The good phase continued till 2008, but the global economic meltdown in that year once again put the Argentine economy backwards, with the GDP shrinking to a meagre 0.8% in 2009. The economy however rebounded strongly in 2010, but slowed again in 2011 even as the government continued to rely on expansionary fiscal and monetary policies. The government then expanded state intervention in the economy throughout 2012. During 2014, the government continued its expansionary fiscal and monetary policies and foreign exchange and imports controls. In July 2014, Argentina and China agreed on an $11 billion currency swap; the Argentine Central Bank has received the equivalent of $3.2 billion in Chinese yuan, which it counts as international reserves. The new Government under President Macri has taken some major steps to liberalize the economy. His administration has lifted capital controls; floated the peso, negotiated debt payments with holdout bond creditors, and removed export controls on some commodities.

GDP: $884.2 billion (2015 est.); $862.9 billion (2014 est.)

Real Growth Rate: 2.5% (2015 est.); -2.5% (2014 est.)

Industries: Food processing, motor vehicles, consumer durables, textiles, chemicals and petrochemicals, printing, metallurgy, steel

Total Exports: $56.76 billion (2015 est.); $68.41 billion (2014 est.)

Export Commodities: Soybeans and derivatives, petroleum and gas, vehicles, corn, wheat

Major Markets: Brazil 17%, China 8.6%, US 5.9% (2015)

Total Imports: $57.18 billion (2015 est.); $62.43 billion (2014 est.)

Import Commodities: Machinery, motor vehicles, petroleum and natural gas, organic chemicals, plastics

Major Suppliers: Brazil 22.4%, US 16.3%, China 15.5%, Germany 5.1% (2015)

DEFENCE & SECURITY

As a major country in the Latin American region, Argentina has yielded great power and dominance over other smaller nations in the continent. Once it competed with Brazil for regional leadership in most part of 20th century. In the present geostrategic stage, Argentina has pursued a policy of common defence cooperation strategy with its neighbours by expanding its regional ties and a cooperative approach towards international security. The country has been a major contributor in several international peacekeeping missions led by the United Nations. It also regularly participates in global security forums.

Argentina does not face any major internal disturbances. Being a part of the 'Southern Cone', it also does not face any external threat from a hostile power. Any prospect of a major inter-state conflict in the continent has also largely dissipated in recent years as a result of greater economic and security integration among the states.

At the international level, its long-standing conflict with the United Kingdom over the Falkland Islands (called The Malvinas by Argentina) in the South Atlantic Ocean has resulted in tensions between the two nations from time to time. The centuries-old dispute over the chain of islands has once again flared up after over three decades of the Falklands War which took place in 1982. The archipelago, located near Argentina, is a self-governing British Overseas Territory with the UK looking after the region's foreign affairs and defence and maintains military presence in the islands by deploying fighter jets, warships and troops. Argentina, claiming sovereignty over the islands, has decided to bolster up its military by acquiring new weapons and platforms. The large-scale military modernisation drive is believed to be aimed at containing Britain's oil drilling plans off the disputed Islands' coast. The two countries are also engaged in territorial spat over the South Georgia and the South Sandwich Islands in South Atlantic Ocean.

Argentina maintains a strong, well-organised military force consisting of the Army, Navy and Air Force. The armed forces are dedicated to protect national sovereignty and maintain territorial integrity while also carrying out international peacekeeping and humanitarian missions. They also stand for the country's continuous presence in the Antarctica.

KEY DEFENCE PERSONS

DEFENCE MINISTER: Julio Cesar Martinez

CHIEF OF JOINT STAFF OF THE ARMED FORCES: Brig. Gen. Bari del Valle Soda

CHIEF OF STAFF OF THE ARMY: Gen. Diego Luis Suner

CHIEF OF STAFF OF THE NAVY: Rear Adm. Marcelo Eduardo Srur

CHIEF OF STAFF OF THE AIR FORCE: Brig. Enrique V Amrein

Internal Conflict

Argentina grapples with the problems of human trafficking, contraband smuggling, illegal immigration from porous borders with neighbouring Bolivia and Paraguay, organised crime and terrorism within the country.

External Conflict

Argentina is not engaged in armed conflict with any country at present, but in the past it has fought a full-scale war with the United Kingdom over territorial sovereignty of the Falkland Islands in the South Atlantic Ocean. The chain of islands, in close proximity to Argentina which it calls The Malvinas, is presently administered by Britain as one of its overseas territories. The island hosts a small population, mostly of British origin.

The dispute over the Falklands dates back to 1833 when Britain reasserted control over the archipelago off Argentina's southeast coast. Argentina has stalked claim over the territory which it says it inherited from the Spanish crown in the early 1800s. The two countries have fought a bitter war over the Falklands in 1982. In recent times, the dispute has once again triggered tension between the two sides. While Britain, with plans to drill oil in the island territory, has dispatched its warships to guard the region, Argentina, on its part, has introduced new rules requiring all ships travelling to the Falklands through its waters to have a permit. Argentina has also raised the issue in the United Nations. A referendum in the self-governing Islands in 2013 resulted in more than 99 percent of voters favouring to remain a British Overseas Territory.

Argentina and UK are also sparring over the South Georgia and the South Sandwich Islands in South Atlantic Ocean.

Another territorial dispute between the two countries includes a sizable landmass of Antarctica, parts of which are also claimed by Chile.

THREAT PERSPECTIVE

UK: The major military threat Argentina faces today is that from the United Kingdom over the Falkland Islands dispute. As tensions mount in recent times over sovereignty claims involving the chain of islands having an area of 12,200 square kilometres, Britain has rallied its warships and weapons by conducting military exercises and started drilling oil in the contested maritime region. The Falklands War in 1982 had resulted in Britain reaffirming its control over the territory after Argentina surrendered. Argentina, on its part, has decided to modernise and empower its military by acquiring new weapons and platforms to counter any potential threat from the UK.

STRATEGIC RELATIONS

Chile: Once hostile to each other for most part of the 19th and 20th century over territorial claims, Argentina and Chile today enjoy relatively warm relations ranging from trade and commerce to defence and strategy. The two neighbours resolved their territorial disputes by signing the Treaty of Peace and Friendship in 1984 which was superseded by the Maipu Treaty of Integration and Cooperation in 2009. The two countries have conducted joint military exercise in the past.

Brazil: The two South American countries share close relations spanning from trade and economy to education, culture, defence and strategy. Strategic partnership between Argentina and Brazil began to revive in the 1990s. In 1988, the two sides inked the landmark Treaty for Integration, Cooperation and Development covering politics, trade, investment, education, defence, technology and space among other aspects. This historic treaty later paved the way for the creation of the regional bloc – the Southern Cone Common Market (MERCOSUR) – three years later. Today, Argentina and Brazil are jointly working on a number of military programmes, including the Embraer KC-390 military transport aircraft and the UNASUR 1 basic jet trainer.

US: Argentina has shared strong economic and strategic ties with the US since the 1980s, especially after it transited from military dictatorship to democracy. The former supported and assisted the US during the Gulf War in 1990 following which it was accorded the status of major non-NATO ally of the US in the Latin American region in 1999. Argentina has been greatly influenced by American military strategy and planning in the past. The US has supplied majority of defence equipment to the Argentine military.

China: Argentina and China share strong bilateral trade ties. China is Argentina's third-largest trading partner after the MERCOSUR bloc and the European Union, and is one of its main destinations for food exports. In 2014, the two countries elevated their relationship to strategic level with China pledging to invest hugely in building infrastructure including hydroelectric power, ships and railways for Argentina. China has also supplied military technology to Argentina. In 2011, the two countries signed a deal to build the Chinese Z-11 light helicopters in Argentina under license, and mass production for the Argentine military began in 2013. The two sides are also discussing other military deals, including OPVs for the Argentine Navy and APCs for the Army.

Russia: Bilateral relationship between Argentina and Russia has mostly revolved around political and economic spheres. In the military sphere, Russia plans to supply helicopters and transport aircraft to Argentina.

It also plans to install a GLONASS monitoring station in Argentina. In 2014, the two sides signed an intergovernmental agreement on cooperation in the use of nuclear energy. Under the deal, Russian atomic energy corporation Rosatom has proposed to construct two new nuclear power units in Argentina. In April 2015, the two countries announced increasing military collaboration during the then Argentine President Cristina Fernandez's Moscow visit.

Multilateral Alliances

MERCOSUR: As a founding member of the MERCOSUR regional trading bloc along with Brazil, Uruguay and Paraguay, Argentina has made efforts towards greater economic, social and political integration among the South American nations. The MERCOSUR, founded in 1991, later included Venezuela as a full member. The regional block, modelled on the European Union (EU), however, has faced differences among member nations over several issues, including adopting a common currency and trade policy.

UNASUR: The Union of South American Nations (UNASUR), founded in December 2004, is a regional grouping formed by 12 South American countries that are united by shared history, religion, common culture and language. Over the years, UNASUR has made many significant achievements including in the sphere of security and defence, democratic stability, reducing crime, energy and financial integration, and institution building in the Latin American region. Its endeavours have also resulted in resolving the constitutional crises in Ecuador and Paraguay, and settled the dispute between Venezuela and Colombia. Being a part of the group, Argentina has benefited by partnering with member nations, especially Brazil, in joint military projects.

ECLAC: The Economic Commission for Latin America and the Caribbean (ECLAC) is a United Nations-led regional commission with the primary goal to encourage economic cooperation. It includes 44 member states -- 20 in Latin America, 13 in the Caribbean and 11 from outside the region. Argentina is a member state.

DEFENCE CAPABILITIES

ARMY: The land forces of Argentina, called the *Ejército Argentino*, is organised into three divisions and one Army Rapid Deployment Force. The Army operates with the primary mission of safeguarding the sovereignty and territorial integrity of the country and its citizens. The Argentine Army operates with the purpose to "defend the interests of the nation, contributing to its scientific, technological, economic and social development, and cooperate to achieve the general welfare of its inhabitants." A fully professional force, it also contributes to international peacekeeping missions led by the United Nations. In its present status, the Argentine Army lacks manpower, modern equipment and infrastructure for combat missions. Budgetary constrain too has restricted it from being entirely modernized.

Personnel: 50,000+

Equipment

Category	Name	In Service
MBT	TAM (Tanque Argentino Mediano)	230
Light Tank	AMX-13 (being withdrawn)	30
	SK-105 Kurassier	110
	Patagon	4
APC/MIC V/ IFV	TAM (VCTP-VCTM-VCPC)	160
	AMX (VCTP-VCDT-VCPC)	30
	M-113	288
	M-557	18
Artillery	155mm AMX F-3	24
	155mm VCA-TAM	17
	155mm CITEFA Model 77 (L33)	109
	105mm Oto Melara M56	70
	105mm M-101 howitzer	60
	155mm M114	48
Mortar	120mm Thompson-Brandt	300
	81mm Thompson-Brandt	350
	60mm Thompson-Brandt	300
MRL	127 mm SAPBA-1	2
	105 mm Pampero	4
ATGW	Mathogo	
	BGM-71 TOW	
ATW	105 mm Model 1968	150
	89 mm Instalaza M65	
	M72 LAW	385
SAM	Strela-2/ Strela-2M	
	Roland	3
AA gun	40mm Bofors L-70	
	35mm K-63 twin	8
	30mm HS 831	20
	20mm GAM-BO1	200

Army Aviation

Category	Name	In Service
Helicopter	AS.332L Super Puma	3
	UH1H Iroquois (being upgraded to Huey II version)	40+
	SA-315B Lama	2
	Bell 205, 206	4+5
Fixed wing aircraft	OV1D MOHAWK	7
	C-212-200	3
	Merlin IIIB/IV A	4
	Twin Otter	2
	Beechcraft B-80 Queen Air	1
	Sabreliner 75A	1

Army aviation continued...

Category	Name	In Service
Fixed wing aircraft	Cessna 207	4
	T-41 Mescalero	5

New Procurements/ Upgrades

✦ Modernisation of TAM MBTs, announced in 2010, is going on. A new contract worth US$ 111 million has been signed with Israel in June 2015 to upgrade 74 of these tanks.

✦ A total of 14 Guarani 6x6 armoured vehicles are on order from Brazil.

✦ Plans are on to acquire over 100 of the Chinese VN1 wheeled APCs and licence-produce them domestically.

✦ The entire inventory of M-113A1 APCs is being progressively upgraded to A2 standard.

✦ Delivery of 20 LVC CP-30 multiple rocket launcher systems is going on. The MRL is capable of firing the 105mm Pampero and the 127mm SAPBA-1 rockets.

✦ 20 Agusta-Bell AB 206 JetRanger helicopters have been purchased from Italy after long years of negotiation under a $3 million deal signed in 2016.

NAVY: The Argentine Navy (*Armada Argentina*) operates with the primary mission to "ready, train and support the means of naval power in the nation to help ensure effective and efficient employment under the joint action." Its secondary missions include participation in peace operations and / or multinational coalitions under mandate of International Organisations, search & rescue, supporting the activity in Antarctica, and humanitarian assistance among other tasks. To achieve efficiency and effectiveness of maritime missions, the Navy has built up C4IS capabilities along with amphibious and underwater prowess. However, the Armada Argentina is in a parlous state at present as majority of its warships are not operational. Budgetary constraints have also plagued the maritime fleet's maintenance and modernisation.

Major naval bases: Puerto Belgrano, Mar del Plata, Espora, Rio Grande, Zarate

Personnel: 20,033

Equipment

Category	Name	In Service
Destroyer	Almirante Brown (Meko 360 H2)	4
Submarine	Santa Cruz (TR 1700)	2
	Salta (Type 209/1200)	1
Frigate	Drummond (Type A 69)	3
	Espora (Meko 140)	6
Patrol Vessel	Punta Mogotes class	2
	Intrépida class (FAC)	2
	Baradero class	4
Mine hunter	Neuquen	2
Amphibious Force	Hercules	1

Navy continued...

Category	Name	In Service
Amphibious Force	LCVP Landing Craft	16
	LCM 6	4
Auxiliary ship	Survey and Research ships	4

Naval Aviation

Category	Name	In Service
Maritime patrol aircraft	King Air 200M Cormoran	4
	S-2T Turbo Tracker	4
	P3B Orion	4
Helicopter	S-61 Sea King D-4	5
	Aérospatiale SA 316 Alouette III	5
	Agusta ASH-3H Sea King	2
	Bell 205 UH-1H	6
	Eurocopter AS 555MN Fennec	3
Trainer	EMB-326GB	7
	Beechcraft T-34C-1 Mentor	9

Marines

Category	Name	In Service
APC/MIC V/IFV	ERC-90	12
	LVTP-7	11
	LARC-5	13
	Panhard VCR/TT	24
	Hummers	24
Artillery	105mm Oto Melara	12
	155mm M114	6
Mortar	81mm	20
SAM	RBS-70	6
AA gun	40mm L70	4

New Procurements/ Upgrades

✦ A contract has been awarded to state-owned INVAP S.E. in September 2014 to modernise all four Meko 360 destroyers and two TR-1700 submarines to extend their operational lifespan.

Meko 360 destroyer *Argentine Navy photo*

✦ Plans are on to upgrade the diesel-electric submarines into nuclear-powered ones and equip them with conventional weapons.

✦ Four 1,800 tonne offshore patrol vessels are on order from German shipbuilder Fassmer with construction expected to start soon in an Argentine shipyard.

✦ Discussion is going on with China to acquire five 1,800-ton P-18N corvettes, which will serve the Argentine Navy as "Malvinas" class OPVs.

✦ Negotiation is going on with France to acquire three retired A 69 class of corvettes. The corvettes are of the same class as the Drummond-class of platforms presently operational in the Argentine Navy. The platforms were acquired from France in the 1980s.

✦ The P-3B Orion MPAs, received from US surplus in the 1990s, are being upgraded with the capability to launch MBDA AM-39 Exocet anti-ship missiles. Argentinean firms have also started service-life extension programme for these aircraft to keep them operational for another 15 years.

✦ The Naval Aviation Command has issued an operational requirement to carry out modernisation of its T-34 C1 Turbo Mentor jet trainers.

AIR FORCE: The Air Force of Argentina, called the *Fuerza Aérea Argentina* (FAA), operates with the primary objective of defending the national air space and protecting national interests, including in Antarctica. It contributes to UN peacekeeping missions. Once reckoned as a strong force, the FAA has lost its prowess following the Falklands War in 1982. Today, it is grappling with obsolete and inadequate air platforms and equipment which has limited its combat capability. While the older-generation frontline fighters (French Mirage) have been retired, acquisition of new warplanes as their replacement is pending owing to budgetary constraints. The Air Force is urgently needing supersonic fighters' fleet for which negotiations are going on with a number of countries but no final decision has been taken yet. The FAA's modernisation plans have been hindered due to financial and political issues. The Government is planning to reorganise the Air Force and its units by streamlining personnel and reshuffling their rank and file.

Major air bases: El Palomar, Tandil, Moreno (Buenos Aires), Parana (Entre Rios Province), Reconquista (Santa Fe Province), El Plumerillo (Mendoza Province), Villa Reynolds (San Luis Province), Comodoro Rivadavia (Chubut Province)

Personnel: 13,200

Equipment

Category	Name	In Service
Fighters/ Attack/ Reconnaissance	A-4 AR Fightinghawk	3
	FMA IA 58 Pucará	31
Transport/ Tanker	Fokker F-28 Fellowship 1000C	7
	KC-130/ C-130 Hercules	2+5
	DHC-6 Twin Otter	9
	Saab 340B	4

Air Force continued...

Category	Name	In Service
Trainer	Beech T-34A mentor	29
	EMB-312 Tucano	25
	IA-63 Pampa II	8
Helicopter	Bell 212 Twin Huey	7
	Bell UH-1H Iroquois	
	Aérospatiale SA-315B Lama	2
	Hughes 500D Defender	10
	Hughes 369	1
UAV	Cabure	
Weapon	R 530 (AAM)	
	R 550 Magic	

New Procurements/ Upgrades

✦ Acquisition of new frontline fighters is under consideration with the Air Force evaluating several aircraft at present. While discussions were previously going on with China to buy the JF-17 or J-10 fighters, now the FAA is weighing other options as well. It is evaluating to buy the Italian company Alenia Aermacchi's M-346 Master advanced jet trainer/light fighter as frontline platforms. France has also shown its willingness to supply a squadron of Mirage F-1 supersonic fighters. Likewise, Korean Aerospace Industries (KAI) FA-50 Fighting Eagle is also being considered by the FAA.

✦ The US State Department in August 2016 has approved the possible Foreign Military Sale of up to 24 Beechcraft T-6C+ Texan II turboprop trainer/light attack aircraft in response to Argentina's request for the jets.

✦ The state-owned Fabrica Argentina de Aviones (FAdeA) has been contracted to supply the Air Force with 40 new Pampa III aircraft. The indigenously developed jet made its maiden flight in Sept. 2015.

✦ Plans are on to upgrade the IA-58 Pucara light strike aircraft into IA-58H configuration by 2020.

✦ The Learjet 35A aircraft is being modified by the Avcon Industries of Argentina for electronic attack configuration.

✦ Eight P2002JF single seat jet trainers have been leased from Italy's Tecnam in September 2016.

P2002JF jet trainer

✦ Plans are on to procure six KC-390 military transport aircraft being built by Brazilian aircraft maker Embraer. The new planes are set to replace the C-130 Hercules of the FAA.

- ✦ Five Hercules C-130 military transport aircraft are being refurbished under the US Foreign Military sales programme. The first aircraft has been delivered in April 2016. All aircraft are scheduled to be delivered by 2019.
- ✦ Incorporation of 3DLA primary radars for the air space control and vigilance system is planned. Six such radars are on order from INVAP S.E.
- ✦ Nine Twin Otter aircraft are being refurbished by Canada's Viking Air Limited under a $15 million contract.
- ✦ The Air Force is planning to acquire new medium transport aircraft to replace the Fokker F-27s which retired in 2016. Purchase of Airbus Military C-295s or Alenia C-27J Spartan is under consideration.
- ✦ A new basic trainer aircraft, designated IA-100, is being developed domestically.
- ✦ MoU has been signed for the procurement of 12 Airbus Helicopters H125 helicopters as a replacement for the SA315B Lama choppers. The new choppers will be used for SAR and other non-military missions.
- ✦ Four Bell 412EP helicopters are being acquired from the US.
- ✦ Development of a new indigenously-built medium altitude long-endurance UAV, Vigia 2A, is going on.

Major Armaments Developers & Producers

Once home to a brimming, well-diversified defence industry and meeting the requirements of its armed forces, Argentina in the last few years has initiated steps to revive the industry which faced shutdown following privatisation and lack of political attention after the country's transition from military dictatorship to democracy. The country is also encouraging foreign investors to invest in its defence sector with an emphasis on technology transfer and partnership in military programmes. Some major military manufacturers in the country include the following.

Fábrica Militar de Armas Portátiles (FMAP) in Rosario, Santa Fe, produces a wide range of small arms and ammunition.

Fabricaciones Militares (FM) in Rio Tercero, Cordoba is a manufacturer of heavy infantry weapons, including anti-tank weapons and light mortars.

Instituto de Investigaciones Cientificos y Technicas de las Feurzas Armadas or The Institute for Scientific and Technological Research for the Defense (CITEFA) in Buenos Aires operates as federal research and development agency. It has developed artillery guns, multiple rocket launchers, anti-tank missiles among others.

The Río Santiago is Argentina's leading shipyard and has built several major ships for the Navy.

Tandanor is the country's largest shipyard based in Buenos Aires. It has been entrusted with the task of building the four new 1,800 ton offshore patrol vessels for the Argentine Navy.

The **Fabrica Argentina de Aviones (FAdeA)** is the leading aircraft manufacturer of Argentina. Operating as a subsidiary of US defence major Lockheed Martin until March 2009, the FAA was nationalised in August 2009 and now operates as a fully state-owned aviation company. It is presently developing the IA-73 'Pampa' turboprop trainer jet for joint purchase by members of the Union of South American Nations (UNASUR). Plans are also on to develop a light strike version of the plane, to be called the 'Pampa NG'. The company is also upgrading the IA 58 Pucará ground attack fighter jets of the Argentine Air Force. It is also building a prototype of the Chinese CZ-11 helicopter, known locally as "Pampero."

DEFENCE DEALS

Joint Venture Programmes

Argentina has joined hands with a number of South American countries to co-develop military platforms and systems. It is partnering with China to produce a light utility helicopter and also planning to co-develop OPVs and armoured vehicles for its military.

Brazil

KC-390: Argentina's FAdeA has partnered with Brazil's Embraer in the development and production of the Brazilian KC-390 military transport aircraft. The FAdeA is supplying structural parts for the new airplane which was unveiled in Sao Paulo in October 2014. As an industrial partner in the programme along with the Czech Republic and Portugal, Argentina has pledged to buy six KC-390s for its air force.

Gaucho infantry vehicle: The two countries jointly developed and produced the Gaucho light infantry vehicle in the 2000s. While Brazil developed the engine and gearboxes for the wheeled all-terrain vehicle, Argentina designed its tubular frame and independent suspension. Presently, the vehicle is in service with the Argentine Army.

Venezuela

Gradicom missile: The two South American countries are developing the new missile system, reportedly with the help of Iran. Argentina's Fabricaciones Militares (FM) and Venezuela's Compañía Anónima Venezolana de Industrias Militares (CAVIM) have signed a joint development contract in 2012 to develop the medium range missile system.

China

CZ-W11 helicopter: Partnering with China's National Aero-Technology Import and Export Corporation, Argentina has successfully developed and produced the CZ-W11 ultra-light helicopter for its military. The new chopper, called the Pampero in Argentina, is presently undergoing flight tests and will be inducted in the Argentine military in near future.

UNASUR

Unasur 1: Seven South American countries, including Argentina, have partnered to develop a new basic trainer. The new trainer jet is called Unasur 1 which is an adaptation of the acronym for the Union of South American Nations (UNASUR). The multinational aviation project was announced in April 2013. Besides Argentina, the other partner nations in the project include Brazil, Ecuador, Venezuela, Chile, Colombia and Uruguay – all members of the UNASUR. The Unasur 1 will be developed with a 1,590kg maximum take-off weight and a flying speed of 388km/h. It will feature a piston engine for primary training and a turbine engine for basic and primary training.

Arms Import & Export

Argentina has acquired most of its military equipment from the US in the recent past. It has also bought arms from Brazil, France, Germany and Spain. Presently, defence import sources for the country is diversifying to include China and Russia from which new military equipment are being procured.

Export of military equipment from the country is nil. The moribund domestic defence industry is being revived to take part in joint defence projects with few other countries, mostly from South America, with export plans. Argentina has also initiated efforts in recent times to build its own indigenous UAVs and trainer jets.

DEFENCE BUDGET

Argentina is planning a 50 percent increase in its defence spending but has left open-ended the amount that will eventually go into the military regeneration programme. Defence expenditure is largely driven by modernisation plans, participation in peacekeeping missions and a dispute with the UK regarding the sovereignty of the Falkland Islands. But a major share of the total defence spending goes towards personnel costs. The defence budget in 2014 was ARS35.1 billion (USD$ 5.7 billion). The proposed defence budget for the year 2015 was US$ 5.9 billion as over 30 new acquisition or modernisation programmes for the Army, Navy and Air Force were under consideration.

CONCLUSION

The leading countries in the South American continent, which is relatively peaceful compared to other regions of the world, are focused on defence modernisation plans and strengthening their overall military capability mainly to tackle internal security issues and safeguard territorial integrity as also to protect the vast natural resources around their vicinity. While Brazil is marching ahead in the race, Argentina has initiated steps to catch up. The military build-up of Argentina is primarily aimed at preparing it to face any immediate or imminent threat from Britain over the long-standing Falkland Islands dispute. Hence, a number of new acquisition as well as modernisation programmes are on the cards. However, continuing economic woes has hindered any immediate action in this regard. Presently, the foremost defence priorities for Argentina are the Falklands, the Argentine Antarctica and continuing its military participation in international peacekeeping missions besides safeguarding national sovereignty.

CONTACT DETAILS

Ministry of Defence
Azopardo 250,
C1107ADB, Buenos Aires
Tel: (+5411) 4346-8800
Website: www.mindef.gov.ar

ARMENIA
(Capital: Yerevan)

INTRODUCTION
Area: 29,743 sq km
Population: 3,051,250 (July 2016 est.)
Coastline: 0 km (landlocked)
Maritime claims: None

KEY POLITICAL PERSONS

PRESIDENT & COMMANDER-IN-CHIEF: Serzh Sargsyan

PRIME MINISTER: Karen Karapetyan

KEY DEFENCE PERSONS

DEFENCE MINISTER: Vigen Sargsyan

CHIEF OF GENERAL STAFF: Lt. Gen. Movses Hakobyan

ECONOMY

Armenia's economy has undergone a major change since independence. Sustained growth, ambitious reforms, and external inflows of capital and remittances have created a market-oriented environment. The Government helped by the ceasefire that has been in effect in Nagorno-Karabakh since 1994, has been able to carry out wide ranging economic reforms. New sectors, such as precious stone processing and jewellery making, information and communication technology, and even tourism are beginning to supplement more traditional sectors such as agriculture in the economy. The steady economic progress has earned the country increasing support from international institutions such as the IMF, World Bank, and the European Bank for Reconstruction and Development (EBRD). With a per-capita GDP of US$ 3,870 (GNI, 2013), Armenia is a lower middle-income country. Russia is Armenia's biggest trade partner. Economic ties with Russia remain close, especially in the energy sector. Armenia joined Russia in the Eurasian Economic Union upon the bloc's launch in January 2015 and the WTO in January 2003.

GDP: $25.32 billion (2015 est.); $24.58 billion (2014 est.)

Real Growth Rate: 3% (2015 est.); 3.5% (2014 est.)

Industries: Diamond processing, metal-cutting machine tools, forging and pressing machines, electric motors, tires, knitted wear, hosiery, shoes, silk fabric, chemicals, trucks, instruments, microelectronics, jewellery, software, food processing, brandy, mining

Total Exports: $1.626 billion (2015 est.); $1.698 billion (2014 est.)

Export Commodities: Pig iron, unwrought copper, nonferrous metals, diamonds, mineral products, foodstuff, energy

Major Markets: Russia 15.2%, China 11.1%, Germany 9.8%, Iraq 8.8%, Georgia 7.8%, Canada 7.6%, Bulgaria 5.3%, Iran 5.3% (2015)

Total Imports: $2.78 billion (2015 est.); $3.754 billion (2014 est.)

Import Commodities: Natural gas, petroleum, tobacco products, foodstuff, diamonds, pharmaceuticals, cars

Major Suppliers: Russia 29.1%, China 9.7%, Germany 6.2%, Iran 6.1%, Italy 4.6%, Turkey 4.2% (2015)

INTERNATIONAL DISPUTES

The dispute over the break-away Nagorno-Karabakh region and the Armenian military occupation of surrounding lands in Azerbaijan remains the primary focus of regional instability. The Organisation for Security and Cooperation in Europe (OSCE) continues to mediate over the dispute. The border with Turkey remains closed over the Nagorno-Karabakh dispute.

DEFENCE

The Armed Forces of the Republic of Armenia comprise two services - the Army and the Air Force. Being a landlocked country, Armenia has no Navy. The General Staff is responsible for operational command of the Armenian Military. The Army is functionally divided into Active and Reserve Forces. The Active Forces mainly have peacekeeping and defensive duties while the Reserve Forces consists of Enhancement Forces, Territorial Defence Forces, and Training Grounds. They deal with planning and reservist preparation, armaments and equipment storage, training of formations for active forces rotation or increase in personnel.

The de-facto independent region of Nagorno-Karabakh maintains its own armed forces – NK Defence Army - under central command and military structure distinct from the Armenian Army.

An Armenian helicopter during a military exercise. *An MoD photo*

ARMY

Personnel: 40,783

Equipment

Category	Name	In Service
MBT	T- 72	144
	T-54/55	
AIFV/APC	BMP-1/2	173
	BRM-1K	
	BTR-60/-70/80	
	MT-LB	
Artillery	D30 122mm	69
	D-20	62
	D-1	
	2A36 152mm (towed)	
	2S1 122mm	10
	2S3 152 mm (SP)	28
Mortar	M120 120mm	12
SSM	Scud-B	
	9M79 Tochka-U	8+
MLR	BM-21 122mm	47
	WM80	4
	Smerch 300mm	
ATGW	9P148	8
	9P149	14

New procurements/Upgrades

✦ Under a deal worth US$ 200 million revealed in February 2016, new anti-tank missiles/rockets, air-defence systems and other equipment are being acquired from Russia. The deal also includes upgrade of T-72 MBTs of the Army.

✦ Iskander-E short-range ballistic missiles and Buk surface-to-air missiles have been acquired from Russia under the deal.

AIR FORCE

Personnel: 1,208

Equipment

Category	Name	In Service
Fighter	Su-25	15
Transport	IL-76	3
Trainer	L-39	10
Helicopter	Mi-24P	8
	Mi-24K	2

Air Force continued...

Category	Name	In Service
Helicopter	Mi-24R	2
	Mi-8MT	10
	Mi-9	2
	Mi-2	7
AA gun	ZSU-23-4	
	ZSU-23	
SAM/Anti-aircraft system	SA-2	
	SA-3	
	SA-4	
	S-60	3
	S-300	

New procurements/Upgrades

No major procurement plans are under consideration at present.

Defence Expenditure

Total defence spending: $500 million (2015)

Estimated defence spending in terms of GDP: Around 4%

Defence Production and R&D

With the help of Russia, Armenia has set up certified centres and joint ventures for the repair and maintenance of weapons and military equipment. The republic is able to carry out repairs and service equipment produced by companies in the Russian military and industrial complex on site. The joint enterprise KAMAZ-Armenia is responsible for the repair and maintenance of the Armenian Defence Ministry's fleet of trucks. There are plans for the production of certain types of ammunition and armoured vehicles and to set up assembly plants for the production of cars based on the KAMAZ chassis. In June 2013, Armenia and Russia signed an agreement on military and technical cooperation. In October 2016, the two sides made a commitment to further develop joint military-technical co-operation, including through formation of joint ventures, mainly in the fields of air defence technology. Russian military industrial support for Armenia comes against a background of tensions between Armenia and Azerbaijan over the disputed Nagorno-Karabakh region.

Defence Procurement

Armenia is modernising its military equipment. In 2010, the Government approved a five-year plan to modernise its armed forces under which the government had put forth a plan to acquire new long-range weapons and other hardware.

CONTACT DETAILS

Ministry of Defence

Bagrevandi 5, 0044

Telephone: + 37410 294699

E-mail: modpress@mil.am, press@mil.am

STRATEGIC INFORMATION

AUSTRALIA

(Capital: Canberra)

INTRODUCTION

Area: 7,741,220 sq km

Population: 22,992,654 (July 2016 est.)

Coastline: 25,760 km

Maritime claims: Territorial sea: 12 nm

Contiguous zone: 24 nm

Exclusive Economic Zone: 200 nm

Continental shelf: 200 nm or to the edge of the continental margin

GEOPOLITICAL IMPORTANCE

Australia, officially the Commonwealth of Australia, is a country in the Southern Hemisphere comprising the mainland of the Australian continent, the island of Tasmania and numerous smaller islands in the Indian and Pacific Oceans. Neighbouring countries include Indonesia, East Timor and Papua New

Guinea to the North, the Solomon Islands, Vanuatu and New Caledonia to the Northeast and New Zealand to the Southeast. The largest country in Oceania, Australia, has emerged as a maritime nation in the making and is an important player in the Asia-Pacific region. The country's strategic positioning at the confluence of the Indian and Pacific Oceans will have an immensely positive influence on regional developments and stability in Asia.

POLITICAL OVERVIEW

The Parliament consists of the Queen, represented by the Governor-General, and two Houses - the House of Representatives and the Senate. The Parliament passes legislation. Proposed laws have to be agreed to by both Houses of Parliament to become law. The two Houses have equal powers, except that there are restrictions on the power of the Senate to introduce or directly amend some kinds of financial legislation.

The Queen - Queen Elizabeth II - constitutional function is to appoint the Governor-General, and in doing this the Queen acts as advised by the Australian Prime Minister. The Governor-General has a wide range of powers, exercised under the authority of the Australian Constitution. The functions and roles of the Governor-General include appointing ambassadors, ministers and judges, giving Royal Assent to legislation, issuing writs for elections and bestowing honours. The Governor-General is also Commander-in-Chief of the Australian Defence Force.

The Prime Minister is the head of the Government. The Cabinet, consisting of senior Ministers presided over by the Prime Minister, is the Government's pre-eminent policy-making body. Major policy and legislative proposals are decided by the Cabinet. The Prime Minister selects Ministers for Cabinet positions. Ministers are selected by the Prime Minister. There are currently 30 Ministers. The maximum number can be increased by legislation. About 19 or so senior Ministers administer the major departments and are, usually, members of Cabinet. Other Ministers are responsible for particular areas of administration within a major department, or may be in charge of a small department. Up to 12 Members and Senators are appointed by the Prime Minister as Parliamentary Secretaries to assist or represent Ministers in their administrative responsibilities.

The judiciary is the legal arm of the Australian Government. The Constitution vests the judicial power of the Commonwealth - the power to interpret laws and to judge whether they apply in individual cases - in the High Court and other federal courts. The High Court is established by the Constitution. Other federal courts are created by legislation of the Parliament. Judges are appointed by the Governor-General acting on the advice of the Prime Minister and Cabinet.

KEY POLITICAL PERSONS

HEAD OF STATE: Queen Elizabeth II

GOVERNOR GENERAL & COMMANDER-IN-CHIEF: Gen. Peter John Cosgrove

PRIME MINISTER: Malcolm Bligh Turnbull

ECONOMY

The Australian economy is one of the largest mixed market economies in the world. Its economy is dominated by its service sector, comprising 68% of GDP. The country is a significant exporter of natural resources, energy, and food. Australia's abundant and diverse natural resources attract high levels of foreign investment and include extensive reserves of coal, iron, copper, gold, natural gas, uranium, and renewable energy sources. Australia is a member of the APEC, G20, OECD and WTO organisations. It has entered into free trade agreements with ASEAN, Chile, New Zealand, Singapore, Thailand, and the United States.

GDP: $1.489 trillion (2015 est.), $1.454 trillion (2014 est.)

Real Growth Rate: 2.5% (2015 est.), 2.6% (2014 est.)

Industries: mining, industrial and transportation equipment, food processing, chemicals, steel

Total Exports: $184.4 billion (2015 est.), $240.8 billion (2014 est.)

Export Commodities: Coal, iron ore, gold, meat, wool, alumina, wheat, machinery and transport equipment

Major Markets: China 32.2%, Japan 15.9%, South Korea 7.1%, US 5.4%, India 4.2% (2015)

Total Imports: $208.4 billion (2015 est.), $240.5 billion (2014 est.)

Import Commodities: Machinery and transport equipment, computers and office machines, telecommunication equipment and parts; crude oil and petroleum products

Major Suppliers: China 23%, US 11.2%, Japan 7.4%, South Korea 5.5%, Thailand 5.1%, Germany 4.6% (2015)

DEFENCE & SECURITY

Australia's most basic strategic interest remains the defence of Australia against direct armed attack. The prime task of the Australian Defence Forces (ADF) is defending the country's sea lanes and infrastructure, Australia's security, stability and cohesion of its immediate neighbourhood, which the country share with Indonesia, Papua New Guinea, East Timor, New Zealand and the South Pacific island states. The ADF is engaged in the stability of the wider Asia-Pacific region and particularly Southeast Asia - assisting countries in the region to respond to threats to their security. The ADF also participates in peacekeeping missions contributing limited forces to operations throughout the world.

The Government is working with communities to build resilience to violent extremism. Cyber security is one of the Australian Government's highest national security priorities. The Australian Signals Directorate (ASD, formerly DSD) provides foreign signals intelligence to the Australian Defence Force and Australian Government to support military and strategic decision-making. The government launched its Cyber Security Strategy in 2009 with the aim to promote a secure, resilient and trusted electronic operating environment that supports Australia's national security.

KEY DEFENCE PERSONS

MINISTER FOR DEFENCE: Marise Ann Payne

CHIEF OF THE DEFENCE FORCE: Air Chief Marshal Mark Binskin

CHIEF OF ARMY: Lt. Gen. Angus Campbell

CHIEF OF NAVY: Vice Admiral Tim Barrett

CHIEF OF AIR FORCE: Air Marshal Gavin Leo Davies

Internal Conflict

Terrorism

Australia faces the threat of a domestic strike by transnational terrorists against innocent civilians. A number of non-state armed groups are reportedly active on Australian territory. The National Counter-Terrorism Plan is maintained by the Australia-New Zealand Counter-Terrorism Committee. It is the primary public document on Australia's national counter-terrorism policy and arrangements. It outlines the responsibilities, authorities and the mechanisms to prevent acts of terrorism within Australia and to manage the consequences if they do occur. The country has signed memorandums of understanding on counter-terrorism with Indonesia, the Philippines, Malaysia, and Thailand. The agreements promote increased bilateral co-operation between intelligence and law enforcement agencies and defence officials of Australia and the signatory countries. The ADF maintains two elite Tactical Assault Groups, the Special Operations Engineer Regiment as well as a high readiness group in each Army Reserve brigade and the 1st Commando Regiment to meet its counter-terrorism responsibilities.

External Conflict

The Australian Defence Force personnel is involved in global military operations such as Afghanistan, Israel/Lebanon, Egypt, Iraq. The ADF currently has several forces deployed to the Middle East as part of 'Operation OKRA' to combat the Daesh (also known as ISIL) terrorist threat in Iraq and Syria. Australia's contribution is being closely coordinated with the Iraqi government, Gulf nations and a broad coalition of international partners. About 780 ADF personnel have deployed to the Middle East in support of Operation OKRA. These personnel make up the Air Task Group (ATG), the Special Operations Task Group (SOTG) and Task Group Taji (TG Taji).

Territorial Disputes

Australia asserts land and maritime claims to Antarctica and this claim is recognised by France, New Zealand, Norway and UK, but it is not recognised by the United Nations, US, Russia or by most other countries. In 2007, Australia and Timor-Leste agreed to a 50-year development zone and revenue sharing arrangement and deferred a maritime boundary. But dispute still continues over the best means to begin production at the Greater Sunrise natural gas field in the Timor Sea. Indonesian groups challenge Australia's claim to Ashmore Reef. Australia has closed parts of the Ashmore and Cartier reserve to Indonesian traditional fishing.

THREAT PERSPECTIVE

- ✦ Australia continues to face a persistent threat of terrorism, espionage, foreign interference, including cyber security threats. The escalating security crisis in Iraq and Syria poses an increasing threat to Australia. The Government is particularly concerned about Australians who travel to conflict zones and return to Australia with skills and intentions acquired from fighting or training with proscribed terrorist groups. The Government is currently working closely with state, territory, and local governments, as well as the private sector and the community, in building an effective, nation-wide counter-terrorism capability, and by contributing to regional and global efforts to counter-terrorism.
- ✦ Challenges from slow economic growth, social and governance challenges, population growth and climate change in the South Pacific region contributes to Australia's instability.
- ✦ Increasing security threats in cyberspace and space. The cyber threat to Australia represents a real and present risk to the Australian Defence Force's (ADF) warfighting capability, and national security and economic prosperity both in peacetime and during armed conflict.

STRATEGIC RELATIONS

Australia shares enduring relationships with United States and United Kingdom and growing ties with New Zealand, Indonesia and Japan in pursuing common strategic and defence interests, particularly in countering global and regional terrorist threats. Australia enjoys close and productive relations with China and India and also has valuable longstanding defence ties with Malaysia, Singapore, Thailand, the Philippines and Papua New Guinea, and developing further ties with many other countries in the Asia - Pacific.

United States: Australia has long been a close friend and security ally of the United States. Australia has a formal alliance with the U S known as the Australia, New Zealand, United States Security (ANZUS) Treaty. The treaty binds Australia and the United States to consult on mutual threats, and, in accordance with respective constitutional processes, to act to meet common dangers. It was invoked by Australia for the first time in response to the terrorist attacks of 11 September 2001. The agreement aims to maintain the security architecture of the Asia-Pacific region.

India: Defence cooperation between India and Australia has expanded significantly in recent years with the signing of MoU on defence cooperation as well as a Joint Declaration on Security Co-operation. The two nations continue to build relations between the defence forces through regular personnel and training exchanges. Enhancing their maritime cooperation, the first formal bilateral naval exercise (AUSINDEX) was held off the coast of Visakhapatnam in 2015. They have also decided to hold AUSINDEX biennially, and the next iteration will take place in Australia in 2017. In February 2016, Australia's Chief of Navy, a Royal Australian Navy ship and a detachment of the RAN Band participated in the Indian Navy's International Fleet Review. India is Australia's seventh-largest trading partner and the fifth-largest export market.

Japan: Australia and Japan have established good bilateral relationships based on mutually complementary economic relations. In recent years, the two countries have strengthened political and security cooperation, which has made them a strategic partner in the Asia-Pacific region. As both countries are allies with the US in the Asia-Pacific region, the two countries have promoted the trilateral relations. Japan is a vital and long-standing export market for Australian business, a reliable supplier of high-quality finished goods, and a key source of investment to the Australian economy. The Japan-Australia Economic Partnership Agreement (JAEPA) entered into force on 15 January 2015.

China: Australia and China are actively engaged economically, cultural and politically which spans different organisations such as APEC, East Asia Summit and the G20. China is Australia's largest trading partner. Further strengthening and deepening their economic relationship is a major priority for both countries, with both governments committed to sustaining the impressive trade and investment performance achieved in the past two decades. The two countries have agreed on an unspecified programme of enhanced military cooperation, including exchanges, training and service-to-service engagement.

DEFENCE CAPABILITIES

The Australian military is an all-volunteer force. Both women and men can enlist in the Australian Defence Force. The ADF comprised of the Australian Army, the Royal Australian Navy (RAN), and the Royal Australian Air Force (RAAF). The role of Commander-in-Chief is vested in the Governor-General, who appoints a Chief of the Defence Force from one of the armed services on the advice of the government. The Chief of Defence (CDF) commands the ADF.

The government in the Defence White Paper 2016 has outlined its programme to upgrade the Defence Force. The defence force will be better equipped to face growing unease in the Asia-Pacific region due to military build-up, including disputes over the South China Sea. While the defence budget will be increased by $29.9 billion over the next decade, defence personnel will be increased to 62,400, according to the White paper. An impressive array of new naval and aerial capabilities will be added to the ADF.

The future force will be more capable of conducting independent combat operations to defend Australia. The Government is strengthening Defence's intelligence, surveillance and reconnaissance capabilities to provide the forces with comprehensive situational awareness. The Defence cyber workforce and systems is being strengthened to deter and defend against the threat of cyber-attack.

The land force will be equipped with new personal equipment for soldiers and a new generation of armoured combat reconnaissance and infantry fighting vehicles, as well as new combat engineering equipment. A new long-range rocket system will further enhance fire power, and armed medium-altitude unmanned aircraft will enhance surveillance and protection for the land force.

Modernising of the maritime capabilities will be a key focus. The submarine force will be increased from 6 to 12 regionally superior submarines. The surface naval capability will include three Hobart Class Air Warfare Destroyers and a new class of nine future frigates supported by new replenishment vessels. There are also plans to induct more capable offshore patrol vessels, new manned and unmanned aircraft and a new large-hulled multi-purpose patrol vessel, the Australian Defence Vessel Ocean Protector. The ADF's capacity for amphibious operations will be strengthened by the introduction of new weapons and equipment for the amphibious ships. New light helicopters will be acquired to support Special Forces operations.

The ADF will be equipped with a potent air combat and air strike capability centred around the F-35A Lightning II and the E/A-18G Growler that builds on its current fleet of F/A-18 Super Hornet, Wedge tail Airborne Early Warning and Control and air-to-air refuelling aircraft while the air lift capability will be increased to comprise 8 heavy lift C-17A Globemasters with additional heavy lift aircraft to be considered in the longer term, 12 upgraded C-130J Hercules, 10 C-27J Spartans and 10 CH-47F Chinook helicopters.

ARMY: The Australian Army is mainly a light infantry force commanded by the Chief of Army. The Army is implementing a number of projects as part of the military modernisation programme. The Army's most recent modernisation programme has been Plan Beersheba. Army is now commencing detailed planning to modernise Army beyond Beersheba. Other modernisation projects include Land 400, Land 200 and Land. The government in its Defence White Paper 2016 has stated that Australia's upgraded military force will include new missile armed unmanned aircraft (drones) for troop protection and a new long range rocket and artillery program.

The Army's operational headquarters, Land Command, is located at Victoria Barracks in Sydney. The Australian Army's three regular brigades are based at Robertson Barracks near Darwin, Lavarack Barracks in Townsville and Gallipoli Barracks in Brisbane.

Personnel: 29,366

Equipment

Category	Name	In Service
MBT	M1A1IM Abrams	59
APC	M113AS4	431
	ASLAV (Piranha 8X8)	257
	Bushmaster	1153
Artillery	105 mm M2A2	240
	155 mm M198	36
	105 mm Hamel L118/119	109
	M777A2	54
Mortar	81 mm F2	296
Anti Tank GW	Milan	
	Javelin	
ATK weapon	84mm Carl Gustav	577
SAM	Rapier BM1 systems	18
	Bofors RBS-70 (with MK3 Bolide missiles)	30

New Procurements/Upgrades

The 2016 Defence White paper has proposed to acquire the following equipment for the land forces:

- ✦ Upgraded M1 Abrams battle tanks
- ✦ Attack helicopters
- ✦ New missile armed unmanned aircraft (drones) for troop protection
- ✦ New long range rocket and artillery program
- ✦ 7 MQ4C Triton unmanned surveillance aircraft (drones)
- ✦ The DoD has shortlisted the AMV 35 to compete in a 12-month Risk Mitigation Activity for the Army's Land 400 programme

Army Aviation

Category	Name	In Service
Helicopter	S-70A Blackhawk	35
	CH-47D Chinook	7
	CH-47F Chinook	10
	MRH-90	33
	ARH Tiger	22
UAV	RQ 7B Shadow 200	

New Procurements/Upgrades

- ✦ Delivery of 40 MRH-90s underway to replace the Blackhawks
- ✦ Three new CH-47Fs Chinooks have been delivered in addition to the existing fleet of seven

The AMV 35. *Image credit: BAE Systems*

An Australian Army soldier of Battlegroup Tiger watches for the 'enemy' while a New Zealand Army soldier reloads his weapon while clearing the township of Iron Knob, South Australia, during Exercise Hamel on 9 July 2016. *Image credit: Australian DoD*

Australian Army Bushmaster protected mobility vehicles assembled and ready to deploy into the 'battlespace' on a mission during Exercise Hamel in Cultana training area, South Australia, on 6 July 2016. *Image credit: Australian DoD*

NAVY: The Royal Australian Navy (RAN) is one of the largest and most sophisticated naval forces in the Pacific region, with a significant presence in the Indian Ocean and worldwide operations in support of military campaigns and peacekeeping missions. The Navy provides maritime forces that contribute to the ADF's capacity to defend Australia, contribute to regional security, support global interests, shape the strategic environment and protect national interests. The RAN is commanded through Naval Headquarters in Canberra. The professional head is the Chief of Navy, who holds the rank of Vice-Admiral.

In the next two decades, Australia will embark on one of the most extensive and ambitious shipbuilding programs to modernise the Royal Australian Navy, according to the Defence White Paper 2016. The Government has announced continuous shipbuilding programs for major surface combatants and minor war vessels. The Navy will double the submarine fleet from six to 12, including new regionally superior submarines to replace the outdated Collins class, nine new anti-submarine warfare frigates to replace the current fleet of Anzac-class ships, 12 new offshore patrol vessels and 7 PA-8 Poseidon spy planes.

Major naval bases: Canberra ACT, Sydney NSW, Garden Island WA, Nowra NSW, Jervis Bay NSW, Westernport VTC, Cairns QLD, Darwin NT.

Personnel: 14,070

Equipment

Category	Name	In Service
Frigate	Anzac class	8
	Adelaide class	4
Submarine	Collins class	6
Mine warfare force	Huon class	4
	MSA	2
	CDT	2
Replenishment ship	HMAS Success	1
	HMAS Sirius	1
Patrol Vessel	Armidale class	13
Amphibious Force	HMAS Adelaide III LHD	1
	HMAS Canberra LHD	1
	Tobruk class LSH	1
	Choules LSD	1
Survey Vessel	Leeuwin class	2
	Paluma class	4

New Procurements/Upgrades

✦ Australia has signed a deal with France for 12 submarines to be built by French defence major DCNS.

✦ The Navy has commissioned HMS Adelaide III, the second one of the two Landing Helicopter Dock (LHD) ships being built for the RAN. The LHDs are the largest warships ever to be built for the RAN.

HMS Adelaide III. *Image credit: Royal Australian Navy*

✦ Australia plans to build 12 next generation of submarines to replace its ageing Collins Class fleet set to be retired from the mid-2020s.

✦ The Government has announced First Pass Approval for both the Offshore Patrol Vessel and the Future Frigates projects. The first Offshore Patrol Vessels will begin construction in 2018 and the Future Frigates in 2020.

✦ Three Hobart Class air warfare destroyers (AWD) are being built for RAN. The lead ship, HMAS Hobart, is slated for completion in June 2017, followed by the HMAS Brisbane in September 2018, and the HMAS Sydney in March 2020.

✦ The upgrade of the Anzac class frigates with an advanced Anti-Ship Missile Defence system on the remaining seven ships will be completed by 2017.

✦ The Government plans to procure 7 PA-8 Poseidon spy planes bringing total to 15 and tactical drones to assist war ship operations.

✦ Austal has been selected as the preferred tenderer to build up to 21 steel-hulled replacement Pacific Patrol Boats in Henderson, beginning in 2017.

Pacific patrol boat replacement design. *Image credit: Austal*

Naval Aviation

Personnel: 14,120

Equipment

Category	Name	In Service
Helicopter	Sikorsky S-70B-2 Seahawk	14
	MH-60R Romeo	14
	Squirrel AS-350B squirrel	13
	Agusta A109E	3
	MRH-90 Taipan	6

New Procurements/Upgrades

✦ RAN has taken delivery of 14 of the 24 MH-60 Romeo helicopters on order.

AIR FORCE: The Royal Australian Air Force (RAAF) is the air force branch of the ADF. The RAAF provides air and space power for Australia's security. The RAAF is the second-oldest independent and permanent air force in the world after Britain's Royal Air Force. The RAAF has taken part in many of the 20th century's major conflicts including the World Wars, the Korean War and the Vietnam War. More recently the RAAF participated in the 2003 invasion of Iraq. Air Force Headquarters is located in Canberra.

The government in its Defence White Paper 2016 has stated its plans to acquire 72 F-35 Joint Strike Fighters, which will begin to enter service by the early 2020s. These aircraft will reportedly be equipped with the Norwegian-developed Naval Strike Missile and Joint Strike Missile. The special forces will receive new helicopters and 12 E/A-18 Growlers. Two C-17 Globemasters will be added to the Air Force and also new air to air, air to surface and anti-ship missile weapons.

Major air bases: Amberly, Edinburgh, Richmond, Williamtown, Tindal, Pearce, Darwin, Townsville, Wagga, Fairbairn, East Sale, Learnmonth and Curtin in WA and Sherger QLD are bare bases.

Personnel: 14,076

Equipment

Category	Name	In Service	On Order
Fighter	F/A-18A	55	
	F/A-18B	16	
	F/A-18F Super Hornet	24	
	F-35 Lightning II	2	70
Reconnaissance/ AEW&C	EA-18G Growler		12
	Boeing 737-700 AEW&C	6	
Transport/ Tanker	C-27J Spartan	10	
	C-17 Globemaster III	8	
	C-130J Super Hercules	12	
	Bombardier Challenger 604	3	
	Boeing 737 BBJ	2	
	King Air 350	3+5	
	KC-30A (A30 MRTT)	5	2
Trainer	Hawk 127	33	
	PC-9/A	60	
Patrol	AP-3C Orion	18	
UAV	IAI Heron	1	

New Procurements/Upgrades

✦ The first PC-21, registered as A54-001, will be handed over to the Air Force in June 2017 after completion of testing and verification work in both Switzerland and Australia.

RAAF's Pilatus PC-21 aircraft taking its first initial production test flight at their factory in Stans, Switzerland. *Image credit: Lockheed Martin*

✦ The purchase of a first batch of 14 F-35A Lightning II has been approved, and two planes are being delivered. The government plans to procure a total of 72 aircraft.

✦ The RAAF is expected to take delivery of the first EA-18G Growler airborne electronic attack aircraft in 2017. The second RAAF Growler has made its first flight while 10 other aircraft are in various stages of assembly.

✦ RAAF is expected to take delivery of the first of the eight P-8A Poseidon aircraft in 2017 and all the aircraft will be fully operational by 2021. The P-8A program is set to provide Australia's future manned maritime patrol and response aircraft capability, replacing in part the AP-3C Orion aircraft.

P-8A Poseidon maritime surveillance aircraft.
Image credit: Royal Australian Air Force

✦ The MoD will purchase two additional KC-30A multi-Role Tanker Transport aircraft for the RAAF with expected delivery in 2018.

✦ RAAF has received its eighth and final C-17A Globemaster III aircraft from Boeing completing the C-17A fleet requirement of the country.

✦ The Government plans to acquire Triton Unmanned Aerial Vehicle (UAV), new air to air, air to surface and anti-ship missile weapons, integrated air and missile defence systems.

Major Armament Producers

The Government has committed to forming a new partnership with Australian defence industry to ensure Defence gets the equipment, systems and personnel it needs on time and on budget. The Defence Materiel Organisation, the Australian Government agency responsible for acquisition of equipment for the Australian Defence Organisation, has been disbanded and its core responsibilities for capability delivery have been transferred to the new Capability Acquisition and Sustainment Group. The new Group will work in partnership with defence industry and, together with the CDIC, will ensure the ADF has the necessary capabilities to fulfil its missions and operations.

Australian defence firms design, manufacture and maintain quality military equipment capable of operating in extreme conditions such as high heat and humidity and cold Antarctic waters. Specialist products include command and control systems, multi-sensor data fusion, signal processing, underwater systems, phased array radar, navigation and positioning aids, logistics support systems and self-monitoring propulsion systems. Specialist services include ship-building, construction and repair of major weapons, systems integration and the provision of land, maritime and air services to support peacetime and operational defence requirements. The defence firms have supplied Australia's defence forces with sophisticated capability solutions for deployments in Iraq, Afghanistan, the Solomon Islands, South-East Asia and East Timor.

DEFENCE DEALS

Joint Venture Programmes

Australia and Japan have signed an accord in 2014 to deepen defence-orientated science and technology co-operation between the two countries. The Transfer of Defence Equipment and Technology agreement was signed in Canberra by Australian Prime Minister Tony Abbott and his visiting counterpart Shinzo Abe. Australia also signed an agreement with India to work together in the aviation and aerospace industries. A Memorandum of Understanding (MOU) was signed between Aviation Aerospace Australia (AAA) and the Society of Indian Aerospace and Technologies Industries (SIATI).

Arms Export

Although small by world standards, Australia maintains a significant sector of industry dedicated to or involved in the manufacture and sale of armaments and military equipment. The country exports arms such as aircraft, armoured vehicles, artillery, radar systems, missiles, and ships and related goods. Australia's biggest customers are the USA, Canada, New Zealand, the UK and Israel. Other customers include Indonesia, Singapore, Philippines, South Africa, India, Egypt, Papua New Guinea, Saudi Arabia, Thailand and Pakistan. The Department of Defence is responsible for administering controls on the export of defence and dual-use goods, and the granting of authorisations to export, in the form of permits and licenses. Within Defence, this role is performed by the Defence Export Control Office (DECO).

Arms Import

Australia is now the world's sixth-largest importer of arms, according to the Stockholm Institute for Peace Research (SIPRI). The arms import and spending on military equipment surged in line with ambitious targets to update the armed forces' fleet of fighter jets, submarines, frigates and armoured vehicles. Its acquisitions increased by 65% between 2010 and 2014, compared with the four previous years, it said.

In recent years Australia, has bought combat helicopters from France, German armoured personnel carriers, radar systems from Sweden, Howitzer artillery guns from Britain, air-refuelling tanker aircraft from Spain, as well as fighter aircraft, helicopters, military transport aircraft, Shadow drones, Hellfire anti-tank and Sidewinder air-to-air-missiles from the US. Other weapons on order includes - F-35 Joint Strike Fighter aircraft, Poseidon surveillance planes, amphibious assault ships, Triton surveillance drones, air-warfare destroyers and submarines.

DEFENCE BUDGET

Total defence spending: $32.3 billion (2016), $31.9 billion (2015)

Estimated defence spending in terms of GDP: 1.88%

The government will provide defence with $32.3 billion in 2016-17 and $142.9 billion over the forward estimates.

The 2016 Defence White Paper has set out comprehensive long-term plans to add an impressive array of new capabilities to the ADF with most of these are focused on the ADF's naval and aerial capabilities. The naval modernization programme includes the sea-1000 programme, the replacement programme six Collins-class submarines. The Air Force is expected to be equipped 72 F-35 joint strike fighters, which will begin to enter service by the early 2020s.

CONCLUSION

Australia, a key democratic partner of the United States, is one of the major political and economic powers in Southeast Asia. The government through its 2016 Defence White has stated its commitment to build a highly capable armed force to protect the security and prosperity and respond to strategic challenges over the coming decades. An ADF with higher levels of preparedness will be able to better respond to strategic developments that threaten Australia's interests where the Government requires a military response. The Government will continue to strengthen the alliance with the United States and deepen their long-standing network of international defence partnerships and work together to meet common threats and challenges.

CONTACT DETAILS

The Department of Defence
Russell Offices
Department of Defence,
Canberra ACT 2600, Australia
Tel: +61-6-265-9111
www.defence.gov.au

AUSTRIA

(Capital: Vienna)

INTRODUCTION

Area: 83,871 sq km
Population: 8,711,770 (July 2016 est.)
Coastline: 0 (Landlocked)
Maritime claims: None

KEY POLITICAL PERSONS

PRESIDENT & COMMANDER-IN-CHIEF: Van der Bellen

FEDERAL CHANCELLOR: Christian Kern

KEY DEFENCE PERSONS

MINISTER OF DEFENCE: Hans Peter Doskozil

CHIEF OF DEFENCE STAFF: Gen. Othmar Commenda

ECONOMY

Led by large service and industrial sectors and a highly-developed agricultural sector, Austria is one of the leading market economies of Europe and ranks as the 25th largest economy in the world. Historically tied to the German economy, Austria has also established closer ties with other European economies after becoming a member of the European Union in 1995 and expanded its growth potential. During the global economic downturn in 2008, the country witnessed economic contraction for a brief period and later made a recovery. The Austrian Parliament has approved an austerity package consisting of a mix of expenditure cuts and new revenues that aims to bring public finances into balance by 2016. Early signs pointed towards a slight improvement in 2016, driven by low interest rates on government debt. Currently, the budget deficit stands at 2.7% of GDP and public debt has reached a post-war high of 84.2% of the GDP.

GDP: $404.3 billion (2015 est.); $400.8 billion (2014 est.)

Real Growth Rate: 0.9% (2015 est.); 0.4% (2014 est.)

Industries: Construction, machinery, vehicles and parts, food, metals, chemicals, lumber and wood, paper and paperboard, communications equipment, tourism

Total Exports: $142.9 billion (2015 est.); $166 billion (2014 est.)

Export Commodities: Machinery and equipment, motor vehicles and parts, paper and paperboard, metal goods, chemicals, iron and steel, textiles, foodstuff

Major Markets: Germany 29.4%, US 6.4%, Italy 6.1%, Switzerland 5.7%, France 4.4%, Slovakia 4.2% (2015)

Total Imports: $140 billion (2015 est.); $164 billion (2014 est.)

Import Commodities: Machinery and equipment, motor vehicles, chemicals, metal goods, oil and oil products; foodstuff

Major Suppliers: Germany 41.5%, Italy 6.3%, Switzerland 6%, Czech Republic 4.2% (2015)

INTERNATIONAL DISPUTES

Austria has no known international dispute, barring its opposition to the Temelin nuclear power plant built by the Czech Republic near the Austrian border.

DEFENCE

ARMY

Personnel: 15,000+

Equipment

Category	Name	In Service
MBT	Leopard 2A4	34
APC, IFV	Ulan	112
	Pandur I	71
	Dingo 2 (protected vehicle)	30+
Artillery	M-109A5OE 155mm	30
Mortar	L16 81 mm	424 (Total)
	M-2/M-30 107 mm	
	M-60	
	M-86 120mm	
ATGW	Bill 1	285

New Procurements/ Upgrades

- ✦ 32 BvS10 armoured vehicles (APCs) will be procured from BAE Systems under a deal inked in 2016.
- ✦ Carl-Gustaf man-portable weapon systems have been ordered from SAAB with deliveries scheduled to take place by 2016-17.
- ✦ 12 Pandur APCs are being upgraded for use on peacekeeping mission in Kosovo.
- ✦ Six new Dingo 2 mine-resistant, ambush-protected (MRAP) vehicles are on order from Krauss-Maffei Wegmann.
- ✦ Delivery of 150 Iveco Light Multirole Vehicles is going on. A total of 150 such vehicles, designated as "Husars" in Austrian service, are on order since 2009.
- ✦ Six Tracker mini unmanned aerial vehicles have been delivered by Airbus Defence in 2015.

AIR FORCE

Personnel: 3,000

Equipment

Category	Name	In Service
Fighter/Attack	Eurofighter Typhoon	15
Transport	C-130K Hercules	3
	Skyvan 3M	2
	PC-6B	10
Helicopter	S-70 Blackhawk	9 (Option for 3 more)
	Bell OH-58B Kiowa	11
	AB-212	23 (being upgraded)
	SA-316 Alouette III	24
Trainer	Saab 105-OE	21
	Pilatus PC-7	12
Air Defence System	Mistral SAM	24
	35mm twin anti-aircraft gun (ZFIAK 85)	25
	20mm Fliegerabwehrkanone 65/68 (anti-aircraft gun)	
	Skyguard 98	30

New Procurements/ Upgrades

- ✦ The entire fleet of Eurofighter Typhoon fighters has been upgraded with enhanced hardware and software including new radio systems, GPS and an identification friend-or-foe (IFF) transponder.
- ✦ The first of three upgraded C-130K Hercules transport aircraft has been delivered to the Air Force by the Marshall Aerospace and

Defence Group in August 2013. The other two C-130Ks are scheduled for delivery by 2015-16.

✦ AugustaWestland has received a contract worth €63 million to upgrade Austria's entire fleet of AB212 helicopters. A total of 23 such helicopters are being upgraded under the deal.

✦ An RAT31DL/M air defence radar system has been ordered from Selex ES. The new radar will reinvigorate Austria's existing inventory of five earlier RAT31S medium-range radars, procured in the 1980s. It will be operational by 2016.

✦ A request has been made to buy three more Sikorsky UH-60 Black Hawk (S-70) utility helicopters from the US. Upgrade of the 9 other Black Hawks is under consideration.

Defence Expenditure

Austria has one of the lowest defence budgets among the European Union nations. Its main defence suppliers include Germany, Italy and few other European countries.

Total defence spending: €2.1 billion (2016)

Defence spending in terms of GDP: 0.57%

Defence Production and R&D

The country's defence industry is primarily producing armoured vehicles, aerospace, ammunition, electronics and naval systems. The indigenous Pandur II armoured fighting vehicle has been sold to countries like Czech Republic and Portugal. Some of the defence equipment have also been sold in Africa, Middle East and few European countries.

Defence Procurement & Future Plans

Defence procurement in Austria has not been very robust as it has been hindered by low defence expenditure. The country has initiated reforms which are directed towards reducing its armed forces personnel strength and diverting the funds to acquire new defence equipment. Following a referendum in January 2013, Austria decided to retain military conscription.

In October 2014, the Defence Ministry announced major restructuring plans for the Austrian military in order to meet the falling defence expenses. Under the plans, a large number of the Army's equipment, including MBTs, armoured vehicles, artillery and other weapons have been retired or sold. The restructuring also involve curtailment of a number of military positions, around 1,400, which are to be gradually cut via retirements and a recruitment freeze. Several military stations have also been shut down.

CONTACT DETAILS

Ministry of Defence

Rossauer Lande 1, A-1090 Vienna

Website: www.bundesheer.at

AZERBAIJAN

(Capital: Baku)

INTRODUCTION

Area: 86,600 sq km

Population: 9,872,765 (July 2016 est.)

Coastline: Landlocked (713km coastline on the Caspian Sea)

Maritime claims: None

KEY POLITICAL PERSONS

PRESIDENT& COMMANDER-IN-CHIEF: Ilham Aliyev

PRIME MINISTER: Artur Rasizade

KEY DEFENCE PERSONS

DEFENCE MINISTER: Gen. Zakir Hasanov

DEFENCE INDUSTRY MINISTER: Jamalov Yavar Talib

CHIEF OF GENERAL STAFF: Col. Gen. Najmeddin Sadykhov

CHIEF OF NAVY: Vice Adm. Sahin Sultanov

CHIEF OF AIR FORCE: Maj. Gen. Ramiz Tahirov

ECONOMY

Oil remains the most prominent product of Azerbaijan's economy. Oil exports through the Baku-Tbilisi-Ceyhan Pipeline, the Baku-Novorossiysk, and the Baku-Supsa pipelines remain the main economic driver. The Southern Gas Corridor between Azerbaijan and Europe is expected to open up another source of revenue from gas exports. Over the last five years' cotton, natural gas and agriculture products have contributed to the country's economic growth. Agriculture is a meaningful component of Azerbaijan's non-oil economy and has significant potential for boosting export revenues for the country.

GDP: $169.4 billion (2015 est.); $167.6 billion (2014 est.)

Real Growth Rate: 1.1% (2015 est.); 2.8% (2014 est.)

Industries: Petroleum and natural gas, petroleum products, oilfield equipment; steel, iron ore, cement, chemicals and petrochemicals, textiles

Total Exports: $15.59 billion (2015 est.); $28.26 billion (2014 est.)

Export Commodities: Oil and gas 90%, machinery, cotton, foodstuff

Major Markets: Italy 26.3%, Germany 13.2%, Indonesia 7%, France 6.8%, Czech Republic 6% (2015)

Total Imports: $9.774 billion (2015 est.); $9.332 billion (2014 est.)

Import Commodities: Machinery and equipment, oil products, foodstuff, metals, chemicals

Major Suppliers: Russia 19.9%, Turkey 16.5%, UK 8.6%, Germany 6.6%, Italy 6.3%, US 4.1% (2015)

INTERNATIONAL DISPUTES

✦ Azerbaijan is still technically at war with its neighbour Armenia, over the disputed Nagorno-Karabakh region. The two countries were at war from 1988 -1994, and diplomatic relations remain strained.

✦ A long-standing Caspian Sea border dispute between Azerbaijan and Turkmenistan continues and neither side have budged to resolve the matter.

✦ Azerbaijan, Kazakhstan, and Russia have ratified the Caspian seabed delimitation treaties based on equidistance while Iran continues to insist on a one-fifth slice of the sea.

DEFENCE

The armed forces have three services – Army, Navy and Air Defence Force.

ARMY

Personnel: 57,135

Equipment

Category	Name	In Service
MBT	T-90	100
	T-72	376
	T-55 (reserved)	35
AIFV/APC	BMP-1	26
	BMP-2	39
	BMP-3	56
	BRM-1K	4
	BTR-60/-70/-80	24
	MT-LB	196
	Matador	15
	Marauder	15
Artillery	D30 122mm	254
	D20 152mm	24
	M46 130mm	36
	2A36 152mm towed	16
	2S19 MSTA-S 152mm SP	18
	2S1 122mm SP	66
	2S7 203mm SP	12
	2S3 152mm SP	16
Mortar	120mm (including 18 SP)	267
	Vena SP	18
MRL	BM-21 122mm	44
	9A52 Smerch 300mm	12
	IMI LYNX 122mm	16
	Grad-LAR	
	160mm	
	306mm	
	TOS-1A Buratino	
SSM	9K79 Tochka (3 launchers)	
ATGW	Kornet	
	Spike	

New Procurements/Upgrades

✦ 36 Firtina 15mm tracked howitzers are on order from Turkey.

✦ An agreement has reportedly been signed with Israel in 2016 to buy the Iron Dome missile defence system.

NAVY

Personnel: 1,786

Equipment

Category	Name	In Service
Submarine (midget)	Triton-2M	
Mine warfare force	Sonya class	1
Light forces	Petya-2 class	1
	Vodnik	1
Patrol boat	S301 class	2+4
Landing ship	Polnochny class	4
Miscellaneous	Hydrographic ship	3

New Procurements/Upgrades

✦ Azerbaijan plans to increase its capacity in the Caspian Sea and to have modern ships by 2020. New base for the Navy as well as divisions will be constructed in the northern and southern directions. The Navy plans also to reinforce its fleet with Kh-35 submarine launched anti-ship missiles.

AIR FORCE

Personnel: 7,944

Equipment

Category	Name	In Service	On Order
Fighter	Mig-29	12	
	MiG-29UB	3	
	Su-24	34	
	Su-25	5	
Trainer	L-29	28	
Helicopter	Mi-17-1B	34	
	Mi-8	11	
	Mi-24G	24	12
	Mi-35M	24	24
SAM	S-300 Favorit		
	BUK-MB		
UAV	Hunter, Aerostar		

New Acquisition/Procurements

✦ A total of 10 AN-178 military transport planes are to be purchased from Ukrainian aircraft manufacturer Antonov in a deal signed in 2015. First two aircraft are expected to be delivered in 2018. Azerbaijan is planning to assemble the planes domestically.

Defence Expenditure

Total defence expenditure: US$ 4.8 billion (2015)

Estimated defence spending in terms of GDP: 3%

Defence production and R & D

Azerbaijan's defence industry has emerged as an autonomous entity with a growing defence production capability. The country's indigenous defence products include 'Istiglal-T' the modernisation version of 14.5-mm calibre 'Istiglal' sniper rifle, 7.62-mm 'Yalguzag' sniper rifle, tank machine gun, HP-7.62 offensive machinegun, universal machinegun 'Zafar', 'Zafar-K' and 'Inam' pistols, drone models, aviation training bombs, ammunition of various calibres, vests, gas masks and others. The ministry is cooperating with the defense sectors of Ukraine, Belarus and Pakistan. Azerbaijan produces 40mm revolver grenade launchers, 107mm and 122mm MLRS systems, Cobra 4×4 vehicles and joint modernization of BTR vehicles in Baku.

Defence Procurement

As part of military modernisation programme Azerbaijan's future defence procurement plans include acquiring two submarines, a mine vessel, and transport ship. The military is interested in T-50 training planes, K-9 self-propelled artillery vehicles and drones.

Russia and Israel are key suppliers of equipment to Azerbaijan. Elbit Systems, Israel's largest non-state-owned defence firm, has been operating a local subsidiary, Elbit Systems of Azerbaijan, since 2009.

Prominent deals in recent years include the Cardom self-propelled recoiling mortar and advanced upgrades to the Russian-made T-72 tanks. From 2012 Baku has purchased at least $1.6 billion worth of UAVs, radars and air defence missiles from Israel Aerospace Industries (IAI) and plans to buy IAI-produced spy satellites. Azerbaijan has acquired Spike anti-tank missiles and targeting systems from Israeli company Rafael.

Ministry of Defence Industries of Azerbaijan (MODIAR) and Roketsan Company of Turkey have signed an agreement for joint production of 107 and 122 mm calibre missiles. The engines for these missiles will be produced by Roketsan and other parts in Azerbaijan.

Azerbaijan is purchasing UMTAS and OMTAS anti-tank missile complexes from Turkey. These missiles can be used both on different vehicles and above ground. OMTAS medium-range anti-tank missile have a range of 4,000 m and UMTAS long-range anti-tank missile 8,000 m. Turkey will also deliver 36 self-propelled howitzers under T-155 Firtina 2011 contract signed between Azerbaijan and Turkey. Three Ukrainian anti-tank guided missiles have been delivered to the Azerbaijan military.

Azeri Arms Parade

CONTACT DETAILS

Ministry of Defence
40, Metbuat Ave. Baku,
AZ 1141, Azerbaijan Republic
Tel.:/Fax: + (994 12) 539 24 53
www.mdi.gov.az
E-mail: info@mdi.gov.az
Ministry of Defence Industry
MEGA Company production
Tel.:+(99412) 539 24 53

BAHRAIN
(Capital: Manama)

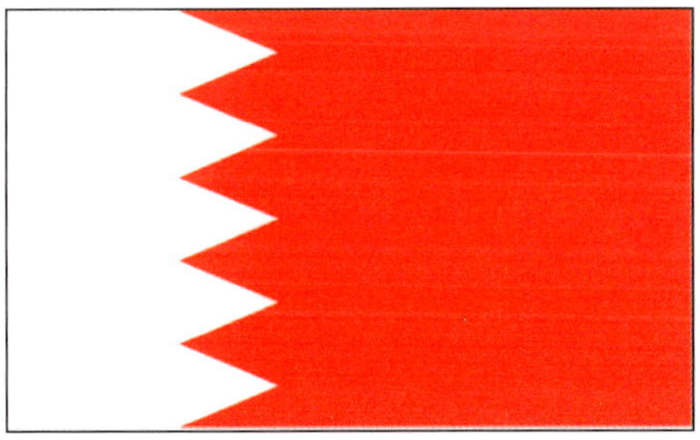

INTRODUCTION

Area: 760 sq km

Population: 1,378,904 (July 2016 est.)

Coastline: 161 km

Maritime claims: Territorial sea: 12 nm
Contiguous zone: 24 nm
Continental shelf: extending to boundaries to be determined

KEY POLITICAL PERSONS

HEAD OF STATE & SUPREME COMMANDER: King Hamad bin Isa Al-Khalifa

PRIME MINISTER: Khalifa bin Salman Al Khalifa

CROWN PRINCE, DEPUTY SUPREME COMMANDER & FIRST DEPUTY PM: Salman bin Hamad Al Khalifa

KEY DEFENCE PERSONS

DEFENCE MINISTER: Maj. Gen. Yusuf bin Ahmed bin Hussain Al-Jalahma

COMMANDER-IN-CHIEF OF DEFENCE FORCE: Field Marshal Shaikh Khalifa bin Ahmed Al Khalifa

CHIEF OF STAFF OF DEFENCE FORCE: Lt. Gen. Dhiab bin Saqr Al-Nuaimi

ECONOMY

Bahrain is one of the most diversified economies in the Persian Gulf. Bahrain's economy continues to depend heavily on oil. As part of its diversification plans, Bahrain implemented a Free Trade Agreement (FTA) with the US in August 2006, the first FTA between the US and a Gulf state. Petroleum production and refining account for more than 60% of Bahrain's export receipts, 70% of government revenues, and 11% of GDP (exclusive of allied industries). Other major economic activities are production of aluminium - Bahrain's second biggest export after oil - finance, and construction. Bahrain competes with Malaysia as a worldwide centre for Islamic banking. Unemployment, especially among the young, is a long-term economic problem Bahrain struggles to address. In 2009, to help lower unemployment among Bahraini nationals, Bahrain reduced sponsorship for expatriate workers, increasing the costs of employing foreign labour. The global financial crisis caused funding for many non-oil projects to dry up and resulted in slower economic growth for Bahrain. Other challenges facing Bahrain include the slow growth of government debt as a result of a large subsidy programme, the financing of large government projects, and debt restructuring, such as the bailout of state-owned Gulf Air.

GDP: $64.8 billion (2015 est.), $62.82 billion (2014 est.)

Real Growth Rate: 3.2% (2015 est.), 4.5% (2014 est.)

Industries: Petroleum processing and refining, aluminium smelting, iron pelletization, fertilizers, Islamic and offshore banking, insurance, ship repairing, tourism

Oil - proved reserves: 100 million bbl (1 January 2016 es)

Natural gas - proved reserves: 92.03 billion cu m (1 January 2016 est.)

Total Exports: $14.2 billion (2015 est.), $20.75 billion (2014 est.)

Export Commodities: Petroleum and petroleum products, aluminium, textiles

Major Markets: Saudi Arabia 3.6%, UAE 2.4%, US 2.2% (2015)

Total Imports: $8.848 billion (2015 est.), $13.32 billion (2014 est.)

Import Commodities: Crude oil, machinery, chemicals

Major Suppliers: Saudi Arabia 29.1%, US 9.5%, China 7.6%, Japan 6.6%, Australia 5.1%, India 4.9% (2015)

INTERNATIONAL DISPUTES

Bahrain has no known international disputes.

DEFENCE

ARMY

Personnel: 8,500

Equipment

Category	Name	In Service
MBT	M60 A3 Patton	180
APC/MICV /IFV	YPR-765	25
	M-113 A2	110
	AT-105 Saxon	10
	Panhard M3	110
	AML-90	22
Artillery	M110 203mm self-propelled	13
	M-198 A1 155mm	18
	L-118 105mm	8
Mortar	81 mm	180
	120 mm	
Anti Tank GW	BGM-71C TOW	2000
	Javelin	60
RCL Gun	120 mm Mobat	6
SAM	RBS 70	160
	FIM-92A Stinger	400
	Crotale	
	MIM-23 Hawk	8
MRL	M270 MLRS	9
	SR5	

New Procurements/ Upgrades

No major procurement plans are under consideration at present.

NAVY

Personnel: 1,770 (including Coast Guard)

Equipment

Category	Name	In Service
Frigate	Sabha Class Frigate	1
Corvette	Al Manama class	2
Missile Patrol Craft	Al Fateh Class	4
	Al Riffa class	2
	Al Jarim Class	2
Landing craft	42mt landing craft	4
SSM	Harpoon	
SAM	SM-1MR	

RBNS Sabha (FFG 90) participates in Exercise Arabian Shark. *Image Credit: US Navy*

Naval Aviation

Category	Name	In Service
Aircraft	SA 365F DAUPHINs	2
	BO-105s	2

New Procurements/ Upgrades

No major procurement plans are under consideration at present.

AIR FORCE

Personnel: 1,500

Equipment

Category	Name	In Service
Fighter	F-16C	15
	F-5 E	8
Helicopter	AH-1E/F Cobra	22
	AB-212	18
	UH-60M	8
Trainer	F-5F	4
	F-16D	4
	Hawk-129	6
	T67 M260	3
	TAH-1P	8
	Bo 105	4

New Procurements/ Upgrades

No major procurement plans are under consideration at present.

A BO-105 helicopter of Bahrain flies inland near a causeway in the aftermath of Operation Desert Storm. *Image Credit: US DoD*

Defence Expenditure

Total defence spending: $1.365 billion (2014)

Defence spending in terms of GDP: 4.2% (2014), 4.1% (2013)

Bahrain has doubled its defence spending. Bahraini has ordered defence imports worth USD 80 million in 2016. United States has been the key military supplier to Manama whose military spending is fuelled by the threat from neighbouring Iran. Among other regional and global organisations, Bahrain is a member of the United Nations, International Monetary Fund, World Bank, and World Trade Organisation.

Defence Production and R&D

Bahrain has not stretched its defence production facilities and it produces very less equipments. The nation had unveiled its first military armoured vehicle 'Faisal' in 2009. Bahrain Defence Force's technical maintenance unit designed and assembled the vehicle, with the aid of a European company with experience in this field.

The Bahrain Ship Repairing & Engineering Company (BASREC) and the Arab Shipbuilding & Repair Yard (ASRY) have both received contracts from the US Navy to repair and overhaul warships in recent years.

Defence Procurement

The army's modernisation and expansion plans have been helped by the willingness of NATO states to cascade surplus equipment. The US has been the dominant supplier of military equipment to Bahrain. The nation also receives a large section of military equipment aid both from Iraq and Iran. On May 2012, the United States has said it is partially resuming sales of military supplies to key Gulf Arab ally Bahrain but maintaining a freeze on certain wire-guided missiles and vehicles. The equipment and services were released for the purpose of helping Bahrain maintain its external defence capabilities. On April 2012, a Congressional Research Service report said Washington had not discontinued arms sales and anti-terrorism aid to Bahrain, but some of the supplies remained on hold or were at reduced levels following the unrest.

The Government of Bahrain plays a key role in the Gulf's security architecture and is an important member of the US-led anti-ISIL coalition. US assistance enables Bahrain to continue to obtain the equipment and training it needs to provide for its own defence and to operate alongside U.S. air and naval forces.

CONTACT DETAILS

Defence Forces HQ

C/o Ministry of Defence

P.O Box 245, HQ Bahrain

West Rifa's

Bahrain

Tel: +973-665599

BANGLADESH
(Capital: Dhaka)

INTRODUCTION

Area: 148,460 sq km

Population: 156,186,882 (July 2016 est.)

Coastline: 580 km

Maritime claims: Territorial sea: 12 nm

Contiguous Zone: 18 nm

Exclusive Economic Zone: 200 nm

Continental shelf: up to the outer limits of the continental margin

KEY POLITICAL PERSONS

PRESIDENT: Abdul Hamid

PRIME MINISTER & DEFENCE MINISTER: Sheikh Hasina Wajed

KEY DEFENCE PERSONS

CHIEF OF ARMY STAFF: Gen. Abu Belal Muhammad Shafiul Huq

CHIEF OF NAVAL STAFF: Adm. Nizamuddin Ahmed

CHIEF OF AIR STAFF: Air Marshal Abu Esrar

ECONOMY

The developing market economy of Bangladesh has registered an annual GDP growth of 6-7% in the past few years. While the country's service sector contributes majority share of the GDP, the agriculture sector, employing around 45% of the population, makes a contribution of around 20%. The country mainly produces rice. Poor infrastructure, over population, corruption coupled with political instability has often hindered the country's growth trajectory. Steady garment export growth combined with remittances from overseas Bangladeshis which totaled about $15 billion and 8% of GDP in 2015 are the largest contributors to Bangladesh's current account surplus and rising foreign exchange holdings.

GDP: $577 billion (2015 est.); $536.5 billion (2014 est.)

Real Growth Rate: 6.4% (2015 est.); 6.3% (2014 est.)

Industries: Jute, cotton, garments, paper, leather, fertilizer, iron and steel, cement, petroleum products, tobacco, pharmaceuticals, ceramics, tea, salt, sugar, edible oils, soap and detergent, fabricated metal products, electricity, natural gas

Natural Gas - Proved Reserves: 264.6 billion cu m (1 January 2014 est.)

Crude Oil - Proved Reserves: 28 million bbl (1 January 2015 est.)

Total Exports: $31.74 billion (2015 est.); $29.92 billion (2014 est.)

Export Commodities: Garments, knitwear, agricultural products, frozen food (fish and seafood), jute and jute goods, leather

Major Markets: US 13.9%, Germany 12.9%, UK 8.9%, France 5%, Spain 4.7% (2015)

Total Imports: $37.63 billion (2015 est.); $37.41 billion (2014 est.)

Import Commodities: Cotton, machinery and equipment, chemicals, iron and steel, foodstuff

Major Suppliers: China 22.4%, India 14.1%, Singapore 5.2% (2015)

INTERNATIONAL DISPUTES

✦ On 7 June 2015, India and Bangladesh signed a historic agreement to simplify their 4,000-km land border and clarify the identities of 52,000 living in enclaves, over four decades after the two neighbours first tried to untangle complex territorial rights set down in 1713. Under the deal, signed in the presence of Prime Minister Narendra Modi and his Bangladeshi counterpart Sheikh Hasina in Dhaka, the two countries have decided to swap 200 tiny enclaves, most of them close to the official border.

✦ On 14 May 2012, the International Tribunal for the Law of the Sea (ITLOS) delivered its judgment in the dispute concerning delimitation of the maritime boundary between Bangladesh and Myanmar. The tribunal decided on an adjusted equidistance line as the boundary between the two countries.

DEFENCE

ARMY

Personnel: 150,000+

Equipment

Category	Name	In Service
MBT	T-59/ T-69	350
	T-62 (light tank)	40
APC	MT-LB	19
	BTR-80, BTR-70	100+
	FAHD	60
	RN-64	
Artillery	M56A1/ L10A1 105mm	150+
	Type 83	110
	D30 122mm	14
	Type 59-1 130mm	62
Mortar	29A1 81mm	11
	Type-57/87/M32 82mm	477
	Type 52 120mm	23
RCL	M40 106mm	246
SAM	HN-5A Manpad	8
	QW-2s	43
AA Gun	Type 55/74 37mm	114
	Type 4 59 57mm	34

Army Aviation

Category	Name	In Service
Aircraft	Cessna 152	4

Army aviation continued...

Category	Name	In Service
Aircraft	Cessna 208B	1
	CASA CN 235-300	2 (on order)
Helicopter	Eurocopter AS365 Dauphin	2
	Bell 206	

New Procurements/Upgrades

✦ A total of 44 MBT-2000 tanks and three armoured recovery vehicles are being procured from China.

✦ Plans are on upgrade the entire fleet of T-59 main battle tanks with new engine, improved guns and other systems and transform them as modern combat platform.

✦ Two CASA CN 235-300 transport aircraft are on order from Airbus.

✦ One C295W medium transport aircraft has been ordered from Airbus in 2016 with delivery scheduled for late 2017.

NAVY

Personnel: 15,000+

Equipment

Category	Name	In Service
Frigate	Bangabandhu (Ulsan class)	1
	Umar Farooq (Salisbury class)	1
	Osman (modified Type 053H)	1
	Type 41 Leopard class	2
Corvette	Castle class	2
	Durjoy class	2
Light Force	Kapatakhaya class OPV	5
	Padma class	5
	Madhumati Patrol boat	1
	Nirbhoy	1
	Salam/Barkat class	2
	Karanphuli class	2
	Type 021 class missile FAC	4
	Durbar class	5
	Shaheed Daulat class gun FAC	4
	Titas class	4
Mine Warfare Force	Shapla class minesweepers	
	Sagar minesweepers	1
Amphibious Force	Shah Poran class LCU	2
	LCT	3
	LCVP	3

Naval Aviation

Category	Name	In Service
Aircraft	Dornier 228 NG	2
Helicopter	AW-109 Power	2
	Z-9C	5

New Procurements/Upgrades

✦ Two Type 035G class diesel-electric attack submarines have been officially handed over to the Bangladesh Navy in November 2016 at China's Liaoning province. China had reportedly received $203 million in 2013 to provide the submarines to Bangladesh which has been building a permanent submarine base and plans to induct the two platforms in the Navy by 2017.

Chinese-built Type 035G 'Ming' class submarine *An internet photo*

✦ Two new C13B advanced corvette class warships, named BNS Shadhinota and BNS Prottoy, have been received from China in 2015. Each warship with a displacement of 1,300 tonnes and a maximum speed of 25 knots, is 90 metres long and measures 11 metres across the beam. The warships are armed with 76 mm and 30 mm naval guns as well as ship-to-ship and ship-to-air missiles.

✦ Two Hamilton-class cutters have been received from the US Coast Guard.

✦ Construction of eight new X12 high-speed patrol boats has started at the country's DEW shipyard. The boats are being built as part of a technology transfer arrangement with Indonesia's PT Lundin shipyard.

✦ Five Z-9C naval helicopters are on order from China for anti-submarine warfare and search and rescue missions.

Patrol boat BNS Nirbhoy *A Navy photo*

AIR FORCE

Personnel: 14,000

Equipment

Category	Name	In Service
Fighter Aircraft	F-7M	12
	F-7BG	15
	MiG-29	8
	L-39 ZA Albatross	8
Transport	C-130B, C-130E	4+ 4 (on order)
	An-32	3
Helicopter	Mi-17	20
	Bell 212	11
	Bell 206L	2
	AW139	2
	Mi-171Sh, Mi-171E	6+1 (on order)

Air Force continued...

Category	Name	In Service
Trainer	T-37B	12
	PT-6	12 (On order)
	Mushahk	10
	K-8W	4

New Procurements/ Upgrades

✦ Bangladesh has inked a major defence deal worth $1 billion with Russia in January 2013 to buy fighter jets, helicopters, armoured vehicles, anti-tank missiles, automatic grenade launchers, radar equipment among others. Russia will issue the $1 billion loan to Bangladesh that will be used to buy Russian weapons and military hardware.

✦ A deal with Russian arms exporter Rosoboronexport has been signed to procure 24 Yak-130 jet trainers under the larger defence deal of $1 billion. Six of the aircraft have been delivered in 2015.

✦ Four Chinese-made K-8W fighter trainers have been inducted in service in 2014. Plans are on to buy five more of the aircraft.

✦ AgustaWestland has in December 2015 delivered two AW139 multirole medium helicopters under a deal signed in 2014.

✦ Plans are on to acquire F-7BGI fighter jets from China.

✦ A request has been made to the US Government to acquire four C-130E transport aircraft through the Foreign Military Sale programme. The proposed sale would cover the regeneration, overhaul, modification and logistics support of the second-hand Hercules planes.

✦ Six Mi-171Sh multi-role (combat-transport) helicopters and one Mi-171E medium-lift are being delivered by Russia's Rosoboronexport.

✦ The FM-90 anti-aircraft short range missile systems, procured from China, have been inducted.

Defence Expenditure

Defence budget: Tk 221.3 billion (US$2.8 billion - 2016-17); Tk 183.8 billion (US$2.3 billion - 2015-16)

Defence spending in terms of GDP: 1.4% (2016-17)

Defence Production and R&D

Bangladesh has produced semi-automatic 7.62 mm SKS rifles and associated ammunition at its Ghazipur factory. With the assistance of China, it has produced variations of the AK series assault rifle. Bangladesh Ordnance Factory (BOF) had obtained license and technology for the local production of the G-3 rifle in 2002 as part of expanding operations. The country is now trying to modernise its factory to build small arms and ammunition. Few shipyards in the country are capable of repairing and refurbishing naval vessels. The Government-owned Khulna Shipyard has built the Padma-class OPVs and also plans to build new corvettes for the Navy.

Defence Procurement

Bangladesh has acquired defence hardware worth Taka 15,000 crore ($2 billion) since 2010 as part of its initiatives to modernise its armed forces. The country has procured 2,600 armoured vehicles, 44 third-generation tanks and 15 APCs for its Army; 16 ships, including five maritime patrol vessels and two corvettes, two maritime helicopters and missiles for its Navy; and short range air defence systems and 16 fourth-generation fighter planes for its Air Force among other platforms and equipment. The Government has also decided to acquire submarines to boost its maritime power. The acquisitions are being made under a 10-year military modernisation drive.

CONTACT DETAILS

Ministry of Defence
Ganabhaban Complex, Sher-E-Bangla Nagar, Dhaka-1207
Ph: 9113006, 9113319
Website: www.mod.gov.bd
Email: info@mod.gov.bd, modgob@bttb.net

BELARUS
(Capital: Minsk)

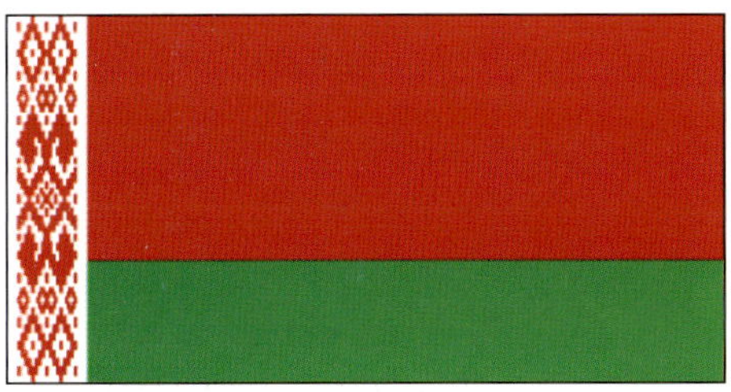

INTRODUCTION

Area: 207,600 sq km
Population: 9,570,376 (July 2016 est.)
Coastline: 0 km (landlocked)
Maritime claims: None

KEY POLITICAL PERSONS

PRESIDENT: Aleksandr Lukashenko

PRIME MINISTER: Andrei Kobyakov

KEY DEFENCE PERSONS

DEFENCE MINISTER: Lt. Gen. Andrei Ravkov

CHIEF OF THE GENERAL STAFF: Maj. Gen. Oleg Belokonev

ECONOMY

Belarus had been one of the more prosperous states as part of the former Soviet Union. After independence, the economy suffered a decline. President Lukashenko has opposed the privatisation of state enterprises, and foreign investors have largely stayed away. Around three quarters of the economy is under government control. Over 80 per cent of agricultural land is state-owned. The country has received several loans from the IMF and Russia since the global financial crisis of 2008. According to the IMF, real GDP growth was 2.1 per cent in 2013. Belarus is dependent on Russia to meet its energy needs. Peat, the country's most valuable mineral resource, is used for fuel and fertilizer and in the chemical industry. Belarus also has deposits of clay, sand, chalk, dolomite, phosphorene, and rock and potassium salt. Forests cover about a third of the land, and lumbering is an important sector. Its economy faces challenges including high external debt servicing payments, a growing trade deficit, stagnant economic growth, and low foreign reserves. In December 2013, Russia announced a new loan for Belarus of up to $2 billion for 2014. Notwithstanding foreign assistance, the Belarusian economy continued to struggle under the weight of high external debt servicing payments and trade deficit. In mid-December 2014, structural economic shortcomings were aggravated by the devaluation of the Russian ruble and triggered a near 40% devaluation of the Belarusian ruble. Belarus entered 2015 with stagnant economic growth and reduced hard currency reserves, with under one month of import cover.

GDP: $167.7 billion (2015 est.); $174.5 billion (2014 est.)

Real Growth Rate: -3.9% (2015 est.); 1.6% (2014 est.)

Industries: Metal-cutting machine tools, tractors, trucks, earthmovers, motorcycles, televisions, synthetic fibres, fertilizer, textiles, radios, refrigerators

Total Exports: $26.19 billion (2015 est.); $35.42 billion (2014 est.)

Export Commodities: Machinery and equipment, mineral products, chemicals, metals, textiles, foodstuff

Major Markets: Russia 39%, UK 11.2%, Ukraine 9.5%, Netherlands 4.3%, Germany 4.1% (2015)

Total Imports: $28.33 billion (2015 est.); $38.06 billion (2014 est.)

Import Commodities: Mineral products, machinery and equipment, chemicals, foodstuffs, metals

Major Suppliers: Russia 56.6%, China 7.9%, Germany 4.6% (2015)

INTERNATIONAL DISPUTES

Belarus has no known internal disputes. However, Poland as a member state that forms part of the EU's external border, has implemented strict Schengen border rules to restrict illegal immigration and trade along its border with Belarus.

DEFENCE

The armed forces of Belarus consist of the Ground Forces and the Air and Air Defence Forces which are under the command of the Ministry of Defence. Being a landlocked country, Belarus has no navy. Belarus has conducted military reforms which have reshaped its armed forces as a relatively effective force. Belarusian army is equipped with Soviet-era arms and holds regular combat readiness exercises.

ARMY

Personnel: 10,188

Equipment

Category	Name	In Service
MBT	T-72	1331
	T-80	61
AIV/APC	BMP-2	1162
	BMP-1	101
	BMD-1	119
	BRM-1	161
	BTR-70	355
	BTR-80	191
	BTR-D	9
	MT-LB	61
Artillery	2A36 152mm	50
	2A65 152mm towed	136
	2S1 122mm	232
	2S19 152mm	13
	2S3 152mm	162
	2S5 152mm	120
	2S7 203mm	24
	D-20 152mm	58
	2S9 120mm SP	54
	D-30 130mm	46
Mortar	120mm 2S12	65
SSM	OTR-21 Tochka	36
MRL	122mm BM-21	210
	300mm 9A52	40
	220mm 9P140	75
ATGM	9P148	275
	9P149	80
SAM	S-300	
	Tor-M2E	

New Procurements/Upgrades

✦ Induction of the 301mm Polonez rocket launch system is being planned. The locally manufactured weapon is believed to be a Belorussian version of the Chinese A200 MRL.

AIR FORCE & AIR DEFENCE FORCE

Personnel: 14,041

Equipment

Category	Name	In Service
Fighter	Yak-130 Mitten	4
	Su-27	20
	MiG-29	38
	Su-25	30
Transport	An-26	3
	IL-76	3
Trainer	L-39	10
Helicopter	Mi-8	38
	Mi-24K	4

Air Force continued...

Category	Name	In Service
Helicopter	Mi-24	17
	Mi-24R	4

New Procurements/Upgrades

✦ Russia has agreed to supply an unspecified number of Su-30 fighter aircraft to the Belarusian Air Force under an initial agreement signed in February 2016. The new fighters will replace the MiG-29 fleet of aircraft.

✦ The air force has received five Yak-130 'Mitten' advanced jet trainer/light attack aircraft ordered in 2012. It has also ordered an additional four Yak-130 jets from Russia.

✦ An order has been placed for 12 Mil Mi-8MTV-5 Hip helicopters. The Air Force is expected to take delivery of six Mi-8MTV helicopters from Russia by the end of 2016 or early 2017.

✦ A contract has been signed in 2016 to upgrade Mi-24 attack helicopters.

✦ A fourth battalion of S-300 PS air defence system has been delivered by Russia in 2016.

✦ Belarus and Russia plan to jointly develop a new short-range air defence system to replace the existing Strela-10M systems.

Defence Expenditure

Estimated defence spending in terms of GDP: 1.3%

Defence Production and R&D

The State Military Industrial Committee coordinates activities of more than 200 entities, which are licensed to develop, produce, improve and maintain military products. Eighty of them are licensed to conduct foreign trade activities in relation to products for military purposes. Currently the Committee includes 24 main defence companies.

Defence Procurement

Belarus and Russia plan to jointly develop a new short-range air defence system. The new system is aimed at replacing the existing Strela-10M systems which is currently in service with the Belarusian armed forces. The work is being carried out in the framework of the Russia-Belarus Union State.

Belarus and Turkmenistan have signed an agreement to jointly produce unmanned aerial vehicles (UAVs). It has also signed a defence cooperation agreement with Indonesia which will lead to research & development and joint production of defence hardware.

CONTACT DETAILS

Ministry of Defence

1, Kommunisticheskaya St., Minsk, 220034, Belarus

Tel/Fax: +375-17-2971581

E-mail: modmail@mod.mil.by

Website: http://www.mil.by/en/

BELGIUM

(Capital: Brussels)

INTRODUCTION

Area: 30,528 sq km

Population: 11,409,077 (July 2016 est.)

Coastline: 66.5 km

Maritime claims: Territorial sea: 12 nm

Contiguous zone: 24 nm

Exclusive Economic Zone: Geographic coordinates define outer limit

Continental shelf: Median line with neighbours

KEY POLITICAL PERSONS

CHIEF OF STATE & COMMANDER-IN-CHIEF: King Philippe

PRIME MINISTER: Charles Michel

KEY DEFENCE PERSONS

MINISTER OF DEFENCE: Steven Vandeput

CHIEF OF DEFENCE STAFF: Gen. Marc Compernol

LAND COMPONENT COMMANDER: Maj. Gen. Jean-Paul Deconinck

MARITIME COMPONENT COMMANDER: Rear Adm. Wim Robberecht

AIR COMPONENT COMMANDER: Maj. Gen. Frederik Vansina

ECONOMY

The modern, well-developed economy of Belgium has benefited from a robust infrastructure, mainly transport network and diversified industrial base. The country, though lacks in natural resources, imports raw material and semi-finished goods, re-processes them and exports to other countries. Belgian economy is well integrated with other European economies. The service sector of the country contributes a majority share of its GDP followed by the industrial sector.

GDP: $494.1 billion (2015 est.); $487.4 billion (2014 est.)

Real Growth Rate: 1.4% (2015 est.); 1.3% (2014 est.)

Industries: Engineering and metal products, motor vehicle assembly, transportation equipment, scientific instruments, processed food and beverages, chemicals, base metals, textiles, glass, petroleum

Total Export: $259.9 billion (2015 est.); $320.2 billion (2014 est.)

Export Commodities: Machinery and equipment, chemicals, finished diamonds, metals and metal products, foodstuff

Major Markets: Germany 16.9%, France 15.5%, Netherlands 11.4%, UK 8.8%, US 6%, Italy 5% (2015)

Total Import: $259.6 billion (2015 est.); $325.2 billion (2014 est.)

Import Commodities: Raw materials, machinery and equipment, chemicals, raw diamonds, pharmaceuticals, foodstuffs, transportation equipment, oil product

Major Suppliers: Netherlands 16.7%, Germany 12.7%, France 9.6%, US 8.7%, UK 5.1%, Ireland 4.7%, China 4.3% (2015)

INTERNATIONAL DISPUTES

Belgium has no known international disputes. The country is one of the founding members of the North Atlantic Treaty Organisation (NATO) which is headquartered at its capital Brussels.

DEFENCE

ARMY

Personnel: 12,000+

Equipment

Category	Name	In Service
APC	Mowag Piranha III	242
	Pandur	41
	Dingo 2	220
Armoured Car	IVECO Light Multirole Vehicle	440
Artillery	LG1 Mark II 105 mm towed howitzer	14
Mortar	120mm 120 RT	48
	81 mm M1	39
	60 mm M19	50+
ATGW	Milan	

Army continued...

Category	Name	In Service
SAM	Mistral	45
Helicopter	NH-90 TTH	4

New Procurements/Upgrades

✦ All Pandur APCs are to be upgraded by RUAG Defence with improved ballistic protection kits.

NAVY

Personnel: 1,400+

Equipment

Category	Name	In Service
Frigate	F930 Leopold I	1
	F931 Louise-Marie	1
Patrol vessel	P902 Liberation (river patrol vessel)	1
Mine sweeper	Tripartite class	6
Auxiliary ship	A962 Belgica (oceanographic research vessel)	1
	960 Godetia (command & logistic support ship)	1
	A999 Barbara (hovercraft)	1
Helicopter	SA316B Alouette III	3
	NH-90 NFH	4

New Procurements/Upgrades

✦ Acquisition of two new frigates is being planned to replace the Leopold I class of ships.

✦ Acquisition of six new minesweepers is being considered to replace the existing ones.

AIR FORCE

Personnel: 5,000+

Equipment

Category	Name	In Service
Fighter/ Attack	F-16A/B Falcon	59
Transport/ Surveillance	Falcon 20E-5	2
	Falcon 900B	1
	Embraer ERJ 135LR	2
	Embraer ERJ 145LR	2
	C-130H	11
	Airbus A400M	
Helicopter	Agusta AW109	20
	Sea King Mk.48	3
UAV	RQ-5 Hunter	12

New Procurements/Upgrades

✦ 34 new combat aircraft are to be acquired to replace the aging fleet of F-16 fighters. The Defence Ministry has issued an RFI to five foreign government agencies -- two American and three European – in this regard. The five contending fighters include F-35 Lightning II, F/A-18F Super Hornet, Dassault Rafale, Saab JAS-39 Gripen, and Eurofighter Typhoon.

✦ Seven Airbus A400M heavy transport aircraft are on order to replace the C-130H planes.

✦ The Government has requested for AIM-9X-2 Sidewinder air-to-air missiles from the US at an estimated cost of $68 million under foreign military sales deal.

Defence Expenditure

Defence spending in terms of GDP: 0.7%

Major suppliers of defence equipment to Belgium include Netherlands, Germany and other European countries. Some Middle East and Latin American countries have bought defence equipment from Belgium.

Defence Production and R&D

Belgium's defence industry is engaged in the production of small arms, ammunition, and military electronics equipment.

Defence Procurement

A Governmental Committee under a Strategic Defence Plan approved in December 2015 has proposed to maintain a military manpower of 25,000 active personnel by 2030. The plan also calls for 9.2 billion euros to be spent on major equipment investment programmes for the Army, Navy and Air Force. These would include 34 new combat aircraft, two frigates, six mine-hunters and six drones. All new acquisitions are to be made by 2030.

Following the signing of a Benelux air policing agreement on 4 March, 2015, Belgium, Luxembourg, and the Netherlands have decided to jointly defend the airspace of all three countries from military threats and also against renegade aircraft. Belgium along with Spain have also taken over NATO's Baltic Air Policing mission in Sept 2016 from Hungary and Germany for a four-month period.

CONTACT DETAILS

Ministry of Defence

Quartier Reine Elisabeth

Block 4, Rue d'Evere 11140 Evere

Brussels

Website: www.mil.be/def

BOLIVIA
(Capital: La Paz)

INTRODUCTION

Area: 1,098,581 sq km

Population: 10,969,649 (July 2016 est.)

Coastline: 0 km (landlocked)

Maritime claims: None

KEY POLITICAL PERSONS

PRESIDENT: Juan Evo Morales Ayma

VICE PRESIDENT: Alvaro Garcia Linera

KEY DEFENCE PERSONS

DEFENCE MINISTER: Reymi Luis Ferreira Justiniano

CHIEF OF ARMED FORCES: Gen. Juan Gonzalo Duran Flores

CHIEF OF GENERAL STAFF: Gen. Gina Reque Teran

COMMANDER GENERAL OF ARMY: Gen. Luis Orlando Arinez Bazzan

COMMANDER GENERAL OF NAVY: Vice Adm. Jose Manuel Puente Guarachi

COMMANDER GENERAL OF AIR FORCE: Gen. Celier Aparicio

ECONOMY

Bolivia is a resource-rich country with strong growth attributed to captive markets for natural gas exports – to Brazil and Argentina. Gas accounts for roughly 50% of Bolivia's total exports and funded more than half of its 2015 budget. In 2005, the government passed a controversial hydrocarbons law that imposed significantly higher royalties and required foreign firms then operating under risk-sharing contracts to surrender all production to the state energy company in exchange for a predetermined service fee. After higher prices for mining and hydrocarbons exports produced a fiscal surplus in 2008, the global recession in 2009 slowed growth. President Evo Morales passed an investment law and promised not to nationalise additional industries in an effort to improve Bolivia's investment climate. The global decline in oil prices in late 2014 exerted downward pressure on the price Bolivia receives for exported gas and may result in lower GDP growth rates and losses in government revenue.

GDP: $74.39 billion (2015 est.), $70.98 billion (2014 est.)

Real Growth Rate: 4.8% (2015 est.), 5.5% (2014 est.)

Industries: Mining, smelting, petroleum, food and beverages, tobacco, handicrafts, clothing, jewellery

Oil - proved reserves: 209.8 million bbl (1 January 2016 est.)

Natural gas - proved reserves: 295.9 billion cu m (1 January 2016 est.)

Total Exports: $8.197 billion (2015 est.), $12.15 billion (2014 est.)

Export Commodities: Natural gas, soybeans and soy products, crude petroleum, zinc ore, tin

Major Markets: Brazil 28.1%, Argentina 16.9%, US 12.1%, Colombia 6.3%, China 5.3%, Japan 4.7%, South Korea 4.3% (2015)

Total Imports: $9.069 billion (2015 est.), $9.935 billion (2014 est.)

Import Commodities: Petroleum products, plastics, paper, aircraft and aircraft parts, prepared foods, automobiles, insecticides

Major Suppliers: China 17.9%, Brazil 16.5%, Argentina 11.8%, US 10.6%, Peru 6.2%, Japan 5.2%, Chile 4.6% (2015)

INTERNATIONAL DISPUTES

Chile and Peru rebuff Bolivia's reactivated claim to restore the Atacama corridor, ceded to Chile in 1884, but Chile offers instead unrestricted but not sovereign maritime access through Chile for Bolivian natural gas.

DEFENCE

ARMY

Personnel: 23,000 (1900 conscripts)

Equipment

Category	Name	In Service
MBT	SK-105 Kurassiers	34
APC/MICV /IFV	EE-9 Cascavel	24
	M113	60
	EE-11 Urutu	24
	M9 Half-track	23
	Mowag Roland	20
	V100 Commandos	15
Artillery	M1 75mm pack Howitzers	36
	M101 105mm	6
	122mm Type 54	18
Mortar	M29 81mm	250
	M120 120mm	
	M30 107mm	
RCL	90mm	50
	160mm	30
AA Gun	2x37mm Type 65	16
	2x20mm Oerlikon K20	80
SAM	HN-5 MANPAD	50

Army Aviation

Category	Name	In Service
Aircraft	Beechcraft King Air C90	1
	CASA C-212 Aviocar	2
	Cessna 206 Stationair	4
	Cessna 421B Golden Eagle	1
	Emb810 Seneca	1

New Procurements/ Upgrades

No major procurement plans are under consideration at present.

NAVY

Personnel: 4,500 (including 1,000 Marines)

Equipment

Category	Name	In Service
Patrol Craft	PR-51 Class	1
	Capitan Bretel	6
	Lake	4
	Boston Whaler	32
Other Ship	Piranha assault boats Mk.1	8
	Tankers	2
	Transport vessel	1
Aircraft	Cessna 206 Stationair	1
	Cessna 402	1

New Procurements/ Upgrades

No major procurement plans are under consideration at present.

AIR FORCE

Personnel: 4,000

Equipment

Category	Name	In Service
Aircraft	K-8 Attack/Trainer	6
	Canadair CT-33 Attack/Trainer	14
	Pilatus PC-7 Attack/Trainer	3
	Foxtrot 4	2
	Robinson R44 Trainer	6
	Learjet 25	2
	Z 242L	2
Helicopter	AS350 Utility	2
	Alouette SA 316	1
	AS332 Cougar	1
	EC145	3
	UH-1H	14
Transport	C-130B	3
	Jetstream 31	2
	King Air 90/200/350	5
	Xian MA60	(2 on order)
UAV	RemoEye-006	1

New Procurements/ Upgrades

No major procurement plans are under consideration at present.

Training aircraft Zlin 242 L of the Bolivian Air Force (FAB). *Image Credit: Bolivian Ministry of Defence.*

Defence Expenditure

Total defence spending: $377 million (est.)

Defence spending in terms of GDP: 1.5% (est.)

Foreign suppliers: Venezuela, Brazil, United States, Austria, People's Republic of China, South Korea, Russia, Peru, Argentina & Mexico.

Defence Production and R&D

The country has plans to manufacture small-arms ammunition, explosives, light vehicles and light vessels. Several small local, government owned shipyards have been constructed to assembly, design and produce small, aluminium river boats. A boat building yard is constructed at Puntiti, Cochabamba. 'Kojak' a buggy-type light attack vehicle has been made by Bolivian army's logistics battalion.

One scout version is equipped with a MAG and an Instalanza or RPG-7V. There is an anti-tank guided weapon version armed with two HJ-8A missiles. Construction has been limited to about 10 of the scout version and at least three of the anti-tank version, with a total requirement of up to 60 vehicles.

Aviotec Ltd. is the only company to produces light aircraft and helicopters for the local and US export market.

Russian and Bolivian Defence Ministers Sergei Shoigu and Reimi Ferrera have signed an agreement on military cooperation. Bolivian defence industry receives aircraft repair and maintenance assistance from Russian government situated at the department of Cochabamba as part of an engagement policy with Bolivia.

Defence Procurement

The Bolivian defence ministry has made a new modernise plan for its forces, which includes the establishment of a Supply, Replacement and Equipment Fund (FARE), which will draw from multiple taxes to fund procurement. Bolivian Defence Minister Walker San Miguel had introduced the 15-point plan in January 2010. Bolivian Navy has procured second-hand equipments for its commercial and civilian section from the US or South American civil markets and its naval inventory is relatively adequate within its present limited operational parameters. The nation receives various new materials from China, Venezuela, Brazil and Argentina. Various helicopters, training aircraft, armour and logistical support equipment are donated to the nation. Bolivia has already received a credit line for up to USD100 million worth of weapons and logistical equipment from Russia in 2009. Bolivia has also signed a 20-year military cooperation agreement with Russia, which may allow Bolivia's credit line to be increased up to USD 300 million in coming years. The Bolivian government awarded USD 216 million worth contract to the French company Thales to install a system of 13 radars for military defence and air traffic control in order to effectively combat drug trafficking and contraband.

The Bolivian Army or Ejército Boliviano.
Image Credit: Bolivian Ministry of Defence

CONTACT DETAILS

Ministry of National Defence
Av.Pedro Salazar esquina 20de Octubre
Plaza Avaroa, 2502 La Paz
Bolivia
Tel: +59122379240, 377131, 371186
Fax: +59122377135
Telex: entel 2320
Armed Forces HQ
Gran Quartel General de Miraflores La Plz
Bolivia
Tel: +59122370120, 378180 through 82
Telex: 5384 comanfab

BOSNIA AND HERZEGOVINA
(Capital: Sarajevo)

INTRODUCTION

Area: 51,197 sq km
Population: 3,861,912 (July 2016 est.)
Coastline: 20 km

KEY POLITICAL PERSONS

CHAIRMAN, PRESIDENTIAL COUNCIL: Bakir Izetbegovic

PRIME MINISTER: Fadil Novalic

KEY DEFENCE PERSONS

MINISTER OF DEFENCE: Marina Pendes

CHIEF OF JOINT STAFF: Gen. Ante Jelec

ECONOMY

The inter-ethnic warfare in Bosnia and Herzegovina has hindered its continuous economic growth with production levels plummeting and unemployment rate soaring from time to time. One of Bosnia's main economic challenges in 2010 was to reduce spending on public sector wages and social benefits to meet the IMF's criteria for obtaining funding for budget shortfalls. The country receives substantial amounts of reconstruction assistance and humanitarian aid from the international community. It relies heavily on West Europe for trade and credit. Building up a strong, competitive private sector, strengthening the fiscal system and carrying out public administration reforms, greater integration into the European Union and obtaining the membership of the World Trade Organisation (WTO) are some of Bosnia's top economic priorities. The country has taken steps to reduce huge public spending and increase revenue collection.

GDP: $40.53 billion (2015 est.); $39.43 billion (2014 est.)

Real Growth Rate: 2.8% (2015 est.); 1.1% (2014 est.)

Industries: Steel, coal, iron ore, lead, zinc, manganese, bauxite, aluminium, motor vehicle assembly, textiles, tobacco products, wooden furniture, ammunition, domestic appliances, oil refining

Total Exports: $3.95 billion (2015 est.); $4.489 billion (2014 est.)

Export Commodities: Metals, clothing, wood products

Major Markets: Slovenia 16.5%, Italy 15.9%, Germany 12.1%, Croatia 11.5%, Austria 11.1%, Turkey 5.2% (2015)

Total Imports: $8.173 billion (2015 est.); $9.982 billion (2014 est.)

Import Commodities: Machinery and equipment, chemicals, fuels, foodstuff

Major Suppliers: Croatia 19.3%, Germany 13.9%, Slovenia 13.8%, Italy 10.9%, Austria 5.7%, Hungary 5.2%, Turkey 4.5% (2015)

INTERNATIONAL DISPUTES

Sections along the Drina River remain in dispute between Bosnia and Herzegovina and Serbia.

DEFENCE

ARMY

Personnel: 10,000+

Equipment

Category	Name	In Service
MBT	M60A3	45
	AMX-30B2	32
	M-84	16
	T-55	150 (in reserve)
Armoured vehicle	M113A2/A3	150
	BVP M-80	90

Army continued...

Category	Name	In Service
Armoured vehicle	BOV	50+
	M1145 HMMWV	25
	AMX 10P	35
	IVECO	
ATGM	AT-4 Spigot	
	AT-3 Sagger	
	HJ-8	
MRL	BM-21 Grad	29
	M-63	5
	M-77	7
Howitzer	D-30	86 (+35 in reserve)
Helicopter	UH-1H	2
	Mi-8	4
	Mi-17	3
	Soko Gazelle Gama	0

New Procurements/ Upgrades

No major procurement plans are under consideration at present.

AIR FORCE

Personnel: 2,500+

Equipment

Category	Name	In Service
Trainer	Lola Utva 75	9
Utility Helicopter	UH-1H	14
	Mi-8/17	14
	Soko Gazelle Gama	4
	Mi-34	1
MANPAD & SAM	FIM-92 Stinger	50
	9K38 Igla	20
	Bofors 40mm L/70	Two batteries
	ZU-23-2, AAA	150
	9K31 Strela-1	15
	2K12 Kub	
	P-40 radar	12

New Procurements/ Upgrades

No major procurement plans are under consideration at present.

Defence Expenditure

Estimated defence spending in terms of GDP: 1.1%

Military suppliers to Bosnia and Herzegovina are Russia, US, China, Turkey, Italy and Germany.

Defence Production and R&D

While in the past, Bosnia and Herzegovina has produced and exported ammunitions and other equipment to a number of countries including the US, presently it does not have any facilities to manufacture defence equipment. Some ammunition plants in the country are currently producing explosives for military and commercial purposes.

Defence Procurement

No major procurement plans are under consideration by the Government at present. While the focus would be to dispose of the existing inventory of obsolete weapons and artillery, the unified Armed Forces is set to adopt a light infantry-based force model mainly for international operations along with disaster relief missions, combating terrorism and other organised crime within the country. The all-volunteer force will constitute of 10,000 professional military personnel and 5,000 reserve force.

CONTACT DETAILS

Ministry of Defence
Hamdije Kresevljakovica 98
71 000 Sarajevo, BiH
Phone: +387 33 285 500
Fax: +387 33 206 094
E-mail: info@mod.gov.ba

BOTSWANA
(Capital: Gaborone)

INTRODUCTION

Area: 581,730 sq km
Population: 2,209,208 (July 2016 est.)
Coastline: 0 km (Landlocked)
Maritime claims: None

KEY POLITICAL PERSONS

PRESIDENT & COMMANDER-IN-CHIEF: Seretse Khama Ian Khama

VICE PRESIDENT: Mokgweetsi Eric Masisi

KEY DEFENCE PERSONS

MINISTER OF DEFENCE: Shaw Kgathie

COMMANDER OF THE DEFENCE FORCES: Lt. Gen. Placid Segokgo

ECONOMY

Botswana has maintained one of the world's highest economic growth rates since independence in 1966. Diamond mining has fuelled much of the expansion and currently accounts for one quarter of GDP, approximately 85% of export earnings, and about one-third of the government's revenues. Tourism is the secondary earner of foreign exchange and many Batswana engage in subsistence farming and cattle raising. Through fiscal discipline and sound management, Botswana transformed itself from one of the poorest countries in the world to a middle-income country with a per capita GDP of $17,700 in 2015. In October 2015, the Government announced a stimulus plan to boost the economy through projects in agricultural production, construction, manufacturing, and tourism development. In 2016, Botswana entered its fourth year of drought, detrimental to Botswana's small, but vital agriculture sector.

GDP: $34.84 billion (2015 est.); $34.96 billion (2014 est.)

Real Growth Rate: -0.3% (2015 est.); 3.2% (2014 est.)

Industries: Diamonds, copper, nickel, salt, soda ash, potash, coal, iron ore, silver; livestock processing; textiles

Total Exports: $6.274 billion (2015 est.); $8.514 billion (2014 est.)

Export Commodities: Diamonds, copper, nickel, soda ash, meat, textiles

Total Imports: $7.09 billion (2015 est.); $7.991 billion (2014 est.)

Import Commodities: Foodstuff, machinery, electrical goods, transport equipment, textiles, fuel and petroleum products, wood and paper products, metal and metal products

INTERNATIONAL DISPUTES

Botswana has no known international disputes.

DEFENCE

Botswana Defence Force consists of three separate service components: Ground Forces Command, Air Arm and Defence Logistics Command. The country has no navy. The military doctrine is to maintain a small tactical, trained force capable of defending itself, and also for rapid deployment anywhere in the continent. The BDF has been successfuly deployed as part of the UN peacekeeping operations in both Somalia and the Darfur region of Sudan.

ARMY

Personnel: 20,000

Equipment

Category	Name	In Service
Tank	SK-105 Kurassier (light)	52
	Scorpion (light)	24
Armoured Vehicle	Piranha III	
Artillery	105mm light guns towed	36
	Soltam 155mm towed	
	Carl Gustav 84mm RCL	
ATGW	Milan, Tow	
SAM	Javelin	
	SA 7	

New Procurements/Upgrades

✦ Acquisition of 45 Piranha 8x8 armoured vehicles is under consideration or a deal might have been singed. No official account for the same is available.

AIR ARM COMMAND

Personnel: 2,500

Equipment

Category	Name	In Service
Fighter	CF-5A	9
	CF-5D	3
Transport/ iaison	C-130B	3
	CN-235	2
	C-212	2
	BN-2 Islander	10
	Cessna 02	5
Trainer	Pilatus PC-7 Mk II	5
Helicopter	Bell 412	4
	AS.350 Ecureuil	4
VIP transport	Gulfstream IV	1
	King Air 200	1
	Bell 412	1
UAV	Hermes 450	

New Procurements/Upgrades

✦ Negotiations are going on with Sweden to buy SAAB Gripen multi-role fighters.

Defence Expenditure

Total defence spending: $467 million (2012), $344 million (2011)

Defence spending in terms of GDP: 2.8% (2012)

Defence Production and R&D

Botswana has no defence industry base. There are some repair facilities for the maintenance of the combat vehicle. The Indian Air Force (IAF) imparts training to the Botswana Defence Force (BDF). The professionalism and contribution made by the IAF Training Team has been acknowledged and appreciated by the Government of Botswana. There have been regular exchanges of high military officials between the two countries. A large segment of BDF officers also received training from the United States.

Piranha armoured vehicle

CONTACT DETAILS

Ministry of Defence Justice and Security

Private Bag 00384, Gaborone

TEL: (+267)3698200/300, FAX: (+267)3933034

Toll Free: 0800 600 971

Email Public Relations: mdjs@gov.bw

STRATEGIC INFORMATION

BRAZIL

(Capital: Brasilia)

INTRODUCTION

Area: 8,515,770 sq km

Population: 205,823,665 (July 2016 est.)

Coastline: 7,491 km

Maritime claims: Territorial sea: 12 nm

Contiguous zone: 24 nm

Exclusive Economic Zone: 200 nm

Continental shelf: 200 nm or to edge of the continental margin

GEOPOLITICAL IMPORTANCE

Brazil is the largest country in the South America region bounded by the Atlantic Ocean on the east. The country is bordered on the north by Venezuela, Guyana, Suriname and the French overseas region of French Guiana; on the northwest by Colombia; on the west by Bolivia and Peru; Colombia to the northwest; on the southwest by Argentina and Paraguay and on the south by Uruguay. Brazil borders all other South American countries except Chile and Ecuador. Semi arid along northeast coast, mountains, hills, and rolling plains in the southwest, including Mato Grosso with Midwestern savannahs; the country has world's largest wetland area and coastal lowland. It also encompasses a number of oceanic archipelagos, such as Fernando de Noronha, Rocas Atoll, Saint Peter and Paul Rocks, and Trindade and Martim Vaz.

POLITICAL OVERVIEW

The Federal Republic of Brazil is one of the world's largest democracies. The political and administrative organisation of Brazil comprises the federal government, the states, the federal district and the municipalities. The form of government is that of a democratic republic, with a presidential system. The President appoints the Ministers of State, who assist in government. Executive power is exercised by the President, advised by a cabinet. Legislative power is vested upon the National Congress, a two-chamber legislature comprising the Federal Senate and the Chamber of Deputies. Judicial power is exercised by the judiciary, consisting of the Supreme Federal Court, the Superior Court of Justice and other Superior Courts, the National Justice Council and the regional federal courts. The Federal Constitution is the supreme law of Brazil.

KEY POLITICAL PERSONS

PRESIDENT & COMMANDER-IN-CHIEF: Michel Temer

VICE PRESIDENT: Position Vacant

ECONOMY

Brazil has the seventh largest economy by nominal GDP in the world, and seventh largest by purchasing power parity. The Brazilian economy is characterised by moderately free markets and an inward-oriented economy. Brazil is one of the world's fastest growing major economies. It has a mixed economy with abundant natural resources. The country is characterised by large and well-developed agricultural, mining, manufacturing, service sectors and is expanding its presence in world markets. Brazil has moderately free markets and an inward-oriented economy. Significant steps taken since the 90s toward fiscal sustainability, as well as measures taken to liberalise and open the economy, have considerably enhanced the country's competitiveness fundamentals, providing a better environment for private-sector development.

GDP: \$3.199 trillion (2015 est.), \$3.327 trillion (2014 est.)

Real Growth Rate: -3.8% (2015 est.), 0.1% (2014 est.)

Industries: Textiles, shoes, chemicals, cement, lumber, iron ore, tin, steel, aircraft, motor vehicles and parts, other machinery and equipment

Oil - proved reserves: 15.31 billion bbl (1 January 2015 est.)

Natural gas - proved reserves: 388.7 billion cu m (1 January 2014 est.)

Total Exports: \$190.1 billion (2015 est.), \$224.1 billion (2014 est.)

Export Commodities: Transport equipment, iron ore, soybeans, footwear, coffee, automobiles

Major Markets: China 18.6%, US 12.7%, Argentina 6.7%, Netherlands 5.3% (2015)

Total Imports: \$172.4 billion (2015 est.), \$230.7 billion (2014 est.)

Import Commodities: Machinery, electrical and transport equipment, chemical products, oil, automotive parts, electronics

Major Suppliers: China 17.9%, US 15.6%, Germany 6.1%, Argentina 6% (2015)

DEFENCE & SECURITY

Brazil constitutes the largest armed force in Latin America. It consists of the Brazilian Army *(Exercito Brasileiro),* the Brazilian Navy *(Marinha do Brasil),* the Brazilian Marine Corps & Brazilian Naval Aviation (*Corpo de Fuzileiros Navais*), and the Brazilian Air Force (*Forca Aerea Brasileira*). Brazil has an established national security strategy to modernise the country and populate the vast central and western areas of Brazil. The Brazilian National Defence Policy is primarily directed by the Brazilian Constitution. Brazil's national defence policy is outlined in the National Defence Policy, the National Mobilization Policy and the National Defence Strategy. The country's armed forces have been participating in various UN peacekeeping operations such as the United Nations Stabilisation Mission (MINUSTAH) in Haiti. Brazil's new National Defence Policy focuses on three principal elements: reorganisation of country's armed forces, restructuring the Brazilian defence industry while promoting economic development; and revising the policies governing the composition of the armed forces, including revisiting the issue of compulsory military service. The five Brazilian strategic areas are: Amazon, Midwest, Prata river Basin, Central Brazil, and the Northeast. The Amazon is the Brazilian top priority strategic area. The armed forces have expanded their presence in the Amazon under the Northern Corridor (*Calha Norte*) programme, with the aim of protecting its natural resources from any action coming from outside the region.

KEY DEFENCE PERSONS

MINISTER OF DEFENCE: Raul Belens Jungmann Pinto

CHIEF OF ARMY: General Eduardo Villas Boas

CHIEF OF NAVY: Adm. Bacellar Eduardo Leal Ferreira

CHIEF OF AIR FORCE: Lt. Gen. Nivaldo Luiz Rossato

Internal Conflict

Brazil has been involved in numerous armed actions inside its territory and also along its borders with its neighbours. Currently, Brazil has good relationships with all of its South American neighbours, and hasn't been to war with any of them since the 19th century. Brazil's main internal security concern is violent crime. The crime rate remains high in most urban centres, including the cities of Brasilia, Rio de Janeiro and Sao Paulo, and is also growing in rural areas within those states. Individuals with ties to criminal entities and traffickers operate along all the Brazilian borders. These organizations are involved in the trafficking of illicit goods and drugs. On Aug 2012, around 10,000 Brazilian troops were deployed in the southern border region, in a show of force against illicit traffickers of all kinds who operate in the region. The border reinforcements are part of Operation Agatha which the government initiated on their country's borders with Bolivia, Argentina, Paraguay and Uruguay. Operation Agatha seeks to combat trans-border and environmental crimes along the borders of the Western Amazon region. Brazilian Navy and Air Force are also deployed on a larger scale and working directly in the operation. In previous stages of the operation, Brazil seized more than 2.3 tonnes of illicit drugs, along with 302 boats used by traffickers and has also confiscated 60 firearms and other weapons. It also destroyed various clandestine airstrips.

According to an UNHCR report, the communities formerly controlled by drug dealers are now in the hands of militias. These militias are made up of police officers, prison guards and firefighters and control up to a third of favelas in Rio de Janeiro. In 2011, several operations were carried out targeting police officers accused of collusion with militia groups. The militias are still active and a security threat in the region. But these criminal gangs operating in Brazil are not normally considered as non-state armed groups.

External Conflict

Brazil hasn't had its territory invaded since year 1865 during the Paraguayan War. Also, Brazil has no contested territorial disputes with any of its neighbours. Few as mentioned is given below:

- ✦ Uncontested boundary dispute between Brazil and Uruguay over Braziliera/ Brasiliera Island in the Quarai/ Cuareim River leaves the tri-point with Argentina in question.
- ✦ Smuggling of firearms and narcotics continues to be an issue along the Uruguay-Brazil border.
- ✦ Colombian-organised illegal narcotics and paramilitary activities penetrate Brazil's border region with Venezuela.

Brazil faces external security threats in the form of drug-trafficking, arms smuggling and guerrilla activity mostly volatile in the regions of Amazon. Colombia's FARC, a leftist insurgency, finances itself largely through illicit activities, which are often located along the Brazilian border. Brazil is planning to double its border security by 2014. The country will increase the number of security personnel along the Bolivian, Peruvian and Colombian borders. The move is focused on curbing out illegal arms and drugs trafficking along Brazil's 16,000 kilometre border.

THREAT PERSPECTIVE

Brazil is by far the largest country in Latin America and enjoys generally good relations with its ten South American neighbours. Brazil has resolved all its border-related disputes peacefully and generally do not interfere in the affairs of other countries. Currently, there is no immediate or concrete threat to the country but border security and sovereignty concerns continues to be a driving factor for Brazil to have a foolproof national security strategy which is important to protect the country's aquifers, its strategic minerals, its biodiversity, oil and gas.

Brazil has some concerns with regard to the triple frontier area. The Tri-Border Area of Argentina, Brazil, and Paraguay has long been used for arms smuggling, money laundering, and other illicit purposes which are all potential funding sources for terror outfits. Brazil has therefore focused their efforts in the areas of Sao Paulo; the Tri-Border Areas of Brazil, Argentina, and Paraguay, and Brazil, Peru, and Colombia; and along the Colombian and Venezuelan borders in order to counter terrorist threats as also safeguard its territory from any terrorist attacks. The country is a member of the Financial Action Task Force (FATF) and the Financial Action Task Force on Money Laundering in South America.

STRATEGIC RELATIONS

US: The United States and Brazil are working together on key global, multilateral, and regional issues. Both the countries have traditionally enjoyed cooperative, active relations encompassing a broad political, economic and military agenda. The military ties between Brazil and US have increased in recent years mainly in the areas of defence, counterterrorism and counternarcotics. Brazilian and US personnel have worked together following the Haiti earthquake. The Brazilian troops have also participated in a number of UN missions around the world. In April 2010, the United States and Brazil signed a Defence Cooperation Agreement designed to promote cooperation in areas such as research and development, technology security, and acquisition of defence products and services. This was followed by a General Security of Military Information Agreement, signed in November 2010, which will facilitate the sharing of classified defence and military information. Additional areas of defence cooperation include information exchanges, combined military training, and joint military exercises.

India: Brazil and India are members of the BRICS group of emerging powers (along with China, Russia and South Africa) and Brasilia views its multi-faceted partnership with New Delhi, including in the defence field, as a "strategic priority". Both countries are having excellent bilateral cooperation in the area of defence, particularly in the joint development of high-tech military aircraft. Both Brazil and India have co-operated in the multilateral level on issues such as international trade and development, environment, reform of the UN and the UNSC expansion. Brazil and India are also involved in the IBSA initiative. The strategic partnership established in 2006 between India and Brazil has deepened with both countries cooperating closely in BRICS, IBSA, G4, G20, BASIC and the wider multilateral context of the United Nations.

China: Brazil and China share a healthy relation in the field of strategic alliance development. As one of the BRICs countries, Brazil has a very good relationship with China and it always welcomes investments as well as products from China. Brazil considers China an important trade partner. The ties between the two armies have also grown stronger through mutual visits and personnel training. China became Brazil's largest trading partner in 2009. Chinese exports to Brazil collapsed in July 2016 in the latest dramatic sign of the deepening recession in Latin America's biggest economy.

Russia: Brazil shares an important alliance with the Russian Federation, with partnerships in areas such as space and military technologies, and telecommunications. Both the countries have seen a significant improvement in recent years, by increasing commercial trade and cooperation in military and technology areas.

Argentina: Brazil and Argentina are both close and share strong integration and partnership between each other. Both the countries are engaged in several joint venture projects in the military field, such as the Gaucho armoured vehicle and the Embraer KC-390 military transport aircraft. Argentina and Brazil have close cooperation in the field of space science. Brazil has been a strong supporter of the Argentine claim over the Falkland Islands. Both account for each other's largest export and import market. Argentine foreign policy has given special emphasis in deepening the strategic alliance with Brazil in all its aspects.

Multilateral Alliances

Brazil is a member of international organisations including the United Nations, Organisation of American States (OAS), Inter-American Development Bank, G-20, International Monetary Fund, World Bank, World Trade Organisation, BRIC, sub-regional Mercosur and Union of South American Nations. Brazil is a founding member of the OAS and the Inter-American Treaty of Reciprocal Assistance (Rio Treaty).

DEFENCE CAPABILITIES

ARMY: The Army's mission is to preserve and guarantee the defence of Brazil. The Brazilian Army has fought in several international conflicts, mostly in South America during the 19th century. In the 20th century it participated on the Allied side in World War I and World War II, as well as in UN peacekeeping missions. The Brazilian Army is trying to renew its equipment and making a redistribution of its barracks in all the Brazilian Regions, prioritising the Amazon. The Brazilian Army is creating an Expeditionary Force (F EXPD) to provide permanent support for the country's participation in foreign missions.

Personnel: 219,663

Equipment

Category	Name	In Service
MBT	Leopard 1A5	250 (30 reserve)
	Leopard 1A1	128
	M60 Patton A3 TTS	91
	M41 (light)	287
APC/MICV /IFV	EE-9 Cascavel (6x6)	600 (To be replaced by VBR-MR 8×8 in 2020)
	EE-11 Urutu (6x6)	223 (To be replaced by VBTP-MR)
	VBTP-MR Guarani (6x6)	250
	M113-BR/A2	584 & 12
	Guarani 6x6	38 (48 on order)
	M577A2	34
	AV-VB4 RE	1
	AV-VBL	32
	Iveco LMV 4x4	(1,464 on order)
Artillery	M114A1 155mm howitzer	92
	M108 155mm self-propelled	72 (To be decommissioned)
	M109 155mm	76 (To be upgraded to M109A5+)
	L118 105 mm howitzer	54
	M101 105 mm howitzer	320
	M102 105 mm howitzer	19
	Oerlikon 35mm	38
Mortar	M30 107 mm	217
	K6A3 120 mm	86

Army continued...

Category	Name	In Service
MRL	Astros 2020	38 launchers vehicles
	AV-SS 12/36	100
RCL	Carl Gustav 84 mm	300
Anti Tank GW	Eryx 136 mm	12
	AT4 84 mm	1,500
	ALAC 84 mm	(2,000 on order)
	MSS-1.2 130 mm	(400 on order)
	MILAN	20
SAM	RBS 70	16
	Pantsir-S1	(1 battery on order)
	9K38 Igla SA24	24
	9K38 Igla SA18	120
AA Gun	40 mm, 35 mm, 57 mm, 90 mm	240
	Gepard 1A2	36

Army Aviation

Category	Name	In Service
Helicopter	EC-725 Caracal	5 (16 on order)
	AS532 Cougar	8
	HM-1 Pantera	33
	AS550	33
	UH-60L	4

New Procurements/ Upgrades

✦ BAE Systems has been awarded a USD 54 million contract to provide 32 upgraded M109A5+ self-propelled howitzers to the Brazilian Army.

✦ The Brazilian Army will receive 50 Armoured Fighting Vehicles external link (AFVs) from the US Army, with M577A2 command and control vehicles forming the majority of this batch.

✦ Saab has signed a contract on deliveries of the RBS 70 VSHORAD (very short range air defence system) to the Brazilian Army. The order has a value of approximately MSEK 80 and comprises deliveries of man-portable launchers, missiles and associated equipment.

✦ The army has awarded USD 25 million contract to Elbit Systems of Israel's Brazilian subsidiary Ares Aeroespacial for the procurement of 12.7/ 7.62 mm remote controlled weapon stations. AEL Sistemas S.A won a USD 15 million contract to supply the Brazilian army with unmanned turrets for armoured vehicles.

✦ Brazil has signed a contract for the purchase of 86 Guarani armoured personnel carriers from Iveco, Italy. The initial Guarani vehicles, worth approximately 240 million Brazilian reals, will be used by infantry and cavalry units of the army. Seven armoured vehicles of the 16-vehicles pilot batch project will be provided to the Army.

✦ A programme had been launched for the procurement of the 4x4 light armoured vehicles with GAUCHO and CHIVUNK model up to 1200 nos. are proposed to be procured.

✦ A contract has been signed for 50 EC-725 Super COUGAR helicopter to be manufactured locally 16 for army aviation, 16 for navy and 18 for the air force.

✦ BAE Systems, in partnership with the Brazilian Army are upgrading 150 M113 armoured personnel carriers for Brazil through a foreign military sales contract worth USD 41.9 million. The Brazilian Army will upgrade the M113B vehicles to the M113A2 Mk1 configuration.

Cavalaria. *Image Credit: Exército Brasileiro*

NAVY: The Brazilian Navy is the largest navy in Latin America. It is equipped with a 32,800-tonne aircraft carrier, the NAe São Paulo (formerly Foch of the French Navy), British-built frigates, locally built corvettes, coastal diesel-electric submarines and many other river and coastal patrol craft, among other vehicles. The main branches of the Brazilian Navy are: Naval Operations Command, Surface Fleet Command, Submarine Fleet Command, Naval Aviation Command, 1st Fleet Division Command, 2nd Fleet Division Command & Marine Corps General Command.

Naval Bases: Val de Cães, Almte. Ary Parreiras, Natal, Ladário Fluvial, Rio Negro, Rio Grande, Almte. Castro e Silva, Ilha das Flores, Ilha do Governador, Rio Meriti, Navy Arsenal of Rio de Janeiro, Rio de Janeiro, Aratu.

Personnel: 60,000 (including 15,000 Marines)

Equipment

Category	Name	In Service
Aircraft Carrier	Clemenceau class (Sao Paulo)	1
Submarine	Type 209 class	5
	Scorpene class	(To be delivered between 2017-2021)
Frigate	Brasil class (Training)	1
	Niteroi class (Multi-purpose)	6
	Type 22 (Anti-submarine)	2
Corvette	Inhauma class	3
	Barroso class	1
Patrol Vessel	Amazonas class	3
	Imperial Marinheiro class	1
	Macae class	3
	Grajau class	12

Navy continued...

Category	Name	In Service
Patrol Vessel	Bracui class	4
	Roraima class	3
	Pedro Teixeira class	2
	Piratini class	6
Amphibious Force	Round Table class	2
	Mattoso Maia class	1
	Foudre class	1
Minesweeper	Aratu class	6
Auxiliary ship		40
Missile	Exocet ASM, MAN-1-SUP ASM, MAN-1-SUB ASM, Aspide SAM, Sea Wolf SAM, Simbad SAM	

Naval Aviation

Category	Name	In Service
Fighter/ Attack	AF-1 & AF-1B Skyhawk	3 & 1 (7 under upgrade)
Tanker	KC-2/C-1 Trader	(4 on order)
Helicopter	Bell 206	17
	S-70B	6 (2 on order)
	Mk. 21/ Mk. 21A	12
	EC725	7
	AS332	5
	AS355	9
	AS350	19
Trainer	AF-1A	1
	AF-1C	1
UAV	FT-100 Horus	1
Weapon	AIM-9H Sidewinder, MAA-1A/1B Piranha, Sea Skua, Python, Derby	

Marines

Category	Name	In Service
APC/MICV/ IFV	SK-105 Kurassier	17 (22 on order)
	M113	30 (42 planned)
	AV-VBL 4x4	3
	AAV7A1	52 (78 planned)
	Piranha IIIC 8X8	30 (42 on order)
Artillery	M114A1	6
	L118 105 mm	18
	K-6A3 120 mm	6
	ASTROS II	6
	M29 A1 81 mm	100
	Bofors 40 mm	6
Missile	Mistral	24
	RBS 70	12
	Pantsir-S1	
UAV	Carcara	42
	Horus FT-100	1

New Procurements/ Upgrades

✦ Helibras to provide the first armed H225M helicopter to the Brazilian Navy. The helicopter was developed specifically for naval missions and features an APS-143 patrol radar, chaff and flare countermeasures and is armed with two missiles.

✦ The US Government has cleared a proposal to sell AGM-84L Harpoon Block II missiles and associated equipment worth USD 169 million to Brazil under the Foreign Military Sales deal. The Brazilian Government had requested a possible sale of 16 AGM-84L Harpoon Block II missiles, 4 CATM-84L Harpoon Block II captive air training missiles, containers, spare and repair parts, support and test equipment, personnel training and training equipment among others.

✦ Elbit Systems of America has received a subcontract from the Brazilian Navy for the upgrade of four Grumman C-1A aircraft. The prime contract is held by Marsh Aviation Company of Mesa, Arizona. The subcontract, valued at USD 106 million, will be performed by Elbit Systems of America over a five-year period.

✦ Brazilian Navy's Naval Aviation has signed a contract with AugustaWestland for a major midlife upgrade of eight Lynx Mk21A helicopters. The contract, valued in excess of USD 160 million, includes replacement of the aircraft's engines with the CTS800-4N product from LHTEC, navigation, displays suite and mission avionics.

✦ Kongsberg has secured a contract from the Brazilian Navy to supply Penguin anti-ship cruise missiles and associated equipment. The contract is valued at worth EUR 33 million (USD 42 million). The missiles will be deployed on the Brazilian Navy's new maritime helicopters.

✦ French shipbuilder DCNS has commissioned its new shore-based integration facility for the combat systems to be installed on the Brazilian Navy's Scorpene conventional submarines (SSKs).

✦ Brazil has begun building the first S-BR Scorpene class submarine designed by Europe's leading shipbuilder DCNS. The programme covers the design and construction of four Scorpene conventional-propulsion submarine submarines, design and construction assistance with the non-nuclear portion of Brazil's first nuclear-powered submarine and support for the construction of a naval base and a naval shipyard. The first of the four Scorpene-class conventional-propulsion submarines is scheduled to enter active service in 2017.

✦ 12 AF-1 and 3 AF-1A SKYHAWK are to be refurbished and modernised by Embraer.

✦ Plan to purchase 6 S-2T TURBO Tracker (3 transport and 3 AEW) to operate from SAO PAULO.

✦ The Brazilian Navy has received the first pair of S-70B Seahawk multi-mission maritime helicopters. The country is procuring six S-70Bs via the US Foreign Military Sales programme.

✦ The Brazilian Navy has awarded a USD 208.0386 million contract to BAE Systems for the supply of three Ocean Patrol Vessels and ancillary support services. The contract also contains a Manufacturing License to enable further vessels of the same class to be constructed in Brazil.

✦ Brazilian Navy to receive two new ships and modernisation of the Arsenal de Marinha do Rio de Janeiro shipyard by South Korea's POSCO Daewoo under a USD 1 billion contract.

BNS Rademaker (F49), a type 22 frigate of the Brazilian Navy. *Image Credit: US Navy*

AIR FORCE: The Brazilian Air Force is the aerospace branch of the Brazilian armed forces and is managed by the "Aeronautics Command" (Comando da Aeronáutica - COMAer). The COMAer was created in 1999, and replaced the Ministry of Aeronautics. Now, the COMAer is one of the three armed forces assigned to the Ministry of Defence (Ministério da Defesa). The COMAer is led by the "Aeronautics Commander" (Comandante da Aeronáutica). He is nominated by the President and reports directly to the Minister of Defence.

Air Bases: Anápolis, Belém, Boa Vista, Brasília, Campo Grande, Canoas, Florianópolis, Fortaleza, São Paulo, Manaus, Natal, Porto Velho, Recife, Galeão, Santa Cruz, Afonsos, Salvador, Santa Maria, Santos, Campo de Provas Brigadeiro Velloso.

Personnel: 77,454

Equipment

Category	Name	In Service
Fighter/ attack/ Reconnaissance aircraft	F-5EM	43
	A-1/A-1M	47
	A-29A	31
	F-39E	(28 on order)
	E-99 AWACS	5
	R-99	3
	Learjet 35	6
	EMB 110	4
	EMB 110	11
	P-3AM	8
Transport Aircraft/ air refuelling	KC-130H/M	2
	KC-767	(3 on order)
	EMB 120	20
	EMB 110	50
	C-130E/H/M	12
	CASA C-295	12 (3 on order)
	Cessna 208	32
	KC-390	(30 on order)

Air Force continued...

Category	Name	In Service
Helicopter	UH-1H	24
	Bell 206	1
	EC725	5
	AS332	5
	UH-60L/M	16
	Mi-35	12
Trainer	AMX-T	8
	A-29B	61
	EMB 312	104
	F-5FM	2
	F-39F	(8 on order)
UAV	Hermes 450	4
	Hermes 900	1
SAM	9K38 Igla	20 (26 on order)
AAM	MAA-1A Piranha, MAA-1B Piranha, A-Darter, Derby, Python-3, Python-4	
Anti-Ship missile	AGM-84 Harpoon, MAN-1	
Anti-tank missile	9M120 Ataka-V, 9K114 Shturm	

New Procurements/ Upgrades

✦ Harris Corp., along with private information and technology firm CPqD, was awarded a two-year contract to provide a monitor and control system as well as on-site operations support for the Brazilian Air Force air traffic control network.

✦ The Brazilian Ministry of Defence has signed a USD 5.4 billion contract with Saab for the supply of 36 Gripen Next-Generation (NG) fighter aircraft to the Brazilian Air Force. The two contracting parties, SAAB and the Brazilian Ministry of Defence through the Aeronautics Command (COMAER), also signed an Industrial Co-operation contract under which the former will deliver "substantial technology transfer" to the Brazilian industry. Delivery of the Gripen NGs to the Brazilian Air Force will be undertaken from 2019 to 2024.

✦ Helibras, the Brazilian subsidiary of Airbus Helicopters, delivered the first two upgraded AS365 Panther K2 rotorcraft to the Brazilian Army, as part of a contract inked between Helibras and the Brazilian Army Aviation Command (AvEx) to upgrade the Panthers to enhance their lifespan by another 25 years. The contract entailed modernizing a total of 34 Panthers.

✦ Additional six UH-60L BLACKHAWK are on order to US.

✦ 53 AMX A-1A, A-1B are refurbished and extensive modernisation in a programme led by Elbit (Israel).

✦ 9 airframes of P-2 ORION are to be refurbished and extensively upgraded by EADS.

✦ F-5E/ F aircraft are under extensive upgrade to F-5M in programme led by Elbit Systems.

✦ Brazilian Air Force has taken delivery of an A-29 Super Tucano and two upgraded F-5M fighter aircraft from Embraer. Embraer also received another contract to upgrade 11 additional F-5 jets.

✦ Rockwell Collins' Talon radio selected by Embraer for Brazilian Air Force I-X Project to provide Talon radios for six Embraer Legacy 500 aircraft.

AMX A-1 fighter bombers. *Image Courtesy: CRUZEX 2013, Ministério da Defesa*

Nuclear Weapons Programme

Brazil has never developed either chemical or biological weapons. In the 1980s, Brazil began a secret nuclear weapons programme in response to Argentina's programme, under the military regime and was supplied with nuclear materials and equipment by US, France and West Germany. However, with the decline of military rule, Brazil also ended the weapons of mass destruction and acceded to the Nuclear Non-Proliferation Treaty in 1998. Brazil has also ratified the Geneva Protocol on 28 August 1970, the Biological Weapons Convention on 27 February 1973, and the Chemical Weapons Convention on 13 March 1996.

Nevertheless, the country still possesses the key technologies needed to produce nuclear weapons. Brazil's nuclear capabilities are the most advanced in Latin America. Reports indicate that Brazil has the capabilities, technological knowhow and is in a position to produce nuclear weapons within a short period. Angra Nuclear Power Plant is Brazil's sole nuclear power plant which consists of Angra I and Angra II. Angra III is currently under construction

The Brazilian Navy is currently developing a nuclear submarine fleet. In 2008, France agreed to transfer technology to Brazil for the joint development of the nuclear submarine hull. On the other hand, Brazil which has signed many treaties has also declared itself a nuclear-weapon-free zone by signing the Treaty of Tlatelolco in 1967. The country is fighting against the nuclear proliferation by being an active participant in the International Atomic Energy Agency and the Nuclear Suppliers Group. Brazil is also the founding member of Brazilian–Argentine Agency for Accounting and Control of Nuclear Materials. The safeguards agency plays an active role in the verification of the peaceful use of nuclear materials that could be used, either directly or indirectly, for the manufacture of weapons of mass destruction.

Major Armaments Producers

Domestic suppliers: Embraer, Avibras, NUCLEP, CBC, IMBEL, Taurus, Helibras, EMGEPRON, Agrale, Mectron, Aero Bravo, Iveco Brazil, MAN Latin America, Odebrecht, Industria Aeronautica Neiva, Troller, INACE, Usiminas, XMobots, Condor S/A

Foreign suppliers: United States, France, Germany, Spain, Canada, Colombia, Russia, Israel, United Kingdom, Sweden, Italy, China

Brazil has firm defence R&D facilities for the development of the aircrafts, artillery, missiles infantry weapons, submarines and space research. The country has joint ventures with Italy for the development of the multirole aircraft.

Major defence development by Brazil

✦ :SS-80 long range artillery rocket (300mm)

✦ MAA-1 piranha air to air missile

✦ AV-SF-70 Skyfire 70mm rocket launcher which is also supplied to many nations

✦ Tupi class submarines

Aircraft/ helicopter/ aerospace equipment manufacturers

✦ **Helibras or Helicópteros do Brasil S.A:** Helicopters of Brazil, Inc. is a Brazil-based helicopter manufacturer which is a wholly owned subsidiary of Eurocopter, a division of EADS.

✦ **Aero Bravo**: It is a Brazilian aircraft manufacturer founded in 1993 in Belo Horizonte. The firm manufactures light aircraft of its own design and under license.

✦ **Indústria Aeronáutica Neiva**: It is a subsidiary of Embraer which produces airplanes and aircraft components. Its main product is the Embraer EMB 202 Ipanema, the most employed agricultural aircraft in Brazil.

✦ **Advanced Composites Solutions (ACS)**: The company constructs light aircraft. ACS is the production of the ACS-100 Sora, a two seat light sport aircraft. ACS is also involved in the development and integration of Unmanned Aircraft Systems through its subsidiary Flight Solutions.

✦ **Embraer:** Embraer S.A. (BM&F Bovespa: EMBR3 / NYSE: ERJ) is a giant aerospace conglomerate of Brazil that produces commercial, military, and executive aircraft and provides aeronautical services.

The Brazilian Aerospace and Defence industry is a strong segment in the country´s economy presenting an impressive growth and worldwide projection. EMBRAER leads this industry and as recognized as one of the four major aerospace manufacturers of commercial aircraft and with a growing presence in the defence and business aviation markets.

ABIMDE- Brazilian Defence and Security Industries Association, congregates the companies that manufacture army stuff, with the objective to sponsor, promote and represent their common interests and objectives, aiming at the social and economic enlargement of the country.

DEFENCE DEALS

Joint Venture Programmes

Brazilian defence firm Embraer and Israel's AEL Sistemas, a subsidiary of Elbit Systems, have formed a joint venture company, Harpia Sistemas, to develop, produce and market unmanned aerial vehicles (UAVs). The two defence firms had announced their partnership to jointly develop, produce and market UAVs and related components. The Brazilian Air Force has signed a contract worth 1.4 million reals with Denel do Brasil to prepare facilities in the industrial park in the city of Sao José dos Campos in the state of Sao Paulo for the production of A-Darter missiles. This new technology, the result of a joint development between Brazil and South Africa, has already reached the end of the test phase, and is ready to move into the next phase, large-scale production. This missile will equip the modernised version of the A-1 fighter aircraft as well as the future F-X2 fighter aircraft. The production of the new missile will involve several Brazilian companies, including Mectron, Avibras and Opto Electronics, which have benefited from a transfer of technology in areas such as optics, navigation, sensors and image processing. Jaragua launched a joint venture with the Italian firm Oto Melara to produce cannons in Brazil and set up a maintenance centre for Latin America.

Arms Import

ST Kinetics, arm of ST Engineering has won a contract worth USD 20.9 million from the Brazilian Navy for the supply of 5 units of 40mm L70 naval gun and an associated Integrated Logistics Support (ILS) package. The contract was signed between the Brazilian Navy and Allied Ordnance of Singapore (Pte) Ltd (AOS), a wholly owned subsidiary of ST Kinetics. Brazil has also an agreement with Colombia, valued at 10 million dollars, for the purchase of patrol vessels to monitor the Amazon region.

Sweden's Saab JAS 39E Gripen won the multi-billion dollar contract to supply Brazilian Air Force with 36 new fighter aircraft by edging out French and US contenders in the competition. Saab's Gripen NG was in competition with the Rafale made by France's Dassault company and Boeing's F/A-18 fighter from US for the FX-2 air force replacement programme. The total value of the contract is estimated at USD 4.5 billion including Gripen NG, sub-systems for Gripen NG, an extensive technology transfer package, a financing package as well as long term bi-lateral collaboration between the Brazilian and Swedish Governments. The first aircraft to arrive in 2018. The contract is to be complete by 2023. The Gripen NG is capable of performing an extensive range of air-to-air, air-to-surface and reconnaissance missions. Munitions include various missiles, laser-guided bombs, and a single 27 mm Mauser BK-27 cannon. The countries which operate Gripen aircraft are Sweden, South Africa, Hungary, Czech Republic, Thailand and the UK Empire Test Pilot School (ETPS). Switzerland has also selected Gripen as a future fighter jet.

Arms Export

Brazil has witnessed several new developments

in its aerospace industry as well as in other defence sectors. The Brazilian helicopter manufacturer, Helibras recently won the contract to assemble a number of Eurocopter EC 725 Super Cougars in its new facility at Itajubá.

Saab AB announced that they will be investing in the Brazilian aero-structure integrator Akaer.

DEFENCE BUDGET

Total defence spending: $31.9 billion (2015)

Estimated defence spending in terms of GDP: 1.39% (2015)

Brazil's national defence policy is outlined in the National Defence Policy, the National Mobilisation Policy and the National Defence Strategy.

Brazil has cut defence procurement to USD 5.4 billion in 2015 leading some projects to get delayed or suspended. Brazil's foreign deployments will also be reduced, with the current number of troops deployed to the United Nation's Stabilisation Mission in Haiti (MINUSTAH) to be cut to 850 by 2016. Overall federal government spending is to be cut by BRL69.9 billion.

The Brazilian government has approved a decree to cut a further USD 400.9 million in military spending. According to a report by SIPRI, Brazil at present spends about 1.6% of GDP on defence; other countries, including Russia and the United States, spend more than 4%. Brazil is planning to spend closer to 2% of GDP on defence in the coming years in order to help the defence industry of the country that could supply the weapons and equipment necessary to fulfil the government's defence goals. Most of the weapon systems in service with Brazilian armed forces are of US or EU origin. Russian weaponry is limited up to SAMs and recently ordered Mi-35 attack helicopter. According to the International Institute for Strategic Studies which appeared in news media, Brazil's defence budget reached USD 36.6 billion in 2011, the world's 10th largest. But most of this budget has been postponed due to delay in deals such as contract for fighter aircraft.

The defence ministry had last year approved [law n°] MP 544/ 11 which creates a special tax regime for the national defence industry. The law is aligned with the Greater Brazil Plan, which aims to increase the competitiveness of domestic industry. In 2008, Brazil's Ministry of Defence released the National Strategy of Defence (NSD), a strategic review of the country's defence posture and objectives. The NSD emphasises the development of a robust domestic defence industrial capability to both compete with external markets in order to increase their production scale and create the technological capacity to gradually rule out the need to purchase imported services and products.

ABIMDE, which groups 170 companies, hopes to invest USD 120 billion over the long term, planning boost annual export sales from USD 1.7 billion to USD 4 billion. The government is also planning to increase the exports of its locally made defence products by shifting the focus more on its industrial and manufacturing sectors.

CONCLUSION

Brazil's international relations are based on Article 4 of the Federal Constitution, which establishes non-intervention, self-determination, international cooperation and the peaceful settlement of conflicts as the guiding principles of Brazil's relationship with other countries and multilateral organisations. Brazil's emerging economic growth has led the country to embark on a more ambitious and long term defence plan to modernise its armed forces and also to build a strong domestic defence industry over the next ten years. Brazil, with largest armed forces in Latin America maintains a stable and friendly relationship with neighbouring countries and has never engaged in armed conflict with any other nation. The country is vast and geographically diverse and is home to the Amazon River and rainforests. Brazil has large reserves of many natural resources and has recently discovered substantial oil reserves. Brazil's armed forces with no serious external or internal threats, are searching for a new role. They are expanding their presence in the Amazon under the Northern Corridor (*Calha Norte*) programme. The country's defence procurements are largely focused on the protection of such resources, including the Amazon system from illegal mining, deforestation and drugs and arms trafficking. The defence ministry focuses on providing its domestic industry with technology procured from foreign OEMs (original equipment manufacturers). The technology procured has assisted the development of the domestic aeronautical sector and continues to provide the Brazilian defence forces with attack helicopters, light attack aircrafts and air cargo transport systems. The recent government incentives to modernize the forces and develop a strong, export-oriented military industrial complex has helped Brazil's defence industry to gain momentum and focus more on invest and growth. Friendship between India and Brazil will stand both the countries in good stead.

CONTACT DETAILS

Ministry of Defence
Esplanada dos Ministérios Bloco Q
CEP 70049-900 Brasília
Distrito Federal
BrazilTel: (+55 61) 33 12 40 00
Websites: www.defesa.gov.br
http://www.exercito.gov.br/
www.mar.mil.br/
www.fab.mil.br/

BRUNEI
(Capital: Bandar Seri Begawan)

INTRODUCTION

Area: 5,770 Sq Km

Population: 436,620 (July 2016 est.)

Coastline: 161 km

Maritime claims: Territorial sea: 12 nm

Exclusive Economic Zone: 200 nm or to median line

KEY POLITICAL PERSONS

SULTAN & PRIME MINISTER: Sir HASSANAL Bolkiah (Also Head of Defence & Finance)

KEY DEFENCE PERSONS

COMMANDER OF ARMED FORCES: Maj. Gen. Mohd Tawih bin Abdullah

CHIEF OF THE LAND FORCE: Brig. Gen. Aminan bin Pengiran Haji Mahmud

CHIEF OF ROYAL BRUNEI NAVY: First Adm. Norazmi Bin Pengiran Haji Muhammad

CHIEF OF ROYAL BRUNEI AIR FORCE: Colonel (U) Shahril Anwar bin Haji Ma'awiah

JOINT FORCE COMMANDER: Brig. Gen. Haji Hamzah bin Haji Sahat

ECONOMY

Brunei has a small well-to-do economy, which depends on revenue from natural resource extraction, and also encompasses a mixture of foreign and domestic entrepreneurship, government regulation, welfare measures, and village tradition. Brunei is a founding member of the 12-nation Trans-Pacific Partnership (TPP) free trade agreement negotiations as well as the Regional Comprehensive Economic Partnership (RCEP) and, with the nine other ASEAN members, will form the ASEAN Economic Community in 2015. Per capita GDP is among the highest in Asia, and substantial income from overseas investment supplements income from domestic production. Crude oil and natural gas production account for about 90% of its GDP. Brunei encourages foreign investment in the domestic economy through various incentives, marketing the country as an opportunity for investors in new industries and economic activities.

GDP: $33.22 billion (2015 est.), $33.29 billion (2014 est.)

Real Growth Rate: --0.2% (2015 est.), -2.3% (2014 est.)

Industries: Petroleum, petroleum refining, liquefied natural gas, construction, agriculture, transportation

Natural gas - proved reserves: 390.8 billion cu m (1 January 2016 est.)

Oil - proved reserves: 3901.1 billion bbl (1 January 2016 est.)

Total Exports: $7.235 billion (2015 est.), $11.11 billion (2014 est.)

Export Commodities: Crude oil, natural gas, garments

Major Markets: Japan 35.9%, South Korea 14.8%, Thailand 10.8%, India 9.8%, NZ 5.6%, Australia 5% (2015)

Total Imports: $3.359 billion (2015 est.), $3.668 billion (2014 est.)

Import Commodities: Iron and steel, motor vehicles, machinery and transport equipment, manufactured goods, food, chemicals

Major Suppliers: Singapore 27.9%, China 25.3%, Malaysia 12.4%, UK 10.6%, South Korea 4.9% (2015)

INTERNATIONAL DISPUTES

✦ According to Letters of Exchange signed in 2009, Malaysia in 2010 ceded two hydrocarbon concession blocks to Brunei in exchange for Brunei's sultan dropping claims to the Limbang corridor, which divides Brunei.

✦ Brunei claims a maritime boundary extending as far as a median with Vietnam, thus asserting an implicit claim to Louisa Reef.

DEFENCE

ARMY

Personnel: 3,200+ 500 reserve

Equipment

Category	Name	In Service
Light Tank	FV101 Scorpion	15
APC	FNSS AIFV	6+
	M113	80
	V150 Commando	80
	AIFV	40
	VAB	35
	Anoa	40
Artillery	155 mm	6
	L16	20
	L118 105 mm	10
Mortar	M81 80 mm	60
RCL	M-67 90 mm	
	M40 106 mm	75

New Procurements/Upgrades

No major procurement plans are under consideration at present.

Brunei and US forces in CARAT 2011.

NAVY

Personnel: 1,200

Equipment

Category	Name	In service
Patrol vessel (Offshore and Inshore)	Waspada Class	3
	Perwira Class	3
	Ijhtihad class	4
	Darussalam Class	4
Amphibious Force	Serasa class	2
	Teraban class	2
Assault Boat	Fast assault boat	26

New Procurements/Upgrades

Brunei ordered the Darussalam-class offshore patrol vessels from Lurssen. The first two vessels were delivered in January 2011. The second batch of two ships was delivered in 2014.

AIR FORCE

Personnel: 1300

Equipment

Category	Name	In Service
Fighter	FA-50	On order
Transport	CN-235	1 (3 on order)
	C-130J	On order
Trainer	Pilatus PC-7	4
	Bell 206	2
Helicopter	UH-60	10
	Bell 214ST	1
	Bo-105	6
	S-70A	4
	S-70i	(12 on order)

Air Defence

Two squadrons are for air defence and are equipped with 10 Rapier and Mistral SAM systems.

New Procurements/Upgrades

✦ Brunei has expressed interest in the FA-50.

✦ Sikorsky Aircraft, a subsidiary of United Technologies Corp. formally signed a contract with the Brunei Ministry of Defence to provide 12 S-70i BLACK HAWK helicopters with associated spare parts, training and ground support equipment. The contract contains an option for 10 additional helicopters.

✦ BAE Systems will continue delivering military aviation training to the Royal Brunei Air Force, with the signing of a contract to train six students across three courses.

Defence Expenditure

A major share of military budget is spend on payment for the weapon system and there maintenance.

Total defence spending: $0.44 billion (est.)

Estimated defence spending in terms of GDP: 3.1% (est.)

Foreign Suppliers: Australia, Belgium, Canada, France, Germany, Indonesia, Italy, Netherlands, Singapore, Sweden, United Kingdom, United States

Brunei's Ministry of Defence has increased their defence budget to USD 409 million USD for the F.Y. 2016-2017, an increase of 4.7%. Brunei has slashed defence spending by 25.26 per cent for the 2015/2016 fiscal year as the country's revenue is projected to decline from falling global oil prices. The USD 537.490 million budget for the Ministry of Defence (MinDef) was presented at the 11th Legislative Council (LegCo) session in 2015. Previously, the defence budget was USD 719.15 million. The ministry is planning to formulate an effective defence strategy to maintain national stability and sovereignty bolster diplomatic relations as well as readiness to address varying threats, humanitarian and natural disasters.

Defence Production and R&D

There is no major defence R&D policy formed by state majority of the defence budget is being spend on the procurement of new systems.

Defence Procurement

Brunei Army/Navy/Air Force have not done any major procurements in recent years.

CONTACT DETAILS

Ministry of Defence - MINDEF,
Directorate of Logistics - DOL
Kementerian Pertahanan
Bolkiah Garrisen
3510 Negara
BruneiTel: (+673 2) 38 63 33
(+673 2) 38 23 60
Fax: (+673 2) 38 68 08
(+673 2) 33 70 72
E-mail: dirlog_mindef@brunet.bn

Public Relations Unit
Ministry of Defence
Bolkiah Garrison BB3510
Brunei Darussalam
Tel: +6732-2386371 / 2386372 / 2386373
Fax: +673-2381934 / +673-2383277
Email: pru@mindef.gov.bn
Website: www.mindef.gov.bn

BULGARIA
(Capital: Sofia)

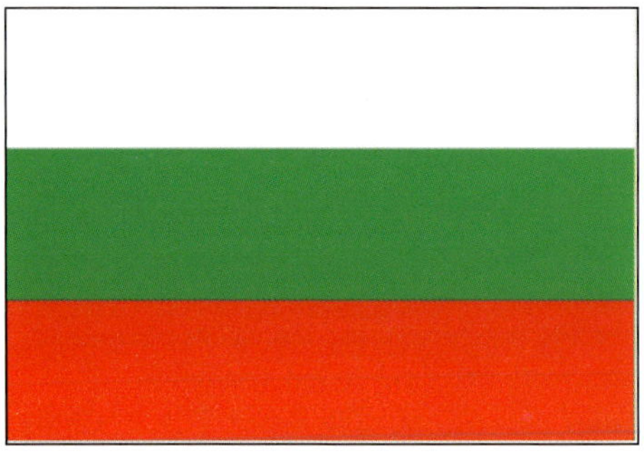

INTRODUCTION

Area: 110,879 sq km
Population: 7,144,653 (July 2016 est.)
Coastline: 354 km
Maritime claims: Territorial sea: 12 nm
Contiguous zone: 24 nm
Exclusive Economic Zone: 200 nm

KEY POLITICAL PERSONS

PRESIDENT: Rumen Radev

PRIME MINISTER: Boyko Borissov (Outgoing)

KEY DEFENCE PERSONS

DEFENCE MINISTER: Nikolay Nenchev

CHIEF OF DEFENCE: Gen. Constantin Popov

COMMANDER OF LAND FORCES: Maj. Gen. Andrey Iliev Botsev

COMMANDER OF NAVY: Rear Adm. Mitko Alexandrov Petev

COMMANDER OF AIR FORCE: Maj. Gen. Tsanko Ivanov Stoykov

ECONOMY

Bulgaria has experienced rapid economic growth in recent years reaching estimated GDP of $123.9 billion (PPP 2014). The World Bank has described Bulgaria as an industrialised upper-middle-income country. The strongest sectors contributing to the economy are energy, mining, metallurgy, machine building, agriculture and tourism. Corruption in the public administration, a weak judiciary, and the presence of organised crime remain significant challenges.

GDP $133.9 billion (2015 est.); $130.3 billion (2014 est.)

Real Growth Rate: 3% (2015 est.); 1.5% (2014 est.)

Industries: Electricity, gas, water; food, beverages, tobacco; machinery and equipment, base metals, chemical products, coke, refined petroleum, nuclear fuel

Total Exports: $24.62 billion (2015 est.); $27.94 billion (2014 est.)

Export Commodities: Clothing, footwear, iron and steel, machinery and equipment, fuel

Major Markets: Germany 12.5%, Italy 9.2%, Turkey 8.5%, Romania 8.2%, Greece 6.5%, France 4.2% (2015)

Total Imports: $26.74 billion (2015 est.); $31.63 billion (2014 est.)

Import Commodities: Machinery and equipment; metals and ores; chemicals and plastics; fuels, minerals, and raw materials

Major Suppliers: Germany 12.9%, Russia 12%, Italy 7.6%, Romania 6.8%, Turkey 5.7%, Greece 4.8%, Spain 4.8% (2015)

INTERNATIONAL DISPUTES

Bulgaria has no known international disputes.

DEFENCE

The Land Forces of Bulgaria are functionally divided into Deployable and Reserve Forces. The Forces are in a process of continued restructuring since 2004. Under the most recent reform, brigades were reduced to regiments, while several garrisons and brigades were disbanded. The Bulgarian Air Force is one of the oldest air forces in Europe and the world. In recent times it has been actively taking part in numerous NATO missions and exercises in Europe. The Navy has traditionally been the smallest component of the Bulgarian military.

ARMY

Personnel: 14,480

Equipment

Category	Name	In Service
MBT	T-72	80
AIFV/APC	BMP-23	70
	BMP-1	90
	BTR-60C	20
	MT-LB	100
Artillery	D20 152mm towed howitzer	24
	2SI 122mm SP howitzer	48
MRL	BM-21 122mm	24
ATGW	9P148 Konkurs	18
	AT-5 Spandrel	
AA Gun	ZU-23-2 23mm	84
SAM	SA-8	24

New Procurements/Upgrades

No major procurement plans are under consideration at present.

NAVY

Personnel: 3,610

Equipment

Category	Name	In Service
Frigate	Drazki class (ex-Belgian Wielingen class)	3
	Smeli (ex-USSR KONI class)	1
Patrol craft	ex-USSR Pauk class	2
Light forces	Mulniya missile FAC	1
Mine warfare force	Tsibar mine hunter (ex-Belgian Tripartite type)	1
	ex-USSR Sonya class minesweeper	3
	ex-USSR Vanya class	2
	Inshore minesweepers	3
Landing craft	Vidra class LCU	1
Auxiliaries	AOTL	2
	ARS	1
	YDT	2
	AGSC	2
	ADG/AX	1

Naval Aviation: Three AS.565MB Panther helicopters.

New Procurements/Upgrades

✦ The Government in 2016 has approved the acquisition of two new multi-role patrol vessels.

AIR FORCE

Personnel: 6,840

Equipment

Category	Name	In Service
Fighter	Su-25K/KB	14
	MiG-29A	12
	MiG-29UB	4
	MiG-21Bis	2 (in reserve)
Transport	C-27J	3
	An-30	1
	PC-12M	1
	An-2T	1
	L-410	2
Trainer	L-39ZA	6
	PC-9M	6
Helicopter	Mi-24D/V	6
	AS.532 AL	12
	Mi-17	6
	Bell 206	6
SAM	S-300	
	S75-Dvina	
	S-125 Pechora	
	S-200	
	2K12 KUB	

New Procurements/Upgrades

✦ At least 10 Russian-built MiG-29 fighters will be upgraded in 2016-2017 to extend their lifespan.

✦ Acquisition of second-hand fighter aircraft, likely from Europe or the USA or Israel, is under consideration. Previously planning to buy new fighter jets, the country now has changed its acquisition criteria and is planning to go for second-hand warplanes. The fighters are to ultimately replace the MiG-29 fleet.

Defence Expenditure

Defence spending in terms of GDP: 1.3% (2015)

The Government is aiming to increase military spending to 2 percent of GDP by 2024.

Defence Production and R&D

The Bulgarian defence industry is divided into two main parts. The first is under the control of the Ministry of Industry and comprises two large sub-groups - Metalchim, which deals with arms and ammunition and Electron in military electronics. The second part is under the control of the Ministry of Defence (MoD) and comprises the TEREM enterprise, dealing with repair works and defence equipment manufacturing. Defence production exports are mainly undertaken by the MoD and four defence trade agencies - Kintex, Teraton, Arimex and Elmet Engineering. Plans are on to privatise state-owned weapons manufacturers.

Defence Procurement

In 2016, the Government has approved three major projects worth over 2.4 billion leva (around 1.2 billion euro) to modernise the Navy and Air Force. The projects include buying fighter jets and multi-role patrol ships, as well as extending the lifespan of MiG-29 fighters. The new acquisitions are aimed at decreasing Bulgaria's dependency on Russia in getting weapons and military hardware.

CONTACT DETAILS

Ministry of Defence

3 Dyakon Ignatiy Street
1092 Sofia
Phone: +359 2 92 20 922
Fax: +359 2 987 96 93
Website:
https://www.md.government.bg/en/

CAMBODIA
(Capital: Phnom Penh)

INTRODUCTION

Area: 181,035 sq km
Population: 15,957,223
Coastline: 443 Km
Maritime claims: Territorial sea: 12 nm
Contiguous zone: 24 nm
Exclusive Economic Zone: 200 nm
Continental shelf: 200 nm

KEY POLITICAL PERSONS

HEAD OF STATE & COMMANDER-IN-CHIEF: Norodom Sihamoni

PRIME MINISTER: Hun Sen

KEY DEFENCE PERSONS

MINISTER OF DEFENCE: Gen. Tea Banh

ARMY CHIEF: Gen. Meas Sophea

NAVY CHIEF: Vice Adm. Tea Vinh

AIR FORCE CHIEF: Lt. Gen. Soeung Samnang

ECONOMY

Cambodia's foreign policy focuses on establishing friendly borders with its neighbours, as well as integrating itself into regional and global trading systems such as ASEAN & WTO. The Cambodian Government has been working with bilateral and multilateral donors, including the Asian Development Bank, the World Bank and IMF, to address the country's many pressing needs. In the last decade, Cambodia saw rapid progress in its economy. Cambodia's per capita income is rapidly increasing but is low compared to other countries in the region. Most rural households depend on agriculture and its related sub-sectors. As per data provided by IMF, the annual average GDP growth for the period 2001–2010 was 7.7% making it one of the world's top ten countries with the highest annual average GDP growth. China is the largest source of foreign aid and FDI. The service sector is heavily concentrated in trading activities and catering-related services. Rubber exports increased about 50% in 2011 due to continued demand for raw rubber, particularly from China, Malaysia, and Vietnam. GDP contracted slightly in 2009 as a result of the global economic slowdown but climbed more than 6% in 2010 and 6.7 in 2011, driven by tourism and renewed exports. The global financial crisis is weakening demand for many Cambodian exports, and construction is declining due to a shortage of credit. The long-term development of the economy remains a daunting challenge. The major economic challenge for Cambodia over the next decade will be fashioning an economic environment in which the private sector can create enough jobs to handle Cambodia's demographic imbalance.

GDP: $54.21 billion (2015 est.), $50.7 billion (2014 est.)

Real Growth Rate: 6.9% (2015 est.), 7.1% (2014 est.)

Industries: Tourism, garments, construction, rice milling, fishing, wood and wood products, rubber, cement, gem mining, textiles

Total Exports: $8.453 billion (2015 est.), $7.445 billion (2014 est.)

Export Commodities: Clothing, timber, rubber, rice, fish, tobacco, footwear

Major Markets: US 23.1%, UK 8.8%, Germany 8.2%, Japan 7.4%, Canada 6.7%, China 5.1%, Vietnam 5%, Thailand 4.9%, Netherlands 4.1% (2015)

Total Imports: $11.92 billion (2015 est.), $10.67 billion (2014 est.)

Import Commodities: Petroleum products, cigarettes, gold, construction materials, machinery, motor vehicles, pharmaceutical products

Major Suppliers: Thailand 28.7%, China 22.2%, Vietnam 16.4%, Hong Kong 6.1%, Singapore 5.7% (2015)

INTERNATIONAL DISPUTES

✦ Cambodia is concerned about Laos' extensive upstream dam construction

✦ Cambodia and Thailand dispute sections of boundary. In 2011 Thailand and Cambodia resorted to arms in the dispute over the location of the boundary on the precipice surmounted by Preah Vihear temple ruins, awarded to Cambodia by ICJ decision in 1962 and part of a planned UN World Heritage site.

✦ Cambodia accuses Vietnam of a wide variety of illicit cross-border activities. Progress continuing on a joint development area with Vietnam is hampered by an unresolved dispute over sovereignty of offshore islands.

DEFENCE

ARMY

Personnel: 80,000

Equipment

Category	Name	In Service
MBT	T-55	250
	Type 62/63	130
	Type 59	200
	PT-76	40
APC/MICV/IFV	BRDM -2	30+
	BTR-60	250
	BMP-1	100
	OT-64	40
	M113	20

Army continued...

Category	Name	In Service
Artillery	76 mm (Zis-3)	20+
	100 mm (T-12)	20
	122 mm Type 81	80
	130 mm M-46	100
	R-70	50
	107 mm Type 63	200
Mortar	160mm	
	81mm	
MRL	122 mm BM-21	
ATW	100mm & 82mm B-10	
AA gun	37mm Type 65/74	100+
	23mm ZU-23-2	100+
	57mm S-60	100+
	14.5mm ZPU-1/-2/-4	100
	12.7mm	
MANPADS	SA-7	
	HN-5	
	FN-12/16	

New Procurements/ Upgrades

No major procurement plans are under consideration at present.

NAVY

Personnel: 2500 (Marines included)

Equipment

Category	Name	In Service
Patrol Craft	Stenka class	4
	Koh Chlam class	2
	Zhuk class	2
	Shmel class	2
	T4 class	2
River patrol Craft	Riverine patrol craft	3

New Procurements/ Upgrades

✦ Cambodia is planning to buy two 140-metre heavy warships from China.

Royal Cambodian Marines embark aboard the amphibious dock landing ship USS Tortuga (LSD 46) to attend opening ceremonies for the first of two Cambodian phases of Cooperation Afloat Readiness and Training (CARAT) 2010. *Image Credit: US Navy*

AIR FORCE

Personnel: 2500

Equipment

Category	Name	In Service
Aircraft	L-39ZA Albatros	
Transport	MA60	1
	An-24	1
	Y-12	2
	A320	1
Helicopter	Mi-17	5
	Mi-8	4
	Mi-26	2
	Harbin Z-9	9
	AS355	1

New Procurements/ Upgrades

No major procurement plans are under consideration at present.

Royal Gendarmerie of Cambodia: The Royal Gendarmerie of Cambodia is a paramilitary unit deployed in all provinces to keep law and order and internal security in Cambodia.

Defence Expenditure

Total defence spending: $345 million (est.)

Defence spending in terms of GDP: 2% (est.)

The European Union (EU) signed an agreement with the government to provide Cambodia with radiation detection equipment and counterterrorism technology to detect illicit trafficking or criminal use of nuclear and radiological materials at the country's border checkpoints. The deal will see radiation detection equipment and alarm communication systems get installed at the New River Port in Phnom Penh and both Phnom Penh and Siem Reap International Airports.

Defence Production and R&D

Cambodia has some repair facilities for the maintenance of combat vehicles. The nation does not possess any defence industry base for producing equipment.

Defence Procurement

Cambodia has bought 100 tanks and armoured vehicles from Eastern Europe to improve the nation's military capacity.

General Tea Banh, Minister of National Defence. *Image Credit: Cambodia Defence Ministry*

CONTACT DETAILS

Ministry of National Defence

Russian Federation Blvd, Corner Street 171

12156 Phnom Penh, Cambodia

Tel/Fax: +855-(0) 23-883274

E-mail: minadet@camnet.com.kh

Website: http://www.mod.gov.kh

CAMEROON

(Capital: Yaounde)

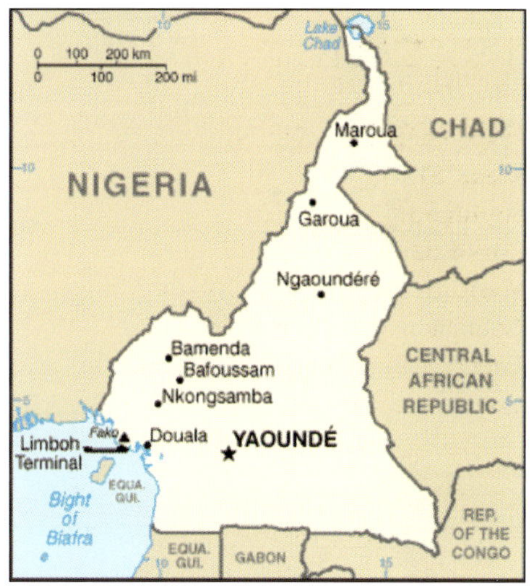

INTRODUCTION

Area: 475,440 sq km

Population: 24,360,803

Coastline: 402 km

Maritime claims: Territorial sea: 12 nm

Contiguous zone: 24 nm

KEY POLITICAL PERSONS

PRESIDENT & COMMANDER-IN-CHIEF: Paul Biya

PRIME MINISTER:
Philemon Yang

KEY DEFENCE PERSONS

MINISTER DELEGATE AT THE PRESIDENCY IN CHARGE OF DEFENCE: Joseph Beti Assomo

CHIEF OF DEFENCE STAFF: Maj. Gen. Rene Claude Meka

CHIEF OF STAFF OF THE ARMY: Maj. Gen. Baba Souley

CHIEF OF STAFF OF THE NAVY: Rear Adm. Jean Mendoua

CHIEF OF STAFF OF THE AIR FORCE: Brig. Gen. John Calvin Momha

ECONOMY

Cameroon has one of the best-endowed primary commodity economies in sub-Saharan Africa. The reason is its modest oil resources and favourable agricultural conditions. International oil and cocoa prices have a significant impact on the economy. Since 1990, the government has embarked on various IMF and World Bank programmes designed to spur business investment, increase efficiency in agriculture, improve trade, and recapitalise the nation's banks. The IMF is pressing for more reforms, including increased budget transparency, privatisation, and poverty reduction programmes. Subsidies for electricity, food, and fuel have strained the budget. New mining projects - in diamonds, for example - have attracted foreign investment, but large ventures will take time to develop.

GDP: \$72.64 billion (2015 est.), \$68.61 billion (2014 est.)

Real Growth Rate: 5.9% (2015 est.), 5.9% (2014 est.)

Industries: Petroleum production and refining, aluminium production, food processing, light consumer goods, textiles, lumber, ship repair

Natural gas - proved reserves: 135.1 billion cu m (1 January 2016 est.)

Oil - proved reserves: 200 million bbl (1 January 2016 est.)

Total Exports: \$5.756 billion (2015 est.), \$6.679 billion (2014 est.)

Export Commodities: Crude oil and petroleum products, lumber, cocoa beans, aluminium, coffee, cotton

Major Markets: China 16.7%, India 15.7%, Spain 6.2%, Belgium 6.1%, France 6.1%, Portugal 5.6%, Netherlands 5%, Italy 5% (2015)

Total Imports: \$6.5 billion (2015 est.), \$6.915 billion (2014 est.)

Import Commodities: Machinery, electrical equipment, transport equipment, fuel, food.

Major Suppliers: China 27.9%, Nigeria 13.9%, France 10.9%, Belgium 4.1% (2015)

INTERNATIONAL DISPUTES

- The United States has handed over six peacekeeping security vehicles to the Cameroonian armed forces in order to help combat Boko Haram, a terrorist organisation originating in Nigeria.
- Sovereignty dispute between Equatorial Guinea and Cameroon over an island at the mouth of the Ntem River.
- Joint Border Commission with Nigeria reviewed a 2002 ICJ ruling on the entire boundary and bilaterally resolved differences, including a June 2006 Greentree Agreement that immediately ceded sovereignty of the Bakassi Peninsula to Cameroon with a full phase-out of Nigerian control and repatriation of residents in 2008.
- Cameroon and Nigeria agreed on maritime delimitation in March 2008.

DEFENCE

ARMY

Personnel: 15,000

Equipment

Category	Name	In Service
APC/MICV/IFV	M-8	10
	FERRET	7
	VBL M-11	4
	COMMANDO V-150	40
	M-3 halftrack	8

Army continued...

Category	Name	In Service
Artillery	M101 105 mm	16
	SOLTAM 155 mm	4
	M116	5
MRL	122mm	20
Anti Tank GW	MILAN	5
	HOT	5
AA Gun	Type 63 37 mm	18
	35 mm twin	18
	Type 58 14.5 mm	16

New Procurements/ Upgrades

- Russia and Cameroon have signed a defence cooperation agreement aimed at boosting Cameroon's military as it combats Boko Haram militants. Russia will provide military equipment, training of Cameroonian military personnel and humanitarian aid to refugees and internally displaced people affected by the Boko Haram conflict.
- In 2014, Cameroon unveiled a variety of new Chinese hardware, including Type 07P infantry fighting vehicles and PTL-02-type tank destroyers.

US Soldiers with 173rd Infantry Brigade Combat Team (Airborne), move a casualty in a simulated patrol with Cameroon Army during Central Accord 2014 (CA 14), Mar. 18, 2014. *Image Credit: US Army*

NAVY

Personnel: 1,500

Equipment

Category	Name	In Service
Patrol Vessel	BAKASSI class (P48S type)	1
	L'AUDACIEUX class missile FAC (P48 type)	1
	ALFRED MOTTO class	1
	Swiftship	20
	Boston Whaler	3
Auxiliary Craft	LCM	2
	7m RHIB	2

New Procurements/ Upgrades

No major procurement plans are under consideration at present.

AIR FORCE

Personnel: 1800

Equipment

Category	Name	In Service
Combat	Atlas Impala	6
	MB-326	5
	Dassault/Dornier Alpha Jets	6
Transport/ Utility	C-130 Hercules	3
	IAI Arava 202	1
	Piper PA-23 Aztec	1
	Gulfstream III	1
	CN-235	1
Training	Humbert Tetras	7
	Dassault/Dornier Alpha Jets	6
Helicopter	SA 330 Puma	1
	Alouette II	1
	Mi-17	2
	Bell 206	
	Bell 412	

Cameroon became the first country outside of Nigeria to launch coordinated air strikes against Boko Haram.

New Procurements/ Upgrades

✦ Insitu Inc., Bingen, Washington, is being awarded $9,396,512 for firm-fixed-price delivery order against a previously issued basic ordering agreement for the procurement of one ScanEagle unmanned aircraft system for the government of Cameroon under the FMS programme. The system consists of analogue medium wave infra-red ScanEagle unmanned air vehicles, launch and recovery equipment, ground control stations, Insitu video exploitation systems and ground support equipment.

Defence Expenditure

Total defence spending: $376 million (est.)

Estimated defence spending in terms of GDP: 1.7% (est.)

Cameroon's government has tabled a 2016 budget of 4,200 billion CFA francs (USD 6.9 billion), up 13.5% on last year. Cameroon has increased its defence budget as it attempts to deal with Boko Haram militants carrying out raids in its territory. Cameroon is part of an 8,700-strong task force including troops from Chad, Niger, Nigeria and Benin that has pledged to destroy Boko Haram.

Defence Production and R&D

Cameroon does not undertake any defence production activities.

Defence Procurement

Cameroon Army and Air Force have not done any major procurement in recent years. The Cameroon Navy on the other hand, has modernised and increased its capabilities with the acquisition of a number of small Rodman patrol craft and the retirement of some small older craft.

CONTACT DETAILS

Ministry of Defence
PO Box 1162, Yaoundé
Cameroon
Tel: (+237 2) 222 69 20
(+237 2) 222 19 31
(+237 2) 223 40 45
Fax: (+237 2) 223 59 71

STRATEGIC INFORMATION

CANADA
(Capital: Ottawa)

INTRODUCTION

Area: 9,984,670 sq km

Population: 35,362,905 (July 2016 est.)

Coastline: 202,080 km

Maritime claims: Territorial sea: 12 nm

Contiguous zone: 24 nm

Exclusive Economic Zone: 200 nm

Continental shelf: 200 nm or to the edge of the continental margin

GEOPOLITICAL IMPORTANCE

Canada is a North American country consisting of ten provinces and three territories. Located in the northern part of the continent, it extends from the Atlantic Ocean in the East to the Pacific Ocean in the West and northward into the Arctic Ocean. It is the world's second largest country by total area, after Russia. Canada's common border with the United States to the South and northwest is the longest in the world. With the opening of the Arctic Ocean for sea trade Canada forms a part of the Arctic Council and is gradually investing in that region.

POLITICAL OVERVIEW

Canada is a federal state that is governed as a parliamentary democracy and a constitutional monarchy with Queen Elizabeth II as its head of state. The Queen's representative, the Governor General of Canada, carries out most of the federal royal duties in Canada.

KEY POLITICAL PERSONS

HEAD OF STATE & COMMANDER-IN-CHIEF: Queen Elizabeth II

GOVERNOR GENERAL: David Lloyd Johnston

PRIME MINISTER: Justin Trudeau

ECONOMY

Canada is the world's 11th largest economy as of 2015, with a nominal GDP of approximately USD 1.79 trillion. Canada has the eleventh largest economy in the world, is one of the world's wealthiest nations, and is a member of the Organization for Economic Co-operation and Development (OECD) and Group of Eight (G8). As with other developed nations, the Canadian economy is dominated by the service industry, which employs about three quarters of Canadians. Canada enjoys a substantial trade surplus with the US, which absorbs about three-fourths of Canadian exports each year. Canada is the US's largest foreign supplier of energy, including oil, gas, uranium, and electric power. Canada achieved marginal growth in 2010 and 2011 and plans to balance the budget by 2015. In addition, the country's petroleum sector is rapidly becoming an even larger economic driver with Alberta's oil sands significantly boosting Canada's proven oil reserves, ranking the country third in the world behind Saudi Arabia and Venezuela.

GDP: $1.634 trillion (2015 est.), $1.616 trillion (2014 est.)

Real Growth Rate: 1.1% (2015 est.), 2.5% (2014 est.)

Industries: Transportation equipment, chemicals, processed and unprocessed minerals, food products, wood and paper products, fish products, petroleum, natural gas

Oil - proved reserves: 171 billion bbl (1 January 2015 est.)

Natural gas - proved reserves: 1.889 trillion cu m (1 January 2014 est.)

Total Exports: $411 billion (2015 est.), $478.3 billion (2014 est.)

Export Commodities: Motor vehicles and parts, industrial machinery, aircraft, telecommunications equipment; chemicals, plastics, fertilizers; wood pulp, timber, crude petroleum, natural gas, electricity, aluminium

Major Markets: US 76.7% (2015)

Total Imports: $428.7 billion (2015 est.), $473.9 billion (2014 est.)

Import Commodities: Machinery and equipment, motor vehicles and parts, crude oil, chemicals, electricity, durable consumer goods

Major Suppliers: US 53.1%, China 12.2%, Mexico 5.8% (2015)

DEFENCE & SECURITY

The Department of National Defence (DND) is the department within the government of Canada with responsibility for all matters concerning the defence of Canada. In addition to the civilian components of the department, this also includes Canada's military, known as the Canadian Forces. DND is led by the Minister of National Defence and is headquartered at National Defence Headquarters (NDHQ) in Ottawa. The Canadian Forces (CF) are the unified armed forces of Canada and consists of the sea, land, and air environmental commands called the: Royal Canadian Navy (RCN), Canadian Army, and Royal Canadian Air Force (RCAF), which together are overseen by the Armed Forces Council, chaired by the Chief of the Defence Staff. The Commander-in-Chief is the Canadian monarch, Elizabeth II, represented by the governor general. Canada's First Defence Strategy is to increase personnel levels to 70,000 active and 30,000 reserves sometime in the next 20 years, with an increase to 69,300 by 2014/2015. The reserves should reach 28,000 by 2014/2015. The Canada First Defence Strategy (CFDS) entails military recruitment, procurement and improvement strategy of the Canadian government to enhance the overall effectiveness of the Canadian Forces. The strategy also aims to enforce Arctic sovereignty with the Royal Canadian Mounted Police and the Canadian Coast Guard. The purpose of the Canada First Defence Strategy is to give Canada a modern military with clearly defined missions and capabilities. The strategy focuses on many key military functions and operations and on improving the equipment and fleets that the Canadian Forces operates.

KEY DEFENCE PERSONS

MINISTER OF DEFENCE: Harjit Singh Sajjan

CHIEF OF THE DEFENCE STAFF: Gen. J.H. Vance

COMMANDER OF THE ARMY: Lt. Gen. P.F. Wynnyk

COMMANDER OF THE NAVY: Vice Adm. M.F.R. Lloyd

COMMANDER OF THE AIR FORCE: Lt. Gen. M.J. Hood

COMMANDER FOR JOINT OPERATIONS COMMAND: Lt. Gen. Stephen Bowes

Internal Conflict

Terrorism: Terrorism in Canada primarily consists of fundraising for terrorist attacks outside of the country. It includes fundraising, lobbying through front organisations, providing support for terrorist operations, procuring weapons and material, coercing and manipulating immigrant communities in Canada, facilitating transit to and from the US and other countries, and other illegal activities.

External Conflict

- Canada has managed maritime boundary disputes with the US at Dixon Entrance, Beaufort Sea, Strait of Juan de Fuca, and the Gulf of Maine, including the disputed Machias Seal Island and North Rock.
- Canada and the United States dispute regarding the division of the Beaufort Sea and the status of the Northwest Passage is being resolved deliberately and both are cooperating on the survey of Arctic continental self.
- US works closely with Canada to intensify security measures for monitoring and controlling legal and illegal movement of people, transport, and commodities across the international border.
- Sovereignty dispute with Denmark over Hans Island in the Kennedy Channel between Ellesmere Island and Greenland.
- Commencing the collection of technical evidence for submission to the Commission on the Limits of the Continental Shelf in support of claims for continental shelf beyond 200 nautical miles from its declared baselines in the Arctic, as stipulated in Article 76, paragraph 8, of the United Nations Convention on the Law of the Sea.

War on Terrorism: Canada joined the United States in the war on terror after the attacks on the World Trade Centre and Pentagon.

Afghanistan War: Canada's goal is to help Afghans rebuild Afghanistan into a viable country that is better governed, more stable and secure, and never again a safe haven for terrorists. To this end, Canadians have assisted in improving security, diplomacy, human rights and development. After 11 September 2001 attacks, Canada announced that it will contribute air, land and sea forces to Operation Enduring Freedom. On July, 2011 Canada terminated its combat mission in Kandahar province. Further on 15 December 2011, the last rotation of troops returns to Canada from Kandahar after completing the closeout of military operations in Kandahar Province as part of Mission Transition Task Force (MTTF) from July to December 2011. Canada's participation is currently in the NATO Training Mission–Afghanistan (NTM-A), which delivers training and professional development support to the national security forces of Afghanistan: the Afghan National Army (ANA), the Afghan Air Force (AAF), and the Afghan National Police (ANP).

Iraq War: Canada did not join the invading coalition, but still participated in the conflict in Iraq, joining a number of non-belligerent nations in helping to rebuild the country post-invasion. This included the training of Iraqi police and army officers, and contributing approximately USD 300 million towards this effort. Canadian Special Forces troops in northern Iraq are involved in several with Islamic State extremists. In November 2015, Prime Minister Justin Trudeau announced the withdrawal of Canada's fighter aircraft from the fight against ISIL, but their surveillance and transport and refuelling aircraft would remain in the area.

NATO: Over the years, Canada has sent over 120,000 troops as part of UN peacekeeping missions, and it has the 2nd highest peacekeeping fatality with 114 fatalities. Currently, Canadian troops are part of the NATO-led mission in Afghanistan, ISAF. The Canadian Forces have also participated in NATO-led missions in Libya and Syria.

Operation REASSURANCE in Europe: Operation REASSURANCE refers to the military activities undertaken by the Canadian Armed Forces (CAF) to support NATO assurance measures through the provision of military capabilities for training, exercises, demonstrations and assigned NATO tasks.

International operations, Ukraine, Operation UNIFIER: Operation UNIFIER is Canada's contribution to help the Ukrainian Armed Forces (UAF) promote security in the region through capacity building, training, and assisting the development of a Western-style (NATO) patterned military. Approximately 200 CAF personnel will be deployed to Ukraine until 31 March 2017 on both a sustained and periodic basis depending on the type of training being conducted.

International operations, Middle East, Operation IMPACT: Operation IMPACT is the CAF contribution to the Middle East Stabilization Force - the multinational military coalition against the Islamic State of Iraq and Syria (ISIS) in the Republic of Iraq and in Syria. As of 10 August 2015, Air Task Force-Iraq has conducted 1325 sorties: CF-188 Hornet fighters have conducted 853 sorties; the CC-150T Polaris has conducted 226 sorties and delivered over 13.681 million pounds of fuel to coalition aircraft; and CP-140 Aurora aircraft have conducted 246 reconnaissance missions.

Continental operations, Northwest Territories, Operation NUNAKPUT: The primary goals of Op NUNAKPUT were asserting sovereignty over Canada's northernmost regions, as well as maintaining interoperability between the Canadian Rangers and other government departments (OGD's) in order to maximise effectiveness in response to safety and security issues in the North.

Continental operations, Saskatchewan, Operation LENTUS 15-02: CAF members deployed to northern Saskatchewan to support local and provincial firefighting efforts following a Request for Assistance made by the Province of Saskatchewan on 4 July. During Operation LENTUS 15-02, CAF members conducted 12 days of fire line operations, from 8-20 July 2015, in the vicinities of Montreal Lake and La Ronge.

THREAT PERSPECTIVE

Although Canada has not often been targeted specifically for a terrorist attack, it is vulnerable to terrorism. Extremists from environmental and animal-rights groups are willing to use dangerous and violent tactics in the fight for their cause. Graduates of terrorist training camps in countries such as Afghanistan and Pakistan reside in Canada or continue to seek access to Canada.

To effectively combat terrorist activities, enhanced participation in joint initiatives and the sharing of information have become critical. Also crucial is the growing collaboration between agencies that seek to prevent attacks and those seeking evidence to support prosecution.

STRATEGIC RELATIONS

United States: Canada and the United States are currently the world's largest trading partners, share the world's longest border, and have significant interoperability within the defence sphere. American defence arrangements with Canada are more extensive than with any other country. The Permanent Joint Board of Defence, established in 1940, provides policy-level consultation on bilateral defence matters. The United States and Canada share North Atlantic Treaty Organization (NATO) mutual security commitments. American and Canadian military forces have cooperated since 1958 on continental air defence within the framework of the North American Aerospace Defence Command (NORAD). Canada has been involved in international responses to the threats from Daesh/ISIS/ISIL in Syria and Iraq, and is a member of the Global Coalition to Counter Daesh.

India: Indo-Canadian relations are the longstanding bilateral relations between India and Canada. Being fellow members of the Commonwealth of Nations, Canada and India share high commissioners rather than ambassadors. Canada and India maintain a dialogue on regional security and global strategic issues of common interest through the annual Canada-India Strategic Dialogue and through regular meetings of the Canada-India Joint Working Group on Counter-Terrorism, as well as annual Foreign Policy Consultations. Canada and India maintain a bilateral dialogue on anti-terrorism, including an annual meeting of the Canada-India Strategic Dialogue, as well as regular meetings of the aforementioned Canada-India Working Group on Counter-Terrorism.

Japan: Canada and Japan are partners in numerous international groups and organizations including the G7, G20, APEC, the ASEAN Regional Forum, and the OECD. Japan and Canada have been in good relations and have frequent high level exchanges. Japan and Canada have been cooperating in many fields. The Japan-Canada/ Canada Japan Inter-Parliamentary Friendship Group meets every year. Talks between the two governments include peace and security, housing, fisheries, finance, and telecommunications.

Multilateral Alliances

Canada is a member of the G7, G8, G20, NATO, OECD, WTO, Commonwealth of Nations, Francophonie, OAS, APEC, and the United Nations.

DEFENCE CAPABILITIES

ARMY: The Canadian Army is the land component of the combined Canadian Forces. Canada's Army is the largest section of the Canadian Forces with 35,000 soldiers. The Regular Force units have 19,500 full time soldiers; the Army Reserve has 16,000 part-time, volunteer soldiers. All are supported by 4,100 civilian employees. The Regular Force has three Mechanized Brigade Groups composed of units stationed in Western Canada, Ontario, Quebec and Atlantic Canada. The Army is equipped with the latest in modern weapons. The Coyote armoured reconnaissance vehicle and the LAV III infantry combat vehicle provide significant surveillance and combat capability. Both Canadian-built vehicles are equipped with state-of-the-art target acquisition systems, night-fighting capability, and enhanced crew protection. The Armoured Regiments fight with the Leopard 2A6M tank. Artillery Regiments have recently been equipped with the new M777 ultra-light howitzer. Specially tasked Air Force units provide dedicated Griffon CH 146 Helicopter capabilities to the Army.

Major Army bases: Edmonton, Alberta; Shilo, Manitoba; Petawawa, Ontario; Kingston, Ontario; Montreal, Quebec; Valcartier, Quebec; Gagetown, New Brunswick and numerous training facilities.

Personnel: 49,500 (17,000 reserve)

Equipment

Category	Name	In Service
MBT	Leopard 2A4+/2A4M	60
	Leopard 2A6M	20
APC/IFV	M113A3/MTVL	289
	Cougar	40
	Grizzly AVGP	274
	Husky AVGP	5
	LAV III	651
	Coyote Reconnaissance Vehicle	203
	Nyala	75
	Buffalo A2	19
	Textron TAPV	500
	Bison	199

Army continued...

Category	Name	In Service
Artillery	105 mm LG1 Mark II	28
	105 mm C3 Close Support Gun	96
	105 mm Field Gun C1	19
	155 mm M777 Ultra Light Howitzer	37
Mortar	60 mm M19	
	81 mm L16	
Anti-Tank GW	TOW 2	40
	ERYX	435
	ITAS	10
Recoilless rifle	Carl Gustav M2	935
	Carl Gustav M3	227
SAM	ADAT	16
UAV	RQ-21 Blackjack	5
	IAI Heron	
	RQ-11 Raven	

New Procurements/ Upgrades

✦ Canadian Army has awarded USD 5.5 million contract to provide a new technology to adapt a C16 automatic grenade launcher system for use in realistic live training environments.

✦ The Department of National Defence is considering the purchase of the Long-Range Precision Rocket System (LRPRS). The Canadian Forces will acquire in total 17 multiple rocket launchers by the year 2018.

✦ The Government of Canada has awarded a contract, valued at USD 10.16 million to Wajax Power Systems of Québec City, Quebec, for the repair and overhaul of engines for the Leopard 2 family of vehicles over three years. The contract includes two one-year extension options, which could bring the total contract value to USD 22.09 million.

✦ The Canadian Defence Ministry has selected Textron Systems Canada to supply hundreds of next-generation Tactical Armoured Patrol Vehicles (TAPV) to the country's Army. The Defence Ministry announced the contract on June 8 under which Textron will supply 500 of the 4x4 advanced wheeled combat vehicles to the regular and reserve forces of the Canadian Army. The contract is valued at USD 603.4 million CAD (USD 585 million) and includes an additional five year in-service support contract of USD 105.4 million CAD (USD 102 million).

✦ The Close Combat Vehicle (CCV) will provide the Canadian Forces with a medium-weight armoured vehicle that is both very well protected and tactically (off-road) mobile. The CCV project will involve the procurement of 108 CCV, to include Infantry Fighting Vehicle, Forward Observation Officer, Engineer Reconnaissance, and Tactical Command configurations. The project scope also includes the option to procure up to 30 additional vehicles, as well as the development and implementation of an in-service support contract.

✦ Rheinmetall has been awarded two contracts, worth a total of €87 million, for equipping the Canadian armed forces. The first contract, worth €70 million, involves supplying the Canadian Army with a total of 304 automatic grenade launcher systems and accompanying ammunition. The launchers will be deployed under the C16 - Automatic Grenade Launcher System designation. The order also includes 250,000 rounds of 40mm practice and service ammunition, to be delivered in several lots. The second contract, worth €17 million, is to modernise and overhaul 42 Leopard main battle tanks taken over from the Dutch armed forces by the Canadian Army. The tanks will be refitted to meet Canadian Army standards and to enable integration into existing C4I structures.

✦ The Canadian Department of National Defence announced that USD 5 billion would be spent to enhance, replace and repair the army's armoured vehicles. LAV III armoured vehicles (AV) will be upgraded with improved protection and automotive components. Of the USD 5 billion announced, approximately 20% of it will be used to upgrade LAV III models. The upgrade will extend the LAV III life span to 2035. The remaining USD 4 billion is to be spent on a "new family of land combat vehicles".

✦ Canada's M113 armoured personnel carrier fleet will be upgraded to various configurations and retained until 2020. Modifications include more powerful engines, upgraded suspension, bolt-on steel armour plates and/ or steel cage armour, and improved armament consisting of either a Cadillac-Gage 1 metre turret or a remote weapon station.

✦ Rheinmetall Canada has been awarded an important contract by Textron Systems Canada Inc. to provide the Canadian Army with new and improved Tactical Armoured Patrol Vehicles.

✦ The FME (Force Mobility Enhancement) project includes the acquisition of a fleet of Armoured Engineer Vehicles that will provide crucial support to the Leopard 2 tanks, LAV III, and future fleets, such as the CCV and the TAPV. The project will also deliver tactical mobility implements capable of being used on the fleet of Leopard 2 Main Battle Tanks.

✦ On January 24, 2013, the Government of Canada announced the on-schedule delivery of Canada's first upgraded Light Armoured Vehicle III, in London, Ontario.

✦ General Dynamics Land Systems Canada, the producer of the LAV 6.0, awarded DRS Sustainment Systems, Inc. an eight-year, firm fixed-fee contract valued at approximately USD 100 million for the production of masted surveillance suites and accompanying operator control stations, as well as program engineering and service support.

Soldiers from the 1st Royal Canadian Regiment (1 RCR) return to their Light Armoured Vehicle 6.0 (LAV 6.0) after completing a sweep of an enemy during Exercise Storming Bear on 1st November 2016, at 4th Canadian Division Support Base Petawawa, Ontario. *Image Courtesy: Corporal Kris Reeve, Royal Canadian Army*

NAVY: The Royal Canadian Navy (RCN) is the naval force of Canada. The navy is composed of approximately 8,500 regular and 5,100 reserve sailors, along with 5,300 civilian personnel. The Canadian Navy fleet, roughly balanced between the Atlantic (Halifax, N.S.) and Pacific (Esquimalt, B.C.) coasts, has 33 destroyers, frigates, replenishment ships submarines and maritime coastal defence vessels, plus many more auxiliary and support vessels. Canada's 33 warships, submarines, and coastal defence vessels are divided more or less evenly between the Atlantic and Pacific coasts. They serve a number of roles; from major international operations, sovereignty operations and patrols, to single ship support of multinational deployments. The Naval Reserve, headquartered in Québec City, is made up of 24 Naval Reserve Divisions (of up to 150-200 personnel each) located across Canada. The Royal Canadian Navy's Atlantic Fleet, known as Maritime Forces Atlantic (MARLANT), is headquartered and home-ported at CFB Halifax in Halifax, Nova Scotia and is supported by CFS St. John's in Newfoundland. The Royal Canadian Navy's Pacific Fleet, known as Maritime Forces Pacific or MARPAC, is headquartered at CFB Esquimalt in British Columbia.

Personnel: 13,600 (5,100 reserve)

Equipment

Category	Name	In Service
Destroyer	Iroquois Class DDG	1
Submarine	Victoria Class	4
Frigate	Halifax Class FFG	12
Patrol Vessel	Kingston Class	12
Training vessel	Orca Class	8
Auxiliary ship	Queenston class	(2-3 Planned)
	Harry DeWolf-class OPV	6 (Planned)
Weapon	Mk 141 Harpoon (SSM), Mk 41(GMVLS), Phalanx 20mm (CIWS), Mk 32 Torpedo, Bofors SAK 57 mm	

Naval Aviation

Category	Name	In Service
Helicopter	CH-124 SeaKing (ASW)	41

New Procurements/ Upgrades

✦ The Arctic/ Offshore Patrol Ship (AOPS) Project has been established in order to deliver to the government of Canada a naval ice-capable offshore patrol ship to assert and enforce sovereignty in Canada's waters including the Arctic. The AOPS Project will deliver to the Canadian Forces six to eight fully supported Arctic/ Offshore Patrol Ships. The estimated cost of acquiring these ships is USD 3.1 billion, with approximately USD 4.3 billion provided for operations and maintenance over their 25-year lifespan.

✦ Two contracts have been awarded - USD 549 million to Halifax Shipyards of Nova Scotia and USD 351 million to Victoria Shipyards Company Limited of British Columbia for the modernisation and life extension of 12 Canadian-built HALIFAX class multi-role patrol frigates. Later, Lockheed Martin Canada was awarded the combat systems integration contract. The final ship upgrade will be completed in 2017.

✦ The JSS project represents a total investment by the Government of Canada of approximately USD 2.6 billion. Canadian shipyard, selected as part of the National Shipbuilding Procurement Strategy, will be engaged to complete the design of and build the Joint Support Ships. The JSS will replace the Navy's current Auxiliary Oiler Replenishment vessels that are now more than 40 years old and nearing the end of their service lives. The new ships will provide core replenishment, limited sealift capabilities, and support to forces ashore.

✦ The Government of Canada has reached agreements with Irving Shipbuilding Inc. and Vancouver Shipyards Co. Ltd. This charts the course for construction of Canada's combat and non-combat surface fleets under the National Shipbuilding Procurement Strategy.

The Canadian patrol frigate HMCS Halifax (FFH 330) transits the Caribbean Sea Jan. 18, 2010. *Image Courtesy: US Navy*

AIR FORCE: The Royal Canadian Air Force (RCAF) is responsible for all aircraft operations in the Canadian Forces, enforcing the security of Canada's airspace and providing aircraft to support the missions of Royal Canadian Navy and Canadian Army. The RCAF is a partner with the United States Air Force in protecting continental airspace under the North American Aerospace Defence Command (NORAD). The RCAF also provides all primary air resources to the National Search and Rescue Programme.

Personnel: 14,500 (2,600 reserve)

Equipment

Category	Name	In Service
Fighters/attack/ Reconnaissance	CF-188 Hornet	77
Transport/ Tanker	CC-130 Hercules	27
	CC-144 Challenger	4
	CC-114U Challenger	2
	CT-142 Dash 8	4
	CC-138 Twin Otter	4
	CC-150 Polaris	5
	CC-115 Buffalo	6 (To be replaced by 2018)
	CC-177 Globemaster	5
	CC-130J Super Hercules	17 (To be replaced by 2018)
Helicopter	CH-124A Sea King	27
	CH-146 Griffon	85
	CH-149 Cormorant	14
	CH-47D Chinook	6
Trainer	CT-114 Tutor	25 (To be replaced by 2020)
	CT-156 Harvard II	24
	CT-155 Hawk	19
Maritime Patrol	CP-140 Aurora	16
	CP-140A Arcturus	2
Missile	AGM-65G Maverick (AGM), AIM-120 (AAM), AIM-9 (AAM), AIM-7 (AAM)	

New Procurements/ Upgrades

✦ Calian Technologies Ltd. announced that Amtek, a Calian subsidiary, has been awarded a contract renewal valued at USD 15 million for the Royal Canadian Air Forces' (RCAF) Airworthiness Programme, Gatineau, Quebec.

✦ MDA has signed a contract amendment for CAUSD 3 million with Canada's Department of National Defence to provide upgrades to the ground systems supporting the radar surveillance systems developed by MDA and used on Canada's fleet of CP-140 Aurora aircraft. The CP-140 Aurora fleet is Canada's Intelligence, Surveillance and Reconnaissance aircraft.

✦ The Royal Canadian Air Force has formally accepted the 17th Lockheed Martin CC-130J Super Hercules. The first CC-130J was accepted in June 2010. Lockheed Martin has delivered more than USD 350 million in Industrial and Regional Benefits (IRB) to Canadian industry as a result of the Government of Canada's procurement of the CC-130J aircraft and In-Service Support.

✦ Canadian Forces Air Command will be replacing CH-124 Sea Kings with a fleet of new maritime helicopters. The force is acquiring a fleet of 28 new, fully-equipped CH-148 Cyclone helicopters bundled with a long-term in-service support programme.

✦ The Canadian government has said it will choose one of four fighter jets in competition to replace its aging fleet of CF-18s. The F-35 has been touted as a technological wonder and the ultimate stealth attack aircraft able to evade radar detection.

✦ Canada signed a USD 1.15 billion contract with Boeing to purchase 15 CH-47Fs, entering service with the Canadian Forces after its planned withdrawal from combat operations in Afghanistan.

✦ Lockheed Martin has been awarded a five-year contract extension worth approximately USD 504.3 million by the Government of Canada to provide In Service Support (ISS) for the Royal Canadian Air Force's (RCAF) fleet of 17 C-130J Super Hercules aircraft (CC-130J).

CF-188 Hornet. *Image Courtesy: Royal Canadian Air Force*

Major Armaments Producers

L-3 Communications MAS, CAE Meggitt Training Systems Canada, Colt Canada, Textron Systems Canada, Kongsberg Protech Systems, Canada, Rheinmetall Defence Canada, Irving Shipbuilding Inc., General Dynamics Land Systems Canada, Raytheon Canada Limited, Seaspan Marine Corporation, Thales Canada, Boeing Canada

Defence Production and R&D

Defence Research and Development Canada (DRDC) is an agency of the Canadian Department of National Defence that responds to the scientific and technological needs of the Canadian Forces. Defence R&D Canada has an annual budget of USD 350 million. DRDC is organised around nine research centres located across the country with headquarters in Ottawa. Each research centre focuses on a particular set of scientific and operational requirements. DRDC supports innovation and S&T as an engine for generating solutions. DRDC accesses knowledge, builds networks and leverages capacity across a broad range of partners, nationally and internationally creating innovative partnerships for new solutions to challenging problems.

General Dynamics Canada has been awarded a contract to develop and demonstrate an automated computer-network defence capability to enhance the security of the Canadian Department of National Defence's (DND) networks. The ARMOUR project, conceptualised by Defence Research and Development Canada (DRDC), will develop and integrate advanced scientific and technological solutions to the increasingly challenging and complex problem of defending DND networks, protecting their mission-critical information. ARMOUR will integrate leading-edge network cyber assessment tools to proactively deal with cyber vulnerabilities and mitigate attacks in real time, as well as automatically generate optimised courses of action for potential future threats. Demonstrations will be used to validate the functionality of the resulting system.

DEFENCE DEALS

Arms Export

Canada remained the sixth-largest arms exporter in 2015-2016 and the bar has been raised high since 2013-2014. A key priority of Canada's foreign policy is the maintenance of peace and security. The Government of Canada guarantees that Canadian military exports are not prejudicial to peace, security or stability in any region of the world or within any country. Canadian Arms Sales are governed by the country's Export and Imports Permits Act. Sales with the United States are also specifically regulated by the 1959 Defence Production Sharing Arrangement. Canada jumped from fifteenth to twelfth largest exporter of military hardware in the world. According to data from the Stockholm International Peace Research Institute (SIPRI), Canada was the 12th largest exporter of military hardware in the world in 2010. The country exports to several governments engaged in human rights violations, like the Philippines, Israel, Saudi Arabia, China, Libya and Tunisia. Between 2006 and 2009 Canada exported some USD 1.4 billion worth of military gear, according to the Report on the Export of Military Goods from Canada for 2007-2009, published by the Department of Foreign Affairs. Canada has become the world's second-largest exporter of defence equipment to the Middle East after United States.

DEFENCE BUDGET

Total defence spending: $13.82 billion (2016-2017)

Estimated defence spending in terms of GDP: 0.97% (2015)

The Canadian Armed Forces said it will continue to play an important role in the coalition fight against the Islamic State of Iraq and the Levant (ISIL), and in North Atlantic Treaty Organization (NATO) assurance measures in Central and Eastern Europe. Budget 2016 also reaffirms Canada's contribution to support Ukrainian forces through military training and capacity-building in 2016-17, in coordination with other countries providing similar training assistance. The Government will seek the input of Canadians, experts, allies and partners, and Parliament on the strategic environment for the Canadian Armed Forces, the roles for the Canadian Armed Forces, and the Canadian Armed Forces size, structure and capabilities.

Budget 2016 proposes to reallocate funding of USD 3.716 billion for large-scale capital projects from the 2015-16 to 2020-21 period to future years. Total funding available to National Defence for large-scale capital projects is USD 84.3 billion over 30 years out to 2044-45, and USD 2.8 billion per year ongoing starting in 2045-46, on an accrual basis. Budget 2016 proposes to provide USD 200.5 million over two years on a cash basis, starting in 2016-17, to National Defence to undertake infrastructure projects at Canadian Forces Bases and other defence properties across Canada. Investments include USD 77.1 million on projects to support readiness for Canadian Armed Forces military operations, USD 67.4 million on projects to support the Reserve Force, USD 6.0 million on projects to support northern operations and USD 50.0 million on projects to support military personnel and their families.

Defence Procurement

The activities of the Canadian Forces and the Department of National Defence (DND), like those of every other federal government organisation, are carried out within a framework of legislation that is approved and overseen by Parliament. The Canadian Forces and the Department of National Defence have complementary roles to play in providing advice and support to the Minister of National Defence and in implementing the decisions of the Government on the defence of Canada and of Canadian interests at home and abroad. In broad terms:

- The Deputy Minister has responsibility for policy, resources, interdepartmental coordination and international defence relations; and
- The Chief of the Defence Staff has responsibility for command, control and administration of the Canadian Forces and military strategy, plans and requirements.

The Government of Canada is delivering on its "Canada First" commitment to strengthen Canada's multi-role, combat-capable defence force. The Department of National Defence is purchasing equipment and related support services for the Canadian Forces.

Tank Replacement Project *(Budget: USD 650 million; Status: 100 tanks purchased)*: The Government of Canada continues to award several contracts under the Tank Replacement Project generating economic benefits for Canadian industry. The heavily protected direct fire capability of a Main Battle Tank is an invaluable tool in the arsenal of any military. The Government of Canada will launch a competition to industry for the in-service support for the tanks required over the next 30 years.

Integrated Soldier System Project *(Budget: USD 250 million; Status: Delivery expected in 2017)*: The Integrated Soldier System Project (ISSP) will acquire and support over 4000 soldier-wearable communication suites, complete with required accessories, support equipment, contract management, training, logistic and engineering support (collectively referred to as the Integrated Soldier System). Defence Ministry announced the awarding of the contract for the Integrated Soldier Systems Project on July 27, 2015. The system will increase soldiers' battlespace awareness using modernised and secure communication and navigation innovation.

Medium Range Radar Project *(Budget: USD 250 million; Status: Delivery expected in 2017)*: The plan is to have 10 systems that will provide the Canadian Armed Forces with a means to detect hostile projectiles and track airborne targets.

Medium Support Vehicle System Project *(Budget: USD 725 million)*: In June 2015, a contract was awarded for the purchase of 1537 standard military pattern trucks.

Halifax-class Modernisation / Frigate Life Extension *(Budget: USD 4.3 billion; Status: On track to meet final operational capability in 2018)*: The contract is for upgrading fleet of 12 modernised and upgraded Halifax-class frigates. Each modernisation and refit period is expected to take approximately 18 months. The final ship upgrade is anticipated to be completed in the 2017 timeframe. Following acceptance of HMCS Toronto, in late Summer 2016, corresponding sea trials and project close-out activities are expected to be completed by January 2018.

Victoria-class Submarines *(Budget: USD 450 million; Status: Completion of final vessel anticipated in 2017)*: The project will extend the life of fleet of four Victoria-class submarines.

Arctic/Offshore Patrol Ships *(Budget: USD 3.5 billion; Status: Construction to begin in 2015)*: The AOPS project will deliver six ice-capable offshore patrol ships that will conduct sovereignty and surveillance operations in Canada's Exclusive Economic Zone, including in the Arctic. The Royal Canadian Navy will also use the AOPS to support other units of the Canadian Armed Forces (CAF) in the conduct of maritime-related operations and to support other government departments in carrying out their mandates, as required. On September 1, 2015, construction of the first Arctic and Offshore Patrol Ship, the future HMCS Harry DeWolf, began. On August 25, 2016, the Arctic and Offshore Patrol Ship project marked another significant milestone with the cutting of steel of the future HMCS Margaret Brooke, the second ship in the class.

Canadian Surface Combatant *(Budget: USD 26.2 billion; Status: First delivery anticipated in 2025)*: The CSC project will renew the Royal Canadian Navy surface combat fleet by replacing the capabilities provided by the destroyers (Iroquois-class) and the multi-role patrol frigates (Halifax-class). The Canadian Surface Combatant project is still in its early stages and construction is not expected to begin until the early 2020s.

Queenston-Class Joint Support Ship *(Budget: USD 2.6 billion; Status: Construction to begin in fiscal 2016-17)*: The JSS project will deliver two ships with an option for a third. The capabilities required of the JSS are crucial to the Canadian Armed Forces (CAF). These ships will enable a Naval Task Group to remain at sea for extended periods of time.

CP-140 Aurora *(Budget: USD 2 billion; Status: Six aircraft have been modernised and life extended)*: The project was initiated to extend the life of 14 CP-140 Aurora aircraft to 2030 with fully modernised platforms.

Maritime Helicopter Project *(Budget: USD 3.2 billion; Status: First six of fleet of 28 delivered)*: The Maritime Helicopter Project is a program that involves the delivery of 28 state-of-the-art, combat-capable CH-148 Cyclone helicopters, associated logistical and in-service support, spare parts, as well as modifications to the Halifax-class ships and construction of a new training facility equipped with a full training suite of flight, mission and maintenance simulators.

CF-188 Hornet: The modernisation programme extends the life of the CF-188 Hornet fighter fleet. The CF-18 has recently been put through a two-phase modernisation programme, a comprehensive mid-life upgrade to ensure that the Canadian Forces have a modern and interoperable fighter fleet.

CONCLUSION

Defending Canadians from threats to their safety and well-being is a critical role for government. To deliver on this core responsibility, the Government is committed to rebuilding the Canadian Forces into a first-class, modern military. Canadian defence policy today is based on the Canada First Defence Strategy. Based on that strategy, the Canadian military is oriented and being equipped to carry out six core missions within Canada, in North America and globally. Canada needs and benefits from combat-capable maritime, land and air forces able to fulfill a broad range of missions and tasks. While Canada faces no direct conventional military threat, there remain direct and indirect threats to the national security which includes drugs, organised crime, illegal immigration, terrorism and the uncertainty caused by the growing proliferation of missiles carrying weapons of mass destruction, according to Canadian Forces. The global economic downturn has put pressure upon all Western governments to reduce costs, and the Defence departments of Canada's allies have all had to trim costs and find efficiencies. Canada is no different in this regard. The force development plans and mission projections for the Canadian Forces (CF) and the ongoing Revolution in Military Affairs (RMA) have far-reaching implications for the future role of Canada's Military Intelligence also. Till now, approximately 41,000 Canadian forces personnel have served in the Afghanistan mission; USD 7.5 billion have been spent on combat operations. Since the Korean War, it was the forces' first combat experience. These experiences in Afghanistan and in other operations, such as counter-piracy missions off the Horn of Africa, have had an important impact on the armed forces. The Canadian Forces are a vital instrument of national defence and sovereignty and a key implement for the achievement of Canadian national goals at home and abroad. A strong and modern military, designed specifically to meet Canada's security and foreign policy needs, will serve Canada's pride and Canada's interests. It is, therefore, incumbent on Canadian government to ensure that Canada's military forces are well-funded, equipped to the highest standards, and recruited and trained to fight alongside the best, against the best. The future of Canadian Forces may look less promising as money is going to be tight. And hence, the nation must continue to transform the force, and focus upon maximising their resources to deliver more operational output. The Canadian Forces contribute to the conduct of Canadian defence diplomacy through a range of activities, including the deployment of Canadian Defence Attachés, participation in bilateral and multilateral military forums, ship and aircraft visits, military training and cooperation, and other such outreach and relationship-building efforts. The Canadian government aim is to strengthen key military capabilities through focused investments supported by long-term funding in order to deliver a balanced, multi-role, combat-capable force that will give Canada Armed Forces the necessary flexibility to respond to a full range of challenges in the years ahead.

Coyote Armoured Vehicles with groups of military personnel. *Image Courtesy: Royal Canadian Army*

CONTACT DETAILS

Department of National Defence
Maj.Gen.George R.Pearkes Bldg
101 Colonel By drive
Ottawa, Ontario K1A 0K2
Canada
Tel: +1-613-995-2534
Fax: +1-613-992-4739
www.forces.gc.ca

(Government Procurement Agency)
Public Works & Government Services Canada
16A1, 102 Corporate Communications, Portage III
E-mail: questions@pwgsc.gc.ca
Tel: +1-800-622-6232-1

Defence Research and Development Canada
www.drdc-rddc.gc.ca
Partnering with Defence R&D Canada
corp.bdo-bda@drdc-rddc.gc.ca
Defence R&D Canada Contacts Abroad
Canadian Defence Liaison Staff (Washington)
wshdc-outpack@dfait-maeci.gc.ca
Canadian Defence Liaison Staff (London)
ldn@dfait-maeci.gc.ca
Canadian Defence Liaison Staff (Paris)
paris@dfait-maeci.gc.ca

STRATEGIC INFORMATION

CHILE

(Capital: Santiago)

INTRODUCTION

Area: 756,102 sq km

Population: 17,650,114 (July 2016 est.)

Coastline: 6,435 km

Maritime claims: Territorial sea: 12 nm

Contiguous zone: 24 nm

Exclusive Economic Zone: 200 nm

Continental shelf: 200/350 nm

GEOPOLITICAL IMPORTANCE

Chile is one of South America's most stable and prosperous nations recognised as an emerging economy and a key regional player. The country lies on a long, narrow strip of land between the Andes Mountains to the East and the Pacific Ocean to the West. To the North is the Atacama Desert, the driest desert in the world, and to the South are the ice-fields and glaciers of Chilean Patagonia. Chile shares frontiers with Argentina, Peru and Bolivia. Its distinctive shape - 4,300 kilometres long and on average 175 kilometres wide - makes it the longest country in the world, North to South, with the fifth lengthiest coastline at over 78 thousand kilometres.

Chile belongs to a number of organisations, including the United Nations, Organisation of American States, Community of Democracies, Asia-Pacific Economic Cooperation forum, Organisation for Economic Cooperation and Development, International Monetary Fund, World Bank, and World Trade Organisation. Chile is also a member of the Pacific Alliance, Union of South American Nations (UNASUR), and Community of Latin American and Caribbean States (CELAC).

POLITICAL OVERVIEW

Chile is a republic with Executive, Legislative and Judicial branches. The executive branch consists of the President, and the legislative branch features a bicameral legislature composed of the High Assembly and the Chamber of Deputies. The judicial branch consists of the Constitutional Tribunal, the Supreme Court, the Court of Appeals and Military courts.

The President of Chile is both head of state and head of government. Executive power is exercised by the government. Legislative power is vested in both the government and the two chambers of the National Congress. The Judiciary is independent of the executive and the legislature of Chile.

The country is divided into 15 regions for administration. These regions are again divided into provinces that are administered by appointed governors. The provinces are further divided into municipalities that are governed by elected mayors. Elections are due every four years. A President cannot run for office for two consecutive terms.

President Michelle Bachelet has announced the launch of a process to draw up a new constitution for Chile. A proposal for a new constitution is expected to be put to Congress in 2017. The Constitution for Chile, drawn up under the military government of General Augusto Pinochet has been amended.

KEY POLITICAL PERSONS

PRESIDENT: Michelle Bachelet Jeria

MINISTER OF FOREIGN AFFAIRS: Heraldo Munoz Valenzuela

ECONOMY

Chile is considered to be one of the fastest growing economies in Latin America. It is the first country in Latin America to join the Organisation for Economic Cooperation and Development (OECD). Chile has signed free trade agreements (FTAs) with several countries including the European Union, South Korea, New Zealand, Singapore, Brunei, China, Japan and Mexico. Copper accounts for 60% of Chile's exports and 15% of gross domestic product. However, GDP growth fell to 1.9% in 2014 and 2.1% in 2015, as a result of the slowdown in the mining sector and the decline in copper prices and private consumption, according to the World Bank. GDP growth is expected to recover slowly in 2017-2018 with the increase in copper prices and levels of private investment.

GDP: $422.4 billion (2015 est.), $413.9 billion (2014 est.)

Real Growth Rate: 2.1% (2015 est.), 1.8% (2014 est.)

Industries: Copper, lithium, other minerals, foodstuffs, fish processing, iron and steel, wood and wood products, transport equipment, cement, textiles

Total Exports: $61.82 billion (2015 est.), $75.68 billion (2014 est.)

Export Commodities: Copper, fruit, fish products, paper and pulp, chemicals, wine

Major Markets: China 24.4%, US 12.3%, Japan 10%, South Korea 6.2%, Brazil 5.4% (2014)

Total Imports: $56 billion (2015 est.), $67.91 billion (2014 est.)

Import Commodities: Petroleum and petroleum products, chemicals, electrical and telecommunications equipment, industrial machinery, vehicles, natural gas

Major Suppliers: China 23.4%, US 18.8%, Brazil 7.8%, Argentina 4% (2015)

DEFENCE & SECURITY

Going beyond the traditional scope of defence, which is the protection of sovereignty, independence and the land and sea of the country, the government of Chile has stressed on the need to look at the international security context and cooperate with other countries. Besides joining international humanitarian and peacekeeping operations, such as the UN mission in Haiti, Chile has also signed defence bilateral and multilateral defence agreements worldwide. Security problems such as drug trafficking, organised crime and piracy has been brought under the broad umbrella of national defence in the National Defence and Security Strategy (ENSYD) released by the government.

The Carabineros and Policia de Investigaciones have been tasked with safeguarding internal security. The Carabineros, although a part of the armed forces, has now come under the Ministry of the Interior *Ministerio del Interior* from the Ministry of Defence *Ministerio de Defensa Nacional.*

Chile's military is well-equipped. Their main objectives are to secure the country from external threats; protect the country against terrorism, drug traffickers, and organized came; and to provide quick and effective responses to natural disasters. Chile collaborates with the United States on security issues related to drug trafficking and with Bolivia in counter-narcotics operations on their shared border. The country also works with Argentina and has a joint military force trained in humanitarian work.

KEY DEFENCE PERSONS

DEFENCE MINISTER: Jose Antonio Gomez

CHIEF OF THE JOINT STAFF: Gen. Arturo Merino Nunez

CHIEF OF ARMY: Gen. Humberto Oviedo Arriagada

CHIEF OF NAVY: Admiral Enrique Larranaga Martin

CHIEF OF AIR FORCE: Gen. Jorge Robles Mella

Internal Conflict

Chile's indigenous Mapuche people, the Coordinadora Arauco-Malleco (CAM), since its formation in 1998, has demanded the return of ancestral territory, jurisdictional autonomy and cultural identity. The Mapuche have waged violent campaign to win back their land. Protests have ranged from marches, hunger strikes and the occupation of public buildings to the setting up of road blocks, the occupation of disputed land, arson and the sabotage of machinery and equipment. In recent years, the country has also experienced high levels of civil unrest with students carrying out protests calling for education reforms. There have been violent clashes between police and protesters.

External Conflict

Chile has long-running territorial disputes with Peru and Bolivia. Chile fought against the countries of Bolivia and Peru in the War of the Pacific (1879-1883) in which the two countries lost significant amounts of territory to Chile. Peru lost what is now the Chilean region of Tarapaca, and Bolivia lost what is now Antofagasta, as well as its access to the Pacific Ocean. Diplomatic relations between Chile and Bolivia broke off as tension continued due to Bolivia's demand for recovery of its outlet to the Pacific Sea.

THREAT PERSPECTIVE

Chile faces threats from the globalisation of terrorism, separatist Mapuche community, from the ever-increasing drug trade, organised crime, civil attacks, and arms proliferation. Territorial issues continue to fester among Chile, Argentina, Bolivia, and Peru. Internal instability in neighbouring countries also poses a threat to Chile.

STRATEGIC RELATIONS

United States: Chile has strong political, economic and trade relations with US. The two nations defence relationship is strong and vibrant based on common values and shared interests. Both countries are working with like-minded partners to promote stability and security regionally and globally. Chile is making important contributions to regional security by training Guatemalan pilots and incorporating Salvadoran and Honduran infantry platoons into Chile's battalion serving in the UN Stabilization Mission in Haiti. Bilateral trade has more than doubled since the US-Chile Free Trade Agreement (FTA) entered into force in 2004. The United States is Chile's second-largest trading partner, and the largest foreign investor in Chile.

China: Chile and China have upgraded their bilateral relations to a strategic partnership implementing high level cooperation mechanisms in the political and economic areas. Chile was the first South American country to establish diplomatic ties with China and also the first country from the region to sign an FTA with China. The signing of Chile - China FTA has brought about an important increase in bilateral trade. China's economic and trade relations with Chile are growing strongly becoming the largest source of imports and largest destination for Chilean exports globally.

India: Chile and India have good bilateral relations. Over the past few years there has been a steady increase in exchanges between the Armed Forces of the two countries. The signing of a Memorandum of Understanding in Defence in Santiago, and the exchange of Defence Attaches are important signposts on the road to institutionalizing this relationship. The government of India has also given its approval for expansion of India - Chile Preferential Trade Agreement (PTA) between India and Chile.

Canada: Canada and Chile continue to expand their bilateral defence collaboration, which has grown in scope and intensity in recent years, particularly since the signature of a Memorandum of Understanding on Defence Cooperation in April 2012. The two nations share many common defence goals, including the desire to increase the coherence and effectiveness of the Inter-American Defence Board. Both countries seek to cooperate in dealing with hemispheric defence and security issues, and to contribute to the enhancement of security in the hemisphere. To that end, Armed Forces from Canada and Chile have successfully initiated trilateral defence cooperation to support capacity-building in Central America. The two countries also work together in support of regional trade integration in the context of the Trans-Pacific Partnership negotiations. As a founding member, Chile was a strong supporter of Canada achieving observer status to the Pacific Alliance.

DEFENCE CAPABILITIES

Chile possesses the most modern and capable armed forces in Latin America. The Armed Forces are subject to civilian control exercised by the President through the Minister of Defence. The President appoints service chiefs to four-year terms and has the authority to remove them. The joint chief of staff is the sole authority in charge of planning the use of armed forces, as well as training. He is also responsible for Chilean troops in international operations, and in the case of natural disasters or emergencies. The Chilean national police *Carabineros* are incorporated into the Defence Ministry. Military service is obligatory in Chile.

ARMY *(Ejercito de Chile)*: The Chilean army is organised into seven divisions, a special operations brigade and an air brigade. Its main mission is to ensure national sovereignty, maintain the territorial integrity and protect the people, institutions and vital resources of the country, in the face of any external threat or aggression.

Personnel: 33,529 (includes 8,512 conscripts)

Equipment

Category	Name	In Service
MBT	Leopard 2A4	131
	Leopard 1V	20APC/IFV
	M-113AA2 (including derivative versions)	477
	Piranha(6X6,8X8)	179
	Marder	173
	YPR-765	169
Artillery	Soltam M86/M71 155mm	48
	M56 105mm pack howitzer	104
	M101 105mm	89
	M109A3 155mm SP	48

Army continued...

Category	Name	In Service
Mortar	81 mm	744
	120mm	276
	120mm SP	78
RCL	M40A1 106mm	221
Anti Tank GW	Spike	62
Weapon	Mistral (SAM)	12
	MIM-72 Chapparal (SAM)	
	AIM-9D Sidewinder (SAM)	

Army Aviation

Category	Name	In Service
Aircraft	CN-235	3
	C-212-300	2
	Cessna C-208 Caravan	3
	Cessna 680	1
Helicopter	AS.532AL Cougar	9
	AS.330L Puma	2
	AS.350B	4
	AS.355F	1
	SA.330L Puma	4
	MD-530F (attack)	10

New Procurements/Upgrades

✦ A Request for Proposal has been circulated for a fixed-wing basic trainer aircraft.

Chilean Army soldiers. *Image credit: Ejercito de Chile*

Katary II exercise of the 3rd Armoured Brigade "La Concepcion". *Image credit: Ejercito de Chile*

NAVY *(Armada de Chile)*: The presence of the long coast in Chile has led to formation of the Chilean Navy. The Navy is responsible for protecting more than 4,300 km of coastline and 4.5 million square kilometres of maritime territory, including the Exclusive Economic Zone and the Continental Shelf. In addition to defending against external attacks, the Navy protects the nation from such threats as drug trafficking, piracy, and terrorism. It also maintains security, promotes national development and supports national interests abroad.

Major naval bases: Pacific zone: Iquique, Easter Island, Valparaíso, Talcahuano, Puerto Montt. Atlantic zone: Strait of Magellan: Punta Arenas, in the Beagle Channel: Puerto Williams. Antarctica: Captain Arturo Prat Base.

Personnel: 26,000

Equipment

Category	Name	In service
Submarine	Thomson class (type 209)	2
	O'Higgins class (Scorpene type)	2
Frigate	Almirante Latorre class	2
	Almirante Cochrane class	3
	Almirante Williams	1
	Almirante Blanco Encalada	2
Patrol vessel	Casma class missile FPB	3
	Piloto Pardo class OPV	3+1
	Contramastre Micalvi OPC	6
Amphibious Force	Sargento Aldea LSDH	1
	Batral - type LST	2
	LSM	2
Auxiliaries	Fleet oilers	2
	Transport ship	1
	Oceanographic ship	1
	Icebreaker	1
	Buoy ship	1
	Training ship	1

A rendering of the Vard design for the Antarctic icebreaking vessel. *Image credit: Vard Marine*

New Procurements/Upgrades

✦ The Navy is planning on upgrading its three Type-23 frigates with US anti-air missile systems such as the Raytheon Evolved Seasparrow Missiles (ESSMs), Evolved Seasparrow Telemetry Missiles, Lockheed Martin Mk 41 Vertical Launch Systems (VLS).

✦ Chile is planning to construct a Crocodile-class 250 submarine. The vessel, introduced as a 'light SSK' is intended to be used in a coastal-littoral denial role, outfitted with both wire-guided torpedoes and anti-ship missiles.

✦ Vard Marine, a Fincantieri company will design and build an Antarctic icebreaking vessel for the Chilean Navy under a contract by the Chilean ASMAR Company. Delivery of the vessel is scheduled in 2021.

Marines

Personnel: 2,125

Equipment

Category	Name	In Service
Light Tank	Scorpion	15
Armoured Car	Mowang	25
	HMMWV	8
Artillery	KH-178 105mmm Howitzer	4
	G-5 155mm towed howitzer	14
Mortar	M29 81 mm	8
ASM/SSM	Trailer mounted MM38 Exocet missile	2

New Procurements/Upgrades

✦ Chile plans to buy surplus US stock of a dozen AAV7 armoured amphibious vehicles for its new sea-borne Amphibious Expeditionary Brigade.

Naval Aviation

Category	Name	In Service
Aircraft	C295 ASW	2
	P-3ACH Orion	3
	C-295 Persuader	3
	CASA 212	3
	P-111	1
	Cessna O-2A Skymaster	7
Helicopter	Bo-105	5
	HH-65 Dauphin	4
	PC-7	7
	SA.365F Dauphin II	4
	AS.532SC Cougar	5

New Procurements/Upgrades

✦ An additional three to four Super Pumas are being sought.

AIR FORCE *Fuerza Aerea de Chile (FACh)* is considered to be one of the most professional and capable air arms in Latin America. The country's defence modernisation programme has benefited the FACh and moulded its inventory into an effective and balanced force.

The Service has five air brigades headquartered in Iquique, Antofagasta, Santiago, Puerto Montt, and Punta Arenas. Its mission is to defend the country through the control and use of the air space, the participation in surface warfare and support to the national and friendly forces.

Major air bases: Los Condores, El Bosque, Pudahuel, Quintero, El Tepual, Chabunco, Cerro Moreno and Santiago. The Air Force also operates an airbase on King George Island, Antarctica and at Quintero, near Valparaiso.

Personnel: 8,143 (including 349 conscripts)

Equipment

Category	Name	In Service
Fighter	F-16 MLU/A/B	36
	F-16 C/D	**10**
	F-5E/F	14
Transport /Air refuelling/ tanker	KC-130R	1
	C-130H	2
	C-130B	1
	Beech 99A	2
	Beech 200 King Air	1
	Boeing 737-300/500	2
	Boeing 707 AEW	1
	Boeing 767-300 ER	1
	C-212	**3**
	C-130H/B	3
	Citation CJ-1	3
	KC-135E Stratotankers	3
	O2-OAa	6
Trainer	A-29 super tucano	**12**
	T-36 Pillan	**19**
	T-35 Pillan	**7**
	A-36 Halcon	**8**
VIP transport	BK-117	1
	Gulfstream IV	1
Helicopter	UH-1H	18
	DHC-6 Twin Otter	12
	S-70 Black Hawk	1
	Bo.105 CB-4	1
	Bell 412	**9**
	Bell 206 B/B2	3
UAV	Hermes 900	3
Weapon	AIM-7 Sparrow (AAM)	
	AIM-9 Sidewinder (AAM)	
	IRIS-T(AAM)	
	AIM-120 AMRAAM (AAM)	
	Python-4(AAM)	
	AGM-45 Shrike (AGM)	
	AGM-65 Maverick (AGM)	
	AGM-88 HARM (AGM)	
	AGM-84 Harpoon	
	AGM-119 Penguin	

Air Defence

The Air Defence operates AAA systems (Miguel FCS), Blowpipe, Mistral SAMs and Mygale system. Each Mygale has Samantha truck-mounted radar and three ASPIC truck-mounted four-cell launchers for Mistral missiles.

New Procurements/Upgrades

- ✦ The FACh has taken delivery of the first of the two KC-130R Hercules refuelling/transport aircraft.
- ✦ Chile has inked a deal with Brazil to participate in the design and development of the new Embraer KC-390s in view of procurement plans to buy six transport aircraft.
- ✦ The FACh has released a tender for six to eight platforms to reinforce its rotary-and fixed-wing medium-lift capabilities with Airbus Helicopters, AgustaWestland, Rosoboronexport, and Korean Aerospace Industries (KAI) set to bid their H215, AW149, Mi-171 Hip, and Surion respectively.

Chilean Air Force peacekeeping mission team from Haiti returns to Chile. *Image credit: FACh*

Defence Production And R&d

The Chilean defence industry was developed to address the military needs as a result of the US embargo on sales to Chile of new weapons and spare parts. Chile is a major regional producer of defence equipment including small arms and ammunition; infantry support weapons, armoured and soft-skinned vehicles, ballistic rocket systems, vessels, command and control systems, aircraft and artillery. However, it depends largely on imports for its armed forces modernisation plans. Chilean state-owned companies continue to dominate as military suppliers and in the maintenance and repair of existing systems. FAMAE, ENAER and ASMAR are the most important arms companies in the country.

Major Armament Producers

FAMAE (Fábricas and Maestranzas del Ejército) is a government-owned weapons producer. The Company specializes in research projects and weapon system modernisation. It is responsible for modifying and maintaining the tanks, aircraft and other military devices that reach the Chilean Army.

ENAER (Empresa Nacional de Aeronáutica), the national aeronautics company design and builds military aircraft, including basic trainers and light jet fighters.

ASMAR shipyard is one of the region's largest and most capable naval production and overhaul facilities, having produced naval vessels up to 2,000 tonnes.

DTS Technology Development and Systems Ltd., specializes in information technology, electronics and telecommunications, providing solutions for the military, specifically for the Air Force of Chile, through the development of major military projects in areas of Command and Control and Electronic Warfare.

DESA is a system integrator located specializing in real time systems focused at defence applications -- ground and naval fire control systems, command and controls systems, tactical network etc.

Linktronic is involved in the design and production of high technology electronic equipment with particular emphasis on Air and Coastal Surveillance Radars, Defence Systems and Computers.

Detroit Chile is a Chile-based company engaged in the marine operations, as well as the production of engines.

SISDEF Ltd. develops systems engineering solutions for defence, security and marine and industry.

The largest import trading partners are currently the US, Spain, Germany and the Netherlands. The US has emerged as the main import trading partner with a substantial increase in share, from 16% during 2006 - 2010 to 30% in 2010 alone. Foreign Original Equipment Manufacturer (OEMs) enters the market through defence exhibitions, joint ventures and direct commercial sales.

DEFENCE DEALS

Joint Venture Programmes

The Chilean Ministry of Defence encourages joint ventures and technology-sharing agreements to enable domestic firms to enhance their capabilities and also promote diplomatic relations between the countries. Chile has signed a Declaration of Intent with Brazil for participation in the development of the new KC-390 military transport aircraft, carried out by the Brazilian aerospace firm Embraer.

Arms Imports

Chile is one of Latin America's largest arms importers. As part of its military modernisation programme Chile has made various acquisitions including F-16s fighters, Embraer EMB-314 Super Tucano trainers, 412EP tactical helicopters, Hermes 900 UAV and Cirrus SR22T intelligence, surveillance and reconnaissance (ISR) aircraft, and KC-135E Stratotanker aircraft. Chile also acquired surplus aircraft from the Netherlands as part of the Amstel I and II programmes. Currently, the FACh is in the process of procuring medium transport aircraft and helicopters, Scorpene class diesel-electric submarines from France, Leopard I-V tanks from Germany, Type 23 frigates from the UK. A Chilean defence delegation which visited India recently has evinced keen interest in the BRAHMOS missile system for induction into its navy.

Arms Exports

Chile's domestic defence industry is still in its nascent stage, and the country relies on imports from various countries to fulfil its weapons procurement needs. The country has capability in vehicles, ships, aircraft, and ammunition. FAMAE provides a number of ground-based systems to include armour, rockets, and anti-aircraft weapons. However key opportunities identified for defence industry include fighter aircraft, landing platform dock (LPD)-type amphibious assault ships, lighter armoured vehicles such as MRAPs, 8x8 armoured personnel carriers and 4x4 armoured security vehicles, AV-8B Harrier aircraft, unmanned aerial vehicles (UAVs) and light attack and reconnaissance aircraft.

DEFENCE BUDGET

Total defence spending: $4.2 billion (2011 est.), $6,579 million (2010)

Defence spending in terms of GDP: 2.1% (2012)

Chile's Defence Expenditure has decreased from US$4.2 billion in 2011 to US$2.6 billion in 2015. The armed forces receive 10 percent of the nation's annual copper exports. Codelco, Chile's state-owned copper miner and the world's largest producer of the metal. In 2014, Codelco contributed US$1.08bn to the armed forces' budget under the copper law and US$1.16bn in 2013, according to media reports. The government is likely to introduce a new bill in another attempt to repeal the so-called copper law.

Defence Procurement

In line with the decision to reform its defence procurement system, the Chilean Government has created a new body Defence Investments *Consejo de Inversiones en Defensa, or CID* whose task is to assess the military forces' procurement projects. It includes the undersecretaries of defence and armed forces, the heads of the three services and the chief of the defence joint staff. The CID began functioning within the Ministry of Defence in March 2011. It has an advisory role in assessing the armed forces' procurement plans. Assessment criteria are to consider the contribution of the projects to the military services' capabilities - as required by their tasks and roles - as well as projects' financial viability. The council has powers to ask the services to revise or modify their projects. The assessment process will end with the issuing of a report that contains recommendations from the CID to the minister of defence, who will then make final decisions in consultation with the chief of government.

CONCLUSION

Chile today is one of South America's most stable and prosperous nations and the Chilean army the most advanced military force in the region. Chile fully implements the principle of international cooperation and has developed an intense multilateral diplomacy by actively participating in the United Nations missions related to peace and security in the world. The country also actively participates in combined forces exercises in third world countries to prevent disruption to global security.

CONTACT DETAILS

Ministry of National Defence
Zenteno 45 4 Floor
Phone: +56 1202
E-mail: mdn@defensa.cl
www.defensa.cl

Chilean Army
www.ejercito.cl/home

Chilean Navy
www.armada.cl

Chilean Air Force
www.fach.cl

STRATEGIC INFORMATION

CHINA
(Capital: Beijing)

INTRODUCTION

Area: 9,596,960 sq km

Population: 1,373,541,278 (July 2016 est.)

Coastline: 14,500 km

Maritime claims: Territorial sea: 12 nm

Contiguous zone: 24 nm

Exclusive Economy Zone: 200 nm

Continental shelf: 200 nm or to the edge of the continental margin

GEOPOLITICAL IMPORTANCE

China is a major Asian power that occupies large landmass and ranks third in the world in terms of area. It is bordered by 14 countries that include India, Pakistan, Afghanistan, Nepal, Bhutan, North Korea, Russia, Tajikistan, Kazakhstan, Kyrgyzstan, Laos, Mongolia, Vietnam and Myanmar. Surrounded by Korea Bay, Yellow Sea, East China Sea and South China Sea, China has a varied terrain ranging from mountains and deserts to plateaus and sea coasts. It is a more enclosed state than any other major power in the world. From geostrategic standpoint, China occupies and plays an important role in Asia, especially in East Asia, South Asia and Southeast Asia. The vast size of the country along with its huge population has made China a rapidly rising market economy and it has replaced the USA as the largest economy in the world in terms producing goods and services as well as in terms of purchasing power parity (PPP). The country's rapid fiscal rise has given way to the swift expansion of its geostrategic and military influence not only in the Asian continent but at the global sphere as well.

POLITICAL OVERVIEW

The People's Republic of China (PRC) functions as a single-party socialist republic. It is ruled by the Communist Party of China (CPC) which is the only political party in the country and influences the state and society. The unitary government led by the CPC controls the state, military and media. State power within the People's Republic of China is exercised through the CPC, the Central People's Government and their provincial and local counterparts. The unicameral legislature of the country, called the National People's Congress (NPC), is controlled by the Communist Party. The political power in China today is not centred in the hands of a single leader, as it was a few decades ago, but is more of an institutionalised affair controlled by a group of leaders. The PRC heralded a new regime in March 2013 when Xi Jinping took charges as the new President and Chairman of the powerful Central Military Commission (CMC) and Li Kequiang became the new Prime Minister of the country. In April 2016, President Xi assumed the designation of Commander-in-Chief of the Joint Battle Command Center of the CMC.

KEY POLITICAL PERSONS

PRESIDENT & CHAIRMAN OF CENTRAL MILITARY COMMISSION:
Xi Jinping

VICE PRESIDENT:
Li Yuanchao

PRIME MINISTER:
Li Keqiang

ECONOMY

China is a leading global economic power. Until 2013, it was the second largest economy in the world after the US. According to the data released by the International Monetary Fund (IMF) in 2014, China had overtaken the US as the largest economy in the world in terms producing goods and services worth $17.6 trillion as against the US's good and services production worth $17.4 trillion. The IMF's updated data on world economy released in October 2014 also listed China as the world's biggest economy in terms of purchasing power parity (PPP).

Being the world's most populous nation, China's market economy has witnessed rapid growth after the country braced up for reforms and adopted an open market system in the 1980s that attracted large-scale foreign investments. The country is the world's largest exporter and the second largest importer of merchandise goods. China is also the world's second-biggest oil consumer – an important factor in driving the country to meet its energy needs by expanding its geopolitical and military footprint across the world. On the whole, the Chinese economy has gained importance over the past decades and is significantly influencing the global economic sphere. The country is a member of leading international financial institutions including the World Bank, International Monetary Fund (IMF), World Trade Organisation (WTO), Asia-Pacific Economic Cooperation (APEC) and the G-20.

Facing a slowdown in its overall economic growth in the last few years amidst the ongoing global financial crisis, China has taken several stimulus measures, including huge investments in infrastructure and other government projects as well as funding its banking sector with fresh liquidity to revive the economy. The country is attempting a transition from an export-led economy to a consumer and services-led economy. In the first quarter of 2016, the country recorded a GDP growth of 6.7%.

GDP: $19.39 trillion (2015 est.); $18.14 trillion (2014 est.)

Real Growth Rate: 6.9% (2015 est.); 7.3% (2014 est.)

Industries: World leader in gross value of industrial output; mining and ore processing, iron, steel, aluminium, and other metals, coal; machine building; armaments; textiles and apparel; petroleum; cement; chemicals; fertilizers; consumer products, including footwear, toys, and electronics; food processing; transportation equipment, including automobiles, rail cars and locomotives, ships, and aircraft; telecommunications equipment, commercial space launch vehicles, satellites

Natural Gas - Proved Reserves: 4.945 trillion cu m (1 January 2016 est.)

Crude Oil - Proved Reserves: 25 billion bbl (1 January 2016 est.)

Total Exports: $2.143 trillion (2015 est.); $2.244 trillion (2014 est.)

Export Commodities: Electrical and other machinery, including data processing equipment, apparel, radio telephone handsets, textiles, integrated circuits

Major Markets: US 18%, Hong Kong 14.6%, Japan 6%, South Korea 4.5% (2015)

Total Imports: $1.576 trillion (2015 est.); $1.809 trillion (2014 est.)

Import Commodities: Electrical and other machinery, oil and mineral fuels; nuclear reactor, boiler, and machinery components; optical and medical equipment, metal ores, motor vehicles; soybeans

Major Suppliers: South Korea 10.9%, US 9%, Japan 8.9%, Germany 5.5%, Australia 4.1% (2015)

DEFENCE & SECURITY

China is a powerful Asian country that has evolved as a leading regional and global military power in the 21st century. Safeguarding its territorial integrity while maintaining domestic and regional stability has driven the country towards a rapid expansion of its military might. While strongly defending its territory, China has also adopted an assertive policy to strategic issues in the Asian continent. The military organisation of the country, called the People's Liberation Army (PLA), is controlled by the Central Military Commission (CMC) which is the supreme defence policy making body. The Chairman of the CMC is the Commander-in-Chief of the PLA. The active forces of PLA consist of the Army, Navy, Air Force, PLA Rocket Force (erstwhile Second Artillery Corps), military schools and national defence scientific research institutions.

China has rapidly modernised its armed forces in the aftermath of the two Gulf Wars fought in 1990 and 2003. Once following the "limited local war" strategy evolved from Mao Zedong's "protracted people's war", the PLA of today has adopted the "people's war under hi-tech conditions" tactics as part of its modernisation drive that began in 1993. The new warfare tactics rely more on advanced technology than on troops. The "people's war" strategy, based on large-scale land and Guerrilla warfare methods, was designed to confront mainly Japan and the erstwhile Soviet Union. However, the doctrine became redundant following the disintegration of Soviet Union in 1991. Shifting from this warfare methodology to a modern, hi-tech condition was necessitated after the rise of the US as the supreme world power.

Of late, the PLA has followed the doctrine of "active defence" relying on state-of-the-art military hardware, modern high-tech weapons, precision guided munitions, sophisticated command & control systems coupled with a strict military training regime and largescale use of information technology. The new military doctrine along with structural changes in its armed forces, development of pre-emptive strike capability, change in warfare methodology, C^4I^2SR, ground, sea and aerospace-based surveillance and information technology have become the core of present-day Chinese armed forces. By making efforts to acquire advanced military capabilities, China has been steadily transforming its posture to that of a belligerent power. The country has especially ramped up its efforts to build and deploy strategically critical missile systems capable of not only defending its territory but also launch offensive strikes against potential enemy installations. Advanced and extensive researches have been undertaken in the past few decades to develop anti-ship ballistic missile systems (ASBM), ballistic missile defence (BMD) capabilities and even anti-satellite weapons to carry out space war. These potential weapons are being developed as part of China's "anti-access and area-denial" (A2AD) strategy. In 2010, China conducted ground-based midcourse interceptor missile test which demonstrated its capability to track down and destroy midcourse ballistic missiles, with many defence experts believing that the country was nearing to achieve BMD capability. In January 2013, PRC conducted another such test to exhibit that it is increasing its potential in missile interception capabilities. The latest test firing of the anti-missile system took place in July 2014 during which a land-based mid-course missile interceptor successfully traced, tracked and destroyed a ballistic missile flying in outer space.

In 2007, the country successfully tested a ground-based direct-ascent anti-satellite (ASAT) weapon system to destroy one of its defunct weather satellites, which established its ability to attack satellites and space assets in low-Earth orbit. The successful ASAT mission demonstrated China's prowess to launch space warfare missions and was viewed by many countries as a serious threat to their vital space-based systems. In May 2013, the country launched a rocket into space which later re-entered Earth's atmosphere above the Indian Ocean. According to military experts, the test was a new interceptor weapon that could be used to destroy satellite in orbit. Another test of a similar weapon, designated Dong Neng-3 (DN-3), was conducted in October 2015.

In yet another new military technical breakthrough in January 2014, China test flew a hypersonic glide vehicle (HGV) capable of cruising at a top speed of Mach 10. The HGV test, marking a step beyond China's ASBM programme, featured a slower, shorter-range manoeuvrable re-entry vehicle (RV). The ultra high speed vehicle was launched into space by an intercontinental ballistic missile. The HGV test put China at the second slot after the US in having successfully tested hypersonic technology. A second test flight of the new system was carried out in August 2014 followed by one more test reported in December 2014. In June 2015, the Chinese Defence Ministry confirmed the fourth test of the hypersonic nuclear delivery vehicle followed by a second test in August 2015 and yet another test in November 2015, which was the sixth test launch of the new weapon system. The most recent test of the glider vehicle took place in April 2016.

Another frontier that has come to occupy centre-stage in China's overall military strategy is the strengthening of its asymmetric warfare capabilities involving cyberspace. China has accelerated its activities in cyberspace domain by developing new capabilities to not only defend its information technology network but even launch offensive operations against its adversaries. Acknowledging the potentiality of information technology in military arena which can have wider implications from both tactical as well strategic perspectives, China has been swiftly building up its information warfare capability to launch computer network-based war that could paralyse and upset the entire civilian and military apparatus of a country by taking control of its information & information-based systems. Development of electronic countermeasure weapons, cyber espionage and carrying out attacks on internet network are becoming the potential tools for China to launch cyberspace-based operations.

The most vital aspect of modern Chinese military is the rapid growth of its maritime power. As the country expands its economic horizons and realises the increased importance of sea lanes as vital trading routes, it is adopting a strategy of "sea control" and "sea denial" to ascertain its supremacy over water resources within its territory and beyond. The country's navy has primarily focused on strengthening its anti-surface and anti-air warfare capabilities while building up a credible at-sea nuclear deterrent.

The potentiality of the PLA's air power too has assumed significance with the country building and deploying new-generation stealth warplanes/ bombers, combat helicopters, drones, long-range transport aircraft, refuelling tanker aircraft along with advanced weapon systems to carry out integrated operations alongside the Army and Navy. The steady enhancement and expansion of the PLA's military aviation capabilities is clearly marked by the fact that the country announced the establishment of an air defence identification zone (ADIZ) over a swath of the disputed East China Sea in December 2013. The move was aimed at putting the Air Force at the forefront in the eventuality of a war and thereby supporting the country's expanded maritime power projection. In recent months, China has also invoked its right to set up an air defence zone in the hotly disputed South China Sea, more so after an International Court ruled in favour of Philippines over a disputed region in the SCS.

According to a Pentagon report, China's air force is "pursuing modernisation on a scale unprecedented in its history and is rapidly closing the gap with Western air forces across a broad spectrum of capabilities including aircraft, command and control, jammers, electronic warfare and data-links."

In a recent decision taken by the CPC, China has announced to optimize the size and structure of its army, while also focusing on optimizing the function and institution settings of the Central Military Commission. In September 2015, Beijing announced that its 2.3 million strong People's Liberation Army, the world's largest, would be cut by 300,000 troops by 2017. The country also announced to strengthen the joint operation command authority under the CMC, and the theatre joint operation command system.

In yet another sweeping reform announced in November 2015, the country for the first time integrated the PLA area commands overseeing India and Pakistan, thereby junking its Soviet-era model for a US-style joint command system to fortify the ruling Communist Party's control. Of the seven military area commands – at Beijing, Nanjing, Chengdu, Jinan, Shenyang, Lanzhou and Guangzhou – the Chengdu military command was in charge of security along India's Eastern sector in the Tibet region including Arunachal Pradesh while the Lanzhou military command looked after the Western sector, including the Kashmir region and Pakistan.

As per the new strategic zone plan unveiled by President Xi, both the Chengdu and Lanzhou get integrated into a strategic command region, making it perhaps the biggest areas for the PLA. The overhaul is aimed at moving away from an army-centric system towards a Western-style joint command in which the Army, Navy and Air Force are equally represented.

In early 2016, the PLA was reformed again so as to include five new theatre commands by replacing the seven previously existing military regions. The Defence Ministry said that the new military commands, under the administration of the CMC, are based on the functions and structure of the military regions (MRs) they will replace, with improved mechanisms for command and logistics. The PLA will establish a Transitional Work Office to ensure a smooth transition from the old MRs to the new theatre commands. As part of the overall military reforms initiated by President Xi Jinping, who also heads the decision-making Central Military Commission and also assumed the role of Commander-in-Chief of the Joint Battle Command Centre of the CMC in 2016, the PLA has dissolved the four existing general headquarters and instead set up 15 new organisations including the Headquarters of the Joint Force. The restructuring of the CMC has been done to consolidate its authority and further centralise the structure of the People's Liberation Army.

The urge to grow economically and militarily has driven China to exert its influence in and around the resource-rich areas of Asia-Pacific region, particularly in the South China Sea and East China Sea which are major trading routes and are also believed to treasure huge amounts of oil and natural gas along with fishing and mineral resources. The Asian giant has intensified efforts in expanding its footprint in the SCS, ECS and even in the Indian Ocean by building up new harbours and other infrastructural setups, thereby adding to the worry of the neighbouring littoral states which are at loggerheads with Beijing. China's growing power projection and aggressive stance over regional and littoral issues in recent times has alarmed the international community, though the country maintains that it follows a "non-offensive strategy" with a "need to expand" while safeguarding its rights and interests. The growing territorial disputes in the Asia-Pacific region, with China being at the centre, have flared up regional tensions and invited global attention and concern. Moreover, the secretive military plans and pursuits by the Communist giant have projected a blurred picture of its real ambitions and intentions before the entire world.

In its latest Defence White Paper, issued in May 2015, Beijing has laid emphasis on a more "active defence posture" and a greater Chinese naval presence farther from the country's shores. It has also emphasized on the importance of building up small, multi-functional and modular units of its military that could take on different purposes for joint operations.

KEY DEFENCE PERSONS

VICE CHAIRMAN OF CENTRAL MILITARY COMMISSION: Gen. Fan Changlong

VICE CHAIRMAN OF CENTRAL MILITARY COMMISSION: Gen. Xu Qiliang

DEFENCE MINISTER: Gen. Chang Wanquan

CHIEF OF JOINT STAFF DEPARTMENT OF CMC: Gen. Fang Fenghui

COMMANDER OF PLA NAVY: Adm. Shen Jinlong

COMMANDER OF PLA AIR FORCE: Air Force Gen. Ma Xiaotian

COMMANDER OF PLA ROCKET FORCE: Gen. Wei Fenghe

Internal Conflict

The major internal areas of concern for China have been the separatists from Tibet, Muslim Uighurs in Xinjiang and Mongolians in Inner Mongolia.

External Conflict

From time to time, China has attempted to assert its supremacy and project its influence in the entire Asian continent, including in the energy-rich Central Asian region consisting of Kazakhstan, Kyrgyzstan, Tajikistan, Turkmenistan and Uzbekistan with which it is steadily establishing strong trade ties and clinching new deals for oil and gas. The resource-rich waters of South China Sea (SCS) and East China Sea (ECS) have also remained the prime areas of conflict in the region. The SCS, a major shipping lane, is also believed to be rich in oil and gas reserves. It is claimed by several littoral countries in the area including Brunei, Malaysia, the Philippines, Taiwan and Vietnam. A group of uninhabited islands in the ECS has also become a major source of regional tension between China and Japan in recent times.

Territorial disputes: China is engaged in border disputes, both land and maritime, with many of its neighbouring nations in the Asia-Pacific region.

India: The two neighbours, sharing a land boundary of 4,057-km, are locked in a dispute over the Himalayan border region and even fought a brief war over the same in 1962. The border dispute between the two Asian countries which historically roots back to 19th century when India was under British rule, has mainly involved two areas – the northeast border areas of Kashmir that includes the Aksai Chin region lying in the northern section of India on China's south-western border; and the North-East Frontier Agency (NEFA, which later became the Union Territory of Arunachal Pradesh) in the north-eastern portion of India, on China's southern border. While the China-administered Aksai Chin region in the Karakoram Mountains, covering an area of around 43,000-sq kms, which remained officially un-demarcated for several years, is claimed by India as a part of its state of Jammu & Kashmir, China has been a claimant of India's north-eastern state Arunachal Pradesh, covering an area of around 90,000-sq km. Even as India has regarded the McMahon Line in Arunachal region as the effective Sino-Indian boundary (which was agreed to by Britain and the then independent state of Tibet as part of the Shimla Accord inked in 1914), the same has been disputed by the Chinese government.

On the maritime front, the Brahmaputra River, one of the major rivers in Asia that flows through India, China and Bangladesh, has remained another source of Sino-India conflict since decades. The river that originates in Tibet and enters India in the state of Arunachal Pradesh is an important source of irrigation, trade and transportation. Recent attempts by China to build as many as five dams -- at Zangmu, Dagu, Jiacha, Jiexu and Lalho -- on the river to divert water for irrigating a large landmass covering the Gobi Desert have caused friction between the two neighbours.

Though negotiations have been going on between the two sides for over two decades now to resolve the long-standing border dispute, no concrete results have come up to date.

Bhutan: China stakes claim over a total of 764-sq km of Bhutanese land covering the North West (269-sq km) and Central parts (495-sq km) of the country. The north-west part constitutes Doklam, Sinchulung, Dramana and Shakhatoe in Samste, Haa and Paro districts. The Central parts are the Pasamlung and the Jakarlung valley in Wangdue Phodrang district of Bhutan.

Japan: China and Japan are claiming a group of uninhabited islands in the East China Sea which has become a major irritant in their bilateral relations in recent times, leading to frequent skirmishes. Both China and Taiwan lay claim over the islands, which are part of what Japan calls the Senkakus, and China calls the Diaoyu group. The dispute has further flared up after Japan's decision in September 2012 to buy the chain of islands from a private Japanese family which the country recognises as its "owner". This has led to China dispatching a number of its naval vessels to enter the disputed waters to conduct patrol. Chinese aircraft have also stepped up activity over the region. In 2014, both the Asian powers scrambled their military planes against each other several times to project their dominance over the disputed areas. China has continued to send its Coast Guard and other ships to sail near the islands, despite Japan's strong protests over the same.

Vietnam: While land border conflict between China and Vietnam has been resolved, territorial water dispute between the two neighbours over a number of islands in the South China Sea, in particular the Spratly Islands and Paracel Islands, has resulted in regional tension. The Vietnamese Navy has conducted military drills in the region and held military training exchanges with the US Navy in a sign of renewing its defence ties with Washington to counter the Chinese assertiveness over the SCS. Recent Chinese moves to explore the disputed waters of the SCS for oil and gas have led to intensified maritime confrontations between the two sides. Beijing has also started constructing lighthouses in the Spratly Islands.

Philippines: The resource-rich South China Sea is again at the centre of China-Philippines territorial spat. The Philippines is locked in a tense dispute with China over the Scarborough Shoal -- a group of islets in the South China Sea. Even as the Chinese ships took control of the islets in 2012, China in 2014 announced plans to begin regular patrols of the South China Sea, known in Manila as the West

Philippines Sea. The Philippines, which had earlier expressed its intent to reinvigorate its military ties with the US over the dispute, and signed deals to buy new fighter aircraft, military helicopters and ex-US warships that could be used in the maritime disputes, however, has taken a sudden U-turn, visibly miffed at Washington's failure to help protect Manila's maritime rights in the region. There has been a major shift in Philippines' policy towards the US in the past few months, including scaling down in joint military exercises between the two allies in the disputed region.

Meanwhile, in a recent development in July 2016, an international tribunal backed by the United Nations has ruled in favour of Philippines in the South China Sea maritime dispute, while stating that China has no legal basis to assert its rights over the resources within its "nine-dash line," which extends hundreds of miles to the south and east of its island province of Hainan and covers some 90% of the disputed waters. Immediately after the verdict, China has released a White Paper denouncing the verdict while insisting that Beijing has claims over the strategic region "for 2,000 years."

Malaysia: A small group of islands in the Spratlys in South China Sea is the area of contention between Malaysia, China and few other littoral states in the region. China's sovereignty claims over the entire South China Sea have further accentuated the multi-national dispute.

THREAT PERSPECTIVE

US: While from political and military perspective, China has achieved its strategic goals by strongly defending its territory and often asserting it beyond the internationally-demarcated areas, the most pertinent threat it faces today is from the United States. The presence of US Navy submarines, aircraft carriers and warships in the Pacific Rim area has been the cause of constant bickering between China and the US as the former regards it as American hegemony and intrusion in the Asia-Pacific region. The US's interventions in Taiwan and in the contested maritime regions of the SCS and ECS have made Beijing even more belligerent in recent years. The autonomously-administered Taiwan, which China claims as part of its territory and intends to unify it with the mainland even by force if need be, has been the foremost reason for US-China confrontations. The US's military support to Taiwan has remained a contentious issue between the two leading world powers. China has vehemently opposed any kind of US military sales to the island state for it fears that this could empower Taiwan against China. Besides, the strong military relations the US enjoys especially with Japan and South Korea have further intensified the race to assert supremacy in the region. The growing US-India relations in recent years has also made China jittery.

With the intent to retain its supremacy, the US has announced a major strategic shift towards the highly-tensed Asia-Pacific region by deciding to deploy 60 percent of its naval assets, including as many as six aircraft carriers, in the area. Washington has also decided to deploy a second X-band missile defence radar in Japan – a move believed to have been aimed at containing China. The US has also inked a new defence pact with Australia under which 2,500 US Marine Corps are set to be deployed in Northern Australia by 2017. Besides, two key space systems including a US Air Force C-band space surveillance radar and a space telescope, are also being stationed in Australia. Even after all these strategic moves by Washington, China remains undeterred and continues to maintain a belligerent position in and around the Asian continent.

Japan: The foremost regional threat China faces in Asia is from Japan. Tokyo, which once regarded the erstwhile Soviet Union as its main adversary, has shifted focus towards its economically and militarily growing neighbour China. Japan is home to US military bases and some 47,000 US troops are stationed in the country. It also conducts regular military exercises with the US. Strongly supported by the US, Japan has significantly strengthened its overall military capability, especially its air and naval power, in the last two decades. While it is making its ground forces more mobile for quick response against any contingency, especially in southwest Japan, its navy has deployed the US-supplied Aegis air defence systems to defend the country, and the warplanes in the air force are being armed with advanced air-to-air missiles. The country is adding more teeth to its naval and air prowess by acquiring new class of submarines, destroyers, helicopter carriers, advanced combat jets, spy planes and related equipment. It is also in race alongside China and Russia to develop a new stealth fighter jet to ascertain its air superiority in the Pacific. Amidst the ongoing maritime skirmishes, Japan has also added a large warship – a helicopter carrier – in its naval fleet. The 19,500-ton warship, named Izumo, is the largest warship in Japan's possession after World War II. The Sino-Japan relationship has ebbed to its lowest in recent years over the Senkaku/ Diaoyu Islands with the fighter planes and naval vessels of two sides confronting each other head-on many times.

Taiwan: China perceives the military rise of Taiwan with American backing as another major threat against it. The autonomously-ruled island state, which China claims as part of its own territory, has embarked on an ambitious military modernisation plan. Its major defence acquisition plans include Perry class frigates, a large number of F-16 fighters, combat helicopters, Patriot air defence systems from the US and new indigenously-built anti-ship missile systems, stealth warships and submarines among others. The country is also developing new-class of missile boats and stealth corvettes along with anti-ship cruise missiles to defend its territory.

India: China is the world's third largest country in terms of geographical area and the most populous nation; India, the seventh largest in terms of landmass and second in terms of population. Hence, the two countries occupy centre-stage in the entire Asian continent. With their rapid economic rise along with military progress, India and China are set to play a crucial role in the global stage in the foreseeable future.

An air of misgiving and mutual competitiveness has prevailed over bilateral relationship between the two Asian neighbours having two completely separate political setups, policies and ideologies. With its deep-seated roots in democracy and transparent policies, India has maintained the posture of a "soft power" and toed the strategy of peaceful coexistence with all its neighbours while favouring an all-round development of the region. The Communist China, on the other hand, has followed strict internal controls and secretive, opaque policies in several spheres, including in its foreign and military dealings. The 7th Defence White Paper of China released in 2011 listed Tibet as one of its three core areas. The Tibetan Autonomous Region (TAR) has been at the heart of centuries-old Sino-Indian contentious border relations. At a time when China has taken an aggressive stance over its territorial claims in Asia, it has not spared the Indian state of Arunachal Pradesh, especially the Tawang district having historical affiliations with Tibet, and has been reasserting its claim over it time and again.

China views India as a major rival to its power and stature in the Asian continent. At the global front, China has maintained an ambivalent position over India's aspiration to become a permanent member of the United Nations Security Council (UNSC). In recent times, New Delhi's repeated efforts of becoming a member of the Nuclear Suppliers Group (NSG) has also been scuttled by Beijing. Even as New Delhi is renewing its defence ties with Washington and also with Tokyo, China has felt the unease over the same.

From the military perspective, after the 1962 Sino-India war, India has gone on to strengthen its overall military capabilities – a move being cautiously but warily watched by China. India's growing maritime prowess, in particular with its Navy's 'Blue Water' ambitions, has troubled China which too is pursuing similar aspirations in the 21st century. Rising Indian defence spending and acquisition of advanced weapon systems and platforms has also raised Chinese eyebrows and even found mention in its Defence White Papers.

China is viewing India as its main rival in the Indo-Pacific region while acknowledging the ramifications of an economically and strategically growing India in the entire Asian continent. Even as New Delhi is making a cautious approach to keep any kind of confrontation with Beijing at bay and diplomatic efforts continue to instil mutual trust between the two sides, India, at the same time, is adding more capabilities to its armed forces to counter any "surprising and unexpected move" from its militarily growing Western neighbour. Acquisition of new frigates, corvettes, submarines, aircraft carriers, multi-role fighters, stealth warplanes, combat drones, new missile systems and other weapons are galvanising the military might of India to face possible hostilities from China.

South Korea: Though South Korea considers North Korea as its major military threat, Seoul is also expanding its air and naval power to counter a rising China in the Asia-Pacific. The country is building new KDX-class of warships equipped with US-made Aegis air defence systems and acquiring submarines. Procurement of new-generation stealth fighters, Unmanned Aerial Vehicles and missile systems are also on the cards.

STRATEGIC RELATIONS

Pakistan: China has attached great importance to its strategic ties with Pakistan and consequently, established strong military relations by supplying weapons and armaments to Islamabad. Over the years, the two countries have become "all-weather allies", further expanding the ambit of bilateral cooperation. Both have embarked on joint production of defence weapons and equipment. Pakistan's nuclear weapons programme is assisted by China. Chinese PLA is working in Pak-occupied-Kashmir (PoK) to construct a new road from China to a part of Gwadar – a port in Balochistan province of Pakistan. The port, located at the peak of Arabian Sea and the entrance of Persian Gulf, is expected to provide a key transit route to China to enter and strengthen its foothold in the Middle East. In a recent development, China has taken over operational control of the strategically located port and decided to deploy its naval ships along with those of Pakistan's to safeguard the port. The port will be connected with China's western province of Xinjiang through rail and road links. China has also extended space cooperation with Pakistan and launched a geostationary communications satellite, PakSat-1R, designed by China Academy of Space Technology, for it in the past.

Iran: Not quite open in its military dealings with other nations, China has maintained secret alliance with Iran by supplying weapons and military hardware, especially ballistic and cruise missile technology to it. Among the other Middle Eastern countries, Iran is believed to have received maximum Chinese assistance in missiles and related technology.

North Korea: The secret strategic tie-up between China and North Korea is aimed at developing missiles and nuclear arsenal. The former has supplied an unknown number of cruise missiles and related technology to the DPRK along with training in rocket engine design, reverse engineering, small warhead design, airframe expertise and other assistance.

Myanmar: China, Myanmar relations date back to 1949 when the latter became one of the first countries to recognise the sovereignty of the People's Republic of China. Though bilateral relation witnessed a downside turn in later years, it again assumed importance in 1988 when Myanmar faced international sanctions following a crackdown on pro-democracy protests. China during that phase supplied arms and aid to the country. The two countries today share good trade relations and China is engaged in several infrastructural development projects in Myanmar. China realises the strategic importance of Myanmar to gain access to the Indian Ocean and has accordingly, focused on improving its political, diplomatic as well as military ties with that country. It has supplied battle tanks, fighter jets, naval craft and other weapons to Myanmar. The Chinese military presence in Myanmar's Coco Islands, located in close proximity to India's Andaman Islands, has often been reported but denied by both Beijing and Nay Pyi Taw.

Russia: China, in the past, has overtly depended on the Soviet-era military technology and weapons. China and Russia have shared strong bilateral trade relations since long and Moscow is also a leading oil supplier to Beijing. The fast-changing geopolitical equations of the 21st century have brought the two powerful countries further closer as they are seeking to renew their strategic alliance so as to reinvigorate their position in the Asia-Pacific, especially after the US's decision to step up its military engagement in the region. The two countries are influential members in the multilateral Shanghai Cooperation Organisation (SCO) forum. Recently, the two sides have held joint military exercises and are viewing issues, both regional and global, in a united manner. In May 2014, Moscow and Beijing signed a 30-year, $400 billion landmark deal to pipe natural gas from Russia's Far East to China. The deal was clinched amidst worsening relations between Russia with Western countries over the Ukraine crisis.

Israel: China and Israel have pledged to strengthen bilateral military ties in recent times. China shares close ties with Israel's defence industry and has received sophisticated defence technology from it in the past. Israel-China military relations date back to the 1970s. Beijing has developed one of its most modern fighter jets, the J-10, from an Israeli design.

Indonesia: With an aim to expand its reach in its extended neighbourhood, China in the past few years has made efforts to renew its ties with Indonesia. While the two countries share political, economic and cultural relations, China has been making efforts to expand it further so as to include strategic alliance with Indonesia. The two Asian countries have pledged to boost their military technological cooperation and extend it further to include joint military exercises, training and maritime security missions as well as technological and defence production.

Thailand: Off late, China has made efforts to enhance its strategic alliance with its Southeast Asian neighbour Thailand which occupies an important geographic position in the continent. Bilateral relation between the two countries is growing stronger with cooperation in the fields of infrastructure, energy and trade. China also supplies military equipment to Thailand. The two sides have conducted joint military training exercises. They have also jointly worked on developing multiple rocket launcher systems in the past and also signed a deal to develop an advanced version of the weapon. In a recent development, Thai and Chinese defence ministries also agreed to hold a joint military exercise involving their air forces for the first time in November 2015.

Seychelles: China's relations with the island country in the Indian Ocean over the last three decades were confined to political, social and cultural spheres. In 2011, it was reported that China was inclined to open a naval base in Seychelles – its first military base outside its territory and in the Indian Ocean Region – to "seek supplies and recuperate" facilities for its Navy. The Chinese Government, however, has denied it.

Sri Lanka: China has endeavoured to strengthen its ties with Sri Lanka in recent times. The two countries have decided to upgrade their relationship to strategic level. The renewed ties between them include military exchanges, including recent submarine dockings at Lankan ports, besides trade and economic relations. China has also been investing in infrastructure development projects in the South Asian island nation. It has also agreed to provide defence technology as well as training to the Lankan army personnel. Sri Lanka, on its part, has committed to join the Maritime Silk Road (MSR) project being planned by China. Beijing has also agreed to support Colombo in developing capabilities in satellite communication, space technology and maritime industries. In November 2012, China launched Sri Lanka's first communications satellite, SupremeSAT-I.

Bangladesh: In order to further expand its strategic reach in South Asia, China has sought to firm up its relations with Bangladesh. The two countries' militaries have pledged to strengthen defence ties. Bangladesh has received massive military aid, military training, technology and equipment from China in the past. The two countries signed the defence cooperation agreement in 2002 and 2004 to strengthen their strategic ties. In 2016, Bangladesh has acquired two Type 035G attack submarines from China.

Djibouti: China has emerged as a new economic partner of Djibouti, the tiny country on the Horn of Africa. In a most recent development, Beijing has signed a 10-year agreement with the African nation to set up a naval logistics base there. The new military base in Djibouti, situated near the strategic maritime chokepoint connecting the Red Sea to the Gulf of Aden, assumes significance amidst China's growing strategic prowess. In February 2014, the two countries had signed a military agreement allowing the Chinese Navy to use the Djibouti port.

Multilateral Alliances

SCO: The Shanghai Cooperation Organisation (SCO), established in 2001, with China, Russia, Kazakhstan, Kyrgyzstan, Tajikistan, and Uzbekistan as its member nations, has emerged as a potential strategic block in Asia. China played an instrumental role in setting up the multilateral organisation. The group has accorded observer status to Afghanistan, Iran, Mongolia and Belarus, while Sri Lanka and Turkey have been granted the dialogue partner status at the organisation. India and Pakistan have been accorded full-time member nation status in 2015 and are set to join the group in 2017. While emphasising on political and economic coordination among its member nations, the alliance has also focused on strategic and security issues in the continent. Many analysts view the SCO as a potential rival block to Western alliances, especially the US-led NATO in the long run.

BRICS: The forum was established in 2006 with Brazil, Russia, India and China (BRIC) as its members. South Africa later joined the block in 2010. The alliance has primarily focused on political, economic and financial cooperation among the member nations.

APEC: The Asia-Pacific Economic Cooperation (APEC) came into existence in 1989 to promote trade and economic cooperation among the Asia-Pacific nations. China joined the forum in 1991. The 21-member nation APEC has solely focused on free and open trade and investment, promoting and accelerating regional economic integration and encouraging economic and technical cooperation among the member countries to build a dynamic and harmonious Asia-Pacific region.

DEFENCE CAPABILITIES

ARMY: The People's Liberation Army (PLA) is the largest active standing army in the world. The PLA Ground Force (PLAGF) has an estimated strength of over 1.8 million personnel and an additional 800,000 in reserve. The 8th Defence White Paper of China, titled "*The Diversified Employment of China's Armed Forces*", released in 2013 for the first time revealed the number of PLAGF troops at 850,000. The ground forces were previously deployed in seven Military Regions (MR) across China, being based at Shenyang, Beijing, Jinan, Nanjing, Guangzhou, Chengdu and Lanzhou. In 2016, the country announced major military reforms under which five new theatre Commands -- the Eastern, Western, Southern, Northern and Central -- have been established and will focus on joint combat and modern warfare. Additionally, two new leading organs, namely the PLA Rocket Force and the PLA Strategic Support Force, have also been established as part of building a modern military. While the Rocket Force has replaced the Second Artillery Force (SAF) and "would strengthen nuclear deterrence and nuclear counter-attack capabilities, intensify the construction of medium and long range precision strike power, and reinforce the strategic check-and-balance capability," the Strategic Support Force would be a new type of command force to maintain national security and would become an important growth point of the PLA's combat capabilities.

The ground force is continuously being restructured to deal with future land warfare operations and move from regional defence to trans-regional mobility. The PLAGF is accelerating the development of army aviation troops, light mechanised units and special operations forces, and enhancing building of digitalized units, gradually making its units smaller, modular and multi-functional in organization so as to enhance their capabilities for air-ground integrated operations, long-distance manoeuvres, rapid assaults and special operations. Induction of advanced battle tanks, armoured vehicles, howitzers, guns, other small arms along with the integration of information systems with weapons and platforms have been undertaken to strengthen the PLA Army.

Personnel: 1.8 million+ (850,000 as per 8th Defence White Paper. To be reduced by 2017)

Equipment

Category	Name	In Service
MBT	Type 99A, A2	800+
	Type 98G	250
	Type 96, 96A, 96G	2500+
	Type 69/79	300+
	Type 59D/D1	
	Type 54/55	
Light Tank	Type 63/63A	800
	ZTD-05	200+
IFV/AIFV/APC/ATV	Type 97	50+
	Type 89	300
	Type 63	2000
	ZBL09 (Type 09)	100
	Type 05 (APC & 105mm Tracked)	400
	Type 04	200
	ZBD-03 (Type 03)	40
	Type 90/92	600+
	Type 86	600+
	Type 85	
Artillery	Type 05 155mm SP	100
	AH4 155mm lightweight howitzer	
	PLZ-45 155mm SP	80+
	Type 89 122mm SP	500
	Type 83 152mm SP	500+

Army continued...

Category	Name	In Service
Artillery	U/I ARTY 130mm SP	
	WZ-551 122mm SP	
	Type 85 122mm SP	
	Type 70 122mm SP	200
	Type 54 122mm SP	
	WZ-551105mm SP	
	Type 88 155mm towed	150
	Type 66 152mm	1,400
	Type 54 152mm	
	Type 59-1 130mm field gun	1,000
	Type 83 122mm	
	Type 60 122mm	
	Type 54 122mm	6000
Mortar	YW-381 120mm SP	
	WZ-551 120mm SP	
MLRS	PHL03 300mm	
	WS-1/2/3/6 SP	
	A-100 300mm	
	Type 83 273 mm SP	
	Type 82 130 mm, Truck	
	Type 70 130 mm, APC	
	Type 63 130 mm, Truck	
	Type 90,122 mm, Truck	
	Type 89,122 mm, APC	
	Type 83 122 mm, Truck	
	Type 81,122 mm Truck	
AT Gun	120mm/125mm SP	
	Type 86, 120 mm SP	
ATGM	HJ-8	
	HJ-9	24
	HJ-10	
	HJ-12	
AA gun	Type 59, 100mm radar, towed	
	Type 56, 85mm, radar, towed	
	Type 59, 57mm radar/optic, towed	
	Type 88, Twin-37mm optic, tracked	
	Type 63, Twin 37mm optic, towed	
	Type 90, Twin 35mm optic, tracked	55+
	Type 87, Twin 25mm optic, tracked	
SAM	HQ-61A	
	HQ-7/7B	
	HQ-64, LY-60	
	HQ-6	
	QW-1/QW-2 (MANPADs)	
SSM (conventional)	DF-11	~200
	DF-15	~100

Army continued...

Category	Name	In Service
SSM (convent-ional)	DF-21 (CSS-5)	~100
	DF-31	15
	DH-10	300+
	HY-2	
	HY-4	

Army Aviation

Category	Name	In Service
Helicopter	HC-120 (EC120)	100+
	Mi-17	20
	Mi-171, Mi-171E	50 (34 on order)
	Mi-17-V5	35
	Mi-6	2
	S-70 (Black Hawk)	19
	Z-9 (utility)	64
	Z-8A/B (to be retired)	12
	Z-9A, Z-9WZ, Z-9WA & other variants (attack)	250+
	Z-10 (attack)	50+
	Z-19 (reconnaissance/ attack)	10+
UAV	CH-1	
	ASN-15	
	ASN-105	
	ASN-104	

New Procurements/Upgrades

✦ Induction of new MBTs is going on. A new variant of Type 96 MBT has been unveiled by NORINCO in 2016 and is designated as Type 96B. The advanced third-generation tank is likely to replace the older-generation Type 59 and Type 69 MBTs of the PLAGF.

✦ The Army is introducing new tracked AIFVs into service.

✦ The Z-10 attack helicopter, unveiled in the 2012 Zhuhai Air Show, is being progressively inducted in all PLAGF aviation units. Designed primarily for anti-tank missions, the new rotorcraft can deploy the HJ-8, HJ-9 and HJ-10 missiles. In air combat role, the gunship is capable of deploying the TY-90 missile along with the larger PL-5, PL-7 and PL-9 AAMs.

✦ Z-20, a new medium-lift helicopter having a maximum take-off weight of 10-tonne, is under development and is expected to enter production by 2018.

✦ A new state-of-the-art HJ-2 anti-tank missile, developed by the China North Industries Group (NORINCO), is being introduced into the Army.

✦ A new 125mm cannon, termed as the world's fastest anti-tank gun, has been delivered to the Army in June 2015 for trial. The multi-purpose tank cannon reportedly has a muzzle velocity of 2,000 meters/second – almost Mach 6 – for armoured piercing fin stabilized discarding sabot rounds (APFSDS) making it the fastest tank gun in the world. The gun is likely to be used for the next generation of Chinese main battle tanks.

✦ NORINCO has developed a new 155 mm/39 calibre lightweight gun-howitzer, called AH-4.

NAVY: The People's Liberation Army Navy (PLAN), which was once a subordinate force to the PLA Ground Force and played a limited role during the Cold-war era, has transformed into a major naval power today. China began modernising its naval forces in the 1990s in an effort to build up a "Blue Water Navy" that would take up operations far and beyond its territorial waters. The Navy is being modernised to follow the strategy of "open water defence" – a shift from its earlier strategy of "offshore defence." The modernisation programme, involving acquisition and deployment of a large number of frigates, corvettes, destroyers, submarines, aircraft carriers, naval combat jets, helicopters and unmanned aircraft along with a broad range of weapons, especially ballistic and cruise missiles, has established China as a major naval power in the world today. The PLAN currently possesses over 60 destroyers and frigates, and nearly 70 submarines, both nuclear and conventional. The Navy plans to induct several new-generation submarines into its naval fleet in the coming years and build more aircraft carriers to form Carrier Battle Groups. Pursuing this ambitious goal to build aircraft carrier battle groups and deploy them in the strategically vital South China and East China Seas to deter the presence of US naval forces as well as other regional powers, China in September 2012 commissioned its sole refurbished Ukrainian aircraft carrier, the Liaoning. The country has also started building more aircraft carriers and support ships indigenously to bolster its global military power projection capabilities.

The latest military whitepaper released in May 2015 has called for a greater role for the PLA Navy in particular by emphasising the maritime domain as one of four "critical security domains" where China must increase its capabilities. The whitepaper states the following: In line with the strategic requirement of offshore waters defence and open seas protection, the PLA Navy will gradually shift its focus from "offshore waters defence" to the combination of "offshore waters defence" with "open seas protection," and build a combined, multi-functional and efficient marine combat force structure. The PLAN will enhance its capabilities for strategic deterrence and counter-attack, maritime manoeuvres, joint operations at sea, comprehensive defence and comprehensive support.

With the acquisition of modern platforms, weapons and systems, China is building up a most diverse and powerful naval force in the entire Asian continent to stake its claim over the resource-rich East China Sea and South China Sea while expanding its strategic reach beyond the territorial waters. With an aim to secure strategic oil routes in the Arabian Sea and Indian Ocean, and lay claim over vital energy resources all across Asia and beyond, China has also made its intention clear in recent times that it wants to build naval bases in foreign nations as well.

The PLAN also conducts complex multidiscipline warfare training throughout the year besides taking part in multilateral naval exercises, including in the RIMPAC. It has expanded its reach and operated in the Mediterranean, increased intelligence collection deployments in the Western Pacific, and even has deployed submarines in the Indian Ocean in recent times.

The PLAN has also fulfilled a long-held dream of gaining access to the "First Island Chain" in the Pacific Ocean. The "First Island Chain" refers to the first major archipelagos off the East Asian continental mainland, including the Japanese archipelago, Ryukyu Islands, Taiwan and the northern Philippines. The Chinese naval vessels have sailed through these major East Asian archipelagos which are closely guarded by the US and Japan. The Navy, for the first time, also showcased its fleet of nuclear submarines during a military exercise in West Pacific in October 2013.

The People's Liberation Army Navy is divided into three fleets – the East China Sea (Beihai) fleet, the South China Sea (Nanhai) fleet and the North China Sea (Donghai) fleet headquartered at Qingdao, Zhanjiang and Ningbo respectively. Each fleet has under its command fleet aviation, support bases, flotillas, maritime garrison commands, aviation divisions and marine brigades. In terms of equipment and war operational responsibilities, the PLAN is composed of submarine, surface vessel, naval aviation, Marine Corps and coastal defence arms.

Major naval bases: Lushun, Huludao, Qingdao, Shanghai, Zhoushan, Wenzhou, Xiamen, Guangzhou, Zhanjiang, Yulin, Yalong, Luda, Dalian, Hong Kong, Whampoa.

Personnel: 270,000 (235,000 as per 8th Defence White Paper)

Equipment

Category	Name	In Service	On Order
Aircraft Carrier	Liaoning (STOBAR)	1	
	Type 001A class (indigenously built)		2+
Training Ship	ATS Shichang (multi-role aviation)	1	
Destroyer	Luhai class (Type 051B)	1	
	Luhu Class (Type 052)	2	
	Luyang I Class (Type 052B)	2	
	Luzhou Class (Type 051C)	2	
	Luyang II Class (Type 052C)	6	

Navy continued...

Category	Name	In Service	On Order
Destroyer	Luyang III Class (Type 052D)	4	8
Submarine	SSBN Jin Class (Type 094)	4	2
	SSBN Xia Class (Type 092)	1	
	SSB Qing (Type 032)	1	
	SSN Shang (Type 093/093G)	2	4
	SSN Han (Type 091)	3	
	SSN Type 095		14
	SS Kilo Class (Project 877EKM & 636)	2 + 10	
	SS Song Class (Type 039)	13	
	SS Yuan Class (Type039A, B, C, AG/ Type 041)	15	5
	SS Ming Class (Type 035)	15	
Frigate/ Corvette	Jiangkai (Type 054)	6	2
	Jiangkai II (Type 054A)	22	3
	Jianghu Class (Type 053)	10+	
	Jiangwei II Class (Type 053H3)	10	
	Jiangdao (Type 056/56-A) [corvette]	30	10 or more
Patrol Boat	Houbei class (Type 022)	70+	
	Houjian (Type 037II)	4	
	Houxin Class (Type 037IG)	25	
	Hainan (Type 037)	40 (being phased out)	
	Shanghai III class (Type 062I) gunboat	10+	
LPD	Yuzhao (Type 071)	4	2 (or more)
LCAC	Jingsah II Class (Type 726)	30	
	Zubr class	1	3
LCM	Yuchin Class (reserve) (Type 068/ 069)	30	
LCU	Yunnan class (Type 067)		
LST	Yuting II (Type 072A)	9	6
	Yuting (Type 072 II/ III)	15+	
	Yukan (Type 072)	3	
LSM	Yuhai class (Type 074)	20	10
	Yudao class (Type 073II/ 073III)		
	Yunshu class (Type 073A)		

Navy continued...

APA	Qiongsha Class (attack transport)	4	
AH	Shichang (Multirole Aviation)	2	
Patrol Hydrofoil	Huchuan Class	20	
Mine sweeper, Mine hunter	Wolei / Bulieijian Class(minelayer)	1	
	Wochi class (Type 81/81-A)	10	
	Wozang-class (Type 082B)	4	
	Type 010 (Sov T-43) (in reserve)	40	
	Lianyun Class (minesweeper-coastal)	50+	
Auxiliary ship	Oiler/ Cargo	50	
	Fuchi-class (Type 903/ 903A) replenishment ship	8	
	Type 908 replenishment ship	1	
	Survey (Ocean/ Hydro) Ship	35+ 3	13
	Hospital Ship	4	
	Reconnaissance / Surveillance/ Intelligence	21+	
	Space Event Ship (Yuan Wang class)	7	
Weapons	ASM C802A C602 C704 C705		
	SAM HQ-61 HQ-7 HQ-9 HQ-16 Shtil		
	Torpedo Yu-1/Yu-1A Yu-2 Yu-3 Yu-4 Yu-6 53-65KE		

New Procurements/ Upgrades

✦ Type 095, a new variant of the Type 093 Shang-class nuclear-powered submarine (SSN), is under development to replace the older generation Han-class of SSNs. The new improved submarines would feature vertical launch system and may be able to carry the YJ-18 anti-ship cruise missile. A total of 14 such subs are being planned to be inducted from 2018 onwards. The first Han-class platform has been retired in 2016 with the remaining two likely to be retired by 2020.

✦ Yuan-class (Type 039A/Type 041) SSKs with AIP technology are being progressively inducted into service.

✦ Two or more Jin class (Type-094) SSBNs are under construction. Each SSBN is expected to carry up to 12 JL-2 SLBMs having an estimated range of over 8,000-km. The new subs with the long-range missiles will give China its first credible sea based nuclear deterrent capability.

✦ The Type-096 class SSBN, currently under development stage, is expected to become the PLAN's next-generation missile submarine. Based on the Type-094 class SSBN, the Type-096 class of subs are likely to be armed with the longer range JL-3 ballistic missiles. The PLA Navy is expected to deploy as many as eight Type-096 class subs by 2020.

✦ Aircraft carrier Liaoning is ready to undertake combat operations, the Defence Ministry has announced in November 2016. The sole aircraft carrier of China was commissioned in September 2012 and since then completed extensive sea trials, including combat training, test flight of ship-borne J-15 fighters, ship-deck test, training of commanders for ship-borne fighters' landing, and flight training of test pilots. The 60,000 tonne carrier, a refurbished Ukrainian warship, is capable of carrying over 30 fixed-wing aircraft and helicopters in its deck.

✦ Work is advancing on two indigenous aircraft carriers, one being built at the Dalian shipyard and another at the Jiangnan shipyard near Shanghai. The first indigenous carrier is expected to enter water in early 2017.

✦ A new class of large guided missile destroyer, designated Type 055, is under construction. The 10,000-ton powerful warship is believed to be capable of carrying over 100 anti-ship, anti-air and ship-to-ground attack cruise missiles. The platform is likely to be used to escort indigenous aircraft carriers of the PLAN which are going to be inducted in service in the coming years.

✦ All six Luyang II Class Type 052C destroyers have been commissioned into service.

✦ Fourth Type 052D (Luyang II class) destroyer has been commissioned into service in July 2016. The new warships are an advanced version of the Type-52C class of destroyers, and are being equipped with two 32-unit vertical launch systems capable of launching HQ-9B air-defence missiles, anti-ship and anti-submarine missiles. The PLAN plans to induct 12 such DDGs in service.

✦ The sole Luhai class (Type 051B) destroyer has completed a major modernisation programme and is now equipped with HHQ-16 SAMs and Type 1130 CIWS, replacing the older weapon systems. The warship's air and surface search radars have also been upgraded.

✦ Type 056A "Jiangdao" class of stealth corvettes are being progressively commissioned into service. The new warships, replacing the Type 037 patrol boats, flaunt a sleek design and will primarily be deployed on escort missions and for anti-submarine warfare operations. The 40th platform of the class has been launched in October 2016.

✦ Type 054A (Jiangkai II) class frigates, capable of undertaking long-range surveillance, air defence and anti-submarine warfare operations, are being inducted progressively. Over 20 such warships are likely to join the Navy.

✦ Four Yuzhao-class (Type 071) landing platform docks have been commissioned into service. The Navy plans to operate six such LPDs.

✦ The first of four Zubr-class Landing Craft Air Cushion (LCAC) hovercraft ordered from Ukraine was delivered in April 2013. The second LCAC is under construction, with a further two to be licence-built in China.

✦ A new space event ship of the Yuan Wang-class has been commissioned into service in July 2016 to track China's space-related developments. The Yuan Wang-class ships supplement China's network of land-based tracking stations.

✦ Two new Type 903A (Fuchi-class) fleet replenishment ships have joined service in 2016.

Marine Forces

Personnel: 12,000

Equipment

Category	Name	In Service
MBT	Type 63A	150+
	Type 60	
APC	Type 63	60
Artillery	Type 83, 122mm	
	Type 54, 122mm	
MLRS	Type 63, 107mm	
ATGW	HJ-73	
	HJ-8	
SAM	HN-5	
	JL-2	

Naval Aviation

Personnel: 30,000+

Equipment

Category	Name	In Service	On Order
Bomber/ Anti-ship strike	Su-30Mk2	22	
	H-6	50	
	JH-7/JH-7A/B	50+	
	J-10A/B	30+	
Fighter/ Ground attack	J-11B/BS	50	
	J-7	30	
	Q-5	40+	
	J-8	40+	
Maritime Patrol/ ISR/ ASW	Be-6	10	
	SH-5	6	
	Y-8X		
	Gaoxin-6		
Trainer	JL-9, JJ-9		
Helicopter	Z-8 Super Frelon	12	
	Z-9	50	
	Ka-27	3	

Naval aviation continued...

Category	Name	In Service	On Order
Helicopter	Ka-28	5	9
Transport	Y-7		
	Y-8		
Amphibi-ous	AG-600		

New Procurements / Upgrades

✦ Gaoxin-6, an anti-submarine aircraft developed and manufactured by Aviation Industry Corp of China and based on the Y-8 transport plane, has been commissioned into service and deployed with the PLAN's North Sea Fleet in 2015.

✦ China has formed its first aircraft carrier-borne aviation force under the PLAN. The force comprises carrier-borne fighter jets, jet trainers and ship-borne helicopters to carry out anti-submarine, rescue and vigilance tasks.

✦ A new stealth warplane, designated J-31 (also FC-31), made its debut flight in October 2012. The J-31 is the second radar evading combat aircraft after the J-20 fighter, unveiled by China in January 2011. The new aircraft is believed to have been designed to operate from the PLA Navy's Liaoning aircraft carrier and also for use by the Air Force. China publicly unveiled the new stealth jet for the first time at the Zhuhai Air Show in November 2014. The fighter made its first international debut at the 2015 Dubai Air Show in November.

✦ The Navy unveiled its new carrier-based fighter jet J-15. The new naval fighters have been flight tested from on-board the refurbished Soviet-era aircraft carrier Liaoning. The indigenously developed J-15 is capable of carrying anti-ship, air-to-air and air-to-ground missiles, as well as precision-guided bombs. Three variants of the new naval fighter are being developed – an advanced trainer, a multi-role combat and an electronic warfare variant.

✦ A new anti-submarine warfare helicopter, designated Z-18 (F & J), is undergoing flight tests. The rotorcraft has been developed from the Changhe Aircraft Industry Group (CAIG) Z-8 which is a derivative of the French SA-321 Super Frelon.

✦ The TA600 (also called AG600) amphibious aircraft has been officially unveiled by the state-owned Aviation Industry Corporation of China (AVIC) in July 2016. It is the world's largest amphibious aircraft designed to be used for passenger transport, maritime patrol, and search & rescue missions.

CHINA COAST GUARD: In the wake of rising maritime disputes with a number of its neighbours, China, in July 2013 operationalised a new, unified Coast Guard, enabling it to be armed with weapons. The previous configuration of the coast guard attached it to the police, fisheries law enforcement and Customs' anti-smuggling maritime police. The new China Coast Guard (CCG) has integrated the functions of marine surveillance with the existing coast guard. The new agency is now dealing with conflicts in the disputed waters as per Chinese law and the divisions that were previously not allowed to be equipped with weapons, are now armed. The CCG reportedly has 11 squadrons and more than 16,000 personnel.

AIR FORCE: The People's Liberation Army Air Force (PLAAF), which was once assigned the task of defending the country by carrying out air defence missions and protect the ground forces, has rapidly evolved as a more modern aggressive air power capable of conducting offensive aerial missions against targets even beyond its territory. The air force's modernisation programme began in the 1990s when the country acquired fourth-generation fighter aircraft along with support aircraft and advanced weapons. The country's economic growth further propelled its defence indigenisation programme which now includes development of new trainer jets, combat helicopters, advanced stealth fighters, unmanned aerial systems, early warning and electronic warfare support aircraft among others for the PLAAF. Since 2012, the Air Force has made increasing efforts to boost its overall capabilities by acquiring a wide range of modern air platforms. In the wake of the continuing territorial disputes in the South China Sea and East China Sea regions, the PLA Air Force has also conducted military exercises in the West Pacific several times in 2016 by dispatching a wide variety of its aircraft including strategic bombers, fighter jets, early warning aircraft and refuelling aircraft. It has also conducted aerial patrols over the East China Sea Air Defence Identification Zone (ADIZ) set up in 2013.

According to the 2015 defence whitepaper, the PLAAF has shifted its strategy from "territorial air defence" to a strategy of both "offense and defence". Besides, it is also focused on building up an air-space defence force structure that can meet the requirements of informationized operations. The PLAAF will boost its capabilities for strategic early warning, air strike, air and missile defence, information countermeasures, airborne operations, strategic projection and comprehensive support, states the 2015 Defence White Paper.

The PLAAF is primarily composed of aviation, ground air defence, radar, airborne and electronic countermeasures (ECM) arms. It is made up of air teams, regiments (air ports and air stations), divisions, and regional air forces (air armies). In terms of equipment and war operational responsibilities, it mainly consists of air units, anti-aircraft units, surface-to-air missile units, paratroops and radar units.

Personnel: 400,000+

Major air bases: Shahezhen, Xijiao, Datong, Zunhua, Chengdu, Jing Hong/Gasa, Luliang, Jiugucheng, Luyang and Zhengzhou. China has also operational airfields in Tibet

Autonomous Region - at Lhasa, Kashgar, Korla, Yankaw, Hotan, Cherchen (Qiemo) and Gardzong

Equipment

Category	Name	In Service	On Order
Fighter/ Ground attack/ Interceptor aircraft	Su-27SK	36	
	Su-30 MKK, Su-30 MKK2	105+100	
	J-11B/BS	120+	100+
	J-7	250	
	J-8	160+	
	J-8II	60	
	Q-5	50+	
	J-10	100+	250+
Bomber	H-6/ 6H/ 6M	120	
	H-6K	40	
	JH-7A/B	100+	
Surveillance/ Reconnaiss-ance aircraft	Y-9G		
	JZ-6/7/8	100	
	HZ-5, HZ-7	40	
AWACS /AEW aircraft	KJ-200		
	KJ-2000	6	
	KJ-500		
ELINT/ SIGINT aircraft	Tu-154M/D	4	
	Y-9JB		
Transport, Tanker aircraft	Y-12	8	
	Y-11	15	
	Y-9	5	
	Y-8	48	
	Y-7	40	
	Boeing 737-200	8	
	CL-601	5	
	Il-76MD	20+	16
	Il-78	4	
	Tu-154M	15	
Air-refuelling aircraft	H-6U	12	
Trainer aircraft	Su-27UBK	26	
	Hongdu L-15 Falcon		
	CJ-6		
	HJ-5		
	JJ-7	50+	
	JL-9	5	35
	JL-8	10	40
Helicopter	Z-6 (Mi-8)	40+	
	Mi-17		
	Mi-171	32	52
	Z-8	30+	
	Z-9 (A/B & W)	40+	
	Bell 214ST	4	

Air Force continued...

Category	Name	In Service	On Order
Helicopter	As332 Super Puma (VIP transport)	6	
	Gazelle/Hot	8	
	S-70 Black Hawk	20 (being retired)	
	Z-10, Z-11, Z-19 (attack)		
SAM	HQ-1/2		
	S-300	6 regiments	
AAM	PL-2, PL-5, PL-7, PL-8, PL-9, PL-11, PL-12		

New Procurements/Upgrades

✦ A new variant of J-11 fighter has been flight tested in April 2015. The new warplane, designated J-11D, is being produced by the Shenyang Aircraft Corporation. It features a phased-array radar and an air refuelling system and is armed with the PL-10 air-to-air missile.

✦ Test flight of new J-20 advanced fifth-generation stealth fighter is under progress with the PLAAF planning to deploy the jet by 2020. The new short-range air-to-air missile PL-10 is undergoing trials to arm the new fighter.

✦ A limited number of new multi-role attack aircraft, designated J-16, has reportedly been inducted in service in 2016. The new fighter is believed to have been derived from the Russian Su-30MKK fighter. It has been designed for both the PLA Air Force and Navy.

✦ The first four Su-35 super-manoeuvrable fighters from Russia have been delivered in December 2016. A deal worth US$2 billion was inked in November 2015 to buy 24 Su-35 fighters. Delivery of the warplanes will continue till 2018.

✦ Induction of advanced H-6K strategic bombers has started. The new bombers, with extended range, are capable of carrying long-range cruise missiles.

✦ Development of a new strategic bomber, called H-20, with long-range strike capability was announced in 2016.

✦ The new KJ-500 airborne early warning and control (AEW&C) aircraft has started entering service. The platform is based on the Shaanxi Y-9 four-turboprop transport aircraft. Development of another improved variant, the third-generation KJ-3000 AEWC, has also started.

✦ Russia's Rosoboronexport has started delivering the 10 Il-76MD military transport aircraft.

✦ The L-15 advance jet trainer has completed development phase and is ready to enter production phase. It completed its first pilot flight in January 2016. Armed version of the jet trainer has started flight testing in 2016. The new supersonic training and light attack jet is being built to train the pilots of China's fourth-gen warplanes, including the J-10, J-11 and JF-17.

✦ Deployment of JL-9, a new third-generation supersonic jet trainer, has started in 2015.

✦ The Y-20 military transport aircraft has been officially commissioned into service in July 2016 after the PLA Air Force received the first plane of the series. The indigenously-built large military transport plane made its maiden test flight in January 2013. Bearing resemblance with the US' C-17 Globe master and Russia's Il-76 freighter aircraft, it is China's largest home-grown military transport plane. An Airborne Early Warning and Control (AEW&C) variant of the aircraft could be developed in future.

✦ An agreement has been signed with Ukraine's Antonov to jointly develop An-225 strategic airlifter.

✦ Russia is delivering the Mi-171E multipurpose helicopters under a 2012 contract for 52 such choppers.

✦ A new variant of the Z-11 attack helicopter has been unveiled in 2016, designated as Z-11WB.

✦ A new variant of the Wing Loong II unmanned combat aerial vehicle (UCAV) is under development and reportedly started flight tests.

✦ The CH-4 medium altitude long-range unmanned combat aerial vehicle (UCAV) has entered service.

✦ A deal has been announced to procure the advanced S-400 Triumf air defence missile systems from Russia. Delivery of the long-range anti-aircraft missiles is expected to begin in 2017-end and will be completed by 2020.

STRATEGIC FORCE COMMAND: The strategic force command or the strategic missile force was created to control China's ground-based strategic missile arsenal, including its nuclear ballistic and conventional missiles. Previously called the PLA Second Artillery Force (PLASAF), it has now been renamed as the **PLA Rocket Force (PRF)** under new largescale military reforms announced in 2016. The PRF has now been upgraded into a "full-fledged" new service on par with the PLA Army, Air Force and the Navy, indicating that it would now have several arms and special troops besides having academies, research institutes and logistic support system under its command. As the core force of China's strategic deterrence, the PRF, unlike the erstwhile PLASAF, is expected to command all three legs of China's nuclear triad and could deploy its nuclear assets on land, sea and air in the coming years. The new force thus could eventually emerge as the first independent service with land, sea and air nuclear forces in the world. The PRF is also expected to separate its nuclear and conventional wings. While the nuclear missiles would be responsible for strategic nuclear deterrence, the conventional ones would be used in the first round of strikes. The PRF would strengthen nuclear deterrence and nuclear counter-attack capabilities, intensify the construction of medium and long range precision strike weapons, and reinforce the strategic check-and-balance capability of the country.

SPACE FORCE COMMAND: Acquisition of advanced space-based technology for civil and military purposes has played a crucial role in China's steady ascent as a rising power, outperforming its other Asian counterparts, including Japan and India. China has avidly pursued the goal to reach outer space and exploit its potential for carrying out both civilian as well as military operations. It has also emphasised on the need to effectively integrate its air and space prowess for launching offensive and defensive operations. From orbiting a large number of communications, navigation, surveillance and reconnaissance satellite systems to developing and testing an anti-satellite (ASAT) weapon, China in recent years has taken giant strides in its space capabilities. The successful ASAT test in January 2007, indeed, made China the first country to hit down a satellite in outer space and raised global concern over a future arms race in space. Such space-based capabilities of the country have been affiliated with the Central Military Command which controls and overviews China's space programme. The country, to date, has also placed in orbit more than 100 satellites of which 40 are dedicated for military purposes.

Beijing has also beefed up efforts to put in place an independent global satellite navigation network by 2020 as a rival to the US's Global Positioning System (GPS). In this direction, it has orbited over a dozen 'Beidou' or Compass navigation satellites so far. The network, which would eventually consist of over 30 such satellites, could be used for both civilian as well as military purposes.

Major Armaments Producers

Land equipment manufacturers

China North Industries Group (Norinco) is a leading producer of most of China's tanks, armoured vehicles and equipment.

Shipbuilders/ naval equipment manufacturers

China State Shipbuilding Corporation (CSSC) is a leading shipbuilding conglomerate consisting of shipyards, naval equipment manufacturers and research institutes.

China Shipbuilding Industry Corporation (CSIC) is another major shipbuilding conglomerate mainly involved in warships design, development, repairing of domestic and foreign naval vessels. In 2016, the conglomerate announced a plan to consolidate some of its shipbuilding activities in a bid to boost efficiencies.

Dalian Shipbuilding Industry Company (DSIC) is the largest shipbuilding company of China operating under CSIC.

Guangzhou Shipyard International (GSI) operates under CSSC. It is the largest modern integrated shipbuilding facilities in Southern China.

Jiangnan Shipyard located in Shanghai designs, develops and repairs both military and civilian ships.

Aerospace equipment manufacturers

China Aerospace Science and Technology Corporation (CASTC) is the leading missile weaponry manufacturer. (It is presently working on a new type of conventional missile with a range of 4000 km which could engage targets in land, sea, air, space.)

China Changfeng Mechanics and Electronics Technology Academy works under CASTC to design, develop and produce air and missile defence systems.

China Aviation Industry Corporation (CAIC) is the aircraft manufacturing consortium. It has been split into two groups – AVIC I which is primarily producing fighters, bombers and other large aircraft, and AVIC I II which is manufacturing trainer jets, transport aircraft, helicopters and other small planes.

Chengdu Aircraft Industrial Corporation (CAC) is a manufacturer and supplier of combat jets and aircraft parts.

Changhe Aircraft Industries Corporation is a helicopter manufacturer.

China National Aeroengine Corporation (CAREC) is an aero engines' producer for manned aircraft.

China National Aerotechnology Import–Export Corporation (CATIC) is a leading trading agency that deals with the import and export of defence products.

Harbin Aircraft Manufacturing Corporation (HAMC) manufactures helicopters. It operates under the Harbin Aircraft Industries Group that produces helicopter, light and general-purpose aircraft, and regional aircraft in China.

Shenyang Aircraft Corporation (SAC) is a civilian and military aircraft manufacturer.

DEFENCE DEALS

Joint Venture Programmes

Pakistan

Al-Khalid (MBT-2000): China and Pakistan jointly developed the main battle tank in the 1990s. The MBTs are operational in the Pakistan Army.

Hongdu JL-8 jet trainer: The Hongdu Aviation Industry Corporation (HAIC) of China and Pakistan Aeronautical Complex (PAC) have jointly developed the Hongdu JL-8 (export variant Karakorum-8 or K-8) basic two-seat jet trainer with light attack capabilities.

JF-17 Thunder (FC-1) fighter: China has developed the multi-role fighter aircraft jointly with Pakistan. The two countries plan to export the fighter to other countries.

Indonesia

Missile: Indonesia and China have agreed to strengthen their defence cooperation, including joint production of C-702 missiles. The two countries signed an MoU in 2011 that stipulates joint military procurement by both the countries in a government-to-government plan and a technological transfer with regard to certain types of weapons production, as well as joint-development and joint-marketing of certain types of military weapons.

Thailand

MRLS: Thailand and China have agreed to jointly develop multiple rocket launchers with a guidance system. The two sides signed an agreement in 2012 under which the Thai Defence Technology Institute will work with China to develop the new multiple rocket launchers called "DTI-1G (Guided)" which will have a greater range than existing systems. In an earlier joint deal, Thailand and China had developed the DTI-1 system, which had a range of between 60 and 180 km.

Arms Export

China has emerged as the world's third largest arms exporter after the US and Russia, according to the data released by the Stockholm International Peace Research Institute (SIPRI) in 2015. China surpassed Germany, France and UK to supply weapons and equipment to foreign countries between 2010 and 2014. China's arms sales account for nearly 6% of the global share and Pakistan remains the top-most receiver of Chinese defence hardware. Beijing has also exported arms and weapons to several other countries including Iran, Libya, Saudi Arabia and Kuwait in the Middle East, Algeria, Egypt and Nigeria in Africa, Bangladesh, Malaysia, Myanmar, North Korea and Sri Lanka in Asia, and Argentina, Ecuador, Venezuela and Bolivia in South America. It is also making efforts to establish bilateral military collaborations with Indonesia, Israel, Brazil, Switzerland and some other Latin American countries, like Peru.

Arms Import

While in the past, China has relied heavily on Russian technology for building majority of its defence equipment including aircraft, missiles and naval vessels, the country, in the last two decades, has vociferously attempted to be self-reliant in military sphere. Before the imposition of international arms embargo as a result of the Tiananmen Square crackdown in 1989, China was also importing weapons from France, Germany, Canada, UK, Israel and Egypt. The country, which was once the world's top arms importer until 2009, now occupies third place in arms imports, owing to the rapid and robust development of its domestic defence industry. With an agenda of modernising and strengthening its armed forces by inducting indigenous platforms, weapons and systems, China has focused on new technological innovations and breakthroughs and successfully produced a wide range of new-generation advanced military hardware, even though many of these find their origin either in the Western or Russian designs and technology.

DEFENCE BUDGET

China is the top-most defence spender in Asia and ranks second in the global chart after the US. The country's defence budget has maintained a double digit growth for over two decades. In 2012, its defence budget crossed the $100 billion mark for the first time. Keeping pace with its rapid economic rise, China has been spending a visible share of its GDP – officially around 1.4% -- on military developments, though according to defence analysts, it could be as high as 3% as the country's defence budget does not include military R & D spending. An all-round military modernisation that includes development of new-generation fighter aircraft, large transport aircraft, combat helicopters, unmanned aerial systems, submarines, warships, aircraft carriers, new weapons and other systems along with existing military hardware upgrade, military training and human resource development have led to a sharp rise in China's annual defence budget. For the year 2016, the country has announced to spend Yuan 954 billion ($146 billion) on its military – an increase of 7.6% over the previous year's budget allocation.

Year	Allocation	% of GDP	% increase over previous year
2016	Yuan 954 billion ($146 billion)	-	7.6
2015	Yuan 886.9 billion ($142 billion)	1.34	10.1
2014	Yuan 808.2 billion (US$ 132 bn)	1.31	12.2
2013	Yuan 720.2 billion (US$ 117 bn)	1.3	10.7
2012	Yuan 670 billion (US$ 106 bn)	1.28	11.2
2011	Yuan 601 billion (US$ 91.5 bn)	1.4	12.7

NAVAL WARFARE SYSTEMS

The naval warfare capabilities of China have improved steadfastly over the last few years with the country swiftly building up a modern, dynamic and powerful Navy capable of showing its influence and dominance in the near and far-sea regions. The naval modernisation programme of China has essentially focused on transforming its Navy into a blue-water maritime force having the capability to control the high seas besides its littoral waters.

Accordingly, the naval modernisation programme of the PLAN has included the development and deployment of a wide range of modern maritime platforms, including aircraft carriers, submarines, destroyers, frigates as well as advanced weapons such as anti-ship ballistic and cruise missiles besides maritime aerial platforms and C4ISR systems for future network warfare missions.

The PLAN, indeed, is building up a most diverse and powerful naval force in the entire Asian continent to not only lay claim over the resource-rich and highly contested South China Sea and East China Sea regions but also expand its military and strategic reach beyond these territorial waters. While many obsolete maritime platforms have been phased out, the Navy continues to induct many advanced ones having improved, state-of-the-art capability. The PLAN could deploy as many as 270 naval platforms by the year 2020.

To carry out its other secondary roles such as humanitarian/disaster response operations or anti-piracy missions, the Chinese Navy has also procured numerous auxiliary/ support vessels.

All major naval platforms, combat as well as support, and weapon systems currently in possession of the PLA Navy are detailed below. Also included are some of the futuristic platforms/ systems/weapons which are likely to be inducted in naval service in near future.

COMBAT PLATFORMS

The PLA Navy presently deploys some 200 major combat vessels. The Chinese naval combatant has been modernised with the induction of larger, multi-role maritime platforms capable of undertaking blue water operations. These combat platforms consist of the following **surface** and **sub-surface ships**.

Surface Ships

PLAN's surface ships include its **aircraft carrier, destroyers, frigates, corvettes** and **missile/ patrol boats,** besides the **amphibious force/vehicles**.

Aircraft Carrier

China took the initial step towards becoming a carrier-capable Navy when it commissioned its first aircraft carrier **Liaoning** in September 2012. The 60,000-tonne Kuznetsov-class former Soviet-era ship built partially was acquired from Ukraine in 1998. It was later completely refurbished in China. The aircraft carrier undertook a series of sea trials and training programmes following its induction. It was deployed in the South China Sea in 2013 to test its equipment. The induction of the aircraft carrier also enabled the aviation wing of the PLAN to activate its sea-based flying prowess when the J-15 naval fighter successfully conducted conventional (Ski-jump) take off and landings on-board the carrier in November 2012. The vessel can carry up to 30 or more fixed wing fighter jets and helicopters. The PLA-Navy maintains that Liaoning is operationalised as a test platform for carrying out scientific research, training and other experiments even as the Asian country has embarked on the ambitious project of building two or more such large warships to form Carrier Battle Groups in future.

Displacement: 46,000 tons (min) - 67,500 tons (max)
Length: 305m (overall); 270m (waterline)
Beam: 75m (overall); 35m (waterline)
Draught: 9m
Propulsion: Steam turbines (8 boilers, 4 shafts)
Speed: 32 knots
Operational Range: 3,850 nautical miles (7,130 km) at 32 knots
Endurance: 45 days
Countermeasure: Decoy/ chaff (24-tube)
Sensors: 3D air/surface search radar; active phased array radar; dipping sonar
Flight deck: Ski-jump
Fixed wing aircraft: J-15/J-15S (trainer) & J-10
Helicopter: Z-9, Z-18, Ka-31 & Ka-27
Armament: HQ-10 SAM; Type 1130 (30mm) CIWS; ASW rocket launchers
Crew: 2,000

China is also building at least two indigenous aircraft carriers designated as **Type 001A** class – one at its Dalian shipyard in Liaoning province and the other, **Type 002**, at the Jiangnan shipyard near Shanghai. The construction of the first warship has reached an advanced stage with the flight deck being installed which is believed to be a ski-jump jet one for short take-off and landing (STOL) operations, like that of the Liaoning carrier. The exact specification of the new class of carriers is not available. According to some Chinese officials, the new ships would bear similarity to the Liaoning aircraft carrier but will use more advanced weapons, systems and technology.

Destroyers

The Chinese Navy in the last few years has steadfastly inducted several modern destroyers into its fleet. The total destroyers operational in the PLAN fleet are estimated to be around 30. These include the **Luda class (Type 051 - retired in 2016), Luhu class (Type 052), Luhai class (Type 051B), Luyang I class (Type 052B), Luzhou class (Type 051C), Luyang II class (Type 052C)** and **Luyang III class (Type 052D).** In addition, the Navy has also started building a new **Type 055** guided missile destroyer.

Luda class (Type 051): The warships were designed and built between the 1970s and 1990s as the first generation of indigenous Chinese guided missile destroyers. The warships were influenced by some Soviet-designed DDGs, most likely by the Kotlin-class. Different variants of the ship were built later, including the Luda II and Luda III class with improved design and other features. The warships were refurbished in the 1990s and fitted with new advanced missile systems and guns among other upgrades. A total of 16 Luda class destroyers were built and inducted in the Chinese Navy. In October 2016, the PLAN decommissioned the last platform of the class.

Displacement: 3,670 tonnes
Length: 132m
Beam: 12.8m
Draught: 4.6m
Propulsion: Steam turbines
Speed: 32 knots
Operational Range: 2,970 nautical miles (5,500 km) at 18 knots
Countermeasure: ECM
Sensors: 3D air/surface search radar; navigation radar; medium-frequency hull-mounted sonar
Armament: YJ-82/83 AShM; HQ-7 SAM; 130mm gun; 37mm AA gun; ASW rocket launcher; torpedo launcher; naval mines
Crew: 280

Luhu class (Type 052): The new destroyers succeeded the first-generation Luda-class DDGs. The Luhu class warship was built as the first modern multi-role guided missile destroyer of China incorporating many improvements over its predecessor. It used many foreign equipment and systems, including the US-made gas turbine engine. The first Luhu class destroyer was commissioned into service in 1994. Presently, two such warships are operational in the PLAN. Both platforms have been upgraded in 2011.

Displacement: 4,800 tonnes
Length: 144m
Beam: 16m
Draught: 5.1m
Propulsion: Gas turbines; Diesel engine
Speed: 31 knots
Operational Range: 4,500 nautical miles (8,000 km) at 15 knots
Countermeasure: ECM X band jammer; ESM; 15 barrel chaff/decoy launcher
Sensors: Long rage air search radar; air/surface search radar; navigation radar; medium-frequency hull-mounted radar
Armament: YJ-83 AShM; HQ-7 SAM; 100mm gun; 30mm CIWS; ASW torpedo tube; ASW rocket launcher
Crew: 260

Luhai class (Type 051B): Only one such destroyer was built in 1996 primarily as an experimental vessel to test various technologies involved in advanced warship design. The sole Luhai class DDG, Shenzhen, was the largest and heaviest surface warship of the PLAN when it entered service in 1999. The destroyer remains in service with the Navy and has recently completed a major modernisation programme at a Chinese shipyard. The warship is now capable of firing the HHQ-16 medium range air defence system, replacing the HHQ-7 short-range SAM and is also armed with two Type 1130 close-in weapon systems which have replaced the previously mounted 37mm AA gun. The destroyer's surface and air search radars have also been upgraded.

Displacement: 6,000 tonnes
Length: 153m
Beam: 16.5m
Draught: 6m
Propulsion: Steam turbines
Speed: 31 knots
Operational Range: 12,100 nautical miles (22,000 km) at 15 knots
Countermeasure: ECM jammer; 15-barrel chaff/decoy launcher; RWR
Sensors: 3D air search radar; air/surface search radar; navigation radar; medium-frequency sonar; towed-array sonar
Armament: YJ-83 AShM; HHQ-16 SAM; 100mm gun; 30mm CIWS; torpedo tube; ASW rocket launchers
Crew: 250

Luyang I class (Type 052B): The Luyang I class was built as a successor of the Luhai class warship. Two such modern multi-role stealth destroyers were built in 2002-03 and commissioned into service in 2004. Both vessels used Russian technology extensively, including improved air defence system. They share similarities with the Russian-built Sovremenny-class destroyer in terms of size and performance.

Displacement: 6,500 tonnes
Length: 155m
Beam: 17m
Draught: 6m
Propulsion: Diesel & gas turbines
Speed: 30 knots
Countermeasure: ECM jammer; 18-barrel chaff/decoy launchers
Sensors: 3D air/surface search radar; medium-frequency active/passive search & attack sonar
Armament: YJ-83 AShM; SA-N-12 Grizzly SAM; 100mm dual gun; 30mm CIWS; ASW torpedo tube; ASW rocket launchers
Crew: 280

Luzhou class (Type 051C): Based on the hull design of the Type 051B Luhai class DDG, the new Luzhou class warships were built in 2004-05 and inducted in service in 2006-07. Two such guided missile destroyers were built as an improvement over their predecessor the Type 052B class of DDGs and deployed the Russian S-300FM long-range air defence missile systems. The two Type 051C destroyers are deployed with the PLAN's North China Sea Fleet.

Displacement: 7,100 tonnes
Length: 155m
Beam: 17m
Draught: 6m
Propulsion: Steam turbines
Speed: 30 knots
Countermeasure: ECM interceptor & jammer; 15-barrel chaff/decoy launchers
Sensors: 3D air search radar; surface search radar; over-the-horizon targeting radar
Armament: S-300FM SAM; YJ-83 AShM; 100mm gun; 30mm CIWS; ASW torpedo tube
Crew: 280

Luyang II class (Type 052C): The warships were built based on the Type 052B hull design and propulsion. Two destroyers of this class were built and commissioned into service between 2003 and 2005. Four more such warships were built in 2012. The new class of air defence destroyers mostly used Chinese systems, unlike the Type 052B ones which used a mix of Russian and Chinese weapons and systems. The warships are operated by the South China Sea fleet and East China Sea fleet.

Displacement: 6,600 tonnes
Length: 155m
Beam: 17m
Draught: 6m
Propulsion: Steam turbines
Speed: 30 knots
Countermeasure: ECM interceptor & jammer; 15-barrel chaff/decoy launchers
Sensors: 3D air search radar; surface search radar; over-the-horizon targeting radar
Armament: S-300FM SAM; YJ-83 AShM; 100mm gun; 30mm CIWS; ASW torpedo tube
Crew: 250

Luyang III class (Type 052D): Developed as a successor to the Type 052C class of warships, the new Lyuang III class are the latest multi-role guided missile destroyers being steadily inducted into the PLAN service. The first DDG of the series was commissioned in 2014. A total of 12 such DDGs are being constructed for the Navy. The new destroyers are primarily designed to provide anti-air area defence to a carrier battle group or amphibious ready group. They have often been compared to the US's Arleigh Burke-class of destroyers having the Aegis combat system.

Displacement: 7,500 tonnes
Length: 156m
Beam: 18m
Draught: 6.5m
Propulsion: Diesel & gas turbines
Speed: 30 knots
Sensors: Active phased array radar; air search radar; VHF search radar
Armament: CY-5 (HN-5 series) ASW missile; HQ-9 SAM; HQ-16 MR-SAM; HQ-10 SR-SAM; YJ-18 or YJ-83 AShM; CJ-10 LACM; 130mm dual purpose gun; 30mm CIWS; torpedo tube; rocket launcher
Crew: 280

Type 055: The new warship presently under construction as a successor to the Type 52D class has been dubbed as Asia's largest battle ship. Construction of the new guided missile destroyer has reportedly started in 2016 but not much details about its exact specifications are available. According to a US military report, the new surface ship could be classified as a guided missile cruiser keeping in mind its huge size with an estimated displacement of 10,000 tons and a length of 175 meter. It is likely to be armed with medium- and long-range air defence missiles, anti-submarine missiles along with land-attack and anti-ship cruise missiles. The battleship could accommodate up to 100 vertical launch systems capable of holding as much number of missiles and weapons. The warship could also deploy the X-ban and S-band phased array radars among other improved electronic systems.

Frigates

The PLAN has inducted several new indigenously built guided missile frigates (FFGs) in recent years. While phasing most of the older generation FFGs built in the 1970s and 80s, China has concurrently focused on building and inducting new-generation frigates. The FFGs active in the PLAN service at present include the **Jianghu class (Type 053), Jiangwei I class (Type 053 H2G), Jiangwei II class (Type 053H3), Jiangkai I class (Type 054) and Jiangkai II class (Type 054A)**.

Jianghu class (Type 053): The older-generation frigates began entering service with the Chinese Navy in the 1970s. The warships are believed to have been designed from the Soviet-built Riga-class frigates. China built several variants of the Type 053 class to conduct different missions, including air defence and surface warfare. More than 30 such frigates were built in total of which few are presently operational with the PLAN and the Chinese Coast Guard. Some of the frigates have been sold to other countries as well.

Displacement: 1,700 tonnes - 2,000 tonnes
Length: 106m - 112m
Beam: 10m - 12m
Draught: 3m - 4m
Propulsion: Diesel engines
Speed: 30 knots
Operational Range: 4,000 nautical miles (7,408 km) at 18 knots
Sensors: Air/surface search radar
Armament: SSM; 100mm gun; ASW rocket launcher (different variants deployed different armaments)
Crew: 200

Jiangwei I class (Type 053 H2G): Of the many variants of the Type 053 family of frigates, the Jiangwei I or Type 053 H2G series entered service in the 1990s as the Chinese Navy's first multi-role frigates. Four such warships were built which were later transferred to the Chinese Coast Guard. The vessels had improved design and systems over their predecessors.

Displacement: 2,400 tonnes
Length: 112m
Beam: 12.4m
Draught: 4m
Propulsion: Diesel engines
Speed: 30 knots
Operational Range: 5,000 nautical miles (9,260 km) at 16 knots
Countermeasures: ECM interceptor & jammer; RWR
Sensors: Air/surface search radar; Navigation radar
Armament: YJ-83 AShM; HQ-61 SRSAM; 37mm AA gun; 100mm dual gun; ASW rocket launcher; ASW torpedo launcher
Crew: 170

Jiangwei II class (Type 053H3): Built as follow-on warship of the Type 053 H2G class, the Jiangwei II class multi-role FFGs entered service in the 1990s and 2000. Ten such frigates were built and are active in the PLAN service. The primary mission for the frigate is coastal patrol besides anti-air warfare and anti-submarine warfare roles.

Displacement: 2,400 tonnes
Length: 114m
Beam: 12.4m
Draught: 4m
Propulsion: Diesel engines
Speed: 27 knots
Operational Range: 5,000 nautical miles (9,260 km) at 16 knots
Countermeasures: ECM interceptor & jammer; RWR; LWR; Electronic signals intelligence system
Sensors: 2D air/surface search radar; Navigation radar; Early warning radar; Medium frequency sonar
Armament: YJ-83 AShM; HQ-7 SAM; 37mm AA gun; 100mm dual gun; ASW rocket launcher; ASW torpedo launcher
Crew: 170

Jiangkai I class (Type 054): The frigates succeeded the Jiangwei II class of ships and were inducted into service in 2005. Two such modern multi-role guided missile frigates were built featuring stealthy hull design and French-supplied electronics and weapon systems. Both ships of the class are presently operational with the PLAN.

Displacement: 3,600 tonnes
Length: 134m
Beam: 16m
Draught: 6m
Propulsion: Diesel engines
Speed: 27 knots
Operational Range: 6,500 nautical miles (12,038 km) at 18 knots
Countermeasure: ECM; ELINT; RWR
Sensors: Air/surface search radar; Navigation radar; Medium frequency active/passive sonar; SATCOM
Armament: YJ-83 AShM; HQ-7 SAM; 100mm gun; 30mm CIWS; Rocket launcher; ASW torpedo launcher
Crew: 250

Jiangkai II class (Type 054A): The previous Jiangkai I class of frigates later gave way to the advanced Jiangkai II class ships having improved sensors and weapons but retaining the same stealthy hull design. The first frigate of the class was inducted in service in 2007-08. The new-generation frigates, capable of undertaking long-range surveillance, air defence, anti-surface and anti-submarine warfare operations, are progressively being inducted. The PLAN is presently operating 22 such FFGs with two or more at different stages of construction.

Displacement: 4,000 tonnes
Length: 134m
Beam: 16m
Draught: 6m
Propulsion: Diesel engines
Speed: 27 knots
Operational Range: 8,000 nautical miles (14,816 km)
Countermeasures: ECM; ELINT; RWR
Sensors: 3D air/surface search radar; Surface search radar; Navigation radar; Medium frequency active/passive sonar; SATCOM
Armament: YJ-83 AShM; HQ-16 SAM; 76mm gun; 30mm CIWS; ASW rocket launcher; ASW torpedo launcher
Crew: 165

Corvettes

The Navy is operating the indigenous Jiangdao class of stealth multi-role corvettes.

Jiangdao class (Type 056): The PLA Navy has been progressively inducting the Jiangdao class (Type 056) of stealth corvettes into service since the last few years. The modern multi-role vessels are replacing the Type 037 patrol boats as well as some older-generation frigates. Flaunting a sleek design, they are capable of carrying out escort missions, anti-submarine warfare, anti-air warfare, anti-surface warfare roles as well as search & rescue operations. The PLAN is likely to operate a large fleet of Type 056 class of corvettes, possibly around 60. Thirty such platforms are believed to have been inducted so far.

Displacement: 1,500 tonnes
Length: 90m
Beam: 12m
Draught: 4m
Propulsion: Diesel engines
Speed: 30 knots
Operational Range: 2,000 nautical miles (3,700 km) at 18 knots
Countermeasures: Jammer; Decoy launcher
Sensors: Air/surface search radar; Navigation radar; SATCOM
Armament: YJ-83 AShM; HQ-10 SAM; 76mm gun; 30mm cannon; Torpedo launcher
Crew: 70

Missile Boats / Patrol Craft

The Chinese Navy presently operates nearly 100 missile patrol craft. While a large number of the Cold War-era missile boats have been phased out from service, the PLA Navy has inducted many modern missile patrol boats to augment its coastal defence and area denial capability even though it has transformed itself into a "blue water" maritime force. The patrol craft presently active in the Chinese Navy include the **Houbei class (Type 022),** few improved variants of the **Hainan class (Type 037)** and some **Shanghai-III class (Type 062I)** gunboats. Many older-generation Shanghai class and Hainan class patrol boats have been decommissioned.

Houbei class (Type 022): The advanced missile boats were built in 2000s and commissioned into service from 2004 onwards. Over 70 such boats are presently operational with the PLAN. The Houbei class incorporates stealthy, catamaran-shaped hull design with wave-piercing technology. Each boat carries anti-ship cruise missiles. The missile craft has been designed to patrol China's coastal areas and operate within its littoral region.

Displacement: 225 tonnes
Length: 42.6m
Beam: 12.2m
Draught: 1.5m
Propulsion: Diesel engines; Waterjet propulsion
Speed: 36 knots
Operational Range: 200 km
Sensors: Surface search radar; Navigation radar
Armament: YJ-83 AShM; HN-2 LACM; 30mm high speed gun
Crew: 12

Houxin class (Type 037IG): Developed as the missile-armed variant of the Type 037 class of submarine chasers, the Houxin class craft entered service from 1991 onwards. Over 20 such vessels are operational with the Navy at present. Some of them have been exported to Myanmar. The missile boats are used in conjunction with multiple missile boats to send a swarm of missiles towards their intended target and then egress from the area immediately.

Displacement: 478 tonnes
Length: 62m
Beam: 7.2m
Draught: 2.24m
Propulsion: Shafts
Speed: 32 knots
Operational Range: 1,700 nautical miles (3,200 km) at 18 knots
Sensors: Surface search radar; Navigation radar
Armament: YJ-series AShM; 37mm AAA gun; 14.5mm anti-aircraft machine gun
Crew: 60

Houjian class (Type 037II): The missile boats succeeded the Houxin class vessels. They were built between 1991 and 2001 and incorporated diesel engine propellers. Four such ships are presently active in the PLAN service.

Displacement: 520 tonnes

Propulsion: Diesel engines

Speed: 30 knots

Operational Range: 1800 nautical miles (3,333 km) at 18 knots

Sensors: Surface search radar; Navigation radar

Armament: YJ-83 AShM; HN-2 LACM; 30mm high speed gun

Crew: 75

Shanghai-III class (Type 062I): Developed as the improved variant of the original Shanghai-I class (Type 062) of gunboat, the Shanghai-III class vessels are larger and armed with heavier guns. Some 12 such boats are operational in the PLA Navy and are primarily used for coastal and inland patrolling. Many of these gunboats have been exported to other countries as well.

Displacement: 170 tonnes

Length: 41m

Beam: 5.3m

Draught: 1.8m

Propulsion: Diesel engines & shafts

Speed: 25 knots

Operational Range: 655 nautical miles (1,210 km) at 17 knots

Sensors: Surface search radar

Armament: 37mm gun; 23mm heavy machine gun

Crew: 40

Amphibious Force

China continues to build up a modern amphibious force to carry out far sea operations. Induction of several modern amphibious ships mainly in the East China Sea fleet and South China Sea fleet is going on even as many older-generation ones have been phased out from service. The PLAN has included some 30 tank landing ships/amphibious transport docks and 22 medium landing ships among others, while purchase contracts have been signed with Ukraine to acquire assault hovercraft, according to a US Congressional report released in 2016. Acquisition of the amphibious ship force "signal China's intent to develop an expeditionary and over-the-horizon amphibious assault capability," states the report.

The amphibious vessels of PLAN include **Yuzhao-class (Type 071)** landing platform dock (LPD), **Yuyi-class (Type 722 & 726 - also called Jinsha II class)** & **Zubr class** landing craft air cushion (LCAC), **Yukan class (Type 072), Yuting class (Type 072 II & III), Yuting II class (Type 072A)** landing ship, tank (LST), **Yuhai class (Type 074), Yudao class (Type 073II/073III)** & **Yunshu class (Type 073A)** landing ship, medium (LSM), **Yunnan class (Type 067)** landing craft, utility (LCU), and **Yuch'in class (Type 068/069)** landing craft, mechanized (LCM).

Yuzhao-class LPD: The first Yuzhao-class landing platform dock built by Shanghai-based Hudong-Zhonghua Shipyard entered service with the PLA Navy in 2007. The amphibious ship has been designed to enable the PLA/ Marine forces to operate far from home waters, and enhance the Navy's ability to rapidly mobilise troops in any amphibious assault. It can carry helicopters, about 20 armoured fighting vehicles, boats and landing craft as well as up to 1,000 soldiers. The Navy has so far inducted four such warships while two or more are likely being constructed.

Displacement: 19,000 tonnes

Length: 210m

Beam: 28m

Draught: 7m

Propulsion: Diesel engines

Speed: 25 knots

Operational Range: 10,000 nautical miles (19,000 km) at 18 knots

Countermeasure: Decoy/ chaff

Sensors: Surface search radar; Air search radar; Navigation radar

Armament: 76mm gun; 30mm CIWS

Crew: 1,000

Helicopter: Z-8 (4)

Yuyi-class LCAC: It was produced in the 1970s as a medium-sized landing craft, also designated as Type 722 (Jinsha I) class). Another improved variant of this type was later developed and designated as Jinsha II class (Type 722 II). The landing craft air cushion (LCAC) featured improved propulsion system and entered service in 1989. The Yuyi-class is the most advanced variant of the series which was developed in 2008-09 and entered service thereafter. It is also designated as **Type 726**. The landing craft, capable of carrying a maximum cargo of up to 60 ton that could include a Type 96 or Type 99 Main Battle Tank, is carried by the Type 071 LPD.

Displacement: 150 tonnes

Length: 30m

Beam: 16m

Propulsion: Turbine engines

Speed: 60 knots

Operational Range: 165 nautical miles (306 km) or more

Armament: 14.5mm machine gun; Grenade launcher

Zubr class LCAC: The Soviet-built hovercraft has been acquired from Ukraine by the PLAN in 2013. A total of four such craft are to be inducted in the PLAN service of which two will be licence-built in China. The Zubr-class landing craft is the world's largest hovercraft, designed for shore landing assault units. It has a cargo area of 400 square meters capable of carrying up to 150 ton cargo and a fuel capacity of 56 tons. The landing craft can carry up to three main battle tanks or 10 armoured fighting vehicles or eight armoured fighting carriers or eight amphibious tanks or 500 amphibious landing troops.

Displacement: 555 tonnes (full load)

Length: 57m

Beam: 16m

Draught: 1.6m

Propulsion: Gas turbines

Speed: 60 knots

Operational Range: 300 nautical miles (555 km) at 55 knots

Countermeasures: ECM; Chaff; ESM Radar

Sensors: Navigation radar

Armament: MANPAD launchers; 30mm CIWS; Rocket launchers

Crew: 500

Yukan class LST: Built in the late 1970s as China's first indigenous large landing ships, the Yukan class or Type 072 class of vessels replaced some ex-US LSTs in service with the Chinese Navy. Three such craft are presently active. The LST is capable of carrying 500 tons of cargo which includes up to 10 tanks and 100 troops.

Displacement: 4,000 tonnes (full load)

Length: 120m

Beam: 15m

Draught: 2.9m

Propulsion: Diesel engines

Speed: 18 knots

Operational Range: 3,000 nautical miles (5,600 km) at 14 knots

Sensors: Navigation radar

Armament: MANPAD; 37mm naval gun

Crew: 100

Yuting class LST: The more capable Yuting class (Type 072-II) of landing ships were built in the 1990s to replace the Yukan class of LSTs. An improved variant, the Type-072III, was built later with enhanced sealift capability and other features, including a flight deck for helicopter take off/landing. Around 15 Yuting class LSTs are operational in the PLAN service at present. The LSTs are capable of carrying a cargo of up to 500 ton which includes 250 troops or 10 tanks.

Displacement: 4,800 tonnes (full load) [Type 072-II]; 7,000 tons [Type 072-III]

Length: 120m

Beam: 16.4m

Draught: 2.8m

Propulsion: Diesel engines

Speed: 14 knots [Type 072-II]; 18 knots [Type 072-III]

Operational Range: 3,000 nautical miles (5,600 km) at 14 knots

Armament: 57mm naval gun; 37mm naval gun

Crew: 250

Yuting II class LST: The latest variant of the LST is being progressively inducted in the PLAN service since 2005. Built as the successor of the Yuting class of LST, the Yuting II class (Type 072A) features a long deck well and a helipad without hanger. Around 10 such vessels have joined service with some more on order. It can carry cargo of up to 500 ton which include 10 armoured vehicles or 250 fully armed troops.

Displacement: 4,800 tonnes (full load)

Length: 120m

Beam: 16.4m

Draught: 2.8m

Propulsion: Diesel engines

Speed: 21 knots

Operational Range: 3,000 nautical miles (5,600 km) at 14 knots

Armament: 37mm naval gun

Crew: 250

Yuhai class LSM: The ships were built in the 1990s as a successor to the Yuling-class (Type 079) of landing ship, medium (LSMs) which were built in the 1960s and later retired. The Yuhai class (Type 074) featured improved equipment along with an enlarged hull design. It is capable of carrying loads up to 100 ton which include two main battle tanks or six light tanks or 350 fully armed troops. Around 12 such LSMs are likely to be active in the PLA Navy.

Displacement: 800 tonnes (full load)

Length: 58m

Beam: 10.4m

Draught: 2.7m

Propulsion: Diesel engines

Speed: 18 knots

Armament: 25mm AAA gun

Crew: 350

Yudao class LSM: The older generation of this class of ships (Type 073) were built in the 1960s. The later variants incorporating many improved featured began rolling out in the 1980s and 1990s. While two Type 073II class of the LSMs are in active service, one of its successor, the Type 073III (also called Yudeng class), is also operational. The latest Type 073III variant is capable of carrying a cargo of 250 ton that includes five medium tanks or 10 light tanks or 500 fully equipped troops.

Displacement: 1,000 tonnes (Type 073II); 1,800 tonnes (Type 073III)

Length: 77.4m (Type 073II); 87m (Type 073III)

Beam: 10.4m (Type 073II); 12.6m (Type 073III)

Draught: 2.5m (Type 073II); 2.25m (Type 073III)

Propulsion: Diesel engines; Shaft

Speed: 15 knots (Type 073II); 17 knots (Type 073III)

Operational Range: 1,000 nautical miles (1,900 km) at 16 knots (Type 073II); 1,500 nautical miles (2,800 km) at 14 knots (Type 073III)

Armament: 25mm AAA gun & 37mm AAA gun (Type 073II); 37mm AAA gun & Type 81H 122 mm rocket (Type 073III)

Crew: 60 (Type 073II); 74 (Type 073III)

Yunshu class LSM: The ships with larger displacement and more advanced equipment and system over their predecessors entered service in 2001-02. Also called **Type 073A**, around 10 such LSMs are presently active in the Chinese Navy. The LSM has the same load carrying capacity as that of its predecessor (250 tons).

Displacement: 2,000 tonnes (full load)

Length: 87m

Beam: 12.6m

Draught: 2.25m

Propulsion: Diesel engines; Shaft

Speed: 17 knots

Operational Range: 1,500 nautical miles (2,800 km)

Armament: 37mm AAA gun; 122mm rocket

Crew: 70

Yunnan class LCU: The utility landing craft was built in the late 1960s as an enlarged version of the Yuch'in class LCM. The LCU, also known as Type 067, is capable of carrying a medium tank or two armoured fighting vehicles or a company of infantry. Over 100 such craft were built of which few are believed to remain in service with the PLAN while some have been converted to buoy tender/cable layer.

Displacement: 135 tons

Length: 28.6m

Beam: 5.4m

Draft: 1.5m

Propulsion: Diesel engines; Shafts

Speed: 12 knots

Operational Range: 500 nautical miles (925 km) at 10 knots

Armament: 25mm AAA gun; 14.5mm heavy machine gun

Crew: 12

Yuch'in class LCM: The craft entered service between 1960s and 1970s and are also knowns as Type 068/069 landing craft, mechanized. The small-sized craft can be maintained and stored on land and are still in service with the Navy in limited numbers. Few of them have been converted into survey craft.

Displacement: 58 tons

Length: 24.8m

Beam: 5.2m

Draft: 1.3m

Propulsion: Diesel engines; Shafts

Speed: 11.5 knots

Operational Range: 450 nautical miles (830 km)

Armament: 14.5mm heavy machine gun

Crew: 12

Sub-Surface Ships

The sub-surface platforms active in the PLA Navy include **submarines** and **unmanned underwater vessels**.

Submarines

The sub-surface force of the Chinese Navy and its capabilities has witnessed a substantial increase in the past few years with progressive induction of new generation submarines, both conventional as well as nuclear. Deployment of advanced weapons and systems in these platforms, mostly built indigenously, has made them more lethal to conduct underwater operations. While many older generation subs have been phased out, steady induction of advanced ones has consolidated China's undersea capabilities further. The number of submarines active in the PLAN service are estimated to be 60 with some more undergoing construction. These platforms include **conventional attack submarines (SS), nuclear-powered attack submarines (SSN), cruise missile carrying nuclear submarines (SSGN)** and **ballistic missile carrying nuclear submarines (SSBN).**

Conventional Attack Submarines (SSK/SSI)

The conventional attack submarine fleet of China includes the **Kilo-class** (Project 877EKM & 636) hunter/killer (SSK), **Ming class** (Type 035) SSK, **Song class** (Type 039) SSK and **Yuan class** (Type 039A) SSI (with AIP technology).

Kilo class SSK: The PLA Navy maintains two earlier generation Russian-origin Kilo class (Type 877EKM) and 10 improved Kilo class (Project 636) diesel-electric attack submarines. The submarines were acquired between 1995 and 2005. While the Type 877EKM vessels are designed for anti-submarine and anti-surface ship warfare missions, the improved class of the ship is capable of carrying out general reconnaissance and patrol missions besides the primary ASW and ASuW roles of its predecessor. China has designed few indigenous submarines later (Type 039A/Yuan class) having hull design resembling that of the Kilo-class submarines.

Type 877EKM

Displacement: 2,300 tonnes (surfaced); 4,000 ton (submerged) at full load

Length: 72m

Beam: 9.9m

Draught: 6.5m

Propulsion: Diesel electric (Two 1,000 kW diesel generators, one 5,500hp propulsion motor, one fuel-economic motor); 1 shaft

Speed: 10 knots (surfaced); 17 knots (submerged)

Operational Range: 6,000 miles (9,700 km) when snorting at 7 knots; 400 miles (640 km) when submerged at 3 knots

Endurance: 45 days

Countermeasure: Electronic support measures; Radar warning receiver

Sensors: MGK-400 sonar; Active radar

Armament: Strela-3 or Igla SAM; Club-S LACM/AShM; 533mm torpedo tubes

Crew: 52

Project 636

Displacement: 2,300 tonnes (surfaced); 4,000 tons (submerged) at full load

Length: 74m

Beam: 9.9m

Draught: 6.5m

Propulsion: Diesel electric (Two 1,000 kW diesel generators, one 5,500hp propulsion motor, one fuel-economic motor); 1 shaft

Speed: 11 knots (surfaced); 20 knots (submerged)

Operational Range: 7,500 miles (12,000 km) when snorkelling at 7 knots; 400 miles (640 km) when submerged at 3 knots

Endurance: 45 days

Countermeasure: Electronic support measures; Radar warning receiver

Sensors: MGK-400EM digital sonar; Active radar

Armament: Strela-3 or Igla SAM; Club-S LACM/AShM; 533mm torpedo tubes

Crew: 52

Ming class SSK: The conventional submarines were built indigenously in the 1970s as a redesign of the Soviet-origin Romeo-class (Type 033) underwater platforms. The Ming class (Type 035) was improved further in later years with its other variants such as Type 035A, Type 035G and Type 035B entering production and induction. About a dozen Type 035G variants were built in 1990-1999, and commissioned into service. Two of these subs were decommissioned in 2015 for exporting to Bangladesh. Fifteen of the Ming class platforms (9 Type 035G, 4 Type 035B & 2 Type 035A) are believed to be operational in the PLA Navy at present.

Displacement: 1,580 tonnes (surfaced); 2,110 tons (submerged)

Length: 76m

Beam: 7.6m

Draught: 5.1m

Propulsion: Diesel engines (2); shafts (2)

Speed: 15 knots (surfaced); 18 knots (submerged)

Operational Range: 8000 nautical miles (14,800 km) when snorting at 8 knots; 330 nautical miles (600 km) when submerged at 4 knots

Countermeasures: Electronic support measures; Radar warning receiver

Sensors: Medium-frequency sonar; low frequency sonar; I-band surface search radar

Armament: 533mm torpedo tubes (for launching Yu-3 torpedoes)

Crew: 55

Song class SSK: Built as the successor of Ming-class SSKs, the Song-class (Type 039) was the first fully independently built conventional submarine of China. Construction of the first and sole platform of the class started in 1991 and it was commissioned into service in 1999 after overcoming several design and performance issues. Many improvements were later incorporated in the platforms built thereafter between 1999 and 2005 which were designated as **Type 039G.** The new class reduced the acoustic signature and enhanced the underwater performance of the submarine. Over 10 such platforms are presently operational in the PLAN.

Displacement: 1,700 tonnes (surfaced); 2,250 tons (submerged)

Length: 75m

Beam: 8.4m

Draught: 5.3m

Propulsion: Diesel-electric engine (1); Shaft (1)

Speed: 15 knots (surfaced); 22 knots (submerged)

Countermeasures: Electronic support measures; Radar warning receiver

Sensors: Medium-frequency active/passive sonar; low-frequency sonar; I-band surface search radar

Armament: YJ-82 AShM; 533mm torpedo tubes (for launching Yu-3, Yu-4, Yu-6 torpedoes); Naval mines

Crew: 60

Yuan class SSI: The conventional attack submarines with Air Independent Propulsion (AIP) technology were the first such platform built in China in the 2000s to replace the older-generation Ming class SSKs. The use of AIP technology made the Yuan class (Type 039A/Type 041) SSIs quieter and stealthier. Progressive induction of this new class of platforms from 2006 onwards has fortified the conventional submarine force of China. Around 15 such vessels built in four different variants (Type 039A/B/C & AG) incorporating design and other technical modifications are presently operational in the PLA Navy while some more are under construction. Few of them are being exported to other countries, including Pakistan.

Displacement: 2,000 tonnes (surfaced); 3,200 tons (submerged)

Length: 75m

Beam: 8.4m

Draught: 6.7m

Propulsion: Diesel electric; shaft (1) with AIP

Speed: 16 knots (surfaced); 20 knots (submerged)

Operational Range: 8,000 nautical miles (14,800 km) at 16 knots

Endurance: 60 days

Countermeasures: Radar warning receiver

Sensors: Surface search radar; Sonar

Armament: YJ-82 AShM; 522mm torpedo tubes (for Yu-6 torpedoes)

Crew: 60

Nuclear powered attack submarines (SSN)

The PLAN operates the **Han class (Type 091)** and **Shang class (Type 093/093G)** of nuclear-powered attack submarines (SSNs). A new **Type 095** class of SSNs are also under development and likely to be deployed by 2018.

Han class: The submarines were built as Chinese Navy's first-generation indigenous SSNs. The first Han class (Type 091) platform was laid down in 1970 and commissioned into service in 1974. The SSN was produced based on technologies of the 1950s and 1960s and had many shortcomings, including noisy nuclear reactor and inability to launch missiles while submerged. Few more platforms of the class with major modifications in terms of weapons and equipment were built subsequently and the last vessel entered service in 1990. Three Han class submarines are presently operational in the PLAN.

Displacement: 4,500 tonnes (surfaced); 5,500 tons (submerged)

Length: 98m

Beam: 10m

Draught: 7.4m

Propulsion: Nuclear reactor (1)

Speed: 12 knots (surfaced); 25 knots (submerged)

Sensors: Sonar

Armament: 533mm torpedo tubes; C-801 (YJ-82) AShM

Crew: 75

Shang class: Believed to have been built in the 1980s or 1990s as a successor to the Han class SSNs, the Shang class (Type 093) platforms took several years to enter PLAN service. Only two such platforms were inducted. Development of these nuclear-powered attack submarines faced several technical issues. Another improved variant, **Type 093G**, was developed later and possibly could have been operationalised. It is capable of deploying advanced anti-ship cruise missiles and long-range land attack cruise missile. The Type 093G has often been classified as an **SSGN (cruise missile carrying nuclear submarine).**

Displacement: 7,000 tonnes (submerged)

Length: 107m

Beam: 11m

Draught: 7.5m

Propulsion: Pressurised water nuclear reactor

Speed: 30 knots (submerged)

Sensors: Sonar

Armament: 533mm torpedo tubes; YJ-18 AShM; CJ-10 LACM

Crew: 100

Type 095: The platforms are being developed as a third-generation nuclear attack submarines and a successor to the Shang class of subs. The submarines will have improved hull structure and other advanced features in terms of technology and material that would make them quieter and stealthier. They will also have more advanced nuclear reactors and sensors and would be able to launch anti-ship cruise missiles. Some 14 such platforms are being planned to be inducted starting from 2018 onwards. Detailed specifications of the new class of submarines is not known yet.

Ballistic missile carrying conventional submarine (SSB)

The PLA Navy operates only one SSB platform called the **Qing class (Type 032)**. It has been developed as an experimental platform to test China's evolving underwater technologies and capabilities.

Qing class: Pegged as the world's largest ballistic missile carrying conventional submarine, the Qing class SSB's development started in 2005 and it began sea trial in 2012 before being commissioned into service in the same year. The submarine is believed to have been influenced from the Soviet-era Golf class submarine. It is presently used for testing China's evolving underwater technologies, such as test firing of SLBMs and SLCMs among others.

A Chinese blogpost photo

Displacement: 3,800 tonnes (surfaced); 6,628 tons (submerged)

Length: 92.6m

Beam: 10m

Draught: 6.85m

Propulsion: Diesel electric; shaft (1), AIP

Speed: 14 knots or above (surfaced)

Armament: 5mm torpedo tubes; 2 SLBM VLS (for JL-2 SLBM); 4 cruise missile VLS (for YJ-18/CJ-20 AShM)

Crew: 100

Ballistic missile carrying nuclear submarine (SSBN)

The SSBNs active in the Chinese Navy include the **Xia class (Type 092)** and **Jin class (Type 094)** of platforms. A new-generation SSBN, designated **Type 096**, is also under development and likely to be deployed by 2020.

Xia class: A derivative of the Han-class (Type 091) SSN, the Xia class (Type 092) is China's first ballistic missile carrying nuclear submarine (SSBN) built in the 1980s and presently active in the PLAN. The first platform of this class entered service in 1987 and it was armed with the JL-1 submarine-launched ballistic missiles. The submarine has undergone several technological upgrades, including its combat system. It is capable of operating in the regional waters only. A second platform of the class was reportedly built in 1982 but was lost in an accident, though no official record has ever mentioned this.

Displacement: 6,500 tonnes (surface); 8,000 tons (submerged)
Length: 120m
Beam: 10m
Draught: 8m
Propulsion: Pressurised water nuclear reactor (1); Steam turbines (2); Shaft (1)
Speed: 22 knots (submerged)
Sensors: Sonar
Armament: JL-1/1A SLBM; 533mm torpedo
Crew: 100

Jin class: The advanced SSBN was developed as a successor to the Xia class of submarines and the first platform was launched in 2004. It is believed to have joined the Navy in 2010. While four Jin class (Type 094) SSBNs are presently operational, two or more are believed to be under construction. The submarines, classified by the Pentagon as "China's first credible sea-based nuclear deterrent", are capable of firing the JL-2 intercontinental range ballistic missile. The SSBN can carry up to 12 JL-2 missiles.

Displacement: 9,000 tonnes (surface); 11,000 tons (submerged)
Length: 135m
Beam: 12.5m
Propulsion: Nuclear reactor; Shaft (1)
Speed: 20 knots or more (submerged)
Armament: JL-2 SLBM; 533mm torpedo
Crew: 100 or more

Type 096: Presently under development, the Type 096 class is being built as the PLAN's next-generation missile submarine. Based on the Type-094 class SSBN, the Type-096 class subs are likely to be armed with the longer range JL-3 ballistic missiles. The new SSBN can fire up to 24 JL-3 SLBMs. The PLA Navy is expected to deploy as many as eight Type-096 class subs by 2020. Detailed specifications of the new class of submarines is not known.

UUV: The unmanned underwater vehicles (UUVs) being developed by China are likely to be used for military applications. Several research institutions and universities are working on developing unmanned systems for both civilian and maritime roles. Some of the UUV prototypes developed by China include the **Zhishui 3** UUV, **Haiyan** glider-type UUV, **Qianlong-1** autonomous underwater vehicle (AUV) and an UUV armed with torpedoes among others.

SUPPORT PLATFORMS

The PLA Navy uses wide-ranging support platforms to carry out littoral as well as blue water operations. Most of these support platforms have been indigenously developed and stationed in naval bases of all three fleets of the Navy. The main support vessels of the Chinese Navy include **fleet replenishment ships, training ships** and **patrol hydrofoil**.

Fleet Replenishment Ships

The first supply ships of China were built in the 1980s to accompany visiting warships for refuelling them. Few tankers were also acquired from Russia in the 1990s. Later, China independently developed several fleet replenishment ships which include the **Type 905**, **Type 908** and **Type 903/903A** class of vessels.

Type 905: The ships were built as China's first-generation of replenishment oilers. Two vessels joined the Navy in 1980 and used to support the Yuan Wang class of tracking ships built to track satellite launch or missile launch events. A total of four Type 905 (Fuqing-class) ships were built of which two are presently active in the PLAN. The fleet oilers are capable of carrying oil fuel, food, munitions, and spare parts of up to 12,000 ton. It also has a helicopter landing pad to support operations of a medium utility helicopter.

Displacement: 22,000 tonnes
Length: 168m
Beam: 21.8m
Draught: 9.4m
Propulsion: Diesel engine (1); Shaft (1)
Speed: 18 knots
Operational Range: 18,000 nautical miles (29,000 km)
Armament: 37mm naval gun
Crew: 130

Type 908: The second-generation replenishment ships were built to succeed the Type 905 class ships. The first ship originally built for the Soviet Navy was acquired by China in 1993. It was then completely overhauled in Chinese shipyard to operate as a multi-product replenishment oiler. The vessel joined the PLA Navy in 1996 and can simultaneously fuel three warships. It can carry cargo up to 10,000 ton and also has a hanger and a flight deck for helicopter operations. A second unit built later was sold to the Thailand Navy.

Displacement: 37,000 tonnes
Length: 189m
Beam: 25.3m
Draught: 10.4m
Propulsion: Diesel engine (1); Shaft (1)
Speed: 16 knots
Armament: 30mm CIWS
Crew: 125

Type 903: The first ship of the class entered service in 2005. An improved variant, Type 903A, was developed later having higher displacement. Progressive induction of these ships, also called **Fuchi-class,** is going on since then with the PLAN inducting the seventh and eighth vessels of the class in 2016. The new ships are capable of providing astern and alongside replenishment at sea for several different types of ships simultaneously. They can also receive helicopters for vertical replenishment operations, and are fitted with armament for self-defence against ground and aerial threats. The cargo carrying capacity of the ship is more than 11,000 tons which includes fuel oil, food, water and ammunition.

Displacement: 20,000 – 23,400 tonnes
Length: 178m
Beam: 24.8m
Draught: 8.7m
Propulsion: Diesel engine (2); Shaft (2)
Speed: 20 knots
Operational Range: 10,000 nautical miles (18,500 km) at 14 knots
Armament: 37mm naval gun
Crew: 130

Training Ships

The PLA Navy presently operates two training ships, **Type 679 (Zheng He)** and **Type 0891A** (**Shichang**).

Type 679: The ship, designated Zheng He, is China's first indigenously built training ship. It was commissioned into service in 1987. The ship has a large hospital on board to accommodate cadet classes. More than forty different classes can be trained on board. The ship has circumnavigated the globe in 2012.

Displacement: 5,550 tonnes
Length: 132m
Beam: 15.8m
Draught: 4.8m
Propulsion: Diesel engine
Speed: 17 knots
Operational Range: 10,000 nautical miles (18,500 km) at 17 knots
Endurance: 30 days
Armament: 30mm naval gun; 57mm naval gun; Rocket launchers
Crew: 170

Type 0891A: Built as China's first multi-role aviation ship, it was commissioned into service in 1997. The training ship, designated Shichang, is primarily used for helicopter operations. The large flight deck of the ship enables multiple helicopter operations. The vessel also has a hospital module and is used for medical treatment facility.

Displacement: 10,000 tonnes
Length: 125m
Beam: 19m
Draught: 10.6m
Propulsion: Diesel engine
Speed: 17 knots
Crew: 170

Patrol Hydrofoils

The PLA Navy has since the 1960s extensively used the Huchuan (Type 025) class of fast attack hydrofoil-torpedo boats. The boats with high speed and stealth capabilities have been used extensively for coastal defence purposes in littoral waters. Although they no longer form a part of the Navy's frontline maritime combat platform, few of these vessels are still in service. Different variants of the boat have been developed with improvements mainly in their armament.

Displacement: 45 tons (full load)
Length: 22.5m
Beam: 3.8m
Draught: 3.6m
Propulsion: Diesel engines (3); Shafts (3)
Speed: 50 knots
Sensors: Surface search radar
Armament: 533mm torpedo tube; 14.5mm heavy machine gun; mines & mine guns
Crew: 15

NAVAL AVIATION

The PLA Navy Air Force (PLANAF) operates a wide range of aircraft and helicopters with the primary role to provide air defence cover to the Navy's warships and submarines. It also conducts aerial patrol and reconnaissance missions in the country's coastline. The aviation wing of the PLAN is also capable of conducting anti-ship and anti-submarine warfare missions besides transport operations. Most of the aircraft used by the PLANAF are also operated by the PLA Air Force. The aircraft operated by the PLANAF include **fighter/bomber/ground-attack aircraft**, **maritime patrol aircraft**, **anti-submarine warfare aircraft**, **amphibious aircraft**, **utility (trainer/transport) aircraft** and **helicopters** of different roles. Commissioning of the Navy's maiden aircraft carrier has also led to the development of the country's first indigenous carrier-borne fighter jet. Three variants of the new naval fighter, **J-15**, are being developed – an advanced trainer, a multi-role combat and an electronic warfare variant.

Fighter/ Bomber/ Ground Attack Aircraft

The fighters, bombers and attack aircraft operational in the Navy's aviation wing include the following.

J-7: The fighter/ground attack aircraft developed in the 1960s remain operational in the PLANAF in limited numbers. The single-seat, single-engine, lightweight fighter jet with supersonic speed and short-range attack capability is based on the Soviet-era MiG-21 fighter. Several variants of the fighter have been developed with improved avionics and other features. The PLANAF primarily uses the warplanes to carry out air defence operations.

Length: 14.88m
Wingspan: 8.32m
Height: 4.11m
Weight: 5,300 kg (empty); 9,100 kg (max at take-off)
Propulsion: 1 turbojet

Speed: Mach 2.0 (max)
Range: 2,200 km
Combat Radius: 850 km
Armament: PL-2/5/7/8/9, Magic R.500 & AIM-9 AAMs; Unguided bombs; 50mm & 90mm rocket pods
Crew: 1

J-8: Of the different variants of the J-8 interceptor fighters developed by China, the PLA Navy uses the improved J-8II series, most likely the J-8B/D fighters having aerial refuelling probe and upgraded avionics. It also uses few J-8Fs – the reconnaissance version of the original J-8 single-seat, supersonic heavy fighter developed in the 1960s.

Length: 21.52m
Wingspan: 9.34m
Height: 5.41m
Weight: 9,820 kg (empty); 18,880 kg (max at take-off)
Propulsion: 2 turbojets
Speed: Mach 2.4 (max)
Range: 2,000 km
Combat Radius: 600 km - 1,000 km
Armament: PL-2/7AAMs; Unguided bombs; Rocket pods
Crew: 1

J-10: The delta-wing, multi-role fighter developed indigenously in by China the 1990s is operational in the PLANAF. The warplane is capable of carrying out aerial combat as well as ground attack operations. The PLA Navy uses the initially built **J-10A** variant of the jet.

Length: 15.5m
Wingspan: 9.75m
Height: 5.43m
Weight: 9,750 kg (empty); 19,000 kg (max at take-off)
Propulsion: 1 turbofan

Speed: Mach 2.2 (max)
Range: 2,000 km
Combat Radius: 550 km
Armament: PL-8/9/11/12 AAMs; C-801/802 AShM; 90mm unguided rocket pods; laser-guided bombs; 23mm gun
Crew: 1

J-11B/BS: The fourth-generation fighter having the capability to conduct air superiority and ground attack missions has been inducted in the PLANAF. The Navy's aviation wing operates the J-11B/BS/BSH variants of the highly manoeuvrable fighter. The J-11 is based on the Soviet-origin Su-27 multi-role fighter but is fitted with Chinese engine, avionics and weapons.

Length: 21.9m
Wingspan: 14.7m
Height: 5.9m
Weight: 16,400 kg (empty); 33,000 kg (max at take-off)
Propulsion: 2 turbofans
Speed: Mach 2.35 (max)
Range: 3,530 km
Combat Radius: 1,500 km
Armament: PL-8/9/12 & R-27/R-73/R-77 AAMs; Unguided rocket; Cluster bombs; 30mm gun
Crew: 1

J-15: Developed as China's first indigenous carrier-borne air superiority fighter, the J-15 is undergoing flight trials from the Liaoning aircraft carrier of the PLAN. The new fighter is believed to be a derivative of the Russian-made Su-33 naval fighter jet but incorporates Chinese avionics, engines and weapon systems mostly used in the J-11B fighters. The J-15 is capable of carrying anti-ship, air-to-air and air-to-ground missiles, as well as precision-guided bombs. Three variants of the J-15 are being developed – an advanced trainer, a multi-role combat and an electronic warfare variant, for use by the PLANAF as well as the PLA Air Force.
Length: 21.9m
Wingspan: 14.7m
Height: 5.9m
Weight: 17,500 kg (empty); 33,000 kg (max at take-off)
Propulsion: 2 turbofans
Speed: Mach 2 (max)
Range: 3,500 km
Combat Radius: 1,500 km
Armament: PL-9/PL-12 & R-77/ R-73 AAMs; AIM-9 AAM; Anti-ship missile; Anti-radiation missile; Rockets & bombs; 30mm gun; ECM pod
Crew: 1 - 2

H-6: The strategic bomber built as a derivative of the Soviet-era Tu-16 fighter bomber is used by the PLA Navy's aviation wing. The conventional medium-range bomber is capable of carrying both nuclear and non-nuclear warheads. Many variants of the bomber have been developed in China, and the PLANAF is believed to have procured the H-6D configuration capable of carrying anti-ship cruise missiles. The bomber is acquired by the PLAN to attack carrier battle groups of the US Navy. It is capable of conducting all-weather, day & night operations.

Length: 35m
Wingspan: 33m
Height: 10.36m
Weight: 37,200 kg (empty); 79,000 kg (max at take-off)
Propulsion: 2 turbojets
Speed: Mach 0.9 (max)
Range: 6,000 km
Combat Radius: 1,800 km
Armament: C-601/602/301 AShMs; Guided bombs
Crew: 4

JH-7/JH-7A/B: The multi-role fighter/ bomber is operational in the PLANAF since the 1990s. New variants of the bomber with improved features have been developed and inducted mainly to carry out anti-ship missile strike missions. The latest JH-7A variant with more powerful airframe, upgraded avionics and more missile carrying capability entered PLANAF service in 2004. Another more improved variant, JH-7B, is presently under development and it has more powerful engine, new avionics, aerial refuelling capabilities and higher combat radius. The bombers are capable of all-weather, day and night operations.
Length: 22.32m
Wingspan: 12.8m
Height: 6.22m
Weight: 14,500 kg (empty); 28,500 kg (max at take-off)
Propulsion: 2 turbofans
Speed: Mach 1.75 (max)
Range: 3,700 km
Combat Radius: 1,800 km
Armament: YJ-82 AShM; PL-8/9 AAM; C-704/705 AGM; YJ-91 anti-radiation missile; 23mm gun; 57mm/90mm unguided rockets; Unguided and laser-guided bombs
Crew: 2 pilots + weapons operator

Su-30Mk2: The Russian-built long-range multi-role naval strike fighter has been acquired by the PLANAF in 2004. The heavy, all-weather fighter jet is a derivative of the Su-30MKK aircraft and incorporates maritime strike capability along with upgraded weapons and better C4ISTAR (command, control, communications, computers, intelligence, surveillance, target acquisition and reconnaissance) capabilities. The warplane is armed with the Kh-31 series of airborne cruise missiles. The fighter offers enhanced combat capabilities against aerial, ground and sea-based targets. It is also used for reconnaissance and training missions. The aircraft also features in-flight refuelling system which further increases its operational range.

Length: 21.9m
Wingspan: 14.7m
Height: 6.4m
Weight: 17,000 kg (empty); 34,500 kg (max at take-off)
Propulsion: 2 turbojets
Speed: Mach 2 (max)
Range: 3,000 km – 5,600 km
Armament: Kh-31A & Kh-35E AShM; Kh-31P & Kh-39ME AGM; R-27, R-73 & R-77 AAMs; aerial bombs & cluster bombs; unguided rockets; 30mm gun
Crew: 2

Q-5: The ground attack aircraft built in China is used by the PLA Navy. The Q-5B variant of the aircraft capable of carrying anti-ship missile is used by the PLANAF. The aircraft was introduced in the PLA service in 1970.
Length: 15.7m
Wingspan: 9.7m
Height: 4.3m

Weight: 6,375 kg (empty); 11,830 kg (max at take-off)
Propulsion: 2 turbojets
Speed: Mach 1.12 (max)
Range: 2,000 km
Combat Radius: 400 km – 600 km
Armament: PL series AAM; AShM; 23mm gun; Unguided rockets and bombs
Crew: 1

Maritime Patrol/ Reconnaissance/ ASW Aircraft

The maritime patrol and reconnaissance aircraft used by the PLA naval aviation include the following.

SH-5: The Chinese-built maritime amphibian aircraft entered service in the 1980s. Few such aircraft having different configurations were built and inducted into service. The PLANAF uses the aircraft for a wide range of roles, including for aerial firefighting, anti-submarine warfare, transport, search & rescue besides maritime patrol and reconnaissance missions.

Length: 38.9m
Wingspan: 36m
Height: 9.8m
Weight: 25,000 kg (empty); 45,000 kg (max at take-off)
Propulsion: 4 turboprops
Speed: Mach 0.45 (max)
Range: 4,760 km
Armament: C-101 AShM; ASW torpedo; Bombs; Naval mines
Crew: 8

Y-8X: Based on the Y-8 turboprop transport aircraft, the Y-8X maritime patrol and anti-submarine warfare aircraft entered service in 1984. Besides maritime patrol, surveillance and ASW operations, the aircraft is also used for search & rescue missions. Few of the aircraft are also equipped for electronic and signal intelligence missions. The Y-8 can conduct day or night operations during all weather conditions.

Length: 34.02m
Wingspan: 38m
Height: 11.16m
Weight: 35,500 kg (empty); 61,000 kg (max at take-off)
Propulsion: 4 turboprops
Speed: Mach 0.53 (max)
Range: 5,620 km
Armament: Torpedoes & mines
Crew: 3

Gaoxin-6: Developed as China's first anti-submarine patrol aircraft, the Gaoxin-6 was first unveiled in 2011 and joined service in 2015. The new fixed-wing, four-engine aircraft built indigenously is based on the older-generation Y-8 MPA/transport aircraft. The long-range anti-submarine aircraft is also capable of carrying out anti-ship operations besides maritime patrol and reconnaissance missions.

Propulsion: 4 turboprops
Speed: Mach 0.53 (max)
Range: 6,000 km
Armament: AShMs; Bombs & Torpedoes
Crew: 10

AG600: The first prototype of the AG600 was unveiled as the world's largest amphibious aircraft in 2016. The aircraft has been indigenously designed and developed by China to conduct numerous operations, including forest firefighting and maritime search & rescue. It is likely to play a key strategic role in the disputed South China Sea and East China Sea regions as well.

Length: 37m
Wingspan: 38.8m
Weight: 53,500 kg (max at take-off)
Propulsion: 4 turboprops
Speed: Mach 0.53 (max)
Range: 5,000 km or more

Trainer aircraft

JL-9: The two-seat fighter trainer jet is operational in the PLANAF since 2013. The JL-9G variant is specifically designed for the Navy's aviation wing as an aircraft carrier-based jet trainer having ski-jump take off and arrested landing capability.

Length: 14.5m
Wingspan: 8.32m
Height: 4.1m
Weight: 5,000 kg (empty); 9,800 kg (max at take-off)
Propulsion: 1 turbojet
Speed: Mach 1.5 (max)
Armament: 23mm gun; Rocket pods
Crew: 2

Helicopter

The PLANAF operates the following helicopters of different roles.

Z-8: The triple engine, multi-role helicopter developed in China as a derivative of the European SA 321 Super Frelon helicopter entered service in 1989. Different variants of the helicopter have been developed, including for SAR role, transport and logistics support missions. The PLANAF primarily uses the Z-8A ship-borne variant of the rotorcraft which is able to land on water.

Z-18: An upgraded version built to replace the Z-8, the Z-18 single-engine, medium lift transport helicopter has entered service with PLANAF. The new rotorcraft is capable of being operated from the Liaoning aircraft carrier. An ASW/ anti-ship variant and an AEW variant of the helicopter have also been developed for PLAN.

Z-9: The utility helicopter is a licence-built version of the Eurocopter AS365 Dauphin. The rotorcraft entered service with the Chinese military in 1994. It has many variants and the PLAN uses the helicopter for ASW and SAR operations. The Z-9 has been deployed on some of the Navy's frigates and also on-board the Liaoning aircraft carrier.

Ka-27/ 28: The Russian-built naval helicopters are operational in the PLAN to carry out anti-submarine warfare operations. They are operated by the PLA Navy's Luyang I (Type 052B) class of DDGs. The mission of the helicopters is to detect, track and destroy submerged submarines at a depth of 500m. The Ka-28 also has day and night operations capability.

Ka-31: The Russian-built rotorcraft designed for airborne early warning (AEW) and control operations is deployed in the PLANAF. Its primary mission is to detect long-range airborne threats including fixed-wing aircraft and helicopters. It also performs surveillance, target tracking and transmission of target data to the command post. It is deployed on the Liaoning carrier.

WEAPONS

The primary weapons used by the PLA Navy for naval warfare missions include missiles, torpedoes and naval guns. While the use of naval guns played a key role in the early days of the PLAN's evolution, development and deployment of a wide range of cruise and ballistic missiles as well as ASW torpedoes in the past few decades has expanded the Navy's operational horizon beyond its coastal waters.

Missiles

The Navy uses indigenously developed ballistic and cruise missiles of different ranges, speed and payloads for carrying out anti-ship, land-attack and anti-air warfare operations. These missiles include the **DF-21D** ASBM, **CJ-10/DH-10** LACM, next-gen **HN-2000/ DH-2000** long-range LACM, **YJ-8** series of AShM, **YJ-82** LACM/AShM, **JL-2** SLBM, **HN-1/2/3** LACM, and **HQ-9/10** air defence missiles.

DF-21D: Developed as an anti-ship ballistic missile (ASBM) having the capability to target moving aircraft carriers, the new weapon has been termed a "carrier killer." The mobile, land-based missile has an estimated range of 1,450-km. It is believed to be capable of employing manoeuvrable re-entry vehicle (MaRV) payloads with a terminal guidance system.

CJ-10/ DH-10: It is a family of long-range land-launched subsonic cruise missiles deployed with the PLA since 2008-09. Over 500 such missiles are estimated to be in the PLA's arsenal, including in the Navy. The naval variant of the missile is designated YJ-100. Another new missile, designated **DH-2000**, is under development for deployment on attack submarines and with anti-carrier strike capabilities.

C-801 (YJ-8): It has been developed as a second generation multi-role, multi-platform anti-ship missile. The subsonic missile, developed in the 1980s, was originally designed for launch from surface-based platforms. It was later modified and developed for launch from submarines, with the designations of **YJ-1** and **YJ-81** respectively. More advanced variants such as the **C-802** and **C-803** were developed subsequently.

JL-2: It is a submarine-launched ballistic missile having a range of 8,000 km. The weapon is deployed on board the Type-094 Jin-class SSBNs. The JL-2 was developed as an improved variant over the original JL-1 missiles that were deployed on the older generation Type 092 Xia class nuclear submarines. **JL-3,** another more advanced version of JL-2, is also under development to arm the futuristic Type 096 class SSBNs. The JL-3 missile will have a higher range than its predecessor.

HN-2: It has been developed as a short, medium and long range ship and submarine-launched land-attack cruise missile variant, designated as **HN-2B** and **HN-2C** respectively. The sub-launched variant has a reported range of 1,400 km. China is also believed to have developed the **HN-3B** submarine-launched variant having a maximum range of 2,200 km.

Torpedoes

The torpedoes presently deployed in the PLA Navy submarines for carrying out anti-submarine warfare and anti-surface warfare operations include the Yu series of indigenously developed ones such as the **Yu-3** acoustic homing torpedo, **Yu-4** passive acoustic homing torpedo, **Yu-5** wire-guided torpedo, **Yu-6** heavyweight torpedo and **Yu-7** lightweight torpedo.

Yu-3: The 533mm acoustic homing anti-submarine torpedo, believed to China's first indigenously designed one, entered service in the 1980s. Several modifications were later incorporated to improve the weapon which was designed to be launched from both nuclear and non-nuclear submarines.

Yu-4: It is based on the Russian built SAET-50 anti-surface warfare passive acoustic homing torpedo. The electric propelled, acoustic homing heavyweight torpedo is inducted in two variants – the passive acoustic-homing variant Yu-4A and the active/passive acoustic-homing variant Yu-4B. Designed to engage surface ships, the torpedo is carried by the PLAN's Song-class SSKs.

Yu-5: The wire-guided anti-submarine torpedo was developed in the 1990s for equipping conventional submarines. It is the first indigenously built wire-guided ASW for conventional submarines. The torpedo is initially guided by wire guidance for most of the time, and at the final stage, the guidance is changed to active / passive acoustic homing guidance.

Yu-6: The new generation heavyweight torpedo is the most advanced one designed to take on both submarines as well as surface ships. The torpedo entered service in 2005 and is carried by the Type 039 (Song class) and Type 039A (Yuan class) of conventional submarines, and also likely by the Type 093 (Shang class) SSNs and Type 094 (Jin class) SSBNs.

Yu-7: The anti-submarine warfare torpedo entered service in the 1990s and is deployed on most of the Navy's surface combatants for anti-submarine warfare. The weapon is a reverse-engineered variant of the American-built Mk 46 Mod. 2 light ASW torpedo. The 324-mm calibre torpedo is designed to be launched from surface combatant torpedo tubes to attack enemy submarines. Another improved variant, designated Yu-11, is believed to have been developed later as a lightweight torpedo for deployment on-board helicopters.

Naval Guns

Type 730 CIWS: The seven barrelled 30mm Gatling gun is designed to provide defence against anti-ship missiles, fixed/rotary wing aircraft, ships and other small craft, and floating mines. The Chinese built CIWS is deployed on most of the PLAN destroyers, and some frigates. It is capable of firing 5,800 rounds per minute for up to a range of 3-km. An advanced variant, designated **Type 1130** with 11 barrels, is arming the Liaoning aircraft carrier.

Type 76: The twin 37mm naval gun is indigenously developed for anti-aircraft and anti-surface operations. Improved variants of the AA gun were developed later of which the Type 76A is deployed on many frontline warships and the Type 76F is arming the auxiliary vessels of PLAN.

Type 79: The twin barrel 100mm gun was developed as China's first large calibre, fully automatic gun. It entered service in the 1980s. An improved variant, Type 79A, was later developed with a firing capacity of 18 rounds per minute for a range of 22km. The naval artillery is deployed on all modern destroyers and frigates of the PLA Navy.

H/PJ38: The new single barrel 130mm naval gun entered service between 2005 and 2007. It is believed to have been reverse engineered from the Russian-built AK 130 twin 130mm naval gun. The H/PJ38 is considered the more powerful and advanced naval artillery over all other smaller calibre guns presently in PLAN service. The gun is deployed on the Navy's Type 052D class of destroyers.

CONCLUSION

While evolving as a global economic powerhouse, China has simultaneously pursued a large-scale, comprehensive military modernisation agenda to rapidly ramp up force projection capabilities in Asia and beyond in the 21st century world. It has laid down an ambitious roadmap to make rapid military strides by 2020 and of winning "informationised" wars by 2050. By flexing its military muscles, China appears to be posing a serious security challenge to several regional powers, including India and Japan, and also to the US. Its "anti-access/ area-denial" strategy to keep away other countries from the vital natural resources in the Asia-Pacific region has raised concern among the international community. A rising China with an aggressive posture also appears to be steadily tilting the balance of power in the entire Asian continent in its favour. By countering the US presence in the Asia-Pacific and asserting its supremacy in the waters of South China and East China seas, the Asian giant has gone loud and clear in stating that it will confront any challenge to maintain its territorial sovereignty even while dictating terms in the region and beyond to safeguard its economic and strategic interests.

The J-11 fighter jets *Image credit PLA DailyChina MoD*

CONTACT DETAILS
Ministry of Defence
Beijing, China
Tel.: 010-66720068
Website: www.mod.gov.cn/ eng.mod.gov.cn/

COLOMBIA
(Capital: Bogota)

INTRODUCTION

Area: 1,138,914 sq km

Population: 47,220,856 (July 2016 est.)

Coastline: 3,208 km (Caribbean Sea 1,760 km, North Pacific Ocean 1,448 km)

Maritime claims: Territorial sea: 12 nm

Exclusive Economic Zone: 200 nm

Continental shelf: 200 m depth or to the depth of exploitation

KEY POLITICAL PERSONS

PRESIDENT & COMMANDER-IN-CHIEF: Juan Manuel Santos Calderon

VICE PRESIDENT: German Vargas Lleras

KEY DEFENCE PERSONS

MINISTER OF DEFENCE: Luis Carlos Villegas Echeverri

GENERAL COMMANDER OF THE ARMED FORCES: Gen. Juan Pablo Rodríguez Barragán

CHIEF OF ARMY: Gen. Alberto Jose Mejia Ferrero

CHIEF OF NAVY: Vice Adm. Leonardo Santamaria Gaitan

CHIEF OF AIR FORCE: Maj. Gen. Carlos Eduardo Bueno Vargas

ECONOMY

Colombia is the world's fourth largest coal exporter and Latin America's fourth largest oil producer. Declining oil prices have resulted in a drop in government revenues. Colombia's real GDP has grown more than 4% per year for the past four years, continuing almost a decade of strong economic performance. Colombia has a free market economy with major commercial and investment ties to the United States. Colombia depends heavily on oil exports, making it vulnerable to a drop in oil prices. Colombia is the third largest Latin American exporter of oil to the US. Economic development is stymied by inadequate infrastructure, weakened further by recent flooding. The SANTOS Administration's foreign policy has focused on bolstering Colombia's commercial ties and boosting investment at home. The government has encouraged exporters to diversify their customer base beyond the United States and Venezuela; traditionally Colombia's largest trading partners. Columbia has signed FTAs with a number of other countries, including Canada, Chile, Mexico, Switzerland, the EU, Venezuela, South Korea, Turkey, Japan, and Israel. Foreign direct investment - notably in the oil sector - reached a record $10 billion in 2008 but dropped to $7.2 billion in 2009, before beginning to recover in 2010, and reached a record high $13 billion in 2011. The business sector remains concerned about Venezuela's trade restrictions on Colombian exports, an appreciating domestic currency, and the pending US Congressional approval of the US-Colombia Trade Promotion Agreement. Colombia's economy has experienced rapid increase over the past three years despite a serious armed conflict. The economy continues to grow in part because of austere government budgets, focused efforts to reduce public debt levels, an export-oriented growth strategy, an improved security situation in the country, and high commodity prices. Colombia estimate a peace deal may bolster economic growth by almost 2%. Tourism in Colombia is an important sector in the country's economy.

GDP: $667.4 billion (2015 est.), $647.5 billion (2014 est.)

Real Growth Rate: 3.1% (2015 est.), 4.4% (2014 est.)

Industries: Textiles, food processing, oil, clothing and footwear, beverages, chemicals, cement; gold, coal, emeralds

Oil - proved reserves: 2.3 billion bbl (1 January 2016 est.)

Natural gas - proved reserves: 134.7 billion cu m (1 January 2016 est.)

Total Exports: $ $38.12 billion (2015 est.), $56.92 billion (2014 est.)

Export Commodities: Petroleum, coal, emeralds, coffee, nickel, cut flowers, bananas, apparel

Major Markets: US 27.5%, Panama 7.2%, China 5.2%, Spain 4.4%, Ecuador 4% (2015)

Total Imports: $52.04 billion (2015 est.), $61.55 billion (2014 est.)

Import Commodities: Industrial equipment, transportation equipment, consumer goods, chemicals, paper products, fuels, electricity

Major suppliers: US 28.8%, China 18.6%, Mexico 7.1%, Germany 4.2% (2015)

INTERNATIONAL DISPUTES

✦ In December 2007, the ICJ allocated San Andres, Providencia and Santa Catalina islands to Colombia under a 1928 Treaty but did not rule on 82 degrees W meridian as the maritime boundary with Nicaragua. It has managed a dispute with Venezuela over maritime boundary and Venezuelan-administered Los Monjes Islands near the Gulf of Venezuela.

✦ Managed dispute with Venezuela over maritime boundary and Venezuelan-administered Los Monjes Islands near the Gulf of Venezuela.

✦ Colombia, Honduras, Nicaragua, Jamaica and the US assert various claims to Bajo Nuevo and Serranilla Bank.

✦ Colombia is involved in an armed conflict which is taking place between guerrilla fighters (Revolutionary Armed Forces of Colombia (FARC) & National Liberation Army (ELN)) and the government, as well as paramilitary groups, which have been linked to the authorities. The United States has been heavily involved in the conflict since its beginnings. On 23 June 2016, the Colombian government and the FARC rebels signed a historic ceasefire deal, bringing them closer to ending more than five decades of conflict. The Colombian government and the FARC on November 24 signed a revised peace deal and the revised agreement was be submitted to Congress for approval. The House of Representatives unanimously approved the plan on November 30, a day after the Senate also gave its backing.

DEFENCE

ARMY

Personnel: 235,538

Equipment

Category	Name	In Service
APC/MICV/IFV	EE-9 Cascavel	174
	EE-11 Urutu	100
	M8 Greyhound	8
	M113	120
APC/MICV/IFV	M1117	39
	BTR-80	20
	RG-31 Nyala	4
	ISBI	16
RCL Gun	106mm M40	69
Artillery	M116 howitzer	70
	M101 howitzer	50
	105 mm GIAT LG1	20
Mortar	81mm	130
	120 mm	120
	155 mm	
	60 mm	
	M30	80
Anti-Tank GW	BGM-71 TOW	18
	Nimrod	8
	Spike	
AA Gun	M1A1 40mm	30
	Bofors 40mm/L70	29

Army Aviation

Category	Name	In Service
Aircraft	Commander 1000	2
	Beechcraft King Air 90	1
	Convair 580	1
	Piper PA-31 Navajo	2
	Rockwell 685 Commander	2
	An-32	2
	CASA C-212 Aviocar	5
	Piper PA-34 Seneca	5
Helicopter	UH-1H	31
	Bell Huey II	27
	Kaman K-MAX	5
	Mil Mi-17	23
	UH-60L	57
UAV	RQ-11 Raven	

New Procurements/ Upgrades

✦ The Colombian Army has contracted Sikorsky Aircraft to buy two additional S-70i BLACK HAWK helicopters. They will augment five S-70i helicopters in the Special Forces of the Colombian Army's Air Assault Division. The Colombian Army has already inducted five new Sikorsky S-70i multi-mission helicopters into its fleet of BLACK HAWK aircraft.

✦ The government is negotiating for the procurement of the at least 50 Leclerc Main Battle Tanks (MBTs) and support vehicles that are surplus from French Army.

NAVY

Personnel: 35,502

Equipment

Category	Name	In Service
Submarine	Type 209/1200	2
	Type SX-506	2
Frigate	Almirante Padilla class	4
Corvette	Donghae class	1
Offshore Patrol Vessel	Fassmer-80	3
	Auxiliary vessels	14
	Boston Whaler Patrol craft	15
	Riverine support patrol vessel	8

Naval Aviation

Category	Name	In Service
Aircraft	Cessna 208 Caravan	2
	Aero Commander 690	5
	C-212-100	1
	Cessna 208	2
	Super King Air	1
	CN-235-200	3
	Piper PA-28 Cherokee	3
	Piper PA-34 Seneca	4
Helicopter	MBB Bo 105 CB	2
	AS 555 Fennec	2
	Bell UH-1N Twin Huey	5
	Bell 412	

Colombian Armada submarine A.R.C. Tayrona enters Mayport Basin during a visit sponsored by US Naval Forces Southern Command (NAVSO). *Image Credit: US Navy*

New Procurements / Upgrades

✦ LRAD has been awarded a contract by the Colombian Navy to deliver 300X systems in support of its coastal defence patrol boats.

✦ Cassidian Optronics improves the mission effectiveness of the 209 class submarines of the Colombian Navy by installing state-of-the-art sighting systems. After the already ordered refurbishment of an attack periscope, the Colombian Navy has now ordered a SERO 250 search periscope from Cassidian Optronics for a 209 Class submarine.

✦ Two existing U209/1200 Class submarines, ARC Pijao and ARC Tayrona, are being modernised at the state-owned ship at Cartagena. These will be joined by another two German U206A submarines.

✦ The ARC is working on additional medium and long-term defence programmes, including the development and acquisition of a number of Coastal Patrol Vessels (Fassmer CPV-40), and the research & development of an indigenous corvette or frigate-class vessel ("Plataforma Estrategica de Superficie"), planned towards 2018-2020, as per reports.

✦ The Colombian Navy has commissioned a new landing craft utility (LCU).

AIR FORCE

Personnel: 15,500

Equipment

Category	Name	In Service
Fighter	IAI Kfir	24
	EMB 314	24
	OV-10	5
	AC-47T	6
	A-37	8
Trainer	T-41D	10
	T-34M	10
	T-37B/C	17
	AT-27	14
	T-90 Calima	2
Tanker/ Transport	707-373C	1
	KC-767	1
	C-212-300	3
	CASA CN-235-200	3
	CASA C-295M	4
	EMB 110P1A	2
	Arava	1
	C-130 Hercules	7
	PA-23 Aztec	1
	PA-31T Cheyenne	1
	PA-34 Seneca	3
	Cessna 208	1
Helicopter	Bell 205	26
	Bell 212 Twin Huey	12
	MD 500 Defender	16
	UH-60 Black Hawk	90
	Bell 206	32
Reconnaissance and Intelligence	SA 2-37	6
	C-26A	5
	Citation SR-560	6
	Cessna 208	5
	Super King Air 300	3

New Procurements / Upgrades

No major procurement plans are under consideration at present.

Defence Expenditure

Total defence spending: $9.1 billion (est.)

Defence spending in terms of GDP: 3.5% (2015)

Domestic suppliers: Indumil, Cotecmar

Foreign suppliers: United States, Israel, South Africa, Spain, Brazil, Belgium, Germany, France, UK, Sweden, Canada, Portugal, Russia & Romania.

Colombia continues to spend a large percentage of its GDP to keep its armed forces trained and armed. According to the Colombian Ministry of Defence, the 2016 budget will continue to focus on strengthening the country's internal security and the strategic forces of the army. The Colombian armed forces have a budget of USD 10.3 billion, which is equivalent to 13.9 percent of the total Colombian budget for 2016. This figure has only gone up by USD 600 million dollars since 2015. The country mainly plans for the modernisation of its armed forces and equipments. The additional defence spending mainly focuses on the expansion, training programmes, small arms purchases and ammunition, with little going to expensive non-domestic procurements. In 2012, Canada and Colombia signed a memorandum of understanding to increase defence cooperation, on the sidelines of the International Security Forum in Halifax. The two countries, which already cooperate on military training and exercises, will work together on issues of military justice as well as share information about fighting home-made bombs.

Defence Production and R&D

In 2006, Grupo Social y Empresarial de Defensa - GSED (Social and Entrepreneurial Defence Group) is the organisation built by the Colombian Defence ministry by involving 18 of the State-run enterprises and institutions that were defence-oriented. GSED has become the second largest Colombian State-run conglomerates, employing over 8,900 people by 2010. Colombia has produced a number of wheeled APC prototypes, armoured cars, river patrol boats, tugs, medium-sized river support ships, light aircrafts, mortars and so on. The GSED companies were involved in several high-tech research and development contracts, including developing sensors and data links, unmanned aerial vehicles (UAVs), 70-mm rockets, and mine-resistant boots for infantrymen and mine detection systems. In 2006, the Corporacón de Ciencia y Tecnología de la Industria Naval, Marítima y Fluvial (COTECMAR), reported sales of $39 million. By 2009 these sales had gone up to $106 million. This included major work on 39 of the Navy's boats and ships. The manufacturing of BTR-80 APC's has also started in the COTECMAR inventory.

Defence Procurement

Former Colombian President Uribe had announced the launch of the Democracy Consolidation Plan in 2007, which is an armed forces expansion and modernisation plan funded by some $3.73 billion obtained from a new tax elaborated by his administration and applied to a number of high performing Colombian enterprises. The defence ministry has also negotiated for the procurement of the at least 50 Leclerc Main Battle Tanks (MBTs) and support vehicles that are surplus from French Army, in February 2010. At present, a wide variety of equipment is on order from France, Germany, Israel and Russia in addition to US defence aid under bilateral anti-drug support programmes.

Colombia Fuerzas Comando 2016. *Image Credit: Colombian Army*

CONTACT DETAILS

Ministry of National Defence
Centro Administrativo Nacional
Avenida Eldorado
Santa Fe de Bogota, D.C
Colombia
Tel: +5712222445
Fax: +57-12221874
Telex: 40816

Defence Forces HQ
Centro Administrativo Nacional
Avenida Eldorado, Carrera 52
Santa Fe de Bogota, D.C
Tel: +5712669300

CROATIA
(Capital: Zagreb)

INTRODUCTION

Area: 56,594 sq km

Population: 4,313,707 (July 2016 est.)

Coastline: 5,835 km (mainland 1,777 km, islands 4,058 km)

Maritime claims: Territorial sea: 12 nm

Continental shelf: 200 m depth or to the depth of exploitation

KEY POLITICAL PERSONS

PRESIDENT & COMMANDER-IN-CHIEF: Kolinda Grabar-Kitarovic

PRIME MINISTER: Andrej Plenkovic

KEY DEFENCE PERSONS

MINISTER OF DEFENCE: Damir Krsticevic

CHIEF OF GENERAL STAFF: Gen. Mirko Sundov

COMMANDER OF THE ARMY: Lt. Gen. Mato Ostovic

COMMANDER OF THE NAVY: Commodore Predrag Stipanovic

COMMANDER OF THE AIR FORCE: Brig. Gen. Miroslav Kovac

ECONOMY

Croatia has a high-income economy. International Monetary Fund data shows that Croatian nominal GDP stood at $57.371 billion in 2013. On 1 July 2013 Croatia joined the EU. The economy of Croatia is a service-based economy, with the service sector accounting for 67% of the total GDP. Food processing and chemical industry take significant portions of industrial output and are responsible for significant portion of exported goods. Industrial Sector represents 27% of Croatia's total economic output, and agriculture represents 6%. The country is currently preparing for membership in the European Union, its most important trading partner. The new government has announced a more flexible approach to privatisation, including the sale in the coming years of state-owned businesses that are not of strategic importance. While macroeconomic stabilisation has largely been achieved, structural reforms lag. The industrial sector with exports of over €1 billion annually is dominated by shipbuilding which accounts for over 10% of exported goods. IMF data projects that Croatian nominal GDP stands at $52 billion, or $12,405 per capita for year 2017.

GDP: $91.1 billion (2015 est.), $89.62 billion (2014 est.)

Real Growth Rate: 1.6% (2015 est.), -0.4% (2014 est.)

Industries: Chemicals and plastics, machine tools, fabricated metal, electronics, pig iron and rolled steel products, aluminium, paper, wood products, construction materials, textiles, shipbuilding, petroleum and petroleum refining, food and beverages, tourism

Oil - proved reserves: 71 million bbl (1 January 2016 est.)

Natural gas - proved reserves: 24.92 billion cu m (1 January 2016 est.)

Total Exports: $11.91 billion (2015 est.), $12.95 billion (2014 est.)

Export Commodities: Transport equipment, machinery, textiles, chemicals, foodstuffs, fuels

Major Markets: Italy 13.4%, Slovenia 12.5%, Germany 11.4%, Bosnia and Herzegovina 9.9%, Austria 6.6%, Serbia 4.9% (2015)

Total Imports: $19.28 billion (2015 est.), $21.39 billion (2014 est.)

Import Commodities: Machinery, transport and electrical equipment; chemicals, fuels and lubricants; foodstuffs

Major Suppliers: Germany 15.5%, Italy 13.1%, Slovenia 10.7%, Austria 9.2%, Hungary 7.8% (2015)

INTERNATIONAL DISPUTES

✦ Dispute remains with Bosnia and Herzegovina over several small sections of the boundary related to maritime access that hinders ratification of the 1999 border agreement.

✦ The Croatia-Slovenia land and maritime boundary agreement, which would have ceded most of Pirin Bay and maritime access to Slovenia and several villages to Croatia, remains unratified and in dispute.

✦ Slovenia has objected to Croatia's claim of an exclusive economic zone in the Adriatic Sea.

✦ In 2009, however Croatia and Slovenia signed a binding international arbitration agreement to define their disputed land and maritime borders, which led to Slovenia lifting its objections to Croatia joining the EU. Slovenia continues to impose a hard border Schengen regime with Croatia.

DEFENCE

ARMY

Personnel: 16,000

Equipment

Category	Name	In Service
MBT	M-84A4 Sniper	72
	M-84D	(under development)
	M-95 Degman	4
APC/MICV /IFV	BVP M-80A	124
	Patria AMV	126
	Iveco LMV	10
	M1151	90
	M-ATV	150
	RG-33	37
	International MaxxPro MRAP	30
Artillery	105mm M56	40
	122mm D-30 RH M94	36
	130mm M-46H1	32
	152mm M84	16
	CITER 155mm L33 Gun	8
	155mm M114	18
	Panzerhaubitze 2000	10+
	122mm 2S1 Gvozdika SP	9
Mortar	60mm M57	69
	82mm M96	69
	120mm M75	43

Army continued...

Category	Name	In Service
MRL	128mm RAK-12	8
	122m APR-40	36
	122mm M-96 Tajfun	4
	122mm M-91 Vulkan	8
Anti-Tank GW	Spike ER	(on order)
	9K115-2 Metis-M	54
	9K111 Fagot	120
Anti-Tank GW	RL90 M95, AT4 RPGs	
	9K11 Malyutka	200
AA Gun	BOV 20/3 SPAAG	54
	L/70	7
SAM	10CROA1	10
	9K38 Igla	60
	Strela 2	513

New Procurements/ Upgrades

✦ Duro Dakovic has signed a contract on the acquisition and equipment of the Armed Forces' combat armoured vehicles with PROTECTOR M151 12.7mm calibre weapon stations. The contract is worth HRK 112 million and Duro Dakovic is supposed to deliver the combat armoured vehicles to the Ministry in course of 2016-2017.

✦ Maintenance and modernisation of the infantry combat vehicle BVP M-80, including modernisation with new anti-armour systems and equipping the Army with Man Portable Short Range Air Defence System (MANPADS).

✦ The US government decided on donating 212 Mine Resistant Ambush Protected Vehicles (MRAP), comprising 162 combat vehicles M-ATV, 20 MRAP HAGA (heavily armed ground ambulance) vehicles and 30 MaxxPro MRAP vehicles to the Republic of Croatia.

✦ Croatian Army has ordered M151 12.7mm calibre Protector remote weapon stations (RWS) from Kongsberg.

✦ Increasing fire power and effectiveness of the artillery systems by procuring fire management systems, reconnaissance radar systems and other related advanced systems.

✦ Procurement of 3 S-10 self-propelled missile systems and modernisation of existing systems.

✦ Modernising of existing IGLA SAM system.

✦ Croatian MoD agreed on an amendment with Patria for the procurement of 126 Patria AMV 8x8 Modular APC/IFV.

✦ Croatia to procure 94 Iveco LMV (Light Multi role Vehicles). The army might purchase additional LMVs once more funds become available. Around 250-300 LMVs are needed.

✦ Modernisation of M-84A and M-84A4 Sniper MBTs and upgrade to M-84D and M-95 Degman standard. Programme calls for an upgrade and modernisation of 75 existing tanks and procurement of additional 29 new tanks over next 4 years due to a plan which requires Croatia to operate at least 104 modern tanks.

NAVY

Personnel: 2000

Equipment

Category	Name	In Service
Missile boat	Kralj class	2
	Helsinki class	2
	Koncar class	1
Patrol boat	Mirna class	4
Landing craft	Silba class	2
	Type 11	2
	Type 22	1
	Type 21	1
Minehunter	Korcula class	1
Weapon	RBS-15	40+
	MOL Mobile anti-ship missile launchers with RBS-15 (Coastal batteries)	3

New Procurements/ Upgrades

✦ The Croatian Navy has received the autonomous underwater vehicles Remus 100 - a donation from by the United States. The Remus -100 are also used in search of sunken objects at depths up to 100m.

✦ Procurement of resources for chemical and biological detection; detection and dose measurement of radiation; means for NBC contamination.

✦ 10 new patrol boats, locally built, 40 meters in length.

✦ 4 new corvettes - 88–120 meters in length.

✦ Overhaul of existing 2 King class fast attack crafts, including new engines.

✦ Upgrading RBS-15 and MOL missiles to latest standard.

During a military exercise in the Adriatic Sea, the Croatian Navy has successfully test-fired a Saab RBS15 anti-ship missile from a naval vessel in November 2015. The missile was fired from the Kralj-class missile boat Dmitar Zvonimir. It was Croatia's first firing of an RBS15 from a warship since 1994. *Image Credit: SAAB Group*

AIR FORCE AND AIR DEFENCE

Personnel: 2000

Equipment

Category	Name	In Service
Fighters/attack/ Reconnaissance	MiG-21bis	9
	MiG-21UM	4
Transport/Tanker	An-32B	2

Air Force continued...

Category	Name	In Service
Helicopter	Mi-171Sh	10
	OH-58D	11
	Mi-8T	9
	Bell 206B-3	8
Trainer	Pilatus PC-9M	20
	242L	5
UAV	Skylark I	6
	Hermes 450	2

New Procurements/ Upgrades

✦ The Minister of Defence of Croatia has decided to select Elbit Systems' proposal to upgrade Croatia Air Force's MiG-21 aircraft.

✦ Procurement of 10-12 Mi-171Sh transport helicopters and all associated spare parts and equipment.

✦ Procurement of 5 Zlin 242L basic trainers.

✦ Procurement of Advanced short to medium range NATO SAM systems and modernisation of existing Russian-made IGLA SAMs.

✦ Croatia plans to modernise its air force with the introduction of 18-24 multirole fighters. In the competition are JAS 39C/D Gripen, Eurofighter Typhoon, F-16C/D and MiG-29M/M2.

✦ Acquisition of up to 16 OH-58D Kiowa Warrior scout/attack helicopters from US Army.

Croatian Air Force. *Image Credit: Ministry of Defence of Croatia.*

Defence Expenditure

Total defence spending: USD 4.75 billion (2015)

Estimated defence spending in terms of GDP: 1.21 (2015)

Domestic suppliers: Đuro Đaković (armoured vehicles), HS Produkt (small arms).

Foreign suppliers: Sweden, Switzerland, Italy, Finland, United States, Norway, Germany.

The Croatian Ministry of Defence is planning on reforms and drafting of the National Security Strategy and the concept of the Homeland Security. The Ministry has increased the defence budget, as the first contribution of the new Croatian Government to the defence sector. According to the modernisation plan, 2015-2024 of Croatia MoD, the Croatian Armed Forces are set to receive vitally needed new equipment, including a battalion of self-propelled howitzers, new infantry fighting vehicles, at least eight brand new helicopters and a squadron of fighter aircraft. The Croatian military budget for the past 6–7 years was kept below 2% of GDP compared to the 1990s when defence expenditure was a significant part of the Croatian budget due to the country's war of independence.

The Ministry of Defence on 4 April 2016 signed 200 mil kuna worth contracts with 27 Croatian manufacturers of weapons and military equipment, stipulating the provision of clothes, footwear, weaponry, repair and overhaul services, catering and insurance services for the members of the Croatian Armed Forces. Croatian defence industry offers products compatible with NATO standards, and the contracts exemplify a co-operation whereby the Ministry promotes Croatian companies and supports their development. The worth of the contracts signed with 27 weapons and military equipment producers, stipulating the provision of nearly 300,000 of various products, is cca 200 mil kuna, which is a significant sum, and will serve as an important incentive to the national economy. According to a press release by Croatian Ministry of Defence, the contract on the provision of combat uniforms has been signed with the Association of Producers of Weapons and Military Equipment led by Kroko Prizvodnja i razvoj d.o.o, which comprises Cateks d.d. Cakovec, Odjeca d.o.o. Zagreb, Sesir d.o.o. Zagreb, URIHO Ustanova Zagreb and VARTEKS PRO d.o.o. Varazdin and Hemco d.o.o. Dakovo as subcontractors. The contracts entail the provision of cca 300,000 various products.

Defence Production and R&D

Croatia has a relatively well-developed arms industry that is highly competitive internationally, with significant annual arms exports. In 2014 Croatia arms exports have reached USD 257 million, the majority of exports being to NATO allies and Australia. Kraljevica Shipyard is a shipbuilder at Kraljevica, on the Adriatic coast of Croatia. The shipyard was founded in 1729 and is claimed to be the oldest continuously operational shipyard in the world. The government of Croatia has repeatedly attempted to privatise the shipyard, in November 2010. In 2008, the Croatian government announced that five of the country's shipyards would be privatised, in accordance with EU regulations for gaining membership. According to the new ten year plan of the Croatian Armed Forces (The Croatian Armed Forces Long-Term Development Plan 2006-2015), the defence force will develop and cooperate with expert scientific and research institutions in Croatia and abroad as well as with the producers of military equipment and weaponry. The Institute of Defence Research and Defence Systems Development will be responsible for research and development of the armed forces' needs. The Institute for Research and Development of Defence Systems (IROS) is where the needs of national defence and security meet the scientific and technological potentials of the Republic of Croatia. The Institute is organised as an administrative unit of the Ministry of Defence which carries out, coordinates and guides scientific activity in the field of defence systems as well as scientific-research and development activities in other fields which are important for the Croatian Armed Forces. The Institute carries out research activities in the field of defence and security, as well as in the interdisciplinary fields of advanced defence technologies, defence systems, management of human resources and many others.

Defence Procurement

Domestic production and procurement remains a top most priority with the Croatian government. The country has established a series of defence procurement priorities for each branch of service. At present, the ongoing procurement is in line with the reforms in progress. Croatia Army/Navy/Air Force has not done any major procurement last year.

CONTACT DETAILS

Ministry of Defence of the Republic of Croatia

Phone: +385 (1) 4567-111
e-mail: infor@morh.hr
Sarajevska cesta 7
10 000 Zagreb

Public Relations and Information Department

Sarajevska cesta 7 (building no. 15)
10 000 Zagreb
Tel.: +385 (1) 4832-433
Fax: +385 (1) 4832-905
E-mail: infor@morh.hr

Institute for Research and Development of Defence Systems (IROS)

Ilica 256b, HR-10000 Zagreb, Croatia
Telephone: + 385 1 37 84 690
Fax: + 385 1 37 84 878
E-mail: lea.gradecak@morh.hr
URL: http://iros.morh.hr/en

Croatian soldiers engage opposing forces during exercise Combined Resolve IV at the US Army's Joint Multinational Readiness Center in Hohenfels, Germany, May 30, 2015

STRATEGIC INFORMATION

CUBA
(Capital: Havana)

INTRODUCTION

Area: 110,860 sq km

Population: 11,179,995 (July 2016 est.)

Coastline: 3,735 km

Maritime claims: Territorial sea: 12 nm

Contiguous zone: 24 nm

Exclusive Economic Zone: 200 nm

GEOPOLITICAL IMPORTANCE

Cuba (the Republic of Cuba) is the most populous and largest island nation in the Caribbean. It is situated between the Caribbean Sea and the North Atlantic Ocean. United States (150 km) and the Bahamas lies to the north of Cuba, Mexico is to the west, the Cayman Islands and Jamaica are to the south, and Haiti and the Dominican Republic are to the southeast. Isla de la Juventud is the main island of Cuba which includes several archipelagos.

POLITICAL OVERVIEW

Cuba is constitutionally defined as a "socialist state guided by the principles of José Julián Martí Pérez, and the political ideas of Vladimir Lenin, Friedrich Engels, and Karl Marx." The constitution describes the Communist Party of Cuba as the "leading force of society and of the state". Cuban government has the executive power and is represented by the Council of State and the Council of Ministers. Cuba has an elected national legislature, the National Assembly of People's Power. The People's Supreme Court is the highest judicial body. The President of Cuba, who is also elected by the Assembly, serves for five years and there is no limit to the number of terms of office. Fidel Castro, who has stepped down in 2008 as Cuba's President, has run the country for 49 years, since 1959 when the communist revolutionary came to power after leading the successful Cuban Revolution. Under his leadership Cuba was converted into a one-party socialist state, with industry and business being nationalised under state ownership and socialist reforms implemented in all areas of society. He adopted Marxism-Leninism as his guiding ideology. By following the path of communism and support from Soviet Union Fidel Castro transformed Cuba economically and socially. With the collapse of the Soviet Union in 1991, Castro had to struggle while Cuba was led into its economic "Special Period". In order to come out of the economic decline Castro formed many alliances with Latin American countries. Subsequently, Cuba under his rule made impressive domestic strides which include universal free healthcare and a much-admired education system, which has produced doctors for the developing world, but the economic growth is still poor. In 2006, amidst failing health, Castro transferred his responsibilities to Vice-President Raul Castro, who then became President in 2008.

KEY POLITICAL PERSONS

PRESIDENT & COMMANDER-IN-CHIEF: Raúl Modesto Castro Ruz

FIRST VICE PRESIDENT: Miguel Díaz-Canel Bermúdez

ECONOMY

The economy of Cuba is centrally planned and controlled by the Government of Cuba. It follows socialist principles in terms of production and employment. The country has recently seen significant foreign investment and participation of private sector. The government controls labour force, capital investment and sets most prices and rations goods. Cuba faced a severe economic downturn in 1990 following the withdrawal of former Soviet subsidies. President Raul Castro after taking over Fidel Castro initiated many significant changes in order to drive Cuba's stagnant, state-run economy. The Cuban sugar economy is the principal agricultural economy in Cuba. The country has authorised the first joint venture in the sugar industry in a bid to boost efficiency, modern technology to revitalise the sector. Tourism in Cuba attracts over 2 million people a year, and is one of the main sources of revenue for the island. However, in recent years Cuba's tourism has decreased due to the economic recession, escalating foreign investment conflicts and fears, and internal economic restrictions. Recently, Cuba and Vietnam hold forums and discussed new opportunities in the areas of tourism and construction. Both are planning to expand economic relations. Cuban cigars rolled from tobacco leaves remain one of the country's leading exports. Recently, US has eased the trade embargo (sanctions which were imposed by US on Cuba since 1962) lifting some travel restrictions and allowing Cuban Americans to send unlimited remittances back home and has also granted a special permit for sending a weekly shipment of humanitarian aid to the Communist-run island.

GDP: $128.5 billion (2014 est.), $126.9 billion (2013 est.)

Real Growth Rate: 1.3% (2014 est.), 2.7% (2013 est.)

Industries: Petroleum, nickel/cobalt, pharmaceuticals, tobacco, construction, steel, cement, agricultural machinery, sugar

Oil - proved reserves: 124 million bbl (1 January 2015 est.)

Natural gas - proved reserves: 70.79 billion cu m (1 January 2014 est.)

Total Exports: $3.903 billion (2015 est.), $5.149 billion (2014 est.)

Export Commodities: Petroleum, nickel, medical products, sugar, tobacco, fish, citrus, coffee

Major Markets: Canada 17.7%, Venezuela 13.8%, China 13%, Netherlands 6.4%, Spain 5.4%, Belize 4.7% (2015)

Total Imports: $13.48 billion (2015 est.), $13.1 billion (2014 est.)

Import Commodities: Petroleum, food, machinery and equipment, chemicals

Major Suppliers: Venezuela 31.8%, China 17.6%, Spain 10%, Brazil 4.8% (2015)

DEFENCE & SECURITY

The Cuban Revolutionary Armed Forces (*Fuerzas Armadas Revolucionarias - FAR*) comprises of Revolutionary Army (*Ejercito Revolucionario, ER*, which includes Territorial Militia Troops (*Milicia de Tropas de Territoriales, MTT*)); Revolutionary Navy (*Marina de Guerra Revolucionaria, MGR*, which includes Marine Corps); Revolutionary Air and Air Defence Forces (*Defensas Anti-Aereas y Fuerza Aerea Revolucionaria, DAAFAR*); and Youth Labour Army (*Ejercito Juvenil del Trabajo, EJT*). The military, intelligence and internal security forces are managed by the Ministry of the Revolutionary Armed Forces (MINFAR) and Ministry of Interior (MININT). The armed forces is the most powerful and influential institution in Cuba with top military generals playing crucial roles in the functioning of the entire system. The Cuban military is currently prepared for two major missions: the defence of Cuba against foreign aggression such as US and the provision of internal security.

KEY DEFENCE PERSONS

MINISTER OF THE REVOLUTIONARY ARMED FORCES: Gen. Leopoldo Cintra Frias

Internal Conflict

Currently there are no reports of non-state armed groups or private military companies in Cuba.

External Conflict

There is currently no armed conflict involving Cuba.

THREAT PERSPECTIVE

USA: Cuba and the United States restored diplomatic relations on July 20, 2015, which had been severed in 1961 during the Cold War. The countries' respective "interests sections" in one another's capitals were upgraded to embassies on July 20, 2015. Since the Cuban Revolution of 1959, relationship between Cuba and United States has deteriorated substantially and been marked by tension and confrontation. Both the countries do not share any formal diplomatic relations and US has maintained a commercial, economic, and financial embargo on Cuba which makes it illegal for US corporations to do any business with them. The United States Interests Section of the Embassy of Switzerland in Havana, Cuba or USINT Havana represents US interests in Cuba. Similarly, the Cuban Interests Section in Washington, DC is the diplomatic mission of the Republic of Cuba to the United States. These respective Interests Sections function as de facto embassies for the two countries. US, at present, have begun implementing a less strict policy towards Cuba. US stated that the country is ready for dialogue and lifting the trade embargo completely if Cuba has political change such as institution of democratic reforms and improving human rights. On July 2012, Cuban President Raul Castro said that his government is willing to mend differences with the US government and sit down to discuss anything, as long as it is a conversation between equals. The US continues to operate a Naval Base at Guantanamo Bay which is leased to US and only mutual agreement or US abandonment of the facility can terminate the lease. The American presence in Guantanamo is a sore point for Havana, which demands the base be shut down. According to a US State Dept. annual report (2011), US have kept Cuba on the blacklist of alleged state sponsors of terrorism for the 30th year running. As per the report, the Cuban govt. has taken no action against the terrorist groups and have provided safe haven to members of illegal armed groups, ETA (armed Basque nationalist and separatist organisation), FARC (Columbia), and other US fugitives. The Government of Cuba responded strongly to the accusations and said that it has maintained a public stance against terrorism and terrorist financing. Available reports indicate that there was no evidence on Cuba giving weapons or paramilitary training to militants. The report also added that, Cuba did not sponsor counter terrorism initiatives or participate in regional or global operations against terrorists in 2010. Cuba, on the other hand, denounced US counterterrorism efforts throughout the world, portraying them as a pretext to extend US influence and power. As per some media reports, Cuba has worked with Venezuela and other countries to intercept and arrest terrorists. In Oct 2013, Cuba has again, for the 22nd consecutive year, asked the United Nations to condemn the United States economic embargo against the island. On 14 Dec 2014, President Obama announced plans to restore diplomatic relations in major policy shift. Obama said the United States will open an embassy in Cuba and relax some of the restrictions on commerce and travel between the United States and Cuba. On 20 March 2016, President Barack Obama visited Cuba, the first by a sitting president in over 80 years.

STRATEGIC RELATIONS

Russia: After 1959, Cuba was increasingly dependent on former Soviet Bloc for economic and military aid. Castro regime built a strong armed force with the help of Soviet equipment and military support. But after the dissolution of the Soviet Union, the assistance was evaporated with deterioration of the relationship between Russia and Cuba and it resulted in the steady decline of the Cuban economy from which it has only partially recovered. With changing times and new governance, the relations improved between both the countries. The two governments also signed many economic agreements for increasing joint cooperation with each other including sending huge amount of humanitarian aid to Cuba. Cuba also politically supported Russia's position in the 2008 South Ossetian war. Russia has also offered Cuba its space technology and would help the latter build its own space centre. An agreement has also been signed in 2009 to explore Cuba's offshore oil deposits, which are believed to be substantial. Russia is also in talks with Cuba to set up naval bases there. Both the countries currently maintain very strong diplomatic relationship. The relations have also been substantially enhanced in the Russian-Cuban military cooperation which has reached a new level. In July 2014, Vladimir Putin also visited Cuba, where he touted a decision to wipe clean 90 percent of the island's USD 35 billion debt to Moscow and announced deals to invest in Cuba's offshore oil industry. The Russian Defence Ministry is considering the question of renewing Russia's presence in the bases in Cuba, according to media reports.

Venezuela: Cuba shares a very warm and significant relationship with Venezuela. Both the governments have taken positive steps to promote investment as well as technical and educational cooperation between the two countries. Cuba and Venezuela have signed many agreements to work together in the area of oil, energy, electricity, agriculture, service, medical, infrastructure, and business joint ventures. Close military ties with Cuba are also helping Venezuela to transform its military as Cuba has decades of experience in military equipment, training & strategy, guerrilla warfare and in counter insurgency operations.

Bolivia: Relationship between Bolivia and Cuba has increased considerably, and Cuba has sent doctors and teachers to Bolivia. Both the countries have also signed agreements on bilateral cooperation and mutual judicial assistance. Bolivia and Cuba had series of visits last year to boost bilateral military relations and continuous exchange between the armed forces. Cuba and Bolivia have enhanced their bilateral relations significantly since Bolivian President Evo Morales took office in 2006, and the two countries have signed a number of cooperative agreements, mostly in the fields of health and education.

China: With the decline of support from former Soviet Union, Cuba had to look for new major allies. China emerged as a new key partner for Cuba thus lifting its diminishing economy. The relations continued to grow between the two countries with focus on trade, credits and investments. China is Cuba's second largest trading partner after Venezuela. China and Cuba have signed many agreements for easing bilateral trade such as opening factories producing local goods, selling wide range of Chinese-made items from bicycles to locomotives and developing oil resources with the aim of providing a boost to Cuba's national infrastructure. Both the countries have recently agreed to promote relations between the two armed forces. Cuba is willing to enhance exchanges with the Chinese military and strengthen bilateral cooperation in personnel training and other areas. China is also working with Cuba to strengthen friendship and enhance cooperation for the sake of national unity and territorial integrity. Cuba and China in August 2016 signed new agreements aimed at deepening bilateral cooperation in a number of fields.

Organisation of American States (OAS): Cuba, despite being a founding member, was excluded from participation in the OAS in January 1962. Cuba was technically still a member but was barred from sending representatives to the OAS. Later in June 2009, the Ministers of Foreign Affairs of the

Americas present for the OAS's 39th General Assembly passed a vote to annul Cuba's suspension from the OAS. The 2009 resolution states that the participation of the Republic of Cuba in the OAS will be the result of a process of dialogue initiated at the request of the Government of Cuba, and in accordance with the practices, purposes, and principles of the OAS. Cuba, though welcomed the decision as a historically significant, the country has confirmed that it will not return to the OAS. The move was published in an editorial by Granma, official newspaper of the Central Committee of the Cuban Communist Party.

Multilateral Alliances

Cuba is a founding member of the Bolivarian Alliance for the Peoples of Our America (ALBA), which is an international cooperation organisation based on the idea of social, political, and economic integration between the countries of Latin America and the Caribbean. Cuba is also a member of United Nations (UN) but is not contributing personnel to UN peacekeeping operations. The country is also a member of the Non-Aligned Movement (NAM) and the Association of Caribbean States (ACS). Ties between the nations of the Caribbean Community (CARICOM) and Cuba have remained cordial.

DEFENCE CAPABILITIES

ARMY: Revolutionary Army (*Ejercito Revolucionario, ER*) consists primarily of armour and artillery units. There are three armies: the eastern, central, and western. The army composed of ground units structured into brigades, regiments and infantry, tank, artillery and air defence battalions, the battalions of regular troops and the Territorial Troops Militia.

Personnel: 145,000 (Include 60,000 conscripts)

Equipment

Category	Name	In Service
MBT	T-62	380
	T-54/55	800
	PT-76	50
	T34 (coastal defence)	400
APC/MICV/ IFV	BRDM-1	50
	BRDM-1	100
	BMP-1	120
	BTR 40/ 50/ 60/ 152	
	76 mm	
	85 mm	
	122 mm	1200
	130 mm	
	152 mm	
Mortar	82mm M-41/ 43	1000
	120 mm M-38/ 43	

Army continued...

Category	Name	In Service
MRL	122mm BM-21	175
	BM-14	
	FROG - 4/ 7	70
RCL	57mm	
Anti Tank GW	AT-1, AT-3	
	9K111	
	D-44, T-12	
	SU-100 (tank destroyer)	100
AA Gun	ZU-23-2	
	ZPU-4	
	ZSU-23-4	
	ZSU-57-2	
	S-60	
	M-1939	
SAM	SA-6	12
	SA-7	
	SA-8	16
	SA-9	59
	SA-13	41
	SA-14	
	SA-16	
	S-75	145
	S-125	59

New Procurements/ Upgrades

No major procurement plans are under consideration at present.

Cuban soldiers during a march past.

NAVY: Revolutionary Navy (Marina de Guerra Revolucionaria) is the naval force of Cuba.

Personnel: 13,500 (8,500 conscripts)

Equipment

Category	Name	In Service
Submarine	Delfin class (midget)	4
Frigate	Rio Damuji class	2
Corvette	Soviet PAUK type	1
Light Force	OSA I-II missile FAC	6
Mine warfare force	Yevgenya class inshore	5
	Sonya coastal	3
Patrol boat	Pauk II	1
SSM	P-15 Termit (Coastal defence)	

Naval Aviation

Category	Name	In Service
Aircraft	Ka-28	4
	Mi-4	4
	MiG-29	6

New Procurements/ Upgrades

No major procurement plans are under consideration at present.

AIR FORCE: Revolutionary Air and Air Defence Forces (Defensas Anti-Aereas y Fuerza Aerea Revolucionaria, DAAFAR) is the air force of Cuba.

Active Naval & Air bases: Cabanas, Holguin, Havana – Playa Baracoa Airport, Havana- Jose Marti International Airport, La Coloma Airport.

Personnel: 22,000 (11,000 conscripts)

Equipment

Category	Name	In Service
Fighter/Attack/Bomber	MiG-21	12
	MiG-23	24
	MiG-29	3
Transport/Tanker	An-26	2
Trainer	Aero L-39	25
Helicopter	Mi-8/17	10
	Mi-35	4

New Procurements/ Upgrades

No major procurement plans are under consideration at present.

Major Armament Producers

Domestic suppliers: Union de Industrias Militares

Foreign suppliers: Russia, China, North Korea, Kazakhstan, Mongolia, Spain, Soviet Union (Former)

Union of Military Industry (Unión de Industria Militar - UIM), started in the year 1998, is the small industry owned by the Cuban government and run by the military of the country. The main task of the organisation is to provide repair, maintenance and upgrade for the military equipments.

The organisation is made up of seven research and development centres, 12 military-industrial enterprises and two scientific technological centres. Marta Abreu de Las Villas Central University at Villa Clara; Doctor Carlos J. Finlay Central Military Hospital; José Martí Military Technical Institute and Simulator Research and Development Centre (SIMPRO) are the research centers involved in UIM. There are 12 military-industrial enterprises. There is some limited production of infantry weapons, such as AKM assault rifles, automatic grenade launchers and RPGs, munitions, land mines and support equipment. The military-industrial complex works for the repair of the weaponry and technology for the ground, air and naval units of the Revolutionary Armed Forces. They are also involved in the modernisation of the forces in order to keep pace with the world technological development. The organisation has series of factories devoted for production of light weapons for infantry, ammunition, mines and other equipment of diverse nature. There are also large workshops, repair bases specialising in tanks, artillery, aviation, navy, radio communications, transport, radars, missile systems, metallurgical products and others.

DEFENCE DEALS

Cuba used to receive huge support in the form of military and financial aid from former USSR. The support was supplied to all the three branches of armed forces. After 1990s, Cuban military power was sharply reduced by the loss of Soviet subsidies. Cuba was totally dependent on indigenous production for the maintenance of existing inventory. In terms of development, Cuban defence industry is only capable of producing small arms like the AKM, RPG-7V and sniper rifles, plus munitions which includes large maintenance and repair capabilities. The army unit of Cuban Revolutionary Armed Forces is equipped with all Russian-made equipment, but most of the weaponry has crossed their time limit and are ageing. There is also shortage of workable spare parts and the country lacks hard currency to buy more specially materials for the tanks, motor transport, and costly munitions anti-tank, anti-aircraft, towed and self-propelled artillery stocks. The scenario has compelled Cuban government to stretch itself to other resources available and strengthening relations with nations believing in their ideology. In Sept 2012, a Cuban senior general agreed to further deepen military cooperation with China during a visit to Beijing. He said that Cuba was willing to enhance exchanges with the Chinese military and strengthen bilateral cooperation in personnel training and other areas. Cuba already share strong relationship with Venezuela.

DEFENCE BUDGET

Total defence spending: Exact defence expenditure figures are unavailable.

Estimated defence spending in terms of GDP: 3.8%

The accurate statistics of Cuban defence budget is not disclosed by the country since 2003. The last estimated budget of the nation was said to be USD 1.9 billion for 2009. Since 2001, military budgets may have increased by up to 400 per cent compared to the dark days of the mid-1990s. Cuban military was deprived of logistic support, arms and ammunitions and financial aid since the disintegration of USSR. The defence expenditure before was around USD 1.4 billion but later dropped to USD 600 million following the collapse of Soviet Union. This sudden change had a deep impact in the economy and the state of Cuban military equipment. Currently, though the exact data is not available, Cuba has increased its defence spending and in 2004 also conducted military exercise at heavy cost. The Cuban army at present remains well trained and professional in nature. However, lack of replacement parts for its existing equipment has increasingly affected operational capabilities. The increase in defence expenditure since 2001 which was very low in the 90s is due to the broadening of horizons of the armed forces and also participating in the commercial markets. Cuba in order to increase the funding of its military operations and controlling the economy is concentrating more on tourism, agriculture which includes sugarcane productions, trade and other varied activities.

CONCLUSION

The armed forces have long been the most powerful institution in Cuba. The military controls 60 percent of the economy through the management of hundreds of enterprises in key economic sectors. Fidel Castro with a rebel armed forces led Cuba's communist revolution to victory in 1959. But with the collapse of Soviet Union in 1991, Castro's rule was severely tested as it went through major downgrade in terms of military and financial aid. It led to a period of huge economic crisis which is known in Cuba as the Special Period. However, the period radically transformed Cuban society and the economy, as it necessitated the successful introduction of sustainable agriculture, decreased use of automobiles, and revamped industry, health, and diet throughout the country. Cuba which was comparatively isolated in the 1990s, has renewed its bilateral cooperation with several South American countries, most notably Venezuela and Bolivia. The country has also developed a growing relationship with the People's Republic of China and Russia. Cuba's armed force, though facing severe shortages of fuel and replacement of parts for its existing equipment, still has the capabilities to give strong resistance to any regional power. In 2006, the power was transferred from Fidel Castro to Raul Castro and with that many changes took place in the political arena of the country including foreign relations. Raul Castro's reforms include the right to buy and sell private property, but so far there has been no migratory reform. The process of reform might have stepped up a gear but is moving in snail pace. With much developments going on all around the region, transition is inevitable in Cuba. The Cuban Revolutionary Armed Forces which controls most of Cuban economy will play a key role in the transition. FAR has long been the strong, influential and competent institution in Cuba. The military, with a strong revolutionary tradition, still plays a pivotal role as a decisive power in the functioning of the country. Raul Castro therefore has significant challenges to deal with including striking a balance between improving of economy and maintaining political and military control.

CONTACT DETAILS

Ministry of the Revolutionary Armed Forces
Plaza de la Revolucion
Havana, Cuba
Tel: +537-511151
http://www.cubagob.cu/

CYPRUS
(Capital: Nicosia)

INTRODUCTION

Area: 9,251 sq km
Population: 1,205,575 (July 2016 est.)
Coastline: 648 km
Maritime claims: Territorial sea: 12 nm
Contiguous zone: 24 nm
Continental shelf: 200 m depth or to the depth of exploitation

KEY POLITICAL PERSONS

PRESIDENT: Nikos Anastasiadis

KEY DEFENCE PERSONS

DEFENCE MINISTER: Christoforos Fokaidis

PERMANENT DEFENCE SECRETARY: Christos Malikkides

COMMANDER IN CHIEF: Lt. Gen. Georgios Mpasiakoulis

CHIEF OF STAFF: Brig. Gen. Georgios Karagiannis

ECONOMY

The area of the Republic of Cyprus under government control has a market economy. The World Bank has classified Cyprus as a high-income economy. The country entered the Euro zone on 1 January 2008 and adopted the Euro as its official currency replacing the Cypriot pound. A major economic crisis hit Cyprus in 2013. In order to boost the economy , the government has introduced a series of measures including attract investment through modernising legislation, promoting development projects, diversifying tourism, introducing tax incentives and speeding up licensing procedures. Recently the discovery of natural gas resources has further upgraded the strategic potential of the island. Tourism, financial services, and real estate have traditionally been the most important sectors. The economy of Northern Cyprus is about one-fifth the size of the economy of the government-controlled area, while GDP per capita is around half. The economy mainly revolves around the agricultural sector and government service. The northern economy experienced growth estimated at 2.8% in 2013 and 2.3% in 2014 and is projected to grow 3.8% in 2015.

GDP: $28.06 billion (2015 est.); $27.62 billion (2014 est.)

Real Growth Rate: 1.6% (2015 est.); -2.5% (2014 est.)

Industries: Tourism, food and beverage processing, cement and gypsum production, ship repair and refurbishment, textiles, light chemicals, metal products, wood, paper, stone and clay products

Total Exports: $2.759 billion (2015 est.); $3.751 billion (2014 est.)

Export Commodities: Citrus, potatoes, pharmaceuticals, cement, clothing

Major Markets: Greece 10.9%, Ireland 10.2%, UK 7.2%, Israel 6% (2015)

Total Imports: $6.286 billion (2015 est.); $7.473 billion (2014 est.)

Import Commodities: Consumer goods, petroleum and lubricants, machinery, transport equipment

Major Suppliers: Greece 25.7%, UK 9.1%, Italy 8%, Germany 7.5%, Israel 5.5%, China 4.8%, Netherlands 4.1% (2015)

INTERNATIONAL DISPUTES

Hostilities in 1974 divided the island into two de facto autonomous entities - the internationally recognized Cypriot Government and a Turkish-Cypriot community (north Cyprus). The UN Peacekeeping Force in Cyprus (UNFICYP) maintains the buffer zone between north and south. Cyprus entered the European Union on 1 May 2004 still divided, with the EU's body of legislation and standards (*acquis communitaire*) suspended in the north. Turkey protests Cypriot Government creating hydrocarbon blocks and maritime boundary with Lebanon in March 2007.

DEFENCE

CYPRIOT NATIONAL GUARD

The Cypriot National Guard also known "National Guard" is the combined arms military force of the Republic of Cyprus. This force consists of Air, Land, Sea and Special Forces. Military service is mandatory for males.

Personnel: 12,000 (50,000 reserve)

Equipment

Category	Name	In Service
MBT	T-80U/UK	80
	AMX-30	82
AIFV/APC	BMP-3	43
	VAB	126
	Leonidas I	168
	EE-9 Cascavel	124
Artillery	F1 155mm towed	12
	Zuzana 155mm SP	12
	M56 105mm	72
	100mm AT guns	20
MRL	M-68 128mm towed	18
	BM-21 122mm	4
Mortar	RT61 120mm	112
ATGW	Milan	45
	Hot	70
SAM	Skyguard/Aspide system	12
	TOR-M1	6
	Mistral	30
	SA-10	

New Procurements / Upgrades

No major procurement plans are under consideration at present.

Air Service

Category	Name	In Service
Helicopter	Mi-35P	11
	SA-342 Gazelle/Hot	4
	AW139 (SAR)	2

New Procurements / Upgrades

No major procurement plans are under consideration at present.

Naval Service

Personnel: 320

Equipment

Category	Name	In Service
Patrol craft		4
ASM	MM40 Exocet	24

New Procurements / Upgrades

✦ Plans to purchase two patrol boats from Israel have been approved.

✦ Cyprus also plans to purchase another two 50m-long vessels from an EU country. Negotiations for it is going on.

Defence Expenditure

Total defence spending: $347 million (2015)

Estimated defence spending in terms of GDP: 1.5%

Defence Production and R&D

An indigenous defence industry, Saft Nife Middle East Ltd. has been set up. Levels of research and development have remained nominal.

Defence Procurement

Priorities were to mechanise its infantry forces with modern equipment, and to build up its holdings of Main Battle Tanks. Other objectives included acquisition of combat helicopters, artillery systems and an advanced air defence system. Russia has become an important supplier of arms.

CONTACT DETAILS

Ministry of Defence
4, Emmanuel Roides Street,
1432 Nicosia
Tel: +357 22807500
E-mail: defence@mod.gov.cy
Website: http://www.mod.gov.cy

CZECH REPUBLIC
(Capital: Prague)

INTRODUCTION

Area: 78,867 sq km

Population: 10,644,842 (July 2016 est.)

Coastline: 0 km (Landlocked)

Maritime claims: None

KEY POLITICAL PERSONS

PRESIDENT & COMMANDER-IN-CHIEF: Milos Zeman

PRIME MINISTER: Bohuslav Sobotka

KEY DEFENCE PERSONS

DEFENCE MINISTER: Martin Stropnicky

CHIEF OF GENERAL STAFF OF ARMED FORCES: Gen. Josef Becvar

LAND FORCES COMMANDER: Maj. Gen. Stefan Kaleta

AIR FORCE COMMANDER: Brig. Gen. Jaromir Sebesta

ECONOMY

The Czech Republic is one of the most stable and prosperous economies among the post-Communist states of Central and Eastern Europe. Maintaining an open investment climate has been a key element of Czech Republic's transition from a communist, centrally planned economy to a functioning market economy. As a member of the European Union, with an advantageous location in the centre of Europe and a well-qualified labour force, the Czech Republic is an attractive destination for foreign investment. The country's service sector is the largest contributor to the GDP followed by industry. Prior to its EU accession in 2004, the Czech government harmonised its laws and regulations with those of the European Union. The conservative Czech financial system remained relatively healthy throughout 2009. Nevertheless, the real economy contracted by 4.1% in 2009, mainly due to a significant drop in external demand as the Czech Republic's main export markets fell into recession.

The economy made a recovery in the following years but was adversely affected by the Euro Zone crisis coupled with falling household spending and declining exports. The economy slowly recovered in the second half of 2009 and registered weak growth in the next two years. In 2012 and 2013, however, the economy fell into a recession again, due both to a slump in external demand in the EU and to the government's austerity measures, returning to weak growth in 2014, and stronger growth in 2015.

GDP: $332.5 billion (2015 est.); $319 billion (2014 est.)

Real Growth Rate: 4.2% (2015 est.); 2% (2014 est.)

Industries: Motor vehicles, metallurgy, machinery and equipment, glass, armaments

Exports: $131 billion (2015 est.); $146.6 billion (2014 est.)

Export Commodities: Machinery and transport equipment, raw materials and fuel, chemicals

Major Markets: Germany 32.4%, Slovakia 8.4%, Poland 6%, UK 5.1%, France 5.1%, Austria 4.4% (2014)

Imports: $122.5 billion (2015 est.); $135.9 billion (2014 est.)

Import Commodities: Machinery and transport equipment, raw materials and fuels, chemicals

Major Suppliers: Germany 30%, Poland 9%, China 8.3%, Slovakia 6.6%, Netherlands 5%, Austria 4.1% (2015)

INTERNATIONAL DISPUTES

✦ Dispute continues over a nuclear power plant in Temelin bordering Austria.

DEFENCE

ARMY

Personnel: 5,500+

Equipment

Category	Name	In Service
MBT	T-72	123 (30 operational)
APC/IFV	Pandur II & BMP-2	200+
	Dingo-2	20
Artillery	SPGH-M77 Dana 152mm self-propelled howitzer	
Mortar	M1982 Pram-L 120mm towed	50+
	SPM-85 Pram-S 120mm self-propelled	8
ATGW	Fagot	
	Spike	
Air Defence Systems	RBS-70 Bofors	
	2K12 Kub-M2	
	9K35 Strela-10M	
	9M32 Strela-2	
	9K33M (OSA-AKM)	

New Procurements/ Upgrades

✦ Plans are on to replace the BMP-2 tracked infantry vehicles.

✦ Delivery of small arms, including CZ 805

Bren attack rifles, CZ 75 Phantom pistols and CZ Scorpion sub-machine guns, are going on.

- ✦ Acquisition of 12 new multi-role helicopters is under consideration with a selection likely to be finalized by 2017.
- ✦ Javelin anti-tank missiles are on order from the US.

AIR FORCE

Personnel: 6,000+

Equipment

Category	Name	In Service
Fighter/ Attack/ Reconnaissance	JAS 39 Gripen	14
	L-159 ALCA	24
Transport	L-410, Yak-40, CL-601 Challenger, A-319 CJ, CASA C-295M	15
Trainer	Aero L-39 Albatros	9
Helicopter	Mi-35/24V (attack)	17
	Mi-8, Mi-17/ Mi-171S (transport)	25
	W-3A Sokol (utility)	10
UAV	Raven RQ-11B	2

New Procurements/ Upgrades

- ✦ The Defence Ministry has announced to extend the country's lease of SAAB JAS 39 Gripen fighters with Sweden until 2027. The 14 Gripen jets – two JAS 39D two-seat jets and 12 JAS 39C jets – were acquired from Sweden under a 10-year agreement signed in 2004. All fighters will undergo a weapons system upgrade to enhance their ground-attack capability under a contract signed with Sweden's FXM.
- ✦ Plans are on to retire the 17 Russian-made Mil Mi-24 'Hind' attack helicopters and most of the fleet of 25 Russian-made Mil Mi-8/17 'Hip' medium transport helicopters in the coming years.
- ✦ Acquisition of new light helicopters, up to 30-35, is being considered.
- ✦ All 10 Polish-made PZL Swidnik W-3A Sokol light helicopters are undergoing mid-life extension upgrades.
- ✦ Acquisition of two new transport aircraft is being considered to replace the Yak-40 planes.
- ✦ A contract has been signed with Russia for upgrading the Mi-8 fleet of helicopters. The upgrades will start from 2016-end.

Defence Expenditure

Total defence spending: CZK 43.5 billion (2015); CZK 41.99 billion [US$ 1.64 billion] (2014)

Defence spending in terms of GDP: 1.09% (2015)

The country's annual defence spending is designed to increase so as to reach CZK 71.5 billion by 2020. Major suppliers of military hardware to Czech Republic are Russia, US and Europe.

Czech Republic signed an agreement with Vietnam in August 2014 to collaborate in military technologies, training, maintenance and repair, and wider defence industrial activity. It is one of the biggest suppliers of military hardware to Vietnam.

Defence Production and R&D

The defence industry, which once thrived during the era of Soviet Union, could not survive following the collapse of former Czechoslovakia in 1993. The defence industry of Czech Republic today is a small fraction of what was once a competitive sector in the economy. The Czech government intends to privatise the defence sector.

Defence Procurement

With the armed forces' inventory replete with Soviet-era equipment, the Czech Republic has not done any major procurement in recent times. The country has been cooperating with Slovakia in producing some defence equipment. Efforts are on to dispose the obsolete weapons and platforms and acquire new systems in the coming years. As part of a modernisation drive, the country plans to increase recruitment of soldiers so that within five years, the Czech Armed Forces could comprise of approximately 27,000 professional soldiers and 5,000 active reservists.

CONTACT DETAILS

Ministry of Defence

Tychonova 1, CZ 160 01 Prague 6

Czech Republic

Tel. +420 973 201 111

E-mail: info@army.cz

Website: http://www.army.cz

DEMOCRATIC REPUBLIC OF CONGO (Capital: Kinshasa)

INTRODUCTION

Area: 2,344,858 sq km

Population: 81,331,050 (July 2016 est.)

Coastline: 37 km

Maritime claims: Territorial sea: 12 nm

Exclusive Economic Zone: Since 2011, the DRC has a Common Interest Zone agreement with Angola for mutual development of off-shore resources

KEY POLITICAL PERSONS

PRESIDENT & COMMANDER-IN-CHIEF: Joseph Kabila

PRIME MINISTER: Samy Badibanga

KEY DEFENCE PERSONS

MINISTER OF DEFENCE: Crispin Atama Tabe

CHIEF OF GENERAL STAFF: Lt. Gen. Didier Etumba Longila

ECONOMY

The Democratic Republic of the Congo is home to a vast potential of natural resources and mineral wealth. Agriculture is the mainstay of the economy. The economy of the third largest country in Africa relies heavily on mining. However, much economic activity occurs in the informal sector and is not reflected in GDP data. The economy was adversely affected by the global economic slowdown in 2008, but later recovered and achieved a GDP growth of 6 to 7 percent in 2010-11. Renewed activity in the mining sector, the source of most export income, has boosted Kinshasa's fiscal position and GDP growth in recent years. The country marked its thirteenth consecutive year of positive economic expansion in 2015.

GDP: $62.87 billion (2015 est.); $58.35 billion (2014 est.)

Real Growth Rate: 7.7% (2015 est.); 9.2% (2014 est.)

Industries: Mining (copper, cobalt, gold, diamonds, coltan, zinc, tin, tungsten), mineral processing, consumer products (textiles, plastics, footwear, cigarettes), metal products, processed foods and beverages, timber, cement, commercial ship repair

Total Exports: $10.35 billion (2015 est.); $12.32 billion (2014 est.)

Export Commodities: Diamonds, copper, gold, cobalt, wood products, crude oil, coffee

Major Markets: China 43.5%, Zambia 25%, South Korea 4.9%, Belgium 4.8% (2015)

Total Imports: $10.46 billion (2015 est.); $11.96 billion (2014 est.)

Import Commodities: Foodstuff, mining and other machinery, transport equipment, fuel

Major Suppliers: China 20.6%, South Africa 17.7%, Zambia 12.3%, Belgium 6.9%, Zimbabwe 5.1%, India 4.7% (2015)

INTERNATIONAL DISPUTES

✦ Members of Uganda's Lord's Resistance Army forces continue to seek refuge in Congo's Garamba National Park as peace talks with the Uganda government evolve.

✦ The location of the boundary in the broad Congo River with the Republic of the Congo is undefined except in the Pool Malebo/Stanley Pool area; Uganda and DROC dispute Rukwanzi island in Lake Albert and other areas on the Semliki River with hydrocarbon potential.

✦ A Boundary commission continues discussions over the Congolese-administered triangle of land on the right bank of the Lunkinda River claimed by Zambia near the DROC village of Pweto.

DEFENCE

ARMY

Personnel: 10,000

Equipment

Category	Name	In Service
MBT	T 72	20
	T-55	12
	Type 59	10+
	Type 62	40+
Armoured vehicles	AML-90	40
	AML-30/60	50+
APC	BMP-1	20
	M-113	10
	M-3	70
	YW-531	10
Artillery	M116 75mm	20+
	Type-56 85mm	20
	M1938/D-30 122mm	20
	Type-60 122mm	10+
	Type-59 130mm	8
Mortar	81mm	
	120mm	
	M-30 107mm	
MRL	BM-21 122mm	6
RCL	M18 57mm	
	M20 75mm	
	M40A1 106mm	
AA gun	ZPU-4 14.5mm	
	M1939/Type-63 37mm	
	L/60 40mm	

New Procurements/Upgrades

✦ Ukraine's Ukroboronprom has signed a contract to deliver 50 T-64BV-1 main battle tanks worth about 100 million hryvnia (US $11.5 million) to an unnamed customer, reportedly believed to be the DRC.

NAVY

Personnel: 2,000+ (including marines)

Equipment

Category	Name	In Service
Patrol Craft	Ex-Chinese Shanghai II class	2

New Procurements/Upgrades

No major procurement plans are under consideration at present.

AIR FORCE

Personnel: 1,200

Equipment

Category	Name	In Service
Combat	Su-25	8
	MB-326GB	8
	MB-326K	6
Transport	DHC-5D Buffalo	3
	C-130H Hercules	5
	C-47	6

Air Force continued...

Category	Name	In Service
Transport	BN-2A Islander	1
Helicopter	SA-330 Puma	4
	SA-332 Super Puma	1
	Alouette III	7
	Mi-2 Hoplite	3
	Mi-8	4
	Mi-24	
	Mi-35	
Trainer	Cessna 150	12
	SF-260 MZ	10
	T-6	10
	Cessna 310	3

New Procurements/Upgrades

No major procurement plans are under consideration at present.

Defence Expenditure

Estimated defence spending in terms of GDP: 1.2%

Defence Production and R&D

No defence production is undertaken in Democratic Republic of the Congo.

Defence Procurement

No confirmed data or information is available on the current acquisition programmes by the DRC military.

A number of countries, including Ukraine, China, US, France, Egypt and South Africa have supplied arms and equipment to DRC in the past after the United Nations Security Council, in 2008, decided to lift its arms embargo imposed on the African nation since 2003 due to continued armed conflict in the country. However, in the wake of continuing armed violence, the UNSC has been renewing the embargo from time to time, and it is now been extended until 1 July 2016.

CONTACT DETAILS

Office of the Prime Minister
Gombe Avenue
Hotel du Conseil Executif
Kinshasa
Tel: (+243 12) 308 92
(+243 12) 309 79

DENMARK
(Capital: Copenhagen)

INTRODUCTION

Area: 43,094 sq km

Population: 5,593,785 (July 2016 est.)

Coastline: 7,314 km

Maritime claims: Territorial sea: 12 n

Contiguous zone: 24 nm

Exclusive Economic Zone: 200 nm

Continental shelf: 200 m depth or to the depth of exploitation

KEY POLITICAL PERSONS

HEAD OF STATE: Queen Margrethe II

PRIME MINISTER: Lars Lokke Rasmussen

KEY DEFENCE PERSONS

MINISTER OF DEFENCE: Claus Hjort Frederiksen

CHIEF OF DEFENCE FORCES: Gen. Peter Bartram

CHIEF OF ARMY: Maj. Gen. Hans Christian Mathiesen

CHIEF OF NAVY: Rear Adm. Frank Trojahn

CHIEF OF AIR FORCE: Maj. Gen. Max Arthur Thorso Nielsen

ECONOMY

A highly modern market economy, Denmark benefits from a high-tech agricultural sector and state-of-the-art industry driven by world-leading firms in pharmaceuticals, maritime shipping and renewable energy. It also depends on foreign trade to a great extent. The country's enjoys a fiscal position in the European Union. Denmark is a net exporter of food and energy and enjoys a comfortable balance of payments surplus. The country is experiencing a lackluster economic recovery, having still not regained the GDP level of 2008. GDP contracted in 2012 and 2013, followed by real growth of 1.3% in 2014, and 1.2% in 2015. The government has projected 1.9% growth in 2016, while private sector estimates are about 1% growth. A historically low level of unemployment rose with the economic downturn but the labour market has strengthened since 2013, and unemployment stood at about 4.5% in early 2016, based on the national measure. Productivity growth was significantly below the OECD average in 2012-2014.

GDP: $258.7 billion (2015 est.); $255.7 billion (2014 est.)

Real Growth Rate: 1.2% (2015 est.); 1.3% (2014 est.)

Industries: Iron, steel, nonferrous metals, chemicals, food processing, machinery and transportation equipment, textiles and clothing, electronics, construction, furniture and other wood products, shipbuilding and refurbishment, windmills, pharmaceuticals, medical equipment

Total Exports: $95.97 billion (2015 est.); $111.4 billion (2014 est.)

Export Commodities: Machinery and instruments, meat and meat products, dairy products, fish, pharmaceuticals, furniture, windmills

Major Markets: Germany 17.8%, Sweden 11.6%, US 8.4%, Norway 6.3%, UK 6.3%, Netherlands 4.4%, China 4.2% (2015)

Total Imports: $85.02 billion (2015 est.); $101.3 billion (2014 est.)

Import Commodities: Machinery and equipment, raw materials and semi-manufactures for industry, chemicals, grain and foodstuff, consumer goods

Major Suppliers: Germany 20.4%, Sweden 12.3%, Netherlands 8.1%, China 7.3%, Norway 6.1%, UK 4.4% (2015)

INTERNATIONAL DISPUTES

- Dispute with Iceland, UK, and Ireland on Denmark's sovereignty claim over the Faroe Islands situated between the Norwegian Sea and the North Atlantic Ocean continues.
- Sovereignty dispute with Canada over Hans Island in the Kennedy Channel between Ellesmere Island and Greenland.

DEFENCE

ARMY

Personnel: 8,400

Equipment

Category	Name	In Service
MBT	Leopard 2A5	57
APC/ IFV/ Other Vehicles	M113	200+ (to be retired)
	CV90 35 MkIII	45
	Piranha IIIC	90
	Eagle IV	30
Artillery	M109 155 mm self-propelled howitzer	12
Mortar	60mm	90
	120mm	20
RCL	Carl Gustaf	774

Army Aviation

Category	Name	In Service	On Order
Helicopter	Fennec AS550	8 (4 in reserve)	

Army aviation continued...

Category	Name	In Service	On Order
Helicopter	EH-101 Merlin	14	
UAV	RQ11B Raven		12

New Procurements/ Upgrades

✦ Plans are on to replace the existing M109A3 155mm self-propelled howitzers with new-generation weapons.

✦ The Defence Ministry has decided to acquire a total of 309 Piranha 5 armoured personnel carriers at a cost of US$600 million to replace the existing fleet of M113 APCs. Delivery of the new vehicles will commence in 2018 and continue till 2023.

✦ 14 Heavy Tactical Recovery Vehicles (HTRV) have been acquired from Rheinmetall MAN Military Vehicles.

NAVY

Personnel: 1,125

Equipment

Category	Name	In Service
Frigate	Iver Huitfeldt-class	3
	Absalon class (support ship)	2
	Thetis-class	4
Patrol vessel	Knud Rasmussen class	2 + 1 on order
	Agdlek class	1
	Diana class	6
	Flyvefisken class (mine-clearing & combat roles)	2
Auxiliary ship	Icebreakers	3
	ECO-101, Mette & Gunnar Thorson-class environmental control vessels	7
	Svanen-class sail training ships	2

Naval Aviation

Category	Name	In Service
Helicopter	Lynx	8

New Procurements/ Upgrades

✦ The first MH-60R Seahawk helicopter, being procured from the US Navy, has been delivered in June 2016. The Defence Ministry is procuring nine MH-60R Seahawk helicopters from the US. The new ship-based rotorcraft are to replace the Danish Navy's Lynx helicopters. All nine helicopters are slated to be delivered by mid-2018.

MH-60R Seahawk

AIR FORCE

Personnel: 2,500

Equipment

Category	Name	In Service
Fighter	F-16A/B	45 (30 operational)
Transport	C-130J Hercules	4
	CL-604 Challenger	3
Trainer	MFI-17 Supporter	27
Helicopter	AS 550 Fennec	8
	EH101 Merlin	14
	Lynx	8

New Procurements/ Upgrades

✦ In June 2016, the Denmark Government officially agreed to buy 27 F-35A joint strike fighters to replace its fleet of aging F-16s. The Lockheed Martin-made fighter will be acquired at an estimated cost of 20 billion Danish crown (US$3.07 billion).

Defence Expenditure

Total defence spending: DKK 20.9 billion (2015)

Estimated defence spending in terms of GDP: 1.2%

Being a NATO member, Denmark receives majority of its military equipment from the international block, mostly from the US and also from other Nordic countries.

Defence Production and R&D

The Danish defence industry has been primarily associated with the production of electronics, avionics, communications and radar systems. Terma and Systematic are Denmark's two leading defence manufacturers in the country.

Defence Procurement

The Danish Government has not announced any fresh procurement plans for its defence forces in recent years. Its main focus has been on consolidating the reform measures started in the last five years and train the personnel to use the new equipment before making them operational. Procurement of new armoured vehicles and other communication equipment for the Danish Army has been finalised. For the Navy, the Defence Ministry has announced procurement of new ship-borne helicopters from the US to replace the Lynx helicopters.

The country also plans to replace the Air Force's aging F-16 fighters' fleet with new advanced warplanes even as the MoD has recommended extending the lifespan of the F-16s to keep them operational for some more years. Under a national defence reforms plan spanning between 2013 and 2017, Denmark has also decided to reduce annual conscription from 5,000 to 4,200 and close some military bases. The country is aiming to restructure its Army, Navy and Air Force by making them leaner and agile so as deploy them for international missions alongside national operations.

Iver Huitfeldt Class Frigate

CONTACT DETAILS

Ministry of Defence

Holmens Kanal 42

1060 Copenhagen K

Phone: + 45 7281 0000

E-mail: fmn@fmn.dk

Website: www.fmn.dk

DOMINICAN REPUBLIC
(Capital: Santo Domingo)

INTRODUCTION

Area: 1,138,914 sq km

Population: 10,606,865 (July 2016 est.)

Coastline: 3,208 km (Caribbean Sea 1,760 km, North Pacific Ocean 1,448 km)

Maritime claims: Territorial sea: 6 nm Contiguous zone: 24 nm

Exclusive Economic Zone: 200 nm Continental shelf: 200 nm or to the edge of the continental margin

KEY POLITICAL PERSONS

PRESIDENT & COMMANDER-IN-CHIEF: Danilo Medina Sanchez

VICE PRESIDENT: Margarita Cedeno de Fernandez

KEY DEFENCE PERSONS

MINISTER OF ARMED FORCES: Maximo William Munoz Delgado

CHIEF OF STAFF OF ARMY: Maj. Gen. Braulio Alcantara A. Lopez

CHIEF OF STAFF OF NAVY: Vice Adm. Miguel Enrique Pena Acosta

CHIEF OF STAFF OF AIR FORCE: Maj. Gen. Louis Napoleon Payan Diaz

ECONOMY

Dominican Republic is an upper middle-income developing country primarily dependent on agriculture, mining, trade, and services. The Dominican Republic has long been viewed primarily as an exporter of sugar, coffee, and tobacco, but in recent years the service sector has overtaken agriculture as the economy's largest employer, due to growth in telecommunications, tourism, and free trade zones. The economy is highly dependent upon the US, the destination for nearly 60% of exports. Remittances from the US amount to about a tenth of GDP, equivalent to almost half of exports and three-quarters of tourism receipts. The country suffers from marked income inequality. The poorest half of the population receives less than one-fifth of GDP, while the richest 10% enjoys nearly 40% of GDP. High unemployment and underemployment remains an important long-term challenge. The Central America-Dominican Republic Free Trade Agreement (CAFTA-DR) came into force in March 2007, boosting investment and exports and reducing losses to the Asian garment industry. The growth of the Dominican Republic's economy rebounded in 2010-11 from the global recession, and remains one of the fastest growing in the region. The Dominican Republic has registered the fastest economic growth in Latin America during 2014-15 with low, stable inflation. The Dominican Republic belongs to a number of the same international organisations, including the United Nations, Organisation of American States, International Monetary Fund, World Bank, and World Trade Organisation.

GDP: $149.7 billion (2015 est.), $139.9 billion (2014 est.)

Real Growth Rate: 7% (2015 est.), 7.3% (2014 est.)

Industries: Tourism, sugar processing, ferronickel and gold mining, textiles, cement, tobacco

Total Exports: $9.523 billion (2015 est.), $9.899 billion (2014 est.)

Export Commodities: Ferronickel, sugar, gold, silver, coffee, cocoa, tobacco, meats, consumer goods

Major Markets: US 42.5%, Haiti 16.5%, Canada 8.1%, India 4.8% (2015)

Total Imports: $16.86 billion (2015 est.), $17.27 billion (2014 est.)

Import commodities: Foodstuffs, petroleum, cotton and fabrics, chemicals and pharmaceuticals

Major suppliers: US 42%, China 9.2%, Venezuela 5.6%, Trinidad and Tobago 4.5%, Mexico 4.4% (2015)

INTERNATIONAL DISPUTES

There is no ongoing armed conflict involving the Dominican Republic.

DEFENCE

ARMY

Personnel: 45,800

Equipment

Category	Name	In Service
MBT	AMX13 light	2
	M41A1 light	6
APC/MICV /IFV	AMLs	20
	V150 Commandos	8
	M3A1 106mm	35
Artillery	M101 105mm Howitzers	6
	105mm M-45 Bofors	20
	Reinosa 105mm/26 howitzers	12
Mortar	M-29 81mm	124
	M-30	4
	M-2 60mm	
	ECIA 120 mm	24
RCL	M40A1 106mm	
AA Gun	Hispano-Suiza HS.404	4
	40mm Bofors L/70	

Army Aviation

Category	Name	In Service
Helicopter	Bell OH-58 Kiowa OH-58A	9
	Robinson R22	4
	Robinson R44	2

New Procurements/ Upgrades

No major procurement plans are under consideration at present.

NAVY

Personnel: 4,000

Equipment

Category	Name	In Service
Patrol boat	USCG 180 Class	1
	USCG 133 White Class	2
Interceptor	32 Defender Class	8
Auxiliary vessel	Ocean Tug ATR 165 Class	1
	Damen Stan Tug 2608 Class	2
Landing Craft	LCU-1600 Class	1
Rescue craft	Damen Stan Patrol 1500 Class	3

Image Credit: Armada de Republica Dominicana (Navy of the Dominican Republic)

Naval Aviation

Category	Name	In Service
Aircraft	SA316Bs	2
	Cassena T41Ds	5
	CH-146 KIOW A	1

New Procurements/ Upgrades

No major procurement plans are under consideration at present.

AIR FORCE

Personnel: 5,500

Equipment

Category	Name	In Service
Fighter	Super Tucano A-29B	8
Trainer	ENAER T-35	6
	Sikorsky S-333	1
	C-172 Skyhawk	2
	R-44, R-22	
Transport	C-212-400	3
Helicopter	Bell 412	
	UH-1H Huey II	8
	OH-58A	10
	OH-6A	1
	Schweizer 333	3

New Procurements/ Upgrades

✦ The Dominican Republic Air Force (FARD) would be the first military operator of the Tecnam P2006T SMP (Special Missions Platform) aircraft. A contract was signed between the Air Force and Tecnam in this regard. The US Government will provide and install a range of special surveillance equipment onto its first Tecnam P2006T SMP aircraft to support CTOC operations on the Caribbean region.

Tecnam P2006T SMP. *Image Credit: Tecnam*

Defence Expenditure

Total defence spending: $322 million (est.)

Estimated defence spending in terms of GDP: 0.7% (2015)

Domestic suppliers: ELBO, EAB, EBO

Foreign suppliers: United States, Germany, France, Belgium, Brazil, Spain

Defence Production and R&D

The army has expressed an interest in building a small-calibre munitions production facility. Besides this, there has been little significant military procurement in the past 10 years, although the Dominican Army needs to replace its obsolete equipment. To be capable of fulfilling its role, it would require modern armoured fighting vehicles, artillery pieces and both anti-armour and air defence weapons.

Defence Procurement

There do not appear to be any plans for either modernisation or procurement in the foreseeable future, and with a military procurement budget of around $15 million there will be little available for anything more than essential spares and ammunition.

CONTACT DETAILS

Ministry of Armed Forces

Avda. 27 de Febrero esq. Luperon
Santo Domingo
Dominician Republic
Tel: +8095305149
Fax: +809-531-1309

ECUADOR
(Capital: Quito)

INTRODUCTION

Area: 283,561 sq km

Population: 16,080,778 (July 2016 est.)

Coastline: 2,237 km

Maritime claims: Territorial sea: 200 nm

Continental shelf: 100 nm from 2,500-m isobath

KEY POLITICAL PERSONS

PRESIDENT & COMMANDER-IN-CHIEF: Rafael Correa Delgado

VICE PRESIDENT: Jorge Glas Espinel

KEY DEFENCE PERSONS

MINISTER OF DEFENCE: Ricardo Patino

CHIEF OF JOINT COMMAND OF THE ARMED FORCES: Vice Adm. Oswaldo Zambrano Cueva

CHIEF OF ARMY: Brig. Gen. Luis Ayala Miguel Angel Castro

CHIEF OF NAVY: Vice Adm. Isaac Sarzosa Angel Aguirre

CHIEF OF AIR FORCE: Lt. Gen. Cesar Abdon Merizalde Pavon

ECONOMY

Ecuador is substantially dependent on its petroleum resources, which have accounted for more than half of the country's export earnings and one-fourth of public sector revenues in recent years. In 1999-2000, Ecuador suffered a severe economic crisis, with GDP contracting by more than 6%. Poverty increased significantly, the banking system collapsed, and Ecuador defaulted on its external debt later that year. In March 2000, the Congress approved a series of structural reforms that also provided for the adoption of the US dollar as legal tender. This stabilised the economy and positive growth returned in the years that followed, helped by high oil prices, remittances, and increased non-traditional exports. President Rafael Correa, who took office in January 2007, defaulted on Ecuador's sovereign debt in December 2008, refusing to make payment on $3.2 billion in international bonds, representing over 80% of Ecuador's private external debt. Economic policies under the CORREA administration - including an announcement in late 2009 terminating 13 bilateral investment treaties - generated economic uncertainty and discouraged private investment. The Ecuadorian economy slowed to 0.4% growth in 2009 due to the global financial crisis and to the sharp decline in world oil prices and remittance flows. Growth picked up to a 3.6% rate in 2010 and 6.5% in 2011. China has become Ecuador's largest foreign bilateral lender since Quito defaulted in 2008, allowing the government to maintain a high rate of social spending. In 2014, oil output increased slightly and production remained steady in 2015, although prices will likely remain lower than in previous years. Ecuador currently faces the challenge of adapting its economic structure to the new international context to achieve strong growth in the medium term and to protect key social advances made during the oil boom.

GDP: $183.4 billion (2015 est.), $183.3 billion (2014 est.)

Real Growth Rate: 0% (2015 est.), 3.7% (2014 est.)

Industries: Petroleum, food processing, textiles, wood products, chemicals

Oil - proved reserves: 8.832 billion bbl (1 January 2016 est.)

Natural gas - proved reserves: 10.99 billion cu m (1 January 2016 est.)

Total Exports: $19.05 billion (2015 est.), $26.6 billion (2014 est.)

Export Commodities: Petroleum, bananas, cut flowers, shrimp, cacao, coffee, wood, fish

Major Markets: US 39.5%, Chile 6.2%, Peru 5.1%, Vietnam 4.3%, Colombia 4.3% (2015)

Total Imports: $20.7 billion (2015 est.), $26.66 billion (2014 est.)

Import Commodities: Industrial materials, fuels and lubricants, nondurable consumer goods

Major Suppliers: US 27.1%, China 15.3%, Colombia 8.3%, Panama 4.9% (2015)

INTERNATIONAL DISPUTES

✦ The Colombian non-state armed group FARC (*Fuerzas Armadas Revolucionarias de Colombia*) is present near the border with Colombia.

DEFENCE

ARMY

Personnel: 30,000

Equipment

Category	Name	In Service
MBT	Leopard 1	40+
	T-55AM	64
	AMX-13 Light Tank	100
APC/MIC V/IFV	Engesa EE-9	32
	EE-3 Jararaca	10
	EE-11 Urutu	30
	AMX-VCI	80
	Panhard ERC-90	
	BTR-3U	
	BTR-60	6
Artillery	M2A1 105mm	24
	M198 155mm	12
	MkF3 155mm SP	12
Mortar	M29 81mm	412
	Soltam M-66 160mm	12
	M30 107mm	
MRL	RM-70 122mm	8
RCL	M67 90mm and M40A1 106 mm	400
AA Gun	Bofors 40mm	30
	M167 20mm	10
	20 mm M163 Vulcan	70
	ZSU-23-4	34
Weapon	Spike, MAPATS, HJ-8, Kornet-E, HOT, MILAN, 9M133 Kornet	
SAM	SA-8, Mistral, Blowpipe, HN-5A, Strela 2, 9K38 Igla, MIM-72	

Army Aviation

Category	Name	In Service
Aircraft	T-41 Mescalero	2
	Cessna 182	4
	Nanchang CJ-6	2
Transport	Pilatus PC-6	1
	Maule M-7	3
	Beechcraft Super King Air	1
	IAI Arava	4
	C-212-400 Aviocar	2
	CASA CN-235	2
	DHC-5D Buffalo	1
Helicopter	Mil Mi-17	8
	SA342L Gazelle	10
	Eurocopter AS332	6
	Eurocopter AS350	2

New Procurements/ Upgrades

No major procurement plans are under consideration at present.

NAVY

Personnel: 9,000

Equipment

Category	Name	In Service
Submarine	Type 209/1300	2
Frigate	Condell class	2
Corvette	Tipo 550 class	6
Patrol Vessel	Fast attack craft	3
Auxiliary Ship	Auxiliaries	6
Missiles	MBDA Exocet, MBDA Aspide (SAM)	

Naval Aviation

Category	Name	In Service
Aircraft	Super King Air	1
	CASA CN-235	1
Helicopter	206B	3
	TH-57A	3
	230T	1
	430	2
Trainer	T-34C-1	3
	T-35B	4
UAV	Heron I	2
	Searcher III	4

The Ecuadorian corvette BAE El Oro (Tipo 550 class) sits pier side in Panama City, Panama before beginning at sea exercises during PANAMAX 2005. *Image Credit: US Navy*

New Procurements/ Upgrades

No major procurement plans are under consideration at present.

AIR FORCE

Personnel: 5,000

Equipment

Category	Name	In Service
Fighters/attack/ Reconnaissance	IAI Kfir	13
	Atlas Cheetah	14
Trainer	EMB 314 Super Tucano	15
	T-35	14
	Atlas Cheetah	2

Air Force continued...

Category	Name	In Service
Trainer	DA20C1	6
	Bell 206	5
Transport	C-130	3
	Boeing 737	1
	Boeing 727	1
	C-295	4
	Hawker Siddeley HS 748	2
	de Havilland Canada DHC-6 Twin Otter	3
Helicopter	HAL Dhruv	withdrawn
AAM	R550 Magic-II, Python MK-IV, Python MK-III, Shafrir MK-II, V-4 R-Darter-BVR	

An Ecuadoran Air Force Kfir aircraft takes off during the joint US/Ecuadoran Exercise BLUE HORIZON'86.

New Procurements/ Upgrades

✦ Eurocopter delivered the first two AS550 C3 Fennec helicopters to the Ecuadorian Ministry of Defence under a contract signed in 2010 for seven of these helicopters and two AS350 B2s, which were delivered in December 2011. The contract contemplates delivery of two Fennecs per year until 2015. The new Fennecs will perform a wide range of missions including rescue, reconnaissance, drug enforcement and border patrol from their base at Porto Viejo.

Defence Expenditure

Total defence spending: $2.3 billion (est.)

Defence spending in terms of GDP: 2.7% (est.)

Foreign suppliers: United States, France, Germany, United Kingdom, Israel, Brazil, South Africa, Russia, Chile, China, Venezuela, Spain, Canada, Argentina

Ecuadorian Armed Forces is equipped with western equipment, with some from Russia and China acquired for emergency upgrades. Ecuador receives military assistance from US, Venezuela and Italy. The administration at the Defence Ministry launched a deep restructuring programme under the name of "PATRIA-I". It involves the modernisation of military equipment, improvement of planning and operations within the Ecuadorian territory. "PATRIA-I" was completed by 2011.

Defence Production and R&D

Ecuador has indigenously produced 9mm PAME-90 submachine guns. Its defence industry mainly relies on the external sources.

Defence Procurement

Ecuador's Defence Minister has announced that the plan to strengthen the armed forces would continue despite of the global financial crisis and that the government would invest more into the armed forces. The total defence spending in 2009 was $1,287 million. This hike in defence spending translated into the acquisition of a variety of systems including a fleet of 24 EMB-314 Super Tucano combat aircraft for $270 million, 2 Mi-17 helicopters for $22 million, IAI Heron and Searcher UAV's for $23 million and radars from China for $60 million.

CONTACT DETAILS

Ministry of National Defence
Exposicion No. 28
Quito, Ecuador
Tel: 593-2 295-1951

STRATEGIC INFORMATION

EGYPT
(Capital: Cairo)

INTRODUCTION

Area: 1,001,450 sq km

Population: 94,666,993 (July 2016 est.)

Coastline: 2,450 km

Maritime claims: Territorial sea: 12 nm

Contiguous zone: 24 nm

Exclusive Economic Zone: 200 nm

Continental shelf: 200 m depth or to the depth of exploitation

GEOPOLITICAL IMPORTANCE

Egypt is bordered by the Mediterranean Sea to the North, the Gaza Strip and Israel to the Northeast, the Red Sea to the East, Sudan to the South and Libya to the West. The predominantly Sunni Muslim country is the cultural and historic centre of the Arab world and has provided a stabilizing influence in volatile West Asia and made peace with Israel after numerous wars. The Suez Canal in Egypt connects the Mediterranean Sea to the Red Sea through the Isthmus of Suez. 12% of all international trade and 22% of the world's total container traffic goes through the Suez Canal. The Sumed oil pipeline located at Suez provides an alternative to the Suez Canal for transporting oil from the Persian Gulf region to the Mediterranean. Egypt's geo-strategic location, size and population, its intellectual achievements, its control of the Suez Canal and its position towards Israel have made Egypt a strategically important country.

POLITICAL OVERVIEW

In the aftermath of 13 June 2013 protests in Egypt former Muslim Brotherhood President, Mohammed Morsi, was ousted from power in a coup by former head of the Egyptian Armed Forces, Abdel Fattah al Sisi who later won the Presidential elections held on 8 June 2014. As a result of the new constitution approved by a referendum in January 2014, Egypt held parliamentary elections between 18 October and 22 November 2015. The two-phase elections saw the regime backed 'For the love of Egypt' coalition win all 120 party-based seats. The country's political transition culminated in the election of a President and a Parliament.

The new constitution adopted in 2014, like the constitution drafted under Morsi, is based on the Egyptian Constitution of 1971. The new constitution strengthens the country's three key institutions - the military, the police and the judiciary. The new constitution sets up a President and Parliament. The President is elected to a four-year term and may serve two terms. The parliament may impeach the president. The military retains the ability to appoint the national Minister of Defence. Under the Constitution, political parties may not be based on "religion, race, gender or geography". The document guarantees an absolute freedom of expression that is subject to broad exceptions. It also gives more rights to women and disabled people, and removes certain Islamist-leaning clauses inserted under

Morsi, while maintaining the principles of Islamic sharia as the main source of legislation.

KEY POLITICAL PERSONS

PRESIDENT& SUPREME COMMANDER: Abdel Fattah El-Sisi

PRIME MINISTER: Sherif Ismail

ECONOMY

Political uncertainty has caused Egypt's economic growth to slow significantly. However, the government has undertaken important measures to restore economic stability with the political transition which began in June 2013 ended in December 2015. The government has announced a wide range of reforms, introducing new taxes, increasing selected taxes, and reducing energy subsidies. Egypt has been benefiting from the large-scale exceptional financing from the Gulf. Saudi Arabia, UAE, and Kuwait have committed around US$20 billion to Egypt through a mix of central bank deposits, cash grants, in-kind grants, and project aid. This has helped the authorities to stabilize the economy and to partially meet the country's energy and food needs. According to the World Bank, economic growth rebounded to 4.2 percent in 2014 to 15, double the growth during the previous four years.

GDP: $1.048 trillion (2015 est.), $1.006 trillion (2014 est.)

Real Growth Rate: 4.2% (2015 est.), 2.2% (2014 est.)

Industries: Textiles, food processing, tourism, chemicals, pharmaceuticals, hydrocarbons, construction, cement, metals, light manufactures

Total Exports: $19.03 billion (2015 est.), $25.27 billion (2014 est.)

Export Commodities: Crude oil and petroleum products, fruits and vegetables, cotton, textiles, metal products, chemicals, processed food

Major Markets: Saudi Arabia 9.1%, Italy 7.5%, Turkey 5.8%, UAE 5.1%, US 5.1%, UK 4.4%, India 4.1% (2015)

Total Imports: $57.17 billion (2015 est.), $64.45 billion (2014 est.)

Import Commodities: Machinery and equipment, foodstuff, chemicals, wood products, fuels

Major Suppliers: China 13%, Germany 7.7%, US 5.9%, Turkey 4.5%, Russia 4.4%, Italy 4.4%, Saudi Arabia 4.1% (2015)

DEFENCE & SECURITY

The Egyptian military, one of the biggest armies in the Middle East, has fought several major wars since its independence. The military with experience in battling militancy has also engaged in war on Islamic State of Iraq and the Levant (ISIL) and the Sinai insurgency. Egypt faces serious security challenges including terrorism and Islamic insurgency and its priority is ensuring stability at home. Many terrorist attacks have been linked to Islamic extremism. The breakdown of the Libyan state, crumbling of border outposts, and rise of warring militant factions all threaten Egypt's internal stability. The Sinai Peninsula is the epicentre of an insurgency initiated in early 2011 as a fallout of the 2011 Egyptian Revolution. Security officials face resilient jihadist insurgents in the peninsula. In order to improve its own defences and to address mounting challenges, the government has tried to forge a closer security relationship with Libya and its neighbours. The governments of Egypt and Libya have signed a military cooperation agreement while North African countries have also formed an advisory committee to help Tripoli address its political and security concerns. On the international scene Egypt is a longstanding and committed contributor to United Nations peacekeeping operations.

KEY DEFENCE PERSONS

COMMANDER IN CHIEF & DEFENCE MINISTER: Col. Gen. Sedki Sobhi

CHIEF OF STAFF OF ARMED FORCES: Lt. Gen. Mahmoud Hegazy

COMMANDER IN CHIEF OF NAVY: Rear Admiral Ahmed Khaled Hassan Saeed

COMMANDER OF AIR FORCE: Air Marshal Younes Hamed

COMMANDER OF AIR DEFENCE FORCES: Maj. Gen. Ali Fahmi

Internal Conflict

Egypt continues to face internal conflict with increasing violence and terrorism. Muslim Brotherhood President, Mohammed Morsi, was deposed in a military coup in 2012 and the parliament dissolved. Abdel Fattah el-Sisi, who announced the overthrow of Morsi, became the leader of Egypt following an election victory in 2014. Although the political situation in Egypt has stabilized since Abdel Fattah el-Sisi became the leader and officially took power, protests have continued despite the government's crackdown. An ongoing insurgency led by Ansar Beit al-Maqdis in the Sinai Peninsula working with the Islamic State of Iraq and the Levant has also become increasingly active. The militant group known as the Sinai Province has been linked to a number of deadly attacks. The group's base, North Sinai, has been under a state of emergency since October 2014 when 33 security personnel were killed in an attack claimed by the group. It has been involved in suicide bombings, drive-by shootings, assassinations and beheadings.

External Conflict/Territorial Disputes

Sudan claims but Egypt de facto administers security and economic development of Halayib region north of the 22nd parallel boundary. The genesis of the disputes dates back to as early as 1958 after Sudan gained independence from being ruled jointly by Britain and Egypt.

Saudi Arabia claims Egyptian-administered islands of Tiran and Sanafir. Tiran Island is the narrowest section of the Straits of Tiran, which is an important sea passage to the major ports of Aqaba in Jordan and Eilat in Israel. The island is currently only inhabited by military personnel from Egypt and the Multinational Force and Observers.

THREAT PERSPECTIVE

Terrorism

Militant groups pose a major challenge to Egypt. Terrorism in Egypt is mainly concentrated in the Sinai region. Thousands of Egyptian militants have joined ISIS in Iraq and Syria. Security agencies estimated the number of militants in Sinai at around 8,000. The Egyptian security forces have suffered considerable losses in attempts to wipe out these terrorist groups. Egyptian naval vessels have been attacked by terrorists and a Russian plane was brought down over the Sinai desert in October 2015, with a bomb being confirmed as being responsible for the crash.

Egypt also face security threat with unchecked influx of terrorist elements and materiel from Libya. Egypt share a 720-mile long western border with the neighbour where Islamist and jihadist alliance led by Ansar al-Sharia - a group with ties to Islamic State has taken control of Benghazi and declared an "Islamic Emirate."

Israel is viewed as a threat notwithstanding an Egyptian-Israeli peace treaty and Egypt's role in stabilising West Asia. It has an enormous arsenal of conventional forces that surpasses in efficiency and capability all the Arab states put together. Such an imbalance poses a threat to the weaker side and is thus considered a source of regional instability.

STRATEGIC RELATIONS

United States: Egypt and the United States are strategic partners and share goals of peace and stability throughout the Middle East, North Africa and beyond. The two countries maintain close military-to-military relations. The two nations have cooperated on counter-terrorism, the Israeli-Palestinian peace process, peacekeeping operations in Africa, deepening regional economic integration and other shared endeavours.

European Union: Egypt's relation with the EU is based on the 2004 Association Agreement, and cooperation is led by a roadmap called the Action Plan. The Euro-Mediterranean Partnership (EMP) is the primary framework guiding Euro-Mediterranean relations and encompassing components and programmes dealing with democracy, while the European Neighbourhood Policy (ENP) addresses reform issues. Egypt is a major trading partner for the EU in the Southern Mediterranean region.

Russia: Relations between Egypt and Russia dates back to the period of the Union of Soviet Socialist Republics (USSR). Relations between the two countries have improved significantly in recent years. Both countries have strengthened their military and trade ties. There have been exchange of high-level visits between the two nations. In 2014, a preliminary deal was reached between both countries to buy arms worth $3.5 billion from Russia. Trade between the two countries stood at $3 billion in 2015.

India: India and Egypt have good bilateral relations and both countries are keen to enhance cooperation in various fields. A number of high level visits have also given a boost to relations. The two nations have decided to enhance bilateral energy and defence ties and take steps to provide better climate and opportunities for promotion of investments. India is among the largest importers of Egyptian products.

China: China and Egypt have upgraded their relationship to a "comprehensive strategic partnership". Egypt was the first country both in the Arab world and Africa to recognise the newly founded People's Republic of China in 1956. Relations between the two nations have remained sound since then regardless of the changes on both sides and the revolution in Egypt. Trade between Egypt and China reportedly amounts to $12bn per year.

Saudi Arabia: Egypt and Saudi Arabia are both highly influential countries in the Arab world, with Egypt being the most populous Arab country and Saudi Arabia being a member of the G20. They shoulder a big responsibility to realize the Arab solidarity and achieve the aspired goals for Arab nations from Atlantic Ocean to Gulf. The relations between the two countries is growing with increasing number of mutual visits from both sides at high levels.

DEFENCE CAPABILITIES

The Egyptian Armed Force consist of the Army, Navy, Air force and the Air Defence force. The supreme commander of the armed forces is the President and the Supreme Council is headed by the Commander-in-Chief and Defence Minister and his deputy, the Chief of Staff of the Armed Forces. Members of the council comprised of the leaders of the five main military branches. The armed forces enjoy considerable power within the Egyptian State. The armed forces have de facto control over all undeveloped non-agricultural land in Egypt. Equipment from the former Soviet Union is being progressively replaced by more modern equipment. President El-Sisi has made building up his country's military a top priority and Egypt has begun the military modernization programme acquiring new equipment and also enhancing its operational combat capabilities. Egypt has one of the biggest navies in the Middle East.

ARMY: The Egyptian army is one of the oldest armies in history and the largest service branch within the Egyptian Armed Forces. The Army considers itself to be the backbone of the regime and the guarantor of national stability.

Personnel: 300,000

Equipment

Category	Name	In Service
MBT	M-1A1	1005+125
	Ramses II	500 (includes upgraded version)
	T-62	500
	M-60A3	1,450
AIFV/ APC	BMP-1	4939 (total)
	YPR-765	
	M-113	
	BMR-600	
	FAHD	
	Commando Scout armoured car	
	M-106A2/M-125A2 mortar carriers	
	M577A2 command posts	
	M-901 ITV tank destroyer	
	YPR-765 Prat tank destroyer	
	M-981 FISTV	
	BRDM 1/-2	

Army continued...

Category	Name	In Service
Artillery	122mm	1200
	130mm	
	152mm	
	180mm	
	M109A2/A3 155mm SP	680
	122mm SP	124
Mortar	120mm	100
	160mm	100
	240mm	200
MRL	BM-11	145
	BM-21	60
	M-88s (all 122mm)	175
	MLRS	26
RCL	82mm	500
	107mm	400
ATK Gun	57mm	300
	76mm	300
	100mm	300
AA Gun	ZSU-23-4	350
	ZSU-57-2 SP	
	M 167A2 Vulcan	
ATGW	TOW	520
	Milan II	220
	Swingfire	200
Weapon	AL SAQR (SA-7) (SAM)	
	Avenger system (SAM)	75
	Patriot-3 missile (SAM)	
	RIM 116B Block (SAM)	
	Scud B/C (SSM)	
	Sakr Eye (SAM)	
	Sakr-36 (SSM)	
	AGM-114 Hellfire (ASM)	
	Frog-7	
	AL-Badr	

New Procurements/Upgrades

✦ A deal was signed with the US to supply the Egyptian Army with Abrams fleet up to 1,130.

✦ Egypt has ordered 356 AGM-114K/R3 Hellfire II missiles.

NAVY: The Egyptian Navy's missions include protection of more than 2,000 kilometres of coastline of the Mediterranean Sea and the Red Sea, defence of approaches to the Suez Canal, and support for army operations. The Egyptian navy's operational capabilities include four improved Romeo-class diesel submarines and Type 209/1400 submarines.

Major naval bases: Mersa Matruh, Port Sa'id, Port Tawfiq, Adabiyya, Ras al Tin, Bur Safaga.

Personnel: 16,000

Equipment

Category	Name	In Service
Frigate	Tahya Misr	1
	Sharm el-Sheikh class (ex-US Perry class)	4
	Damyat class (ex-US Knox class)	2
	EL Suez class (ex-Spanish Descubierta class (training frigate)	2
	Zaffer class (Chinese Jianghu 1 type	2
	Tarig class (training frigate)	1
Submarine	Romeo type (ex-Chinese)	4
	Type-209/1400	1
Patrol craft	'23 July' class (ex-German type 148 class)	5
	Ramadan class FAC (missile)	6
	Ex-USSR Osa class FAC	
	Ezzat fast missile craft (Ambassador IV type)	4
	MRP-20 type fast interceptor craft	
	Timsah class	12
	Crestitalia type	6
Amphibious force	ex-Polish Polnochy class LCT	3
	ex-USSR Vydra class LCU	9
	ex-USSR SMB1 class LCU	4
	LCM	10
Mineswee-per	ex-US Osprey class (Al Shareen class)	2
	ex-USSR Yurka class	4
	ex-USSR T-43 type (Assiout class) MCMV	3
	ex-USSR T-301 class	2
	Swiftship type	6
	SRN-6 hovercraft	3
Auxiliaries	Underway replenishment ship (ex-German Glucksburg)	1
	Ammunition transport ship (ex-German Odenwald)	1
Weapon	Harpoon	
	MK-44 Guided Missile Round Pack	
	Exocet	

Missile corvette Ahmed Fadel at the Alexandria naval base. *Image credit: Egyptian MoD*

New Procurements/Upgrades

✦ The Navy has commissioned into service a missile corvette called Ahmed Fadel donated by Russia. The vessel R-32 (832) has been delivered to the naval base in Alexandria with supersonic 3M80 Moskit anti-ship missiles.

✦ The first Gowind 2500 corvette is under construction in Lorient. The vessel being built by DCNS is the first of a series of four units that will be delivered to Egypt before 2019. The delivery of the first vessel is slated for 2017.

✦ French Company DCNS has delivered the two Mistral-class helicopter carriers - the LHD Gamal Abdel Nasser and the LHD Anwar El-Sadat.

✦ Egypt has taken delivery of the first four Type 209 attack submarines from Germany. The submarine is 64-metre long vessels armed with torpedoes and capable of firing land- and ship-striking missiles from on deck.

Egyptian Navy's new submarine in Hamburg, Germany. Image credit: Egyptian official spokesperson *Facebook page*

✦ US shipbuilder Swiftships has been awarded a contract to build four 28-metre-long coastal patrol vessel kits for Egypt, which will assemble the craft in Alexandria. The 28 metre Coastal Patrol Craft (CPC) will include 4.7-metre rigid inflatable boats and forward looking infrared systems.

✦ The US State Department has approved the sale of Boeing UGM-84L Encapsulated Harpoon Block II submarine-launched anti-ship missiles to be used by Egypt's new Type 209 submarines.

Naval Aviation

Category	Name	In Service
Helicopter	SA-342L Gazelle	9
	SH-2G Super Seasprite	12

New Procurements/Upgrades

✦ Egypt is to get the first of 46 navalised Ka-52 Alligator helicopters in 2017 to equip its new French-built Mistral-class amphibious assault ships.

Coastal Defences

Coastal defences continued...

Category	Name	In service
Artillery	100mm	2 brigades
	130mm	
	152 gun	
Artillery	Otomat batteries with 2 twin truck-mounted launch units each	3
	P-15 Termit launcher	

AIR FORCE: The Egyptian Air Force (EAF) primary role is the air defence of the nation, involving international search-and-rescue operations in the desert, the Mediterranean Sea, and the Red Sea. The EAF also extends support to national security organizations engaged in the war against terrorism - the Sinai Insurgency, the Second Libyan Civil War and the Intervention in the Yemen.

Personnel: 25,000+

Major air bases: Almaza, Cairo West, Bilbeis, Beni Suwayf, Luxor, al-Minya, ras Banas, Hurghada, In Shas, Fayid, Gianaclis, Tanta, Al Mansura, Gabel el Basur, Mersa Matruh.

Equipment

Category	Name	In Service
Fighter	Rafale	6
	F4E Phantom II	25
	F-6	36
	Mirage 5D/E	48
	F-16A	34
	F-16B	12
	F-16C	138+
	F-16D Block 52	195
	F-7	54
	MiG-21	84
	Mirage 2000	16
AWAC	E-2C Hawkeye/ Hawkeye 2000	6
	Beechcraft 1900 Elint	4
Transport	DH-5D	4 (3 regiment s)
	C-130H	19
	C-295	12+
	Falcon 20	2
	Gulfstream IV-SP	2
	Gulfstream III	1
	Gulfstream G400	2
	An-74TK	3
Trainer	Alpha Jet	70
	K-8	120
	EMB-312 Tucano	54
	L-39	10
	L-59E	48
	IL-14	4
	Rockwell Commander 114B	6
Other aircraft	KC-135 tanker	3
	Boeing 707 tanker/transport	3

Air Force continued...

Category	Name	In Service
Helicopter	AH-64D Apache	46
	UH-60L Blackhawk	24+
	Commandos	15
	Mi-4	12
	Mi-6	10
	Mi-8 assault	42
	Mi-8T	15
	SA.342K Gazelle (including 44 with HOT)	65
	CH-47D	18
	UH-12	18
	AS-61	2
	S-70A21 (VIP)	2
	AW139 SAR	2

New Procurements/Upgrades

✦ The Air Force has taken delivery of the second batch of three Rafale multirole combat fighters. The first three Rafales were received last year. Egypt and France has signed a contract for 24 Rafales, including 16 two-seaters and eight single-seater fighter jets.

✦ The first batch of MiG-29 (or MiG-35) Fulcrum multi-role fighters for the Air Force are under production. Russia has reportedly agreed on a contract to supply 46 MiG-29 to replace Egypt's aging fleet of MiG-21s and J-7s.

✦ Egypt has received the final four F-16 fighter jets completing the delivery of the aircraft totalling 195.

✦ Egypt to buy from US, AN/AAR-57 common missile warning system (CMWS) for AH-64E Apache, UH-60 Blackhawk, and CH-47 Chinook helicopter.

AIR DEFENCE FORCES: The Egyptian Air Defense Forces (EADF) is the Anti-aircraft warfare branch of the Egyptian Armed Forces, responsible for protecting the Egyptian airspace against any hostile air attacks. The ADF's capabilities include antiaircraft guns, rocket and missile units, interceptor planes, and radar and warning installations. Its air-defence system is one of the largest and most complex systems in West Asia.

Personnel: 80,000

Equipment

Category	Name	In Service
SAM	SA-2	320
	SA-3	180
	SA-6	76
	Tor M1/M2	
	Buk M1/M2	
	Hawk PIP III batteries with 72 systems	1

Air Defence Force continued...

Category	Name	In Service
SAM	Chaparral batteries (Sentinel radars)	12
	Crotale batteries with 36 systems	14
	Skyguard/Amoun	13
AA gun	20mm	250
	23mm	250
	37mm	250
	Twin 35mm (Amoun system)	250
	57mm	350
	85mm	350
	100mm	300

Egypt's ballistic missile arsenal is reportedly composed of short-range missiles. The country currently deploys R-300 Elbrus, Project T (Scud-B-100) short range ballistic missiles (SRBM); R-70 Luna-M artillery rockets; and Sakr-80 artillery rockets. It does not possess any intercontinental ballistic missiles (ICBMs) and it is not a member of the Missile Technology Control regime (MTCR). Cairo is a state party to the treaty on the Non-Proliferation of Nuclear Weapons (NPT). It is also one of the few states to have used chemical weapons in wartime (North Yemen Civil War (1962-1970), period of Egyptian involvement 1963-1967). Egypt acceded to the Geneva Protocol on December 6, 1928, but has remained outside the Chemical Weapons Convention (CWC).

DEFENCE PRODUCTION AND R&D

The Ministry of Military Production supervises the National Organisation for Military Production (NOMP), which was set up in 1949 to make equipment for the armed forces. The NOMP acts as a holding company for the military production companies in Egypt, and has its own legal structure. The organisation oversees 17 companies that stretch across electronic, chemical, engineering and mining industries with the bulk of its companies in the engineering area. In addition to NOMP, Egypt has two state agencies, the Arab Organisation for Industrialisation (AOI) and National Service Products Organisation (NSPO), which run a number of companies that manufacture dual-purpose products.

Egypt boasts one of the most advanced defence industries in the Arab world, supported by a robust industrial complex, particularly in the production of vehicles. Egypt assembles a variety of trainer aircraft, alongside armoured fighting vehicles (AFVs), artillery pieces, surface-to-air missiles and M1A1 Abrams tanks. Military goods produced in Egypt include: small calibre and heavy ammunition, mortars, mines, grenades and other explosives, antitank rockets, rocket motors, radars and electronic equipment, smoke and pyrotechnic devices, rifles, pistols (Beretta licensee) and machine guns, jet trainer aircraft (Alpha and Tucano), armoured personnel carriers, Alpha jet engines, field and aircraft communications equipment, Gazelle helicopters and engines, gyroscopes, weapon sights, binoculars, periscopes, tanks, MLRs, and artillery pieces. In addition to manufacturing small arms and ammunition, Egypt had begun producing or assembling more advanced weapons systems through licensing and joint venture agreements with companies based in the United States and Western Europe.

Major Armament Producers

Arab Organisation for Industrialisation (AOI) supervises nine military factories which produce civilian goods as well as military products. AOI is owned by the Government of Egypt and administered by a Supreme Committee, the chairman of which is the Egyptian president and which includes several cabinet ministers and is considered a division of the Ministry of Military Production.

National Service Products Organisation operates three companies that manufacture military and civilian products.

Arab-American Vehicle Co. (AAVCO): AAV Company is a joint Egyptian and American venture. The company produces military jeeps and other light vehicles.

Arab-British Defence Co. (AB-DCO): AB-DCO is a joint Arab-British project company. The company's product includes ATGM and missiles.

DEFENCE DEALS

Joint Venture Programmes

M1A1 Abrams tank: The United States and Egypt will resume co-production of M1A1 Abrams tanks, weapon systems and accessories after a deal formalizing the collaboration between the Egyptian Tank Plant and US defence equipment manufacturer General Dynamics Land Systems. The bulk of the tank order will be produced locally at Egyptian Tank Plant 200. Other joint venture programmes with the US includes the M88A2 coproduction programme and the HAWK rebuild programme.

Arms Export

Egypt achieved partial self-sufficiency in production of arms and began to develop an export market in the Middle East and Africa. Egypt's arms export record has been its sale of military-grade jeeps. Two military-grade models -the Jeep TJL and the Jeep J8 - are exclusively produced at the AAV factory, and have previously been exported to countries including Libya. Egypt is also making efforts to export weapons that contained technology of US origin, including potential sales of M1A1 tanks to Iraq, ammunition to Saudi Arabia, and the provision of technical support for Turkey's arsenal of Hawk Missiles. Apart from light arms manufacture, Egypt's military industry worked only on repair and maintenance of imported weapons.

Arms Import

Egypt plans to continue import of weapons from the United States whose companies like Lockheed Martin, Boeing, General Electric, Raytheon, General Dynamics and BAE Systems all do big business with Cairo selling fighter jets, tanks, radars, artillery and other hardware to Egypt. Boeing sells Cairo CH-47 Chinook transport helicopters while General Dynamics Land Systems is responsible for licensing the manufacture of M1A1 Abrams main battle tanks in Egypt.

Egypt has also continued its build-up of French-made defence capabilities with a deal to acquire two helicopter carriers originally destined for Russia before their delivery was halted by Paris in retaliation for Moscow's aggression in the Ukraine. The purchase of the two Mistrals is the latest in a string of French-supplied equipment orders from Egypt in the recent past. Egypt has ordered four Gowind corvettes and 45 MM-40-3 Exocet missiles to arm the vessels, a FREMM frigate from French shipyard DCNS and 24 Rafale fighters from Dassault. and an agreement has been signed with Russia to supply 50 Kamov Ka-52 Alligator attack helicopters.

DEFENCE BUDGET

Total defence spending: ~US$7.85 billion (2015), LE31 billion (2014)

Defence spending in terms of GDP: 1.76% (2014)

Egypt's military budget is largely unknown. The defence expenditures on the economy were difficult to assess because the military did not make this information public. Egypt remains the second largest recipient of US Foreign Military Financing aid of annual $1.3 billion after arms freeze was lifted.

The Egyptian government plans to secure a €3.37 billion (US $3.76 billion) loan from France to fund the acquisition of weapons and military equipment, including Navy vessels and fighter jets, as part of the country's ongoing force modernisation programme. The government has been spending massively on arms ordering vast amounts of military equipment from France and Russia, from frigates to fighter jets.

CONCLUSION

Egypt is among the strongest military powers in the Arab world. Modern Egypt is considered to be a regional and middle power, with significant cultural, political, and military influence in North Africa, the Middle East and the Muslim world. The country is central to trade and international relations. Twelve per cent of all international trade goes through the Suez Canal in Egypt and 22% of the world's total container traffic. A destabilised Egypt would disrupt the world's energy security. Egypt remains strategically important to the US and to the EU while Israel's national security rests largely on its treaty with Egypt.

An Egyptian Crotale P4R low-altitude surface-to-air missile system is positioned in the desert during Operation Desert Shield.

CONTACT DETAILS

Ministry of Defence and Military Production

23 July St., AlKobba bridge, Cairo

Ph.: 24032158

Fax: 22916227

E-mail: mmc@afmic.gov.eg

Website: http://www.mmc.gov.eg

ESTONIA

(Capital: Tallinn)

INTRODUCTION

Area: 45,228 sq km

Population: 1,258,545 (July 2016 est.)

Coastline: 3,794 km

Maritime claims: Territorial sea: 12 nm

Exclusive Economic Zone: Limits fixed in coordination with neighbouring states

KEY POLITICAL PERSONS

PRESIDENT: Kersti Kaljulaid

PRIME MINISTER: Taavi Roivas

KEY DEFENCE PERSONS

MINISTER OF DEFENCE: Margus Tsahkna

COMMANDER OF DEFENCE FORCES: Lt. Gen. Riho Terras

CHIEF OF STAFF: Col. Martin Herem

COMMANDER OF NAVY: Captain Juri Saska

COMMANDER OF AIR FORCE: Col. Jaak Tarien

ECONOMY

A well-developed modern market economy, Estonia has benefited from its trade ties with other European nations including Finland, Sweden, and Germany. The economy also benefits from strong electronics and telecommunications sectors and sound fiscal policies adopted by the government. While the economy was adversely affected during the global economic slowdown in 2008, it bounced back and registered a GDP growth of over 7% in 2011. Estonia became a member of the European Union in 2004 and adopted the Euro as its national currency in 2011 which boosted foreign investments in the country. Least affected by the eurozone crisis, Estonia achieved fastest economic growth in the EU by adopting a number of austerity measures and tax reforms. Strong exports and foreign investments also helped the Estonian economy maintain a lead over its other European neighbours. After two years of robust recovery in 2011 and 2012, the Estonian economy faltered in 2013 with only 1.6% GDP growth, mainly due to continuing recession in much of the EU. GDP growth in 2014 was 2.9% but dropped to 1.2% in 2015 due to lower demand in key Scandinavian export markets. GDP growth is expected to be about 2.2% in 2016.

GDP: $37.55 billion (2015 est.); $37.15 billion (2014 est.)

Real Growth Rate: 1.1% (2015 est.); 2.9% (2014 est.)

Industries: Engineering, electronics, wood and wood products, textiles; information technology, telecommunications

Total Exports: $12.24 billion (2015 est.); $15.18 billion (2014 est.)

Export Commodities: Machinery and electrical equipment 34%, food products and beverages 9%, mineral fuels 9%, wood and wood products 10%, metals 7%, furniture 9%, vehicles and parts 6%, chemicals 5% (2015 est.)

Major Markets: Sweden 18.8%, Finland 16%, Latvia 10.4%, Russia 6.7%, Lithuania 5.9%, Germany 5.2%, Norway 4.1% (2015)

Total Imports: $13.19 billion (2015 est.); $16.51 billion (2014 est.)

Import Commodities: Machinery and electrical equipment 28 %, mineral fuels 11%, food and food products 10%, vehicles 9%, chemical products 8%, metals 8% (2015 est.)

Major Suppliers: Finland 14.5%, Germany 11%, Lithuania 9%, Sweden 8.5%, Latvia 8.3%, Poland 7.4%, Russia 6.1%, Netherlands 5.5%, China 4.8% (2015)

INTERNATIONAL DISPUTES

✦ Border dispute with Russia over parts of the Narva region has not been resolved completely even after the two sides signed a border treaty in February 2014.

DEFENCE

ARMY

Personnel: 2,000+

Equipment

Category	Name	In Service
APC/ IFV	Sisu XA-180	56
	Sisu XA-188	60
	CV-90	44 (on order)
Armoured car	Mamba Alvis-4	9
Artillery	D-30 122mm howitzer	42
	FH-70 155mm howitzer	24
	M-40 105mm howitzer	38
Mortar	B455 81mm	40
	NM95 81mm	10
	M252 81mm	80
	2B11 120mm	14
	M-41D 120mm	165
RCL	M40A1 106mm	30
	Pv-1110 90mm	130
ATW	84mm Carl Gustav	
	148mm & 180mm MAPATS	
ATGW	115mm MILAN-2	
SAM	90mm Mistral	25

New Procurements/Upgrades

✦ Delivery of Sisu XA-188 APCs from Netherlands is going on.

✦ Delivery of 44 CV9035 NL infantry fighting vehicles (IFVs) from the Netherlands has started in October 2016. The country will also purchase 37 CV90 IFV vehicles from Norway which will be rebuilt as armoured support vehicles.

✦ The US government has approved the sale of Javelin anti-tank launchers and missiles under Foreign Military Sales programme. Delivery of the weapon has started.

✦ Delivery of Mistral M3 short-range air defence missiles is going on.

NAVY

Personnel: 200+

Equipment

Category	Name	In Service
Mine Warfare	Admiral Cowan class	3
Auxiliary ship	Lindormen class (diver and support vessel)	1 (+1 in reserve)
Training ship	Rihtniemi class	1

New Procurements/Upgrades

✦ Thales UK has been awarded a contract to provide an additional four years of technical support and maintenance to the Admiral Cowan class mine sweepers. The new contract extends the support programme until December 2016.

AIR FORCE

Personnel: 300+

Equipment

Category	Name	In Service
Transport	An-2	2
Trainer	L-39	2
Helicopter	Robinson R44	4

New Procurements/Upgrades

✦ Two Sherpa C-23B light transport aircraft are being acquired from the US Army reserves.

✦ Two Ground Master 403 medium-range air defence radar systems have been inducted in service, completing the military air surveillance network for the country.

Defence Expenditure

Defence spending: €456 million (Proposed budget for 2016); €412 million (2015)

Defence spending in terms of GDP: 2.1% (estimated for 2016)

As a NATO member, the Estonian Government had targeted to gradually increase the defence spending to 2% of GDP. The defence budget, which had hovered near the 2% mark between 2004 and 2010, reached the 2% target in 2012 and continues to remain intact for last two years. For 2016, the Government has announced plans to increase defence budget by 9%, making it 2.1% of GDP.

Defence Production and R&D

In recent times, the Estonian defence industry has focused on cyber security, protection against improvised explosive devices (IEDs) and chemical weapons. In the research and development sphere, the country has successfully produced the anti-improvised explosive device IRIS. An Unmanned Ground Vehicle system has been developed while effort is on to develop light armoured panels. The allocation of funds for R&D projects has increased significantly since 2001. The country has adopted a 10-year National defence development programme spanning from 2013 to 2022 under which it decided to procure weaponry, equipment and ammunitions worth €111.5 million in 2014-15.

Defence Procurement & Future Plans

The Defence Ministry has released National Defence Development Plan 2013-2022, replacing the Military Defence Development plan 2009-2018. Under the new plan, Estonia plans to procure modern third-generation anti-tank missile systems, 155mm self-propelled howitzers, tanks and medium-range air-defence systems for its land forces. The plan also outlines equipping its 1st infantry brigade with modern infantry vehicles and APCs. The 2nd infantry brigade will achieve full operational capability by 2022. For its air wing, the development plan foresees an increase of Estonia's air surveillance and defence capability and development of its Amari air base. The Air Force will be completely integrated with NATO's air and missile defence system and provide uninterrupted surveillance information for Estonia and all of NATO.

CONTACT DETAILS

Ministry of Defence
Sakala 1, 15094, Tallinn
Tel.: +372 717 00 22
Website:
http://www.kaitseministeerium.ee/en

Estonian Defence Forces
Juhkentali 58
EE15007, Tallinn
Website: http://www.mil.ee/en/defence-forces

ETHIOPIA
(Capital: Addis Ababa)

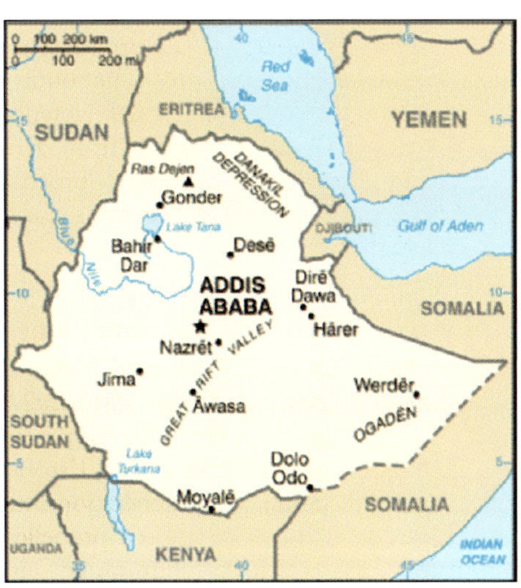

INTRODUCTION

Area: 1,104,300 sq. km
Population: 102,374,044 (July 2016 est.)
Coastline: 0 km (Landlocked)
Maritime claims: None

KEY POLITICAL PERSONS

PRESIDENT: Mulatu Teshome

PRIME MINISTER & COMMANDER-IN-CHIEF OF THE NATIONAL DEFENCE FORCES: Hailemariam Desalegn

KEY DEFENCE PERSONS

MINISTER OF DEFENCE: Siraj Fegessa Shereffa

CHIEF OF STAFF OF ARMED FORCES: Gen. Samora Yenus

ECONOMY

The Ethiopian economy is primarily based on agriculture which employs 85% of the country's population and contributes over 40% of the GDP. The Government holds all land in the country and gives it on long-term lease to tenants. While the agricultural and manufacturing sectors have attracted foreign investments in recent times, the country has not opened its banking and insurance sectors to foreign investment. The Government is making efforts to implement economic reforms, including privatisation of state enterprises. Even though the economy has registered a good GDP growth, with the country becoming the twelfth fastest growing economy in the world in 2012, a burgeoning population has made the country one of the poorest in the world. Ethiopia's economy continues on its state-led Growth and Transformation Plan. The country has achieved high single-digit growth rates through government-led infrastructure expansion and commercial agriculture development. Ethiopia remains a one-party state with a planned economy. In the fall of 2015, the government finalized and published the current 2016-2020 five-year plan, known as the Growth and Transformation Plan (GTP II).

GDP: $162 billion (2015 est.); $147 billion (2014 est.)

Real Growth Rate: 10.2% (2015 est.); 10.3% (2014 est.)

Industries: Food processing, beverages, textiles, leather, chemicals, metals processing, cement

Total Exports: $2.935 billion (2015 est.); $3.283 billion (2014 est.)

Export Commodities: Coffee, khat, gold, leather products, live animals, oilseeds

Major Markets: Switzerland 14.3%, China 11.7%, US 9.5%, Netherlands 8.8%, Saudi Arabia 5.9%, Germany 5.7% (2015)

Total imports: $15.87 billion (2015 est.); $15.86 billion (2014 est.)

Import Commodities: Food and live animals, petroleum and petroleum products, chemicals, machinery, motor vehicles, cereals, textiles

Major Suppliers: China 20.4%, US 9.2%, Saudi Arabia 6.5%, India 4.5% (2015)

INTERNATIONAL DISPUTES

✦ Border dispute with Eritrea over the town of Badme

✦ Conflict with Somalia over the un-demarcated former British administrative line having little meaning as a political separation to rival clans within Ethiopia's Ogaden and southern Somalia's Oromo region.

✦ An initial deal has been signed between Ethiopia, Egypt and Sudan in 2015 to end a long-standing dispute over sharing of the Nile waters and the building of Africa's biggest hydroelectric dam, in Ethiopia. Egypt has been opposing the project on the ground that it could diminish its share of Nile River waters, which provides almost all of the desert nation's water needs.

DEFENCE

ARMY

Personnel: 180,000

Equipment

Category	Name	In Service
MBT	T-72	50+
	T-62	75
	T-54/55	
APC/IFV	BRDM-1	70
	BMP-1	20
	BTR-40/60/152	90
	WZ551	
Artillery	Howitzer	300
	122 mm	360
MLRS	BM-21 122mm	
SAM	SA-2/3/7	

New Procurements/Upgrades

✦ A contract has been signed with Ukraine's state arms exporter Ukrspecexport SC to supply 200 T-72 main battle tanks. Delivery of the MBTs has begun in Sept 2013.

✦ Ukrspecexport SC has also been awarded contract to upkeep the existing fleet of T-72s operational in the Army.

AIR FORCE

Personnel: 3,500

Equipment

Category	Name	In Service
Fighter	Su-27	15
	Su-25	10
	MiG-21	15
	MiG-23	12
Transport	C-130	2
	An-26	1
	C-47	8
	Yak-40	1
Helicopter	Mi-6	10
	Mi-8	12
	Mi-24 (attack)	10
	Mi-14 (anti-submarine)	2
	SA330 Puma	1

New Procurements/Upgrades

No major procurement plans are under consideration at present.

Defence Expenditure

Defence Budget: ETB7.5 billion ($380.5 million - 2014-15)

Estimated defence spending in terms of GDP: 0.6%

Ethiopia receives majority of military equipment from the US, Israel, Italy and Russia.

Defence Production and R&D

The country does not undertake any defence production activities.

CONTACT DETAILS

Ministry of Defence

Churchill Road

PO Box 1373, Addis Ababa,

Ethiopia

FIJI

(Capital: Suva)

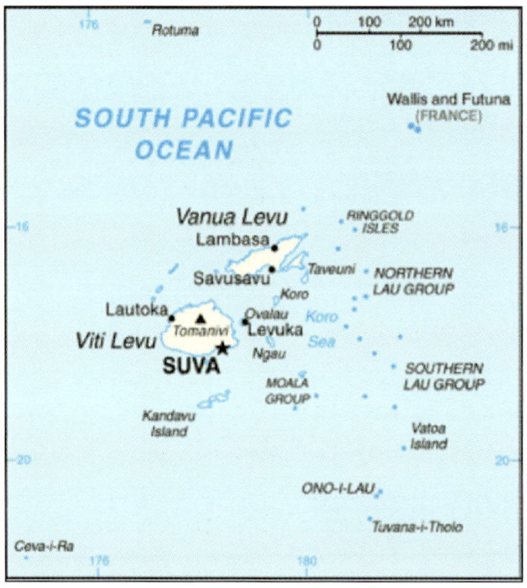

INTRODUCTION

Area: 18,274 sq km

Population: 915,303 (July 2016 est.)

Coastline: 1,129 km

Maritime claims: Territorial sea: 12 nm

Exclusive Economic Zone: 200 nm

Continental shelf: 200 m depth or to the depth of exploitation; rectilinear shelf claim added

KEY POLITICAL PERSONS

PRESIDENT & COMMANDER-IN-CHIEF: Maj. Gen. (Retd.) Jioji Konousi Konrote

PRIME MINISTER: RAdm. (Retd.) Josaia Voreqe Bainimarama

KEY DEFENCE PERSONS

MINISTER FOR DEFENCE & NATIONAL SECURITY: Ratu Inoke Kubuabola

COMMANDER REPUBLIC OF FIJI MILITARY FORCES (RFMF): Rear Adm. Viliame Naupoto

ECONOMY

Fiji is endowed with forest, mineral, and fish resources. It is one of the most developed of the Pacific island economies though still with a large subsistence sector. Fiji is a member of the WTO. Fiji's current account deficit peaked at 23% of GDP in 2006, and has been improving since. Fiji's sugar remains a significant industry and a major export. The sugar industry reforms since 2010 have improved productivity and returns. Urbanisation and expansion in the service sector have contributed to recent GDP growth. Fiji is a member of the WTO.

GDP: $8.048 billion (2015 est.), $7.716 billion (2014 est.)

Real Growth Rate: 4.3% (2015 est.), 5.3% (2014 est.)

Industries: Tourism, sugar, clothing, copra, gold, silver, lumber, small cottage industries

Total Exports: $1.233 billion (2015 est.), $1.152 billion (2014 est.)

Export Commodities: Sugar, garments, gold, timber, fish, molasses, coconut oil

Major Markets: US 13.4%, Australia 10.2%, Samoa 6.7%, Tonga 5.9% (2015)

Total Imports: $2.283 billion (2015 est.); $2.403 billion (2014 est.)

Import Commodities: Manufactured goods, machinery and transport equipment, petroleum products, food, chemicals

Major Suppliers: China 16.2%, South Korea 15.7%, NZ 14%, Australia 13.4%, Singapore 8.7%, France 7% (2015)

INTERNATIONAL DISPUTES

Fiji is not currently engaged in an armed conflict.

DEFENCE

ARMY

Personnel: 3,400

Equipment

Category	Name	In Service
Artillery	25-pdr gun	4
	105 mm howitzer	5
Mortar	81 mm	10

Army Aviation

Category	Name	In Service
Helicopter	Eurocopter AS355F2	1
	Eurocopter SA365N	1

New Procurements / Upgrades

No major procurement plans are under consideration at present.

NAVY

Personnel: 300 (approx.)

Equipment

Category	Name	In Service
Patrol Vessel	Pacific class	3
	Levuika class	2
	Ex-Israeli Dabur Class	4

New Procurements / Upgrades

No major procurement plans are under consideration at present.

Defence Expenditure

Total defence spending: 93.6 million (2016)

Estimated defence spending in terms of GDP: 1.1% (2015)

According to Government of Fiji, a total budget of USD 93.6 million is provided to the RFMF in 2016. the Republic of Fiji Military Force is to ensure at all times the security, defence and well- being of Fiji and all Fijians. It also makes a significant contribution to United Nations Peacekeeping operations in troubled areas around the world. The RFMF's naval division provides surveillance of Fiji's maritime zone and operates a costal radio station for all vessels within Fijian waters.

The US Department of Defence offers defence-related support in two main areas: Foreign Military Financing (FMF), which the government uses to buy military equipment from the US and the International Military Education and Training (IMET) programme, which funds training programmes provided by Washington.

Defence Production and R&D

Fiji does not undertake any defence production activities.

Defence Procurement

The Republic of Fiji Military Forces (RFMF) acquire small amounts of equipment mainly from Australia, South Korea, New Zealand, Singapore, UK and US and some is received in the form of military co-operation/aid. Currently, changes are taking place to meet the military's operational requirements and government demands in areas such as infrastructure development.

The Republic of Fiji Military Forces (RFMF)

CONTACT DETAILS

President Office Location

Government House
Berkley Crescent
P.O. Box 2513, Government Buildings
Suva
Phone: 3314 244
Fax: 3301 645

PM's Office Location

4th Floor, New Government Buildings,
Suva, Postal Address:
PO Box 2353, Govt Bldgs, Suva
Phone: 3211 201
Fax: 3306 034
pmsoffice@connect.com.fj

Military Force HQ

Queen Elizabeth Barracks
Maddocks Road
P.O Box 102, Fiji
Phone: +6793385222
www.rfmf.mil.fj

FINLAND

(Capital: Helsinki)

INTRODUCTION

Area: 338,145 sq km

Population: 5,498,211 (July 2016 est.)

Coastline: 1,250 km

Maritime claims: Territorial sea: 12 nm (in the Gulf of Finland - 3 nm)

Contiguous zone: 24 nm

Continental shelf: 200 m depth or to the depth of exploitation

KEY POLITICAL PERSONS

PRESIDENT: Sauli Niinisto

PRIME MINISTER: Juha Sipila

KEY DEFENCE PERSONS

DEFENCE MINISTER: Jussi Niinisto

CHIEF OF DEFENCE FORCES: Gen. Jarmo Lindberg

COMMANDER OF THE ARMY: Lt. Gen. Seppo Toivonen

COMMANDER OF THE NAVY: Rear Adm. Veijo Taipalus

COMMANDER OF THE AIR FORCE: Maj. Gen. Kim Jaameri

ECONOMY

Finland has a highly industrialized, largely free-market economy with per capita output almost as high as that of Austria, Belgium, the Netherlands, and Sweden. Trade is important with exports accounting for over one-third of GDP in recent years. Finland is strongly competitive in manufacturing - principally the wood, metals, engineering, telecommunications, and electronics industries. Finland excels in hi-tech exports such as mobile phones. Except for timber and some minerals, Finland depends on imports of raw materials, energy, and some components for manufactured goods. Because of the climate, agricultural development is limited to maintaining self-sufficiency in basic products. Forestry, an important export earner, provides a secondary occupation for the rural population. Finland had been one of the best performing economies within the EU in recent years and its banks and financial markets avoided the worst of global financial crisis. However, the world slowdown hit exports and domestic demand hard in 2009, with Finland experiencing one of the deepest contractions in the euro zone. A recovery of exports, domestic trade, and household consumption stimulated economic growth in 2010-12 but continued recession within the EU dampened the economy in 2012-14. The recession affected general government finances and the debt ratio, turning previously strong budget surpluses into deficits, losing its coveted triple-A credit rating, and on pace to breach EU debt limits in 2015. Finland's main challenge will be to stimulate growth while faced with weak export demand in the EU and its own government austerity measures.

GDP: $224.9 billion (2015 est.); $224.5 billion (2014 est.)

Real Growth Rate: 0.2% (2015 est.); -0.7% (2014 est.)

Industries: Metals and metal products, electronics, machinery and scientific instruments, shipbuilding, pulp and paper, foodstuff, chemicals, textiles, clothing

Total Exports: $61.29 billion (2015 est.); $75.6 billion (2014 est.)

Export Commodities: Electrical and optical equipment, machinery, transport equipment, paper and pulp, chemicals, basic metals, timber

Major Markets: Germany 13.9%, Sweden 10.1%, US 7%, Netherlands 6.6%, Russia 5.9%, UK 5.2%, China 4.7% (2015)

Total Imports: $58.5 billion (2015 est.); $74.43 billion (2014 est.)

Import Commodities: Foodstuff, petroleum and petroleum products, chemicals, transport equipment, iron and steel, machinery, textile yarn and fabrics, grains

Major Suppliers: Germany 17%, Sweden 16%, Russia 11%, Netherlands 9.1%, Denmark 4.1% (2015)

INTERNATIONAL DISPUTES

✦ Various groups in Finland advocate restoration of Karelia and other areas ceded to the Soviet Union, but the Finnish Government asserts no territorial demands.

DEFENCE

ARMY

Personnel: 24,000+ (including conscripts)

Equipment

Category	Name	In Service
MBT	Leopard 2A4	120
APC/IFV	BMP-2	110
	CV90	100
	BTR-50PK	50
	XA-180/102/203	600+
	BTR-60PB	100+

Army continued..

Category	Name	In Service
Artillery	Towed pieces	700+
Mortar	XA 361 - 120 mm	24
	Krh-TeKa - 120 mm	27
MRLS	298 RsRakH 06 (M270)	22
	Rakh 89 122mm	30
ATGW	TOW	
	Spike	
SAM	ITO-90	10
	ITO 2005, ITO 2005 M	16

Army Aviation

Category	Name	In Service
Helicopter	MD 500D/MD 500E (Utility)	7
	NH90 TTH	20
UAV	Ranger	11
	Orbiter	55 systems

New Procurements/Upgrades

✦ Defence Ministry in January 2014 cleared a contract worth EUR199.9 million ($271 million) to acquire 106 used Leopard 2A6 main battle tanks from the Netherlands. Delivery of the MBTs has started in 2015 and would be completed by 2019.

✦ The state-owned Patria company has been awarded a contract on the lifecycle update of the XA-180 Pasi armoured vehicles. The contract involves the overhaul of a pre-series vehicle and a series of 70 vehicles in 2015-2017 alongside an option to modernise 210 vehicles by the end of 2021. The first upgraded APC was delivered in November 2014.

✦ The Government has requested a possible sale of 40 Guided Multiple Launch Rocket Pods from the US.

✦ Lockheed Martin has received $45.3 million contract to upgrade the Army's entire fleet of M270 Multiple Launch Rocket Systems. The Defence Ministry has also approved the purchase of 90 M31A1 and 150 M30A1 missiles for the 22 M270 MLRS.

✦ Acquisition process for mini unmanned air vehicles for the army is going on. Five bidders have responded to the RFI which was issued in 2009.

✦ Delivery of all the 20 NH90 TTH helicopters has been completed in 2015.

NAVY

Personnel: 2,000+

Equipment

Category	Name	In Service
FAC	Hamina class	4
	Rauma class	4
Patrol vessel	Kiisla class	2
Amphibious ship	LCUs, LCPs	40+
Mine sweeper	Kuha class	6

Navy continued..

Category	Name	In Service
Mine sweeper	Kiiski class	7
	Hameenmaa class (minelayer)	2
	Pohjanmaa class (minelayer)	1 (to be sold)
	Pansio class	3
Auxiliary ship	Kemio class	1
	Kampela class	3
	Valas class	5
	Hauki class	8
	Hila class	4
	Jehu class	3

New Procurements/Upgrades

✦ Under the Laivue 2020 (Squadron 2020) project approved the Government, four new corvettes will be procured to replace the three minelayers of the Pohjanmaa- and Hameenmaa-classes, as well as the four light fast attack craft of the Rauma-class. The new vessels will be built between 2019 and 2024, with initial operational capability 2025.

✦ Three Jehu-class of 200-tonne combat/ landing boats have been delivered to the Navy in June 2015. Finland has ordered 12 of these fast, manoeuvrable vessels from Marine Alutech Oy Ab.

✦ Patria has upgraded and delivered four Rauma class missile FACs in 2014.

AIR FORCE

Personnel: 3,000+

Equipment

Category	Name	In Service
Fighter	F/A-18 C/D	62
Tanker & Transport	Learjet 35	3
	CASA C-295	3
	Fokker F-27	1
	C-17 Globemaster	3 (shared with NATO members)
Trainer	Hawk Mk.51/51A (trainer)	49
	Hawk Mk.66 (attack)	16
	L-70 Vinka	28
	L-90 Redigo	7
	Pilatus PC-12	6

New Procurements/Upgrades

✦ New-generation fighter jets will be procured to replace the entire fleet of 62 F/A-18 Hornet fighters. The Defence Ministry has issued an RFI to the UK, France, Sweden and the USA in April 2016 for the procurement programme. The tender for the Hornet replacement solution is due to be issued in 2018, followed by final contract award by 2021.

✦ The Government is also considering to acquire armed drones along with the new fighters.

✦ Under the US Foreign Military Sales programme, Lockheed Martin has been contracted to support the integration of the long-range Joint Air-to-Surface Standoff Missile (JASSM) onto the Finnish Air Force's F-18C/D fighters.

✦ Planning to refurbish the F-18 fighters and keep it operational till 2020, Finland has sought Mid-Life Upgrade (MLU) equipment kits from the US under the Foreign Military Sales programme.

✦ A deal has been reached to procure 28 used Grob G115E trainer aircraft from the UK-based Babcock International Corporation for about 6.6 million euros. The aircraft will be delivered in 2016-17.

✦ 12 Ground Master 403 long-range air defence radars are being delivered to enhance the country's air surveillance capability.

Defence Expenditure

Total defence spending: €2.9 billion (2016); €2.7 billion (2015)

Defence spending in terms of GDP: 1.37% (2016)

Leading arms suppliers to Finland are the US, Germany, Russia and Sweden.

Defence Production and R&D

Finland's domestic defence industry is ruled by Patria Industries, which is majority-owned by the Government. The defence industry is primarily involved in the production of land-based systems, including armoured personnel carriers, self-propelled mortars and air-defence systems among others. However, the majority of land and aerospace systems are imported from other countries. The Finnish government is also exploring options to cooperate with European nations for the production and maintenance of defence equipment.

Defence Procurement

Finland, in the past, has acquired several Soviet-era defence equipment from Russia. However, strained relations with Russia over the ongoing Ukraine crisis has made the Nordic country inch closer towards NATO. Finland has partnered with European countries and has been making defence procurements from both Europe and the USA. Future military procurements, mainly for the Navy and Air Force, have been outlined recently. The country is planning to curtail the number of defence personnel by 200 while increasing budgetary allocation by additional 30 million euros to acquire new military hardware.

F-18 Hornet of Finnish Air Force

CONTACT DETAILS

Ministry of Defence
Etelainen Makasiinikatu 8
P.O. Box 31, FIN-00131,
Helsinki
Finland
Tel.: +358 295 16001
Fax: +358 9 1608 8244
E-mail: kirjaamo@defmin.fi
Website: http://www.defmin.fi

STRATEGIC INFORMATION

FRANCE
(Capital: Paris)

INTRODUCTION

Area: Total 643,801 sq km; 551,500 sq km (metropolitan France)

Population: 66,836,154 (July 2016 est.)

Coastline: Total: 4,853 km; 3,427 km, metropolitan France, (has numerous islands under its control in the Pacific Ocean and Indian Ocean).

Maritime Claims: Territorial sea: 12 nm

Contiguous zone: 24 nm

Exclusive Economic Zone: 200 nm (does not apply to the Mediterranean)

Continental shelf: 200 m depth or to the depth of exploitation

GEOPOLITICAL IMPORTANCE

France is the largest western European country with several overseas territories and islands located on other continents. Metropolitan France extends from the Mediterranean Sea to the English Channel and the North Sea and from the Rhine to the Atlantic Ocean. It is bordered by Belgium, Luxembourg, Germany, Switzerland, Italy and Monaco, with Spain and Andorra to the Southwest. The country is linked to the United Kingdom by the Channel Tunnel and has land borders with Suriname and Brazil through French Guiana, as well as with the Kingdom of the Netherlands. France is an important player in European geopolitics - it holds the key to the entirety of Europe both in and out of the United Nations. Along with Germany France is a major player of the European Union. It continues to be a major power in the world with its well-developed economy, membership in European organisations, strong military posture and political influence.

POLITICAL OVERVIEW

France's current system of government, known as the Fifth Republic, is based on a constitution that was adopted by popular referendum in 1958. The current constitution has greatly increased the powers of the President making it the most powerful position in the French political system. Since the passing of the 2008 Constitutional reform, the maximum number of terms a president can serve has been limited to two. The office of President under the Third and Fourth Republic was a largely ceremonial one. The presidential term is five years. He is directly elected in a two-stage voting system. A candidate who receives more than 50% of the vote in the first round is elected. However, a second round is held if no candidate receives 50%. The French Parliament consists of two chambers – the National Assembly (lower house) and the Senate (upper house). The Prime Minister is head of the government. He is nominated by the majority party in the National Assembly and appointed by the President for an indefinite term. The President names the prime minister, presides over the cabinet, commands the armed forces, and concludes treaties. The country's highest appellate court is called the Cour de Cassation and the six chief judges are appointed by the President. The Constitutional Court, a creation of the Fifth Republic, holds the power of judicial review.

KEY POLITICAL PERSONS

PRESIDENT:
Francois Hollande

PRIME MINISTER:
Bernard Cazeneuve

ECONOMY

France has long been part of the world's wealthiest and most developed national economies. Its economy is the fifth largest in the world by nominal GDP. It has the second-largest economy in Europe, after Germany. The government dominates major sectors of the economy as a large shareholder in many semi-public enterprises. Tourism is one of the many contributors to its economy with France being the most visited country in the world. The country has substantial agricultural resources, a large industrial base, and a highly skilled work force. France has been very successful in developing dynamic telecommunications, aerospace, and weapons. According to SIPRI, France is also the fourth largest weapons exporter in the world.

GDP: $2.666 trillion (2015 est.), $2.632 trillion (2014 est.)

Real Growth Rate: 1.3% (2015 est.), 0.6% (2014 est.)

Industries machinery, chemicals, automobiles, metallurgy, aircraft, electronics; textiles, food processing; tourism

Total Exports: $510.5 billion (2015 est.), $584.5 billion (2014 est.)

Export Commodities: Machinery and transportation equipment, aircraft, plastics, chemicals, pharmaceutical products, iron and steel, beverages

Major Markets: Germany 15.9%, Spain 7.3%, US 7.2%, Italy 7.1%, UK 7.1%, Belgium 6.8% (2015)

Total Imports: $537.5 billion (2015 est.), $631.1 billion (2014 est.)

Import Commodities: machinery and equipment, vehicles, crude oil, aircraft, plastics, chemicals

Major Suppliers: Germany 19.5%, Belgium 10.7%, Italy 7.7%, Netherlands 7.5%, Spain 6.8%, US 5.5%, China 5.4%, UK 4.3% (2015)

DEFENCE & SECURITY

The priority mission for the armed forces is protection of the country's territory and population. According to the Defence White Paper 2013, the armed forces will retain the capacity to set up command structures for the land, sea and air components on the level of an army brigade or equivalent. Intelligence capabilities will be developed as a priority, with a greater effort to modernise human resources and upgrade the technical imaging and electromagnetic interception resources, whether space-based, air, naval or land. For the first time, the armed forces model includes military cyber defence capabilities, in close liaison with intelligence and defensive and offensive planning, in preparation for or support of military operations. The Special Forces will be reinforced, in terms of personnel, command capabilities and their capacity to coordinate with the intelligence services; the joint dimension of the COS (special operations command) will also be strengthened.

The White Paper said the Atlantic Alliance is an essential component of the defence and national security strategy, therefore, France intends to capitalise on its full participation in the military structures of NATO.

✦ The Defence White Paper has introduced a change and set forth four guiding principles for the new armed forces model:

✦ Strategic autonomy, which is key to freedom of decision and action and to France's capacity to take the initiative in operations it may deem necessary, likewise to wield influence in a multinational coalition in which it might be engaged.

✦ Capacity to engage in coercive and initial-entry operations in a theatre of war, crisis management operations in all the varying forms they make take today and in actions required to protect security interests over the whole spectrum of the most probable military missions.

✦ Differentiation of forces, which consists of distinguishing between them as a function of the types of missions they are required to perform – deterrence, protection, crisis management, coercion and war operations.

✦ Pooling, which will give the capacity to allocate a core of multi-role, scarce and technologically high-level capabilities to several missions (protection, deterrence and intervention); this principle also entails pooling some of the most sophisticated technical resources between the different intelligence services.

KEY DEFENCE PERSONS

DEFENCE MINISTER: Jean-Yves Le Drian

CHIEF OF DEFENCE STAFF: Gen. Pierre de Villiers

CHIEF OF STAFF OF THE ARMY: Gen. Jean-Pierre Bosser

CHIEF OF STAFF OF THE NAVY: Admiral Christophe Prazuck

CHIEF OF STAFF OF THE AIR FORCE: Gen. Andre Lanata

DIRECTOR GENERAL OF THE GENDARMERIE: Gen. Richard Lizurey

NATIONAL ARMAMENT DIRECTOR: Laurent Collet-Billon

External conflict

France is part of the United States led military campaign against the jihadists in Syria and Iraq. The two countries have pledged to work together on countering the threat from the Islamic State. France has intensified its military campaign in Syria to retaliate against Islamic State, which claimed responsibility for the terror attacks in the French capital in mid-November 2015, closely coordinating with the U S military. France also drafted a UN Security Council resolution urging UN members to "take all necessary measures" in the fight against Islamic State and al-Nusra Front. On 20 November 2015, the UN Security Council unanimously passed the French-British drafted-sponsored resolution.

Territorial disputes: For decades, a territorial dispute had simmered between Canada and France because of the tiny French islands of St Pierre and Miquelon which lie just south of Newfoundland. Madagascar claims the French territories of Bassas da India, Europa Island, Glorioso Islands, and Juan de Nova Island. While Comoros claims Mayotte, Mauritius claims Tromelin Island. There is a territorial dispute between Suriname and the French overseas department of French Guiana. France also asserts a territorial claim in Antarctica (Adelie Land) and both France and Vanuatu claim Matthew and Hunter Islands, east of New Caledonia.

Peace Operation: France is an active participant in the United Nations peacekeeping missions. The French armed forces are present in a number of conflict zones around the world. France supported the UN peacekeeping operations such as UNTSO, UNIFIL, MINURCAT, MINURSO, MONUSCO, UNMIL, UNOCI, and MINUSTAH.

THREAT PERSPECTIVE

Terrorism: Terrorist threat in the country remains high. A series of coordinated terrorist attacks consisting of mass shootings, suicide bombings, and hostage-taking occurred in Paris and in the northern suburb of Saint-Denis. The Islamic State of Iraq and Syria (ISIS) claimed responsibility for the attacks in which 129 people were killed in mid-November, 2015. Following the latest attack, France's President Francois Hollande declared that France was at war and has called on the United States and Russia to join a global coalition to destroy Islamic State.

France with a colonial history is determined to combat terrorism in all its forms with its partners, under the auspices of the United Nations, within the European Union, in the G8, and under other regional cooperation frameworks. In 2005, the EU adopted a European strategy to fight against terrorism structured around four principles - prevention, protection, pursuit and reaction.

The Defence White Paper 2013 stated that the level of threat and the climate of uncertainty characterising France's international environment since 2008 have not diminished and hence must take the following phenomena into consideration. Overall there is no external threat to France.

Threats related to power: The rise in military spending particularly in Asia; power politics (Russia, China); regional destabilisation (Middle East); proliferation of weapons of mass destruction and the consequences thereof; cyber attacks instigated by countries as also the US which tapped the mobile phone of the French President.

Risks related to weakness: Certain weak or failed States can become a threat. This strategic phenomenon is taking on a new dimension. When certain States prove incapable of exercising their responsibilities, the risks and threats that crystallise in their territory can rapidly spill over and undermine the security situation.

Threats and risks intensified by globalisation: They include threats and risks to the ever-increasing flow of goods, merchandise and people resulting from economic globalisation; the risks to maritime security with the increase in piracy; the risks of terrorism, the exponential increase in risks to digital infrastructure through cyber attacks and potential threats in outer space. All these security threats are intensified by globalisation.

Cyber threat: France has announced plans to bolster its cyber-defence capabilities having been a victim of cyber attack. The new white paper highlights cyber threats as a significant concern, stating that the world of cyberspace is now its own battlefield and in response, France will develop both its offensive and defensive capabilities. These capabilities include intelligence, technical capabilities and the ability to determine the origin of a threat.

STRATEGIC RELATIONS

India: France and India have extensive strategic co-operation. The strategic partnership includes the civil nuclear cooperation, Defence cooperation, Counter-terrorism cooperation and Space cooperation. France and its defence industry actively contribute to the "Make in India" programme in the defence sector. The first conventional submarine, Scorpene, which started being built in India in 2008 with transfer of technology and support from DCNS, began sea trials in 2015. An agreement has been signed on India's acquisition of 36 Rafale fighter jets. The two nations are important partners in space technology and applications. Indian Space Research Organisation (ISRO) and its French counterpart Centre National de Etudes Spatiales (CNES) also have a rich history of cooperation and collaboration. The annual joint military exercises have grown more elaborate.

United States: France and the United States continue to cooperate closely on many issues, most notably in combating terrorism, efforts to stem the proliferation of weapons of mass destruction (WMD), and on regional problems, including in Africa, Lebanon, and Kosovo. The two nations are among the five permanent members of the UN Security Council (P5). Trade and investment between them are strong. US exports to France include industrial chemicals, aircraft and engines, electronic components, telecommunications, computer software, computers and peripherals, analytical and scientific instrumentation, medical instruments and supplies, and broadcasting equipment. The US is the top destination for French investment and the US is the largest foreign investor in France. They have a bilateral convention on investment and a bilateral tax treaty addressing, among other things, double taxation and tax evasion.

United Kingdom: France and UK have signed a Defence and Security Cooperation Treaty with an aim to develop co-operation between British and French Armed Forces, for sharing and pooling of materials and equipment including through mutual interdependence, the building of joint facilities, mutual access to each other's defence markets, and industrial and technological co-operation. Key planks include the creation of a 10,000-strong joint expeditionary force involving 5,000 troops from each country in all three services, training and exercising together and prepare for high-intensity combat, rescue and peacekeeping missions; and the development of an integrated strike force to be deployed on either nation's aircraft carriers.

China: France was the first among Western powers to establish diplomatic relations with China. It views China as its very important strategic partner. In 2015 French Prime Minister, Manuel Valls, and the Chinese premier, Li Keqiang, visited each other's country. They also signed a contract for the sale of 75 Airbus A330 aircraft. The two countries are permanent members of the United Nations Security Council and the world's two largest economies. Both the countries have decided to expand nuclear cooperation besides deepening their cooperation in the aerospace industry and participation of French manufacturers in China's C919 aerospace programme.

Israel: France was one of the first countries to establish diplomatic relations with Israel on 11 May 1949, and subsequently played an active role in helping to consolidate the newly formed State by contributing to its defence effort. The political dialogue between France and Israel has greatly improved and in recent years' trade between the two countries has remained stable with some major investments or joint ventures. Bilateral relations in the cultural, scientific and economic fields and in tourism are currently expanding fast. Cooperation has also been observed in the high-tech fields of defence.

Germany: France and Germany have united after World War II in pursuing political objectives to guarantee that there would never be another war between the two nations. Consequently, the European Union was created with the integration of other European countries. The partnership between Europe's two largest economies has been vital to the functioning of the EU and the 17- nation euro zone. The new currency Euro, a French-German confection, was launched in world money markets on 1 January 1999. The two countries have always been important players in furthering the ideals of European integration. Germany is the fifth-leading foreign investor in France and in 2009 became the leading country of origin in terms of foreign investments and creators of jobs in France.

DEFENCE CAPABILITIES

The French military is one of the most capable forces in Europe. The French Armed Forces encompass the French Army, the French Navy, the French Air Force and the National Gendarmerie. The president of the republic heads the armed forces. He is the supreme authority for military matters and is the sole official who can order a nuclear strike. The French military doctrine is based on the concepts of national independence, nuclear deterrence and military self-sufficiency. The armed force structure is suited for expeditionary employment. The force has its own indigenous ballistic missile submarines, maintains its own air and land-based nuclear deterrent. It is equipped with LeClerc MBT, VBCIs, an integrated infantry combat system (FELIN), self-propelled howitzers (CAESER), and attack helicopters (Tiger). It maintains its own nuclear-powered aircraft carrier, combat aircraft including the Rafale and Mirage 2000 fighters and AWACS aircraft. France military modernisation programme - the Scorpion programme - is aimed at transforming the French Army in order to meet the operational challenges of tomorrow. It brings together all the French Army platform programmes: EBRC - Armoured Reconnaissance and Combat Tank, VBMR - Multirole armoured vehicle, modernised Leclerc heavy tank, Felin, Scorpion Combat and Information System (SIC-S) and communication means.

ARMY: The French army is the largest component of the French Armed Forces. The Army comprised of the Troupes de Marine, the Armoured Cavalry Branch, the Artillery, the Military engineers, the Infantry, which includes the Chasseurs Alpins, specialist mountain infantry, Maintenance Matériel, Logistics (Train), Signals (Transmissions), and Commissariat.

The Defence White Paper 2013 seeks to overhaul the armed forces to create a more mobile army and to boost intelligence resources and Special Forces. According to the document the land forces will rely on a land-based operational force of around 66,000 deployable troops including, apart from back-up and operational support, seven joint brigades: two brigades trained for coercive combat, three multi-purpose brigades, two light brigades; 200 heavy tanks, 250 median tanks, 115 utility helicopters, 140 reconnaissance and attack helicopters and some 30 tactical drones.

Personnel: 113,083

Equipment

Category	Name	In Service
MBT	Leclerc	254 (200 modernised)
Armoured car /AIFV /APC	ERC-90	110
	AMX-10RC	256
	VBCI	630
	VPC	110
	VAB	3,200
	AMX-10P/PC	110
	Buffalo	5
	Nexter Aravis	15
	BVs10 mKII	53
Artillery	155mm & AUF -1155mm SP howitzer	157
	155mm TRF-1 towed	
	Ceasar 155mm truck-mounted	577
Mortar	RTF1 120mm	364
MRL	MLRS (converted for unitary warhead rockets)	13
Anti-tank weapon	AT4CS	
	APILAS	
ATGW	Javelin	76
	Milan	947
	Hot (on VAB)	135
	Hot (on helicopters)	110
	ERYX	700
AA gun	53T2 20mm	525
UAV	Crecerelle	12
	CL-289	54
	Sperwer	18
Weapon	Mistral (SAM)	336
	Crotale (mobile SAM)	

New Procurements/Upgrades

- The DGA has awarded Nexter Systems the Leclerc tank modernisation contract to deliver 200 modernised Leclerc MBTs and 18 modernised DCL armoured recovery vehicles from 2020.
- France will acquire Jaguar EBRC 248 (Engins Blindés de Reconnaissance et de Combat (armoured reconnaissance and combat vehicles) to replace the Army's AMX 10RC and Sagaie light tanks, and the VAB HOT tank destroyers. Its main armament will comprise a 40mm automatic gun and MMP missiles. Delivery is scheduled for 2019.
- Under the Scorpion Programme France will acquire 1,722 Véhicules Blindés Multi Rôles (armoured multi-role vehicles, VBMR), called Griffon, to replace the current Véhicules de l'Avant Blindé (VAB) wheeled armoured vehicles. Delivery of the first VBMR is expected in 2018.
- The Army is scheduled to receive the first three of 95 modified VBCI 8x8 armoured vehicles. The vehicles will be 29-tonne VCI (Véhicule de Combat d'Infanterie) IFV

variants modified to a gross vehicle weight (GVW) of 32 tonnes with improved protection against improvised explosive devices. Delivery is expected to complete in 2017.

Army Light Aviation: The French Army Light Aviation was established for observation, reconnaissance, assault and supply duties. It operates numerous helicopters in support of the French Army.

Personnel: 2,200

Equipment

Category	Name	In Service
Combat /transport helicopter	Tigre HAP	39
	Tigre HAD	3+
	Puma SA.330	84 (to be withdrawn)
	Gazelle SA.342	147
	EC-725 Caracal	8
	Fennec AS.555	15
	NH-90TTH Caiman	8+
Utility Aircraft	Cougar AS.532	18
	Cessna F406	2
	PC-6	5
	TBM-700	8

New Procurements/Upgrades

✦ The DGA has amended the Tigre contract with a firm order for 24 attack helicopters and a retrofit of 36 escort units. Expected delivery of the first retrofit of the escort to the attack model is 2017.The existing fleet of escort units will be upgraded with laser-guided Hellfire II missiles and a more powerful engine. The first three of the attack, or hélicoptère appui-destruction (HAD), Block 2 version has been delivered.

✦ France has ordered six additional NHIndustries NH90 troop transport helicopters for its army, increasing the service's eventual inventory to 74. Deliveries of the rotorcraft are due to take place between 2017 and 2019.

NHIndustries NH90 troop transport helicopter. *Image credit: Airbus Helicopters*

NAVY: The blue water French Navy has five branches - the Force d'Action Navale, the Forces Sous-marines, the Aviation Navale, the Fusiliers Marins including Commandos de Marine and the Gendarmerie maritime. France aims to continue sailing blue water Navy that is able to project maritime power on the waves, silently below and by striking from the air. The French Navy's area of operation includes the North Atlantic, the Mediterranean Sea, and the Indian Ocean.

The French submarine fleet consists of nuclear-propelled attack (SSNs) and ballistic missile submarines (SSBNs). Its SSBN force is the nucleus of the country's strategic deterrent, and maintains a continuous at-sea presence. By 2025, the navy plans to be equipped with all the capabilities needed to complete a full range of missions, from defence to security. Current important naval programme includes the FREMM multi-mission warship programme and the Barracuda class nuclear attack submarine programme.

Personnel: 36,658

Major naval bases: Brest, Toulon, Ile Longue and Cherbourg in Metropolitan France, Fort de France, Degrad des Cannes, Port des Galets, Nouméa and Papeete.

Equipment

Category	Name	In Service
Aircraft carrier	Charles de Gaulle	1
Submarine	Triomphant class with 16 M-45 SLBM	4
	Rubis class SSN	6
Destroyer/ Frigate	Horizon class	2
	Aquitaine class	2
	Georges Leygues class	5
	Floreal class	6
	Cassard class	2
	La Fayette class	5
Amphibio-us force	Mistral class LHD	3 + 1
	Batral class assault ship	5
	LCM (Landing craft mechanised)	20
	LCT (Tank Landing Craft)	11
	LST (Landing Ship, Tank)	5
OPV	D'estienne D'orves class	9
Patrol craft	L'Audacieuse class	4
	Grebe	1
	Sterne	1
	Arago	1
	Leopard PCC	8
	Flamant class	4
	L'Adroit OPV	1
Mine warfare force	Eridan class minehunter	13
	Vulcain class support ship	4
	Antares class sonar towing vessel	4
Auxiliaries	Durance class tanker	3
	Hydrographic ship	3
	Oceanographic ship	2

Navy continued...

Category	Name	In Service
Auxiliaries	Trial ship	10
	Command ship (Indian Ocean)	3
Missile tracking ship	A601 Monge	1
Electromagnetic research ship	A759 Dupuy de Lôme	1
Weapon	Exocet missile (ASM)	
	Aster missile (SAM)	
	Crotale missile (SAM)	

✦ The DGA has ordered two more BSAH offshore support and rescue vessels for the Navy. The previous order for two BSAH vessels will be delivered in 2018 and the other two in 2019.

✦ The first of the five new Fregate de Taille Intermediaire (FTI) ship is planned for delivery in 2023 to replace the La Fayette class frigates. The FTI will join the navy's eight FREMM multirole frigates, six in anti-submarine warfare configuration and two fitted for air defence, and two Horizon class anti-air warfare frigates.

DCNS' FTI concept design for the French Navy. *Image credit: Luca Peruzzi*

✦ Order placed for third Barracuda-type nuclear-powered attack submarine. The navy plans to acquire six Barracuda nuclear-powered hunter-killer submarines to replace the Rubis-class attack subs.

✦ Multi-mission OPVs are expected to enter service around 2024. The boats will replace a fleet of some 15 vessels of the P400, Aviso 69 and fisheries protection boats.

Naval Aviation: The Aviation navale is a combination of carrier squadrons and naval patrol air force.

Major naval air bases: Hyeres (main), Landivisiau, Lann-Bihoue and Lanveoc-Poulmic

Equipment

Category	Name	In Service
Interception aircraft	Rafale M F3	39
AEW	E-2C Hawkeye (shipborne)	3
MP/ASW	Atlantique 2	22
Surveillance aircraft	Falcon 50M	4
	Falcon 200	5

Naval aviation continued...

Category	Name	In Service
Surveillance aircraft	EMB 121AN	6
ASW	LYNX	15
ASuW	AS565SA Panther	16
Transport/ SAR	EC225 Super Puma (SAR)	2
	AS365N Dauphin SAR	3
	Alouette III	6
	Falcon DA	5
	MER	10
	Alquette II	4
	Rallye	8
	CAP 10	7

New Procurements/Upgrades

✦ 48 production-standard Rafale M have been ordered in view of an operational fleet of 40 aircraft by 2020.

✦ 15 Atlantique 2 MP/ASW aircraft will be modernised with re-deliveries in 2018-2023.

✦ 27 NFH-90 helicopters are on order.

✦ A MoU has been signed with the UK, confirming joint orders for a future helicopter launched anti-ship missile.

AIR FORCE: The French Air Force is the fourth largest air force in terms of aircraft in NATO, and the second largest in the European Union. The Minister of Defence is responsible for execution of military policy. The Chief of Staff of the Armies (CEMA) acts as advisor to the Minister of Defence. He is responsible for the preparation and logistic support of the air force. The CEMAA is assisted by the air force staff and by its subordinate services. The Air Force operates aircraft of various types including combat, transport, liaison, training and helicopters.

The 2013 Defence White Paper has stated that the air forces will deploy the airborne component of the deterrence mission. They will have the capacity for initial entry in war operations, deep- penetration strikes and support for ground manoeuvres, on the one hand, and to conduct crisis management missions or missions to protect national air space or approaches, on the other. Their transport capability will be upgraded and they will make a strong contribution to surveillance and intelligence in crisis or engagement zones.

Major air bases: Avord, Bordeaux, Cazaux, Cognac, Creil, Dijon, Evreux, Istres, Luxeuil, Monte-de-Marsan, Nancy, Orange, Orleans, Paris, Saint-Dizier, Salon, Solenzara, Toulouse, Tours,Villacoublay, BA104 (UAE). They have airbases on islands in the Pacific and Indian Ocean. Further their friendly relations with African countries, where they had colonized enables them to use their air bases.

Personnel: 42,276

Equipment

Category	Name	In Service	On Order
Fighter tanker	KC-135FR/KC-135R	14	
	Rafale B	42+	
	Rafale C	45+	
	Mirage 2000B/C	50	
	Mirage 2000-5F	34	
	Mirage 2000N	65	
	Mirage 2000D	80	
AEW	E-3F AWACS	2	
Transport/ VIP	A400M Atlas	7	50
	C-130Hs & C-130H-30	14	
	C-135FR	11	
	KC-135R	3	
	C-160 Transall	51	
	A-340	2	
	A-310	3	
	A-319CJ	2	
	Falcon 900	2	
	Falcon 7X	1	
	CN-235	27	
	Falcon 50	4	
	TBM 700	15	
	Twin Otter	5	
	Socata TBM	15	
Trainer aircraft	Alpha jet	85	
	EMB-121 Xingu	23	
	Jodel-D-140	18	
	Walter Extra 300	3	
Helicopter	Puma	24	
	Super Puma	5	
	Cougar	5	
	EC725 Caracal (CSAR)	11	
	AS555 - Fennec	35+	
UAV	Harfang	4	
	Reaper	3	
Weapon	Crotale NG	12	
	SAMP-T	8	

New Procurements/Upgrades

✦ France has ordered four more Lockheed Martin C-130 Hercules transport aircraft. The first two aircraft will be C-130J Super Hercules transports, and the second two in the KC-130J tanker/transport version. The first two aircraft will be delivered by the end of 2017 and in early 2018 and the two-tanker aircraft in 2019.

✦ The DGA has awarded Dassault Aviation and MBDA a contract for the mid-life upgrade of the Mirage 2000D fleet to allow them to operate alongside the Rafale combat aircraft beyond 2030. The upgrade includes the fitting of an automatic cannon for ground attack as well as the replacement of their previous-generation Magic self-protection missiles by the MICA.

Mirage 2000D strike fighter. *Image credit: French Air Force*

✦ A total of 180 Rafale fighters are on order with delivery of 150 aircraft scheduled for 2019. Dassault Aviation will develop a new standard of the Rafale fighter aircraft, designated "F3 R". The new configuration will integrate Meteor long-range air-to-air missile, a new-generation laser pod for carrying out air-to-ground strikes and the latest version of AASM laser-guided bomb besides new electronic components for the fighter.

✦ The MoD has received its ninth A400M Atlas military transport aircraft. Two more aircraft are expected to be delivered before the end of 2016. A total of 174 aircraft are currently on order.

The newest A400M transport for France. *Photo courtesy Airbus DS*

✦ 12 Airbus A330 MRTT tanker-transport aircraft (Phoenix) are on order to replace the KC-135. The first two aircraft will be delivered in 2018-2019.

✦ France is planning to bring a total of 12 Predator B/MQ-9 Reaper RPA into service by 2019.

✦ DoD has ordered two KC-130J for France for 2020 delivery.

The Gendarmerie National is an armed force attached to the Ministry of the Interior and employed on a day-to-day basis for internal security missions. The gendarmerie nationale is in charge of public safety, with police duties among the civilian population. It also contains a military force and a Special Forces component. It is mandated to fulfill national security duties and duties in support of its parent ministry.

STRATEGIC NUCLEAR FORCES

- ✦ The Strategic Nuclear Force command controls all the air force's nuclear assets. This force controls the following units:
- ✦ The FOST Navy command with 4SSBNs with a total of 48 SLBMs: 16 M4s with a maximum of six.
- ✦ TN70-TN71 150kton warheads each and 32 M45s with a maximum of 6TN 75 150kton warheads each.
- ✦ The FAS Air Force Command with a total of 60 Mirage 2000Ns armed with the ASMP and ASMP-A nuclear- tipped missile, plus11 C-135R + 3 KC-135R tankers.
- ✦ Two Ship borne Super Etendard squadrons with a total of 30 aircraft.

Major Armament Producers

SAFRAN is a French conglomerate involved in defence, aerospace propulsion and equipment, and security.

The Thales Group products include radars, secure communications, drones, electronic missile systems.

Dassault Aviation is a French aircraft manufacturer of military, regional and business jets, a subsidiary of Dassault Group.

EADS: Co-manufacturer of MBDA missile systems, NH90 helicopters for troop transfer and Tiger combat helicopters, Airbus A400M military transport planes, Harfang drones, Astrium anti missile defence system.

Nexter: Manufacturers Leclerc tanks, VBCI armoured vehicles, Caesar wheeled self-propelled guns, Aravis tanks, munitions. They have recently developed the 155 mm (52 calibre) towed gun.

DCNS is a naval defence company. The group designs, builds and supports surface combatants, submarines, systems and equipments. Products include conventional and nuclear powered submarines and FREMM frigates.

DEFENCE DEALS

Joint Venture Programmes

India

Dassault Reliance Aerospace: India's Reliance Group and France's Dassault have set up a Joint Venture called 'Dassault Reliance Aerospace' which will focus on research and development of unspecified defence projects in India. The defence partnership was formed as part of a fighter jet deal agreed between the two nations. India signed a contract to buy 36 Rafale fighter jets. Dassault, which is building the aircraft, agreed under the terms of the deal to invest about 50 percent of the value of the contract in India.

Scorpene submarine: Under a technology transfer agreement with France, India is currently building Scorpene submarines at Mumbai's Mazagon Docks Limited (MDL). The first of six Scorpene diesel-electric attack submarines (SSKs), INS Kalvari, has been completed. MDL is building five other the DCNS-designed vessels under licence at its facilities in Mumbai.

Maitri SRSAM: India and France plans to jointly develop a Short-Range Surface to Air Missile. Talks are on for joint development and production of surface-to-air missile systems between the Defence Research and Development Organisation (DRDO) and MBDA. The MoU to co-develop the SR-SAM as part of a 'Maitri' project was signed during French President Francois Hollande's visit to India in February, 2013.

United Kingdom

UCAS: The UK and France have announced a £1.54bn project to build a prototype of the Future Combat Air System. The next-generation UAV programme builds on the successful £120m joint feasibility study undertaken after the last UK- France Summit in 2014. With full scale development of prototypes set to begin in 2017. The end result should be a UCAS capability that can perform sustained surveillance, target designation, intelligence gathering and strike missions in hostile territories. Six industry partners will work on the project from the British and French industrial bases: BAE Systems, Dassault Aviation, Thales France, Leonardo Airborne and Space Systems (formally Selex ES), Rolls-Royce and Safran.

Missile: France and the UK have signed the Demonstration and Manufacture (D&M) contract for the Future Anti-Surface Guided Weapon (Heavy)/Anti Navire Léger (FASGW(H)/ANL) anti-ship missile programme. Designed to equip the Royal Navy's (RN's) new Wildcat HMA.2 helicopter, and the French Navy's NH90 Caiman and Panther maritime helicopters, FASGW(H)/ANL is a 100kg-class anti-ship missile intended to address targets from fast attack craft up to corvette size.

Italy

FREMM frigate: The FREMM European multi-mission frigate is a joint programme between France and Italy. It will build 21 FREMM frigates for the French Navy and the Italian Navy. The prime contractors for the FREMM programme are Armaris of France and Orizzonte Sistemi Navali of Italy.

Drone programme: Italy, France and Germany have signed an agreement to develop a European drone programme for reconnaissance and surveillance. The defence ministers of the three countries have pledged a two-year study to lay the basis for a European drone to be operating by 2025. Airbus, Dassault and Alenia Aermacchi are likely to develop the drones.

Brazil

ProSub: Brazil under the ProSub programme plans to build four diesel-electric attack submarines and a single much larger nuclear-powered submarine with the help of France. The first conventionally-powered submarine is expected to be completed in 2017. These submarines being developed with French assistance are based on the French Scorpene class. The nuclear submarine is slated to begin construction in 2016 and to enter operational service 2025.

Europe

nEUROn: France is involved in the European programme for an unmanned combat air vehicle (UCAV) technology demonstrator. The nEUROn programme is designed to validate the development of complex technologies representing all mission systems: high-level flight control and stealth, launching real air-to-ground weapons from an internal bay, integration in the C4I environment, innovative industrial collaboration processes, etc. The nEUROn is being developed with international cooperation, led by the French company Dassault Aviation.

Arms Export

France signed arms export contracts worth around US $17.5 billion in 2015, with Dassault Aviation, MBDA, DCNS and Airbus at the top of the list for the foreign deals. The figure was driven by the sale of 48 Rafale fighter jets to Egypt and Qatar, which brought in €11.5 billion, with the latter holding options for 12 more units. Airbus sales to South Korea of helicopters and A330 multirole tanker transport aircraft put sales to Asia second to that of the Middle East. Saudi Arabia bought €100 million of French arms.

French arms exports rose 18 percent in 2014, the country's best export performance for 15 years, according to the Ministry of Defence. France sold 9.1 billion of weapons in that year. These large contracts represented a total of 4.79 billion euros - a gain of 71 percent compared to the previous year. France's main markets over the period 2010 to 2014 were the Middle East (38 percent), followed by Asia (30 percent). After that came Europe (13 percent), the Americas (11 percent) and Africa (four percent). The top French client over the period was Saudi Arabia, which snapped up 12 billion euros' worth of weapons over the period, followed by India (six billion euros), Brazil (under six billion), the United Arab Emirates (four billion), the United States and Morocco.

France's key customers include Singapore, Greece, NATO partners, North Africa region and Francophone countries. Its major military industries have produced the Rafale fighter, the Charles de Gaulle aircraft carrier, the Exocet missile and the Leclerc tank among others. The most exported French defence equipment includes ships, aircrafts and missiles. India has decided to buy 36 of the Rafale jets in a fly-away condition under a government-to-government (G2G) contract.

DEFENCE BUDGET

Total defence spending: €32 billion (2016), €32.3 billion (2015)

Defence spending in terms of GDP: 2.2% (2014)

France increased its 2016 defense budget by $671 million to $32 billion according to a parliamentary bill. The budget has been increased from roughly $18.7 billion to over $19 billion for military equipment, covering a number of deliveries slated for next year, including nine Rafale fighters, three A400M transport aircraft, five Tiger combat helicopters, and six NH90 transport helicopters, along with a frigate, missiles for the frigate and a Barracuda attack submarine, and 25 heavy vehicles.

Defence Procurement

The Direction générale de l'armement (DGA) is the French Government Defence procurement agency responsible for the programme management, development and purchase of weapon systems for the French military. The DGA covers three distinct missions within the Ministry of Defence. It is responsible for future weapons systems in terms of conducting targeted research and innovation projects. It equips the armed forces in the context of European industrial and technological policy. The DGA supports and promotes French defence industry exportation.

France plans to procure four Lockheed Martin C-130 Hercules transport aircraft. The first two aircraft will be delivered between the end of 2017 and early 2018, with the two refuelling versions in 2019. A330 Multi-Role Tanker Transport (MRTT) aircraft for the Air Force with 12 to be delivered between 2018 and 2025. The air force will also receive more Talios laser designation pods. The French Army light aviation (ALAT) will receive seven more Airbus Helicopters Tiger combat helicopters and six more NHIndustries NH90 Caïman utility helicopters.

For the Navy, new ships will be purchased while a fourth B2M multi-mission vessel will be added to the three previously planned. Six FREMM multi-mission frigates will be delivered before 2019. France's intermediate-sized frigate programme will also be accelerated, with the first delivery scheduled for 2023.

France has also ordered 200 air-to-air Meteor missiles to equip its Rafale fighters. The first missiles will be delivered in 2018. The DGA has contracted Swedish Security Company Saab to supply the Roquette Nouvelle Generation, (Roquette NG) next-generation shoulder-launched weapon system for the French armed forces. The Roquette NG is a general-purpose weapon system which will be employed by all three branches of the French armed forces.

NAVAL WARFARE SYSTEMS

France has built up and maintained a blue water naval capability for many centuries to not only defend its national territory and protect the vital sea lanes of communication, but also to safeguard its overseas islands in the Indian Ocean and Pacific Ocean regions. The four major commands of the French Navy include Force d'Action Navale (Naval Action Force), Forces Sous-Marines (Submarine Forces), Aviation Navale (Naval Air Wing) and Fusiliers-Commandos (Marine Commandos). The Navy presently maintains a fleet of over 100 surface combat platforms including an aircraft carrier, 10 submarines and around 200 fixed-wing aircraft and helicopters. While it is steadily phasing out some of its older-generation platforms and systems, induction of many new warships, aircraft and weapon systems have been undertaken to retain the country's maritime power. The Navy has also reached a deal with the UK for shared use of aircraft carriers in future.

All major combat and support platforms currently in possession of the French Navy have been listed below. Some of the future maritime platforms, systems and weapons which are set to join service in near future have also been detailed.

COMBAT PLATFORMS

The combat platforms of French Navy include **surface** and **sub-surface warfare ships**. The Navy maintains over 100 such platforms. The surface fleet of the Navy is primarily tasked with anti-surface, anti-submarine and anti-air warfare missions. It also deploys forces through amphibious means. The sub-surface platforms, including nuclear-powered submarines, are used for sea control/sea denial operations, intelligence gathering missions, and also for strategic nuclear deterrence of the country.

Surface ships

The surface warfare ships of French Navy include an **aircraft carrier**, **destroyers**, **frigates**, **corvettes**, **missile/patrol craft** and **amphibious platforms** consisting of **LHDs**, **LPDs**, **LSTs** and **LCMs**.

<u>**Aircraft Carrier**</u>

CHARLES DE GAULLE (Nuclear-Powered) – 01 in service

Charles de Gaulle is the French Navy's only nuclear-powered aircraft carrier in service. It is the only nuclear-powered carrier in the world other than those of the US Navy. It is the tenth French aircraft carrier came into service so far. The platform was designed and constructed in French naval dockyards and commissioned into active service in 2001. So far, the flagship carrier has played an important role in creating French domination across the world.

By USN - U.S. Navy official website,

Displacement: 43,182 t (full load)
Length: 261 m
Flight deck: 261m X64.4 m
Beam: 64.36 m
Draught: 9.43 m
Speed: 27 knots
Capacity: 800 commandos, 500 tonnes of ammunitions
Armament: MBDA Aster 15 surface-to-air missile; Mistral short range missiles; 8 × 20 mm cannons
Aircraft carried: Rafale M: 26 aircraft
E-2C Hawkeye: 02 aircraft
SA365 Dauphin: 02
Ec725 Caracal AS532 Cougar: 02

<u>**Destroyers**</u>

FORBIN Class - 02 in service

The Forbin class are air defence destroyers designed and constructed in DCNS Dockyards. First of these vessels entered service in Oct 2014. These multirole warships are specifically designed to provide:

1.Protection of a naval task force (naval, amphibious or even civilian) against aircraft threats, including modern anti-ship missiles.

2. Command and coordination task of air operations from the sea, including aircraft from NATO allies.

© Marine nationale / Johann Peschel

Displacement: 7,150 t (full load)
Length: 142m
Beam: 20m
Draught: 8m
Speed: 29 knots (GT); 18 knots (Diesel)
Range: 13000 Km@18 knots, 7000@ 25 knots
Endurance: More than a month
Armament:

SAM: MBDA PAAMS/Aster 15, Aster 30, Mistral
Anti-ship: 8 × Exocet MM40 Block 3 (Mid-ship)
Anti-submarine: 2 × torpeo tubes
Guns: 2 × 76 mm super rapid guns
2 × 20 mm gun
Aircraft carried: 1 Nh90 helicopter

AQUITAINE Class - 04 in active service; 08 are planned for the Navy by 2020

The Aquitaine class of warships was launched under the Frigate European Multi-Mission (FREMM) Programme. This class of vessels were designed by France's DCNS and Italy's Ficantieri in a joint programme to replace the existing destroyers and frigates within the French and Italian navies. The platforms are also being marketed extensively by both the countries in the international defence market. They have so far been exported to Moroccan and Egyptian navies.

© *Marine nationale / Guillaume Izard*

Displacement: 6,096 t (full load)
Length: 137m
Beam: 19.7m
Draught: 5.4m
Speed: 27.5 knots
Range: 6000 nautical mile @15 knots
Armament:
*SAM,:*MBD PAAMS/Aster 15, Aster 30
Anti-ship: 8 × Exoce MM40 Bloc3 (Mid-ship), SCALP- MdCN LACM (to be fitted in near future)
Anti-submarine: 2 × torpedo tubes
Guns: 2 × 76 mm super rapid gun
2 × 20 mm gun & 2 HMG
Aircraft carried: 1 NH90 helicopter

CASSARD Class - 02 in Service

The Cassard class are multi-role destroyers which can be deployed for anti-aircraft operations such as escort of aircraft carriers or LHDs as a part of naval task force. Two of these vessels are currently in service with French Navy. First of these vessels was inducted in 1988 and is due to be retired from active service in the coming decade. The platforms were constructed by DCNS at its Lorient dockyards. Originally four of such vessels were planned but due to budget limitation, only two were completed and joined service.

Type F 70 AA Cassard - © *Marine nationale*

Displacement: 5,080 t (full load)
Length: 139m
Beam: 15m
Draught: 6.5m
Speed: 29.5 knots
Range: 8000 nautical mile @17 knots
Complement: 250 (25 officers)
Armament:
SAM: Standard SM-1MR medium range, Mistral
Anti-ship: 8 × MM40 Exocet
Anti-submarine: 2 × fixed torpedo tubes
Guns: 1 × 100 mm Dual Purpose gun
2 × 20 mm anti-aircraft guns
4 × 12.7 mm machine guns
Aircraft carried: 1 AS 565SA helicopter

GEORGES LEYGUES & Modified GEORGES LEYGUES Class - 05 in Service

First of this class entered service in 1982, mainly designed to counter Soviet submarine forces. The last of these vessels was commissioned in 1990. The platforms are primarily designed for anti-submarine roles. Equipped with powerful underwater sensors, the warships have been the backbone of French Navy's anti-submarine forces. The first three vessels of the class share similarities, while the last two are different and also called the modified GL class. The advancements carried out in the modified GL class are in terms of weapons, sensors and superstructure of the ship which includes additional decks for bridge.

Type F 70 Montcalm© *Marine nationale / Tonard Jacques*

Displacement: 4,900 t (full load)
Length: 139m
Beam: 14m
Draught: 5.9m
Speed: 30 knots
Range: 8000 nautical mile @15 knots
Armament:
SAM: Crotale, Mistral short range
Anti-ship: 4 × MM38/MM40 Exocet
8 × MM38/MM40 Exocet (modified GL class)
Anti-submarine: 2 × fixed torpedo tubes
Guns: One 100mm Dual Purpose gun
2 × 20 mm guns, 4 × 12.7 machine guns
Aircraft carried: 1-2 LYNX Westland (ASW role) helicopters

Frigates

LA FAYETTE Class - 05 in Service

They are the most advanced frigates currently in service with the French navy. Designed and constructed by DCNS, they are deployed at the Toulon port. The first vessel of the class entered service in 1996, while last one was inducted in 2001. These ships can be deployed for multiple roles enlisting from maritime patrol and surveillance, air defence and maritime superiority missions. They are the first ships designed by DCNS involving modular design and interoperability concept to be used with other NATO ships as well. The platfroms have also been ordered by Saudi Arabia, Taiwan and Singapore.

Displacement: 3800 t (full load)
Length: 115m
Beam: 15.4m
Draught: 5.8m
Speed: 25 knots
Range: 7000 nautical mile @15 knots
Armament:
SAM: Crotale
Anti-ship: 8 × MM40 Exocet
Anti-submarine: 2 × torpedo tubes
Guns: 1 × 100 mm dual purpose gun,
2 × 20mm anti-aircraft guns
Aircraft carried: 1 AS 565SA/NH90 helicopter

Image courtesy: Marine nationale

FLORÉAL Class - 06 in Service

The Floréal class are light surveillance frigates of the French Navy. The first platform was commissioned in 1992, while the last one was inducted in 1994. The ships were built by Chantiers de l'Atlantique in Saint Nazaire, France along with Lorient dockyards (DCNS). The ships are deployed across the world to enforce French Navy's domination in different regions. The platforms are fully equipped to carry out various missions, including protection of EEZ, humanitarian relief, evacuation, navigation, combat escort etc. Similar class of ships have been delivered to the Royal Moroccan Navy as well.

Image courtesy Jean-Michel Roche

Displacement: 3000 t (full load)
Length: 93.5m
Beam: 14m
Draught: 4.5m
Speed: 20 knots
Range: 10,000 nautical mile @15 knots
Armament:
SAM: MISTRAL
Anti-ship: MM38 Exocet
Anti-submarine: 4 × fixed torpedo tubes
Guns: 1 × 100 mm dual purpose gun
2 × 20mm anti-aircraft guns
Aircraft carried: 1 AS 565MA/AS 332F Super Puma helicopter

Corvettes/ OPVs

D'ESTIENNE D'ORVES Class - 09 in Service

These ships are now only used for patrol and anti-submarine operations in French waters. The ships have been outfitted with old ASM and SAM capabilities and only provide coverage for ASW role in addition to patrol and SAR duties at present.

Image courtesy: Marine nationale

Displacement: 1270 t - 1350 t (full load)
Length: 81m
Beam: 10m
Draught: 5.5m
Speed: 25 knots
Range: 4500 nautical mile @15 knots
Armament:
SAM: MISTRAL
Anti-submarine: 4 × fixed torpedo tubes
Guns: 1 × 100 mm gun, 2 × 20mm guns
4× HMG

Bâtiment multi-mission OPVs - 02 in service; 04 in total order

This is a new class of ship being built for the French navy. The multi-purpose ships have ocean going capability and can perform a variety of missions such as law enforcement and logistics missions, policing, as well as combating illegal fishing, trafficking besides SAR duties.

Image courtesy: Marine nationale

Displacement: 2300 t (full load)
Length: 65m
Beam: 14m
Draught: 4.20m
Speed: 15 knots
Crew: 40

LHD

MISTRAL Class - 03 in service

The Mistral class is one of the world's most modern amphibious assault ships or helicopter landing dock (LHD) in service of the French Navy. Three of such ships were built by France's DCNS and STX. The platforms, referred to as "projection and command ships" (French: bâtiments de projection et de commandement or BPC), have been designed to carry out multiple roles such as helicopter carrier ship, a hospital ship, amphibious assault ship, command & control centre etc.

First platform of the class was commissioned in 2006, followed by two other ships in next two years. The LHDs were also ordered by the Russian Navy, but the deal was scrapped later on and the ships were sold to Egypt.

Image courtesy: Marine nationale

Displacement: 22,000 t (full load)
Length: 199m
Beam: 32m
Draught: 6.2m
Speed: 19 knots
Range: 11,000 nautical miles @15 knots
Capacity: Up to 900 troops and 16 armoured vehicle/ 13 MBTs, 4 LCUs, 2 LCACs
Complement: 160 (18 officers)
Armament:
SAM: MISTRAL
Gun: 2 × 30mm, 4× HMG (12.7mm)
Aircraft carried: 16 × NH90 / Super PUMA/COUGAR/TIGRE attack helicopters

BATRAL CLASS - 02 in service

The small landing ships in service with the French Navy since the 1970s are due to be retired from service by 2017. Three 1500-tonne Bâtiments Multimission (B2M) OPVs would replace these older-generation platforms.

Image by: Jean-Michel Roche

Displacement: 1600 t (full load)
Length: 80m
Beam: 13.5m
Draught: 2.4m
Speed: 15 knots
Range: 4500 nautical miles @13 knots
Capacity: up to 180 troops 12 vehicle and 350 t combat load.
10 LCU/ 20 APC- amphibious
Complement: 52 (7 officers)
Armament:
Gun: 2 × 20mm, 2× HMG (12.7mm)
Aircraft carried: 1 × Lynx/ PANTHER

SUPPORT PLATFORMS

The support platforms of the France Navy include the following ships.

Replenishment Vessel/Tanker

DURANCE Class - 04 in Service

The Durance class fleet replenishment vessels are meant to supply fuel, ammunition and other supplies to fleet ships in order to sustain them for longer duration sea operations without going to the shore. First of this class of platforms entered service in 1980, while the last one was inducted in 1990. These are large displacement ships equipped to operate at high

© *Marine nationale*

sea conditions. All Durance class platforms are equipped with 40mm bofors guns to provide self defence capability while operating in hostile waters.

Displacement: 17,800 tonnes (full load)
Length: 157.2m
Beam: 21.2m
Draught: 8.65m (average); 10.80m full load
Speed: 19 knots
Range: 9,000 nautical miles
Armament: 1 Bofors 40mm guns; 6 12.7mm M2 Browning machine guns; Mistral short-range SAM
Aviation facilities: Alouette III helicopter

Special Purpose ship

Dupuy de Lome (Electromagnetic Intelligence gathering ship)

Dupuy de lome is an intelligence gathering ship capable of collecting important surveillance data beyond the enemy line. The ship has been in active service with the French Navy from 2005 onwards. Currently the Navy operates one such platform which is under the command of Military Intelligence Directorate (DRM). The DRM defines the ship's missions and provides collected information to different wings of the French armed forces. The ship has been equipped with latest measures to maintain itself even with minimal number of crew.

Photo: Jean-Michel Roche

Displacement: 3,100 t (3,600 t full load)
Length: 101.75m
Beam: 15.85m
Draught: 4.9m
Speed: 16 knots
Range: 6,300 km

Sub-surface ships

The sub-surface platforms operational in the French Navy include **ballistic missile submarines (SSBNs)** and **nuclear-powered attack submarines (SSN)**.

Ballistic missile carrying nuclear submarines (SSBN)

LE TRIOMPHANT Class - 04 in service

The Le Triomphant class is the French Navy's ballistic missile carrying nuclear-powered submarine (SSBN). The underwater platforms replaced the L'Inflexible M4 class of SSBNs. The submarines were designed and built at DCNS's Cherbourg shipyard. The first of this class of submarine was launched in July 1993 and entered service in 1997 while the keel for the fourth and last, Le Terrible (S619), was laid down in October 2000. The submarine was launched in March 2008 and commissioned in September 2010. The vessel is equipped with 16 vertically launched M45 ballistic missiles with 6 MRVs, each capable of carrying 150kt of nuclear warhead. The range of the missile is 6,000-km. The submarine can also carry the improved M51 missiles, weighing more than 50t and carrying warheads with 12 MIRVS, having an increased range of 8,000-km.

Sources: © Marine National

Displacement: 12,640 tons (surfaced); 14,335 tons (submerged)
Length: 138m
Beam : 12.50m
Draught: 10.60m
Speed: over 25 kts (46 km/h)
Range: Unlimited distance; 20–25 years
Test depth: Over 400m
Complement: 15 officers & 96 sailors
Armament: 16 M45 or M51 SLBMs with TN 75 warheads
533mm torpedo tubes & Exocet SM39 SLCM

Nuclear powered attack submarines (SSN)

RUBIS Class – 06 in service

They are the first-generation of nuclear-powered attack submarines in the French Navy service. The first platform was inducted

Image courtesy: Marine nationale

in service in 1983 and the last one was commissioned in 1993. This class is set to be replaced by the next-generation Barracuda-class of SSNs presently under development.

Displacement: 2400 tons (surfaced); 2670 tons (submerged)
Length: 73.6m
Width: 7.6m
Maximum depth: 300 meters
Propulsion: 1 nuclear reactor, 1 turbo-reducer group of 7 MW
Crew: 2 crews of 70 sailors
Speed: > 20 knots
Armament: 4 torpedo tubes of 533mm or SM-39 Exocet missiles

BARRACUDA Class – 06 being built

Six Barracuda class nuclear-powered attack submarines are being built for the French Navy. The submarines will replace the existing Rubis-class of SSNs. The first Barracuda platform is expected to be delivered to French Navy in 2018. The 5000+ tonne vessel will be equipped with MBDA MdCN Scalp land attack cruise missiles in addition to latest anti-submarine torpedoes.

Artistic impression

NAVAL AVIATION

Fighter Aircraft

RAFALE M - 35 in service (26 are carrier borne)

It is the carrier-borne variant of the Rafale multi-role fighter which entered service in 2001. For carrier operations, the single-seat M model has a strengthened airframe and it weighs about 500 kg more than the Rafale C single-seat land-based version of the fighter. The Rafale M is the only non-US fighter type cleared to operate from the decks of US carriers, using catapults and their arresting gear. Manufactured by Dassault Aviation, it is an omni-role fighter used for a variety of missions.

Image courtesy: Marine nationale

Wing span: 10.90m
Length: 15.30m
Height: 5.30m
Weight: Overall empty weight 10 t (22,000 lbs) class
Max. take-off weight: 24.5 t (54,000 lbs)
Store stations:
Total: 14
Heavy-wet: 5
Performance:
Max. Thrust: 2 x 7.5 t
Limit load factors -3.2 g / +9 g
Max. Speed: M = 1.8 / 750 knots
Service ceiling: 50,000 ft

Aircraft for MSR / ASuW / AEW role

Dassault Falcon 50MI/MS/10MER

Eight of these aircraft are in service. The aircraft are primarily used for SAR and maritime surveillance roles in the Atlantic Ocean region. Falcon 10 maritime surveillance and patrol aircraft are also being used by the French Navy for transport missions.

F10 aircraft

Falcon F-50 - *Image courtesy: Marine nationale*

Length: 18.52 m
Wingspan: 18.86 m
Height: 6.98 m
Empty weight: 9889 kg
Max. take-off weight: 18,008 kg
Power plant: 3 × Honeywell TFE731-40 turbofan engines
Maximum speed: 1,015 km/h
Cruise speed: 850 km/h
Range: 6000 km
Service ceiling: 14,936m
Crew: 2
Capacity: 8-9 passengers

ATLANTIQUE-II

Twenty Two (22) such aircraft, presently in service, are being used for maritime surveillance, anti-submarine warfare and other specific roles such as communication intelligence and electronics intelligence etc.

Image courtesy: Marine nationale

Length: 31.62 m
Wingspan: 37.42 m
Height: 10.89 m
Empty weight: 25,700 kg
Loaded weight: 45,000 kg
Max. take-off weight: 46,200 kg
Power plant: 2 × Rolls-Royce 2-shaft turboprops
Maximum speed: 648 km/h
Cruise speed: 315 km/h
Ferry range: 9,075 km
Endurance: 18 hours
Service ceiling: 9,145 m
Armament: Up to 3,500 kg (7,700 lb), including torpedoes, depth charges, mines and/or buoys
Crew: 12

E-2C HAWKEYE - 03 in service; 01 on order

The Gurmann E-2C is a carrier bound aircraft used for AEW role. The Hawkeye is one of the most advanced airborne early warning (AEW) aircraft in the world that combines advanced sensors and command & control capabilities. The aircraft is capable of detecting and tracking aerial targets at more than 500 km away. The aircraft undertakes missions such as detection and tracking of all airborne targets; safety of a naval force against aerial and surface threats; support to aerial missions of interception and assault against naval and land targets; and to work as an information and data hub within the carrier group. Three such aircraft are presently in service with the French Navy and one more is on order.

Length: 17.60 m
Wingspan: 24.56 m
Height: 5.58 m
Max. take-off weight: 26,083 kg
Maximum speed: 350 knots (648 km/h)
Cruise speed: 256 knots (474 km/h)
Ferry range: 1,462 nmi (2,708 km)
Endurance: 6 hr
Service ceiling: 34,700 ft
Crew: 5

Helicopter

Eurocopter EC725/COUGAR/EC225

The EC225 is used for surveillance/search and rescue missions. It also ensures the training of commandos and provides support to the strategic forces.

Image courtesy:Marine nationale

Length: 19.5 m
Height: 4.97 m
Empty weight: 5,256 kg
Gross weight: 11,000 kg
Max take-off weight: 11,200 kg
Power plant: 2 × Turbomeca Makila 2A1 turboshaft engines
Cruising speed: 270 km / h (320 km / h maximum)
Range/Endurance: 4 hours
Capacity: 16 people/6 stretchers + 4 people seated/3.8t in suspended loads
Equipment: Modern Avionics with Screen
Main rotor diameter: 16.2 m
Crew: 2

Nh90 CAIMAN MARINE - 27 on order with deliveries going on

The medium-sized, twin-engine, multi-role military helicopters have been on order for the French Navy with deliveries going on. The naval helicopter is designed to perform a wide range of missions from anti-submarine, to anti-surface warfare, search & rescue, maritime surveillance and control and special operations including counter-terrorism and anti-piracy.

Image courtesy: Marine nationale

Length: 16.13 m
Weight: 10,600 kg
Rotor diameter: 16.30 m
Height: 5.23 m
Maximum speed: 300 km/h
Range: 1,000 km (NHS/NHC)
Service ceiling: 6 km
Capacity: 2 pilots + 20 seated troops/ medevac role up to 12 stretchers
Armaments: 2×MU-90 Compact Torpedo

DAUPHIN SA365 – 08 in service

The medium-weight multipurpose twin-engine helicopter is primarily being used for SAR roles. The rotorcraft is also deployed onboard the Navy's aircraft carrier Charles de Gaulle.

Image courtesy: Marine nationale

Length: 13.68m (44 ft 11 in)
Height: 3.97 m (13 ft 0 in)
Empty weight: 2,380 kg (5,247 lb)
Max take-off weight: 4,300 kg (9,480 lb)
Power plant: 2 × Turboméca Arriel 2C turboshaft
Main rotor diameter: 11.94 m (39 ft 2 in)
Maximum speed: 306 km/h (190 mph; 165 kn)
Service ceiling: 5,865 m (19,242 ft)
Rate of climb: 8.9 m/s (1,750 ft/min)
Crew: 1 or 2 pilots
Capacity: 10 troops
Armament: Guns: Door mounted 7.62 mm machine gun
RTE MK2/S anti-ship missile (possible for integration
Mk46 or Whitehead A.244/S torpedoes

WESTLAND LYNX Mk4 - 18 in service

Out of 21 such aircraft originally received from the UK, currently 18 are in service with the French Navy. The helicopters are deployed onboard the six Georges Leygues class of destroyers. They are used for training roles.

Image courtesy: Marine nationale

Length: 15.241 m
Rotor diameter: 12.80 m
Height: 3.7 m
Empty weight: 3,291 kg
Max. take-off weight: 5,330 kg
Power plant: 2 × Rolls-Royce Gem turboshaft
Maximum speed: 324 km/h (201 mph)
Range: 528 km (328 miles) with standard tanks
Capacity: 8 troops
Payload: 1,480 kg
Crew: 2 or 3
Armament Naval : 2 x torpedoes or 2 x depth charges

ALOUETTE III-SA 319B

Some 23 helicopters are in service with the Navy. Most of the platforms are being used for training, test and evaluation trials and SAR roles.

Image courtesy: Marine nationale

Length: 10.03 m
Main rotor diameter: 11.02 m
Height: 3.00 m
Empty weight: 1,143 kg (2,520 lb)
Gross weight: 2,200 kg (4,850 lb)
Power plant: 1 × Turbomeca Artouste IIIB turboshaft
Maximum speed: 210 km/h (130[5] mph)
Cruising speed: 185 km/h (115 mph)
Range: 540 km (335 miles)
Service ceiling: 3,200 m (10,500 ft)
Rate of climb: 4.3 m/s (850 ft/min)
Crew: 2
Capacity: 5 passengers

WEAPONS

Mistral: The short range surface-to-air missile system provides air defence coverage to surface ships. The weapon is capable of engaging low flying aircraft, helicopters & UAVs. Mistral is deployed on most platforms of the French Navy, from aircraft carrier to patrol boats.

MBDA photo

Length: 1.86m
Weight: 19.7kg
Diameter: 0.09m
Speed: Mach 2.6
Range: 6 km (Max); 0.5 km (Min)
Reaction time: 3-5 sec
Rate of Fire: 2 shot /min
Kill probability: 70%
Crew: 3

Crotale: The short-range anti-air missile is used for air defence roles. It is used to intercept anti-ship missiles and aircraft. The weapon has been deployed on the French Navy destroyers and frigates.
Length: 2.35m
Weight: 76kg
Diameter: 0.165m
Speed: 1200 m/s
Range: Up to 16 km

Aster 15: The short to medium range surface-to-air missile is equipped on several modern warships of the French Navy. The weapon provides air defence coverage to the naval platforms. It is designed to engage a wide range of aerial threats, including supersonic anti-ship cruise missiles, aircraft and UAVs.
Length: 4.2 m
Weight: 310 kg
Diameter: 0.18m
Maximum speed: Mach 3
Range: In excess of 30 km

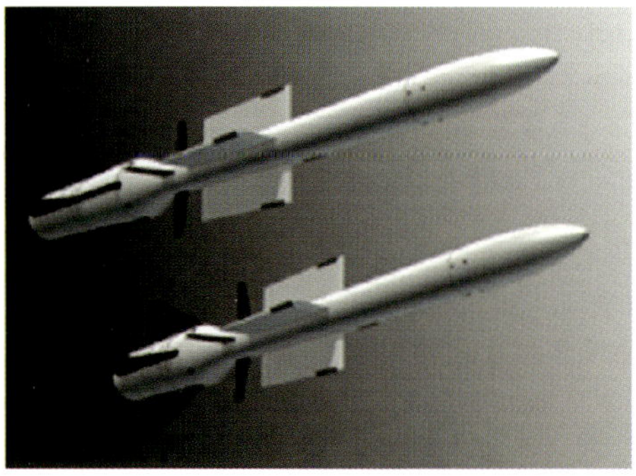

Aster 15. *MBDA photo*

Aster 30: It is the longer-range variant of Aster 15 and is used to provide air defence coverage to naval platforms.

MBDA photo

Exocet: The anti-ship cruise missile developed by MBDA in the 1970s is an all-weather, multi-target weapon fitted in the French Navy's surface ships and submarines. Advanced variants of the missile have been developed later on. The weapon can be used for littoral warfare and coastal land attack operations.

MBDA photo

CONCLUSION

France is among the top countries that can claim to be major powers in the world today. It plays an influential global role as a permanent member of the United Nations Security Council, NATO, the G-8, the G-20, the EU and other multilateral organisations. It ranks as the world's fifth largest national economy, second biggest in Europe behind Germany. France is the only US ally to muster a full-spectrum defence capability across land, air and sea, and with nuclear weapons of its own.

CONTACT DETAILS

Ministry of Defence
14, Rue St-Dominique
75700 Paris,
France
Postal Address: 00452 ARMEES
Tel: +33-1-44424180
Fax: +33-1-44424171
Website: www.defense.gouv.fr

General Armaments Commission
DGA – Delegation Generale pour l'Armement
14, Rue St-Dominique
75700 Paris,
France
Postal Address: 00457 ARMEES
Tel: +33-1-42193011 Fax: +33-1-42194427

GEORGIA
(Capital: **Tbilisi**)

INTRODUCTION

Area: 69,700 sq km
Population: 4,928,052 (July 2016 est.)
Coastline: 310 km
Maritime claims: Territorial sea: 12 nm
Exclusive Economic Zone: 200 nm

KEY POLITICAL PERSONS

PRESIDENT: Giorgi Margvelashvili

PRIME MINISTER: Giorgi Kvirikashvili

KEY DEFENCE PERSONS

DEFENCE MINISTER: Levan Izoria

CHIEF OF GENERAL STAFF: Maj. Gen. Vakhtang Kapanadze

ECONOMY

Georgia's main economic activities include the cultivation of agricultural products such as grapes, citrus fruits, and hazelnuts; mining of manganese and copper; and output of a small industrial sector producing alcoholic and non-alcoholic beverages, metals, machinery, aircraft and chemicals. The economy according to the World Bank achieved robust growth between 2003-2012, averaging 6.1 percent annually following structural reforms that stimulated capital inflows and investment. Reforms helped improve the business environment, strengthened public finances, upgraded infrastructure facilities and liberalized trade. However, the economy started to slow down since the last quarter of 2012 and GDP registered 1.7 % growth in the first three quarters of 2013. The country has a sizable hydroelectric capacity, a factor that has become an increasingly important component of its energy supplies and policies. The government published its 2020 Economic Development Strategy in early 2014 and Georgia also signed an association agreement with the European Union paving the way to free trade.

GDP: $35.68 billion (2015 est.); $34.72 billion (2014 est.)

Real Growth Rate: 2.8% (2015 est.); 4.6% (2014 est.)

Industries: Steel, machine tools, electrical appliances, mining (manganese, copper, and gold), chemicals, wood products, wine

Total Exports: $3.043 billion (2015 est.); $3.995 billion (2014 est.)

Export Commodities: Vehicles, ferro-alloys, fertilizers, nuts, scrap metal, gold, copper ores

Major Markets: Azerbaijan 10.9%, Bulgaria 9.7%, Turkey 8.4%, Armenia 8.2%, Russia 7.1%, China 5.7%, US 4.7%, Uzbekistan 4.4% (2015)

Total Imports: $7.363 billion (2015 est.); $8.29 billion (2014 est.)

Import Commodities: Fuels, vehicles, machinery and parts, grain and other foods, pharmaceuticals

Major Suppliers: Turkey 17.2%, Russia 8.1%, China 7.6%, Azerbaijan 7%, Ireland 5.9%, Ukraine 5.9%, Germany 5.6% (2015)

INTERNATIONAL DISPUTES

✦ Russia's military support and subsequent recognition of Abkhazia and South Ossetia independence in 2008 continue to sour relations with Georgia.

DEFENCE

The Georgian armed forces comprised of the Land Forces, Air Force and a paramilitary organisation Georgian National Guard. The Land forces consist of five infantry and two artillery brigades with additional brigades or battalions attached. In 2010, the Air Force was merged into the Army and was renamed the Army Air Section. The Navy was abolished in 2009 and was incorporated into the Coast Guard. The National Guard of Georgia was established on 20 December 1990. It represents the first Georgian armed formation, which became the base of the foundation for modern Georgian Armed Forces.

ARMY

Personnel: 36,000

Equipment

Category	Name	In Service
MBT	T-72, T54/55	
	Didgori	50
	BMP-2	120
	BMP-1	80
	Nurol Ejder	80

Army continued...

Category	Name	In Service
MBT	BTR-70/80	125
	Otokar cobra	200+
	Wolf	50
Artillery	D30 122mm towed	25
	D20 152mm towed	25
	DANA 152mm SP	40
	2S3 152mm SP	30
	2S7 203mm SP	10
Mortar	82mm	900
	120mm	
ATGW	Metis-M	
	Konkur	
	Fagot	
SAM	OSA-AKM	
	BUK-M1	
	Топ	
	Spyder	
	Strela	
	Igla	

Army Air Wing section Equipment

Category	Name	In service
Aircraft/ Helicopter	Su-25	
	L-39	
	Mi-24	3
	UH-1H	8
	Mi-8	2
	Mi-2	1

New Procurements/Upgrades

✦ New air defence systems from France are expected to be delivered in 2017. A deal had been signed with France in 2015 to acquire the air-defence systems and missiles. The type of weapon has not been disclosed.

Defence Expenditure

Total defence spending: $368 million (2016); $322 million (2015)

Defence spending in terms of GDP: 2.2% (2016)

Defence Production and R&D

Georgian military research and development is severely constrained by the national defence budget. The country retains scientists with considerable expertise from the Soviet military industrial complex and these have targeted foreign technical partnerships to revive Georgia's advanced defence production capabilities. The State Military Scientific-Technical Centre DELTA was established in 2005 to co-ordinate military research and development.

Defence Procurement

New air defence systems are being procured.

Georgian troops during a military exercise

CONTACT DETAILS

Ministry of Defence
N20 General Kvinitadze Str
0112 Tbilisi,
Georgia
Tel: (+995 032) 2723535
(+99532) 911963
Fax: +995-32-910645

Ministry of the Interior
Ulitsa Bolshaya Alleya 10
380014, Tbilisi,
Georgia

STRATEGIC INFORMATION

GERMANY
(Capital: Berlin)

INTRODUCTION

Area: 357,022 sq km

Population: 80,722,792 (July 2016 est.)

Coastline: 2,389 km

Maritime claims: Territorial sea: 12 nm

Exclusive Economic Zone: 200 nm

Continental shelf: 200 m depth or to the depth of exploitation

GEOPOLITICAL IMPORTANCE

Germany, the seventh largest European country in terms of area and the largest in terms of population, is situated partly in Western Europe and partly in Central Europe. It is surrounded by nine countries – Denmark in the north, Poland and Czech Republic in the east, Austria in the southeast, Switzerland in the south, France in southwest, Luxembourg, Belgium and Netherlands in the west. To its northeast lies the Baltic Sea and to its northwest, the North Sea.

After remaining a loosely-controlled territory for several centuries lacking sovereignty and unity, Germany emerged as a strong united empire under the leadership of Otto von Bismarck in the latter half of 18th century. Like the other European nations, it also followed imperialist policies and acquired overseas colonies in Asia and Africa. Germany's aggressive military posture and its intent to exercise control in entire Europe resulted in World War I in 1914. The country was eventually defeated by the Allied Powers and was forced to pay heavy reparations for causing the devastating war. Later, the rise of Nazi regime led by Adolf Hitler and his audacious territorial expansionist designs in the 1930s again saw Germany occupy the centre-stage in the post- World War I period which subsequently led to World War II in 1939. Hitler's attempt to occupy the entire European continent, including the Soviet Russia, failed at the end and Germany was again defeated in the war. After its defeat, the country was divided into two blocks – East Germany and West Germany, controlled by two Cold War adversaries – the former Soviet Republic and the US, respectively. It was only after the fall of the Berlin Wall in 1989 that the country was reunited again. Since then, it has assumed an important position in the political, economic and military affairs of Europe and to certain extent, in the globe.

POLITICAL OVERVIEW

The Federal Republic of Germany, comprising of 16 states, functions as a Parliamentary representative democracy with the President as the head of the state and the Chancellor as the head of the government. While the President is bestowed with nominal powers and responsibilities, the Chancellor carries out executive duties along with the Cabinet. The legislative power is exercised by the country's bi-cameral Parliament constituting of the Bundestag (Federal Diet) and Bundesrat (Federal Council). The federal structure of Germany grants certain degree of autonomy to all its states, with each having its own constitution. The states function as parliamentary republics, parallel to that of the central structure, and are ruled by a cabinet and a unicameral legislative body, called Landtag (State Diet). The independent judiciary of Germany is based on the civil law system with affiliations to the erstwhile Roman law, having its core principles being codified into a referable system and used as the primary source of law.

KEY POLITICAL PERSONS

PRESIDENT:
Joachim Gauck

CHANCELLOR:
Angela Merkel

ECONOMY

Germany is the largest economy in Europe and thus, regarded as the economic powerhouse of the continent. The country benefits from a highly skilled labour force along with a well-developed infrastructure sector. It is the third largest exporter in the world and mainly exports automobiles, machinery, chemicals, computer and electronic goods, electrical equipment, pharmaceuticals, metals, transport equipment, foodstuff and textiles. While the service sector contributes around 70% of the total national GDP, the share of industry and agriculture is 29.1%, and 0.9% respectively. The German economy was adversely affected by the global economic slowdown in 2008-2009, though it survived the Eurozone crisis that followed later. Being the single largest national economy, Germany, due to its strong industrial output and demand for exports, resisted the regional economic slump which affected most other Eurozone countries. The economy, however, registered a slowdown in later half of 2011. In the 2013-14 period too, the country's exports slowed, leading to an overall economic slow-down caused by global factors as also domestic issues of rising energy prices following the Government's decision to shut down all nuclear power plants by 2022. Germany plans to replace nuclear power largely with renewable energy, which accounted for 27.8% of gross electricity consumption in 2014, up from 9% in 2000. Domestic consumption, bolstered by low energy prices and a weak euro, are likely to drive German GDP growth again in 2016. The economy at present suffers from low levels of investment, and a government plan to invest 15 billion euros during 2016-18, largely in infrastructure, is intended to spur needed private investment.

GDP: $3.86 trillion (2015 est.); $3.804 trillion (2014 est.)

Real Growth Rate: 1.5% (2015 est.); 1.6% (2014 est.)

Industries: Among the world's largest and most technologically advanced producers of iron, steel, coal, cement, chemicals, machinery, vehicles, machine tools, electronics, automobiles, food and beverages, shipbuilding, textiles

Natural Gas - Proved Reserves: 116 billion cu m (1 January 2014 est.)

Crude Oil - Proved Reserves: 226.8 million bbl (1 January 2015 est.)

Total Exports: $1.309 trillion (2015 est.); $1.48 trillion (2014 est.)

Export Commodities: Motor vehicles, machinery, chemicals, computer and electronic products, electrical equipment, pharmaceuticals, metals, transport equipment, foodstuff, textiles, rubber and plastic products

Major Markets: US 9.6%, France 8.6%, UK 7.5%, Netherlands 6.6%, China 6%, Italy 4.9%, Austria 4.8%, Poland 4.4%, Switzerland 4.2% (2015)

Total Imports: $1.017 trillion (2015 est.); $1.18 trillion (2014 est.)

Import Commodities: Machinery, data processing equipment, vehicles, chemicals, oil and gas, metals, electric equipment, pharmaceuticals, foodstuff, agricultural products

Major Suppliers: Netherlands 13.7%, France 7.6%, China 7.3%, Belgium 6%, Italy 5.2%, Poland 5%, US 4.7%, Czech Republic 4.5%, UK 4.2%, Austria 4.2%, Switzerland 4.2% (2015)

DEFENCE & SECURITY

The defence and security aspects of present-day Germany have been largely influenced and shaped by its history. While it held the centre-stage in the two World Wars of the 20th century with its offensive posture and expansionist goals, a unified Germany, after the end of Cold War, has focused more on its internal development, peace and stability and has reoriented its armed forces - the Bundeswehr - and its military doctrine, accordingly. The Bundeswehr operates as a defensive military force. And being firmly tied to the NATO for maintaining its own security and territorial integrity, Germany has adhered to the military doctrine of collective European defence and security cooperation. Safeguarding its interests in the European and transatlantic context forms the basis of Germany's present-day defence policy. However, the German armed forces, unlike their other leading European counterparts like the British and French military, have lacked a clear-cut strategy and capability of engaging in overseas multinational military missions and thus, assume the role of a static, regional force rather than an expeditionary one. This makes Germany a more continental power, rather than a global power, with limited strategic and security objectives.

Germany does not face any direct conventional military threat from any country and has continued to focus on safeguarding its national interests while jointly working with its NATO partners, including the US. There are no border disputes with any of its neighbours. The country until recently was free from any major internal problems except the issues of illicit drug trade and organised crime, but is now battling the increasing threat of Islamic terrorism and extremism.

In 2016, Berlin reversed its earlier policy of troops cut and announced to raise the number of its armed forces for the first time since the Cold War in response to growing international challenges, including Islamist terrorism, instability in Africa, civil wars in Syria and Libya, Russia's new-found assertiveness and ongoing Ukraine crisis. In a new Defence Whitepaper released in July 2016, the Government announced major policy shifts which confirm Germany's growing willingness to play an important role in the global strategic affairs at large and also lead the European nations to collectively combat the emerging security challenges in the continent. The country's overall military force had been reduced from around 240,000 to 177,000 troops. Now, the Government has announced to add some 20,000 more soldiers to the military over the next seven years. Alongside, an increase in the annual defence expenses as decided by the Government will bolster the military capabilities of Germany in the coming years.

KEY DEFENCE PERSONS

FEDERAL MINISTER FOR DEFENCE & COMMANDER-IN-CHIEF: Ursula von der Leyen

CHIEF OF STAFF, BUNDESWEHR: Gen. Volker Wieker

CHIEF OF ARMY: Lt. Gen. Jorg Vollmer

CHIEF OF NAVY: Vice Adm. Andreas Krause

CHIEF OF AIR FORCE: Lt. Gen. Karl Muellner

THREAT PERSPECTIVE

Terrorism: The most significant threat Germany faces today is radical Islamic terrorism. The Islamic extremists have targeted the country in the past few months with many cases of attacks occurring in public places. The Angela Merkel Government's open-door migrant policy that allowed an estimated 900,000 asylum seekers from Syria, Iraq, Afghanistan and other such strife-torn regions to enter Germany is being blamed as the root cause for the series of bungled militant attacks by refugees inspired by the Islamic State (ISIS). The Government has issued a high level of security alert after the spate of attacks. The country has joined hands with US-led NATO forces to fight the ISIS.

STRATEGIC RELATIONS

US: Berlin and Washington strengthened their bilateral relations after the end of Cold War when a unified Germany came into existence. Both being primary NATO members, and with Germany positioned at the centre of European affairs, bilateral cooperation in the areas of defence and security assumed significance alongside their political, trade and economic ties. Washington played a key role in Germany's integration into the NATO and other leading European institutions. Germany underlines the importance of the US as its essential security partner and the two sides regularly conduct joint military wargames and engage in defence exchanges. As a NATO member, Germany has contributed troops in the US-led mission in Afghanistan, though it abstained from the Iraq war.

France: Germany's decades-old relationship with France is based on the Élysée Treaty (Treaty of Friendship) the two sides inked in January 1963 which put an end to centuries-old rivalry, enmity and war between the two neighbours. Both countries have strived for greater European integration for decades. Strategic and military cooperation between the two sides dates back to the 1950s when they agreed to jointly develop weapons and armaments which included missiles, transport aircraft, trainer jets, battle tanks and even nuclear weapons. This relationship further intensified when they agreed to implement the defence clauses of the Élysée Treaty in 1982 and have since then expanded the purview of bilateral defence ties. The two countries have, as recently as 2012, singed a military agreement which includes renewed cooperation in new defence purchases, joint defence projects and exploring possible cooperation on missile defence and related areas. The two countries' Air Forces have also signed a partnership agreement for joint training programme for the Airbus Military-built A400M military transport aircraft aircrew and mechanics. In October 2016, the two countries signed an agreement to create a joint air transport squadron that will operate the Lockheed Martin C-130 Hercules aircraft. The new Franco-German air transport squadron is expected to be operational by 2021.

Russia: Even while remaining a strong NATO member, Germany has maintained strong bilateral relations with Russia. The multi-faceted German-Russian relationship has often been defined as a strategic partnership between the two sides. The two countries have been especially cooperating in the fields of trade, investment, energy and technology. A major share of Germany's natural gas needs is met by supplies from Russia and Germany, on its part, has provided technological expertise to Russia. Moreover, Berlin has made efforts over the years for Moscow's inclusion into Europe's economic and security apparatus. The ongoing Ukraine crisis, however, has put the relationship under strain.

Israel: With Israel, Germany shares long-standing military relations and has supplied sophisticated defence technology in the past. Berlin is regarded as one of Israel's closest strategic allies after the US. The two countries have supplied defence technology to each other and also cooperated on jointly developing weapon systems. While Germany has supplied its Dolphin-class submarines to the Israeli Navy, Israel on its part has provided anti-tank missiles to German armed forces.

India: Indo-German relations covering bilateral trade, economic and cultural exchanges exist for several decades. The framework for strengthening their political, scientific and strategic cooperation was laid down towards the end of 20th century. The security and strategic aspects of bilateral ties were further highlighted when the two sides adopted the "Agenda for German Indian Partnership in the 21st Century" in New Delhi in 2000. Military cooperation between them was strengthened with the singing of a new defence and security agreement in 2006. Germany has supplied weapons and equipment to Indian armed forces and is ranked as the fifth largest arms exporter to India. India has acquired tanks, submarines and aircraft from Germany in the past and the two countries' navies have also been engaged in joint naval manoeuvres.

MULTILATERAL ALLIANCES

NATO: West Germany joined NATO on 6 May 1955 when the Western Alliance's second round of enlargement took place. East Germany, which had joined the rival Warsaw Pact, came into NATO fold after the unification of Germany in 1990. Since then, the multinational military block has provided a vital security cover to Berlin that helped Germany to grow as a powerful European nation. Berlin has taken part in several NATO-led combat and humanitarian missions since then, including the 1999 Kosovo conflict, 2001 Afghanistan war and the Libya War in as recently as 2011, though it later pulled out of the Libya conflict. Presently, Germany is assisting NATO in Syria in the fight against the Islamic State (ISIS).

In another development, Germany will host the command centre of the NATO's European missile defence system which will be located at the Headquarters of Allied Air Command at Ramstein. A decision in this regard was taken in February 2012 at a meeting of the defence ministers of NATO member states in Brussels. The German defence minister also announced that the country's Patriot systems would become a part of the missile defence system.

EU: Being at the heart of European affairs, Germany has enjoyed a key position in the European Union. As a founding member of the EU of which it became a part in 1952, Germany has played crucial role in the larger European integration and adoption of Euro as a common currency in the continent. During the global financial crisis and the Euro Zone crisis, the country stood out strongly and maintained a healthy GDP growth as compared to its other European allies. Berlin, however, has been reluctant to take any leadership role in the Union's financial or security affairs.

DEFENCE CAPABILITIES

ARMY: The German ground forces, called the Heer, forms the largest part of the German armed forces, the Bundeswehr. It is divided into infantry, artillery and armoured forces. The Government in 2010 announced a reforms plan under which military conscription was abolished from 1 July 2011. The reforms aim to make the ground forces more flexible, well-trained, agile and capable of undertaking overseas military missions with its allied partners in a more efficient way. The Government's earlier decision to reduce the number of some existing land platforms has now been reversed and new procurements are lined up to add more platforms and systems.

Personnel: 69,000

Equipment

Category	Name	In Service	On Order
MBT	Leopard 2A5/ 2A6	225	
APC/IFV	Fennek	217	
	MTW M113	594	
	Marder 1A3	500+	
	Dingo 1 & 2	395	85 (Dingo 2)
	Puma	20	330
	Boxer	272	131
	Mungo	400+	31
	Fuchs	898	
	BV 206	183	87
	Wiesel 1 & Wiesel 2	251	
	Duro	113	100 (Duro III)
	Eagle	198	451 (Eagle IV & V)
Artillery	PzH 2000 155mm (self-propelled howitzer)	89	
Mortar	120mm	87	
MRL	M270	55	
ATGW	Milan	1500	
	PARS 3 LR		680
SAM	Stinger		
AA gun	Gepard 1 A2	85	

Army Aviation

Category	Name	In Service	On Order
Light transport	C-212-300	2	
	Short 330	2	
	OV-10	75	
	Beech 99	1	
	T-41	23	
	King Air	1	
	Jetstream 41	2	
	ERJ-135	2	
Helicopter	Bell UH-1D	65	
	NH90 TTH	23	59
	CH-53	66	
	Bo105	142	
	EC 135	15	
	Tiger (attack)	13	27

Army continued..

Category	Name	In Service	On Order
UAV	Aladin	115	
	KZO	5	
	Luna X	7	
	Mikado (Mini UAV)		

New Procurements/Upgrades

✦ Delivery of Puma AIFVs has started from June 2015. The Marder IFVs are being phased out and replaced by the Puma. Delivery of all 350 Puma vehicles is scheduled to be completed by 2020.

Puma AIFV

✦ The number of Leopard 2 MBTs will be increased from 225 to 328 following the Ukraine crisis. 20 of the existing tanks are being upgraded to new improved 2A7 variant. The first 2A7 has been delivered by Krauss-Maffei Wegmann in December 2014.

✦ The Defence Ministry is evaluating options to develop a new-generation MBT as a successor to the Leopard 2 MBTs which are likely to remain operational until 2030.

✦ Delivery of 85 Dingo 2 vehicles is going on.

✦ 131 Boxer wheeled armoured vehicles have been ordered from Artec GmbH for €476 million. Deliveries will commence from 2017.

✦ 70 Eagle IV vehicles are on order in addition to 218 previously ordered. 176 Eagle Vs have been ordered between 2013 and 2014.

✦ 87 BV 206 transport vehicles are on order.

✦ Rheinmetall will modernise 25 more Fuchs/Fox 1 armoured transport vehicles for the Army under a €36 million contract. The company had previously been contracted to upgrade a total of 152 Fuchs to the new 1A8 version and by February 2013, 123 of the APCs were delivered.

✦ Serial production of PARS 3LR anti-tank guided missile has started after completion of test launches. Dielhl Defence has received the contract to produce 680 PARS 3LRs which will arm the Tiger attack helicopters.

Tiger attack helicopter

NAVY: The naval wing of the Bundeswehr functions with the primary mission to defend and protect Germany's maritime territory and secure its sea lanes. Being closely integrated in the NATO system, the Navy also operates with the allied forces and contributes in combat and humanitarian missions overseas. It operates a number of submarines, frigates, corvettes, aircraft and helicopters to carry out its missions. While plans are on to phase out some of the aging platforms from service, acquisition of new naval weapons and platforms is also under consideration. The modernisation programme for the navy is being undertaken to improve its capabilities in carrying out joint operations internationally.

Major naval bases: Wilhelmshaven, Glucksburg, Kiel

Personnel: 15,448

Equipment

Category	Name	In Service	On Order
Sub-marine	Type 212A (U212) class	5	1
Frigate	Bremen-class (F-122)	4	
	Brandenburg-class (F-123)	4	
	Sachsen class (F-124)	3	
	Type F125		4 (to replace Bremen -class)
Corvette	Braunschweig class (K130)	5	
	Korvette K131		6
FAC	Gepard-class (Type 143A)	8	
Mine sweeper	Frankenthal class mine hunter	9	
	Kulmbach class mine hunter	2	
	Ensdorf class mine sweeper	5	
Auxiliary ship	Barbe class (Type 520) utility landing craft	2	
	Elbe class (Type 404) supply boat	6	
	Oste class (Type 423) electronic surveillance ship	3	
	Westerwald (Type 760A) ammunition transport	1	
	Walchensee class (Type 703) fleet oiler	2	
	Rhön class (Type 704) fleet oiler	2	
	Berlin class (Type 702)	2	1
	Fehmarn class	1	
	Wangerooge class	2	
Weapon	(ASM) AGM84, Hapoon, MM38 Exocet, RBS15, MK3		
	(AGM) Sea Skua		
	(SAM) RIM-116 RAM,RIM-7-Sea Sparrow, RIM-162 Evolved Sea Sparrow, SM-2		

Naval Aviation

Category	Name	In Service	On Order
Fighter/ ASW/ Reconnai-ssance	P-3C Orion (to be upgraded)	8	
	Dornier DO 228 LM (maritime pollution control)	2	
Helicopter	Sea Lynx Mk 88 A	22	
	Sea King Mk 41	21	
	Camcopter S-100		6

New Procurements/Upgrades

✦ Delivery of the sixth and last Type 212(A) class conventional submarine built by Howaldtswerke-German Werft GmbH (HDW) is scheduled for 2016-17.

✦ Four Type F125 frigates are on order with scheduled deliveries between by 2019-2020. First frigate of the series, Baden Wurttemberg, is expected to be delivered by mid-2017. ThyssenKrupp Marine Systems laid down the keel for the third frigate of the series in June 2014 and it was christened "Sachsen-Anhalt" in March 2016.

✦ Four F123 frigates are being upgraded by Thales Deutschland under a 2010 contract. The first ship of the class has been refurbished and undergoing sea trials.

✦ Five new platforms under MKS 180 Multi-Role Combat Ships (MRCSs) programme is being planned to be built.

✦ Five additional K130 class of corvettes are being planned to be built.

✦ Upgrade of the Lockheed-built P-3C Orion MPAs has been approved by the US Govt under Foreign Military Sales programme at a cost of $250 million. The upgrades will include installation of new hardware and software. A contract has been signed in this regard with Lockheed Martin, and Airbus Defence and Space GmbH in July 2015. Delivery of the first upgraded aircraft is scheduled for November 2016.

✦ Procurement of 18 Sea Lion-variant of NH90 helicopters has been approved by the Defence Ministry under a US$1.6 billion deal announced in March 2015.

✦ Airbus Helicopters has been awarded a contract to upgrade the entire fleet of Sea King Mk 41 helicopters.

✦ 6 Camcopter S-100 helicopter UAVs are on order.

✦ The Navy has received the new RBS15 Mk3 anti-ship missiles from Diehl Defence. The missiles are being deployed on the Navy's K130-class corvettes.

AIR FORCE: Luftwaffe or the German Air Force operates with the primary mission to protect and defend the national air space. It also contributes to the military stability of Europe by undertaking joint operations with the allied forces of NATO. The Luftwaffe is planned to be restructured with a focus on four priority areas – air-surface integration (ASI), missile defence, military use of outer space and unmanned aircraft systems. While the ASI will give thrust on developing the Air Force's integrated mission capabilities by synergising with the ground and naval wings of the country's military as also with the allied forces, strengthening the Luftwaffe's missile defence apparatus will be a key element in the NATO's integrated missile defence programme with an initial thrust on providing short-range defence against missiles. To firm up its network centric capabilities, the Air Force plans to extensively use data and services provided by the SAR-Lupe reconnaissance satellite system and the SATCom Bw 2 satellite communication system launched by the country in the last few years. Acquisition and use of a wide range of advanced unmanned aerial vehicle systems will further boost up its reconnaissance and surveillance capabilities.

Major air bases: Wittmundhafen, Neuburg, Büchel, Schleswig, Ramstein, Erding, Holzdorf, Spangdahlem, Frankfurt

Personnel: 33,000

Equipment

Category	Name	In Service	On Order
Fighter/ Attack/AWS	Eurofighter Typhoon	107	36
	F-4F Phantom II	51	
	PA-200 Tornado	89	
Transport/ Tanker/ Surveillance	Airbus A319 CJ	2	
	Airbus A310	4	
	Transall C-160D	43	
	Challenger 601	6	
	Airbus A400M	3	50
	Airbus A340-313		2
Trainer/ Helicopter	T-6 Texan II	60+	
	T-38A Talon	35+	
	NH-90		42
	Cougar AS 532 U2	3	
	Bell UH -1D	64	
UAV	Heron-1	3	
Air Defence System	Patriot SAM	29 (to be cut to 14)	
	Mantis	2 systems	
	Stinger		
	Taurus KEPD 350 ASM	600	
	AGM-88B HARM		
AAM	IRIS-T		
	AIM-120 AMRAAM		
	AIM9-L Sidewinder		
	Meteor (under development)		

New Procurements/Upgrades

✦ The Defence Ministry has slashed the acquisition of Eurofighter Typhoons from 177 to 143. **Initially,** the country had planned to acquire over 250 Eurofighters, but steadily reduced the number to 180. In a Feb. 2014 decision, the Defence Ministry further decided to cancel the order for the second tranche of 37 jets.

- ✦ The number of Airbus A400M heavy lift transport aircraft to be retained with the Air Force has been cut down from 53 to 40. While the country will buy 53 planes, it will try to sell 13 of them to a third party and eventually maintain a fleet of 40 A400Ms. The Air Force has taken delivery of three A400Ms so far.
- ✦ The Air Force is weighing options to acquire up to 10 Lockheed Martin-built C-130J military transport aircraft from the US in the wake of delayed delivery of Airbus A400M transport planes.
- ✦ 42 NH-90 Tactical Transport Helicopters (TTHs) are on order.
- ✦ 15 H45M multi-role helicopters have been ordered from Airbus Helicopters.
- ✦ The Euro Hawk drone programme, under which the country was to acquire five modified Northrop Grumman-made Global Hawk UAVs, has been shelved by the Government over safety issues raised by the European authorities. Germany had received the first drone in 2011 and already spent €508 million (US $657 million) under the project. The remaining four high altitude long endurance (HALE) drones were to be delivered between 2015 and 2017.
- ✦ Acquisition of up to 16 armed drones was being planned and talks were going on with Israel and the US for the Heron and Reaper UAVs respectively. In January 2016, a decision was announced to lease between three and five Heron TP UAVs from 2018.
- ✦ Discussions are going on with European partners France and Italy to develop a European medium altitude long endurance (MALE) UAV. The new pan-European UAV, MALE2020, is expected to be operational by 2025.
- ✦ The SAATEG (system for imagery reconnaissance deep in the area of operations) interim medium-altitude long-endurance unmanned air system, consisting of Israel-made Heron 1 UAVs and ground stations, is being reduced from 16 to 22.
- ✦ Signing of a contract for the Medium Extended Air Defence System (MEADS) which was chosen over Raytheon's Patriot system (to replace the current Patriot system fielded in the 1980s) has been delayed. The German Defence ministry was to sign the contract with European MBDA and US's Lockheed Martin for the missile defence system.
- ✦ The Patriot air defence systems are being reduced from 29 to 14.
- ✦ The last of six ThalesRaytheonSystems (TRS)-built Ground Master 400 (GM 400) long-range air defence radar system has been inducted in service in September 2015. The radars will provide enhanced air surveillance throughout Germany.

DEFENCE PRODUCTION

Germany has a well-developed defence industry with some 200 companies catering to the domestic as well as international defence market. Avionics and aircraft engine, armoured vehicles, missile systems and naval platforms are some of the leading areas of defence production in the country. The arms production is primarily confined to private sector which accounts for 85% of all military research and development, procurement, and maintenance. Germany also collaborates with its European allies in defence production. Indeed, nearly 70 percent of Germany's major military equipment is manufactured as part of international projects in collaboration with several NATO countries.

In recent times, when the domestic demand for military equipment is facing a cut down due to budget constraints, German military exports has witnessed an upward trend, though some of the major export deals have faced political resistance and are yet to be finalised. The country's cutting-edge defence technology and weapon systems have found a place in the Middle East, Asia, Latin America, US and Europe. Exports to European nations, however, are categorised as military transfers and not military export. In 2013, total military exports increased by €1.1 billion to around €5.8 billion with major defence export deals announced for Saudi Arabia, Algeria, Indonesia and Qatar. In 2014, however, the total weapons exports approved by the German government dropped to €6.5 billion ($7.3 billion), a decline of €1.8 billion compared to the year 2013, according to the government's weapons exports report released in June 2015. Ships and submarines for South Korea, Singapore, Saudi Arabia and Israel made up the lion's share of exports in this category.

Arms exports by Germany almost doubled in 2015, with the country approving arms sales worth €7.9 billion in that year, a jump from €3.97 billion in the previous year. Major export deals for the year included four tanker aircraft for the UK, and MBTs and howitzers for Qatar. Other countries that have received arms & equipment from Germany in recent years include Israel, Saudi Arabia, Algeria, India, Kuwait and South Korea.

Major Armaments Producers

Rheinmetall Defence is one of the Europe's leading manufacturers and suppliers of combat systems, electronic solutions and wheeled vehicles for the Army.

Krauss-Maffei Wegmann GmbH is a leading designer, developer and producer of armoured vehicles, tanks and weapon systems for ground forces.

ThyssenKrupp Marine Systems is a consortium of naval platforms producers. The group is involved in the production of surface ships and submarines. The Kiel-based Howaldtswerke-German Werft (HDW) - a leading shipbuilder - is part of the group.

Diehl BGT Defence GmbH is an arms manufacturer and produces a wide range of missile systems for army, navy and air force. The Diehl Defence Land Systems is involved in repairing and modernisation of various military vehicles.

Joint Ventures/ Collaboration Programmes

Europe

Eurofighter Typhoon: Germany is one of the four partners in the consortium manufacturing the Eurofighter combat jets along with Spain, Italy and the UK. EADS Germany is the partnering company in the programme.

A400M: Germany is one of the seven NATO member nations in the Airbus A400M military transport plane programme led by European conglomerate EADS.

Eurocopter: Leading European helicopter manufacturer Airbus Helicopter, which was initially created with the merger between the helicopter divisions of Aerospatiale-matra (France) and DaimlerChrysler Aerospace (Germany), today is a subsidiary of larger European defence consortium EADS. The company, which designs, develops and supplies a wide range of rotor craft, is at present composed of three entities - the parent company Eurocopter, the German subsidiary Eurocopter Deutschland, and the Spanish subsidiary Eurocopter Espana.

France

UAV: Germany and France signed an agreement in September 2012 to jointly develop a medium-altitude, long-endurance (MALE) unmanned aerial vehicle system.

Poland

APC: Rheinmetall Defence subsidiary Rheinmetall MAN Military Vehicles (RMMV) has signed a partnership agreement with Poland's Polska Grupa Zbrojeniowa (PGZ) to develop an amphibious wheeled armoured personnel carrier (APC). The two countries will sell the new APC in the international market.

DEFENCE BUDGET

Keeping in view the growing international security challenges, Germany has decided to allocate more funds for defence expenses. Reversing the earlier trend of reduction in its defence budget which put Berlin at a much lower rank as compared to other European countries, including the UK and France, and also way below the NATO stipulated mark of 2 per cent of GDP, the Government has announced to steadily increase the annual defence budget. Keeping in view the ongoing Russia-Ukraine crisis, the country in 2015 pledged to increase its defence spending by 6.2 per cent over the next five years. Accordingly, the defence budget for 2016 was €34.2 billion, an increase of 4.2 per cent over 2015. For 2017, the proposed defence budget figure stands at €36.61 billion.

Previously planning to restructure the Bundeswehr with a curtailment in a number of new military acquisitions, Germany now plans to reform and expand the military with an increase in troops' level as well as modernisation of existing weapons and platforms in addition to some new procurements. It has also committed to widen its engagement with the NATO while making efforts to spend many more billions in the coming years to meet the NATO standard of 2 percent GDP on defence expenses.

Year	Allocation	% of GDP
2016	€ 34.29 billion	1.2
2015	€ 32.94 billion	1.16
2014	€ 32.44 billion	1.14
2013	€33.3 billion	1.5
2012	€31.87 billion	1.5
2011	€31.55 billion	1.4
2010	€31.1 billion	1.3

CONCLUSION

Due to historical factors, Germany has remained a territorial power rather than an expeditionary power – a position more comfortably accepted by the entire European fraternity. Berlin's allegiance to the NATO has provided it with the necessary security cover. The NATO connection has, indeed, helped Germany to fully integrate into the European community. While still trading a cautious path in its military approaches, Germany is making efforts to prepare its armed forces to take up expeditionary roles in the future – an urgent need arising out of the rapidly evolving geostrategic equations wherein the US is steadily shifting its focus away from Europe to Asia. The revised military strategy of Germany has taken into consideration the ongoing crisis in Ukraine and resurgence of Russia as a major military power, the prolonged civil wars in Syria and Libya, and the increasing threat of Islamic terrorism to Europe. Berlin has announced to align its defence expenses to meet the NATO standards. Future military procurements by the country would focus less on conventional warfare and more on flexibility, manoeuvrability, deployment and most importantly, force protection capabilities. Germany, together with its other powerful European allies, is gearing up to play a crucial role in ensuring the European continent's safety and security in the coming years.

CONTACT DETAILS

Ministry of Defence
Stauffenbergstrasse 18
10785 Berlin
Germany
Tel.: 030 1824-22276

Websites
Ministry of Defence: www.bmvg.de
Army:www.deutschesheer.de/portal/a/heer
Navy:www.marine.de/portal/a/marine
Air Force:
www.luftwaffe.de/portal/a/luftwaffe

GHANA
(Capital: Accra)

INTRODUCTION

Area: 238,533 sq km
Population: 26,908,262 (July 2016 est.)
Coastline: 539 km
Maritime claims: Territorial sea: 12 nm
Contiguous zone: 24 nm
Exclusive Economic Zone: 200 nm
Continental shelf: 200 nm

KEY POLITICAL PERSONS

PRESIDENT: Nana Akufo-Addo

VICE PRESIDENT: Kwesi Amissah-Arthur

KEY DEFENCE PERSONS

MINISTER OF DEFENCE: Dr Benjamin Bewa-Nyog Kunbuor

CHIEF OF DEFENCE STAFF: Air Marshal Michael Samson-Oje

CHIEF OF ARMY STAFF: Maj. Gen. Obed Boamah Akwa

CHIEF OF NAVY STAFF: Rear Adm. Peter Kofi Faidoo

CHIEF OF AIR STAFF: Air Vice Marshal Maxwell Mantserbi-Tei Nagai

ECONOMY

Ghana is one of the top-ten fastest growing economies in the world, and the fastest growing economy in Africa. The country is well endowed with natural resources. The domestic economy continues to revolve around agriculture, which accounts for more than a third of GDP. Gold, cocoa production and individual remittances are major sources of foreign exchange. Ghana became an oil exporter in 2010. Tullow Oil Plc (TLW) pumps about 100,000 barrels of crude a day at the Jubilee field. Expansion of Ghana's oil industry has boosted economic growth, but the recent oil price crash has reduced Ghana's 2015 anticipated oil revenue and the government failed to curb election-related spending resulting in rising wage bill while revenue fell. Ghana has signed a $920 million extended credit facility with the IMF in April, 2015 to help it address its growing economic crisis. The IMF fiscal targets will require Ghana to reduce the fiscal deficit by cutting subsidies, decreasing the bloated public sector wage bill, strengthening revenue administration, and increasing revenues.

GDP: $115.4 billion (2015 est.); $111.1 billion (2014 est.)

Real Growth Rate: 3.9% (2015 est.); 4% (2014 est.)

Industries: Mining, lumbering, light manufacturing, aluminium smelting, food processing, cement, small commercial ship building, petroleum

Total Exports: $10.36 billion (2015 est.); $13.22 billion (2014 est.)

Export Commodities: Oil, gold, cocoa, timber, tuna, bauxite, aluminium, manganese ore, diamonds, horticultural products

Major Markets: India 25.2%, Switzerland 12.2%, China 10.6%, France 5.7% (2015)

Total Imports: $13.47 billion (2015 est.); $14.6 billion (2014 est.)

Import Commodities: Capital equipment, refined petroleum, foodstuff

Major Suppliers: China 32.6%, Nigeria 14%, Netherlands 5.5%, US 5.4% (2015)

INTERNATIONAL DISPUTES

✦ Maritime border dispute continues between Ghana and Cote d'Ivoire.

DEFENCE

Ghana Armed Forces (GAF) comprised of the Army, Navy and Air Force supervised by the Ghanaian Ministry of Defence and the Chief of Defence Staff. The army is divided into two brigades - Northern Command (Kumasi) and Southern Command (Accra). There are two operational commands of the Navy - the Western Naval Command at Sekondi and the Eastern Naval Command at Tema. Currently modernisation programme of the GAF is being carried out. The plan calls for the GAF to be reorganized, re-equipped and restrained with an emphasis on the protection of the country's oil sites.

ARMY

Personnel: 16,000

Equipment

Category	Name	In Service
APC	Piranha	50
	Tactica	20
	Ratel	39
	Sisu XA-185	
	WZ523	50+
	Hunter light strike vehicles	12

Army continued...

Category	Name	In Service
MRL	Type 83 122mm	
Mortar	81mm	
	107mm	
	120mm	
ATK weapon	Carl Gustav	50
	RPG-7	

New Procurements/Upgrades

No major procurement plans are under consideration at present.

NAVY

Personnel: 2,000

Equipment

Category	Name	In Service
Patrol boat	Warrior class	2
	Blika class	4
	Achimota class (FPB57 type)	2
	Dzata class (FPB45 type)	2
	Defender class inshore response boat	7
	Stephen OTU ex-South Korean Chamsuri class	1
	Anzone class transport (ex-USCG Balsam class)	1

New Procurements/Upgrades

No major procurement plans are under consideration at present.

AIR FORCE

Personnel: 1,910

Equipment

Category	Name	In Service
Fighter	MB-339s	2
	MB-326 F/K	4
Transport	F-28	1
	B-N2 Islander	1
	C-212	2
	F-27	4
	Falcon 900	1
	C-295M	2
Trainer	K-8	4
	L-29	12
Helicopter	Bell 412	1
	A-109	2
	SA-318 Alouette III	2
	Mi-17	10

New Procurements/Upgrades

✦ Five A-29 Super Tucano advanced trainer/light attack aircraft have been ordered from Brazilian aircraft maker Embraer with delivery expected in 2016-2017. The jet trainers will be deployed for advanced training, border surveillance and internal security missions.

✦ Four new Harbin Z-9EH helicopters have been delivered by China in September 2015.

✦ Negotiations are going on to acquire one Mil Mi-35 attack helicopter from Russia.

✦ Additional Mi-17 helicopters are on order from Russia.

✦ Delivery of one more Airbus C295M transport aircraft is awaited.

Airbus C-295M transport plane of Ghana Air Force *An internet photo*

Defence Expenditure

Total defence spending: $190 million (2016)

Estimated defence spending in terms of GDP: 0.4%

Defence Production and R&D

Ghana does not undertake any defence production activities.

Defence Procurement

The Air Force is acquiring new aircraft and helicopters to enhance its capabilities.

CONTACT DETAILS

Ministry of Defence
Directorate of Public Relations
General Headquarters
Burma Camp, Accra
Director of Public Relations: Col Eric Aggrey-Quashie
Telephone: +233-302-775665
Fax: +233-302-772241
Web: http://www.gaf.mil.gh/

GREECE

(Capital: Athens)

INTRODUCTION

Area: 131,957 sq km

Population: 10,773,253 (July 2016 est.)

Coastline: 13,676 km

Maritime claims: Territorial sea: 12 Nm

Contiguous Zone: 24 Nm

Exclusive Economic Zone: 200 nm

KEY POLITICAL PERSONS

PRESIDENT & COMMANDER-IN-CHIEF: Prokopis Pavlopoulos

PRIME MINISTER: Alexis Tsipras

KEY DEFENCE PERSONS

MINISTER OF DEFENCE: Panagiotis Kammenos

CHIEF OF DEFENCE: Adm. Evangelos Apostolakis

CHIEF OF ARMY STAFF: Lt. Gen. Basil Tellidou

CHIEF OF NAVY STAFF: Vice Adm. George Giakoumakis

CHIEF OF AIR FORCE STAFF: Lt Gen Christos Vaitsis

ECONOMY

The economy of Greece mainly comprises the service sector (85.0%) and industry (12.0%), while agriculture makes up 3.0% of the national economic output. Greece has a capitalist economy with a public-sector accounting for about 40% of GDP and with per capita GDP about two-thirds that of the leading euro-zone economies. The Greek government-debt crisis started in late 2009. It was the first of five sovereign debt crises in the eurozone - later referred to collectively as the European debt crisis. The economy went into recession in 2009 as a result of the world financial crisis, tightening credit conditions, and Athens' failure to address a growing budget deficit. The economy contracted by 2.3% in 2009, 3.5% in 2010, and 6.0% in 2011. Austerity measures reduced the deficit to 11% of GDP in 2010 and about 9% in 2011. Eroding public finances, inaccurate and misreported statistics, and consistent underperformance on reforms prompted major credit rating agencies in late 2009 to downgrade Greece's international debt rating, and has led the country into a financial crisis. European leaders and the IMF agreed in October 2011 to provide Athens a second bailout package of $169 billion. The second deal however, calls for Greece's creditors to write down a significant portion of their Greek government bond holdings. In exchange for the second loan Greece has promised to introduce an additional $7.8 billion in austerity measures during 2013-15. These massive austerity cuts are lengthening Greece's economic recession and depressing tax revenues. Uncertainty regarding Greece's future in the Eurozone has dampened investor confidence and lowered growth projections for 2015. Greece exited its six-year recession in the second quarter of 2014, but the challenges of securing political stability and debt sustainability remain.

GDP: $286 billion (2015 est.), $286.6 billion (2014 est.)

Real Growth Rate: -0.2% (2015 est.), 0.7% (2014 est.)

Industries: Tourism, food and tobacco processing, textiles, chemicals, metal products; mining, petroleum

Oil - proved reserves: 10 million bbl (1 January 2016 est.)

Natural gas - proved reserves: 991.1 million cu m (1 January 2016 est.)

Total Exports: $27.5 billion (2015 est.), $35.6 billion (2014 est.)

Exports Commodities: Food and beverages, manufactured goods, petroleum products, chemicals, textiles

Major Markets: Italy 11.2%, Germany 7.3%, Turkey 6.6%, Cyprus 5.9%, Bulgaria 5.2%, US 4.8%, UK 4.2%, Egypt 4% (2015)

Total Imports: $46.62 billion (2015 est.), $65.17 billion (2014 est.)

Imports Commodities: Machinery, transport equipment, fuels, chemicals

Major Suppliers: Germany 10.7%, Italy 8.4%, Russia 7.9%, Iraq 7%, China 5.9%, Netherlands 5.5%, France 4.5% (2015)

INTERNATIONAL DISPUTES

✦ Greece and Turkey continue discussions to resolve their complex maritime, air, territorial, and boundary disputes in the Aegean Sea.

✦ Greece rejects the use of the name Macedonia or Republic of Macedonia.

✦ The mass migration of unemployed Albanians still remains a problem for developed countries, chiefly Greece and Italy.

DEFENCE

ARMY

Personnel: 90,000

Equipment

Category	Name	In Service
MBT	Leopard 2 A6 HEL	170
	Leopard 2A4	183
	Leopard 1A5/GR	501
	M-48A5/MOLF	390
	M-60A3 TTS	70
APC/MICV /IFV	VBL	242
	BMP-1	382
	G-127 Leonidas	439
	M113A1s/A2	1,867
Artillery	M-56 105mm	18
	M101 155mm	329
	M114 155mm towed	195
	PzH-2000	24
	M109A1/A2/A3/A5s 155mm	195
	M110A2 203mm SP	147

Army continued...

Category	Name	In Service
ATK	Carl Gustaf	2000
ATGW	TOW	361
	MILAN	290
	AT-4 Spigot	262
	Kornet-E	196
AA Gun	20mm	204
	Zu-23-2 23 mm	523
SAM	I-Hawk PIP IIIs	42
	TOR-1Ms	21
	OSA-AK (SA-8)	31
	Stingers	1,516

Army Aviation

Category	Name	In Service
Helicopter	U-17A/Bs	25
	C-12C (VIP)	1
	C12/APs	2
	Bell 205s/UH-His	100+
	CH-47D Chinook	15
	AH-64A/D Apache	18
	AB-206s	14
	NH-300C	17
	B-212	1
UAV	SAGEM Sperwer	10+

New Procurements/ Upgrades

✦ Bofors Carl Gustaf AB, a company within the Celsius Group, has been awarded a contract valued at SEK 170 million from Greece for ammunition to be used in the 84 mm Carl Gustaf-system. This contract involves direct deliveries to the Hellenic Army of armour penetrating ammunition as well as deliveries of component kits to the Hellenic Arms Industry (EBO) for further production in Greece.

✦ Greece has signed a contract with Russia for the supply of spare parts for air defence systems "Tor-M1" and "Osa-AKM".

Greek infantry loading the M40 recoilless rifle

Hellenic Army artillery in parade

NAVY

Personnel: 25,000

Equipment

Category	Name	In Service
Submarine	Glavkos class	3
	Poseidon class	4
	Okeanos class	1
	Papanikolis class	4
Frigate	Elli class	9
	Hydra class	4
OPV/IPV/ Missile Boat	Rouseen class	5 (2 on order)
	Laskos missile FAC	4
	Kavaloudis class missile FAC	5
	Votsis class	3
	Niki class	2
	Kasos class	2
	Armatolos class	2
	Machitis class	4
	Tolmi class	2
	Anromeda class	4
	Stamou class	2
Amphibious force	Samos class LSTs	5
	Zubr class	4
	Naxos class	4
Mine Warfare Force	Evniki class	2
	Evropi class	2
Main Auxiliary	Prometheus fleet	1
	Axios class	2

Naval Aviation

Category	Name	In Service
Helicopter	S70B6	11
	AB-212 ASW	8
	ALOUETTE III	2
	P-3B Orion	6

New Procurements/ Upgrades

Modernisation of the Greek military's P-3B Orion maritime patrol aircraft has started with the Hellenic Navy. The mid-life upgrade of four P-3B Orion aircraft will be carried out under a USD 141.9 million contract awarded by the US Navy to Lockheed Martin in February 2016.

The Greek frigate HS Psara, F-454, receives fuel during an underway replenishment with the fast combat support ship, USS Seattle, AOE-3. *Image Credit: US Navy*

Hellenic Navy

AIR FORCE

Personnel: 28,500

Equipment

Category	Name	In Service
Fighter/ Attack	Lockheed F-16 C/D	150+
	Mirage 2000	46
	F-4 Phantom II	10
	Embraer R-99 AEW	3
Transport	C-130 Hercules	15
	C-27J Spartan	8
	Embraer ERJ-135ER	2
	Gulfstream V	1
Trainer	T-41 Mescalero	44
	T-6 Texan II	25
	T-2 Buckeye	35
Helicopter	AW109	3
	AS332 Super Puma	12
	Agusta-Bell AB205	13
	Agusta-Bell AB212	4
	Bell 47	7
SAM	Patriot batteries	6
	Skyguard/Sparrow squadrons	6
	SU-300 PMU1	2
	TOR-M1	4
	Crotale NG/GR	9
AAM	AIM-120, AIM-9M, MICA, IRIS-T	
UAV	HAI Pegasus	10+

New Procurements/ Upgrades

✦ Kongsberg Defence & Aerospace and Raytheon of the US have signed a contract with the Greek Armed Forces to upgrade their HAWK air defence system. The total contract is valued at about NOK 1 billion.

✦ Greece will be receiving Boeing CH-47D Chinooks. Boeing has completed contract negotiations with the U.S. Army for the purchase of seven new Boeing CH-47D Chinook heavy-lift helicopters that will be delivered to the Greek Army under a Foreign Military Sale programme. The value of the FMS programme is in excess of USD 200 million.

Defence Expenditure

Total defence spending: $2.7 billion (2016)

Estimated defence spending in terms of GDP: 2.6% (2015)

Domestic Suppliers: ELBO, EAS (EBO, Pyrkal), Miltech, KEA, EAB, Barracuda,

Elefsis, Hellenic Shipyards, Kioleides, EODH

Foreign Suppliers: Brazil, Canada, France, Germany, Italy, Russia, United Kingdom, United States.

Despite its economic distress, Greece shows no sign of cutting back its considerable military budget. Military spending of Greece went from $10.1 billion in 2008 to $7.5 billion today. The country is still the top EU military spender in terms of ratio to GDP (2.72%).

Defence Production and R&D

Greece has a well-developed naval construction industry but it lacks sophisticated production techniques in the battlefield and aerospace sectors. Over the decades, Greece has sought to be as self-sufficient as possible in defence production. Due to the event of hostilities with its neighbour and fellow NATO member Turkey, Greece's arms suppliers, mainly the US and Western European states, would impose embargoes on the supply of military equipment. Nonetheless, despite the development of an arms industry, Greece remains largely dependent on imports. According to Stockholm International Peace Research Institute (SIPRI) data, Greece was ranked the seventh highest importer of major conventional weapons in the world in between the period 1998-2002 and later it was ranked 5th highest arms importer in 2004-2008. Due to the country's poor financial condition, the major arms build-up (launched in 1997, covering combat aircraft, MBTs and other equipment) has slowed down. The nation has faced an unprecedented expenditure while hosting the Athens 2004 Olympic Games. It has also affected on the allocation of the Greek budget, and had generating procurement difficulties for the country.

Defence Procurement

The Armed forces are being steadily modernised and expanded, mainly as a result of national concern over Turkey's larger military forces. Emphasis is being placed on introducing new technology and increasing military flexibility. Communications facilities are also being improved. Greece launched a major arms build-up in 1997, covering combat aircraft, MBTs, submarines and other equipment. However, acquisition and modernisation plans ran into some difficulties in 2001, when the government decided in March to cut USD4.74 billion from its five-year procurement plan and postpone the acquisition of 60 Eurofighter Typhoon fighter aircraft until after 2004. The nation has a new EMPAE (Unified Mid-Term Program and Modernization of the Armed Forces), which covers the ten-year timeframe of 2006-2015. It will analysis the overall framework of Greek defence procurement. The Greek defence procurement is used as a main platform to achieve the goals like reducing dependency on foreign sources, expanding the role of the Greek defence forces as a technology developer and cooperating with other European countries towards a more competitive European defence industry.

CONTACT DETAILS

Ministry of Defence
Mesogeion 227-231
Holargos15451, Greece
Tel.: +30 210 6598607
Fax: +30 210 6443832

National Defence General Staff
Stratopedo Papagou
15501 Holargos
Athens, Greece
Tel.: +30-210-6552050, -51

STRATEGIC INFORMATION

INDIA
(Capital: New Delhi)

INTRODUCTION

Area: 3,287,263 sq. km
Population: 1,266,883,598 (July 2016 est.)
Coastline: 7,517 km
Maritime claims: Territorial sea: 12 nm
Contiguous zone: 24 nm
Exclusive economic zone: 200 nm
Continental shelf: 200 nm or to the edge of the continental margin

GEOPOLITICAL IMPORTANCE

The largest country in South Asia in terms of landmass and population, India is bordered by Pakistan in the west, Myanmar and Bangladesh in the east, China, Nepal and Bhutan in the northeast. In its south-east lies the Bay of Bengal and in south-west, the Arabian Sea. To its south lies the Indian Ocean which is a vital sea trading route and connects the Atlantic and Pacific Oceans. India maintains an important position in the Indian Ocean Region through its chain of Andaman and Nicobar Islands in the east and through Lakshadweep islands in the west. Sri Lanka, Maldives and Indonesia are India's maritime neighbours.

Since centuries, India with its rich economic, cultural and intellectual strengths has remained an attractive destination for the entire world. Being the largest country in South Asia, it has influenced the regional power play by being actively involved with its immediate neighbours in more than one way to maintain peace and prosperity in the entire South Asian region. Its predominant geostrategic location in South Asia and proximity to the Indian Ocean has given it the additional edge to play a pivotal role in the regional sphere. With a rapidly growing market economy in the backdrop of an ever-expanding population coupled by the steady military rise has placed India as a major Asian powerhouse alongside China and Japan.

POLITICAL OVERVIEW

India is the largest democracy in the world and functions as a multi-party system with a Parliamentary form of government. The Constitution of India defines the country as a sovereign socialist secular democratic republic. It has made provision for a Parliamentary system of governance with a federal structure to dilute power between the Centre and the States. The Union Government comprises of three branches: Executive, Legislature and Judiciary. The Executive includes the President, Prime Minister and his council of ministers. The Legislature constitutes of the two houses of Parliament – Lok Sabha (lower house) and Rajya Sabha (upper house), the President, Prime Minister and the Council of Ministers. The Judiciary includes the Supreme Court, High Courts and all other subordinate courts in the country. While the President is the nominal head of the state and the supreme commander of the armed forces, the actual executive power rests with the Prime Minister and his Council of Ministers.

KEY POLITICAL PERSONS

PRESIDENT & COMMANDER-IN-CHIEF: Pranab Mukherjee

VICE PRESIDENT: Hamid Ansari

PRIME MINISTER: Narendra Modi

ECONOMY

India has transformed itself into a modern, developing economy where agriculture, industry and services sector are primarily contributing to the national GDP. A major share of the GDP in the past came from agriculture, fisheries, handicrafts and other small enterprises. To expand economic development and catapult domestic growth, India embraced economic reforms in 1991 and opened up its markets for foreign trade and investments. The liberalisation phase, which included industrial deregulation, privatisation of state-owned enterprises, and reduced control over international trade and investment, spearheaded India's economic growth and transformed the economy into one of the leading open-market economies in the world. As per the World Bank's 2014 data, India displaced Japan as the world's third largest economy in terms of purchasing power parity (PPP). It ranks as the 7th largest economy in the world in terms of GDP as per 2015 World Bank data. India has also overtaken China as the fastest growing major economy in the world in 2016, expanding by 7.6 percent in 2015-16 period as per the national statistics. To expand its economic reach beyond its boundaries and immediate neighbourhood, New Delhi has in recent years has initiated efforts for major entrepreneurial and financial forays into a number of countries in Asia, Africa, Europe and South America. Securing vital natural resources, like minerals, oil & natural gas, have particularly become important to sustain the country's rapid all-round growth and development. A number of state-owned as well as private-sector enterprises are making efforts to capture a pie of the highly competitive international markets for unhindered national growth.

GDP: $7.965 trillion (2015 est.); $7.421 trillion (2014 est.)

Real Growth Rate: 7.3% (2015 est.); 7.2% (2014 est.)

Industries: Textiles, chemicals, food processing, steel, transportation equipment, cement, mining, petroleum, machinery, software, pharmaceuticals

Natural Gas - Proved Reserves: 1.489 trillion cu m (1 January 2016 est)

Crude Oil - Proved Reserves: 5.675 billion bbl (1 January 2016 est)

Total Exports: $272.4 billion (2015 est.); $328.4 billion (2014 est.)

Export Commodities: Petroleum products, precious stones, machinery, iron and steel, chemicals, vehicles, apparel

Major Markets: US 15.2%, UAE 11.4%, Hong Kong 4.6% (2015)

Total Imports: $409.2 billion (2015 est.); $472.4 billion (2014 est.)

Import commodities: Crude oil, precious stones, machinery, fertiliser, iron and steel, chemicals

Major suppliers: China 15.5%, UAE 5.5%, Saudi Arabia 5.4%, Switzerland 5.3%, US 5.2% (2015)

DEFENCE & SECURITY

As the largest country in the subcontinent in terms of geographical area and population, India has played a central role in South Asian. It has steadfastly emerged as an economic powerhouse and alongside strengthened its defence forces to counter both conventional and asymmetric threats. The country, with its huge population and wide-ranging religious, cultural, regional, linguistic and social diversity, has faced varied security challenges, both internal and external, which have posed threat to its sovereignty and integrity from time to time. Even though it is the largest country in South Asia, India has more often witnessed a volatile neighbourhood which has directly or indirectly hindered its own growth and progress in the Asian continent, as well as in the global stage. Internal political upheavals in almost all of its surrounding countries now and then have spilled into India and directly or indirectly influenced its political and strategic policies and position. With its strong democratic credentials, India, however, has maintained a peaceful posture towards all its neighbouring nations and endeavoured to maintain friendly relations with them so as to create an integrated South Asia. Yet, New Delhi has often faced mistrust and misconception of being a dominant neighbour and even perceived as a threat by few of its smaller neighbours.

The unresolved border dispute with two of its immediate neighbours – China and Pakistan – has resulted in full-scale wars with them in the past. After the 1962 war with China, three wars with Pakistan in the past and the localised Kargil conflict in 1999, India has diligently pursued a massive military modernisation programme to safeguard its sovereignty and territorial integrity while preparing itself to counter any external aggression in future. Acquisition of advanced military platforms, weapon systems and equipment for the Army, Navy and Air Force along with upgradation of existing ones have been the top-most priority for the country. India also plans to revise its decades-old military doctrine to meet any modern, asymmetric, fourth-generation warfare threats. Phased induction of a large number of modern battle tanks, frigates, destroyers, submarines, aircraft carriers, multi-role fighters, stealth warplanes, combat drones and sophisticated weapons among others are emboldening the Indian Armed Forces in a big way, and preparing them to meet any "two-front" conflict scenario with Pakistan and China in future. To prepare itself for the modern-day network-centric warfare missions, India has also launched an integrated defence communication network enabling the Indian Army, Air Force, Navy and the Special Forces Command to share situational awareness for a faster decision-making process. The Defence Communication Network (DCN), a strategic, highly secure and scalable system, has a pan-India reach – from Ladakh to the North East to island territories.

KEY DEFENCE PERSONS

DEFENCE MINISTER: Manohar Parrikar

MINISTER OF STATE FOR DEFENCE: Subhash Ramrao Bhamre

CHIEF OF ARMY STAFF: Gen. Bipin Rawat

CHIEF OF NAVAL STAFF & CHAIRMAN, CHIEFS OF STAFF COMMITTEE: Adm. Sunil Lanba

CHIEF OF AIR STAFF: Air Chief Marshal BS Dhanoa

Internal Conflict

Terrorism: India has been fighting cross-border terrorism since its independence. Terrorism-related violence in the country has stemmed mainly from the territorial dispute with Pakistan. Armed insurgency and proxy war waged by secessionists and separatists over the Kashmir issue has often taken violent proportions and resulted in large-scale death and destruction across India. Terrorist attacks in many parts of the country have raised international concern.

Left Wing Extremism: Another major internal security challenge for India over the decades has emanated from Naxalism or Left-wing extremism. It is estimated that Naxalite groups are present in 200 of India's 602 districts covering a third of the country's landmass. In some of the districts, mostly in central and eastern parts of the country having vast, mineral-rich forests, the Naxalites collect their own taxes and deliver their own system of justice. Security analysts describe the Naxalite-affected areas as a 'Red Corridor' running from eastern to southern India. The Maoist problem in India's north-eastern states has also raised its head over several factors, including the region's geographical distance from the rest of the country and the social, economic and political inequality of its natives, which has been further exacerbated by their ethnic and linguistic diversity. Several armed groups in the region, some demanding autonomy while others seeking complete independence, have indulged in violence in the region for many years from now.

External conflict

Territorial disputes: India has land boundary disputes with two of its immediate neighbours – China and Pakistan. There is also a maritime boundary dispute of Sir Creek with Pakistan and disputes with China over construction of dams on the Brahmaputra River.

Pakistan: The flashpoint between the two neighbours since independence is Kashmir over which they have fought three wars in the past and the localised Kargil military conflict as recently as in 1999. As of 2010, India controls approximately 43% of the region, including most of Jammu, the Kashmir Valley, Ladakh, and the Siachen Glacier. Pakistan occupies approximately 37% or 78,000 sq km of Kashmir, known as Pakistan occupied Kashmir (PoK) and the Northern Areas of Gilgit-Baltistan. The rest of the territory, around 5,180 sq km, is under the control of China, illegally ceded to it by Pakistan under the China-Pakistan Boundary Agreement of 1963. Pakistan has been strongly contesting New Delhi's position that Jammu & Kashmir is an integral part of India.

Another territorial dispute with Pakistan involves the Siachen Glacier located in the eastern Karakoram range in the Himalaya Mountains. The glacier is the highest battlefield on Earth with both India and Pakistan maintaining permanent military presence in the area. The conflict in the region originates from the 1972 Shimla Agreement that did not clearly mention who controlled the glacier. India occupied the entire glacier and its tributary glaciers in 1984 while Pakistan presently controls areas just a few kilometres to the west.

India is also entangled in a maritime dispute with Pakistan which involves the Sir Creek estuary at the mouth of the Rann of Kutch in the Arabian Sea. The strip of water divides the Kutch region of Indian state of Gujarat with the Sindh Province of Pakistan. Both countries seek a technical resolution of the issue.

China: India and China share a land border of 4,057-km. India's decades-old boundary dispute with its north-eastern neighbour has revolved around the Aksai Chin region administered by China. The full-scale Sino-India war in 1962 resulted in China occupying Aksai Chin, covering an area of 37,244 sq km. Pakistan later illegally ceded another 5,180 sq km of the region to China. India is the claimant of the area as the eastern most part of its state of Jammu & Kashmir. Though an uninhabited high altitude desert, Aksai Chin has remained a strategically vital region, giving China direct access to Pakistan via the Pak-occupied-Kashmir (PoK). China has built a 2,143-km highway running through the disputed land border area (covering some 150-km) that connects Tibet with its Xinjiang prefecture. In recent years, the country has also completed repaving the road which was built between 1951 and 1957.

The other region of contention between the two countries is the Indian state of Arunachal Pradesh over which China stakes claim. China considers Arunachal Pradesh (formerly North East Frontier Agency), covering an area of 90,000 sq km, as part of its territory and calls it 'South Tibet', a claim vehemently rejected by India, which says the state is its integral part. China also does not recognise the McMahon Line as the border in the North East.

The Brahmaputra River, one of the major rivers in Asia that flows through India, China and Bangladesh, has remained another source of Sino-India conflict since decades. The river that originates in Tibet and enters India through the state of Arunachal Pradesh is an important source of irrigation, trade and transportation. Recent attempts by China to build about four dams on the river to divert water in order to irrigate a large landmass covering the Gobi Desert have caused tensions between the two neighbours.

Even though discussions at political and diplomatic level continue between the two sides to find a solution to the long-standing boundary dispute and build mutual trust and confidence, no concrete results have been achieved so far.

THREAT PERSPECTIVE

Pakistan: India's relations with Pakistan remain strained over numerous issues. These include border dispute on Jammu & Kashmir as also terrorist activities originating from the Pakistani soil over which the Pakistan government shows less concern and often denies involvement. Pakistan views India as its major military adversary and has accordingly developed its military capability to match India's strategic outreach. It has also adopted covert tactics against India and engaged in proxy war over the Kashmir issue. Backed by China, which has been supplying sophisticated nuclear and other weapons and technology to it, Pakistan is strengthening its military might to take on India. Further, they threaten to use nuclear weapons to stall a provocative offensive by India.

China: Even while steadily emerging as the leading economic powerhouses in Asia, India and China have also vied to strengthen and expand their military clout to remain influential in the region. The two neighbours have fought a full-scale war in 1962, and China's aggressive posture in recent times over international territorial claims, including its reassertion over Arunachal Pradesh, has further heightened regional tensions. With a belligerent posture, China has been robustly engaged in developing massive road and rail infrastructure along the disputed land border region, thereby enabling swift mobilisation of its troops there – a move being cautiously watched over by India. According to the 2011-2012 annual report released by the Indian Defence Ministry, China has built a network of rail lines and more than 58,000 km of roads in the Tibet Autonomous Region (TAR) along with nuclear missile silos. Beijing has also announced the completion of a highway linking the Medog county in Tibet (located near the border of Arunachal) with the rest of China. It has also made six of its airfields - Gongar, Linzhi, Pangta, Hoping, Gar Gunsa and the most recently added Kashgar - in the TAR fully operational. Regular military manoeuvres involving Chinese fighter jets along with other advanced weapons and platforms in the Himalayan region have also increased in recent times. A 2010 Pentagon report stated that China has moved its nuclear-capable advanced CSS-5 (DF-21) ballistic missiles closer to borders with India. Beijing is also engaged in rapid infrastructure development in the disputed Pak-occupied-Kashmir (PoK) region despite India's strong opposition.

To further widen its geopolitical and military reach and secure vital energy resources across Asia, China has also been trying to expand its foothold in the Indian Ocean Region (IOR) and beyond. According to a US Defence Department report published in 2003, China over the years is steadily building up a "String of Pearls" in South Asia – a farfetched maritime strategy to not only encircle India but also gain access to the energy resources in entire Asia. The "String of Pearls" refers to Beijing's efforts to build sea ports and airfields, enhance political, diplomatic and military ties with Asian littoral states that extend from the South China Sea through the Strait of Malacca, across the Indian Ocean, and on to the Arabian Gulf. Its recent efforts to help building the Gwadar port in Pakistan, the Chittagong port in Bangladesh and the Hambantota port in Sri Lanka are viewed to be a part of the larger "String of Pearls" strategy. The recent dockings of two Chinese Navy submarines – one at Sri Lanka's Colombo port and another at Pakistan's Karachi port – are a further confirmation that such a strategy is being implemented by Beijing.

Even as India is making a cautious approach to keep any kind of confrontation with China at bay while acknowledging the rapid military rise of that country in its immediate and extended neighbourhood, Indian Armed Forces are acquiring new capabilities and strengthening themselves to remain prepared for any eventuality. In 2009, India decided to deploy additional troops along the border in Arunachal Pradesh. The additional troops, consisting of some 35,000 personnel (downsized from previous number of 90,000 troops) would be deployed as the Army's new Mountain Strike Corps, expected to be fully operational by 2018-19. The new military wing will have units spread all across the 4,057-km LAC from Ladakh to Arunachal Pradesh. The Indian Army has also identified strategically important infrastructure requirements along the Line of Actual Control (LAC) and they are being developed in a phased manner.

The IAF has deployed its force multiplier assets such as the C-130J Super Hercules special operations aircraft at the Panagarh air base in West Bengal which would be the headquarters of the new Mountain Strike Corps. The air base, renamed as Air Force Station Arjan Singh, went operational in May 2016.

The Air Force too has refurbished its helipads and airfields in the north-eastern sector and deployed frontline combat aircraft and advanced weapon systems to defend and protect the territory from any external aggression. In a subtle show of strength, the IAF has also landed its C-130J airlifter at the Daulat Beg Oldie airstrip near the LAC in the Ladakh area of Jammu and Kashmir for the first time in August 2013. It has also re-activated its strategic Advanced Landing Ground (ALG) of Mechuka in West Siang district of Arunachal Pradesh in May 2016 followed by the Pasighat ALG in that state in August 2016. The new ALG is capable of operating all types of aircraft, including the Su-30MKI, as well as helicopters.

The Navy, for its part, has strengthened its military presence in the frontline islands of Andaman and Lakshadweep chains by setting up new naval bases there so as to secure the crucial sea lanes in the IOR. New Delhi in recent few years has also given more emphasis to its "Look East" policy and making renewed efforts to strengthen its relations, both strategic and economic, with East Asian and South East Asian countries. India today forms a part of the East Asia Summit and has strong relations with the ASEAN.

STRATEGIC RELATIONS

Russia: India's close relationship with Russia - the erstwhile Soviet Union - dates back to the 1950s when the former was making efforts to become a self-reliant nation after gaining independence from the British rule in 1947. From industrial development to advancement in military technology, space science and nuclear energy, the Soviet Union played a vital role in shaping up modern India. It also provided strong political support to India in international forums. The strategic relationship between the two countries reached its zenith when the Indo-Soviet Friendship Treaty was signed in 1971 which incorporated security clauses in its ambit along with mutual cooperation in other areas. The two countries enjoyed strong strategic ties until the disintegration of the USSR in 1991 which gave way to a new uni-polar world order with the United States emerging as the sole superpower. While both India and Russia continued to share deep bilateral bond, they also began to explore new geostrategic alliances in the emerging world order. Nevertheless, the decades-old friendship between the two sides remains crucial even today. India continues to remain one of the top-most clients of the Russian defence industry, receiving critical weapon systems and technology from Moscow, in spite of opening up its arms market to the US and many other Western powers. The two countries are also exploring avenues for joint space missions in future and cooperating in developing India's nuclear technology for energy security.

Israel: India established diplomatic ties with Israel in 1991. The two countries have become strategic partners thereafter. Israel, with a state-of-the-art defence industry, has supplied many sophisticated military technologies to India in the past. India's defence trade with Israel includes a range of military equipment including modernisation of tanks and fighter aircraft, supply of UAVs, missiles, command and control systems and ammunition. This multi-billion dollar India-Israel defence market is set to grow further in the coming years. In the last few years, Israel has emerged as the one of the largest defence suppliers to India after the US and Russia.

Bhutan: The two South Asian neighbours have enjoyed close economic, political and strategic ties since decades. Under the Treaty of Friendship and Cooperation signed between them in 1949, the two countries agreed for non-interference in each other's internal affairs. Bhutan, under the treaty, also agreed to let India guide its foreign policy and both nations would consult each other closely on foreign and defence affairs. India has provided economic assistance to Bhutan and helped in the Himalayan nation's all-round development. Bhutan's transition to democracy in 2008 has further strengthened Indo-Bhutanese ties.

Sri Lanka: India's centuries-old relationship with its closest neighbour in the subcontinent, Sri Lanka, has included historical, cultural, intellectual, linguistic and religious aspects. Their alliance has further expanded to cover economic, political and strategic cooperation in last few decades. India has been actively involved in infrastructure development and modernisation of projects in Sri Lanka which is strategically positioned in the Indian Ocean region. Bilateral defence ties between the two neighbours have strengthened over the years with India providing military training to the Lankan armed forces. The two sides also conduct joint military exercises on a regular basis as the commonality of concern for both countries include the safety and security of their sea lanes of communication in the Indian Ocean region. India is also building two OPVs for the Lankan Navy.

Afghanistan: India and Afghanistan have since long shared close cultural and historical ties. The Treaty of Friendship between the Government of India and the Royal Government of Afghanistan was inked on 4 January 1950 to strengthen bilateral relations. In the past decade, India has generously assisted Afghanistan in its overall development. The two countries signed the Strategic Partnership Agreement in New Delhi on 4 October 2011 to open the doors to training, equipping and capacity-building in defence sector. The accord also seeks to impart a long term commitment to their multifaceted bilateral relations and to actively develop them in political, economic, trade, scientific, technological, cultural and other fields in the years ahead. This is the first Strategic Partnership Agreement India has signed within its South Asian neighbourhood. Afghanistan is also keen to continue its defence cooperation with India. Reversing its earlier position of refraining from supplying any military and related equipment to the war-torn Afghanistan, New Delhi has now decided to supply defence hardware to it and handed over four Mi-25 attack helicopters in December 2015. Kabul is seeking more defence supplies from India.

Maldives: The geographic position of Maldives in the Indian Ocean region makes the island country strategically vital for India. Bilateral relations between the two nations have involved high-level visits, cooperation in defence and security arena, trade and investment and cultural ties. India has supplied a naval platform, surveillance aircraft and radars to the island nation in the past to secure it from sea piracy and terrorism as also to safeguard its own coastal borders from any sea-borne attack. Maldivian military personnel also receive training in India. New Delhi has also expressed interest in building a naval port at Uthuru Thila Falhu in the Kaafu atoll of the island.

Singapore: Bilateral relations between India and Singapore have revolved around political, economic, cultural and strategic areas. Singapore has played an important role in reconnecting India with the countries of South East Asia since the adoption of New Delhi's "Look East" policy in the early 1990s. The two countries have signed a Defence Cooperation Agreement in 2003 under which they carry out joint military exercises.

Seychelles: India and Seychelles share strong trade and economic relations along with cooperation in the fields of science & technology, health care and tourism. The most vital aspect of bilateral ties between the two sides is their deep-rooted strategic partnership which has evolved over the years with India providing defence equipment and platform to the island country in the Indian Ocean. India has provided a comprehensive maritime security solution including provisioning of military hardware, training, consultancy as well as hydrographic assistance to the Seychelles Government in the past. In November 2014, New Delhi gifted a Fast Attack Craft to the Seychelles Coast Guard to carry out anti-piracy missions. Indian Navy has already supplied an FAC to Seychelles in February 2005. India is also setting up coastal surveillance radars for the Seychelles Coast Guard. In 2016, the Seychelles Government announced that it has provided a piece of land to India in the country's Assumption Island to build a naval base in the Indian Ocean region.

Mauritius: The century-old multi-faceted relationship between India and Mauritius has revolved around historical, cultural, democratic values and intimate ancestral ties. The strategically located island nation had drawn inspiration from India's freedom movement and it gained independence in 1968. In recent times, India has forged extensive defence relations with Mauritius to safeguard vital sea lanes of communication from pirates. New Delhi has provided helicopter, an offshore patrol vessel, fast interceptor boats and two FACs to the government of Mauritius to bolster its counter-piracy capabilities. Indian naval ships are also conducting regular surveillance and joint patrolling missions in the vast Exclusive Economic Zone of Mauritius.

Vietnam: The decades-old Indo-Vietnam relation has included political, economic and strategic ties between the two Asian countries. Vietnam has also played a crucial role in India's "Look East" policy. The ambit of India's strategic alliance with Vietnam has further expanded in recent times with the two countries agreeing to strengthen their defence cooperation in the backdrop of new evolving geostrategic scenario in the Asian continent. Defence relations between the two countries have included high level bilateral visits, training of personnel, upgrading military capabilities, assistance in defence production and joint exercises. India, for the first time, has also provided a $100 million credit line to Vietnam which is being utilised by Hanoi for the procurement of Offshore Patrol Boats for its Border Guards. The South East Asian country has also evinced keen interest in India's defence research programmes and the two sides have identified the major areas for cooperation that includes upgrade of Soviet legacy systems, upgradation of Thermal Sights and Fire Control Systems for the BMP vehicles, T-54 and T-55 tanks, upgrade of MI-17/ Mi-8 helicopters, shipbuilding programmes, missile systems from India and software defined radios from Vietnam. India is also training the Vietnam Navy personnel in - among other things - operating the Russian-origin Kilo-class submarines. So far it has trained over 500 personnel of the Vietnamese Navy. New Delhi has also agreed to train Vietnam's pilots in flying the Russian-built Sukhoi-30 fighter aircraft from 2017 onwards. The Vietnamese pilots will arrive in India to be trained by the Indian Air Force fighter pilots. India and Vietnam are also working on an agreement to launch Vietnam's satellites into space. India is keen on establishing a satellite tracking, data reception and processing centre in the Ho Chi Minh city.

Thailand: India and Thailand share a maritime boundary in the Andaman Sea. Bilateral relations between the two countries have revolved around trade and economic exchanges, cultural ties and cooperation in the field of defence that includes joint military exercises and training of armed forces. The two countries signed a bilateral Memorandum of Understanding on Defence Cooperation in January 2012. In June 2016, they have also agreed to forge a closer partnership in maritime cooperation. India also plans to take up infrastructure development projects with Thailand and one such project is the construction of India-Myanmar-Thailand Trilateral highway, connecting the three Asian countries by road.

Myanmar: India and Myanmar share a land border of 1,650-km and a common maritime boundary in the Bay of Bengal. The geographic position of Myanmar, which connects South Asia with South East Asia, has strategic significance for India. The two neighbours have primarily been cooperating in the fields of trade and commerce. Myanmar has received military hardware from India in the past. India has also taken up various developmental tasks, including infrastructure build up, in Myanmar. Navies of the two countries have taken part in multilateral exercises in the past. India has recently agreed to Myanmar's request for assistance in building OPVs, which are likely to be constructed at an Indian shipyard. It is also considering to train Myanmarese pilots to fly the Russian-origin Mi-35 attack helicopters.

Indonesia: As part of India's "Look East" policy and the renewed "Act East" policy of the present dispensation, Indonesia has received increased attention from New Delhi with bilateral cooperation covering economic, political, security and military aspects. The bilateral ties further strengthened with the signing of the 'Joint Declaration on Establishing a Strategic Partnership' in 2005 between the two sides. The two countries have been closely coordinating in defence and security fields and the navies of both sides regularly conduct coordinated maritime patrols (CORPAT) in the Indian Ocean Region. Their armies have also hold joint military training exercises. During their biennial defence dialogue held at Jakarta in Oct 2012, India had agreed to train and support the Indonesian Air Force in operating its fleet of Sukhoi fighter jets acquired from Russia.

Malaysia: The two Asian countries establishing strategic partnership in 2010. The bilateral ties involve political, regional and international cooperation besides the defence & security aspect. While conducting regular military exchanges, the two countries in November 2015 have also agreed to set up a "SU-30 Forum" for cooperation in training, maintenance, technical support and safety-related issues, building on the assistance by India for successful completion of the SU-30 MKM (being operated by the Malaysian Air Force) training programme by Indian pilots in Malaysia. The two sides have also announced joint collaboration on projects of mutual interest in the defence sector.

Iran: India and Iran have shared strong cultural and historical relations since several decades. India has always acknowledged the strategic importance of Iran in the Asian continent. Though Indo-Iranian ties have passed through several ups and downs in the past few years due to changing regional and global geostrategic equations, nevertheless, the two countries have valued each other as important trading and economic partners. Iran has remained a crucial energy supplier for India and meets major part of its crude oil demand. The two countries have especially been making efforts to improve bilateral cooperation in economic, political, energy and security areas. The two sides, during Prime Minister Narendra Modi's Tehran visit in May 2016, signed a historic deal to develop the Chah Bahar port in Iran. The deal was signed after a meeting between Prime Minister Modi and Iran's President Hassan Rouhani. The port in South East Iran and having direct road links to Afghanistan border will offers huge trade opportunities for India as also for Iran in Central Asia and Afghanistan. Apart from the bilateral agreement between India and Iran, a trilateral transit and trade corridor deal was also signed by India, Iran and Afghanistan which would ensure easy movement of goods between the three countries. Further, with the possibility of improvement in Iran-US relations following the historic nuclear deal inked with Tehran in 2015, negotiations are likely to commence on the long-pending Iran-Pakistan-India proposed gas pipeline project.

Japan: While bilateral relationship between the two Asian powers remained confined only to the economic sphere in the past, in recent times, the two countries have made sincere efforts to expand it further so as to include political and strategic tie-ups. Keeping in mind the changing regional and global geostrategic equations, India and Japan have also decided to hold joint military exercises involving their defence forces. In the first-ever bilateral maritime engagement, the two countries' navies conducted the 'JIMEX 12' naval exercise in June 2012. The two countries are also participating in the multi-lateral maritime drill "Malabar".

Mongolia: Bilateral ties between India and Mongolia date back to 1955 when New Delhi established diplomatic relations with Ulan Bator. The two countries also share deep cultural and historic ties for many centuries. In the year 2015, India upgraded its bilateral ties with Mongolia to "Strategic Partnership" during Prime Minister Narendra Modi's visit, the first-ever visit by an Indian Premier, to the Central Asian nation. New Delhi also pledged US$ 1 billion credit line to Mongolia for infrastructure development in that country.

US: Recalibrating its position vis-à-vis the world's superpower, India has in recent times inched closer to establish strong security and defence ties with the US. The strategic partnership between the two countries is witnessing a dramatic transformation of sorts wherein the US is acknowledging India as its "security partner of choice" in the broader Indo-Pacific region. The resurgence of China, which is brandishing its ever-expanding military clout, has also brought New Delhi and Washington further closer to counter the Asian giant and its increasing belligerence in the Asian continent. High-level joint military exercises are being regularly conducted between the Indian and US armies, navies and air forces in both the countries. India has also been conducting joint maritime exercise with the US in the Indian Ocean. The two sides have inked technology exchange agreement and most importantly, India in August 2016, has signed the Logistics Exchange Memorandum of Agreement (LEMOA) with the USA. The agreement has paved the way for the two nations to become logistical allies, enabling both the militaries to use each other's assets and bases for repair and replenishment of supplies. The US has also agreed to elevate defence trade and technology sharing with India to a level commensurate with its closest allies and partners. Washington has also explored the ever growing Indian defence market to export its cutting-edge defence equipment and technology. It has, in fact, emerged as one of the leading suppliers of defence hardware to India in the past few years.

EU: India and the European Union have built-up a longstanding relationship since the 1960s and have been coordinating on a broad range of areas, including politics, trade, economy, science & technology, education and strategic sphere. In 2004 India became one of the EU's "Strategic Partners". In recent years, some of the leading powers of the EU, including France, Germany and Italy, have evinced interest in the hugely growing Indian defence sector.

Multilateral Alliances

Besides maintaining bilateral relations with its neighbouring and other countries, India has also been engaged in several multilateral forums which include SAARC, BIMSTEC, IBSA and BRICS.

SAARC: As one of the founding members of the South Asian Association for Regional Cooperation (SAARC), India has endeavoured to promote political, economic, social and technological cooperation among the member nations in the block which include Afghanistan, Bangladesh, Bhutan, Maldives, Nepal, Pakistan and Sri Lanka. The multilateral forum, founded in 1985 with its headquarters in Kathmandu, Nepal, has so far accorded observer status to Australia, China, the European Union, Iran, Japan, South Korea, Mauritius, Myanmar and the USA. The forum has been engaged in promoting regional trade and economy, and enhancing mutual cooperation in several fields including agriculture, education, energy, science and technology and social development.

BIMSTEC: The Bay of Bengal Initiative for Multi-Sectoral Technical and Economic Cooperation (BIMSTEC) is a regional block comprising of India, Bangladesh, Sri Lanka, Thailand, Myanmar, Nepal and Bhutan. The multilateral forum with the South Asian and South East Asian nations has 13 priority areas - trade and investment, technology, energy, transport and communication, tourism, fisheries, agriculture, cultural cooperation, environment and disaster management, public health, people-to-people contact, poverty alleviation, counter-terrorism and transnational crimes. The block, originally named Bangladesh, India, Sri Lanka, and Thailand Economic Cooperation (BISTEC), was formed in 1997 to primarily promote economic and trade cooperation among the member nations. Myanmar, Nepal and Bhutan subsequently joined the forum which was then renamed as Bay of Bengal Initiative for Multi-Sectoral Technical and Economic Cooperation.

IBSA: With a view to promote South-South cooperation, India, Brazil and South Africa joined to form IBSA in 2003, bringing together three developing countries from three continents of the globe – Asia, South America and Africa. The trilateral block has provided a platform to its member nations to engage in cooperation in several areas, including political, trade and economic, social, agricultural, and defence and technological relations. The group has also been holding a biennial joint naval drill, called India-Brazil-South Africa Maritime (IBSAMAR) exercise since 2008.

BRICS: The forum was formed in 2006 with Brazil, Russia, India and China (BRIC) as its founding members. South Africa joined the block in 2010 which was then renamed BRICS. The alliance was primarily focused on political, economic and financial cooperation among the member nations. Over the years, the block has expanded its purview of cooperation so as to include evolving issues including international terrorism, climate change, food & energy security, millennium development goals and strategic partnership.

SCO: India has been accorded an observer status in the six-nation Shanghai Cooperation Organisation (SCO). The grouping, with China, Russia, Kazakhstan, Kyrgyzstan, Tajikistan, and Uzbekistan as its founding members, agreed to include India along with Pakistan as full-time members during its annual summit held at Ufa, Russia in July 2015. Both India and Pakistan are expected to join the SCO as full members at its next annual summit going to be in Astana, Kazakhstan in June 2017. Joining the SCO would pave the way for New Delhi to strengthen its connectivity with the resource-rich Central Asian nations. India has renewed efforts in this direction with its "Connect Central Asia" policy. While emphasising on political and economic coordination among its member nations, the SCO is steadily emerging as an influential regional block in the Eurasian space. The organisation is also focused on the rapidly evolving strategic and security issues in the region.

Arctic Council: The eight-member Arctic Council is a high-level intergovernmental forum set up to address the issues faced by the Arctic governments and the indigenous people of the Arctic region. Canada, Denmark, Finland, Iceland, Norway, Russia, Sweden, and the US are the Council's permanent members. India, along with 11 other non-Arctic countries including China, have been admitted into the council as permanent observers in 2013. New Delhi has expressed its desire for increased cooperation in the council in view of the geo-strategic importance of the Arctic Sea and the region's important role in governing climate change that has global consequences.

DEFENCE CAPABILITIES

ARMY: With a strength of nearly 1.2 million active personnel and around a million reserve force, the Indian Army is the second largest army in the world after China and the largest standing volunteer army in the world. It is the largest wing of the Indian defence forces and is equipped with huge conventional and nuclear arsenal to defend and protect national sovereignty and integrity. As India has faced security challenges from both state and non-state actors from time to time, the Army, operating from the remotest and most difficult terrains of the country, has carried out the most complex, delicate and sensitive operations to protect the nation from any internal or external threat. It also takes up humanitarian missions across the nation at the times of national disasters or adversities.

To add more teeth to its existing capabilities, Indian Army is in the process of acquiring new arms, weapons and platforms, including howitzers, missiles, combat vehicles, helicopters, UAVs, transport, surveillance and fighter aircraft among other small arms. Besides strengthening itself to counter enemy threat, Indian Army has also been actively taking part in the United Nations-led peacekeeping missions in several strife-torn countries worldwide.

Personnel: 1.2 million (+1 million reserve)

Equipment

Category	Name	In Service	On Order
MBT	Arjun Mk-I	124	124
	Arjun Mk-II		118
	T-90S/M	500+	1000+
	T-72 M1, M2	1900+	
	T-55	900	
IFV, APC, ARV	BMP-1, BMP-2	2000+	362 (BMP-2)
	Casspir	250+	
	OT-64 SKOT/ OT-62 TOPAS	200+	
	VT-72B	200	
	WZT-3	550	
Artillery/ Weapon	155mm light field gun	180	
	105mm light field gun	700+	
	FH77B 155mm howitzer	410	
	M46 130mm howitzer (upgraded to 155mm standard)	300	
	M777 howitzer		145
	122mm 2S1 howitzer		
	81mm mortar	500	
	106mm RCL		
	Smerch 300mm	38	
	Pinaka MRL	80	2 Regiments
	BM-21 MRL	100+ (being withdrawn)	
	57mm ATK gun		
	100mm ATK gun		
	AT3 Sagger ATGM	500	
	Milan ATGM		
	Snapper ATGM		
	Spike		8,356
	Inver ATGM		10,000
	Nag ATGM		443

Army continued...

Category	Name	In Service	On Order
Artillery/ Weapon	9M113 Konkurs ATGM		
	9K22 Tunguska SPAAG		
	40mm L-70 AAG	1000	
	ZSU-23-2	800	
	SA-6 SAM	180	
	SA-8 SAM		
	SA-13 SAM	50	
	SA-18 SAM		
	SA-9 SAM		
	Akash SAM		2 Regiments
	Prithvi SSM	40+	
	BRAHMOS SSM		

Army Aviation

Category	Name	In Service	On Order
Helicopter	Chetak	60	
	Cheetah	120	
	Cheetal		22
	ALH Dhruv	50	149
	Kamov Ka-226T LUH		200
	LCH		
	Rudra (attack)	20	40
UAV	Searcher I & II		
	Nishant	3	
	Heron		15

New Procurements/Upgrades

✦ Procurement of advanced T-90 tanks is going on. The Army plans to induct over 1500 such tanks, mostly built in India under license from Russia, in its inventory by 2020. The Defence Ministry in November 2016 cleared a proposal to acquire 464 of the modernized T-90MS MBTs from Russia. The tanks will be manufactured by India's OFB under the "Make-In-India" programme.

T-90MS tank

✦ The Army will get about 362 BMP-2 series of infantry fighting vehicles built by the Ordinance Factory Board. Over 900 existing BMP-2s will also be upgraded with new fire control systems, anti-tank missiles and third generation night vision capability. Reliance Defence has joined hands with a Ukrainian company to carry out the upgrade of the IFVs.

✦ The Government has invited proposals from 10 Indian public and private defence firms for building a future infantry combat vehicle (FICV). The new vehicle will replace the BMP-2 IFVs. Over 2,600 FICVs are being planned to be built under the Make-In-India programme.

✦ Two regiments of BRAHMOS supersonic cruise missile system have been fully operationalised in the Army while a third regiment for carrying out mountain warfare operations is being raised. Government in 2016 has cleared the induction of fourth BRAHMOS regiment for Army. The advanced missile with "trajectory manoeuvre and steep-dive capabilities" will be deployed in the country's Eastern sector for mountain warfare missions.

✦ The indigenously-built Akash surface-to-air missile system has been inducted in the Army in May 2015. The Army's initial order is for two Akash regiments, each comprising of 125 missiles. Both regiments are expected to be fully operationalised by 2016-17.

✦ Delivery of the indigenously-developed 155mm/45-calibre Dhanush howitzers has started in July 2016. The Defence ministry has cleared a proposal to induct 114 Dhanush systems in the Army. The artillery gun is being produced by the Ordnance Factory Board in Jabalpur on the basis of transfer of technology of the Swiss Bofors guns.

✦ A proposal has been cleared to buy 145 ultra-light howitzers, M-777, from the US for a price of worth US $700 million. While 25 of the guns will be procured in completed form through the Foreign Military Sales (FMS) programme, the remaining ones will be built locally. BAE Systems, the manufacturer of the M-777, has announced to tie up with India's Mahindra Defence for assembling, integrating and testing the howitzers in India.

✦ Procurement of 814 mounted gun systems (MGS) for an estimated $3 billion will be made over the coming years. The 155 mm/52 calibre guns will be bought under the "buy and make" Indian category. While the first 100 MGSs can be imported ready-built from a foreign firm having a joint venture with an Indian company, the remaining 714 are to be manufactured in India. Negotiations are advancing between India's L&T and a South Korean firm to acquire the K-9 self-propelled howitzers, which would be manufactured by L&T as K-9 Vajra.

✦ The Defence Ministry has cleared a proposal to acquire domestically built 130mm Arjun Catapult artillery systems. A total of 40 such systems will be acquired at a cost of worth Rs. 820 crore.

✦ An agreement has been signed with Russia to produce the 80 km-range Smerch rockets. India's Ordnance Factory Board has signed the MoU with Russia's Rosoboronexport and Splav 'Spa' to manufacture five versions of Smerch rockets based on transfer of technology from Russia.

✦ Two more indigenous Pinaka multi-barrel rocket regiments for the Army will be acquired at a cost of Rs. 3,300 crores. Each regiment will come with 18 Pinaka launchers and associated systems.

✦ 10,000 Invar anti-tank guided missiles will be procured from Russia for the Army's T-90 main battle tanks. Bharat Dynamics Ltd has received a Rs 3,000-crore contract to licence-produce the missiles at its facilities.

✦ A total of 8,356 Spike third generation anti-tank guided missiles and 321 launchers will be procured from Israel at an estimated cost of Rs. 3,200 crore.

✦ The Army is considering to buy quick reaction air defence missile systems to protect its assets and troops from enemy aircraft and unmanned aerial vehicles. Israel's Spyder QR-SAM has emerged a frontrunner for the likely procurement.

✦ The Defence Ministry has cleared a proposal to procure around 22 Cheetal helicopters for the Army. The HAL-developed rotorcraft are expected to be deployed in Ladakh and Siachen regions to support the Army's high-altitude operations.

✦ An RFI has been issued in October 2014 to Indian companies for an unspecified number of reconnaissance and surveillance helicopters (RSHs) for the Army to replace the aging fleet of Cheetah and Chetak helicopters.

✦ A proposal has been cleared by the Defence Ministry in November 2016 to acquire 15 Light Combat Helicopters (for Army & Air Force).

✦ Russia and India have agreed on joint production of Ka-226T military utility helicopters. India plans to produce at least 140 of the multi-role rotorcraft locally.

✦ 15 Heron UAVs are being procured from Israel at a cost of Rs 1,200 crore.

✦ The Defence Ministry in November 2016 has cleared the procurement of 598 unmanned aerial vehicles.

NAVY: The Indian Navy operates with the primary vision and mission "to create and sustain a three dimensional, technology enabled and networked force capable of safeguarding the country's maritime interests on the high seas and projecting combat power across the littoral." Besides safeguarding the country's territorial waters, the Navy also works to strengthen maritime cooperation with other nations and assists in humanitarian missions. Although smaller than the other two wings of the Indian armed forces, the Navy is sizable and well-equipped, a reflection of India's focus on regional power projection. Once a "brown water" navy, Indian Navy has since the 1990s made efforts to strengthen its overall capabilities in order to transform itself into a "blue water" navy to carry out long range littoral as well as high sea operations. This transition has become necessitated in the wake of the fast changing geo-strategic landscape where Asia has emerged as the new power block amidst a raging China flaunting its military might. Of particular significance for Indian Navy in recent times is the Indian Ocean Region (IOR) which has witnessed a spurt in activities, especially by China.

To bolster its capabilities, the Indian Navy is undergoing a full scale modernisation drive by acquiring new maritime platforms including aircraft carriers, submarines, frigates, destroyers, corvettes, naval fighter jets, surveillance aircraft, helicopters and modern combat systems. The new maritime assets, equipped with cutting-edge weapons and armaments, promise to transform Indian Navy as a "world class" maritime force.

The Naval Fleet headquarters is based at New Delhi, while naval commands are located at Mumbai (Western Command), Vishakhapatnam (Eastern Command) and Kochi (Southern Command). The naval aviation HQ is based at Arakonam, while the Marines also have a joint command HQ based at Port Blair on the Andaman Islands in the Bay of Bengal.

Major naval & naval air bases: Mumbai, Kolkata, Kochi, Vishakapatnam, Port Blair, Kavaratti, Karwar, Porbandar, Okha, Panaji, Arakkonam, Ramanathapuram

Personnel: 100,000+ (Including Marine Commandoes, Naval Air & Coast Guard personnel)

Equipment

Category	Name	In Service	On Order/In Production
Aircraft Carrier	INS Vikramaditya	1	
	INS Vikrant		1 (+1 or more planned to be built)
	INS Viraat	1 (Set to be decommi-ssioned)	
Destroyer	Delhi class	3	
	Rajput class	5	
	Kolkata class (Project-15A)	3	
	Project-15B		4
Submarine	Shishumar class	4	
	Sindhughosh class	9	
	INS Chakra (Akula II class)	1	1
	Arihant class SSBN	1	5
	Scorpene (Project 75)		6
	Project 75-I		6
Frigate	Talwar class	6	
	Brahmaputra class	3	
	Godavri class	2	

Navy continued...

Category	Name	In Service	On Order/In Production
Frigate	Shivalik class	3	
	Project 17A		7
Corvette	Khukri class	4	
	Kora	4	
	Veer class	10	
	Abhay class	4	
	ASW Kamorta class	2	2
FAC	Bangaram class	4	
	Car Nicobar class	11	
	INS Tarmugli (advanced water-jet FAC of Car Nicobar class)	1	3
	Trinket class	1	
	Super Dovra II class	5	
OPV	Sukanya class	6	
	Saryu class	4	
Minesw-eeper	Pondicherry class & Karwar class	6	
Amphibi-ous Force	LPD (INS Jalashwa)	1	4
	Kumbhir class LST	4	
	Magar class LST	2	
	Shardul class LST	3	
	LCU	6	8
Fleet tanker	Deepak class	2	
	Jyoti class	1	
	Aditya class	1	
Torpedo recovery vessel	Astrabahini	1	
Research / Survey vessel	Sagardhwani	1	
	Sandhayak class	7	
	Makar class	1	5
Training/ Sailing ship	Tir	1	
	Tarangini class	2	
Weapon	K-4/ K-15 SLBM		
	Dhanush, (ballistic missile)		
	(AShM/ LACM) BRAHMOS, Klub KH-35, P-20		
	(AAM) R-77		
	(SAM) Barak-1, Shtil, 9K38 Igla (MANPAD)		

Navy continued...

Category	Name	In Service	On Order/In Production
Weapon	(Torpedo) A244-S, APR-3E, SET-65E/53-65KE Type 53-65, TEST 71/76, Varunastra		

Naval Aviation Equipment

Category	Name	In Service	On Order
Naval Fighter	MiG 29K/ KUB	41	4
	Naval LCA		40
Maritime patrol/ Utility	Tu-142M Bear	8	
	IL-38	5	
	P-8I	8	4
	Dornier Do 28	18	
Trainer	Hawk 132 AJT	11	6
	HJT-16 Kiran	20	
Helicopter	Ka-28	10	
	Ka-31	14	
	UH-3H Sea King	32	
	AgustaWestland Sea King Mk 42	20	
	Chetak (Aloutte III) utility	55	
	ALH Dhruv	6	16
UAV	Searcher II		
	Heron		

New Procurements/ Upgrades

✦ The Russian-built aircraft carrier INS Vikramaditya, commissioned in the Navy in November 2013, is stationed at the Karwar naval base in Karnataka. The 44,570-ton warship is the second aircraft carrier of India after INS Viraat. The carrier, with a length of 284 metres, is capable of deploying MiG-29K naval combat aircraft along with Kamov 31 and Kamov 28 anti-submarine warfare and maritime surveillance helicopters.

✦ Construction of the 37,500 tonne first indigenous aircraft carrier, INS Vikrant, is progressing steadily. The warship successfully undocked from the Cochin Shipyard in June 2015. Designed by Directorate of Naval Design and built by the state-run Cochin Shipyard Ltd., INS Vikrant was launched in August 2013. The warship is expected to go for extensive trials in 2016 and inducted into the Navy by 2017-18.

✦ All three Kolkata-class (Project-15A) guided missile destroyers have been inducted in service. INS Kochi, the second platform was commissioned into service in September 2015 followed by the third and final ship of the class, INS Chennai, in November 2016. The first ship of the class, INS Kolkata, had joined service in August 2014. Each warship is capable of deploying 16 BRAHMOS supersonic cruise missile systems besides other weapons. BRAHMOS has been successfully test fired from both INS Kochi and INS Kolkata. The warships are designed by Navy's design bureau and constructed by Mazagon Dockyards Ltd.

The third and final Project-15A guided missile destroyer INS Chennai

✦ "Mormugao", the second of four new P15-B class of stealth destroyers, has been launched in Sept 2016. The first and lead ship of the class, "Visakhapatnam", was launched in April 2015. The new warships are being built indigenously at Mumbai's Mazagon Docks as a follow-on project to the Kolkata class DDGs. One of the most powerful weapon systems on board the new warships would be BRAHMOS missile.

✦ INS Arihant, India's first indigenously-built nuclear submarine, has been commissioned into service in August 2016. The 6,000-tonne underwater platform is armed with K-4 and K-15 ballistic missiles (SLBMs). A total of six such nuclear-powered submarines are planned to be built.

✦ A deal has been signed with Russia in October 2016 to lease a second Akula-II class nuclear submarine. The Navy is already operating INS Chakra, the first platform of the class.

✦ INS Kalvari – the first of six Scorpene diesel-electric attack submarines has been floated out in April 2015 with its commissioning scheduled for 2017. Second platform of the class, Khanderi, has been launched in Jan. 2017.

✦ Six submarines, under Project 75 India (P-75-I), will be acquired in addition to the six Scorpene submarines (Project 75) currently under development. All the six conventional submarines, equipped with AIP technology and land-attack missile capability, will be built in India under the "Make-In-India" programme.

✦ Four Kilo-class conventional submarines are to be upgraded by Russia in association with an Indian partner under a Rs. 5,000 crore programme.

✦ Two Shishumar class (Type 209/ 1500) submarines will be upgraded under a $38.4 million contract signed in June 2016 with Germany's Thyssenkrupp Marine Systems.

✦ Plans are on to acquire four Grigorovich-class (Project 11356M) guided missile frigates (advanced variant of the six Talwar-class frigates Indian Navy currently operates) from Russia under the Make-In-India programme wherein a Russian shipyard will form partnership with an Indian shipyard to build the warships in India.

✦ Seven new frigates are planned to be built under the Project-17A programme -- a follow-on project to the newly-inducted Project-17 Shivalik-class frigates. The new warships, with advanced stealth features and indigenous weapons and sensors, will be built by MDL, Mumbai and GRSE, Kolkata.

✦ INS Kadmatt, the second ship of Project 28 (Kamorta class) anti-submarine warfare corvette, has been commissioned into service in January 2016. The remaining two ships of the series, Kiltan and Kavaratti, are scheduled for delivery by 2017.

✦ The first GRSE-built Mk IV Landing Craft Utility (LCU) vessel has been delivered to the Navy in September 2016. GRSE has been contracted to build eight such LCUs at a cost of Rs 2,170 crore.

✦ Plans are on to acquire five fleet support ships. An RFI will be issued to public and private sector shipyards for the vessels which will be used by the Navy to support the operations of its aircraft carrier fleet.

✦ The State-run GSL is developing 12 mine countermeasure vessels (MCMV).

✦ INS Astradharini, an indigenously built torpedo launch and recovery vessel, has joined the Navy in October 2015.

✦ INS Tarasha, the fourth water-jet FAC designed by Kolkata's Garden Reach Shipbuilders & Engineers Ltd (GRSE), has been launched in June 2016. First platform of the class has been inducted.

✦ The first naval MiG-29 squadron, comprising of 16 advanced carrier-based MiG-29K fighter jets, was commissioned into the Navy in May, 2013. The Navy has so far received 41 of the naval jets while delivery of the remaining four aircraft is expected to be completed by 2016-17. The supersonic fighters, armed with advanced anti-aircraft and anti-ship missiles, precision bombs and a sophisticated weapon delivery system, will be operating from the Navy's aircraft carriers INS Vikramaditya and INS Vikrant.

✦ Two naval variant of indigenously-built Light Combat Aircraft (LCA) are undergoing flight tests. Navy has plans to induct 40 such aircraft so far.

✦ The Defence Ministry has issued a global Request for Information (RFI) to procure new amphibious aircraft for the Navy. Discussions are going on with Japanese firm ShinMaywa for its US-2i planes.

✦ US's Sikorsky has been shortlisted to supply 16 S-70B Seahawk naval multi role helicopters for anti-submarine, anti-surface and Special Commando operation roles. The Defence Ministry is yet to finalise the deal.

✦ Eight P-8I maritime surveillance aircraft ordered from the US have been inducted in service. Four more such planes have been ordered from Boeing with deliveries slated to start from 2020.

✦ The tender to procure 56 naval utility helicopters from abroad at an estimated cost of Rs 9,000 crore has been scrapped as the Defence Ministry has decided to get them manufactured in India by local firms with foreign collaboration.

✦ The first squadron of HAL-built Dhruv advanced light helicopter (ALH) joined service as Indian Naval Air Squadron (INAS) 322 in Nov. 2013. A new order for 16 more such rotorcraft has been cleared by Defence Ministry in June 2014.

✦ Six Russian-built Kamov-28 anti-submarine warfare helicopters will undergo mid-life upgrade by being equipped with new sensors and other systems while four of them will undergo engine overhauling in Russia. A contract worth nearly $300 million has been awarded to Russia's Rosoboronexport to modernize the anti-submarine helicopters.

✦ Negotiations are at an advanced stage with the US to procure 22 unarmed Predator B Guardian drones for maritime surveillance purposes.

✦ The second and third Unmanned Aerial Vehicle squadrons consisting of Searcher and Heron UAVs have been commissioned at Porbandar and Ramanathapuram respectively.

✦ Varunastra, an indigenously developed anti-submarine warfare torpedo by DRDO's Naval Science and Technological Laboratory, has been inducted in the Navy in June 2016.

AIR FORCE: Indian Air Force (IAF), the air arm of Indian defence forces, is primarily tasked to defend and secure national airspace and conduct aerial warfare during conflict situations. The IAF coordinates with other wings of the Indian armed forces to carry out disaster relief operations and humanitarian assistance. It takes part in joint wargames with the Army and Navy and also holds aerial drills with the air forces of friendly nations. Over the years, the IAF has grown from a "tactical force" to one with "transoceanic" reach. At a time when air power around the world has assumed increasing significance in deciding the outcome of future conflicts, the IAF has embarked on a comprehensive modernisation drive and reoriented itself to acquire multi-role capability of platforms and equipment, along with multi-skill capability of its personnel. Induction of fourth and fifth generation fighter aircraft, new transport planes, refuelling tankers, trainer jets, helicopters, unmanned aerial platforms, advanced combat systems, radars and network centric warfare systems along with upgradation of existing fleet of aircraft and weapons has been taken up with an aim to expand and enhance the IAF's operational capability and prepare it to meet the evolving challenges posed by the security scenario in the region.

Personnel: 1,70,000+

Major air bases: The IAF is divided into five major air commands – Western Air Command, Eastern Air Command, Central Air Command, Southern Air Command and South Western Air Command. It operates from over 50 air bases all across the country.

Equipment

Category	Name	In Service	On Order
Fighter, Attack	Su 30 MKI	200+	272
	Rafale		36
	Mirage 2000	48 (being upgraded)	
	MiG-21	200+ (being phased out)	
	MiG-27	50+ (being phased out)	
	MiG-29	60 (being upgraded)	
	Jaguar	114	
	LCA Tejas		42 (Initial order; more to be inducted in future)
	FGFA		100+
Transport	IL-76	26 (to be upgraded)	
	An-32	104 (being upgraded)	
	Avro	56 (to be phased out)	
	Boeing 737		
	Dornier	40+	
	C-130J	5	7
	C-17 Globemaster III	10	3
	Embraer (VIP transport)		
Tanker, Trainer	IL-78	7	
	Mid-air refuellers		6
	HJT-16 Kiran	190	
	HJT-36 Sitara		70+
	HTT-40		
	Hawk Mk. 132 AJT	40+	66 (plan for more aircraft)
	PC-7 MkII	75	38
AEWAC	A-50 EI Phalcon	3	
	EMB 145 AEW&C		3
Helicopter	Mi-8 (to be phased out)	80	
	Mi-17 & Mi-17V5	58 + 151	48 (proposed)
	Mi-26		
	Mi-25/ Mi-35	20	
	LCH		65
	AH-64D Apache		22

Air Force continued...

Category	Name	In Service	On Order
Helicopter	Cheetah	24	
	Chetak	48	30
	Cheetal		12
	ALH Dhruv	40	
	Chinook CH47D		15
UAV	Searcher II	100+	
	Heron		
	Harop (UCAV)		
Surveillance Equipment	Aerostat	2	4
Weapon	S-13 (rocket)		
	SA-125 Pechora SAM		
	9K38 Igla MANPAD SAM		
	9K33 Osa-Akm SAM		
	S-300 SAM		
	SPYDER SAM		
	Akash SAM	2 squadrons	6 squadrons
	Land attack BRAHMOS		2 squadrons
	Air-launched BRAHMOS		
	R-60 AAM		
	R-73, R-77 AAM		
	R-550 AAM		
	Prithvi-II SRBM		
	Mica AAM		500

New Procurements/Upgrades

✦ A contract worth 7.87 billion Euro has been signed with France in September 2016 for 36 Rafale fighter aircraft equipped with latest missiles and weapons along with multiple India-specific modifications. The contract follows an inter-governmental agreement inked between India and France in January 2016 to acquire the medium multi-role fighters from French firm Dassault Aviation. Delivery of the warplanes will start in 36 months and would be completed in 66 months from the date the contract was signed.

✦ The first squadron of the indigenous Tejas Light Combat Aircraft (LCA) has been commissioned into service on 1st July 2016 with the state-run Hindustan Aeronautics Ltd. (HAL) handing over the first two aircraft to the Indian Air Force. The Government in November 2016 has cleared the procurement of 83 Tejas LCAs. The LCAs will be inducted as a replacement for the aging fleet of MiG-21 and MiG-27 fighters. The IAF has so far committed to acquiring and forming up to six squadrons (consisting of around 120 fighters) of Tejas in the coming years.

✦ The Defence Ministry is looking for the acquisition of medium, multi-role, single engine fighter jets from a foreign vendor with "Make-In-India" clause associated with the programme. The new fighters would be replacing the IAF's aging fleet of MiG-21 and MiG-27 fighters.

✦ Discussion is going on with Russia for the joint development and production of the fifth-generation fighter aircraft with stealth capabilities. IAF plans to acquire over 100 FGFAs.

✦ The Government in February 2015 has cleared the purchase of 38 additional Pilatus basic trainers for the IAF. Delivery of 75 Pilatus PC-7 Mk-II jets, previously ordered in May 2012, has been completed with the delivery of the last aircraft in Nov. 2015. IAF has estimated its total requirement of 181 basic trainer aircraft, some of which will be built in India.

✦ HAL-made indigenous jet trainer HTT-40 has made its inaugural flight in June 2016. The Government has announced to order adequate number of the new planes for the Indian Armed Forces as well as for export to make it commercially viable. The programme aims to achieve operational clearance by 2018.

✦ Plans are on to order more of the BAE Systems-built Hawk advanced jet trainers which are being licence-produced by HAL in India. A combat variant of the jet trainer is also in the making jointly by HAL and BAE.

✦ The Defence Ministry has issued a tender worth over Rs 12,000 crore to replace the IAF's entire fleet of Avro planes with 56 new aircraft which will be produced by an Indian private sector firm in partnership with a foreign firm. While the initial 16 aircraft will be supplied in fly-away condition, the remaining 40 would be manufactured in India.

✦ Government has cleared the IAF's proposal to upgrade its entire fleet of Russian-origin IL-76 transport aircraft with new avionics and re-engining the fleet of IL-78 aerial refuelling aircraft at a cost of Rs. 4,300 crore.

✦ Europe's Airbus Military has been selected as the preferred bidder to supply six A330 multi-role tanker transport (MRTT) aircraft to the IAF after a long-drawn selection process. The Defence Ministry in March 2015 cleared a proposal worth $818 million to modify two of the A330s to develop an indigenous AWACS platform. A final decision on acquiring the remaining four tankers is yet to be taken.

✦ India has ordered 10 C-17 strategic airlifters from the US and proposed to acquire three more such aircraft in future.

✦ The Defence Ministry has cleared the proposal to acquire six more C-130J Super Hercules aircraft from the US in a deal worth over Rs 4,000 crore. The new airlifters, to be procured through the Foreign Military Sales (FMS) route, will add to the existing fleet of five such planes being operated by the IAF. One more C-130J is also on order for the one lost in an air crash in March 2014.

✦ An RFI has been issued to procure nine "special mission" aircraft. The RFI specifies that two of the nine aircraft should be certified to perform signals intelligence (SIGINT) role, given that the mission package for SIGINT commonly requires permanent modifications to the airframe. The remaining seven aircraft are to be configured by the prime contractor for multi-mission role, supporting aerial survey, target towing, communications jamming (COMJAM) and flaring.

✦ An Inter-governmental Agreement has been signed in October 2016 to purchase S-400 Triumf air defence missile systems from Russia.

✦ The Government has cleared funds worth $1.2 billion for testing and integration of BRAHMOS supersonic cruise missiles on the IAF's frontline Su-30MKI fighters.

✦ A contract worth $964 million has been signed with Russia's RAC MiG to upgrade the IAF's fleet of MiG-29 fighters. French firms Thales and Dassault Aviation have been awarded $2.4 billion to upgrade Mirage-2000 aircraft, two of which have been delivered in March 2015.

✦ HAL has been contracted to rejig the Jaguar fighters and it has signed an MoU with BAE Systems in this regard.

✦ Ukraine's Antonov is progressively delivering the upgraded An-32 transport planes.

✦ A £250 million contact with MBDA-UK has been signed for the procurement of advanced short range air-to-air missiles (ASRAAM) for the Jaguar fighters.

✦ A Euro 950 million ($1.1 billion) deal has been cleared to procure 500 MICA air-to-air missiles from European firm MBDA for the IAF's Mirage 2000 aircraft fleet.

✦ Delivery of Mi-17V5 multi-role helicopters by Russia has been completed in 2016. India had placed orders for 151 such choppers. The IAF has proposed to acquire 48 more such choppers in near future.

✦ 22 AH-64D Apache Longbow attack helicopters and 15 Chinook CH-47D heavy-lift transport helicopters will be procured from US defence major Boeing under a deal worth around $3 billion.

✦ An RFP has been issued by the IAF to defence aerospace PSU HAL for buying 12 Cheetal helicopters which will be used for carrying out operations in high altitude areas in the northern borders.

Predator drone *A General Atomics photo*

✦ Negotiations are going on with the US to buy up to 100 armed Predator C Avenger drones.

✦ The Harop unmanned combat aerial vehicles (UCAVs) are being procured from Israel under a 2010 deal.

STRATEGIC FORCE COMMAND

The Strategic Forces Command of the Indian armed forces is responsible for the command and control of the country's tactical and strategic nuclear arsenal. Established in 2003, the SFC forms part of India's Nuclear Command Authority (NCA) which comprises the Political Council chaired by the Prime Minister and an Executive Council headed by the National Security Advisor. The SFC functions under the leadership of a Commander-in-Chief of the rank of Air Marshal (or its service equivalent). The Command works with experts at the Department of Atomic Energy (DAE) and the DRDO. It has the sole responsibility of initiating the process of delivering nuclear weapons and warheads only after securing explicit approval from the NCA as India remains committed to the "no first use" of nuclear weapons doctrine.

Potential weapons: India's nuclear doctrine envisages building and maintaining a credible minimum deterrent posture with a "second strike capability". The country has successfully built up a "credible and invulnerable" deterrent nuclear triad with retaliatory strike capability through weapon systems from land, air and sea. The potential weapons controlled by the SFC include the Agni and Prithvi family of missiles. Agni is a series of surface-to-surface ballistic missile family developed under India's Integrated Guided Missile Development Programme (IGMDP) in the 1990s. The indigenously-built strategic weapon system is capable of carrying both conventional and nuclear warheads. Prithvi is a short-range surface-to-surface ballistic missile. The single-stage, liquid-fuelled tactical weapon is capable of carrying nuclear warhead. India has also developed a submarine-launched ballistic missile for its indigenously-built nuclear submarine.

Variants

✦ Agni-I is a short to intermediate range road mobile ballistic missile. The single-stage solid-propelled missile has an operational range between 700-km and 1200-km.

✦ Agni-II is a medium range ballistic missile (MRBM). The two-stage solid-propelled missile can hit targets 2,000 km away.

✦ Agni-III is an intermediate-range ballistic missile (IRBM) developed as the successor to Agni-II. The missile has a range of over 3,000 km.

✦ Agni-IV is a modification of Agni II series. It was initially developed as Agni-II Prime. The intermediate range missile has a strike range of about 4,000-km.

✦ Agni-V is an intercontinental ballistic missile (ICBM) having a range of 5,000-km. It is the first canistered road-mobile ballistic missile in India's military arsenal.

✦ Prithvi-I is a short-range tactical missile with a range of 150-km.

✦ Prithvi-II is a single stage liquid-fuelled short range ballistic missile. The surface-to-surface missile has a strike range of 250-km to 350-km.

✦ Dhanush has been developed as the naval variant of Prithvi ballistic missile system.

India has also successfully developed its first submarine-launched ballistic missile (SLBM) for the indigenously-built nuclear submarine (SSBN) INS Arihant. Development of the underwater-launched ballistic missile in two variants – the longer-range K-4 (up to 3,500 km) and medium range K-15 (up to 700km) – has enabled India in completing its nuclear triad under which it now has the capability to strike from air, land and beneath the sea. With the new DRDO-designed missiles in its arsenal, India has joined the elite club of nations including US, Russia, UK, France and China who have developed such weapon systems for their armed forces.

INTEGRATED SPACE CELL

India officially announced the formation of an Integrated Space Cell (ISC) under the aegis of the Integrated Defence Services Headquarters in June 2008. The ISC has been created to counter growing threat to national space assets and will act as a single window for integration among the Armed Forces, the Department of Space and the Indian Space Research Organisation (ISRO). The establishment of a joint aerospace command is under consideration since early 2000s and the present Government is deliberating on the issue as part of articulating a national space policy in future.

Major Armaments Developers & Producers

DEFENCE RESEARCH & DEVELOPMENT ORGANISATION (DRDO)

Established in 1958, the state-controlled DRDO is the premier defence research agency of India. It has a network of 52 dedicated defence laboratories all across the country which are engaged in developing critical defence technologies covering a wide array of disciplines ranging from aeronautics, armaments, electronics, combat vehicles, missile systems, engineering systems, electronic warfare systems, radar systems, laser technology, unmanned vehicles, military engineering, life sciences, advanced materials, composites, underwater sensors/weapons and platforms, warship technology to instrumentation, advanced computing, networking and simulation for the Indian armed forces. The defence organisation by its unflinching efforts has designed and produced independently as also with foreign collaboration a whole range of state-of-the-art weapon systems, platforms, sensors and allied equipment for all three wings of the Indian defence forces. Despite of the strict technology denial regime imposed on India by developed nations for several years, DRDO, after starting from the scratch, has made immense contributions to Indian military science and built up the country's defence apparatus in a major way.

In its initial years, DRDO primarily focused on developing basic military equipment and related services like small arms, explosives, machine gun, mountain gun and field gun, military psychology for selection of personnel, physiology for application by troops serving in harsh terrains and electronics systems. Later, it moved to the next stage by successfully developing radars and sonars, flight simulators and electronic warfare systems. In the third phase of its growth, the research agency focused on designing and developing major platforms and weapon systems like battle tanks, combat vehicles, nuclear submarines, supersonic fighter jets, ballistic missiles, radar & sonar systems and other equipment.

Today, DRDO has achieved a distinct place in designing, developing and integrating modern military platforms and systems and positioned India as one of the technologically advancing countries not only in Asia but in the world.

DRDO has tied up with about 40 academic institutions, 15 national Science and Technology agencies, 50 public sector undertakings and over 250 private sector enterprises to carry out research and development work and fulfilling its foremost objective of making India self-sufficient in defence production.

The defence organisation is presently working on several key projects that include indigenous aircraft carrier, nuclear submarine, guided missile destroyers, frigates, state-of-the-art weapon and systems for the Navy; advanced main battle tanks, combat vehicles and artillery for the Army; light combat aircraft, advanced medium combat aircraft and airborne early warning and control (AEW&C) system for the Air Force. It is also working on developing micro UAVs, an unmanned combat aerial vehicle (UCAV) and Medium Altitude Long Endurance (MALE) UAV for the armed forces. The premier defence research institution is also designing and developing a wide array of sonar and radar systems.

According the Defence Ministry, DRDO has also been jointly working with a number of other countries, including Russia, Israel, Germany, France, USA, Belarus, Brazil and Singapore, to design and develop new sophisticated military platforms and systems.

Contribution in Aerospace Technology & Systems

LCA: The successful development of Light Combat Aircraft (LCA) Tejas for the Indian defence forces is a major feat achieved by the DRDO after several years of research work. The indigenous fighter jet is developed by DRDO's Aeronautical Development Agency (ADA) in partnership with Hindustan Aeronautics Ltd. The project also involves other DRDO & CSIR laboratories, public and private sector industries and academic institutions of the country. The single-seat supersonic all-weather multi-role fourth-generation light fighter is being developed for the IAF and Navy in two variants -- Mk I and Mk II. The lightweight jet having a tailless, compound delta wing design, is powered by a single engine.

The Mk I variant for the IAF has been developed and is powered by the General Electric F-404 turbofan engine. It is capable of flying at Mach 1.6 (about 2,000 km/h) and can carry out air-to-air, air-to-ground and air-to-sea combat operations. The more powerful Mk II variant will be equipped with the GE F414 engine. The LCA will carry a wide variety of weapons including the Beyond Visual Range (BVR) missiles, air-to-air and air-to-surface missiles, conventional and laser-guided bombs, air defence guns, counter-counter measures and drop tanks to take on surface targets over land or at sea. Designed to replace the IAF's MiG-21 fighter jets, the first squadron of Tejas has been inducted in the Indian Air Force in 2016.

In its naval configuration, the LCA has been designed with Ski-jump Take Off but Arrested Recovery (STOBAR) operations. It has high landing loads than its Air Force variant. A dedicated shore-based test facility (SBTF) has been set up at INS Hansa in Goa to carry out the fighter's ski-jump sorties. Two naval prototypes are undergoing flight trials at present. Developed as the only carrier-borne aircraft in light category in the world, the Naval LCA will carry a wide variety of operational weapons and equipment like the Beyond Visual Range (BVR) missile, conventional and laser-guided bombs, air defence guns, counter-counter measures and drop tanks to take on surface targets over land or at sea.

AEW&C: To provide the IAF with an indigenous "eyes in the sky" capability, India has taken up the project of developing a state-of-the-art Airborne Early Warning and Control (AEW&C) system. Under the programme, DRDO's Bangalore-based Centre for Airborne Systems (CABS) laboratory has developed the internal and external mission systems for integration on-board the Brazilian Embraer-built EMB 145I aircraft. The CABS have developed over 300 mission system items, including the dummy Active Antenna Array Unit having an Active Electronically Scanned Array (AESA) radar for integration on the aircraft. The AEW&C System is mounted on top of the aircraft's fuselage. The indigenous AEW&C system is a multi-sensor system providing for all aspects of airborne early warning & control in present-day defence scenario. It is designed to detect, identify and classify threats present in the surveillance area and act as a command and control centre to support air operations. It can "look" around

270 degrees within a short time and has a range of 350 km. Three AEW&C systems are presently under development with the IAF gearing up to induct two.

AWACS: DRDO has been entrusted with the task of developing an indigenous Airborne Warning and Control System (AWACS) with the capability to penetrate "longer distances" into the enemy territory by using radars and electronic warfare systems without actually venturing into the region physically. The AWACS, which will be different from the AEW&C system, will have a coverage area of about 360 degrees and it will fly at a higher altitude. It will remain in the sky for longer duration and will have better visibility. The project is slated to be completed by 2020. In March 2015, the Government has approved the development of two new AWACS platforms based on the European A330 aircraft.

UAVs: India has undertaken drone development programme for its military since the past few years. The remotely-controlled pilotless aircraft are being developed in various configurations to meet the needs of the Army, Navy and Air Force. DRDO has played a key role in developing these systems listed below.

Rustom: A Medium Altitude Long Endurance (MALE) UAV, Rustom is being developed by DRDO's Bangalore-based Aeronautical Development Establishment (ADE) in two variants – Rustom I and Rustom-II. The remotely-operated aircraft is capable of undertaking various military missions like reconnaissance and surveillance, target acquisition, target designation, communications, battle damage assessment and gathering signal intelligence. The UAV has been designed for the Indian Army, Navy and Air Force. While Rustom-I, weighing around 700kg, has been undergoing flight tests since 2010, the more advanced Rustom-II has undergone its maiden flight in November 2016. The two-ton UAV has a payload capacity of 350-kg and an endurance of over 24 hours. It has an altitude ceiling of 35000 feet and an operational range of 250-km.

Nishant: Designed and developed by DRDO's ADE lab, Nishant's role includes battlefield reconnaissance in day and night, surveillance, target tracking and correction of artillery fire. The electro optical, electronic intelligence and communication intelligence payload on-board the UAV make it suitable for a range of operations both during wartime and for counter-insurgency operations. The 380-kg drone can be launched from a hydro pneumatic launcher, without needing a runway and is recovered by parachute. It has an endurance level of five to eight hours and a payload capacity of 45-kg. The tactical UAV has an altitude ceiling of 3,600 km and an operational range of 150 km. The Army has started taking delivery of the new drones from DRDO.

Lakshya: The pilotless target aircraft Lakshya is one of the successful projects undertaken by ADE. The system is manufactured by HAL. The indigenously-built micro-light UAV is a reusable aerial target system which can realistically simulate an enemy aircraft for air-to-air and surface-to-air weapon system operators. Remotely operated from ground, Lakshya is used for the training of gun and missile crew and air defence pilots for all three Services. The subsonic single-engine UAV has a maximum speed of 857km/h and a range of 150 km. It can attain a height of 29,528 ft. It has been developed in two variants – Lakshya I and Lakshya II. DRDO has already delivered the aircraft with ground control and associated systems to the Armed Forces.

UCAV: Keeping pace with the leading global powers' effort to develop unmanned combat aerial vehicles, India too has put forth its endeavour to build one. Three DRDO laboratories – the Aeronautical Development Agency (ADA), the Aeronautical Development Establishments (ADE) and the Defence Avionics and Research Establishment (DARE) – have joined hands to design and develop the stealth UCAV – to be called 'AURA' (Autonomous Unmanned Research Aircraft). The combat drone, weighing around 15 tonne, will fly at an altitude of 3,000 feet with combat payloads which would include missiles, bombs and precision guided munitions. It will have short take-off and landing capability on prepared runway. DRDO has not divulged much detail about the new stealth drone as yet which, according to some defence analysts, could take several years for induction in the Indian armed forces.

Rotorcraft: DRDO has played a crucial role in the design and development of rotary-wing aircraft in HAL-led projects like advanced light helicopter (ALH) 'Dhruv', light combat helicopter (LCH) 'Rudra' and armed variant of ALH Dhruv, ALH Weapons System Integrated (ALH-WSI).

Contribution in Naval Technology & Systems

In the naval research domain, DRDO has designed and developed state-of-the-art underwater sensors, weapons and material. Three DRDO laboratories – the Naval Physical Oceanographic Laboratory (NPOL), the Naval Material Research Laboratory (NMRL), and the Naval Science & Technological Laboratory (NSTL) are exclusively dedicated to R&D activities in the field of naval technology. The NPOL is a key contributor in designing and developing a wide array of sonar systems like "HUMVAD" hull-mounted sonar, "Humsa" hull-mounted panoramic sonar, "Panchendriya" and "USHUS" submarine sonars, "Nagan" towed array sonar, "Mihir" airborne sonar for Indian Navy platforms.

Similarly, NSTL has indigenised a wide range of technologies in the field of underwater weapons by developing sophisticated torpedoes like "Varunastra" heavy weight torpedo, "Shyena" light torpedo, "Takshak" thermal torpedo), decoys, targets, counter measures, mines and fire control systems. Research work has also led to the creation of the required infrastructure needed for developing and testing such weapons and systems. The NMRL has made key contributions in the field of naval materials technology like high capacity fuel cells, electrochemistry & electrochemical processes, polymer and polymer matrix, composites, elastomers. The Impressed Current Cathodic Protection (ICCP) technology developed by the laboratory to supplement the protection provided by paints to underwater structures against sea water corrosion is widely used by the Navy.

DRDO is also playing a critical role in producing several indigenised components and subsystems for Navy's warships and submarines as well as equipping them with advanced weapon systems.

Contribution in Land Systems

DRDO has proved its dexterity in the field of combat vehicle and engineering by developing a wide range of multi-disciplinary systems including bridge layer tank, amphibious floating bridge & ferry system, anti-tank ammunition, mine field marking equipment, mortar carrier vehicle, armoured engineering reconnaissance vehicle and armoured amphibious dozer among others. A major achievement in this arena is the development of Arjun MBT for the Army. The MBT has been developed as a world-class combat vehicle and DRDO is presently working on an advanced variant of the MBT, to be called Arjun Mk II which will incorporate a total of 93 improvements, including its fire-power, laser protection suite and improved armoured protection for the vehicle. Besides, it is also developing a Futuristic Main Battle Tank (FMBT) for the Army which will feature stealth technology and more firepower. The Combat Vehicle Research and Development Establishment (CVRDE) is the leading lab dedicated for land systems development projects.

Contribution in Missile Technology

The missile technology programme of India led by DRDO has achieved great heights with the successful development of a number of strategic and tactical missile systems. The foremost among them is the Agni series of ballistic missiles. Developed under the Integrated Guided Missile Development Programme (IGMDP) of India, the Agni family today boasts of several variants – Agni I, Agni II, Agni III, Agni IV and the most recent Agni V. Other missile systems designed and developed by the research agency include Akash surface-to-air missile, Prithvi ballistic missile, Astra BVRAAM, BRAHMOS supersonic cruise missile (in partnership with Russia), and the K-4 & K-15 SLBM among others. It has also successfully developed the ballistic missile defence (BMD) system involving a supersonic interceptor missile, called the Advanced Air Defence (AAD) missile, and the Prithvi ballistic missile (PAD) as the target missile. The research wing has set an ambitious goal of developing a hypersonic cruise missile and an advanced Agni-VI ICBM with MIRV capability to deliver several payloads at different targets.

ORDNANCE FACTORIES: The state-run defence firm is a conglomerate of 39 factories spread across India and is engaged in the production of a range of armaments including machine guns, mortar guns, naval guns, ammunition, military vehicles, armoured vehicles like main battle tanks and NBT recce vehicles and related equipment for the Indian armed forces.

BHARAT DYNAMICS LTD: BDL manufactures munitions, torpedoes and missile systems for Indian armed forces. The first missile that entered production with BDL was the Prithvi surface-to-surface missile.

BHARAT ELECTRONICS: This public-sector enterprise produces a whole range of defence equipment, including electronic warfare systems, naval systems, communication systems, radars and weapon systems, for the Indian defence forces.

BEML LTD.: The company, partly owned by the Government, produces a number of defence equipment including TATRA vehicles, missile launchers, missile transporters, armoured recovery vehicles and light recovery vehicles among others. The company has recently initiated the process of setting up an Aerospace Manufacturing Division at SEZ Park near Bangalore International Airport, Devanahalli, which will pave the way for BEML to enter into aircraft and helicopter manufacturing sector. The company has also set up an Aerospace Manufacturing Division at Mysore and certified to design aircraft, helicopter and engine parts.

HINDUSTAN AERONAUTICS LTD.: India's leading public-sector aerospace company and one of Asia's largest, HAL has built up comprehensive skills in design, manufacture and overhaul of fighters, trainers, helicopters, transport aircraft, engines, avionics and system equipment.

COCHIN SHIPYARD LTD.: The CSL is the largest shipbuilder in the country capable of building and repairing large-sized naval vessels in India. The shipyard is presently building the indigenous aircraft carrier (IAC) for Indian Navy.

GARDEN REACH SHIPBUILDERS & ENGINEERS LTD.: The Government-run GRSE has emerged as a leading shipyard of India, building a wide range of naval vessels, including frigates, corvettes, landing craft, fast interceptor boats and survey ships.

GOA SHIPYARD LTD.: The state-run GSL is involved in designing and building a wide range of sophisticated naval vessels for varied applications in defence and commercial sectors with special expertise in building modern patrol vessels of steel and aluminium hull structure. It also undertakes repairing and modernisation of existing naval platforms. The shipyard's defence repertoire includes OPVs, interceptor craft, missile boat, fast attack craft, sail training ship and landing craft for Navy.

HINDUSTAN SHIPYARD LTD.: The Government-owned HSL is one the premier shipbuilding organisations in the country catering to the needs of shipbuilding, ship repairs inclusive of sophisticated and state-of-art submarine retrofitting and construction of offshore vessels. The shipyard has built 150 ships and repaired nearly 1800 ships. Previously functioning under the Shipping Ministry, HSL was transferred to the Defence Ministry in 2009 to meet India's security needs and build strategic platforms for the Indian Navy.

MAZAGON DOCK LTD.: One of India's leading shipyards, the state-run MDL is engaged in the constructing a wide range of naval platforms, including warships and submarines, as well as offshore platforms. It also undertakes ship repairing work.

LARSEN & TOUBRO LTD.: The leading private sector company in technology, engineering, construction and manufacturing has ventured into the production of defence systems for armed forces. The company has formed alliances with leading global defence firms like EADS of Europe to develop and market defence products in the fields of electronic warfare, radars and military avionics. It is also setting up a shipyard at Kattupalli in Tamil Nadu which would enable India to compete globally in building large-sized warships, car carriers and submarines.

RELIENCE DEFENCE AND ENGINEERING LTD: Formerly known as Pipavav Defence, the private sector entity has now been christened as Reliance Defence and Engineering Ltd (RDEL) in 2016 after its take- over by the Reliance Group. The company, with the capability of building a range of naval platforms, has already forayed into the defence sector and plans to build warships, submarines and aircraft carriers for Indian Navy in partnership with foreign firms. Indian Navy has awarded it a contract to build offshore patrol vessels for it.

ABG SHIPYARDS LTD.: The country's largest private ship-building company, ABL is a manufacturer and service provider for a variety of ships, including bulk carriers, diving support vessels, anchor handling supply ships, anchor handling tugs, and vessels for the defence forces like High speed Interceptor Boats and Pollution Control Vessels. The shipbuilder has recently received a Rs 485-crore contract from the Indian Defence Ministry for building a cadet training vessel for the Indian Navy.

Joint Venture Programmes

Russia

BRAHMOS: World's only supersonic cruise missile system BRAHMOS is a joint venture programme between India and Russia. The two countries partnered to set up the JV company, BrahMos Aerospace, in 1998 to jointly design, develop, produce and market the BRAHMOS universal weapon system. The cruise missile, with a range of 290-km and speed of 2.8 Mach, has been operationalised in the Indian Army and Navy and is being prepared for flight trial by the Indian Air Force from its frontline Su-30MKI fighter. The success of BrahMos JV between India's DRDO and Russia's NPOM has paved the way for more such future joint defence projects between the two allies with greater Indian industry participation.

After successfully developing and delivering BRAHMOS in different configurations, BrahMos Aerospace has now initiated work on a novel concept to design and develop a smaller and lighter variant of BRAHMOS to be called BRAHMOS-NG (Next-Gen). Under the new project, plans are on to significantly reduce the size and weight of the 3-ton missile so as to fit it onto a wider number of new-generation military platforms. Keeping pace with rapid advances in military technology systems, BRAHMOS-NG will be designed as a weapon of choice for the Indian defence forces' newly-acquired advanced platforms and also for export in the international market. A hypersonic BRAHMOS-II(K) is also being planned to be developed in the next 10 years or so having a very high speed of Mach 5-7. India and Russia in 2016 have also agreed to increase the existing range of BRAHMOS from 290-km to up to 600-km.

FGFA: An inter-governmental agreement signed between India and Russia in October 2007 entails that India's Hindustan Aeronautics Limited (HAL) and Russia's Sukhoi Corporation will jointly develop the futuristic platform under the Fifth-Generation Fighter Aircraft (FGFA) project, to be initially based on Sukhoi's Perspective Aviation Complex of Frontline Aviation (PAK-FA). The FGFA project is pegged to be the biggest ever in the bilateral Indo-Russian defence cooperation. A final agreement is yet to be inked between the two partnering nations in this regard.

Israel

LR-SAM: DRDO has joined hands with Israel Aerospace Industries (IAI) to develop the Long Range Surface to Air Missile (LR-SAM) system under an offsets contract worth Rs. 2,606 crore signed in 2009. The missile is being developed as an advanced naval air defence system to arm Indian Navy's new P-15A Kolkata class guided missile destroyers. The first successful test firing of the missile, having a range of 70-km, took place on November 10, 2014 in Israel. State-run Bharat Electronics Ltd. (BEL), which has inked MoU with IAI in December 2012 for cooperation in the programme under the aegis of DRDO, will function as the lead integrator and produce major sub-systems for the missile while IAI will continue to act as the design authority and produce sub-systems as main sub-contractor of BEL.

MR-SAM: DRDO and IAI have jointly developed the Medium-Range Surface-to-Air Missile (MR-SAM) for the Indian Air Force. The missile, designed as an air defence weapon for the IAF, is undergoing test firings ahead of its induction expected by 2017. The weapon is also being developed for the Indian Army.

EW System: India and Israel have set up a joint venture to develop advanced electronic warfare (EW) systems for their air forces' fighter aircraft. The proposed joint venture has been signed between India's Defence Avionics Research Establishment (DARE), Bangalore and the Elisra Group, Bene Beraq, Israel.

France

Scorpene: India's Mazagon Dockyards Limited (MDL) and French shipbuilder DCNS have partnered to build six Scorpene submarines for the Indian Navy. The agreement has incorporated Transfer of Technology to MDL.

SR-SAM: India's DRDO and Paris-based missile manufacturer MBDA have in the past discussed ways to jointly work on developing a Short Range Surface to Air Missile (SRSAM) system for the armed forces. The final agreement and contract for the project is yet to be officially concluded.

Arms Import & Export

Even as it is striving to achieve indigenisation in defence technology, India has significantly relied on arms import from foreign countries to meet the requirements of its armed forces. The country depends on foreign suppliers for more than 70 per cent of its defence acquisition needs. According to the Stockholm International Peace Research Institute (SIPRI), India has emerged as the world's largest defence importer overtaking China. For the period 2011-2015, India occupied top slot in arms imports. Defence imports by the country accounted for 14 per cent of the total global arms imports, the latest SIPRI report states. While Russia maintains a strong lead as the top supplier, purchases from US are also increasing fast.

A large share of the defence imports by India has been devoted to acquiring combat vehicles, warships, submarines, advanced fighter aircraft, helicopters, transport planes and other combat weapons from countries including Russia, Israel, US, UK, France and Germany. According to a 2012 Defence Ministry data, India spent Rs.1.98 lakh crore in the past few years in procuring defence equipment for its armed forces. The country targets to spend over $150 billion in the next five years on major defence procurement programmes.

Attaining self-reliance in defence design and defence production, especially at a time when its armed forces are undergoing large-scale modernisation and transformation, has become the prime focus area for the Government of India. The country is making steady progress in this direction and has been successful in developing critical defence technologies covering wide areas including weapons and armaments, electronics, combat vehicles, missile systems, engineering systems, electronic warfare systems, radars, military engineering, warship technology, aeronautics, avionics, unmanned aerial systems etc. To further accelerate the pace of indigenisation in defence production, the Government has announced several major policy decisions in the past two years, including increasing the FDI limit in defence sector from 26% to 49%. The Defence Ministry has also geared up its efforts in boosting up indigenous defence production by selecting private sector Indian defence firms to collaborate with foreign firms in developing and producing cutting-edge military platforms and equipment for the Indian military through transfer-of-technology method. The Government's ambitious "Make-In-India" programme is aimed at largescale indigenisation of defence products and systems by the domestic defence industry in the coming years.

At the exports front, defence exports by the country, mostly critical components in the aerospace sector, have witnessed a steady rise. According to official records, India exported defence equipment worth Rs 512.48 crore in 2011 and equipment worth Rs 446.75 crore in 2012. In 2013, the figure reached Rs 686.27 crore. Defence exports worth Rs. 166.67 crore were made till September, 2014. A major military export deal executed in 2014 was the handing over of a 1.3-ton OPV, "Barracuda", to Mauritius. Built by the state-owned GRSE, the vessel was the first-ever warship delivered to a foreign country by an Indian shipyard. In 2015, the first nine months too witnessed a more than three-fold rise in the value of defence exports by private companies which stood at Rs 441.06 crore as against Rs 132.17 crore in all of previous year. The total value of exports undertaken by both private and state-run defence manufacturers during this period was Rs 695.70 crore, according to official figures released by the Government. The Defence Ministry has set a target to increase defence exports by India to $2 billion in the next two years.

In the past few years, Indian Defence public sector undertakings have exported products to several key countries including the US, UK, Canada, China, Israel, Germany and France, according to the Indian Defence Ministry. The countries in the Indian sub-continent to which defence products have been exported include Bangladesh, Bhutan, Myanmar, Nepal and Sri Lanka. With its expanding indigenous military industrial base, India is now eyeing to grab a fair share of the lucrative international defence market and has initiated efforts to sell military hardware, including missiles, fighter aircraft, helicopters, naval vessels and other equipment to several countries.

DEFENCE BUDGET

Defence spending by India over the past few years has witnessed an upward trajectory, mainly because of an all-round modernisation of its armed forces. A large share of the defence budget also goes for the huge manpower as India maintains one of the largest armed forces in the world in terms of personnel strength. The annual defence allocation has remained around 2 per cent of the country's GDP in the past few years. As the country is rising economically, maintaining a military edge in the Asian region as well as in the world has necessitated the increased defence spending. The allocated Defence Budget for 2016-17 is Rs.3,40,921.98 crore (US$ 52 billion) – an increase of 13.1% over previous year's allocation of Rs. 2,46,727 lakh crore. With this, India has become the world's fourth largest spender on defence, going up two positions higher compared to previous year.

Official Budgetary Allocation

Year	Allocation	% of GDP	% Increase over previous year
2016-2017	Rs. 3.41 lakh crore ($52 billion)	2.26	13.1
2015-2016	Rs. 2.46 lakh crore ($40.4 billion)	1.75	7.74
2014-2015	Rs. 2.29 lakh cr ($38.35bln)	1.81	12.5
2013-2014	Rs. 2.03 lakh cr ($37.4 bln)	1.79	5.31
2012-2013	Rs 1.93 lakh cr ($41 bln)	1.90	17

NAVAL WARFARE SYSTEMS

The maritime power of India has witnessed a steady transformation in the last few years with the country building up a large, influential Navy capable of not only protecting its territorial sovereignty and integrity but also expanding its horizons beyond the littoral waters as part of its strategic power projection. Once known as a "brown water" coastal defence force, the Indian Navy has aspired to transform itself into a formidable "blue water" fleet in the wake of rapidly changing geostrategic positions wherein the Asian continent has emerged as the new global power block.

With this new defence rebalancing, the role of Indian Navy as a major maritime power in the Indian Ocean Region (IOR) has significantly increased. Its maritime reach is also widening further to expand India's strategic foothold in the areas ranging from the Persian Gulf and the Horn of Africa to the Strait of Malacca and far beyond.

Even though smaller than the other two wings of the Indian Armed Forces in terms of troops' size and platforms, the Indian Navy is nevertheless a strong, well-equipped force. Swift induction of top-of-the-line maritime assets along with advanced weapons and systems in recent years has emboldened the Navy to undertake near-sea missions as well as far-off ocean operations effectively. While some of the platforms have become obsolete and hence, need replacement, the Navy has ambitiously put forth its goal to acquire over 100 new platforms, including aircraft carriers, nuclear & conventional submarines, destroyers and frigates over the next 10 years or so. It also plans to double the size of its manpower in the next 15 years. Alongside, the Navy is continuously updating its operational procedures and naval doctrines so as to keep pace with the rapidly evolving regional and global geostrategic equations.

The entire existing maritime combatant of Indian Navy, including its warships, submarines, aircraft and weapon systems, is comprehensively detailed below. Also listed are some of the futuristic platforms /systems/weapons which are going to join operational service in near future.

COMBAT PLATFORMS

As one of the powerful and leading navies in the Asian continent, the Indian Navy maintains a combat platform constituting of **surface ships** such as aircraft carrier, destroyers, frigates, corvettes, fast attack craft/patrol boats and offshore patrol vessels. The **sub-surface** fleet of the Navy consists of submarines of various kinds.

Surface Ships

The Navy at present maintains some 100 surface ships for various operational missions and deployments. These active platforms include **aircraft carrier, destroyers, frigates, corvettes, fast attack craft/patrol boats** and **offshore patrol vessels**.

Aircraft Carrier

India holds the distinct capability of operating aircraft carriers since 1961 – a capability no other Asian country had at that time. **INS Vikrant**, originally built as a Majestic-class CATOBAR (Catapult Assisted Take Off but Arrested Recovery) carrier for the British Navy, could not be completed and was later acquired by India in 1957. Construction on the warship was completed in 1961 and it was inducted in Indian Navy in the same year. The aircraft carrier played a pivotal role in the 1971 Indo-Pak war. It was decommissioned from service in 1997.

India is presently constructing its first indigenous aircraft carrier, bearing the same name, **INS Vikrant**, to keep the prestigious legacy of the erstwhile carrier alive. The new warship is being readied for sea trials following which it is slated for induction by 2018.

A second aircraft carrier, **INS Viraat**, joined the Navy in 1987 as the first STOVL carrier operating the Sea Harrier aircraft. Soon after the acquisition of INS Viraat, the first carrier INS Vikrant was also converted from a CATOBAR carrier to a STOVL (Short Take-off and Vertical Landing) carrier. **INS Viraat** has been taken out of service in July 2016 and awaiting decommissioning. The sole aircraft carrier operational in the Indian Navy at present is **INS Vikramaditya**.

INS Vikramaditya: The refurbished aircraft carrier joined Indian Navy as its largest warship in November 2013. The Soviet-built Kiev-class VTOL (Vertical Take-off and Landing) missile cruiser carrier was active in the Russian Navy until 1996 when India decided to acquire it and signed an initial contract in this regard in 2004. The warship was later modernised completely in Russia, including its super structure, flight deck, weapons suite and sensors among other installations, to convert it into a "short take-off but arrested recovery" (STOBAR) aircraft carrier as per the requirements of the Indian Navy. The entire modernisation programme took nearly a decade to complete. The large warship was operationally deployed in May 2014 along with its fleet of MiG-29K naval fighter aircraft. The carrier has also made goodwill visits to foreign ports besides taking part in the International Fleet Review in Visakhapatnam in February 2016. Induction of the new warship has transformed Indian Navy as a major maritime force having strategic power projection capability in the 21st century. The warship is presently undergoing a maintenance refit during which it will be equipped with air defence systems and weapons.

Displacement: 44,500 tonnes (Max)

Length: 284m (overall)

Beam: 60m (overall)

Draught: 10.2m

Propulsion: Steam turbines (8 boilers, 4 shafts)

Speed: > 30 knots

Operational Range: 13,500 nautical miles (25,000 km) at 18 knots

Endurance: 45 days

Countermeasure: Decoy, ECM

Sensors: Long range air surveillance radars

Flight deck: Ski-jump & arrested gear

Fixed wing aircraft: MiG-29K/KUB

Helicopter: Ka-28, Ka-31, Sea King Mk 42, ALH Dhruv/Chetak

Armament: Barak-8 LR-SAM (to be fitted); Barak-1 SAM; AK-630 (30mm) CIWS

Crew: 1,600

INS Vikrant: The construction of India's first Indigenous aircraft carrier, INS Vikrant, is progressing steadily with the Navy hopeful to deploy it in the coming years. Construction of the new STOBAR carrier started at the Cochin Shipyard in 2006 and its keel was laid down in 2009. The warship, designed by Directorate of Naval Design of the Indian Navy, was launched in August 2013.

Displacement: 38,000 tonnes (min) - 42,000 (max)

Length: 262m (overall)

Beam: 60m (overall)

Draught: 8.4m

Propulsion: Shafts (2)

Speed: 28 knots

Operational Range: 8,000 nautical miles (15,000 km)

Sensors: AESA radar

Flight deck: Ski-jump & arrested gear

Fixed wing aircraft: Mig-29K, Naval LCA

Helicopter: Ka-31, Sea King Mk 42, ALH Dhruv

Armament: Barak-8 LR-SAM; Barak-1 SAM; AK-630 (30mm) CIWS

Crew: 1,400

India also intends to construct another aircraft carrier, to be named **INS Vishal**, which is likely to be the country's first **nuclear-powered supercarrier**. The detailed blueprint of the new futuristic carrier is yet to be finalised.

Destroyer

The Navy presently operates a total of 10 guided missile destroyers while five more such large warships are going to be deployed in the coming years. The DDGs' fleet includes **Delhi class (3)**, **Rajput class (5)** and **Kolkata class (2)** while construction of one more Kolkata class and **four** new **Project-15(B)** class of warships is going on. Some of the older-generation destroyers have been decommissioned.

Delhi class: The large warships were the first to be built in India under Project-15 programme at the Mazagon Dock Ltd with Russian assistance and having some European technology as well alongside Indian systems.

The lead ship of the class, INS Delhi, was commissioned into service in 1997, followed by the second ship in 1999 and third one in 2001. The warships are armed with Russian and Israeli weapon systems. The Government in 2015 has cleared a proposal to install BRAHMOS supersonic cruise missiles on-board the destroyers.

Displacement: 6,200 tonnes
Length: 162m
Beam: 17.4m
Draught: 6.5m
Propulsion: Gas turbine plant (2); Shaft (2)
Speed: 28 knots
Operational Range: 5,000 nautical miles (8,000 km)
Countermeasure: Chaff launcher; Jammer
Sensors: E-band surface search radar; I-band navigation radar; D-band air search radar; sonar
Armament: Kh-35 AShM; Barak SAM; AK-100 (100mm) gun; AK-630 (30mm) CIWS; 533mm torpedo tubes; ASW rocket launcher
Crew: 350

Rajput class: The Soviet-built guided missile destroyers joined Indian Navy between 1980 and 1988. The warships, primarily based on the Soviet-designed Kashin-II class of destroyers, have incorporated major changes in their electronics and combat suite among other features so as to fit in Indian Navy's requirements. The destroyers have been designed for anti-aircraft and anti-submarine warfare missions for carrier task force. They are the first surface ships of Indian Navy to deploy BRAHMOS supersonic cruise missile system on-board. All five Rajput-class destroyers are presently operational in the Eastern Naval Command.

Displacement: 4,974 tonnes
Beam: 15.8m
Draught: 5m
Propulsion: Gas turbines (2); Shaft (2)
Speed: 35 knots
Operational Range: 4,000 nautical miles (6,500 km) at 18 knots
Countermeasure: ECM; Chaff
Sensors: 3D air/surface search radar; I-band navigation radar; C-band air search radar; sonar
Armament: BRAHMOS ASCM; P-15 Termit AShM; M-1 Volna SAM; Barak SAM; 76.2mm gun; AK-630 (30mm) CIWS; 533mm torpedo tubes; ASW rocket launcher
Aircraft carried: Ka-28; HAL Chetak
Crew: 320 (35 officers)

Kolkata class: The new indigenously built large warships have been developed as a follow-on class of the Project-15 Delhi class of destroyers under **Project-15(Alpha)** class. The lead ship of the class, INS Kolkata, joined the Navy in August 2014 followed by the second ship of the class, INS Kochi, in September 2015. Both warships are deploying BRAHMOS weapon system. Another destroyer of this class, INS Chennai, is awaiting induction. Designed by the Navy's bureau, the Kolkata class DDGs have been constructed at the Mazagon Docks Ltd. The new ships incorporate modern design concepts for improved survivability, stealth, sea-keeping and manoeuvrability besides other advancements in weapons and sensors over their predecessor, the Delhi class of DDGs. The Kolkata-class also has a flight deck and enclosed hangar to embark up to two medium-lift helicopters such as Dhruv ALH and Sea King Mk 42 helicopter.

Displacement: 6,800 tonnes
Length: 163m
Beam: 17.4m
Draught: 6.5m
Propulsion: Gas turbines; Diesel engines
Speed: 30 knots
Operational Range: 8,000 nautical miles (15,000 km) in 18 knots
Countermeasure: Decoys
Sensors: S-band AESA multi-role radar; D-band air search radar; towed array sonar
Armament: BRAHMOS ASCM; Barak-8 LR-SAM; 76mm main gun; AK-630 CIWS; 533mm torpedo tubes; ASW rocket launcher
Crew: 400

Project-15B (BRAVO) class: The Bravo class of stealth destroyers are being built as a successor to the Alpha class of (Project-15A) of DDGs. The lead ship of the class, INS Visakhapatnam, was launched into the sea in April 2015 with an expected commissioning by 2018. The second ship, to be named Mormugao, was launched in June 2016. All four ships of the class are being planned to be inducted by 2024. While featuring the same hull design as that of the Project-15A class of warships, the new BRAVO-class will have enhanced capability to operate in nuclear, biological and chemical atmosphere. Once operational, they will become the largest destroyers in the Indian Navy's fleet. The warships will also deploy BRAHMOS missiles as prime strike weapon.

Displacement: 7,334 tonnes
Length: 163m
Beam: 17.4m
Draught: 6.5m
Propulsion: Gas turbines; Diesel engines
Speed: > 30 knots
Operational Range: 4,000 nautical miles (7,400 km) at 14 knots
Countermeasure: Decoy
Sensors: S-band AESA multi-role radar; D-band air search radar; towed array sonar
Armament: BRAHMOS ASCM; Barak-8 LR-SAM; CIWS; Torpedo tubes; ASW rockets
Crew: 300

Frigates

The Navy uses few older-generation guided missile frigates (FFGs) along with some modern ones that have inducted in the last few years. Most of the frigates have been designed and built in India while incorporating some Russian and Western technology. A total of 14 frigates serve the Navy at present which include the **Godavari class**, **Brahmaputra class**, **Talwar class** and **Shivalik class**. Design and development of a new **Project-17A** class of advanced stealth frigates has been approved with construction expected to start soon. Alongside, negotiations are going on with Russia to acquire four new Grigorovich-class of frigates.

Godavari class: The older-generation guided missile frigates are the first significant indigenous warship design and development initiative taken by the Indian Navy. Three such warships were built in the 1980s at Mumbai's Mazagon Dock Ltd of which two remain operational at present. The warships used

Indian, Russian and Western combat systems and technology. The frigates are designed to carry two utility/transport/ASW helicopters (Sea King or HAL Dhruv/ Chetak).

Displacement: 3,850 tonnes

Length: 126.5m

Beam: 14.7m

Draught: 4.5m

Propulsion: Steam turbines

Speed: 27 knots

Operational Range: 4,500 nautical miles (8,300 km) at 12 knots

Countermeasure: ESM/ECM decoys; Chaff

Sensors: Air search radar; Air/surface search radar; Navigation radar

Armament: P-15 Termit AShM; Barak-1 SAM; AK-630 (30mm) CIWS; Torpedo tubes

Crew: 313

Brahmaputra class: The frigates were designed and built as an upgraded variant of the Godavari class of warships under **Project-16A** class. They had similar hull design and dimension but incorporated various advanced systems, sensors and armament, both Indian and foreign, for enhanced capability. Three such FFGs were built at GRSE, Kolkata, and inducted into service between 2000 and 2005. All three frigates remain operational and are primarily deployed in anti-surface warfare and anti-submarine warfare operations.

Displacement: 3,850 tonnes

Length: 126.5m

Beam: 14.5m

Draught: 4.5m

Propulsion: Steam turbines

Speed: 27 knots

Operational Range: 4,500 nautical miles (8,300 km) at 12 knots

Countermeasure: EWS/ESM; Decoy; Chaff

Sensors: Air search radar; Air/surface search radar; Navigation radar; Search & attack sonar; Towed array sonar

Armament: Kh-35 AShM; Barak SAM; 76mm OTO Melara gun; AK-630 (30mm) CIWS; 324mm ASW torpedo tubes

Aircraft carried: Sea King (2) or HAL Dhruv/Chetak (2)

Crew: 313

Talwar class: The modern stealth warships, jointly developed by Russia and India under Project 1135.6, are modified variants of the Soviet-origin Krivak-III class frigates. While first three frigates of the class joined Indian Navy in 2003-04, three more follow-on ships were ordered in 2006. The new frigates incorporated advanced design features, systems and weapons, including BRAHMOS missiles. They were inducted in 2012-13. The state-of-the-art Talwar-class FFGs are designed to carry out a wide range of tasks, primarily hunting down enemy submarines and large surface ships. They also support naval forces during air, surface and sub-surface missions.

Displacement: 4,000 - 6,200 tonnes

Length: 125m - 163m

Beam: 15.2m - 17.4m

Draught: 4.2m

Propulsion: Gas turbines

Speed: > 30 knots

Operational Range: 4,850 nautical miles (8,000 km) at 14 knots

Countermeasure: EWS; Decoy

Sensors: Air search radar; Surface search radar; Navigation radar

Armament: Klub/ BRAHMOS AShM; SA-16/Shtil SAM; A190 (100mm) gun; AK-630 (30mm) CIWS; Kashtan CIWS; 533mm torpedo tubes; ASW rocket launcher

Aircraft carried (1): Ka-28/ Ka-31/ Dhruv

Crew: 350

Shivalik class: The modern, multi-role frigates are the first-of-their-kind be built in India using stealth technology. The warships, constructed at Mazagon Dock Ltd as an upgrade of the Talwar class FFGs, are one of the leading surface combatants of Indian Navy in the 21st century. Three such frigates have been commissioned into service between 2010 and 2012. The FFGs are armed with Indian, Russian and Israeli weapons, including BRAHMOS ASCM/LACM system.

Displacement: 5,300 - 6,200 tonnes

Length: 142.5m

Beam: 16.9m

Draught: 4.5m

Propulsion: Diesel engines & gas turbines

Speed: > 30 knots

Operational Range: 5,000 nautical miles (9,000 km) at 18 knots

Countermeasure: EWS

Sensors: 3-D air/surface search radar; hull-mounted sonar; towed array sonar

Armament: BRAHMOS AShM; Shtil & Barak-1 SAM; 76mm OTO Melara gun; AK-630 (30mm) CIWS; Torpedo launchers; ASW rocket launchers

Aircraft carried: HAL Dhruv (2) or Sea King Mk 42B (2) or Ka-31 (2)

Crew: 260

Project-17A (Alpha) class: The new ships will be built as a follow-on class of the Shivalik-class of FFGs. Seven such warships, with advanced stealth features, weapons and sensors, will be built by MDL, Mumbai and GRSE, Kolkata. They would be longer and broader than their predecessor. The likely specifications of the new warship are as follows.

Displacement: 6,670 tonnes

Length: 149m

Beam: 17.8m

Draught: 5.22m

Propulsion: Diesel engine

Sensors: Sonar

Armament: BRAHMOS ASCM/LACM; Barak-8 LR-SAM; 127mm or 76mm OTO Melara gun; ASW Torpedo tubes

Aircraft carried: Sea King/ Dhruv (2)

Corvettes

The Indian Navy operates around 30 corvettes of different class for various missions. These include the **Khukri class**, **Kora class**, **Veer class** and **Abhay class** while induction of new **Kamorta-class** of anti-submarine warfare corvettes is going on.

Khukri class: Four such corvettes, partly built in India, are in service with the Navy since 1989. The diesel engines of the warships were assembled in India.

Displacement: 1,350 tonnes

Length: 91.1m

Beam: 10.5m

Draught: 3.07m

Propulsion: Diesel engine

Speed: 25 knots

Operational Range: 4,000 nautical miles (7,400 km) at 16 knots

Countermeasure: ESM; Decoy; Chaff

Sensors: Air search radar; Air/ Surface search radar; Navigation radar

Armament: P-15 Termit AShM; Strela-2M SAM; 76mm gun; AK-630 CIWS
Crew: 80

Kora class: The indigenous guided missile corvettes, designed by Navy's design bureau under Project-25A, were built by GRSE and outfitted at MDL. Four such warships were inducted between 1998 and 2004. The warships are primarily used as surface combatant of Indian Navy.

Displacement: 1,350 tonnes
Length: 91.1m
Beam: 10.1m
Draught: 3.07m
Propulsion: Diesel engines (2)
Speed: 25 knot
Operational Range: 3,450 nautical miles (6,400 km) at 16 knots
Countermeasure: EWS; Chaff; Decoy
Sensors: Air/surface search radar; Navigation radar
Armament: Kh-35 AShM; Strela-2M/ Igla-1E SAM; 76mm gun; AK-630 CIWS
Crew: 134

Veer class: Ten such corvettes are presently active in the Indian Navy. They were commissioned between 1987 and 2002.

Displacement: 455 tonnes
Length: 56m
Beam: 10.5m
Draught: 2.5m
Propulsion:
Speed: 32 knots
Armament: P-15 Termit AShM; Strela-2 SAM; 76mm gun; AK-630 CIWS
Crew: 40

Abhay class: The ASW corvettes are customized variants of the Soviet Pauk-class of corvettes. The class is primarily designed for coastal patrol and anti-submarine warfare operations. Four such vessels are operational since 1989.
Displacement: 485 tonnes
Length: 57m

Beam: 10.2m
Propulsion: Diesel engine
Speed: 32 knots
Sensors: Towed array sonar
Armament: Strela-2M SAM; 76mm gun; 533mm torpedo tubes; ASW rocket launchers
Crew: 32

Kamorta class (ASW): The new-generation stealth ASW corvettes built indigenously under Project-28 by GRSE are being progressively inducted in the Navy. Two such warships have joined the Navy in 2014 and 2016 while two more are set to join service in coming years. The advanced warships are primarily designed to carry out ASW missions along with anti-surface warfare (AsuW) and anti-air warfare (AAW) operations.

Displacement: 3,500 tonnes
Length: 110m
Beam: 14m
Propulsion: Combined diesel and diesel (CODAD)
Speed: 30 knots
Operational Range: 3,450 nautical miles (6,000 km) at 25 knots
Countermeasure: EWS; Decoy
Sensors: 3D surveillance radar; bow mounted sonar; towed-array sonar
Armament: Barak SAM; 76mm gun; AK-630 CIWS; Torpedo tubes; ASW rocket launcher
Aircraft carried: Sea King Mk 42B
Crew: 200

Fast Attack Craft

The FACs or small fighting ships used by the Navy mainly for coastal defence and surveillance purposes include the **Super Dovra II class**, **Trinket class**, **Bangaram class**, **Car Nicobar class** and the newly-inducted **water-jet FACs**.

Super Dovra II class: The Israeli-built FACs, also known as T-80 class, joined Indian Navy between 1998 and 2004. The high-speed vessels are mainly used for coastal patrol and coastal defence purposes.

Displacement: 60 tonnes
Length: 25.4m
Beam: 5.5m
Draught: 1.1m
Propulsion: Diesel engine
Speed: 40 knots
Operational Range: 700 nautical miles (1,300 km) at 14 knots
Sensors: --
Armament: 20mm cannon; Machine gun
Crew: 2 officers, 12 sailors

Trinket class: One such vessel, built by GRSE, is presently operational in the Navy which was commissioned into service in 2000. The vessel is used to carry out fisheries protection, anti-poaching, counter-insurgency and search-and-rescue operations in coastal areas and in the EEZ. Three other such craft have been transferred to India's maritime neighbouring countries.

Displacement: 210 tonnes
Length: 38m
Beam: 7.5m
Propulsion: Diesel engines
Speed: 30 knots
Armament: 30mm gun
Crew: 32

Bangaram class: Built as an improved variant of the Trinket class of FACs, the Bangaram class vessels are designed of interdiction tasks against fast moving surface vessels and search/rescue operations in coastal waters. The vessels are also capable of generating a limited naval quick reaction at short notice. The GRSE-built ships are more advanced in many ways, especially in terms of their ergonomic design and operational capabilities over their predecessor. Four such

ships are operational in the Navy since 2006.
Displacement: 260 tonnes
Length: 46m
Beam: 7.5m
Propulsion: Diesel engine
Speed: 28 knots
Armament: 30mm gun
Crew: 44

Car Nicobar class: The lightly armed ships built by GRSE succeeded the Bangaram class of FACs. The advanced vessels are designed to carry out coastal patrol, search & rescue and anti-piracy operations in India's EEZ. Ten such platforms have been inducted in the Navy between 2009 and 2011. The last four ships of the class were built as water jet-propelled FACs having higher speed, agility and swift response capability.

Displacement: 325 tonnes
Length: 49m
Beam: 4m
Propulsion: Diesel engine; Water Jet (last four units)
Speed: 27 knots - 35 knots
Operational Range: 1,700 nautical miles (3,200 km) at 14 knots
Sensors: Navigation radar; Surface search radar; Sonar
Armament: Igla SAM; 30mm gun
Crew: 42

Water-jet FAC: The new variants follow the Car Nicobar class of FACs and incorporate latest communication equipment and radars besides water jet propulsion for higher speed, manouverability and effective surveillance. The indigenously designed Water Jet Fast Attack Craft (WJFAC) are being built by GRSE, the first of which has been inducted in May 2016. Three more such vessels are under construction.
Displacement: 325 tonnes
Length: 49m
Beam: 4m
Propulsion: Water Jet & Diesel engine
Speed: 35 knots
Operational Range: 2,000 nautical miles (3,700 km) at 14 knots
Armament: Igla SAM; 12.7mm heavy machine gun; Medium machine gun; 30mm gun
Crew: 45

Offshore Patrol Vessel

The OPVs operational in the Indian Navy include the following.

Sukanya class: The large, lightly armed OPVs built by a South Korean firm and later on by Hindustan Shipyard, are operational in the Navy from 1990 onwards to carry out offshore patrolling of India's exclusive economic zone. Six such ships are presently in active service. With a large hull design, the OPVs can also be heavily armed and upgraded to light frigates if needed. Two vessels of the class have also been used as test beds for the installation of Dhanush ship-based ballistic missile launch system.

Displacement: 1,890 tonnes
Length: 101m
Beam: 11.5m
Draught: 3.4m
Propulsion: Diesel engine; Shaft
Speed: 21 knots
Operational Range: 7,000 nautical miles (13,000 km) at 15 knots
Sensor: Radar with automatic radar plotting aid (ARPA)
Sensors: Surface search radar; Navigation radar
Armament: 40mm Bofors anti-aircraft gun; 12.7mm machine gun; Dhanush ballistic missile (in two units)
Aircraft carried: HAL Chetak
Crew: 140

Saryu class: The advanced patrol ships indigenously built by Goa Shipyard Ltd are capable of maritime patrol, ocean surveillance, monitoring and safeguarding of shipping lanes. Four Saryu class OPVs have joined the Navy in 2013-14.

Displacement: 2200 tonnes
Length: 105m
Beam: 12.9m
Draught: 3.6m
Propulsion: Diesel engine
Speed: 25 knots
Operational Range: 6,000 nautical miles (11,000 km) at 16 knots
Armament: 76mm Oto Melara gun; AK-630 CIWS
Aircraft carried (1): ALH Dhruv or Chetak or Sea King
Crew: 120

Amphibious Force

To ramp up its amphibious warfare capability and maintain forward position in the strategically important Indian Ocean Region in addition to carrying out large-scale humanitarian operations, India started acquiring multi-role maritime support platforms such as LPDs, LSTs and LCUs from 1987 onwards. The amphibious vessels operated by the Navy include the **Austin-class (ex-US)** landing platform dock (LPD), **Shardul-class** & **Magar class** landing ship, tank (LST), **Kumbhir class** landing ship, medium (LSM) and **MK-3 class** landing craft, utility (LCU). In addition, eight new MK 4 class of LCUs are to be inducted while four new indigenous LPDs are to be built in the coming years.

Austin-class LPD: The amphibious warfare ship, formerly named USS Trenton, was acquired from the US Navy under a contract signed in 2005. It was commissioned into service in 2007 as INS Jalashwa. The large landing platform dock, capable of carrying mechanised landing craft, helicopters and troops, was refurbished to meet Indian Navy's requirements. The Navy also purchased six refurbished Sikorsky UH-3 heavy lift helicopters to operate from the LPD. In 2012, the warship underwent its second refurbishment during which it was fitted with indigenous sensors and other components.

Displacement: 16,600 tonnes
Length: 174m
Beam: 30.4m
Draught: 6.7m
Propulsion: Steam turbines, shaft & boilers
Speed: 20 knots
Sensors: Navigation radar
Armament: Phalanx CIWS; 25mm gun; 50 calibre AA gun
Aircraft carried: UH-3 Sea King helicopter
Crew: 407

Magar class LST: The first-generation of amphibious warfare vessels were built at the Hindustan Shipyards Ltd. and fitted out at GRSE, Kolkata. Two such landing ship, tanks (LSTs) were built and inducted in service between 1987 and 2009 and are stationed at the Navy's Eastern Command. Each warship can carry main battle tanks, APCs, two medium-lift helicopters and up to 500 troops.

Displacement: 5,665 tonnes
Length: 120m
Beam: 17.5m
Draught: 4m
Propulsion: Diesel engine
Speed: 15 knots
Operational Range: 3,000 nautical miles (5,600 km) at 14 knots
Countermeasure: EW suite
Sensors: Navigation radar
Armament: 40mm Bofors AA gun; 22mm multiple-barrel rocket launchers
Aircraft carried: Sea King Mk42
Crew: 140

Shardul-class LST: The new class of ships evolved from the Magar class platforms. Built by GRSE, the first Shardul class landing ship entered service in 2007 followed by the second ship in 2008 and third in 2009. The LSTs incorporate indigenous content of over 90% with state-of-the-art equipment. They can carry 500 troops, one medium lift helicopter, main battle tanks and APCs/military vehicles.

Displacement: 5,600 tonnes
Length: 125m
Beam: 17.5m
Draught: 4m
Propulsion: Diesel engine
Speed: 16 knots
Operational Range: --
Countermeasure: Decoy
Armament: 30mm AA gun; MANAPDs; Rocket launchers
Aircraft carried: Sea King or ALH Dhruv
Crew: 156

Kumbhir class LSM: The medium amphibious warfare vessels of the Navy are a derivative of the Polish-designed Polnocny class of landing ships. Eight such LSMs were acquired between 1974 and 1986 of which four are presently in active service. The ship is capable of carrying a cargo of up to 200 ton, including MBTs, combat vehicles, 200 troops and one utility helicopter.

Displacement: 1,200 tonnes
Length: 84m
Beam: 17.5m
Propulsion: Diesel engine
Speed: 18 knots
Operational Range: 1200 nautical miles at 16 knots
Countermeasure:
Sensors: Navigation radar
Armament: 30mm gun; MANPAD; 40mm rocket launchers
Crew: 108

MK-3 class LCU: The older-generation utility craft are used to transport equipment and troops to the shore. The LCUs are capable of transporting tracked or wheeled vehicles and troops from amphibious assault ships to beachheads or piers. The ships were acquired by Indian Navy in the 1980s and four of them remain operational.
Displacement: 500 tonnes
Length: 58m
Beam: 8.2m
Draft: 1.6m
Propulsion: Diesel engine
Speed: 12 knots
Operational Range: 1,800 km at 8 knots
Sensors: Navigation radar
Armament: 40mm Bofors AA gun; Naval mines
Crew: 270

MK 4 class LCU: The new landing craft are being built by GRSE for various maritime roles including beaching operations, humanitarian and disaster relief operations, and evacuation from distant islands. The first ship of the class was launched in 2014 and delivered to the Navy in September 2016. The LCU is capable of accommodating 56 crew members and 160 fully equipped soldiers. It can carry a freight of 145 tonnes.

Displacement: 830 tonnes
Length: 63m
Beam: 11m
Draft: 2.2m
Propulsion: Diesel engine
Speed: 15 knots
Operational Range: 1,500 nautical miles (2,780 km) at 12 knots
Armament: 30mm naval gun
Crew: 56

Sub-Surface Ships

The undersea fleet of Indian Navy consists of conventional and nuclear submarines. The Navy is presently operating 14 underwater platforms while awaiting the induction of over a dozen more in the coming years. The submarines are used for a wide range of tasks, including nuclear deterrence, fleet protection, reconnaissance, hunting down surface-based and underwater military installations/targets and other covert operations.

Submarines

The active submarine platforms of Indian Navy include the **Sindhughosh (Kilo) & Shishumar class** of **conventional attack submarines (SSKs)** and **Chakra nuclear-powered attack submarine (SSN).** India is also advancing towards the induction of the country's first indigenously built **ballistic missile carrying nuclear-powered submarines (SSBN) INS Arihant** in near future. Besides, the first of six **Scorpene class conventional attack submarine with AIP technology (SSI)** is slated for commissioning by 2016-end.

Conventional Attack Submarine (SSK)

Sindhughosh class SSK: The Russian-origin Kilo-class (Type 877EKM) diesel-electric submarines were acquired under a contract between Russia's Rosvooruzhenie and Indian Ministry of Defence. The first platform of the class entered service in 1986. Of the 10 such submarines commissioned between 1986 and 2000, nine remain in service at present. Four of the platforms are set to undergo a mid-life upgrade programme at a Russian shipyard in association with an Indian partner. The remaining submarines have received mid-life upgrades in Russian and Indian shipyards.

Displacement: 2,300 tonnes (surface); 3,000 tons (submerged) at full load
Length: 72.6m
Beam: 9.9m
Draught: 6.6m
Propulsion: Diesel electric (Two 2,722 kW diesel-electric generators, one 4,400 kW motor,

two 152 kW auxiliary motors, one 97 kW economic speed motor); 1 Shaft

Speed: 10-12 knots (surfaced); 17-25 knots (submerged)

Operational Range: 6,000 miles (9,700 km) when snorting at 7 knots; 400 miles (640 km) when submerged at 3 knots

Endurance: 45 days

Countermeasure: Electronic support measures

Sensors: USHUS integrated sonar

Armament: Strela-3 SAM; Klub-S LACM/AShM; 533mm torpedo tubes

Crew: 53

Shishumar class SSK: The conventional attack submarines were acquired from Germany's Howaldtswerke-Deutsche Werft (HDW) under a contract signed in 1981. While two of the submarines (Type 209/1500) were assembled at the HDW facility at Kiel in Germany, the third and fourth vessels were license-built at Mumbai's Mazagon Dock Limited. The submarines were commissioned between 1986 and 1994 and are in active service at present.

Displacement: 1,660 tonnes (surface); 1,850 ton (submerged)

Length: 65m

Beam: 6.5m

Draught: 6m

Propulsion: Diesel electric (four diesel engines of 1,790 kW each); 1 shaft

Speed: 11 knots (surfaced); 22 knots (submerged)

Operational Range: 11,000 nautical miles (20,300 km) at 10 knots when surfaced; 400 nautical miles (740 km) at 4 knots when submerged

Endurance: 50 days

Armament: Wire guided torpedoes; naval mines; Harpoon AShM (to be fitted under upgradation)

Crew: 36

Scorpene class SSI: The new attack submarines with Air Independent Propulsion (AIP) system are being built at MDL, Mumbai in partnership with French shipbuilder DCNS under a 2005 contract. Six such platforms being built under Project-75 are slated to be inducted in the Navy by 2029. The first submarine of the class, INS Kalvari, is undergoing extensive sea trials and is likely to be inducted in service by 2016-end. The state-of-art conventional attack submarines will have superior stealth and the ability to launch precision strike weapons while underwater or on surface. The vessels will be used for missions such as area surveillance, intelligence gathering, anti-submarine warfare (ASW), anti-surface warfare (ASuW), and minelaying operations.

Displacement: 2,000 tonnes

Length: 61.7m

Beam: 6.2m

Draught: 5.8m

Propulsion: Diesel electric; Batteries; AIP

Speed: 20 knots (submerged)

Operational Range: 6,500 nautical miles (12,000 km) at 8 knot (surfaced); 550 nautical miles (1,020 km) at 5 knot (submerged)

Endurance: 50 days (50+21 days with AIP)

Sensors: Surveillance radar; S-CUBE integrated, modular sonar suite

Armament: SM-39 Exocet AShM; 533 torpedo tubes

Crew: 31

Nuclear powered attack submarines (SSN)

INS Chakra: The platform is an Akula-II class (originally designated "K-152 Nerpa") nuclear attack submarine leased from Russia for a 10-year duration. The submarine joined Indian Navy in April 2012 as INS Chakra, making India the sixth country in the world to operate a nuclear-powered hunter-killer submarine and giving the Navy the operational flexibility in blue-water operations. The platform is primarily being used to train the navy personnel in operating future nuclear-powered submarines of India. In October 2016, India has reached a deal with Russia to lease another Akula-II class platform.

Displacement: 8,140 tonnes (surface); 12,770 tons (submerged)

Length: 110m

Beam: 14m

Draught: 9m

Propulsion: Pressurised water nuclear reactor; Auxiliary electric engines (2); Auxiliary diesel engines (2)

Speed: 20 knots (surfaced); 35 knots (submerged)

Endurance: 100 days

Armament: RK-55 Granat nuclear cruise missile; Klub-S AShM; 533mm & 650mm torpedo tubes

Crew: 73 (31 officers)

Ballistic missile carrying nuclear submarine (SSBN)

INS Arihant: The platform has been built as India's first indigenous nuclear-powered ballistic missile submarine. It was jointly designed and developed by the Indian Navy, Bhabha Atomic Research Centre (BARC) and Defence Research and Development Organisation (DRDO) originally under a project code-named "advanced technology vessel" (ATV) which started in 1998. The submarine, launched at the Eastern Naval Command in Visakhapatnam in 2009, has joined the Navy in August 2016 after completing extensive sea trials during which it has test fired SLBMs. Construction of two more such platforms is advancing. The Navy plans to induct a total of six Arihant-class SSBNs in service.

Displacement: 6,000 tonnes

Length: 110m

Beam: 11m

Draught: 10m

Propulsion: Pressurised water nuclear reactor

Speed: 12 knot - 15 knot (surfaced); 24 knot (submerged)

Sensors: USHUS integrated sonar

Armament: K-15/ K-4 SLBM; AShM & LACM; 533mm torpedo tubes

Crew: 95

SUPPORT PLATFORMS

The support platforms operational in the Indian Navy include **fleet replenishment ships**, **training ships**, **torpedo recovery vessel, diving support ship, research vessel** and **survey ships.** The vessels are primarily used for carrying out various missions along the country's littoral waters.

Fleet Replenishment Ship

With an aim to expand its future blue water operations, Indian Navy has acquired the **Deepak class** of fleet replenishment tankers from Italy's Fincantieri. The Navy also operates two other vessels – **INS Jyoti** and **INS Aditya** – for fleet replenishment purposes.

Deepak class: The two large vessels of the class – INS Deepak and INS Shakti – were inducted into service in 2011 to help strengthen the Navy's combat support operations at sea and expand the reach of the country's maritime power. The platforms can be used for transportation and supply of ammunitions, materials and provisions, as well as for humanitarian assistance and disaster relief operations with their modern medical health facilities. Each ship is equipped with storage facilities for carrying over 15,000 ton of liquid cargo, including diesel oil, aviation fuel and fresh water. It also offers large storage space to accommodate up to 500 ton of solid cargo, including food stores and ammunition. The vessel can carry troops and a 10-ton helicopter on its flight deck as well. The two fleet tankers have become the mainstay support ships of Indian Navy in the 21st century.

Displacement: 27,000 tonnes
Length: 175m
Beam: 25m
Draught: 19.3m
Propulsion: Diesel engine
Speed: 20 knots
Operational Range: 10,000 nautical miles (18,500 km) at 16 knots
Armament: AK-630 (30mm) CIWS
Aircraft carried: Sea King & ALH Dhruv
Crew: 250

INS Jyoti: The Soviet-built Komandarm Fedko-class replenishment oiler joined service in 1996. The ship, built to be a Project 15966M merchant tanker, was modified and acquired by the Indian Navy. It is the second largest ship of Indian Navy after aircraft carrier INS Vikramaditya. The platform is primarily used for fleet replenishment and sustaining blue water operations by the Navy. It can carry a cargo of up to 28,000 ton and has a helicopter deck. The ship has also been armed with close-in weapons for self-defence.

Displacement: 35,900 tonnes
Length: 178m
Beam: 25
Draught: 12m
Propulsion: Steam turbine
Speed: 15 knots
Operational Range: 12,000 nautical miles (22,000 km) at 15 knots
Sensors: Navigation radar
Armament: CIWS
Crew: 100

INS Aditya: The indigenously built ship was commissioned into service in 2000. It can carry up to 12,000 tons of liquid cargo, and 5,000 tons of solid cargo. Though smaller in size when compared to its predecessor INS Jyoti, the vessel has more advanced capabilities, including the ability to permanently carry a medium-lift transport/utility helicopter.

Displacement: 24,612 tonnes
Length: 172m
Beam: 23m
Draught: 7.5m
Propulsion: Diesel engines; Shaft
Speed: 20 knots
Operational Range: 8,640 nautical miles (16,000 km) at 16 knots
Armament: Igla SAM; 30mm gun
Crew: 197

Training / Sailing Ships

INS Tir: The indigenously built training ship was commissioned in 1986. It is capable of accommodating up to 235 personnel on board. The vessel also has a flight deck for carrying a Chetak or Dhruv helicopter.

Displacement: 3,200 tonnes
Length: 105.9m
Beam: 13.2m
Draught: 4.8m
Propulsion: Diesel engine; Shaft
Speed: 18 knots
Operational Range: 6,000 nautical miles (11,000 km) at 12 knots
Sensors: Surface search radar
Armament: Bofors 40mm AA gun; 30mm auto-cannon
Crew: 235

INS Tarangini: The first sail training ship of the Indian Navy joined service in 1997. The vessel, built at Goa Shipyard, became the first Indian naval ship to circumnavigate the globe in 2003-04 during which it carried out training for naval cadets, including those from the overseas. The ship also sails extensively across the Indian Ocean Region for the purpose of providing sail training experience to the officer cadets of Indian Navy.

Displacement: 420 tonnes
Length: 54m
Beam: 8.5m
Draught: 4.5m
Propulsion: Diesel engine
Speed: 10 knots
Crew: 60

INS Sudarshini: The ship built by Goa Shipyard as a follow-on vessel of INS Tarangini was commissioned in 2012. The ship has made a six-month-long naval voyage South East Asia in 2012-2013.

Displacement: 513 tonnes
Length: 54m
Beam: 8.5m
Draught: 4.5m
Propulsion: Diesel engine
Speed: 10 knots
Crew: 60

Torpedo Recovery Vessel

INS Astradharini: The indigenously developed torpedo recovery vessel joined the Navy in October 2015. The advanced warship, having a catamaran hull design, can operate at high seas. It has a large deck area with torpedo launchers for deploying and recovering various kinds of torpedoes during test trials. It can undertake multiple trials during a single sortie. The vessel is equipped with modern power generation, navigation and communication systems. INS Astradharini is primarily being used to carry out technical trials of underwater weapons and systems developed by DRDO-led labs.

Length: 50m
Speed: 15 knots
Crew: 42

Diving support ship

INS Nireekshak: The diving support vessel, built by MDL, Mumbai, was acquired by the Navy and joined service in 1989. The ship was later modified to operate as an interim submarine rescue vessel after it was refitted with diving bell and other rescue equipment removed from a former Russian submarine rescue vessel. The platform is in active service with the Navy.

Displacement: 3,600 tonnes
Length: 70.5m
Beam: 17.5m
Propulsion: Diesel engine; Shaft
Speed: 12 knots
Crew: 63

Research/ Survey vessels

INS Sagardhwani: Commissioned in 1994, the multi-purpose Marine Acoustic Research Ship (MARS) has been used extensively for conducting deep-sea experiments by DRDO's NPOL laboratory scientists. The DRDO-designed naval platform was built at GRSE, Kolkata. The vessel is designed with floating floors for scientific laboratories, anti-vibration mountings for machinery and equipment, balloon launching container and wind weather radar among other state-of-the-art equipment. It also has a helicopter landing pad.

Displacement: 2,000 tonnes
Length: 85m
Beam: 12.8m
Propulsion: Diesel engines; Shaft
Speed: 16 knots
Operational Range: 6,000 nautical miles (11,000 km) at 16 knots
Sensors: Navigation radar
Crew: 82

Sandhayak class: Seven such indigenously-built survey vessels commissioned between 1985 and 2002 serve the Navy at present. Equipped with an array of surveying, navigational and communication systems, the ships are designed to undertake shallow coastal and deep oceanic hydrographic survey and collect oceanographic and geophysical data. They can also assist in times of war and natural calamities as troop transporters. The platform also has a helicopter deck.

Displacement: 1,930 tonnes
Length: 87.8m
Beam: 12.8m
Draught: 3.5m
Propulsion: Diesel engines
Speed: 16 knots
Armament: 40mm Bofors AA gun
Crew: 178

Makar class: The lead ship of the class was commissioned into service in September 2012 as the Navy's first indigenously-built catamaran hull hydrographic survey vessel. Built by the Alcock Ashdown Gujarat Ltd, the platform is the first of its kind to join the Navy. The primary role of the ship is to undertake hydrographic surveys, required for production of nautical charts and publications aimed at improving navigational safety at sea. It is equipped with an array of latest oceanographic and land survey equipment. Five more such ships are slated to join the Navy in future.

Displacement: 500 tonnes
Length: 53m
Beam: 16m
Draught: 2.2m
Propulsion: Diesel engines; Bow thrusters
Speed: 18 knots
Operational Range: 3,000 nautical miles
Sensors: Advanced Electronic Positioning System; Multi-beam Swath Sounding System
Crew: 50

NAVAL AVIATION

The aviation wing of Indian Navy operates over 200 aerial platforms ranging from carrier-borne fighter jets to maritime patrol aircraft, ASW aircraft, and rotorcraft of different roles besides utility planes, jet trainers and unmanned aerial vehicles. The fixed-wing aircraft operated by Indian Navy include **MiG-29K/KUB fighter/trainer aircraft**, **TU-142M maritime reconnaissance & ASW aircraft, Il-38 and P-8I maritime patrol aircraft**, **Dornier 228 utility aircraft**, **Kiran HJT basic trainer aircraft** and **Hawk advanced trainer aircraft.** Induction of the naval variant of the indigenously developed **Light Combat Aircraft 'Tejas'** is also being planned in near future.

The rotorcraft operational in the Navy include **Kamov-28, Kamov-31, UH-3H Sea King (Sikorsky), Sea King Mk 42 (AgustaWestland), ALH Dhruv, Chetak (utility).** The Navy also has inducted the **Searcher II** and **Heron** unmanned aerial vehicles (UAVs).

Fighter Aircraft

MiG-29K/KUB: The Russian-built naval fighters and their trainer variants have been acquired under a contract signed in 2004. Induction of the combat jets is going on since 2010. Delivery of all 45 fighters/trainers is scheduled to be completed by 2016-end. The state-of-the-art, all-weather, air dominance supersonic fighters are operating from the Navy's aircraft carrier INS Vikramaditya. The MiG-29K squadron has been commissioned into frontline service to form the sword arm of the Indian Navy. The air-to-air refuelling capability enables the fighter to extend its range to perform power projection and air space dominance roles.

Length: 17.3m
Wingspan: 11.99m
Height: 4.4m
Weight: 24,500 kg (max at take-off)
Propulsion: 2 turbofans
Speed: Mach 2+
Range: 3,000 km (with 3 drop tanks)
Combat radius: 850 km to 1,300 km
Armament: R-77 RVV-AE AAMs; Kh-35E AShM; KAB 500KR/OD TV guided bombs; S-8KOM rockets
Crew: 1 (2 for KUB trainer)

LCA Tejas: Indigenously designed and developed by DRDO, Tejas is a single-seat supersonic all-weather multi-role fourth-generation light fighter. The naval variant of the light combat aircraft has been designed with Ski-jump Take Off but Arrested Recovery (STOBAR) operations from the Navy's indigenous aircraft carriers. It has high landing loads as compared to the Air Force variant. A dedicated shore-based test facility (SBTF) has been set up at INS Hansa in Goa to carry out the fighter's ski-jump sorties. Two naval prototypes are undergoing flight trials at present and are expected to be inducted in the Navy shortly. Developed as the only carrier-borne aircraft in light category in the world, the Naval LCA will carry a wide variety of operational weapons and equipment like the Beyond Visual Range (BVR) missiles, conventional and laser-guided bombs, air defence guns, counter-counter measures and drop tanks to take on surface targets over land or at sea.

Length: 13.25m

Wingspan: 8.20m

Height: 4.4m

Weight: 6,560 kg (empty); 13,500 kg (max at take-off)

Propulsion: 1 turbofan

Speed: Mach 1.8

Range: 3,000 km

Combat radius: 500 km

Armament: R-73/ R-77 AAM; Astra BVRAAM; Kh-59 ASM; 23mm twin barrel gun

Crew: 1

Maritime Patrol/ Reconnaissance/ ASW Aircraft

The maritime patrol and reconnaissance aircraft used by the Indian naval aviation include the following platforms.

P-8I MPA: India acquired the US's Boeing-built aircraft under a contract signed in 2009. Delivery of the long-range, multi-mission maritime patrol aircraft started in 2013 and the Navy has so far inducted eight P-8Is in service while a new order has been placed to get four more such aircraft. The aircraft is designed to carry out anti-submarine warfare, anti-surface warfare, intelligence, maritime patrol, and surveillance and reconnaissance missions. Acquisition of the aircraft has given the Navy the required reach and flexibility to undertake extensive surveillance missions as well as respond swiftly and effectively to contingencies in India's areas of interest, especially in the increasingly militarised Indian Ocean Region.

Length: 39.47m

Wingspan: 37.64m

Height: 12.83m

Weight: 62,730 kg (empty); 85,820 kg (max at take-off)

Propulsion: 2 turbofans

Speed: Mach 0.73 (907 km/h)

Range/ Combat Radius: 2,222 km

Armament: AGM-84L Harpoon AShM; MK-54 lightweight torpedoes; Rockets

Crew: 9

IL-38 MPA: The Russian-built maritime patrol and anti-submarine warfare (ASW) aircraft joined the Navy in 1977. A mid-life upgrade of the entire fleet of five such planes was completed in 2009 and the aircraft were re-designated as IL-38SD. The IL-38 can be deployed for surveillance, search and rescue, maritime reconnaissance and anti-submarine warfare operations. The aircraft is capable of detecting and intercepting surface vessels and submarines.

Length: 40.18m

Wingspan: 37.4m

Height: 10.17m

Weight: 35,500 kg (empty); 66,000 (max at take-off)

Propulsion: 4 turboprops

Speed: Mach 0.52 (645 km/h)

Range: 7,500 km

Armament: AShM; Torpedoes; Free-fall bomb

Crew: 7-8

TU-142M: The aircraft acquired from Russia were commissioned into service 1988. Eight such planes are in service and have undergone

an upgrade with the last upgraded one delivered in 2014. The aircraft have been used for long range reconnaissance, ASW and ESM/ EW missions in the Indian Navy.

Length: 49.5m

Wingspan: 50m

Height: 12.12m

Weight: 185,000 kg (max at take-off)

Propulsion: 4 turboprops

Speed: Mach 0.74 (925 km/h)

Range: 12,800 km (max)

Combat Radius: 6,500 km

Armament: Kh-35 AShM; AM-23 automatic cannon; Torpedoes

Crew: 11

Trainer aircraft

Kiran HJT: The two-seat intermediate jet trainer built by Hindustan Aeronautics Limited has been in service 1973. The Navy operates 20 such jets which are used in basic training and armament training roles for naval pilots.

Length: 10.6m

Wingspan: 10.7m

Height: 3.63m

Weight: 2,965 kg (empty); 4,950 kg (max at take-off)

Propulsion: 1 turbojet

Speed: Mach 0.56 (695 km/h)

Range: 615 km

Armament: Bombs and rockets

Crew: 2

Hawk AJT: The British BAE-made fourth-generation advanced jet trainers licence-built by HAL joined the Navy in 2013. Out of 17 such aircraft to be procured, 11 have been inducted so far. The Hawk 132 is equipped with advanced navigation systems and has the capability to deploy an array of weaponry including air-to-air missiles, air-to-ground rockets, bombs and guns. It can also carry two extra fuel tanks under the wing, which extends its considerable strike range further. The jets are used for training of combat pilots in the Navy and bridging the gap between basic flying training and advanced fighter flying.

Length: 12.43m
Wingspan: 9.94m
Height: 3.98m
Weight: 4,480 kg (empty); 9,100 kg (max at take-off)
Propulsion: 1 turbofan
Speed: Mach 0.84 (1,028 km/h)
Range: 2,520 km
Armament: AIM-9 Sidewinder AAM/ ASRAAM/ A-Darter AAM; Air-to-ground rocket; Precision-guided bombs; Gun
Crew: 2

Helicopter

The Indian Navy operates the following helicopters for different missions.

Kamov-28: The submarine-hunting helicopters were acquired from Russia in the 1980s. Ten such rotorcraft are operational in the Navy to carry out ASW missions. They are deployed on the Rajput class destroyers, Talwar class frigates and also are designated to perform ASW role for aircraft carrier INS Vikramaditya. A contract has been signed with Russia for mid-life upgrade of the helicopters.

Kamov-31: The airborne early warning (AEW) helicopter is designed for radar surveillance of air and water spaces, early detection of surface ships and low-flying targets at long distances. Russia delivered nine Ka-31s to India in 2004 followed by five more in 2013. The rotorcraft is operating from the Navy's aircraft carrier, destroyers and frigates.

Sea King Mk 42: The helicopters manufactured by AgustaWestland, UK, were

acquired in the 1970s. The advanced, multi-role helicopters have been deployed on aircraft carrier, destroyers and frigates of the Navy. The rotorcraft is used for anti-submarine warfare missions and also for search & rescue operations.

UH-3H Sea King: The American-built ASW/utility/transport helicopters are part of the Indian Navy and some 30 of them remain in service at present. The versatile UH-3Hs can be used for troop carriage, search & rescue, casualty evacuation cargo operations, para-jumping and combat free-fall of special forces among other roles.

ALH Dhruv: The advanced light helicopter, indigenously designed and developed by HAL, joined Indian Navy in 2002. The first ALH squadron was commissioned in 2013 at INS Garuda, Kochi. The ALH is being employed for a variety of missions including advanced search and rescue, special heli-borne operations, armed patrol, sniper operations and VVIP carriage.

Chetak: The light utility helicopter is a licence-built variant of the French-made Aerospatiale Alouette III helicopter. Carried by both combat and non-combat ships, the Chetaks have primarily been used for training, transport, casualty evacuation, communications and liaison roles.

UAV

The Navy operates two UAV squadrons consisting of the Israeli-made **Searcher Mk-II** and **Heron** drones. The units operate from Kochi, Kerala and Porbandar, Gujarat. The unmanned aerial vehicles are used for maritime reconnaissance and provide surveillance cover to high-value assets in the coastal waters. While the Searcher Mk-II is a third-generation UAV, Heron is a larger variant of Searcher. The Searcher carries a modern and sophisticated electro-optic camera and a communication intelligence payload. Heron is equipped with a maritime patrol radar. Both UAVs have long endurance capabilities, can operate before and after sunset and beam real time live pictures of maritime targets.

Heron UAV

WEAPONS

The weaponry deployed in the maritime platforms of Indian Navy include ballistic and cruise missiles for different roles, torpedoes and naval guns. While most of the weapons are of Soviet-origin, induction of new-generation advanced weapons, including indigenous ones in recent times, have given a major fillip to the Navy's warfare capability.

Missiles

The Navy uses a wide range of advanced ballistic and cruise missile systems for multifarious missions. The ballistic weapons include the **K-4/ K-15** SLBM and **Dhanush** SSM. The cruise missiles used for anti-ship/ land-attack operations include the supersonic **BRAHMOS** system, **Klub**, **Kh-35**, **P-15 Termit** and **Harpoon**. For air-to-air operations, the Navy uses the **R-73/ R-77** AAMs, while for air defence roles, it has deployed SAMs such as the Israeli-made **Barak** and Russian-made **Shtil, 9K 22 Tunguska, 9K33 Osa, S-125 Neva/Pechora, 9K32 Strela-2** (MANPAD) and **9K38 Igla** (MANPAD) systems.

K-4/ K-15 SLBM: While the K-15 'Sagarika' has been developed as a medium-range nuclear-capable ballistic missile, the **K-4** is an advanced variant of the K-15. The SLBMs have been designed and developed by DRDO to equip India's indigenously-built nuclear submarine INS Arihant and the follow-on

nuclear submarines of the Navy. The canisterised naval weapons have a range of over 700-km and 3,500-km respectively. Once deployed, the SLBMs will form India's nuclear triad and provide a solid second-strike weapon capability. Developmental trials of the weapon are on.

Dhanush: The indigenously-built short-range ballistic missile has both anti-ship and land attack capabilities. The SSM is capable of carrying a conventional or a nuclear warhead. The first successful test firing of the weapon took place in September 2001 and it has been inducted in service.

BRAHMOS: The supersonic cruise missile having both anti-ship and land-attack capability has been deployed on all frontline destroyers and frigates of the Navy since 2005. The weapon can be launched in both vertical and inclined mode from a static or moving warship. It also has salvo firing capability wherein more than one missile can be launched simultaneously from a warship towards different targets. The missile's underwater variant has also been successfully test fired in 2013. Yet another advanced variant which will have near vertical dive and surround capability is being developed and could be used as a potential "carrier killer" weapon in future.

Klub: The Russian-made missile has been developed in anti-ship, land-attack, air and submarine launch variants. Indian Navy was the first customer of the ship-borne missile system. The Navy presently uses the Klub-S and Klub-N variants which are deployed on the Kilo-class (Sindhughosh) submarines and Talwar class frigates respectively. The sea-skimming missile has a range of 220-km, and it can cruise at a supersonic speed at terminal phase.

Kh-35: The Russian-made tactical missile is fitted on the Navy's Delhi-class destroyers, Brahmaputra-class frigates and Kora-class corvettes. Another variant, the Kh-35E, is an air-launched missile having an extended range of 260-km is deployed on the MiG-29K naval fighter. The anti-ship variant of the weapon also arms the TU-142M maritime reconnaissance/ASW aircraft of Indian Navy.

P-15 Termit: The Soviet-origin short range anti-ship missiles are presently arming few of the Navy's surface ships. The Navy uses the P-20 variant of the weapon having an improved guidance system. The missile has a range of 80-km and is primarily being used as a coastal defence system.

Harpoon: The US-built anti-ship missiles are being procured to arm the Shishumar-class submarines and P-8I maritime patrol/ASW aircraft. The new Block-II variant of the subsonic weapon, which will be acquired by India, has a range of around 250-km.

R-73: The short-range air-to-air missile has been acquired from Russia. India's indigenously developed Tejas light combat aircraft's naval variant will carry the weapon on-board after being inducted into service.

R-77: The Russian-made medium-range active radar homing AAM are deployed on the MiG-29K carrier-borne fighter. The fighter deploys the advanced RVV-AE variant of the weapon having an estimated range of 80-km. India is the first export customer of the missile.

Barak: The Israeli **Barak-1** short range air defence missile system has been deployed on over 10 Indian Navy warships. The weapon is designed to be used as a ship-borne point-defence missile system against aircraft, anti-ship missiles, and UAVs. The **Barak-8**, another advanced variant jointly developed by Israel and DRDO, is designated as Long-Range Surface-to-Air (LR-SAM) system. It is slated to be deployed on the Navy's combat ships.

Shtil: The radar-homing guidance missile having a range of 25-km is used to protect the warships against cruise missiles, helicopters, ballistic missiles and even shore and sea-based targets. The Russian-made weapon has been deployed on board the three Russian-built Talwar class frigates and their follow-on class, the three Teg class frigates besides the Shivalik class frigates.

9K38 Igla: The Soviet-origin MANPADs are arming some of the corvettes, FACs and support ships of Indian Navy.

Torpedoes

The Navy has been using a wide range of ASW torpedoes, including the Italian **A244-S** lightweight torpedo, Russian **APR-3E** light acoustic homing torpedo and the German **AEG SUT-66 Mod-1** wire-guided, active/passive homing torpedo. Efforts have been made by DRDO's Naval Science and Technology Laboratory to produce indigenous torpedoes such as the advanced light torpedo **Shyena** and heavyweight ASW torpedo **Varunastra**.

A244-S: The lightweight, fire-and-forget torpedo is employed for anti-submarine warfare. Launched from surface vessels, it can locate the target by means of an acoustic seeker. The weapon is deployed on Brahmaputra class frigates.

APR-3E: The airborne light ASW acoustic homing torpedo is designed to engage submarines. It is designed to be carried by various fixed wing and rotary wing platforms, including the Tu-142 ASW aircraft, Il-38 MPA/ASW aircraft and Ka-28 helicopter.

AEG SUT-66 Mod-1: The wire-guided, active/passive homing torpedoes are arming the Navy's German-built Shishumar class SSKs.

Shyena: The first indigenously developed 324mm torpedo of India has been inducted in the Navy in 2012. The lightweight ASW torpedo with passive/active homing is designed to engage submarines with a speed of 33 knots.

Varunastra: The indigenously-built advanced heavyweight ASW torpedo joined the Navy in June 2016. The electric torpedo is capable of targeting quiet and stealthy submarines, both in deep and littoral waters in intense counter-measure environment. The 533-mm weapon can be launched from Kolkata, Delhi, Teg, Talwar and Kamorta classes of warships.

Naval Guns

AK-100: The Russian-built 100mm naval cannon are installed on several Indian naval ships. The weapon can fire at a maximum rate of 60 rounds per minute.

Oto Melara 76mm: The Italian-built naval artillery has been designed to be used for various roles such as short-range anti-missile point defence, anti-aircraft, anti-surface, and ground support system. The gun is equipping small warships like frigates and FACs of the Indian Navy.

Bofors 40mm: The Swedish-built naval anti-aircraft gun are arming few of the Navy's offshore patrol vessels, amphibious ships and support vessels.

AK-630 30mm CIWS: The Soviet-origin fully automatic Gatling gun designed to provide defence against anti-ship missiles and precision guided weapons, fixed or rotary wing aircraft, ships and other small craft are installed on the Navy's all frontline destroyers, frigates and corvettes. They are also deployed on aircraft carrier INS Vikramaditya and also on the futuristic INS Vikrant. The six-barrelled CIWS can fire at a maximum rate of 5000 rounds per minute.

Kashtan M CIWS: The modern air defence gun/missile system is designed to provide self-defence against air-launched precision guided weapons such as anti-ship missiles and aircraft/ helicopters. It can also engage small-sized sea targets. Talwar class frigates are armed with the Russian-built CIWS.

Phalanx CIWS: The weapon is deployed on INS Jalashwa LPD of Indian Navy. The US-made radar-guided Gatling gun is designed to defeat anti-ship missiles and other close-in air and surface threats.

CONCLUSION

The multifarious wars with its two of its immediate neighbours in the past, a volatile subcontinent and the swiftly changing geostrategic landscape in Asia and the world has led India to strengthen its overall defence capabilities and stand out prominently in the global stage in recent times, even though the country has never projected itself as a threatening, expansionist and belligerent power either in its immediate neighbourhood or at the global stage. While sincerely toeing the idea of peaceful coexistence with its neighbours, India has at the same time staunchly dealt with any kind of transgression against its sovereignty and territorial integrity. Since independence, India, while maintaining its own strategic autonomy, has pursued a non-aligned policy and propagated a multi-polar world order. With its rising economic clout, it has been making all-out efforts to maintain regional peace, balance and stability while driving growth, development and prosperity at the home front. The adoption of "Look East" policy by India in 1991 has further strengthened its cause of greater engagement, both economically and strategically, with its East Asian and South East Asian neighbours. The country has also expanded its horizons to reach out to the world community to secure its national interests as well as for greater global integration.

CONTACT DETAILS

Ministry of Defence
South Block
New Delhi-110011
India
Website: www.mod.nic.in
Army: http://indianarmy.nic.in/
Navy: http://indiannavy.nic.in/
IAF: http://indianairforce.nic.in/

INDONESIA
(Capital: Jakarta)

INTRODUCTION

Area: 1,904,569 sq km
Population: 258,316,051 (July 2016 est.)
Coastline: 54,716 km
Maritime claims: Territorial sea: 12 nm
Exclusive Economic Zone: 200 nm

KEY POLITICAL PERSONS

PRESIDENT: Joko Widodo

VICE PRESIDENT: Muhammad Jusuf Kalla

KEY DEFENCE PERSONS

MINISTER OF DEFENCE: Ryamizard Ryacudu

CHIEF OF ARMED FORCES: Gen. Gatot Nurmantyo

CHIEF OF ARMY: Gen. Mulyono

CHIEF OF NAVY: Admiral Ade Supandi

CHIEF OF MARINE CORPS: Maj. Gen. RM Trusono

CHIEF OF AIR FORCE: Marshal Agus Supriatna

ECONOMY

Indonesia has a mixed economy in which both the private sector and government play significant roles. The country is a member of G-20 major economies. Its economy is the largest in Southeast Asia and is one of the emerging market economies of the world. The government plays a significant role in its market economy by owning more than 164 enterprises and administers prices on several basic goods, including fuel, rice, and electricity. During the global financial crisis, Indonesia outperformed its regional neighbours and joined China and India as the only G20 members posting growth in 2009. Fitch and Moody's upgraded Indonesia's credit rating to investment grade in December 2011. Indonesia still struggles with poverty and unemployment, inadequate infrastructure, corruption, a complex regulatory environment, and unequal resource distribution among regions. Indonesia was the 25th biggest exporting country in the world in 2014, moving up fifth places from the previous five years. It is the 16th largest economy in the world by nominal GDP and is the 8th largest in terms of GDP (PPP).

GDP: $2.842 trillion (2015 est.), $2.712 trillion (2014 est.)

Real Growth Rate: 4.8% (2015 est.), 5% (2014 est.)

Industries: Petroleum and natural gas, textiles, apparel, footwear, mining, cement, chemical fertilizers, plywood, rubber, food, tourism

Oil - proved reserves: 3.693 billion bbl (1 January 2016 est.)

Natural gas - proved reserves: 2.875 trillion cu m (1 January 2016 est.)

Total Exports: $148.4 billion (2015 est.), $175.3 billion (2014 est.)

Export Commodities: Oil and gas, electrical appliances, plywood, textiles, rubber

Major Markets: Japan 12%, US 10.8%, China 10%, Singapore 8.4%, India 7.8%, South Korea 5.1%, Malaysia 5.1% (2015)

Total Imports: $135.1 billion (2015 est.), $168.3 billion (2014 est.)

Import Commodities: Machinery and equipment, chemicals, fuels, foodstuffs

Major Suppliers: China 20.6%, Singapore 12.6%, Japan 9.3%, Malaysia 6%, South Korea 5.9%, Thailand 5.7%, US 5.3% (2015)

INTERNATIONAL DISPUTES

✦ Three stretches of land borders with Timor-Leste have yet to be delimited, two of which are in the Oecussi exclave area, and no maritime

✦ Indonesia challenges Australia's claim to Ashmore Reef. Australia has closed parts of the Ashmore and Cartier Reserve to Indonesian traditional fishing and placed restrictions on certain catches.

✦ Land and maritime negotiations are continuing with Malaysia and disputed areas include the controversial Tanjung Datu and Camar Wulan border area in Borneo and the maritime boundary in the Ambalat oil block in the Celebes Sea.

✦ Indonesia and Singapore continue to work on finalizing their 1973 maritime boundary agreement by defining unresolved areas north of Indonesia's Batam Island.

✦ Indonesia and Vietnam are continuing the Exclusive Economic Zone negotiations. In 2011, the two countries agreed to work together to reduce illegal fishing along their maritime boundary.

DEFENCE

ARMY

Personnel: 428,517

Equipment

Category	Name	In Service
MBT	Leopard 2	100+
	AMX-13	125
	Scorpion	80
APC/IFV	AMX-VCI MICV	78
	Ferret	55
	Commando Scout	28
	AMX-10P	50
	Marder	
	M113	200+
	Stromer	20
	Saracen	60
	Pindad Badak	
	Pindad Anoa	
	BTR-40	130
	VAB	50
	VBL	18
	APS-3 ANOA	150
	P2 protected carrier	
Artillery	FH 2000 155mm	5+
	Astros II 300mm	
	M109 SP	
	105mm	40
	105LG 105mm	20
	122mm	
	76mm gun/howitzer	50
Mortar	81 mm	
	120mm	
MRL	BM14 140mm	24
RCL	90mm	
	106mm	
SAM	Rapier/Blindfire	24
	Mistral	
	Starstreak	
	RBS-70	
AA Gun	23mm ZU-23-2	
	35mm Oerlikon	
	57mm	

Army Aviation

Category	Name	In Service
Aircraft	BN-2	1
	Cessna 310P	1
	Aero Commander 680	2
	Cessna 185	2
	Wilga	18
	Cessna 207	2
	CN-212-200	7
	DHC-5D Buffalo (VIP)	3
Helicopter	Mi-35	8
	AH-64	8
	AS550	4
	Bell- 205	15
	Mi-17	16
	Bell 412	31

New Procurements / Upgrades

✦ UMS Skeldar to provide Indonesian Army with new security control measures to patrol world's second longest coastline. Incorporating three F-330 fixed wing systems, which can be deployed in applications including convoy support, route surveillance, reconnaissance and direct support, the intensive training programme has been designed to provide operatives with all aspects of UAV management from ISR through to UAV piloting.

✦ The US Army awarded Lockheed Martin a USD 80.6 million production contract for the Modernised Target Acquisition Designation Sight/Pilot Night Vision Sensor (M-TADS/PNVS) on the AH-64 Apache helicopter. This Lot 9 contract includes production of eight targeting and pilotage systems and spares for the US Army and nine systems for the Indonesian Army.

✦ Elbit Systems will provide 300 integrated situational-awareness and targeting helmets for AH-64E Apache Guardian attack helicopter pilots of the government of Indonesia.

✦ The Army received up to 100 refurbished Leopard 2A6 main battle tanks (MBTs) from Germany.

✦ Orders have been placed for 22 TARANTULA 6x6 armoured cars.

✦ Renault SHERPA Light Scout 4x4 protected vehicles are on order.

✦ Orders have been placed for unspecified numbers of Polish WR40 LANGUSTA artillery rocket systems.

✦ 300 Panser Anoa 6x6 APC including the P2- armoured vehicles are on order.

✦ The US Defence Security Cooperation Agency sold 8 AH-64D APACHE Block III LONGBOW Attack Helicopters through Foreign Military Sale to the Government of Indonesia of and associated equipment, parts, training and logistical support. The estimated cost is USD 1.42 billion.

✦ Rheinmetall formally transferred its

Leopard 2A4 RI simulators to the Indonesian armed forces following a successful site acceptance test on 17 March 2016. The Leopard Gunnery Skills Trainer (LGST) and Driver Training Simulator (DTS) are systems specifically designed for training Leopard 2A4 tank crews, primarily commanders, gunners and drivers.

✦ FNSS of Turkey and PT. Pindad of Indonesia have advanced on their Joint Development Modern Medium Weight Tank (MMWT) project. The conceptual design of MMWT has been completed and reveals during Indo Defence 2016. Turkey and Indonesia initiated the project to develop MMT in 2014. The production of the first prototype has already started.

AMX-VCI during Army exercises in Sumatra. *Image Credit: Dinas Penerangan TNI-AD*

Unveiled at the Indo Defence show in Jakarta, the Modern Medium Weight Tank is a joint concept study by Turkey's FNSS and Indonesia's PT Pindad as an alternative to main battle tanks. *Image Credit: FNSS.*

NAVY

Personnel: 74,000

Equipment

Category	Name	In Service
Submarine	Nagapasa class	3
	Cakra class	2
Frigate	Martadinata class	2
	Ahmad Yani class	6
Corvette	Diponegoro class	4
	Bung Tomo class	3
	Ex-Parchim class	10+
	Fatahillah class	3
Patrol craft/Fast Attack Craft	Mandau class missile FAC	4
	Kakap class SAR	4
	Clurit class	5+
	Singa class FAC	4
	Andau class	4

Navy continued...

Category	Name	In Service
Patrol craft/Fast Attack Craft	Sibarau class	6
	Todak class missile FAC	4
	Siada class	8
	Waspada class	2
	Pandrong class	2
Mine warfare force	Pulau Rengat class	2
	Ex-Soviet T-43 class	2
	Kondor class	9
Amphibious vessel	Tanjung Dalpele LSD	1
	Makassar class LPD	4
	Teluk Semangka class LST	6
	Teluk Gelimanuk class	12
	Teluk Sirebong class	2
	LCU	12
	Coastal vessel	22
Support vessel	ARUN fleet oiler	1
	Transport/hospital ship	1
	Driver support ship	1
Anti-ship Missile	RBS-15, Exocet, C-802, Harpoon	

Naval Aviation

Category	Name	In Service
Aircraft	N24A/N22B Nomad	6 (being replaced)
	C-212	6
Transport	Aero Commander 100	4
	PA-38	5
	CN-235	5
	DHC-5D Buffalo	2
Helicopter	NBO-105C	5
	AS565 Panther	3 (8 on order)
	Bell 412	4
	WASP (ASW)	8
	NAS.332B/L Super Puma (ASW with AM.39 Exocet missile)	30

The Indonesian frigate KRI Karel Satsuit Tubin (KST 356) operates with USCG Cutter Sherman (WHEC 720) during a maneuvering event as part of the at-sea phase of exercise Cooperation Afloat Readiness and Training (CARAT). *Image Credit: US Navy*

New Procurements/Upgrades

✦ Airbus Helicopters has delivered the first three of the 11 AS565 MBe Panther helicopters to Indonesia, ahead of contract schedule, at a ceremony at Airbus Helicopters' headquarters in Marignane, in the presence of representatives from Indonesia's Ministry of Defence and the Indonesian Navy. The delivery of the remaining helicopters will run through to 2018.

✦ PT PAL has launched the first of two SIGMA 10514 Perusak Kawal Rudal (PKR) guided-missile frigates designated for the Indonesian Navy (TNI AL). PT PAL built these ships in collaboration with Damen Schelde Naval Shipbuilding (DSNS).

✦ A light frigate based on the SIGMA 10514 design has been formalised to be built with the help of DSNS technical assistance.

✦ Daewoo Shipbuilding and Marine Engineering (S Korea) will deliver three submarines each weighing 1,400 tonnes to Jakarta by the first half of 2018.

✦ The new Type 209 submarine will replace CAKRA class.

✦ Indonesia ordered 11 AS565 Helos for Navy.

✦ A fifth DIPONNEGORO class corvette is under construction in national yards.

✦ Three CN-235-200 MP aircraft are on order with an option for 3 more.

✦ A requirement is being formulated for a naval helicopter to operate from on board the DIPONEGORO class corvettes.

✦ The first of two SIGMA 10514 PKR naval frigates for the Indonesian Navy has successfully completed its sea trials as of 7 September 2016. The two vessels have been/are being been built using a collaborative modular process operating simultaneously at Damen Schelde Naval Shipbuilding (DSNS) in the Netherlands and PT PAL shipyard in Indonesia. The first Sigma 10514 is on schedule for delivery at the end of January 2017.

✦ The consortium made up of Indra and Navantia has been awarded a contract to update the combat system of the Indonesian navy's KRI-362 Malahayati corvette, in a contract worth 18 million dollars, or some 15.7 million euros, with the shipbuilder PT PAL. The two companies will partner to equip the

The first three AS565 MBe Panther helicopters have been handed over to PT Dirgantara Indonesia, who will outfit them to the customer specifications, primarily anti-submarine. Final delivery is expected for 2018. *Image Credit: Airbus*

Indonesian Navy with the latest electronic technology used by the Spanish Navy. Modernisation of the combat system includes renovation of sensors and fire control systems, integrating these via a modern combat management system.

Marines

Personnel: 15,000

Equipment

Category	Name	In Service
APC	AMX-10P	19
	AMX-10PAC 90	40
	BTR-50, BTR-80	
	PT-76 Tank	50+
	P2 Protected carrier	
	BVP-2 AIFV	32
	BMP-2, BMP-3	80
Transport	NC-212-200	3

New Procurements / Upgrades

✦ The MoD has signed a contract with Russia for an additional 37 BMP-3F amphibious infantry fighting vehicles for the Marine Corps.

AIR FORCE

Personnel: 35,850

Equipment

Category	Name	In Service
Fighter	F-16	24
	EMB 314 Super Tucano	8
	F-5E/F	6 (To be retired)
	KAI T-50	10
	Hawk 209	15
	Hawk 109	5
	Su-27 SKM	5
	Su-30 Mk2	7
	Su-30MK	2
Transport	F-28	3
	CN-235	13
	C-130B/H & KC-130B	34
	NC-212	10
	C-130-H-30	8
	L-100-30	2
	Boeing 707	1
	AS.332 Super Puma	6
Maritime Patrol	Boeing 737 surveiller	3
	C-130H-MP	2
	NC-212-200	3
	CN-235-220	1
Helicopter	NAS.332 Super Puma	16
	Bell 206	2
	MDH-500	12

Air Force continued...

Category	Name	In Service
Helicopter	EC-725	1
	NBO-105C	12
	EC-120B	11
	SA-330J	7
	AS332 Super Puma	6
Trainer	HAWK Mk.53	6
	AS-202 BRAVO	40
	SF-260W	15
	KT-1B	12
Missile	Aim-9, Aim-120, AMRAAM, MAA-1A, AA-10, AA-2, AGM-65, AS-14/17/13	

New Procurements / Upgrades

✦ Leonardo-Finmeccanica's SAGE Electronic Support Measures (ESM) system has now entered service on the Indonesian Air Force's first modified CN-235 Maritime Patrol Aircraft (MPA). The modification programme was led by PT Dirgantara Indonesia (PTDI) and Integrated Surveillance and Defence (ISD) in the US. SAGE is a digital sensor for aircraft used for intelligence, surveillance and reconnaissance (ISR) missions.

✦ European consortium Eurofighter GmbH has offered its Eurofighter Typhoon multi-role fighters to Indonesia with the promise that a final assembly facility would be set up in the Southeast Asian country if Jakarta chooses the warplane.

✦ Indonesia has signed a USD 1.3 billion deal with South Korea to jointly develop Seoul's next-generation fighter aircraft KF-X/IF-X. Under the deal signed with Korea Aerospace Industries (KAI), Indonesia's defence ministry will invest about USD 1.3 billion in the Korean Fighter Experimental (KF-X/IF-X) programme.

✦ The Air Force's plan to purchase 24 second-hand F-16C/D Block 25 fighters from US surplus is being finalised with deliveries already started.

✦ Six more Su-30 Mk2 are on order.

✦ Eight EMB-314 Super TUCANOs are ordered from Brazil.

✦ Sixteen KAI T-50 advanced trainers are on order.

✦ Four C-130H transport aircraft are being transferred from RAAF surplus.

An Indonesian Air Force C-130 at Langkawi International Airport flightline on March 16, 2015. *Image Credit: US Air Force*

✦ Nine C-29 transports are on order to replace the F-27s.

✦ Local assembly line for Super PUMA Mk III helicopters with production of planned 18 aircraft underway.

✦ Contract has been signed to supply 16 supersonic T-50 Golden Eagle trainer aircraft by Korea Aerospace Industries (KAI).

Defence Expenditure

Total defence spending: USD 8.05 billion (2016); USD 7.3 billion (2015)

Defence spending in terms of GDP: 0.89% (2015)

Domestic Suppliers: PT Pindad, PT PAL, LAPAN, IAe, PT Lundin Industry Invest, PT DAHANA, PT SRITEX, PT Sentra Surya Eka Jaya (SSE), CV Maju Mapan, PT Fista Bahari Internusa, PT Sari Bahari Malang, PT Palindo Marine Shipyard, PT Len Industri (Persero), PT CMI Teknologi, PT. Dok dan Perkapalan, (DKB) Kodja Bahari & PT. Tesco Indomaritim.

Foreign Suppliers: France, United Kingdom, Russia, United States, Germany, China, South Korea, Brazil, Spain, Canada, Switzerland, Turkey, Australia, Netherlands, Sweden & Poland

Indonesia Government and Ministry of Defence in 2016 has finalised a high defence budget in order to modernise and upgrade its military facilities and equipment. Parliament approved an increase to the defence ministry's budget this year to USD 8.3 billion, up nearly 10 percent from the earlier 2016 budget estimate

F.Y.	Budget in USD (est.)
2017	USD 8.3 billion
2016	USD 7.3 billion
2015	USD 8.05 billion

Defence Production and R&D

Indonesia is committed for the re-establishment of its local defence industry base which was once well-developed and produced a wide range of military equipment. From the recent scenario, the country has preferred local assembly of helicopters, ship building and their refurbishment for developing its own defence industrial base but as of now, a majority of defence budget is spent on procuring defence equipment from other countries. The country was planning to introduce a defence offsets policy in 2011 with an objective to strengthen its own defence industry by transfer of technology and local manufacturing of imported weapon systems.

Defence Procurement

The major defence procurement drive that Indonesia is planning in the next five years includes its Army and the Air Force in particular. While the air force has recently inducted the new Sukhoi fighters in its fleet, it plans to procure more such fighters in future. Besides, procurement of new transport aircraft, multi-role and ASW helicopters and modernisation of existing aircraft for both the army and air force are also being planned. Indonesia has also announced its plan to join South Korea in the development of the new KF-X fighters.

CONTACT DETAILS
Department of Defence and Security
Jl. Merdeka Barat 13-14 Jakarta Pusat
KOTAK POS 2005 Jakarta 10020
Indonesia
Tel.: ++62-21-366184, -374408
Email: postmaster@dephan.go.id

STRATEGIC INFORMATION

IRAN
(Capital: Tehran)

INTRODUCTION

Area: 1,648,195 sq km

Population: 82,801,633 (July 2016 est.)

Coastline: 2,440 km (Iran also borders the Caspian Sea - 740 km)

Maritime claims:

Territorial sea: 12 nm

Contiguous zone: 24 nm

Exclusive Economic Zone: Bilateral agreements or median lines in the Persian Gulf

Continental shelf: Natural prolongation

GEOPOLITICAL IMPORTANCE

Iran is the largest country in West Asia. Ethnically, the people are Persian and by religion, most are Shi'ite Muslims. Iran is a country of particular geopolitical significance because of its location. The country is bordered on the north by Armenia, Azerbaijan and Turkmenistan. It is bordered on the East by Afghanistan and Pakistan, on the South by the Persian Gulf and the Gulf of Oman, on the West by Iraq and on the Northwest by Turkey. As Iran is a littoral state on the Caspian Sea; which is an inland sea, Kazakhstan and Russia are also Iran's direct neighbours to the North. The Islamic Republic is also the reference point of the Shiite galaxy, the minority branch of the Muslim world whose members live along the coast of the Persian Gulf. This Shiite crescent is perceived as a serious threat by Saudi Arabia, a Sunni country, and by other Gulf States. Iran's clerical leadership believes that the Islamic Republic plays a central role in world affairs as the standard bearer of revolutionary Islam, the defender of the interests of Muslims throughout the world. The country is endowed with oil reserves, the third largest in the world. Iran's four strategic maritime areas are the Gulf of Oman, the Strait of Hormuz, the Persian Gulf and the Caspian Sea. The Strait of Hormuz, on the south-eastern coast of Iran, is an important route for oil exports from Iran and other Persian Gulf countries. At its narrowest point the Strait of Hormuz is 21 miles wide, yet an estimated 17 million bbl/d flowed through it in 2011.

POLITICAL OVERVIEW

Iran's current constitution was introduced following the 1979 revolution that established the Islamic Republic of Iran. The Supreme leader is the highest ranking political and religious authority in Iran. He appoints the chiefs of posts such as the commanders of the armed forces, chief judges, prosecutors as well as six of the Islamic jurists who sit on the 12-member Guardian Council. The Assembly of Experts, an 86 - member chamber monitors the highest religious leader's performance. The President is the second highest-ranking government official elected by popular vote to a four-year term. He selects the Council of Ministers - the cabinet – which must be approved by parliament. He sets the country's economic policies and oversees the management of national affairs. He also chairs the Supreme National Security Council. The Guardian Council is an unelected body of 12 clerics and Islamic jurists. The supreme leader directly appoints half of those serving on the council while the others are approved by the Iranian parliament. The council has the constitutional authority to veto parliamentary decisions and vet electoral candidates. The Parliament known as the Majlis is the national legislative body, with 290 publicly elected representatives. Elections for the parliament are held every four years and all potential candidates for the legislature must first be approved by the Guardian Council. The Revolutionary Guard was set up in the wake of the Islamic Revolution to protect the new political leaders and bodies and uphold the spirit of the revolution. The commanders of the Revolutionary Guard are selected by the supreme leader. The Expediency Council members created in 1988 wields influence through its role as national policy adviser to the supreme leader. Its members include heads of the three government branches, the clerical members of the Guardian Council and various other members appointed by the supreme leader for three-year terms. The Supreme National Security Council - the president heads the council which includes the speaker of parliament, head of judiciary, chief of the combined general staff of the armed forces, key cabinet ministers and commanders of the regular military and Revolutionary Guard. The president guides the council in executing the supreme leader's foreign policy directives. The Judiciary nominates six lay members of the Guardian Council who are usually lawyers. Public courts deal with civil and criminal offenses. Separate "revolutionary" courts try other categories of offenses such as crimes against national security or offenses that threaten the Islamic republic. A Special Clerical Court deals with crimes allegedly committed by clerics and occasionally by lay people. Decisions from the revolutionary courts or the clerical court are final and cannot be appealed.

KEY POLITICAL PERSONS

SUPREME LEADER OF ISLAMIC REVOLUTION & COMMANDER-IN-CHIEF OF ARMED FORCES: Ayatollah Sayyid Ali Hosseini Khamenei

PRESIDENT: Hassan Rouhani

ECONOMY

Iran's economy is dominated by oil and gas production, with ten per cent of the world's proven oil reserves and 15 per cent of its gas reserves. It is the world's eighteenth largest by purchasing power parity (PPP) and twenty-nine by nominal gross domestic product. The country is a member of Next Eleven because of its high development potential. Most of the country's exports are oil and gas, accounting for a majority of government revenue in 2010. However, fiscal and monetary constraints, following the expansion of international sanctions in 2012 on Iran's Central Bank and oil exports, significantly reduced Iran's oil revenue. Iran's economy contracted for the first time in two decades during both 2012 and 2013, and grew only slightly in 2014. The limited sanctions relief for Iran provided under the Joint Plan of Action of November 2013, helped to forestall the decline in the economy in 2014. Iran is ranked as an upper-middle income economy by the World Bank. After Iran had dismantled major parts of its nuclear programme, economic sanctions were lifted from the country in 2016.

GDP: \$1.371 trillion (2015 est.), \$1.371 trillion (2014 est.)

Real Growth Rate: 0% (2015 est.), 4.3% (2014 est.)

Industries: Petroleum, petrochemicals, fertilizer, caustic soda, textiles, cement and other construction materials, food processing (particularly sugar refining and vegetable oil production), ferrous and non-ferrous metal fabrication, armaments

Oil - proved reserves: 157.8 billion bbl (1 January 2016 est.)

Natural gas - proved reserves: 34.02 trillion cu m (1 January 2016 est.)

Total Exports: \$64.6 billion (2015 est.), \$86.47 billion (2014 est.)

Export Commodities: Petroleum 80%, chemical and petrochemical products, fruits and nuts, carpets, cement, ore

Major Markets: China 22.2%, India 9.9%, Turkey 8.4%, Japan 4.5% (2015)

Total Imports: \$52.42 billion (2015 est.), \$65.08 billion (2014 est.)

Import Commodities: Industrial supplies, capital goods, foodstuffs and other consumer goods, technical services

Major Suppliers: UAE 39.6%, China 22.4%, South Korea 4.7%, Turkey 4.6% (2015)

DEFENCE & SECURITY

Since the Iran–Iraq war, Iran's main policy goal has been to be as independent as possible of alliances and foreign arms supplies. Accordingly, the country has avoided military alliances, although it has reached military supply agreements with a number of countries. Iran has two parallel land forces, the regular Artesh (Army), and the Army of the Guardians of the Islamic Revolution, also known as the Pasdaran (IRGC). The regular military defends Iran's borders and maintains internal order. The Pasdaran control Iran's strategic missile and rocket forces. The country has a dedicated force to train and equip non-state actors like Hezbollah, Hamas and Shiite extremists in Iraq. Iran has given the highest priority to the procurement and development of various types of missiles and reportedly has the largest and most diverse inventory of long-range artillery rockets and ballistic missiles in the Middle East. Its acquisition of long-range missiles and development of its own liquid - and solid - fuelled missiles has also given it a strike capability. The country has declared that it is a chemical weapons power and may have a biological weapons programme. Iran is a member of International Atomic Energy Agency, Organisation for the Prohibition of Chemical Weapons, United Nations and World Trade Organization (observer status). The country is also a signatory to the Comprehensive Nuclear Test Ban Treaty, Biological Weapons Convention, Chemical Weapons Convention, International Atomic Energy Agency Safeguards Agreement, Partial Test Ban Treaty, and Treaty on the Non-Proliferation of Nuclear Weapons.

KEY DEFENCE PERSONS

MINISTER OF DEFENCE: Cdr. Brig. Gen. Hossein Dehghan

CHIEF OF STAFF OF THE ARMED FORCES: Maj. Gen. Mohammad Hossein Bagheri

COMMANDER-IN-CHIEF OF ISLAMIC REVOLUTIONARY GUARD CORPS (IRGC): Maj. Gen. Mohammad Ali Jafari

COMMANDER-IN-CHIEF OF ARMY: Maj. Gen. Atallah Salehi

COMMANDER-IN-CHIEF OF NAVY: Rear Adm. Habibollah Sayari

COMMANDER-IN-CHIEF OF AIR FORCE: Brig. Gen. Hassan Shah Safi

COMMANDER OF IRAN AIR DEFENCE FORCE (IRIADF): Brig. Gen. Farzad Esmaili

Internal Conflict

Supreme Leader Ayatollah Ali Khamenei and the clergy have been dominating power politics in the country. Islamic law is applicable particularly with regard to women. Over a period of time elections have been conducted and there has been a flip flop of relaxations returning back to strictness with regard to implementation of Islamic laws on daily life. The actual power lies in the hands of the Supreme Leader while the executive functions are performed by the President despite the rise of large numbers of liberal and moderate political leaders. The overall life in the country remains strictly governed by Islamic rules. Iran also faces internal political tensions between rival factions especially supporters of Supreme Leader and of the President. The political conflicts have two dimensions - one is the broader conflict over the direction of the country, and the other involved decisions and disagreements within the conservative ranks of people who are known as "principlists" in Iran – involving significant institutions of the Islamic republic and the relations between the executive branch and the parliament, the judiciary and executive branch. The broader conflict in Iran has been somewhat managed but not resolved through severe repression of the Green Movement, which arose during the disputed 2009 election and the imprisonment of key reformist political leaders.

External Conflict

Nuclear Programme: Iran's nuclear programme has been a matter of contention with the international community. Iran has reportedly undertaken research and experiments geared to developing a nuclear weapons capability. Highly enriched uranium can be used to make a nuclear bomb, but uranium enriched to lower levels can be used for energy purposes. The International Atomic Energy Agency (IAEA) was unable to confirm that Iran's activities are exclusively for peaceful purposes. Meanwhile the six world powers – US, Russia, France, UK, China, and Germany want Iran to curb its nuclear programme in return for the lifting of economic sanctions. A series of meetings were held for a framework deal over the nuclear programme. On 2 April 2015, the talks came to a conclusion and an agreement on a framework deal was announced in which Iran agreed to dismantling parts of its nuclear programme. In return, the P5 have agreed to lift sanctions imposed on the Islamic republic. Officials from the EU and the UN nuclear watchdog agency confirmed in January 2016 that Iran has completed the necessary steps in a deal to restrict its nuclear programme, meaning international economic sanctions are lifted. After the lifting of economic sanctions against Iran, the country has joined the UK, United States, France, Germany, China, Russia and the EU in the field of peaceful users of nuclear energy.

Saudi Arabia: Iran and Saudi Arabia have been locked in a cold war for decades. The rivalry has been fuelled by sectarian tensions - Iran has a predominantly Shiite Muslim population, while Saudi Arabia is mostly Sunni - but also centres on their respective ambitions to exercise political and economic power throughout West Asia. The sectarian tension is exacerbated by the fact that hundreds of thousands of Shi'a Muslims live in Saudi Arabia's Eastern Province - the province in which the bulk of Saudi Arabian petroleum reserves are located. Both countries are major oil & gas exporters and have clashed over energy policy. Saudi Arabia currently has no diplomatic relations with Iran since January 2016.

Afghanistan: A long-standing dispute between Iran and Afghanistan has centred on Iran's claim to a portion of the Helmand's waters. Iran has raised voice against Afghanistan's limiting flow of dammed tributaries to the Helmand River in periods of drought. Conflicting territorial claims and disputes over navigation rights between Iran and Iraq beyond the mouth of the Shatt al Arab in the Persian Gulf was also the main factor Iran–Iraq War. Iranian government is strongly against the presence of US military in Afghanistan. The trade relations between Iran and Afghanistan has increased since 2001 with over USD 800 million exports achieved in 2008. Iran is also planning to boost industrial cooperation with its neighbour in near future.

Terrorism: Iran has been accused as the most active state sponsor of terrorism - funding, providing equipment, weapons, training and giving sanctuary to terrorists for the Lebanese Hezbollah and Palestinian groups such as Hamas and Palestinian Islamic Jihad. The US State Department first designated Iran a state sponsor of terrorism in 1984 and the designation has been continuing each year. According to one US State Department report said, Iran continued its terrorist-related activity in 2014, including support for Palestinian terrorist groups in Gaza, Lebanese Hezbollah, and various groups in Iraq and throughout the Middle East. The United States has long-standing concerns over Iran's nuclear programme, sponsorship of terrorism, and human rights record. The United States and the international community have imposed comprehensive sanctions against Iran to compel Iran to engage seriously in discussions with the international community and address concerns over its nuclear programme and other issues.

Cyber Warfare: From its past bitter experience with events such as the Stuxnet cyber-worm and the 2009 Iranian election protests due to internet activism, Iran views cyberspace warfare as an effective tool to inflict serious damage on an enemy with military and technological superiority. The Iranian cyberspace system comprises a large number of cyber organisations. One central organisation with a primarily defensive orientation is the Cyber Defence Command, operating under Iran's Passive Defensive Organisation, affiliated with the General Staff of the Armed Forces. According to the Institute for National Security Studies (INSS) Iran has made significant gains in three key areas - creating a defence envelope against cyber attacks on critical infrastructures, neutralising cyber operations by opposition elements and regime opponents, keeping western ideas and content out of Iranian cyberspace. Iran is gaining prominence as the sixth member of the cyber superpower club.

Territorial Disputes: Festering dispute over three Gulf islands claimed by both Iran and United Arab Emirates. Iran and the UAE are at loggerheads over the islands of Abu Musa, Greater Tunb and Lesser Tunb, which lie in the strategic Strait of Hormuz in the Gulf. Iran took control of them in 1971 when colonial-era Britain withdrew from that part of the Gulf. The UAE claims ownership in line with an agreement signed with Britain, and has garnered support from other Gulf Arab states and its ally the United States.

THREAT PERSPECTIVE

Israel often threatened an air strike against Iran's nuclear facilities. The two nations are engaged in a series of proxy conflicts and covert operations against each other. Iran support terrorist groups such as Hamas and Hezbollah that seek the Jewish state's destruction. Although the controversial missile defence plans of the Bush administration was abandoned, in 2009 US President Barack Obama unveiled a new anti-missile system, known as the European Phased Adaptive Approach (EPAA). Iran views the NATO assets in Turkey as a threat to its missile capabilities. On July 14, 2015, the P5+1, the European Union, and Iran reached a Joint Comprehensive Plan of Action (JCPOA), aimed at preventing the Islamic republic from building a nuclear weapon in return for the lifting of sanctions that have isolated the country.

STRATEGIC RELATIONS

Russia: Iran and Russia view each other as integral to its own national security, internal stability, and territorial integrity. They are united by their joint opposition to US foreign policy in the Middle Eastern region and beyond. They are partners in supporting the Assad regime in Syria, and they have common interests in the Caspian Sea and Caucasus region. The Islamic Republic of Iran Navy has made its base at Bandar Abbas available to the Russian Navy as a friendly and secure port where Russian Navy ships can refuel, resupply, and make repairs. Russia is one of the few arms and atomic reactor manufactures willing and able to sell them to Iran. Similarly, Iran is one of the few customers for arms and atomic reactors willing and able to buy them from Russia. Russia has assisted Iran with its conventional military, including training and defensive weapons that could serve to protect against air strikes targeting Iran. Both countries also involved militarily. RSII coalition the Russian-led operation against ISIS (ISIL) involves coordinating efforts with regional governments, including those of Syria, Iran and Iraq. Iran and Russia have reached significant progress in all spheres of bilateral ties as well as in international issues.

China: China and Iran have developed a broad and deep partnership centred on China's energy needs and Iran's abundant resources as well as significant non-energy economic ties, arms sales and defence cooperation. China was a key player in reviving and developing Iran's nuclear programme in the 1990s. The nuclear facility in Isfahan is the brainchild of Chinese-Iran nuclear cooperation. The country also played a crucial role in improving Iran's ballistic missile and naval capabilities. The two nations are seeking greater cooperation on counter-terrorism as well as more conventional military cooperation. In 2016, Iran and China have signed an agreement to boost defence-military cooperation and fight terrorism.

North Korea: Pyongyang and Tehran have gradually expanded their military, economic and diplomatic cooperation. The two nations have signed an agreement on bilateral support for developing ballistic missiles. Though the two countries appear isolated internationally because of their nuclear programmes, experts said that there are differences in their nuclear perceptions. North Korea has conducted nuclear tests seeking to have strong nuclear deterrence against enemy forces that pose a threat to its regime. Iran has been developing nuclear programmes, arguing that they are purely for peaceful, civilian purposes. Much of the trade relations between Iran and North Korea have included missile sales, mainly from North Korea to Iran.

India: India and Iran have traditionally shared good ties. Relations between the two countries warmed in the 1990s when India collaborated with Iran to support the Afghan Northern Alliance against the Taliban. There are significant trade ties, particularly in crude oil imports into India and diesel exports to Iran.

Iran is the second largest supplier of crude oil to India, supplying more than 425,000 barrels of oil per day, and consequently India is one of the largest foreign investors in Iran's oil and gas industry. India-Iran enjoys economic and commercial ties covering many sectors. However, the trade relations have traditionally been buoyed by Indian import of Iranian crude oil resulting in overall trade balance in favour of Iran. The India-Iran bilateral trade during the fiscal 2014-15 was USD 13.13 billion. India and Iran is also involved in the International North-South Transport Corridor (INSTC) which is the ship, rail, and road route based multi modal transportation for moving freight between India, Russia, Iran, Europe and Central Asia.

Pakistan: Iran and Pakistan maintained their bilateral relationship in an atmosphere of Islamic brotherhood and as good neighbours, with mutual acceptability. The conflict in Afghanistan in the 1980s brought the two closer as each had an interest in the conflict. Pakistan whole-heartedly supported Iranian viewpoint on the issue of its nuclear programme and maintained that Iran has the right to develop its nuclear programme within the ambit of Non-Proliferation Treaty (NPT). Both countries are founding members of the Economic Cooperation Organization (ECO). Pakistan tries to maintain balanced relations between Saudi Arabia, US and the EU, for which Iran used to warn against and raised many concerns. The two countries have conducted joint operations against terrorists and drug traffickers in the border regions.

Afghanistan: Iran has also played an active role in Afghanistan's reconstruction since 2001. Iran was a key factor in the overthrow of the Taliban and has since helped revive Afghanistan's economy and infrastructure.

Germany: Germany and Iran have a special relationship. The relationship between the countries continued to grow even more after the 1979 Islamic revolution and the overthrow of the Shah. From 1979 to this day most Western countries that were considered US allies, including the US, were shut out of the Iranian market except for Germany. Germany remains one of Iran's biggest trading partners. As per some media reports, Germany has recently agreed to support Iran by providing USD 1.27 billion credit line to help finance a key rail project, the biggest foreign financing the country has been able to secure after the removal of sanctions.

DEFENCE CAPABILITIES

Iran's military has a powerful mix of capabilities for both regular and Islamic Revolutionary Guard Corps (IRGC) forces to defend its territory, intimidate neighbours and threaten the flow of oil and shipping through the Gulf. It has a dedicated force to train and equip non-state actors like Hezbollah, Hamas and Shiite extremists. The country has built up its strike capability by acquiring long-range missiles from North Korea and by developing its own liquid- and solid-fuelled missiles. The country has a substantial inventory of missiles capable of reaching targets throughout the region, including United States military bases and Israel, and it continues to develop more sophisticated missiles. The Iranian armed forces consist of the Islamic Republic of Iran Army, Islamic Republic of Iran Navy, Islamic Republic of Iran Air Force, and the Islamic Republic of Iran Air Defence Force. It has declared that it is a chemical weapons power, and may have a biological weapons programme. The country has also acquired the technology to produce fission nuclear weapons and has enriched uranium to levels where it is clear it can eventually produce fissile material.

ARMY: Iran has two parallel land forces with some integration at the command level: the regular Artesh (Army), and the Army of the Guardians of the Islamic Revolution, also known as the Pasdaran (IRGC). The regular army is responsible for guarding the independence and territorial integrity of the country and maintaining order. The army, which falls under the control of the Supreme Leader, must be committed to Islamic ideology. In a show of its military strength Iran has held several major defence drills in the recent past.

Personnel: 350,000

Equipment

Category	Name	In Service
MBT	Zulfiqar 1/2/3	250+
	M-60	135
	Chieftain	100
	T-62	85
	T-72S	300+
	T-72Z	225
APC/IFV	Boragh	100
	M-113A1	200
	BMP-1	200
	BTR40/50/60/152	320
	Scorpion (light)	50
	EE9 Cascavel	
	Ferret	
	Greyhound	
	Fox	
Artillery	75mm	
	85mm	2000
	M-101 105mm	
	122mm	540
	130mm towed	22
	M-107 175mm SP gun	
	M114 towed	
	M109A1 SP 155mm	200
	M115 towed	
	M110 203mm SP	

Army continued...

Category	Name	In Service
Artillery	RAAD (Thunder) 1 122mm SP	
	RAAD (Thunder) 2 155mm SP	
	GHN45	
	G5 155mm towed gun/Howitzer	50+
	FH77B	18
	HM41 155MM	
MRL	BM21	
	Hadid	64
	Fadjr-3 (240mm)	
	Dadjr-5 (333mm)	
Mortar	81mm	
	107mm	
	120mm	
RCL	5mm	
	75mm	
	106mm	
AA gun	23mm	
	35mm	
	40mm	
	57mm	1800+
	85mm towed	
	ZSU-23-4	
	ZSU-57-2 SP	
ATGW	Entac	
	SS11/12	
	Dragon	
	Tow	130
	Konkur	
Weapon	1-Hawk (SAM)	
	SA-7 (SAM)	
	Stinger (SAM)	
	Misgah-1 (SAM)	
	Shahab Thaqeb (SAM)	
	Ra'ad	
	Tor M-1 (SAM)	29
	RBS-70 (SAM)	

Army Aviation

Category	Name	In Service
Aircraft	F-14	30+
	Cessna 185	28
	Cessna 310	6
	Cessna O-2A	10
	Falcon 20	2
	Aero Commander 690	5
	PC-6B	16
Helicopter	AH-1J Twin Cobra	12
	Bell 214	

Army Aviation continued...

Category	Name	In Service
Holicoptor	Mi-17	50
	Mi-8	
	Boeing CH-47C	40
	AB2/206	84
	NAS.332 Super Puma	7

New Procurements/Upgrades

✦ Iran in August 2016 has released images of its first domestically built new long-range Bavar 373 missile defence system. The system was designed to intercept cruise missiles, drones, combat aircraft and ballistic missiles. The new Bavar 373 has Iran's first vertical launcher, using Sayad 3 missiles that were first tested in September 2014.

✦ In 2015, shortly before the conclusion of an international agreement on Iran's nuclear programme, Moscow re-authorised the delivery of the S-300 system. Iran's army said in May 2015 that it was now equipped with the S-300 system, though further parts are due over the coming months.

NAVY : Iran has two naval forces: The Islamic Republic of Iran navy or IRIN and the Islamic Revolutionary Guard Corps Navy or IRGN. The IRIN is the naval branch of Iran's Artesh, the traditional military force that existed prior to the 1979 revolution. The Iranian Revolutionary Guard Corps Navy (IRGCN) emerged after the Islamic revolution during the Iran-Irag war in the 1990s. The regular navy's inventory consists of surface ships, including frigates and corvettes, and submarines. The IRIN operates mainly in the Persian Gulf and Gulf of Oman but also as far afield as the Red Sea and the Mediterranean Sea. The Revolutionary Guards naval force has a large inventory of small fast attack craft, and specializes in asymmetric, hit-and-run tactics. Both navies maintain large arsenals of coastal defence and anti-ship cruise missiles and mines.

Major naval bases: Bandar Abbas, Busher, Shabahar, Hengham Island, Kharg Island, Khorramshar. The IRGC Naval Wing operates another five bases.

Personnel: 16,000

Equipment

Category	Name	In Service
Submarine	Taregh class (Kilo type)	3
	Ghadir class midget	16+
Frigate	Alvand class (Vosper Mk5 type)	3
	Jamaran class	2+1
Patrol Craft	ex-North Korean Chaho class	3
	ex-Chinese Hudong class FAC	10
	Kaman class FAC (Combattante II type)	8

Navy continued...

Category	Name	In Service
Patrol Craft	ex-Iraqi Osa II type missile FAC	1
	ex-Iraqi Bogomol type missile FAC	1
	ex-Chinese Hegu type FAC	10
	ex-USCG patrol boat (Keywan)	1
	Parvin class (PGM71)	3
	Armed Launches	20
	22t	100
	Tondar class hovercraft	2
Mine warfare force	ex-US Bluebird class minesweeper (Schahrokh)	1
	ex-US Cape class	1
	Harishi class	1
Amphibious force	Hengham class LSL	4
	Iran AJR class LST	2
	Iran Hormuz class LST	3
	Rotork craft type	6

Naval Aviation

Category	Name	In Service
Aircraft	F-27	4
	Strike Commander	6
	Falcon 20E	4
	P-3F Orion	2
Helicopter	AB205	15
	AB206	10
	RH53D	4
	SH3D Sea King	10
	AB.212	6
	M-17	5

New Procurements/Upgrades

✦ The Navy has taken delivery of a destroyer named Damavand which will be used in the Caspian Sea. The ship is 90 metres long, weighs 1,300 tonnes and is armed with cruise missile systems, torpedos and electronic warfare equipment.

✦ The Navy is also developing a new destroyer 'Sahand' the third destroyer produced as part of the Mowj (Wave) project.

✦ The Fateh (Conqueror) submarine, the country's newest semi-heavy submarine which weighs nearly 500 tonnes, is undergoing trials.

✦ A new indigenously designed and built diesel-electric submarine 'Besat' class is currently under construction.

✦ The second Jamaran-class vessel that Iran has built domestically is undergoing trials.

✦ Construction of the Ghadir class of midget submarine continues.

✦ The Kayvan and Parvin class patrol boats have been converted to carry anti-ship missiles.

✦ Anti-ship missiles – Nasr 1, Noor, Mehrab and Ghader - are in production.

A starboard quarter view of the Iran Navy's second Russian-built Kilo class diesel-powered attack submarine en route to Iran.
Image Credit: US Defenceimagery.mil

AIR FORCE: The Islamic Republic of Iranian Air Force (IRIAF) came into being in the early 1980s when the former Imperial Iranian Air Force was renamed. The air force is organised on geographical lines and has three regional commands, namely the Western Area Command (WAC), the Southern Area Command (SAC) and the Eastern Area Command (EAC).

Major air bases: Ahwaz, Bushire, Chahbahar, Dezful, Doshen-Tappeh, Galeh-Marghi, Hamadan, Isfahan, Mashad, Mehrabad, Shiraz, Tabriz, Zahidan, Abu Musa.

Personnel: 30,000

Equipment

Category	Name	In Service
Fighter	MiG-29	15
	Su-24	25
	F-14A Tomcat	20+
	F-5E/F Tiger II	50
	Saegheh	12
	F-4D/E	40
	RF-4E	6
	F-6	20
	Chengdu F-7	15+
	SU-25	13
Transport	Boeing 747	9
	Boeing 707	14
	C-130H Hercules	15
	F-27	13
	Falcon 20	2
	Aero Commander	2
	DHC-2	7
	Y-7	04
AEW	IL-76 Mainstay	2
	DA-20A Falcon	1-3
Maritime patrol	P-3F Orion	1
Trainer	MF1-17 Mushshak	23
	F-33A	45
	U-22	49

Air Force continued...

Category	Name	In Service
Trainer	EMB.312 Tucano	
	PC-7	35
	TB21	6
	TB200	6
Helicopter	AB-205	6
	AB-206	72
	HH-34F	10
	AB-212	5
	CH-47C	24
	Bell 214C	22
	S-61	2
SAM	Rapier	200 units
	Hawk	
	Sayyad	
	Sayyad-2	
	Shahab Thaqeb	
	Mersad	

New Procurements/Upgrades

- ✦ Iran has unveiled the second generation of its indigenously developed Saeghe (Thunderbolt) combat aircraft.
- ✦ 14 Xian Y-7 transport aircraft are being delivered to the Air Force.
- ✦ The third semi-heavy Homa helicopter has carried out flight tests and is in the final stages of development.
- ✦ Iran also unveiled a new home-made helicopter named Sorena.
- ✦ Several models and types of UAVs have been officially presented like the Fotros, Sgahed, Yasser and Rad etc.
- ✦ Iran has expressed its interest in the purchase of an order of Su-30SM fighter aircraft from Russia.

AIR DEFENCE FORCE

The Air Defence Force controls all of Iran's military land-based air defence. Iran has attained self-sufficiency in producing important military equipment and systems. The country has designed and manufactured a variety of indigenous missiles, including Sayyad-2, Khalij-e-Fars (Persian Gulf), Mehrab (Altar), Ra'd (Thunder), Qader (Mighty), Nour (Light) and Zafar (Triumph).

ISLAMIC REVOLUTIONARY GUARD CORPS

The Islamic Revolutionary Guard Corps (IRGC), or Pasdaran in Persian, was formed by former Supreme Leader Ayatollah Khomeini in the aftermath of the 1979 Islamic Revolution. The IRGC is a combined arms force with its own ground forces, navy, air force, intelligence, and Special Forces. It has served not only as a territorial defence force but as an ideological army, merciless toward political opposition, and as the main tool in the regime's export of revolution. Over the years, the corps' power has increased militarily, politically, and economically. The Revolutionary Guards controls the country's strategic missile forces, mounts foreign and domestic intelligence operations, and is responsible for protecting the regime; the Guards has sole jurisdiction of patrolling the Iranian capital.

Iran's elite Revolutionary Guards said in October 2016 that they had developed a "suicide drone" capable of delivering explosives to blow up targets at sea and on land. The new drone is primarily for maritime surveillance and has not been designed to be armed with missiles. But it can carry heavy payloads of explosives for combat missions to launch suicide attacks. The Revolutionary Guards have also claimed to have produced a new attack drone, the Saegheh (Thunderbolt), by reverse-engineering a US Central Intelligence Agency RQ-170 Sentinel drone that was captured in December 2011.

DEFENCE PRODUCTION AND R&D

Iran's military industries are run by the Ministry of Defence's Defence Industrial Organisation and the Revolutionary Guard's military production authority. After the 1979 Islamic revolution and the start of the Iran–Iraq War, economic sanctions and an international arms embargo coupled with a high demand for military hardware forced Iran to rely on its domestic arms industry for repair and spare parts. The Islamic Revolutionary Guards Corps was put in charge of re-organising the domestic military industry. Under its command, Iran's military industry was dramatically expanded. Over the years, considerable investment has been made in new equipment and facilities for research, development and production. In terms of human resources, Iran has a sophisticated coterie of engineers and technicians with the capability to work on sophisticated, high-tech projects. Iran has specialized in the "reverse-engineering" of Western equipment. Iran continues to conduct research on military electronics, battlefield command and communications equipment, and integrated control systems. Key product lines include missiles, ammunition, armaments, vehicles, communications, marine, chemical and aviation. The most significant growth area has been in the missile industries. Iran has started the mass production of its domestically-manufactured anti-warship cruise missile dubbed Qadir. Several Qadir missiles were reportedly handed over to the IRGC's naval forces after it was unveiled in 2014. Other equipment being produced includes Anti-Tank Guided Weapons (ATGW) and long-range artillery rockets. Iran has also produced several domestically developed unmanned aerial vehicles that can be used for reconnaissance and combat operations.

Major Armament Producers

Domestic suppliers: Defence Industries Organisation, Iran Aviation Industries Organisation, Aerospace Industries Organisation, Iran Electronics Industries, Marine Industries Organisation

Defence Industries Organisation is a conglomerate of companies run by the Islamic Republic of Iran whose function is to provide the military of Iran with the necessary manufacturing capacity and technical abilities. It produces tanks, rockets, bombs, guns, armoured vehicles, etc.

Iran Electronics Industries is a state-owned subsidiary of Iran's Defence Industries Organisation. IEI was established as a government company in 1972 in order to develop military materials for the Iranian Military. Its products include missile electronics, radar, and electronic warfare.

Aviation Industries Organisation acts as a policy maker and coordinator to promote an indigenous Iranian aeronautical industry by providing and assisting the Iranian aircraft industries with needed technologies, knowledge and parts.

Aerospace Industries Organisation is a subsidiary of Defence Industries Organisation responsible for managing Iran's missile programme.

Marine Industries Organisation produces Ships, hovercrafts and submarines.

DEFENCE DEALS

Arms Import

Iran imported most of its weapons from the United States and Europe before the Islamic revolution. However due to economic sanctions and a weapons embargo put on Iran by the United States it was forced to rely on its domestic arms industry for weapons and spare parts. Despite the steady development of its domestic defence industry Iran continues to rely on some imported equipment from Russia, China, and North Korea. Iran acquired the Type 877 Kilo-class submarines from Russia. Iran's most important recent arms import was the Russian Tor-M1 (SA-15) medium-range SAM systems. Iran and Russia also signed a contract in 2007 for the supply of five S-300PMU-1 battalions but the contract was annulled and in 2015 the ban on the supply of S-300 systems to Tehran was lifted. According to media reports, Russia has begun delivery of the S-300 air defence missile systems to Iran. In August 2016, Iran confirmed that Russia had delivered main parts of the system, adding that the missile system would be completely delivered to Iran by the end of 2016. As per Iranian state media reports, the country has deployed the Russian-supplied S-300 surface-to-air missile defence system around its Fordow underground uranium enrichment facility.

Arms Export

Iran's defence industries claims to offer a wide variety of armaments, spanning from watercraft to armoured vehicles, from rockets to small arms ammunition, and from artillery to missiles. Due to nuclear programme, the United Nations Security Council placed sanctions on Iran forbidding it from exporting any form of weapons. Despite the sanctions, Iran reportedly sold some military equipment to countries such as Sudan, Syria, and North Korea. According to the arms-transfer database maintained by the Stockholm International Peace Research Institute (SIPRI), Iran exported at least USD 200 million worth of arms and ammunitions between 2010 and 2014. Military equipment exported by Iran includes small arms and ammunition, communications equipment, night-vision goggles, thermal scopes, high-power sniper rifles, long-range mortars, man-portable air-defence systems (MANPADS), and artillery rockets.

DEFENCE BUDGET

Total defence spending: $10.2 billion (2015)

Defence spending in terms of GDP: 2.5% (2015)

Iran's defence spending has in 2016 has increased by almost 40% to reach USD 10.3 billion, according to a draft budget proposal that went to parliament this week, putting defence expenditure growth during Hassan Rouhani's presidency. The defence budget allocated to the Revolutionary Guards in the draft budget for 2017 rose by 53%, reaching some USD 6.9 billion. According to various news reports, the government of Iran, in 2016, has allocated USD 1.5 billion to strengthen the defence of the country. The military budget for the armed forces has increased from USD 15.6 billion to USD 17.1 billion. The supreme leader of the Islamic Republic of Iran, Ayatollah Ali Khamenei, has ordered the allocation of at least 5% of the public budget to defence in order to upgrade the defence capabilities of the country by expanding its missile technology, expanding weapons production capacity and producing defence equipment in order to defend against all forms of external threats.

CONCLUSION

Iran's military is described as the Middle East's most powerful military force. Iran holds a position of immense strategic importance for the three major powers in the world today as well as a regional importance among the nations of the Persian Gulf. The country has managed to build up its military strength despite sanctions. In recent years Iran has designed and manufactured different weapons and military systems in aerospace, naval, aerial, ground, electronic and optic fields. It has large arsenal of anti-ship cruise missiles and developed a wide-range of unmanned aerial vehicles while the Navy has been increasing its presence in international waters. Having signed the nuclear deal framework agreement with world powers - the P5+1 - Iran is likely to gain economic advantage on sanctions relief, which it could use to increase its military strength and seek to play dominant role in the Middle east region. The Iranian Government makes a show of military force on Islamic Republic of Iran Army Day with parades every 18 April, often demonstrating new defence technologies. The defence technologies, many of which produced in the country, due to economic sanctions and arms embargo on Iran is the backbone of Iran's defence systems such as domestically produced tanks, armoured personnel carriers, radar systems, guided missiles, marine systems, military vessels and fighter aircraft. The Islamic Revolutionary Guards were put in charge of creating what is today known as the Iranian military industry with investments from Iran's Ministry of Defence.

Iran's strategic UAV dubbed "Fotros." *Image credit: Iran Ministry of Defence*

CONTACT DETAILS

Ministry of Defence and Logistics
E-mail: vds@isiran.com
Website: http://www.mod.ir/

Aerospace Research Institute
15th Street, PO Box 834-14665
Tehran, Iran
Tel: (+98 21) 88 36 60 30 39
Fax: (+98 21) 88 36 20 11
e-mail: info@ari.ac.ir
Web: www.ari.ac.ir

IRAQ
(Capital: Baghdad)

INTRODUCTION

Area: 438,317 sq km

Population: 38,146,025 (July 2016 est.)

Coastline: 58 km

Maritime claims: Territorial sea: 12 nm
Continental shelf: not specified

KEY POLITICAL PERSONS

PRESIDENT: Fouad Massoum

PRIME MINISTER & COMMANDER-IN-CHIEF: Haider al-Abadi

KEY DEFENCE PERSONS

DEFENCE MINISTER & CHIEF OF STAFF (IRAQI JOINT FORCES): Lt. Gen. Othman Al-Ghanimi

COMMANDER OF THE AIR FORCE: Lt. Gen. Anwar Hamad Amin

COMMANDER SPECIAL FORCES: Maj. Gen. Fadhil Jalil al-Barwari

ECONOMY

The Iraqi economy is dominated by the oil sector, which provides more than 90% of government revenue and 80% of foreign exchange earnings. The country maintained macroeconomic stability in 2013, however, the lower than expected oil revenues and increased spending pressures arising from the difficult security situation weighed on the overall fiscal performance, according to the IMF. Government revenues decreased during the second half of 2014 as global oil prices fell by more than 30%. Iraq's contracts with major oil companies have the potential to further expand oil exports and revenues, but Iraq will need to make significant upgrades to its oil processing, pipeline, and export infrastructure to enable these deals to reach their economic potential, according to experts. Unemployment remains a problem throughout the country. Reducing corruption and implementing reforms such as bank restructuring and developing the private sector would be important steps in this direction. Iraq is making efforts to enact laws and developing the institutions needed to implement economic policy. The government of Iraq is eager to attract additional foreign direct investment, but it faces a number of obstacles, including a tenuous political system and concerns about security and societal stability.

GDP: $534 billion (2015 est.); $546.9 billion (2014 est.)

Real Growth Rate: -2.4% (2015 est.); -0.4% (2014 est.)

Industries: Petroleum, chemicals, textiles, leather, construction materials, food processing, fertilizer, metal fabrication/processing

Total Exports: $54.67 billion (2015 est.); $83.98 billion (2014 est.)

Export Commodities: Crude oil 84%, crude materials excluding fuels, food and live animals

Major Markets: China 22.6%, India 21.1%, South Korea 11.2%, US 7.8%, Italy 6.7%, Greece 6% (2015)

Total Imports: $43.84 billion (2015 est.); $45.2 billion (2014 est.)

Import Commodities: Food, medicine, manufactures

Major Suppliers: Turkey 20.7%, Syria 19.6%, China 19.2%, US 4.8%, Russia 4.4% (2015)

INTERNATIONAL DISPUTES

✦ Iraq's lack of a maritime boundary with Iran prompts jurisdiction disputes beyond the mouth of the Shatt al Arab in the Persian Gulf.

✦ Turkey has expressed concern over the autonomous status of Kurds in Iraq.

DEFENCE

The Iraqi Armed Forces have been rebuilt after the 2003 invasion which toppled the Saddam Hussein regime with the help of United States. Plans are on to reach a total of two armoured and five mechanized divisions. Existing infantry divisions are being progressively reorganized and re-equipped. A new Army Aviation Command has been formed to command helicopter forces. Iraq has announced plans for complete restructuring the army.

ARMY

Personnel: 230,000

Equipment

Category	Name	In Service
MBT	T-72	77
	T-55	50
	M1A1 Abrams	146
AFV /APC	BMP-1	150
	M113	1026
	MTLB	1200
	BTR-80	115
	FUCH	20
	BTR-4	150
Mortar	120mm	(55 batteries)
	81mm	(108 platoons)
Artillery	M109A4/A5 155mm SP howitzer	50
	M198 155mm towed howitzer	128
	152 mm D-20 towed howitzer	18
SAM	Avenger	40
	Pantsir S1	
	Igla - Manpad	

New Procurements/Upgrades

✦ Additional 175 M-1 Abrams tanks are on order.

✦ 280 BMP-1 tracked IFVs are being received from German/Swedish surplus.

✦ 250 MRAP protected carriers have been donated by the US in January 2015.

✦ 200 additional M1151A1 HMMWV are on order.

✦ 40 Pantsir S1 SP air defence systems are being delivered.

✦ Negotiations are on with Russia for the purchase of BTR-82 wheeled APCs.

✦ Negotiations are going on for the purchase of up to 200 M2A2 Bradley tracked IFVs from US Army surplus.

✦ The Army Aviation Command has taken delivery of an initial consignment of 128 mm air-launched rockets.

NAVY

Personnel: 4,000 (including 900 marines)

Equipment

Category	Name	In Service
Patrol craft	Al Basrah class 60m OSV	2
	Fatah-class OPV (Fincantieri Saettia Mk type)	4
	140t 'Predator' type patrol boat	5
	AL FAW-class patrol	6

Navy continued...

Category	Name	In Service
Patrol craft	PB-301	12
	'Defender'-type RIHB fast small boat	26

New Procurements/Upgrades

No major procurement plans are under consideration at present.

Marines

Two battalions of Marines Regiment have been established. The first battalion consists of 12 platoons, six of which execute point of defence of the oil platforms, while the other six are trained to conduct vessel board, search and seizure (VBSS operations). The second battalion also consists of 12 platoons tasked with port security and other ground security duties.

AIR FORCE

Personnel: 6,000

Equipment

Category	Name	In Service
Aircraft	Su-25	5
	F-16	4+
	Sama CH2000	8
	SB7L-360 Seeker	6
	King Air 350	6 +1
Trainer	Lasta 95	20
	Cessna AC-208B Caravan	13
	T-6A Texan II	15
Helicopter	UH-1HP Huey II	16
	Mi-171E	22
	Mi-35	40+
	Mi-28NE	15+
	Bell 206-B3	10
Transport	C-130E	3
	An-32	6
	C-130J	6
UAV	Scan Eagles	10

New Procurements/Upgrades

✦ The Air force has taken delivery of batch of four US F-16 fighter aircraft ordered in 2011 and 2012.

✦ An order has been placed with Pakistan to buy MFI-395 Super Mushshak trainer aircraft.

✦ A deal has been signed to buy 24 T-50IQ multi-role fighters from South Korea's KAI.

✦ Iraq has requested US to supply 12 Bell 412 EP helicopters.

✦ Eurocopter has been awarded a contract for 24 EC-635 helicopters.

✦ The US Defence Department has cleared the sale of two more Cessna AC-208 attack aircraft.

Cessna AC-208 aircraft of Iraqi Air Force *An internet photo*

Defence Expenditure

Total defence expenditure: $5.4 billion (2014)

Estimated defence spending in terms of GDP: 9%

Defence Production and R&D

Iraq has made considerable efforts in armoured fighting vehicle development, combat aircraft conversion and repair, aerial weapon production, artillery system and rocket development and small arms production. It has signed a deal with the Czech Republic to build an Iraq-based tank repair and upgrade facility for its existing T-72 main battle tanks.

Defence Procurement

Iraq seeks to achieve 'strategic independence' between 2016 and 2020. The military aim is to be able to defend its international borders without external support. It is keen to diversify the purchase of defence equipment away from excessive reliance on the United States. Current procurement plans include Main Battle Tanks. Iraq procured 140 upgraded US M1A1 Abrams tanks in an 815-million-dollar deal. Between 2015 and 2025, a total of eleven divisions are likely to be fully mechanised, requiring Iraq to procure and operate a further 1,800 main battle tanks, giving it a total of around 2,200. Ukraine is delivering at least 156 BTR-4E wheeled armoured personnel carriers upgraded with US communications equipment to equip two divisions, while Bulgaria is pursuing a sale of five hundred MT-LB armoured personnel carriers. Iraq has signed a 2.3-billion-dollar deal for 42 Pantsir-S1 missile systems, enough to equip seven army divisions.

CONTACT DETAILS

Ministry of Defence

Bab Al-Muadam, Baghdad, Iraq

Tel: (+964 1) 888 90 71

Web: https://www.mod.mil.iq/

STRATEGIC INFORMATION

ISRAEL

(Capital: Jerusalem)

INTRODUCTION

Area: 20,770 sq km

Population: 8,174,527 (July 2016 est.)

Coastline: 273 km

Maritime claims: Territorial sea: 12 nm

Continental shelf: to depth of exploitation

GEOPOLITICAL IMPORTANCE

Israel - the only Jewish majority state - became a nation on 16 May1948. Israel is positioned between three continents or sub-continents, Africa, the Arab Peninsula (sub-continent) and Europe. The country is bordered by Egypt on the West, Syria and Jordan on the East, and Lebanon on the North. Its maritime plain is extremely fertile. The southern Negev region, which comprises almost half the total area, is largely a desert. The Jordan River flows from the North through Lake Hule (Waters of Merom) and Lake Kinneret (also called Sea of Galilee or Sea of Tiberias), finally entering the Dead Sea 1,349 ft (411 m) below sea level. Israel contains the occupied areas of Jerusalem from which it has not withdrawn after 1967 war. Currently Jerusalem is the capital and borders Bethlehem which is located in Palestine.

POLITICAL OVERVIEW

Israel has a parliamentary democracy system with partial separation of powers between the judicial branch on the one hand, and the legislative and executive branches on the other. The President is elected by the Knesset (Parliament) to a seven-year term. The 120 members of the Knesset are elected in nation-wide elections from party lists. The prime minister appoints a cabinet with the Knesset's approval. Israel does not have a constitution, but it does have a number of basic laws that govern how the government is structured, rights of individuals and similar matters. Scheduled general elections are held regularly four years after the last elections. In addition, a government may fall either by a motion of the government dissolving the Knesset, or on a no confidence motion of parties in the Knesset. To limit the number of such motions and stabilise the government, a no confidence motion requires 61 votes to be carried out, and the initiators must name a candidate whom they will back to form a government. The major political parties are - Kadima, Likud, Israel Beitenu, Labour, Shas, United Torah Judaism (UTJ), Meretz, National Union, Jewish Home, United Arab List, Hadash, Balad.

The Judicial branch is an independent branch of the government. The judicial structure consists of three main types of courts - civil, religious, and military. There are special courts for labour, insurance, traffic, municipal, juvenile, and other disputes. Judges for all courts are appointed by the Judicial Selection Committee. The committee is composed of nine members - two cabinet members (one being the Minister of Justice), two Knesset members, two members of the Israel Bar Association, and three Supreme Court justices (one being the President of the Supreme Court). The committee is chaired by the Minister of Justice.

KEY POLITICAL PERSONS

PRESIDENT: Reuben Rivlin

PRIME MINISTER: Benjamin Netanyahu

KEY DEFENCE PERSONS

DEFENCE MINISTER: Avigdor Lieberman

CHIEF OF GENERAL STAFF: Lt. Gen. Gadi Eizenkot

COMMANDER OF NAVY: Maj. Gen. Eli Sharvit

COMMANDER OF AIR FORCE: Maj. Gen. Amir Eshel

ECONOMY

Israel ranks high among the world's economies in terms of its technological readiness, venture capital availability, and the quality of its research organisations. A number of internationally recognised Israeli companies have been bought by top business conglomerates in multi-billion dollar transactions. The number of new start-ups is very high due to the extraordinary innovative talent in Israel, coupled with the availability of highly skilled manpower. Israel is a leading world diamond manufacturing and trading centre. In 2008 diamond exports amounted to $9.4 billion. The Defence industry in Israel is also a strategically important sector and a large employer within the country. It is a major player in the global arms market and is the 11th largest arms exporter in the world as of 2012. There are over 150 active defence companies based in the country. Much of the exports are sold to the United States and Europe. Israel is considered to be the leading UAV exporter in the world. According to the Stockholm International Peace Research Institute, Israeli defence companies were behind 41% of all drones exported in 2001-2011. In 2013, Israel's economy grew by 3.3%, the lowest rate in ten years, according to Globes, Israel's business news, with 25% of the growth attributed to the discovery of natural gas - the Tamar and Leviathan fields. Reduced investment due to uncertainties caused by the Gaza conflict in summer 2014 has reduced GDP growth to about 2% during 2014.

GDP: $281.9 billion (2015 est.), $274.8 billion (2014 est.)

Real Growth Rate: 2.6% (2015 est.), 2.6% (2014 est.)

Industries: High-technology products (including aviation, communications, computer-aided design and manufactures, medical electronics, fibre optics), wood and paper products, potash and phosphates, food processing, beverages, and tobacco, caustic soda, cement, construction, metals products, chemical products, plastics, diamond cutting, textiles, footwear

Oil - proved reserves: 13.95 million bbl (1 January 2016 est.)

Natural gas - proved reserves: 199 billion cu m (1 January 2016 est.)

Total Exports: $56.29 billion (2015 est.), $63.76 billion (2014 est.)

Export Commodities: Machinery and equipment, software, cut diamonds, agricultural products, chemicals, textiles and apparel

Major Markets: US 27.5%, Hong Kong 8%, UK 6.1%, China 4.9% (2015)

Total Imports: $59.49 billion (2015 est.), $70.36 billion (2014 est.)

Import Commodities: Raw materials, military equipment, investment goods, rough diamonds, fuels, grain, and consumer goods

Major Suppliers: US 13%, China 9.3%, Switzerland 7.1%, Germany 6.1%, Belgium 5.3%, Italy 4% (2015)

DEFENCE & SECURITY

Israel's military priority is to ensure the existence of Israel and the security of its citizens. Its national security strategy is founded on the premise that Israel cannot afford to lose a single war. Israeli strategy begins with the maintenance of a credible deterrent posture, which includes the willingness to carry out pre-emptive strikes. Should deterrence fail, Israel would seek to prevent escalation, and determine the outcome of war quickly and decisively. Israel must prevent the enemy from entering its territory, and must try to quickly transfer the battle to enemy territory.

The state's security doctrine was laid down by the first Prime Minister, David Ben-Gurion, in the 1950s. The doctrine was founded based on three principles - deterrence, early warning and military superiority. In 2006, a committee recommended adding a fourth concept to Israel's security concept, that of defence. The document proposes the addition of four new components to the security doctrine - Prevention and Pre-emption, Alliance with the United States, Regional alliances and "Adaptation." Prevention and pre-emption refer to a policy that Israel has already adopted, using all the means at its disposal in order to pre-empt threats of non-conventional weapon use against it. The document emphasized the importance of the special relationship between Israel and the US as a central component of Israel's security doctrine.

However, in recent years Israel has adopted a new military doctrine relying on high tech defence, quick reaction to counter terrorism and special operations forces. This change in Israeli strategic military doctrine is expected to result in a major reduction in force for Army, Navy and Air force officers and units. According to military analysts, the overhaul would focus on countering threats from guerrilla armies with rockets embedded in civilian areas, such as Hezbollah and Hamas. Instability in Egypt and Syria has prompted Israel to build up forces against cross-border terrorist attacks from small militias. Israel will also focus on cyber warfare and confronting Iran which it accuses of seeking to build a nuclear weapon.

External Conflict

Palestinian-Israeli conflict: The main sticking points preventing peace between Israelis and Palestinians is that Palestinians want Israel to withdraw to pre-1967 borders. The West Bank and the Gaza Strip are Israeli-occupied with current status subject to the Israeli-Palestinian Interim Agreement. The Palestinians in the West Bank and eastern Jerusalem have lived under Israeli occupation since 1967. The settlements that Israel has built in the West Bank are home to nearly 500,000 people and are deemed to be illegal under international law, although Israel disputes this. The West Bank has enormous strategic importance to any country wishing to invade Israel. Israel therefore insists on guarantees that the Palestinian state would not allow a foreign army to enter its borders. Israel evacuated its settlers from the Gaza Strip in 2005 and withdrew its forces, ending almost four decades of military occupation. However, after the militant Islamic group Hamas seized control of Gaza in June 2007, Israel intensified its economic blockade of the Strip. At the end of 2008 it launched a major military assault on Gaza to halt cross-border rocket attacks. Peace process began with the Palestinians after years of an uprising known as the *Intifada*. Despite the handover of Gaza and parts of the West Bank to Palestinian control, a final agreement is yet to be reached. The latest round of peace negotiations began in July 2013 and was suspended in 2014. In recent times, there has been an escalation of violence between Israel and Palestinians. In view of the deadly surge in violence, world leaders have increased their calls to resume the peace process.

Arab-Israeli conflict: The Arab-Israeli conflict began as a struggle over land known internationally as Palestine. It is a small area approximately 10,000 square miles. The land was divided into - the State of Israel, the West Bank and the Gaza Strip - following the Arab Israeli war of 1948. The nature of the conflict has shifted over the years from the large scale regional Arab-Israeli conflict to a more local Israeli–Palestinian conflict. Peace agreements were signed between Israel and Egypt in 1979, and Israel and Jordan in 1994. A ceasefire currently stands between Israel and Syria, as well as with Lebanon (since 2006). The conflict between Israel and Hamas-ruled Gaza, which resulted in the 2009 ceasefire is also included as part of the Israeli-Palestinian conflict and hence the Arab-Israeli conflict.

Golan Heights: Israel unilaterally annexed the Golan Heights in 1981. Lebanon claims the Shab'a Farms area of Golan Heights and Syria claims the entire region. Israel seized the Golan Heights from Syria during the Six-Day War. Israel also annexed Shebaa Farms as part of the Golan Heights, a move not recognized by the international community. Syria tried to retake the Golan Heights during the 1973 Middle East war. Both countries signed an armistice in 1974 and a UN observer force has been in place on the ceasefire line since 1974. Syria wants to secure the return of the Golan Heights as part of any peace deal. However, public opinion in Israel does not favour withdrawal.

THREAT PERSPECTIVE

Geographically, Israel is a tiny state. The state has been plagued by the threat of both low-intensity conflict and full-scale war throughout its entire history. Major threats facing the country include the growing number of foreign fighters joining jihadist groups in the Middle East, cyber warfare, the Iranian nuclear and missile threat, the transforming of the Mediterranean Basin into a launch pad for terrorism in Europe, and threats to maritime and energy security. Israel is rattled by Iran's arsenal of missiles capable of striking the Jewish state, its frequent talk of Israel's annihilation, and its alliances with anti-Israel militants like Hamas in the Gaza Strip and Hezbollah based in Lebanon, who control tens of thousands of rockets. Syria reportedly funds and arms terrorist organizations like the Hamas and Hezbollah, as proxies to attack Israel. In light of increased tensions the Israeli military has increased the number of joint exercises in the recent past. The drills focus on a range of combat scenarios, including underwater missions, air-based missions and the deployment of infantry units.

STRATEGIC RELATIONS

United States: Israel and the US maintain a strategic alliance which represents a clear interest for both sides. Israel is one of the United States' two original major non-NATO allies in the Middle East. In both military and non-military spheres, Israeli technological achievements - often developed with US support - are shared with America. Israel is the largest annual recipient of US foreign assistance. Almost all US aid to Israel is now in the form of military assistance. The US-Israel bilateral relationship is strong, anchored by over $3 billion in Foreign Military Financing annually. In addition to financial support, the US participates in a high level of exchanges with Israel, to include joint military exercises, military research, and weapons development. Through the Joint Counterterrorism Group and a semi-annual Strategic Dialogue, the US and Israel have enhanced their cooperation in fighting terrorism. The United States is Israel's largest single trading partner.

India: Israel has consolidated defence ties with India into a strategic relationship. India is the largest customer of Israeli arms in the world. The first major military deal between the two countries was the sale of Israeli EL/W-2090 AEW radars. India and Israel have signed many important defence deals and also embarked on extensive space cooperation. Both the countries have held numerous joint anti-terror training exercises. India imports critical defence technologies from Israel. There are regular exchanges between the armed forces and defence personnel. There is ongoing cooperation on counter-terrorism issues, including through a Joint Working Group on Counter-Terrorism. ISRO and the Israel Space Agency revived their contacts and held bilateral meetings in 2014 and 2015. Cyber has emerged as one of the areas in which both countries have started to collaborate. Israel currently has plans for fresh joint ventures and technology transfer in developing weapons systems and ensuring implementation of Make in India initiative in the key sector.

Azerbaijan: Israel and Azerbaijan have developed strong strategic ties. Relations between the two nations have strengthened in the political, military, and technology spheres. Azerbaijan is the biggest supplier of oil to Israel while Israel sells sophisticated arms including missile systems and drones. Israel's drone planes are as much in demand in Azerbaijan. Israel also sells its Azeri partner armoured troop carriers, multiple rocket launchers, Tavor rifles, and ammunition.

Germany: Israel and Germany have significant and long-standing military cooperation. Germany has supplied Israel with Dolphin class submarines while Germany utilizes the Israeli-designed Spike Anti-Tank Missile. Israel continues to be a top recipient of German military technology. Germany and Israel are reportedly jointly developing a nuclear warning system, dubbed Operation Bluebird. They have also begun counter-terrorism cooperation in the wake of the terrorist attack at the 1972 Munich Olympics.

Myanmar: Myanmar has become one of Israel's strongest allies. Military ties between Myanmar and Israel date back to the 1950s, with Israel over the years selling military hardware and providing technology and training. Myanmar is reportedly in the process of purchasing Super Dvora patrol boats from Israel. Israel also sold Myanmar air-to-air missiles.

DEFENCE CAPABILITIES

The Israel Defence Forces (IDF) is one of the best battle trained armed force in the world. The IDF consists of the Ground Forces, Air Force and Navy. The IDF is headed by the Chief of General Staff, the Ramatkal, subordinate to the Defence Minister of Israel. Israel follows a conscript model for its armed forces. Everyone in Israel is a soldier. Israel is unique in that military service is compulsory for both men and women. The 'Nachshol' IDF's first company consisting entirely of female combat soldiers was established in 2006 whose mission is to patrol and gather field intelligence in real time.

The country's major military operations include the 1948 War of Independence, 1951-1956 Retribution operations, 1956 Sinai War, 1964 -1967 War over Water, 1967 Six- Day War, 1967-1970 War of Attrition, 1968 Battle of Karameh, 1973 Operation Spring of Youth, 1973 Yom Kippur War, 1976 Operation Entebbe, 1978 Operation Litani, 1982 Lebanon War, 1982-2000 South Lebanon conflict, 1987-1993 First Intifada, 2000-2005 Second Intifada, 2002 Operation Defensive Shield, 2006 Lebanon War, 2008-2009 Gaza War, 2012 Operation Pillar of Defence.

The IDF possesses top-of-the-line weapons and computer systems. The IDF uses several technologies developed in Israel such as the Merkava main battle tank, Achzarit armoured personnel carrier, high tech weapons systems, the Iron Dome missile defence system, Trophy active protection system for vehicles, and the Galil and Tavor assault rifles.

Israel has deployed a variant of the Samson RCWS, a remote-controlled weapons platform, which can include machine guns, grenade launchers, and anti-tank missiles on a remotely operated turret, in pillboxes. The country has also developed observation balloons equipped with sophisticated cameras and surveillance systems used to thwart terror attacks. The IDF also possesses advanced combat engineering equipment which includes the IDF Caterpillar D9 armoured bulldozer, IDF Puma CEV, Tzefa Shiryon and CARPET minefield breaching rockets, and a variety of robots and explosive devices.

Israel also possesses a sizeable arsenal of short and medium-range ballistic missiles, land, sea, and air-launched cruise missiles, and is working towards a layered and comprehensive missile defence capability. The country's missile defence system includes the Arrow system, the Iron Dome system, David's Sling. Israel has also worked with the US on development of a tactical high energy laser system against medium range rockets called Nautilus or THEL. It is believed that Jericho intercontinental ballistic missile is capable of delivering nuclear warheads.

ARMY: The IDF Ground Forces is responsible for the general organisation and structure of field forces; infantry, paratroopers, artillery, armour, engineering and field intelligence. The primary duty of the branch is the development, acquisition and maintenance of combat facilities. The branch also takes responsibility for planning and the administration of manpower, training and military exercises. As part of a larger reform of its ground forces the IDF has made a series of major organisational and structural changes in their Armoured Corps including disbanding its six reservist tank brigades. Under the new structure, every armoured battalion will consist of two instead of three active-duty tank companies; a reserve tank company equipped with the same type of tank, and would get an additional supporting infantry company. The infantry companies will be composed of three platoons that specialise in observation, reconnaissance and operating Keshet mortar carriers.

Personnel: 130,000

Equipment

Category	Name	In Service
MBT	Merkava Mk1/Mk2/Mk3/Mk4	2,500+
	M60 (Magach 5/6/7)	1,700
APC	M113 (all versions)	2500+
	Namer	100+
	Achzarit	300+
	Nagmachon & Nakpadon	
	ZE'EV	150
	David (urban warfare roles)	
	Wolf Armoured Vehicle	
Artillery	D30s (122mm)	100
	Soltam M68/71 (155mm) towed	300+
	M107 175mm (SP)	175
	L33 M-50 155mm (SP)	120
	M-109 155mm (SP)	600
	M-110 203mm (SP)	72
	81mm	900
Mortar	120mm	
	160mm	
MRL	122mm	
	M270	30
	135mm	
	240mm	
	290mm	
	MRLS	48
ATGW	TOW	

Army continued...

Category	Name	In Service
ATGW	TOW	
	MAPATS	
	SPIKE	
	TAMUZ	
AA gun	20mm missile/gun	24
	30mm & 40mm	
	FIM-92 Stinger (SAM)	500
Weapon	Machbet	400
	Pereh Guided Missile Carrier	

New Procurements/Upgrades

✦ Israeli Ministry of Defence Director General has instructed the Directorate of Production and Procurement (DOPP) to purchase hundreds of additional Trophy active protection systems (APS), manufactured by Rafael. The new order will ensure that every new Merkava 4 tank and Namer APC will be equipped with the Trophy system, providing significant additional protection for every tank and APC.

✦ The new Namer heavy APC is in series production with a planned total of 600 vehicles. By 2017 Israel plans to acquire a total of 170 Namer APCs.

✦ The Keshet 120mm mortar system is in continued procurement.

NAVY: The Israeli Navy officially known as the Sea Corps (Hel Yam) plays an important role in the IDF's warfare capabilities. The Navy is building up a fleet of sophisticated submarines and a special submarine course is being imparted to sailors to operate the navy's next generation submarines. The entire combat surface fleet is being upgraded with new German-built, Israeli-equipped Sa'ar-6 corvettes and new radars and electronic warfare (EW) systems integrated in existing Sa'ar-5 and Sa'ar 4.5 ships. The navy's modernisation plans calls for upgrading the smaller Sa'ar 4.5s with the ELM-2258 ALPHA, an advanced lightweight rotating and scanning phased array radar that will enhance performance of the platform's Barak-1 ship defence system. The new Sa'ar-6 EEZ defence platforms will feature the same EW suite and the ELM-2248 Adir, a multifunction surveillance, track and guidance radar; and the Barak-8, an anti-missile and air-defence system. INS Lahav is already operational with the new Adir radar and Barak-8. Other modernisation programs aimed at fortifying the Navy's combat surface fleet against an array of missile and air breathing threats include development of an infrared staring sensor, a new countermeasures control system and an upgrade of shipboard communications capabilities.

Major naval bases: Ashdod, Eilat, Haifa, Atlit

Personnel: 9,500

Equipment

Category	Name	In Service
Submarine	Dolphin class	5+1
Corvette	Sa'ar 5 class	3
Missile Boat	Sa'ar 4.5 class	8
Patrol Boat	Super Dvora Mk III/II	14
	Shaldag Mk I/II	7
	Dabur	
Unmanned Vehicle	Protector USV	
	Orbiter Mini	
Weapon	Harpoon (ASM)	
	Popeye	
	Gabriel	
	Barak 1 & Barak 8	
	Typhoon	
	Rafael overhead weapon station	
	Samson remote controlled weapon station	

New Procurements/Upgrades

✦ MoD has signed a contract for four Sa'ar class corvettes. New Sa'ar 72-class corvettes are under construction at Israel Shipyards.

✦ The sixth and final Dolphin class submarine is under construction with delivery scheduled for 2019.

✦ Three Super Dvora Mk3 FPBs are on order.

Naval Aviation

Category	Name	In Service
Helicopter	SA-366G (ship borne)	2
	AB-206B	6
	AS565MA (shipborne on the Eilaths)	5
	C-130 Hercules	2
	UH-1N Twin Huey	15+

Sa'ar 5 class missile corvettes of the Israeli Navy cruise off the shore of Israel during a training exercise. *Image Credit: Israel Defense Force*

AIR FORCE: The Israeli Air Force officially known as the Air Corps (Hel Avir) is the aerial arm of Israel Defence Forces. The IAF has made a series of historic structural changes to increase the force's operational capabilities. The Chief of IAF Air Group has been divided into two positions - Chief of Aerial Operations and Chief of the Air Group. The Operations Division will now come under the command of the Chief of Aerial Operations and will consist of three colonels who will oversee all operations divided into the following categories - Strike Operations, Defensive Operations and joint operations with the IDF Ground Forces.

The Air Force operates surface-to-air missile and anti-aircraft artillery units. Their major role has been interception of surface-to-surface missiles and rockets fired into Israel. The IAF has begun operating the 'Iron Dome' anti-rocket missile system. The government has also approved the establishment of a second F-35 stealth fighter's squadron. This is the first fifth generation aircraft to be integrated in the IAF. The first of the 14 stealth fighters of the second squadron are scheduled to land in Israel in 2019 and will join the 19 JSFs of the first squadron which will be operational by then. Plans are also on to develop Israel-specific configuration for 31 JSFs.

The IAF's Air Defence Division has established a founding team, marking the start of the absorption process of the David's Sling Weapon System (DSWS) also known as 'Magic Wand'. The DSWS is a surface-to-air missile system designed to intercept medium-to-long-range rockets and cruise missiles, fired at ranges from 40km to 300km. The system is developed to work alongside the Iron Dome, Patriot and Arrow missile systems as the future of Israel's aerial defence formation. The system is expected to enter operational service as early in 2016. Israel has officially established its first Lockheed Martin F-35I squadron. A second F-35I squadron is planned in the future, and the air force is already examining the possibility of standing up a third unit equipped with the fifth-generation fighter.

Major air bases: Ramat David, Sdot Micha, Hatzerim, Ovda, Hatzor, Tel Nof, Ramon, Nevatim, etc

Personnel: 30,000

Equipment

Category	Name	In Service
Fighter	F-15I	25
	F-15A/B	46
	F-15C/D	25
	F-16A/B	60
	F-16C/D	137
	F-16I	102
	F-35I	1st 2 received in Dec 2016
Tanker	KC-130H	5
	KC-707	3
AWE&C/EW	Gulfstream G550	
Transport	C-130J Super Hercules	
	Arava	10

Air Force continued...

Category	Name	In Service
Transport	C-130E/H	16
	Do28	10
	King Air B200	3
Trainer	A-4 Skyhawk	60
	M-346 Master	
	G-120	
	Beechcraft T-6 Texan II	
Helicopter	AH-64A	29
	AH-64D-1	18
	AH-1 G	30
	Bell 206	38
	S-70A-50 Peace Hawk	25
	CH53A/D	40
	UH-60A Blackhawk	34
	CH-53 Sea Stallion	
Liaison	Cessna U-206C	20
	Beech Queen Air	18
UAV	Heron	
	Eitan	
	Harpy	
	Harop	
	Hermes 450	
	Skylark	
Weapon	MIM-23 Hawk (SAM)	
	MIM-104 Patriot (SAM)	
	Arrow (ABM)	
	Shafrir missile	
	Python (AAM)	
	Popeye (ASM)	
	AGM-65 Maverick (ASM)	
	AGM-45 Shrike (ASM)	
	AGM-114 Hellfire (ASM)	
	AIM-120 AMRAAM (AAM)	
	AGM-65 Maverick (AGM)	
	AIM-9 Sidewinder (AAM)	
	MIM-72 Chaparral (SAM)	
	Delilah cruise missile	
	Iron Dome	
	David's Sling (SAM)	
	Jericho II ICBM	
	Jericho III ICBM	

New Procurements/Upgrades

✦ The IAF is receiving a total of 50 F-35 fighter aircraft within the next few years. The first two are being flown from America and are landing on December 12, 2016. The others will be shipped by sea, including a test plane which will arrive in 2019. All maintenance and testing of the plane will be done in Israel, and the planes are projected to be operational in about one year.

✦ The IAF has received an additional C-130J Super Hercules tactical transport aircraft. The new C-130J Shimshon, equipped with self-defence systems, has joined with the IAF's Elephants squadron - the first C-130J squadron in the air force to operate the advanced transport plane.

✦ The IAF's Sikorsky CH-53 life extension project has been completed. The helicopter entered the workshop of the Aerial Maintenance Unit as part of "Yas'ur 2025" project, aimed at extending the helicopters' operational lifespan to reach 2025 at the minimum.

✦ The IAF has taken delivery of the first two of 30 M-346 jet trainers ordered from Italy's Alenia Aermacchi. The aircraft will replace TA-4 aircraft. Another six aircraft are completing assembly at Venegono facility near Milan and a further five are undergoing structural part assembly.

✦ Unspecified numbers of Hermes 900 and Hermes 450 UAVs are on order.

Elbit Systems announced that it has completed the delivery of the Ground Based Training System (GBTS) center for the M-346 "Lavi" trainer aircraft. This comprehensive training solution was selected by the Israeli Air Force (IAF) and the Directorate of Production and Procurement (DoPP), to effectively and affordably qualify its pilots and Weapon Systems Operators (WSOs) to progress directly to fourth and fifth generation fighter aircraft (F-16, F-15 and F-35).

M-346 of Israeli Air Force. *Image credit: Alenia Aermacchi*

✦ IAI Heron, an Unmanned Aerial Vehicle developed by the Malat (UAV) division of Israel Aerospace Industries. Image Credit: US Air Force

AIR DEFENCE

Iron Dome short-range air defence system

Spyder short and medium range mobile air defence system

David's Sling medium-to long-range rocket SAM

Arrow anti-ballistic missile system

DEFENCE PRODUCTION AND R&D

Innovative military technologies have been viewed as strategically crucial for Israel which has been in a perpetual state of conflict with its neighbours. The country invests heavily in research and development and with the participation of the research institutes, the defence industry and the army it has produced a high-tech weapons factory.

The defence industrial sector in Israel contains around 150 firms, divided into three categories - the large state owned or government controlled defence companies such as Israel Aerospace Industries (IAI), Israeli Military Industries (IMI) and Rafael. The second group contains medium-sized firms, all in the private sector, which relies on defence production for their viability but also have large-scale civilian production particularly in the production of telecommunication equipment. The third group is small and medium sized enterprises which each produce a narrow range of products mainly geared to the defence sector. The industry has a high concentration of well-qualified scientists, engineers, and technicians.

Israel's arms industry is the acknowledged world leader in the area of unmanned aerial vehicles (UAVs), manufacturing variants for both intelligence-gathering examples - Heron, Hermes and Searcher Mk III and unmanned combat air vehicles (UCAVS) like Harop and Jaguar. Its defence industries have also manufactured weapons and vehicles such as the Merkava battle tank series, the Kfir fighter aircraft, and various small arms such as the Galil and Tavor assault rifles, and the Uzi submachine gun. Much of the IDF's electronic systems (intelligence, communication, command and control, navigation etc.) are Israeli-developed, including many systems installed on foreign platforms (esp. aircraft, tanks and submarines), as are many of its precision-guided munitions.

Major Armament Producers

IAI: Israel Aerospace Industries is Israel's prime aerospace and aviation manufacturer, producing aerial systems for both military and civilian usage. It also designs, develops and manufacturers naval and ground systems, electronic warfare and radar equipment and missiles.

IMI: Israel Military Industries (IMI) manufactures assault weapons - from the classic Uzi sub-machine gun to the Tavor assault rifle - heavy ammunition, aircraft and rocket systems, armoured vehicles like the Merkava tank, and integrated security systems.

The Rafael Arms Development Authority: Rafael designs and develops a wide range of high technology based defence systems for air, land, sea and space applications.

Elbit Systems: The Company is involved in manufacturing electronic equipment, artillery guns, radio communications systems, unmanned aerial vehicles, remote weapon systems, radar and naval systems.

Aeronautics Defence Systems Ltd. specializes in providing comprehensive defence solutions.

Israel Weapons Industries is an Israeli firearms manufacturer.

Astronautics C.A Ltd is a leading provider of airborne, naval and ground forces defence systems and solutions.

Israel Shipyards is one of the largest shipbuilding and repair facilities in the eastern Mediterranean.

Automotive Industries is an Israeli Automaker and major supplier of the Israeli Security Forces.

Plasan develops, manufactures and assembles custom-built vehicle armour systems and chassis up-armour designs as well as Add-On Armour Protection Kits for lightweight military tactical trucks and APCs, commercial vehicles, and a major supplier of personnel protection armour.

Defence firms in the private sector: Cyclone Aviation, Urdan Industries makers of the Merkava tank; Magal Security Systems, BVR Technologies, the Elul Group and RSL Electronics.

DEFENCE DEALS

Joint Venture Programmes

United States: David's Sling is being jointly developed by the Israeli defence contractor Rafael Advanced Defence Systems and the American defence contractor Raytheon, designed to intercept medium-to long-range rockets and cruise missiles. David's Sling is meant to replace the MIM-23 Hawk and MIM-104 Patriot in the Israeli arsenal. The US and Israel have also jointly developed Missile Defence Systems such as the Iron Dome, the Arrow and the Patriot.

India: The Barak 8 surface-to-air missile (SAM) is being jointly developed by Israel Aerospace Industries, India's Defence Research and Development Organisation (DRDO), Israel's Administration for the Development of Weapons and Technological Infrastructure, Elta Systems, Rafael and other companies. On 10 November 2014 the Barak 8 was successfully test fired in Israel with all integrated operational components for both the marine & land system.

India and Israel have also agreed to jointly develop a medium-range surface-to-air missile (MRSAM) system for the Indian Army. The land version of MRSAM would be an extension of the ongoing Air Force MRSAM project, which is expected to begin induction by 2017. The Army mobile MRSAM systems will be jointly developed by India's state-owned defence research agency, DRDO and Rafael and Israel Aerospace Industries (IAI) of Israel.

Arms Export

Israel is one of the world's major exporters of military equipment ranking 6th in the world in 2014. Three Israeli companies- Elbit Systems, Israel Aerospace Industries and Rafael- are among the world's top 100 arms-producing and military service companies. Much of the Israeli defence equipment exports are sold to the United States and Europe and other major regions include Southeast Asia and Latin America. India is also major country for Israeli arms exports and has remained Israel's largest arms market in the world. Israel is considered to be the leading UAV exporter in the world.

According to the Ministry of Defence export deals worth USD 5.66 billion were signed with Israeli defence industry companies in 2014. The 2013 total had been USD 6.5 billion, itself a USD 1 billion drop from 2012 figures. The 12.9 percent drop stemmed from a reduction in defence budgets and acquisition programmes in the United States and Europe, according to the MoD report. Most of the acquisitions in 2014 were for improvements to planes and aerial systems, as well as sales of electronic warfare and signals intelligence systems, armament, aerial defence systems and radar, and drones (UAVs). Meanwhile Israeli exports of weapons to African countries rose by 40 percent in 2014 compared with the previous year. Israeli security industries signed deals worth $318 million in Africa, compared with USD 223 million in 2013, which was the previous all-time high. Arms deals with African countries have risen steadily on an annual basis since 2009 (USD 77 million), with only a slight dip in 2012 ($107 million) compared with 2011 (USD 127 million), according to data published by the MoD

Arms Import

In Sept. 2016, the US and Israel has signed USD 38 billion worth military aid package under which it will buy advanced planes and weaponry and boost the missile defence shield of the Israeli military, displaying unprecedented commitment to Israel. The new 10-year MoU on security assistance was signed by the US Administration with the Government of Israel. As per the MoU, the US commits to providing Israel USD 38 billion in military aid over ten years, including USD 33 billion in foreign military financing funds and an unprecedented commitment of USD 5 billion for missile defence. This additional funding will, among other things, allow Israel to update the lion's share of its fighter aircraft fleet, including the acquisition of additional F-35s and F-15s. It will also enable Israel to substantially enhance the mobility of its ground forces.

Israel import arms mostly from the United States. However, in recent years, Israel has reportedly acquired Russian-made weapon systems. Most of the arms used by the Israel Defence Forces are manufactured locally or acquired with aid funds from the US. For the past several years US aid to Israel has come to USD 3.1 billion a year. In July 2014 Israel imported 600 air-to-air AIM-9 Sidewinder missiles, together with about 50 training missiles, replacement parts and technical support, in a deal valued at around USD 544 million. Also, reportedly there was the sale of six V-22 aircraft-to-aircraft missiles at a cost of more than USD 1.1 billion. Meanwhile, Israel has negotiated the purchase of two F-35 squadrons, totalling 30+ jets. Israel recently confirmed that it will buy 17 more F-35 fighter jets from the United States. The purchase, approved by the country's security cabinet, will bring the total number of the advanced aircraft in the Israeli fleet to 50.

Israel has also bought Germany submarines from HDW shipyards in Kiel. Israel has purchased six submarines so far, three of which are in use by the Israel Navy. Italy also sold 30 M-346 training aircraft to Israel. The first of these are already making their first flights. As part of the deal, Israel sold Italy several defence products, an Israeli-made satellite and a deterrent aircraft produced by Israel Aerospace Industries in a deal estimated at $2 billion dollars. Emerging economies such as India, Brazil and Chile have also increased their military trade and cooperation with Israel.

DEFENCE BUDGET

Total defence spending: $18.7 billion (2015); $23.2 billion (2014)

Defence spending in terms of GDP: 5.44% (2015)

According to Israel's Treasury 2017-18 budget proposal, which included reforms, tax cuts, and a surge in spending. The proposed budget for 2017 and 2018 raises the deficit target to 2.9 percent of GDP in each of those years, well above the present targets of 2.5 percent in 2017 and 2.25 percent in 2018. The government adopted a new multiyear framework for the defence budget, which sets the budget level for each of the next five years, and delineates the conditions under which the budget would be changed. The 2017 general budget stands at USD 94.1 billion, and was put at USD 98.8 billion for 2018.

Israel is spending the highest percentage of its gross domestic product on defence according to Forbes. Israel reportedly spent 5.2% of its GDP in 2014 on arming the IDF. Israel receives just over $3 billion a year in American aid under a 10-year deal signed in 2007. The agreement stipulates that over 70 percent of that aid has to be spent buying US military hardware. Among the purchases to be covered by the new deal are supplies of the V-22 Osprey tilt-rotor aircraft as well as up to 75 fifth-generation F-35 joint strike fighters.

Defence Procurement

The Israel Ministry of Defence is responsible for procurement, contractual, legal, financial and economical aspects of IDF activities. The Procurement and Production Directorate (PPD) handles procurement and oversees the manufacture of systems and products, maintenance services for the military and civilian defence systems, and ministry agencies' overseas procurement activities. The PPD oversees the development of the country's military-industrial infrastructure. The PPD also administers the maritime and air transport of equipment and goods acquired abroad for Ministry of Defence and other government ministries, as well as equipment and goods exported from Israel. The PPD is composed of five executive and several staff units. The IDF also has several large internal research and development departments, and it purchases many technologies produced by the Israeli security industries including IAI, IMI, Elbit Systems, Rafael, and many more.

CONCLUSION

The Israel Defence Forces emphasises close relations between the Army, Navy, and Air Force. The IDF possesses top-of-the-line weapons and computer systems. Israel has also developed its own independent weapons industry. Despite its war-torn history Israel has made remarkable advancements particularly in the area of innovative defence technologies. Israeli defence companies rank as some of the largest in the world, contributing significantly to Israeli industry and economy. Israel ranks as the most powerful military in the Middle East, according to a report by Business Insider. The country has a qualitative edge over the region - this edge includes space assets, advanced fighter jets, drones, and nuclear weapons. As the immediate threat of full-scale conventional war has receded, the Israel Defence Forces (IDF) has devoted more and more resources to countering the threats posed by low-intensity conflict and weapons of mass destruction. Israel has built a comprehensive standing army and furnished it with an arsenal of highly advanced military hardware. The government, which owns three major defence firms, also encouraged the formation of private companies to equip the IDF. The development of a sophisticated defence industry inevitably led to exports, which today account for a majority of its revenues and allows the country's defence industry to compete against some of the largest companies in the world for foreign contracts, in addition to producing many of the arms needed for Israel's own defence. Israel has also emerged as an important factor with decades of experience in fighting terrorism and existing technology and products to offer. Many countries are now turning towards Israel to buy equipments to fight terrorism.

CONTACT

Ministry of Defence

Post office Box: 7093,

Hakirya,

Tel-Aviv 61070, Israel

Phone number: 97236977458

Website: www.mod.gov.il

STRATEGIC INFORMATION

ITALY

(Capital: Rome)

INTRODUCTION

Area: 301,340 sq km

Population: 62,007,540 (July 2016 est.)

Coastline: 7,600 km

Maritime claims: Territorial sea: 12 nm

Continental shelf: 200 m depth or to the depth of exploitation

GEOPOLITICAL IMPORTANCE

Situated in Southern Europe, Italy, with a large coastline, is bordered by Switzerland and Austria in the north, France and Monaco in the northwest, Slovenia and San Marino in the northeast. The country takes the shape of a peninsula and extends to the central Mediterranean Sea. It is surrounded by the Tyrrhenian Sea on the west, the Ionian Sea on the south, and the Adriatic Sea on the east.

Italy enjoys a unique geostrategic position in the European continent. Its proximity to the Balkan states of Albania, Bosnia and Herzegovina, Bulgaria, Croatia, Greece, Macedonia, Montenegro, Romania, Serbia and Slovenia in the south-eastern Europe, and to the littoral states of North Africa – Morocco, Algeria, Libya and Egypt – makes the country strategically important. Sicily, Sardinia and the surrounding minor islands administered by Italy help the country occupy a vital position in the Mediterranean Sea. The country also has a number of gulfs, bays and ports that are suitable for being used as strategic military bases.

POLITICAL OVERVIEW

Erstwhile a monarchy, Italy today functions as a parliamentary democratic republic with the President as the ceremonial head of the state. The three branches of the government include the executive, the legislature and the judiciary. While the executive powers are exercised by the Council of Ministers led by the Prime Minister, the legislative power is vested with the two houses of Parliament – the Chamber of Deputies and the Senate of the Republic. The judiciary, constituting of all courts of law, functions independently.

KEY POLITICAL PERSONS

PRESIDENT & COMMANDER-IN-CHIEF: Sergio Mattarella

PRIME MINISTER: Paolo Gentiloni

ECONOMY

With a diversified economy, which primarily includes industry, agriculture and services sector, Italy is the fourth largest economy in Europe in terms of GDP and fifth largest in terms of purchasing power parity. The country's economy is largely dependent on manufacturing and service sector which contribute a major share of the GDP. Italy is a member of the European Union, World Trade Organisation and Organisation for Economic Co-operation and Development (OECD). The Euro zone crisis adversely affected the country's economic growth. Though it eased to some extent in 2013, yet the GDP remained 8% below its 2007 pre-crisis level. Public debt has increased steadily since 2007, topping 135% of GDP in 2015, but investor concerns about Italy and the broader euro-zone crisis eased in 2013, brining down Italy's borrowing costs on sovereign government debt from euro-era records. In 2014, economic growth and labour market conditions continued to deteriorate with overall unemployment rising to 12.7% and youth unemployment around 40%, but Italy began to recover in 2015 with marginal growth and a slight reduction in unemployment.

GDP: $2.175 trillion (2015 est.); $2.159 trillion (2014 est.)

Real Growth Rate: 0.8% (2015 est.); -0.3% (2014 est.)

Industries: Tourism, machinery, iron and steel, chemicals, food processing, textiles, motor vehicles, clothing, footwear, ceramics

Natural Gas - Proved Reserves: 59.43 billion cu m (1 January 2014 est.)

Crude Oil - Proved Reserves: 544.5 million bbl (1 January 2015 est.)

Total Exports: $450.1 billion (2015 est.); $518.3 billion (2014 est.)

Export Commodities: Engineering products, textiles and clothing, production machinery, motor vehicles, transport equipment, chemicals; food, beverages and tobacco; minerals, nonferrous metals

Major Markets: Germany 12.3%, France 10.3%, US 8.7%, UK 5.4%, Spain 4.8%, Switzerland 4.7% (2015)

Total Imports: $391.2 billion (2015 est.); $455.2 billion (2014 est.)

Import Commodities: Engineering products, chemicals, transport equipment, energy products, minerals and nonferrous metals, textiles and clothing; food, beverages, and tobacco

Major Suppliers: Germany 15.4%, France 8.7%, China 7.7%, Netherlands 5.6%, Spain 5%, Belgium 4.7% (2015)

DEFENCE & SECURITY

After remaining a colonial power for over half a century and influencing the two World Wars in the 19th century, Italy joined the international military alliance North Atlantic Treaty Organisation (NATO) as one of the founding members in 1949. The country spent heavily in building up a strong military force. Today, Italy commands a prominent position in the entire European continent as well as in global military affairs. Though it does not face any direct military threat from any foreign country, Italy's geostrategic position makes it particularly vulnerable to sea piracy, proliferation of weapons, illegal immigration, drug trade, human trafficking and other forms of organised crime. Islamic terrorism has posed another major security challenge for the country in recent times. To counter these asymmetric threats, the country has maintained a strong and well organised military force. The two primary roles assigned to the Italian armed forces include territorial control/ protection operations and peacekeeping operations. Italy's allegiance to NATO is of greater importance for its national security and also helps it to standardise and upgrade its own military capabilities. Italy's military strategy and policies are primarily influenced by those of the NATO. The country has been engaged in several multilateral operations and peacekeeping missions led by the international military alliance. It also regularly takes part in joint military exercises conducted by NATO. With the rise of the Islamic State and its terror acts which have emerged as the most serious security threat to the entire European continent in recent times, Italy has pitched for a stronger European defence mechanism to fight the menace collectively.

Under a large-scale defence reforms plan announced by the Defence Ministry a year ago, the country planned to cut down the strength of its armed forces by 40,000 personnel over the next 20 years (from 1,90,000 to 1,50,000 by 2024), shut many of its military bases and reduce its defence equipment and platforms by selling them off to other countries. Such a widespread reduction is aimed at making the Italian armed forces a leaner force in consonance with NATO and European standards of building up a flexible, expeditionary military capable of undertaking operations outside the country more efficiently.

In its new Defence Whitepaper, released in April 2015 after a gap of 13 years, the country aimed at taking a key role in the Mediterranean military affairs even as it has grappled with illegal immigration from the neighbouring strife-torn Libya.

KEY DEFENCE PERSONS

MINISTER OF DEFENCE: Roberta Pinotti

CHIEF OF DEFENCE STAFF: Gen. Claudio Graziano

CHIEF OF STAFF OF ARMY: Gen. Danilo Errico

CHIEF OF NAVY: Adm. Valter Girardelli

CHIEF OF AIR FORCE: Gen. Enzo Vecciarelli

Internal Conflict

Italy's major security challenges are regional in nature and originate in the country's vicinity. Its geostrategic position with a large coastline makes it vulnerable to illegal immigration, illicit drug trade, human trafficking, weapons proliferation, organised crime and terrorism-related activities. To counter these internal and transnational threats, Italy maintains a large police force.

External Conflict

Italy does not face direct external military threat from any country. In the last decade, it has taken part in US-led and NATO-backed military interventions in a number of countries including Afghanistan, Iraq and Libya. While its role in the Afghanistan conflict is confined to peacekeeping and nation building, Italy has carried out combat operations against Libya in 2011. In Iraq, its military role began after the war ended and remained confined mainly to stabilization of Coalition forces. The country also provided transit routes – land and maritime – to the participating nations for deploying troops in Iraq, although it did not permit use of its territory for carrying out direct attacks against the West Asian country.

THREAT PERSPECTIVE

Terrorism: Like other parts of Europe, Italy is facing the threat of Islamic terrorism in recent times. Terrorism has posed a serious security challenge to the country, especially to its capital city Rome. The country remains vigilant over any purported terror attacks and has tightened its overall security apparatus to counter any kind of violent act by the Islamic extremists.

STRATEGIC RELATIONS

US: Bilateral strategic ties between Italy and the US roots in their allegiance to the NATO. Under longstanding bilateral agreements flowing from NATO membership, Italy hosts US military forces at Vicenza and Livorno (Army); Aviano (Air Force); and Sigonella, Gaeta, and Naples – home port for the US Navy Sixth Fleet. The US has over 10,000 military personnel – army, navy, marine and air force – stationed in Italy.

EU: As an important member of the European Union (EU), Italy has traditionally supported the idea of greater European integration. Its affiliation to the EU also makes it less vulnerable to the peripheral threats it faces from its surrounding regions. While economic cooperation is the main driving force for the multinational block, it has also been working on several military projects to cater to the European defence market as well as for importing to foreign countries. Italy plays a key role in several such projects, including the Eurofighter Typhoon fighter aircraft, nEUROn unmanned combat aerial vehicle (UCAV) programme and MBDA-led missile systems programme.

Russia: Since decades, Italy has made efforts to build strong relations with Russia, despite being a NATO member. The two sides share deep economic and energy ties. They have explored ways to develop multidimensional and strategic relations in recent times. Italy also entered the Russian defence market in 2010 when Moscow acquired the design, technology, and local assembly rights over the Iveco M65 LMVs. Moscow has also expressed interest in involving Italian shipbuilder Fincantieri in developing Russian shipyards. The two countries had initiated work on a joint project to design and develop a next-generation diesel submarine, called S-1000, for exporting to international customers. It has however been stalled over the ongoing Ukraine crisis. The project was started in 2004.

Turkey: Italy has supported the idea of Turkey's inclusion as a member of the NATO as well as in the EU. The two countries share long-standing economic, political and strategic relations.

DEFENCE CAPABILITIES

ARMY: The Italian Army, which was traditionally been assigned the task of defending national sovereignty, has focused more on enhancing internal stability in recent times in the wake of peripheral dangers emanating from its porous borders. It has also been taking part in several peacekeeping missions led by the United Nations. Being a NATO member, Italy's Army personnel have been part of multinational combat operations from time to time. The Army has aimed at becoming a lighter-weight, digital fighting force in the 21st century.

Personnel: 104,000+

Equipment

Category	Name	In Service
MBT	Ariete	200
	Leopard 1A5	120 (in reserve)
Armoured Car	B1 Centauro	240
APC/ IFV	M113	250+
	VCC-1/ VCC-2	2,000+
	Bv 206	105
	Puma	540
	Cougar	6
	VAB NBC	15
	Dardo	200
	VBM Freccia	249
	VTLM Lince	1260
Artillery	M109L SP howitzer	130+
	PzH 2000 SP howitzer	70
	FH-70 towed howitzer	160
Mortar	120mm F1	200
	60mm M6-111	300

Army continued...

Category	Name	In Service
MLRS		22
ATGW	TOW (to be withdrawn)	244
	Milan	2000+
	Spike MR/LR	990 (with 90 launchers)
AA Gun	Sidam 25	275
SAM	MIM-23 Hawk	60+
	SPADA 2000	50
	SAMP/T	6

Army Aviation

Category	Name	In Service	On Order
Transport	Dornier Do 228	3	
	P180 Avanti	3	
Helicopter	A129 Mangusta (attack)	60	
	Agusta A109	12	
	Bell 205	42 (to retire)	
	Bell 206	30 (to be replaced)	
	Bell 412	23	
	CH-47C Chinook	17 (being phased out)	
	ICH-47F Chinook	2	14 (+4 option)
	NH90 TTH	31	29
UAV	Raven RQ-11A/11B	58	

New Procurements/ Upgrades

✦ Fund has been allocated in 2016 for the development of a new wheeled tank called Centauro II. The Army is expected to procure up to 150 of the new lightweight tanks.

✦ Additional 381 Freccia armoured vehicles are to be procured after a Parliamentary approval in 2015.

✦ The AW129C Mangusta attack helicopters are to be upgraded to new 'G19' configuration featuring new observation, targeting and weapon system. Initial contract is to upgrade 32 helicopters with option for 16 more.

✦ AgustaWestland has received a contract to deliver 16 ICH-47F Chinook transport helicopters. The first two choppers have been delivered in October 2014. Delivery of the remaining Chinooks is scheduled for completion by 2017.

✦ The SAMP/T missile defence systems, deployed in the Army base in Mantova, have reached Full Operational Capability.

NAVY: The Italian Navy, which was the fourth largest navy in the world during World War II, has transformed itself into a modern naval power. The geostrategic position of Italy near the Mediterranean Sea has rendered it necessary to build and project a strong naval force. The Italian Navy, called the "Marina

Militare", maintains a fleet of advanced battleships, including aircraft carriers, submarines, frigates and destroyers. The Navy operates with a mission "to support and to contribute to the safety of navigation and to the National Defence, to promote the study of all sea related matters and the protection of the marine environment." It has also taken part in several NATO-led peacekeeping operations.

Personnel: 30,000+

Major naval bases: Taranto, La Spezia, Naples

Equipment

Category	Name	In Service	On Order
Aircraft Carrier (STOL)	Cavour	1	
	Giuseppe Garibaldi	1	
Destroyer	Orizzonte class	2	
	Durand de la Penne class	2	
Submarine	Todaro class (U212A Type)	2	2
	Sauro III class	2 (To be retired)	
	Sauro IV class	2	
Frigate	Maestrale class	8	
	Lupo class	17	
	Bergamini class FREMM	2	8
Corvette	Minerva class	8	
Patrol vessel	Cassiopea class	4	
	Esploratore class	4	
	Comandanti class	4	
	Multipurpose Offshore Patrol Ship		7
Amphibious assault dock	San Giorgio class	3	
Mine sweeper	Lerici class 1 series	4	
	Lerici class 2 series	8	
Auxiliary ship	Etna class	3	
	Stromboli class	2	
Weapon	Aspide (SAM)		
	Aster 15 & Aster 30 (SAM)		
	Otomat (ASM)		
	(Torpedo) MU-90 Mk 32 A-184 Black Shark		

Coast Guard

Category	Name	In Service
OPV	Diciotti-class	2
APC, IFV	AAV-7	18
	VCC 1/VCC 2	70
Mortar	120mm	4

Coast Guard continued...

Category	Name	In Service
Mortar	81mm	8
ATGW	Milan	12
	TOW	3
	Panzerfaust	25
SAM	Stinger	4

Naval Aviation

Category	Name	In Service	On order
Fighter, Attack, AWS	AV-8B Harrier II	16	
	F-35A/ B Lightening II		90 (navy and air force)
Transport, Surveillance /MPA	P-180 Avanti	3	
	ATR-72 ASW	4	
Helicopter	Aw101 (ASW/ ASuW)	24	
	EH101	20	
	AB-212	20	
	NH-90 TTH & NFH		10+46

New Procurements/Upgrades

✦ Fincantieri launched the third Todaro class (U212A Type) submarine for the Navy in October 2014. Named "Pietro Venuti", it is the first of two new units of the U212A "Todaro" class submarines on order. The second vessel, "Romeo Romei", was launched in July 2015. Delivery of the two new subs, which will replace the Sauro III class submarines, is scheduled for 2016-17.

✦ Five FREMM multi-role frigates have been commissioned into service so far. The sixth platform of the class has been launched in December 2015 with delivery scheduled for 2017. A total of 10 FREMM frigates are to be acquired by the Navy to replace the existing Maestrale class and Lupo class of platforms.

✦ Fincantieri and Finmeccanica have received a contract worth over 1.1 billion euros for the construction of a multi-purpose amphibious unit (a Landing Helicopter Dock - LHD) for the Navy. The industry partners will also develop one logistic support ship with additional funding.

✦ The first NH90 Step B naval helicopter (with full operational capability) has been delivered to the Navy in Nov. 2013. Italy has ordered a total of 56 NH90 helicopters, 46 of which are the NFH (NATO Frigate Helicopter) variant and 10 TTH variant. Five earlier aircraft delivered in the interim Step A configuration will be upgraded next.

✦ Seven Multi-Purpose Offshore Patrol Ships (PPAs) will be acquired with the first delivery expected by 2021.

AIR FORCE: Primarily tasked with defending the country's airspace and preventing and neutralising any aerial threat, the Italian Air Force has also supported peacekeeping and humanitarian missions outside its territory. The Air Force, in coordination with the other two wings of the Italian military, ensures homeland security. It also acts as an advanced and technological organisation which drives the scientific and industrial progress of the country.

Personnel: 45,000

Major air bases: Amendola, Aviano, Ghedi, Sigonella, Trapani, Vicenza

Equipment

Category	Name	In Service	On Order
Fighter/ Attack	AMX/ AMX-T Ghibli	60+	
	F-16	28	
	Eurofighter Typhoon	60+	36
	Tornado	60+	
	F-35A/ B Lightening II		90 (for Air Force & Navy)
Transport, Tanker, Surveillance	C 27J Spartan	12	
	C-130J/C-130J-30 Hercules II	21	
	P180 Avanti	15	
	Airbus A319 (VIP transport)	3	
	Falcon (VIP transport)	7	
	Boeing KC-767A (aerial refuelling)	4	
	ATR 72 MP		4
Trainer	MB-339A/ MB-339CD	200+	
	SF.260EA	30	
	M-346 Master	6	12
Helicopter	HH-139A	10	
	HH 101A (AW101)		12 (+3 in option)
	AB-212	32	
	NH-500E	49	
AEW	Br-1150 Atlantic	8	
	G-550 Conformal		2
UAV	RQ-1 Predators	10+	
	MQ-9 Reaper	6	
Weapon	(AAM) AIM-9 Sidewinder AIM-132 ASRAAM AIM-120 AMRAAM IRIS-T		
	Storm Shadow (AGM)	200+	
	JDAM		
	Paveway (laser guided bomb)		

New Procurements/Upgrades

- A contract has been awarded to Alenia Aermacchi in Nov. 2013 to convert six C-27J Spartan cargo aircraft into multi-mission gunship configuration called the MC-27J Pretorian. The first three C-27Js will be converted into gunship configuration by 2016. The modifications will involve new sensors, communications and cannon. Alenia has partnered with US' defence and aerospace company ATK to carry out the conversion.
- The first G-550 Conformal Airborne Early Warning & Control System (CAEW) has been delivered in December 2016. Two such platforms are being acquired from Israel Aerospace Industries (IAI) under a contract signed in 2012.
- Delivery of the HH-101A (AgustaWestland AW101) medium twin-engine helicopters is going on. The new choppers will replace the HH-3F Pelicans.
- The Government has reiterated its stand over the purchase of 90 F-35 Joint Strike Fighters (JSF) amidst calls from the Parliament to halve the total order for the multi-role fighters to 45. The Government in May 2015 said that it maintains the funding of over 10 billion Euro for the acquisition programme through 2027. Italy has already cut down its earlier order of a total of 131 F-35 (69 conventional take-off and landing F-35As and 62 short take-off and vertical landing variant F-35Bs) for its Air Force and Navy to 90 amidst financial woes. The first F-35 for the Italian Air Force assembled in Italy underwent its maiden flight in September 2015 and was delivered to the Air Force in December 2015.
- The country has decided to cut its original orders of 121 Eurofighter Typhoons by 25 to save worth $2.6 billion. The air force will now receive 96 Eurofighters.
- Panavia GmbH has been awarded contract to upgrade 25 Tornado fighter aircraft with advanced communication, navigation, armament and defence systems. In February 2013, the company received a new contract from the NATO Eurofighter and Tornado Management Agency (NETMA) to integrate two new weapons -- the advanced anti-radiation guided missile (AARGM) and the small-diameter-bombs (SDBs) -- on the Tornado RET 7 and RET 8 configuration.
- Delivery of M-346 advanced jet trainers from Alenia Aermacchi is going on. The Air Force has ordered 18 of the trainers.
- A team of Alenia Aeronautica, Avio and Lockheed Martin has been awarded $204 million contract to provide technical and logistical support services to IAF's fleet of C-130J airlifters.
- General Atomics Aeronautical Systems has been contracted to provide weapons (HELLFIRE missiles, bombs, M36-E8 Captive Air Training Missiles, laser guided bombs) and other logistical support for the MQ-9 Reaper UAVs.

DEFENCE PRODUCTION AND R&D

Major Armaments Producers

Finmeccanica: The second largest industrial group and the largest of the hi-tech industrial groups based in Italy, Finmeccanica works in the strategic fields of defence, aerospace and security. The defence conglomerate in March 2016 decided to rename itself as "Leonardo". It is one of Europe's leading defence systems company and its subsidiaries include Alenia Aermacchi, AgustaWestland, SELEX Sistemi Integrati and Oto Melara among others. The conglomeration is partially owned by the Italian government and produces aerospace systems, helicopters, trainer and fighter aircraft, UAVs, missile, naval artillery and armoured vehicles in cooperation with other leading European defence firms.

Fincantieri: The leading shipbuilder in Italy and one of the largest in Europe, Fincantieri constructs both commercial ships and military naval vessels. It also undertakes repair and modernisation work of the naval vessels.

DEFENCE DEALS

Joint Venture Programmes

Europe

EADS: The largest conglomerate of defence and aerospace companies in Europe, the EADS group includes Airbus, Airbus Military, Cassidian and Eurocopter which produce civilian and military aircraft, defence equipment including land, naval and aerial systems, and a range of helicopters respectively. Italy is a part of the mega conglomerate.

Eurofighter Typhoon: Italy has partnered with Germany, Spain and UK to produce Eurofighter Typhoon - Europe's largest military collaborative programme. Finmeccanica subsidiary Alenia Aermacchi joined hands with other European firms in manufacturing the fourth-generation multi-role fighter aircraft.

nEUROn UCAV: Six European countries - France, Sweden, Italy, Spain, Greece and Switzerland – are jointly developing an unmanned combat aerial vehicle (UCAV) called nEUROn. The programme was launched by the French defence ministry in 2003 with the aim to design and develop an advanced combat unmanned aerial vehicle capable of fighting futuristic warfare and perform stealth missions. Sweden's Saab, Italy's Alenia Aermacchi, Spain's EADS-CASA, Greece's HAI and Switzerland's RUAG later joined the programme led by French major Dassault Aviation.

Missile Systems: MBDA - the leading European missile manufacturer - is a partnership company between BAE Systems, EADS (Europe's largest defence and aerospace organisation) and Italy's Finmeccanica. The company produces a wide range and family of missile systems for European nations as well as for foreign supply.

France

FREMM Frigate: The FREMM programme is a joint Italian-French defence industrial co-operation project to build a new generation multi-mission frigates. Italy's Fincantieri and France's DCNS have been partnering in the project which has been pegged as Europe's largest naval programme.

Germany

Type 212A class submarine: In joint partnership with Germany, Italy is building two additional Type-212A Todaro-class submarines for its Navy. The new vessels are slated to be delivered by 2017. The Italian Navy is presently operating two such submarines (Type-212) which were delivered in 2006 and 2007.

US

F-35 Joint Strike Fighter: The Joint Strike Fighter, pegged as the most expensive weapons programme in the US defence history, is being developed as a joint programme between nine partnering nations. Italy is one of the allies in the project led by US defence major Lockheed Martin.

Russia

S-1000 Submarine: Under a deal inked in 2004, Italy's Fincantieri and Russia's Rubin Central Design Bureau are developing the S-1000 new-generation conventional submarine. The two sides had renewed efforts to upgrade the submarine's previous design and make it smaller so as to market it in Middle East and Southeast Asia. However, the ongoing Ukraine crisis has stalled the project.

Arms Export

Led by a well-developed defence industry, Italy is one of the major suppliers of military equipment in the international market. It has also partnered with leading European countries to manufacture and supply military hardware. In past few years, Italian defence exports have witnessed an upward trajectory. The country's leading defence firms have collectively clinched deals, both civilian and military, with countries like Saudi Arabia, UAE, Qatar, Egypt, Turkey, Germany, US, UK, India, Israel, Singapore and Australia worth several million Euros. According to the Stockholm International Peace Research Institute (SIPRI), Italy ranks 8th in the world in military exports. It accounted for three percent of the share of international arms exports between 2010 and 2014.

Arms Import

Majority of military imports made by Italy have been sourced from European countries and the US. Israel has also supplied some defence hardware to the country. While maintaining self-sufficiency in defence production, Italy has acquired sonar and radar systems, missiles, combat vehicles and unmanned aerial vehicles from other countries. The US has supplied four KC-767 advanced aerial refueling tanker aircraft to Italy besides UAVs.

DEFENCE BUDGET

Military spending by Italy has remained quite obscure as the annual defence budget announced by the Defence Ministry does not include spending on procurement, defence R&D and global peacekeeping operations which are sourced from the Ministry of Economic Development. Defence expenditure by the country has maintained an uneven graph over the last five years. While witnessing an increase in 2008 over the previous fiscal, the defence budget declined in the subsequent year and then increased again till 2011. In 2012, the budget was expected to reach at a mere 0.84% of GDP far less than NATO's stipulated defence spending of 2% of GDP for all member countries.

According to the Italian Defence Ministry figures, in the period 2004-2012, the Italian GDP grew by 15% in monetary terms (from €1.391 billion to €1.622 billion). The budgetary resources allocated to defence function passed in the same period were 1.01% of GDP in 2004 to 0.84% of GDP in 2012 - a reduction of 16%. The Euro-zone crisis has forced Italy to adopt austerity measures that includes overall spending cuts and as a result, total defence expenditure as a percentage of GDP is expected to come down to 1.1% by 2016. Even while the country's defence spending continues to focus on military modernisation programmes and UN peacekeeping missions, plans to reduce its overall military strength reflects its reduced defence spending as part of austerity measures. The defence budget for FY 2015-16 stood at 13.19 billion euro, a decline over the previous year's budget. The funding, however, increased slightly in 2016-17 with the country allocating €13.36 billion.

Budgetary allocation

Year	Allocation	% of GDP
2016-17	€13.36 billion	
2015-16	€13.19 billion	
2014-15	€13.58 billion	0.91%
2013-14	€14.4 billion	1%
2012-13	€13.6 billion	0.84%
2011-12	€20.5 billion	1.27%
2010-11	€17.6 billion	1.1%

CONCLUSION

The 21st century Italy, though subdued than what it was during the two World Wars, has nevertheless maintained a military prominence not only in Europe but also in the global stage largely due its allegiance to the NATO. Its involvement in some NATO-led combat operations, including the recent intervention in Libya, has provided a scope for its military to actively exercise and showcase its prowess outside its territory while at the same time, gauge its operational readiness to defend and protect the homeland. The country, led by a strong domestic defence industry, has focused on modernising its armed forces by acquiring new-age military hardware and equipment despite the ongoing financial crunch in Europe. As the country's new Defence Whitepaper states, Italy is now willing to take a lead role in the military affairs of the Mediterranean region.

Eh101 helicopters of Italian Navy-An MoD photo

CONTACT DETAILS

Ministry of Defence
Palazzo Baracchini
Via XX Settembre 8-00187 Rome
Tel.: (+39) 06-46911
Website: www.difesa.it

Army
Army Palace
Via XX Settembre, 123 / A - 00187 Rome
Tel.: (+39) 800 299 665
Website: http://www.esercito.difesa.it/

Navy
Office for Relations with the Public (URP)
Palace Marina
Square Marina, 4 - 00196 Rome
Tel.: (+39) 06-36804442
Website: http://www.marina.difesa.it/

Air Force
Office for Relations with the Public (URP)
Palazzo Aeronautica
University Avenue, 4 - 00185 Rome
Tel.: (+39) 06-49866626
Website:
http://www.aeronautica.difesa.it/

STRATEGIC INFORMATION

JAPAN
(Capital: Tokyo)

INTRODUCTION

Area: 377,915 sq km

Population: 126,702,133 (July 2016 est.)

Coastline: 29,751 km

Maritime claims:

Territorial sea: 12nm

Contiguous zone: 24 nm

Exclusive Economic Zone: 200nm

(Included the Senkaku group of islands by purchasing it from the owners in September 2012).

GEOPOLITICAL IMPORTANCE

Japan, a country of islands, is situated in East Asia. Located in the Pacific Ocean, it lies to the East of the Sea of Japan, China, North Korea, South Korea and Russia, stretching from the Sea of Okhotsk in the North to the East China Sea and Taiwan in the South. Philippines also share maritime boundary with the country. The four largest islands (from North to South) are Hokkaido, Honshu, Shikoku, and Kyushu. The Ryukyu Islands, including Okinawa, are a chain to the south of Kyushu. Together they are often known as the Japanese Archipelago.

POLITICAL OVERVIEW

Japan is a constitutional monarchy with a parliamentary government. The Emperor has limited powers and is defined as "the symbol of the state and the unity of the people". Power is held primarily by the Prime Minister of Japan and other elected members of the Diet, while sovereignty is vested in the Japanese people. The judiciary is independent.

KEY POLITICAL PERSONS

HEAD OF STATE: Emperor Akihito

PRIME MINISTER: Shinzo Abe

ECONOMY

The economy of Japan is the third largest in the world after the United States and the People's Republic of China and is the world's second largest developed economy. Japan's reservoir of industrial leadership and technicians, well-educated and industrious work force, high savings and investment rates, and intensive promotion of industrial development and foreign trade produced a mature industrial economy. Japan has few natural resources, and trade helps it earn the foreign exchange needed to purchase raw materials for its economy. Japan is the world's third largest automobile manufacturing country, has the largest electronics goods industry, and is often ranked among the world's most innovative countries. Japan ranks 27th of 189 countries in the 2014 Ease of doing business index and has one of the smallest tax revenues of the developed world. Japan is a leading nation in scientific research, particularly in fields related to the natural sciences and engineering. The Japanese electronics and automotive manufacturing industry is popular worldwide.

GDP: $4.843 trillion (2015 est.), $4.817 trillion (2014 est.)

Real Growth Rate: 0.5% (2015 est.), 0% (2014 est.)

Industries: Among world's largest and most technologically advanced producers of motor vehicles, electronic equipment, machine tools, steel and nonferrous metals, ships, chemicals, textiles, processed foods.

Crude oil - proved reserves: 541.6 million bbl (March, 2015 est.)

Natural gas - proved reserves: 20.9 billion cu m (1 January 2014 est.)

Total Exports: $622 billion (2015 est.), $699.1 billion (2014 est.)

Export Commodities: motor vehicles 14.9%; iron and steel products 5.4%; semiconductors 5%; auto parts 4.8%; power generating machinery 3.5%; plastic materials 3.3% (2014 est.)

Major Markets: US 20.2%, China 17.5%, South Korea 7.1%, Hong Kong 5.6%, Thailand 4.5% (2015)

Total Imports: $627.3 billion (2015 est.), $799 billion (2014 est.)

Import Commodities: petroleum 16.1%; liquid natural gas 9.1%; clothing 3.8%; semiconductors 3.3%; coal 2.4%; audio and visual apparatus 1.4% (2014 est.)

Major Suppliers: China 24.8%, US 10.5%, Australia 5.4%, South Korea 4.1% (2015)

DEFENCE & SECURITY

The Japan Self-Defence Forces or JSDF, occasionally referred to as JSF or SDF, are the unified military forces of Japan. The Self Defence Force is divided into three branches: Ground Self Defence Force (Army); Maritime Self Defence Force (Navy); Air Self Defence Force (Air Force). The purpose of Self Defence Forces is to preserve peace, public order and Japan's independence and safety. The Prime Minister is the commander-in-chief of the Self Defence Forces and is advised by the Chief of Staff of the Joint Staff Council consisting of all the three services. Military authority runs from the Prime Minister to the cabinet-level Minister of Defence of the Japanese Ministry of Defence. The defence policy of Japan is pursued under the Constitution and is based on the "Basic Policy on National Defence" adopted by the National Defence Council and the Cabinet in May 1957. The "Basic Policy on National Defence" first states the promotion of efforts for peace such as international collaboration and the establishment of a basis for national security through the stabilisation of livelihood of the people, etc., and then the buildup of efficient defence capability and Japan-US security arrangements as the basis of Japan's defence. The objective of national defence is to prevent direct and indirect aggression, but if invaded, to repel such aggression, and thereby, to safeguard the independence and peace of Japan based on democracy. For most of the post-war period the JSDF was confined to the islands of Japan and not permitted to be deployed abroad. In recent years they have been engaged in international peacekeeping operations. Recently, Japan has stressed the need to step up its military presence in its annual defence white paper. The document says Japan is coping with "increasingly severe security environment," and has named the DPRK, China, and Russia as major security threats. The country made an overhaul to its defense status, including a decision on exercising the right to collective self-defence, a shift away from post-war Pacifism. Prime Minister Shinzo Abe in the year 2015 won a crucial vote in Parliament for legislation that would give Japan's military limited powers to fight in foreign conflicts for the first time since World War II. A total of 11 security-related bills were approved.

KEY DEFENCE PERSONS

MINISTER OF DEFENCE: Tomomi Inada

CHIEF OF JOINT STAFF OFFICE: Adm. Katsutoshi Kawano

CHIEF OF GROUND SELF-DEFENCE FORCE: Gen. Toshiya Okabe

CHIEF OF MARITIME SELF-DEFENCE FORCE: Adm. Tomohisa Takei

CHIEF OF AIR SELF-DEFENCE FORCE: Gen. Yoshiyuki Sugiyama

Internal Conflict

There are no internal security problems in Japan.

External Conflict

- The sovereignty dispute over the islands of Etorofu, Kunashiri, and Shikotan, and the Habomai group, known in Japan as the "Northern Territories" and in Russia as the "Southern Kuril Islands," occupied by the Soviet Union in 1945, now administered by Russia and claimed by Japan, remains the primary sticking point to signing a peace treaty formally ending World War II hostilities.
- Japan and South Korea claim their control

over islands known as Dokdo in Korean and Takeshima in Japanese. On August 2012, Japan delivered a diplomatic document to South Korea proposing that the two countries should take their territorial dispute over a group of small islets to the primary judicial entity of the United Nations. South Korea responded by saying that Japan's suggestion to take the issue to the International Court of Justice has no value because no territorial dispute exists and thus there is no need for negotiations. According to reports, South Korea has maintained a presence on the islets since 1952 when its coast guard was dispatched to the rocks nearly equidistant from the Korean peninsula and the main Japanese island of Honshu.

✦ Japan is in dispute with both China and Taiwan for their claims to the uninhabited islands of the Senkaku-shoto (Diaoyu Tai) and Japan's unilaterally declared exclusive economic zone in the East China Sea, the site of intensive hydrocarbon prospecting.

THREAT PERSPECTIVE

China: Japan and China are both strategically important neighbours, separated only by a strip of water. It is therefore important for them to become constructive partners in the international community. However, recent actions over the Senkaku Islands dispute have brought relations between these East Asian neighbours to a standstill. The islands, known as Senkaku to the Japanese and Diaoyu to the Chinese, are also claimed by Taiwan. On 26 August 2012, Japan announced that the government had budgeted 2.05 billion yen (USD 26 million) for purchasing the islands from their private Japanese owners. And soon on 10 September 2012, the Japanese government signed a contract with a private owner to buy three of the Senkaku Islands, amid intensified opposition from China, which claims territorial rights over those uninhabited islets. The Senkaku issue has sparked and renewed tensions between Tokyo and Beijing. The Chinese government has submitted a nautical chart with baselines of the territorial sea on Senkaku Islands to the United Nations. China has also launched a programme to increase the use of unmanned drones to conduct marine surveillance as it tries to increase its presence around uninhabited East China Sea islands at the centre of a dispute with Japan. The islands, which are also claimed by Taiwan, sit near an important shipping lane and nearby waters are believed to harbour valuable mineral resources. Recently, Prime Minister Shinzo Abe of Japan said that he believed his country and China had taken a major step forward in repairing relations, and that from now on, for the sake of the Asia-Pacific region, there should be frequent dialogue between them. Prime Minister Shinzo Abe and Chinese President Xi Jinping, in Nov 2016, agreed to improve bilateral ties ahead of the 45th anniversary of their normalization of relations.

North Korea: The relationship between Japan and North Korea is not established formally but diplomatic discussions are on between both the governments on issues such as North Korea's nuclear programme. North Korean missile tests are also a concern for Japan because the missiles sometimes travel over Japanese territory and through its airspace. According to the annual white paper of Defence of Japan, North Korean military behaviour has increased tension over the Korean Peninsula, and constitutes a serious destabilising factor for the entire East Asian region. There have also been several confrontations between the two nations over North Korean clandestine activity in Japan besides the abductions including drug smuggling, marine poaching, and spying. Relations between the two countries are severely strained and marked by tension and hostility.

Russia: Relations between the two nations are hampered by the long-running Northern Territories (Kuril) dispute. The dispute over the Southern Kuril Islands deteriorated Russia-Japan relations when the Japanese government published a new guideline for school textbooks on 16 July 2008 to teach Japanese children that their country has sovereignty over the Kuril Islands. Relation between Russia and Japan is improving and getting closer. Recently, Prime Minister Shinzo Abe of Japan and President Vladimir V. Putin of Russia discussed concluding a Japanese-Russian peace treaty in time for the 70th anniversary of the end of World War II in 2015. Russia and Japan are again holding consultations between their foreign and defence ministries, which will improve their bilateral ties and intensify trade & economic cooperation.

STRATEGIC RELATIONS

USA: The United States and Japan have firm and very active political, economic and military relationships. Japan relies on the United States for its national security to a high degree as the later is former's closest ally. Japan and US being world's top economic powers, both also rely on close economic ties for their wealth. The Treaty of Mutual Cooperation and Security between US and Japan declares that both nations will maintain and develop their capacities to resist armed attack in common and that each recognises that an armed attack on either in territories administered by Japan will be considered dangerous to the safety of the other. US has positioned close to 50,000 troops in Japan, the headquarters of the US 7th Fleet and more than 10,000 Marines to undertake military operations. Japan's limited intelligence gathering capability and personnel are focused on China and North Korea, as the nation primarily relies on the US NSA. US and Japan in 2015 enhanced their military relationship by strengthening ties between the two nations on cyber, space and industrial programmes.

India: The friendship between India and Japan has a long history rooted in spiritual affinity and strong cultural and civilizational ties. Both the countries have close military relations. They have shared interests in maintaining the security of sea-lanes in the Asia-Pacific and Indian Ocean, and in cooperation for fighting international crime, terrorism, piracy and proliferation of weapons of mass destruction. The two nations have frequently held joint military exercises and cooperate on technology. India and Japan concluded a security pact on 22 October 2008. Both the countries have also decided to enhance cooperation in anti-piracy operations off the coast of Somalia in the Gulf of Aden and launched a maritime dialogue mechanism. In Dec 2015, the two countries signed a pact regarding the transfer of defence equipment and technology after the annual summit between Prime Minister Narendra Modi and his Japanese counterpart Shinzo Abe. Prime Minister Narendra Modi, in November 2016, during the meeting, India and Japan signed the "Agreement for Cooperation in Peaceful Uses of Nuclear Energy

Australia: Japan describes Australia as its second most important security partner. Japan and Australia have established good bilateral relationships based on mutually harmonising the economic relations. Japan is one of Australia's major economic partners. In recent years, the two countries have also strengthened political, defence and scientific cooperation, which has made them a strategic partner in the Asia-Pacific region. As both countries are allies with the US in the Asia-Pacific region, the two countries have promoted the trilateral relations. The Australia-Japan-United States Trilateral Strategic Dialogue is a key security policy mechanism for Japan. Australia and Japan also have a strong history of cooperation in United Nations peacekeeping missions. Australia and Japan have been maintaining dialogue on humanitarian assistance and disaster relief, maritime security, cyber security, and peacekeeping activities, in order to not only deepen practical cooperation, but also prepare for future challenges.

South Korea: Japan and South Korea are close neighbours. South Korea and Japan are planning to sign a pact on sharing military intelligence. The General Security of Military Information Agreement is aimed at better coping with North Korea's missile threats. In June 2012, Japan and South Korea planned to have military talks on accords to share intelligence and provide each other with fuel and medical support. The accords were aimed at strengthening defence cooperation by sharing important intelligence, mostly on North Korea, and assisting each other's military with fuel and medical supplies during peacekeeping operations abroad. Seoul and Tokyo are important trading and diplomatic partners, but the possibility of such a military treaty is a sensitive topic in South Korea, where many people still harbour strong resentment against Japan's 35-year occupation until the end of World War II. Therefore, as a result of a political firestorm in South Korea's cabinet on the issue, the country was forced to postpone signing of the military agreement with Japan. And now the fate of the accord is uncertain. At present, sharing and protecting of sensitive military data about North Korea and China and missile defences between Japan and South Korea take place through Washington.

United Nations Security Council: Japan, a member of G8, APEC, "ASEAN Plus Three", participant in the East Asia Summit, continues to hold many international cooperation's within the United Nations since 1956. Japan, has participated in many areas such as UN peacekeeping / peacebuilding, international humanitarian relief, nuclear-non-proliferation/ disarmament, and also earned a unique status as a non-nuclear weapon state and one of the world's largest economies. The country is the second-largest contributor to the UN budgets among the Member States, bearing 12.5% of the total budget. It is one of the G4 nations seeking permanent membership in the Security Council. Japan is also a strong advocate of a Reform of the United Nations Security Council (UNSC), in a joint campaign with Germany, India, and Brazil. Japan is striving to gain the permanent seat in the chamber.

DEFENCE CAPABILITIES

ARMY: The Japan Ground Self-Defence Force, or JGSDF, is the armed force of Japan. The largest of the three services of the Japan Self-Defence Forces, the Ground Self-Defence Force operates under the command of the chief of the ground staff. Regionally the JGSDF is organised into five armies, the Northern Army, North Eastern Army, Eastern Army, Central Army, and Western Army. JGSDF divisions and brigades are combined arms units with infantry, armoured, and artillery units, combat support units and logistical support units. The JGSDF has two reserve components: rapid-reaction reserve component

Personnel: 150,863 (8,075 reserve)

Equipment

Category	Name	In Service
MBT	Type 10	65
	Type 90	300
	Type 74	280
APC/MICV /IFV	Type 89	100+
	Type 16	4
	Type 60	30
	Type 73	340
	Type 87 Recon. & Warning Vehicle	100
	Type 82 (Command & Comm.)	230
	Type 96	300
	Komatsu LAV	1,665+

Army continued..

Category	Name	In Service
Artillery	Towed FH70 155mm	480
	Type 75 155 mm self-propelled	140
	Type 99 155 mm self-propelled	20
	M110 203mm howitzers	90
Mortar	Type 96 120mm self-propelled	400
	120mm rifled towed	380
	M2 107mm	90
	M6C-210 60 mm	
	L16 81mm	
MRL	Type 75 130mm	260
	M270	99
RCL Gun	Type 60 106mm	100
Anti-Tank GW	Type 01	1,073
	Type 64	160
	Type 79	250
	Type 87 Chu-MAT	50
	Chu-MPM	46
	Carl Gustaf	
	Panzerfaust-3	
SSM	Type 88	
	Type 12	
SAM	I-HAWK system	520
	Type 81	57
	Type 91	
	200+	
	Type 93	90
	Type 03 Chu-SAM	
	FIM-92A Stingers	80
AA gun	35mm twin	30
	Type 87 self-propelled	68 (200 planned)
SSM (Coastal Defence)	Type 88 (SSM-1)	48 systems (each with six cell)

Army Aviation

Category	Name	In Service
Aircraft	Beechcraft Super King Air LR-2	7
	MU-2 LR-1	2
Combat Helicopter	AH-1S Cobra	88
	Tilt-rotor aircraft (V22)	4
	CH - 47JA	(6 on order)
	AH-64D Apache Longbow	10
	OH-1	38
Liaison & light transport Helicopter	EC225 Super Puma (VVIP)	3
	UH-1H/J	120
	CH-47J/JA	58

Army Aviation continued...

Category	Name	In Service
Liaison & light transport Helicopter	OH-6D	106 (to be replaced)
	UH-60 Blackhawk	34
	TH-480B	12
	V-22B	17
UAV	ScanEagle, R-MAX, RQ-16 T-Hawk	

New Procurements / Upgrades

✦ BAE Systems has been awarded a USD 10.8 million foreign military sales contract to provide two Assault Amphibious Vehicles for Japan. The firm-fixed-price contract will allow BAE to provide the material and technical engineering to build, integrate, test and deliver one AAV command variant and one AAV recovery variant.

✦ The United States has sent a second missile defence radar system to Japan. The new Army Navy/Transportable Radar Surveillance system (AN/TPY-2) aims to enhance sensor coverage for ballistic missile defence of Japan and the US homeland.

✦ Japan is planning to procure amphibious landing craft for the Ground Self-Defence Force for the first time.

✦ The Ministry of Defence has signed a contract with Eurocopter to acquire an EC225, a heavy twin-engine helicopter from Eurocopter's Super Puma family.

✦ Enstrom Helicopter has delivered the eighth TH-480B helicopter to the Japanese Ground Self-Defence Force (JGSDF), bringing the total number of aircraft delivered under the programme to ten. The company is contracted to deliver 30 of the aircraft to the JGSDF, where the fleet will be used for training purposes. The TH-480B will replace the OH-6 as the JGSDF's primary training helicopter.

✦ Lockheed Martin Mission Systems & Sensors, Moorestown, New Jersey, is being awarded a contract for Advanced Electronic Guidance and Instrumentation System (AEGIS) in-service combat systems engineering; computer programme maintenance: annual inspection and regular overhaul execution support: in-country support: staging: replacement of legacy cathode ray tube character readout devices for Japan.

✦ Leonardo-Finmeccanica will provide the latest Identification Friend or Foe (IFF) technology for Japan Ground Self Defence Force (JGSDF). The deal, signed with a Japanese prime contractor, will see Leonardo supplying its SIT-422/5J interrogator for the JGSDF.

✦ Fuji Heavy Industries Ltd. (FHI) has signed a contract with the Ministry of Defence for the initial project of developing a prototype of the next-generation utility helicopter (UH-X) for the Japan Ground Self-Defence Force (JGSDF). FHI will co-develop with Bell Helicopter Textron Inc. an advanced variant of the 412EPI, the latest model of the Bell 412 series.

✦ Bell Boeing, a strategic alliance between Bell Helicopter, a Textron company, and Boeing, was awarded a US Navy contract for five Bell Boeing V-22 Osprey tiltrotor aircraft to be delivered to Japan, marking the first sale of the aircraft through the US government's foreign military sales programme. The contract for the Block C aircraft includes support, training, and equipment.

AH-1S Cobra of Japan Ground Self-Defence Force

The V-22 Osprey is a joint service multirole combat aircraft utilising tiltrotor technology to combine the vertical performance of a helicopter with the speed and range of a fixed-wing aircraft. *Image Courtesy: Boeing.*

NAVY: The Japan Maritime Self-Defence Force, or JMSDF, is the naval branch of the Japan Self-Defence Forces, tasked with the naval defence of Japan. The JMSDF has a large fleet and its main tasks are to maintain control of the nation's sea lanes and to patrol territorial waters.

Naval Bases: Maritime Staff Office (Tokyo), Self Defence Fleet Headquarters (Yokosuka), main bases (Yokosuka, Kure, Sasebo, Maizuru and Ominato), other bases (Yoichi, Hakodate, Kobe, Shimonoseki, and Katsuren) and air bases (Hachinohe, Ominato, Simofusa, Tateyama, Atsugi, Komatsujima, Tokushima, Iwakuni, Ozuki, Oomura, Kanoya, Naha, Matzuru).

Personnel: 45,364 (11,000 for naval aviation)

Equipment

Category	Name	In Service
Helicopter Destroyer (DDH)	Hyuga class	2
	Izumo-class	2
	Shirane class	1 (to be replaced by Izumo class)
Guided Missile Destroyer (DDG)	Kongo class	4
	Hatakaze class	2
	Atago class	2
Destroyer (DD)	Takanami class	5
	Asagiri class	8
	Hatsuyuki class	2
	Murasame class	9

Navy continued...

Category	Name	In Service
Destroyer (DD)	Akizuki class	4
	Asahi-class	1 (to be commissioned in 2018)
Frigate/ Destroyer Escorts (DE)	Abukuma class	6
Submarine	Soryu class	7 (5 on order)
	Oyashio class	10
Patrol vessel	Hayabusa class	6
Landing ship	Osumi class	3
	1-Go class	2
	Yura class	2
Landing craft	YF 2121 class	11
	YF 2150 class	2
	Landing Craft Air Cushion	6
Mine warfare ship	Uraga class	2
	Yaeyama class	1
	Uwajima class	2
	Sugashima Class	12
	Enoshima class	3
	Ieshima-class	2
	Hirashima class	3

Naval Aviation

Category	Name	In Service
Aircraft	P-3C Orion	70
	U 36-A Learjet 35	4
	NAMC YS-11T	3 (to be replaced by KC-130R)
	US-2	5
	EP-3C	4
	UP-3D	3
	Fuji T-5 (trainer)	40
	Beechcraft King Air	31
	Kawasaki P-1	5 (to replace P-3C Orion)
Helicopter	UH-60J	19
	SH-60J	40
	SH-60K	48
	MH-53E	8 (to be replaced)
	MCH-101	4
	MD 500	5
	CH-101	2
	Eurocopter EC 135 (TH-135)	15

New Procurements / Upgrades

✦ Airbus Helicopters Japan has signed a contract with Japan Maritime Self-Defence Force (JMSDF) for Performance-based Logistics (PBL) benefiting from Airbus Helicopters' new customer fleet availability service offer called HCare. The comprehensive agreement covers maintenance, parts supply, as well as repair of 15 TH-135 training helicopters currently in operation at the JMSDF.

✦ JMSDF received final 3 of 13 TH-135 helicopters from Airbus Helicopters Japan. The choppers will allow and step-up training activities of its naval pilots.

✦ Japan has unveiled a large warship - its largest ever since the World War II. The 19,500-ton warship -- a helicopter carrier -- is named 'Izumo' and will be deployed with the country's Maritime Self-Defence Force (MSDF). The ship, having a flight deck of 250 meters long, is designed to carry up to 14 helicopters. The warship can also operate the US Osprey tilt-rotor aircraft as well but is not designed for any fixed-wing aircraft. Another destroyer of the Izumo size is under construction.

✦ Northrop Grumman Corporation has delivered the first of four AQS-24A airborne mine-hunting vehicles to the Japanese Maritime Self-Defence Force (JMSDF) for deployment on Japan's new MCH-101 helicopter platform. The AQS-24A is a high-speed mine-hunting system that is primarily towed from the MH-53E helicopter and has been adapted easily to the JMSDF version of the EH-101 aircraft.

✦ Northrop Grumman will deliver the first of four Airborne Laser Mine Detection Systems (ALMDS) to the JMSDF to operate in conjunction with the AQS-24A. ALMDS is a laser-based, light detection and ranging sensor system that detects, classifies and localizes near-surface mine-like objects from above the waterline and is complementary to the AQS-24A.

✦ 11 AW101 AgustaWestland helicopters are to be procured for minesweeping and transportation role. Two variants of MCH-101 & CH-101 will be procured.

✦ Japan is planning to build a 19,500-tonne aircraft carrier capable of housing helicopters. Japan already has two helicopter carriers - the Hyuga deployed in March 2009 and the Ise deployed in March 2011 - but the planned new vessel will be bigger. A keel-laying ceremony for the first 22DDH helicopter carrier for the Japan Maritime Self-Defence Force (JMSDF) was held at IHI Marine United's (IHIMU's) Yokohama on 27 January 2012. The 22DDH will cost around USD 1.04 billion. It can carry 11 helicopters.

✦ New class of Destroyers are planned to replace the Hatsuyuki class ships; total 19 new ships are being planned.

✦ Japan has awarded Northrop Grumman a series of follow-on contracts by the Japanese Maritime Self-Defence Force (JMSDF) to supply three additional AQS-24A airborne mine hunting systems. The contracts, which also include airborne electronics and ground-based support equipment, follow an initial contract awarded in October 2011 for Northrop Grumman to deliver Japanss first

AQS-24A. The JMSDF will use all four complete AQS-24A airborne mine hunting sonar systems on its new MCH-101 helicopters.

✦ The Japan Maritime Self-Defence Force has purchased Northrop Grumman's four helicopter-mountable, laser mine detection systems. The Airborne Laser Mine Detection System (ALMDS) will help protect its coastline and the daily maritime traffic coming in and out of the country's ports.

✦ The Akizuki class destroyer (19DD) of JMSDF will replace the five Hatsuyuki Class destroyers in service. Currently the first vessel is in service.

✦ JMSDF already commissioned three submarines in the Soryu class submarine (16SS). The Soryu Class submarine can be armed with UGM-84 Harpoon anti-ship missiles and 6 torpedo tubes 533mm.

✦ JMSDF will be receiving 10 more EC 135 from Eurocopter Japan.

✦ USD 490 million contract paves way for Aegis Ballistic Missile Defense on new US Navy destroyer, two Japanese Atago-class destroyers and three South Korea KDX-III destroyers. The seventh and eighth Aegis ships will join Japan's fleet. Aegis provides four of Japan's Kongo-class destroyers and two Atago-class destroyers with advanced sea, air and undersea threat detection capabilities.

Japan Maritime Self Defence Force (JMSDF) submarine Hakuryu (SS-503)

The Japan Maritime Self-Defence Force helicopter destroyer JS Hyuga (DDH-181) is underway in the Pacific Ocean as Sea Hawk helicopters from the Chargers of Helicopter Anti-Submarine Squadron (HS) 14 fly in formation alongside the ship. *Image Courtesy: US Navy.*

AIR FORCE: The Japan Air Self-Defence Force, or JASDF, is the aviation branch of the Japan Self-Defence Forces responsible for the defence of Japanese airspace and other aerospace operations. The JASDF carries out combat air patrols around Japan, while also maintaining an extensive network of ground and air early warning radar systems. The JASDF also provides air support for ground and sea operations of the JGSDF and the JMSDF and air defence for bases of all the forces. The JASDF is not allowed to have strategic bombers for that would go against the self defence–only policy.

Air Bases: Kasuga, Hamamatsu, Komaki, Niigata, Shizuhama, Chitose, Naha, Hofu, Komatsu, Iruma, Misawa, Ashiya, Tsuiki, Gifu, Miho, Kasuga, Miho

Personnel: 46,940

Equipment

Category	Name	In Service
Combat	F-35A	6 (42 on order)
	Mitsubishi F-2 A/B (fighter/fighter trainer)	63
	F-4 Phantom II (fighter/reconnaissance)	71
	F-15 Eagle (fighter/fighter trainer)	154
Transport/ Aerial refueling/ Aerial Surveillance	C-130H	15
	Kawasaki C-1 (EW)	1
	Boeing 747-400	2
	Boeing KC-767	4
	KC-130H	1
	Gulfstream IV	5
	Boeing KC-767 (refueling)	4
	Hawker 800	30
	Kawasaki C-1/C-2	30
	NAMC YS-11 (EW)	7
	U-125A BAe	32
AWACS	E-2 Hawk Eye	13
	Boeing E-767	4
Helicopter	UH-60 Mitsubishi	35
	CH-47J	16
Trainer	Kawasaki T-4	200
	Mitsubishi F-2B	20
	Hawker T-1 Jayhawk	13
	NAMC YS-11	2
	Fuji T-3	48
Weapon	AIM-9 Sidewinder	
	Mitsubishi AAM-3	
	Mitsubishi AAM-4	
	AIM-7 Sparrow	
	ASM-1 and ASM-2 anti-ship missiles	
	J/AAQ-2 FLIR	

Air Defence

✦ PAC-2, PAC-3 (MIM-104 Patriot) systems 6 group with 24 batteries

✦ Type 81 SAM systems about 36 of such units are still in service

✦ Type 91 SAM more than 350 systems.

✦ M167 Vulcan Air Defence System (VADS).

New Procurements/ Upgrades

✦ Northrop Grumman Corporation has received a US Navy contract modification worth USD 285,975,244 for non-recurring engineering and recurring support to configure the first Japanese Air Self Defence Force (JASDF) E-2D Advanced Hawkeye.

✦ The Japan Ministry of Defence has announced its selection of the Lockheed Martin F-35 Lightning II (a total of 42 aircraft) as the Japan Air Self Defence Force's (JASDF) next generation fighter aircraft. Further, on June 2012, Japan signed a formal agreement with the United States to buy an initial four F-35 fighters built by Lockheed Martin and other equipment for 60 billion yen (USD 756.53 million).

✦ Japan will be acquiring 6 Lockheed Martin KC-130R aircraft and 30 Rolls Royce T-56-A-16 engines from US Defence Security Cooperation Agency through Foreign Military Sale. The contract will also provide regeneration, overhaul, modifications and support for 6 KC-130R aircraft and associated engines Also included are 6 non-EDA spare T-56-A-16 engines, 6 AN/APS-133 Radars, 9 AN/ APX-119 Transponder Systems (6 installed and 3 spares), transportation, aircraft ferry support, repair and return, spare and repair parts, support equipment, tools and test equipment, technical data and publications, personnel training and training equipment, US Government and contractor engineering, technical, and logistics support services, and other related elements of logistics support. The estimated cost is USD 170 million.

✦ Japan is looking to join the United States, China and Russia with a stealth fighter that can be ready for a prototype (dubbed "Shinshin," or "Spirit") test flight in just three years, significantly upping the ante in the intensifying battle for air superiority in the Pacific. The country has put 39 billion yen (USD 473 million) into the project since 2009.

✦ RF-4 Phantoms are to be phased out and the equivalent number of F-15 will receive the SAR pod.

✦ PAC-2 systems are under up gradation to PAC-3 systems.

✦ Japan has acquired UH-60J helicopters to replace current search and rescue helicopters in the Japanese Air Self Defence Force.

✦ Japan has announced a request for proposals for its next-generation FX fighter aircraft programme to replace F-4EJs.

Gen. Yoshiyuki Sugiyama, Japan Air Self Defence Force Chief of Air Staff, address the ceremony audience as Japan's first F-35A aircraft is revealed at the Lockheed Martin's production facility in Fort Worth, Texas, Sept. 23 2016. *Image Courtesy: Lockheed Martin photo by Beth Steel*

Anti-Ballistic Missile Programme

Japan decided to join in the American anti-ballistic missile (ABM) defence programme in 1998, soon after North Korea's Taepodong-1 missile launch over northern Japan. In August 1999, Japan, Denmark and the US governments signed a Memorandum of Understanding (MoU) of joint research and development on the Aegis Ballistic Missile Defence System. In 2003, the Japanese government decided to deploy two types of ABM system, sea-based Aegis and land-based Patriot Advanced Capability 3 (PAC-3) ABM. On April 2012, Japan deployed Patriot missiles at three military facilities in the greater Tokyo region and the defence ministry dispatched three Aegis destroyers carrying sea-based interceptor missiles, reportedly to the East China Sea, as a precautionary measure to boost its defences against a planned North Korean rocket launch.

The Aegis Ballistic Missile Defence System (Aegis BMD) is a United States Department of Defence's Missile Defence Agency programme developed to provide defence against ballistic missiles. In 2010, Japan, in partnership with the US, has successfully conducted an Aegis Ballistic Missile Defence (BMD) intercept flight test off the coast of Hawaii. The JMSDF has equipped three vessels for Long Range Surveillance and Tracking (LRST) and engagement: JS Kongo, JS Chokai, JS Myoko, and in 2010 the JS Kirishima.

Major Armaments Producers

The Japanese defence industry is the major supplier of the nation's own Self-Defence Forces. The country has a strong base of defence R&D and production the major contribution in the defence development & production are from Mitsubishi Heavy Industries (MHI) and Kawasaki, apart from these industries there is consortium of the industries which also work for defence equipment development & production.

Domestic suppliers: Mitsubishi Heavy Industries, Mitsubishi Electric, NEC, Kawasaki Heavy Industries, Toshiba, Fujitsu, Fuji Heavy Industries, IHI Corporation, Komatsu Limited, Japan Steel Works, Hitachi Ltd., Daikin Industries, Oki Electric Industry, ShinMaywa, Howa, Sumitomo Heavy Industries.

Foreign suppliers: United States, United Kingdom, Germany, Italy, Switzerland, France, Russia, Sweden.

DEFENCE DEALS

The country is going to expand its defence capabilities in the entire horizon, as per the policy change the procurement of the arms is going to be raise in the coming decade.

Joint Venture Programmes

Lockheed Martin, Mitsubishi Electric Corporation, SAMPA Kogyo K.K. and Mitsubishi Corporation have formed a joint venture company that will serve the emerging combat system engineering and lifetime support requirements of the Japan Maritime Self-Defence Forces (JMSDF). The joint venture company, MLS Corporation, represents a full service partnership arrangement between industry and government, and leverages Japanese and US industry expertise in life cycle support and integration services to benefit the JMSDF. Through the MLS partnership, Lockheed Martin will transfer its proven models to Japan for application across JMSDF surface ship programmes.

Arms Export

Japan has recently relaxed ban on military exports to make its defence manufacturers more competitive. According to media reports, the country has cancelled its self-imposed decades-old ban on export of military equipment. The move will open new corridors for defence contractors. The ban was issued in 1967 when Japan decided not to sell defence products to communist nations and those involved in international conflicts. Later, the country extended the ban on exports with defence industries barred from engaging in any development or production of weapons with any country other than United States. With the relaxation, Japan can now indulge in joint development and production of weapons with other countries and also provide military weapons and support for humanitarian missions. Though the ban has been lifted, Japan will limit its exports to strategic allies like the USA. The move is intended to bolster the domestic arms industry and reduce national defence spending. Under the new standards, Japan will be able to transport military equipment for missions of peace-building and international cooperation. The new rules will allow Japan to develop and produce arms jointly with the US and European countries and to export military equipment for peaceful and humanitarian purposes such as UN peace-keeping operations. Tokyo will also be allowed to provide defensive equipment, such as helmets and bullet-proof vests, to the limited number of countries in which its Self-Defence Forces are deployed.

Arms Import

The Japan Ministry of Defence (MoD) has released the five-year mid-term defence programme which lists major equipment purchases and envisions a total budget of approximately 23, 490 billion yen (USD 267.6 billion) to be spent over the next five years. The US-Japan security alliance is the cornerstone of relation between US contractors and Japan defence market. Major US defence contractors, as well as other US defence makers of defence-related technologies and equipment, play important role in the Japanese defence market, participating through product licensing and direct sales.

The Equipment Procurement and Construction Office (EPCO), under Japan Ministry of Defence (MoD) conducts all procurement of necessary equipment for executions of self-defence force missions (firearms, guided weapons, telecommunications, ships, aircraft, vehicles, machinery, ammunitions, foods, fuel, textile, other necessary materials) and designated primary materials by Minister of Defence for the services. EPCO is a procurement agency that executes the big budget of about 1.3 trillion yen which is equivalent of almost 1/4 of whole defence related costs.

DEFENCE BUDGET

Total defence spending: $42 billion (2016)

Estimated defence spending in terms of GDP: 1% (2016)

Japan's defence ministry has requested USD 50.2 billion defence budget for F.Y. 2017, which is an increase of 2.3% from 2016. The request includes proposals to develop and potentially purchase new anti-ballistic missiles that can be launched from ships or land, and to upgrade and extend the range of the country's current land-based missile defence systems which also includes buying of an additional submarine and new fighter aircraft.

Defence Related Expenditures (F.Y 2016-17 – (Unit: ¥ 100 million)

Exchange rate for FY 2017: USD = JPY ¥ 110

Source: MOD/Budget, Japan

	FY 2016 Budget	FY 2017 Budget Request
Defence related expenditures	48607 (50541)	49735 (51685)
Personnel & Provisions expenses	21273	21551
Material Expenses	27135 (29069)	28184 (30134)
Obligatory outlay expenses	17187 (18377)	17958 (19165)
General material expenses	9948 (10692)	10226 (10969)
Future Obligations concerning new contracts	20800 (22875)	22997 (25052)

Budget Breakdown Details (F.Y 2016-2017) – (Unit: ¥ 100 million)

Source: MOD/Budget, Japan

	FY 2016 Budget	FY 2017 Budget Request
Defence expenditure	48607	49735
Ministry of Defence	48607	49735
Ground Self Defence Force (GSDF)	17489	17754
Maritime Self Defence Force (MSDF)	11954	12166
Air Self Defence Force (ASDF)	11196	11570
(Ministry of Defence Head Office)	47152	48027
Acquisition, Technology and Logistics Agency	1263	1504

Details of Material Expenses (Contract Base) (F.Y 2016-2017) – (Unit: ¥ 100 million)

Item	FY 2016	FY 2017
Research & Development	1211	1391
Equipment Acquisition	5203	6154
Aircraft Acquisition	4232	3922
Shipbuilding, etc.	1567	1281
Facility Improvements, etc.	1541	1867
Maintenance, etc.	11656	13160
Other	803	813

Details of Obligatory Outlay Expenses (F.Y 2016-2017) – (Unit: ¥ 100 million)

Item	FY 2016	FY 2017
Research & Development	780	977
Equipment Acquisition	3789	3951
Aircraft Acquisition	1893	2984
Shipbuilding, etc.	1647	1661
Facility Improvements, etc.	1102	1214
Maintenance, etc.	7440	6646
Other	108	86

Major R&D programmes for F.Y. 2017

✦ Develop a long-range ship-to-air missile capable of countering enemy aircraft, etc. in order to strengthen the air defence capability of destroyers.

✦ Develop the Type-12 surface-to-ship missile (modified) with upgraded functions and performance, including a longer range compared with existing missiles and a new air-to-ship missile for patrol aircraft in order to enhance the capability to counter enemy surface vessels, etc.

✦ Research intended to realize continuous operation of information and communication infrastructure of the MOD/SDF in the event of cyber-attacks.

✦ Conduct research on autonomous surveillance technology and a sensor system for unmanned underwater vehicles that will significantly enhance the underwater surveillance capability.

✦ Conduct research on an innovative electromagnetic acceleration system intended to extend the range and increase the destructive power of shells by increasing the shell speed to a level that would be impossible to achieve for conventional artillery.

✦ Conduct research concerning a whole system design using a simulation model and the reduction of the size of high-power engines in order to enhance sea-surface mobility.

✦ Conduct research on a simulator that enables vehicle simulations under simulated onshore, water's edge and off-shore environments.

F.Y. 2017 Budget - security of seas and airspace

✦ Capability improvement for fixed-wing patrol aircraft (P-3C) (¥500 million)

✦ Life extension of fixed-wing patrol aircraft (P-3C) (3 aircraft: ¥1.8 billion)

✦ Life extension of patrol helicopters (two SH-60Ks and two SH-60Js) (4 helicopters: ¥4.7 billion)

✦ Life extension of imagery intelligence gathering aircraft (OP-3C) (1 aircraft: ¥700 million)

✦ Acquisition of ship-based multipurpose helicopter

✦ Capability Improvement of Airborne Warning And Control System (AWACS) (E-767) (2 aircraft: ¥22 billion)

✦ Acquisition of Unmanned Aerial Vehicles (Global Hawk) (¥17.3 billion)

✦ Life extension of destroyers (life extension work for 5 destroyers and parts procurement for 6 destroyers: ¥5.6 billion)

✦ Construction of a submarine (3,000 tonnes) (1 submarine: ¥76 billion)

✦ Life extension of submarines (life extension work for 3 submarines and parts procurement for 6 submarines: ¥3.8 billion)

✦ Building of an ocean minesweeper (1 vessel: ¥17.8 billion)

✦ Building of an ocean surveillance ship (1 ship: ¥23.4 billion)

CONCLUSION

Japan will build up its defence capabilities while focusing on the further enhancement of joint functions in order to seamlessly and dynamically fulfil its defence responsibilities, which include providing effective deterrence and response to a variety of security situations, supporting stability in the Asia-Pacific region, and also improving the global security environment. The security environment surrounding Japan has become increasingly severe, with various challenges and destabilising factors becoming more tangible and acute. Japan is striving to develop appropriate defence capabilities to protect the life and properties of its nationals and to defend the territorial land, sea, and airspace of Japan. At the same time, it is strengthening the Japan–US Alliance with the United States, which shares basic values and interests with Japan. The peace and security of Japan is ensured through developing seamless defence measures by coupling Japan's own defence capabilities with the Japan-US security arrangements. Japan and the United States are revising their mutual defence guidelines for the first time in nearly two decades to respond to China's military expansion and increase Japan's role in regional defence. US and Japan are pursuing a wider partnership that requires enhanced capabilities and greater shared responsibilities. The revision, the first since 1997, has come at a time of heightened Japan-China tensions over islands claimed by both countries in the East China Sea, as well as continuing concern about North Korea's missile and nuclear weapons development. The revised guidelines will factor in policy changes under Prime Minister Shinzo Abe that enable Japan to shoulder more responsibility for its own and regional defence, and relieve some of America's military burden in the Asia-Pacific region. Japan is gradually building up its forces in the region. This strategic shift of raising the prowess of its armed forces and equipment are seen as a significant step to defend and respond strongly to China's and North Korea's military growth in East Asia. Japan and India have been increasing cooperation in the field of maritime security. It is important to further strengthen our cooperation. Both India and Japan have signalled a keenness to beef up their defence ties in recent months to counter an increasingly assertive China. The country, in an effort to protect its territories and waters, is strengthening the naval and air force by acquiring submarines, ships, next generation fighter aircraft and adding new amphibious infantry vehicles. Japan's increasing military power is however limited keeping its economic conditions and shrinking military expenditures in perspective. The nation relies to a large degree for its defence on its alliance with the United States, which has a significant number of fighters and other aircraft, along with some 50,000 troops, stationed around the Japanese archipelago. According to defence analysts, the current strategy for Japan should be to remain closer to US as an alliance partner while strengthening itself as a potent force. Japan is already started focusing on making its military presence more visible in the region, in a move to discourage China from extending its reach into the waters controlled by Japan. In July 2014, in a historic shift in Japan's post World War II defence policy, Abe's Cabinet approved a new interpretation of Japan's pacifist constitution. The reinterpretation allows the military to defend the US and other allies under what is known as collective self-defence. Japan also adopted defence guidelines last December that make southern island defence a priority, notably the Japanese-controlled disputed islands, which are called Senkaku in Japan and Diaoyu in China.

CONTACT DETAILS

Ministry of Defence

5-1 Honmura-cho, Ichigaya,
Shinjuku-ku, Tokyo 162-8801
Phone: 03-5366-3111
Website: http://www.mod.go.jp
E-mail: infomod@mod.go.jp

Equipment Procurement and Construction Office (EPCO)

http://www.epco.mod.go.jp/en/

JORDAN
(Capital: Amman)

INTRODUCTION

Area: 89,342 sq km
Population: 8,185,384
Coastline: 26 km
Maritime claims: Territorial sea: 3 nm

KEY POLITICAL PERSONS

HEAD OF STATE & COMMANDER-IN-CHIEF: Abdullah II bin al-Hussein

PRIME MINISTER & MINISTER OF DEFENCE: Hani Mulki

KEY DEFENCE PERSONS

CHAIRMAN OF JOINT CHIEF OF STAFF: Lt. Gen. Mahmoud Abdel-Halim Freihat

COMMANDER OF THE AIR FORCE: Maj. Gen. Mansour Salem Jobour

ECONOMY

The economy of Jordan is among the smallest in the Middle East, with insufficient supplies of water, oil, and other natural resources underlying the government's heavy reliance on foreign assistance. Jordan relies on natural gas for 10% of its domestic energy needs. Jordan used to depend on Iraq for oil until the Iraq invasion in 2003 by the United States. The country lacks in coal reserves, hydroelectric power; large tracts of forest or commercially viable oil deposits. Jordan is also classified as an emerging market. Since King Abdullah II's accession to the throne in 1999, liberal economic policies have been introduced, such as such as opening the trade regime, privatizing state-owned companies, and eliminating most fuel subsidies, which in the past few years have spurred economic growth by attracting foreign investment and creating some jobs. This has resulted in a boom lasting for a decade continuing even through 2009. Jordan is now one of the freest and most competitive economies in the Middle East. In 2011 the government approved two economic relief packages and a budgetary supplement, largely to improve the living conditions for the middle and poor classes. Jordan's finances have also been strained by a series of natural gas pipeline attacks in Egypt, causing Jordan to substitute more expensive heavy fuel oils to generate electricity. Amman likely will continue to depend heavily on foreign assistance to finance the deficit in 2012. Jordan has the third freest economy in the Middle East and North Africa, behind only Bahrain and Qatar.

GDP: $82.73 billion (2015 est.), $80.71 billion (2014 est.)

Real Growth Rate: 2.5% (2015 est.), 3.1% (2014 est.)

Industries: Clothing, fertilizers, potash, phosphate mining, pharmaceuticals, petroleum refining, cement, inorganic chemicals, light manufacturing, tourism

Oil – proved reserves: 1 million bbl (1 January 2016 est.)

Natural gas - proved reserves: 6.031 billion cu m (1 January 2016 est.)

Total Exports: $7.829 billion (2015 est.), $8.385 billion (2014 est.)

Export Commodities: Clothing, fertilizers, potash, phosphates, vegetables, pharmaceuticals

Major Markets: US 21%, Saudi Arabia 16.5%, Iraq 10.3%, India 8.7%, UAE 4.8%, Kuwait 4.4% (2015)

Total Imports: $18.04 billion (2015 est.), $20.35 billion (2014 est.)

Import Commodities: Crude oil, machinery, transport equipment, iron, cereals

Major Suppliers: Saudi Arabia 15.4%, China 12.8%, US 6.2%, Germany 4.7%, UAE 4.2% (2015)

INTERNATIONAL DISPUTES

A 2004 agreement settled border dispute with Syria pending demarcation.

DEFENCE

ARMY

Personnel: 90, 000

Equipment

Category	Name	In Service
MBT	Al Hussein Challenger 1	300+
	Khalid	350
	M60A1/A3	342
	Scorpion	80
APC/ IFV	M113 A2	1260
	Ratel	324
	Spartans	100
	Saracen	120
	Saladin	130
	Ferret	140
Artillery	105 mm. M102	54
	155 mm. M59	17
	155 mm. M114	38
	203 mm. M114A2	128
	105 mm. M52	35
	155 mm. M44	29
	155 mm. M109	250+
	203 mm. M110A2	100
	Norinco 120mm. Howitzer	20+
Mortar	120 mm.	
	107 mm.	
	81 mm.	
Anti-Tank GW	TOW launchers	330
	DRAGON	310
	Javelin	30
AA Gun	40 mm. M42	200
	Vulcan	100
	ZSU234 Shilka	16
SAM	Redeye	270
	Hawk	532
	SA6, SA8, SA9, SA13, SA14, Starburst	16

New Procurements/ Upgrades

✦ Paramount Group has been awarded multi-million dollar contract for production of combat vehicles in Jordan. The contract for the production of the first 50 Mbombes was signed by the CEO of KADDB, Major-General Omar Al Kaldi and the Chairman of Paramount Group, Ivor Ichikowitz at IDEX 2015.

✦ Jordan has requested for the procurement of JAVELIN Anti-Tank Guided Missiles, Fly-to-Buy Missiles, JAVELIN Command Launch Units (CLUs) with Integrated Day/Thermal Sight and associated parts, equipment, training and logistical support for a complete package worth approximately $388 million from the US.

✦ 12 HIMARS artillery rocket system on order.

✦ A military cooperation agreement between the Jordan Armed Forces-Arab Army and the Kenyan armed forces was signed in 2015, stipulating exchanging expertise in defence, training and anti-terrorism fields.

✦ Raytheon Co., Ft. Wayne, Indiana, was awarded a USD 13,846,183 firm-fixed-price, foreign military sales contract (Jordan) with options for materials and services for the Jordan Armed Forces Automated Command and Control (J4ISR-C2) Tactical System (JCTS) project.

King Abdullah II, during a visit to the King Abdullah II Design and Development Bureau (KADDB), accompanied by Chairman of the Joint Chiefs of Staff, Gen. Mishaal Zaben. *Image Credit: KADDB*

NAVY

Personnel: 480

Equipment

Category	Name	In Service
Patrol Craft	Al Hussein Class	3
	Faysal Class	4
	Abdullah Class	6
	Al Hashim	3
	Bremse Class	2
	Coastal craft	5

New Procurements/ Upgrades

No major RFPs/Contracts were signed in recent years.

AIR FORCE

Personnel: 12,500

Equipment

Category	Name	In Service
Fighter	F 16A	12
	AT-802	4
	AC-235	1
	AC-295	1
Transport	CASA C-212 Aviocar	2
	An-28	1
	C-208	4+
	C-130H	4
Helicopter	AS332	10
	EC-635	12
	UH-60L	8
	Challenger 604	2
	L-1011 Tristar	1
	S-70A	5
	SA-316C	1
	AH-1SW Cobra	21
Trainer	C-101 Aviojet	13
	Slingsby T-67 Firefly	16
	TB 20 Trinidad	2
	MD 500D	8
Weapons	MANPADS, MIM-23B/E, MIM-104 Patriot, 9K33 Osa, 9K35 Strela-10, SA-22 Pantsir-S1E	

New Procurements/ Upgrades

✦ King Abdullah II Design and Development Bureau (KADDB) has awarded a contract to ATK to modify two of the country's CASA-235 transport aircraft into highly-capable and cost-effective special mission aircraft.

✦ The Government of Jordan has signed a Letter of Intent with Boeing for the AH-6i Light Attack/Reconnaissance helicopters which will enhance border security.

Jordanian F-16 Fighting Falcon aircraft with Squadron One flies an air refuelling mission over Shaheed Mwaffaq Air Base, Jordan, Oct. 17, 2011, during Eager Tiger Falcon Air Meet 2011. *Image Credit: US Air Force*

Defence Expenditure

Total defence spending: $2.5 billion (2015)

Estimated defence spending in terms of GDP: 4.31% (2015)

Domestic Suppliers: King Abdullah Design and Development Bureau.

Foreign Suppliers: USA, UK, France, China, Russia, South Africa, South Korea, Taiwan, Iraq, Ukraine.

Annual imports: $300 million (est.)

Annual exports: $72 million (est.)

Defence Production and R&D

The Jordanian Armed Forces own much of the country's defence industry and are closely involved in research, development, production and marketing of its products. The *King Abdullah II Design and Development Bureau*, KADDB was established to provide an indigenous capability for the supply of scientific and technical services to the Jordanian Armed Forces (JAF). KADDB was also created to provide a one-stop solution for the supply of defence and commercial equipment optimised for Middle East requirements.

Jordan Light Vehicle Manufacturing LLC (JLVM) is a joint venture between KADDB and the Jankel Group Ltd. of the UK to design, develop, produce, and market military vehicles.

Jordan Ammunition Manufacturing and Services Co (JORAMMO) caters for the ammunition needs of the Jordanian Armed Forces as well as other armed forces in the region and worldwide. The Laboratory has a full capability to test and evaluate different type of Small Arms, Light Weapons Ammunition, Armour Materials and Ballistic Resistance Systems for Military and Law Enforcement applications according to various international testing standards. KADDB's numerous products in the field of defence industries includes Al-Washaq ATV Model, the Stallion 4x4 LAV Model, in addition to the 8x8 wheeled Armoured Vehicle Model.

Defence Procurement

The Directorate of Procurement is part of General Headquarter of Jordan Armed Forces. It was established with the mission of executing the whole purchases of Jordan Armed Forces requirements and needs. The Directorate of Procurement carries out sales surplus and unserviceable equipment, Vehicles, Weapons, spare parts and other types of goods. The main duties of the Directorate of Procurement are:

✦ Executing the policy of General Headquarters of the Jordan Armed Forces.

✦ Producing and developing the rules and regulations related to procurement in cooperation with concerned government department.

✦ Specifying the best and optimal sources of purchase whether locally or abroad.

✦ Signing, implementing and pursuing of contracts regarding buying, items, Services Etc. with the concerned parties.

✦ Implementing the laws, regulations and customs instructions regarding purchasing process.

✦ Authenticating the latest references, contract, catalogues, specifications and standardization.

✦ Classifying and arranging local and foreign purchasing resources.

CONTACT DETAILS

Ministry of Defence
PO Box 79
Amman
Jordan

King Abdullah II Design and Development Bureau
928125 Amman, 11190
Jordan
Tel: +962 6 4603230
Fax: +962 6 5627203 - +962 6 4626804, (962) 6 5001186
E-Mail: info@kaddb.mil.jo

Directorate of Defence Procurement
P. O. Box: 926680
Tel: (+962) 6 5000108
Fax: (+962) 6 5001186
Vehicles, Weapons & Ammunitions
vehicle@jafdop.mil.jo
Communications & Electronic Systems.
commsys@jafdop.mil.jo
Quartermaster, Food, and Medical Supplies.
qmaster@jafdop.mil.jo
Shipping & Insurance.
shipin@jafdop.mil.jo
Military Customs & Clearance.
customs@jafdop.mil.jo
Loans & Eastern Supply.
loans@jafdop.mil.jo
Central Tenders Committee
ctc@jafdop.mil.jo

KAZAKHSTAN
(Capital: Astana)

INTRODUCTION

Area: 2,724,900 sq km

Population: 18,360,353 (July 2016 est.)

Coastline: 0 km (Landlocked) Note: Kazakhstan borders the Aral Sea, now split into two bodies of water (1,070 km), and the Caspian Sea (1,894 km)

Maritime claims: None

KEY POLITICAL PERSONS

PRESIDENT & SUPREME COMMANDER OF THE ARMED FORCES: Nursultan A Nazarbayev

PRIME MINISTER: Bakytzhan Sagintayev

KEY DEFENCE PERSONS

DEFENCE MINISTER: Col. Saken Zhasuzakov

COMMANDER OF LAND FORCES: Maj. Gen. Murat Bektanov Karibayevich

COMMANDER OF NAVY: Vice Adm. Zhandarbek Zhanzakov

COMMANDER OF AIR DEFENCE FORCES: Lt. Gen. Ormanbetov Nurlan Sekenovich

ECONOMY

Kazakhstan, geographically the largest of the former Soviet republics, excluding Russia, possesses substantial fossil fuel reserves and other minerals and metals, such as uranium, copper, and zinc. It also has a large agricultural sector featuring livestock and grain. The government realizes that its economy suffers from an overreliance on oil and extractive industries and has embarked on an ambitious diversification programme, aimed at developing targeted sectors like transport, pharmaceuticals, telecommunications, petrochemicals and food processing. Kazakhstan's vast hydrocarbon and mineral reserves form the backbone of its economy. The country is landlocked and depends on Russia to export its oil to Europe. In 2010, Kazakhstan joined Russia and Belarus to establish a Customs Union in an effort to boost foreign investment and improve trade. The Customs Union evolved into a Single Economic Space in 2012 and the Eurasian Economic Union (EEU) in January 2015. The economic downturn of its EEU partner, Russia, and the decline in global commodity prices have contributed to an economic slowdown in Kazakhstan, which is experiencing its slowest economic growth since the financial crises of 2008-09.

GDP: $458.2 billion (2015 est.); $453 billion (2014 est.)

Real Growth Rate: 1.2% (2015 est.); 4.3% (2014 est.)

Industries: Oil, coal, iron ore, manganese, chromites, lead, zinc, copper, titanium, bauxite, gold, silver, phosphates, sulphur, uranium, iron and steel; tractors and other agricultural machinery, electric motors, construction materials

Total Exports: $46.29 billion (2015 est.); $80.24 billion (2014 est.)

Export Commodities: Oil and oil products, natural gas, ferrous metals, chemicals, machinery, grain, wool, meat, coal

Major Markets: China 15.1%, Russia 12.3%, France 9.2%, Germany 7.9%, Italy 6.7%, Greece 4.1% (2015)

Total Imports: $33.65 billion (2015 est.); $43.55 billion (2014 est.)

Import Commodities: Machinery and equipment, metal products and foodstuff

Major Suppliers: Russia 32.9%, China 25.9%, Germany 4.2% (2015)

INTERNATIONAL DISPUTES

Creation of a seabed boundary with Turkmenistan in the Caspian Sea remains under discussion; field demarcation of the boundaries commenced with Uzbekistan in 2004 and with Turkmenistan in 2005; ongoing demarcation with Russia began in 2007; demarcation with China was completed in 2002; Kyrgyzstan has yet to ratify the 2001 boundary delimitation with Kazakhstan.

DEFENCE

Kazakhstan had its first military parade in its history on 7 May 2013, which took place at Otar military base. The Ground Forces have four regional commands: Regional Command Astana, Regional Command South at Taraz, Regional Command East at Semipalatinsk, Regional Command West at Aktobe, as well as the Air Defence Forces, the Airmobile Forces with four brigades, and the Artillery and Missile Forces. Kazakhstan planned to switch to a fully professional army by 2016.

ARMY

Personnel: 45,000

Equipment

Category	Name	In Service
MBT	T-72	650
	T-62	280
AIFV/APC	BMP1/2	508
	BTR-60/70/80 (many stored)	292
	BTR-82	12
	BTR-4	
Artillery	D-30 122mm	200
	M-46 130mm	200
	D-20 152mm (towed)	100
	2S1 12mm	75
	2S3 152mm SP	150
MRL	BM21 122mm	
	Uragan 220mm	

New Procurements/Upgrades

✦ Kazakhstan plans to manufacture100 BTR-4 wheeled APCs through a license production agreement with Ukraine.

✦ An agreement with Turkey covers local production of the Otokar Cobra 4x4 tactical vehicle.

NAVY

Personnel: 3,000

Landlocked Kazakhstan operates a small navy on the Caspian Sea - the world's largest inland body of water - base at Aktau. The Navy is currently undergoing a major force restructuring. The Ministry of Defence has set up a directorate of naval forces to coordinate development of military activities in the Caspian Sea. Russian state arms export agency Rosoboronexport has been selected as the lead contractor to develop a littoral fleet and modern naval infrastructure for the Kazakh Navy's Caspian Sea fleet. In 2012 Kazakhstan launched its first domestically produced ship, a missile boat called 'Kazakhstan.'

Equipment: Su-27 fighter jets, 1 Kazakhstan class missile boat, 4 Almaty class vessels, 1 Dauntless class vessel, 5 Guardian class vessels, 2 Project 1400M Grif class patrol craft , 2 Türk vessels.

New Procurements/Upgrades

✦ Acquisition of corvettes is under consideration. Agreements have been signed with German and Turkish shipbuilders in 2016 to build the corvettes for the Kazakh Navy. A Russian shipyard could also provide the corvette platforms.

✦ A modern hydrographic survey ship is on the Navy's wish list.

✦ A new small patrol boat has been launched by the country's Zenit Uralsk Shipyard in 2016.

AIR FORCE

Personnel: 10,000

Equipment

Category	Name	In Service
Fighter	Su-30SM	4
	MiG-31	
	MiG-29	40
	MiG-27	
	Su-27	12
	Su-24	20+
	Su-25	14
	MiG-23	20
Transport	Airbus C295	4
	An-26	12
	An-72	10
	An-12	
Helicopter	EC145	2+
	UH-1H	6
	Mi-8	40
	Mi-17	20
	Mi-24	30
	Mi-26	10
	Ka-32A11BC	2
SAM	S-300	1 battery

New Procurements/Upgrade

✦ The air force has taken delivery of four Su-30SM fighters from Russian JSC Corporation Irkut.

✦ Two Su-30 multi-role fighters and two Mi-17 transport helicopters are slated to be delivered by Russia.

✦ Modernisation of Su-25 fleet of fighters is going on with the first refurbished aircraft expected to be delivered in 2017.

✦ A deal is expected to be signed for four Russian Mi-35M combat helicopters.

✦ Acquisition of heavy-lift military transport aircraft is under consideration.

✦ Russia to deliver five battalions of S-300 air defence systems.

✦ The Air Force has taken delivery of two additional EADS C-295 light turboprop aircraft with plans to buy eight more.

✦ Orders have been placed for purchase of 20 Eurocopter EC725 medium transport helicopters.

✦ The air force plans to purchase a total of 45 EC145 helicopters.

✦ A second batch of two SAR+ four Medevac helicopter is on order.

✦ An agreement has been signed for local assembly of 20 BC725 Caracals.

✦ Local assembly of Skylark and Hermes UAVs, produced by Israel's Elbit Systems, is being planned. An agreement with Israel is expected to be signed in this regard.

✦ Two Wing-Loong combat drones have been delivered by China in March 2016.

✦ A MoU has been signed with ThalesRaytheonSystems (TRS) to develop and supply fully digital 3D air surveillance radars, Ground Master 400 (GM400).

Defence Expenditure

Total defence spending: $2 billion (2015-16)

Estimated defence spending in terms of GDP: 1.3%

Defence Production and R&D

Kazakhstan has announced ambitious plans to become the leading regional exporter of arms in Central Asia by transforming its defence industry capabilities in order to export mainly artillery systems. Its military is armed primarily with Russian and Soviet equipment but Astana is trying to diversify its sources for armaments from other countries so as not to be too dependent on one. A joint venture has been set up with the European helicopter manufacturer to produce EC 145 helicopters for the Kazakh military and other government agencies. South Africa's Paramount Group and Kazakhstan have entered a deal to jointly produce armoured vehicles. The project will see the production of 120 to 360 military and non-military vehicles a year. Kazakhstan has also signed a military technical cooperation with Israel.

Defence Procurement

The government has signed a decree approving a state programme to modernise the Kazakh military. The defence ministry is actively introducing modern and advanced military systems and platforms. Both the Navy and the Air Force are acquiring new platforms, weapons and systems as part of the modernisation drive. The 2011 Military Doctrine had outlined the priority areas for military procurement as: manufacture and repair of aircraft, motor vehicles, missile and artillery systems, automated C2 as well as communications and other specialist military equipment.

CONTACT DETAILS

Ministry of Defence

Astana, Dostyk 14

Office: +7(7172) 72-15-84, +7(7172) 72-13-85

E-mail: mod@mod.gov.kz

KENYA
(Capital: Nairobi)

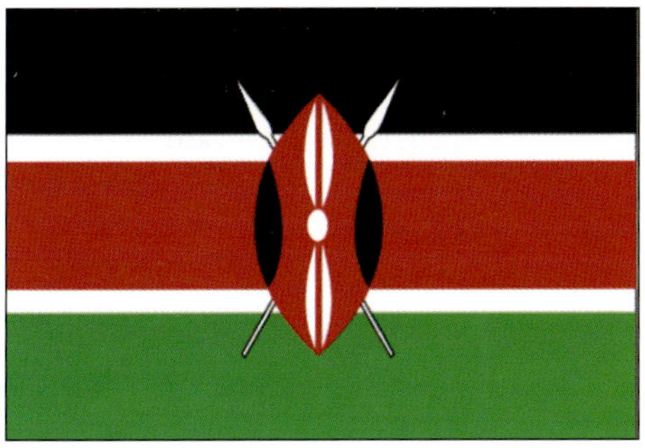

INTRODUCTION

Area: 580,367 sq km

Population: 46,790,758 (July 2016 est.)

Coastline: 536 km

Maritime claims: Territorial sea: 12 nm

Exclusive Economic Zone: 200 nm

Continental shelf: 200 m depth or to the depth of exploitation

KEY POLITICAL PERSONS

PRESIDENT & COMMANDER-IN-CHIEF: Uhuru Muigai Kenyatta

DEPUTY PRESIDENT: William Ruto

KEY DEFENCE PERSONS

DEFENCE SECRETARY: Raychelle Omamo

CHIEF OF DEFENCE FORCES: Gen. Samson Mwathethe

ARMY COMMANDER: Lt. Gen. Robert Kibochi

NAVY COMMANDER: Maj. Gen. Levi Franklin Mghalu

AIR FORCE COMMANDER: Maj. Gen. Samuel Thuita

ECONOMY

The market-based mixed economy of Kenya has registered positive growth in the last few years despite being marred by corruption. It is the most industrially developed country in East Africa and an attractive destination for foreign trade in the continent. While the services sector is the major contributor to the GDP, agriculture and industry too have a visible share in it.

GDP: $142.2 billion (2015 est.); $134.6 billion (2014 est.)

Real Growth Rate: 5.6% (2015 est.); 5.3% (2014 est.)

Industries: Small-scale consumer goods (plastic, furniture, batteries, textiles, clothing, soap, cigarettes, flour), agricultural products, horticulture, oil refining; aluminium, steel, lead; cement, commercial ship repair, tourism

Total Exports: $5.982 billion (2015 est.); $6.174 billion (2014 est.)

Export Commodities: Tea, horticultural products, coffee, petroleum products, fish, cement

Major Markets: Uganda 11.2%, US 8.3%, Tanzania 8.1%, Netherlands 7.4%, UK 6%, Pakistan 4.2% (2015)

Total Imports: $15.56 billion (2015 est.); $17.61 billion (2014 est.)

Import Commodities: Machinery and transportation equipment, petroleum products, motor vehicles, iron and steel, resins and plastics

Major Suppliers: China 30%, India 15.5%, UAE 5.7%, US 4.8%, Japan 4.7% (2015)

INTERNATIONAL DISPUTES

The boundary that separates Kenya's and Sudan's sovereignty is unclear in the "Ilemi Triangle," which Kenya has administered since colonial times.

DEFENCE

ARMY

Personnel: 19,000

Equipment

Category	Name	In Service
MBT	T-72	70
	Vickers Mk3	80
Armoured Car	Saladin	3
	Ferret	16
	AML90	30
	AML60	20
	Shorland	8
	Saracen	5
APC	UR-416	38
	Panhard M3	8
	BRM SP	85
Artillery	2S7 203mm SP	25
	NORA 152mm SP	18
	122mm towed	48
	105mm light gun	40
	L5 105mm pack	
Mortar	L16 81mm	20
	120mm	8
ATGW	Swingfire	14
	Milan	40
SAM	Mistral/ASPIC	
	BUK	
AA gun	TCM20 20mm	50
	ZPU-4	70
ATK weapon	Carl Gustav 84mm	56
Helicopter	Harbin Z9WA	4

New Procurements/Upgrades

No credible information is available about latest procurements by the Army. According to reports, it has acquired some weapons and equipment (artillery and APCs) from Serbia in 2014.

NAVY

Personnel: 1,400

Equipment

Category	Name	In Service
Fleet	Nyayo class large missile FAC	2
	Ex-Spanish P101 type patrol boat	5
	Shupavu class OPV	2
	LSM	1

New Procurements/Upgrades

No major procurement plans are under consideration at present.

AIR FORCE

Personnel: 2,500

Equipment

Category	Name	In Service
Fighter	F-5E/F Tiger II	20
Transport	Fokker 70 (VIP)	1
	Piper NAVAJO	1
	DHC-5D Buffalo	7
	Piper Chieftain	1
	Harbin Y-12	11
	DASH 8	3
Trainer	Tucano	11
Helicopter (attack & utility)	MD 500M	10
	SA.330 Puma	3
	IAR330	6
	Mi-17	

New Procurements/Upgrades

- ✦ Five refurbished UH-1 Huey II utility helicopters will be received from Bell Helicopters under a foreign military sales programme worth US$ 52.24 million cleared by the US Defence Department in September 2016.
- ✦ One ScanEagle UAV is to be delivered by the USA in 2016-17.

Defence Expenditure

Estimated defence spending: Sh75.8 billion ($819 million - 2014-15)

Estimated defence spending in terms of GDP: 2.4%

Major defence suppliers to Kenya are the US, Russia and Europe.

Defence Production and R&D

Kenya has very limited defence production and research facilities. Majority of its military equipment are procured from outside Africa. The state-owned Kenya Ordnance Factories Corporation (KOFC) is primarily involved in manufacturing small arms and ammunition.

Defence Procurement

The Kenyan Army, Navy and Air Force have inducted several new equipment and platforms in the past decade as part of their modernisation drive. The country, however, maintains secrecy over its defence procurement programmes. In the past over two years, Kenya is believed to have acquired armoured vehicles, main battle tanks, helicopter gunships, multiple rocket launchers among other arms and equipment for its military.

CONTACT DETAILS

Ministry of Defence

The Permanent Secretary
Ministry of State for Defence
Ulinzi House,
Lenana Road
P O Box 40668 - 00100,
Nairobi
Tel: 2721100
Fax: 2737322
E-Mail: info@mod.go.ke
Website: http://www.mod.go.ke

KOSOVO
(Capital: Pristina)

INTRODUCTION

Area: 10,887 sq km

Population: 1,883,018 (July 2016 est.)

Coastline: 0 km (Landlocked)

Maritime claims: None

KEY POLITICAL PERSONS

PRESIDENT:
Hashim Thaci

PRIME MINISTER:
Isa Mustafa

KEY DEFENCE PERSONS

MINISTER OF THE KOSOVO SECURITY FORCE: Haki Demolli

COMMANDER OF KOSOVO SECURITY FORCE: Lt. Gen. Rahman Rama

ECONOMY

After gaining independence in 2008, Kosovo's relatively new economy has grown significantly by transitioning into a modern free market system from a centralist and controlled economy. Foreign investments into the country are mostly from Europe, including Germany, Austria and Switzerland, among others. It also receives remittances from citizens working in neighbouring Nordic and other countries. Imports are higher than exports and the country has been making efforts to bridge the huge trade deficit. Kosovo still remains the poorest country in Europe even though it has been least affected by the Euro Zone financial crisis mainly due to the lack of its full-scale integration with the larger European and global market. To further accelerate economic growth, the Government has taken initiatives to privatise state-owned enterprises. In June 2009, Kosovo joined the World Bank and International Monetary Fund. In order to help integrate Kosovo into regional economic structures, the United Nations Interim Administration Mission in Kosovo (UNMIK) signed the country's accession to the Central Europe Free Trade Area (CEFTA) in 2006. Serbia and Bosnia previously had refused to recognise Kosovo's customs stamp or extend reduced tariff privileges for Kosovo products under CEFTA, but both countries resumed trade with Kosovo in 2011. The official currency of Kosovo is the euro, but the Serbian dinar is also used illegally in Serb enclaves. Kosovo's tie to the euro has helped keep core inflation low.

GDP: $17.54 billion (2015 est.); $16.87 billion (2014 est.)

Real Growth Rate: 4% (2015 est.); 1.2% (2014 est.)

Industries: Mineral mining, construction materials, base metals, leather, machinery, appliances, foodstuff and beverages, textiles

Total Exports: $349 million (2014 est.); $408 million (2013 est.)

Export Commodities: Mining and processed metal products, scrap metals, leather products, machinery, appliances, prepared foodstuff, beverages and tobacco, vegetable products, textile and textile articles

Major Markets: Italy 25.8%, Albania 14.6%, Macedonia 9.6%, China 5.5%, Germany 5.4%, Switzerland 5.4%, Turkey 4.1% (2012 est.)

Total Imports: $2.687 billion (2014 est.); $3.398 billion (2013 est.)

Import Commodities: Foodstuff, livestock, wood, petroleum, chemicals, machinery, minerals, textiles, stone, ceramic and glass products and electrical equipment

INTERNATIONAL DISPUTES

Boundary dispute with Serbia continues as Serbia and a number of other countries do not recognise Kosovo as an independent sovereign state.

DEFENCE

The Kosovo Security Force (KSF), formed in March 2008 with the initiatives of the NATO-led Kosovo Force (KFOR) and the Kosovo Protection Corps (KPC), assumed full operational capability in July 2013. Constituted as a lightly armed force responsible for civil protection operations and assisting civil authorities in responding to natural disasters and other emergencies, the KSF has been trained to NATO standards. The security force has been assigned the task of conducting non-military security functions such as search and rescue operations, explosive ordnance disposal, control and clearance of hazardous materials, fire-fighting and other humanitarian assistance tasks.

The Kosovo Government, in March 2014, proposed to create a 5,000-strong army to "protect sovereignty" of the ethnic Albanian territory. The future army will also have 3,000 reservists with a yearly budget of €65 million ($89.5 million) and it is expected to be fully operational after 2019. The proposal, however, has been opposed by Serbia.

Existing Personnel: 4,500 (+2,000 reserve force)

Defence Expenditure

Defence Budget: €45 million (2015)

Estimated defence spending in terms of GDP: 0.9%

Defence Production and R&D

Kosovo has not undertaken any defence production activities and no defence production facilities exist in the country.

Defence Procurement

No major procurement plans are under consideration at present.

CONTACT DETAILS
(NATO-Kosovo Forces: KFOR)
Website: www.nato.int/kfor/

KUWAIT
(Capital: Kuwait City)

INTRODUCTION

Area: 17,818 sq km

Population: 2,832,776 (July 2016 est.)

Coastline: 499 km

Maritime claims: Territorial Sea: 12 nm

KEY POLITICAL PERSONS

CHIEF OF STATE & SUPREME COMMANDER OF ARMED FORCES: Sabah Al-Ahmad Al-Jaber Al-Sabah

PRIME MINISTER: Jaber Mubarak Al-Hamad Al-Sabah

KEY DEFENCE PERSONS

DEPUTY PRIME MINISTER & MINISTER OF DEFENCE: Mohammad Al-Khaled Al-Hamad Al-Sabah

CHIEF OF GENERAL STAFF: Lt. Gen. Mohammad Khaled Al-Kheder

COMMANDER OF GROUND FORCES: Maj. Gen. Khaled Al-Saleh Al-Sabah

NAVY CHIEF: Maj. Gen. Khaled Ahmed Abdullah

AIR FORCE CHIEF: Maj. Gen. Abdullah Yaqoob Al-Foudari

ECONOMY

Kuwait has a geographically small, but wealthy, relatively open economy with self-reported crude oil reserves of about 102 billion barrels - about 9% of world reserves. Petroleum accounts for nearly half of GDP, 95% of export revenues, and 95% of government income. Kuwaiti officials have committed to increasing oil production to 4 million barrels per day by 2020. Kuwait, however, has done little to diversify its economy, in part, because of this positive fiscal situation. Nonetheless, the government in May 2010 passed a privatisation bill that allowed the government to sell assets to private investors, and in January 2011 passed an economic development plan that pledged to spend up to $130 billion in five years to diversify the economy away from oil, attract more investment, and boost private sector participation in the economy. Many of the projects, however, did not materialise because of the uncertain political situation. In 2015, Kuwait, for the first time in 15 years, realized a budget deficit after decades of high oil prices. Kuwaiti authorities have tried to reduce the deficit by decreasing spending on subsidies for the local population, but with limited success. Despite Kuwait's dependence on oil, the government has cushioned itself against the impact of lower oil prices, by saving annually at least 10% of government revenue in the Fund for Future Generations.

GDP: $289.9 billion (2015 est.); $286.7 billion (2014 est.)

Real Growth Rate: 1.1% (2015 est.); 0.6% (2014 est.)

Industries: Petroleum, petrochemicals, cement, shipbuilding and repair, water desalination, food processing and construction materials

Natural gas - proved reserves: 1.798 trillion cu m (1 January 2014 est.)

Crude oil - proved reserves: 104 billion bbl (1 January 2015 est.)

Total Exports: $55.32 billion (2015 est.); $104.5 billion (2014 est.)

Export Commodities: Oil and refined products, fertilizers

Major Markets: South Korea 14.5%, China 12.1%, India 12.1%, Japan 10.4%, US 7.6%, Pakistan 5.9%, Singapore 4.3% (2015)

Total imports: $27.34 billion (2015 est.); $26.99 billion (2014 est.)

Import Commodities: Food, construction materials, vehicles and parts, clothing

Major Suppliers: China 13.2%, US 9.6%, Saudi Arabia 7.7%, Japan 6.5%, Germany 5.1%, France 4.3%, India 4.2% (2015)

INTERNATIONAL DISPUTES

Kuwait and Saudi Arabia continue to negotiate a joint maritime boundary with Iran.

DEFENCE

ARMY

Personnel: 10,000+

Equipment

Category	Name	In Service
MBT	M1A2 Abram	236
	M84	150
	Chieftain	45 (storage)
APC, AFV	Desert Warrior	254
	BMP-2	40

Army continued...

Category	Name	In Service
APC, AFV	BMP-3	55
	M113	60
	Ferret	80
Artillery	Norinco PLZ45 155mm SP	51
	M-109 155mm SP	23
	GCT 155mm SP	15
	F3 155mm SP	18
MLRS	Smerch 300mm	25
ATGW	TOW	118

New Procurements/Upgrades

✦ 12 Fox nuclear, biological, and chemical (NBC) reconnaissance vehicles have been ordered from Rheinmetall with deliveries slated to begin from 2017.

✦ An initial agreement has been signed to buy 120 Sherpa Light armoured 4x4 vehicles made by France's Renault Trucks Defence.

NAVY

Personnel: 1,500

Equipment

Category	Name	In Service
FAC (missile & gun)	Al Sanbouk class (Type TN C45)	1
	Istiqlal class (Type FPB 57)	1
	Um Al Maradim class (CMN PB-37BRL)	8
Patrol Craft	ASI 315 Type	2
	NAJA 12 Type	12
	Seagull	4
	In-shore patrol craft	57
	Al Tahaddy class LMC	2
	Haditah Class LC	2

New Procurements/Upgrades

✦ A total of eight landing craft and high-speed protection vessels have been ordered from Abu Dhabi Shipbuilding (ADSB) under a US$71 million contract. The platforms include two 64-metre landing craft, one 42-metre landing craft and five 16-metre composite fast landing craft.

✦ Plans are on for mid-life upgrade of the eight P37 fast patrol boats. The Navy is also planning to order a range of ship-launched missiles to equip these boats.

✦ Delivery of 10 MkV-C fast Interceptor boat on order from the US is going on. The vessels are being procured to conduct coastal patrol, surveillance and interdiction missions.

✦ The Navy has ordered one DSV (diving support vessel). Contract details have not been disclosed.

AIR FORCE

Personnel: 2,500

Equipment

Category	Name	In Service
Fighter	F/A-18C	32
	F/A-18D Hornet	8
Transport	L100-30 Hercules	2
	C-17 Globemaster	1
Trainer	Tucano	8
	Hawk Mk. 64	10
Helicopter	AS.332 Super Puma	5
	SA 330 Puma	5
	SA342K Gazelle	13
	AH-64 Apache Longbow	16
SAM	I-Hawk battery	1
	Skyguard	10
	Patriot	
	Starburst	50

New Procurements/Upgrades

✦ A final deal has been signed with Eurofighter GmbH in April 2016 for 28 Eurofighter Typhoon jets at a cost of €7.957 billion (US $9.062 billion). The Eurofighter consortium will deliver 22 single-seat and six twin-seat (all third tranche configuration) Eurofighters. Delivery is expected to start from 2019 onwards.

Eurofighter Typhoon jets

✦ The US Government in Sept 2016 has cleared a proposal to sell F/A-18 E/F Super Hornet fighter jets to Kuwait. Kuwait had made a request for acquiring 28 of the jets in April 2015.

✦ A deal worth US$ 1.1 billion has been signed with France for the purchase of 30 Airbus-built H225M Caracal helicopters and associated weapons. The deal was signed in August 2016. The Kuwait Air Force and National Guard will fly the helicopters for combat search and rescue, and naval operations.

✦ The US plans to sell a total of 60 Patriot missile systems to Kuwait. The existing Patriot systems are also being upgraded.

✦ Kuwait has sought 300 AGM-114R3 Hellfire II missiles, along with spare and repair parts, support and test equipment, training equipment and personnel training, from the US under the foreign military sales programme.

✦ A potential sale of 80 AIM-9X-2

Sidewinder Block II missiles and associated equipment for the Kuwait Air Force has been notified by the DCSA.

✦ Lockheed Martin has received a $245 million contract from the US Government under FMS for three KC-130J tanker aircraft for Kuwait. Additionally, the Aviation Training Consulting, LLC, (ATC) has received $25 million to provide training and instruction to the Kuwait Air Force in support of the new KC 130Js.

✦ Plans are on to buy a second C-17 Globemaster III strategic airlifter from the US.

✦ Kuwait has requested continuation of logistics support, contractor maintenance, and technical services in support of its fleet of F/A-18 C/D fighters, including avionics software upgrade and engine component improvement services under a deal worth $200 million under FMS.

Defence Expenditure

Estimated defence spending in terms of GDP: 3.7%

The Kuwait Parliament in January 2016 has announced to allocate an additional US$ 10 billion for the defence budget over the next 10 years to finance the ongoing military procurement programmes, including the purchase of new fighter jets, tanks and air defence systems.

The US is a major military supplier to the country.

Defence Production and R&D

Kuwait does not undertake any defence production activities.

Defence Procurement

Kuwait opted for completely restructured armed forces after the Iraqi invasion and the subsequent war in 1990-91. The Government signed a 10-year defence cooperation agreement with the US that allowed US military equipment to be stored in its territory and joint military operations. On 11 February 1992, Kuwait signed a bilateral defence cooperation agreement with Britain. Kuwait has undertaken a comprehensive all-round upgrade programme by acquiring new weapons and platforms for all three wings of its defence forces.

CONTACT DETAILS

Ministry of Defence

P.O Box 1170

Safat 3012, Kuwait

Tel: +9654848300, 4819288

KYRGYZSTAN
(Capital: Bishkek)

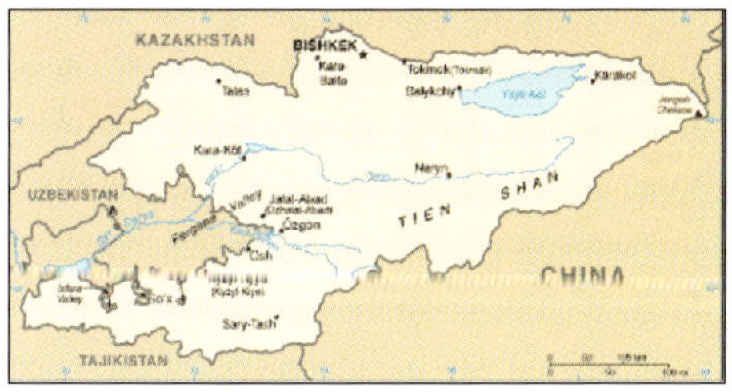

INTRODUCTION

Area: 199,951 sq km

Population: 5,727,553 (July 2016 est.)

Coastline: 0 km (landlocked)

Maritime claims: None

KEY POLITICAL PERSONS

PRESIDENT & COMMANDER-IN-CHIEF OF THE ARMED FORCES: Almazbek Atambayev

PRIME MINISTER: Sooronbay Jeenbekov

KEY DEFENCE PERSONS

CHAIRMAN, STATE COMMITTEE FOR DEFENCE AFFAIRS: (Vacant)

CHIEF OF GENERAL STAFF: Raimberdi Duishenbiyev

ECONOMY

Kyrgyzstan's economic performance has improved in the last few years. A series of reform-oriented governments since the political crises of 2010 have sought to restore economic and social stability. The economy depends heavily on gold exports - mainly from output at the Kumtor gold mine. The agricultural sector accounts for about a quarter of the country's GDP. The energy sector is one of the largest in the Kyrgyz economy. The bulk of the country's current generating capacity is hydropower. The economy also depends heavily on remittances from Kyrgyzstani migrant workers, primarily in Russia. To address governance issues in mining, the Kyrgyz government started implementing the Extractive Industries Transparency Initiative in 2004. Kyrgyzstan was progressive in carrying out market reforms, such as an improved regulatory system and land reform. Kyrgyzstan was the first Commonwealth of Independent States (CIS) country to be accepted into the World Trade Organization. In December 2014 Kyrgyzstan agreed to join the Eurasian Economic Union. Kyrgyz leaders hope the country's August 2015 accession to the Union will bolster trade and investment, but slowing economies in Russia and China, low commodity prices, and currency fluctuations continue to hamper economic growth.

GDP: $20.29 billion (2015 est.); $19.61 billion (2014 est.)

Real Growth Rate: 3.5% (2015 est.); 4% (2014 est.)

Industries: Small machinery, textiles, food processing, cement, shoes, sawn logs, refrigerators, furniture, electric motors, gold, rare earth metals

Total Exports: $1.61 billion (2015 est.); $1.794 billion (2014 est.)

Export Commodities: Gold, cotton, wool, garments, meat, tobacco; mercury, uranium, hydropower; machinery; shoes

Major Markets: Switzerland 26%, Uzbekistan 22.6%, Kazakhstan 20.8%, UAE 4.9%, Turkey 4.5%, Afghanistan 4.5%, Russia 4.2% (2015)

Total Imports: $3.648 billion (2015 est.); $5.204 billion (2014 est.)

Import Commodities: Oil and gas, machinery and equipment, chemicals, foodstuff

Major Suppliers: China 56.4%, Russia 17.1%, Kazakhstan 9.9% (2015)

INTERNATIONAL DISPUTES

Kyrgyzstan is yet to ratify the 2001 boundary delimitation with Kazakhstan; disputes in Isfara Valley delay completion of delimitation with Tajikistan; delimitation of 130 km of border with Uzbekistan is hampered by serious disputes over enclaves and other areas.

DEFENCE

The Armed Forces of the Republic of Kyrgyzstan, originally formed from former Soviet forces of the Turkestan Military District, includes the Army, Air Force, Air Defence Forces, the Northern and Southern Groups of Forces, Interior Troops, Agency of

National Security and Border Troops. Kyrgyzstan has not developed its air capability. The air defence forces have received aid from Russia. The Russians also help patrol Kyrgyz airspace as part of the Joint CIS Air Defence System.

ARMY

Personnel: 12,000

Equipment

Category	Name	In Service
MBT	T-72	150
AIFV/APC	BTR-60 - 53	53
	BTR-70	25
	BTR-80	10
	BMP-1	230
	BMP-2	90
	BRDM-2	30
Artillery	100mm BS-3 towed	18
	122mm howitzer D-30	72
	122mm howitzer M-30	35
	152mm D-1	16
	2S9 Anona SP	12
	2S1	18
	M120	48
	2S12 120mm Sani mortar	6
AA gun	100mm T-12	18
MRL	BM-21 GRAD	15

New Procurements/Upgrades

No major procurement plans are under consideration at present.

AIR FORCE

Personnel: 1,000

Equipment

Category	Name	In Service
Aircraft	L-39 Albatross	30
Helicopter	Mi-8	15
	Mi-17	15
	Mi-24	12
VIP transport/liaison	An-26	2
	Tu-154	2
	Boeing 737	2

New Procurements/Upgrades

No major procurement plans are under consideration at present.

Defence Expenditure

Total defence spending: $2.458 billion (2013)

Estimated defence spending in terms of GDP: 3.74%

Defence Production and R&D

Kyrgyzstan does not have any defence production capability.

Defence Procurement

Kyrgyzstan's Ministry of Defence lacks funds to procure new military platforms or weapons. The country relies on foreign aid and military assistance from Russia. Arms transferred by Russia include small arms, mountain guns and mortars, several attack and transport helicopters and individual combat kits. Other Commonwealth of Independent States (CIS) also supply Kyrgyzstan with military equipment. Bishkek has also received military supplies from China, Turkey, the US and Germany. India and Kyrgyzstan have signed an agreement to bolster defence cooperation and hold annual joint military exercises.

CONTACT DETAILS

State Committee for Defence

Bishkek

Tel., Fax: (312) 66-18-04

E-mail: ud@mil.gov.kg

LAOS
(Capital: Vientiane)

INTRODUCTION

Area: 236,800 sq km

Population: 7,019,073 (July 2016 est.)

Coastline: 0 km (Landlocked)

Maritime claims: None

KEY POLITICAL PERSONS

PRESIDENT & COMMANDER-IN-CHIEF: Bounnhang Vorachith

PRIME MINISTER: Thongloun Sisoulith

KEY DEFENCE PERSONS

DEFENCE MINISTER: Lt. Gen. Chansamone Channhalat

CHIEF OF GENERAL STAFF OF ARMY: Lt. Gen. Suvon Luongbunmi

ECONOMY

Laos is a landlocked country and remains as one of the poorest countries in Southeast Asia. However, its economy is rapidly growing as the government began to decentralise control and encouraged private enterprise. According to the Asian Development Bank, foreign investment in mining and hydropower remains robust. The government is pursuing poverty reduction and education for all children as key goals. The country has become a rising regional player in its role as a hydroelectric power supplier to neighbours such as China, Vietnam and Thailand. Laos began a one-year chairmanship of ASEAN in January 2016. The country is in the process of implementing a value-added tax system. The government appears committed to raising the country's profile among foreign investors and has developed special economic zones replete with generous tax incentives, but a small labour pool remains an impediment to investment.

GDP: $37.62 billion (2015 est.); $34.97 billion (2014 est.)

Real Growth Rate: 7.6% (2015 est.); 7.5%

(2014 est.)

Industries: Mining (copper, tin, gold, and gypsum); timbre, electric power, agricultural processing, rubber, construction, garments, cement, tourism

Total Exports: $2.928 billion (2015 est.); $2.662 billion (2014 est.)

Export Commodities: wood products, coffee, electricity, tin, copper, gold, cassava

Major Markets: Thailand 30.4%, China 27%, Vietnam 17.6% (2015)

Total Imports: $4.058 billion (2015 est.); $4.271 billion (2014 est.)

Import Commodities: Machinery and equipment, vehicles, fuel and consumer goods

Major Suppliers: Thailand 60.9%, China 18.6%, Vietnam 7.3% (2015)

INTERNATIONAL DISPUTES

✦ Border conflict with Thailand over islands in Mekong River remains unresolved

✦ Cambodia and Vietnam are concerned about Laos' extensive upstream dam construction.

DEFENCE

ARMY

Personnel: 34,000+

Equipment

Category	Name	In Service
MBT	T-54/55	30
	PT-76	25
APC	M8	25
	BTR-40/60	30
	BTR-152	40
	M-113	
Artillery	75mm	
	105mm	
	M1938 & D30 122 mm	40
	M46 130mm	10
Mortar	81 mm	20
	155 mm	20
	107 mm	
RCL	75mm, 106mm	
SAM	SA-7	
	SA-3	
AA gun	37mm	
	23mm twin	

Army Aviation: 4 Cessna U-17A

NAVY (riverine section of the Army): 600 personnel & 40 patrol craft

New Procurements/Upgrades

No major procurement plans are under consideration at present.

AIR FORCE

Personnel: 3,500+

Equipment

Category	Name	In Service
Fighter	MiG-21	21 (most grounded)
Transport	AN-24	7
	AN-2	6
	An-74	1
	YAK-40	2
	Y7-100C	
Helicopter	Mi-8	10
	Mi-6	1
	Mi-17	10
	Ka-32T	6
	Z-9A	6
Trainer	MiG-21U	4

New Procurements/Upgrades

✦ Russian Helicopters has signed a contract to repair the Mi-17 helicopters of the Laotian Air Force.

Defence Expenditure

Exact defence expenditure data is unavailable.

Estimated defence spending in terms of GDP: 0.2%

Defence Production and R&D

Laos does not undertake any defence production activities.

CONTACT DETAILS

Ministry of National Defence

Phon Kang Road

Vientiane, Laos

Website: www.mod.gov.la/

LEBANON

(Capital: Beirut)

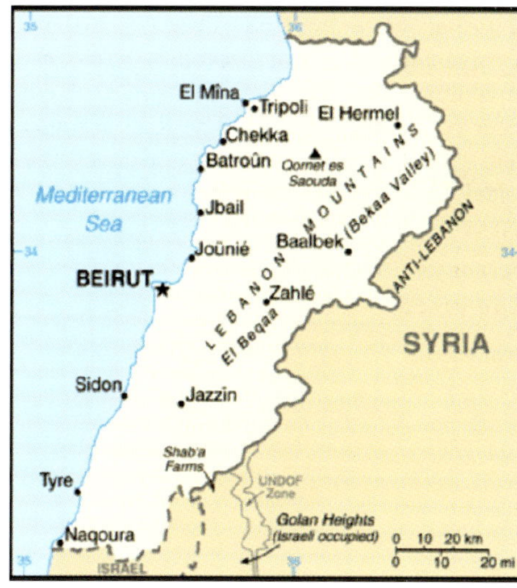

INTRODUCTION

Area: 10,400 sq km

Population: 6,237,738 (July 2016 est.)

Coastline: 225 km

Maritime claims: Territorial sea: 12 nm

KEY POLITICAL PERSONS

PRESIDENT: Michel Aoun

PRIME MINISTER: Saad Hariri

KEY DEFENCE PERSONS

MINISTER OF DEFENCE: Samir Moqbil

COMMANDER-IN-CHIEF: Gen. Jean Kahwaji

CHIEF OF STAFF OF THE ARMED FORCES: Maj. Gen. Hatem Mallak

COMMANDER OF NAVY: Rear Adm. Majed Alwan

COMMANDER OF AIR FORCE: Gen. Ghassan Chahine

ECONOMY

Lebanon has a free-market economy and a strong laissez-faire commercial tradition. The Lebanese economy is service-oriented; the main growth sectors include banking and tourism. The urban population in Lebanon is noted for its commercial enterprise. Although Lebanon is ideally suited for agricultural activities in terms of water availability and soil fertility, as it possesses the highest proportion of cultivable land in the Arabic-speaking world, it does not have a large agricultural sector. Attracting only 12% of the total workforce, agriculture is the least popular economic sector in Lebanon. The government does not restrict foreign investment; however, the investment climate suffers from red tape, corruption, arbitrary licensing decisions, high taxes, tariffs, and fees, archaic legislation, and weak intellectual property rights. The collapse of the government in early 2011 over its backing of the Special Tribunal for Lebanon and unrest in neighbouring Syria slowed economic growth to 1.5% after four years of 8% average growth. In September 2011, the Cabinet endorsed a bill that would provide $1.2 billion in funding to improve Lebanon's downtrodden electricity sector, but fiscal limitations will test the government's ability to invest in other areas, such as water.

GDP: $83.06 billion (2015 est.), $82.23 billion (2014 est.)

Real Growth Rate: 1% (2015 est.), 2% (2014 est.)

Industries: Banking, tourism, food processing, wine, jewellery, cement, textiles, mineral and chemical products, wood and furniture products, oil refining, metal fabricating

Total Exports: $3.551 billion (2015 est.), $4.1 billion (2014 est.)

Export Commodities: Jewellery, base metals, chemicals, miscellaneous consumer goods, fruit and vegetables, tobacco, construction minerals, electric power machinery and switchgear, textile fibres, paper

Major Markets: Saudi Arabia 12.1%, UAE 10.6%, Iraq 7.6%, Syria 7.1%, South Africa 6.6% (2015)

Total Imports: $16.71 billion (2015 est.), $19.16 billion (2014 est.)

Import Commodities: Petroleum products, cars, medicinal products, clothing, meat and live animals, consumer goods, paper, textile fabrics, tobacco, electrical machinery and equipment, chemicals

Major Suppliers: China 11.5%, Italy 7.1%, Germany 6.8%, France 6%, US 5.7%, Russia 4.6%, Greece 4.4% (2015)

INTERNATIONAL DISPUTES

✦ Lacking a treaty or other documentation describing the boundary, portions of the Lebanon-Syria boundary are unclear with several sections in dispute.

✦ Since 2000, Lebanon has claimed Shab'a Farms area in the Israeli-occupied Golan Heights.

✦ The roughly 2,000-strong UN Interim Force in Lebanon has been in place since 1978.

DEFENCE

ARMY

Personnel: 80,000

Equipment

Category	Name	In Service
MBT	T-54/T-55	218
	M48A5	92
	M60A3	10
APC/IFV	M113	1,073
	Panhard AML-90	70
	VAB	73
Artillery	105 mm M101-A1	141
	122 mm D-30	
	130 mm	
	155 mm M114-A1	
	155 mm M109-A3	12
Mortar	120 mm	43
Recoilless rifle	106 mm	
Anti-Tank GW	Milan, TOW	
AA Gun	20 mm	131
	23mm	
	30mm	

New Procurements/ Upgrades

✦ Russia's Defence Minister Sergei Shoigu said his country is ready to provide military aid to Lebanon.

The Lebanese army has taken delivery of 48 MBDA Milan short-range anti-tank missiles which are the first instalment of a USD 3 billion arms package financed by Saudi Arabia and supplied by France. France will provide training services from mid-2015 to mid-2018.

✦ Advanced Technology Systems Co., McLean, Virginia, was awarded a firm-fixed-price, multi-year, foreign military sales (FMS) contract with a maximum value of $26,666,806 for the procurement of tetra trunked radio communication systems for the Lebanese Armed Forces. The Army Contracting Command, Aberdeen Proving Ground, Md., is the contracting activity.

✦ The Lebanese Army signed a contract with West Virginia as Clark International Logistics LLC. has secured a $5.1-million contract with the US Army to repair and refurbish armoured personnel carriers.

✦ Lebanon received 12 M109 155mm self-propelled howitzer from Jordan and 400 Tow 2 in 2015.

Lebanese Army Commando Regiment's M-113 armed with ZU-23

An M-60A3 main battle tank moves along a street in Germany during Central Guardian, a phase of Exercise Reforger '85. *Image Credit: US Department of Defence*

NAVY

Personnel: 5,500

Equipment

Category	Name	In Service
Patrol Vessel	Tabarja Class	1
	Amchit Class	1
	Tracker Class	7
	MK2 combat support	27
	Fast patrol boats	15
Landing Craft	EDIC-III Class	2

New Procurements/ Upgrades

✦ Negotiations to supply Lebanon with three fast attack craft (FACs) are continuing.

✦ The UAE promised to donate three more patrol boats the Lebanese Navy, the type and equipment are still unknown.

✦ The United States promised to deliver one 42m blue-water CSC-137 Class-1 patrol craft.

AIR FORCE

Personnel: 2,500

Equipment

Category	Name	In Service
Aircraft	AC-208B Combat Caravan	3
	Hawker Hunter	3
	Scottish Aviation Bulldog T1	3

Air Force continued...

Category	Name	In Service
Aircraft	Embraer A-29A	6 on order
Helicopter (attack & utility)	AB-205/UH-1H	27
	Aérospatiale SA-342L Gazelle	8
	IAR 330SM / Aérospatiale SA-330 Puma	11
	AgustaWestland AW139	1
	Mil Mi-24	6
	Robinson R44	4
	Sikorsky S-61N Mk II	3
UAV	RQ-11B Raven	12

New Procurements/ Upgrades

✦ The Lebanese Air Force has selected A-29 Super Tucano for Close Air Support role. The aircraft sale is part of a larger, more comprehensive package, including infrastructure improvements, which will be fulfilled by other parties not involved in the Embraer/SNC partnership. The contract includes logistics support for aircraft operation as well as a complete training system for Lebanese Air Force pilots and mechanics.

✦ UAE officially announced approval to provide Lebanon with 10 Aérospatiale Puma transport helicopters.

✦ The country signed a contract with Eurotech to revamp 13 Gazelle helicopters of the original and ex-UAE delivered in addition to equip the ex-UAE Puma helicopters.

✦ Lebanon decided to replace previously offered MiG-29 jets with Mi-24 attack helicopters.

✦ Bell Helicopter Textron Inc., Fort Worth, Texas, was awarded an USD 18,935,563 modification (0005) to foreign military sales contract W58RGZ-11-G-0011 for the procurement of three UH-1H (Huey II) aircraft in support of the government of Lebanon.

✦ Lebanon will receive the first batch of OH-58D light attack helicopters from US

✦ The Air Force placed an order for 6 Embraer Super Tucano.

Defence Expenditure

Total defence spending: USD 1.275 billion (est.)

Estimated defence spending in terms of GDP: 4.75% (2015)

Lebanon's military spending remains far behind most countries in the region, allocating only 2.8 percent of its GDP to military expenditures. Recently, the Lebanese government approved a five-year plan to arm the Lebanese military.

Foreign suppliers: United States, France, United Kingdom, Russia, Saudi Arabia.

Defence Production and R&D

There are no defence production capabilities in Lebanon.

Defence Procurement

Lebanese Army/Navy/Air Force has not done any major defence procurements in recent years.

CONTACT DETAILS

Ministry of Defence
Yarzé
Beyrouth Lebanon
Tel: (+961 5) 45 24 00
Fax: (+961 5) 95 87 57
(+961 1) 44 47 62
E-mail: ministry@leb-army.gov.lb

Lebanese Armed Forces
Yarzeh, Lebanon
Website: www.lebarmy.gov.lb

LIBYA
(Capital: Tripoli)

INTRODUCTION

Area: 1,759,540 sq km

Population: 6,541,948 (July 2016 est.)

Coastline: 1,770 km

Maritime claims: Territorial sea: 12 nm

KEY POLITICAL PERSONS

PRESIDENCY COUNCIL HEAD & NOMINATED PRIME MINISTER OF GOVT OF NATIONAL ACCORD: Fayez Serraj

KEY DEFENCE PERSONS

DEFENCE MINISTER: Al-Mahdi Al-Barghathi

ECONOMY

Libya's economy is almost entirely dependent on the nation's energy sector, which generates about 65% of its GDP and 96% of government revenue. Income from the sale of crude oil and natural gas, coupled with a small population, give Libya one of the highest nominal per capita GDPs in Africa, but Libya's leaders have hindered economic development by, for the most part, failing to use these financial resources to invest in national infrastructure. Libyan sales of oil and natural gas collapsed during the Revolution of 2011, rebounded in 2012 and 2013, but then fell sharply in late 2013 and throughout 2014 due to major protest disruptions at Libyan oil ports and around the country. In early 2015, armed conflict between rival forces for control of the country's largest oil terminals caused a decline in Libyan crude oil production, which never recovered to more than one-third of the average pre-Revolution highs of 1.6 million barrels per day. The state sector is large and growing, with the majority of Libyan workforce receiving government salary in 2014.

GDP: \$92.8 billion (2015 est.); \$99.12 billion (2014 est.)

Real Growth Rate: -6.4% (2015 est.); -24% (2014 est.)

Industries: Petroleum, petrochemicals, aluminium, iron and steel, food processing, textiles, handicrafts, cement

Natural gas - Proved Reserves: 1.549 trillion cu m (1 January 2014 est.)

Crude oil - Proved Reserves: 48.36 billion bbl (1 January 2015 est.)

Total Exports: $10.86 billion (2015 est.); $13.81 billion (2014 est.)

Export Commodities: Crude oil, refined petroleum products, natural gas, chemicals

Major Markets: Italy 32.1%, Germany 11.3%, China 8%, France 8%, Spain 5.6%, Netherlands 5.4%, Syria 5.3% (2015)

Total Imports: $11.24 billion (2015 est.); $20.43 billion (2014 est.)

Import Commodities: Machinery, semi-finished goods, food, transport equipment, consumer products

Major Suppliers: China 14.8%, Italy 12.9%, Turkey 11.1%, Tunisia 6.5%, France 6.1%, Spain 4.6%, Syria 4.5%, Egypt 4.4%, South Korea 4.3% (2015)

INTERNATIONAL DISPUTES

Libya claims more than 32,000 sq km in south-eastern Algeria and about 25,000 sq km in the Tommo region of Niger in a currently dormant dispute.

DEFENCE

ARMY

Personnel: No exact data available

Equipment: While the Libyan Army maintained a large inventory of mostly Soviet-supplied equipment, including main battle tanks, armoured fighting vehicles, small arms and artillery, it lost majority of them during the NATO-led war in 2011. No definitive data is available on the Army's present equipment level other than the recent new procurements.

New Procurements/ Upgrades

✦ 10 Khrizantema-S tank destroyers ordered from Russia were inducted in the Army, taking its total number to 14.

✦ 20 PUMA armoured vehicles were delivered by Italy in 2013.

✦ Jordan has delivered 49 NIMR wheeled armoured military vehicles.

✦ Delivery of 287 Humvee multi-role vehicles from the US is in progress.

NAVY

Personnel: 2,000+

Equipment

Category	Name	In Service
Frigate	Koni class	1
Corvette	Nanuchka class	1
Minesweeper	Ex-USSR Natya class	1

New Procurements/Upgrades

✦ Libya is planning to procure new warships for the Navy.

AIR FORCE

Personnel: 2,000+

Equipment

Category	Name	In Service
Fighter	MiG-23	3
	MiG-21	2
	Mirage F1 AD/ED	2
	J1 Jasterb	
Transport	An-26	2
	An-32	2
Helicopter	Mi-8	
	Mi-24/ Mi-35	3
Trainer	L39 Albatros	2

New Procurements/ Upgrades

✦ The Air Force is planning a major overhaul of its fleet by acquiring new fighter jets, transport aircraft and helicopters. Potential acquisitions could include two squadrons of Rafale fighter jets and a number of F1-Mirage jets from France, Eurofighter Typhoons from the UK and C-130 Hercules transport planes and Chinook helicopters from the US. The US notified in June 2013 the possible sale of two C-130 planes to the country.

✦ Under a defence cooperation agreement signed with France in February 2012, Paris is to renovate Libya's Dassault Mirage F1 fighter jets and train its personnel.

Defence Expenditure

Estimated defence spending in terms of GDP: 1%

Libya has procured arms and equipment mostly from the erstwhile Soviet Union and France in the past. In recent times, the US is eying a share of the country's defence market. Libya has also requested the US to impart personnel training for 6,000 to 8,000 troops at an estimated cost of $600 million.

Defence Production and R&D

Libya does not undertake any defence production activities.

Defence Procurement

After most of its defence equipment was destroyed in the 2011 NATO-led war, the new regime in Libya intends to restructure and modernise the armed forces by acquiring new platforms and weapon systems for the Army, Navy and Air Force. Many of the previously signed deals during Muammar Gaddafi's regime have been scrapped by the present Government. Acquisition of new armoured vehicles for the Army is going on while for the air force, purchase of new airlifters, helicopters and fighter jets are being planned. The country plans to spend some $4.7 billion for new military acquisitions. Continuous political instability, however, has marred these military procurement programmes in recent times.

CONTACT DETAILS

Ministry of Defence
Tel.: 00218214800734
Fax: 00218214800168
E-mail: info@defense.gov.ly
Website: http://www.defense.gov.ly/

LUXEMBOURG
(Capital: Luxembourg)

INTRODUCTION

Area: 2,586 sq km

Population: 582,291 (July 2016 est.)

Coastline: 0 km (Landlocked)

Maritime claims: None

KEY POLITICAL PERSONS

CHIEF OF STATE & COMMANDER-IN-CHIEF: Grand Duke Henri

PRIME MINISTER:
Xavier Bettel

KEY DEFENCE PERSONS

MINISTER OF DEFENCE: Etienne Schneider

CHIEF OF DEFENCE STAFF: Gen. Romain Mancinelli

ECONOMY

The small, stable, high-income economy of Luxembourg – benefiting from its proximity to France, Belgium, and Germany – has historically featured solid growth, low inflation, and low unemployment. The industrial sector, initially dominated by steel, has become increasingly diversified to include chemicals, rubber, and other products. Growth in the financial sector has more than compensated for the decline in steel. The country continues to enjoy an extraordinarily high standard of living - GDP per capita ranks third in the world, after Liechtenstein and Qatar, and is the highest in the EU. Luxembourg's GDP contracted 3.6% in 2009, rebounded in 2010-12, fell again in 2013-14, but recovered in 2015. Unemployment has remained below the EU average despite having increased from a historically-low rate of 4% in the 2000s to 7% in 2014. The country continues to enjoy an extraordinarily high standard of living - GDP per capita ranks among the highest in the world and is the highest in the euro zone. Luxembourg has one of the highest current account surpluses as a share of GDP in the euro zone, and it maintains a healthy budgetary position and the lowest public debt level in the region.

GDP: \$56.02 billion (2015 est.); \$53.43 billion (2014 est.)

Real Growth Rate: 4.8% (2015 est.); 4.1% (2014 est.)

Industries: Banking and financial services, iron and steel, information technology, chemicals, glass, telecommunications, cargo transportation, food processing, metal products, engineering, aluminium, tires, tourism, biotechnology

Total Export: \$17.81 billion (2015 est.); \$24.57 billion (2014 est.)

Export Commodities: Machinery and equipment, steel products, chemicals, rubber products, glass

Major Markets: Germany 22.1%, Belgium 16.7%, France 16.6%, UK 4.7%, Italy 4.6%, Netherlands 4% (2015)

Total Import: \$20.22 billion (2015 est.); \$24.81 billion (2014 est.)

Import Commodities: Commercial aircraft, minerals, chemicals, metals, foodstuff, luxury consumer goods

Major Suppliers: Belgium 27.6%, Germany 22.9%, China 11.7%, France 9.5%, US 8.4%, Netherlands 4.2%, Mexico 4.1% (2015)

INTERNATIONAL DISPUTES

Luxembourg has no known international disputes.

DEFENCE

ARMY

Personnel: 400

Equipment

Category	Name	In Service
Vehicle	Dingo 2	48
	HMMWV Hummer (armoured)	30
	HUMMER (TOW Carrier)	6
	MLST	
	MAN 4T truck	
Artillery	105mm howitzer	
ATGW	BGM-71 TOW	6

Army Aviation

Category	Name	In Service
Aircraft	E-3A Sentry (NATO)	17
	Boeing 707	2
	A400M	

New Procurements/Upgrades

No major procurement plans are under consideration at present.

AIR FORCE

Luxembourg does not have an air force, though the 17 NATO AWACS aircraft are for convenience registered as aircraft of Luxembourg. The country has ordered one A400M heavy transport aircraft from Airbus which will be delivered in 2019 and would be jointly used by Luxembourg and Belgium. It also plans to operate two Airbus A330 MRTT tanker/transport aircraft jointly with Netherlands. The two countries have inked a Memorandum of Understanding (MoU) in this regard and a contract with Airbus has been signed in July 2016. The two aircraft will be a NATO property.

Defence Expenditure

Estimated defence spending in terms of GDP: 0.6%

Defence Production and R&D

Luxembourg does not undertake any defence production activities.

Defence Procurement

Being one of the smallest military forces in the NATO group of countries with a very limited defence budget, Luxembourg has not done any major military procurement in the past. The country, without a Navy and Air Force, has procured equipment for its Army, mainly bought from the US. Belgium and Luxembourg have jointly formed the Office of Defence Cooperation, Belgium-Luxembourg (ODC BELLUX), with an objective to support the US and other NATO militaries through the acquisition of defence goods and services and by participating in international armaments cooperation activities.

CONTACT DETAILS

Ministry of Foreign Affairs
(Department of Defence)
6, Rue I'anvien Athenee
L-1144 Luxembourg
Armed Forces HQ
38-44, Rue Goethe
L-1637 Luxembourg
PO Box 1873
L-1018 Luxembourg
Tel: 26 848-1
Fax: 26 84 56 01
E-mail: info@armee.lu

STRATEGIC INFORMATION

MALAYSIA
(Capital: Kuala Lumpur)

INTRODUCTION

Area: 329,847 sq km

Population: 30,949,962 (July 2016 est.)

Malay 50.4%, Chinese 23.7%, indigenous 11%, Indian (Tamils) 7.1%, others 7.8%

Coastline: 4,675 km

Maritime claims: Territorial sea: 12 nm

Exclusive Economic Zone: 200 nm

Continental shelf: 200 m depth or to the depth of exploitation; specified boundary in the South China Sea

GEOPOLITICAL IMPORTANCE

Malaysia is a Southeast Asian country located partly on a peninsula of the Asian mainland and partly on the northern third of the island of Borneo. Land borders are shared with Thailand, Indonesia, and Brunei, and maritime borders exist with Singapore, Vietnam, Brunei and the Philippines. It has coastlines on the South China Sea and the Straits of Malacca.

POLITICAL OVERVIEW

Malaysia is a federal constitutional monarchy. It consists of thirteen states and three federal territories. The head of state is the 'Yang di-Pertuan Agong', commonly referred to as the king and the Prime Minister of Malaysia is the head of government.

KEY POLITICAL PERSONS

HEAD OF STATE & COMMANDER-IN-CHIEF: Sultan Abdul Halim Mu'adzam Shah

PRIME MINISTER: Mohd Najib bin Tun Abdul Razak

ECONOMY

Malaysia, a middle-income country, has transformed itself from a producer of raw materials into an emerging multi-sector economy. Malaysia is a relatively open state-oriented and newly industrialised market economy. The Najib administration is continuing efforts to boost domestic demand and reduce the economy's dependence on exports. International trade, facilitated by the adjacent Strait of Malacca shipping route, and manufacturing are key sectors of the country's economy. Malaysia is an exporter of natural and agricultural resources, the most valuable exported resource being petroleum. Malaysia has recently recorded significant increase in export of manufactured and agricultural products to almost all its markets. The country is one of the world's largest exporters of semiconductor devices, electrical goods, and information and communication technology products. However, the present government is mainly focusing on boosting domestic demand and reducing the economy's dependence on exports. In an effort to diversify the economy and make it less dependent on exported goods, the government has pushed to increase tourism to Malaysia. Knowledge-based services are also expanding. In 2010, Malaysia launched the New Economic Model (NEM) with an aim to achieve high-income status by 2020 and to move farther up the value-added production chain by attracting investments in Islamic banking, high technology industries, biotechnology, and services.

GDP: $815.6 billion (2015 est.), $777.2 billion (2014 est.)

Real Growth Rate: 5% (2015 est.), 6% (2014 est.)

Industries: Peninsular Malaysia - rubber and oil palm processing and manufacturing, petroleum and natural gas, light manufacturing, pharmaceuticals, medical technology, electronics and semiconductors, timber processing; Sabah - logging, petroleum and natural gas production; Sarawak - agriculture processing, petroleum and natural gas production, logging

Oil Proved Reserves: 3.6 billion bbl (1 January 2016 est)

Natural Gas Proved Reserves: 1.183 trillion cu m (1 January 2016 est)

Total Exports: $175.7 billion (2015 est.), $207.5 billion (2014 est.)

Export Commodities: Semiconductors and electronic equipment, palm oil, petroleum and liquefied natural gas, wood and wood products, palm oil, rubber, textiles, chemicals, solar panels

Major Markets: Singapore 13.9%, China 13%, Japan 9.5%, US 9.4%, Thailand 5.7%, Hong Kong 4.7%, India 4.1% (2015)

Total Imports: $147.7 billion (2015 est.), $172.9 billion (2014 est.)

Import Commodities: Electronics, machinery, petroleum products, plastics, vehicles, iron and steel products, chemicals

Major Suppliers: China 18.8%, Singapore 12%, US 8.1%, Japan 7.8%, Thailand 6.1%, South Korea 4.5%, Indonesia 4.5% (2015)

DEFENCE & SECURITY

The Malaysian Armed Forces (*Angkatan Tentera Malaysia, ATM*) have three branches, the Royal Malaysian Navy, the Royal Malaysian Army, and the Royal Malaysian Air Force. The military of Malaysia is responsible for assisting civilian authorities to overcome all international threats, preserve public order, assist in natural disasters and participate in national development programmes. Malaysia's Armed Forces are responsible for the protection of the country against internal and external threats. They serve in a wide variety of situations such as assisting civil authorities in addressing domestics threats, maintaining public security, providing aid following the onset of natural disasters, and assisting in national development programmes.

KEY DEFENCE PERSONS

DEFENCE MINISTER: Datuk Seri Hishammuddin Hussein

CHIEF OF DEFENCE FORCES: Gen. Tan Sri Dato' Sri Zulkifeli bin Mohd Zin

CHIEF OF ARMY: Gen. Datuk Raja Mohamed Affandi Raja Mohamed Noor

CHIEF OF NAVY: Admiral Dato 'Seri Panglima Ahmad Kamarulzaman

CHIEF OF AIR FORCE: Dato' Sri Haji Roslan Bin Saad

Internal Conflict

Terrorism: Malaysia has experienced terrorism since 1948 during the battle against communists. Malaysian Special Operations Force is a multi service force tasked with a counter-terror mission. Malaysia maintains a high level of security and has not suffered incidents of terrorism for several years, but was vulnerable to terrorist activity and continued to be used as a transit and planning site for terrorists. Malaysia has implemented series of measures in the name of counter-terrorism to detect and prevent terrorism, and to minimise damage from such terrorist acts should they occur. These measures involve all levels of society, including military, police, border and infrastructure security, civil defence, medical readiness, and psychological preparedness. Malaysia also participates actively in international counter-terrorism schemes. It is working closely with intelligence agencies of the United States, Britain, China, Pakistan and India to combat international terrorism. Malaysia is a member of the Asia/ Pacific Group on Money Laundering, a Financial Action Task Force-style regional body. The country has an established Financial Intelligence Unit and capacity to report suspicious transactions. In January 2016, the Malaysian police said the country was on the highest security alert. Malaysian authorities in June 2016 faced a terrorist attack by individuals with links to Daesh (formerly referred to as ISIL).

External Conflict

Since independence, Malaysia, bordered by Thailand, Indonesia, Singapore and Brunei, has been involved in territorial disputes and overlapping maritime claims with almost all its neighbours. Malaysia's territorial and maritime disputes are located in the Gulf of Thailand, the Andaman Sea, the Straits of Malacca, the Straits of Singapore, the South China Sea, the Sulu Sea and the Celebes Sea. Some of these disputes were resolved through bilateral and multilateral treaties.

✦ Disputes continue over deliveries of fresh water to Singapore, Singapore's land reclamation, bridge construction, and maritime boundaries in the Johor and Singapore Straits.

✦ Land and maritime negotiations with Indonesia are ongoing, and disputed areas include the controversial Tanjung Datu and Camar Wulan border area in Borneo and the maritime boundary in the Ambalat oil block in the Celebes Sea.

✦ While the 2002 "Declaration on the Conduct of Parties in the South China Sea" has eased tensions over the Spratly Islands, it is not the legally binding "code of conduct" sought by some parties.

✦ Malaysia was not party to the March 2005 joint accord among the national oil companies of China, the Philippines, and Vietnam on conducting marine seismic activities in the Spratly Islands.

✦ In 2008, ICJ awarded sovereignty of Pedra Branca (Pulau Batu Puteh/ Horsburgh Island) to Singapore, and Middle Rocks to Malaysia, but did not rule on maritime regimes, boundaries, or disposition of South Ledge.

✦ Separatist violence in Thailand's predominantly Muslim southern provinces prompts measures to close and monitor border with Malaysia to stem terrorist activities.

✦ Philippines retain a dormant claim to Malaysia's Sabah State in northern Borneo.

✦ As per Letters of Exchange signed in 2009, Malaysia in 2010 ceded two hydrocarbon concession blocks to Brunei in exchange for Brunei's sultan dropping claims to the Limbang corridor, which divides Brunei.

✦ Another serious non-military security challenges confronting Malaysia is piracy. The Straits of Malacca, situated between Indonesia and Malaysia, was long considered to be the world's most dangerous waters for pirate attacks. Malaysia, Singapore and Indonesia have increased their patrols in the Straits of Malacca and Singapore Straits to combat piracy. Malaysia has established the Maritime-Enforcement Co-ordination Centre (MECC) to co-ordinate anti-piracy activities and surveillance. Pirate activity in the region, has thereby declined significantly since 2005 due to increased military patrols and vessel security, according to the International Maritime Bureau (IMB).

THREAT PERSPECTIVE

Philippines: Both Malaysia and Philippines, have a long history of close economic and political ties; the relationship is often overshadowed by a series of territorial disputes such as claim over the Sabah state in northern Borneo and ongoing disputes over ownership of the Spratly Islands. The Philippine-Malaysia dispute over Sabah was, and still is, a moot diplomatic issue. The Philippines has a claim over the Sabah state in northern Borneo though this is currently not being actively pursued. Malaysia and the Philippines both claim a portion of the disputed Spratly Islands. The archipelago lies off the coasts of the Philippines and Malaysia, about one third of the way from there to southern Vietnam. Malaysia's claims are based upon the continental shelf principle, and have clearly defined coordinates. The Philippines base their claims of sovereignty over the Spratly Islands on the issues of Res nullius and geography. Malaysia along with Philippines and Indonesia is planning to conduct joint military drills for their army personnel starting January 2017 to fight terror group Abu Sayyaf and secure the Sulu Sea from rampant piracy. The joint training will take place in North Kalimantan, Malaysia's Tawao Island and the Philippines' Bongao Island.

Thailand: Relationship between Malaysia and Thailand has been soured by territorial issues. The Malaysia–Thailand border consists of both a land boundary across the Malay Peninsula and maritime boundaries in the Straits of Malacca and the Gulf of Thailand/South China Sea. Malaysia lies to the south of the border while Thailand lies to the north. There are two stretches of the Malaysia–Thailand border which are subject to dispute. The first involves the land border in the Bukit Jeli (Jeli Hill) at the headwaters of the Golok River and the second involves the continental shelf boundary in the Gulf of Thailand. Both disputes have not resulted in any aggression between the two countries. Both Malaysia and Thailand participate in the Cooperation Afloat Readiness and Training (CARAT) annually together with other ASEAN countries.

Indonesia: The border between Indonesia and Malaysia consists of both a land border separating the two countries' territories on the island of Borneo as well as maritime boundaries along the length of the Straits of Malacca, in the South China Sea and in the Celebes Sea. The border in the Celebes Sea is subject to dispute between the two countries. Part of the dispute was settled by the judgement of the International Court of Justice in the Sipadan and Ligitan Case in 2002 and is now awaiting delimitation between the two countries. The two countries however still have overlapping claims over the continental shelf which Indonesia refers to as Ambalat, a sea block in the Celebes Sea which is currently under Indonesia sovereignty. The Ambalat Sea block is located off the coast of the Indonesian province of East Kalimantan and south-east of the Malaysian state of Sabah.

Singapore: Singapore and Malaysia have a broad and multi-faceted relationship. Bilateral trade, investment, and tourism ties are robust. Singapore got separated from the Federation of Malaysia in 1965. Accordingly, certain factors such as geography, history, politics, ideology, economy, culture and ethnicity of both the nations are inextricably linked. These factors sometimes have created tensions between both countries. Since the independence of Singapore, several disputes have surfaced between Singapore and Malaysia. This includes, among other things, the dispute over the revision of the price of water under the 1961 and 1962 Water Agreements and the ownership of Pedra Branca, an island off the coast of Johor. The Pedra Branca dispute was a territorial dispute between Singapore and Malaysia over several islets at the eastern entrance to the Singapore Strait. The dispute began in 1979 and was largely resolved by the International Court of Justice (ICJ) in 2008, which opined that Pedra Branca belonged to Singapore and Middle Rocks belonged to Malaysia. On May 2010, Malaysia and Singapore announced that they have resolved their long-standing disputes over land and water. Malaysia decided to relocate its railway station, close to one of the two bridges linking the two countries, freeing up land in the city-state for redevelopment. Singapore decided it would not seek to extend a water agreement dating back to 1961. Singapore will also hand over the waterworks it operates in Johor to the Malaysian state government when the current agreement lapses. Singapore and Malaysia are full members of the

Commonwealth of Nations. Both countries are part of the Five Power Defence Arrangements (FPDA) along with New Zealand, Australia and the United Kingdom. Both sides along with Indonesia help each other when threatened with threats posed by Jeemah Islamiyah (JI).

STRATEGIC RELATIONS

United States: United States and Malaysia enjoy strong security cooperation. Both the countries share a strong military-to-military relationship with numerous exchanges, training, joint exercises, and visits. The United States and Malaysia cooperate closely on security matters, including counter terrorism, maritime domain awareness, and regional stability. Security cooperation and training builds capabilities among Malaysia's armed forces and coast guard, allowing it to take on an expanded international role, including peacekeeping operations. On February 4, 2016, the United States and Malaysia joined the rest of the 12 Trans-Pacific Partnership (TPP) negotiating partners to sign the TPP, an Asia-Pacific trade agreement.

China: Malaysia and China share close bilateral ties and cordial regional cooperation. Friendship and cooperation between China and Malaysia have achieved significant progress in recent years. Leaders of both countries exchanged visits frequently and cooperation in all areas deepened continuously. The two countries have also agreed to further strengthen military cooperation between the two countries in a bid to maintain peace and stability in the region. During a recent defence and security consultation in September 2012, the two sides reached consensus in a broad range of issues to strengthen exchange and cooperation between the PLA and Malaysian Armed Forces. Both sides agreed to maintain high level military exchanges between the armed forces, make good use of the defence and security consultation mechanism, strengthen exchange and communication in training cooperation and training systems, while deepening cooperation in non-conventional security issues. Malaysia and China signed a series of agreements in Nov 2016 which cover among other things energy and defence coordination.

Russia: Russia and Malaysia share strong ties since 1967 when diplomatic relationships were established between Malaysia and former USSR. Both countries maintain political dialogue and regular meetings on current bilateral issues and international concerns. Malaysia is also one of the largest MTC (military-technical cooperation) partners of Russia in Southeast Asia. The Russian-Malaysian Intergovernmental Joint Commission on cooperation in defence, defence technologies, and defence industry is a working body through which both nations maintain their relationships. Malaysia is one of the main Russia's trading partners in South East Asia. Palm oil is supplied to Russia in accordance with an agreement on extending Malaysian credit facility signed in Moscow in July 2002 as well as through private channels.

India: India and Malaysia have traditionally been close and friendly. India-Malaysia defence relations have also steadily grown over the years. A MoU on Defence Cooperation was signed in 1993. Malaysia-India Defence Cooperation meetings at the level of Defence Secretary are held regularly; the ninth meeting of the MIDCOM was held in 17 January 2010. The IAF Training Team deployed in Malaysia trained Malaysian pilots on the SU-30SKM aircraft for two-and-a-half years from February 2008. Indian naval ships regularly make port calls in Malaysia. Malaysia participates in the biennial MILAN event regularly. Malaysia is one of the significant investors in India. It is estimated that, if the Mauritius route is also included, the investment of Malaysia in India could be as high as USD 7 billion.

Brunei: The close relationship between Malaysia and Brunei has seen some changes in the past few years. Both countries have recently agreed to step up bilateral cooperation in the oil and natural gas sector. Significant progress has also been made on joint demarcation and survey of the land boundary with survey going on for the complicated process of border demarcation. Regarding South China Sea developments, both countries have proposed the idea of a cooperation package which includes cooperation in trade, investment, defence, finance, and transportation. Such joint cooperation will further develop Malaysia-Brunei relations. The two countries share a land border on the island of Borneo.

Australia: Malaysia and Australia's defence relationship is based on practical cooperation including the Malaysia-Australia Joint Defence Programme, an ongoing Australian presence at the Royal Malaysian Air Force (RMAF) Base at Butterworth, and common membership of the Five Power Defence Arrangements (FPDA). Australia and Malaysia cooperate closely on a range of security issues, and have signed a MoU on co-operation to combat international terrorism. MAJDP is the cornerstone of the bilateral defence relationship under which both countries have defence officers posted in each other's country, involved in training and professional exchanges.

Pakistan: The relationship between Pakistan and Malaysia is growing stronger. Both countries enjoy close and cordial relations. Pakistan and Malaysia have a convergence of views on world affairs that translates into cooperation and support for each other, at various international stage. Both are members of Organization of Islamic Conference (OIC) and the Commonwealth of Nations. Both countries have signed a number of agreements and Memoranda of Understanding to enhance bilateral cooperation. The two countries have also signed agreements and MoUs for cooperation in defence, education, science & technology, and a number of economic and commercial agreements on palm oil, etc.

Multilateral Alliances

Malaysia is an active member of various international organisations, including the Commonwealth of Nations, the United Nations, the Organisation of Islamic Cooperation (OIC), Asia-Pacific Economic Cooperation (APEC) forum and the Non-Aligned Movement (NAM). It was a founding member of the Association of Southeast Asian Nations (ASEAN). Malaysia is a frequent contributor to UN and other peacekeeping and stabilisation missions, including recent deployments to Lebanon, Timor-Leste, Philippines, Indonesia, Pakistan, Sierra Leone, Sudan, Western Sahara, Nepal, and Kosovo. Malaysia has also deployed a medical unit to Afghanistan. Malaysia is also an active member of Five Power Defence Arrangements (FPDA) along with United Kingdom, Australia, New Zealand and Singapore.

DEFENCE CAPABILITIES

ARMY: The Royal Malaysian Army (*Tentera Darat Malaysia)* is the land component of the Malaysian Armed Forces. The Malaysian Army (RMA) is responsible for the safeguarding the nation against land-based threats.

Personnel: 80,000 (60,000 reserve)

Equipment

Category	Name	In Service
MBT	PT91M	48
	FV101 Scorpion	30
APC/ MICV/IFV	ACV-300 Adnan	250
	K-200 KIFVs	111
	AV8	250
	Condor 4x4	310
	Sibmas	180
	AV4	20
	Stormer	25
Artillery	G5 Mk3 155mm towed	28
	FH70 155mm towed	14
	OTO Melara Mod 56 105mm	111
	Astros II MLRS	55
	M109 howitzer	30
Mortar	82 mm 2B14 Podnos	
	81 mm L16	
RCL	106mm	150
Anti-Tank GW	ERYX	
	9K115 Metis-M	
	AT-4, M-40, M72 LAW	
	INGWE	216
SAM	Rapier 2000	12
	FN-6	

Army continued...

Category	Name	In Service
SAM	9K38 Igla	70
	Anza 72 mm	95
	Starstreak	
	Starbrust	
AA gun	12.7mm	70
	Bofors 40mm	24
	Oerlikon 35mm	28

Army Aviation

Category	Name	In Service
Helicopter	A-109 LOH	11
	MD 500 Defender	4
	S-61A-4 Nuri	12

ACV 300 Adnan on display

New Procurements/ Upgrades

✦ MD Helicopters Inc. will supply six MD 530G scout attack helicopters to Malaysian Army, completing the full fleet delivery by the end of Q1 2017.

✦ Denel has signed a €340 million contract with Malaysia to supply a range of turret and integrated weapon systems to be fitted onto 8x8 armoured infantry fighting vehicles.

✦ ACV-300 ADNAN tracked vehicle are on order.

✦ 85 URO Vamtac 4X4 high mobility tactical vehicle are being delivered.

✦ Letter of intent has been signed for development of 8X8 APC by local DefTech, which will be based on the FNSS AV8 PARS design.

✦ 12 6X6 VABs are being acquired from French surplus to replace the existing CONDORs.

✦ Malaysia has signed a contract for the initial order of 20 vehicles AV-4 4x4.

✦ Thales has signed a contract with Global Komited, a company within The Weststar Group, to supply the Malaysian Armed Forces with ForceSHIELD, the Thales integrated Advanced Air Defence system. In addition to the STARStreak missiles, the system comprises CONTROLMaster 200 radar and weapon coordination systems, RAPIDRanger and RAPIDRover mobile weapon systems and the Next Generation Lightweight Multiple Launcher, as well as associated communications.

NAVY: The Royal Malaysian Navy (Tentera Laut Diraja Malaysia) is the naval arm of the Malaysian Armed Forces. The Royal Malaysian Navy (RMN) protects Malaysia's coastlines, territorial waters, and economic zones from potential trespass or illegal activity. The Navy Future Fleet programme is component of the second batch of Lekiu class frigates, Scorpene class submarines, New Generation Patrol Vessels (NGPV), Multi-Purpose Support Ship (MPSS) and maritime patrol aircraft. The ultimate goal is to build a six vessels squadron of each class by year 2020. The Special Forces arm of the Royal Malaysian Navy is known as PASKAL (Pasukan Khas Laut or Naval Special Forces). In peacetime, the unit is tasked with responding to maritime hijacking incidents as well as protecting Malaysia's numerous offshore oil and gas platforms. On 15 April 2009, PASKAL was renamed KD Panglima Hitam.

Naval Bases: TLDM Lumut, Perak, TLDM Tanjung Gelang, Pahang, TLDM Sepanggar, Sabah, TLDM Tanjung Gerak, Langkawi, Kedah, TLDM Tanjung Pengelih, Johor, TLDM Sandakan, Sabah, TLDM Semporna, Sabah, TLDM Sejingkat, Sarawak, TLDM Bintulu, Sarawak.

Personnel: 12,000 (1,200 reserve)

Equipment

Category	Name	In Service
Submarine	Scorpene class	2
Frigate	2nd generation PV	(6 ships on order)
	Lekiu class	2
Corvette	Kasturi class	2
	Laksamana class	4
OPV	Kedah class (Meko A100)	6
	Littoral Mission Ship	(On order)
Missile Boat	Perdana	4
	Handalan	4
	Jerung	6
	CB-90	17
	Kris	2
	Sri Tiga	2
Mine Hunter	Mahamiru class	4
Amphibious Force	Sri Indera Sakti	1
	Mahawangsa	1
Auxiliaries	Bunga Mas Lima, Bunga Mas Enam, Mega Bakti	
Missile	Exocet, Sea Skua, Aspide, Otomat, MICA, Sea Wolf, Naval Strike Missile	

Naval Aviation

Category	Name	In Service
Aircraft (Maritime Patrol)	Beech King A 200	4
Helicopter	Super Lynx 300 (with sea SUKA anti-ship missile)	6
	AS 555SN Fennec	6

The Royal Malaysian Navy frigate KD Lekir (FF 26) leads the Ticonderoga-class guided missile cruiser USS Bunker Hill (CG-52) during a passing exercise. *Image Courtesy: US Navy photo by Mass Communication Specialist 2nd Class James R. Evans/Released*

New Procurements/ Upgrades

✦ Malaysia has agreed to purchase four littoral mission ships from China, two of which will be built in China and two in Malaysia.

✦ Daya Materials JV received 49.4 million euro (RM198mil) contract for the ventilation system for six new vessels of the Malaysian Navy. The contract includes design, supply, installation, testing and commissioning of the heating, ventilation and air-conditioning system for the six vessels under the navy's littoral combat ship programme.

✦ Boustead Naval Shipyard awarded RM9 billion contract by Malaysian Ministry of Defence to build six Second Generation Patrol Vessel (SGPV) as part of the Malaysian Littoral Combat Ship (LCS) programme.

✦ A requirement has been defined for a new multi-role support vessel as a replacement of Sri Inderapura LST.

✦ Malaysia has signed a letter of intent for two British-built Jebat-class frigates.

✦ Malaysia announced that it has signed to purchase six new Gowind class corvettes at an estimated price of RM9 billion. It is expected that the ships will be operational by 2017.

✦ Selex ES, a Finmeccanica company, has signed a contract with AIROD SDN BHD for the supply and support of equipment for the Government of Malaysia, Royal Malaysian Navy Super Lynx MK100 helicopter.

AIR FORCE: The Royal Malaysian Air Force (Tentera Udara Diraja Malaysia) maintains and operates a capability that is ready and able to project aerospace power to secure Malaysia's airspace and protect her national interests.

Air Bases: Kuala Lumpur Air Base, Subang Air Base, Butterworth Air Base, Kuantan Air Base, Gong Air Base Kedak, Labuan Air Base, Kuching Air Base, College of the Air Force Air Base, Jugra Air Base, Hill Air Force Base Lunchu, Hill Air Force Base Ibam.

Personnel: 15,000 (3,000 reserve)

Equipment

Category	Name	In Service
Fighter	Su-30MKM	18
	MiG-29N/NUB	12
	F-5	15 (to be retired)

Air Force continued...

Category	Name	In Service
Fighter	F/A-18D	8
	BAE Hawk	20
	MB-339	18
Transport	C-130H	10
	C-130MP	1
	KC-130 Tanker	4
	A400M	3
	CN-235	8
	Falcon 900	1
	F-28	1
	Cessna 402B	10
Helicopter	SA316B/ SA319B Alouette III	30
	S-70 Black Hawk	6
	EC120 Colibri	1
	SH-3 Sea King	17
	A109C	1
	AS-61N1	2
	EC725 Cougar	12
Trainer	PC-7 MkI	30
	PC-7 MkII	17
	MD3-160	20
UAV	CTRM Aludra, ScanEagle, CTRM Eagle ARV	
Missile	AIM-7 Sparrow, AIM-120 AMRAAM, AIM-9 Sidewinder AA-12 Adder, AGM-65 Maverick, AS-14 Kedge, AS-13 Kingbolt	

A Royal Malaysian Air Force Sukhoi Su-30MKM flys over the South China Sea during a bi-lateral exercise promoting the longstanding military-to-military cooperation efforts between the U.S. Navy and the Royal Malaysian military. *Image Courtesy: US Navy.*

New Procurements/ Upgrades

✦ Malaysia is planning to buy 18 combat aircraft to replace its ageing fleet of Russian-made MiG-29s and have shortlisted Eurofighter, Gripen, Rafale, Super Hornet, Sukhoi SU-30MKM. The Air Force will fly its MiG-29N/NUB fighter aircraft until 2020 or longer and will receive upgrades to keep them up to date.

✦ Eurocopter has signed a cooperation contract with Malaysian Ministry of Defence to deliver 12 EC725 helicopters to the Royal Malaysian Air Force.

✦ Russia has signed a contract to supply the RVV-AE air-to-air missiles to the Malaysian Air Force. The contract is worth USD 35 million.

✦ The Malaysian Government has ordered four Airbus Military A400M new generation airlifters. The Royal Malaysian Air Force (RMAF) will be receiving them by 2016.

✦ Sukhoi has signed a USD 100 million deal with the Malaysian Defence Ministry, for technical maintenance and spare parts for the RMAF's 18 SU-30MKMs that were delivered to Malaysia between 2007 and 2009 under a USD 900-million contract signed in 2003.

✦ Malaysia is evaluating offers from several companies to replace the MiG-29N jet fighters with multi-role combat aircraft for the country's Air Force. The ministry is also looking into the package offered by the companies, including the reciprocal offer by the manufacturing companies, such as technology transfer and payment duration.

The Malaysian Air Defence Ground Environment Sector Operations Centre III (MADGE SOC III) programme provides the Royal Malaysian Air Force with significant upgrades to their legacy capabilities. The contract, originally for the addition of a single command and control operations centre and long-range air defence radar, was expanded to include multiple operations centres with either Sector Operations Centre or Air Defence Operations Centre functionality derived from the Sentry® command and control software. Sentry is the basis for the Battle Control System, which is used operationally by the U.S. Air Force and Canadian Air Force. The MADGE SOC III program continues a long history of cooperation between ThalesRaytheonSystems and the Royal Malaysian Air Force, which dates back to the installation of their first automated command and control air defence system.

Major Armaments Producers

Malaysia decided to undertake defence industrialisation for both economic and military reasons. Since the 1970s, Malaysia had defence programmes aimed at improving military capabilities with procurement of weapon systems and in general, military modernisation. In the sector of defence R&D, Malaysia mostly relies on partnerships with international contractors. The Malaysian Defence Industry Council (MDIC) was formed in August 1999 to ensure coordinated and orderly development of the defence industry sector in Malaysia. It was later renamed as Malaysian Defence, Security and Enhancement Industry Council (MIDES). MIDES also discusses common issues and challenges encountered by defence, enforcement and security related companies. Domestic companies are now engaged in joint venture projects such as shipbuilding, manufacturing small arms & explosives, aerospace sector and assembling of military vehicles for the armed forces. The government has given more focus on indigenisation of defence industry in order to attain the goal of self-reliance. The government has been actively engaging & marketing the Malaysian Defence Industry through exhibitions. International exhibitions such as LIMA and DSA hosted by Malaysia periodically have geared up the defence industry to new heights. The Langkawi International Maritime and Aerospace (LIMA) Exhibition is one of the largest defence and civil showcases in the Asia Pacific, regularly attracting the global maritime and aerospace industry to Langkawi. DSA is Asia's largest defence and security exhibition. Thus, the Government of Malaysia is fully committed to developing the nation's competitiveness by enhancing its economic and technological capabilities.

Domestic suppliers: Astronautic Technology, Airod, SME Ordnance, Malaysian Marine and Heavy Engineering, Malaysian International Shipping Corporation & Boustead Naval Shipyard Sdn Bhd, Boustead Heavy Industries, SME Ordnance.

Foreign suppliers: United States, Russia, United Kingdom, Germany, Italy, Switzerland, France, Sweden, Canada, Poland, South Africa, China, South Korea, Spain, Turkey, Austria, Australia, Belgium, Brazil, European Union, Pakistan, Japan, Italy, France.

DEFENCE DEALS

Malaysian Defence Ministry has signed RM9.3 billion worth of contracts, memoranda of understanding (MoU), letters of acceptance (LOA) and letters of Intent (LOI) were signed at LIMA 2015. Major contracts included MoU, LOA and LOI signed by the Defence Ministry for the supply of articles, services and technical assistance pertaining to optronic avionics and communication equipment of the Royal Malaysian Air Force (RMAF) SU-30MK with Thales Optronique. Contracts were also signed for the supply of parts and services for life rafts and life jacket owned by the Royal Malaysian Navy (RMN). Petronas signed an MoU to handover drone system to Malaysian Armed Forces. Other major contracts signed during the exhibition held in 2015 were an LOA with MTU Services (M) Sdn Bhd for the supply and delivery of spare parts, maintenance and training of MTU engines, sub-assemblies, components and electronics monitoring system for the RMN. An LOA was also signed with Zetro Services Sdn Bhd for the maintenance of air defence radar system and associated support equipment as well as an LOI for the maintenance of ground electronics equipment for the air traffic control system and accessories. Destination Marine Services Sdn Bhd also signed an LOI with the Malaysian Maritime Enforcement Agency (MMEA) for the supply and delivery of six units of patrol vessels. The Defence Ministry also signed an LOA with Boustead Naval Shipyard Sdn Bhd for the supply and delivery of spare parts, maintenance, integrated logistics support and training for the 17th patrol vessel squadron for the RMN. BHIC Bofors Asia Sdn Bhd signed an LOA for the supply and delivery of spare parts, services, maintenance and training of Bofors cannon 57mm for RMN, while Thales Raytheon System inked an LOA for the maintenance contract for the Malaysian Air Defence Environment Sector Operation Centre 3 (MADGE SOC 3).

European defence firm Thales has teamed up with Malaysia's Global Komited Sdn Bhd, a wholly owned subsidiary of the Weststar Group of Companies, to develop, promote, market, and distribute a wide range of ground-based air defence systems for the Malaysian Armed Forces. The partnership agreement was inked on the inaugural day of the Defence Services Asia (DSA) exhibition on April 14 in Kuala Lumpur. The event marked the appointment of Global Komited Sdn Bhd as a key industrial partner agent for Thales in the region. The GK-M1 is capable of mounting a lightweight multiple missile launcher that can fire the Thales-built STARStreak high velocity missiles. The STARStreak is the latest addition to Thales-developed Ground-based Air Defence systems. The new missile will enable the Malaysian Armed Forces, particularly the Royal Malaysian Air Force, to replace the previously commissioned Starburst missile system, according to Thales.

BAE Systems is planning to expand its existing Malaysian joint venture with Boustead Heavy Industry Corporation, BHIC Bofors Asia Sdn Bhd (BHIC Bofors), which already has an industrial facility in Kota Kinabalu. This expansion will include manufacture and supply of gun systems for the Royal Malaysian Navy (RMN) and the wider Southeast Asian region. Through the joint venture, BHIC Bofors will produce new guns locally, using an increased number of parts from local suppliers. The Malaysian supplier base will be expanded through a 'vendor development programme' which includes technology transfer from BAE Systems Bofors' factory in Karlskoga, Sweden. BAE Systems has supplied Malaysia with Bofors naval guns for decades and today there are about 140 in service. BAE Systems' Malaysian joint venture, with BHIC Bofors Asia currently supports the RMN with spares, maintenance, through-life support, training, lifetime extension, test, installation and setting to work on ships. Eurofighter Typhoon is a leading contender in the Malaysian Multi-Role Combat Aircraft competition.

Based on agreements concluded from 1989 onwards (and in particular the Defence Agreement signed in 1993 by Defence Ministers), the French-Malaysian cooperation has been focusing on three key points: strategic issues, defence and armament co-operation. France is one of the main military partners of Malaysia as far as the supply of military equipment and defence services—such as training, assessments and technical assistance—is concerned. Since 2002, Malaysia has chosen to develop a closer co-operation with France to create a submarine force. Malaysia ordered two Scorpene units along with an ex-French Navy Agosta boat for training purposes. The second submarine entered service with the Royal Malaysian Navy in 2010. Malaysia govt. has also chosen French military helicopters and missiles, military electronics and some important equipment in new heavy tanks, new fighting aircraft and the tactical radios of the armed forces.

Malaysia is a key Rolls-Royce customer, with over 100 Rolls-Royce military engines in service across a number of platforms. The Royal Malaysian Navy (RMN) was the first defence force to take delivery of the Super Lynx 300 helicopter in 2003, powered by the T800 engine from LHTEC, a joint venture between Rolls-Royce and Honeywell. The RMN now operates six of the highly sophisticated Super Lynx helicopters at sea from its two frigates, the KD Jebat and KD Lekiu. The Royal Malaysian Air Force (RMAF) is one of the most experienced operators of the Adour Mk871-powered Hawk trainer aircraft and also operates the Aermacchi MB339 trainer, powered by the Viper engine and the VVIP Global Express powered by the BR710.

Raytheon awarded an initial contract to Malaysian-based Contraves Advanced Devices Sdn. Bhd. for the production and supply of components for the Evolved Sea Sparrow Missile (ESSM). This is the first contract of its type awarded to a Malaysian company and is aligned with US and Malaysian strategic initiatives as well as the Malaysian Economic Transformation Programme. Raytheon has proposed ESSM to Malaysia for its upcoming Second Generation Patrol Vessel programme for the Royal Malaysian Navy. Selection of ESSM for the SGPV programme would significantly enhance the capabilities of the Royal Malaysian Navy and provide for interoperability with the US Navy and with the Australian, Japanese and Canadian navies in the Pacific theatre. Raytheon has also issued requests for quotation to Malaysian industry for additional components for the ESSM, the MK56 ESSM vertical launching system, MK73 illuminator, and test equipment in support of the Malaysian SGPV programme as well as other international production programmes. ESSM is the world's most advanced ship self-defence and local area defence surface-to-air missile system.

Arms Imports

The Malaysian defence arm requirements are supplied by the Russians and the US. The current trend indicates the domination of the Russian weapons systems in the Malaysian defence market. Under the Tenth Malaysia Plan (2011-15), Malaysia's armed forces plan to acquire three LPD-type multi-role support ships, additional combat aircraft (either additional Su-30MKKs or F-18E/Fs), AEW aircraft, and new armoured vehicles. Other equipment likely to be ordered within the next one to two years includes new ATGWs and a battlefield management system integrated with command post AFV variants.

On November 2016, Malaysia and China signed a major defence deal and promised significant cooperation in the South China Sea, which is seen worldwide as a potential strategic shift by Malaysian Prime Minister Najib Razak. Both the countries signed nine agreements covering defence, business and other areas of mutual interests with special focus on naval cooperation.

Arms Exports

Malaysia External Trade Development Corporation (Matrade) has secured RM761 million (USD 248 million) from export of defence equipment to neighbouring countries- Cambodia and Vietnam, according to Matrade. Cambodia recorded potential sales of RM629.5 million while Vietnam concluded potential sales of RM131.5 million. Items for sale included unmanned aerial vehicles; fast interceptor crafts; fire, rescue and specialised vehicles; simulation training equipment and software; communications and command systems; maintenance, repair and operations as well as design services. Eight companies participated in its first ever specialised marketing mission on the defence industry to Phnom Penh, Cambodia and Hanoi, Vietnam in 2012. The mission's objective was to expand on trade and economic collaboration between Malaysia, Cambodia and Vietnam specifically in the area of defence related products and services, at the same time further strengthening existing business networks with the two nations.

DEFENCE BUDGET

Total defence spending: $3.61 billion (FY2017)

Defence budget in terms of GDP: 1.54% (2015)

Malaysia's defence budget is modest compared to many countries. It is based on a percentage of annual government expenditure and it is not pegged to GDP or GNP. Defence allocation, therefore, depends on affordability of the government. If reflects the interplay of the military strategic objectives and national resource constraints.

Malaysia has slashed its defence budget to 12% from last year, i.e.; a total of USD 3.14 billion has been announced by the Defence Ministry for F.Y. 2017 from total USD 59.50 billion allocated for the new financial year, which is only 5.76% from the whole budget. The Malaysian Armed Force will procure six Second-Generation Patrol Vessel - Littoral Combat Ships for the Royal Malaysian Navy, deliveries of AV8 Gempita 8x8 armoured vehicles, A400M transport aircraft, MD 530G helicopters, refits for Scorpene submarines, SOC 3 radar system and Infra Kota Samarahan and upgrades to C-130, Hawk and S-61A-4 aircraft. According to Defence Minister Datuk Seri Hishammuddin Hussein, slash in the 2017 defence budget will not affect the 48 ongoing projects to be continued next year. The Ministry of Defence received an allocation of RM15.1 billion, out of which RM11.7 billion will be spend for management expenses while RM3.37 billion is meant for development expenditure.

Malaysia's defence policy outlines the three main bases, the country's strategic interests, the principles of defence and defence concepts. It emphasises the need to maintain the strategic importance of a stable and secure environment. In Malaysia, defence spending is hard to assess because of the country's rather opaque budgeting system for development spending. Malaysia's budgeting system for defence is based on a formal five-year planning structure and process which establishes force-manpower-ringgit planning levels to guide the formulation of annual budget – the five-year plans or development expenditure and the annual estimates or operating expenditure. The five-year plans projects spending trends into the future and is used as the primary planning instrument for development or acquisition of new capabilities. In 2012, from the total allocation of RM251.6 billion, the Defence Ministry got RM15.2 billion. Management allocation is RM11.9 billion while development received RM3.2 billion. The Malaysian Armed Forces received USD 4.2 billion to tighten border security in the Budget 2014 which also includes USD 765 million for the procurement of six OPVs, four cargo aircraft, and main battle tanks. The forces have also received USD 2.4 million to continue ESSCOM (Eastern Sabah Security Command) operations.

CONCLUSION

Malaysia's security perspective is influenced by the geographical, political, military and economic realities of the region and its strategic location in Southeast Asia. The defence of these national interests is crucial to Malaysia's sovereignty, territorial integrity and economic well-being. Other than this, the country also faces security challenges from terrorism, illegal immigrants and piracy. Malaysia's defence policy is an extension of its foreign policy. It is a capability-driven policy as opposed to a threat-dependent. It simply means that while it does not foresee immediate threats to sovereignty and territorial integrity, the possibility of threats arising in the future cannot be totally ignored. The main security challenge which Malaysia currently faces is safeguarding and defending the country's national interests which include political stability and peace and harmony of the citizens. From the defence point of view, Malaysia, as an independent sovereign nation, is growing its defence capabilities by building up armed forces and tying strong relations with strategic neighbours in order to defend itself from any external interference and armed attack. Malaysia which shares common land boundaries with Thailand, Brunei and Indonesia, as well as maritime boundaries with Thailand, Brunei, Indonesia, Singapore, the Philippines, Vietnam, China and Taiwan is entwined in complex interdependence with its neighbours with strategic interests in economic, political, cultural and defence arenas. The complexity is further heightened with the constant presence and influence of extra regional powers such as US, Russia, Japan and China. In order to safeguard the security of the country and strengthening the socio-economic and politics, Malaysia must promote and contribute for a stable regional security environment and have closer regional co-operation by resolving territorial disputes and focus more on shared economic interests in the region.

CONTACT DETAILS

Ministry of Defence
Wisma Pertahanan,
Jalan Padang Tembak,
50634, Kuala Lumpur
MalaysiaTel: (+60 3) 26 92 13 33
(+60 3) 20 71 57 46
Fax: (+60 3) 26 91 45 76
Website: www.mod.gov.my

MEXICO
(Capital: Mexico City)

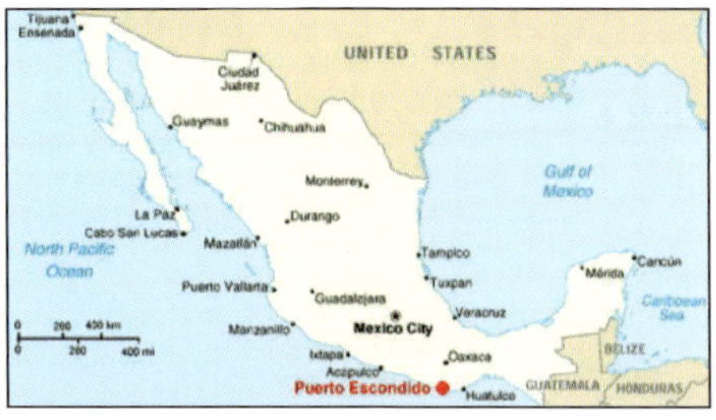

INTRODUCTION

Area: 1,964,375 sq km
Coastline: 9,330 km
Population: 123,166,749 (July 2016 est.)
Maritime claims: Territorial sea: 12 nm
Contiguous zone: 24 nm
Exclusive Economic Zone: 200 nm
Continental shelf: 200 nm or to the edge of the continental margin

KEY POLITICAL PERSONS

PRESIDENT & COMMANDER-IN-CHIEF: Enrique Pena Nieto

KEY DEFENCE PERSONS

SECRETARY OF NATIONAL DEFENCE: Gen. Salvador Cienfuegos Zepeda

CHIEF OF STAFF OF NATIONAL DEFENCE: Maj. Gen. Arturo Granados Gallardo

SECRETARY OF NAVY: Adm. Vidal Francisco Soberon Sanz

COMMANDER OF AIR FORCE: Lt. Gen. Carlos Antonio Rodriguez Mungia

ECONOMY

Mexico has become the US' second-largest export market and third-largest source of imports. The USD 1.3 trillion economy of Mexico has become increasingly oriented toward manufacturing in the 21 years since the North American Free Trade Agreement (NAFTA) entered into force. The economy of Mexico is the 13th largest in the world in nominal terms; and the 11th by purchasing power parity, according to the World Bank.

The economy contains rapidly developing modern industrial and service sectors, with increasing private ownership. As an export-oriented economy, more than 90% of Mexican trade is under free trade agreements (FTAs) with more than 40 countries, including the European Union, Japan, Israel, and much of Central and South America. GDP posted positive growth of 5.4% in 2010 and 3.8% in 2011, with exports - particularly to the United States - leading the way. The administration continues to face many economic challenges, including improving the public education system, upgrading infrastructure, modernizing labour laws, and fostering private investment in the energy sector.

GDP: $2.227 trillion (2015 est.), $2.172 trillion (2014 est.)

Real Growth Rate: 2.5% (2015 est.), 2.3% (2014 est.)

Industries: Food and beverages, tobacco, chemicals, iron and steel, petroleum, mining, textiles, clothing, motor vehicles, consumer durables, tourism

Oil - proved reserves: 9.7 billion bbl (1 January 2016 est.)

Natural gas - proved reserves: 432.9 billion cu m (1 January 2016 est.)

Total Exports: $381 billion (2015 est.), $397.7 billion (2014 est.)

Export Commodities: Manufactured goods, oil and oil products, silver, fruits, vegetables, coffee, cotton

Major Markets: US 81.1% (2015)

Total Imports: $395.6 billion (2015 est.), $400.4 billion (2014 est.)

Import Commodities: Metalworking machines, steel mill products, agricultural machinery, electrical equipment, car parts for assembly, repair parts for motor vehicles, aircraft, and aircraft parts

Major Suppliers: US 47.3%, China 17.7%, Japan 4.4% (2015)

INTERNATIONAL DISPUTES

✦ The US has intensified security measures to monitor and control legal and illegal personnel, transport, and commodities across its border with Mexico.

✦ Belize and Mexico are working to solve minor border demarcation discrepancies arising from inaccuracies in the 1898 border treaty.

DEFENCE

ARMY

Personnel: 700,000

Equipment

Category	Name	In Service
APC/ IFV/ Light Tank	AMX-VCI	409
	M2A1	34
	Buffalo	22
	BDX	95

Army continued...

Category	Name	In Service
APC/ IFV/ Light Tank	ERC 90	
	VBL M11	40
	HWK11	40
	VCR/TT	40
	Mowag Roland	25
Artillery	105 mm M2A1	46
	105 mm M101	
	155mm M-114	
	105 mm. M90	24+
Mortar	120 mm.	34
	81 mm.	330
	60 mm.	1,561
Anti-Tank GW	Milan	6

New Procurements/ Upgrades

✦ On May 16, 2014, the Mexican government sent a request to the U.S. State Department for the purchase of 3,335 M1152 HMMWVs. The Mexican government also sent a purchase request to the U.S. in late June for the sale of 18 UH-60M Black Hawk helicopters, according to a United States DSCA news release.

NAVY

Personnel: 56,000

Equipment

Category	Name	In Service
Destroyer	Quetzalcoatl class	1
	Manuel Azueta class	1
Frigate	Allende Class	4
	Nicolas Bravo Class	2
Patrol Vessel/ Patrol Craft	Huracan Class	2
	Oxaca Class	6
	Durango Class	4
	Uribe Class	6
	Valle Class	12
	Azteca Class	20
	Democrata Class	2
	Yaqui Class	4
	Point Class	5
	Olmeca Class	13
	Polimar Class	8
	Acurio A	6
	Acurio B	2
	Inshore patrol	4
	River Patrol	5
Amphibious Force	Transport Ship	2
	LST	3
	Auxiliary vessel	20
	Rio Tuxpan hydrographic vessel	1

Naval Aviation

Personnel: 1,200

Equipment

Category	Name	In Service
Aircraft	C212-200 MPA Aviojet	7
	An-32	6
	E-2C Hawkeye	3
	Dash-8 Q200	1
	Cessna 180	2
	Cessna 310	3
	Cessna 150	2
	Cessna 337	3
	Cessna 402	4
	DHC-5	1
	L-90 TP RediGo	12
	King Air 90	1
	CN-235	2
Helicopter	MD Explorer	8
	MBB BO-105	11
	SA.319B Alouette III	4
	MD500E	5
	AS-555 Fennec	4
	AS 565 Panther	2
	Mi-8	20
	Mi-17V1	19
	R-44 Clipper	1
	R22 Mariner	2
	Schweizer 300	4

A Mexican Navy MBB BO-105CB helicopter lands on deck of the US Navy (USN) Guided Missile Cruiser USS YORKTOWN (CG 48) during the 43rd annual UNITAS exercise.

Marines

Personnel: 10,000

Equipment

Category	Name	In Service
APC/IFV	Pegaso VAP 3550	25
	BTR-70	
	VAP 3560/1	
Artillery	105 mm. M56	8
MRL	Firos-6	
Mortar	60 mm.	100
	81 mm.	
Recoilless rifle	106 mm.	
Patrol Vessel	Pollaris/ Pollaris II Class FAC	40

New Procurements/ Upgrades

✦ Rockwell Collins Simulation and Training Solutions, Cedar Rapids, Iowa, was awarded a USD 12,087,181 multi-year, firm-fixed-price, foreign military sales contract (Mexico) for a transportable Blackhawk Operations Simulator for the Mexican navy. One bid was solicited with one received. Work will be performed in Mexico, with an estimated completion date of Sept. 11, 2018.

✦ Mexico has signed a contract with Airbus Defence and Space for the acquisition of two additional Airbus C295 medium transport aircraft for the Mexican Navy. The Mexican Navy, which is already flying four C295, six CN235 and two C212 aircraft, will operate these two new aircraft on a diverse range of transport missions for military, civil protection and humanitarian purposes.

✦ Mexican Navy will be receiving three UH-60M aircraft and three auxiliary power units inlet barrier filter kits for the UH-60M from the Sikorsky Aircraft Corporation under a $35 million firm-fixed-price contract.

✦ The Mexican Navy has awarded a contract to Damen Shipyards Group for the fourth Damen Stan Patrol 4207 type boat.

AIR FORCE

Personnel: 11,770

Equipment

Category	Name	In Service
Fighters/ attack/ Reconnaiss-ance/bomber	F-5E	8
	F-5F	2
	Cessna C-182 Skylane	75
	EMB-145 AEW&C	1
	EMB-145MP	2
Transport/ Tanker	C-130S/C-130K/C-130E/L-100-20	14
	Boeing 727	3
	An-32	2
Attack/ Transport Helicopter	Mi-8	11
	Mi-17	21
	Mi-26	2
	MD-530F	24
	Bell 212	22
	S-70	12
	CH-53	2
	Bell 412EP	5
Trainer	PC-7	37
	T-6C+	6

New Procurements/ Upgrades

✦ Sikorsky Aircraft Corp. awarded a $203.6 million modification to a US Army contract to procure 18 green UH-60M Black Hawk helicopters for use by the Mexican Air Force.

✦ Eurocopter, an EADS company has signed an agreement with the Mexican Ministry of Defence for the supply of six additional EC725 helicopters.

✦ Five Bell 412EP ("enhanced performance") helicopters are on order from the US Defence Security Cooperation Agency. Mexico will use them to fight drug traffickers, gangs, and other criminal activities in its ongoing "Cartel War." The estimated value of the contract is over $93 million.

✦ Beechcraft Corporation has announced a follow-on sale of its Beechcraft T-6C+ military trainer to the Mexican Air Force (FAM). The T-6C+, an enhanced version of the T-6 military trainer aircraft, is capable of carrying external stores and delivering practice weapons for training purposes.

✦ Mexico has requested the US for the sale of two C-130J-30 aircraft and associated equipment, parts, training and logistical support for an estimated cost of $412 million.

✦ Cascade Aerospace has acquired its first international military C-130 customer with the successful signing of a contract with the Canadian Commercial Corporation (CCC) to undertake modernisation and sustainment activities on two Mexican Air Force C-130K Hercules aircraft. This contract was established under the auspices of a Memorandum of Understanding between the Canadian and Mexican governments.

✦

Mexico is one of the few countries which has capabilities to manufacture nuclear weapons but has renounced them and pledged to only use its nuclear technology for peaceful purposes following the Treaty of Tlatelolco in 1968. In the 1970s Mexico's national institute for nuclear research successfully enriched weapons-grade uranium which is used in the construction of nuclear weapons. However, in April 2010, Mexico agreed to turn over its weapons-grade uranium to the United States. This agreement hasn't been approved by the Mexican Congress and the uranium remains in Mexico.

Defence Expenditure

Total defence spending: 11.6 billion (2014 est.)

Estimated defence spending in terms of GDP: 0.68% (2015)

Mexico has the second largest defence budget in Latin America, spending about 0.5% GDP on its military.

Major foreign suppliers: United States, Russia, Spain, Israel, Germany, France, Switzerland, UK, Italy, Sweden, Brazil, Denmark, Norway, South Africa and Colombia.

Defence Production and R&D

Mexico has a small but long-established defence industry. The Mexican defence industry is fairly active producing "soft" items: manuals, uniforms, etc. The "Industria Militar" (owned by the government) produces uniforms and other supplies, and small arms, including G3 machine guns. No equipment such as tanks, aircraft, etc., is manufactured and there are no known plans to produce this type of equipment. Mexico's major shipyards at Tampico on the Gulf of Mexico and Salina Cruz on the Pacific Ocean have been involved in the construction of patrol craft and auxiliary vessels. The largest programme, involving the manufacture of Azteca-class patrol craft, was carried out under a licensed production agreement with a British firm.

Defence Procurement

The Mexican Armed Forces are currently under-equipped as much of their inventories are ageing. To meet the present requirements of the forces, Mexico needs a significant procurement programme.

CONTACT DETAILS

Secretariat of National Defence
Boulevard Manuel Avila Camacho
Avenida Industria Militar
Col. Lomas de Sotelo
11640 Delegacion Miguel Hidalgo
Distrito Federal
Tel: (+52 21) 22 88 00
Website: www.sedena.gob.mx

MONGOLIA
(Capital: Ulaanbaatar)

INTRODUCTION

Area: 1,564,116 sq km

Population: 3,031,330 (July 2016 est.)

Coastline: 0 km (Landlocked)

Maritime claims: None

KEY POLITICAL PERSONS

PRESIDENT & COMMANDER-IN-CHIEF: Tsakhiagiyn Elbegdorj

PRIME MINISTER: Erdenebat Jargaltulga

KEY DEFENCE PERSONS

MINISTER OF DEFENCE: B. Bat-Erdene

CHIEF OF GENERAL STAFF: Lt. Gen. Dulamsurengiin Davaa

ECONOMY

Mongolia's economy is traditionally based on agriculture and livestock. The country has extensive mineral deposits - copper, coal, molybdenum, tin, tungsten, and gold. In recent years Mongolia's economy has expanded rapidly. According to the Economist Intelligence Unit, it was one of the top performers in 2013, with economic growth of 11.7 per cent. The main driver of this rapid economic expansion has been the mining industry's development. The main export commodities are copper, gold and coal. Mongolia is now one of the major coal exporters to China, briefly overtaking Australia. During the next five years major mining projects of copper - Oyu Tolgoi project and coal Tavan Tolgoi project - is expected to reach their full potential, likely further increasing economic expansion. The country's economy had grown more than 10% per year between 2011 and 2013 - largely on the strength of commodity exports and high government spending - before slowing to 7.8% in 2014 and 2.3% in 2015.

GDP: $36.07 billion (2015 est.); $35.26 billion (2014 est.)

Real Growth Rate: 2.3% (2015 est.); 7.9% (2014 est.)

Industries: Construction and construction materials; mining (coal, copper, molybdenum, fluorspar, tin, tungsten, and gold); oil; food and beverages; processing of animal products, cashmere and natural fibre and manufacturing

Total Exports: $4.619 billion (2015 est.); $5.774 billion (2014 est.)

Export Commodities: Copper, apparel, livestock, animal products, cashmere, wool, hides, fluorspar, other nonferrous metals, coal and crude oil

Major Markets: China 84%, Switzerland 9% (2015)

Total Imports: $3.512 billion (2015 est.); $4.746 billion (2014 est.)

Import Commodities: Machinery and equipment, fuel, cars, food products, industrial consumer goods, chemicals, building materials, cigarettes and tobacco, appliances, soap and detergent.

Major Suppliers: China 39.9%, Russia 28.4%, Japan 6.4%, South Korea 6.2% (2015)

INTERNATIONAL DISPUTES

Mongolia has no known international disputes.

DEFENCE

The Mongolian Armed Forces consists of three branches - General Purpose Forces, Border Defence Forces, and Internal Security Forces. This peace-time structure could be changed and reorganised in case of war or a war-like situation. The Armed Forces maintains an Air Forces Defense Command under the command of the General Staff. The country has the goal of developing a full-fledged air force in future.

ARMY

Personnel: 20,500

Equipment

Category	Name	In Service
MBT	T 54/55	
MICV/APC	BMP-1	400
	BRDM-2	130
	BTR-40	200
	BTR-60	
	BTR-80	450
Artillery	76 mm	300
	100mm	
	130mm	
	152mm Howitzer	
ATGW	Snapper	
	Sagger AT-3	
AA gun	37mm, 57mm	

New Procurements/Upgrades

No major procurement plans are under consideration at present.

AIR FORCE

Personnel: 3,000

Equipment

Category	Name	In Service
Fighter	MiG-21	10
Transport	An-24	6
	An-2	15
	An-26	1
	An-32	1
Trainer	Yak-18	6
Helicopter	Mi-8	16
	Mi-24	10
	Mi-17 (attack)	
SAM	SA-2	1 battalion

New Procurements/Upgrades

- Mongolia plans to buy three C-130J transport airplanes manufactured by Lockheed Martin.

Defence Expenditure

Estimated defence spending in terms of GDP: 1.1%

Defence Production and R&D

Mongolia does not undertake any defence production activities.

Defence Procurement

The US pledged to increase its military training and aid to Mongolia, currently valued at about $3 million a year. During his visit to Mongolia in April 2014, US Secretary of Defence Chuck Hagel and Mongolian Defence Minister Dashdemberal Bat-Erdene signed a joint statement promising to strengthen military cooperation and increase joint training programmes. Mongolia and Russia have also signed a defence co-operation agreement. Mongolia has taken delivery of some new hardware, including BTR-80 armoured personnel carriers, S-125-2M Pechora-2M surface-to- air missile systems and two Mi-171 'Hip transport helicopters. The Ministry of Defence intends to buy five new MiG-29 'Fulcrum' fighter aircraft. Russia and China are the main suppliers of arms to Mongolia. Turkey also began military assistance after signing an agreement with the country in 2002. India has also extended its defence cooperation agreement with Mongolia by another ten years. India has imparted training to Mongolian military officers in Indian training establishments. Military officers from both the countries have also been visiting each other's bases for joint exercises and academic exchanges.

CONTACT DETAILS

Ministry of Defence

Website: http://www.mod.gov.mn/

MOROCCO
(Capital: Rabat)

INTRODUCTION

Area: 446,550 sq km

Population: 33,655,786 (July 2016 est.)

Coastline: 1,835 km

Maritime claims: Territorial sea: 12 nm

Contiguous zone: 24 nm

Exclusive Economic Zone: 200 nm

Continental shelf: 200 m depth or to the depth of exploitation

KEY POLITICAL PERSONS

HEAD OF STATE, COMMANDER-IN-CHIEF & CHIEF OF GENERAL STAFF:
King Mohammed VI

PRIME MINISTER:
Abdelilah Benkirane

KEY DEFENCE PERSONS

MINISTER DELEGATE TO THE P.M. FOR NATIONAL DEFENCE:
Abdellatif Loudiyi

INSPECTOR GENERAL OF ROYAL ARMED FORCES (FAR):
Gen. Bouchaib Arroub

ECONOMY

Morocco has become a major player in African economic affairs, and is the 5th-largest African economy by GDP (PPP). The World Economic Forum placed Morocco as the most competitive economy in North Africa, in its African Competitiveness Report 2014-2015. Morocco's economy is considered a relatively liberal economy governed by the law of supply and demand. Additionally, Morocco was ranked the 1st African country by the Economist Intelligence Unit' quality-of-life index, ahead of South Africa in 2011. High food and fuel prices strained the government's budget and widened the country's current account deficit. Key economic challenges for Morocco include fighting corruption, reducing government spending, reforming the education system and judiciary, addressing socioeconomic disparities, and building more diverse, higher value-added industries.

GDP: $273.5 billion (2015 est.), $261.8 billion (2014 est.)

Real Growth Rate: 4.5% (2015 est.), 2.4% (2014 est.)

Industries: Phosphate rock mining and processing, food processing, leather goods, textiles, construction, energy, tourism

Oil - proved reserves: 160 bbl/day (2015 est.)

Natural gas - proved reserves: 1.444 billion cu m (1 January 2016 est.)

Total Exports: $18.48 billion (2015 est.), $20 billion (2014 est.)

Export Commodities: Clothing and textiles, electric components, inorganic chemicals, transistors, crude minerals, fertilizers (including phosphates), petroleum products, citrus fruits, vegetables, fish

Major Markets: Spain 22.1%, France 19.7%, India 4.9%, US 4.3%, Italy 4.3% (2015)

Total Imports: $32.74 billion (2015 est.), $40.68 billion (2014 est.)

Import Commodities: Crude petroleum, textile fabric, telecommunications equipment, wheat, gas and electricity, transistors, plastics

Major Suppliers: Spain 13.9%, France 12.4%, China 8.5%, US 6.5%, Germany 5.8%, Italy 5.5%, Russia 4.4%, Turkey 4.3% (2015)

INTERNATIONAL DISPUTES

✦ Claims and administers Western Sahara whose sovereignty remains unresolved.

✦ Morocco protests Spain's control over the coastal enclaves of Ceuta, Melilla, and Penon de Velez de la Gomera, the islands of Penon de Alhucemas and Islas Chafarinas, and surrounding waters. Both countries claim Isla Perejil (Leila Island).

✦ Discussions have not progressed on a comprehensive maritime delimitation, setting limits on resource exploration and refugee interdiction, since Morocco's 2002 rejection of Spain's unilateral designation of a median line from the Canary Islands.

✦ Algeria's border with Morocco remains an irritant to bilateral relations, each nation accusing the other of harbouring militants and arms smuggling.

✦ The National Liberation Front's assertion of claim to Chirac Pastures in south-eastern Morocco is a dormant dispute.

DEFENCE

ARMY

Personnel: 185,800

Equipment

Category	Name	In Service
MBT	T-72B/BK	148
	M60/ M60A3TTS	427
	M48A5	224
	Kurassier	105
	M1A1SA	222
	VT-1A	150
APC/MICV/ IFV	AMX13	54
	M113	1,300
	UR416	55
	VAB VCI/VTT	395
	Ratel IFV	60
	EBR75	16
	AMX10RC	110
	AMX10P	10
	AML90	140
	AML60	35
	ERC90 Lynx	20
Artillery	L118	36
	M101	18
	105 mm. M101A1	40
	130 mm. M46	18
	M114	22
	155 mm. FH70	30
	155 mm. M109	248

Army continued...

Category	Name	In Service
Artillery	155 mm. Mk F3	80
	203 mm. M-110 A2 SP	60
	M198 155 mm.	35
	M1950 155 mm.	35
Mortar	81 mm.	20
	120 mm.	91
Recoilless rifle	75 mm.	
	106 mm.	
	M72 Law	
	Strim89	
Anti-Tank GW	Milan	
	Tow	
	Dragon	
	HOT	
	Sagger	
AA Gun	20 mm.	100
	Twin 200 mm. mounts	42
	ZPU232	160
	M163 Vulcan	40
SAM	SA-7	200+
	Chaparral	59
	Tunguska M-1	12
	Crotale	30+

Army Aviation

Category	Name	In Service
Helicopter	Alouette II	4
	Gazelle	3
	A-109	6

New Procurements/ Upgrades

✦ Harris Corp., Rochester, New York, was awarded a USD 405,500,000 firm-fixed-price, foreign military sales (Morocco), indefinite-delivery/indefinite-quantity contract for the Single Channel Ground and Airborne Radio Systems (SINGARS) including ancillary items, spare parts, installation kits, training and fielding support services. One bid was solicited with one received, with an estimated completion date of April 21, 2021.

✦ The Kingdom of Morocco has requested the US government for enhancement and refurbishment of 200 M1A1 Abrams MBTs and associated parts, equipment, logistical support and training. The cost of the contract is $1.015 billion. The possible sale will also provide 150 AN/VRC-87E and 50 AN/VRC-89E Exportable Single Channel Ground and Airborne Radio Systems (SINCGARS), 200 M2 Chrysler Mount Machine Guns, and 400 7.62MM M240 Machine Guns. The possible sale also includes 12,049,842 Ammunition Rounds (including 1400 C785 SABOT, 1800 CA31 HEAT, and 5400 AA38 SLAP-T), 200 M250 Smoke Grenade Launchers, support equipment, spare and repair parts, personnel training and training equipment, publications and technical data, communication support, US Government and contractor technical assistance, and other related logistics support. This package of M1A1 tank enhancements will contribute to the modernisation of Morocco's tank fleet, enhancing its ability to meet current and future threats.

✦ The Kingdom of Morocco has requested a possible sale of 600 TOW 2A anti-tank guided missiles and associated equipment from the United States. The State Department has made a determination approving a possible Foreign Military Sale to Morocco for TOW 2A, Radio Frequency (RF) Missiles (BGM-71E-4B-RF), M220A2 TOW Launchers and associated equipment, parts, training and logistical support for an estimated cost of $157 million. Also included with this request are Missile Support Equipment; Government-Furnished Equipment; Technical Manuals/Publications; Spare Parts; Tool and Test Equipment; Training; U.S. Government Technical Support/Logistical Support; Contractor Technical Support; and other associated equipment and services.

NAVY

Personnel: 42,000

Equipment

Category	Name	In Service
Frigate	SIGMA class	3
	Mohammed V Class	1
	Hassan II	1
	Mohammed VI FREMM Class	1
Corvette	Lieutenant-Colonel Errahmani	1
Patrol Vessel	Rais bargach Class OPV	5
	El Khattabi Class ex-FAC	4
	Okba Class FAC	2
	El Lahiq	4
	LV Rabhi	6
	El Wacil Class/ Erraid Class	
Auxiliary ship/ Amphibious Force	Sidi Mohammed Ben Abdallah	1
	Batral Type	3
	Dakhla logistics support	1

Naval Aviation

Category	Name	In Service
Aircraft	Britten-Norman Defender	14
Helicopter	Eurocopter AS565 MA	3

New Procurements/ Upgrades

✦ Moscow and Rabat may sign an agreement on the supply of the Russian Amur-1650 diesel-electric 4th-generation Project 677E submarine to Morocco.

✦ DCNS has signed a contract for a multi-mission FREMM frigate for the Royal Moroccan Navy. Morocco is the first international customer for this new-generation vessel.

✦ Morocco signed a $1.2 billion contract with Schelde Naval Shipbuilding for two Light frigate SIGMA 9813 and one Light frigate SIGMA 10145 which are modified versions of the existing SIGMA Class design. On September 2011, Damen Schelde Naval Shipbuilding (DSNS) has handed over the first of three SIGMA Class Frigate 'Tarik ben Ziyad' to the Royal Moroccan Navy after successful sea trials. On Sept 2012, the Royal Moroccan Navy took delivery of the SIGMA Class Frigate, Allal Ben Abdellah, the last ship of a series of three built by DSNS.

✦ The Navy has received Stan Tug 2208 from Dutch shipbuilder Damen. Damen is currently building five Interceptor 1503 vessels for Morocco.

AIR FORCE

Personnel: 13,500

Equipment

Category	Name	In Service
Fighters/ attack/ Reconnaissance/ bomber	Mirage F-1CH	10
	Mirage F-1 MF2000	15
	F-5E	22
	Dassault Falcon 20 ELINT / ESM	2
	F-16C/D Block 52+	24
Transport/ Tanker	KC-130H	2
	C-130H	13
	CN-235	6
	Gulfstream II/III	2
	King Air 200/300/350	8
	Alenia C-27J	3
	Boeing 707	1
	Citation V	2
Attack/ Transport Helicopter	SA341	23
	SA 330 Puma	26
	Bell 205	47
	Bell 206	5
	Bell 212	4
	CH47C Chinook	3
Trainer	Super King Air 100	4
	Bell 206	17
	T-6C Texan II	24
	Alpha Jet	24
	F-5F	4

New Procurements/ Upgrades

✦ Lockheed Martin has been awarded a nearly $841.9 million contract by the US government to deliver 24 F-16 Block 52 fighters under a previously announced deal with Morocco. Morocco is paying $35 million per F-16 Block 52, which include advanced countermeasures, electronic warfare and support equipment.

✦ Hawker Beechcraft Corporation (HBC) has been awarded a contract to provide 24 of its new Beechcraft T-6C trainer aircraft to the Royal Moroccan Air Force. The value of the contract is $185.3 million. The aircraft will replace T-34s and T-37s. On July, 2010, HBC rolls out the first of 24 Beechcraft T-6C military trainers ordered by the Royal Moroccan Air Force (RMAF).

✦ A contract was signed between Finmeccanica unit Alenia Aeronautica and Morocco for the supply of four C-27J Spartan. The value of the contract is $163.8 million.

✦ Morocco has received three CH-47D Chinook helicopters and associated parts, equipment and logistical support from United States in 2015.

CH-47D Chinook helicopter

ROYAL MOROCCAN GENDARMERIE

Personnel: 24,000

The Moroccan Royal Gendarmerie is the Gendarmerie body of Morocco, and comes under the authority of the Ministry of Defence. The service aim of the gendarmerie is to ensure especially administrative, judicial and military policing activities directly and to help to the competent authorisations with the envisaged laws. It is a public force designed to guarantee public security and public order and the implementation of laws.

Category	Name	In Service
Helicopter	SA315	7
	SA342 Gazelle	5
	Eurocopter AS355	3
	Sikorsky S-70 S-70A-26	2
	SA 330 Puma C/F/H	7
	Eurocopter EC145	2
	Eurocopter AS332 AS332/ EC225	2
	Eurocopter EC135	3

Defence Expenditure

Total defence spending: USD 3.37 billion (est.)

Estimated defence spending in terms of GDP: 3.26% (2015)

Moroccan defence spending will rise from US$3.8 billion in 2014 to US$4.5 billion in 2018 because of the procurement of military aircraft, armoured vehicles, radar systems, diesel-electric submarines and patrol ships.

Major foreign suppliers: France, Spain, Russia, China, Romania, Egypt and United States.

Defence Production and R&D

Morocco's small aerospace sector is entirely based at Casablanca's Muhammad V Airport and has civil and military applications. There are no defence production facilities in Morocco.

Defence Procurement

Moroccan procurement has been limited by financial constraints and the European Code of Conduct on arms sales, which some member states interpret as prohibiting weapons sales to Morocco due to the Western Sahara dispute. Morocco allocates around $180 million per year for military capital spending, including procurement and defence construction.

CONTACT DETAILS

National Defence Administration

6Bis, Rue Patrice Lumumba

Rabat, Morocco

Tel: +212-7-62731, -60150, -60805

MYANMAR

(Capital: Rangoon)

INTRODUCTION

Area: 676,578 sq km

Population: 56,890,418 (July 2016 est.)

Coastline: 1930 km

Maritime claims: Territorial sea: 12 nm

Contiguous zone: 24 nm

Exclusive Economic Zone: 200 nm

Continental shelf: 200 nm or to the edge of the continental margin

KEY POLITICAL PERSONS

PRESIDENT: Htin Kyaw

VICE PRESIDENT (First & Second): Myint Swe & Henry Van Thio

KEY DEFENCE PERSONS

DEFENCE MINISTER: Lt. Gen. Sein Win

COMMANDER-IN-CHIEF OF THE ARMED FORCES: Senior Gen. Min Aung Hlaing

COMMANDER OF THE ARMY: Vice Senior Gen. Soe Win

COMMANDER OF THE NAVY: Vice Adm. Tin Aung San

COMMANDER OF AIR FORCE: Gen. Khin Aung Myint

ECONOMY

Myanmar is an emerging economy. The government has taken steps toward reforming and opening up the economy by lowering export taxes, easing restrictions on its financial sector, and reaching out to international organisations for assistance. According to Asian Development Bank, growth is supported by rising investment propelled by improved business confidence, commodity exports, rising production of natural gas, buoyant tourism, and credit growth - all complemented by the government's ambitious structural reform programme. The most productive sectors continue to be oil and gas, mining, and timbre. The major agricultural produce is rice which covers about 60% of the country's total cultivated land area. Myanmar's economy grew at 8.5 percent in real terms in 2014/15 but growth is projected to moderate to 6.5 percent in 2015/16 due to floods and slowing investments, according to the World Bank.

GDP: $283.5 billion (2015 est.), $264.9 billion (2014 est.), $243.7 billion (2013 est.)

Real Growth Rate: 7% (2015 est.), 8.7% (2014 est.), 8.4% (2013 est.)

Industries: Agricultural processing, wood and wood products, copper, tin, tungsten, iron; cement, construction materials; pharmaceuticals; fertilizer; oil and natural gas; garments; jade and gems

Total Exports: $9.135 billion (2015 est.), $9.083 billion (2014 est.)

Export Commodities: Natural gas, wood products, pulses and beans; fish; rice; clothing; minerals, including jade and gems

Major Markets: China 37.7%, Thailand 25.6%, India 7.7%, Japan 6.2% (2015)

Total imports: $12.49 billion (2015 est.), $12.09 billion (2014 est.)

Import Commodities: Fabric, petroleum products, fertilizer; plastics; machinery; transport equipment; cement, construction materials; food products, edible oil

Major Suppliers: China 42.2%, Thailand 18.5%, Singapore 11%, Japan 4.8% (2015)

INTERNATIONAL DISPUTES

On 8 May 2014, India and Myanmar signed a Memorandum of Understanding (MoU) on Border Cooperation. Ensuring security along the 1,643 km land border shared by India and Myanmar has been one of the most interminable problems facing the two countries. The entire border length touches Arunachal Pradesh (520 km), Nagaland (215 km), Manipur (398 km) and Mizoram (510 km).

DEFENCE

The Myanmar Armed Forces is officially known as "Tatmadaw." The Myanmar Army has always been the largest service and has always received the lion's share of the country's defence budget. The Air Force was formed on 16 January 1947, while Myanmar was still under British rule. The Navy has expanded in recent years to provide blue water capability and external threat defence role in Myanmar's territorial waters.

ARMY

Personnel: 290,000

Equipment

Category	Name	In Service
MBT	MBT-2000	130
	Type 69II	100
	Type 63	100
	T-72	139
	Type 59D	135
	PTL-02 6x6 tank destroyers	
APC/AIFV	Type 85	250
	Type 90	150
	Ferret	45
Artillery	Type WP52 155mm	
	Soltam 155mm	36
	140mm	
	M1948 76mm	100
	M101 105mm	110
	NORA B-52 155mm	30
Mortar	81mm	
	120mm	

Army continued...

Category	Name	In Service
MRL	Type 63 107mm	30
	122mm	
	240mm	30+
ATK weapon	Carl Gustaf	
SAM	Pechora-2M	14
	SA-19	15
AA gun	Type 80 57mm	
	M1 40mm	10
	Tunguska gun SP systems	
	Type 74 37mm twin	

New Procurements/Upgrades

No major procurement plans are under consideration at present.

Type 59D MBT of the Myanmar Army. *Image credit: Myanmar Army Corps*

NAVY

Personnel: 16,400 (800 marines)

Equipment

Category	Name	In Service
Frigate	Mahar Bandoola class	2
	Haung Zeya	1
Corvette	Anawratha class	2
Patrol craft	55m missile FAC	5
	Houxin class (missile) FPB	6
	ex-Chinese Hainan coastal patrol boat	10
	Chinese Houxin class Missile FPB	6
	INDAW class OPV	1
	PGM-type	12
	PB-90	3
	river patrol craft	40
	river gun boat	16
Landing craft	LCU	4
	ex-US LCM	10

New Procurements/Upgrades

- A new frigate Kyansittha is fitting out in national yards,
- A preliminary agreement has been signed with India for the construction of 5 OPVs with Indian support.

AIR FORCE

Personnel: 9,500

Equipment

Category	Name	In Service
Fighter	NAMC A-5C	48
	Chengdu F-7M Airguard	52
	MiG-29B + MiG-29UB	18
	F-6	54
	F-7M	25
	A-5C	24
	K-8	12
	Super Galeb	6
	K-8W	6
Transport/ trainer	Beechcraft 1900D	2+
	FH-27	3
	ATR 42	1
	F-27	2
	Y-8D/F	6
	PC-6B Turboporter	5
	C-212	2
	Cessna 180	6
	Cessna Citation	1
	BN-2 Islander	2
	Super Galeb	6
	PC-1	15
	PC-9	4
Helicopter	Mi-35	9
	AS365 Dauphin 2	3
	Mil Mi-35P 'Hind E'	1
	NH-43	7
	UH-1	12
	Bell 206	6
	ex-Polish Mi-2	10
	W-3	12
	Mi-17V	13
	SA-316B Alouette III	9
	Bell 206 Jetranger III	3
	Grob G 120TP	20
SAM	SA-3	
	SA-6	
	TOR M1	

New Procurements/Upgrades

✦ On 24 June 2015, the air force has inducted several aircraft and helicopters which includes - Beechcraft 1900D light transport aircraft, at least 10 Grob G 120TP basic trainer aircraft, one Mil Mi-35P 'Hind E' attack helicopters, three Airbus Helicopters AS365 Dauphin 2s and three Bell 206 Jetranger III helicopters, strengthening the defence capability of the armed forces.

✦ Other aircraft inducted on 25 February include six K-8W jet trainers along with a single ATR 42 transport and a pair of Beechcraft 1900Ds - which will also be used for SAR and surveillance missions and serve as trainers for pilots moving onto ATR 42, ATR 72, Y-8 and Airbus transports.

✦ Myanmar has reportedly signed a deal to buy the JF-17 Thunder multirole fighter from Pakistan.

Defence Expenditure

Total Defence Spending: $2.7 billion (2016), $2.4 billion (2015)

Defence spending in terms of GDP: 3.38% (2015)

Defence Production R&D

Myanmar is reportedly planning to acquire technologies from China and Pakistan to build JF-17 multirole combat aircraft at its own aircraft factory. If the Transfer of Technology (ToT) is not possible the government plans to assemble the aircraft in the country. In recent times, Myanmar has developed the Chinese Sky 02A UAV's in-country as Yellow Cat A2 to perform basic surveillance missions.

The Myanmar Defence Industries (DI) produces approximately 70 major products which include automatic rifles, machine guns, sub-machine guns, anti-aircraft guns, complete range of mortar and artillery ammunition, aircraft and anti aircraft ammunition, tank and anti-tank ammunition, bombs, grenades, anti-tank mines, anti-personnel mines such as the M14 pyrotechnics, commercial explosives and commercial products, and rockets and so forth. DI has produced new assault rifles and light machine-guns for the infantry.

Defence Procurement

The government reportedly plans to purchase the jointly produced Sino-Pakistani fighter aircrafts - the FC-1 with prospects of creating the country's own licensed assembly line. The fighter planes will be assembled in Pakistan while the engines will be supplied by Russia. The government also plans to purchase of a large batch of high-precision "air-to-surfaces" missiles and guided bombs.

CONTACT DETAILS

Ministry of Defence

Building 24, Nay Pyi Taw, Myanmar.

Tel: 067 - 404300, Fax: 067 - 404299

Email: admm.myanmar@mptmail.net.mm

adsom.myanmar@mptmail.net.mm

admmewg.myanmar@mptmail.net.mm

NEPAL

(Capital: Kathmandu)

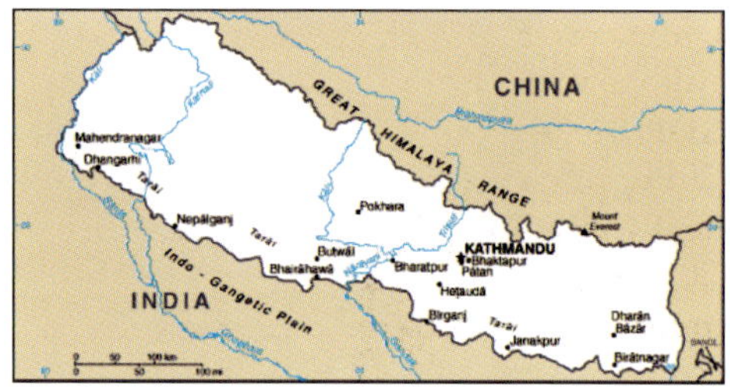

INTRODUCTION

Area: 147,181 sq km

Population: 29,033,914 (July 2016 est.)

Coastline: Landlocked

Maritime claims: None

KEY POLITICAL PERSONS

PRESIDENT & COMMANDER-IN-CHIEF: Bidya Devi Bhandari

PRIME MINISTER: Pushpa Kamal Dahal

KEY DEFENCE PERSONS

DEFENCE MINISTER: Bal Krishna Khand

CHIEF OF ARMY STAFF: Gen Rajendra Chhetri

ECONOMY

Nepal is a landlocked country located in South Asia. Nepal is among the poorest and least developed countries in the world, with almost one-quarter of its population living below the poverty line. Major contribution in the state GDP is from agriculture products. Agriculture employs 76% of the workforce, services 18% and manufacturing and craft-based industry 6%. Nepal has considerable scope for exploiting its potential in hydropower, with an estimated 42,000 MW of feasible capacity. The lack of political consensus in the past several years has delayed national budgets and prevented much-needed economic reform, although the government passed a full budget in 2013 and 2014. Nepal and India signed trade and investment agreements in 2014 that will increase Nepal's hydropower potential.

GDP: $70.09 billion (2015 est.), $67.81 billion (2014 est.)

Real Growth Rate: 3.4% (2015 est.), 5.4% (2014 est.)

Industries: Tourism, carpets, textiles; small rice, jute, sugar, and oilseed mills; cigarettes, cement and brick production

Total Exports: $813.1 million (2015 est.), $1.022 billion (2014 est.)

Export Commodities: Clothing, pulses, carpets, textiles, juice, pashmina, jute goods

Major Markets: India 61.3%, US 9.4% (2015)

Total Imports: $6.511 billion (2015 est.), $7.581 billion (2014 est.)

Import Commodities: Petroleum products, machinery and equipment, gold, electrical goods, medicine

Major Suppliers: India 61.5%, China 15.4% (2015)

INTERNATIONAL DISPUTES

Joint Border Commission continues to work on contested sections of boundary with India, including the 400 square kilometre dispute over the source of the Kalapani river.

India has instituted a stricter border regime to restrict transit of Maoist insurgents and illegal cross-border activities.

DEFENCE

ARMY

Personnel: 95,000

Equipment

Category	Name	In Service
APC/MICV /IFV	BTR-70	135
	Indian Mine Protected Vehicle/APC	70
	Ferret Armoured Car	40
Artillery	105 mm artillery pack	10
	3.7 inch Mountain Howitzer	100
	105 mm (Indian Field Gun)	100
	81 mm & 120mm mortar	1100

Army continued...

Category	Name	In Service
AA Gun	94 mm QF 3.7 inch AA gun	10
	14.5 mm Type 56 AD Guns	30
	40 mm L60 AD Guns	10

New Procurements/ Upgrades

✦ Nepal Army has received two Russian made choppers Mi-17 V-5 as per the procurement agreement made on December 19, 2013. Mi-17V-5 is the latest modification of Mi-17 and can transport heavy cargo, troop, etc.

Nepalese Army Air Corp

Personnel: 500

Category	Name	In Service
Aircraft (Trainer/ Transport)	M28 Sky truck	2
	BN-2T islander	2
	HS-748	1
Helicopter	Mi-17	5
	HAL Lancer	2
	HAL Dhruv	4
	HAL Chetak	3
	HAL Cheetah	2
Helicopter (VIP Transport)	Super Puma (VIP)	1

New Procurements/ Upgrades

✦ India has handed over to Nepal the Advanced Light Helicopter (ALH) Mark III Dhruv which will be used for military and civilian operations. As India is imparting military training and providing physical equipment to the Nepal Army, the helicopter gifted will certainly strengthen the humanitarian assistance and natural disaster relief programme being carried out by Nepal Army.

✦ Nepal Army has received two Russian made choppers Mi-17 V-5. Russia's state-run arms exporter Rosoboronexport's Joint Stock Company delivered the two choppers to the Nepal Army.

The Nepalese Army has opened recruitment process for women since 1961. *Image Credit: Nepalese Army*

Defence Expenditure

Total defence spending: $207 million (est.)

Estimated defence spending in terms of GDP: 1.51% (2015)

Major foreign suppliers: India, China, United States, North Korea.

Defence Production and R&D

Nepal does not possess any defence R&D base. There are certain facility for repair and maintenance of the combat vehicles.

Defence Procurement

Nepal is planning the procurement of the T-72 tanks, APCs, artillery guns & sighting equipments from the India. Nepal is looking for the procurement of the about 100 main battle tanks in two phases. No firm plans for the procurement has been announced.

CONTACT DETAILS

Ministry of Defence

Singa Durbar

Kathmandu,Nepal

Phone Number: 4211289

Fax: 4211294

Website: www.mod.gov.np

NETHERLANDS

(Capital: Amsterdam)

INTRODUCTION

Area: 41,543 sq km

Population: 17,016,967 (July 2016 est.)

Coastline: 451 km

Maritime claims: Territorial sea: 12 nm

Contiguous zone: 24 nm

KEY POLITICAL PERSONS

HEAD OF STATE: King Willem-Alexander

PRIME MINISTER: Mark Rutte

KEY DEFENCE PERSONS

MINISTER FOR DEFENCE: Jeanine Hennis-Plasschaert

CHIEF OF DEFENCE STAFF: Gen. Tom Middendorp

COMMANDER OF ARMY: Lt. Gen. Leo Beulen

COMMANDER OF NAVY: Lt. Gen. Rob Verkerk

COMMANDER OF AIR FORCE: Lt. Gen. Dennis Luyt

ECONOMY

The Netherlands has the 6th largest economy in the European Union and 17th largest economy in the world, and ranks 10th in GDP (nominal) per capita. The Netherlands is the fifth-most competitive economy in the world, according to the World Economic Forum's Global Competitiveness Report. The country is a founding member of the European Union, the OECD and the World Trade Organisation. The Netherlands has a prosperous and open economy, which depends heavily on foreign trade. In 2008, the government nationalized two banks and injected billions of dollars of capital into other financial institutions, to prevent further deterioration of a crucial sector. A highly mechanised agricultural sector employs only 2% of the labour force but provides large surpluses for food-processing and underpins the country's status as the world's second largest agricultural exporter. The Netherlands is part of the Eurozone, and as such, its monetary policy is controlled by the European Central Bank. The Netherlands continues to be one of the leading European nations for attracting foreign direct investment and is one of the five largest investors in the United States.

GDP: $832.6 billion (2015 est.), $816.9 billion (2014 est.)

Real Growth Rate: 1.9% (2015 est.), 1% (2014 est.)

Industries: Agro-industries, metal and engineering products, electrical machinery and equipment, chemicals, petroleum, construction, microelectronics, fishing

Oil - proved reserves: 144.7 million bbl (1 January 2016 est.)

Natural gas - proved reserves: 760.9 billion cu m (1 January 2016 est.)

Total Exports: $476.5 billion (2015 est.), $574.7 billion (2014 est.)

Export Commodities: Machinery and equipment, chemicals, fuels; foodstuffs

Major Markets: Germany 24.5%, Belgium 11.1%, UK 9.3%, France 8.4%, Italy 4.2% (2015)

Total Imports: $389.6 billion (2015 est.), $469.4 billion (2014 est.)

Import Commodities: Machinery and transport equipment, chemicals, fuels, foodstuff, clothing

Major Suppliers: Germany 14.7%, China 14.5%, Belgium 8.2%, US 8.1%, UK 5.1% (2015)

INTERNATIONAL DISPUTES

Netherlands has no known international disputes.

DEFENCE

ARMY

Personnel: 21,500

Equipment

Category	Name	In Service
MBT	Leopard 2A7	18+
APC/MICV /IFV	CV9035 MkIII	149
	PzH 2000	50
	XA-188	70
	Luchtmobiel Speciaal Voertuig MWV	160
	Bushmaster	96
	Fennek	365
	Boxer AFV	200
Artillery	155 mm. SP PzH-2000	57
Mortar	81 mm.	40
	120 mm.	130
Anti-Tank GW	SPIKE launchers	300
ATK Weapon	SRAT	336
	VSRAT	834
	AT4	
SAM	FIM-92 Stinger	300+
	NASAMS II	3
	MIM-104 Patriot	3
UAV	ScanEagle, RQ-11 Raven	

New Procurements/ Upgrades

✦ A dozen Bushmasters are going into production in Bendigo for export to the Netherlands, after the country's Ministry of Defence ordered new vehicles from Thales in Australia. The vehicles will complement the 86 Bushmasters previously purchased by the Dutch customer between 2006 and 2009, and will be delivered by 2017.

✦ The Netherlands ordered around 1000 Boxer 8X8 AFV, to replace the M577 and the support variants of the YPR-765 in the Royal Netherlands Army, with the Netherlands having already confirmed the procurement of 200 vehicles in autumn of 2006. Deliveries to continue till 2016-17.

✦ The Netherlands Ministry of Defence procurement authority has placed orders with Rheinmetall for 10 Kodiak armoured engineer (breaching) vehicle.

✦ BAE Systems Hägglunds and the Netherlands MoD have signed a contract for a mid-life upgrade (MLU) of the Dutch fleet of 95 Bv206 vehicles. The upgrade includes replacement of the engine, transmission, tracks, electrical systems and the renovation of other sub-systems.

Leopard 2A6 Main battle tank. *Image Credit: Ministry of Defence (Netherlands)*

Dutch soldiers dismount a CV 90 combat vehicle to conduct an assault in an urban environment during Exercise Allied Spirit I at the US Army Joint Multinational Readiness Center's Hohenfels Training Area, Germany, Jan. 20, 2015. *Image Credit: US Army/ US DoD*

NAVY

Personnel: 10,500

Equipment

Category	Name	In Service
Submarine	Walrus Class	4
Frigate	De Zeven Provincien Class	4
	Karel Doorman Class	2
Patrol Vessel	Holland class	4
Mine Warfare Force	Alkmaar Class Minehunters	10
Amphibious Force	Rotterdam Class LPDs	2
	LCVP	12
	LCU	5
Support Vessel	Amsterdam Class	1
	Karel Doorman	1
	Zuiderkruis Class	2
Helicopter	NH-90	

New Procurements/ Upgrades

✦ The Holland Class offshore patrol vessels (OPVs) are being built by Damen Schelde Naval Shipbuilding for the Royal Netherlands Navy. Netherlands has received its third Holland-class offshore patrol vessel (OPV) from the Damen Schelde Naval Shipbuilding.

✦ Extensive upgrading of the 2 remaining F827 Karel Doorman class, new SEWACO systems and lay-out of the ships. After these upgrades the ships can last till 2020–2025.

✦ Purchase of new joint support ship Zuiderkruis class JSS. The Dutch Ministry of Defence announced that this vessel will be able to combine sea replenishment capabilities with the ability to act as a sea based platform in support of amphibious operations (incl. CH-47 & AH-64 Helicopters).

✦ Extensive upgrading for Walrus class Submarines including new Sonar, weapons upgrades and probably AIP system for near shore operations.

✦ Upgrading of the of Zeven Provincien class LCF frigates with Theatre Ballistic Missile Defence.

✦ The Damen Shipyards Group has signed a contract with the Dutch and Swedish Ministries of Defence for a seven-year, in-service maintenance contract covering a total of five vessels. All of these are tugs purchased from Damen.

✦ Raytheon Missile Systems, Tucson, Arizona, is being awarded a $13,495,233 modification to previously awarded contract for procurement of 19 SM-2 Block IIIA Zumwalt flight-test-rounds for DDG 1000 class Ships (Zumwalt configuration). This contract modification also involves a purchase for the Netherlands as a result of an international agreement between the governments of the United States and the Netherlands.

Navy ships Zr. Ms. Mercuur, Zr. Ms. Walrus and Zr. Ms. Schiedam. *Image Credit: Royal Netherlands Navy*

Royal Netherlands Navy vessels HNLMS Holland and HNLMS Pelikaan for the Haitian coast. I*mage Credit: Ministry of Defence (Netherlands)*

AIR FORCE

Personnel: 9,200

Equipment

Category	Name	In Service
Fighters/ attack/ Reconn-aissance	F-35A Lightning II	2 (35 on order)
	F-16AM/BM Fighting Falcon	60+
Transport/ Tanker	KDC-10	2
	A330 MRTT	
	Gulfstream IV	1
	C-130H-30 Hercules	4
	C-17 Globemaster III	3 (shared with NATO)
Trainer	PC-7 Turbo Trainer	13
	F-16 MLU	14
Helicopter	CH-47D, CH-47F Chinook	11 & 14
	AS 352U2 Cougar	17
	SA316B Alouette III	4
	AH-64D Apache	29
	Agusta-Bell 412	3
	Westland Lynx	21
	NH-90	20

Air Force continued...

Category	Name	In Service
UAV	MQ-9 Reaper	4
	ScanEagle	12
	RQ-11 Raven	70+

New Procurements/ Upgrades

✦ Lockheed Martin completed rapid delivery of 14 Sniper Advanced Targeting Pods (ATP) to the Royal Netherlands Air Force (RNLAF), outfitting a squadron in support of urgent mission requirements. Five of the 14 Sniper ATPs were delivered within 180 days of contract for fielding on the RNLAF's F-16s. Initial training, spares and support equipment were also provided.

✦ Belgium and the Netherlands have signed a Letter of Intent for the multi-national Smart Defence programme for providing air-to-ground precision-guided munitions (PGMs).

✦ The Defence Security Cooperation Agency notified Congress October 2012 of a possible Foreign Military Sale to the Government of the Netherlands for 28 AIM-9X-2 SIDEWINDER Block II All-Up-Round Missiles and associated equipment, parts, training and logistical support for an estimated cost of $60 million.

✦ The Netherlands Air Force wants to replace its F-16 fleet in the next decade. Candidates for the replacement were the Dassault Rafale, Lockheed Martin's F-16 Block 52/60, the Eurofighter Typhoon, the Saab Gripen and Lockheed Martin's F-35. In 2002 The Netherlands signed a MOU (Memorandum Of Understanding) to co-develop the F-35 as a 'Tier 2' Partner. That deal is worth around 800 million US dollars, advanced by the Dutch government on behalf of Dutch industries.

✦ The Royal Netherlands Navy has ordered a total of 20 NH90 NFH aircraft from AgustaWestland, a Finmeccanica company.

✦ The Royal Netherlands Air Force (RNLAF) has taken delivery of its first two CH-47F Chinook heavy-lift helicopters under contract from the country's Defence Materiel Organization (DMO) at Gilze-Rijen air base in Noord-Brabant, Netherlands. The delivery follows a deal signed with Boeing in February 2007 for the supply of six helicopters along with support services to help expand the air force's operational Chinook fleet to 17.

✦ Netherlands is a participant in the NATO (Strategic Airlift Capability Program) with three C-17 Globemaster III transport aircraft.

✦ The Netherlands has confirmed its decision to purchase eight F-35 Lightning II Joint Strike Fighters that will arrive in the country in 2019. In 2013 the Netherlands decided to replace the F-16's with F-35 Lightning II aircraft.

✦ The United States has approved a possible Foreign Military Sale to the Netherlands for MQ-9 Reapers and associated equipment, parts and logistical support for an estimated cost of USD 339 million.

✦ The Netherlands is ready to purchase 2 Airbus tanker/transport aircraft with Luxembourg. The aircraft will be NATO property and will be stationed at Eindhoven Air Base for pooling and sharing. Belgium, Germany, Norway and Poland intend to join the agreement at a later date. The Netherlands is the lead nation in this multinational collaboration project. The expected life span of the fleet is 30 years and the investment budget is between €250 million and €1 billion.

✦ The Netherlands Ministry of Defence awarded AeroVironment contract for Small Unmanned Aircraft Systems and Upgrades Valued at USD 10.3 Million

The F-35 F-001 of RNLAF receives the traditional welcome after landing on Airbase Leeuwarden (May 2016). *Image Credit: Ministry of Defence (Netherlands)*

Defence Expenditure

Total defence spending: $9,839 million (est.)

Estimated defence spending in terms of GDP: 1.18% (2015)

Next year, the Dutch government will make an additional € 200 million available to the Ministry of Defence, i.e.; in 2017, the armed forces will have almost € 8.7 billion available. The armed forces received another EUR 220 million in 2016, increasing to EUR 345 million in 2020. This is the fourth time in a row that more money has been allocated. The increase is in keeping with the cabinet's multi-year vision, prompted by incessant conflicts, as well as the structurally more stringent demands being placed on the armed forces in this respect by organisations such as NATO. Furthermore, the 2016 budget contains greater provisions for funding the deployment of the armed forces in international missions, with a structural increase of EUR 60 million for this purpose, according to Defence Ministry. Supplies will be increased with more spare parts and munitions. Materiel can also be repaired more quickly with extra maintenance capacity. Germany and the Netherlands planning to integrate 43 Mechanised Brigade in Havelte into a German armoured division. Germany is also integrating a tank battalion into the Dutch brigade and the Netherlands is supplying personnel for a company in this German tank battalion. As part of this cooperation, the Netherlands will supply the last 16 Leopard 2A6 tanks. Cooperation with other countries will also be intensified in numerous areas, as per the policy agenda. The Ministry of Defence is moving ahead with plans to make an initial down payment in 2016 for the eight F-35 aircraft that will be delivered in 2019. Defence is also investing in the modernisation of the SMART-L radars on board the air defence and command (LC) frigates. In addition, Defence is also increasing the military's striking power by investing in various capital ammunition types, including precision-guided munitions for the armoured howitzer, MK-48 torpedoes for the submarines and SM-2 guided missiles for the LC frigates. Various projects have been rearranged. The Chinook Replacement and Modernisation and the Defence-wide Replacement of Operational Wheeled Vehicles projects are more in line with the expected or desired realisation. Projects including medium-altitude long-endurance unmanned aerial vehicles (MALE UAV), the light indirect fire weapon system, the infantry fighting vehicle protection packs and the Electronic Warfare Capability Upgrade will be delayed, as reported by the Ministry of Defence.

Defence Production and R&D

Despite its relatively small size, the Dutch defence industry supplied the majority of defence items to the Netherlands. The Netherlands' defence industrial base is best described as one that supports major projects. The Industrial Marketing Association of The Netherlands (NIID) is an independent group financed by Dutch industry and aims at promoting optimal participation of Dutch industry in the field of defence production for both national and foreign forces. NIID operates in close liaison with the Ministeries van Defensie en Economische Zaken (Ministries of Defence-MOD, and Economic Affairs-MEA). They also assist foreign companies with offset obligations in finding potential partners in The Netherlands. NIID also assists small and medium sized Dutch firms in obtaining defence orders, especially as subcontractors.

Defence Procurement

The Defence Materiel Organisation (DMO) is responsible for all materiel within the Defence organisation: from procurement and major maintenance to disposal. The DMO also establishes internal materiel policy. The DMO works closely with the Netherlands Organisation for Applied Scientific Research (TNO), educational institutions and national and international industry on new technological developments. In addition, the DMO is in contact with similar materiel organisations of foreign ministries of defence, in order to enable international coordination. The Directorate of Materiel Policy is the only DMO unit that is part of the Central Staff of the Ministry of Defence. Its tasks include developing materiel logistics policy and policy in respect of scientific research and development (R&D) and supporting the National Armaments Director (NAD). Contributing to the administrative and political decision-making process concerning major materiel projects, managing the materiel logistics process and R&D and supporting the administrative and political leadership in those areas are also among the directorate's tasks. The Directorate of Projects & Procurement (DP&V) is responsible for the process of providing materiel, which consists of procurement and project management. The Directorate manages all procurement activities for the armed forces at a functional level and issues strict guidelines to all procurers. It obtains new materiel and/or services on a project basis and ensures that materiel is maintained by carrying out large-scale and/or complex modification programmes, also on a project basis. The directorate also negotiates contracts for the disposal of surplus Defence goods. The Directorate of Weapon Systems is responsible for ensuring high-quality materiel for the entire armed forces during all stages of its life cycle in a cost-effective way. This is achieved by setting standards in weapon systems management and by working on a project basis. The services of the required in-house, high-quality expertise base are also rendered to the Directorate of Projects & Procurement, the Directorate of Logistic Agencies and the external clients of the Directorate of Weapon Systems.

CONTACT DETAILS

Ministry of Defence
PO Box 20701
NL-2500 ES's-Gravenhage
Netherlands

Ministry of Defence, Press and Information Service
Kalvermarkt 38
PO Box 20701
NL-2500 ES's-Gravenhage
Tel: (+31 70) 318 88 88
Fax: (+31 70) 318 78 88
E-mail: persvoorlichting@mindef.nl
www.mindef.nl

National Agency for Aerospace Programmes - NIVR
PO Box 35
NL-2600 AA Delft
Netherlands

Defence Materiel Organisation
Van der Burchlaan 31,
Frederik Barracks, building 35,
A-L towers
P.O. Box 90822
2509 LV The Hague
Netherlands

Directorate of Materiel Policy
Tel: +31 (0) 70-3161551
Fax: +31 (0)70-3161553

Directorate of Projects & Procurement
Tel: +31 (0)70-3168079
Fax: +31 (0)70-3162577

Directorate of Weapon Systems
Tel: +31 (0)70-3162610
Fax: +31 (0)70-3166810

Directorate of Logistic Agencies
Tel: +31 (0) 70-3164474
Fax: +31 (0) 70-3164275

NEW ZEALAND
(Capital: Wellington)

INTRODUCTION

Area: 267,710 sq km
Population: 4,474,549 (July 2016 est.)
Coastline: 15,134 km
Maritime claims: Territorial sea: 12 nm
Contiguous zone: 24 nm
Exclusive Economic Zone: 200 nm
Continental shelf: 200 nm or to the edge of the continental margin

KEY POLITICAL PERSONS

HEAD OF STATE: Queen Elizabeth II

GOVERNOR GENERAL & COMMANDER-IN-CHIEF: Jerry Mateparae

PRIME MINISTER: Bill English

KEY DEFENCE PERSONS

DEFENCE MINISTER: Gerry Brownlee

CHIEF OF THE DEFENCE FORCE: Lt. Gen. Timothy (Tim) Keating

CHIEF OF ARMY: Maj. Gen. P.T.A.E Kelly

CHIEF OF NAVY: Rear Admiral John Martin

CHIEF OF AIR FORCE: Air Vice-Marshal Mike Yardley

COMMANDER JOINT FORCES NZ: Maj. Gen. Tim Gall

ECONOMY

New Zealand has a modern, prosperous and developed market economy. The economy is greatly dependent on international trade, mainly with Australia, the European Union, the United States, China and Japan. It has only small manufacturing and high-tech sectors, being strongly focused on tourism and primary industries like agriculture (though both sectors are highly profitable). Economic free-market reforms of the last decades have removed many barriers to foreign investment, and the World Bank in 2005 praised New Zealand as being the most business-friendly country in the world, before Singapore. The economy fell into recession before the start of the global financial crisis and contracted for five consecutive quarters in 2008-09. The economy posted a 2% decline in 2009, but pulled out of recession late in the year, and achieved 1.7% growth in 2010 and 2% in 2011. The government plans to raise productivity growth and develop infrastructure, while reining in government spending. New Zealand was ranked sixth in the 2013 Human Development Index and 13th in INSEAD's 2012 Global Innovation Index. New Zealand is heavily dependent on international trade.

GDP: $168.2 billion (2015 est.), $162.7 billion (2014 est.)

Real Growth Rate: 3.4% (2015 est.), 3% (2014 est.)

Industries: Food processing, wood and paper products, textiles, machinery, transportation equipment, banking and insurance, tourism, mining

Oil - proved reserves: 67.2 million bbl (1 January 2016 est.)

Natural gas - proved reserves: 36.9 billion cu m (1 January 2016 est.)

Total Exports: $34.41 billion (2015 est.), $41.96 billion (2014 est.)

Export Commodities: Dairy products, meat, wood and wood products, fish, machinery

Major Markets: China 17.5%, Australia 16.9%, US 11.8%, Japan 6% (2015)

Total Imports: $35.8 billion (2015 est.), $41 billion (2014 est.)

Import Commodities: Machinery and equipment, vehicles and aircraft, petroleum, electronics, textiles, plastics

Major Suppliers: China 19.4%, Australia 11.8%, US 11.7%, Japan 6.6%, Germany 4.7%, Thailand 4.2% (2015)

INTERNATIONAL DISPUTES

✦ Asserts a territorial claim in Antarctica (Ross Dependency).

DEFENCE

ARMY

Personnel: 6,003

Equipment

Category	Name	In Service
APC/MICV/IFV	NZ Light Armoured Vehicle (NZLAV)	105
	Infantry Mobility Vehicle (IMV)	95
	Light Obstacle Blade Vehicle (LOB)	7
	Pinzgauer High Mobility ATV	352
	Recovery Vehicle (LAV-R)	3
Artillery	105 mm. L-118	24
Mortar	81 mm.	50
Anti-Tank GW	Javelin	24
Recoilless Rifle	84 mm. Carl Gustav	68
SAM	Mistral	12

New Procurements/ Upgrades

✦ New Zealand's Defence Ministry announced that Rheinmetall MAN Military Vehicles Australia (RMMVA) has been awarded a contract to supply the country's armed forces with up to 200 vehicles.

✦ The New Zealand Defence Force (NZDF) has awarded Lockheed Martin an initial 10-year contract valued at more than $100 million to provide its Army with warehousing and repair services. The agreement includes the maintenance, repair and overhaul of equipment and supply chain services ranging from soldier store acquisition, such as tents and weapons, to warehouse sustainment.

A NZLAV during a static live firing shoot. *Image Credit: New Zealand Defence Force, Wellington, New Zealand*

NAVY

Personnel: 2,104

Equipment

Category	Name	In Service
Frigate	Anzac class	2
Patrol Vessel	Protector class OPV	2
	Protector class Inshore	4
Logistics and Support	HMNZS Canterbury	1
	HMNZS Endeavour	1
Survey Vessel	HMNZS Resolution	1

Naval Aviation

Category	Name	In Service
Helicopter	SH-2G Super Seasprite	5

The HMNZS CANTERBURY off Samoa coastline. *Image Credit: New Zealand* **Defence Force, Wellington, New Zealand**

New Procurements/ Upgrades

✦ The ANZAC FSU project will upgrade the surveillance, combat and self-defence capabilities of the ANZAC frigates. New Zealand Government provided approval for the Project to commit to contract at a total project cost of $446 million including project management costs, contingency, and introduction into service and capitalisation costs. The project is expected to be completed in 2018.

✦ Upgradation of the Phalanx Close-In Weapon System (CIWS) by Raytheon. The system provides a last line of defence and protection from Anti-Ship Missiles (ASM) and strike aircraft. The first system has been installed on HMNZS Te Mana, which has completed its Harbour Acceptance Testing and will undergo Sea Acceptance Testing in early December.

✦ MBDA's Sea Ceptor has been selected by the New Zealand Ministry of Defence (MoD) as the Local Area Air Defence (LAAD) system for the Royal New Zealand Navy's (RNZN) ANZAC Frigate Systems Upgrade (FSU) project.

✦ Babcock International Group has secured a $NZ300 million partnering agreement with the Government of New Zealand which will see the company deliver through life support services across the New Zealand naval fleet for the next seven years. Under the new contract, Babcock New Zealand will assume the role of Prime Systems Integrator (PSI) for the Royal New Zealand Navy and New Zealand Ministry of Defence (NZMoD) under a five year + two agreement that creates a benchmark for naval support in New Zealand.

AIR FORCE

Personnel: 2,595

Equipment

Category	ame	In Service
Patrol aircraft	Lockheed P-3K Orion	6
Transport/ Tanker	Boeing 757-200	2
	Lockheed C-130H Hercules	5
Helicopter	Kaman SH-2G	8
	AW109	5
Trainer	Super King Air	4
	T-6 Texan II	7

New Procurements/ Upgrades

✦ Eight NHI NH90 helicopters have been purchased by the Ministry of Defence for $771 million, including costs for parts and services.

✦ A contract was signed with Agusta Westland to acquire five A109 LUH (NZ) Training/Light Utility Helicopters, which will replace the Sioux helicopter.

Lockheed P-3K Orion. *Image Credit: New Zealand Defence Force, Wellington, New Zealand*

Defence Expenditure

Total defence spending: $2.87 billion (est.)

Estimated defence spending in terms of GDP: 1.19% (2015)

Foreign suppliers: United States and European Union.

According to the "NZDF Vote Defence Force 2016-17 Official Estimates", the Minister of Defence is responsible for Defence Force appropriations in Vote Defence Force, totalling just over $3,280 million for the

2016/17 financial year.

This includes just over $2,533 million for departmental output expenses covering the following:

✦ a total of just under $2,038 million on Air ($770 million), Army ($809 million) and Navy ($459 million) capabilities prepared for joint operations to provide the Government with a range of military forces to protect and advance the security and interests of New Zealand. These are held at appropriate levels of capability and preparedness to protect New Zealand's territorial sovereignty and to contribute to regional and global security efforts. Most of these forces will also contribute to a range of services to other government departments and the New Zealand community when not committed to operations overseas, and

✦ a total of just over $495 million on Multi-Category Appropriations (MCA) for Advice to the Government, Operations Contributing to New Zealand's Security, Stability and Interests, and the Protection of New Zealand and New Zealanders.

✦ Of the total output expenses, about $14 million is expected to be funded from third-party revenue (Revenue Department and Revenue Other) in 2016/17. The remainder is funded with Revenue Crown via Treasury.

✦ There is also a capital expenditure appropriation of up to $747 million for the purchase of assets.

✦ The Minister of Veterans' Affairs is responsible for Veterans' Affairs appropriations in Vote Defence Force totalling just under $137 million for the 2016/17 financial year.

	2015/16		2016/17
NZDF - Capital Injection (M22)	Final Budgeted $000	Estimated Actual $000	Budget $000
	1,000	1,000	6,000

Source: Minister(s) Responsible for Appropriations at 26 May 2016. Vote Defence Force, New Zealand Defence Force.

Defence Production and R&D

The New Zealand defence and homeland security industry comprises over 250 companies with the capability and capacity to participate in international defence and security markets. New Zealand companies have major defence acquisition and maintenance contracts with Australia, Singapore, the United Kingdom and the United States, and also export to China, Southeast Asia, South America and Europe. Australia is the most significant market destination, accounting for almost half of all exports. The ANZAC frigates project – a joint venture between the governments of New Zealand and Australia signed in 1989 – saw New Zealand companies provide more than $800 million of work to construct 10 frigates. New Zealand Trade and Enterprise estimates that the value of New Zealand exports of defence and security and services is $200 to $250 million per annum. Defence products and services range from VT-Fitzroy servicing of the French Navy's Light Transport Ship Jacques Cartier, to Rakon's quart crystal components. New Zealand's defence and security industry was confirmed by the United States Defence Director Ted Glum in 2009 to be a secure supply chain partner.

Industry contacts

The New Zealand Defence Industry Association has a broad mandate to support New Zealand companies operating in the defence and homeland security sector (www.nzdia.co.nz).

Defence Procurement

✦ **NZ DEFENCE Procurement has three aspects:**

✦ *Capital Equipment:* This is procured by the Ministry of Defence.

✦ *Minor Capital Equipment:* Capital equipment procured by the NZDF. It is generally items or projects below $NZ7 million.

✦ *Operating Requirements:* These are day to day needs to keep the NZDF, its equipment and its people functioning. It includes basic necessities such as fuel, food, clothing etc, and a wide range of other specialist items to meet particular needs.

CONTACT DETAILS

Ministry of Defence
Level 4, Freyberg House
2-12 Aitken Street, Wellington,
New Zealand
Ph: +64 4 496 0999
Fax: +64 4 496 0859
Ministry of Defence
PO Box 12703
Molesworth Street,
Wellington 6144, New Zealand
www.defence.govt.nz

New Zealand Defence Force
Defence House
2-12 Aitken St
Wellington
+64 4 496 0999
www.nzdf.mil.nz

NIGER
(Capital: Niamey)

INTRODUCTION

Area: 1.267 million sq km

Population: 18,638,600 (July 2016 est.)

Coastline: 0 km (Landlocked)

Maritime claims: None

KEY POLITICAL PERSONS

PRESIDENT & COMMANDER IN CHIEF: Mahamadou Issoufou

PRIME MINISTER: Brigi Rafini

KEY DEFENCE PERSONS

DEFENCE MINISTER: Karidjo Mahamadou

JOINT CHIEF OF STAFF: Gen. Seyni Garba

ECONOMY

Niger is a landlocked, sub-Saharan nation, whose economy centres on subsistence crops, livestock, and some of the world's largest uranium deposits. Agriculture contributes nearly 40% of GDP and provides livelihood for most of the population. The UN ranked Niger as the least developed country in the world in 2014 due to multiple factors such as food insecurity, lack of industry, high population growth, a weak educational sector, and few prospects for work outside of subsistence farming and herding. Since 2011 public debt has increased in part from a large loan financing a new uranium mine. The government relies on foreign donor resources for a large portion of its fiscal budget. The economy in recent years has been hurt by terrorist activity and kidnappings near its uranium mines and instability in Mali, and concerns about security have boosted fiscal spending on defence. Future growth may be sustained by exploitation of oil, gold, coal, and other mineral resources. Niger has sizable reserves of oil and oil production. Food insecurity and drought remain perennial problems for Niger, and the government plans to invest more in the agriculture sector, most notably irrigation. The mining sector may be affected by the government's attempt to renegotiate extraction rights contracts to increase royalty rates and reduce tax exemptions. Niger's three-year $131 million IMF Extended Credit Facility agreement for years 2012-15 was extended until the end of 2016, although formal private sector investment needed for economic diversification and growth remains a challenge, given the country's limited domestic markets, access to credit, and competitiveness.

GDP: $19.05 billion (2015 est.); $18.32 billion (2014 est.)

Real Growth Rate: 4% (2015 est.); 7.1% (2014 est.)

Industries: Uranium mining, cement, brick, soap, textiles, food processing, chemicals, slaughterhouses

Total Exports: $1.099 billion (2015 est.); $1.448 billion (2014 est.)

Export Commodities: Uranium ore, livestock, cowpeas, onions

Major Markets: France 53.1%, Nigeria 20.3%, China 13.8% (2015)

Total Imports: $1.888 billion (2015 est.); $2.157 billion (2014 est.)

Import Commodities: Foodstuff, machinery, vehicles and parts, petroleum, cereals

Major Suppliers: France 12%, China 10.5%, Nigeria 9.5%, French Polynesia 9%, Togo 6.1%, Belgium 5.3%, Cote dIvoire 5.3%, US 4.3% (2015)

INTERNATIONAL DISPUTES

- ✦ Libya claims about 25,000 sq km in a currently dormant dispute in the Tommo region.
- ✦ The location of the Benin-Niger-Nigeria tri-point is unresolved.
- ✦ Only Nigeria and Cameroon have heeded the Lake Chad Commission's admonition to ratify the delimitation treaty that also includes the Chad-Niger and Niger-Nigeria boundaries.
- ✦ The border dispute with Burkina Faso was referred to the ICJ in 2010.

DEFENCE

ARMY

Personnel: 5,000+

Equipment

Category	Name	In Service
Armoured Car/ APC	AML-60	12
	AML-90	88
	M-3	14
Mortar	60mm	
	81mm	
	120mm	
RCL	M20 75mm	6
	106mm	6
AA gun	20mm	30+

New Procurements/ Upgrades

No major procurement plans are under consideration at present.

AIR FORCE

Personnel: 300

Equipment

Category	Name	In Service
Fighter	Su-25	2
Transport/ Liaison/ Utility	C130H	1
	Do-28D	1
	Do-228	1
	Boeing 737-200 (VIP)	1
	ULM TETRA	3
	Cessna	2
Helicopter	SA341F Gazelle (armed)	3

Gazelle helicopter

New Procurements/ Upgrades

No major procurement plans are under consideration at present.

Defence Expenditure

Total defence spending: 33 billion CFA francs ($54.7 million - 2015)

Estimated defence spending in terms of GDP: 1%

Defence Production and R&D

Niger does not undertake any defence production activities.

Defence Procurement

No major procurement plans are under consideration at present by the Army, Navy and Air Force.

CONTACT DETAILS

Ministry of Defence

P.O. Box 626

Niamey, Niger

NIGERIA

(Capital: Abuja)

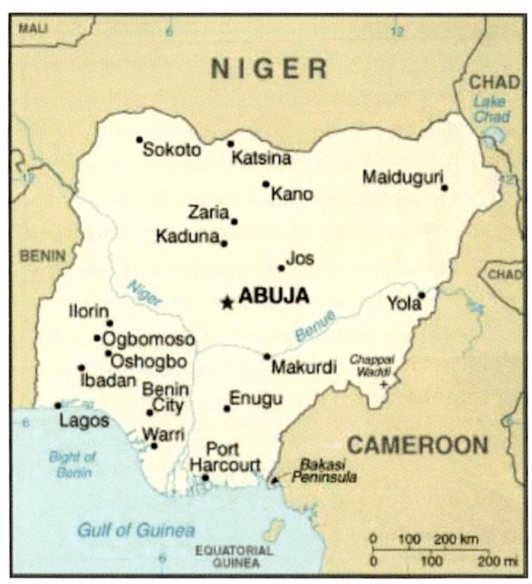

INTRODUCTION

Area: 923,768 sq km

Population: 186,053,386 (July 2016 est.)

Coastline: 853 km

Maritime claims: Territorial sea: 12 nm, Exclusive Economic Zone: 200 nm Continental shelf: 200 m depth or to the depth of exploitation

KEY POLITICAL PERSONS

PRESIDENT & COMMANDER-IN-CHIEF: Muhammadu Buhari

VICE PRESIDENT: Yemi Osinbajo

KEY DEFENCE PERSONS

DEFENCE MINISTER: Mansur Muhammad Dan-Ali

CHIEF OF DEFENCE STAFF: Gen. Gabriel Abayomi Olonisakin

CHIEF OF ARMY STAFF: Maj. Gen. Tukur Buratai

CHIEF OF NAVAL STAFF: Vice Adm. Ibok-Ete Ekwe

CHIEF OF AIR STAFF: Air Marshall S B Abubakar

ECONOMY

Oil-rich Nigeria, long hobbled by political instability, corruption, inadequate infrastructure, and poor macroeconomic management, has undertaken several reform measures over the past decade. As a result, the country has witnessed strong economic growth. While it still remains overtly dependent on the capital-intensive oil sector, which provides 95% of foreign exchange earnings and about 80% of budgetary revenues, the present Government has taken initiatives to diversify the economy. It is taking steps towards developing stronger public-private partnerships for building up national infrastructure. Following an April 2014 statistical "rebasing" exercise, Nigeria has emerged as Africa's largest economy, with 2015 GDP estimated at $1.1 trillion.

GDP: $1.092 trillion (2015 est.); $1.063 trillion (2014 est.)

Real Growth Rate: 2.7% (2015 est.); 6.3% (2014 est.)

Industries: Crude oil, coal, tin, columbite; rubber products, wood; hides and skins, textiles, cement and other construction materials, food products, footwear, chemicals, fertilizer, printing, ceramics, steel

Crude Oil - proved reserves: 37 billion bbl (1 January 2016 es)

Natural gas - proved reserves: 5.111 trillion cu m (1 January 2016 es)

Total Exports: $45.89 billion (2015 est.); $82.59 billion (2014 est.)

Export Commodities: Petroleum and petroleum products 95%, cocoa, rubber

Major Markets: India 18.2%, Netherlands 8.5%, Spain 8.2%, Brazil 8.2%, South Africa 7.8%, France 5.2%, Japan 4.5%, Cote dIvoire 4.2%, Ghana 4% (2015)

Total Imports: $52.33 billion (2015 est.); $61.59 billion (2014 est.)

Import Commodities: Machinery, chemicals, transport equipment, manufactured goods, food and live animals

Major Suppliers: China 25.7%, US 6.4%, Netherlands 6.1%, India 4.3% (2015)

INTERNATIONAL DISPUTES

✦ Land and maritime boundary dispute with Cameroon continues over the potentially oil rich Bakassi peninsula and around Lake Chad.

✦ Location of Benin-Niger-Nigeria tripoint remains unresolved.

DEFENCE

ARMY

Personnel: 100,000+

Equipment

Category	Name	In Service
MBT	Vickers Mks3	100+
	T-55	20+
APC	AML-60/90	100+
	SALADIN	15
	FOX	50
	EE-9 Cascavel	50+
	BTR-3U	42
	Piranha	70
	K7AFA	250+
	MT-LB	60
	Otokar Cobra	100+
Artillery	M-56 105mm	200
	D30/D74 122mm	200
	FH77 155mm	25
	PALMARIA 155mm	25
Mortar	81mm towed	200+
MRLS	BM21 122mm	
RCL	Carl Gustav 84mm	
	M40A1 106mm	
AA Gun	20mm	
	40mm	
	ZSU-23-4 SP	
SAM	Roland	
	Blowpipe	40

Army continued...

New Procurements / Upgrades

✦ A new armoured vehicle is being built indigenously.

✦ 30 mine-resistant ambush-protected vehicles (MRAPs) have been received from China. Nigeria has reportedly ordered 120 MRAPs.

NAVY

Personnel: 5,000

Equipment

Category	Name	In Service
Frigate	Aradu class	1
Corvette	Erin'mi class	2
	Thunder class (cutter) ex-US	2
Light Forces	NNS Andoni FAC	1
	Siri class missile FAC	2
	Ekpe class missile FAC	1
	Manta class 17mm inshore patrol boats	6
	Shaldag type patrol craft	6
	Swiftship type patrol craft	4
	Kyanwa class patrol craft	4
Mine sweeper	Ohue class mine hunter	2
Auxiliary ship	Ambe class LST	2
	Oton class LCU	2

Naval Aviation

Category	Name	In Service
Fighter/ Surveillance /AWS role	AW139	1
	Lynx Mk89	2
	A-109	3

New Procurements/ Upgrades

✦ Two new offshore patrol vessels built by China under the Nigerian navy's P18N project have been delivered in 2015 and 2016.

The new OPV of Nigerian Navy *An internet photo*

✦ China is building a stealth frigate for the Nigerian Navy.

✦ The Navy commissioned the first locally-built warship, NNS Andoni (P100), in 2012. The 100ft long warship, designed and built by Nigerian engineers, is used for anti-piracy missions.

✦ Six Shaldag type patrol craft on order since 2009 are being procured from the Israeli Shipyards.

AIR FORCE

Personnel: 9,000

Equipment

Category	Name	In Service
Fighter	F-7NI	12
	Dornier Alpha	10+
	L39 Albatross	21
Transport/ Surveillance	C130H/H-30	8
	Dornier Sky-Servant	16
	Gulfstream	1
	Do-228	5
	ATR-42 (maritime patrol)	2
	Falcon 90	1
	BBJ 737	1
Helicopter	SA 330 Puma	2
	AS 332 Super Puma	6
	Mi-171	4
	Mi-24	2
	Mi-35	3
	AW101	2
	AW109	
	AW139	2
Trainer	F-7NI	3
	Air Beetle	20

New Procurements/ Upgrades

✦ Plans are on to acquire at least three Pakistan-China-made JF-17 fighter jets for which budgetary allocation has been made in 2016.

✦ The government, in 2014, approved the acquisition of new fighter aircraft and helicopters as well as upgrades for some existing aircraft of the Air Force. The type of aircraft has not been disclosed.

✦ A contract has been signed with Russia's Rosoboronexport for the acquisition of 12 Mi-171Sh transport helicopters and six Mi-35 combat helicopters.

✦ Five C-130 transport aircraft and Do-228 aircraft are being upgraded. Two upgraded Mi-35 helicopters have been delivered.

✦ 17 new F-7NI fighters and 3 FT-7NI trainers have been ordered from China. 12 Chengdu F-7NI fighters and 3 FT-7NI trainers were acquired in 2010.

✦ A contract has been signed with Pakistan in October 2016 to buy 10 Super Mushshak trainer aircraft.

✦ Delivery of A109 (LUH) helicopters ordered from AgustaWestland is going on.

✦ Undisclosed number of Eurocopter EC135 (now Airbus Helicopters H135) and AS365 Dauphin helicopters have been transferred to the Air Force from the Nigerian National Petroleum Corporation (NNPC).

Defence Expenditure

Proposed defence budget for 2016: NGN 294.5 billion; NGN 338.8 billion (2015)

Estimated defence spending in terms of GDP: 1%

Nigeria has imported arms and equipment from several European countries including Germany, Italy and Bulgaria in the past. Present military suppliers include China and Israel.

Defence Production and R&D

The country is primarily producing small arms like rifles and pistols which are used by the Nigerian Police. The state-owned Defence Industry Corporation (DIC), which was set up in 1964 with the objective to make the country self-reliant in defence production and provide all such equipment to the Nigerian military, is overlooking the light weapons production today. Nigeria has explored cooperation with a number of countries in military hardware manufacturing. The country has also entered the naval ship building domain, inducting an indigenously-built warship in 2012 and building few more platforms. In 2013, the country's air force also commissioned into service an indigenous unmanned aircraft built in cooperation with a British institute.

Defence Procurement

Nigeria has mostly relied on foreign countries in acquiring weapons and equipment for its defence forces. Most of its weapons were bought from European countries in the 1980s. In the 1990s, the country, in an effort to give boost to its domestic defence industry, sought military partnership with a number of countries including Romania, Yugoslavia, France and China. Nigeria has also discussed naval cooperation with India in the past. It has announced plans to acquire new weapons and equipment for its military and train them for internal stability while also strengthening their role in international peace efforts.

CONTACT DETAILS

Ministry of Defence

The Defence Headquarters
Area 10, Garki, Abuja, Nigeria
Website: http://www.defence.gov.ng/

STRATEGIC INFORMATION

NORTH KOREA
(Capital: Pyongyang)

INTRODUCTION

Area: 120,538 sq km

Population: 25,115,311 (July 2016 est.)

Coastline: 2,495 km

Maritime Claims: Territorial sea: 12 nm
Exclusive economic zone: 200 nm

(Note: Military boundary line 50 nm in the Sea of Japan and the exclusive economic zone limit in the Yellow Sea where all foreign vessels and aircraft without permission are banned.)

GEOPOLITICAL IMPORTANCE

North Korea or the Democratic People's Republic of Korea (DPRK) is an East Asian country and occupies the northern half of Korean Peninsula. The country is bordered by China and Russia in the north and South Korea in the south. It is surrounded by the Sea of Japan in the East and Yellow Sea and Korean Bay in the West. The Korean peninsula, constituting North Korea and South Korea or Republic of Korea (ROK), has strategically remained an important region for four major global powers – China, Japan, Russia and the US. The peninsula is part of the Northeast Asian region which is gradually gaining importance among the leading continental and sea powers intending to wield greater economic and military clout in the Asia-Pacific. Hence, a unified Korea is being viewed as a strategically important region as it could emerge as a new maritime zone in East Asia. Uniting the two Koreas is regarded to be the key for establishing regional security and stability alongside economic prosperity in the region. The two Koreas, which were undivided until 1948, however, are engaged in unabated military strife to claim sovereignty over the entire peninsular region.

POLITICAL OVERVIEW

North Korea functions as a single-party republic country. It is run by the Workers' Party of Korea (WPK) which has been in power since the establishment of DPRK in 1948. The political system of North Korea is widely viewed as totalitarian dictatorship having Communist credentials. The head of the state is also the Chairman of the National Defence Commission which is the highest military leadership body in the country.

KEY POLITICAL PERSONS

HEAD OF STATE, CHAIRMAN OF NATIONAL DEFENCE COMMISSION & SUPREME COMMANDER OF KOREA PEOPLES REPUBLIC ARMY: Kim Jong-un

ECONOMY

The state-controlled economy of North Korea has witnessed a sharp decline following the collapse of Soviet Union in the 1990s. The country, which once boasted of a modern, industrialised and productive economy, is today marred by chronic financial problems and struggling to feed its people. It relies on international aid to feed the population. The primary reasons for the economic dishevel is DPRK's isolation from international trade and its large-scale military spending by diverting funds needed for investment and production. The country's economic woes have been further compounded by a plethora of sanctions imposed by the US, South Korea and Japan. The Government often highlights its goal of becoming a "strong and prosperous" nation, attracting foreign investment to improve the overall standard of living. In 2013-14, the regime rolled out 20 new economic development zones - now totaling 25 - set up for foreign investors, although the initiative remains in its infancy. Firm political control remains the government's overriding concern, which likely will inhibit changes to North Korea's current economic system.

GDP: $40 billion (2014 est.); $40 billion (2013 est.)

Real Growth Rate: 1% (2014 est.)

Industries: Military products; machine building, electric power, chemicals; mining (coal, iron ore, limestone, magnesite, graphite, copper, zinc, lead, and precious metals), metallurgy; textiles, food processing; tourism

Total Exports: $4.152 billion (2015 est.); $4.366 billion (2014 est.)

Export Commodities: Minerals, metallurgical products, manufactures (including armaments), textiles, agricultural and fishery products

Major Markets: China 75.8% (2015)

Total Imports: $4.819 billion (2015 est.); $5.586 billion (2014 est.)

Import Commodities: Petroleum, cooking coal, machinery and equipment, textiles, grain

Major Suppliers: China 76.4%, Republic of the Congo 5.5% (2015)

DEFENCE & SECURITY

A unified Korean Peninsula, which was ruled by Japan till the end of World War II, split into two parts with North Korea coming under the Soviet command and South under the US. While both the sovereign states intended to rule over the entire region, the Cold War prevailing between the then two superpowers – Soviet Union and the US – further heightened military tension between the two Koreas. But it was North Korea which launched a surprise attack against the South in 1950 with a motive to bring that region under its control. Though the two countries signed the Armistice Agreement in 1953 to end the fight, they have not shunned military conflict altogether since then.

North Korea's decades-long military build-up and acquisition of large number of weapons and equipment has intensified tensions between the two neighbours. The Communist state's unabated attempts to possess nuclear weapons and develop asymmetric warfare capabilities have been a cause of concern for the global powers, especially for the US which strongly supports South Korea militarily. North Korea today maintains one of the largest militaries in the world. It has over one million armed personnel – the fifth largest in the world. Along with the massive conventional armed forces, the country is also acquiring asymmetric warfare capabilities in the form of nuclear, biological and chemical weapons as also long-range ballistic missiles. It follows the Songun or "Military First" doctrine meaning that the armed forces have first priority over national resources. This policy also guides the political and economic life in the Communist country.

DPRK's military doctrine aims at invading South Korea (ROK) and bring the entire peninsula under one regime. Though this offensive doctrine still prevails, a strong US backing to South Korea has deterred the North from waging a full-scale war against its neighbour.

KEY DEFENCE PERSONS

MINISTER OF PEOPLE'S ARMED FORCES: Gen. Pak Yong Sik

CHIEF OF GENERAL STAFF: Gen. Ri Myong-Su

External Conflict

The inter-Korean conflict is the only major external military conflict North Korea is engaged in. The decades-old conflict has often resulted in open confrontations between the two Koreas with each threatening to launch war against the other. The ROK's closer military tie-up with the US and their regular joint military exercises off the Korean coast has also intensified the DPRK-ROK conflict in recent times.

Territorial disputes: DPRK's territorial disputes revolve mainly around some islands in the region.

S Korea: The South-controlled Yeonpyeong Island in the Yellow Sea, which is nearly 12 kilometers from the North Korean border, has often witnessed confrontations between the two Korean navies since the past decade. The two countries are engaged in a maritime border dispute as the North wants the western maritime line or the Northern Limit Line, which separates the two countries, to be drawn farther south.

China: North Korea and China dispute the sovereignty of certain islands in Yalu and Tumen rivers though it has not hampered their bilateral ties. The Baekdu mountain lying in the border between the two countries has become another region of contention in recent times.

Though both countries jointly administer the mountain and its surrounding lake, recent Chinese attempts to bring the region under its command has resulted in friction with the DPRK.

THREAT PERSPECTIVE

South Korea: The ROK is preparing itself to counter the DPRK by acquiring a range of weapons and equipment. Though it maintains a military smaller than that of the North, South Korea is strengthening its conventional warfare power with support from the US. Its strategy is to use its fleet of advanced fighter jets, warships, missiles along with other weapons to deter the North. The country is building new KDX-class of warships equipped with US-made Aegis air defence systems and also acquiring new submarines. Procurement of new-generation stealth fighters, spy drones and missile systems are also being planned. However, ROK lacks the capability to counter the asymmetric threat posed by DPRK, especially from ballistic missiles and nuclear weapons. Accordingly, it is believed that South Korea is also developing nuclear weapons.

US: The United States views North Korea as a rouge state and believes that it can launch a sudden, surprise attack on Seoul and even on Washington. The Asian country's unclear military intentions have therefore strengthened US-ROK military tie up. Being an ROK ally, the US has stationed around 28,500 troops – air, naval and ground forces – in South Korea since the end of Korean War and is pledged to help defend Seoul in the wake of another war. Both countries' navies often conduct joint military exercises off the Korean coast by deploying advanced warships and weapons.

Japan: North Korea's secretive nuclear programme has wearied Japan. Rockets launched by the North have often flown over Japan, prompting the latter to deploy missile interceptors to shoot down any such weapons flying over its airspace. Japan's military alliance with South Korea in recent times has also infuriated DPRK. South Korea has also joined Japan-US military exercises as an observer.

STRATEGIC RELATIONS

China: North Korea is regarded as China's strategic buffer zone in Northeast Asia. The two countries share a 1,400 km-long border. Since China perceives the US as its biggest military threat over the issue of Taiwan, the DPRK acts as a guard post for Beijing against the US troops deployed in South Korea. This allows China to deploy its military away from the North-eastern region and focus more on the Taiwan Strait. North Korea's development of nuclear weapons is viewed to be acting as a shield in the wake of US military intervention in the Asia Pacific region. China's economic support to DPRK has helped the latter in its nuclear armaments development programme. China was also a major supplier of military weapons and equipment to North Korea in the 1970s.

Russia: The military tie-up between the then Soviet Union and North Korea was at its peak in the 1990s when the former was supplying a large number of defence equipment to DPRK that included tanks, submarines, warplanes and missiles among other hardware. The two countries were also engaged in bilateral military wargames involving their navies and air forces. Following the disintegration of Soviet Union in 1991, the military relationship froze. Recently, there has been a thaw in their bilateral defence ties. The two sides are also discussing economic and energy cooperation with each other.

Iran: North Korea for over three decades has helped Iran in building nuclear and ballistic missiles. The clandestine Iran-DPRK military affair of sharing sophisticated weapons technology, mostly provided by China, has continued despite United Nations sanctions against both North Korea and Iran.

Pakistan: North Korea has close relations with Pakistan. It is reported that North Korean missile technologies have been supplied to Pakistan.

DEFENCE CAPABILITIES

ARMY: North Korea maintains one of the largest militaries in the world. The ground force is called the Korean People's Army (KPA). With nearly million-strong active ground personnel, the KPA forms the largest and most formidable branch of North Korean armed forces. The KPA is equipped with a large number of arms and weapons to carry out DPRK's offensive military doctrine of bringing the entire Korean Peninsula under one regime and also the defensive policy of protecting its territory from any threat posed by South Korea. The Army also assists in maintaining internal security of the country.

Personnel: 9, 50,000 (Including reserves, total strength is estimated to be around 4 million)

Equipment

Category	Name	In Service
MBT	Type-59	700
	T-62	800
	T-54	1600
	PT-85	50+
	Pokpung-ho	250+
APC	BMP-1	2500
	BA-64	
	VTT-323	
	YW-531	
	BTR-60	
	BTR-50	
	BTR-40	
	BTR-152	

Army continued...

Category	Name	In Service
Artillery	M-1989 170mm SP	8000
	M-1978 170mm SP	
	M-1977 152mm SP	
	M-1974 152mm SP	
	M-1992 130mm SP	
	M-1991 130mm SP	
	M-1981 130mm SP	
	M-1991 122mm SP	
	M-1985 122mm SP	
	M-1981 122mm SP	
	M-1977 122mm SP	
	M-1992 120mm SP	
	M-1985 towed 152 mm	
	M-1943 towed 152 mm	
	M-1937 ML-20	
	Type 66 towed 152 mm	
	M-46 / Type-59 130 mm	
	D-74 122 mm	
	D-30 122 mm	
	A-19 122 mm	
	M-30 / Type-54 122 mm	
	M-1931 122 mm	
Mortar	82 mm	5000
MLRS/ Vehicle mount rocket launcher	M-1991 240 mm	2500+
	M-1989 240 mm	
	BM-24	
	M-1985	
	BMD-20 200 mm	
	BM-14, 140 mm	
	RPU-14	
	Type-63, 130 mm	
	M-1993	
	M-1992	
	M-1985	
	M-1977	
	BM-21	
	BM-11	
	Type-63	
ATGW	Susong-Po (Sagger)	
SSM	Scud B (SRBM) Hwasong-5	100
	Scud C (SRBM) Hwasong-6	36
	9K52 & 2K6 Luna	24
	Scud-ER	~350
	Nodong-A	~200
	Nodong-B	20+
	Taepodong-1	
	Taepodong-2	~2
SAM	Hwasung-Chong (SA-7)	
	SA-2	800

Army continued...

Category	Name	In Service
SAM	SA-3	30
	SA-5	24
AA gun	KS-19, 100 mm	500
	KS-12, 85 mm	400
	M-1985 SP, 57 mm	
	ZSU-57-2, 57 mm	250
	S-60, 57 mm	
	M-1992,SP,37 mm	
	M-1939, 37 mm	500+
	M-1992,SP,30 mm	
	M-1992,SP,23 mm	
	ZSU-23-(Quad),23 mm	100
	ZU-23,23 mm	1500
	ZPU-4, ZPU-2, ZPU-1	

New Procurements/Upgrades

No official information is available on military procurements made by DPRK.

NAVY: The North Korean Navy has often been designated as a "brown water" navy that operates mostly in the country's surrounding waters. The Navy, called the Korean People's Army Naval Force, acts as a coastal defence force. It deploys small patrol-sized craft capable of defending the country's territorial waters. The Navy possesses some 50 to 60 submarines acquired from Russia and China. The exact number of the Navy vessels is not known due to extreme secretiveness of the country's military programme and acquisitions. The large number of submarines along with other amphibious vessels can covertly insert troops into S Korean territory.

Personnel: 60,000

Major naval bases: Pipagot, Sagon Ni, Najin, Wonsan

Equipment

Category	Name	In Service
Submarine	SSG Soviet Golf class	
	Golf II class	1
	SS PRC Type-031 (Soviet Romeo Class)	22
	SS Whiskey (Soviet)	4
	Sang-O class (Midget)	15
	SSI (Midget)	30
Frigate	FF Soho	1
	FFL Najin	2
Corvette	PG Sariwon	4
	Tral	2
Patrol vessel (Missile)	PTG Soju (OSA)	15 + 8
	PRC Huangfeng	4
	PTG Sohung	6
	PTG Komar	10

Navy continued...

Category	Name	In Service
Patrol Boat (Tarpedo)	PT Shershen	3
	PT Sinnam	
	PT Sin Hung	40
	PT P-4 19.3	
	Ku Song	60
Patrol Craft	PG T Class	5
	PG Mayang	10
	PC Taechong I/II	13
	PC Hainan	6
	PC Chodo	3
	PC/PT/PTG/WPC Chong-Ju	6
	PC SO-142	18
	PC Shanghai II	12
	PB Chongjin	16
	PB Chaho	50+
	PB/PT P-6 PB Shantou	10
LCU	Hantae class	10
LCM	Hungnam class	15
LCVP	Nampo class	100+
ACV	Kongbang class	140+
Amphibious		10
Mine sweeper	MSI Yukto I/II	20+
Auxiliary ship		10+
ASM	Silkworm	
	P-15 Termit	
	KN-01	
	KN-02	
	Kh-35	

Coastal Defence

Category	Name	In Service
Artillery	M-1931/-37	
	SM-4-1	
	M-1992	
	M-1937	
ASM (Coastal Defence)	Silkworm	1 Regiment

New Procurements/Upgrades

✦ An old Soviet-era Golf-II class ballistic missile submarine (SSBN) has been launched after being refurbished. The DPRK plans to use the vessel either for active service or for constructing a similar submarine of its own.

AIR FORCE: The Korean People's Army Air Force operates with the primary mission of defending the country's airspace. It maintains a large but mostly obsolete inventory of Soviet-era fighter planes, bombers, attack helicopters, transport aircraft and other fixed wing planes along with a range of air defence weapons. Some military hardware have also been acquired from China. The air force also provides tactical support to the Army and Navy.

Personnel: 110,000

Major air bases: Chongjin, Haeju, Hwangsuwon, Hyesan, Hyon-ni, Kanch'on, Taetan, Sunchon

Equipment

Category	Name	In Service
Fighter/ Ground Attack	Su-25	10
	MiG 19/Q5	30
	MiG-21	140
	MiG-23	50
	MiG-29	40
	F7B	40
	F6/F5	180+
	A5 Fantan	40
Bomber	H-5	~50
Transport	Y-2	300+
	An-24	6
	Il-18	2
	Il-62M	2-3
	Tu-134	2
	Tu-154	4
Attack/ Transport/ Surveillance Helicopter	Mi-8	20+
	MDH 300/500	27
	MDH300/500(AT)	~60
	Mi-24	50+
	Mi-26	4
	Mi-18	
	Z-5	40+
Trainer	MiG 23	4
	Yak 18	120
	L-39 Albatross	5
AEW	An-24	2
Air Defence System	KN-06	
	SA-2/3	
	SA-5	
	SA-6	
	SA-7	
	SA-17	

New Procurements/ Upgrades

No official information is available on military procurements made by DPRK.

STRATEGIC FORCE COMMAND: Though it is not officially known whether North Korea has built-up a strategic force command, the country certainly possesses a large number of ballistic missiles (according to some estimates over 1,000) and appears to be advancing in its nuclear weapons programme. The country's missile programme has been developed from the Soviet-made Scud tactical ballistic missiles.

North Korea officially declared its nuclear weapons programme in 2003 and detonated its first nuclear device in 2006 followed by more such tests in 2009 and 2013. In 2016, the North conducted two nuclear tests in January and September. In confirming its nuclear weapons programme in 2003, the Asian country nullified the Agreed Framework signed between it and the United States on 21 October 1994 in Geneva.

The country possesses ICBM capability with a range of about 6,000 km. The Taepodong-2, having this range, was test launched for the first time in 2006 and again in 2009, though it failed on both occasions. The Nodong 1 and 2 IRBMs with 1000+ km range are also capable of carrying nuclear warheads.

In April 2012, Pyongyang launched a long-range rocket, which it said, was aimed at putting a satellite into orbit for peaceful research purposes. The 30-metre Unha-3 rocket (believed to be the space launcher version of Taepodong-2 ICBM) had been positioned at a newly-built space centre on the country's north-western Yellow Sea coast but the launch ended in failure with the rocket disintegrating mid-air soon after blastoff. South Korea, the US and a number of other countries viewed the launch as a disguised ballistic missile test.

A second attempt to launch the long-range rocket was accomplished successfully on December 12, 2012. "The launch of the second version of our Kwangmyongsong-3 satellite from the Sohae Space Centre on December 12 was successful," the official Korean Central News Agency (KCNA) announced after the launch took place.

The country, in May 2015, released images indicating the launch of a missile from one of its Navy's submarines. Previously, it was reported that DPRK was on the verge of developing a submarine-launched ballistic missile (SLBM) capable of being launched from the Navy's Golf-class SSBN. A second test of the missile in November 2015 ended in failure. The country has also reportedly acquired the sea-based copy of a Russian cruise missile, believed to be the Kh-35.

In 2016, the country for the first-time test fired an inter-mediate range ballistic missile (IRBM) named Musudan or BM-35. The IRBM is believed to have been deployed with the N Korean military since 2007. In 2016, the missile, having a reported range of 3,500-km, has undergone a series of test firings.

DEFENCE PRODUCTION

The secretive military regime of N Korea does not disclose any details of its defence industry. While in the past it has acquired majority of weapons from Soviet Union and China, DPRK appears to have become steadily self-reliant in defence production. The indigenous defence production includes a whole range of military equipment for the Army, Navy and Air Force including artillery, armoured vehicles, tanks, submarines, warships, helicopters, trainer aircraft, missiles and air defence systems. The country possesses indigenous R&D facilities for development of almost all types of missiles including long range SSM, ICBM, IRBM, SRBM and other variants. Some defence analysts suggest that the missiles can be armed with nuclear warheads.

DEFENCE BUDGET

North Korea allocates quite a sizable amount of its GDP to defence. The annual defence budget is estimated to be between 15 and 20 percent of the GDP in the last few years. For 2016, the estimated military expenditure was 15.8 percent of GDP. DPRK has constantly put up an aggressive and offensive military stance despite the stiff economic sanctions against it. The huge budgetary allocations have also been contributed to strengthen its domestic defence industry. Estimated budgetary figures of last few years are listed in the following table.

Year	Allocation	% of GDP
2011	$5.5 billion	15
2010	No Data available	
2009	76.250 billion won ($8.77 billion)	15.8
2008	71.330 billion won	15.8

CONCLUSION

The extremely secretive and isolated North Korea has been viewed as a steadily rising military threat, particularly by South Korea and the US. The country has maintained a belligerent stance and offensive posture to unify the Korean Peninsula and bring it under its control. With a massive military force and asymmetric warfare capabilities, DPRK poses the greatest threat to the ROK. Its nuclear weapons programme and nuclear proliferation activities have been a major cause of concern for the international community.

North Korean Air Force's MiG 29

CONTACT DETAILS

Ministry of Defence - DPRK
Pyongyang
Korea, North

NORWAY
(Capital: Oslo)

INTRODUCTION

Area: 323,802 sq km

Population: 5,265,158 (July 2016 est.)

Coastline: 25,148 km

Maritime claims: Territorial Sea: 12 nm
Contiguous Zone: 10 Nm
Exclusive Economic Zone: 200 Nm
Continental Shelf: 200 nm

KEY POLITICAL PERSONS

CHIEF OF STATE & COMMANDER-IN-CHIEF: King Harald V

PRIME MINISTER: Erna Solberg

KEY DEFENCE PERSONS

DEFENCE MINISTER: Ine Marie Eriksen Soreide

CHIEF OF DEFENCE: Adm. Haakon Bruun-Hanssen

CHIEF OF ARMY: Maj. Gen. Odin Johannessen

CHIEF OF NAVY: Rear Adm. Lars Saunes

CHIEF OF AIR FORCE: Maj. Gen. Per-Egil Rygg

ECONOMY

Norway has the second-highest GDP per-capita among European countries (after Luxembourg), and the fourth-highest GDP (PPP) per-capita in the world. The Norwegian economy is a prosperous bastion of welfare capitalism, featuring a combination of free market activity and government intervention. The government controls key areas, such as the vital petroleum sector, through large-scale state-majority-owned enterprises. The country is richly endowed with natural resources - petroleum, hydropower, fish, forests, and minerals - and is highly dependent on the petroleum sector, which accounts for nearly half of exports and over 30% of state revenue. Norway is the world's second-largest gas exporter. Its position as an oil exporter has slipped to ninth-largest as production has begun to decline. Norway opted to stay out of the EU during a referendum in November 1994. In anticipation of eventual declines in oil and gas production, Norway saves almost all state revenue from the petroleum sector in the world's second largest sovereign wealth fund, valued at over $500 billion in 2010. After solid GDP growth in 2004-07, the economy slowed in 2008, and contracted in 2009, before returning to positive growth in 2010-11, however, the government budget is set to remain in surplus. Norway is the world's second-largest gas exporter; and seventh largest oil exporter, making one of its largest offshore oil finds. The country has a very high standard of living compared with other European countries, and a strongly integrated welfare system. Norway's modern manufacturing and welfare system rely on a financial reserve produced by exploitation of natural resources, particularly North Sea oil.

GDP: $356.2 billion (2015 est.), $350.7 billion (2014 est.)

Real Growth Rate: 1.6% (2015 est.), 2.2% (2014 est.)

Industries: Petroleum and gas, food processing, shipbuilding, pulp and paper products, metals, chemicals, timber, mining, textiles, fishing

Oil - proved reserves: 5.1 billion bbl (1 January 2010 est.)

Natural gas - proved reserves: 1.922 trillion cu m (1 January 2016 est.)

Total Exports: $102.9 billion (2015 est.), $143.7 billion (2014 est.)

Export Commodities: Petroleum and petroleum products, machinery and equipment, metals, chemicals, ships, fish

Major Markets: UK 22.2%, Germany 17.9%, Netherlands 10.2%, France 6.6%, Sweden 6.1%, Belgium 5%, US 4.5% (2015)

Total Imports: $75.79 billion (2015 est.), $90.89 billion (2014 est.)

Import Commodities: Machinery and equipment, chemicals, metals, foodstuffs

Major Suppliers: Sweden 12%, Germany 11.8%, China 10.9%, UK 6.7%, US 6.6%, Denmark 6% (2015)

INTERNATIONAL DISPUTES

✦ Norway asserts a territorial claim in Antarctica (Queen Maud Land and its continental shelf)

✦ Russia and Norway reached an agreement on aligning the Barents Sea and Arctic Ocean boundaries over EEZ and continental shelf on 15 September 2010; this agreement is pending ratification by the respective national assemblies.

✦ As a founding member of the North Atlantic Treaty Organisation, Norway has been an active participant in NATO since the signing of the treaty in Washington. The Norwegian Delegation to NATO is staffed by personnel from the Norwegian Ministry of Foreign Affairs and the Ministry of Defence.

DEFENCE

ARMY

Personnel: 7,700 (60,000 reserve)

Equipment

Category	Name	In Service
MBT	Leopard 2A4NO	52
AIFV & APC	CV-9030N	103
	Supacat HMT Extenda	(On Order)
	M113	500
	Patria XA series	74
Artillery	M109A3GN 155mm	26
	M125A2	10
MLRS	MLRS	12
ATK Weapon	M72 LAW	
SAM	RBS70 MKII	300
UAV	RQ-11 Raven, PD-100	

New Procurements/ Upgrades

✦ The UK high mobility vehicle specialist, Supacat, has signed a £23 million contract with The Norwegian Defence Logistic Organisation (NDLO) to supply a new fleet of High Mobility Vehicles. Supacat is supplying the HMT Extenda vehicle, the most capable vehicle in its class with the highest levels of mobility, protection, payload and firepower.

✦ BAE Systems has delivered the first CV90 STING, an engineering variant of the CV90 vehicle, for Norway at the Company's Ornskoldsvik facility in Sweden. The contract was awarded in June 2012 to upgrade and deliver a total of 144 CV90 vehicles to the Norwegian Army. The first CV90 IFV, an infantry fighting variant, was delivered. The CV90 contract to Norway includes delivery of five vehicle variants, 74 infantry fighting, 21 reconnaissance, 15 command, 16 engineering, 16 multi-role and two driver training vehicles.

✦ The Norwegian Army has ordered 24 Archer 155mm SP Truck mounted howitzers in a joint programme with Sweden.

✦ BAE Systems and Kongsberg Devotek A/S of Norway will jointly develop a new system of gears and transmissions for the Spitterskyddad Enhets Platform (SEP) 8x8 armoured vehicle.

✦ Navantia will provide technical assistance for five F-310 frigates over the next three years.

✦ The Norwegian Defence Logistics Organisation (NDLO) has agreed to the procurement of a further tranche of 62 Light Armoured Vehicles (LAV) from Iveco Defence Vehicles.

Leopard 2A4 Main battle tank of the Norwegian Army. *Image Credit: Ministry of Defence, Norway.*

NAVY

Personnel: 3,700

Equipment

Category	Name	In Service
Submarine	Ula Class	6
Frigate	Fridtjof Nansen class	4
Corvette	Skjold class	6
Mine warfare force	Oksoy class	4
	Alta Class	3
Auxiliary	Reine class	1

Naval Aviation

Category	Name	In Service
Aircraft	Mk86P3B Orion	2
Helicopter	LYNX	6

New Procurements/ Upgrades

✦ The Norwegian Defence Logistics Organisation (NDLO) forwarded a "Request for Information" (RFI) to a number of prequalified shipyards. These include DCNS (France), Fincantieri (Italy), Navantia (Spain), ThyssenKrupp Marine Systems (representing Kockums AB, Sweden and HDW, Germany) and Daewoo Shipbuilding & Marine Engineering (DSME, South Korea). The purpose of the RFI is to investigate investment cost, life cycle costs, production time, performance and other important aspects related to new submarines that in turn will shape a decision on life extension or fleet replacement.

✦ Kongsberg Gruppen has received a $25 million contract with the Norwegian Defence Procurement Division for the development of a Joint Strike Missile.

✦ Kongserg has recieved a $466.4 million contract from the Norwegian Armed Forces' Logistics Organisation for the serial production of the new Naval Strike Missile (NSM) for Norway's Nansen Class AEGIS frigates, and Skjold Class catamaran-hovercraft fast attack craft.

✦ Kongsberg Defence & Aerospace has signed a contract for the delivery of Sea PROTECTOR Remote Weapon Stations (RWS) for the Royal Norwegian Navy (RNoN) combat vessels. The contract has an overall scope of 133 MNOK.

✦ Analox Military Systems has been awarded a contract to supply submarine equipment to the Norwegian Navy. The contract was signed with Norwegian Defence Logistics Organisation (NDLO) to supply the Norwegian Navy with two carbon monoxide monitoring systems for six ULA-class submarines.

✦ The Norwegian Government decided 25th November to procure five P-8A Poseidon for delivery in 2021 to 2022. The decision taken in the Government has been proposed to the Parliament.

Ula class submarine of the Royal Norwegian Navy. *Image Credit: Ministry of Defence, Norway.*

AIR FORCE

Personnel: 1430 (approx)

Equipment

Category	Name	In Service
Fighter	F16	57
	F-35	4 (more on order)
ASW Maritime Patrol	P3C Orion	4
	P3Ns	2
EW/ AEWCS	Falcon 200	2
Transport	C-130J-30s	4
Helicopter	Bell 412SPs	18
	NH-90	5+
	Sea King Mk4	12
	LYNX Mk86s	6
Trainer	SAAB Supporter	15
SAM	NASAMS	2

New Procurements/ Upgrades

✦ Airbus Defence and Space has signed a four-year framework agreement with the Norwegian Defence Logistic Organisation (NDLO) for the provision of X-band satellite capacity and has committed to deliver up to 30Mhz of capacity annually on the Skynet military satellite constellation. Skynet will augment the Norwegian forces' existing secure communications in maritime, land and aeronautical applications across Northern Europe, the Arctic and for Allied nations missions globally.

✦ Kongsberg has signed a Phase III contract with the Norwegian Defence Logistics Organisation (NDLO) to complete the development of the Joint Strike Missile (JSM) and prepare it for integration on the F-35 Lightning II Joint Strike Fighter (JSF). The contract is valued at NOK 1.1 billion (USD 179 million), and comes on the heels of a bridge phase contract signed on 29 November 2013.

✦ The first Norwegian F-35 was rolled out at Lockheed Martin's production facilities in Fort Worth, Texas in September 2015. Norway has ordered the first two of dozens of F-35 fighter aircraft it plans to buy from the United States as part of its largest-ever government spending project. The total deal for 52 fighter aircraft is worth $10 billion. Norway is planning to acquire up to 52 F-35As in the years leading up to 2025. The F-35 will replace Norway's current fleet of F-16 fighters which date back to the early 1980s.

✦ Norway has requested the purchase of two C-130J-30 from the US Air Force (USAF) baseline aircraft and associated parts at an estimated cost of $300 million in June 2012, to boost its capability in providing intra-theatre lift for its armed forces.

✦ Norwegian Defence Logistics Organisation awards Pratt & Whitney contract to support the Royal Norwegian Air Force's F100-PW-220E engine fleet.

F-16 fighter aircraft of the Royal Norwegian Air Force. *Image Credit: Ministry of Defence, Norway.*

Defence Expenditure

Total defence spending: $7.2 billion (est.)

Estimated defence spending in terms of GDP: 1.52% (2015)

According to a press release by the Government of Norway, the Norwegian Government in Dec 2016 followed up on the new Long Term Defence Plan presented in June with a 2017-budget that involves a significant increase in the funding level for the Norwegian Armed Forces. In total, the Government proposes a NOK 1.9 billion increase year on year, of which NOK 1.14 billion addresses specific measures recommended by the Long Term Defence Plan.

The focus will be on the shortfalls in maintenance, spare parts and ammunition, and improving training levels and readiness, it seeks to create a solid basis for future investments. It will also address the cost growth of military equipment. Among other major proposals, the Government is requesting authorization to order a further 12 F-35 combat aircraft out of the planned total of 52, divided by two batches of six delivered in 2021 and 2022.

The Government also requested an allocation of NOK 8.6 billion to cover existing commitments relating to the acquisition of the F-35 and related equipment and infrastructure. The Norwegian Government also recommends accelerating the purchase of three new helicopter capable Cost Guard vessels to replace the current Nordkapp-class. The Government now plans to present this procurement project for Parliamentary approval in early 2018.

These funding increases serve as an initial step in the process that aims to add NOK 165 billion to the budgets of the Norwegian Armed Forces over the coming 20 years. The Norwegian parliament has agreed upon a new Long Term Plan for the Norwegian Armed Forces. The plan is entitled Capable and

Sustainable and represent a historic defence spending increase. The main priorities in its implementation will be to 1. Strengthen our national defence, 2. Strengthen NATO's ability for collective defence, 3. Contribute to international crisis management and 4. Further develop the total defence concept.

In 2017, the additional funds will go towards reducing a backlog that has built up with regard to materiel maintenance and reserve stocks. Starting in 2018, extra resources will be employed to increase activity in all branches of the military, and in 2019 and 2020 will see a focus on new capabilities to strengthen defence and situational awareness in the High North.

The Norwegian Defence Estates Agency, the Norwegian Defence Materiel Agency, the Norwegian National Security Authority and the Norwegian Defence Research Establishment will receive the allocation letter for the Long Term Plan covering the 2017–2020 period from the Ministry of Defence.

Key figures *(Source: Government of Norway)*:

Total defence budget	Personnel and operations	Infrastru-cture	Procure-ment
NOK 50.9 billion	NOK 35.6 billion	NOK 3.1 billion	NOK 12.1 billion

Defence Production and R&D

Norwegian Defence Research Establishment (FFI) is the prime institution responsible for defence-related research in Norway. It conducts research and development on behalf of the Norwegian Armed Forces and provides expert advice to political and military defence leaders. In particular, its task is to keep track of advances in the fields of science and military technology which might affect the assumptions on which Norwegian security policy and/or defence planning is based.

The Norwegian Defence and Security Industries Association (FSI) is the largest and most important association in Norway advocating the interests of the Norwegian Defence and Security Industry and the primary interlocutor for the government in matters of importance to the industry.

The association comprises more than 100 companies from all parts of Norway. It is a diversified group ranging from the major national defence contractors to one-man businesses with unique niche capabilities built on innovation and advanced technology serving both the military and civilian markets. Norway plays a significant role in global security and the production of defence products. The high growth, which occurred in this sector over the last years, is due to the fact that the Norwegian defence industry is highly competitive and to the constructive relationship that exists between the armed forces and industry.

Kongsberg Gruppen is a Norwegian defence contractor and maritime automation supplier. It is an international technology corporation that delivers advanced and reliable solutions that improve safety, security and performance in complex operations and during extreme conditions. KONGSBERG works with demanding customers in the global defence, maritime, oil and gas and aerospace industries. Kongsberg Defence & Aerospace AS (KDA), one of two operating companies of Kongsberg Gruppen (KOG) of Norway, is a supplier of defence and space related systems and products, mainly anti-ship missiles, military communications, and command and weapons control systems for naval vessels and air-defence applications.

Defence Procurement

The Norwegian Defence Materiel Agency (NDMA) is an important partner for Norwegian industry. The Defence White Paper 38 stip□ ulates that the Norwegian Armed Forces and Norwegian defence industry are strategic partners.

Norway has set up a defence equipment agency called Forsvarsmateriell in 2015, responsible for acquisition and production in the sector. Forsvarsmateriell's foremost mission and raison d'être is to support the Armed Forces and the country's defence capability. The agency is directly under the Ministry of Defence. The main tasks of the agency include material investments and material management in the defence sector. Mette Sorfonden was appointed the Director of the agency on 20 November 2015.

CONTACT DETAILS

Ministry of Defence
Myntgt. 1
P.O. Box 8126 Dep
0032 Oslo 1, Norway
Tel: +4723098000
Fax: +4723092323
E-mail: postmottak@fd.dep.no

Norwegian Armed Forces
P.O. box 800, Postmottak
2617 Lillehammer
Norway
Tel: (+47) 915 03 003
Email: forsvaret@mil.no

OMAN
(Capital: Muscat)

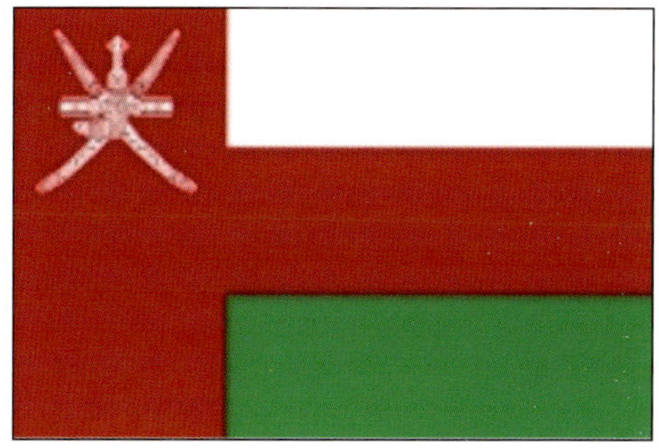

INTRODUCTION

Area: 309,500 sq km

Population: 3,355,262 (July 2016 est.)

Coastline: 2,092 km

Maritime claims: Territorial sea: 12 nm

Contiguous zone: 24 nm

Exclusive Economic Zone: 200 nm

KEY POLITICAL PERSONS

HEAD OF STATE, PRIME MINISTER & MINISTER OF DEFENCE: Sultan Qaboos Bin Said Al Said

DEPUTY PRIME MINISTER: Fahd Bin Mahmoud Al Said

KEY DEFENCE PERSONS

MINISTER RESPONSIBLE FOR DEFENCE AFFAIRS: Sayyid Badr Bin Saud Bin Harib Al-Busaidi

CHIEF OF STAFF OF ARMED FORCES: Lt Gen Ahmed bin Harith al Nabhani

COMMANDER OF ARMY: Maj. Gen. Matar bin Salim bin Rashid Al-Bulushi

COMMANDER OF NAVY: Rear Adm. Abdullah bin Khamis Al Raisi

COMMANDER OF AIR FORCE: Air Vice Marshal Matar bin Ali bin Matar Al-Obaidani

ECONOMY

Oman is heavily dependent on dwindling oil resources, which generate 77% of government revenue. Current GDP per capita has expanded continuously in the past half-century. Because of declining reserves, Muscat has actively pursued a development plan that focuses on diversification, industrialisation, and privatisation, with the object of reducing the oil sector's contribution to GDP to 9% by 2020. By using enhanced oil recovery techniques, Oman succeeded in increasing oil production, giving the country more time to diversify, and the increase in global oil prices through 2011 provided the government greater financial resources to invest in non-oil sectors. Oman authorities are comfortable with short-term budget deficits and have approved an expansionary 2015 budget.

GDP: $171.4 billion (2015 est.), $164.6 billion (2014 est.)

Real Growth Rate: 4.1% (2015 est.), 2.9% (2014 est.)

Industries: Crude oil production and refining, natural and liquefied natural gas (LNG) production; construction, cement, copper, steel, chemicals, optic fibre

Oil - proved reserves: 5.3 billion bbl (1 January 2016 est.)

Natural gas - proved reserves: 688.1 billion cu m (1 January 2016 est.)

Total Exports: $34.43 billion (2015 est.), $53.22 billion (2014 est.)

Export Commodities: Petroleum, re-exports, fish, metals, textiles

Major Markets: China 35.4%, UAE 15.3%, South Korea 6.8%, Saudi Arabia 5.8%, Pakistan 4.2% (2015)

Total Imports: $28.27 billion (2015 est.), $27.89 billion (2014 est.)

Import Commodities: Machinery and transport equipment, manufactured goods, food, livestock, lubricants

Major Suppliers: UAE 29.7%, Japan 10.2%, US 7.5%, China 6.7%, India 6.3% (2015)

INTERNATIONAL DISPUTES

Boundary agreement reportedly has been signed and approved with UAE in 2003 for entire border, including Oman's Musandam Peninsula and Al Madhah exclave.

DEFENCE

ARMY

Personnel: 50,000

Equipment

Category	Name	In Service
MBT	Challenger 2	38
	M60A3/ A1	79
	Scorpion	37
APC/MICV /IFV	AT105 Saxon	15
	Stormer	9
	Piranha 8X8	160
	WZ5516X6	50
	Saladin	38
	V150 Commando	100+
	VBC90	15
	Fahd	49
Artillery	155 mm G6	24
	155 mm FH70	12
	155 mm M109	15
	130 mm	15
	105 mm towed	35
Mortar	81 mm	
	107 mm	
	120 mm	
MRL	122 mm Type 90	10+

Army continued...

Category	Name	In Service
Anti-Tank GW	Milan	50
	TOW	18
	Javelin	30
AA Gun	ZU232	
	35 mm twin GDF-005	10
SAM	SA7	34
	Javelin	30
	Albi/Mistral	
	THAAD anti-missile system	

New Procurements/ Upgrades

✦ The US State Department has made a determination approving a possible Foreign Military Sale to Oman of TOW 2B missiles, related equipment and support. The estimated cost is USD 51 million. The Government of Oman has requested a possible sale of 400 tube-launched optically-tracked wire guided (TOW) 2B Aero, Radio Frequency (RF) missiles (BGM-71F-3-RF) and 7 TOW 2B Aero, RF missile (BGM-71F-3-RF) Fly-to-Buy missiles.

✦ Oman is to receive nine CENTAURO Armoured 8x8 Vehicles with 120 mm guns. These vehicles will replace the VBC90 of the Royal Guards Brigade.

NAVY

Personnel: 4,500

Equipment

Category	Name	In Service
Corvette	Qahir class	2
	Khareef Class	3
Patrol Vessel	Province Class	4
	Al Bushra Class	3
	Al Ofouq	4
HSSV	Al Mubshir	1
Amphibious Force	Nasir Al Bahr	1
	Fulk al Salamah	1
	Al Sultana	1
Missile	Crotale	50
	Exocet MM-40	160
	Harpoon	
	Mica	

New Procurements/ Upgrades

✦ The Royal Navy of Oman has taken delivery of the High Speed Support Vessel (HSSV) RNOV Al Mubshir built by defence prime contractor, Austal. The RNOV Al Mubshir is the first of two 72 metre HSSVs to be delivered to the Royal Navy of Oman under a US$124.9 million contract, Austal said in a statement. The Company commenced construction of the HSSVs in August 2014 at its Henderson, Western Australia shipyard, with the second HSSV scheduled to be delivered later in CY2016, in line with the contract.

✦ The Navy has received the second Khareef-class corvette, Al Rahmani, from BAE Systems. The company was awarded a £400 million contract to design and construct three Khareef-class corvettes.

✦ Royal Navy of Oman is to equip patrol boats and corvettes with Vigile ESM (Electronic Support Measures) systems that include Vigile 200 and Vigile 400 systems. The agreement was signed with Thales.

✦ Raytheon Anschuetz was awarded a contract to provide the company's Integrated Bridge and Navigation Systems for four Offshore Patrol Vessels for the Royal Navy of Oman.

✦ UK defence major BAE Systems has launched the third and last Project Khareef class corvette for the Royal Oman Navy. BAE had received the £400 million project in 2007 to build three Khareef class corvettes for Oman. The deal includes through-life support and services, training for RNO personnel and initial logistics support package for the ships.

✦ Oman Navy has received Al Shamikh Vessel, the first ship of Al Khareef project, which is a vital source that would contribute to the strengthening of military capabilities and capacities of the Sultan's Armed Forces (SAF).

High Speed Support Vessel (HSSV) RNOV Al Mubshir during sea trials off Western Australia, March 2016. *Image Credit: Austal*

AIR FORCE

Personnel: 5,000

Equipment

Category	Name	In Service
Aircraft	Eurofighter Typhoon	12 (On order)
	F-16C Fighting Falcon	14
	Hawk 203/103	10
	CASA C-295	3 (on order)
	Skyvan	3
Transport/ Tanker	Airbus A320 Prestige	2
	C130H	3
	C-130J	4
Attack/ Transport Helicopter	Bell 205	5
	Bell 206	4
	SA330	2
	Puma	3

Air Force continued...

Category	Name	In Service
Attack/ Transport Helicopter	AS332	2
	Super Lynx300	15
	NH90	15
Trainer	Pilatus PC-9	12
	Hawk 128	5
	F-16D Block Fighting Falcon	4

New Procurements / Upgrades

✦ Raytheon is to sign a multi-billion dollar contract with Oman for the sale of its ground-based air defence system. The contract for purchase of National Advanced Surface-to-Air Missile System (NASAMS), valued at $1.28 billion, includes ground support equipment, a full training package, and technical assistance.

✦ The USA Defence Security Cooperation Agency has notified the congress of a possible sale of 18 F-16 Block 50/52 to Oman in a contract worth 3.5 Billion USD. In addition to the new fighters, the contract includes upgrading existing 12 F-16 C/D in the RAFO inventory. In 2011 it was announced that Oman had agreed to buy an additional 12 F-16C/D Block 50s to join the 12 F-16s C/Ds already in service.

✦ Oman to receive two additional C-130J Super Hercules aircraft from Lockheed Martin. Oman ordered its first C-130J in July 2009. In Sept 2012, Sultanate of Oman accepted the first C-130J Super Hercules aircraft. Oman currently operates a fleet of three C-130Hs purchased in the early 1980s.

✦ The Royal Air Force of Oman (RAFO) has signed a contract with Airbus Military to acquire eight C295 tactical aircraft. The contract, signed on Monday, makes provision for five C295s in tactical transport configuration and the remaining three as maritime patrol aircraft (MPA).

✦ ITT Exelis has won a $23.6 million Foreign Military Sales (FMS) contract from Oman to provide 12 airborne electronic warfare defensive systems.

✦ Lockheed Martin to provide additional Sniper Advanced Targeting Pods (ATP) for Oman's F-16 fleet worth $23 million contract value.

Defence Expenditure

Total defence spending: $9.3 billion (est.)

Estimated defence spending in terms of GDP: 14.07% (2015)

Foreign Suppliers: United Kingdom, United States, India, Turkey, Arab League, Russia, Pakistan.

India and Oman in May 2016 vowed to deepen bilateral defence ties as the two countries signed four key agreements in the critical sector with focus on enhancing military cooperation. The agreements on defence cooperation, marine crime prevention, maritime issues and flight safety information exchange was signed.

Oman has kept defence spending at a constant level of 10-12 per cent of GDP, and 30 per cent of total government spending, since the early 1990s. Oman registered the sharpest increase in defence spending in 2012. SIPRI, Stockholm has reported increased military spending throughout the Middle East. The institute said the largest percentage increase worldwide in 2012 was by Oman. In 2012, Oman has ordered offshore patrol vessels, anti-tank guided missiles and fighter-jets.

Defence Production and R&D

Sultan Qaboos bin Said Al Said is very much interested in promoting the development of defence and technology in the country and encourages the involvement of the Omani companies or state organisations. In 2000, Alenia Marconi Systems (AMS) signed an agreement with the Royal Air Force of Oman (RAFO) to jointly set up and manage an air controller training school in the Sultanate.

Defence Procurement

The Gulf Cooperation Council [GCC] was established in an agreement concluded on 25 May 1981 in Riyadh, Saudi Arabia between: Bahrain, Kuwait, Oman, Qatar, Saudi Arabia and UAE. These countries declared that the GCC is established in view of the special relations between them, their similar political systems based on Islamic beliefs, joint destiny and common objectives. The GCC is a regional common market with a defence planning council as well. The geographic proximity of these countries and their general adoption of free trade economic policies are factors that encouraged them to establish the GCC. The GCC States seek to build up their defence forces according to a common conception. In this context, they have unified operational procedures, training, and military curricula. They also endeavour to accomplish compatibility of their military systems.

The Office of the Secretary General of Ministry of Defence Oman main roles and responsibilities include managing and implementing the financial and procurement policy, as well as paying invoices to MOD suppliers. The office provides e-Services to facilitate improved communications with external organisations that deal with the Ministry of Defence.

CONTACT DETAILS

Ministry of Defence

Chief of Staff Sultan's Armed forces

BOX 113 – Post Muscat – Sultanate Of Oman

Tel: 24333361 – 24338734 / Fax: 24333369

http://www.mod.gov.om/

Email: jundoman@mod.gov.om

Office of Deputy Prime Minister for Security and Defence

P.O. Box 113, Muscat, Oman

Tel: +968704096 www.omanet.om

STRATEGIC INFORMATION

PAKISTAN
(Capital: Islamabad)

INTRODUCTION

Area: 796,095 sq km

Population: 201,995,540 (July 2016 est.)

Coastline: 1,046 km

Maritime claims: Territorial sea: 12nm

Contiguous Zone: 24 nm

Exclusive Economic Zone: 200 nm

Continental Shelf: 200 nm or to the edge of the continental margin

GEOPOLITICAL IMPORTANCE

The Islamic Republic of Pakistan is a South Asian country and was once a part of the larger Indian subcontinent. It is bordered by India in the east, Iran and Afghanistan in the west and China in the far northeast. In its south lies the Arabian Sea. The geographic location of Pakistan makes it a vital access point between South Asia and Southwest Asia. The four neighbours of Pakistan – India, China, Afghanistan and Iran – all play important roles in the Asian continent as well as in international sphere. Pakistan, being positioned in the midst of these countries, plays a key role in some or other ways. The Islamic Republic's relative proximity to the resource-rich Central Asian region also gives it a strategic edge. The Arabian Sea in the south connects Pakistan with the Gulf countries. In the present-day world order, Pakistan has attained an important stature and is influencing the international geostrategic sphere for numerous reasons.

POLITICAL OVERVIEW

The Constitution of Pakistan defines the country as a federal republic with the President as head of the state and Commander-in-Chief of armed forces, and the Prime Minister as head of the Government. However, political instability in the country has given way to authoritarian military regimes several times in its history. Hence, the political system of Pakistan has remained volatile in nature with the country ruled either by democratically elected governments or by the military.

KEY POLITICAL PERSONS

PRESIDENT:
Mamnoon Hussain

PRIME MINISTER:
Muhammad Nawaz Sharif

ECONOMY

The developing economy of Pakistan has suffered setbacks due to national and international political turmoil. The country ranks 43rd largest in terms of nominal GDP according to World Bank data. In the last decade, it has witnessed economic inconsistency with GDP rising from 5% to 8% between 2004 and 2007 and sliding down thereafter largely due to global recession. Foreign investments in the country have significantly declined in recent times due to political instability, internal unrest and terrorism. Foreign aid, mostly from the US, and financial assistance from global institutions including the World Bank, IMF and Asian Development Bank, has been helping Pakistan to revive its economy. Falling global oil prices in 2015 contributed to a narrowing current account deficit and lower inflation, despite weak export performance. Pakistan's programme with the IMF – a three-year, $6.7 billion Extended Fund Facility focusing on reducing energy shortages, stabilizing public finances, expanding revenue, and improving the external balance – was slated to conclude in September 2016. While passing most quantitative targets, Pakistan has missed targets on structural reforms and performance criteria throughout the programme.

GDP: $931 billion (2015 est.); $893.1 billion (2014 est.)

Real Growth Rate: 4.2% (2015 est.); 4% (2014 est.)

Industries: Textiles and apparel, food processing, pharmaceuticals, construction materials, paper products, fertiliser, shrimp

Natural Gas - Proved Reserves: 669.4 billion cu m (1 January 2016 est)

Crude Oil - Proved Reserves: 400 million bbl (1 January 2016 est)

Total Exports: $22.73 billion (2015 est.); $24.78 billion (2014 est.)

Export Commodities: Textiles (garments, bed linen, cotton cloth, yarn), rice, leather goods, sports goods, chemicals, manufactures, carpets and rugs

Major Markets: US 13.1%, UAE 9.1%, Afghanistan 9.1%, China 8.8%, UK 5.4%, Germany 4.9% (2015)

Total Imports: $39.29 billion (2015 est.); $42.65 billion (2014 est.)

Import Commodities: Petroleum, petroleum products, machinery, plastics, transportation equipment, edible oils, paper and paperboard, iron and steel, tea

Major Suppliers: China 28.1%, Saudi Arabia 10.9%, UAE 10.8%, Kuwait 5.6% (2015)

DEFENCE & SECURITY

Pakistan, which was once a part of Indian subcontinent, has throughout its existence maintained an aggressive posture towards India. This has resulted in three wars – in 1948, 1965 and 1971. The 1971 War led to the creation of a new nation Bangladesh which was carved out from East Pakistan. Further, there was localised military conflict over Kargil in 1999. A militarily "imbalanced" Pakistan has also adopted covert war tactics against India and engaged in proxy war over Kashmir.

Pakistan has maintained a sizable military force since its creation in 1947 after separation from India. It ranks seventh in the world in terms of active armed personnel. The armed forces of Pakistan constitute of the Army, Navy and Air Force. The country has built up a large conventional armed force, though its size and strength pales in comparison to India's vast military power. By possessing nuclear weapons, Islamabad has tried to add teeth to its military capabilities against New Delhi which still adheres to a "no first use" policy of nuclear weapons.

Drawing lessons from the past Indo-Pak wars, Pakistan, while maintaining that it follows a defensive military doctrine, however, has realised the need for an "offensive defence" strategy to counter the superior conventional military might of India. The Islamic nation has always regarded India as its foremost military rival in the entire Asian region. To overcome the conventional military asymmetry against India, Pakistan has acquired nuclear weapons capability. It has also been increasingly aligned towards the other Asian power China to procure advanced arms and weapons. The two "all-weather allies" also conduct regular joint defence drills. In the past few years, Pakistan and China have also been working on joint development and production of sophisticated weapon systems and military platforms. The two partners have been aggressively making efforts on the strategic containment of India and its rise as an influential global economic and military power.

KEY DEFENCE PERSONS

DEFENCE MINISTER: Khawja Muhammad Asif

CHAIRMAN JOINT CHIEFS OF STAFF COMMITTEE: Gen. Zubair Mahmood Hayat

CHIEF OF ARMY: Gen. Qamar Javed Bajwa

CHIEF OF NAVY: Adm. Muhammad Zakaullah

CHIEF OF AIR FORCE: Air Marshal Sohail Aman

Internal Conflict

Pakistan is grappling with several internal problems, including terrorism, religious, ethnic and sectarian violence, and organised crime.

Terrorism: The foremost internal threat that Pakistan faces today is terrorism. Absence of political stability, alternating military regimes and the country's support to US-led war on terror has aggravated the problem which has become a major cause of concern for the country's internal security. The armed conflict in North West Pakistan has intensified in recent times with militant elements in the region, which were once nurtured by the country, posing a serious security threat. In its latest military doctrine, the Pakistan Army has described home-grown militant groups and internal dangers as the "biggest threat" to the country's security. The Army has introduced changes in its operational priorities for the first time as the new doctrine describes the ongoing guerrilla war in the tribal belt and along the western border and bomb attacks by militant groups as the greatest threat.

Ethnic and religious unrest: The Islamic state has been dealing with religious and ethnic unrest since its creation in 1947. The religious violence in Pakistan has largely been rooted in the country's birth as a theocratic state. The sectarian strife between Shia and Sunni Muslims in the country has often taken violent form. Political, economic and social factors have also played a role in the ethnic, religious and sectarian conflicts in the country.

Organised crime: Pakistan deals with the issues of drug trafficking, money laundering, extortion, fraud and other forms of organised crime which have often been linked to terrorist groups operating in the country.

External Conflict

The external conflict area for Pakistan has solely revolved around India ever since both the nations were separated over six decades ago. While Kashmir has remained the flashpoint in bilateral relations between the two nations, several other factors have also accentuated the Indo-Pak dispute.

Territorial disputes: Pakistan is engaged in land border disputes with two of its immediate neighbours, India and Afghanistan.

India: The central point of territorial dispute between Pakistan and India is Kashmir. While India controls approximately 43% of the region, including most of Jammu, the Kashmir Valley, Ladakh, and the Siachen Glacier, Pakistan has occupied approximately 37% of Kashmir, namely Azad Kashmir and the northern areas of Gilgit and Baltistan. The rest of the territory is under the control of China. Islamabad has been strongly contesting New Delhi's position that Jammu and Kashmir is an integral part of India. The troubled history of Partition and Jammu and Kashmir's accession to India has resulted in bitter conflicts between the two countries from time to time.

Another territorial dispute between Pakistan and India involves the Siachen Glacier located in the eastern Karakoram range in the Himalaya Mountains. The glacier is the highest battlefield on Earth with both India and Pakistan maintaining permanent military presence in the area. The conflict in the region originates from the 1972 Simla Agreement that did not clearly mention who controlled the glacier. India occupied the entire glacier and its tributary glaciers in 1984 while Pakistan presently controls areas just a few kilometres to the west. A ceasefire agreement is effective in the area since 2003.

A maritime dispute between Pakistan and India involves the Sir Creek estuary at the mouth of the Rann of Kutch in the Arabian Sea. The strip of water divides the Kutch region of Indian state Gujarat with the Sindh province of Pakistan. Both countries seek a technical resolution of the issue.

Afghanistan: Pakistan and Afghanistan spar over the Durand Line that separates the two countries. The Durand Line was drawn in 1893 between British government and then Afghan king Abdur Rahman Khan. It brought some tribal regions under the British administration which was then ruling over the undivided India. The line became the border between Pakistan and Afghanistan in 1947 after India's partition. Afghanistan, however, does not recognise it as the Af-Pak border and stakes claim over some of the areas the Durand Line runs through that are inhabited by Pashtun tribes and fall in Pakistan's territory. Pakistan's attempts to convert the Durand Line into international border have been opposed by Afghanistan.

THREAT PERSPECTIVE

India: Pakistan perceives India as its primary military threat. India's conventional military prowess has outclassed that of Pakistan's in all aspects. The Indo-Pak wars in the past have clearly determined the military imbalance between the two countries. India's obvious dominance in this lopsided equation stems from the very fact that while New Delhi has always aspired to rise as a major regional and global powerhouse in Asia and beyond, with military being an important part of it, Pakistan, on the other hand, has single-mindedly focused on building up its defence arsenal only to counter India. The Islamic country lives in the perpetual "pseudo" fear of an Indian invasion of its territory and hence bears an aggressive posture towards New Delhi.

Though India does not pose an "existential threat" to Pakistan and has, in fact, adopted a "Cold Start" military strategy which is meant to deter its powerful military from eliminating the neighbour, nevertheless it is preparing itself to face any asymmetric and fourth-generation warfare tactics raised by both state and non-state actors of Pakistan. The Indian Army, Navy and Air Force are well-equipped and trained to counter any misdemeanour by Pakistan.

India has pursued a military modernisation programme over the years by adding strength to its armed forces. Acquisition of new-generation weapons and platforms for the Army, Navy and Air Force along with modernisation of existing military hardware are being carried out even as New Delhi is evaluating options to revise its military doctrine to meet asymmetric and fourth-generation warfare tactics. Steady induction of a large number of advanced battle tanks, new frigates, submarines, aircraft carriers, multi-role fighters, stealth warplanes, combat drones and new missile systems are bolstering up the military might of India to counter any "two-front" simultaneous war scenario with Pakistan and China.

Conflict between the two countries has intensified in recent times with both sides exchanging fire at their borders for many months from now. India has also conducted a "surgical strike" on the terror launch pads along the LoC following a terror attack at an Indian Army camp at Uri, Kashmir in September 2016.

Afghanistan: Afghanistan, which provides direct access to Pakistan to enter the resource-rich Central Asian region, has been engaged in military skirmishes with the latter, mostly over the border issue. Pakistan's attempts to establish a "proxy regime" in Afghanistan to gain strategic benefits has caused tension between the two countries. The cross-border insurgency issue has led to acrimony between the two neighbours, with each accusing the other of harbouring militants in its territory. Military conflict between Pakistan and Afghanistan has escalated in past few months with the two sides engaged in heavy gunfire exchange at their borders.

STRATEGIC RELATIONS

China: China has remained Pakistan's most important strategic and military ally for the last sixty years. Beijing has supplied several key defence equipment, jointly developed sophisticated weapons systems & platforms, and conducted joint military drills with Pakistani forces from time to time. The Chinese assistance to Pakistan is mostly aimed at building up the latter's indigenous conventional military power with a special thrust on air combat and naval capabilities. Most of Pakistan's nuclear arsenal and missile technology have also been provided by China. The growing defence tie up between the two countries is widely viewed as a strategy to counter India as a rising regional and global power. The conventional military equipment supply contracts China has clinched with Pakistan includes those for main battle tanks, naval frigates, combat jets and fighter bombers which also include transfer of technology. Pakistan has also signed deals to acquire larger warships and submarines from China.

US: The US military assistance to Pakistan dates back to 1950s when the two countries signed the Mutual Defence Assistance Agreement in 1954 and the US later supplied a large number of F-86 Sabre fighters to the Pak Air Force. Military cooperation between the two sides deepened following 9/11 terror attacks on the US. As a key ally of the US in the war against terror, Pakistan received a number of sophisticated platforms and weapons including fighter jets, surveillance aircraft, attack helicopters, naval craft and anti-missile defence systems. In 2006, US arms sales to Islamabad topped $3.5 billion, nearly matching total purchases by Pakistan from the United States during the fifty years prior to 2001. According to a US Congressional report, the total Foreign Military Sales agreements with Pakistan was worth about $5.2 billion for FY 2002-FY 2012. However, Pakistan's continuous and clandestine support to militant groups hiding inside its territory which are hostile to the US has resulted in a dwindling Pak-US relationship in recent times. Washington has sharply cut both military and economic aid to Pakistan in recent years, showing its resentment over the nuclear-armed Islamabad's tacit support to the outlawed terror outfit Taliban in the neighbouring Afghanistan. The total security-related aid from the US for Pakistan in 2015 was worth $0.34 billion.

Iran: The two South Asian countries have maintained close strategic relationship over the last five decades. Defence cooperation between Iran and Pakistan has involved joint development of defence equipment including armoured vehicles, tanks, helicopters and unmanned systems. The two countries have also conducted joint military exercises in the past. Often, the Iran-Pakistan strategic relationship has remained a clandestine affair with the latter supplying nuclear technology to the former which has raised concern in the international community.

Saudi Arabia: The decades-old close strategic and military ties between Pakistan and Saudi Arabia has largely remained an interdependent affair. While Saudi Arabia has provided extensive financial and other aid to Pakistan, in return, it has received military aid and expertise from the latter. Pakistan-Saudi Arabia relation extends to close military cooperation including provision of Pakistan Army Contingents for Saudi security in crisis situations. Pakistan has also provided training teams for Saudi Armed Forces. The most significant aspect of this relationship in recent times, however, has been Pakistan's secret attempts to provide nuclear technology to Saudi Arabia. Recently, the two sides have pledged to enhance strategic relations between their armies and increase joint exercises. Saudi Arabia is also reportedly planning to buy the JF-17 fighter jets jointly produced by China and Pakistan.

Turkey: The two countries, which were once part of the Central Treaty Organisation during the Cold War era, have remained strong partners in the area of military cooperation. Turkey has supplied defence equipment to Pakistan. Presently, both countries are exploring options of collaboration in defence production with Ankara offering its military hardware, including combat vehicles, naval vessels and Unmanned Aerial Vehicles to Islamabad along with assistance in manufacturing the systems. A Turkish firm has bagged a contract from Pakistan to build a fleet tanker for the Pak Navy. Pakistan has also delivered indigenously built parts for the Turkish UAV "ANKA" to Turkish Aerospace Industry (TAI). TAI, which manufactures the UAVs, hopes to export them to Pakistan. The TAI has also carried out avionics and structural modernisation of 41 F-16 fighter jets operated by Pakistan Air Force. Recently, Pakistan has evinced interest in buying Turkey's first indigenously-built main battle tank called "Altay". According to a 2016 news report, Turkish Aerospace Industries is also negotiating with Pakistan to upgrade further F-16 fighters (nearly 74) of the Pak Air Force. Pakistan, on its part, has won a contract in 2016 to 100 Super Mushshak jet trainers to Turkey.

Malaysia: Bilateral strategic and defence cooperation between Pakistan and Malaysia has gained momentum in recent times with the two countries holding regular dialogues to boost their relations. Military relations between the two sides date back to the 1950s and involve joint exercises, training programmes and defence exchanges. The two countries are also exploring joint production of military equipment and systems.

DEFENCE CAPABILITIES

ARMY: The Pakistan Army is the largest wing of Pakistan's armed forces and is a voluntary force. It is primarily tasked with defending the homeland from external aggression. The powerful Army, however, has attempted to control the entire Pakistan from time to time with the intention to usher in effective administration and make the country corruption-free while instilling a sense of unity among the citizens. All military regimes in Pakistan's history have been headed by an Army official.

The modernisation plans of Pakistan Army include upgradation of main battle tanks (MBTs), acquisition of new artillery, attack and transport helicopters along with other hardware while phasing out obsolete ground platforms and systems.

Personnel: 5,50,000 (+ 5,20,000 reserves)

Equipment

Category	Name	In Service
MBT	Al Khalid	375
	T85-IIAP	250 (to be withdrawn)
	Type 69	250
	T-80 UD	330
	Type 59 (Al Zarar)	500+ (being upgraded)
Light tank	T60	100+
APC	M113	800+
	Al Fahd	140
	Talha	500+
	BTR-70	120
Artillery	155mm Panter	12

Army continued...

Category	Name	In Service
Artillery	M198 155mm	120+
	M114 155mm	65
	Type-59 130mm	200
	Type-60 122mm	600
	M101 105mm	300+
	Type-56 85mm	180
	M7 105mm	12
	M109A2	300
	M109A5 155mm	115
	M110 203mm	12
	Norinco SH1 155mm wheeled	85+
Mortar	120mm	
MLRS	BM2 122mm	30+
RCL	75mm	
	106mm	
ATGW	TOW (I & II)	5000
	M901 TUA	24
	Cobra	
	Baktar Shikan	
SSM	Ghauri-1	10 - 15
SAM	HQ-7B (FM-90)	
	RBS 70	410
	CSA1	6+
	Stringer	100
	Anza (MK I /II)	
AA Gun	ZU23 30mm	
	ZU23 35mm	
	37mm	
	40mm	60
	57mm	
	90mm	15

Army Aviation

Category	Name	In Service
Combat/ Utility helicopter	Z-10	3
	Cobra AH-1S/1F	40
	SA.330 Puma	33
	Bell 206	10
	Mi-8	7
	Mi-17	28+
	Alouette III	24
	UH-1	15
	Bell 412EP	26
Liaison & Light Transport	Cessna O-1E	35
	Mushshak	130+
VIP Transport	Cessna 421	2
	Queen Air	
	Y-12	2
	Turbo commander	

New Procurements/Upgrades

✦ Induction of upgraded T-59 'Al Zarar' (Chinese built) MBTs is going on.

✦ Over 300 Al Khalid MBTs are being inducted.

✦ Pakistan and China have signed a deal to manufacture Al Khalid-I – the latest version of the existing Al Khalid MBT. The new combat vehicle will have improved capability and systems.

✦ Development work is going on for yet another advanced variant, Al Khalid-II MBT, which is expected to be ready by 2019.

✦ US's Bell Helicopter has been awarded a contract to develop weapon system for the Army's Cobra AH-1S/1F attack helicopters.

NAVY: The Pakistan Navy is the maritime wing of Pakistan Armed Forces and also constitutes of marine forces and coast guard. The Navy operates a few submarines, destroyers, frigates, corvettes and missile boats along with fixed-wing aircraft and helicopters. The primary task of the Navy is to protect and defend the country's territorial waters and to guard Pakistan's sea frontiers, sea trade and maritime interests. The Navy also operates with the mission to be a technologically driven and sustainable force; maintain qualitative superiority in conventional forces and asymmetric fighting capability to offset the advantage of numerically superior adversaries; and to operate in synergy with other services, especially the PAF and be ready to participate in Joint Operations whenever required. To strengthen its maritime strike capability, the Navy is presently undergoing modernisation as part of the Armed Forces Development Programme 2019 and inducting new platforms, including frigates and surveillance planes. Acquisition of new attack submarines and combat ships have also been lined up as part of the programme.

Major naval bases: Ormara, Pasni, Gwadar, Karachi, Jiwani

Personnel: 20,000+

Equipment

Category	Name	In Service	On Order
Submarine	Khalid class (French-Agosta 90B)	3	
	Hashmat class (French - Agosta)	2	
	Sx404 class (Midget)	3	
Frigate	Zulfiquar class (F-22P)	4	
	Tariq class (ex-UK Type 21)	5	
	Alamgir (ex-USS McInerney)	1	
Missile boat/ FAC	Larkana class	2	
	Quwwat class	2	
	Jalalat class	4	
	Azmat class	3	1
	MRTP-33	2	

Navy continued...

Category	Name	In Service
Mine Sweeper & Mine Hunter	Munsif class (French Tripartite Type)	3
Auxiliary ship	Fleet oiler	2
	Coastal tanker	2
	Inshore tanker	1
	Small tanker/utility ship	2
Weapon	(ASM) C-602 Zarb (new C-602 variant for coastal defence) C-802A Harpoon SM 39 Exocet	
	Babur LACM	
	(Torpedo) Tp 45 (lightweight) Mk-46 Black Shark	

New Procurements/Upgrades

✦ A deal has been announced in April 2015 to buy eight diesel-electric attack submarines from China at an estimated cost of US$5 billion. The possible underwater vessels could be the Yuan-class, also known as the Type 039A or Type 041, attack submarines. The platforms are expected to be inducted in the Navy by 2028.

✦ A contract has been awarded to Turkish company STM in June 2016 to upgrade the three Khalid-class (French Agosta 90B) submarines. The upgrade will be carried out in collaboration with a Pakistani shipyard.

✦ Plans are on to build four more "improved" F-22P frigates with enhanced sensors and weaponry in partnership with China.

✦ The third Azmat-class missile FAC has been launched in Sept 2016. The warship has been built by Karachi Shipyard and Engineering Works (KSEW) with Chinese assistance. The first and second ships of the series have been commissioned in 2012 and 2014.

✦ A new fleet tanker, developed in partnership with Turkey, has been launched in August 2016. The tanker is designed by Turkish firm M/s STM while its complete construction/assembly activities have been undertaken at the Karachi Shipyard and Engineering Works Ltd.

✦ Six new maritime patrol vessels are under construction with Chinese assistance.

Naval Aviation

Category	Name	In Service
ASW/ Surveillance	P-3 Orion	7
	Atlantic	2
	F-27	6
	ATR-72	3

Naval Aviation continued...

Category	Name	In Service
ASW/ Utility Helicopter	Sea King	6
	Alouette III	15
	Z-9C	12

New Procurements/Upgrades

✦ The third ATR-72 turboprop aircraft has been inducted in August 2016. Two such aircraft were previously commissioned in 2013. The aircraft are equipped with collision avoidance system and semi-glass cockpit instrumentation.

✦ 12 Harbin Z-9C anti-surface warfare helicopters have been inducted to equip the new Chinese-built F-22P frigates.

✦ Unspecified number of AgustaWestland AW139 multi-role helicopters have been ordered from Italy's Leonardo-Finmeccanica.

✦ The first squadron of indigenously-built Uqab-II unmanned aerial vehicles (UAVs) has been commissioned.

✦ A Scaneagle UAV has joined operational service in August 2016.

AIR FORCE: The Pakistan Air Force (PAF) operates with the mission to provide aerial defence to Pakistan in synergy with the Army and Navy. The PAF operates over 500 fighter jets along with some reconnaissance, transport and trainer aircraft. Under the Armed Forces Development Programme 2019 initiated by the Government, the PAF is set to undergo modernisation by acquiring new platforms and weapons, including airborne early warning aircraft, mid-air refuellers, UAVs and long-range missiles.

Major air bases: Karachi, Peshawar, Quetta, Kamra, Mianwali, Shorkot, Jacobabad and Chaklala (Rawalpindi) and Sargodha

Personnel: 42,000

Equipment

Category	Name	In Service
Fighter	F-16 A/ B	54
	F-16 C/ D	18
	Mirage IIIEP	16
	F-7P	270+
	F-7PG	
	Mirage IIIRP	3
	Mirage 5PA/ DPA	56
	Mirage 5EF	34
	Mirage IIIEA	33
	Mirage IIIDA	5
	Mirage 5DF	5
	JF-17	80+
	A5III	120+
Transport/ Surveillance / Tanker	C130B/ E	16
	CN-235-200	3 + 1 (VIP)

Air Force continued...

Category	Name	In Service
Transport/ Surveillance/ Tanker	F-27	1
	Falcon 20	1
	Boeing 707	2
	A310	1
	Cessna 172	4
	Y-12	2
	Il-78	4
AWACS /AEW	Falcon 20F	2
	Mi-171	3
	Saab 2000 Erieye	3 (+2 on order)
	ZDK-03	4
Trainer	MFI17B	78
	FT5	25
	T37B	20
	K-8	40+ (total order for 75)
Helicopter	Alouette-III	12
Weapon	(AAM) AIM-9B Sidewinder, PL-5E, PL-9, PL-12 (SD-10), Hatf-8 (Raad)	
	(SAM) Crotale AD systems, Aspide 2000, HQ-2B	
	(AGM) AM 39 Exocet, AGM-65 Maverick, C-802	

New Procurements/ Upgrades

✦ Discussions are underway with Russia for the possible acquisition of Su-35 Flanker-E multi-role fighters.

✦ Turkey has announced to provide 34 of the T-37 jet trainer and light attack aircraft to the Pak Air Force free of cost.

✦ Turkish Aerospace Industries (TAI) has completed the delivery of all 41 upgraded F-16 fighter jets of the PAF with the last four jets handed over in Sept. 2014. TAI is now negotiating to upgrade 74 more F-16 fighters of the PAF.

✦ Delivery of 18 F-16 C/ D Block 52 fighters have been completed after the last two jets of the series were handed over to the PAF in Feb. 2012. Pakistan intended to buy 16 more such fighters from the US. The US Government in Feb. 2016 announced its decision to sell 8 F-16 fighters to the PAF, but the deal could not materialise following Pakistan's failure to pay for the fighters.

✦ 13 F-16 jets acquired from Jordan have been inducted into service in 2014. Pakistan had inked a deal to acquire 13 of the Jordanian surplus F-16A/B fighters. Following its failure to buy the F-16s from US, Pakistan is now considering to acquire six of the used warplanes from Jordan yet again.

✦ A decision is yet to be taken on the PAF's plans to acquire two squadrons (36) of the advanced J-10B multi-role, all-weather fighter aircraft from China at a cost of $1.4 billion, negotiations for which were initiated in 2009.

✦ Series production of JF-17 'Thunder' light fighter aircraft in cooperation with China is underway. PAF has raised four JF-17 squadrons by April 2016 and plans to induct a total of 250 JF-17s in the coming years. Production of the advanced Block II variant of the fighter started in December 2013, with a total of 50 jets planned to be manufactured. The Chinese planes are replacing the Mirage and F-7 jets of the PAF.

✦ Production of yet another advanced variant, JF-17B 'Thunder', has been launched with the PAF intending to start induction of the new fighter by April 2017. The production of the first prototype of JF-17B was initiated by China and Pakistan at an induction ceremony held at Chengdu Aerospace Corporation in China's Sichuan province in April 2016.

✦ Five SAAB 2000 AEW&C aircraft with EriEye radar are on order with deliveries going on.

✦ All four Karakorum Eagle 03 (KE-03/ ZDK-03) airborne early warning and control (AEW&C) aircraft have been received from China, forming the KE-03 AEW&C squadron of the PAF.

✦ Rockwell Collins has been awarded a contract to upgrade the C-130 Hercules transport aircraft.

✦ A deal has been finalised to acquire four Mi-35 'Hind E' attack helicopters from Russia. Negotiations are going on to buy more such gunships from Russia.

✦ The US Defence Department has in April 2015 approved the sale of AH-1Z Viper attack helicopters. Pakistan will receive nine of the combat helicopters by 2018-end.

✦ Purchasing of Turkey-built T-129 attack helicopters is under consideration.

✦ In response to Pakistan's request, the US has transferred 24 Bell 412EP utility choppers to it under Foreign Military Sales programme.

✦ Acquisition of unmanned aerial vehicles from Europe is being planned.

✦ The US Government has cleared a $15.2 million FMS deal to provide ScanEagle UAVs to Pakistan.

STRATEGIC FORCE COMMAND

Pakistan formed the National Command Authority (NCA) on 2 February 2000 to oversee the country's strategic nuclear arsenal and related organisations. The NCA operates with a unified central command structure and comprises of the Strategic Plans Division (SPD) and Strategic Forces Commands of the three military branches – Army, Navy and Air Force. The SPD has an elaborate Security

Division which includes a counter-intelligence network to safeguard the activities of strategic organisations. It also has in place a dedicated multi-layered security apparatus to safeguard strategic assets. The Strategic Forces Command comprises of Army Strategic Forces Command, Naval Strategic Forces Command and Air Force Strategic Command. Pakistan raised the strength of its SPD Security Force to 25,000 in 2013 to provide foolproof security to the country's nuclear weapons and strategic assets. In May 2012, the Navy also inaugurated a new Naval Strategic Force Command (NSFC). The new command will perform a pivotal role in the development and employment of the Naval Strategic Force and would be the custodian of the nation's 2nd strike capability.

Potential weapons

Hatf is a series of ballistic and cruise missiles with different ranges. The missile has been developed as an offensive weapon against enemy attack. The indigenously-built missiles can carry both conventional and nuclear warheads.

Variants

- Hatf-1 is a two-stage solid propelled short range ballistic missile (SRBM) having a range between 70 km and 100 km.
- Hatf-2 (Abdali-1) is a short range surface-to-surface ballistic missile. The weapon has a range of 180 km and can carry nuclear and conventional warheads.
- Hatf-3 (Ghaznavi) is a road mobile, solid propellant, single warhead ballistic missile with a range of 290 km.
- Hatf-4 (Shaheen-1) is a single stage SRBM having a range between 750 km to 900 km. A new improved variant of the missile, known as Shaheen-1A, was tested in April 2012. The range of the missile has not been revealed.
- 5Hatf-6 (Shaheen-2) is a medium range ballistic missile (MRBM) having a range between 2,000 km and 2,500 km.
- Hatf-5 and Hatf-5B (Ghauri-1, Ghauri-2) are MRBMs having operational range of 1300 km and 2,300 km respectively.
- Hatf-7 (Babur) is a nuclear-capable cruise missile having a range of 700 km. The multi-tube cruise missile can carry both conventional as well as nuclear warhead.
- Hatf-8 (Raad) is an air-launched cruise missile having a range of 350 km. The weapon can carry both conventional and nuclear warhead and has been developed exclusively for launch from combat aircraft.
- Hatf-9 (Nasr) is a short-range surface-to-surface ballistic missile capable of carrying nuclear warhead. The missile has a range of 60 km.

Major Armaments Producers

Heavy Industries Taxila (HIT) is a state-controlled defence firm that produces main battle tanks, combat vehicles, artillery and related equipment for the Pakistan Army.

Pakistan Ordnance Factories (POF) is the leading producer of small arms and ammunitions.

Karachi Shipyard and Engineering Works (KSEW) is the leading producer of naval vessels, including warships and submarines.

Pakistan Aeronautical Complex (PAC) is the state-owned aerospace firm engaged in assembly, overhaul and production of aircraft and related components for the PAF.

Air Weapons Complex produces air-borne weapons systems.

National Development Complex has been established as an aerospace and defence agency to undertake research and development of various missile programmes.

Surveillance and Target Unmanned Aircraft (SATUMA) is a private-run firm that produces unmanned aerial vehicles and related equipment.

DEFENCE DEALS

Joint Venture/ Collaboration Programmes

China

Al-Khalid MBT: Pakistan developed the Al-Khalid main battle tank with Chinese assistance in the 1990s. The MBTs are currently operational with the Pakistan Army.

Hongdu JL-8 jet trainer: The Pakistan Aeronautical Complex (PAC) joined hands with China's Hongdu Aviation Industry Corporation (HAIC) to manufacture the Hongdu JL-8 (export variant Karakorum-8 or K-8) single engine two-seat advanced jet trainer aircraft with light attack capabilities. Pakistan's share in the joint venture is 25%. PAF uses the K-8P variant of the jet trainer.

JF-17 Thunder (FC-1) combat aircraft: The multi-role JF-17 Thunder combat aircraft was jointly developed by China and Pakistan in the 1990s. Pakistan announced serial production of the new fighter jet in February 2011. Pakistan Air Force plans to acquire as many as 250 such fighters in future and also intends to sell the jet in the international market. The two countries have announced to jointly export an upgraded variant of the multi-role fighters to foreign customers.

F-22 naval frigate: Pakistan, with China's assistance, has built four F-22 guided missile frigates. The Pakistan Navy has inducted the warships along with the Chinese Harbin Z-9EC helicopters.

Saudi Arabia

Pakistan has expressed interest to form joint ventures with Saudi Arabia to produce defence systems, including tanks, armoured vehicles and missiles. Saudi Arabia has also evinced interest in the JF-17 Thunder with Pakistan reportedly offering the fighter jet along with technology transfer and co-production.

Arms Export

Pakistan, while attempting to attain self-sufficiency in defence production, mainly with Chinese assistance, has been exporting some military parts and weapons to other countries. Defence exports worth $300 million have been made mostly to countries in South Asia, Middle East and Africa in the last few years. In 2015, the country targeted to earn over $100 million in military exports to some 30 countries. Presently, Islamabad is making persistent efforts to export the JF-17 Thunder fighters built in partnership with China.

Arms Import

Pakistan has mostly relied on US and China to acquire military hardware. Other countries, including France, Germany, Switzerland, Italy and Brazil have also supplied defence equipment to Pakistan. According to a Pentagon report, Pakistan is the top-most buyer of a whole range of Chinese defence equipment that includes fighter aircraft, frigates, helicopters, jet trainers, early warning and control aircraft, tanks, air-to-air missiles and anti-ship cruise missiles. In recent years, Pakistan has inclined more towards China in acquiring new military hardware, leaving behind the US.

DEFENCE BUDGET

The Pakistan government has traditionally made defence allocations with the objective of maintaining conventional parity with India. To keep pace with India's rising defence expenditure, Pakistan too has increased its defence allocation over the years. However, its annual defence expenditure is much less when compared to that of India's. Pakistan's association with the US-led war on terror has also pushed up its annual defence allocation as the country's defence budget is supplemented by military aid provided by the US. The country's defence budget for 2015-16 was PKR 781 billion. The allocation was estimated to be around 2.54% of the country's GDP. For 2016-17 period, the Government has announced to spend PKR 860 billion on defence – an increase of 11% over the previous year's allocation.

Official Budgetary Allocation

Year	Allocation	% Increase over previous year
2016-17	Pkr 860.2 billion ($8.2 billon)	11
2015-2016	Pkr 781 billion ($7.8 billion)	11
2014-2015	Pkr 700. 2 billion ($7 billion)	11.64
2013-2014	Pkr 627 billion ($6.36 billion)	10
2012-2013	Pkr 570 billion (revised) ($5.82 bn)	10
2011-2012	Pkr 495 billion ($5.75bn)	12

CONCLUSION

Pakistan has maintained a large military force since its inception to counter the conventional military power of India. The small country has built-up its armed forces to maintain a minimum credible deterrence against India which it regards as its most potent adversary. Toeing the doctrine of "offensive defence" strategy against militarily superior India, Pakistan has focused on acquiring latest weapons and equipment and increased its defence expenditure over the years. Though in terms of size, strength, strategic, military and technological perspective, India overpowers Pakistan, in terms of nuclear prowess though, the latter has perhaps surpassed India by possessing more nuke warheads, according to defence analysts. As per a report by the Council on Foreign Relations (CFR), a leading US think-tank, Pakistan is believed to be pursuing the fastest growing nuclear weapons programme in the world and it could have enough fissile material to produce more than 200 nuclear devices by 2020. The country has deployed or is developing some 11 delivery systems for its nuclear warheads, including aircraft, ballistic missiles and cruise missiles, the report states. Moreover, the absence of a "no first use" nuclear policy by Pakistan (unlike India) has helped the country maintain a credible military deterrence against India which, for its part, fully acknowledges the cost and consequences of a full-scale nuclear fallout in the entire Asian continent.

CONTACT DETAILS

Minister of Defence

Tel: 9271107, 9271108

Fax: 9221596

Website: http://www.mod.gov.pk/

PERU

(Capital: Lima)

INTRODUCTION

Area: 1,285,216 sq km

Population: 30,741,062 (July 2016 est.)

Coastline: 2,414 km

Maritime claims:

Territorial sea: 200 nm

Continental shelf: 200 nm

KEY POLITICAL PERSONS

PRESIDENT: Pedro Pablo Kuczynski

PRIME MINISTER: Fernando Zavala Lombardi

KEY DEFENCE PERSONS

MINISTER OF DEFENCE: Jorge Nieto Montesinos

CHIEF OF JOINT COMMAND OF THE ARMED FORCES: Adm. Jorge Moscoso Ricardo Flores

COMMANDER OF ARMY: Gen. Humberto Ramos Hume

COMMANDER OF NAVY: Adm. Edmundo Luis Enrique DEVILLE del Campo

COMMANDER OF AIR FORCE: Gen Del Aire Dante Arevalo Abate

ECONOMY

Peru is one of the world's fastest-growing economies owing to the economic boom experienced during the 2000s. The country's economic performance has been tied to exports, which provide hard currency to finance imports and external debt payments. The Peruvian Gross Domestic Product has grown to its fastest rate reaching 9.5% (Annual Calculated Increase) in 2010. This paved the road for the Foreign Investment rate to get the highest level ever with more than 50 billion dollars invested in Peru at an accumulated level. The economy of Peru is the 42nd largest in the world. The country is an emerging, market-oriented economy characterized by a high level of foreign trade. The US-Peru Trade Promotion Agreement entered into force on 1 February 2009, opening the way to greater trade and investment between the two economies. Trade agreements with South Korea, Japan, and Mexico also were signed in 2011. Although Peru has continued to attract foreign investment, political disputes and protests impede development of some projects related to natural resource extraction.

GDP: $389.1 billion (2015 est.), $376.9 billion (2014 est.)

Real Growth Rate: 3.3% (2015 est.), 2.4% (2014 est.)

Industries: Mining and refining of minerals; steel, metal fabrication; petroleum extraction and refining, natural gas; fishing and fish processing, textiles, clothing, food processing

Oil - proved reserves: 700 million bbl (1 January 2016 est.)

Natural gas - proved reserves: 414.1 billion cu m (1 January 2016 est.)

Total Exports: $34.16 billion (2015 est.), $39.53 billion (2014 est.)

Export Commodities: Copper, gold, zinc, crude petroleum and petroleum products, coffee, potatoes, asparagus, textiles, fishmeal

Major Markets: China 22.1%, US 15.2%, Switzerland 8.1%, Canada 7% (2015)

Total Imports: $36.99 billion (2015 est.), $40.48 billion (2014 est.)

Import Commodities: Petroleum and petroleum products, plastics, machinery, vehicles, iron and steel, wheat, paper

Major Suppliers: China 22.7%, US 20.7%, Brazil 5.1%, Mexico 4.5% (2015)

INTERNATIONAL DISPUTES

✦ Peru's November 2005 unilateral legislation to shift the axis of their joint treaty-defined maritime boundaries along the parallels of latitude to equidistance lines was rejected by Chile and Ecuador.

✦ Peru rejects Bolivia's claim to restore maritime access through a sovereign corridor through Chile along the Peruvian border.

DEFENCE

ARMY

Personnel: 75,000

Equipment

Category	Name	In Service
MBT	T-55	50
	Al-Khalid	50+
	AMX13	35
APC/IFV	M8 armoured cars	60
	M113A1	300
	UR416	225
	M325	300
	Fiat/OTO 6616	20
	BRDM2	30
	BTR-60	12
	Casspir	20
	V-100 Commando	40
Artillery	105 mm M101	130
	105 mm M2	70
	105 mm M56	50
	155 mm M114	36
	130 mm M46	30
	122 mm towed D30	38
	155 mm M109A2	12
	155 mm MkF3	12
	122 mm 2S1	
Mortar	81 mm, 120 mm, 160 mm	500+
	Fiat/OTO 6614 mortar carriers	
MRL	122 mm. BM21	26
Anti-Tank GW	Spike	244
	Kornet	200
AA Gun	4X23 mm. Shilka	25
	Zu-23-2	30
	Bofors	20
SAM	SA7	400
	SPYDER	
	SA-14	120
	SA-16	200+
	TOR M1	

Army Aviation

Army aviation continued...

Category	Name	In Service
Aircraft	An-32	3
Aircraft	An-28	3
	L410	1
	Beechcraft Queen Air	1
	Il-103	6
Helicopter	SA 315B Lama	6
	Alouette II/III	8
	Mi8/17	50
	Mi-171	15
	Mi26	3
	Agusta A-109	12
Trainer	Cessna 150	3
	Cessna 172	2
	Cessna 182	1
	Enstrom F-28	3

New Procurements / Upgrades

✦ Tonbo Imaging, an India based start-up, has been awarded a USD 100-million contract to manufacture and export night-vision sights for guns to be sold by a US company named Unified Weapons Systems (UWS) to the Peruvian army.

✦ The MBT-2000, built and marketed internationally by China, was recently tried by the Peruvian Army for possible acquisition. It is based on the Chinese Type 90-II, which combines technologies from several Soviet/western tanks.

✦ The Peruvian Defence Ministry and Russian state-controlled arms exporter Rosoboronexport have signed a USD 250 million worth contract for the sale of eight Russian-built helicopters. Russia has completed delivery of all Mi-171Sh military transport helicopters to Peru.

✦ Russia is offering a full-scale model of its T-90S main battle tank (MBT) to the Peruvian Army.

✦ Peru has acquired 27 China-made 90B self-propelled rocket launchers for army artillery units. Each truck carries 40 units of 122-mm rocket launchers with a range of 20-40 km.

Live firing exercise with 90B self-propelled rocket launchers by 1st artillery brigade. *Image Credit: Peruvian Army*

Women in the Peruvian Army. The Third Army Division, has female staff in all garrisons of Arequipa, as in the case of the Sixth Armoured Brigade. *Image Credit: Peruvian Army*

NAVY

Personnel: 25,715

Equipment

Category	Name	In Service
Submarine	Type 209/1200	6
Cruiser	De Zeven Provincien-class	1
Frigate	Carvajal class	3
	Lupo class	4
Corvette	PR-72P-class	6
Patrol Vessel/ Gun Boat	Loreto class	2
	Clavero class	2
	Maranon class	2
Amphibious Force	Terrebonne Parish class	2
Missile	Exocet, Otomat, Aspide, FN-6, Iqla	

Naval Aviation

Category	Name	In Service
Aircraft	An-32	2
	Fokker 60	1
	Beechcraft B200T Super King Air	5
	T-34 C.1 "Turbo Mentor"	5
	Cessna 206	1
	F27-200	3
	De Havilland Canada DHC-6 Twin Otter	1
Helicopter	AB 212 ASW	5
	Bell 206 Jet Ranger	3
	ASH3D Sea King	3
	Mi8T	3
	Enstrom F28F	5

New Procurements / Upgrades

✦ Kaman Aerosystems (Kaman) announced that its Air Vehicles and MRO (AVMRO) division has successfully secured a contract for USD 39.8 million with General Dynamics Mission Systems-Canada to commence work on the implementation phase of the previously announced Peruvian Navy's SH-2G Super Seasprite aircraft programme. The total expected value to Kaman for the combined programme, including this contract and previously issued contracts, now totals USD 50.5 million, total sales of USD 10.6 million had already been recognised as revenue through April 1, 2016, and the balance is expected to be recognized through the remainder of the programme.

✦ The Peruvian Navy will soon be receiving two Newport class tank landing ships from the US Navy's inactive reserve.

✦ The Peruvian Navy will be receiving two multifunctional surveillance radars KRONOS NV from SELEX Sistemi Integrati, a

Finmeccanica company, which will be installed on board the Aguirre Class frigates.

✦ ThyssenKrupp Marine Systems (TKMS) finalised an agreement with Peruvian naval shipyard SIMA to upgrade four submarines.

A Peruvian Agusta AB 212 helicopter approaches the flight deck of the guided missile cruiser USS Thomas S. Gates (CG 51) during a helicopter cross deck training exercise in the Caribbean Sea. *Image Credit: US Navy*

AIR FORCE

Personnel: 15,000

Equipment

Category	Name	In Service
Fighters/ attack/ Reconnaissance	MiG-29	15+
	A-37B	20
	Su-25	15+
	Mirage-2000	11
	C-26 Metroliner III	4
Transport/ Tanker	L-100-20	5
	C-27J	1 (3 on order)
	DHC-6 Twin Otter	10+
	C-130A	1
	C-130D	4
	DC-862CF	1
	C-26A Metro III	4
	737-200	3
	An-32B	6
Helicopter	Bell 212	11
	Bell 412	2
	Bell 206	8
	MBB Bo 105	10
	SA.316B Alouette III	10
	Mi-8 / Mi 17 Hip	35
	AB 214ST	9
	Mi-25 Hind	24
Trainer	Il-103	6
	Z242L	16
	EMB 312 Tucano	20
	MB-339	13
	Cessna 150	2
	Bell 206	2

New Procurements/ Upgrades

✦ Viking Air Limited of Victoria, British Columbia, and the Peruvian Air Force have entered into a multi-year contract for the comprehensive support of twelve Twin Otter Series 400 aircraft operated by the FAP in the Amazonian region of north-east Peru.

✦ The Peruvian Ministry of Defence has selected Alenia Aermacchi, a Finmeccanica Company, to supply two C-27J Spartan tactical airlifters. The contract has a value of around 100 million euro inclusive of a substantial logistic support package for the two aircraft.

✦ Eight Peruvian Air Force MiG-29S/UB ('Fulcrum-C/B') fighter aircraft are currently being upgraded by Russian Aircraft Corporation MiG under a USD 106 million worth contract. Renamed MiG-29SMP (P for Peruvian), the fighters will receive improvements from the MiG-29SMT upgrade package, including new avionics, communications and electronic countermeasures systems, multifunction colour displays in the cockpit and HOTAS (hands-on throttle and stick) controls.

✦ South Korea has signed a $200 million deal to export 20 KT-1 trainer jets to Peru over the next four years.

✦ Korea Aerospace Industries (KAI) has awarded a contract to CMC Electronics to deliver its integrated glass cockpit for installation in the Peruvian Air Force's KT-1P turboprop trainer aircraft.

✦ Four Mil Mi-25 Hind D attack helicopters, modernised in Russia, have joined the Peruvian Air Force in Oct 2012 to boost its combat capability in the fight against terrorism and drug trafficking.

✦ The Peruvian Air Force has received the first of four C-27J Spartan military transport aircraft from Italy's Alenia Aermacchi. The new aircraft is part of a contract signed between Alenia Aermacchi and the Peruvian Air Force (FAP) in December 2013.

Defence Expenditure

Total defence spending: $1,992 million (est.)

Estimated defence spending in terms of GDP: 1.58% (2015)

Major defence suppliers: Russia, United States, France, Italy, Belarus, Spain, China, Ukraine, South Korea, Serbia, Israel, Argentina, Mongolia, Malaysia, India & Cuba.

Peru's defence budget rose to 20 percent, the main reason behind this is internal security. Peru's budget proposal places top priority on national security and defence. The budget provides financing for a new remuneration scale for the Peruvian armed forces and police as well as improved logistics. Giving a boost to its ties with the fastest growing Latin American country, India has signed four agreements with Peru including one on defence cooperation and another on establishment of a Joint Commission. Peru also came out in support of India's claim for permanent membership in an expanded UN Security Council. The two countries also asserted that terrorism cannot be justified in any form for any cause or for any reason.

Defence Production and R&D

The Peruvian Army's weapon and ammunition factory (FAME) was re-organised and established in 2009 as an independent entity in the defence industry sector of the Ministry of Defence. Currently, Peru is concentrating on maintenance and upgrade capabilities. Shipbuilding is also developing there.

Defence Procurement

The Peruvian Armed Forces' existing arsenal is impressive but is extraordinarily heterogeneous in origin and much of it is ageing, in storage, or out of service. Plans to retire most unnecessary equipment will improve overall readiness. Major acquisitions might take place in areas such as armoured weapons and air defence systems and also improvements in command and control assets.

CONTACT DETAILS

Ministry of National Defence
Av La Peruanidad S / N,
Quinones building (Campus Martius) - Jesús María
Lima, Peru
Tel: (+511) 209 8530
www.mindef.gob.pe

Joint Command of the Armed Forces of Peru
Corpancho 289 Nicholas Street,
Santa Beatriz. Lima – Peru
Phone: (511) 3151030
Email: informaciones@ccffaa.mil.pe

FAME
Ex. Hacienda Nievería Km. 3.5 Carretera Cajamarquilla
LURIGANCHO
CHOSICA
PERU
E-mail: ventas@famesac.com
Tel: (+511) 717 8791
Fax: (+511) 356 4119
http://www.famesac.com

Peruvian Army:
www.ejercito.mil.pe
Peruvian Navy:
www.marina.mil.pe
Peruvian Air Force:
www.fap.mil.pe

STRATEGIC INFORMATION

PHILIPPINES
(Capital: Manila)

INTRODUCTION

Area: 300,000 sq km

Population: 102,624,209 (July 2016 est.)

Coastline: 36,289 km

Maritime claims: Territorial Water up to 100 nm from coastline

Exclusive Economic Zone: 200 nm

Continental shelf: to depth of exploitation

GEOPOLITICAL IMPORTANCE

The Philippines is a very important and strategic gateway to Asia from the Pacific. The country is located in South-Eastern Asia, an archipelago between the Philippine Sea and the resource-rich South China Sea, East of Vietnam. Forming part of the Malay Archipelago, it consists of 7,107 islands and islets of which only 2,773 have names and about 500 are larger than a square kilometre. The three major island groups are Luzon, Visayas and Mindanao. More than half of the world's shipping tonnage and 80 percent of crude oil shipments headed to Japan and South Korea passes through the Philippines which lie at a vital maritime crossroads. The nation's large population size and economic potential have led it to be classified as a middle power. Its proximity to other Asian countries makes it an ideal hub for industry and commerce. It is also a founding member of the United Nations, World Trade Organization, Association of Southeast Asian Nations, and East Asia Summit.

POLITICAL OVERVIEW

The Philippines is a republic with a presidential form of government wherein power is equally divided among its three branches: Executive, Legislative, and Judicial. One basic corollary in a presidential system of government is the principle of separation of powers wherein legislation belongs to Congress, execution to the Executive, and settlement of legal controversies to the Judiciary.

The Legislative branch is authorised to make laws, alter, and repeal them through the power vested in the Philippine Congress. This institution is divided into the Senate and the House of Representatives.

The Senate or Senado or upper chamber consists of 24 seats wherein one-half are elected every three years and members elected at large by popular vote to serve six-year terms. They can be re-elected but they are no longer eligible to run for a third consecutive term.

The House of Representatives or Kapulungan ng mga Kinatawan or lower chamber consists of 212 members representing districts plus 24 sectoral party-list members and members elected by popular vote to serve three-year terms. The Constitution prohibits the House of Representatives from having more than 250 members.

The Executive branch is composed of the President and the Vice President who is elected by direct popular vote and serves a term of six years. The Constitution grants the President authority to appoint his Cabinet. These departments form a large portion of the country's bureaucracy. The President functions as both the head of state, the head of government and the Commander-in-Chief of the Armed Forces of the Philippines. The vice-president is second in line to succession should the president resign, been impeached or die in office.

The Judicial branch holds the power to settle controversies involving rights that are legally demandable and enforceable. This branch determines whether or not there has been a grave abuse of discretion amounting to lack or excess of jurisdiction on the part and instrumentality of the government. It is made up of a Supreme Court and lower courts.

The Constitution expressly grants the Supreme Court the power of Judicial Review as the power to declare a treaty, international or executive agreement, law, presidential decree, proclamation, order, instruction, ordinance or regulation unconstitutional.

KEY POLITICAL PERSONS

PRESIDENT & COMMANDER-IN-CHIEF OF THE ARMED FORCES:
Rodrigo Duterte

VICE PRESIDENT:
Leni Robredo

ECONOMY

The Philippine economy has been growing steadily in recent years and it is one of the emerging markets. The International Monetary Fund (IMF) has described it as the 39th largest economy in the world. The Philippines has been named as one of the Tiger Cub Economies together with Indonesia, Malaysia and Thailand. According to HSBC, the Philippine economy will become the 16th largest economy in the world, 5th largest economy in Asia and the largest economy in the Southeast Asian region by 2050. Economic growth has accelerated, averaging 6.0% per year from 2011 to 2015, and competitiveness rankings have improved. The Philippines has not sustained steady growth in foreign direct investment, which continues to lag regional peers. Long term challenges include reforming governance and the judicial system, building infrastructure, improving regulatory predictability, and the ease of doing business, attracting higher levels of local and foreign investments. The Philippines is a participant in the Asia-Pacific Economic Cooperation, Asian Development Bank, the Colombo Plan, Group of 24, G-20, G-77, the World Bank, Next Eleven and the World Trade Organization.

GDP: $741 billion (2015 est.); $693.4 billion (2014 est.)

Real Growth Rate 5.8% (2015 est.); 6.1% (2014 est.)

Industries: Electronics assembly, garments, footwear, pharmaceuticals, chemicals, wood products, food processing, petroleum refining and fishing

Total Exports: $43.28 billion (2015 est.); $49.82 billion (2014 est.)

Export Commodities: Semiconductors and electronic products, transport equipment, garments, copper products, petroleum products, coconut oil, fruits

Major Markets: Japan 21.1%, US 15%, China 10.9%, Hong Kong 10.6%, Singapore 6.2%, Germany 4.5%, South Korea 4.3% (2015)

Total Imports: $64.97 billion (2015 est.); $67.16 billion (2014 est.)

Import Commodities Electronic products, mineral fuels, machinery and transport equipment, iron and steel, textile fabrics, grains, chemicals, plastic

Major Suppliers: China 16.2%, US 10.8%, Japan 9.6%, Singapore 7%, South Korea 6.5%, Thailand 6.4%, Malaysia 4.8%, Indonesia 4.4% (2015)

DEFENCE & SECURITY

The Armed Forces of Philippines (AFP) is a volunteer force. The three major services – Army, Navy, Air Force – are unified under a Chief of Staff who is assisted by a Vice Chief of Staff. Each of the three major branches is headed by an officer. The Philippines does not currently have a modern military posture capable of independent defence of its territorial matters and claims in the Spratly group. Most of the armed forces' equipment is obsolescent or suffers from poor readiness. As a result of the Joint Defence Assessment (JDA) of the Philippines and United States in October 1999, the Philippine Defence Reform (PDR) Programme was formulated. The PDR priority programmes include - Multi-Year Defence Planning System (MYDPS), Improve Intelligence, Operations, and Training Capacities, Improve Logistics Capacity, Professional Development Programme, Improve Personnel Management System, Multi-year Capabilities Upgrade Programme (CUP), Optimization of Defence Budget and Improvement of Management Controls, Centrally Managed Defence Acquisition System Manned by a Professional Workforce, Development of Strategic Communication Capability, Information Management Development Programme.

The Government of the Republic of the Philippines (GRP) is engaged in a concerted effort to modernise its military into a service that is capable of projecting a posture of credible external deterrence. The Armed Forces of the Philippines (AFP) Modernisation Act was passed in 2012 with an aim to equip the AFP with the necessary capabilities to protect the territorial integrity of the state, offset evolving foreign defence challenges, and ensure the attainment of the country's strategic maritime interest, particularly in the South China Sea. The most contentious region covers the Spratlys Island. The transformation will focus on four key areas - the full implementation of the Internal Peace and Security Plan (IPSP); organisational reforms with the goal of ensuring full transparency in military spending; strengthening the AFP's territorial defence capabilities; and the development of a naval force that can protect Manila's maritime interests in the West Philippine Sea.

The Philippine Department of National Defence (DND) framework for bringing reforms is based on an environment of increasing economic prowess and a gradually decreasing threat level over time to address AFP capability gaps, implement capability for seamless interoperability by developing proficiency in the conduct of joint operations, eliminating crisis handling by individual major services, improve effectiveness of internal security operations, enhance capability to counter terrorism and other transnational threats, provide sustainment or long-term viability of acquired capabilities, improve cost-effectiveness of operations, improve accountability and transparency in the DND, increase professionalism in the AFP through reforms in areas such as promotions, assignments, and training.

Manila aims to spend $1.8 billion on new defence technology by 2017. The Philippines has also signed Enhanced Defence Cooperation Agreement with the United States in 2014 which will see the US forces return to the country's bases after a 20-year absence. Philippines' primary security threats are land-based – separatist, communist insurgent and terrorist groups. However, recent national policies have shifted the strategic direction of the AFP towards external, territorial defence as opposed to previous internal issues.

KEY DEFENCE PERSONS

SECRETARY OF NATIONAL DEFENCE: Maj. Gen. Delfin Negrillo Lorenzana

CHIEF OF STAFF OF THE ARMED FORCES: Gen. Eduardo Ano

COMMANDER OF THE ARMY: Lt. Gen. Glorioso Miranda

COMMANDER OF THE NAVY: Vice Adm. Ronald Joseph S. Mercado

COMMANDER OF THE AIR FORCE: Lt. Gen. Edgar Restor Fallorina

COMMANDER OF THE MARINES: Maj. Gen. Andre M. Costales Jr.

Internal Conflict

The Philippines has been confronted with multiple and simultaneous insurgencies. The country has a long history of conflict in the southern regions, mostly in the remote islands of central Mindanao, especially on Basilan and Jolo. The myriad of rebel groups including Muslim separatists, communists, clan militias and criminal groups are all active in the area. Three groups comprise the Muslim separatists - the Moro National Liberation Front (MNLF), the Moro Islamic Liberation Front (MILF) and the Abu Sayyaf. The MILF and the Abu Sayyaf are breakaway factions of the MNLF. The communist insurgency, on the other hand, is propagated by the Communist Party of the Philippines' (CPP) military wing, the New People's Army (NPA).

In March 2014, the government and the Moro Islamic Liberation Front (MILF) signed a peace deal. The deal grants mostly Muslim areas of Mindanao greater political autonomy in exchange for an end to armed rebellion. Other rebel groups have vowed to keep fighting for full independence of the region. The conflict involving the Moros has killed tens of thousands of people and, since 2000, displaced 3.5 million people, according to the Internal Displacement Monitoring Centre (IDMC).

The new Philippines Government has taken a firm stand against drug trafficking in the country which have been termed as "Oplan Double Barrel" and "Oplan Tokhang" – officially, the Philippines' drug war. So far, the Government has ordered stern action against drug traffickers and drug addicts with a mentioning "shoot to kill" policy. In 2016, the country has executed more than 2,600 drug traffickers to fight the crime.

This brutal war to fight the menace of illegal drugs has pitted the Philippines Government against the drug offenders and has been widely condemned. It has also resulted in an unexpected and sharp decay in Philippines' relationship with one of its crucial strategic allies – the US. In the past few months, there has been more cancellation of bilateral military exercises between the two partner nations. Manila has also become extremely critical of Washington for not taking any firm stand in support of the Southeast Asian country against the growing aggression of China in the region.

In the wake of the drug war, the Philippines President Rodrigo Duterte has remarked that he might follow Russia's example and withdraw from the International Criminal Court (ICC), where his critics say he could be charged over the thousands of killings in his war on drugs. In his statement before the annual summit of Asia-Pacific leaders in November 2016, the President said that the United Nations has been unsuccessful in stopping wars (across the world). He further said if China and Russia decide to create a new world order, he would be the first to join them and leave the UN which "is dominated by the U.S."

External Conflict

The Philippines is one of the contenders in the South China Sea dispute. Both the Philippines and China lay claim to the Scarborough Shoal (known as Huangyan Island in China) – a little more than 100 miles (160km) from the Philippines and 500 miles from China. The dispute also involves the Paracels and the Spratlys Islands also claimed by China, Malaysia, Taiwan, and Vietnam. The Philippines invokes its geographical proximity to the Spratly Islands as the main basis of its claim. Malaysia and Brunei are the other claimants of the territory in the South China Sea. Brunei does not claim any of the disputed islands, but Malaysia claims a small number of islands in the Spratlys. The interests of different nations include acquiring fishing areas around the two archipelagos; the potential exploitation of suspected crude oil and natural gas under the waters of various parts of the South China Sea; and the strategic control of important shipping lanes. The sea is also a major shipping route and home to fishing grounds that supply the livelihoods of people across the region.

In July 2016, Philippines won an arbitration case against China over the South China Sea in the International Tribunal for the Law of the Sea (ITLOS). The tribunal ruled that Philippines has exclusive sovereign rights over the West Philippine Sea (in South China Sea) and that China's "nine-dash line" is invalid. Philippines had filed its case in January 2013 following a tense standoff between Chinese and Philippines ships at Scarborough shoal in April 2012. After Manila's initiation of the arbitration case, China had conducted several massive reclamation projects to turn submerged reefs into artificial islands capable of hosting military structures and equipment.

THREAT PERSPECTIVE

The Philippines has been facing external threats to its maritime interests in the South China Sea. None of the other claimant countries pose a threat to Philippine interests the way China does. China reportedly occupies seven features in the Philippine claim, while Malaysia and Vietnam occupy three and 19, respectively. The issue of far greater concern for the Philippines is the fact that five of the nine dotted lines in China's map are clearly on the Philippines' continental shelf and well within 200 nautical miles of its main islands, recognised by China as the sovereign territory of the Republic of the Philippines. China has consistently refused to recognise the jurisdiction of any international body over territorial delimitation and sovereignty-related issues while accusing the Philippines of unnecessarily provoking a crisis by internationalizing what it sees as an essentially bilateral territorial dispute, which should be resolved primarily through diplomatic channels.

STRATEGIC RELATIONS

United States: The Philippines and the US have a long history of bilateral relations and common strategic purposes that date back to joint anti-Japanese operations during World War II.

However, there has been a major shift in Philippines' policy towards the United States in the past few months. Despite the 28 April 2014 agreement on Enhanced Defence Cooperation & 1955 RP-US alliance pact, Philippines Navy and the US military joint patrols and naval exercises in the disputed South China Sea have been put on hold by the new Philippines administration – the first concrete break in Philippines-US defence cooperation after months of increasingly strident comments by President Rodrigo Duterte. Further upping the ante, the country's defence secretary Delfin Lorenzana has also stated that the 107 US troops involved in operating surveillance drones against Muslim militants would be asked to leave the southern part of the country once the Philippines acquires those intelligence-gathering capabilities in near future. President Duterte has made his intention clear to halt the 28 joint military exercises that are carried out with the US forces every year. The US-Philippine amphibious beach landing exercise which took place in October 2016 would be the last in his six-year presidency, Duterte has stated, even as he backs away from what he views as too much dependence on the US.

Australia: The Philippines and Australia have been steadfast partners in trade, development, good governance and security. According to Australia's 2013 Defence White Paper, the Philippines remain an important long-standing defence partner for Australia. It is the Philippines' primary partner, along with the United States, in the implementation of the Coast Watch System programme which seeks to improve the Philippines' maritime domain awareness and border security. The Status of Visiting Forces Agreement (SOFVA) in force since 2012 also provides a comprehensive legal framework for the presence of Australian forces in the Philippines and the latter's forces in the former. Australia is also the largest bilateral grant donor to the Philippines' education sector.

Japan: The Philippines-Japan Strategic Partnership formalised in 2011 originally began as an enhanced economic relationship. A common strategic interest in protecting the sea lines of communication of the two maritime countries was also identified as a foundation of the enhanced relationship. Tokyo has decided to extend a stand-by credit loan for the Philippines' disaster preparedness and offer continuous assistance to the Mindanao peace process where the Philippine government has been negotiating for a comprehensive peace pact with the Moro Islamic Liberation Front. The two nations have also agreed to work together to help the United States maximise it's rebalance to the Asia Pacific. In addition, Manila has expressed willingness to provide Japanese maritime vessels access to some of its naval bases, alongside the US.

South Korea: The Philippines was one of the first countries that extended diplomatic recognition to the Republic of Korea. This was cemented with the Philippines Government's deployment of the Philippine Expeditionary Force to Korea (PEFTOK) to help South Korea against the invasion of the communist North during the Korean War in the 1950s. Today, South Korea is one of the Philippines' most important bilateral partners. Trade between the Philippines and South Korea remains active. The two nations have signed a Memorandum of Understanding on Logistics and Defence Industry Cooperation. Seoul has sealed a deal to export a dozen Korean-made FA-50 fighter jets to Manila. South Korea also hopes to export naval escort vessels to the Southeast Asian nation.

DEFENCE CAPABILITIES

ARMY: The Philippine Army has several regular units dedicated to counter-insurgency and conventional army operations. A reorganization programme is underway to reduce the number of infantry units and boost maritime security and territorial defence. Under the Armed Forces Modernization Act initiated in 2013, the Army aims to procure armoured vehicles, body armour, rocket launchers and other equipment. Recently, the Army has also made considerable changes in its HORIZON-I programme initiated in 2013. The fund, originally allocated for the procurement of a shore-based missile system (SBMS), has been diverted for the priority procurement of infantry equipment such as sniper rifles, night vision equipment, tactical communication equipment, trucks and transport vehicles among others.

Personnel: 67,000 (+30,000 reserves)

Equipment

Category	Name	In Service
Tank	Scorpion (light)	11
APC/AIFV	FNSS	6+
	M113	94
	V150 Commando	70
	AIFV	33
	GKN Simba	135
Artillery	Soltam M71	2
	M114A1 155 Howitzer	4
	M101A1 105 mm	229
Mortar	M81 81 mm	358
RCL	M67 90 mm	189

New Procurements/Upgrades

✦ A total of 114 M113A2 APCs have been handed over to the Army by the US Government through the Excess Defence Articles grant.

✦ The Army has taken delivery of the first batch of 28 upgraded M113A2 APCs from Israel. Fourteen of the M-113s will be configured as fire support vehicles, four as infantry fighting vehicles, six as armoured personnel carriers, and another four as armoured recovery units.

✦ The army has taken delivery of 23 HMMWV armoured ambulances.

✦ An international tender has been issued for the purchase of 12 155mm SP howitzers.

✦ The government plans to buy anti-aircraft guided missiles from Israel. It is also considering the acquisition of Raytheon MIM-23 Hawk SAM.

✦ A new programme is expected to be launched in 2017 for the procurement of a shore-based missile system. Few Israeli firms have demonstrated their systems to the Philippines officials recently.

✦ Php 6.5 billion allocated for the HORIZON-1 programme has been diverted for the procurement of sniper rifles, tactical communication, ballistic protection equipment, tactical transport etc.

NAVY : Under a programme designed to bolster its maritime capability, the Philippines has been acquiring new naval vessels and helicopters to protect its territorial integrity in the face of rising tensions in the South China Sea. The core of the Philippine naval fleet consists of seven Chamsuri class patrol vessels. Two decommissioned Hamilton class Coast Guard frigates remain the largest and most modern vessels in the Navy's fleet. In early 2014, the Philippine government requested a third Hamilton class frigate from the US. The South Korean government has donated a landing craft and a Pohang class corvette.

Major naval bases: Sangley Point (Cavite), Naval Operating Base San Vicente (Aparri, Cagayan), Naval Operating Base Mactan (Cebu), Naval Operating base Ulugan (Palavan), Naval Operating base Batu-Batu (Tawi-Tawi), Naval training Base San Miguel (Zambales)

Personnel: 21,000 (includes 8,500 Marines)

Equipment

Category	Name	In Service
Frigate	Rajah Humabon class (ex-US Cannon class)	1
	Gregorio Del Pilar (ex-USCG Hamilton cutter)	2
Corvette	Rizal class	2
	Cebu class	7
Patrol craft	General Mariano Alvarez (ex-US Cyclone class)	1
	Emilio Jacinto class (ex-RN Peacock class)	3
	PCF 370 class (Halter marine type)	22
	PC 394 class (ex-UCG Cutter type)	2
	PG118 (ex-ROK navy)	1
	Tomas Batilo	6
Amphibious Force	MPAC	3
	LSV	2
	LCU	2
	LST	5
Auxiliary ship		10

Naval Aviation

Category	Name	In Service
Aircraft	BN-2 Islander	6
	Cessna 172	2
	C-23 Sherpa	2
	Fokker F27 Friendship	1
Helicopter	Bo 105 (being withdrawn)	5
	AW109	4

New Procurements/Upgrades

✦ The Navy has inducted two recently decommissioned Australian Navy's Balikpapan-class heavy landing craft (LCH) vessels. The vessels were re-commissioned into the Philippines Navy as BRP Ivatan (AT298) and BRP Batak (AT299).

✦ South Korea's Hyundai Heavy Industries (HHI) has been selected to supply two new frigates to the Philippine Navy.

✦ Three multi-purpose attack craft are on order.

✦ Indonesia's PT Pal shipbuilding has been selected for the construction of two strategic sealift vehicles with delivery scheduled between 2016 and 2017.

✦ The Navy has expressed its intent to acquire submarines for deterrence and undersea warfare and plans to purchase at least one platform by 2020.

✦ Four AW109 Power maritime helicopters have been inducted in service in 2015.

✦ Two Lynx AW-159 anti-submarine and anti-surface warfare helicopters are on order from Italy's Leonardo Finmeccanica with delivery scheduled for 2018.

✦ Japan has leased five of its surplus TC-90 aircraft to the Philippine Navy for conducting air patrolling missions over the disputed Spratly Islands region.

Navy's Del Pilar class frigate

Marines

Equipment

Category	Name	In Service
IFV	V150 Commando	19
APC	V300 with 90 mm gun	12
	LVT	12
	GKN Aqua tracks	2
Gun	40 mm	4
	Twin 12.7 truck mount	4
	20mm AA gun	
	M101 105 mm	23
	L10A1 Otomelara 105mm	5
RCL	M40 A1 106 mm	

New Procurements/Upgrades

✦ Eight Samsung Techwin KAAV7 Amphibious Assault Vehicles are to be ordered to replace the LVTs.

✦ Six M777 155mm lightweights are to be ordered

✦ SMAW assault weapons are on order.

AIR FORCE: The Philippines Air Force is undergoing modernisation under the Capability Upgrade Programme (CUP). The PAF has shortlisted its procurement priorities which include purchase of more advance assets such as AEWACs, ECM aircraft, medium/heavy lift helicopters and advance multi-role fighters.

Major air bases: Basa, Fernando, Mactan Ebuen, Edwin Andrews, Antonio Bautista, Kalayaan Island, Sangley Point, Vilamor

Personnel: 14,389

Equipment

Category	Name	In Service
Combat	S-211	15
	FA-50 Fighting Eagle	2
	OV-10 Bronco	24
	SF-206TP	7
	MG-520	24

Air Force continued...

Category	Name	In Service
Transport	Airbus C295	3
	C-130H/ L-100	3+1
	F-27/F-28	1 each
	Nomad	9
VIP Transport	F-28,F-27,S-70	1 each
	Bell-214	1
	Bell 412	6
Helicopter	PZL-Swidnik W-3 Sokol	7
	Bell 412S	6
	UH-1H Iroquois	25
	UH-1D	7
SAR	S-76	7
	Bell 205	6
Trainer	SF-260F	18
	SF-260M	21
	T4-1D	30
	LC-210	1

New Procurements/Upgrades

✦ First two FA-50 light attack aircraft ordered from South Korea have been delivered in December 2015. The contract for 12 of the Korea Aerospace Industries-built FA-50 fighters was signed in March 2014 with the delivery scheduled to be completed by 2017. The Government plans to acquire more such fighters in future.

The FA-50 fighter of the Philippines Air Force *Air Force photo*

✦ The Air force has taken delivery of the third and final Airbus C295 medium transport aircraft ordered from Airbus Defence and Space.

✦ Two surplus Lockheed Martin C-130T Hercules transport aircraft have been received from the US in 2016.

✦ Plan is on to procure close-air support aircraft to replace the OV-10 Bronco planes presently operational.

✦ The Government has terminated a contract with a US supplier over its provision of 21 refurbished UH-1 Huey utility helicopters to the Air Force.

✦ Eight AW-109E attack helicopters have been inducted in service in 2015 with a plan to acquire more such platforms in future.

✦ The Philippines has ordered 12 T-50 (Golden Eagle) supersonic advanced trainer from KAI.

✦ Bell Helicopters is to supply eight 412 EP helicopters to Philippine DND. Five of the helicopters will be configured for combat utility operations, including relief efforts, whereas the remaining aircraft are scheduled to be modified for VIP transportation.

DEFENCE PRODUCTION AND R&D

The Philippines has indigenous arm manufacturing capability engaged mainly in small arms and ammunition manufacturing. The Self-Reliant Defence Posture (SRDP) programme, initiated in 1974, took the development of a domestic defence industry as its objective. Defence officials contracted SRDP projects with the government arsenal and local manufacturers, encouraging the use of indigenous raw materials and production capacity. Projects included domestic production of small arms, radios, and assorted ammunition. One of the most significant SRDP operations was the manufacturing of the M 16A1 rifle under license from Colt Industries, an American company.

Major Armaments Producers

Domestic suppliers: Government Arsenal, Armscor, FERFRANS, Floro International Corporation, United Defence Manufacturing Corporation, Steelcraft Industrial & Development Corporation.

Foreign suppliers: United States, South Korea, Israel, Italy, Germany, United Kingdom, France

DEFENCE PROCUREMENT

The Armed Forces of Philippines (AFP), Department of National Defence, Commission on Audit and Budget secretary along with other civilian agency are responsible for any defence procurement in the country. The DND has a system of internal auditing with a multi-year planning with yearly planning system. The new procurement is defined by multi-year planning which is implemented on the basis of the strategic importance and need of procurement and budget allocation.

As part of the military modernisation programme the Philippines is planning several projects which include acquisition of fighter jets, naval helicopters, patrol aircraft, frigates, patrol craft, multi-purpose attack vessels and other weapons both for internal security and a subsequent transition to external defence.

Under the multi-year Capabilities Upgrade Programme first phase, the Navy's three Jacinto class patrol vessels underwent upgrades of their command and control systems and installed new M242 Bushmaster 25 mm cannons. Two Patrol Killer Medium class boats (PKM or Chamsuri class) were also acquired from South Korea in 2006. Among the most high profile acquisitions, the AFP purchased three decommissioned Hamilton-class cutters from the US Coast Guard; two are already in the possession of the Philippine Navy. The AFP is also taking delivery of twelve the new FA-50 fighter jets from Korea Aerospace Industries. Other arms purchases since 2013 include 28 M113 APCs from Israel, five Bell-205 helicopters from Germany, eight helicopters from Poland, and 25 Humvees from the United States.

The Department of National Defence (DND) reportedly plans to sign a contract for three IAI Elta EL/M-2288 air defence radars, with one to be positioned in Palawan province to cover the South China Sea. The Philippines will seek four further systems in the next five years.

DEFENCE BUDGET

Proposed defence spending: \$2.9 billion (2017)

Defence spending in terms of GDP: 1.35% (2016)

The Government has proposed an annual defence spending worth PHP134.2 billion (USD2.9 billion) for the year 2017, a hike of nearly 15% over previous year's allocation. The proposed budget however does not include funds for military procurements and only takes into account personnel expenses, maintenance and operations, and capital outlays.

CONCLUSION

The Philippines has taken initial steps to build a more credible military defence against China and other neighbouring countries, not only for the bounty that the Spratlys offers but for its own territorial integrity. Building and maintaining the required support infrastructure such as enhanced air and naval bases, radar facilities and missile systems to support more modern air and naval assets, however, may take years to accomplish. Experts opine that it would take a major shift in government policy to make the required paradigm change. Further, Manila's recent major shift in political stance towards the US has also fuelled speculations with regard to its future policies and priorities. Other than its old-time ally US, the Philippines is now exploring new trading partners, including China and Russia.

CONTACT DETAILS

Department of National Defence

DND Building,
Segundo Ave. Camp General Emilio Aguinaldo Quezon City,
Philippines 1110
Email: info@dnd.gov.ph
Website: http://www.dnd.gov.ph/
Trunkline: (+632) 982 - 5600

POLAND
(Capital: Warsaw)

INTRODUCTION

Area: 312,685 sq km

Population: 38,523,261 (July 2016 est.)

Coastline: 440 km

Maritime claims: Territorial sea: 12 nm

Exclusive Economic Zone: defined by international treaties

KEY POLITICAL PERSONS

PRESIDENT & COMMANDER-IN-CHIEF: Andrzej Duda

PRIME MINISTER: Beata Szydlo

KEY DEFENCE PERSONS

MINISTER OF NATIONAL DEFENCE: Antoni Macierewicz

CHIEF OF GENERAL STAFF OF THE ARMED FORCES: Lt. Gen. Mieczyslaw Gocul

INSPECTOR OF LAND FORCES: Gen. Brig. Wojciech Grabowski

INSPECTOR OF NAVY: Adm. Miroslaw Mordel

INSPECTOR OF AIR FORCE: Gen. Brig. Cezary Wisniewski

ECONOMY

Poland has a strong domestic market, low private debt, flexible currency, and not dependent on a single export sector. The economy of Poland is a high income economy and is the sixth largest in the EU and one of the fastest growing in Central Europe. Poland has made great economic progress over the last decade, and is ranked 21st worldwide in terms of the GDP. The largest component of its economy is the service sector. Since 2004, EU membership and access to EU structural funds have provided a major boost to the economy. Unemployment is 2% more than the EU average. Inflation reached a low of about 2.6% in 2010 due to the global economic slowdown, but climbed to 4.3% in 2011. While the Polish economy has performed well over the past five years, growth slowed in 2013 and picked back up in 2014.

GDP: $1.005 trillion (2015 est.), $970.2 billion (2014 est.)

Real Growth Rate: 3.6% (2015 est.), 3.3% (2014 est.)

Industries: Machine building, iron and steel, coal mining, chemicals, shipbuilding, food processing, glass, beverages, textiles

Oil - proved reserves: 100 million bbl (1 January 2010 est.)

Natural gas - proved reserves: 81.66 billion cu m (1 January 2016 est.)

Total Exports: $190.8 billion (2015 est.), $210.7 billion (2014 est.)

Export Commodities: Machinery and transport equipment 37.8%, intermediate manufactured goods 23.7%, miscellaneous manufactured goods 17.1%, food and live animals 7.6%

Major Markets: Germany 27.1%, UK 6.8%, Czech Republic 6.6%, France 5.5%, Italy 4.8%, Netherlands 4.4% (2015)

Total Imports: $188.4 billion (2015 est.), $215 billion (2014 est.)

Import Commodities: Machinery and transport equipment 38%, intermediate manufactured goods 21%, chemicals 15%, minerals, fuels, lubricants, and related materials 9%

Major Suppliers: Germany 27.6%, China 7.5%, Russia 7.2%, Netherlands 5.9%, Italy 5.2%, France 4.1% (2015)

INTERNATIONAL DISPUTES

As a member state that forms part of the EU's external border, Poland has implemented the strict Schengen border rules to restrict illegal immigration and trade along its eastern borders with Belarus and Ukraine.

DEFENCE

ARMY

Personnel: 55,000

Equipment

Category	Name	In Service
MBT	Leopard 2A4/2A5	230
	PT-91	232
	T-72M	586
APC/IFV	BWP-1	1,297
	Rosomak	487
	Cougar	40
	BRMD-2	376
Artillery	122 mm SP 2S1 Gvozdika	150
	152 mm SP Dana	111
Mortar	120 mm M120	138
	98 mm M98	99
Anti-Tank GW	Spike	96
	9K11	129
	9K111 Fagot	77
	9K113	18
	9K115	7
MRL	122 mm. BM21	205
	RM70	30
	WR-40 Langusta	1
AA Gun	ZSU-23-4 SPAAG	36
	ZSU-23-2 Towed	404
SAM	SA-6 KUB	80
	SA-8 OSA	64
	Poprad	
	SA-7 STRZALA	246
	GROM	336
UAV	Orbiter, FlyEye	

Army Aviation

Category	Name	In Service
Helicopter	W-3A	37
	Mi-2	46
	Mi-8	17
	Mi-17	13
	Mi-24	31

New Procurements/ Upgrades

✦ Poland is planning the delivery of 70 helicopters for the Polish Army. It will also receive an additional five Mi-17 helicopters to strengthen its military contingent in Afghanistan. The contract is worth $106 million.

✦ The Polish Army is to receive a total of 307 Rosomak armoured vehicles from 2014-2019 which will increase the number of Rosomak (Wolverine) combat vehicles in its inventory to 877, providing a major element in the Polish 2022 defence modernisation plan.

Nearly 200 soldiers from the Polish and the United States, on 12-16 January 2016 in Skwierzyna, held a joint training in air and missile defence. *Image Credit: General Command of the Armed Forces*

NAVY

Personnel: 12,000

Equipment

Category	Name	In Service
Submarine	Kilo class	1
	Kobben class	4
Frigate	Oliver Hazard Perry class	2
Mine Warfare Force	Krogulec Class	3
	Gardno Class	10+
	Mamry Class	4
Amphibious Force	Lubin Class	5
	Deba Class	3
Corvette	Kaszub class	1
FAC	Orkan class	3
Support Vessel	Czernicki Class	1
	Heweliusz Class	2
	Nawigator Class	2
	Baltyk Class	1
	Piast Class	2
	Zbyszko Class	2
	Wodnik Class	1
	Iskra Class	1

Naval Aviation

Personnel: 1,300

Equipment

Category	Name	In Service
Aircraft	M-28	5+
Helicopter	Mi-17	2
	Mi-2	1
	SH-2G	4
	W-3RM	5
	W-3T	2
	Mi-14PL	10
	Mi-14PS	2
	Mi-2	1

Kobben class (ORP Kondor) submarine.
Image Credit: Polish Ministry of National Defence

New Procurements/ Upgrades

✦ The Polish Navy to undergo a full modernisation by 2018.

✦ The Polish Navy has acquired RBS15 Mk3 and Norwegian Naval Strike Missiles for vessels and coastal defence units.

✦ The Polish Navy plans to reinforce the Navy's helicopter fleet by around seven ASW and SAR units.

✦ A new submarine is planned to join the fleet by 2017.

✦ The original order of seven Gawron-class corvettes has been reduced to two as it is planned to construct three Kormoran 2 class minehunters instead.

✦ Serving vessels of the Polish Navy will undergo upgrade and overhaul to maintain operational status.

✦ The Polish Defence Ministry has placed an order to build three new mine countermeasure ships and complete one long-stalled patrol vessel.

✦ The Polish Ministry of Defence has signed a contract with the US government to modernise the Navy's Oliver Hazard Perry-class guided-missile frigate, ORP Gen. Kazimierz Pulaski. The deal is worth about $34 million.

AIR FORCE

Personnel: 27,500

Equipment

Category	Name	In Service
Fighters/ attack/ Reconna-issance	MiG-29A	26
	F-16 C/D Block 52+	48
	Su-22M4/Su-22UM3K	48
Transport /Tanker	C-295M	11

Air Force continued...

Category	Name	In Service
Transport/ Tanker	PZL M-28 Bryza	22
	C-130E	5
	Tu-154M	2
	Yak-40	4
	M-28	2
	Bell 412HP	1
Trainer	TS-11 Iskra	54
	PZL SW-4	11
	PZL-130	34
	SA-3 Goa	90
	SA-3	60
	SA-4	18
	SA-45	12

New Procurements/ Upgrades

✦ Italian aircraft maker Leonardo-Finmeccanica has rolled out the first of eight Aermacchi M-346 advanced jet trainers (AJTs) meant for Poland. Poland had ordered the eight Aermacchi M-346 AJTs in 2014. Deliveries of all the aircraft will be completed by November 2017.

✦ Lockheed Martin Corporation has been awarded a $250 million contract to supply Poland's air force with JASSM long-range air-to-surface missiles to arm Poland's 48-strong fleet of F-16 fighter jets. The missiles are scheduled to be operational in 2017.

✦ Poland aims to acquire 16 medium-lift transport helicopters for the land forces, three search-and-rescue (SAR) helicopters for the Air Force and seven in the anti-submarine and SAR variant for the Navy. The purchase is worth more than 2 billion zloty ($610 million). Deliveries are scheduled from 2015 to 2017. The contract will also include the procurement of five flight simulators.

✦ The Polish Air Force has signed a contract with Airbus Military for the acquisition of five C295 transport aircraft.

✦ Poland plans to spend $3 billion on multirole military helicopters for its armed forces.

✦ Lockheed Martin has submitted a bid for a Polish Ministry of Defence tender to establish new integrated aircrew training systems (IATS) in the country.

✦ First upgraded MiG-29A has been delivered to Polish Air Force. The upgradation contract is worth $39.5 million.

✦ BAE Systems has won a contract from Wojskowe Zaklady Lotnicze Nr 2 S.A. to support the Polish Air Force (PLAF) C-130 fleet.

✦ Poland has a $41 billion programme to modernise its armed forces by 2022. Poland is set to push for the involvement of local arms producers in new contracts for rockets, ships, helicopters, tanks and small arms.

✦ Lockheed Martin recently received a $7.9 million Foreign Military Sales contract from the U.S. Air Force to support additional integration of the Joint Air-to-Surface Standoff Missile (JASSM) onto the Polish Air Force F-16C/D aircraft. This second contract for Poland includes support for a broad range of certification and flight activity. Integrated logistics support and program and technical oversight will conclude in 2018.

Defence Expenditure

Total defence spending: $9.3 billion (2016); $10.5 billion (2014)

Estimated defence spending in terms of GDP: 2% (2016); 1.8% (2015)

Domestic suppliers: Polish Defence Holding, WZM S.A., PZL-Świdnik, HSW S.A. & OBRUM

Foreign suppliers: European Union, Israel, Norway, Switzerland, South Korea & United States

Annual Exports: European Union, Nigeria, Libya, Iraq, Malaysia & Vietnam

Total Defence Expenditure in 2016 (Estimated Exchange Rate: 1 USD = 3,66 PLN):

Defence Expenditure	2015 (Budgetary Act)	2016 (Budgetary Act)	2016/ 2015
Part 29 National Defence	32.518,1 (excluding 5.363,3 MPLN for F-16 programme)	35.43,2 (including National Defence: 27.912,3 MPLN)	109,0%
Other Parts	296,9	445,5	109,4%
Total	**32.815,0**	**35.898,7**	**109,4%**

Source: BUDGETARY DEPT. Poland Ministry of Defence

	2013	2014	2015	2016
Capital Invest-ment Expenses	6.157,8 MPLN	8.190,3 MPLN	12.764,2 MPLN	10.247,3 MPLN

Source: BUDGETARY DEPT. Poland Ministry of Defence

Defence Production and R&D

The Polish defence industry comprises 38 manufacturing companies, 12 renovation and manufacturing military enterprises, 10 research-and-development establishments, and three companies specialising in trading arms. Revenue from this sector of the economy amounted to $1.27 billion in 1999, while the sector's corporate assets are estimated at over $1.71 billion. Up to now, out of the 38 companies recognised as strategically important for the country, only two have been successfully privatised.

Defence Procurement

Poland's Ministry of National Defence (MND) created a new 'Armaments Inspectorate' '(Inspektorat Uzbrojenia - IU) in January 2011. The Armament Inspectorate performs analytical-conceptual tasks in the process of the acquisition of the armament for the Polish Armed Forces needs.

CONTACT DETAILS

Ministry of National Defence
Ulitsa Klonowa 1
PL-00909 Warszawa
PolandTel: (+48 22) 684 02 12
(+48 602) 28 14 34
Fax: (+48 22) 684 02 24
E-mail: dpimon@wp.mil.pl
newsroom.dpimon@wp.mil.pl
www.wp.mil.pl
www.mon.gov.pl

Armament Inspectorate
Aleje Niepodległości 218
Warsaw
Tel: (+48 22) 68 461 77
E-mail: rzecznik.iu@mon.gov.pl

PORTUGAL
(Capital: Lisbon)

INTRODUCTION

Area: 92,090 sq km

Population: 10,833,816 (July 2016 est.)

Coastline: 1,793 km

Maritime claims: Territorial sea: 12 nm

Contiguous Zone: 24 nm

Exclusive Economic Zone: 200 nm

Continental Shelf: 200 m depth or to the depth of exploitation

KEY POLITICAL PERSONS

PRESIDENT & SUPREME COMMANDER OF THE ARMED FORCES: Marcelo Rebelo de Sousa

PRIME MINISTER: Antonio Luis Santos da Costa

KEY DEFENCE PERSONS

DEFENCE MINISTER: Jose Alberto Azeredo Lopes

CHIEF OF THE GENERAL STAFF: Gen. Artur Neves Pina Monteiro

CHIEF OF STAFF OF THE ARMY: Gen. Frederico Jose Rovisco Duarte

CHIEF OF STAFF OF THE NAVY: Vice Admiral Luis Manuel Fragoso Fourneaux Apple

CHIEF OF STAFF OF THE AIR FORCE: Gen. Manuel Teixeira Rolo

ECONOMY

Portugal's economy is based primarily on services and industrial production. The country has been a part of the Eurozone since its inception. Some state enterprises have been privatized. The economy grew by more than the EU average for much of the 1990s, but the rate of growth slowed in 2001-08. The economy contracted in 2009, and fell again from 2011 to 2013, as the government implemented spending cuts and tax increases. However, the economy reportedly has been steady, expanding continuously since the third quarter of 2014, with a yearly GDP growth of 1.5% registered in the second quarter of 2015. Austerity measures have contributed to record unemployment and a wave of emigration not seen since the 1960s. According to reports, overall unemployment has decreased from 2013 to the first quarter of 2014 but remains well above pre-crisis levels.

GDP: $289.8 billion (2015 est.), $285.6 billion (2014 est.)

Real Growth Rate: 1.5% (2015 est.), 0.9% (2014 est.), -1.1% (2013 est.)

Industries: textiles, clothing, footwear, wood and cork, paper and pulp, chemicals, lubricants, automobiles and auto parts, base metals, minerals, porcelain and ceramics, glassware, technology, telecommunications; dairy products, wine, other foodstuffs; ship construction and refurbishment; tourism, plastics, financial services, optics

Total Exports: $54.33 billion (2015 est.), $62.72 billion (2014 est.)

Export Commodities: agricultural products, foodstuffs, wine, oil products, chemical products, plastics and rubber, hides, leather, wood and cork, wood pulp and paper, textile materials, clothing, footwear, machinery and tools, base metals

Major Markets: Spain 25%, France 12.1%, Germany 11.8%, UK 6.7%, US 5.2%, Angola 4.2%, Netherlands 4% (2015)

Total Import: $64.49 billion (2015 est.), $75.29 billion (2014 est.)

Import Commodities: agricultural products, chemical products, vehicles and other transport material, optical and precision instruments, computer accessories and parts, semiconductors and related devices, oil products, base metals, food products, textile materials

Major Suppliers: Spain 32.9%, Germany 12.9%, France 7.4%, Italy 5.4%, Netherlands 5.1% (2015)

INTERNATIONAL DISPUTES

Portugal does not recognise Spanish sovereignty over the territory of Olivenza based on a difference of interpretation of the 1815 Congress of Vienna and the 1801 Treaty of Badajoz.

DEFENCE

The Portuguese Army is one of the oldest armies in the world, established in the 12th century. The Government, in April, 2013, approved structural reform of the National Defence and the Armed Forces, referred to as "Defence 2020".

ARMY

Personnel: 18,190

Equipment

Category	Name	In service
MBT	Leopard 2A6	37
	M-60A3/A4	90
Armoured car/ APC/AIFV	V-150 Chaimite	15
	M-113A1	190
	M-113A2	35
	M-577A	50
	V-200	51
	Panhard M11	32
	HMMWV	29
	Pandur II 30mm	22
Artillery	M119 light gun	18
	M101 105mm	8
	M114A1 155mm	25
	Oto Melara 105mm pack	5
	M-109A2	6
	M-109A5	17
Mortar	81mm	162
	107mm M30	52
	120 mm Tampella	49
SAM	Stinger Manpads	20
	Chaparral	31
ATGW	Tow	95
	Milan	85
ATK	Carl Gustav 84mm	152
	90mm	34
	M-40 106mm	51

New Procurements/Upgrades

✦ Portugal is planning to buy 47 new 4x4 tactical communications vehicles to expand its SIC-T network.

✦ Portugal is planning to buy tactical mini-unmanned aerial vehicles (UAVs) for a maximum EUR6 million (USD6.6 million), with payments between 2016 and 2021 through a tender to be run by the NATO Support and Procurement Agency.

NAVY

Personnel: 9,847

Equipment

Category	Name	In Service
Submarine	Tridente class (Type 209PN)	2
Frigate	Bartolomeu Dias class	2
	Vasco Da Gama class	3
	Baptista De Abdrade class (patrol)	3
	Joao Coutinho class (patrol)	4
Patrol craft	Viana Do Castelo class OPV	2
	Cacine class large patrol craft	4
	Centauro class (coastal)	4
	Argos class (coastal)	5
	Rio Minho class (inshore)	1
	Albatroz class (coastal)	2
Amphibious force	Bacamarte LCT	1
Miscellaneous	Replenishment ship	1
	Hydrographic ships	4
	Training ships	4

Naval Aviation: Five Super LYNX 95 ASW helicopters (shipborne).

New Procurements/Upgrades

✦ Portugal looks to procure new replenishment ship within wider naval modernisation programme for the period spanning 2015-26.

✦ Portugal plans to commission multi-purpose Logistic Ship (LPD) in 2017.

✦ The navy plans to acquire an LPD of about 10,000 tons sometime beyond 2020.

✦ There are also plans to procure a new replenishment ship with an overall length of 140 metres, and around 11,000 tons in 2021.

AIR FORCE

Personnel: 6,868

Equipment

Category	Name	In Service
Fighter	F-16 AM 15MLU	25
	F-16 BM 15MLU	5
	F-16A	12
	F-16B	3
Trainer	Alpha JET	7
	TB30 Epsilon	16

Air Force continued...

Category	Name	In service
Trainer	Chipmunk	
	Super Blanik	
Maritime patrol	P-3C Orion	5
Transport/ Support	C-130H	3
	C-130H-30	3
	Falcon-S50	3
	C-295M	12
	C-212-100	2
	C-212-300	2
Helicopter	EH.101	12
	SA.330S Puma	5
	SA.316 Alouette III	14

New Procurements/Upgrades

✦ The Portuguese aerospace company Indústria Aeronáutica de Portugal (OGMA) has begun to upgrade three Lockheed Martin F-16A Block 15 Fighting Falcon fighters for the Air Force (FAP). The three aircraft will be upgraded to the F-16AM Mid-Life Upgrade (MLU) standard by 2017.

✦ The P-3 fleet is undergoing a modernisation programme.

✦ A Declaration of Intent has been signed with Brazil for possible Portuguese participation in the development programme of the KC-390 transport aircraft.

Defence Expenditure

Total defence expenditure: €2.14 billion (2015), €2.138 billion (2014)

Defence spending in terms of GDP: 1.1% (2016 est.); 1.85% (2015)

Defence Production and R&D

The military is currently undertaking several major acquisitions supported by national industries. These include two new submarines with air-independent propulsion that feature communications, information systems and integration. The Ministry of Defence is acquiring armoured vehicles for the army and navy with communications systems provided by home companies. The country produces licensed equipment such as assault rifles and has its own arms producing industry. A declaration of intent between the Brazilian and Portuguese ministries of defence was signed in September 2010. Portugal's membership of the consortiums involved with the production of the NH90 helicopter and KC-390 transport aircraft is expected to increase the country's defence exports.

Defence Procurement

Portugal imports the majority of its defence equipment as its domestic defence industry is relatively under-developed. Ships, aircraft and armoured vehicles account for the largest expenditure. The country's main arms suppliers are the US and Netherlands. Submarines, armoured vehicles and frigates are expected to be imported from the European countries, with transport aircraft coming from the US. All major defence procurement projects are managed by the MoD procurement agency Direcçao-Geral de Armamento e Infra-Estruturas de Defesa (DGAIED). In 2001, the Parliament gave its approval for the nation's Law of Military Programmes (LMP) to launch a plan for re-equipping the Portuguese Armed Forces over the next 35 years. Investment will amount to about € 5.5 billion ($ 5 billion) released in six-year cycles until 2018.

CONTACT DETAILS

Ministry of Defence
Madeira, 1 - 1400-204 Lisbon.
Phone: 21 303 85 20
Fax: 21 301 95 55
E-mail: dscrp@defesa.pt

Public Relations of CEMGFA
Madeira, 1 - 1449-004 Lisbon.
Phone: 21 304 37 89
Fax: 21 304 32 62
E-mail: emgfa_rp@emgfa.pt

STRATEGIC INFORMATION

QATAR
(Capital: Doha)

INTRODUCTION

Area: 11,586 sq km

Population: 2,258,283 (July 2016 est.)

Coastline: 563 km

Maritime claims: Territorial sea: 12 nm

Contiguous zone: 24 nm

Exclusive Economic Zone: as determined by bilateral agreements or the median line

GEOPOLITICAL IMPORTANCE

Qatar, also known as the State of Qatar, is an Arab emirate, in the Middle East, occupying the small Qatar Peninsula on the north-easterly coast of the much larger Arabian Peninsula. Its sole land border is with Saudi Arabia to the south, with the rest of its territory surrounded by the Persian Gulf. A strait of the Persian Gulf separates Qatar from the nearby island state of Bahrain. The country is in a strategic location near major petroleum deposits. Qatar is therefore, one of the richest countries in the region due to the exploitation of large oil and gas fields since the 1940s. Qatar is a peninsula located halfway down the west coast of the Arabian Gulf. Its territory comprises a number of islands including Halul, Sheraouh, Al-Ashat and others.

POLITICAL OVERVIEW

Qatar has an unelected, monarchic, emirate-type government. There are no democratic institutions or elections, and power is assumed on a hereditary basis. The head of state is the Emir, and the right to rule Qatar is passed on within the Al Thani family. The Emir's role is influenced by continuing traditions of consultation, rule by consensus, and the citizen's right to appeal personally to the Emir. The Emir, while directly accountable to no one, cannot violate the Sharia (Islamic law) and, in practice, must consider the opinions of leading notables and the religious establishment. Their position was institutionalised in the Advisory Council, an appointed body that assists the Emir in formulating policy. There is no electoral system. Political parties are banned. But Qatar on the other hand, claims that it is developing into a constitutional monarchy. Suffrage is currently limited to municipal elections, with the voting age set at 18. Elections in 1999 for a 29-member municipal council were the first in which Qatari women were allowed to vote and stand for office. A constitution, providing for democratic reforms, came into force in 2005. Qatar has a discretionary system of law controlled by the Emir, although civil codes are being implemented; Islamic law (Sharia) is significant in personal matters. The first elections for the Consultative Assembly of Qatar are planned for middle to late 2013, according to Qatari Emir Hamad bin Khalifa Al Thani.

KEY POLITICAL PERSONS

HEAD OF STATE, COMMANDER-IN-CHIEF & MINISTER OF DEFENCE: Sheikh Tamim bin Hamad Al Thani

PRIME MINISTER: Abdullah bin Nasser bin Khalifa Al Thani

ECONOMY

Qatar is one of the richest country in the world. Qatar has prospered in the last several years with continued high real GDP growth. GDP was driven largely by the oil and gas sector. It is the richest country in the world. Petroleum is the cornerstone of Qatar's economy and accounts for more than 70% of total government revenue, more than 60% of gross domestic product, and roughly 85% of export earnings. Production and export of natural gas are becoming increasingly important. Long-term goals feature the development of off-shore petroleum and the diversification of the economy. Economic policy is focused on developing Qatar's non-associated natural gas reserves and increasing private and foreign investment in non-energy sectors, but oil and gas still account for more than 50% of GDP, roughly 85% of export earnings, and 70% of government revenues. Qatar has prospered in the last several years with continued high real GDP growth in 2011-2012. Oil and gas likely have made Qatar is the second highest per-capita income country in the world.

GDP: $319.8 billion (2015 est.), $309.7 billion (2014 est.)

Real Growth Rate: 3.3% (2015 est.), 4% (2014 est.)

Industries: Liquefied natural gas, crude oil production and refining, ammonia, fertilizers, petrochemicals, steel reinforcing bars, cement, commercial ship repair

Oil - proved reserves: 25 billion bbl (1 January 2016 est)

Natural gas - proved reserves: 24.53 trillion cu m (1 January 2016 est)

Total Exports: $77.29 billion (2015 est.), $126.7 billion (2014 est.)

Export Commodities: Liquefied natural gas (LNG), petroleum products, fertilizers, steel

Major Markets: Japan 25.4%, India 14.6%, China 8.4%, UAE 6.8%, Singapore 5.6%, UK 5.5%, Thailand 4.2% (2015)

Total Imports: $28.5 billion (2015 est.), $31.15 billion (2014 est.)

Import Commodities: Machinery and transport equipment, food, chemicals

Major Suppliers: China 11.9%, US 11.3%, UAE 9%, Germany 7.7%, Japan 6.7%, UK 5.9%, Italy 4.6%, Saudi Arabia 4.4% (2015)

DEFENCE & SECURITY

The Qatar Armed Forces are the military forces of Qatar. The country plays an active role in the collective defence efforts of the Gulf Cooperation Council; the other five members are Saudi Arabia, Kuwait, Bahrain, the UAE, and Oman. Qatar maintains a modest defence establishment which includes army, navy, and air force. The country lacks strategic depth, have small populations, and small, untrained armies. Qatar's significant oil and natural gas reserves have made the country the potential target for aggression and dependent on outside forces for defence and security. The country has a public security force of about 8,000 men, including a coast guard, national fire fighting force, air wing, marine police, and an internal security force. Qatar also has signed defence pacts with the US, UK, and France. Qatari forces played an important role in the first Gulf War.

KEY DEFENCE PERSONS

MINISTER OF STATE FOR DEFENCE AFFAIRS: Khalid bin Mohammad Al Attiyah

CHIEF OF STAFF OF QATAR ARMED FORCES: Maj. Gen. Ghanem bin Shaheen Al Ghanem

COMMANDER OF NAVY: Maj. Gen. Mohammed Nasser Al Mohannadi

COMMANDER OF AIR FORCE: Maj. Gen. Mubarak bin Mohamed al-Kumait al-Khayarin

Internal Conflict

Qatar is not currently engaged in any armed conflict.

External Conflict

Bahrain: The territorial dispute with Bahrain over the Hawar Islands and the maritime boundary dispute with Bahrain were solved by the International Court of Justice (ICJ) in The Hague. In the June 2001 decision, Bahrain kept the Hawar Islands and Qit'at Jaradah but dropped claims to Janan Island and Zubarah on mainland Qatar, while Qatar retained significant maritime areas and their resources.

Other Issues: Qatar has played an important role in mediating regional conflicts including in Lebanon, where it deployed peacekeepers as part of the UN Mission in Lebanon (UNIFIL) and, more recently, in Darfur, Somalia and Yemen. Qatar also mediated the Eritrea-Djibouti border dispute. The country also chairs the Follow-Up Committee of the Arab Peace Initiative and has in recent months hosted a series of ministerial meetings of the Committee in Doha.

Libyan crisis: Qatar pledged 4-6 fighter aircraft and a cargo plane as part of its contribution to the March 2011 military intervention in Libya following UN Security Council Resolution 1973 of 2011.

ISIS: Qatar, in response to ISIS (or ISIL) militants, has joined many states along with USA and Russia on a coalition against them in Syria and Iraq. The country has done airstrikes against ISIL targets in Syria.

THREAT PERSPECTIVE

Qatar in order to prevent militants traveling to and from Syria and nearby conflict zones has joined hands with international organisations and also taken number of new counterterrorism measures in recent years to secure its critical infrastructure. In March 2005, a car bomb blast occurred at the Doha Players theatre outside the capital, Doha, near a British school. It was the first attack of its kind in the Gulf Arab state, which hosts the US military's Central Command. Since then, the country fears that it would become a regular target for militant attacks. Currently, the threat status remains low in Qatar but still military, security agencies maintain a high alert scenario and readiness to face any impending threat. The country is active on the regional and world arena, having mediated in disputes in the Middle East and Africa. It is pursuing an Afghan peace deal.

STRATEGIC RELATIONS

Gulf Cooperation Council (GCC): Qatar plays an active role in the collective defence efforts of the Gulf Cooperation Council (GCC--the regional organization of the Arab states in the Gulf; the other five members are Saudi Arabia, Kuwait, Bahrain, the UAE., and Oman).

US: Bilateral relations are strong and expanding. Qatar and the United States coordinate closely on regional diplomatic initiatives, cooperate to increase security in the Gulf, and enjoy extensive economic links, especially in the hydrocarbons sector. Qatar has supported US military operations critical to the success of Operation Enduring Freedom and Operation Iraqi Freedom. The two countries have a Defence Cooperation Agreement that provided for US access to Qatari bases, pre-positioning of United States material, and combined military exercises. Qatar is host to the US AFCENT forces and serves as a critical forward-deployed location in the region. The United States provides no development assistance to Qatar. Both the countries have extensive economic ties.

UK: Bilateral relations are strong and there is good co-operation in a number of areas. Qatar and the UK enjoy a special defence relationship underpinned by the Defence Cooperation Arrangement (DCA) signed in May 2006. The spirit of the DCA is to recognise UK support to the Qatari Armed Forces through training and exercises both in Qatar, providing Loan Service Officers and Training Teams to the Emiri Guard, Qatar Military Academy and the Infantry School; and in the UK with many Qatari officers graduating each year from Sandhurst, Dartmouth, the Joint Command and Staff College Shrivenham, and the Royal College of Defence Studies. Britain and Qatar signed a Defence Cooperation Agreement in November 2014.

India: The India-Qatari strategic engagement is based on the Qatari interests to engage India as a big market for its natural gas. Recently, India and Qatar have signed a landmark defence cooperation pact based on defence and security, dealing with key areas like maritime security, intelligence sharing on terrorism, money laundering and transnational crime. India-Qatar Defence Cooperation Agreement, signed during the visit of former PM to Qatar in November 2008 and extended till 2018, is implemented through the Joint Defence Cooperation Committee (JDCC).

Iran: Qatar is investing in missile defence systems to counter the threat across the Gulf from Iran, which has built up its missile arsenal. Stronger relations between Tehran and Doha contribute to peace and stability in the Middle East region. Qatar and Iran share close ties and membership in the Organisation of Petroleum Exporting Countries (OPEC), the Non-Aligned Movement (NAM) and the Organisation of the Islamic Conference (OIC). Qatar and Iran have made several bilateral agreements for to further develop their economic relationship. Iran and Qatar have a close economic relationship, particularly in the oil and gas industries.

Iraq: Following the 1990-91 Gulf War, in which Qatar and Iraq were on opposing sides, Qatar closed their embassy in Baghdad. They reopened the embassy for the first time twenty-five years in mid-2015.

Multilateral Alliances: Qatar, which is a member of NATO's Istanbul Cooperation Initiative (ICI) launched in 2004, has also contributed actively to the NATO-led Operation Unified Protector in Libya. The country was also an early member of OPEC. It is a member of the Arab League and Organisation of Islamic Cooperation (OIC).

DEFENCE CAPABILITIES

ARMY: Qatari Emiri Land Force (QELF) is the largest branch of the Qatar Armed Forces and maintains a modest military force.

Personnel: 8,500

Equipment

Category	Name	In Service
MBT	Leopard 2A7	6 (56 on order)
APC/ MICV/ IFV	EE-9 Cascavel	30
	AMX 10RC	12
	Piranha II 8x8	40
	AMX 10P	30
	AMX-VTT	14
	VAB	160
	Commando vehicle	8
Artillery	155 mm G5	12
	PzH 2000 self-propelled	2 (22 on order)
	155 mm F3	22
Mortar	81 mm	24
	120 mm	15
MLRS	Astros II	4
	HIMARS	6
	122mm BM-21 Grad	
Anti-Tank GW	Milan	630
	Hot	1000
	Swingfire	
Missile	SA-7	
	Patriot PAC-3	10
	Stinger	60
	Starburst	
	Roland MK-II	200

Army continued...

Category	Name	In Service
Missile	Blowpipe	60
	Mistral	500
	MIM-23 Hawk	
	Rapier	250
	THAAD	13

New Procurements/ Upgrades

- ✦ A Vectrus subsidiary has received a USD 26 million contract modification from the US Army for Qatar Base Operations Support Services at Camp As Sayliyah and Camp As Sayliyah.
- ✦ Qatar will be receiving 62 Leopard-2A7 main battle tanks from Germany replacing AMX-30 for the Qatar Land Forces at a cost of around USD 2.46 billion. The tanks would be upgraded to the latest 2A7 standard.
- ✦ Qatar plans to buy US Patriot missile batteries and Apache attack helicopters in an arms deal worth about USD 11 billion. The sale would provide Qatar with roughly ten radars and 34 launchers for Patriot systems designed to knock out incoming missiles, as well as 24 Apache helicopters and Javelin anti-tank missiles.

Members of the Qatari Armed Forces look at different pieces of kit the Royal Marines use whilst in theatre.

NAVY: Qatari Emiri Navy (QAN) has a small number of 1,500-man navy, including its marine police force and coastal defence artillery. The navy headquarters is at Doha there is also a base at Halul island.

Personnel: 1,800 (including Marine Police)

Equipment

Category	Name	In Service
Fast Attack Craft	Barzan Class (Vita Class)	4
	Damsah Class	3
Patrol Vessel	Vosper class	6
	Damen Polycat	6
	MV-45	3
	P-1500 class	5
	DV-15 class	3
	Fast Interceptor	10
Missile	Exocet MM40 Block-III, MM-38, Mistral, Goalkeeper	

New Procurements/ Upgrades

✦ Fincantieri and the Qatari Ministry of Defence, have signed a contract in June 2016 for the construction of seven new generation units included in the national naval acquisition programme. The contract, amounting for Fincantieri to approx. 4 billion euros, envisages the supply of seven surface vessels, of which four corvettes of over 100 meters in length, one amphibious vessel (LPD - Landing Platform Dock), and two patrol vessels (OPV - Offshore Patrol Vessel) as well as support services in Qatar for further 15 years after the delivery of the vessels.

✦ MBDA has been awarded a contract by Qatar to supply a coastal defence system for the Qatari Emiri Naval Forces (QENF). The contract entails the delivery of two different missile systems – the Marte ER (the Extended Range version of the Marte missile) and Exocet MM40 Block 3 missile.

✦ The new Qatar Emiri Naval Forces Base will be the centre for Qatar's maritime security and will be built 14 kilometres from Doha. The new naval base will be operational in 2017.

✦ Qatar plans to procure two 700 tonne corvettes to replace the French-built Qatar Combattante III M Class fast attack craft of 345 tonnes displacement.

✦ The Qatari Emiri Navy will procure a batch of Exocet MM40 Block 3 missiles to equip its fleet of four Vita class Patrol Boats from the US missile developer MBDA.

✦ Nakilat Damen Shipyards Qatar (NDSQ) and Qatar Armed Forces have signed two MoUs for the construction of seven vessels at Qatar's premier shipyard (USD 851 million). The MoUs signed by NDSQ and Qatar Armed Forces concern six 50m-long axe-bow high-speed patrol vessels and one 52m-long diving support vessel for the Qatar Armed Forces.

AIR FORCE: The Qatar Emiri Air Force (QEAF) is the aerial branch of the Qatar Armed Forces. It was formed in 1974, three years after achieving independence from Great Britain in 1971.

Personnel: 1,500

Equipment

Air Force continued...

Category	Name	In Service
Fighters/ attack/ Reconnaissance/ bomber	Dassault Mirage 2000-5	12
	Boeing F-15Ea	(30+ on order)
	Boeing 737 AEW&C	(3 on order)
	Dassault Rafale	(24 on order)
Transport/ Tanker	C-17A Globemaster III	2 (5 on order)
	C-130J Super Hercules	4
	Boeing 747-SP	2
	Airbus A330 MRTT	(2 on order)
	Airbus 300/310/320	1 each
Helicopter	Gazelle SA 342	14
	AH-64 Apache	(24 on order)
	AW139	21
	NH-90	(22 on order)
	UH-60R Sea Hawk	(5 on order)
	Westland Commando	12
	UH-60R Sea Hawk	(6 on order)
Trainer	Alpha Jet-A	6
	Pilatus PC-21	(24 on order)
SAM	I-Hawk Battery	1
	Roland II	6
	Rapier Battery	1
AAM	AIM-9M Sidewinder, Mica-RF, Magic-II R.550, Matra R.530	
AGM	AS-30L	
ASM	AM-39 Exocet	

New Procurements/ Upgrades

✦ The US Government has approved the sale of F-15E Strike Eagle to Qatar Emiri Air Force worth at around USD 4 billion.

✦ Lockheed Martin received a USD 90.6 million foreign military sale contract from the US Army to provide Modernised Target Acquisition Designation Sight/Pilot Night Vision Sensor (M-TADS/PNVS) systems to the Qatar Emiri Air Force.

✦ The Government of Qatar has requested a possible sale of 6 MH-60R Seahawk Multi-Mission Helicopters from the Defence Security Cooperation Agency. Under the USD 750 million contract, the DSCA will also supply communication equipment; support equipment, technical, and logistics support services, and other related elements of logistics support. The MH-60R helicopters will supplement and eventually replace the Qatar Air Force's aging maritime patrol helicopters.

✦ In May 2015, the Air Force signed a deal for 24 Dassault Rafale fighters worth USD7 billion. This deal makes Qatar the third export customer for the fighter after Egypt and India. Qatar joins France, Egypt and India that have ordered the Rafale. Qatari Emiri Air Force will be receiving 24-36 new fighter aircraft soon. The new aircraft would replace Dassault Mirage 2000-5s.

✦ Qatar has signed a 240-million-euro deal with the European Aeronautic Defence and Space Company (EADS) to boost radar-based protection and safety measures for Qatari airspace. The project, to be carried out over three years, will create a monitoring network through placement of monitoring stations and advanced technology on the sea, land and air borders of the state of Qatar.

✦ Qatar will be acquiring a fleet of 24 Pilatus PC-21 turboprop trainers and associated equipment. The Qatar Emiri Air Force selected the PC-21 as their fixed-wing basic and advanced training aircraft, after a thorough evaluation both in Switzerland and in Qatar.

A Qatar Emiri Air Force Dassault Mirage 2000-5 fighter jet takes off as part of a Joint Task Force Odyssey Dawn mission. *Image Courtesy: US Navy*

Major Armaments Producers

There is no domestic production facility in Qatar for defence related products and equipments. Qatar also has no military research and development programmes in progress. The Gulf Cooperation Council [GCC] was established in an agreement concluded on 25 May 1981 in Riyadh, Saudi Arabia between: Bahrain, Kuwait, Oman, Qatar, Saudi Arabia and UAE. These countries declared that the GCC is established in view of the special relations between them, their similar political systems based on Islamic beliefs, joint destiny and common objectives. The GCC is a regional common market with a defence planning council as well. The geographic proximity of these countries and their general adoption of free trade economic policies are factors that encouraged them to establish the GCC. The GCC States seek to build up their defence forces according to a common conception. In this context, they have unified operational procedures, training, and military curricula. They also endeavour to accomplish compatibility of their military systems.

DEFENCE DEALS

Arms Import

Qatar is planning to procure Super Mushshak aircraft from Pakistan which will be used to train pilots. Aeronautical Complex (PAC) and Qatar Emiri Air Force signed an agreement in this regard. Thales has been awarded a contract to supply a military satellite communications (SATCOM) system to the Qatar Armed Forces. Under the contract, the company will supply the systems needed to transmit and receive communications through satellites and process data on the ground. Thales will also deliver a network operations centre that will help the Qatari armed forces plan and manage the complete system.

Qatar has a less significant military and therefore the procurement percentage is mere compare to other Gulf States. Still, the country does 98 percent of arms procurement from the United States. According to a report by

Stockholm International Peace Research Institute (SIPRI), Qatar's defence spending is small compared with that of Saudi Arabia and the UAE, the purchases it did make between 2005 and 2009 were overwhelmingly awarded to US companies. Qatar plans to buy US Patriot missile batteries and Apache attack helicopters in an arms deal worth about USD 11 billion. The sale would provide Qatar with roughly ten radars and 34 launchers for Patriot systems designed to knock out incoming missiles, as well as 24 Apache helicopters and Javelin anti-tank missiles. Qatar was investing in missile defence systems to counter the threat across the Gulf from Iran, which has built up its missile arsenal. Qatar has signed another defence contract with US through Foreign Military Sale worth USD 6.5 billion. The Government of Qatar has requested a possible sale of 2 Terminal High Altitude Area Defence (THAAD) Fire Units, 12 THAAD Launchers, 150 THAAD Interceptors, 2 THAAD Fire Control and Communications, 2AN/ TPY-2 THAAD Radars, and 1 Early Warning Radar (EWR). Also included are fire unit maintenance equipment, prime movers (trucks), generators, electrical power units, trailers, communications equipment, tools, test and maintenance equipment, repair and return, system integration and checkout, spare/ repair parts, publications and technical documentation, personnel training and training equipment, U.S. Government and contractor technical and logistics personnel support services, and other related support elements. The estimated cost is USD 6.5 billion. Qatar is currently preparing for a major purchase of 36 fighter aircraft that would replace the Qatar Emiri Air Force's (QEAF) existing fleet of French-made Dassault Mirage 2000-5s. The contract is likely to be won by US or French firms, as per defence analysts.

Qatar has signed QR460.5 million contract with Aurora for drone sensor integration; QR365 million worth contract with Reiner Stemme Utility Air-Systems for drones; and aQR200 million for a long-distance satellite ground imaging station from Airbus. The Qatar Armed Forces have also signed a contract for three AW139 medium twin helicopters plus a comprehensive logistics support package with AgustaWestland, a Finmeccanica company. These aircraft will be operated by the Qatar Emiri Air Force to perform emergency medical service missions. This latest contract adds to the one for eighteen AW139s signed by the same customer in 2008. Some of the 18 have already been delivered to carry out a range of operations including utility, troop transport, search and rescue, border patrol, special forces operations, law enforcement and homeland security. Qatar is also planning to buy six MH-60R Seahawk multi-mission helicopters for USD 750 million from the Defence Security Cooperation Agency through a possible Foreign Military Sale. The proposed sale of the MH-60R Seahawk helicopters will improve Qatar's capability to meet current and future anti-surface warfare threats. Qatar will use the enhanced capability to strengthen its homeland defence.

Qatar's Ministry of Defence recently signed a three-year, USD 43.92 million contract with UK-based Serco to run courses for mid-level officers, such as majors and lieutenant-colonels, from the Qatari Emiri Navy, Army and Air Force. The Qatari Armed Forces recently during the DIMDEX exhibition announced deals with US weapons manufactures worth USD 7.6 billion, including Apache helicopters, Javelin missiles and PAC 3 Patriot systems and also included 17 Turkish fast patrol vessels and German tankers.

Important defence contracts and MoUs signed during DIMDEX 2016

- A €42 million MoU between the Qatar Emiri Naval Forces and Qatari company Nakilat Damen Shipyards Qatar (NDSQ) for diving support boats
- A €41 million MoU between the Qatari Emiri Naval Forces and Turkish company Yonca Onuc (J.V.) for the purchase of fast interceptor vessels
- A QAR 68 million MoU between the Qatari Emiri Naval Forces and European company Zodiac for the purchase of armed fast attack vessels
- A QAR 200 million MoU between the Qatar Armed Forces and French company Airbus for a long-distance satellite ground imaging station
- A QAR 34.6 million MoU between the Qatar Armed Forces and South Korean company AK & Partners, Inc. for surveillance vehicles
- A QAR 460.5 million MoU between the Qatar Armed Forces and US company Aurora for drone sensor integration
- A QAR 126.8 million MoU between the Qatar Armed Forces and the US company L-3 for MX20D and M25 drone cameras
- A QAR 54.7 million MoU between the Qatar Armed Forces and the US company Textron Systems for Aerosonde MK 4.7 drones
- A MoU between the Qatar Emiri Air Force and US company Lockheed Martin for a QAR 50 million, 3-year C-130 maintenance contract
- A QAR 2.6 billion MoU between the Qatar Emiri Naval Forces and European company MBDA for a coastal battery system
- A QAR 240 million MoU between the Qatar Emiri Naval Forces and MBDA for new Exocet MM40 B3 missiles
- A QAR 95 million MoU between the Qatar Emiri Naval Forces and German company MTU Friedrichshafen GmbH for maintenance and overhaul of the MTU propulsion system
- A MoU between the Qatar Emiri Naval Forces and Qatari company Nakilat for the training of navy officers
- A Euro 32.5 million MoU between Al Zaeem M.B.A.A. Air Academy and French company DCI for fighter and helicopter pilot training
- A QAR 365 million MoU between the Qatar Emiri Air Force and German company Reiner Stemme Utility Air-Systems GmbH for the production of drones
- A MoU between the Qatar Armed Forces and the Polish company WKK for the purchase of 51% of the company's shares, who specialise in the manufacture of composite materials and fuselage production
- A QAR 60 million MoU between the Qatar Armed Forces and French company Thales for Searchmaster radar
- A MoU between the Qatar Armed Forces and the Chinese National Precision Machinery Import and Export Corporation (CASIC) for the provision of support in the fields of production and military cooperation between the drone project committee and CASIC

DEFENCE BUDGET

Total defence spending: $12.6 billion

Estimated defence spending in terms of GDP: 2.3%

The State of Qatar has witnessed rapid economic growth in recent years with a predicted GDP growth forecast to hit 7.8 % in 2015, its fastest rate since 2011. Oil and gas have given Qatar one of the highest per capita incomes in the world and made it one of the fastest-growing economies. Qatar's military spending follows the Middle Eastern trend by investing increasing amounts of GDP to modernise and strengthen its military forces, with $23 billion of deals announced at DIMDEX 2014 with the Qatar Armed Forces. Qatar, during DIMDEX 2016, Qatar penned eight new MoUs totalling QAR 1.29 billion.

Qatar, along with other Middle East GCC nations like Saudi Arabia and UAE are continuing with their massive military spending despite falling oil prices. Qatar is planning to modernise its armed forces and secure its maritime borders by signing agreements for approx. USD 8.95 billion (QR32.58 billion). Qatar is one of the wealthiest countries in the world and has strong bonds with the European Union especially USA and France. Qatar has defence pacts with the US, UK, Russia and France (Circa 1998). There is a little transparency in the military spending. The country does not report their military spending separately from their broader spending on security (including internal security). According to a report by the IMF, the defence budget was around USD 1.91 billion in 2007 and USD 2.19 billion in 2008.

Qatar's procurement and military spending is growing rapidly. There is therefore a greater need for the development of efficient, effective and interoperable logistics systems and strategies. Recent military operations have highlighted both the importance of coalitions as well as the logistics capabilities of Qatar.

CONCLUSION

Qatar has a very high per capita income and is strongly involved in nearby regional conflicts. The Middle East nation plays an active role in the collective defence efforts of the Gulf Cooperation Council; the other five members are Saudi Arabia, Kuwait, Bahrain, the UAE, and Oman. The presence of a large American military base in the country provides the country with a guaranteed source of defence and national security. The country has one of the most creative foreign policies in this unstable region. Qatar is planning to emerge as a world player and also as a strong nation among the rest of the Gulf States. According to media reports, the country has donated hundreds of millions of dollars to victims of Hurricane Katrina, flooding in Pakistan and civil strife in Darfur. It won a bid to host the 2022 World Cup in recognition of its status as an oasis of stability and a global mediator. Qatar, one of the wealthiest nations in the middle-east is rich in oil and natural gas has the world's largest per capita production and proven reserves of both oil and natural gas. The country thus shares a mutual interest with western countries. The risk of civil unrest is minimal and less political protests take place even as anti-Western sentiment is high. The US military has a substantial military presence in Qatar, including the operational headquarters for the US Central Command (CENTCOM) and the Central Air Operational Component at Al-Udeid Air Base. The country, having less population, minimal military power, is fully economy driven with major investments in banking, financial services, wealth and education, as well as in development of industrial plants. From military point of view, she is friendly to Iran and also serves as a base for the US armed forces. Qatar has been historically sensitive to outside military intervention in the Gulf and was eager to bolster regional security measures. Also, it is focusing all its energies on improving cooperation and coordination on mutual defence issues while also continuing to work together in social, cultural, political, and economic spheres with other GCC members. The Qatar Armed Forces is aiming to develop a technically advanced military force which will ensure its readiness to meet the challenges of the maritime domain in the 21st Century.

CONTACT DETAILS

Ministry of Defence and Armed Forces HQ

P.O. Box 37, Doha, Qatar

Tel: +974-4614111

Procurement Directorate – Logistics Department

Tel: 44612969

Tender and Auction Committee

Tel: 44612784

www.gov.qa

REPUBLIC OF CONGO

(Capital: Brazzaville)

INTRODUCTION

Total Area: 342,000 sq km

Population: 4,852,412 (July 2016 est.)

Coastline: 169 km

Maritime claims: Territorial sea: 12 nm

Contiguous zone: 24 nm

Exclusive economic zone: 200 nm

KEY POLITICAL PERSONS

PRESIDENT & COMMANDER-IN-CHIEF: Denis Sassou-Nguesso

KEY DEFENCE PERSONS

MINISTER TO THE PRESIDENCY OF THE REPUBLIC IN CHARGE OF NATIONAL DEFENCE: Gen. Charles Richard Mondjo

CHIEF OF ARMY STAFF: Gen. Francois Olenga Tete

ECONOMY

The economy of the Republic of the Congo is a mixture of agriculture and village handicrafts, an industrial sector based largely on petroleum extraction, support services, and a government characterised by budget problems and overstaffing. The Congo's growing petroleum sector is by far the country's major revenue earner. Oil has supplanted forestry as the mainstay of the economy, providing a major share of government revenues and exports. Natural gas is increasingly being converted to electricity rather than being flared, greatly improving energy prospects. New mining projects, particularly iron ore, which entered production in late 2013, may add as much as $1 billion to annual government revenue. The country has undertaken economic reform measures with support from leading international financial institutions, including the World Bank and International Monetary Fund.

GDP: $29.36 billion (2015 est.); $28.64 billion (2014 est.)

Real Growth Rate: 2.5% (2015 est.); 6.8% (2014 est.)

Industries: Petroleum extraction, cement, lumber, brewing, sugar, palm oil, soap, flour, cigarettes

Natural gas - proved reserves: 90.61 billion cu m (1 January 2016 est.)

Crude oil - proved reserves: 1.6 billion bbl (1 January 2016 est.)

Total Exports: $5.231 billion (2015 est.); $9.12 billion (2014 est.)

Export Commodities: Petroleum, lumber, plywood, sugar, cocoa, coffee, diamond

Major Markets: China 42.1%, Italy 16.9%, US 4.9%, India 4.7%, Portugal 4.2% (2015)

Total Imports: $3.934 billion (2015 est.); $4.939 billion (2014 est.)

Import Commodities: Capital equipment, construction materials, foodstuff

Major Suppliers: China 20.3%, France 14.2%, South Korea 9.8%, US 4.9%, UK 4.4%, Italy 4.1%, India 4.1% (2015)

INTERNATIONAL DISPUTES

Most of the Congo River boundary with the Republic of the Congo is undefined as no agreement has been reached on the division of the river or its islands, except in the Pool Malebo/Stanley Pool area.

DEFENCE

ARMY

Personnel: 10,000

Equipment

Category	Name	In Service
Battle Tank	T-54/55	26
	T-59	14
	PT-76 light	3
	Type 62	8
Armoured Vehicle	BRDM1/2	24
	Panhard M3	
	BTR-152	20
	BTR-50	18
	BTR-60	28
Artillery	75mm	6
	100mm guns	10
	122mm D-30 howitzers	8
Mortar	81mm	
	120mm	
MRL	122mm	
ATGW	AT-3 SAGGER	
AA gun	14.5mm	16
	37mm	40
	57mm	8
ATK gun	100mm	

New Procurements/ Upgrades

No major procurement plans are under consideration at present by the Army.

NAVY

Personnel: 300

Equipment

Category	Name	In Service
Patrol Vessel	Arco class river patrol	4
Tug	Hinda class	1

New Procurements/ Upgrades

No major procurement plans are under consideration at present by the Navy.

Note: Angola, Cameroon, the Democratic Republic of Congo, the Republic of the Congo, Gabon, Equatorial Guinea and Nigeria have reached an agreement for the establishment of a joint force (Gulf of Guinea Guard Force, GGGF) to monitor their shared interests in the Gulf of Guinea, in particular the valuable offshore oil facilities in the region.

AIR FORCE

Personnel: 500

Equipment

Category	Name	In Service
Fighter	MiG-21	12
	L-39 Albatross	4
Transport	An-26	1
	F-28	1
	An-24	5

Air Force continued...

Category	Name	In Service
Transport	Nord 2501	1
Helicopter	Mi-26	1
	Alouette II/III	4
	SA-365 Dauphin	1

New Procurements/ Upgrades

No major procurement plans are under consideration at present by the Air Force.

Defence Expenditure

Estimated defence spending in terms of GDP: 1.7%

Most of the defence equipment and hardware of the Congo military are of Soviet, Chinese and Czech origin.

Defence Production and R&D

No defence production is undertaken in Congo.

Defence Procurement

No military procurement plans are under consideration by the Government at present.

CONTACT DETAILS

Ministry of National Defence

Quartier General

Brazzaville, Congo

Tel.: +242 22 281 4559

ROMANIA
(Capital: Bucharest)

INTRODUCTION

Area: 238,391 sq km

Population: 21,599,736 (July 2016 est.)

Coastline: 225 km

Maritime claims: Territorial sea: 12 nm

Contiguous zone: 24 nm

Exclusive Economic Zone: 200 nm

Continental shelf: 200 m depth or to the depth of exploitation

KEY POLITICAL PERSONS

PRESIDENT & SUPREME COMMANDER OF THE ARMED FORCES: Klaus Iohannis

PRIME MINISTER: Dacian Julien Ciolos

KEY DEFENCE PERSONS

DEFENCE MINISTER: Mihnea Ioan Motoc

CHIEF OF THE GENERAL STAFF: Lt. Gen. Nicolae Ionel Ciuca

CHIEF OF ARMY STAFF: Maj. Gen. Dumitru Scarlat

CHIEF OF NAVY STAFF: Rear Admiral Alexandru Mirsu

CHIEF OF AIR STAFF: Maj. Gen. Laurian Anastasof

ECONOMY

Romania has a developing, upper-middle income market economy. The country has considerable economic potential with over 10 million hectares of agricultural land, diverse energy sources such as coal, oil, natural gas, hydro, nuclear and wind, a substantial manufacturing base and opportunities for expanded development in tourism. The implementation of a bold package of macro-stabilization and structural measures, supported by a multilateral programme with the World Bank, International Monetary Fund (IMF), and the European Commission (EC), helped the country overcome the effects of the global crisis by restoring macroeconomic balances and reviving economic growth, according to the World Bank. Exports remained the engine of economic growth, led by trade with the EU, which accounts for roughly 70% of Romania trade.

GDP: $413.8 billion (2015 est.), $398.9 billion (2014 est.)

Real Growth Rate: 3.7% (2015 est.), 3.0% (2014 est.)

Industries: electric machinery and equipment, auto assembly, textiles and footwear, light machinery, metallurgy, chemicals, food processing, petroleum refining, mining, timber, construction materials

Oil - proved reserves: 600 million bbl (1 January 2016 est.)

Natural gas - proved reserves: 105.5 billion cu m (1 January 2016 est.)

Total Exports: $54.52 billion (2015 est.), $62.16 billion (2014 est.)

Export Commodities: machinery and equipment, other manufactured goods, agricultural products and foodstuffs, metals and metal products, chemicals, minerals and fuels, raw materials

Major Markets: Germany 19.8%, Italy 12.5%, France 6.8%, Hungary 5.4%, UK 4.4% (2015)

Total Imports: $63.12 billion (2015 est.), $70.5 billion (2014 est.)

Import Commodities: machinery and equipment, other manufactured goods, chemicals, agricultural products and foodstuffs, fuels and minerals, metals and metal products, raw materials

Major Suppliers: Germany 19.8%, Italy 10.9%, Hungary 8%, France 5.6%, Poland 4.9%, China 4.6%, Netherlands 4% (2015)

INTERNATIONAL DISPUTES

Romania has a dispute over Ukrainian-administered Zmiyinyy/Serpilor (Snake) Island and Black Sea maritime boundary delimitation. The ICJ ruled largely in favour of Romania in its dispute submitted in 2004 over Ukrainian-administered Zmiyinyy/Serpilor (Snake) Island and Black Sea maritime boundary delimitation. Romania also opposes Ukraine's reopening of a navigation canal from the Danube border through Ukraine to the Black Sea.

DEFENCE

The Romanian Military is modernising its forces. The first short-term stage was completed in 2007. 2015 marked the end of the second stage and in 2025, the long-term stage is to be completed. The stages aim at modernising the structure of the armed forces, reducing personnel as well as acquiring newer and more improved technology that is compatible with NATO standards.

ARMY

Personnel: 53,000

Equipment

Category	Name	In Service
MBT	T-55	299
	TR-580	47
	TR-85MI	58
IFV/APC	MLI-84	20
	B33 TAB Zimbru	69
	TAB-71	369
	TAB-77	150
	TAB-79	414
	TAB Zimbru	58
	MRAP	60
	TABC-79	371
	Piranha IIIC	31
Artillery	M30 120mm howitzer	40
	M81 152mm	315
	M85 152mm	103
	2SI 122mm	6
	89 122mm	18
	89 122mm SP	18
Mortar	82mm	299
	120mm	281
MLRS	APR-40	133
	Larom 160mm	54
ATK gun	M77 100mm	231
	Su-100 SP	23

Army continued...

Category	Name	In Service
ATGW	9P122	12
	9P133	78
	9P148	48
AA Gun	35 mm twin	24
	37mm	18
	Gepard SPAAG	36

New Procurements/Upgrades

No major procurement plans are under consideration at present.

Romanian armoured vehicles go on the offensive during Combined Resolve II at the Hohenfels Training Area, May 26, 2014. *Image Credit: Romanian Land Forces & 7thArmyJMTC*

NAVY

Personnel: 7,000

Equipment

Category	Name	In Service
Frigate	Regela Ferdinand class	2
	Marasesti class	1
Corvette	Tetal class	4
Patrol craft	ex-USSR Tarantul class missile corvettes	3
	Epitrop class torpedo boats	3
Mine warfare Force	Corsar class minelayers	1
	Misca class minesweepers	4
River squadron	Mihail Kogalniceanu class river monitors	3
	Brutar class river monitors	5
	VD 141 class river patrol minesweepers	6
Marine	TABC-79	
	TAB-1M	
Helicopter	PUMA	3

New Procurements/Upgrades

✦ Requirements have been formulated for a new class of multi-purpose corvettes, MCMVs and a logistic ship.

AIR FORCE

Personnel: 13,450+

Equipment

Category	Name	In Service
Fighter	MiG-21 Lancer	39
Trainer	MiG-21 Lancer TS	10
	IAR-99	10

Air Force continued...

Category	Name	In Service
Trainer	IAR-99 SOIM	11
	Yak-52	12
Transport	C-130B	3
	C-130H	2
	C-27J	6
	An-30	35
Helicopter	IAR-330 Socat (attack)	23
	IAR-330 L/M/MEDevac (utility/SAR)	35
	IAR-316B	6
Air defence artillery	S-60 57mm gun	96
SAM	SA-2	7
	Hawk XXI system	4

New Procurements/Upgrades

✦ The Air Force has taken delivery of its sixth and final Alenia Aeronautica C-27J Spartan tactical transport aircraft.

✦ Purchase of 12 Lockheed Martin F-16AM/BM Fighting Falcons from Portugal has been completed

✦ Four new SAM Systems are planned to be produced by 2017 to replace the SA-2s.

✦ Five new Shorad/VShorad integrated systems are planned to be procured by 2016 to replace the S-60 guns.

Defence Expenditure

Total defence spending: $2.1 billion (2014), $58.14 billion (2012)

Defence spending in terms of GDP: 1.39% (2015)

Defence Production and R&D

Romania is carrying out a military modernization programme. New steps have been taken for finding a fundamental solution for the Romanian Defence Industry, such as: reorganizing and resizing the sector according to the real needs of the Romanian Defence Forces and Romania's potential in the external market. Romania has the capability and there are companies that have their production devoted to defence. There are over 25 such companies that need to be integrated in the European defence industry system. RomArm is the government-owned national company for military technology and is heavily involved in nearly all aspects of defence acquisition and development, including armoured vehicles, artillery and small arms.

Defence Procurement

The Ministry of National Defence - Armaments Department is responsible for armaments systems acquisition management, component of the Romanian defence policy in the field of enhancing the country's defence capability and the inter-operability of Romanian armed forces with NATO and EU countries. Being the authority in the acquisition domain for the Ministry of National Defence, the Armaments Department acts towards the harmonisation of the national with international laws by concerning the product and services procurement organisation, establishing the goals of modern technology integration, and optimum resource utilisation and keeping balance between domestic and foreign acquisitions.

CONTACT DETAILS

Ministry of Defence
Izvor Street, nr. 110
Sector 5, cod 050561

Information and Public Relations Directorate
E-mail: presamapn@mapn.ro
Contact: 021-319.60.22

STRATEGIC INFORMATION

RUSSIAN FEDERATION
(Capital: Moscow)

The map of Russia is according to the Federal Constitutional Law dated 21.03.2014 No. 6-FCL/FKZ (Version 31.12.2014)

"On accession of Crimean Republic to Russian Federation and on formation of new constituent entities of Crimean Republic and Sevastopol – the city of Federal status." (Valid with amendments and addenda from 01.01.2015)

INTRODUCTION

Area: 17,098,242 sq km

Population: 142,355,415 (July 2016 est.)

Coastline: 37,653 km

Maritime claims: Territorial sea: 12 nm

Contiguous zone: 24 nm

Exclusive Economic Zone: 200 nm

Continental shelf: 200m depth or to the depth of exploitation

GEOPOLITICAL IMPORTANCE

The Russian Federation is a large country which is carved out of the erstwhile Soviet Union. It occupies most of Eastern Europe and North Asia, stretching from the Baltic Sea in the West to the Pacific Ocean in the East, and from the Arctic Ocean in the North to the Black Sea and the Caucasus in the South. It is bordered by Norway and Finland in the Northwest; Estonia, Latvia, Belarus, Ukraine, Poland, and Lithuania in the West; Georgia and Azerbaijan in the Southwest; and Kazakhstan, Mongolia, China, and North Korea along the Southern border. The country is endowed with the largest reserves of mineral and energy resources and is the largest producer of oil and natural gas in the world. As one of the five permanent members of the United Nations Security Council, Russia has played a major role in maintaining international peace and security.

POLITICAL OVERVIEW

The Russian Federation is a federal presidential republic. The head of the Russian state is the President who is elected after every six years. The President, who is also the Supreme Commander in Chief of the Russian military forces, has the rights to lead international politics of the country. The President is eligible for a second term but constitutionally he is barred from a third consecutive term. The Prime Minister is appointed by the President with the approval of the Duma.

The upper chamber of the Russian parliament, the Federation Council, consists of two members from each of the 83 federal subjects: 21 republics, 46 oblasts, nine krais, two federal cities, four autonomous okrugs and an autonomous oblast. The lower chamber or the State Duma has 450 members who are known as deputies. The state Duma is the more powerful house. All the bills, even those proposed by the Federation Council, must first be considered by the Duma. The federal bodies include federal ministries, federal services and federal agencies. Judges are appointed by the President with the consent of the Federation Council. The Constitutional Court and general courts are headed by the Supreme Court and arbitration courts by the Supreme Arbitration Court.

On 18 March 2014, Russian President Vladimir Putin added the Crimean Peninsula to the map of Russia, a day after Crimea's regional assembly declared independence from Ukraine and applied to join the Russian Federation.

Crimea is home to Russia's Black Sea fleet. The Crimean Peninsula is a multi-ethnic region which until February 2014 was administered by Ukraine. Russia now administers the territory consisting of the Autonomous Republic of Crimea and the city of Sevastopol as two federal subjects within the Crimean Federal District - the Republic of Crimea and the federal city of Sevastopol.

KEY POLITICAL PERSONS

PRESIDENT & SUPREME COMMANDER-IN-CHIEF OF THE ARMED FORCES: Vladimir Putin

PRIME MINISTER: Dmitry Medvedev

ECONOMY

Russia has a high income mixed economy. Market reforms of the 1990s privatised much of Russian industry and agriculture, with notable exceptions in the energy and defence-related sectors. The country has an abundance of natural resources, including oil, natural gas and precious metals, which make up a major share of Russia's exports. Russia also has a large and sophisticated arms industry, capable of designing and manufacturing a wide range of high-tech military equipment, including a fifth-generation fighter aircraft. In 2013, the value of Russian arms exports totalled $15.7 billion. Top military exports from Russia include combat aircraft, air defence systems, ships and submarines. However, declining oil prices, lack of economic reforms, and the imposition of foreign sanctions following Crimea's annexation in 2014 have led to an economic downturn. According to the World Bank, the recession continues in Russia, although the pace of GDP decline has slowed down. The recession, which started in the third quarter of 2014 and persisted in the first half of 2016, has now declined substantially after the government's policy response package. Real GDP shrank by 0.9 percent year-on-year in the first half of 2016 compared to -3.7 percent in 2015.

GDP: $3.718 trillion (2015 est.); $3.862 trillion (2014 est.)

Real Growth Rate: -3.7% (2015 est.); 0.7% (2014 est.)

Industries: Complete range of mining and extractive industries producing coal, oil, gas, chemicals, and metals; all forms of machine building from rolling mills to high-performance aircraft and space vehicles; defence industries (including radar, missile production, advanced electronic components), shipbuilding; road and rail transportation equipment; communications equipment; agricultural machinery, tractors, and construction equipment; electric power generating and transmitting equipment; medical and scientific instruments; consumer durables, textiles, foodstuff, handicraft

Natural Gas - Proved Reserves: 47.8 trillion cu m (1 January 2016 es)

Crude Oil - Proved Reserves: 80 billion bbl (1 January 2016 est.)

Total Exports: $341.5 billion (2015 est.); $497.8 billion (2014 est.)

Export Commodities: Petroleum and petroleum products, natural gas, metals, wood and wood products, chemicals, and a wide variety of civilian and military manufactures

Major Markets: Netherlands 11.9%, China 8.3%, Germany 7.4%, Italy 6.5%, Turkey 5.6%, Belarus 4.4%, Japan 4.2% (2015)

Total Imports: $193 billion (2015 est.); $308 billion (2014 est.)

Import Commodities: Machinery, vehicles, pharmaceutical products, plastic, semi-finished metal products, meat, fruits and nuts, optical and medical instruments, iron, steel

Major Suppliers: China 19.2%, Germany 11.2%, US 6.4%, Belarus 4.8%, Italy 4.6% (2015)

DEFENCE & SECURITY

Russia remains an important international actor both in the Eurasian region and in the world at large even after the disintegration of the Soviet Union. The Russian Armed Forces is one of the largest in the world and possess the second-largest nuclear arsenal. It is in the midst of a major programme of reform, including a gradual transition to an all-volunteer makeup with organisational changes and re-equipment of advanced weapons with the aim of making the country more agile and adaptable in meeting new threats both at home and abroad. Even as the reforms, initiated in the year 2008 continue, efforts are going on to modernize 70 percent of the military platforms and weapon systems by 2020.

The Supreme Commander of the Armed Forces is the President of the Russian Federation. Direct leadership of the armed forces is vested with the Ministry of Defence and the General Staff exercises operational control. The Russian military is divided into the Russian Land Force, the Russian Navy, the Russian Air Force, Russian Aerospace Defence Force, Strategic Missile Force, and the Russian Airborne Troops.

In the wake of the rapidly evolving global geostrategic landscape and its new national security priorities, Moscow has reoriented its Armed Forces to meet four major objectives - - deterring the military and political threats to the security or interests of the Russian Federation; supporting economic and political interests of the Russian Federation; mounting other-than-war enforcement operations; and using military force. The Russian Armed Forces are trained to directly engage in armed conflict, local war, regional war and large-scale war.

Russia's strategic goal seeks to prevent global and regional wars and conflicts. The country is also increasing its security cooperation with other Soviet successor states that belong to the Commonwealth of Independent States (CIS). The updated military doctrine of the country released in 2014 calls for a more aggressive stance toward NATO, boosting Moscow's presence in the Arctic and strengthening its cooperation with India and China. It also exhorts for joint setting up of missile defence systems by Russia and allied countries to counter NATO's efforts to extend its air and anti-missile defence coverage over Europe. The doctrine has also highlighted the deterrent capacity of Russia's conventional weapons (in addition to its nuclear arsenal which can be used only in the event of an existential threat), referring to the country's growing conventional might. As a major nuclear power, Russia has inherited the arsenal of all of the former Soviet states which consists of silo-based as well as rail and road mobile ICBMs, sea-based SLBMs, strategic bombers, strategic aerial refuelling aircraft, and long-range tactical aircraft capable of carrying gravity bombs, standoff missiles, and cruise missiles.

As part of a wide-ranging military modernization programme, structural reforms are being carried out while combat capabilities of almost all branches of the Russian armed forces have increased with the acquisition of new weapon systems and advanced platforms. The Army is planning to replace a wide range of its current armoured vehicles fleet with a family of modular platforms. While new tanks and missiles are being rolled out for the Army, the Navy has been inducting new-generation submarines and surface ships equipped with high-precision weapons. The submarine-based intercontinental ballistic missile Bulava which saw few successful test flights in recent times has also been deployed. The Air Force is in the process of inducting hundreds of new warplanes and special mission aircraft over the next decade. The state-of-the-art warplanes being received by the Air Force include new models such as the Su-34 bombers, Su-35 fighters and Mi-28 helicopter gunships equipped with sophisticated electronics and high-precision missiles. The military also plans activation of a new cyber warfare unit to protect the country against cyber-attacks and to wage cyber warfare. The Kremlin has approved a new "information security doctrine" in December 2016 aimed at bolstering the country's defences against cyberattacks.

The country's Armed Forces are also conducting regular military exercises across the country and also bilateral and multi-lateral drills beyond its territory. Between December 2014 and September 2015, the Russian troops have conducted about 2,000 tactical exercises of different levels. Soldiers are being trained to operate the latest unmanned aerial vehicles. A dedicated unmanned air vehicle unit has been established reportedly in the eastern region of Chukotka.

A separate Arctic command has been formed by the country and the military has also restored long-abandoned Soviet-era airfields and built radar stations in the region. Russia is also in the process of building 10 airfields and 13 air-defence radar stations in the New Siberian Islands, Franz Josef Land, Novaya Zemlya and other parts of Russian Arctic. The ground element of the command will incorporate two Arctic motorised infantry brigades, Air Defence forces consisting of Pantsir-S1 missile and other systems, and the Northern Fleet will become its main striking force.

The Russian Defence Ministry serves as the administrative body of the country's Armed Forces. The country has four military districts (tactical-strategic commands) established as part of the military reforms which started in 2008, replacing the former six districts.

KEY DEFENCE PERSONS

DEPUTY PRIME MINISTER IN CHARGE OF DEFENCE INDUSTRY & MILITARY PROCUREMENT: Dmitry Rogozin

MINISTER OF DEFENCE & GENERAL OF THE ARMY: Sergei Shoigu

CHIEF OF GENERAL STAFF, FIRST DEPUTY DEFENCE MINISTER & GEN. OF THE ARMY: Valery Gerasinov

COMMANDER-IN-CHIEF OF THE LAND FORCES: Col. Gen. Oleg Salyukov

COMMANDER-IN-CHIEF OF THE NAVY: Adm. Vladimir Korolev

COMMANDER OF THE AEROSPACE FORCES: Col. Gen. Viktor Bondarev

COMMANDER OF THE STRATEGIC MISSILE FORCES: Col. Gen. Sergei Karakaev

COMMANDER OF THE AIRBORNE FORCES: Col. Gen. Andrey Serdyukov

Internal Conflict

Russia's internal problems stem mainly from separatist uprisings and attempts to secede. The disintegration of the Soviet Union on 26 December 1991 resulted in political instability leading to ethnic conflict in Russia. The minority groups have been fighting for autonomy to control their own people, language, culture, and religion. Post-Soviet entities, especially Georgia, represent the greatest problem to Russia even today.

Georgia and Russia had a military conflict over the autonomous region of South Ossetia, which had been de-facto independent from Georgia. On 26 August 2008, Russia unilaterally recognised the independence of South Ossetia and Abkhazia. This recognition was strongly condemned by the European Union, the United Sates, North Atlantic Treaty Organisation (NATO) and the Organisation for Security and Co-operation in Europe (OSCE). Russia's military support and subsequent recognition of Abkhazia and South Ossetia independence continue to sour Moscow's relations with Georgia.

The Russia-Chechen conflict is another major internal area of concern for Moscow. The Southern Caucasus region is a strategic pathway from Russia to oil reserves in Central Asia and the Caspian Sea. In 1991, secessionist movements developed in Chechnya were strongly resisted by Moscow. Russia's implacable opposition to Chechen independence was also based on the region's importance as a centre of the oil industry and home to several chemical factories. Even though Russia has established its firm control over the autonomous region after decades of fighting for independence, separatist groups continue low-level guerrilla attacks even today. Besides, jihadist groups, including those aligned with Islamic State terrorist organization, also exist in the region.

Territorial Disputes

Japan: Russia's territorial dispute with Japan over a group of islands in the Pacific Ocean continues. The islands, administered by Russia and called as the Southern Kurils, are being claimed by Japan. These islands of Etorofu, Kunashiri, and Shikotan, and the Habomai group, known in Japan as Northern Territories, were annexed by the then Soviet Union in 1945 following Japan's defeat in World War II. The long-standing dispute has prevented the two countries from signing a formal peace treaty following World War II.

External Conflict

Ukraine: Russia is involved in a conflict with Ukraine over the control of Crimea, a Black Sea peninsula. Prior to the crisis, Crimea was a part of Ukraine as the Autonomous Republic of Crimea and the administratively separate municipality of Sevastopol. On 17 March 2014, the Crimean Parliament declared independence from Ukraine and asked to join the Russian Federation. The next day Russia and the separatist government of Crimea signed a treaty of accession of the Republic of Crimea and Sevastopol into the Russian Federation. However on 27 March 2014, the UN General Assembly passed a non-binding Resolution that declared the Crimean referendum invalid and the incorporation of Crimea into Russia as illegal. On 15 April 2014, the Ukrainian parliament declared Crimea a territory temporarily occupied by Russia.

THREAT PERSPECTIVE

NATO/US: According to the text of Russia's revised military doctrine of 2014, "NATO's military build-up" and the bloc's expansion toward the Russian borders are among the main external threats to Russia's security and territorial integrity. Other external threats listed in the revised doctrine include the development and deployment of strategic missile defence systems, the implementation of the "global strike" doctrine, plans to place weapons in space, as well as the deployment of high-precision conventional weapon systems.

The North Atlantic Treaty Organisation (NATO) – an alliance formed to defend Europe from Russia – is regarded by Moscow as its main military threat. The bloc is seen as one of the instruments of America's anti-Russian policies. NATO's building up of a missile defence shield over Europe with the US deploying interceptor missiles in Eastern European countries, including in Poland and Romania, has been viewed by Russia as a direct threat to its territory. The US has advanced in its plans to deploy air defence missiles, including the Patriot, THAAD and Standard Missile systems in several parts of Europe and on ships in the Black Sea and Mediterranean. The missile shield deployed in Romania has been activated in May 2016, further escalating the tensions with Moscow. NATO also prepares a new deterrent in Poland and the Baltics following Russia's annexation of

Crimea in 2014. Russia, on its part, has stationed anti-aircraft and anti-ship missiles in to its enclave of Kaliningrad, located between Poland and Lithuania.

While the US and NATO officials insist that the shield is designed to counter threats from the Middle East and not Russia, they remain vague on whether the radars and interceptors could be reconfigured to defend against Russia in case of a conflict.

Russia has also vehemently opposed Ukraine's integration with the West which may help NATO expand to its western borders.

Separatism/Terrorism: Russia is surrounded by a fairly unfriendly environment. Its neighbours, mostly the former Soviet Republics, are not keen to see Russia re-emerge as a great world power in the 21st century and become an influential centre of a multi-polar world order as it was once upon a time. Many of its neighbours could hinder Moscow in pursuing its national interests, while trying to weaken its position in Europe, West Asia, Trans-Caucasus, Central Asia and in the Asia-pacific Region.

Islamic terrorism also poses another major threat to the security of Russia with most terrorist activity taking place in Chechnya and Dagestan. The Russian government has banned seventeen terrorist organisations.

STRATEGIC RELATIONS

India: Russia has shared a very deep strategic partnership with India for decades. Over the years, defence ties between the two countries have transformed from a buyer-seller relationship to one involving joint research, design, development, and production of advanced military hardware. The two nations are working together on a number of high-technology military systems, including the highly successful BRAHMOS supersonic cruise missile system. Russia has shared defence technology of a strategic nature with India, including that of the aircraft carriers and nuclear submarines. The two sides also have periodic exchanges of armed forces personnel and military exercises. Moscow has also provided the precision usage of its global navigation system GLONASS to India.

China: The rapidly changing global geostrategic equations have firmed up strategic ties between Russia and China in recent times. Both are permanent members of the UN Security Council and have called for a multi-polar world order. They have often presented a united front against the West and its allies and also stood together on the 2011 Libyan conflict and the ongoing civil war in Syria. Moscow has also allied with Beijing in the South China Sea dispute which has flared up in recent times, especially after an international tribunal's decision favouring Philippines over territorial claims in the region. Both Russia and China have also conducted a joint naval drill in the region in September 2016 following the ruling. Both countries' militaries are also carrying out joint drills in Central Asia and East Asia regularly.

Armenia: After the disintegration of Soviet Union, Armenia has been considered as the only ally of Russia in all of Transcaucasia. The two countries are members of an inter-governmental military alliance called the Collective Security Treaty Organization (CSTO) along with four other ex-Soviet countries – a relationship that Armenia finds essential to its security. Both countries are engaged in vigorous political dialogue. Bilateral relations have been successfully developing on the basis of good traditions and reached the level of strategic partnership, with Russia supplying critical military equipment to Armenia. The two sides have signed an agreement on military and technical cooperation. They have also signed an agreement in 2016 on joint force grouping of the two countries in the Caucasus region

Iran: Russia and Iran have been increasing their military cooperation in recent years. Russia has assisted Iran with its conventional military, including training and defensive weapons that could serve to protect against air strikes targeting Iran. Although Russia has voted in favour of UNSC sanctions against Iran's nuclear programme, it has helped Iran develop the Bushehr facilities and has supported Iran's right for peaceful use of nuclear technology in the International Atomic Energy Agency (IAEA). Moscow has also supplied the S-300 long-range air defence systems to Tehran.

North Korea: Moscow has gained unique and exclusive communications capabilities with Pyongyang based on the development of trust between the two states' leaderships at the highest political levels. Using its "special" relations in the political sphere, Moscow indirectly stimulated the beginning of economic reforms in North Korea and its gradual involvement in the processes of international cooperation in Northeast Asia. There is an exchange of Defence Ministry delegations on the basis of an agreement signed back in 1992.

Kazakhstan: Russia and Kazakhstan have good bilateral relationship. The two nations also work together within the CSTO. Agreements and conventions are periodically renewed which outline the general framework and other aspects of their close cooperation. Russia has invited Kazakhstan to join its missile early warning system and space monitoring network.

Belarus: Russia and Belarus have made great strides in regional military alignment, coordinated air defence and joint military manoeuvres in the past decade. They have decided to take their bilateral relationship further by signing a draft treaty on the exchange of geostrategic information between their militaries. Russia also plans to base fighter jets at a military airbase in Belarus. Moscow has delivered four battalions of S-300 surface-to-air missiles and also agreed to supply Sukhoi-30 fighters to the Belarusian Air Force.

Multilateral Alliances

SCO: The Shanghai Cooperation Organisation (SCO) is an inter-governmental group of Central Asian countries founded by Russia, China, Kazakhstan, Uzbekistan, Kyrgyzstan and Tajikistan. India and Pakistan are set to join the forum as full members in 2017. The organisation aims to promote economic, political and strategic cooperation between its member states. The SCO's main goal has been to serve as a forum to ease tensions in the region and fight against "terrorism, extremism and separatism."

BRICS: The BRICS is an independent international organisation encouraging commercial, political and cultural cooperation between the BRICS nations - Brazil, Russia, India, China, and South Africa. As of 2014, the BRICS nations represented 18 percent of the world economy.

DEFENCE CAPABILITIES

ARMY: The Land Forces of the Russian Federation is primarily tasked with "repelling the aggression of the enemy in continental theaters of military operations, protect territorial integrity and national interests of the Russian Federation." It is the equipped with a wide range of weapons and trained in various methods of warfare, including in joint operations alongside the other wings of the Armed Forces. The Land Forces consists of the following units: Motorised Rifle Troops, Tank Troops, Missile Troops and Artillery, Air Defence Troops, Reconnaissance formations and military units, Engineer Troops, Troops of Radiological, Chemical and Biological Defence, Signal Communications Troops. The Army is becoming more professional and gradually moving away from universal conscription.

Personnel: 300,000+

Equipment

Category	Name	In Service
MBT	T-14 Armata	100+ (on order)
	T-90	400
	T-80	3,500
	T-72	9000
	T-72 B3	250+
Armoured car/IFV/APC	Kurganets-25	
	T-15 Armata	
	BMD4	105 +
	BMD-4M	12
	BMD	2000
	BMP-1/ 2	25,000
	BMP-3	700
	BRDM-2	2000+
	ATOM	
	Boomerang	
	Kurganets-25	

Army continued...

Category	Name	In Service
Armoured car/IFV/ APC	BTR-50/-60/-70/-80	25,000
	BTR-90	
	BTR-82A/AM	300
	BTR-MDM Rakushka	12
	BPM-97	200+
	MT-LB	3000+
	2S25 Sprut-SD	24
	IVECO LMV	358
Artillery	122mm	40,000
	130mm	
	152mm (towed)	
	122mm & 152mm (SP guns & howitzers)	
Mortar	82mm	
	120mm	14,000
	160mm	
	240mm	
	Gun/mortars	1,100
MLR	122mm	
	140mm	
	220mm	
	240mm	7,500
	300mm Smerch (Tornado)	
	TOS-1 Buratino	
RL	RPG 30	
	40mm RPG-7	
	64mm RPG-15	
	73mm RPG-16	
ATK gun	57mm	
	76mm	
	85mm	8,000
	100mm (towed)	
	ASU-57/85 SP gun	
ATGW	3M11 Fleyta	
	3M6 Shmel	
	9M14 Malyutka	
	9K111 Fagot	
	9M113 Konkurs	
	9K114 Shturm	
	9K115 Metis	
	9K112 Kobra	
	9M117 Bastion	
	9M119 Svir	
	9M133 Kornet	
	9M13 Metis-M	
	9M112 Kobra	
	9M120 Atakas	
	9M123 Khrizantema	
Weapons	SS-21 Tochka (SSM)	100+

Army continued...

Category	Name	In Service
Weapons	Iskander-M (SSM)	57
	2K11 Krug/SA-4 (SAM)	500
	2K12 Kub/SA-6 (SAM)	400
	9K33 Osa/SA-8 (SAM)	400
	9K31 Strela-1/SA-9 (SAM)	200
	Buk/SA-11 (SAM)	250
	S-300V/SA-12 (SAM)	100
	9K35Strela-10/SA-13 (SAM)	350
	Tor missile/SA-15 (SAM)	100
	Igla/SA-16/-18/-19 (SAM)	
	Pantsir-S1/SA-22 (SAM)	
	S-400 Triumf/SA-21 (SAM)	

New Procurements/Upgrades

✦ The Defense Ministry has signed an order for a pilot batch of over 100 state-of-the-art T-14 "Armata" main battle tanks. Delivery of the new MBTs is expected to continue till 2025.

T-14 Armata MBT *An internet photo*

✦ Test trial of the new Kurganets-25 combat platform is expected to start in 2017. The new amphibious IFV/APC versions were unveiled at the Victory Day Parade in Moscow in May 2015.

✦ Mass production of the new Boomerang APC, unveiled in 2015, is scheduled to begin in 2017 with deliveries to the Army likely to start in 2019. The heavily armed and amphibious Boomerang 8x8 APCs are set to replace the BTR-80 and BTR-82 family of wheeled APCs as well as other older models currently operational in the Russian army.

✦ The army has taken delivery of at least 250 T-72B3s, the upgraded version of the Soviet-era T-72B MBTs.

✦ The army has taken delivery of 12 BTR-MDM Rakushka APCs. About 2,500 BTR-MD Rakushka tracked APCs are to be procured by 2025.

✦ Nearly 600 BMP-2 IFVs are being upgraded with deliveries going on and expected to be completed by 2018.

✦ A new armoured recovery vehicle (ARV) based on the Kurganets-25 IFV is under development. The ARV will be deployed to support infantry and marine battalions equipped with the new Kurganets-25 and Boomerang combat vehicles, as well as the older BTR-80, BTR-82B, BMP-2 and BMP-3 vehicles.

✦ The MoD plans to buy up to 120 Iskander-M tactical ballistic missile systems to equip at least five missile brigades of Iskander-M complexes.

✦ Army trials of the new Koalitsiya-SV self-propelled howitzer is expected to start in 2017.

✦ The advanced RPG-30 anti-tank rocket launcher has entered service.

Army Aviation Equipment

Category	Name	In Service
Helicopter	Ka-52 Alligator	
	Mi-28N(E) Night Hunter	10
	Mi-24/V/PN (attack)	700
	Mi-8 (transport)	700
	Mi-26 (transport)	12
	Mi-17V-5	10

By 2020, Russia plans to add six more Army Aviation airbases to the eight existing ones. In addition, over 1,000 new helicopters are planned to be inducted in the coming years. In 2016, more than 50 new platforms, including the Kamov Ka-52 "Alligator," the Mi-28NE "Night Hunter" and Mi-8AMTSh "Terminator" helicopters have been inducted in the Army Aviation wing.

NAVY: The Russian Navy, consisting of the Northern Fleet, the Pacific Fleet, the Black Sea Fleet, the Baltic Fleet, the Caspian Flotilla, Naval Aviation, and the coastal troops consisting of the Naval Infantry and the Coastal Missile and Artillery Troops, is primarily tasked with defending and protecting Russia's interests and conduct of operations in the maritime theatre of war. The Navy is capable of attacking enemy land-based facilities with nuclear weapons, destroying enemy naval forces at sea and on bases, disrupting its maritime lines of communication and protecting friendly maritime shipping operations. It can also support the Ground Forces during the operations in the continental theatre of war, insert maritime forces into the area, repel enemy's assaults and accomplish a number of other missions.

The Russian navy is currently undergoing rapid modernization. As part of the naval rearmament programme, the Defence Ministry intends to procure some 100 new warships, including new-generation submarines, destroyers, corvettes and frigates, by 2020 along with new weapons, including advanced ballistic and cruise missile systems. Inclusion of all the new platforms and weapon systems are going to bolster Russia's position as a major maritime power in the world. The country is also building up a highly capable surface combatant.

A new 2015-2020 naval doctrine has also been released which focuses on Russia's strategic objectives along its borders and also in the Atlantic Ocean, the Arctic and the Antarctic.

The document underlines the Pacific, the Caspian Sea and the Indian Ocean as the areas of strategic importance to Russia.

Major naval bases:

Northern Fleet – Severonmorsk HQ, Motovskij Gulf, Polyarny, Severodvinsk, Archangelsk

Baltic Fleet – Baltiysk, Kaliningrad HQ, Kronstad

Black Sea Fleet – Sevastopol HQ, Poti, Novrissisk

Pacific Fleet – Vladivostok HQ, Petrapavlovsk, Sovyetskaya Gavan, Korsakov, Providenie, Magadan, Najin

Personnel: 105,000+

Equipment

Category	Name	In Service
Aircraft Carrier	Admiral Kuznetsov class	1
Cruiser	Pyotr Velikiy (nuclear powered)	1
	RFS Admiral Nakhimov	1
	Moskva Slava class	1
	Marshal Ustinov Slava class	1
	Kerch class	1
Destroyer	Admiral Chebanenko (Udaloy II type)	1
	Udaloy class (+1 reserve)	7
	Sovremenny class (+5 reserve)	5
Submarine/ Research submarine	Borei class SSBN	8 (total order)
	Typhoon class (test boat)	1
	Delta IV class	6
	Oscar II class	4
	Severodvinsk class	1+4
	Sierra I/II class	2
	Akula I/Akula II class	11
	Lada class	1 (+2 under constr-uction)
	Kilo class	9 (+8 reserve)
	Varshavyanka-class (Project 636.3)	6
	Kalitka	1
	Orenberg	1
Frigate	Neustrashimy class	2
	Krivak I/II/III class (+4 reserve)	3
	Tatarstan class	2
Corvette	Steregushchiy class	4+4
	Grisha I/II/III/IV/V class (+ some 20 reserve)	8
	Parchim class (+4 reserve)	8
	Tarantul I/II/III class	10
	Nanuchka I/III class	6
	Dergach class (hydrofoil)	2

Navy continued...

Category	Name	In Service
Patrol craft	Grad Sviyazhsk	2+5
	Murayev class attack hydrofoils	9
	Matka class FAC missile hydrofoils	8
	Babochka class FAC hydrofoils	3
	Pauk I/II (patrol)	24
	Svetlyak class	30
	Stenka class	20
	Astrakan class river gunboats	3
Minesweeper	Evgenya class	12
	Natya class	30
	Sonya class	51
	Gorya class	2
	Lida class	11
Amphibious Force	Ropucha class	19
	Aligator class IST	7
	Polnocny class LCT	4-5
Hovercraft	Pomornik class	10
	Aist class	9
	Lebed class	2
	Utka class (WIG)	1
	Drakon class (WIG)	2
Auxiliaries & Support ship	Boris Chilkin class AOs	4
	Ugra class submarine tender	2
	Don class submarine tender	1
	Don class	2
	Susanin icebreaker	13
	Lama class (missile support ship)	5
	Amga class (missile support ship)	3
Air Defence	Igla-1M	
Weapon	3M-54 Klub	
	K-300P Bastion-P	
	Kh-35 (ASM)	

New Procurements/Upgrades

✦ The Kirov-class nuclear-powered battlecruiser Admiral Nakhimov, currently being overhauled and modernised, is expected to join back the Navy in 2018.

✦ Plans are on to upgrade the Navy's sole aircraft carrier, Admiral Kuznetsov.

✦ The Navy plans to order 12 advanced Leader-class (Project 23560) destroyers to enter service from 2023-2025. The new platforms will be built in two variants – one with nuclear power plant and another with gas turbine power generating units. The design work on the new platforms has started. The destroyers will replace several existing warship types of Russian Navy's inventory, including guided missile cruisers and large antisubmarine warfare ships.

✦ A contract has been signed in 2016 to design and develop a class of fifth-generation nuclear-powered submarines which are expected to be built after 2020. The new platforms would be named as Husky-class.

✦ Additional five Borei class nuclear submarines are under construction. A total of eight such platforms are expected to enter service by 2020.

✦ A new class of fifth-generation conventional submarines under Project Kalina are being developed with air-independent propulsion (AIP) system. The first vessel is slated to be developed by 2018.

✦ The fifth and sixth Varshavyanka-class (Project 636.3) diesel-electric submarines have been delivered in 2016 for induction in the Navy's Black Sea Fleet. A new contract has been awarded to the Admiralty Shipyard in September 2016 to build six more platforms of the same class with deliveries scheduled to be completed before 2021. The new submarines will be deployed in the Pacific Fleet of the Navy.

✦ Acquisition of at least 10 fourth-generation Project 885 Yasen-class multi-role nuclear attack submarines is planned by 2020.

✦ Two more Lada class (Project 677) diesel-attack submarines are under construction with expected delivery scheduled for 2019. The lead sub of the class has been operational since 2010.

✦ The MoD has decided to overhaul the Project 971 (Akula class) of nuclear-powered attack submarines. The refurbished platforms will feature better stealth capabilities and improved electronics among other upgrades.

✦ Six new frigates under Project 22350 Admiral Gorshkov-class are under construction and are slated for induction by 2025. The lead ship of the class is undergoing sea trials and is expected to join the Navy in 2017.

✦ The first two Admiral Grigorovich-class (Project 11356M) frigates have been commissioned into service in March and June 2016 while delivery of a third platforms is expected in the coming months.

✦ A new class of missile corvettes under Project-20386 is being planned to be built. The lead ship of the class has been assigned the name "Dersky" the keel for which has been laid in October 2016.

An illustration of the new Project-20386 class missile corvette *A Russian MoD photo*

✦ Five of Project 21631 Buyan-M class of corvettes have entered service with the Navy. The MoD had previously ordered a total of nine Buyan-M corvettes. A new contract has been signed in 2016 to procure more such platforms. The exact number of the ships to be delivered to the Navy under the new contract has not been specified.

✦ New multi-role corvettes under Project 20385 (Gremyashchy class) are being built with the lead ship of the class expected to be launched in 2017. The new platforms are larger and more advanced variants of the Project 20380 and Project 20381 design (Steregushchiy-class) of corvettes. All the new ships are intended to replace the Navy's Project 1124/1124M (Grisha-class) anti-submarine warfare corvettes.

✦ New class of small missile corvettes under Project 22800 are being built. Induction of the new small corvettes is planned to start from 2017-18, with a total of 18 units to be built.

✦ Three new landing ships - Denis Davydov, Lieutenant Rimsky- Korsakov, and Michman Lermontov have joined the Baltic fleet.

✦ Russia is building an innovative class of Dyugon high-speed landing craft (LCU) under Project 21820 to replace the Ondatra class.

✦ Two landing ships under Project 11711 are being built. The lead ship, Ivan Gren, is expected to join the Navy in 2017.

✦ Construction of a new class of patrol ships under Project 22160 has started. The lead ship of the class, Vasily Bykov, is slated to join the Navy in 2017. A total of six such platforms will be built.

✦ Four new minesweepers under Project 12700 (Aleksandrit class) are being built and will be inducted by 2019.

✦ The Navy has taken delivery of the reconnaissance vessel Yuri Ivanov (Project 18280). The second vessel of the class is under construction. The navy plans to procure a total of four medium intelligence ships.

✦ A new torpedo, Khishchnik ("Predator"), is being designed to replace the Shkval ("Squall") torpedo presently in service with the Navy.

Naval Aviation

Equipment

Category	Name	In Service	On Order
Fighter	MiG-29K	10	10
	MiG-29KUB	4	
	Su-30SM	8	20
	Su-33 (shipborne)	12	
Bomber	Tu-22M Backfire	58	
	Su-24 Fencher	51	
ASW	Tu-142 Bear F	30	
	Be-12 Mail	30	

Naval Aviation continued...

Category	Name	In Service	On Order
ASW	IL-38/38N May	25	
	Mi-14 Haze	20	
	Ka-25 Hormone-A	20	
	Ka-27 Helix	100	
	Ka-31		
Transport	Tu-134	4	
	An-12 Cub	31	
	An-24 Coke	12	
	An-26 Curl	39	
	An-72	2	
	An-140	1	
Recce/EW/ Targeting	Tu-142D	15	
	An-12	7	
	Su-24	6	
	Kamov Ka-27		
	Kamov Ka-52K		
Helicopter	Mi-8	63	
	Mi-14	9	

New Procurements/Upgrades

✦ A new high speed helicopter is under development for the Navy. The advanced rotorcraft would be capable of cruising at 450 km/h and flying a distance of up to 1,500 km.

✦ The Navy has taken delivery of 10 MiG-29K and four MiG-29KUB fighters. The MoD had signed a contract for 20 MiG-29Ks and four MiG-29KUBs.

✦ 8 Su-30SM fighters have been delivered so far. 20 more are on order with deliveries scheduled to be completed by 2018.

✦ The Ilyushin Il-38N, an upgraded version of the Il-38 ASW aircraft, is being delivered and inducted in the Navy's Pacific Fleet. The aircraft is capable of carrying out electronic intelligence (ELINT) missions.

✦ New Kamov Ka-60 utility helicopters are on order.

Coastal Artillery & Rocket Troops: 7,000+ personnel

Naval Infantry

Personnel: 8,000

Equipment

Category	Name	In Service
MBT	T-80	150
	T-72	150
	BRDM	60
APC	BTR-60/-70/-80	300
	MT-LB	300
Artillery	2SI 122mm	120
	2S3 152 SP	60
AA Gun	ZSU-23-4 SPAAG	50
SAM	SA-7/-8/-9/-13	

New Procurements/Upgrades

✦ The new BTR-82 wheeled APC is being introduced.

AEROSPACE FORCES: In August 2015, the Russian Government merged the country's Air Force along with the Aerospace Defence Forces and the Air Defence Troops into a new branch of the Armed Forces known as the Russian Aerospace Forces. As part of its military modernisation programme, Russia plans to acquire more than 600 modern aircraft, including fifth-generation fighters, as well as over 1,000 helicopters and a range of air defence systems.

The country is successfully replacing its older-generation combat aircraft with newer variants. The new fighter platforms introduced in service include the Su-35S multi-role jets and Su-34 fighters/bombers while induction of the advanced MiG-35 warplanes and the fifth-generation T-50 PAK FA fighters are expected from 2017 onwards. Development of new military transport aircraft, refuelling aircraft, AWACS aircraft and other platforms and weapons systems are also advancing with inductions planned in the coming years. The country has also announced a programme to spend about US$9 billion by 2020 on the development and procurement of a wide range of tactical and strategic drones to boost reconnaissance, communications and combat capacities of its armed forces. The Air Force is set to receive its first unmanned combat drone (UCAV) by 2020.

Personnel: 180,000+

Equipment

Category	Name	In Service	On Order
Fighter/ Interceptor	MiG-31BM	24	
	MiG-31	122	
	MiG-29	120	
	MiG-29UB/UBT	30+6	
	MiG-29SMT	29	16
	Su-27/27SM	340	
	Su-30M2	30	
	Su-30SM	30	88 (Total order)
	Su-35S	48	50
Bomber/ Strike/ Ground Attack	Tu-95MS	63	
	Tu-160/160M Blackjacks	16	
	Tu-22M Backfire	145	
	Tu-95SM Bear H6	25+	
	Tu-95SM Bear H16	30	
	Su-34	36	200 (Total order)
	Su-24M/M2 Fencer	400	
	Su-25B/SM	180	

Category	Name	In Service	On Order
Reconnaissance/EW	MiG-25R	40	
	Su-24M2	32	
	Su-24M	70	
AWACS/ AEW&C	A-50 Mainstay	15	
	A-50U	3	
Tanker	IL-78 Midas	20	
Helicopter	Mi-24/-6/-8/-26/-14		
	Mi-28N		60+
	Kazan Ansat		
Transport aircraft	An-12	700	
	An-124	11	
	IL-76MD/TD	120	
	An-24	100	
	An-32	50	
	An-72	30	
	Tu-134s/154	20	
	L410	150	
Trainer	Yak-52	350	
	L-29/L-39	1500	
	Yak-130	70+	109 (total order)
Weapon	SA-10/12 (SAM)		
	S-500 (SAM)		
	SA-21 Growler (SAM)		
	Pantsir M (SAM)		
	S400 (SAM)		
	S-300V4 (ABM)		
	Topol-M (ICBM)		
	Yars missile (ICBM)		
	RT-12M2 Topol-M (ICBM)		
	Iskander-M (BM)		

New Procurements/Upgrades

✦ The T-50 fifth generation stealth fighters are expected to enter service in 2017. Designed and built by the Sukhoi Company, nine prototypes of the new combat plane are presently undergoing flight trials.

✦ The MiG-35 is a 4++ generation jet fighter currently under development. Ten prototypes have been built and are undergoing field trials. The single seat version is designated MiG-35 and the two-seat version is MiG-35D. The Air Force hopes to induct the new fighters in service from 2017 onwards.

✦ Delivery of modernised MiG-31BM 'Foxhound' interceptors is going on. The Air Force will receive more than 130 upgraded MiG-31B interceptor fighters.

✦ Air Force has taken delivery of 48 Su-35S multi-role fighters so far while another order has been awarded to Sukhoi in 2015 to deliver 50 more such fighters.

✦ The Air Force has so far received 30 Su-30SM fighters. A total of 88 such jets are set to be inducted into service by 2018.

✦ Delivery of 16 upgraded MiG-29 fighters (in MiG-29SMT) configuration is going on.

✦ Delivery of Su-34 frontline bombers is going on. The Defence Ministry has placed an order for around 200 Su-34s which are to be delivered by 2020.

Su-34 fighter/bomber *A Sukhoi photo*

✦ Delivery of Yak-130 combat trainers is going on. Orders have been placed for over 100 such jets.

✦ The upgrade of Tu-160 Blackjack strategic bombers fleet to Tu-160M2 standard is expected to be completed by 2018 with official tests due for completion in 2021. The Defense Ministry plans to buy around 50 such platforms.

✦ Delivery of the upgraded Tu-95MS bombers, now designated Tu-95MSM, is going on.

✦ The maiden flight of the first PAK DA next generation long-range bomber is expected to be performed in 2019. PAK DA will be equipped with the newest complex of the radio electronic combat and high-precision weaponry.

✦ Russia is developing the Beriev A-100 airborne early warning and control (AWACS) aircraft based on the Il-76MD-90A (Il-476) transport aircraft.

✦ Development of the Ilyushin Il-96-400TZ air-to-air refuelling aircraft is progressing with two platforms expected to enter service in near future.

✦ The first modernised IL-476 (also Ilyushin Il-76MD-90A) heavy-lift military transport aircraft has been delivered in 2015. A total of 39 such platforms are on order.

✦ Upgrade of Il-76MD transport aircraft to Il-76MDM configuration is going on with the first refurbished aircraft test flown in February 2016. Some 30 such platforms are to be delivered to the Air Force by 2020.

✦ The IL-214T medium-airlift military transport aircraft and the IL-112V light-military transport aircraft projects are currently under development to replace the An-12, An-24 and An-26 fleet of aircraft. The new planes are expected to enter service by 2018.

✦ Work on the development of a new interceptor aircraft has started to replace the existing 122 MiG-31 Foxhound planes by 2028.

✦ Procurement of over 60 Mi-28N attack helicopters is going on which are set to replace the existing fleet of Mi-24 Hind helicopters.

✦ Order has been placed for an initial batch of 24 Mi-28UB trainer variants of the Mi-28N/NE attack helicopters. A total of 60 such platforms with combat capability are expected to be inducted by 2020.

✦ Air Force is in the process of procuring the highly modernised Mi-35 Hind combat helicopters, about 30 Mi-8 and Mi-38 transport helicopters and Mi-26T heavy lift helicopters.

✦ Test flight of the new Yak-152 basic trainer aircraft has started in September 2016. The new trainer jet is likely to replace the Air Force's aging fleet of Yak-52 trainers. The Defence Ministry has ordered 150 of the new trainers.

STRATEGIC MISSILE FORCES

The Strategic Missile Forces (SMF) represents the main component of the country's strategic nuclear force. They are designed for nuclear deterrence of a possible aggression. They control Russia's land-based Inter Continental Ballistic Missiles (ICBMs). The SMF operates missile systems such as the silo-based R-36M2 which carries ten warheads, the UR-100NUTTH, the single warhead mobile RT-2PM Topol, and the upgraded Topol-M called RS-24 Yars. A new super-heavy thermonuclear-armed intercontinental ballistic missile, RS-28 Sarmat, is also expected enter service by 2018-19 and replace the RS-36M family of missiles by 2020. The SMF presently holds an inventory of 400 intercontinental ballistic missiles deployed all across the country.

Personnel: 120,000

Equipment

Category	Name	In Service
ICBM/SLBM	R-36M2 Voyevoda	40+
	RT-2PM2 Topol-M	18
	RS-24 Yars	58
	RS-12M Topol	70+
	UR-100 NUTTH	55+
	R-29 Sineva	
	RS-12M Topol-M	60

New Procurements/Upgrades

✦ Operational deployment of the RS-24 Yars ICBM is underway. The ICBM is a redesigned version of the Topol-M carrying an MIRV payload with four warheads. The SMF is expected to be equipped with 108 RS-24 missile defence systems by 2020.

✦ Russia is building a new railway-mobile missile system called Barguzin based on the Yars ICBM.

✦ The R-26 Rubezh land-mobile system currently under development is expected to become operational in 2016.

✦ RS-28 Sarmat, the future liquid-fuelled, MIRV-equipped, super-heavy thermonuclear ICBM is under development. The RS-28 Sarmat is expected to become operational in 2018.

✦ The new Servodvinsk class SSGNs will be deployed with either conventional warhead or 1 Kton nuclear warhead.

✦ The Kh-101/Kh-102 family of long-range cruise missiles is entering service to replace the Kh-55/Kh-555.

✦ Several S-300 PMU-2, Antey 2500 and S-400 land mobile batteries have been deployed. Delivery of the advanced S-400 air defence systems is going on.

AIRBORNE FORCES

The Airborne Forces (AbF) is primarily tasked "to flank enemy around by air and performing tasks in its rear to disrupt control of troops, seize and destruct ground elements of precision weapons, frustrate advance and deployment of reserves, violation of logistical support and communications, as well as to defend some directions, areas, open flanks, block and destroy air assaults, along with other tasks." In peacetime, the Airborne Forces perform the basic tasks on maintaining combat and mobilisation readiness at a level ensuring their successful use as intended.

The AbF has been the elite force in the Russian Military before World War II. They are not just a light infantry but a very mobile force fully mechanized and with a lot of firepower. The artillery of the AbF includes such systems as the 2S9 Nona, the D-30 howitzer or the 2S25 self-propelled tank destroyer. Its personnel currently totals 50,000 men and include 5 Airborne Divisions and 6 Airborne Brigades. All of them are air-deployable and on a top-level or readiness.

In 2016, the airborne troops carried out tasks concerning increase of their combat capabilities as well as re-equipment of formations and military units with modern armament and military hardware. The AbF formations have received new, modern armament and military hardware, including over 60 BMD-4Ms and BTR-MDMs, about 40 BREM-Ds, 2S9-1 Nona self-propelled artillery systems, Verba MLRS, modern EW stations, about 40 BTR-82A APCs as well as around 6,000 parachute systems and about 100 pieces of different airborne hardware. The troops have also held more than 2,300 combat training activities of different scales.

Major Armaments Producers

Aircraft Manufacturers/Design Bureau

Sukhoi, Mikoyan, Ilyushin, Yakovlev, Beriev.

Helicopter Designers & Manufacturers

Mil Moscow Helicopter Plant, Russian Helicopters, Federal State Unitary Enterprise Splav, Tupolev.

Weapons Manufacturers

Uralvagonzavod (UVZ), KAMAZ, Almaz-Antey, Tactical Missiles Corporation, IZH, Tikhomirov Scientific Research Institute of Instrument Design, Moscow Institute of Thermal Technology, Almaz Scientific Industrial Corporation, Vympel NPO, NPO Mashinostroyenia (NPOM.)

Shipbuilders

Amur Shipyard, Almaz shipbuilding, Yantar Baltic Shipbuilding, Sevmash, Admiralty Shipyard, Northern Shipyard (Severnaya Verf), Zalenodolsk Shipyard, Kalashnikov Concern.

Foreign suppliers: France, Finland, Austria and Italy

DEFENCE DEALS

Joint Venture Programmes

India

BRAHMOS: The world-class BRAHMOS supersonic cruise missile system with land-attack and anti-ship capabilities is a product of India-Russia Joint Venture programme. Currently, the air-launched version of the formidable weapon is advancing for flight trials onboard the Sukhoi-30 fighter. The missile has been inducted in the Indian Army, Navy and Air Force. The Governments of India and Russia have also agreed in principle to export BRAHMOS to a list of mutually agreeable countries for defensive purposes.

FGFA: India and Russia have agreed to work jointly on the development and production of the advanced fifth-generation fighter aircraft (FGFA) project for the Indian Air Force. Discussions are going on between the two Governments over the broad contours of the project with the signing of a work share agreement expected in the coming months or so. India is expected to commit an investment worth $4 billion over the coming years to develop a tailor-made version of the fighter. New Delhi has already spent $300 million in a preliminary design contract for the project with Russia that was completed in June 2013. State-run Hindustan Aeronautics Ltd. (HAL) is the Indian partner in the project.

China

Moscow and Beijing are in the final stages of negotiations on joint construction of a series of helicopters. The new line of helicopters will be built in China with materials and technologies provided by Russia.

Brazil

Russia and Brazil expect to sign a contract for the supply of combined short to medium range surface-to-air missile and anti-aircraft artillery weapons systems - the Pantsir S1 with the possibility of joint-production of sub-assemblies and accessories in Brazil.

Arms Export

Russia has retained its second position in the global arms exports market with a 27% share in 2014 despite sanctions imposed on its following the Ukrainian crisis. According to a report, Russia's current portfolio of arms export orders has grown by US$1 billion, and is now valued at around $50 billion.

Exports were largely unharmed by sanctions, as Russia's major trading partners are non-Western nations with close political ties to Moscow, such as China, India, Algeria and Venezuela. Russian Sukhoi Su-30 fighters, diesel submarines and medium and long-range air defence systems are particularly popular in the Asia-Pacific region. According to a Stockholm International Peace Research Institute (SIPRI) report, although Russian exports increased by 6.2 percent over 2014, the rate was "significantly slower than the 48.4 percent growth rate between 2013 and 2014".

Moscow signed contracts worth more than $4 billion with Iraq supplying the Solntsepyok flamethrower system and Mi-28NE attack helicopters to Baghdad. It has also supplied 63 Mi-17V-5 helicopters to Afghanistan and Mi-171-Sh helicopters to Peru. Russia is holding talks with Indonesia to sell Su-35 aircraft to replace Jakarta's aging fighters' fleet, and is in negotiations with Argentina, Nicaragua, Uruguay, and Peru on possible deliveries of new airplanes like the Yak-130 and MiG-29. According to the SIPRI report, Azerbaijan and Iraq featured on the list of top buyers of Russian armaments while Syria and Venezuela fell off the list. Moscow is now exploring new markets in Southeast Asia, Latin America and Africa to sell its military products.

DEFENCE BUDGET

Adversely hit by declining oil prices, Russia has announced a major curtailment of its annual military expenditure. The draft budget announced for the year 2017 has fixed the defence expenditure at 2,840 billion rubles ($45.15 billion), or 3.3 percent of GDP and a reduction of 1,000 billion rubles ($15.89 billion), or approximately 30 percent over previous year's allocation. The country's defence budget is projected to continue the downward trend – 3 percent of GDP in 2018, with a further drop to 2.8 percent in 2019.

NAVAL WARFARE SYSTEMS

The present-day Russian Navy was formed on the foundation laid by the Soviet Navy. After the disintegration of Soviet Union in the 1990s, the Russian fleet inherited the depleting base infrastructure from Soviet naval forces. Today, Russia is actively restoring the fleet's repair and maintenance system in order to strengthen its presence in the seas. With the renewal of the Russian Navy's activity in the middle of the 2000s, navigation in remote seas and oceanic zones increased. With the dramatic change of geopolitical balance in global maritime activity, Russia's maritime capacity, its economic and military components were revamped to real capability to maintain and protect Russia's interests in the

World Ocean. The country constructed technical base for North Fleet along with bases for the Pacific Ocean and Baltic Fleets. Russian Navy, today, consists of the following branches: submarine forces, surface forces, naval air force, marines and coastal defence troops. It also includes ships and vessels, unconventional units, rear divisions and subdivisions. Key Fleets include the Black Sea Fleet, the Pacific Ocean Fleet, the Northern Fleet, and the Baltic Fleet along with the Russian Caspian Flotilla, Naval Aviation, and the coastal troops. Russian Navy presently comprises over 90 ships 1 rank, more than 70 ships 2 rank and upward of 140 ships 3 rank, including those under repair, in reserve and being constructed.

Surface Ships

Aircraft Carrier: The Aircraft carrier serves as the flagship of the Russian Navy, equipped with a full-length flight deck and facilities for carrying, arming, deploying, and recovering aircraft.

Project 11435, Kuznetsov-class - Admiral Flota Sovetskogo Soyuza Kuznetsov is a Project 1143.5 Heavy Aircraft Carrying Cruiser, the only Russian aircraft carrier. The Admiral Flota Sovetskogo Soyuza Kuznetsov joined Northern Fleet on January 20, 1991. The Kuznetsov-class are used to support and defend strategic missile-carrying submarines, surface ships, and maritime missile-carrying aircraft of the Russian fleet. Russian aircraft carrier Admiral Kuznetsov's main fixed-wing aircraft is the multi-role Sukhoi Su-33 along with the Kamov Ka-27 and Kamov Ka-27S helicopters for anti-submarine warfare, search and rescue, and small transport.

Displacement: 43,000 tonnes

Length: 302.3m

Beam: 72.3m

Draft: 9.14m

Speed: 54 km/h

Propulsion: Steam turbines, 8 turbo-pressurised boilers, 4 shafts, 200,000 hp

Range: 15,700 km at 33 km/h

Endurance: 45 days

Armament: missile launchers (12 Granit anti-ship missiles and 60 Udav-1 ASW rockets), Klinok SAMS (192 missiles, 24 launchers) and Kashtan CIWS (256 missiles)

Aircraft: Capable to carry up to 24 Ka-27 multipurpose helicopters, up to 16 Yak-41M supersonic VTOL multirole aircrafts and up to 12 Su-27K deck-based fighters

Crew: 1960

Admiral Flota Sovetskogo Soyuza Kuznetsov. *(Image Credit: upload.wikimedia.org.)*

Heavy Missile Cruisers: The heavy cruiser is a type of cruiser designed for long range, high speed and an armament of naval guns.

Project 11442, Kirov-Class - The Kirov-class heavy missile cruiser is a class of nuclear-powered warship of the Russian Navy. Only Pyotr Velikiy (Peter the Great) is currently operational. Petr Veliky is a Project 11442 heavy nuclear-powered guided missile cruiser. The country's most powerful nuclear-powered battleship is the flagship of Northern Fleet. In May 2016 the Pyotr Velikiy put to sea for the first time in two years for drills off the coast of Northern Russia. Admiral Nakhimov is projected to re-enter the Russian Navy in 2018. Russia planned to reactivate the remaining two vessels by 2020.

Displacement: 24,300 tonnes

Length: 252m

Beam: 28.5m

Height: 59m

Draft: 9.1m

Propulsion: CONAS; 2 steam turbines and nuclear reactor; 140,000 hp

Speed: 59 km/h

Range: 1,850 km at 56 km/h

Endurance: 60 days for provisions and supplies, 3 years for fuel (unlimited for nuclear reactor)

Armament: AK-130 dual-mount 130-mm multipurpose turret artillery system; 20 SM-233 separate underdeck vertical launchers; 12 launchers (96 missiles) of Rif zonal AD system; 2 eight-tube containers of Klinok (Kinzhal) short range and close-in AD system; 6 3M87 Kashtan (Kortik) combined missile and artillery AD systems; 2 RPK-6M Vodopad 533-mm ASW missile and torpedo systems; RKPTZ-1 Udav-1M antitorpedo system; RBU-12000 and RBU-1000 Smerch-3 anti-submarine rocket launchers; 2 PK-14 coupled 150-mm launchers (decoy launch system)

Aircraft: 2 Ka-27 Helix heavy multipurpose helicopters

Crew: 655

HMS Dragon (foreground) with the Russian Kirov Class battlecruiser 'Pyotr Velikiy'. *(Image Credit: Royal Navy/MOD)*

Cruisers: The cruiser is the largest and most powerful surface combatant after the aircraft carrier.

Project 1164, Moskva - Moskva is a Project 1164 Atlant (Slava) lead ship, Guided Missile Cruiser. The cruiser has been repeatedly deployed in the Mediterranean and paid numerous calls at ports of the region. The ship, formerly known as Slava, was recommissioned as Moskva in April 2000 and replaced the Kynda-class cruiser Admiral Golovko as the flagship of the Black Sea Fleet. On July 22, 2016 Guards Missile Cruiser Moskva was awarded order of Nakhimov.

Displacement: 9,800 tonnes

Length: 186.4m

Beam: 20.8m

Draft: 8.4m

Propulsion: 4 COGOG gas turbines, 2 shafts 90,000 kW

EW & decoys: Rum Tub and Side Globe EW antennas; 2 × PK-2 DL (140mm chaff / flare)

Speed: 59 km/h

Range: 19,000 km at 30 km/h

Endurance: 30 days

Armament: 16 anti-ship missile launchers (16 P-500 Bazalt missiles; after modernization – 16 P-1000 Vulkan missiles); 1 x 2 130-mm gun mount AK-130, 6 sextuple 30-mm gun mounts AK-630; 8 x 8 B-204 launchers of S-300F Rif SAM system (ammunition load is 64 missiles); 2 x 2 launchers of Osa-MA SAM system (ammunition load is 48 missiles); 2 x 5 533-mm torpedo tubes; 2 x 12 RBU-6000 anti-submarine rocket launchers

Aircraft: Ka-27 helicopter

Crew: 510

Moskva is the lead ship of the Project 1164 Atlant class of guided missile cruisers in the Russian Navy. *(Image Credit: en.wikipedia.org)*

Project 1164, Varyag - Varyag is a Project 1164 Atlant (Slava) Guided Missile Cruiser. The warship joined the Pacific Fleet in 1990. Later, the ship was named flagship of the Russian naval task force positioned in the eastern Mediterranean in recent years.

Displacement: 11490 tonnes

Length: 186.4m

Beam: 20.8m

Height: 42.5m

Draft: 8.4m

Speed: 59 km/h

Endurance: 30 days

Armament: 16 Bazalt anti-ship missile launchers (ammunition load is 16 P-500 missiles), 2 torpedo tubes, RBU-6000 anti-submarine rocket launchers, 2 130-mm gun mounts AK-130, 1 30-mm gun mount AK-630, 2 Osa-MA SAM launchers, 8 S-300F Rif

SAM launchers
Aircraft: Ka-25/Ka-27 ASW helicopter
Crew: 480

Russian missile cruiser Varyag and project 1155 large ASW ships in the port of Vladivostok on 7 May 2010 *(Image Credit: United States Navy)*

Project 1164, Marshal Ustinov - Marshal Ustinov is a Project 1164 Atlant (Slava) Guided Missile Cruiser. Marshal Ustinov is assigned to the 43rd Missile Ship Division of the Russian Northern Fleet.
Displacement: 9,800tonnes
Length: 186.5m
Beam: 20.8m
Height: 42.5m
Draft: 7.6m
Speed: 59.264 km/h
Endurance: 30 days
Crew: 476
Armament: 16 Bazalt anti-ship missile launchers; 2 five-container 533-mm torpedo tubes; 2 RBU-6000 antisubmarine rocket launchers; 2 coupled 130-mm gun mounts AK-130; 6 30-mm gun mounts AK-630; 2 coupled Osa-MA SAM launchers, 8 S-300F Rif SAM launchers
Aircraft: Ka-25/Ka-27 ASW helicopter

Destroyers: Destroyers can operate independently or as part of carrier strike groups, surface action groups, amphibious ready groups, and underway replenishment groups.

Project 956, Sovremennyy class - The Sovremennyy-class destroyer is the principal anti-surface warship of the Russian Navy. The ships encompass anti-ship, anti-aircraft, anti-submarine and coastal bombardment capability. The ship was primarily made for anti-ship operations and was designed to complement anti-submarine warfare Udaloy destroyers. Destroyers in this class include Admiral Ushakov, Gremyaschy, Bezboyaznenny, Boevoy, Bistry, Bespokoiny, Nastoichivy, Burniy. Bystryy (715), Gremyashchiy (406), Bespokoynyy (620), Nastoychivyy (610) and Admiral Ushakov (474) are currently in active service.
Displacement: 6,600 tonnes
Length: 156m
Beam: 17.3m
Draft: 5.96m
Endurance: 30 days
Propulsion: 2 GTZA-647 steam turbines 50,000 hp each; 2 fixed-pitch screws
EW & Decoys: 2 PK-2 decoy dispensers (200 rockets)
Speed: 60.6 km/h
Range: 26,000 km at 26 km/h
Armament: 2 coupled gun mounts AK-130/54; 4 30-mm six-barrel gun mounts AK-630; 2 quadruple launchers for P-270 Moskit anti-ship cruise missiles; 2 six-barrel RBU-1000 ASW rocket launchers; 2 Shtil SAM launchers; 2 coupled 533-mm torpedo tubes
Aircraft: 1 Ka-27 helicopter
Crew: 350

A view of the Soviet Project 956 "Sarych" (Sovremenny class) destroyer.
(Image Credit: US DoD)

Project 01090, Kashin-class - The Kashin-class ship was designed to fulfil a variety of roles including anti-air and anti-submarine operations to defend task forces from submarines, low-flying aircraft, and cruise missiles. One ship of the Kashin-class in service with the Russian Navy. Russian destroyer Smetlivy is a Kashin-class guided missile destroyer of the Russian Navy and is now a part of the Black Sea Fleet.
Displacement: 3,400 tonnes
Length: 144m
Beam: 15.8m
Propulsion: 2 × COGAG; 2 shafts; 4 × M8E gas turbines
Speed: 70km/h
Range: 6,480km at 33km/h
Crew: 300
Armament: 2 quadruple launchers of Kh-35 Uran anti-ship missiles; 76-mm twin gun mount AK-726; 2 coupled launchers of Volna SAM system (16 missiles); 5 x 533-mm torpedo tubes (5 torpedoes); 2 ASW rocket launchers RBU-1000

Russian destroyer Smetlivy *(Image Credit: A.Kuzenkov, kchf.ru)*

Project 1155, Udaloy-class - The Udaloy I class are a series of anti-submarine destroyers, currently in the service of the Russian Navy. Ships in this category include Vice-Admiral Kulakov, Marshal Shaposhnikov, Admiral Vinogradov, Admiral Levchenko, Admiral Panteleyev, etc. The Project 1155.1 Udaloy II is equipped with more capable sonars, an integrated air defence fire control system, and a number of digital electronic systems based on state-of-the-art circuits. The destroyer Vice-Admiral Kulakov, which had been retired in 1990, was upgraded to Udaloy II for anti-surface warfare operations.
Displacement: 6,930 tonnes
Length: 163m
Beam: 19.3m
Propulsion: 2 shaft COGAG, 4 gas turbines
Speed: 65km/h
Range: 19,400km at 26km/h
Crew: 300
Armament: SS-N-14 Silex missile, SS-N-22 Sunburn missile; SA-N-9 Gauntlet; Kortik SAM; DP guns; AK-130 130mm guns; 30mm AA guns on Kortik CIWS; 553mm torpedoes; RBU-6000 anti-submarine rocket launchers
Aircraft carried: 2 Ka-27 'Helix' series helicopters

The Russian destroyer Admiral Vinogradov (DDG 572) is underway near the guided-missile cruiser USS Vella Gulf (CG 72) while conducting operations in the Gulf of Aden.

Frigates: Frigates are generally lighter than a destroyer are used to protect other warships and merchant navy ships. Russia has three active class namely Project 11540, Project 1135 and Project 1135.6M.

Project 11540, Neustrashimyy-class - Neustrashimy is a Project 11540 Yastreb Frigate which joined Baltic Fleet on March 14, 1991. Two ships, Neustrashimyy and Yaroslav Mudryy are currently in active service. The ships are used for anti-submarine warfare and incorporates some stealth technology.
Displacement: 3,800tonnes
Length: 129m
Beam: 15.6m
Draft: 8.35m
Propulsion: 2 shaft COGAG (gas turbines)
Speed: 56km/h
Crew: 210
Armament: Anti-ship missiles; Gauntlet SAM; ASW RBU-6000 launcher; 100mm gun, Kashtan CIWS; 533-mm torpedo tubes
Aircraft carried: Ka-27 Helicopter

Project 1135, Krivak-class – The Krivak-class ships were designed as a successor to the Riga class. Ladny is a Project 1135 Frigate. The ship was commissioned in 1981 in the Russian Black Sea Fleet. The Russian Navy Krivak II class Frigate Pytlivyy (Project 1135M) is operational again following maintenance work.

Displacement: 3,300tonnes

Length: 123.5m

Beam: 14.1m

Draft: 4.28m

Propulsion: 2 shaft COGAG; M-8k gas-turbines; M-62 gas-turbines

Speed: 59km/h

Range: 9,251 km at 26km/h

Crew: 200

Armament: URK-5 SSM/ASW missiles; Osa-MA SAM systems; RBU-6000 anti-submarine rockets; 76 mm guns; 533 mm torpedo tubes

Russian frigate Ladny (Project 1135) in Sevastopol Bay *(Image Credit: wikimedia.org)*

Project 1135.6M, Admiral Grigorovich-class - The Admiral Grigorovich class is the latest class of frigates ordered by the Russian Navy for the Black Sea Fleet. The frigates are based on the Talwar-class frigate. 6 ships are planned and 2 are ships are in service of Russian Navy, namely- Admiral Grigorovich and Admiral Essen. Admiral Makarov is scheduled for delivery in the third quarter of this year. The ships are assigned to the Black Sea Fleet.

Displacement: 3,620tonnes

Length: 124.8m

Beam: 15.2m

Propulsion: 2 shaft COGAG; DS-71 cruise gas turbines; DT-59 boost gas turbines

Speed: 56km/h

Range: 8,980km at 26km/h

Crew: 200

Armament: 100mm A-190 Arsenal naval gun; 8 UKSK VLS cells for Kalibr and 3M55 Oniks; Kashtan CIWS; Igla-1E; 533mm torpedo tubes; RBU-6000 rocket launcher

Aircraft carried: Ka-27PL or Ka-31 helicopter

Admiral Grigorovich *(Image Credit: Russian Ministry of Defence)*

Corvettes: Corvettes are small warships used for coastal patrol and as fast attack craft.

Steregushchiy-class - The Steregushchiy class is the newest class of corvette in the Russian Navy. It was commissioned into Baltic Fleet in 2008. The Steregushchiy-class ships are multi-purpose corvettes, designed to replace the Grisha class. The ships are used for littoral zone operations, engagement of enemy submarines and surface ships, and gun support of landing operations. Steregushchiy, Soobrazitelnyy, Boykiy and Stoykiy are currently active in service.

Displacement: 1,800 tonnes

Length: 104.5m

Beam: 11.6m

Draft: 8.6m

Propulsion: 2 shaft CODAD, 4 Kolomna 16D49 diesels

Speed: 50km/h

Range: 7,000km at 26km/h

Crew: 90

Armament: 00mm A-190 Arsenal; Kashtan CIWS-M; 4 Uran Kh-35; Redut VLS cells; AK-630M CIWS; 330mm torpedo tubes; 14.5mm MTPU pedestal machine guns

Aircraft carried: Ka-27 Helicopter

Parchim-class - The Parchim-class corvette are used for coastal anti-submarine warfare. 8 ships are in service with the Baltic Fleet of the Russian Navy.

Displacement: 800tonnes

Length: 72.50m

Beam: 9.40m

Propulsion: 3 shaft M504 diesels

Speed: 45.7km/h

Range: 3,900km at 26km/h

Crew: 80

Armament: 57 mm gun AK-725; 30 mm gun AK-230; SA-N-5 MANPAD; RBU-6000; 400 mm torpedo tubes

Tarantul-class - The Project 1241 or Tarantul-class corvettes are smaller in size compare to Nanuchka-class. Tarantula-II had a better sensor suite/equipment than the ships of the following class Tarantul-I. Tarantul-III has important upgrades such as a new type of propulsion — a CODAG (COmbined Diesel And Gas) system. Project 1242.1 and project 1241.8 Molniya are further developments of the Tarantul family of ships. Currently, the Russian Navy has 6 ships (project 1241.1/1241.7) and 20 ships (under Project 12411/1242.1) are in service.

Displacement: 480 tonnes

Length: 56.0m

Beam: 10.5m

Propulsion: 2 shaft COGAG turbines

Speed: 78km/h

Range: 26km/h

Crew: 50

Sensors and processing systems: Radar: Spin trough, Bass Tilt, Peel pair, Pop group

Armament: P-15 Termit/SS-N-2 Styx; P-270 Moskit/SS-N-22 Sunburn; Kh-35 Uran/SS-N-25; SA-N-5 SAM MANPAD; 76 mm AK-176 AK-630 30 mm gun; CADS-N-1 Kashtan CIWS

Russian missile corvettes project 12411 (Tarantul-III class) *(Image Credit: Wikepedia.org)*

Nanuchka-class - The Nanuchka class corvettes were designed around the P-120 Malakhit anti-ship missile. 11 ships under Nanuchka III (Project 1234.1) are in service with the Russian Navy.

Displacement: 560 tonnes

Length: 59.3m

Beam: 12.6m

Draft: 2.7m

Propulsion: 3 shaft Diesels

Speed: 56km/h

Range: 4,630km at 22km/h

Electronic warfare: Vympel-R2 suite with Foot Ball-A intercept, Half Hat-B intercept, 2 PK-16 decoy RL, 4 PK-10 decoy RL

Crew: 60

Armament: SS-N-9 SSM; SA-N-4 SAM; Twin 57mm AK-257 gun; 76mm AK-176 gun; 30mm AK-630 gun

Grisha-class – The Albatros (Grisha-class) are a series of anti-submarine corvettes. The Grisha-I class was decommissioned in 1979. Grisha-II are in service with the Russian Maritime Border Guard. Grisha-III incorporated several small-scale modifications, including a 30-mm gun and new electronics. About 20 remain in Russian service. Grisha-IV class was a trial version. Grisha-V had some new modifications such as a single 76 mm gun. 28 ships remain in the Russian Navy.

Displacement: 880 tonnes

Length: 71.6m

Beam: 9.8m

Propulsion: 3 shaft, 2 M-507A cruise diesels; boost gas turbine

Speed: 63km/h

Grisha V-class corvette (Soviet-era, Project 1124-M) *(Image Credit: US Navy)*

Range: 7,400km at 19km/h

Crew: 60

Armament: SA-N-4 Gecko surface-to-air missile launcher; 57mm 70-caliber AK-257 dual purpose guns; RBU-6000 anti-submarine rocket launchers; 533 mm torpedo tubes; depth charge racks or mines

Bora-class – Bora is a Project 1239 hover-borne guided missile corvette of the Russian Navy

which are used for marine combat purposes. Project 1239 Bora guided missile air cushion ship is designed to destroy enemy surface ships, fast attack craft, and transports in all water areas of closed sea theatres of operation. Bora and Samum naval ships are assigned to the Black Sea Fleet.

Displacement: 1,050 tonnes

Length: 64m

Beam: 18m

Draft: 3.3m

Propulsion: gas turbines M10-1; diesels M-511A; diesels M-504 6

Speed: 22km/h

Range: 4,600km at 102km/h

Crew: 68

Electronic warfare: Vympel-R2 suite with Foot Ball-A intercept, Half Hat-B intercept, 2 PK-10 decoy RL, 2 PK-16 decoy RL

Armament: quadruple launchers of Moskit ASM; MANPADS Osa-M launcher; 76-mm gun mount AK-176; 30-mm 6-barreled gun mount AK-630

The Bora-class missile corvette of the Russian Navy *(Image Credit: Wikipedia.org)*

Buyan-class - The Buyan-class corvette (Project 21630) was designed to reinforce Caspian Flotilla in littoral sea zone and rivers. The first ship of her class, Astrakhan, was commissioned in 2006. Buyan-M (Project 21631), the missile version of Buyan-class is used for the defence of national economic zones and to engage surface warships. These ships will be armed with missile and artillery weapons, and equipped with EW equipment. Ships namely, Astrakhan, Volgodonsk, Makhachkala, Grad Sviyazhsk, Uglich, Velikiy Ustyug, Zelenyy Dol and Serpukhov are in service. Vyshniy Volochek, Orekhovo-Zuyevo, Ingushetiya and Grayvoron will be commissioned in 2017.

Displacement: 500 tonnes

Length: 62m

Beam: 9.6m

Draft: 2m

Propulsion: 2 shaft CODAD; Zvezda M520; Pumpjet

Speed: 52km/h

Range: 2,800km

Crew: 29-35

Electronic warfare: 2 × PK-10 decoy launchers

Armament: 100 mm A-190; 30 mm AK-630 (AK-630-M2 in 21631); A-215 "Grad-M" (only 21630); UKSK VLS cells with Kalibr-NK system or P-800 Onix (only 21631); 3M47 Gibka (only 21630); Komar (only 21631); DP-65 anti-saboteur grenade launcher; 14.5mm KPV type; 7.62mm PKM type (only 21631)

Volgodonsk (Buyan class) *(Image Credit: Ministry of Defence of the Russian Federation)*

Landing ships: These ships are used to support amphibious operations by carrying vehicles, cargo, and landing troops.

Alligator-class - The Project 1171 (Tapir/Alligator) class are amphibious transport docks. Ships namely, Orsk, Saratov, Nikolai Filchenkov and Nikolai Vilkov are in active service with the 197th Brigade of Landing Ships in the Black Sea Fleet.

Displacement: 3,400 tonnes

Length: 112.8m

Beam: 15.8m

Propulsion: CODAD power plant; 2 diesels; 2 shafts

Speed: 33km/h

Capacity: 1,000 tonnes

Crew: 55

Armament: 122mm UMS-72 Grad-M; SA-N-5 SAM; 57-mm gun mount ZIF-31B; 2M-3M anti-aircraft machineguns; MANPADS launchers; MLRS A-215 Grad-M

Ropucha-class - The Project-775 Ropucha-class landing ships are designed for beach landings.

Displacement: 4,080 tonnes

Length: 112.5m

Beam: 15m

Draft: 3.7m

Propulsion: Diesel engines; propellers

Speed: 33km/h

Capacity: 500 tonnes of armour vehicles & 225 troops

Crew: 87

Armament: AK-257 guns; AK-176; 122 mm rocket launcher A-215 Grad-M; Strela 2(SA-N-5); 30 mm AK-630

Project 775M Ropucha-II class Russian tank landing ship Azov in Sevastopol bay.

(Image Credit: Wikipedia.org)

Zubr-class - The Project-12322 Zubr-class is an air-cushion landing craft and is utilised for sealift marines and tanks to shores. Yevgeniy Kocheshkov and Mordovia are in active service. The rectangular pontoon are the main load carrying part of the ship and provides high strength and buoyancy. The Zubr-class landing craft has a cargo area of 4,300 sq ft and a fuel capacity of 56 tonnes.

Displacement: 415 tonnes

Length: 57m

Beam: 25.6m

Draft: 1.6m

Propulsion: Kuznetsov NK-12MV gas turbines; propellers

Speed: 116.68km/h

Crew: 27

Electronic warfare: EC system & decoys; MS-227 chaff launcher; MP-411 ESM radar

Armament: Strela-3; Strela 2 quad launchers; 30mm AK-630; 140mm Ogon launchers; 122mm retractable rocket launchers; mines

Mordovia and Yevgeniy Kocheshkov (Zubr-class) landing in Kaliningrad Oblast.

(Image Credit: Ministry of Defence of the Russian Federation)

Serna-class – The Project-11770 Serna-class is a class of landing craft being constructed for the Russian Navy.

Displacement: 61 tonnes

Length: 25.6m

Serna-class landing craf*t (Image Credit: Ministry of Defence of the Russian Federation)*

Beam: 5.6m
Draft: 1.52m
Speed: 55.5km/h
Propulsion: M-503A-3 diesels; DG-16/1500 diesel generators
Capacity: 92 troops or 50 tonnes of cargo
Crew: 5

Dyugon-class - The Project 21820 Dyugon-class being constructed for the Russian Navy. Ataman Platov and Michman Lermontov are Dyugon-class landing crafts.
Displacement: 280 tonnes
Length: 46m
Beam: 8.6m
Propulsion: M502A-2D diesel engines
Speed: 65km/h
Capacity: 140 tonnes of cargo
Crew: 6

Michman Lermontov landing in Kaliningrad Oblast
(Image Credit: Ministry of Defence of the Russian Federation)

Ondatra-class - The Project-1176 Akula Ondatra-class are ideal for amphibious operations and littoral combat.
Displacement: 107.3 tonnes
Length: 24m
Beam: 5.2m
Draft: 1.55m
Speed: 21km/h
Propulsion: 3D12 diesels; DGR1A-16/1500-0M4 diesel-generator
Endurance: 2 days
Capacity: 22 troops or 50 tonnes of cargo
Crew: 5

Patrol Boats: The patrol boats are small naval vessel generally designed for coastal defence duties and are commonly found engaged in various border protection roles.

Grachonok-class - The Project-21980 Grachonok-class are used for water region patrol and are designed to combat sabotage and terrorist forces. Ships include Nakhimov, Kadet, Suvorov & Kursant-Kirovets.
Displacement: 139 m
Length: 31.04 tonnes
Draft: 1.85m
Speed: 43km/h
Endurance: 5 days
Crew: 8

Armament: marine machine-gun installation MTPU; DP-65A and DP-64 grenade launchers; Igla-1

Raptor-class - The Project 03160 is a Raptor-class high-speed patrol boats. The boats will be operated by the Russian Navy's Black Sea Fleet. The Russian Navy will receive over 10 Raptor high-speed patrol boats and special purpose tugs by the end of 2018. The leading boat of the class was delivered to the Russian Navy in 2014. The Raptor boats are used during troop landing and search and rescue operations.
Displacement: 23 tonnes
Length: 16.7m
Beam: 4m
Draft: 0.85m
Speed: 88.8km/h
Propulsion: Caterpillar C18 diesels; pump-jets
Crew: 24

Osa-class - The Project-205 Moskit (Osa) are a class of missile boats. Osa-II, used by Russian Navy, are the upgraded Project-205 with tube-shaped missile launchers for the improved P-15U missile.
Displacement: 235 tonnes
Length: 38.6m
Beam: 7.64m
Propulsion: Shafts
Speed: 73km/h
Endurance: 5 days
Crew: 29
Armament: AK-230 twin 30 mm CIWS; P-15 Termit anti-ship missiles

Mine sweeper/Mine Hunter: Mine sweeping vessels are designed for the location of and destruction of naval mines.

Natya-class - The Natya-classs minesweepers are used for ocean mine sweeping.
Displacement: 875 tonnes
Length: 61m
Beam: 10.2m
Propulsion: M-503 Diesel engines
EW & Decoys: AT-2 acoustic sweep; GKT-2 contact sweep; TEM-3 magnetic sweep
Crew: 70
Armament: naval mines; depth charges; MKT-210; 30mm AK-230; RBU1200

Sonya-class - The Sonya-class minesweepers are built as a successor to the Vanya class with new sweeps and more effective sonar.
Displacement: 400 tonnes
Length: 48.8m
Beam: 8.8m
Propulsion: diesel engines
Range: 5,550km
Crew: 40
Armament: 25 & 30mm guns; Sweeps GKT, PEMT-2, ST-2

Russian Navy minesweeper German Ugryumov in 2015
(Image Credit: Ministry of Defence of the Russian Federation)

Gorya-class – The Gorya-class minesweeper are in service with the Russian Navy. The sophisticated mine detection equipment are used for deep ocean sweeping.
Displacement: 900 tonnes
Length: 70m
Beam: 12m
Propulsion: Diesel engines
Speed: 30km/h
Crew: 70
EW & Decoys: AT-2 acoustic sweep; GKT-2 contact sweep; TEM-3 magnetic sweep
Armament: 30 & 76mm guns; SA-N-14 SAM missiles; torpedo tubes

Sub-surface ships

Ballistic missile submarines: A ballistic missile submarine is a submarine deploying submarine-launched ballistic missiles. The Russian Navy has deployed strategic submarines with the Northern Fleet and the Pacific Fleet.

Borei-class - The Borei-class (Project 955) are nuclear-powered ballistic missile submarine operated by the Russian Navy. These submarines which are much smaller than the Typhoon-class are planned to replace the Delta III, Delta IV and Typhoon classes. Yuriy Dolgorukiy, Aleksandr Nevskiy and Vladimir Monomakh, ships in Borei-class, are already inducted with the Navy. In 2011 a Bulava missile was launched for the first time from the Borei-class submarine Yuriy Dolgorukiy. K-535 Yuriy Dolgorukiy is the first Borei-class ballistic missile submarine of the Project 955 in service with the Russian Navy. K-550 Alexander Nevsky is a Russian nuclear ballistic missile submarine of the fourth generation Borei class (Project 955A). Vladimir Monomakh is a Russian ballistic missile submarine of the fourth generation Borei class (Project 955) that became operational in 2015.
Displacement: 14,720 tonnes (surface); 24,000 tonnes (submerged)
Length: 170m
Beam: 13.5m
Propulsion: OK-650B nuclear reactor; AEU steam turbine; shaft and propeller
Speed: 28km/h (surface); 56km/h (submerged)

Endurance: 90 days

Crew: 107

Armament: RSM-56 Bulava SLBMs with 6-10 MIRVed warheads (both 955A Borei II & Project 955); 533 mm torpedo tubes; RPK-2 Viyuga cruise missiles

Alexander Nevsky in Vilyuchinsk

(Image Credit: Ministry of Defence of the Russian Federation)

Typhoon-class - The Project 941 or Akula (Typhoon) are the largest class of submarines ever built by the Soviet Navy in 1980s. Dmitri Donskoy (TK-208) is currently the submarine active in service under Akula-class. It is the largest nuclear submarine in the world. In normal conditions, a Typhoon-class submarine can stay submerged for periods up to 120-180 days.

Displacement: 23,200–24,500 tonnes (surface); 33,800–48,000 tonnes (submerged)

Length: 172m

Beam: 23.3m

Draft: 11m

Propulsion: OK-650 pressurised-water nuclear reactors; steam-turbine nuclear power plants; shafts with 7-bladed shrouded screws

Speed: 41.15 km/h (surface); 50 km/h (submerged)

Endurance: 180 days

Crew: 160

Armament: 9K38 Igla SAM; 533 mm torpedo tubes; RPK-2 Viyuga cruise missiles; Type 53 torpedoes; D-19 launch system; RSM-52 SLBMs

Soviet Typhoon-class ballistic missile submarine *(Image Credit: US Navy)*

Delta IV-class - The Project 667BDRM (Delta IV) class submarines operate in the Northern Fleet. Seven were built and currently six are active in service namely, K-51 Verkhoturye, K-84 Yekaterinburg K-18 Karelia, K-117 Bryansk, K-407 Novomoskovsk and K-114 Tula. The Delta-IV submarines are strategic nuclear missile submarines designed to carry out strikes on military and industrial installations and naval bases.

Displacement: 11,740 tonnes (surface); 18,200 tonnes (submerged)

Length: 167.4m

Beam: 11.7m

Draft: 8.8m

Speed: 25.928km/h (surface); 44.448km/h (submerged)

Propulsion: two pressurised water-cooled reactors powering two steam turbines with two fixed-pitched shrouded propellers

Endurance: 90 days

Crew: 140

Armament: 533-mm bow torpedo tubes; SLBM R-29RM launchers; MANPADS 9K310 Igla-1/9K38 Igla launcher

Delta-III-class - The Project 667BDR (Delta III) is a large ballistic missile submarine operated by the Russian Navy. The Delta III submarines are significantly quieter and have a higher missile section for new longer missiles. The Delta III class are fitted with a new sonar system, the MGK-400 Rubikon.

Displacement: 10,600 tonnes (surface); 13,700 tonnes (submerged)

Length: 155m

Beam: 11.7m

Draft: 8.7m

Propulsion: Two pressurised water-cooled reactors (reactor system OK-700A) powering two steam turbines; five-bladed fixed-pitched shrouded propellers

Endurance: 90 days

Crew: 130

Speed: 25.928km/h (surface); 44.448km/h (submerged)

Armament: RSM-50 R-29R "Vysota" missiles; 533 mm (21 in) torpedo tubes in the bow with 16 torpedoes (SET-65, SAET-60M, 53-65K, 53-65M)

Cruise missile submarines: A cruise missile submarine is a submarine that can launch cruise missiles.

Oscar-class – The Project 949A (Oscar II) are Russian Navy cruise missile submarines. Project 949 (Granit) and Project 949A (Antey) are the largest cruise missile submarines in service with Russian Navy. 11 Project 949A Antey submarines were completed at Severodvinsk, of which five were assigned to the Northern Fleet, with several more in reserve or waiting for repairs. Russia is planning to modernise three Oscar-class submarines to a new Project 949AM standard.

Displacement: 12,500/14,700 tonnes (surface); 16,500/19,400 tonnes (submerged)

Length: 155m

Beam: 18.2m

Propulsion: steam turbines; 190 MW nuclear reactors OK-650V

Speed: 28km/h (surface); 59km/h (submerged)

Endurance: 120 days

Crew: 108

Armament: 533mm & 650mm torpedo tubes; 533 mm and 650 mm weapons; SS-N-15 Starfish ASM; nuclear warheads; SS-N-16 Stallion; Type 40 anti-submarine torpedo or 32 ground mines; P-700 Granit cruise missile

The nuclear submarine Tomsk with cruise missiles on board

(Image Credit: Ministry of Defence of the Russian Federation)

Attack submarines: Attack submarines are designed to seek and destroy enemy submarines and surface ships.

Sierra-I class - The Project 945, Sierra-I class can dive to greater depths, reduce the level of radiated noise and increase resistance to torpedo attacks. B-276 Kostroma and B-239 Karp are the ships active under Russian Navy's Northern Fleet.

Displacement: 7,200 tonnes (surface); 8,300 tonnes (submerged)

Length: 107.16m

Beam: 12.28m

Draft: 8.8m

Propulsion: nuclear reactor OK-650A; steam generators; turbo generators; storage batteries; DG-300; 370 kW low speed engines

Speed: 18.5 km/h (surface); 63.0 km/h (submerged)

Crew: 59

Armament: 650mm & 530mm torpedo tubes; SS-N-21 Sampson; SS-N-15 Starfish; HE Type 40 torpedo; SS-N-16 Stallion; MANPADS 9K310 Igla-1 or 9K38 Igla launchers

Sierra-II class – The Project 945A (Sierra II) class has a longer sail and a larger pod. The ships in this class include B-336 Pskov and B-534 Nizhny Novgorod. The Sierra-II class nuclear-powered attack submarine Pskov returned to service on 28 December 2015, following a four-year refit. Ships in this category include B-336 Pskov & B-534 Nizhny Novgorod.

Displacement: 7,600 tonnes (surface); 9,100 tonnes (submerged)

Length: 110m

Beam: 14.2m

Speed: 18.5km/h (surface); 59.3km/h

Crew: 61

Armament: 650mm & 530mm torpedo tubes;

SS-N-21 Sampson; SS-N-15 Starfish; HE Type 40 torpedo; SS-N-16 Stallion; MANPADS 9K310 Igla-1 or 9K38 Igla launchers

Victor-III class - The Project 671RTMK (Victor III) class nuclear-powered submarine are in service and has an advanced pod which is identified as a hydrodynamic housing for a reelable towed passive sonar array; the system was subsequently incorporated into the Sierra class and Akula-class submarine. These submarines are quieter than the previous versions. Active submarines in this class include Daniil Moskovsky, Tambov, B-138 Obninsk & B-388 Petrozavodsk.

Displacement: 6,900 tonnes (surface); 7,250 tonnes (submerged)

Length: 107.1m

Beam: 10.8m

Draft: 7.8m

Speed: 59 km/h

Propulsion: VM-4-type water-cooled and water-moderated reactors; turbine-geared propulsion unit GTZA-615; OK-300 steam turbines; One shaft line; multi-blade screw; two auxiliary motors

Endurance: 80 days

Crew: 96

Armament: 650 mm torpedo tubes; Type 88R SS-N-16 Stallion cruise missiles; 533mm torpedo tubes; 84RN SS-N-15 Starfish cruise missiles; VA-111 Shkval rocket torpedoes

An aerial port quarter view of the Russian Northern Fleet VICTOR III class nuclear-powered attack submarine underway on the surface *(Image Credit: US Navy)*

Akula-class - The Project 971 (Akula/ Shchuka-B) class nuclear-powered attack submarine incorporates a double hull system. Akula-I are equipped with MGK-500 Skat sonar system. The improved versions of Akula-I are quieter than the original Akulas. The Akula-II are equipped with MGK-540 Skat-3 sonar system. The Akula-III is longer and has a larger displacement compared to the Akula II. K-157 Vepr, K-295 Samara, K-317 Pantera, K-461 Volk, K-331 Magadan, K-328 Leopard, K-154 Tigr, K-419 Kuzbass, K-263 Barnaul, K-391 Bratsk & K-322 Kashalot are all Akula-class (Project 971 Schuka-B) submarines.

Displacement: 8,140 tonnes (surface); 12,770 tonnes (submerged)

Length: 114.3m

Beam: 13.6m

Draft: 9.7m

Propulsion: Nuclear reactor OK-650BZ; turbine

Speed: 19 km/h (surface); 65km/h (submerged)

Endurance: 100 days

Crew: 73

Armament: 533 mm torpedo tubes; 650 mm torpedo tubes; Igla-M surface-to-air missile launcher; Granat/ Kalibr cruise missiles

Akula class submarine underway *(Image Credit: US Department of Defence)*

Yasen-class - The Project 885 Yasen class submarine is the newest Russian nuclear-powered attack submarine. The first submarine of this class called Severodvinsk was delivered to the Russian Navy on June 17, 2014. According to Russia's Naval Doctrine, the Yasen-class submarines will become the main multipurpose nuclear-powered subs in the Russian Navy.

Displacement: 8,600 tonnes (surface); 13,800 tonnes (submerged)

Length: 120m

Beam: 15m

Speed: 37km/h (surface); 52km/h (submerged)

Propulsion: KPM type pressurised water reactor

Endurance: 100 days

Crew: 90

Armament: VLS equipped silos with Oniks; Kalibr-PL anti-ship, anti-submarine and land attack submarine launched weapon; Kh-101 cruise missiles; 650mm & 533mm torpedo tubes

Yasen class submarine

Kilo-class - The Project-877 Kilo-class diesel electric submarine are in naval service for anti-shipping and anti-submarine operations in relatively shallow waters and are equipped with Rubikon MGK-400 & MG-519 Arfa sonar system. The Improved Kilo-class (Project 636) are equipped with improved MGK-400EM, with MG-519 Arfa also upgraded to MG-519EM. The Project 877V (Kilo) class submarines, namely B-871 Alrosa, uses a pump jet propulsion system and are active with the Russian Black Sea Fleet.

Displacement: 2,300–2,350 tonnes (surface); 3,000-3,950 tonnes (submerged)

Length: 70.0–74.0m

Beam: 9.9m

Draft: 6.5m

Propulsion: Diesel-electric propulsion; Diesel generators

Speed: 19-22km/h (surface); 31-46km/h (submerged)

Endurance: 45 days

Crew: 52

Armament: 533mm torpedo tubes; Kalibr / Club land-attack cruise missile, anti-ship missile and anti-submarine missile, SA-N-8 Gremlin or 8 SA-N-10 Gimlet Surface-to-air missiles; mines

An aerial bow-on view of the Russian Northern Fleet KILO class diesel-powered attack submarine underway on the surface *(Image Credit: US Navy)*

Lada-class - The Project 677 Lada class are fourth generation diesel-electric submarines. B-585 Saint Petersburg is a Lada-class diesel-electric submarine with the Baltic Fleet. This class is intended to be the highly-improved version of the Project 636 Kilo class. The submarine is designed for anti-submarine, anti-ship, anti-aircraft warfare including protection of naval bases, seashore and sea lanes, conducting of reconnaissance missions.

Displacement: 1,765 tonnes (surface); 2,700 tonnes (submerged)

Length: 72m

Beam: 7.1m

Propulsion: Electric propulsion motor; diesel Air Independent Propulsion; Storage battery

Speed: 19km/h (surface); 39km/h (submerged)

Endurance: 45 days
Crew: 36
Armament: 533mm torpedo tubes; rocket torpedoes; Igla SAM system

Naval Aviation

Antonov An-24 - The An-24 is a 44-seat twin turboprop transport/passenger aircraft. The performance of An-24 has made it usable and adapted to perform many secondary missions, as well as further development to produce the An-26 tactical transport, An-30 and An-32 with more powerful engines.
Wingspan: 29.2m
Length: 23.53m
Height: 8.32m
Capacity: 50
Crew: 4
Weight: 13,300kg
Power plant: Ivchenko AI-24 Turboprop engines
Cruise Speed: 450km/h
Range: 2,761km

Antonov An-26 - The An-26 is a twin-engine turboprop civilian and military transport aircraft.
Wingspan: 29.20m
Height: 8.58m
Payload: 5,500kg
Capacity: 40
Crew: 5
Weight: 15,020kg
Power plant: Progress AI-24VT turboprops
Cruise Speed: 440km/h
Range: 2,500km

Antonov An-26 'Curl' transport aircraft, taken from a RAF Typhoon aircraft during a QRA (Quick Reaction Alert) intercept *(Image Credit: RAF/MOD)*

Beriev Be-12 - The Beriev Be-12 is a turboprop-powered amphibious aircraft designed for anti-submarine and maritime patrol duties. 9 aircraft are in operation with the Black Sea Fleet and used for search and rescue role.
Wingspan: 29.84m
Length: 30.11m
Height: 7.94m
Weight: 24,000kg
Power plant: Ivchenko Progress AI-20D turboprops
Speed: 530km/h
Range: 3,300km
Crew: 4

Armament: bombs, depth-charges and torpedoes

Ilyushin Il-38 - The Ilyushin Il-38 is a maritime patrol aircraft and anti-submarine warfare aircraft. The aircraft is a version of the four-engine turboprop Ilyushin Il-18.
Wingspan: 37.4m
Length: 40.18m
Height: 10.17m
Weight: 35,500kg
Power plant: Ivchencko/Progress AI-20M turboprop engines
Speed: 645km/h
Range: 7,500km
Crew: 7
Armament: depth-charges, mines, torpedoes and bombs

Kamov Ka-27 - The Ka-27 is a military helicopter used for ferrying and anti-submarine warfare. Russia in 2013 tested the new Kamov Ka-27M with an active electronically scanned array radar. Serial upgrading to the level of combatant helicopters Ka-27M are being planned including eight Ka-27PLs of the Northern Fleet.
Capacity: 4,000kg or 16 troops
Length: 11.3m
Weight: 6,500kg
Speed: 270km/h
Range: 980km
Avionics: Radar, MAD or dipping sonar, sonobuoys
Armament: Torpedoes or Sonobuoys

A Russian KA-27PS (Helix D) helicopter flies past the flight deck of USS Belleau Wood (LHA 3) (not shown) to prepare for landing on the amphibious assault ship during Exercise Cooperation from the Sea '96, off the coast of Vladivostok, Russia *(Image Credit: US Navy)*

Kamov Ka-31 - Ka-31 is a military helicopter used in airborne early warning and control role. One significant feature of the Ka-31 is the large antenna of the early-warning radar.
Length: 12.5m
Rotor Diameter: 14.50m
Height: 5.6m
Power plant: Isotov TV3-117VMAR turboshaft
Speed: 250km/h
Range: 600km
Crew: 2

Mikoyan MiG-29K – The MiG-29K is an all-weather carrier-based multirole fighter aircraft. The aircraft has multi function radar, HOTAS control, RVV-AE AAM, anti-ship and anti-radar missiles and precision guided weapons. The Russian Navy ordered 24 MiG-29Ks for the Admiral Kuznetsov. Deliveries of the MiG-29K for the Russian Navy started in 2010. Later, Russian Naval Aviation ordered 20 MiG-29K and 4 MiG-29KUB. These aircraft were all delivered by early 2016.
Wingspan: 11.99m
Length: 17.3m
Height: 4.40m
Weight: 18,550kg
Power plant: Klimov RD-33MK afterburning turbofans
Speed: 2,200+km/h
Crew: 1 (MiG-29K); 2 (MiG-29KUB)
Armament: 30 mm GSh-30-1; Vympel R-73, Vympel R-27, Vympel R-77; Kh-25ML, Kh-29T, Kh-35U; Kh-25MP, Kh-31P; Kh-31A, Kh-35; Bombs

MiG-29K jet at Zhukovskiy LII air field

Sukhoi Su-25UTG - The Su-25UTG is a two-seater aircraft carrier variant of the Su-25UB designed to train pilots in take-off and landing on a land-based simulated carrier deck. This aircraft is a carrier capable version used to carrier out full deck-landing training aboard the Navy's aircraft carrier. Around 10 of these aircraft are currently operational in the Russian Navy as part of the 279th Naval Aviation Regiment.

Sukhoi Su-27 - The Su-27 is a twin-engine aircraft which was designed for air superiority missions. The Su-27's basic design is aerodynamically similar to the MiG-29, but it is significantly larger. Russia is planning to replace the Su-27 and the MiG-29 with the Sukhoi PAK FA stealth fifth-generation multi-role twin-engine fighter. 53 Su-27s are in service with the Russian Navy as of January 2014.
Wingspan: 14.7m
Length: 21.9m
Height: 5.92m
Weight: 16,380kg
Power plant: Saturn/Lyulka AL-31F turbofans
Speed: 2,500km/h
Range: 3,530km
Armament: 30 mm GSh-30-1 cannon; medium-range AA missiles R-27; short-range heat-seeking AA missiles R-73
Crew: 1

Sukhoi Su-30 - The Su-30 is a twin-engine,

two-seat supermanoeuvrable fighter aircraft used for all-weather, air-to-air and air-to-surface and other multirole missions. Russian Navy received nine Su-30SMs as of 2016, 50 are planned and 28 are on order. These aircraft will be used in the Black Sea Fleet. Su-30SM is considered as 4+ generation fighter aircraft with upgrades based on radar, radio communications systems, ejection seats, weapons, and other aircraft systems.

Wingspan: 14.7m

Length: 21.94m

Height: 6.36m

Weight: 17,700kg

Speed: 2,120km/h

Crew: 2

Armament: GSh-30-1 gun; R-27ER (AA-10C); Kh-31P/A; R-77 RVV-AE; R-73E; R-27ET; Aerial bombs; Kh-29T/L; Kh-59ME

Sukhoi Su-33 - The Su-33 is an all-weather carrier-based twin-engine air superiority fighter aircraft having a powerful undercarriage and structure, folding wings and other features such as upgraded engines and a twin nose wheel, and is air refuellable, all for carrier operations.

Wingspan: 14.70m

Length: 21.19m

Height: 5.93m

Weight: 18,400kg

Power plant: AL-31F3 afterburning turbofans

Speed: 2,300km/h

Range: 3,000km

Crew: 1

Armament: 30 mm GSh-30-1 cannon; R-27R/T/ET/EM; R-73 air-to-air missile; Kh-31A Krypton anti-ship missile; ECM pods, bombs & rockets

A Su-33 on board Admiral Kuznetsov in 1996. U.S. Navy sailors from the USS San Jacinto are visiting the carrier *(Image Credit: US Navy)*

Naval Weapons

P-700 Granit - The P-700 Granit is a Russian naval anti-ship cruise missile. The missile has surface-to-surface and submarine-launched variants. The P-700 was derived from the P-500 Bazalt missile with a turbojet.

Weight: 7,000kg

Length: 10m

Warhead: HE or Nuclear

Engine: Ramjet

Range: 625km

Speed: 16Mach

Launch Platform: Oscar class submarines, Kirov & Admiral Kuznetsov class ships

Kashtan CIWS - The Kashtan close-in weapon system (CIWS) is a modern naval air defence gun-missile system deployed by the Russian Navy and used on aircraft carrier, cruisers and frigates. The system provides defence against anti-ship missiles, anti-radar missiles and guided bombs. Variants include Kortik-M/Kashtan-M.

Weight: 15,500kg

Height: 2,250mm

Barrels: 2 × 6 (guns); 2 × 4 launch tubes

Action: Gas-operated rotary cannon

Warhead: Continuous-rod w/ frag layer

Armament: 9M311K; 9M311-1E; AO-18KD autocannon

Speed: 910m/s

RSM-56 Bulava - The Bulava is a submarine-launched ballistic missile (SLBM) developed for the Russian Navy. These are deployed on the new Borei class of ballistic missile nuclear submarines.

Weight: 36.8 tonnes

Length: 11.5m

Diameter: 2m

Engine: Solid-fuel rocket and Liquid-propellant rocket

Propellant: solid propellant and liquid fuel

Range: 8,000km

Launch Platform: Borei-class submarines; Typhoon-class submarine; Dmitri Donskoi

Bulava SLBM

3M-54 Klub - The Klub (Kalibr) missile system are Russian land-attack cruise missile, anti-ship missile and anti-submarine missile. The 3M54K is a submarine launched missile. 3M54T is deployed in a surface ship with a VLS launched system and a thrust vectoring booster. 3M14K & 3M14T are an inertial guidance land attack variant, deployed by the Russian Navy. The Russian Navy currently uses the 3M14, 3M54, 91R1, 91RT2.

Weight: 1,300 – 2,300 kg (depends on variants)

Length: 6.2m – 8.9m (depends on variants)

Warhead: 5000kg or nuclear

Engine: Solid-Fuel rocket; or Turbojet; or Solid fuel rocket (depends on variants)

Speed: 0.8 - 2.5 - 2.9 Mach (depends on variants)

Range: 50km - 2500km (depends on variants)

Launch Platform: naval ships, submarines, containers, airplanes

K-300P Bastion-P - The K-300P Bastion-P is a Russian mobile coastal defence missile system. These are in service with the Black Sea Fleet & Northern Fleet. The main role of the Bastion-P is to engage various surface ships. The Bastion-P launcher carries the P-800 Oniks anti-ship cruise missiles. Variants include K-300P, K-330s & Bastion E.

Weight: 41 tonnes

Length: 12.8m

Missile Length: 8.6m

Range of fire: upto 300km

Height: 3.6m

Engine: YaMZ-846 diesel

Range: 1,000km

Guidance Systems: active-passive radar seeker head

Launch Platform: coastal installations

P-800 Oniks - The P-800 Oniks is a Russian supersonic anti-ship cruise missile which is carried in flight by aerodynamic lift. Bastion-P complexes are stationed armed with these missiles. Newest class of Russian nuclear-powered attack submarines, Yasen-class submarine, can also launch the missile.

Wingspan: 1.7m

Length: 8.9m

Weight: 3,100kg

Speed: 2.5 Mach

Engine: Ramjet

Warhead: 250kg

Launch Platform: coastal installations, naval ship, and Fixed-wing aircraft

Kh-35 - The Kh-35 is the aircraft-launched version of a Russian subsonic anti-ship missile. The missile (Uran/Bal) can also be launched from helicopters, surface ships and coastal defence batteries with the help of a rocket booster. 112 Kh-35 (3M-24) are delivered to Russia. The Russian Navy has 16 mobile launch systems.

Wingspan: 1.33m

Length: 4.4m (ship); 3.85m (aircraft)

Weight: 620kg (ship); 520kg (aircraft)

Guidance System: Inertial, active radar

Range: 130km

Speed: 0.8Mach

Warhead: 145kg HE penetrator

Launch Platform: Aircraft, ships and boats, coastal

P-270 Moskit - The P-270 Moskit is a Russian ramjet powered anti-ship missile. The Moskit and its various variants can be launched from ship, modified trucks, submarines, and aircraft. The missile can carry conventional and nuclear warheads. Variants include P-80, P-270 Moskit, Kh-41 & 3M-80MVE.

Wingspan: 2.10m

Length: 9.75m

Weight: 4,500kg

Warhead: conventional and nuclear

Range: 120 - 250km

Speed: 3Mach
Launch Platform: naval ships, fixed-wing aircraft, coastal and land

Metel Anti-Ship Complex - The Metel Anti-Ship Complex is a family of anti-submarine missiles. The missile is based on the P-120 Malakhit anti-ship missile. Variants include 60R, 70R, 83R/URPK-3, 84R/URPK-4, 85RU/URPK-5. The missile carries an anti-submarine torpedo.
Length: 7.2m
Weight: 3,930kg
Warhead: ASW torpedo or nuclear depth charge
Propulsion: solid fuel
Speed: 0.95Mach
Range: 10 - 90km (varies)

RBU-6000 - The RBU-6000 Smerch-2 is a 213-mm calibre anti-submarine rocket launcher which contains twelve launch barrels and are remotely directed by the Burya fire control system. It fires RGB-60 unguided depth charges. The RPK-8 system is an upgrade of the RBU-6000 system which can fire the 90R rocket.

Weight: 3,100kg launcher; 110kg RGB-60 projectile; 112.5kg 90R projectile
Warhead: 25kg RGB-60 projectile; 19.5kg 90R projectile
Length: 2m launcher; 1.83m RGB-60 projectile; 1.83m 90R projectile
Range: 350m - 5800m RGB-60 projectile; 600m - 4,300m 90R projectile
Depth: 10 - 500m RGB-60 projectile

A RBU-6000 anti-submarine rocket launcher aboard the Udaloy class guided missile destroyer Admiral Tributs *(Image Credit: www.dodmedia.osd.mil)*

RPK-6 Vodopad/RPK-7 Veter - RPK-6 Vodopad is a 533 mm anti-ship missile. RPK-7 Veter is a 650-mm version of anti-ship missile. Both torpedo-tube launched missiles are powered with solid fuel rocket engine.
Range: 100Km
Payload: 400 torpedo; depth charges
Guidance: Inertial

RPK-9 Medvedka - RPK-9 Medvedka is a modern missile system used for engaging submarines. Each torpedo, with 8 missiles, in the system acts as the warhead along with the launcher.
Missile Weight: 800kg
System Weight: 19.4 tonnes (8 missiles)
Diameter: 400mm
Depth: 15 - 500m

R-29RMU Sineva - The R-29RMU Sineva is a liquid-fuel SLBM capable of deployment as a MIRV. The Delta IV ballistic missile submarines carries these missiles, every sub carries 16 such missiles, each carrying four MIRVs. The R-29RMU is expected to remain in service until 2030.
Weight: 40.3 tonnes
Length: 14.8m
Diameter: 1.9m
Warhead: 4 MIRV or 8 MIRV
Propulsion: Three-stage liquid propellant
Range: 8,300km
Guidance: Astro-inertial
Launch Platform: Submarine

R-29RMU Sineva

R-29RMU2 Layner - Russia's new Layner submarine-launched ballistic missile was inducted into service with the Russian submarine fleet in the beginning of 2014. The Layner missiles is a modification of the Sinyeva. The missile has the same flight performance characteristics as the Sinyeva, but with more modern missile defence suppression systems and a longer range. The Layner can carry twelve nuclear warheads.
Weight: 40 tonnes
Length: 15m
Diameter: 1.9m
Warhead: 12 low yield warheads
Propulsion: Three-stage liquid propellant
Range: 8,300-12,000Km
Guidance: Astro-inertial with Glonass
Launch Platform: Submarines

P-500 Bazalt - The P-500 Bazalt is a turbojet-powered, supersonic cruise missile used by Russian Navy. The missile was later added to both the Echo II class submarine and the Juliett class submarine. A version of the P-500 Bazalt with improved guidance and engines is used on the Slava class cruiser. The P-1000 with increased missile weight has been installed on the Slava class cruiser Varyag. Eight pairs of P-500 canisters are a distinctive feature of Slava class cruisers. The Varyag have been upgraded to the P-1000 Vulkan.
Weight: 4,800Kg
Length: 11.7m
Diameter: 1.2m
Warhead: HE or nuclear
Propulsion: Turbojet
Range: 550Km
Speed: 2.1Mach
Guidance: Semi-active, terminal active radar
Launch Platform: Echo II & Juliett submarines, Kiev & Slava class ships

Russian sailors man the rails as Russian navy missile-cruiser Varyag departs San Francisco Bay *(Image Credit: US Navy)*

CONCLUSION

Russia is engaged in its largest military build-up since the collapse of the Soviet Union in 1991. The country is rising in its influence once again at the global stage even as it is preparing to project power in the military, economic, and political spheres across Eastern Europe, Central Asia, and the North Pacific. Simultaneously, it is strengthening its relations in the Middle East, especially with Iran and Syria, and also with Egypt. Russia is also beginning to bolster its geopolitical presence and military might by expanding the Collective Security Treaty Organization (CSTO). Moscow is carrying out sweeping military reforms that hopes to revitalise almost 70 percent of the country's armed forces by 2020. The rearmament programme focuses on equipping the military with new ships, submarines, airplanes, nuclear missiles and battlefield equipment for soldiers. Plans are also on to completely overhaul the strategic nuclear arsenal of the country during this period. The Crimean Peninsula continues to give Moscow continuing access to the naval base at Sevastopol, which is home to Russia's Black Sea Fleet. Operating from Sevastopol, the Black Sea Fleet provides Russia with the ability to project power in and around the Black Sea, while also serving as a potent symbol of Russian military power. Russia has also launched a military push in the Arctic region to restore its Soviet-era stake in a hotly contested area that contains the world's largest untapped energy resources. The larger Russian goal appears to be building up a strong military force in the 21st century to recover the economic, political, and geostrategic assets lost by the Soviet state in the previous century.

CONTACT DETAILS

Ministry of Defence
19 Znamenka Str., Moscow
Russia-119169
Tel.: 8 (495) 696-71-71
Website: http://eng.mil.ru/

Information Desk of the Ministry of Defence
Phone: 8 (495) 696-88-00

STRATEGIC INFORMATION

SAUDI ARABIA

(Capital: Riyadh)

INTRODUCTION

Area: 2,149,690 sq km

Population: 28,160,273 (July 2016 est.)

Coastline: 2,640 km

Maritime claims: Territorial sea: 12 nm
Contiguous zone: 18 nm
Continental shelf: not specified

GEOPOLITICAL IMPORTANCE

The Kingdom of Saudi Arabia, commonly known as Saudi Arabia is the largest state in the Middle East by land area, constituting about 80 percent of the Arabian Peninsula, and the second-largest in the Arab World. It is bordered by Jordan and Iraq on the north and northeast, Kuwait, Qatar and the United Arab Emirates on the east, Oman on the southeast, and Yemen on the south. It is also connected to Bahrain by the King Fahd Causeway. The largest and most influential nation on the Arabian Peninsula, Saudi Arabia is strategically located between the Red Sea and the Arabian Gulf. Saudi Arabia's geography is dominated by the Arabian Desert and associated semi-desert and shrub land. Its extensive coastlines on the Persian Gulf and Red Sea provide great leverage on shipping (especially crude oil) through the Persian Gulf and Suez Canal. Saudi Arabia's maritime claims include a twelve-nautical-mile territorial limit along its coasts. The Saudis also claim many small islands as well as some sea-beds and sub-soils beyond the twelve-nautical-mile limit. Saudi Arabia sees itself as the leader of the Arab countries. Being a rich state it champions the Middle-east cause in all forums.

POLITICAL OVERVIEW

Saudi Arabia is an absolute monarchy with a Council of Ministers and Consultative Council. Although, according to the Basic Law of Saudi Arabia adopted by royal decree in 1992, the king must comply with *Sharia* (i.e., Islamic law) and the Quran. The Basic Law specifies that the king must be chosen from among the sons of the first king, Abdul Aziz Al Saud, and their male descendants subject to the subsequent approval of religious leaders (the *ulema*). The country has given the *ulema* (the body of Islamic religious leaders and jurists) a direct role in government. The king combines legislative, executive, and judicial functions and royal decrees form the basis of the country's legislation. The king is also the prime minister, and presides over the Council of Ministers, which comprises the first and second deputy prime ministers (usually the first and second in line to the throne respectively) and 23 ministers with portfolio and five ministers of state. The Quran and the *Sunna* (the traditions of Muhammad) are declared to be the country's constitution, but no modern constitution has ever been written for Saudi Arabia, and it remains the only Arab nation where no elections have ever taken place, since its creation. No political parties or national elections are permitted. The royal family dominates the political system. On September 2011, King Abdullah announced that women will have the right to stand and vote in future local elections and join the advisory *Shura* council as full members.

KEY POLITICAL PERSONS

HEAD OF STATE & PRIME MINISTER: King Salman bin Abdulaziz Al Saud

CROWN PRINCE & DEP. PRIME MINISTER: Muhammad bin Nayef bin Abdulaziz Al Saud

ECONOMY

Saudi Arabia has an oil-based economy with strong government control over major economic activities, with roughly 75% of budget revenues and 90% of export earnings come from the oil industry. It possesses 25% of the world's proven petroleum reserves, ranks as the largest exporter of petroleum, and plays a leading role in OPEC (the Organization of Petroleum Exporting Countries). Saudi Arabia has a centrally planned economy. Private enterprises do exist; they are however regulated by the Saudi government. Saudi oil reserves are the largest in the world, and Saudi Arabia is the world's leading oil producer and exporter. Oil accounts for more than 90% of the country's exports and nearly 75% of government revenues. Proven reserves are estimated to be 263 billion barrels, about one-quarter of world oil reserves. The economy has currently benefitted tremendously from the high-oil price environment between 2003 and 2008 and again from 2010 onwards. Saudi Arabia's ninth development plan (2010-2014) aspires to eliminate poverty and increase development in infrastructure, medical services, educational capacity, and residential housing. The plan also aims to increase real GDP by 15% over 5 years and calls for substantial government investment in human resource development, in order to decrease Saudi unemployment from 9.6% to 5.5%.

GDP: $1.683 trillion (2015 est.), $1.628 trillion (2014 est.)

Real Growth Rate: 3.4% (2015 est.), 3.6% (2014 est.)

Industries: Crude oil production, petroleum refining, basic petrochemicals, ammonia, industrial gases, sodium hydroxide (caustic soda), cement, fertilizer, plastics, metals, commercial ship repair, commercial aircraft repair, construction

Oil - proved reserves: 269 billion bbl (1 January 2016 est.)

Natural gas - proved reserves: 8.489 trillion cu m (1 January 2016 est)

Total Exports: $202.3 billion (2015 est.), $342.5 billion (2014 est.)

Export Commodities: Petroleum and petroleum products 90%

Major Markets: China 13.2%, Japan 10.9%, US 9.6%, India 9.6%, South Korea 8.5% (2015)

Total Imports: $155 billion (2015 est.), $158.5 billion (2014 est.)

Import Commodities: Machinery and equipment, foodstuffs, chemicals, motor vehicles, textiles

Major Suppliers: China 13.9%, US 12.7%, Germany 7.1%, South Korea 6.1%, India 4.5%, Japan 4.4%, UK 4.3% (2015)

DEFENCE & SECURITY

The Saudi military consists of the Saudi Army, the Royal Saudi Air Force, the Royal Saudi Navy, the Royal Saudi Air Defence, the Saudi Arabian National Guard – the 'SANG' (an independent military force), and paramilitary forces. In addition, there is a military intelligence service. The armed forces are mainly the responsibility of the Ministry of Defence and Aviation. However, the Saudi Arabian National Guard is independent of the Ministry, and is commanded by King Abdullah's son, Prince Mutaib bin Abdullah. Khalid bin Sultan, son of Crown Prince Sultan bin Abdulaziz, is the Assistant Minister of Defence and Aviation of Saudi Arabia.

KEY DEFENCE PERSONS

MINISTER OF DEFENCE: Deputy Crown Prince Mohammad Bin Salman Abdulaziz Al Saud

CHIEF OF GENERAL STAFF: Gen. Abdurrahman bin Saleh bin Abdullah Al-Bunyan

DEPUTY CHIEF OF STAFF: Lt. Gen. Fayyad bin Hamid bin Raqed Al-Ruwaili

COMMANDER OF THE LAND FORCES: Gen. Eid bin Awwadh Al-Shalawi

COMMANDER OF THE NAVY: Lt. Gen. Abdullah bin Sultan bin Mohammed Al-Sultan

COMMANDER OF THE AIR FORCE: Gen. Mohammed bin Ahmad bin Abdurrahman Al-Shallan

Internal Conflict

Political opposition: The Al Saud dynasty holds a monopoly of power in Saudi Arabia and any form of political parties are banned. This has encouraged the growth of dissident groups and militant Islamists. They face opposition from four sources: Sunni political activism (which itself takes on many forms, from moderate voices to militant Jihadi ideologies), liberal criticism, Shiite minority agitation, and tribal and regional politics.

Terrorism: Saudi Arabia has a growing terrorism problem unleashed by radical Islamic fighters. Their targets include foreign civilians—mainly Westerners affiliated with its oil-based economy—as well as Saudi civilians and security forces. After the terrorist attacks on New York and Washington of 11 September 2001 - carried out mainly by Saudi nationals - the Saudi authorities were further torn between their natural instincts to step up internal security and pressure to allow a greater degree of democracy. Al-Qaeda in the Arabian Peninsula (AQAP) is a militant Islamist organisation, primarily active in Yemen and Saudi Arabia. The group has carried a number of attacks in Saudi Arabia. Saudi security services have waged an active counterterrorism campaign that has largely neutralised terrorist organisations, though sporadic instances of terrorism still occur.

✦ In May 2006, terrorists attempted to attack the major ARAMCO oil-processing facility at Abqaiq.

✦ In February 2007, four French nationals were killed in western Saudi Arabia in a suspected terrorist attack.

✦ In August 2009, an al-Qaeda in the Arabian Peninsula (AQAP) suicide bomber attempted to assassinate a Saudi royal and senior Ministry of Interior official.

✦ The Specialised Criminal Court in 2011 started hearing charges including possessing explosives, missiles, military weapons, and chemical materials and smuggling items.

✦ In 2012 special security police arrested two Saudis and six Yemenis from cells operating in Riyadh and Jeddah.

✦ In 2014, the Saudi interior ministry issued a royal decree branding all 'deviants' as terrorists.

✦ In 2015, a suicide bomber attacked the Imam Ali mosque for which ISIS claimed responsibility.

✦ In 2016, four suicide bombs exploded in three locations in Saudi Arabia. No group has claimed responsibility for the attacks but the ISIS is under the suspicion radar.

Saudi Arabia has undertaken several initiatives to counter extremism and to promote moderation and tolerance among its citizenry. It has built a special counterterrorism force and increased the size, training and professionalism of its security forces. These forces are equipped with cutting-edge technology, and have been deployed aggressively against terrorists that threaten targets inside the Kingdom and around the world. The Kingdom has implemented the use of aerial reconnaissance aircraft to help patrol remote areas to protect against smuggling and infiltration. New barriers, motion detectors and thermal imaging systems have been installed which provide additional coverage, helping to seal the Kingdom's borders. Saudi Arabia has also shared its counterterrorism expertise with its allies. It has put in place one of the world's strictest financial control systems to prevent funds going to support terrorism.

External Conflict

Yemen: Saudi Arabia began a military intervention alongside eight other Arab states and with the intelligence and logistical support of the United States against the Yemini Houthi rebels, at the request of Hadi's government, in 2015. The intervention included a bombing campaign, naval blockade and deployment of ground forces in Yemen. Later, Saudi Arabia and its coalition partners said they would launch a political and peace efforts after Operation Decisive Storm and will be called Operation Restoring Hope. However, the coalition did not rule out using force in order to eliminate enemy forces. The coalition has yet to dislodge the Houthi group from the capital Sanaa. King Salman meanwhile has doubled his country's Yemen aid pledge to USD 540 million, as per news reports. The border between Saudi Arabia and Yemen has long been a source of political dispute, particularly from 1990 - when North Yemen and South Yemen unified - to 2000, when the two countries signed an agreement.

ISIS: The Government of Saudi Arabia is working closely with United States since 2014 to fight against ISIS militants. The country is supporting by providing training and equipping Syrian fighters in this coalition battle against ISIS and its propaganda. The Saudi government's counter-terrorism actions have considerably helped, such as by cutting off money flows to terrorist groups, according to a statement by US Secretary of State John Kerry. Saudi Arabia has also put USD 500 million into the coffers of the UN humanitarian aid agencies in Iraq.

✦ Kuwait and Saudi Arabia continue discussions on a maritime boundary with Iran.

✦ Saudi Arabia claims Egyptian-administered islands of Tiran and Sanafir.

THREAT PERSPECTIVE

Saudi Arabia is one of the key powers in the region given both its position as the world's most important oil producer and also the two holiest sites of Islam, Mecca and Medina, are located in the kingdom. Saudi Arabia, in the initial times, was primarily concerned about the safety and security of royal blood(s) of the kingdom mainly king and the royal family. Their defence systems were concerned chiefly about the safety of the holy places and the security of the borders. Since the discovery of oil in the country, and its overwhelming export to nations around the world, gave rise to new security concerns engulfing Saudi Arabia and added precautionary measures were taken in maritime security, since majority of its oil exports are through waterways in the Persian Gulf.

Iran: Saudi Arabia broke off diplomatic relations with Iran on January 4, 2016. The Iranian government barred its citizens from making the hajj. Iran's relations with Saudi Arabia have been a function of the two countries' geopolitical situation in the Middle East and political rivalries between Tehran and Riyadh. Both Saudi Arabia and Iran have aspiration for Islamic leadership and both countries possess different visions of regional order. Iran, which after the Islamic Revolution strictly followed an anti-US policy, always deemed Saudi Arabia as an agent of the US in the Persian Gulf region that speaks for the US interests. Saudi Arabia's concerns about Iran are mainly associated to its plans of expanding influence to other parts of the Persian Gulf region, especially in post-Saddam Iraq and the quest to build its own nuclear arsenal. Both the countries share a hate-hate relationship. Saudi Arabia announced in January 2016 it was severing its ties to Iran after its embassy in Tehran was firebombed in protest at the kingdom's execution of Shiite cleric Sheikh Nimr al-Nimr. The oil-rich rivals have also been divided over the nearly five-year war in Syria, where Iran is backing the regime, and the conflict in Yemen where a Saudi-led coalition is battling Shiite rebels.

Iraq: Saudi Arabia and Iraq are longstanding regional rivals. Neither holds diplomatic relations with the other. Saudi leaders were relieved when Iraq was defeated, but they recognised that relations with Baghdad had been damaged. Post-war Saudi policy focused on ways to contain potential Iraqi threats to the kingdom and the region. The Saudi leadership opposed the US plan to invade Iraq, and did not join the Coalition. The Kingdom stands behind Iraq, and desires to see a stable, peaceful, united and independent Iraq that is representative of all Iraqis. Saudi Arabia's efforts in diplomacy have been coupled with support for humanitarian relief and assistance to the Iraqi people. The Saudi Arabia-Iraq border is one of the most patrolled borders in the world, and the Kingdom has dramatically increased its presence to stop illegal border crossings. In August 2016, Iraq has called on Saudi Arabia to replace its ambassador to Baghdad over comments he made about Iran's involvement in Iraq. The Saudi embassy in Baghdad only reopened in December 2015 after being closed since the Iraqi invasion of Kuwait.

STRATEGIC RELATIONS

US: Saudi Arabia's unique role in the Arab and Islamic worlds, its possession of the world's largest reserves of oil, and its strategic location make its friendship important to the United States. The United States and Saudi Arabia share common concerns about regional security, oil exports and imports, and sustainable development. Although Saudi Arabia's relations with the United States were strained after 11 September 2001, Saudi Arabia is now one of the United States' strongest partners against terrorism. Saudi Arabia is a strong partner in the campaign against terrorism, providing military, diplomatic, and financial cooperation. Counterterrorism cooperation between Saudi Arabia and the United States increased significantly since 2013. Saudi Arabia and the US have established two Joint Task Forces - one to combat terrorists, another to combat terror financing. Experts from both governments work side-by-side, share real-time information about terror networks. The United States has provided logistical and intelligence support to Saudi Arabia in their military intervention in Yemen, establishing a Joint Planning Cell with Saudi Arabia. The United States has sold Saudi Arabia military aircraft, air defence weaponry, armoured vehicles, and other equipment.

UK: The United Kingdom and Saudi Arabia have long been close allies. Saudi Arabia is the United Kingdom's primary trading partner in the Middle East. There are more than 150 joint ventures between British and Saudi Companies. Both the countries share strong military cooperation with each other. Both the countries are strong strategic partners with shared regional and wider interests. Saudi Arabia is one of UK's most vital partners in global counter-terrorism.

Pakistan: Bilateral relations between the Islamic Republic of Pakistan and the Kingdom of Saudi Arabia are exemplary. Pakistan is the closest non-Arab ally of Saudi Arabia. Saudi Arabia and Pakistan are leading members of the Organisation of Islamic Cooperation (OIC). Pakistan maintains close military ties with Saudi Arabia, providing extensive support, arms and training for the Military of Saudi Arabia. They have had a deep strategic military relationship for decades and today have an unacknowledged nuclear partnership to provide the kingdom with a nuclear deterrent on short notice if ever needed. Pakistan, on the other hand, has received more aid from Saudi Arabia than any country outside the Arab world since the 1960s. In 2015, Pakistan has declined a Saudi Arabian request to join the coalition against Yemen intervention and maintained a neutral diplomatic stance. However, Pakistan said it will come to Saudi Arabia's defence if the country's sovereignty or territorial integrity is threatened.

India: India has always sought to maintain strong ties with Saudi Arabia. Saudi Arabia is the one of largest suppliers of oil to India. Saudi Arabia has supported granting observer status to India in the OIC and has expanded its cooperation with India to fight terrorism. In Sept 2012, India and Saudi Arabia have decided to enhance their military exchanges during the first meeting of their joint committee on defence cooperation held in New Delhi. The Joint Committee was mandated to formulate a programme to develop areas of cooperation between the defence establishments of both countries. Both nations also agreed on joint ventures and the development of oil and natural gas in public and private sectors. Reflecting growing congruence in bilateral ties, India and Saudi Arabia have inked a defence cooperation pact to take their strategic partnership further in areas of security. The defence cooperation pact will allow exchange of defence-related information, military training and education as well as cooperation in areas varying from hydrography and security to logistics. The recent visit of Prime Minister Narendra Modi to Riyadh in 2016 could be seen as a turning point in our growing engagement with the Kingdom of Saudi Arabia, which has taken an upward strategic direction.

Multilateral Alliances

Saudi Arabia, a member of UN, is also a founder member of the Arab League, Persian Gulf Cooperation Council, Muslim World League, and the Organisation of Islamic Cooperation (OIC). The country plays a prominent role in the International Monetary Fund and the World Bank, and in 2005 joined the World Trade Organisation. According to the Saudi Ministry of Foreign Affairs, Saudi foreign policy is focused on co-operation with the Gulf States, the unity of the Arab world, solidarity with Muslim countries, and support for the UN. Saudi Arabia also is an observer to the Organisation of American States.

DEFENCE CAPABILITIES

ARMY: Royal Saudi Land Forces (RSLF) is a branch of the Saudi Armed Forces. The combat strength of the Saudi Army consists of 4 armoured brigades, 17 mechanised infantry brigades, three light motorised rifle brigades, and one airborne brigade. It also has five independent artillery brigades and an aviation command. The primary task of the army is to protect the kingdom's borders against attack by a conventional force.

Personnel: 8,75,000 (including 5,75,000 Reserve & National Guard)

Equipment

Category	Name	In Service
MBT	M1A2S Abrams	440 (more on order)
	M60A3	406
	AMX30	135
APC/MICV /IFV	Al-Fahd	100
	AMX-10P	500
	M2A2 Bradley	400
	BMP-3	250 (on order)
	M113	1100

Army continued...

Category	Name	In Service
APC/MICV /IFV	Panhard M3	100
	TPz Fuchs	150
	HMMWV	15000
	Nyoka Mk2	2000
Artillery	105 mm M101	100
	105 mm M102	100
	155 mm M198	90
	155 mm FH70	35
	M109A2/A5	280
	155 mm. AMX GCT	50
	M198	40
	155 mm. M114	50
	PLZ-45	50+
Mortar	M224	
	107 mm	
	60 mm	150+
	120 mm	110
MLRS	Astros II MLRS	60
Anti-Tank GW	Dragon	4000
	BGM-71 TOW	
	ITOW	2000
	HOT/HOT-2	3200
	Javelin	
	Swingfire	
	MILAN	
IRBM	DF3A	
SAM	Stinger	
	Mistral	600

Army Aviation

Category	Name	In Service
Helicopter	UH-60A	20
	S-70	5+
	Bell 406 Combat Scout	15
	CH-47	
	AH64/64D Apache	90 (29 on order)
UAV	Aeryon Scout, Saqr	

New Procurements/ Upgrades

✦ DynCorp has won a USD 61 million, five-year, foreign military sales contract from the US Army to provide maintenance services for Saudi Arabia's Land Forces Aviation Command.

✦ GD Land Systems has been awarded a contract to convert 15 of Saudi Arabia's M1A1 tanks into the M1A2S model. The contract is an addition to a USD 58 million contract originally awarded in 2008 to design, develop, convert, implement and test a hybrid configuration of the M1A1, M1A2 and M1A2 System Enhancement Package tank variants. The US Army TACOM Lifecycle Management Command awarded the USD 19 million contract on behalf of the Royal Saudi Land Forces.

✦ USA to sell missiles and munitions, including thousand bunker-buster GBU-39 bombs to Saudi Arabia. The sale will also include sophisticated air-launched cruise missiles that can hit targets from a long distance. The weapons are designed for use by US-made F-15 and F-16 fighter aircraft previously purchased by Saudi Arabia. The Saudis will purchase 650 of the Boeing-manufactured SLAM-ERs and 973 Joint Standoff Weapons, made by Raytheon, as well as other missiles. Also includes, hundreds of Standoff Land Attack missiles, or SLAM-ERs, and Joint Standoff Weapons (JSOW). Under the arms package, the Saudis are due to receive USD 6.8 billion in weapons, parts, training and logistical support.

✦ Government of Saudi Arabia has ordered 150 Javelin Guided Missiles from the US Defence Security Cooperation Agency under the Foreign Military Sale (FMS) programme. The total value of the contract is estimated worth USD 71 million. Currently, Saudi Arabia does not have JAVELIN Anti-tank missiles in its inventory.

✦ Saudi Arabia requested 24 UH-60L Blackhawk helicopters. United Technologies subsidiary Sikorsky and General Electric (engines) are the principal contractors. The total value of the contract is estimated by the DSCA is USD 350 million.

✦ Raytheon awarded USD 55 million contract to deliver TOW (Tube-Launched, Optically-Tracked, Wireless-Guided) missiles to Saudi Arabia.

✦ The US State Department has made a determination approving a possible Foreign Military Sale to the Kingdom of Saudi Arabia for ammunition for the Royal Saudi Land Forces (RSLF) and associated equipment, parts and logistical support for an estimated cost of USD 500 million.

✦ Lockheed Martin Corp. Rotary and Missions Systems, Orlando, Florida, was awarded a USD 48,554,905 firm-fixed-price foreign military sales contract to upgrade the M1A2S Advanced Gunnery Training System of Saudi Arabia.

A Brazilian Avibras ASTROS-II SS-30 rocket is launched from an ASTROS SS-30 multiple rocket system mounted on the back of a Tectran 6x6 AV-LMU truck. The display is part of a demonstration of Saudi Arabian equipment taking place during Operation Desert Shield. *Image Courtesy: US Department of Defence*

NAVY: Royal Saudi Naval Forces (RSNF) is the naval force of the Kingdom of Saudi Arabia. It also includes Marine Forces and Special Forces. The Naval headquarters is in Riyadh. The Western Fleet is based in the Red Sea with the main base at Jeddah. The Eastern Fleet is based in the Persian Gulf with the headquarters at Jubail. Other naval facilities were located at Yanbu, Dammam, and Ras al Mishab.

Naval Bases: Jeddah, Jubail, Dammam, Ras Tanura, Ras-al-Mishab

Personnel: 60,500 (+ 12000 Marines)

Equipment

Category	Name	In Service
Frigate	Al Riyadh class	3
	Al Medinah	4
Corvette	Badr Class	4
Patrol Boat	Al Siddiq	9
	Naja 12	39
	Halter Type	17
Mine Warfare Force	Sandown-class	3
Auxiliary ship	Boraida Class (Fleet Tanker)	2
Amphibious Force	Afif Class Landing Craft Utility	4
	LCM	4
SSM	Exocet MM40 Block II	8 systems
	Otomat	8 systems
SAM	Aster 15	
	Crotale	26

Naval Aviation

Category	Name	In Service
Helicopter	MH-60R	18
	AS332	18
	AS565	20+

New Procurements/ Upgrades

✦ Saudi Arabia is planning to procure 5 corvettes from Navantia.

✦ Requirements are being finalised for a new class of multi-purpose frigates to replace the Medina Class, as well as new missile corvettes to replace both the BADRs and the Al Siddiqs.

✦ Saudi Arabia is planning to procure 30 Mark V patrol boats and associated equipment, parts, training and logistical support for an estimated cost of USD 1.2 billion. The country has requested a possible sale of 30 Mark V patrol boats, 32 27mm guns, spare and repair parts, support equipment, personnel training and training equipment, publications and technical documentation, US govt. and contractor engineering, technical, and logistics support services, and other related elements of logistics support.

✦ Booz Allen Hamilton has been awarded a USD 12 million worth US Navy contract modification to provide support services for the Royal Saudi Navy.

AIR FORCE: The Royal Saudi Air Force is the aviation branch of the Saudi Arabian armed forces. The RSAF maintains the third largest fleet of F-15s after the JASDF and the USAF.

Personnel: 60,000

Equipment

Category	Name	In Service
Fighters/ attack/ Reconnaissance/ bomber	F-15E	(84 on order)
	F-15C/D Eagle	70
	F-15S Strike Eagle	68 (to be upgraded to SA standard)
	Eurofighter Typhoon	60 (50+ on order)
	Panavia Tornado IDS	80
	An-178	(20+ on order)
Transport/ Tanker	E-3 Sentry (AWACS)	10
	A330 MRTT	6
	C-130 Hercules	30+
	C-130J Super Hercules	2 (20 on order)
	Gulfstream III/IV	3
	Saab 2000 AEW&C	2
	L-100 Hercules	6
	CN-235	5
	An-132	(2 on order)
Attack/ Transport Helicopter	Bell 212	25
	Bell 205	24
	AS-532 Cougar	12
	AS365 Dauphin	24
	SA-332 Super Puma	13
	AS-61	3
	Alouette III	6
	S-70 Blackhawk	16
	Cougar MkII	11
Trainer	BAe Hawk	29 (22 on order)
	BAe Jetstream	1
	Super Mushshak	20
	F-5B	15
	F-15D	15
	Pilatus PC-9	47
	Pilatus PC-21	40
	SR22	20
	King Air 350	6 (3 on order)
	T-41A	12

New Procurements/ Upgrades

✦ Al-Salam Aircraft Co (AAC) Ltd., Riyadh, Saudi Arabia, has been awarded a not-to-exceed USD 32,500,000 contract action for services required for the teardown and assessment of three F-15 aircraft and the conversion and repair of one F-15S aircraft to the F-15SA configuration.

✦ Pakistan is planning to sell JF-17 Thunder combat jets and trainer aircraft to Saudi Arabia. Pakistan is interested in selling arms to Saudi Arabia, including the JF-17 Thunder jet co-developed with China, the Mushshak trainer aircraft and other equipment.

✦ Lockheed Martin and AEC will establish the Sniper ATP Expanded Repair Capability facility in the Kingdom of Saudi Arabia - the first Sniper support established outside the United States. Designed to support the Royal Saudi Air Force's (RSAF) Sniper ATPs, the Expanded Repair Capability facility will provide both affordability benefits and expedite repair times for the RSAF.

✦ Saudi Arabia will be receiving two KC-130J Hercules tanker-transport aircraft worth USD 181 million dollars from Lockheed Martin. The Foreign Military Sales (FMS) contract, which was awarded by the US Department of Defence (DoD) includes engineering support.

✦ The Royal Saudi Air Force has inducted into service the Airbus Military A330 MRTT new generation tanker/transport aircraft. The aircraft is the first of a batch of three that have now been contractually delivered.

✦ Saudi Arabia has awarded BAE Systems a 1.6 billion pound (USD 2.5 billion) deal to supply Hawk jet trainers, Pilatus PC-21 turboprop and other training devices. The centrepiece of the deal is the provision of 22 BAE Hawk Advanced Jet Trainers and the supply of 55 PC-21 machines built by the Swiss manufacturer Pilatus.

✦ The Royal Saudi Air Force has selected the GE Aviation-made F110-GE-129E engines to power its new F-15 fighter jets it has ordered from the US. The RSAF has placed an order for a total of 193 F110-GE-129E engines that will be fitted on the 84 Boeing-made F-15SA fighters. With the order, the RSAF becomes the largest operator of F110 engines other than the US government. This follows earlier RSAF orders for 156 engines to power 71 re-engined F-15S aircraft.

✦ BAE Systems was awarded a USD 367 million contract to deliver 70 Digital Electronic Warfare Systems (DEWS)/ Common Missile Warning Systems (CMWS) to upgrade F-15S fighter jets in support of the Royal Saudi Air Force. The work is scheduled to be completed by November 2018.

✦ Saudi Arabia to procure USD 60 billion worth of aircrafts, including as many as 84 new F-15 fighters, an upgrade of 70 more, and three types of helicopters- 70 Apaches, 72 Black Hawks and 36 "Little Bird" defenders from the United States. Boeing was awarded USD 11.4 billion by US DoD for the procurement of 84 new F-15 fighter aircraft. Development of the new F-15 fighters is set to be completed by 2020.

✦ Saudi Arabia is expected to buy 150 helicopters -- 30 Mi-35 attack helicopters and up to 120 Mi-17 transport helicopters -- more than 150 T-90S tanks, around 250 BMP-3 infantry fighting vehicles and "several dozen" air defence systems, from Russia under a USD 2 billion contract.

✦ Lockheed martin has signed a USD 40 million foreign military contract for the delivery of Sniper Advanced Targeting Pods (ATPs) to the Royal Saudi Air Force.

✦ Saudi Arabia government signed an USD 7.21 billion agreement with the United Kingdom (UK) government for the purchase of 72 Eurofighter Typhoon multi-role fighter aircraft under Project Salam. The first squadron of 24 already delivered to the Royal Saudi Air Force (RSAF).

✦ The A330 MRTT transport/ tankers have been ordered by the Saudi Arabia. The country finalised an agreement to purchase three A330 MRTT equipped with both an Aerial Refuelling Boom System (ARBS) and two Cobham 905E under-wing refuelling pods.

✦ US defence company Raytheon has received a USD 475 million contract from Saudi Arabia to provide the Paveway laser guided bombs. The advanced versions of the bomb use GPS/ INS guidance capabilities.

✦ The US Department of Defence has announced a USD 71,475,750 contract for a Maryland firm to modify two King Air 350 extended range surveillance planes for the Kingdom of Saudi Arabia.

✦ ManTech International Corp. has been awarded USD 175 million contract to provide technical support to the Royal Saudi Air Force's fleet of F-15s.

✦ PKL Services, Inc. has been awarded a contract by the United States Air Force (USAF) in support of the Royal Saudi Air Force (RSAF), beginning January 2017.

A Royal Saudi Air Force Panavia Tornado F.3 aircraft taxis on a runway as it prepares to take off on a mission during Operation "Desert Storm" on 2 February 1991. *Image Courtesy: US Defenseimagery.mil*

ROYAL SAUDI AIR DEFENCE: Royal Saudi Air Defence is the fourth branch of Saudi Armed Forces. It was a part of the Army until 1981, when it was made independent by Field Marshal Khalid bin Sultan. The RSADF has its headquarters in Riyadh.

Personnel: 36,000

Equipment

Category	Name	In Service
AA Gun	35 mm. twin towed	156
	Vulcan towed	100
	30 mm. AMX30	18
Missile	Improved HAWK	17 batteries
	Shahine I/ Shahine II	151

Royal Saudi Air Defence continued...

Category	Name	In Service
Missile	MIM-104 Patriot	10 system (to receive PAC-3 missiles)
Radar	AN/FPS-117	15
	AN/TPS-43	5

SAUDI ARABIAN NATIONAL GUARD (SANG): The Saudi Arabian National Guard (SANG) is a separate military force of the Kingdom of Saudi Arabia. It is not part of the Saudi Arabian Defence Forces. It is a tribal force forged out of those tribal elements loyal to the Saud Family. It serves both as defence force against external threats and as a security force against internal threats. Its duties include protecting the royal family, guarding against coups, protecting strategic facilities and resources, and protecting the Holy Places of Mecca and Medina. It is always commanded by a high ranking member of the royal family.

Prince Mutaib bin Abdullah bin Abdul-Aziz Al-Saud is the Commander of the Saudi National Guard since 2010. He is also a state minister. He is a son of King Abdullah of Saudi Arabia. He conducted a major USD 3 billion reorganisation of SANG to develop its firepower and artillery.

Personnel: 2,15,000

Equipment

Category	Name	In Service
APC/ IFV	LAV-25	
	LAV-III	15 (LAV 6.0 on order)
	Al-Fahd	80+
	EE-11 Urutu	20
	Piranha II	1000+
Mortar/ Assault	120 mm. AMS	73
	90 mm. LAV Assault Gun	130
Artillery	155 mm. M198	27
AA Gun	M167 Vulcan	
Helicopter	AS532U Super Puma (Jointly with Navy & Coast Guard)	12
	AH-6	(20+ on order)

New Procurements/ Upgrades

✦ DynCorp International, Fort Worth, Texas, was awarded a USD 68,961,578 firm-fixed-price, foreign military sales contract, with options, for integrated maintenance support services for the Saudi Arabian National Guard Aviation.

✦ The US Army has awarded Northrop Grumman Corporation a contract to continue the modernisation and training of the Saudi Arabia National Guard (SANG). The contract has a potential value of USD 550 million.

Major Armaments Producers

Domestic suppliers: Advanced Electronics Company, Military Industries Corporation, Abdallah Al Faris Company for Heavy Industries, Alsalam Aircraft Company, BAE Systems Saudi Arabia, SELEX Galileo, King Abdulaziz City for Science and Technology, Prince Sultan Advanced Technology Research Institute

Foreign suppliers: France, China, Germany, United Kingdom, Russia, Brazil, United States, Spain, Turkey, Austria, Pakistan, Japan, Switzerland, Italy, Brazil, Australia, Canada

The Abdallah Al Faris Company for Heavy Industries, based in Dammam manufactured Al-Fahd Infantry fighting vehicle and the Al-Faris 8-400 armored personnel carrier. After the collapse of Arab Organisation for Industrialisation (AOI), Saudi Arabia established its own small arms industry at 'Al Kharj', producing rifles, machine guns, and ammunition under license. In 1985 a royal decree by King Fahd led to the creation of the 'General Establishment of Military Industries' to oversee and coordinate the kingdom's existing and proposed domestic defence projects. The Al-Fahd Infantry fighting vehicle and the Al-Faris 8-400 armoured personnel carrier, used by Saudi land forces, were manufactured by the Abdallah Al Faris Company for Heavy Industries, based in Dammam. Saudi Arabia, despite budget hurdles, remains a major buyer of defence arms and equipment. The country, like other Gulf States, devotes a larger share of their gross domestic product (GDP) to military spending than the global average. The Gulf Cooperation Council [GCC] was established in an agreement concluded on 25 May 1981 in Riyadh, Saudi Arabia between: Bahrain, Kuwait, Oman, Qatar, Saudi Arabia and UAE. These countries declared that the GCC is established in view of the special relations between them, their similar political systems based on Islamic beliefs, joint destiny and common objectives. The GCC is a regional common market with a defence planning council as well. The geographic proximity of these countries and their general adoption of free trade economic policies are factors that encouraged them to establish the GCC. The GCC States seek to build up their defence forces according to a common conception. In this context, they have unified operational procedures, training, and military curricula. They also endeavour to accomplish compatibility of their military systems.

Prince Sultan Advanced Tech. Research Institute (PSATRI) was established to conduct applied scientific and technical research and development to support the Royal Saudi Air Force and other Saudi military and security users in becoming more effective and efficient regional leaders. PSATRI benefits from its close collaboration with MOD, Private organizations and strong interaction with KSU, local and global universities and research centres.

The King Abdulaziz City for Science and Technology (KACST) is Saudi Arabia's national science agency and home to its advanced laboratories. KACST is promoting the development and investment in the national system of science, technology and innovation, by orienting research outputs towards industrial diversification. KACST has created the Joint Centres of Excellence Program (JCEP) to conduct cutting-edge research ideas via partnerships with leading academic and industrial establishments from around the world.

DEFENCE DEALS

Arms Import

Saudi Arabia is projected to become the fifth biggest arms buyer in 2020 with an estimated defence budget of roughly USD 60 billion. Despite falling oil prices, Saudi Arabia is seeing surge in military spending. The vast majority of Saudi Arabia's military equipment is imported from European and North American suppliers. Canada's General Dynamics has received USD 10 billion contract to sell armoured vehicles to Saudi Arabia. General Dynamics Land Systems is the manufacturer of the LAV III armoured vehicle Canada used in Afghanistan. It also makes similar Stryker armoured vehicles for the United States. The United Kingdom has sold USD8.2billion of arms to Saudi Arabia in five years.

Saudi Arabia's modern high-technology arsenal makes Saudi Arabia among the world's most densely armed nations, with its military equipment being supplied primarily by the US, France and Britain. Saudi Arabia is planning to buy 20 military transport planes and five refuelling aircraft along with related defence equipment, worth an estimated USD 6.7 billion, from the US. The country has requested a possible sale of 20 C-130J-30 Aircraft, 5 KC-130J Air Refuelling Aircraft, 120 Rolls Royce AE2100D3 Engines (100 installed and 20 spares), 25 Link-16 Multi-functional Information Distribution Systems, support equipment, spare and repair parts, personnel training and training equipment, publications and technical data, US Government and contractor technical assistance, and other related logistics support. The proposed sale of C-130J and KC-130J aircraft will provide a modern configuration, improve reliability and enhance the Royal Saudi Air Force's ability to effectively field, support, and employ these aircraft. The Defence Security Cooperation Agency notified the Congress on November 2012 about the details of such a sale deal to Saudi Arabia.

The United States plans to sell Patriot missile batteries to Saudi Arabia worth USD 1.75 billion. The Saudi government had requested the purchase of 202 Patriot Advanced Capability (PAC)-3 missiles -- the most sophisticated version of the Patriot anti-missile weapons -- as well as a flight test target, telemetry kits and other related equipment. The proposed sale will help replenish Saudi's current Patriot missiles which are becoming obsolete and difficult to sustain due to age and the limited availability of repair parts. The

Patriot missiles have an estimated range of up to 300 kilometres and have more advanced radar than the older systems. Raytheon has received a contract to supply Command, Control, Communications, Computers and Intelligence (C4I) system to the Kingdom of Saudi Arabia. The contract, valued worth over USD 600 million, has been awarded by the Saudi Ministry of Defence (MOD).

DEFENCE BUDGET

Total defence spending: $87.2 billion (2016)

Estimated defence spending in terms of GDP: 13.5% (2015)

Saudi Arabia has become the largest importer of arms in the Middle East, ranking second internationally. Saudi Arabia surpassed Russia to become the third-largest spender, mainly due to the fall in the value of the rouble, according to SIPRI news report. Saudi Arabia has increased expenditure on defence and security by 20.5% to USD 80.8 billion for 2014 with which spending on defence and security will increase to 35.4%. According to an earlier defence budget report, Saudi Arabia is ranked at number 7 as the biggest spender for its military. It spends over 10% of GDP on defence, more than double the proportion spent by US. The Saudi government spends billions of dollars to arm the military forces and usually buys the military equipment from the main ally United States. The vast majority of Saudi Arabia's military equipment is imported from European and North American suppliers.

'Al Yamamah' is the name of a series of record arms sales by the United Kingdom to Saudi Arabia. British Prime Minister Margaret Thatcher and Saudi Prince Bandar bin Sultan played a key role in the negotiations for the deal, which in the end totalled more than USD 80 billion. It was the largest arms deal in UK history. The prime contractor has been BAE Systems, formerly British Aerospace. The deal involved massive orders for British Tornado fighter planes, helicopters, tanks and ammunitions, most of which were built and supplied by BAE. Controversy surrounding the deal emerged later on. In the Al-Yamamah arms deal between the UK and Saudi Arabia, BAE Systems was accused of establishing a complex web of international banking arrangement and third parties that was designed to provide the greatest secrecy and obscurity for payments to decision takers, according to media reports.

CONCLUSION

Saudi Arabia has one of the best-funded defence forces in the Middle East. The kingdom spends 25% of its budget, or about nearly USD 88 billion on military and safeguarding its borders. Saudi Arabia maintains a complex diplomatic position between the Middle East and the West. It has consistently sought to promote Arab unity, defend Arab and Islamic interests, and support a peaceful resolution of the Israeli-Palestinian conflict (insisting, however, that Israel must withdraw from the territories occupied in 1967). On the other hand, Saudi Arabia has been a partner with the West in economic endeavours and the war against terrorism. Saudi Arabia's military, the best-equipped in the Gulf region, is a force to be reckoned with. It ranks among the world's most densely armed nations, and it has ambitious plans to further upgrade its arsenal. It has always devoted significant resources to improving its military. Since the Cold War era, it has been militarily aligned with the US. Saudi Arabia imports most of its military arms and equipment. The strategic position of Saudi Arabia is directly related to its impact on the stability of the Gulf, which encompasses the military forces, defence expenditures, arms imports, military modernisation, readiness, and war fighting capability. In general, Saudi foreign policy objectives are to maintain its security and its paramount position on the Arabian Peninsula, defend general Arab and Islamic interests, promote solidarity among Islamic governments, and maintain cooperative relations with other oil-producing and major oil-consuming countries. The country, because of its possession of the world's largest reserves of oil and its strategic location, is having good diplomatic relations with western countries, especially United States. The US is Saudi Arabia's largest trading partner, and Saudi Arabia is the largest US export market in the Middle East.

CONTACT DETAILS

Ministry of Defence and Aviation
Airport Road
11165 Riyadh
Saudi Arabia
Tel: +96614785900/477313
Fax: +96614011336
Website: http://moda.gov.sa/ModaEng/

Armed Forces HQ
C/o Ministry of Defence
Tel: =96614023106

National Guard HQ
Riyadh, Saudi Arabia
Tel: +96614024600, 4024277

SERBIA
(Capital: Belgrade)

INTRODUCTION

Area: 77,474 sq km
Population: 7,143,921 (July 2016 est.)
Coastline: 0 km (Landlocked)
Maritime claims: None

KEY POLITICAL PERSONS

PRESIDENT: Tomislav Nikolic

PRIME MINISTER: Aleksandar Vucic

KEY DEFENCE PERSONS

MINISTER OF DEFENCE: Zoran Djordjevic

CHIEF OF DEFENCE STAFF: Gen. Ljubisa Dikovic

COMMANDER OF ARMY: Lt. Gen. Milosav Simović

COMMANDER OF AIR FORCE & AIR DEFENCE: Maj. Gen. Ranko Zivak

ECONOMY

Serbia has a transitional economy mostly dominated by market forces, but the state sector remains large and many institutional reforms are needed. The economy relies on manufacturing and exports, driven largely by foreign investment. Serbia has made progress in trade liberalisation and enterprise restructuring and privatisation, but many large enterprises - including power utilities, telecommunications, natural gas company, national air carrier, and others - remain in state hands. Serbia has made some progress towards EU membership, signing a Stabilization and Association Agreement with Brussels in May 2008, and with full implementation of the Interim Trade Agreement with the EU in February 2010, gained candidate status in March 2012. In January 2014, Serbia's EU accession talks officially opened. Serbia's negotiations with the World Trade Organization (WTO) are advanced, with the country's complete ban on the trade and cultivation of agricultural biotechnology products representing the primary remaining obstacle to accession. Serbia is also pursuing membership in the WTO, and accession negotiations are at an advanced stage. Structural economic reforms needed to ensure the country's long-term prosperity have largely stalled since the onset of the global financial crisis. Serbia, however, is slowly recovering from the crisis. High unemployment and stagnant household incomes are ongoing political and economic problems. Factors favourable to Serbia's economic growth include a strategic location, a relatively inexpensive and skilled labour force, free trade agreements with the EU, Russia, Turkey, and Central European Free Trade agreement countries; and a generous package of incentives for foreign investments.

GDP: \$97.5 billion (2015 est.); \$96.78 billion (2014 est.)

Real Growth Rate: 0.7% (2015 est.); -1.8% (2014 est.)

Industries: Automobiles, base metals, furniture, food processing, machinery, chemicals, sugar, tires, clothes, pharmaceuticals

Total Exports: \$12.6 billion (2015 est.); \$14.14 billion (2014 est.)

Export Commodities: Iron and steel, rubber, clothes, wheat, fruit and vegetables, nonferrous metals, electric appliances, metal products, weapons and ammunition, automobiles

Major Markets: Italy 16.2%, Germany 12.6%, Bosnia and Herzegovina 8.7%, Romania 5.6%, Russia 5.4% (2015)

Total Imports: \$17.03 billion (2015 est.); \$19.58 billion (2014 est.)

Import Commodities: Machinery and transport equipment, fuels and lubricants, manufactured goods, chemicals, food and live animals, raw materials

Major Suppliers: Germany 12.4%, Italy 10.6%, Russia 9.6%, China 8.5%, Hungary 4.8%, Poland 4.2% (2015)

INTERNATIONAL DISPUTES

✦ Serbia with several other states protested the US and other states' recognition of Kosovo declaring itself as a sovereign and independent state in February 2008; Ethnic Serbian municipalities along Kosovo's northern border challenge the final status of the Kosovo-Serbia boundary.

✦ Serbia has delimited about half of the boundary with Bosnia and Herzegovina, but sections along the Drina River remain in dispute.

DEFENCE

ARMY

Personnel: 12,000+

Equipment

Category	Name	In Service
MBT	M-84/84A	200
APC, IFV	M-80/80A	320
	M-86	49
	BTR-50	
Armoured car	BRDM-2	70+
Artillery	M84 152mm	36
	M46 130mm	36
	122 mm D30	
	2SI 122mm Sp	72
Mortar	M74, M75	
MRLS	M-63/94 128mm	18
	M77 128mm	60
	M87 ORKAN 262mm	4
	Morava 128mm	

Army continued...

Category	Name	In Service
AA Gun, SAM	L-70 Bofors	
	9K35M Strela-10M	
	9K31M Strela-1M	
	9K32M Strela 2M	

New Procurements / Upgrades

✦ Mass production of the indigenously-built Lazar multi-role armoured vehicle and its advanced variants is expected to begin in 2017.

✦ The indigenously developed Morava 128mm multiple rocket launchers (MRL) are being progressively inducted.

✦ 30 T-72 MBTs are to be donated by Russia.

AIR FORCE

Personnel: 4,000+

Equipment

Category	Name	In Service
Fighter, Trainer	MiG-21/21-U (being phased out)	15
	MiG-29/MiG-29UB	4
	Galeb G-2	1
	Galeb G-4	6
Transport, Laison	An-26	14
	An-2TD	1
	Seneca V	1
Helicopter	SA-341H/SA-342L Gazelle	50
	Mi-24, Mi-8, Mi-17	2+8+1
SAM	S-125M NEVA	6
	2K12 KUB	20+
	S-10/S-2M	100+

New Procurements / Upgrades

✦ The Defence Ministry is planning to procure over a dozen new fighter planes and is evaluating options between Russia's Sukhoi Su-30 (also recently the MiG-35 fighter), the US F16 and F18, France's Rafale, Sweden's JAS Gripen, Chinese JF-17 and Eurofighter Typhoon.

✦ Russia will donate 6 of its MiG-29 aircraft to Serbian Air Force.

✦ A total of 15 Utva Lasta (designated V-54 by Serbia) basic trainers are on order with deliveries going on.

✦ Funds have been allocated to upgrade the Supergaleb G4 jet trainers.

✦ Two Mi-17V-5 military helicopters have been delivered by Russia in 2016 as part of a contract signed in 2015.

Mi-17V5 helicopter *Serbia Armed Forces photo*

✦ The S-300 PMU SAM defence systems and Pantsir ground-based air defence systems are being sought from Russia, but no deal has been finalised.

Defence Expenditure

Defence Budget: 55 billion dinars ($480 million) - 2015

Estimated defence spending in terms of GDP: 2%

Defence Production and R&D

Serbia has a well-developed defence industry. Once part of erstwhile Yugoslavia – a leading arms exporter – Serbia, until the 1990s, was a hub of weapons manufacturing units. Several of these arms factories were destroyed by NATO during the Kosovo War in 1999. Serbia again revived defence production and is one of the leading producers and exporters of military equipment in the region. Serbia has eight state owned military plants. The country has exported weapons and military hardware to a number of countries in the Middle East, Africa, Asia, Europe and the US. It has also jointly produced weapon systems with its neighbours Bosnia, Macedonia and Croatia. Russia has also expressed interest in establishing military-technical cooperation with the Balkan nation to jointly work on the production and modernisation of some types of weapons and equipment. The country had also inked a deal with the UAE to jointly produce the ALAS multi-role missile. Serbia's Yugoimport SDPR and UAE's Emirates Advanced Research and Technology Holding had signed contracts in this regard. In December 2013, Serbia and the European Defence Agency (EDA) signed an agreement which would allow the Defence Ministry of Serbia to participate in EDA's projects and R&D programmes.

Defence Procurement

Serbia has revived its defence procurement plans after a decade of inactivity due to lack of funds. While the Army plans to procure a number of modern APCs to strengthen its fleet, acquisition of new fighter jets for the Air Force is under consideration by the Defence Ministry. Plans are on to acquire missile defence systems from Russia.

CONTACT DETAILS

Ministry of Defence

Republic of Serbia

5th Bircaninova Street, Belgrade

Website: http://www.mod.gov.rs/

Armed Forces website: http://www.vs.rs/

Minister of Defence

Tel.: +381 11 3006-311

Fax: +381 11 3006-062

Email: ministar.odbrane@mod.gov.rs

SINGAPORE

(Capital: Singapore)

INTRODUCTION

Area: 697 sq km

Population: 5,781,728 (July 2016 est.)

Coastline: 193 km

Maritime claims: Territorial sea: 3 nm

KEY POLITICAL PERSONS

PRESIDENT & COMMANDER-IN-CHIEF: Tony Tan Keng Yam

PRIME MINISTER: Lee Hsien Loong

KEY DEFENCE PERSONS

DEFENCE MINISTER: Dr. Ng Eng Hen

CHIEF OF DEFENCE FORCE: Lt. Gen. Perry Lim Cheng Yeow

CHIEF OF ARMY: Maj. Gen. Melvyn Ong

CHIEF OF NAVY: Rear Adm. Lai Chung Han

CHIEF OF THE AIR FORCE: Maj. Gen. Mervyn Tan

ECONOMY

Singapore has a highly developed and successful free-market economy. It enjoys a remarkably open and corruption-free environment, stable prices, and a per capita GDP higher than that of most developed countries. Trade and commerce are key parts of the economy in Singapore. About 28% of Singapore's GDP comes from industry. The petrochemical industry in particular is very important for the economy. The country imports a lot of crude oil, using it for refined petroleum products. It also places great emphasis on high-end manufacturing - semi-conductors and consumer electronics, as well as machinery, transport equipment, and ships are typical for this field. The government is also trying to foster future growth sectors such as aerospace, precision engineering, and especially life sciences. The government has also invested in education for decades. Human capital and a skilled workforce contribute to the prosperous economy in Singapore.

GDP: $471.9 billion (2015 est.) , $454.3 billion (2014 est)

Real Growth Rate: 2% (2015 est.), 2.9% (2014 est.),

Industries: Electronics, chemicals, financial services, oil drilling equipment, petroleum refining, rubber processing and rubber products, processed food and beverages, ship repair, offshore platform construction, life sciences, entrepot trade

Total Exports: $377.1 billion (2015 est.) , $437.3 billion (2014 est.),

Export Commodities: Machinery and equipment (including electronics and telecommunications), pharmaceuticals and other chemicals, refined petroleum products, foodstuffs and beverages

Major Markets: China 13.7%, Hong Kong 11.5%, Malaysia 10.8%, Indonesia 8.2%, US 6.9%, Japan 4.4%, South Korea 4.1% (2015)

Total Imports: $294.5 billion (2015 est.) , $360.9 billion (2014 est.)

Import Commodities: Machinery and equipment, mineral fuels, chemicals, foodstuff, consumer goods

Major Suppliers: China 14.2%, US 11.2%, Malaysia 11.2%, Japan 6.3%, South Korea 6.1%, Indonesia 4.8% (2015)

INTERNATIONAL DISPUTES

Singapore has a dispute with Malaysia over deliveries of fresh water. Indonesia and Singapore continue to work on finalisation of their 1973 maritime boundary agreement by defining unresolved areas north of Indonesia's Batam Island. In 2008, ICJ awarded sovereignty of Pedra Branca (Pulau Batu Puteh/Horsburgh Island) to Singapore, and Middle Rocks to Malaysia, but did not rule on maritime regimes, boundaries, or disposition of South Ledge;

DEFENCE

ARMY

Singapore's defence policy is based on the twin pillars of deterrence and diplomacy. The SAF enjoys interactions with other militaries, which range from exchanges of visits and joint exercises to attendance on each other's courses. The country works closely with its ASEAN neighbours and hosts the annual Shangri-La Dialogue (SLD), organised by the International Institute of Strategic Studies, since 2002. Singapore is also part of the Five Power Defence Arrangement. The Singapore Ministry of National Defence (MND) has officially opened the Airborne-Trooper Training Facility (ATF) at Pasir Ris Camp in Singapore, to enable Singapore Armed Forces (SAF) troopers to hone their skills in basic airborne, heli-rappelling insertion and heli-roping insertion anytime and regardless of the weather conditions.

Personnel: 71,600

Equipment

Category	Name	In Service
MBT	Leopard 2A4SGP	196
	AMX-13SM1	(Stored)
AIFV/APC	Terrex AV-81	135
	Bionix I/II	200+
	M113A2 Ultra IFV	950
Artillery	M-71 155mm	35
	FH-88 155mm	50
	FH2000 155mm	50+
	Pegasus 155mm lightweight	54
	SPH-1 Primus 155mm SP	54

Army continued...

Category	Name	In Service
Mortar	160mm	
	120mm	
	81mm	
	60mm	
MLR	Himars	18
ATGW	Milan-2	30
	Spike	
ATK weapon	Matador	
	Carl Gustaf 84mm	

New Procurements/Upgrades

✦ Series production of the Bionix II AIFV continues.

✦ Series production of at least 135 Terrex 8x8 ICV is underway.

NAVY

Personnel: 4,000

Equipment

Category	Name	In Service
Submarine	Challenger class (ex-Sjoormen class)	6
	Archer class (ex-Vastergotland class)	2
Frigate	Formidable class (modified La Fayette design)	6
Patrol craft	Victory class missile corvette	6
	Fearless class ASW/patrol vessel	11
Mine warfare force	Bedok class (Landsort design)	4
	FB 31-42 class inshore patrol vessel	12
Amphibious Force	Endurance class LPD	4
	Perservance LSL	1
	LCM	6
	LCU	30

New Procurements/Upgrades

✦ Singapore has launched the first Littoral Mission Vessel (LMV) 'Independence' delivered to the RSN on 26 May 2016. It is expected to be fully operational by 2017. Singapore had signed the contract for eight LMVs to replace the existing Fearless-class patrol vessels.

The Republic of Singapore Navy's first Littoral Mission Vessel "Independence."
Image Credit: MoD Singapore

✦ Two TKMS Type 218SG submarines are on order with delivery projected from 2020 to start replacing the Challengers.

✦ Eight new 75m patrol corvettes are on order from ST Engineering with deliveries starting in 2016 to replace the Fearless class.

AIR FORCE

Personnel: 8,000

Equipment

Category	Name	In Service
Fighter	F-16C/D	62
	F-5S Tiger II	30
	F-15SG	40+
	RF-5S Tigereye	6
Transport/ Tanker	C-130 H	6
	KC-130B	4
	KC-130H	1
	KC-135	4
Trainer	Northrop F-5	9
	Alenia M-346	12
	EC-120B Colibri	5
	Pilatus PC-21	19
Helicopter	AH-64D Apache Longbow	20
	AS.332 Super Puma	46
	AS 532 Cougar	24
	CH-47D/SD	14
	S-70B Sea Hawk	6
Maritime patrol	F-50 Enforcer	5
UAV	Searcher	
	Heron 1	
	Hermes 450	
SAM	I-Hawk	6
	Mistral	
	IGLA	75
	Rapier Bix	10
	RBS-70	330
AA Gun	35mm AA gun	24

New Procurements/Upgrades

✦ The Air Force Hermes 450 (H-450) UAV has achieved Full Operational Capability (FOC) status. The H-450s is the first advanced UAV to achieve FOC status in the SAF.

✦ The RSAF 150 squadron has received the full fleet of 12 M-346 aircraft ordered in 2010 as part of the Advanced Jet Trainer programme to replace the A-4SU Super Skyhawk. The first aircraft was delivered in 2012.

✦ Singapore plans to acquire the ASTER-30 Surface-to-Air Missile System manufactured by European defence firm MBDA.

Defence Expenditure

Total defence spending: $9.3 billion (2014),

SG$ 12.34 billion (2013)

Defence spending in terms of GDP: 3.22% (2015)

Defence Production and R&D

Economic reforms over the past decades have enabled Singapore to invest heavily in defence and now consistently spend about 5 per cent of its GDP on developing its military. As a result of this investment Singapore Technologies Engineering (ST Engineering) has become one of the world's most important defence firms, with over 100 subsidiaries. ST Engineering has a very close working relationship with the country's armed forces. According to SIPRI, Singapore has sold defence products to Indonesia, Chad, Nigeria, the Philippines, the United Arab Emirates and Brazil since 2000, generating $1.75 billion in 2010 alone.

Defence Procurement

✦ The Defence Science and Technology Agency (DSTA) is responsible for all aspects of defence procurement process such as contract preparations, pricings and risks, identification of contractors, as well as the evaluation of tenders received, and the subsequent contract award. MINDEF adopts three principles in all its defence procurement activities:

✦ Open and transparent tendering system and approval process

✦ Systematic and legally-binding contracting procedures

✦ Elaborate scheme of checks and balances

CONTACT DETAILS

Ministry of Defence
MINDEF Building
303 Gombak Drive
669645 Singapore
Tel: +656760-8828
Fax: +656764-6119
www.mindef.gov.sg

SLOVAKIA
(Capital: Bratislava)

INTRODUCTION

Area: 49,035 sq km
Population: 5,445,802 (July 2016 est.)
Coastline: 0 km (Landlocked)
Maritime claims: None

KEY POLITICAL PERSONS

PRESIDENT & COMMANDER-IN-CHIEF: Andrej Kiska

PRIME MINISTER: Robert Fico

KEY DEFENCE PERSONS

DEFENCE MINISTER: Peter Gajdos

CHIEF OF DEFENCE STAFF: Gen. Milan Maxim

COMMANDER OF LAND FORCES: Maj. Gen. Ondrej Novosad

COMMANDER OF AIR FORCE: Maj. Gen. Miroslav Korb

ECONOMY

Slovakia has made significant economic reforms since its separation from the Czech Republic in 1993. With a population of 5.4 million, the Slovak Republic has a small, open economy, with exports, at about 92% of GDP, serving as the main driver of GDP growth. Slovakia joined the European Union (EU) in 2004 and the Eurozone in 2009. The country's banking sector is sound. Slovakia has led the region garnering FDI, because of its relatively low-cost, highly-skilled labour force, reasonable tax rates, and favourable geographic location in the heart of Central Europe. However, recent increases in corporate taxes, as well as changes to the Labour Code, slow dispute resolution, and ongoing corruption potentially threaten the attractiveness of the Slovak market. Moreover, the energy sector is characterised by high costs, unpredictable regulatory oversight, and growing government interference.

GDP: $161 billion (2015 est.); $155.4 billion (2014 est.)

Real Growth Rate: 3.6% (2015 est.); 2.5% (2014 est.)

Industries: Automobiles; metal and metal products; electricity, gas, coke, oil, nuclear fuel; chemicals, synthetic fibers, wood and paper products; machinery; earthenware and ceramics; textiles; electrical and optical apparatus; rubber products; food and beverages; pharmaceutical

Total Exports: $73.12 billion (2015 est.); $83.16 billion (2014 est.)

Export Commodities: Vehicles and related parts 27%, machinery and electrical equipment 20%, nuclear reactors and furnaces 12%, iron and steel 4%, mineral oils and fuels 5% (2015 est.)

Major Markets: Germany 22.7%, Czech Republic 12.5%, Poland 8.5%, Austria 5.7%, Hungary 5.7%, France 5.6%, UK 5.5%, Italy 4.5% (2015)

Total Imports: $71.09 billion (2015 est.); $79.36 billion (2014 est.)

Import Commodities: Machinery and electrical equipment 20%, vehicles and related parts 14%, nuclear reactors and furnaces 12%, fuel and mineral oils 9% (2015 est.)

Major Suppliers: Germany 19.4%, Czech Republic 17.4%, Austria 9.1%, Hungary 6.3%, Poland 6.3%, South Korea 5.5%, Russia 5.2%, China 4.1% (2015)

INTERNATIONAL DISPUTES

✦ Bilateral government, legal, technical and economic working group negotiations continued in 2006 between Slovakia and Hungary over Hungary's completion of its portion of the Gabcikovo-Nagymaros hydroelectric dam project along the Danube.

✦ As a member state that forms part of the EU's external border, Slovakia has implemented the strict Schengen border rules.

DEFENCE

ARMY

Personnel: 6,000+

Equipment

Category	Name	In Service
MBT	T-72	17
APC, IFV	BMP-1/ 2	200
	Tatrapan T1	
	OT-90	50+
Armoured Car	Aligator	40
Artillery	D-30	8
	155mm ZUZANA	16
Mortar	81mm, 82mm, 98mm	
MRLS	RM 70/85 Modular	20+
ATGW	9P135	90
	9P148	6
	9S428	100

New Procurements / Upgrades

✦ The planned purchase of 31 Rosomak 8x8 armoured personnel carrier vehicles from Poland has been scrapped. The new platforms, which would have been inducted as Scipio infantry fighting vehicles in the Slovakian Army, were to be acquired under a 2015 agreement signed between Slovakia and Poland.

✦ Carl Gustaf 84 mm shoulder-fired recoilless rifles have been ordered from Sweden's SAAB with delivery to begin from 2016 onwards.

AIR FORCE

Personnel: 4,000

Equipment

Category	Name	In Service
Fighter, Trainer	MiG-29	5
	L-39	8
Transport	An-26	1
	L-410	5
Helicopter	Mi-17	9
SAM	2K12KUB	4
	S-300 PMU	1 Battery
	9K38 Igla	

New Procurements / Upgrades

✦ The Defence Ministry is planning to acquire new fighter jets for the Air Force to replace its aging fleet of MiG-29s set to retire by 2016-17. Plans to lease eight Swedish JAS-39 Gripen fighters for a period of 10-15 years has been deferred.

✦ Two C-27J Spartan tactical transport aircraft have been ordered from Italy's Alenia Aermacchi under a contract signed in October 2014. The aircraft are scheduled for delivery in 2016-2017.

✦ Nine Sikorsky UH-60M Black Hawk military helicopters will be acquired from the US to replace the Mil Mi-17 'Hip' utility helicopters. Delivery of the new aircraft is scheduled between 2016 and 2019.

L-39 jet trainer *A Slovak Air Force photo*

Defence Expenditure

Defence spending: €880.6 million (2016)

Defence spending in terms of GDP: 1.12% (2015)

Defence Production and R&D

The Slovakian defence industry is expanding ties with Russia, China and other Asian countries in an attempt to find markets for its products while meeting its own procurement needs. It has also tied up with Poland to jointly produce and market military land platforms. A 2010 European Defence Agency assessment stated that Slovakia has a good research & development base for future high-tech military production.

Defence Procurement

Even as the Government plans to acquire modern platforms including new fighter jets and armoured vehicles for its military to replace the aging inventory of Soviet-era weapons and systems, lack of sufficient funds has delayed any final decision in this regard. The country intends to be self-reliant in defence production and planning.

CONTACT DETAILS

Ministry of Defence
Kutuzovova 8
832 47, Bratislava, Slovakia
Tel: 0960 11 22 33
E-mail: minister@mod.gov.sk
Website: http://www.mosr.sk/

STRATEGIC INFORMATION

SOUTH AFRICA
(Capital: Pretoria)

INTRODUCTION

Area: 1,219,090 sq km

Population: 54,300,704

Coastline: 2,798 km

Maritime claims: Territorial sea: 12 nm

Contiguous zone: 24 nm

Exclusive Economic Zone: 200 nm

Continental shelf: 200 nm or to edge of the continental margin

GEOPOLITICAL IMPORTANCE

South Africa (Republic of South Africa) is located at the southern tip of Africa. It is divided into nine provinces. To the north of the country lie the neighbouring territories of Namibia, Botswana and Zimbabwe; to the east are Mozambique and Swaziland; while Lesotho is an enclave surrounded by South African territory. South Africa is the 25th largest country in the world. Mafadi in the Drakensberg at 3,450 m (11,320 ft) is the highest peak in South Africa. South Africa is a maritime nation, endowed with a double geo-political identity, both land and the sea. It is strategically situated along vital sea routes of the world, the South Atlantic, the Indian Ocean and the Southern Oceans. South Africa's maritime border extends from the Orange River in the West to Punta do Ouro in the East - a coastline of about 3,000 km and along which its marine resources are spread.

The geo-strategic position of the country is also defined by its maritime zones, marine resources, marine ecology and conservation - as well as its maritime trade.

POLITICAL OVERVIEW

South Africa is a parliamentary republic and the President is both head of state and head of government. The President, Deputy President and the Ministers make up the executive branch of the national government. Executive power is exercised by the government. Legislative power is vested in both the government and the two chambers of Parliament, the Council of Provinces and the National Assembly. The judiciary is independent of the executive and the legislature. Government is three-tiered, with representatives elected at the national, provincial and local levels. Since the end of apartheid in the 1990s the African National Congress (ANC) has dominated South Africa's politics. The ANC is the ruling party in the national legislature, as well as in eight of the nine provinces. Nelson Mandela served as president of South Africa from 1994 to 1999, the first ever to be elected in a fully representative democratic election. He joined the African National Congress in 1944 and was engaged in resistance against the ruling National Party's apartheid policies after 1948. His negotiations in the early 1990s with South African President F.W. de Klerk helped end the country's apartheid system of racial segregation and ushered in a peaceful transition to majority rule. Mandela and de Klerk were jointly awarded the Nobel Prize for Peace in 1993 for their role in the ending of apartheid.

KEY POLITICAL PERSONS

PRESIDENT & COMMANDER-IN-CHIEF: Jacob Gedleyihlekisa Zuma

DEPUTY PRESIDENT: Matamela Cyril Ramaphosa

ECONOMY

South Africa has a mixed economy, the second largest in Africa after Nigeria. It also has a relatively high GDP per capita compared to other countries in Sub-Saharan Africa. South Africa is a middle-income, emerging market with an abundant supply of natural resources; well-developed financial, legal, communications, energy, and transport sectors; a stock exchange that is the 18th largest in the world; and modern infrastructure supporting a relatively efficient distribution of goods to major urban centres throughout the region. The economy of South Africa is also the largest in Africa. In a 2010 survey, South Africa was found to have the second most sophisticated financial market and the second-lowest effective business tax rate, out of 14 surveyed countries. Nevertheless, intimidating economic problems remain from the apartheid era - especially poverty, lack of economic empowerment among the disadvantaged groups, and a shortage of public transportation. South Africa's economic policy is fiscally conservative focusing on controlling inflation and attaining a budget surplus. In 2011, after a year of observer status, it officially joined the BRICS group of five emerging-market nations at the summit held in Hainan, China. Principal international trading partners of South Africa - besides other African countries - include Germany, the United States, China, Japan, the United Kingdom and Spain.

GDP: $723.5 billion (2015 est.), $714.4 billion (2014 est.)

Real Growth Rate: 1.3% (2015 est.), 1.5% (2014 est.)

Industries: Mining (world's largest producer of platinum, gold, chromium), automobile assembly, metalworking, machinery, textiles, iron and steel, chemicals, fertilizer, foodstuffs, commercial ship repair

Oil - proved reserves: 15 million bbl (1 January 2016 est.)

Natural gas - proved reserves: 15.01 billion cu m (1 January 2012 est.)

Total Exports: $81.63 billion (2015 est.), $92.75 billion (2014 est.)

Export Commodities: Gold, diamonds, platinum, other metals and minerals, machinery and equipment

Major Markets: China 11.3%, US 7.3%, Germany 6%, Namibia 5.2%, Botswana 5.2%, Japan 4.7%, UK 4.3%, India 4.2% (2015)

Total imports: $84.33 billion (2015 est.), $98.62 billion (2014 est.)

Import Commodities: Machinery and equipment, chemicals, petroleum products, scientific instruments, foodstuffs

Major Suppliers: China 17.6%, Germany 11.2%, US 6.7%, Nigeria 5%, India 4.7%, Saudi Arabia 4.1% (2015)

DEFENCE & SECURITY

The South African National Defence Force (SANDF) is the name of the present-day armed forces of South Africa. SANDF was created in 1994, following South Africa's first post-apartheid national elections and the adoption of a new constitution. The SANDF is subdivided into- the South African Army, the South African Air Force, the South African Navy, Joint Operations Command, Military Intelligence and the South African Medical Service. The commander of the SANDF is appointed by the President from one of the armed services and is accountable to the Minister of Defence. The SANDF continues to be part of the United Nations Peace support and enforcement missions in the DRC. The SANDF plays a key role in peace keeping operations on the African continent as well as providing election observers from time to time. Recent peacekeeping actions by the South African military include the South African intervention in Lesotho in order to restore the democratically-elected government after a coup, as well as extensive contributions to United Nations and African Union peacekeeping operations in the Democratic Republic of the Congo (MONUSCO) and Burundi.

KEY DEFENCE PERSONS

MINISTER OF DEFENCE & MILITARY VETERANS: Nosiviwe Noluthando Mapisa-Nqakula

CHIEF OF NATIONAL DEFENCE FORCE: Gen. Solly Zacharia Shoke

CHIEF OF ARMY: Lt. Gen. Lindile Yam

CHIEF OF NAVY: Vice Adm. Mosuwa Samuel Hlongwane

CHIEF OF AIR FORCE: Lt. Gen. Fabian Zimpande (Zakes) Msimang

Internal Conflict

South Africa is not currently engaged in an armed conflict.

External Conflict

- South Africa has placed its military along the border to apprehend the thousands of Zimbabweans fleeing economic dysfunction and political persecution.

✦ As of January 2007, South Africa also supports large numbers of refugees and asylum seekers from the Democratic Republic of the Congo (33,000), Somalia (20,000), Burundi (6,500), and other states in Africa (26,000).

✦ Managed dispute with Namibia over the location of the boundary in the Orange River.

✦ In 2006, Swazi king advocated resort to the ICJ to claim parts of Mpumalanga and KwaZulu-Natal from South Africa.

THREAT PERSPECTIVE

South Africa introduced apartheid in 1948, as a systematic extension of pre-existing racial discrimination in the country. As a result, the country became increasingly isolated internationally until apartheid was ended and racial equality introduced in 1993.

After South Africa held its first non-racial election in April 1994, most sanctions imposed by the international community in opposition to the system of apartheid were lifted. On 1 June 1994, South Africa rejoined the Commonwealth, and on 23 June 1994, the UN General Assembly accepted its credentials. South Africa served as the African Union's (AU) first president from July 2003 to July 2004.

South Africa after emerging from the international isolation of the apartheid era has become a leading international actor.

Its principal foreign policy objective is to promote the economic, political, and cultural regeneration of Africa, through the New Partnership for African Development (NEPAD); to promote the peaceful resolution of conflict in Africa; and to use multilateral bodies to insure that developing countries' voices are heard on international issues.

South Africa has played a key role in seeking an end to various conflicts and political crises on the African continent, including in Burundi, the Democratic Republic of the Congo, Madagascar, Sudan, Comoros, and Zimbabwe.

South African democracy is not under terminal threat. The country currently has no major defence and security related issues with other nations around the world and is more focused and determined in developing excellent and long-lasting relations with all countries, especially its neighbours in the Southern African Development Community (SADC) and the other members of the African Union.

In 2016, the United States embassy in Pretoria issued a statement warning US citizens about the threat of near-term attacks against places where US citizens assemble in South Africa. The embassy issued the warning after threats of terrorist attacks from ISIS. Later, South Africa disagreed with the assessment of the US and a statement in this regard was also issued the Department of International Relations and Cooperation (DIRCO) and the State Security Agency (SSA). South Africa Govt. noted that the information provided as a basis for the latest terror alerts on South Africa has been found to be very sketchy, dubious, unsubstantiated.

STRATEGIC RELATIONS

Africa: Africa is at the centre of South Africa's foreign policy. South Africa's development is inextricably linked to the development of Africa and the southern African region. Africa faces the challenge of positioning itself to address the marginalisation of the continent by engaging global role-players on socio-economic development and facilitating a fair and just global order. Angola-South Africa relations are quite strong as both fought together during Angolan Civil War and South African Border War. Botswana and South Africa issued a communiqué which said the two heads of state reviewed bilateral, regional and international issues of mutual interest such as the Agreement on the Establishment of a Joint Permanent Commission for Cooperation between Botswana and South Africa. Bilateral political and economic relations between Egypt and South Africa have improved greatly since 1993. The relationship is in good terms with Malawi also where the two countries in 2008 has signed a Memorandum of Understanding designed to enhance the relationship between the two countries through enhanced security cooperation. South Africa–Zimbabwe relations have been generally cordial since the end of apartheid in South Africa, although there have been tensions due to political troubles in Zimbabwe in recent years. Both the nations have recently agreed maintain border control simple and also discussed on progress made on a MoU on migration. Finally, in order to ensure peace, stability and security in the African continent, South Africa should continue to play leading role with initiatives from African Union (AU) and UN in conflict prevention, peacekeeping, peace-building, post-conflict reconstruction and in providing lasting solutions to other outstanding issues in the continent by keeping the commitment to African unity and integration within the framework of the Constitutive Act of the AU.

India: The bilateral relations between the Republic of India and the Republic of South Africa have grown strong since the end of apartheid in South Africa in 1994. Both nations have since developed close strategic, cultural and economic ties. The strategic significance of South Africa for India is undeniable. Both the countries have also developed defence cooperation, trading military equipments and joint exercises and programmes to train their armed forces. A number of bilateral agreements have been concluded between the two countries in diverse areas ranging from economic and commercial cooperation, defence, culture, heath, human settlements, public administration, science and technology and education.

United States: The bilateral relations between South Africa and United States have been more significant and beneficial since the abolition of apartheid and 1994 democratic elections. The two countries share development objectives throughout Africa, and South Africa plays a key economic and political role in the African continent. South Africa is a strategic partner of the United States, particularly in the areas of health, security, and trade. South Africa and US belong to a number of the same international organizations, including the United Nations, International Monetary Fund, World Bank, G-20, and World Trade Organisation.

China: The relationship between China and South Africa was officially established in January 1998. The two countries currently enjoy an increasingly close relationship. Increasingly numerous official visits to each other's countries by their respective officials and rapidly increasing trade between the two countries has drawn them ever increasingly together. Relationship scaled new heights in 2010, with the comprehensive strategic partnership established between the two countries. There are over 140 medium-sized or large Chinese companies in South Africa with a combined investment of USD 13 billion as of 2015-2016.

Russia: The historical links between South Africa and the Russian Federation are strong. Full diplomatic relations were established between South Africa and the Russian Federation on 28 February 1992. Russia and South Africa distinguishes close positions on key international issues such as the strengthening of multilateralism, the role of the UN and international law, the fight against terrorism, non-proliferation, regional conflicts (North Africa, Middle East, Iraq, Afghanistan, the Iranian "nuclear dossier"), and in sub-Saharan Africa. Both the countries cooperate in various international platforms such as the UN, BRICS, the Group of Twenty, the format of Group of Eight plus five, IAEA, and others. Both of them are coordinating efforts on foreign policy issues.

Multilateral Alliances

South Africa is active in the United Nations (UN), the African Union (AU), the Commonwealth of Nations, and International Atomic Energy Agency (IAEA), Inter-State Defence and Security Committee (South Africa) and others. Considered a possible permanent addition to the United Nations Security Council, South Africa was elected by the UN General Assembly to serve on the Security Council in 2007 for the first time ever. South Africa was a non-permanent member of the United Nations Security Council from October 2006 until 2008. The relationship between European Union (EU) and South Africa increased significantly since 1994 when both the countries entered into a broad framework agreement that would cover all spheres of cooperation. South Africa and the EU now cooperate with each other at various levels and in many forums. Other than trade and development cooperation, various dialogue meetings address issues pertinent to South Africa, the EU, Africa and the rest of the world. South Africa is a member of BRICS.

DEFENCE CAPABILITIES

ARMY: The South African Army is an integral part of the new South African National Defence Force. It is now becoming increasingly involved in peacekeeping efforts in southern Africa, often as part of wider African Union operations. The South African Army released its ARMY VISION 2020 guidelines document during 2006. This strategic guide is an attempt to transform the South African military into a professional, effective, well-equipped and well-trained force that is able to deploy quickly. As per the vision, the objective of the army is to defend and protect South Africa and its territorial integrity and vital interests by providing prepared and supported landward combat forces, also to contribute to peacekeeping operations in the continent at large.

Personnel: 40,000 (12,000 reserve)

Equipment

Category	Name	In Service
MBT	Olifant 1A/1B	150
APC/MICV /IFV	Rooikat 76	170
	Ratel ZT3 tank destroyer	55
	Ratel IFV	1000
	Casspir	330+
	Mamba	500
Artillery	GV5 155mm towed	60
	GV6 155mm SP	44
Mortar	M1/M4 60mm	
	M3 81mm	1000
	M5 120mm	25
RCL Gun	M40 105mm	173
MRL	Bateleur 127mm	25
	Valkiri Mk 1 127mm	25
SAM	Cactus (Crotale)	54
	Starstreak	90
Anti-Tank GW	Ingwe	
	MILAN ER	300
AA gun	Oerlikon 35mm	35
	23mm	37
UAV	Vulture	3

New Procurements/ Upgrades

✦ Armscor has awarded a multibillion rand contract to Denel to manufacture over 200 armoured vehicles to the South African National Defence Force (SANDF) over a 10-year period. The 8 x 8 Badger is a best-of-breed vehicle in its class and will contribute to the modernisation of the SANDF. In the future, it will eventually replace the SANDF's existing fleet of armour protected combat vehicles like the Ratel.

✦ Project Citizen: Partial acquisition of a new-generation maintenance, repair and recovery system.

✦ The South African Army Air Defence Artillery has awarded the CSIR's DPSS unit a R24.82 million contract for decision support for a project study on the integration, command, control and sensor needs for Ground Based Air Defence Systems (GBADS).

✦ Development and partial acquisition of an advanced multi-role light artillery gun capability in the form of the 105mm "Light Experimental Ordnance.

✦ The South African Army had awarded Denel Land Systems an R8.4 billion contract to acquire 264 Badgers. A R1.048 billion order to develop a prototype of each was also awarded. The South African 'Badger' variant is equipped with an under-floor armour package from Land Mobility Technologies (LMT) and Denel Land Systems' Modular Combat Turret (MCT). There will be five versions: a standard infantry carrier, a command car, fire support variant, mortar carrier and tank hunter.

✦ Upgrade of 30 Milan launchers to Milan ADT-ER status and acquisition of 300 missiles for R167.4 million. Delivery underway. Up to 100 more launchers can be upgraded.

NAVY: The South African Navy is the navy of the South African National Defence Force. The role of the navy is to prepare for and to conduct naval operations in defence of the Republic of South Africa, its citizens and interests and to carry out peacetime operations in support of other national objectives. Other responsibilities include the maintenance, preservation and the provision of naval services in support of other state departments and authorities, including search and rescue, protection of maritime resources, and diplomatic sea transport support.

Naval Bases: Simons Town (main naval base and dockyard), Naval Station Port Elizabeth, Naval Station Durban.

Personnel: 7,700 (1,000 reserve)

Equipment

Category	Name	In Service
Submarine	Heroine Class	3
Frigate	Valour class	4
Light Force	Warrior class FAC	4
	T Craft Class IPV	3
	Namacurra Class HPB	
Mine warfare force	River Class	4
	CITY Class type 351	4
Auxiliary	SAS Drakensberg	1
	SAS Protea	1
SAM	Umkhonto	16/32
ASM	Exocet MM40 Block 2	8

Naval Aviation

Category	Name	In Service
Helicopter	Atlas Oryx M2	1
	Westland Super Lynx 300 Mk64	4
Aircraft	Douglas Jet Prop DC-3 AMI	5

The South African Navy does not operate any aircraft itself. The aircraft used on ships or supporting the Navy are operated by 22 Squadron of the South African Air Force.

New Procurements/ Upgrades

✦ Saab's local South African Defence and security company, Saab Grintek Defence (SGD) has been awarded a contract by Armscor (IMT- Institute Maritime Technology) for a TactiCall Integrated Communications System, which will be used to enhance the existing communications network at the Navy's Silvermine Maritime Domain Awareness Centre (MDAC).

✦ South Africa is planning to locally build the new Offshore Patrol Vessels (OPVs) to be acquired under Project Biro to replace the remaining strike craft and minehunters. Under Project Biro, the Navy will acquire several offshore and inshore patrol vessels. A request for information for Biro-category vessels was issued last year.

✦ A 45m in-shore patrol vessel is being planned.

✦ South African defence firm Denel has successfully conducted the first live firing of the modified Umkhonto surface-to-air missile (SAM) system from a ground-based launcher recently. The test demonstrated three war-shot missiles successfully destroying low-cost aerial target systems (LOCATS), two at 15 km range and one at 20km. The naval missile has been developed in Block 1 and Block 2 variants and deployed on the SA Navy's Meko A200 class frigates.

South African Navy submarine SAS Charlotte Maxeke cuts through the surface ahead of Type 23 frigate HMS Portland during an exercise of the coast of South Africa. *Image Courtesy: LA(Phot) Caroline Davies/MOD, Royal Navy (defenceimagery.mod.uk)*

SAS Amatola. *Image Courtesy: South African Navy*

AIR FORCE: The South African Air Force (SAAF) is the air force of South Africa, with headquarters in Pretoria. SAAF is involved in regional peace-keeping, disaster relief and maritime patrol operations. Current air combat capabilities are limited to the Gripen multi-role fighter and the Rooivalk combat support helicopter to allow regional deployments while maintaining national air space protection and training obligations.

Air Bases: Bloemspruit, Durban, Waterkloof, Langebaanweg, Overberg, Swartkop, Thaba Tshwane, Ysterplaat, Hoedspruit, Makhado, Port Elizabeth, TEK Base, Valhalla, & Air Force Mobile Deployment Wing.

Personnel: 10,815 (830 reserve)

Equipment

Category	Name	In Service
Fighter	Gripen C & D	26
Helicopter	Oryx - MKI and MKII	40
	Rooivalk MKI	10
	MBB/ Kawasaki BK 117	5
	A109LUH	27
	Westland Super Lynx Mk300	4
Trainer	BAE Hawk Mk120	24
	Pilatus PC-7 MkII	52
Transport	C-130BZ Hercules	7
	Cessna 208B	10
	Super King Air 200/ 300	4
	Pilatus PC-12	1
	CASA C-212-200 / 212-300	4
	Cessna Citation 550	2
Maritime Patrol	C-47TP	10
Weapon	IRIS-T (AAM)	
	A-Darter (AAM)	
	Mokopa (ASM)	
	Lightning III targeting pod	
	Type 159 Rocket Launcher	
	FZ90 70mm FFAR	
	GBU-12 Paveway II	
	Umbani PGM	
Ground defence	RADAR sector control centres	2
	Mobile long range radar	2
	Static radar	2
	Tactical mobile radar	4
	Protection squadrons	12

New Procurements/ Upgrades

✦ South African armaments acquisition agency Armscor has finally awarded Saab a support contract for the South African Air Force's 26 Gripens. The R285 million contract will see Saab support the Gripen fleet to 2016.

✦ AgustaWestland, a Finmeccanica company, has signed a comprehensive service support contract for the South African Air Force's fleet of AW109LUH helicopters. The annual contract includes the provision for it to be extended for up to 5 years.

✦ Thales has been selected by the South African Department of Defence, South African Air Force and Armaments Corporation of South Africa (ARMSCOR) for a five year Through Life Support (TLS) contract of all Thales avionics equipment on-board several fleets of aircrafts. Under the contract, the company will support a large variety of avionics equipment on board the aircraft, including the TopDeck suite for the C-130BZ tactical transport aircrafts, avionics equipment for the Rooivalk combat helicopters, TopFlight avionics systems for the Hawk Lead-In-Fighters and avionics suites for the Super Lynx helicopters.

✦ Denel Aviation will soon deliver the remaining Rooivalk attack helicopter to the South African Air Force (SAAF), allowing the upgraded aircraft to enter operational service. Recently, two more upgraded and operation-ready Rooivalk helicopters were delivered at the beginning of September 2012, with ten now in service with the 16 Squadron at AFB Bloemspruit. The earlier helicopters have already been deployed in joint military exercises conducted by the SANDF and have proven their value in a combat support role.

✦ The Mokopa is a South African air-to-ground missile, designed primarily for use as an anti-tank weapon. It is currently in its final stages of development, and is being integrated onto the South African Air Force's Rooivalk attack helicopters.

Nuclear Weapons Programme

South Africa was the first and only country to build nuclear weapons and then possibly voluntarily dismantle them in 1989. The country was involved in research and development of strategic weapon systems including nuclear, biological, and chemical weapons from 1960s to 1980s. South Africa also acquired the technology to build nuclear weapons. South Africa developed at least six nuclear warheads, which it later acknowledged, along with a variety of missiles and other conventional weapons. Later, after intensive diplomatic efforts by other nations, the South African government dismantled all of its nuclear weapons. The country has been a signatory of the Biological Weapons Convention since 1975, the Nuclear Non-Proliferation Treaty since 1991, and the Chemical Weapons Convention since 1995. The country also banned any further development, manufacture, marketing, import, or export of nuclear weapons or explosives, as required by the NPT. The country joined the Nuclear Suppliers Group (NSG) in 1995 and also played a leading role in the establishment of the African Nuclear Weapons Free Zone Treaty, also known as the Treaty of Pelindaba in 1996. South Africa also signed the Comprehensive Test Ban Treaty (CTBT) in 1996 and ratified it in 1999. The Treaty of Pelindaba came into effect on 15 July 2009 and the African Commission on Nuclear Energy, in order to verify compliance with the treaty, has been established and headquartered in South Africa.

Major Armament Producers

Domestic suppliers: ARMSCOR, Denel, Reutech Radar Systems, Thales, Paramount Group.

Foreign suppliers: AgustaWestland, BAE Systems, Heckler & Koch, IVECO, MAN, Saab AB, Thales, ThyssenKrupp.

Foreign supplier countries: US, France, Switzerland, Israel, UK, Belgium.

In the past, South Africa has attempted to attain a level of self-sufficiency in the design and manufacture of complete armament systems commensurate with the high threat perception prevalent at the time. Since the fall of the Berlin Wall and the negotiated settlement in South Africa which followed shortly thereafter, the strategic environment within the region has undergone a dramatic change. The central strategic challenge is no longer that of deterrence, but of building common regional security.

Armscor (also ARMSCOR), the Armaments Corporation of South Africa, was a government-supported weapons producing conglomerate officially established in 1968, primarily as a response to the international sanctions by the United Nations against South Africa that began in 1963 and were formalised in 1967. The corporation later assumed the role of a procurement agency for the South African National Defence Force (SANDF). ARMSCOR's primary function is to provide a cost-effective service to the SANDF for the execution of its capital acquisition programmes and the logistical and maintenance support that it requires. Armscor maintains strategic capabilities and technologies and promotes the local defence industry. Following the split off of Armscor and its manufacturing subsidiaries, the South African government established Denel, a leading state-owned manufacturer of defence and aerospace systems, in 1991. Denel inherited most of Armscor's production and research facilities. According to a recent budget speech by South African Minister of DoD, Armscor has developed a new strategy which strongly focuses on the SANDF as its primary client. This strategy deals with the rapid acquisition of equipment in support of urgent operational requirements for the SANDF. Armscor is assisting the DOD with developing the Funding Model to support the implementation of the Defence Review 2015.

Denel along with its associated companies has been manufacturing a number of weapons, including artillery, air defence missiles, anti-tank missiles, UAVs and aircraft components for the South African armed forces as well as for the global market.

DEFENCE DEALS

While acquisition of new naval vessels for the navy is over, delivery of new fighter aircraft for the Air Force is going on as part of the modernisation and transformation programme. Procurement of a large number of BADGER Armoured Fighting Vehicles for the Army is also going on.

Development of a new Main Battle Tank under project 'Aorta' and a new generation armoured personnel carrier under project 'Sepula' for the Army are in planning stage. Under project Protector, the government is planning the development and partial acquisition of a mobile ground-based air defence system. For the Navy, acquisition of nine indigenous multi-purpose hull vessels is being planned under project Biro. Under project Millennium, one to three multi-mission strategic support ships are being planned.

Joint Venture Programmes

✦ Russia's Rosoboronexport and South Africa are discussing the possibility of joint development of defence-related products primarily in the area of radar equipment and rocket engines.

✦ South Africa's Denel has confirmed the establishment of a joint venture (JV) with United Arab Emirates (UAE) company Tawazun Holdings. The JV is called Tawazun Dynamics. The joint venture will develop, manufacture and integrate guided systems for conventional air munitions. The UAE's Air Force will be the first customer of Tawazun Dynamics.

✦ The Advanced High Performance Reconnaissance Light Aircraft (Ahrlac), is a light reconnaissance and counter-insurgency aircraft developed in South Africa by AHRLAC Holdings, a joint venture between the Paramount Group and Aerosud. A joint venture between Paramount Group and Aerosud has given South Africa's aviation sector an important boost, creating a world-class Mirage F1 supply, support and upgrade capability for worldwide customers.

Arms Export

India, world's largest arms importer, has signed defence deal worth nearly USD 80 million with South African companies for integrated self-protection systems on its indigenously developed Dhruv Advanced Light Helicopters. The deal, expected to give a huge boost to South African defence industry, was signed between Dhruv's manufacturer Hindustan Aeronautics Limited (HAL) and Saab Grintek Defence for Integrated Defensive Aids Suites (IDAS). The IDAS systems will be produced at its South African subsidiary Saab Grintek Defence in Centurion, north of Johannesburg, and will be delivered between 2015 and 2018.

South Africa is aiming for high position in the world arms market with military exports ascending ahead. Denel, the largest manufacturer of defence equipment in South Africa has made $16.5 million in profits in 2011. Denel currently has nearly USD 744 million in export contracts on its books. Denel's output is primarily directed to meet the requirements of the South African military, which recently purchased state-of-the art Rooivalk helicopters manufactured by Denel. According to reports by SIPRI, South Africa exported major conventional weapons in Sub-Saharan Africa and AU from 2000-2009.

DEFENCE BUDGET

Total defence spending: $3.092 billion (est.)

Estimated defence spending in terms of GDP: 1.11% (2015)

The defence budget for the financial year FY 2016/17 is R47 billion which is approximately 1.05% of South Africa's Gross Domestic Product (GDP). The MTEF allocations indicate that for FY 2017/18, the country's defence budget will decline to 1.03% of GDP and for FY 2018/19 a further decline to 0.98% of GDP. These figures indicate a persistent decline of the defence budget. According to Minister of Defence and Military Veterans, the reduction in real terms of the defence budget allocation over the last few years has forced to reduce the numbers of MSDS from 8955 in 2012 to 3863 in 2015.

The 2016/17 Annual Performance Plan contains 12 outcomes that can be achieved within the current budget allocation.

✦ Sustained ordered defence commitments

✦ Capability sustainment and renewal

✦ Maintained defence facilities

✦ Restructuring of the SANDF

✦ DOD Human Resources Management

✦ Enhancement of Military Discipline

✦ Establishment of relationships between DOD education, training and development programmes and accredited tertiary institutions for military and civilian members

✦ Development of the Defence Funding Model

✦ Establishment of Defence Industry Engagements

✦ Strategic direction for the implementation of Defence Review 2015

✦ Strategic communication intervention

Operation Thusano is directed at the maintenance, repair and preservation of military equipment, with the assistance of a technical team from the Cuban Defence Force. Since the inception of the programme in 2014, a total of 1320 vehicles are now operational and 850 vehicles have gone into preservation. The planned maintenance and repair of vehicle for FY 15/16 would have cost the department about R700 million. We repaired the vehicles for R276 million with a saving of R424 million. It is estimated that the DOD will 10 save approximately R1.75 billion over four years.

Defence Budget – Actual Expenditure 2015-2016

Programme	2015/2016		2015/2016
	Final Appropri-ation	Actual Appropria-tion	(Over)/ Under Expendi-ture
	R'000	R'000	R'000
Administration	4,984,514	4,981,493	3,021
Force Employment	3,616,407	3,602,801	13,606
Land Defence	15,118,951	3,602,801	0
Air Defence	7,085,719	7,085,719	0
Maritime Defence	3,732,748	3,732,748	0
Defence Intelligence	830,060	830,060	0

Source: Defence Vote 19 actual expenditure versus adjusted appropriation, Defence Annual Report 2015/16

CONCLUSION

South Africa and SANDF continues to support regional and continental processes to respond to and resolve crises, promote peace and security, strengthen regional integration, significantly increase intra-African trade and champion sustainable development in Africa. South Africa defence ministry is focusing on national imperatives such as the National Development Plan Vision 2030, Medium-term Strategic Framework 2014–2019, New Growth Path, Industrial Policy Action Plan and importantly the SA Defence Review 2014. The South Africa Defence Review 2014 provides the long-term defence policy for South Africa that informs the defence trajectory to be pursued over multiple Medium-term Strategic Framework periods. It maps out five strategic planning milestones as firm foundations to direct the development of the long term defence strategic trajectory to enable the restoration of South Africa's defence capabilities through an Extended Long-term Development Plan, inclusive of an aligned funding trajectory integrated into the government national planning, budgeting and reporting cycle. South Africa is a major player in the African continent which aspires to shape the peace and security of the region. South Africa's vision for 2025 is to be a successful and influential member of the international community, supported by a globally competitive economy on a continued growth path that has made significant inroads in addressing unemployment, inequality and poverty in South Africa and contributing to the development of the region and continent.

South Africa's longer-term stability is linked to the success of the South African government and its partners in fighting poverty and reducing the toll of the AIDS pandemic. The country is the only African nation with strength and industrial capability to develop a wide variety of military products. South

Africa's influence in the African continent by playing a leading role both on specific issues and within their regions such as taking policy initiatives, building institutions and originating solutions has created unprecedented opportunities for the country in the international arena. The next strategic challenge is for South Africa to utilise this opportunity to take the initiative in shaping a new global order. Hence, South Africa should continue keep strong bilateral relations in order to enhance the strength of South Africa's international positions and influence in multilateral organizations. SANDF, as a national asset, is defending the country's borders and supporting the national security imperatives, foreign policy objectives and its' economic interests.

CONTACT DETAILS

Ministry of Defence
Defence Headquarters
cnr Nossob & Boeing Street
Erasmuskloof
PRETORIA
Tel: (012) 355 6999
Email: info@dod.mil.za

Information Centre
Department of Defence (Defence Corporate Communication)
Private Bag X161
Pretoria 0001
Tel: (012) 355 6321; Fax: (012) 355 6398
E-mail: info@mil.za

STRATEGIC INFORMATION

SOUTH KOREA
(Capital: Seoul)

INTRODUCTION

Area: 99,720 sq km

Population: 50,924,172 (July 2016 est.)

Coastline: 2,413 km

Maritime claims: 12 nm; between 3 nm and 12 nm in the Korea Strait

Contiguous zone: 24 nm

Exclusive Economic Zone: 200 nm

GEOPOLITICAL IMPORTANCE

South Korea occupies the southern portion of the Korean Peninsula. The mountainous peninsula is flanked by the Yellow Sea to the West, and Sea of Japan (East Sea) to the East. Its southern tip lies on the Korea Strait and the East China Sea. The country is a world leader in IT, shipbuilding and vehicle manufacture. Korea participates in various United Nations Peace Keeping Operations and peace activities in Multi-National Forces (MNF).

POLITICAL OVERVIEW

The Republic of Korea is a republic with powers nominally shared among the presidency, the legislature, and the judiciary, but traditionally dominated by the President. The President is the chief of state and is elected for a single term of 5 years with no additional terms being allowed. The President cannot dissolve the National Assembly, but the Assembly can hold the President ultimately accountable to the Constitution by means of an impeachment process. The Prime Minister is appointed by the President and approved by the National Assembly. As the principal executive assistant to the President, the Prime Minister supervises the administrative ministries and manages the Office for Government Policy Coordination under the direction of the President. The 299 members of the unicameral National Assembly are elected for a 4-years term. South Korea's judicial system comprises a Supreme Court, appellate courts, and a Constitutional court. The judiciary is independent under the constitution. The country has nine provinces and seven administratively separate cities – the capital of Seoul, along with Busan, Daegu, Daejeon, Gwangju, Incheon and Ulsan. Political parties include the Grand National Party (GNP), Democratic Party (DP), Liberty Forward Party (LFP), New Progressive Party (NPP), Pro-Park Alliance (PPA), and Renewal Korea Party (RKP).

KEY POLITICAL PERSONS

PRESIDENT & COMMANDER IN CHIEF: Park Geun-hye

PRIME MINISTER: Hwang Kyo-ahn

ECONOMY

South Korea has a market economy which ranks 15th in the world by nominal GDP and 12th by purchasing power parity (PPP). The country has achieved an incredible record of growth and global integration from the 1960s to become a high-tech industrialised economy today. South Korea, which is one of the Asian Tigers, is a member of Organisation for Economic Co-operation and Development (OECD). It is the only developed country so far to have been included in the group of Next Eleven countries. The country adopted numerous economic reforms following the Asian financial crisis. Growth moderated to about 4-5% annually between 2003 and 2007. With the global economic downturn in late 2008, South Korean GDP growth slowed to 0.2% in 2009. In the third quarter of 2009, the economy began to recover, in large part due to export growth, low interest rates, and an expansionary fiscal policy, and growth exceeded 6% in 2010. In 2011, the governments of both the United States and South Korea ratified the Free Trade Agreement which came into effect in March 2012. Between 2012 and 2015, the economy experienced slow growth – 2%-3% per year - due to sluggish domestic consumption and investment. The administration in 2015 faced the challenge of balancing heavy reliance on exports with developing domestic-oriented sectors, such as services. The economy's long term challenges include a rapidly aging population, inflexible labour market, and over dependence on manufacturing exports to drive economic growth.

GDP: $1.849 trillion (2015 est.); $1.802 trillion (2014 est.)

Real Growth Rate: 2.6% (2015 est.); 3.3% (2014 est.)

Industries: Electronics, telecommunications, automobile production, chemicals, shipbuilding, steel

Total Exports: $548.8 billion (2015 est.); $613 billion (2014 est.)

Export Commodities: Semiconductors, petrochemicals, automobile/auto parts, ships, wireless communication equipment, flat display displays, steel, electronics, plastics, computers

Major Markets: China 26%, US 13.3%, Hong Kong 5.8%, Vietnam 5.3%, Japan 4.9% (2015)

Total Imports: $428.5 billion (2015 est.); $524.1 billion (2014 est.)

Import Commodities: Crude oil/petroleum products, semiconductors, natural gas, coal, steel, computers, wireless communication equipment, automobiles, fine chemical, textiles

Major Suppliers: China 20.7%, Japan 10.5%, US 10.1%, Germany 4.8%, Saudi Arabia 4.5% (2015)

DEFENCE & SECURITY

In South Korea, military service is a constitutional duty, and the Law on Conscription applies to all males. The Ministry of National Defence is gradually reducing the service period to relieve citizens' burden and allow more people to fulfil their obligations by assigning service duties fairly. The President is the Commander-in-Chief of the South Korean Armed Forces. The Chairman of the Joint Chiefs of Staff, a 4-star General or Admiral is the Senior Officer of the armed forces and has the operational authority over the armed forces, with directions from the President through the Defence Minister. The Joint Chiefs of Staff enables close mutual support between the country's Army, Navy, and Air Force and the integration of the three Services. The JCS carries out joint and combined operations by executing operational command over joint units and operations commands.

The ROK military is currently undergoing a structural reform to optimize itself for combat missions by tailoring the structures of its command, troop, unit, and force to the mission and function of each unit at each echelon, execution of combat operations, and the operational environment. The command structure would be streamlined to focus on combat missions. The troop structure would be scaled down but will be better trained and equipped. The force structure will be reformed with the focus on strengthening the frontline combat units and attaining victory in battle, according ROK's latest Defence Reform 2020.

The military modernisation is being carried out to respond to the widening fields of threat, mitigate technological and interoperability gaps with the United States forces and eventually attain a self-reliant defence posture. The ROK military has adopted 'Advanced Elite Military' as its vision to achieve the objectives of national security and national defence. To this end, it has identified the following eight policy tenets which it is actively pursuing – establishing a defence posture for comprehensive security; strengthening the ROK-US military alliance and expansion of defence diplomacy and cooperation; providing military support for the advancement in inter-Korea relations; bolstering advanced military capabilities; nurturing highly qualified military personnel and improving the training and education system; enhancing management efficiency; nurturing an attractive and rewarding military; striving to become a defence force that serves the people. Following continuous provocations by North Korea in the West Sea in the last few years, a Northwest Island Defence Command (NWIDC) has been established and the ROK military's response capabilities in the Northwest Islands area have been strengthened significantly.

The ROK military is maximizing the efficiency of its existing military forces by reinforcing the vulnerable areas of the forces that are currently in operation, while at the same time continuously preparing its forces to respond to a wide array of military provocations by North Korea. The ROK military also maintains a 24-hour surveillance and early warning system on the Korean Peninsula and its adjacent islands including North Korea. Additionally, the ROK and the US are jointly identifying and assessing threat indicators such as North Korea's long-range missile launches and nuclear tests.

The country has laid out a major defence reforms plan, to be executed by 2020, under which it intends to transform its military from a manpower-intensive force to a technology-intensive force, from a military-dominated one to a civilian-dominated one, and from a service-oriented force structure to having one with joint operational capability with a long-term vision.

Highlights of Defence Reforms-2020:

- ROK armed forces will build their military structure and force systems to befit modern warfare.
- Civilian workforce in the MND will be expanded to carry out support role, while the military workforce will concentrate on combat & associated tasks.
- Defence management system will be innovated with new methods & technology in armed force would be used in more efficient and cost effective manner.

KEY DEFENCE PERSONS

MINISTER OF DEFENCE: Han Min-koo

CHAIRMAN OF THE JOINT CHIEFS OF STAFF: Gen. Lee Soon-Jin

CHIEF OF STAFF OF THE ARMY: Gen. Jang Jun-kyu

CHIEF OF NAVY: Adm. Um Hyun-seong

CHIEF OF STAFF OF THE AIR FORCE: Gen. Jeong Kyung-Doo

COMMANDER OF MARINE CORPS: Maj. Gen. Lee Sang-hoon

External Conflict

The Northern Limit Line or North Limit Line (NLL) is a disputed inter-Korea maritime demarcation line in the Yellow (West) Sea between the Democratic People's Republic of Korea (DPRK) in the North, and the Republic of Korea (ROK) in the South. This line of military control acts as the de facto maritime boundary between North Korea and South Korea.

South and North Korea have remained technically in a state of war ever since the 1950-53 Korean War which ended in a truce and not a peace treaty. Over the past decades, small-scale skirmishes have flared up repeatedly along their land and sea borders. Tension has been running particularly high in the Korean peninsula after the sinking of the South Korean warship Cheonan in 2010. Tensions escalated between the two sides in 2015 when Seoul resumed loudspeaker broadcasts after a landmine at the border injured two of its soldiers. The tension between the two neighbours have flared up in recent times, especially after repeated nuclear weapons tests carried out by the North.

Territorial disputes

The Liancourt Rocks, called Dokdo in Korean and Takeshima in Japanese, are a group of islets in the Sea of Japan (East Sea) whose ownership is disputed between Japan and South Korea. Although Liancourt Rocks are claimed by both Korea and Japan, the islets are currently administered by South Korea, which has its Korean Coast Guard stationed there.

THREAT PERSPECTIVE

Security challenges for South Korea include transnational and non-military threats such as the proliferation of Weapons of Mass Destruction (WMD), terrorism, cyber attacks, and large scale disasters. North Korea in particular poses a major security threat by developing and increasing its large stockpile of conventional military hardware, while advancing in its nuclear weapons programme, and other weapons of mass destruction. The attack and sinking of a ROK Navy ship and the shelling of Yeonpyeongdo Island in 2010 further escalated the military conflict between Seoul and Pyongyang. The nuclear tests carried out by the North in 2016 have strained ties between the two sides with South Korea conducting regular military exercises in the region jointly with its ally, the US.

STRATEGIC RELATIONS

United States: The US-ROK alliance is stronger than ever and is evolving into a comprehensive strategic alliance beyond a military alliance. The two nations became allies during the Cold War. Washington helped establish South Korea and fought on its UN-sponsored side in the Korean War (1950-1953). The two partners have decided to fully cooperate in dealing with new threats such as in space and cyberspace domains. Both countries remain close trading partners as shown by official figures which rank Korea among the top ten trading partners of the US. After the September 2016 nuclear tests conducted by the North, the US has provided strategic bomber cover over the South for more than a month, followed by multiple joint military exercises. Washington has also supported toughest ever sanctions against North Korea through the UN.

In another major development, White House is also looking towards deploying a Terminal High Altitude Area Defense (THAAD) anti-missile system battery in South Korea as quickly as possible. The missile defence system is expected to be operational within a year of formal approval from the US Government. Once deployed, the weapon could have major strategic implications in the entire East Asian region and even beyond.

Japan: Tokyo has recognised South Korea as the only legitimate government in the whole Korean peninsula based on the Treaty on Basic Relations between Japan and the Republic of Korea. And despite not having any defence relationship with Seoul, Tokyo has nonetheless backed the US's contingency plans to dispatch the United States' armed forces based in Japan to South Korea in the event of a North Korean attack on the South. Japan is South Korea's second-biggest trading partner.

India: ROK and India, the two largest democracies in Asia, have declared their relationship as a strategic one. The two sides have signed agreements, including in trade and defence cooperation. They have decided to enhance their engagement in civil nuclear energy cooperation as well as space activities.

China: The ROK-China relationship has been elevated to a 'strategic cooperative partnership' level, leading to active defence cooperation and exchange. Visits by naval vessels, military sports teams, and military history and training institutions have also been very active. Despite a robust economic relationship, China and South Korea diverge over a number of issues, notably the nuclear issue on the Korean peninsula, fishing in disputed territories as well as a lack of mutual trust in strategic terms.

Russia: South Korea's relations with Russia has been elevated to strategic cooperative partnership. The two sides are strengthening their cooperation to resolve the North Korean nuclear issue and to develop energy resource opportunities. They are stepping up the level of cooperation and exchange in the area of defence and security through senior level exchange visits and cooperation in the defence industry and technology sectors. They have also decided to hold defence strategic dialogues on a regular basis.

European Union: The EU-Korea relationship has evolved over the past few years, based on shared values, common issues of global concern and the increasing role of both partners at world stage. This is demonstrated by the signing of two agreements during 2010 – the new Framework Agreement and Free Trade Agreement (FTA). The agreements address a wide range of international concerns, including non-proliferation of weapons of mass destruction, human rights, cooperation in the fight against terrorism, climate change, energy security and development assistance. On the basis of these two agreements, the EU and South Korea decided at their October 2010 summit to upgrade their relationship to a Strategic Partnership.

DEFENCE CAPABILITIES

ARMY

The ROK Army consists of the Army Headquarters, two Field Army Commands, Operations Command, Capital Defence Command, Special Warfare Command, Army Aviation Operations Command, Army Missile Command, and other supporting units.

Personnel: 495,000

Equipment

Army continued...

Category	Name	In Service
MBT	Type 88 K1/K-1 A1E/K2	1500+
	M-48A2C	
	M-48A3	950
	M-48A5K	
	M47	250 (reserve)
	T-80U	35
MICV/ APC	K-21 IFV	65
	KIFV	1400
	M-577	500
	M-113	
	BMP-3	110
	KM-900	400
	LVTP-7	61
	BV-206	93
	M-8	45
Artillery	M101/KH178 105mm	1000
	M114/KH179 155mm towed howitzers	
	M115 203mm towed howitzers	48
	K55 155mm	1000
	K9 thunder 155mm SP	100
	M110 203mm SP howitzer	
	M107 175mm	100
Mortar	81mm towed	5000
	107mm	
SSM	Hyunmoo 2B (ballistic)/ Hyunmoo 3C (cruise)	
MLRS	MLR 227mm ATACMS Block1A missile	58
	130mm Kooryong	156
RCL	57mm, 75mm, 106mm LAW	
ATGW	TOW, Metis	
AT weapon	Panzerfaust	3
SAM	Javelin	100
	Mistral	35
	IGLA	
	KP-Sam	
	Sin Gung	
AA gun	40mm	40
	Vulcan 20mm	500
	35mm twin	

New Procurements/Upgrades

- Around 100 of the indigenously-built K-2 Black Panther MBTs are expected to be deployed by 2017-18. Over 400 of the new-generation battle tanks are on order to replace the older M-48 tanks.
- Delivery of the advanced K21 IFVs is underway. A total of 450 vehicles are set to replace the older KIFV systems.
- The requirement of 500 K-9 Thunder SP howitzers is being formulated for the Army.
- Series production of the Pegasus (Chun-Mai) mobile SAM and the B1 HO twin 30mm SPAAG is underway.
- Delivery of the KP-SAM Vshorad system is underway.
- A new truck-mounted 105mm howitzer system is being procured.

✦ Raybolt man-portable anti-tank missile systems from LIG Nex1 are on order to replace the TOW ATGMs.

✦ Korean Air Aerospace Division (KAL-ASD) has been ordered to produce 64 advance tactical unmanned aerial vehicles by 2020 for the Army.

✦ Chun-Mu artillery rocket systems (ARS) deliveries are underway. The 239mm tactical MLRS with effective range of 80 km has been developed by Doosan DST.

Army Aviation

Equipment

Category	Name	In Service
Helicopter	Surion	65
	MDH – 500 Defender	150
	Bo - 105	12
	OH-6S	100
	KH-4	5
	AH-1F Cobra (attack)	60
	CH-47 Chinook	15
	UH-60P Black Hawk	120
	DHC-2 Beaver	10
	O-1A	10
	King Air	1
	Mi-172	2

New Procurements/Upgrades

✦ Series production of Surion utility helicopters is underway. Nearly 250 helicopters are on order which will be delivered by 2022.

Surion helicopter *ROK Army photo*

✦ Plans are being finalized for the indigenous development of a new Light Civil Helicopter based on the H-155B1 design for maritime surveillance and transport, to be eventually modified into a Light Armed Helicopter.

✦ The Israeli Heron UAV system is on order.

NAVY

The ROK Navy is focusing on modernisation efforts to guard against North Korean coastal threats and to strengthen its naval capabilities in view of China and Japan's military build-up, as they are locked in an ongoing territorial conflict over disputed islands in the East China Sea. South Korea has deployed more submarines and anti-torpedo weapons as well as an increased number of artillery shells on frontline islands. An independent submarine command has been launched. The command, located in the southern port city of Jinhae, operates 13 diesel-electric submarines – nine 1,200-ton 209-class, and four 1,800-ton 214-class subs – with five more 214 subs being planned to be commissioned by 2020. To beef up anti-submarine operations, the ROK Navy also plans to procure 12 more helicopters. Under the FFX batch I programme, five 2,300-ton Incheon-class frigates have been commissioned. The warships are designed for coastal patrol and anti-submarine warfare missions. Under the second batch programme, up to 12 more frigates are projected to enter service by 2020. The country is also considering building and procurement of nuclear-powered submarines to match and counter the growing nuclear threats emanating from North Korea.

The ROK Navy consists of the Navy HQs, Naval Operations Command, Marine Corps HQs/Northwest Islands Defence Command (NWIDC), and other supporting units.

Major naval bases: Chinhae, Tonghae, Pyontaek, Pusan, Pohang

Personnel: 70,000 (to be reduced to 64,000 under Defence Reforms Plan)

Equipment

Category	Name	In Service
Destroyer	Sejong Daewang (KDX-3) (DDG)	3 (+3 advanced variants to be built)
	Chungmugong Yi Sun Shin (KDX 2)	6
	King Kwang-Gaeto class (KDX-1)	3
Frigate	Ulsan class (being withdrawn)	7
	Incheon class	5
Corvette	Po Hang class (being withdrawn)	
Patrol craft	Yoon Young Ha-Ham class missile FAC	12+
	Chamsuri (being phased out)	50+
Submarine	Son Won IL class (Type 214)	9
	Chang Bogo class (Type 209)	9
	Tolgorae class (midget)	2
	Cosmos type (midget)	7
Mine warfare force	Yangyang- AM	3+4
	Kang Keong class MCMV	6
Amphibious ship	Dokdo class (LHD)	4
	Alligator type	4
	LST type	4
	LSMR type	1
	ex-US LSM type	2
	ex-US LCU type	3
	ex-US LCM	10

Navy continued...

Category	Name	In Service
Auxiliary ship	Chung Jee class logistic support vessel	3
	Soyang class, combat support ship	1
	Submarine rescue vessel	1
	T-Agor type ship	2

New Procurements/Upgrades

✦ ROKN plans to commission three more 7,600-ton KDX-III Aegis destroyers by 2023 to develop a strategic mobile fleet. An initial agreement has been signed with HHI in 2016 for the development of the new warships.

✦ A programme for six new-generation smaller destroyers (under project KDDX) will be launched after 2023.

✦ The seventh and eighth Type 214 attack submarines under the KSS-II programme have been launched in 2016. The navy expects to commission the platforms in 2017 and 2018 respectively.

✦ Plans are to develop up to nine 3,000-ton heavy attack submarines under the improved KSS-III programme. The new platforms could be deployed with submarine-launched ballistic missiles.

✦ Five Incheon-class guided-missile frigates under FFX-I programme have been delivered with a sixth and final ship set for induction in early 2017.

✦ The first vessel of a new class of frigates under the improved FFX-II programme has been launched in 2016 with delivery expected in 2017-18. The new frigate, being built by Daewoo Shipbuilding and Marine Engineering (DSME), is a larger variant of the Incheon-class FFGs. DSME plans to build at least nine such platforms.

Type 214 attack submarine *A RoKN photo*

✦ A total of 42 Yoon Young-ha class of missile FACs in two different variants are being progressively inducted.

✦ Three additional Dokdo class amphibious assault ships are on order.

✦ Three landing ship tank platforms under LST-II programme are being built by HHI with delivery of the first vessel expected in 2017.

✦ RoKN has launched the second Tongyeong-class salvage ship. The ship is being built under the Navy's Auxiliary Towing and Salvage (ATS) II project.

✦ Three Soyang-class of combat support ships are on order from HHI. The first ship of the class has been launched in November 2016 with expected commissioning in 2018.

Naval Aviation

Equipment

Category	Name	In Service
ASW/ Surveillance/ Utility Helicopter	P-3CK/P-3C Orion	16
	F406 Caravan II	5
	500 MD	25
	Super Lynx	22
	AW159 Wildcat	4

New Procurements/Upgrades

✦ KAI and Airbus Helicopters have signed a contract to develop 214 Light Attack Helicopters (LAH) and about 100 Light Civil Helicopters (LCH), based on Airbus AS365B1 Dauphin helicopter, now re-designated H155.

✦ Eight MH-60S helicopters are on order for mine countermeasure roles.

✦ First batch of four AW159 Wildcat helicopters have been delivered in 2016. Total eight AW159s are on order.

AW159 naval helicopter *A Leonardo Finmeccanica photo*

✦ The Navy will have a force of 16 upgraded Lockheed Martin P-3 Orion maritime aircraft by 2018 under an L-3 Communications and Korean Air programme to refit eight aircraft.

MARINE FORCES

The Marine Corps operates as a distinct arm of the South Korean armed forces. The Marine Corps was founded as a reconnaissance force just prior to the start of the Korean War. South Korea plans to develop the Marine Corps into a force that can readily adapt to carry out a variety of tasks based on situations and missions such as multi-dimensional high-speed landing, rapid response, and ground operations among others.

Personnel: 27,000+

Equipment

Category	Name	In Service
MBT	M47	40
APC	LVTP7	100
Helicopter	SA-316B/SA-319B	11
Weapon	Mistral	
	Spike NLOS	

AIR FORCE

The ROK Air Force consists of the Air Force HQs, Air Force Operations Command, and other supporting units. The Air Force Operations Command (AFOC) controls air operations in a centralised manner. The Northern and Southern Air Combat Commands, under the AFOC's centralised command, execute decentralised missions in order to defend the airspace of their respective areas of operation. The Air Defence Artillery Command (ADAC) carries out all-directional air defence missions in preparation against air attacks by enemy aircraft and missiles. The Air Defence and Control Wing executes missions such as air control within the ROK theatre, air surveillance and aircraft identification, and air support operations. The ROK Air Force intends to transform itself into an aerospace force with increased operational reach, air superiority, and precision strike capabilities.

Major air bases: Ch'ongju, Kangnong, Kunsan, Kwangju, Osan, Sunch'on, Suwon, and Taegu

Personnel: 65,000

Equipment

Category	Name	In Service
Fighter/ AWS	RF-16 C/D Block 52	140
	F-16 C/D Block 32	40
	F-15K Slam Eagle	60
	F-5E/F (to be withdrawn)	60
	F-4E	50
Transport	C-130J	4
	C-130H-30	4
	C-130H	8
	CN-235 (including 2 VIP)	20
	Boeing 737 AEW&C	4
ELINT/SIG INT/EW	Hawker 800SIG	4
	800RA	4
Helicopter	AS.332 Super Puma	3
	Bell UH-1D/N (VIP)	5
	Bell 212/412 (SAR)	10
	UH-1H (SAR)	5
	Ka-32 (SAR)	7
	S-92 (VVIP)	3
Trainer	KT-1	85
	T-35 (on lease)	30
	Hawk (T-59)	16
	T-50 Golden Eagle	50
	TA-50	22
	T-50B (being delivered)	
SAM	Patriot PAC-3	18
	Hawk (to be withdrawn)	
	Mistral	

New Procurements/Upgrades

✦ ROK has selected the A330 Multi Role Tanker Transport, powered by Rolls-Royce, as its future air-to-air refuelling and transport aircraft.

✦ Over 100 new 4+ generation fighters are to be acquired by 2025 under the KF-X Project to be built by KAI in partnership with US's Lockheed Martin. The advanced warplanes are to replace the Air Force's ageing fleet of F-4s and F-5s.

✦ ROK has decided to procure 40 of the Lockheed Martin-built F-35 Lightning II Joint Strike Fighter to replace ROKAF's ageing fleet of F-4E fighters.

✦ Deliveries of 60 FA-50 aircraft is going on. The new platforms are replacing the F-5Es.

✦ DAPA has placed an order for four RQ-4 Global Hawk UAVs from the US for delivery between 2016 and 2019.

✦ Two Falcon 2000 aircraft are on order and will be equipped for ELINT roles.

✦ Delivery of Taurus KEPD 350K air-launched stand-off cruise missiles for integration on the F-15K fighters is underway. Around 180 missiles are on order.

✦ Acquisition of two more Boeing E-737 airborne early warning and control (AEW&C) platforms is under consideration in addition to four already in service.

✦ ROKAF has announced plans to acquire four RQ-4 Global Hawk UAVs.

Boeing 737 Peace Eye AEW&C aircraft

Missiles

✦ The state-run Agency for Defence Development (ADD) has carried out the test-launch of ballistic missile Hyunmoo-2B having a range of over 500 km. Under the revised guideline on ballistic missiles, South Korea was allowed to expand the range of missiles from 300 km to 800 km.

✦ South Korea has developed the Hyunmu-3 cruise missile with a range of 1,500-km and the Hyunmu-2 ballistic missile with a range of 300-km.

✦ ROK has requested the US for the supply of 260 AIM-120C-7 AMRAAM missiles, along with containers, missile support and test equipment, provisioning, spare and repair parts, support equipment, personnel training and training equipment, as well as other related elements of programme support.

✦ The Government plans to acquire an Iron Dome anti-missile defence system from Israel in an effort to bolster protection against rockets launched from North Korea.

✦ DAPA has awarded Raytheon Company a $31 million contract for additional AGM-65G-

2 Maverick missiles.

✦ ROK has requested from US the supply of a total of 76 AIM-9X-2 Sidewinder Block II missiles, along with 24 CATM-9X-2 captive air training missiles (CATMs), and eight CATM-9X-2 Block II missile guidance units.

✦ Series production of the new Cheongung SAM is being launched to replace the Hawk system.

✦ DAPA has awarded a contract to Raytheon for additional AGM-65G-2 Maverick missiles for the Air Force.

✦ The MoD has approved the purchase of Patriot PAC-3 missiles and upgrade of South Korea's air defence system.

Nuclear capability: Seoul signed the Non-Proliferation of Nuclear Weapons (NPT) treaty in April 1975. It is a state party to the Comprehensive Nuclear Test Ban Treaty (CTBT), and a member of the Nuclear Suppliers Group and the Zangger Committee.

Biological Weapons: South Korea ratified the Biological and Toxin Weapons Convention (BTWC) in June 1987 and joined the Australia Group in October 1995.

Chemical weapons: South Korea ratified the Chemical Weapons Convention (CWC) in April 1997. According to Nuclear Threat Initiative (NTI) report, South Korea completed the destruction of its entire chemical weapons stockpile in July 2008, becoming only the second CWC member to do.

DEFENCE PRODUCTION

South Korea has progressively built up a formidable defence industry over the past 30 years. The existence of a modern industrial base, consistent financial backing and encouragement from the government, generous technological assistance from the US and Japan, and a well-educated, skilled work force has contributed to the success of the South Korean defence industry. The country's defence sector has developed competences in a number of areas such as manufacturing of armoured vehicles, naval vessels and aircraft. Domestic manufacturers provide approximately 77 per cent of the weapon systems to the country's armed forces. In the reform acquisitions mandated under the Defence Reform 2020 plan, emphasis has been placed on improvement of individual and joint C4ISR capabilities of the armed services. During the next ten years, the ROK Government aims to double the number of workers in its defence industry and increase its annual weaponry exports to US$4 billion. The defence industry develop, design and manufacture a full range of military systems, including imaging satellites, armoured vehicles, artillery, missiles and rockets, C4I systems, fixed- and rotary-wing aircraft, surface combatants and submarines. The Defence Acquisition Programme Administration (DAPA) is responsible for defence industry equipment purchases by the country.

Major Armaments Producers

Korea Aerospace Industries Ltd.: KAI is a South Korean aerospace company involved in the development and production of military aircraft. Major products are T-50 Golden Eagle supersonic jet trainer, KT-1 turboprop trainer, and Korea Utility Helicopter. South Korea's Defense Acquisition Program Administration (DAPA) has signed a contract with Korea Aerospace Industries (KAI) to formalise the company's development of the Korean Fighter Experimental (KFX) aircraft, which is expected to start in 2017.

Samsung Thales: The Company is a leading high-tech company, with its main operations focused on defence. Its core business areas are command, control and tactical communication systems, radar systems, naval combat management systems, electro-optics and fire control systems, aviation/electronics warfare systems, Land C2 systems.

Samsung Techwin: Samsung Techwin Defence Company produces huge artillery system for the South Korean military which includes: K9 Thunder, K10 Ammunition Resupply Vehicle, K77 Fire Direction Center Vehicle, K55A1 Self-propelled Howitzer, K-56 Ammunition Resupply Vehicle. It also constructs Hybrid-Unmanned Vehicle, Hybrid Tactical Vehicle and Unmanned Ground Combat Vehicle.

Hanwha: Hanwha's Defence Products Division is involved in developing and manufacturing various state-of-the-art weapons ranging from precision munitions to precision guided weapon systems, underwater equipment and satellite launch vehicles. The company is also contracted to localise production of Korea's next generation fighter aircraft engine development and manufacturing.

Poongsan: The company has made significant contribution towards defence indigenisation through mass production and localization of ammunitions. The military products include ammunition for small arms, anti-aircraft guns, mid-to large-size arms.

Hanjin Heavy Industries: The company has built a variety of special ships for the Korean Navy such as Dokdo-ham, a mega transporter, corvettes, landing ships, air cushion vehicles and patrol boats.

LIG Nex1: The company has produced a wide range of advanced precision electronic systems including missile, underwater weapon systems, radars, electronic warfare, avionics, tactical communication systems, fire control systems, naval combat systems, and electro-optics.

DEFENCE DEALS

Joint Venture Programmes

Europe: The Airbus Group NV and KAI have signed a contract to produce military helicopters in South Korea. KAI and the European aircraft maker intend to jointly build at least 300 helicopters. All the helicopters will be assembled in South Korea. The two companies are currently producing the utility helicopter, KUH-1 Surion.

Israel: Sharp Elbit Systems Aerospace (SESA), a joint venture entity between Israel's Elbit Systems and ROK's Sharp Aviation K, has been formed in 2013 to provide capabilities in the domains of maintenance, repair and production of advanced avionics for military applications. The company is established to manage research & development of systems and avionics for ROK military's existing and future projects, including the Korean Light Attack Helicopter (LAH) and the Korean Future Fighter (KFX) development programmes.

Indonesia: Indonesia and South Korea have signed an agreement to continue joint development of the Korean Fighter Experimental, Indonesian Fighter Experimental - Ed (KFX/IFX) aircraft. The KF-X/IF-X development cooperation programme is being undertaken in three phases, including the technology development (TD), engineering and manufacturing development (EMD), as well as the production development (PD) phase. The two sides have also agreed to establish a joint programme management office to assist the implementation of the project.

Pakistan: Pakistan and Korea have announced their interest in establishing mutually beneficial cooperation in the field of defence production and agreed to further enhance each other's capabilities by helping through joint ventures. A statement by the Pakistani government has indicated that a potential focus of the collaboration is naval shipbuilding and development of new shipyards.

USA: South Korea has secured technologies from the US to support its programme to develop the Korean Fighter Experimental (KFX) aircraft through defence offset of F-35 deal between the both countries. Technologies include active electronically scanned array (AESA) radars, electro-optical targeting pods, infrared search and track systems, and radio frequency jammers. KAI's T-50 supersonic jet trainer, jointly developed and produced in partnership with US's Lockheed Martin, is one of the contenders in the US military's T-X programme under which the US Air Force plans to replace its older-generation T-38 trainer jets with a new-generation trainer aircraft.

Arms Import

South Korea is progressively bolstering its military strength by purchasing warplanes, anti-missile systems and other weapon systems. The country inked its biggest-ever arms deal worth US$7.04 billion for the procurement of 40 Lockheed Martin-built F-

35 fifth-generation fighter jets. DAPA also announced the purchase of four aerial tankers to enhance its air defence system in addition to upgrading its missile system with PAC-3 interceptors. The ROK Government is also making efforts to boost other military capabilities, such as the development of advanced army tanks, navy frigates, new-generation submarines, surveillance systems such as UAVs as well as launching military satellites. South Korea has also finalised a deal for the US THAAD missile defence system which is expected to be deployed in 2017.

Arms Export & Major Export Deals

South Korea is emerging as one of Asia's leading arms exporting powerhouses with more and more countries deciding to purchase arms from the country. 96 South Korean companies were operating in the defence industry as of 2012. Five of these companies, including Samsung, Korea Aerospace Industries, and LIG Nex1, were named among the Top 100 arms-producing and military services companies in the world in 2013 by the Stockholm International Peace Research Institute.

In 2014, defence exports by ROK reportedly hit a record high at $3.6 billion, which is the largest-ever the country had ever sold in the international market since the Defence Acquisition Program Administration (DAPA) was established in 2006. Major deals included a $1.2 billion contract to build six corvettes for the Royal Malaysian Navy and a $420 million deal with the Philippines for 12 FA-50 fighter jets. In 2013, sale of weapons by Seoul stood at $3.42 billion. The major international defence sales during the year included KAI's contract to sell 24 FA-50 Fighting Eagle armed trainer aircraft to the Iraqi Air Force. Other major export contracts are Thailand's order in August of a DW300H frigate from Daewoo shipbuilding and Marine Engineering (DSME); Colombia and Peru's orders in February and November respectively of patrol vessels made by STX Offshore and Shipbuilding; and Norway's contract, signed in June, to purchase a logistics support vessel from DSME. The country generated $2.4 billion dollars in arms exports over the course of 2011, the bulk of which came from the sale of three Changbogo-class submarines ($1.1 billion) and 16 KAI T-50 Golden Eagle trainer jets ($400 million) to Indonesia. KAI has also entered into an export deal for a total of 20 KT-1Ps worth 210 million dollars with Peru in 2012 through a government-to-government transaction.

In 2015, South Korean defence firms secured export orders worth US$3.5 billion. In 2016, HHI secured a major deal worth US$337 million to build two new frigates for the Philippines Navy.

DEFENCE BUDGET

Proposed defence budget for 2017: KRW 40.3 trillion (US $36.5 billion)

Defence spending in terms of GDP: 2.6% (2015)

Seoul's defence budget has grown rapidly in the recent past with 2017 seeing the highest year-on-year growth since 2011. The state military budget has been set at KRW 40.3 trillion for 2017. In 2016, the country had allocated KRW 39 trillion for defence expenditure.

CONCLUSION

South Korea has emerged from a chaotic history to one of the most dynamic countries in the world. Over the past four decades the country has demonstrated incredible growth and global integration to become a high-tech industrialised economy. The ROK military has also become a great regional naval power in the span of a decade and its military shipbuilding industry and general munitions industries have seen successes abroad. With the country continually advancing in defence technology due to the perceived threat from North Korea, ROK is seen as an emerging maker of high-end military systems and advanced military technologies.

CONTACT DETAILS

Ministry of National Defence
2, Itaewon-ro, Yongsan-gu
Seoul-04383
Korea, South
Tel: 02748-1111

SOUTH SUDAN
(Capital: Juba)

INTRODUCTION

Area: 644,329 sq km
Population: 12,530,717 (July 2016 est.)
Coastline: 0 km (Landlocked)
Maritime claims: None

On 9 July 2011, South Sudan became an independent country called Republic of South Sudan and on 14 July 2011, the country became the 193rd member state of the United Nations. The government has formed various organs which constitute the state.

KEY POLITICAL PERSONS

PRESIDENT & COMMANDER-IN-CHIEF: Salva Kiir Mayardit

VICE PRESIDENT: Taban Deng Gai

KEY DEFENCE PERSONS

DEFENCE MINISTER: Kuol Manyang Juuk

CHIEF OF GENERAL STAFF: Gen. Paul Malong Awan

ECONOMY

Industry and infrastructure in landlocked South Sudan are severely underdeveloped and poverty is widespread, following several decades of civil war with Sudan. Subsistence agriculture provides a living for the vast majority of the population. Property rights are tentative and price signals are missing because markets are not well organized. South Sudan has little infrastructure. The government spends large sums of money to maintain a large army; delays in paying salaries have periodically resulted in riots by unruly soldiers. South Sudan depends largely on imports of goods, services, and capital from the North. Despite these disadvantages, South Sudan has abundant natural resources. It produces nearly three-fourths of the former Sudan's total oil output of nearly a half million barrels per day. The government of South Sudan derives nearly 98% of its budget revenues from oil. The economy of South Sudan undoubtedly will remain linked to Sudan for some time, given the long lead-time and expense required to build another pipeline. In early 2012, South Sudan suspended production of oil because of its dispute with Sudan over trans-shipment fees. This had a devastating impact on GDP, which declined by at least 55% in 2012. In 2013, GDP grew by nearly 30%. However, the outbreak of conflict on 15 December 2013 combined with a further reduction of oil production and exports, meant that GDP growth fell significantly in 2014 and poverty and food insecurity rose. The country is currently burdened by considerable debt, accrued largely in 2012, because of rapidly accumulating arrears and increased military spending. It has received more than $4 billion in foreign aid since 2005, largely from the UK, the US, Norway, and the Netherlands. Long-term challenges for the country include diversifying the formal economy, alleviating poverty, maintaining macroeconomic stability, improving tax collection and financial management and improving the business environment.

GDP: $23.69 billion (2015 est.); $23.73 billion (2014 est.)

Real Growth Rate: -0.2% (2015 est.); 2.9% (2014 est.)

Natural Gas - Proved Reserves: 63.71 billion cu m (1 January 2016 est.)

Crude Oil - Proved Reserves: 3.75 billion bbl (1 January 2016 est.)

INTERNATIONAL DISPUTES

✦ The South Sudan-Sudan boundary represents the 1 January 1956 alignment. A final alignment pending negotiations and demarcation; and the final sovereignty status of Abyei Area are pending negotiations between South Sudan and Sudan.

✦ Periodic violent skirmishes with South Sudanese residents over water and grazing rights persist among related pastoral populations along the border with the Central African Republic.

✦ The boundary that separates Kenya and South Sudan's sovereignty is unclear in the "Ilemi Triangle," which Kenya has administered since colonial times.

DEFENCE

Following the independence of South Sudan in 2011, the Sudan People's Liberation Army (SPLA) became the regular army of the new republic. The SPLA was founded as a guerrilla movement in 1983 and was a key protagonist of the Second Sudanese Civil War. The US Security Council established the United Nations Mission in South Sudan (UNMISS) for an initial period of one year starting from 9 July 2011 in view of the security situation in South Sudan where a series of deadly ethnic clashes had occurred.

The SPLA maintains some military platforms such as the T-72 MBTs, T-55 MBTs, APCs and Mi-17 helicopters along with small arms and weapons to carry out its operations.

Defence Expenditure

No data is available on the country's defence budget.

Defence Production and R&D

South Sudan does not undertake any defence production activities.

Defence Procurement

SPLA has not officially announced any major defence procurements, but it has reportedly procured amphibious vehicles, ATGWs and other small arms and ammunition from China in 2014. The country has also acquired MANPADs, believed to be from China, which have been delivered to it through its neighbouring countries of Kenya and Uganda. With these new acquisitions, South Sudan hopes to modernize and professionalize its armed forces.

CONTACT DETAILS
SPLA Headquarters,
Juba, South Sudan

SPAIN
(Capital:Madrid)

INTRODUCTION

Area: 505,370 sq km

Population: 48,563,476 (July 2016 est.)

Coastline: 4,964 km

Maritime claims: Territorial sea: 12 nm

Contiguous zone: 24 nm

Exclusive Economic Zone: 200 nm (applies only to the Atlantic Ocean)

KEY POLITICAL PERSONS

SUPREME COMMANDER & CHIEF OF STATE: King Felipe VI

PRESIDENT: Mariano Rajoy Brey

KEY DEFENCE PERSONS

MINISTER OF DEFENCE: Maria Dolores de Cospedal Garcia

CHIEF OF DEFENCE STAFF: Gen. Adm. Fernando Garcia Sanchez

CHIEF OF STAFF OF THE ARMY: Gen. Jaime Dominguez Buj

CHIEF OF STAFF OF THE NAVY: Adm. Jaime Munoz-Delgado

CHIEF OF STAFF OF THE AIR FORCE: Air Gen. Francisco Javier García Arnaiz

ECONOMY

Spain's mixed capitalist economy is the 14th largest in the world, by nominal GDP. After almost 15 years of above average GDP growth, the Spanish economy began to slow in late 2007 and entered into a recession in the second quarter of 2008. The economic boom of the 2000s was reversed, leaving over a quarter of Spain's workforce unemployed by 2012. Despite the poor recent performance of the Spanish economy generally, the country's international trade situation has improved. In 2013 it achieved a trade surplus for the first time in three decades. The European Commission criticized Spain's 2016 budget for its easing of austerity measures and its alleged overly optimistic growth and deficit projections. Spain's borrowing costs are dramatically lower since their peak in mid-2012, and despite the recent uptick in economic activity, inflation has dropped sharply, from 1.5% in 2013 to a negative 0.6% in 2015.

GDP: $1.572 trillion (2014 est.), $1.551 trillion (2013 est.)

Real Growth Rate: 3.2% (2015 est.) 1.4% (2014 est.)

Industries: Textiles and apparel (including footwear), food and beverages, metals and metal manufactures, chemicals, shipbuilding, automobiles, machine tools, tourism, clay and refractory products, footwear, pharmaceuticals, medical equipment

Crude oil - proved reserves: 150 million bbl (1 January 2016 est.)

Natural gas - proved reserves: 2.548 billion cu m (1 January 2016 est.)

Total Exports: $277.9 billion (2015 est.), $317.1 billion (2014 est)

Export Commodities: Machinery, motor vehicles; foodstuffs, pharmaceuticals, medicines, other consumer goods

Major Markets: France 15.7%, Germany 11%, Italy 7.4%, UK 7.4%, Portugal 7.1%, US 4.5% (2015) **Total Imports:** $302.6 billion (2015 est.), $345.6 billion (2014 est.)

Import Commodities: Machinery and equipment, fuels, chemicals, semi-finished goods, foodstuffs, consumer goods, measuring and medical control instruments

Major Suppliers: Germany 14.4%, France 11.7%, China 7.1%, Italy 6.5%, Netherlands 5%, UK 4.9% (2015)

INTERNATIONAL DISPUTES

✦ Portugal does not recognise Spanish sovereignty over the territory of Olivenza based on a difference of interpretation of the 1815 Congress of Vienna and the 1801 Treaty of Badajoz.

✦ Morocco protests Spain's control over the coastal enclaves of Ceuta, Melilla, and the islands of Penon de Velez de la Gomera, Penon de Alhucemas, and Islas Chafarinas, and surrounding waters; both countries claim Isla Perejil (Leila Island).

✦ Spain disapproves of United Kingdom plans to grant Gibraltar greater autonomy.

DEFENCE

The Spanish army is one of the oldest active armies - dating back to the late 15th century. Following the suspension of conscription, the Spanish Army has become a fully professionalized volunteer force. The Spanish Navy was responsible for a number of major historic achievements in navigation.

ARMY

Personnel: 80,000

Equipment

Category	Name	In Service
MBT	Leopard 2/E	216
	Leopard 2/A4	100+
Armoured Car	VEC-M1 Pegaso	95
	Centauro B1	84
APC/AFV/IFV	Pizarro	345
	Piranha IIIC	21
	BMR-600	682
	LINCE	395
	RG-31	180
Artillery	155mm/62mm APU SBT	84
	L-118 105 Light Guns	53
	Mountain howitzer 105/14mm	18
	M-109A5 155mm SP howitzer	96
Mortar	120mm on M-106	192
	81 mm (medium mortar)	1200
	120mm on BMR	8
	120 mm towed	265
MLRS	Teruel 140 mm	14
ATGW	Tow	200
	Milan	442
	Spike	260
	Hot	28
ATK weapon	C-90CR M3	
	C-100 Alcotan	
SAM	I-Hawk PIP III	36
	Patriot	8 batteries
	Roland	16
	Aspide	6
	Mistral	105

Army continued...

Category	Name	In Service
SAM	Nasams	8
AA gun	35mm twin	48
UAV	Raven System	27
	Searcher	4

New Procurements/Upgrades

✦ An additional 70 GDSBS 155mm/52cal APU SBT howitzers are on order, including 66 SIAC versions to be delivered by 2013 for the field artillery and 4 V07 versions to join coastal defence assets. The 12 existing systems are being upgraded to V07 standard.

Army Aviation

Category	Name	In Service
Helicopter	CH-47D Chinook	17
	Eurocopter Tiger (Attack)	6
	AS.332B Super Puma	16
	AS-532AL Cougar	22
	EC-135 (UME)	14+8
	Bo-105	10
	AB-212	6

New Procurements/Upgrades

✦ 22 Eurocopter Tiger combat helicopters have been ordered to replace the Bo-105s. The first batch of 6 helicopters has been delivered in the HAP-E configuration.

✦ The MoD plans to upgrade the Chinook from the CH-47D to CH-47F versions. (2016)

✦ A total of 32 NH90 helicopters are on order for the three services.

NAVY

Personnel: 22,300

Equipment

Category	Name	In Service
Submarine	Galerna class	3
	S-80 class	(under construction)
Frigate	Alvaro De Bazan class	5
	Santa Maria class	6
Amphibious force	LCM-1 E landing craft	14
	Galicia class LPD	2
	Juan Carlos I (LHD)	1
Patrol ship	Cabo Fradera class	1
	Descubierta class	4
	Serviola class	4
	Anaga class	3
	Toralla class	2
	Meteoro class	4
	P-114 (coastal surveillance)	2
LCU	LCM1-E class	14
Mine sweeper	Diana BMA (M-11)	1
	M-30 Segura class	6

Navy continued...

Category	Name	In Service
Auxiliary ship	Patino AOR	1
	Cantabria	1

New Procurements/Upgrades

✦ Four Spanish designed new generation submarines with AIP propulsion are under construction to replace the Galernas (S-80 programme Isaac Peral class) with the first vessel expected to be delivered in 2018.

Naval Aviation

Category	Name	In Service
Fighter/Attack/ AWS role	AV-8B	16
Trainer	TAV-8B	1
Transport/ Surveillance	Cessna Citation II/VII	3
Helicopter	AB-212	10
	Hughes 500	0
	SH-3W AEW	3
	SH-60B (Sea Hawk)	12
	SH-3D	7

A port bow view of the Spanish Navy, F 100 Class Frigate, ALMIRANTE JUAN DE BORBON (F 102) underway in the Pacific Ocean during Combat Systems Ship Qualifications Trails on the Pacific Missile Test Centre Range off the coast of California (CA). *Image Credit: US Navy*

An AV-8B Harrier II from the Spanish aircraft carrier Principe de Asturias (R 11) prepares to land after a live fire exercise as part of a passing exercise with ships assigned to Standing NATO Maritime Group Two (SNMG-2). *Image Credit: US Navy*

Marines

Personnel: 5,836

Equipment

Category	Name	In Service
Vehicle	M-60A3 TTS	16
	AAV-7A1	19
	Piranha IIIC	24
	BLR	35
Artillery	M-109A2 (155 mm howitzer)	6
	M-56 105 mm pack howitzer	12
	C-90/CR	
ATGW	Tow 2A	12
	Spike	18
SAM	Mistral	12

AIR FORCE

Personnel: 24,000

Equipment

Category	Name	In Service
Fighter/ Attack	Eurofighter Typhoon	61
	EF-18 (F/A-18 Hornet)	87
	Eurocopter EC120 Colibri	15
Transport/ Surveilla-nce	PA-3A/B Orion	7
	Casa C-212	24
	Casa CN-235	20
	Casa C-295	13
	C-130-Hercules	12
	KC-130H	5
	Falcon 900	5
	Falcon 20	2
	Airbus A-300	4
Trainer	Northrop F-5	19
	Casa C-101-Aviojet	78
	Sikorsky S-76 Spirit	8
	Enaer T-35C	36
	Beechcraft Bonanza F-33C	35
	Northrop F-5M	20
Helicopter	SA.330-Puma	7+
	AS.332 Super Puma	19
	AS.532 Cougar	2
	Eurocopter EC-120 Colibri	17
	Sikorsky S-76	8
	Beechcraft C-90 King Air	4

New Procurements/Upgrades

✦ The training helicopters, the Airbus H120 and the Sikorsky S-76C, are due to be taken out of service from around 2027, with the replacement almost certain to be the H135. (2016)

✦ The Air Force has taken delivery of 51 of the 73 Eurofighter Typhoons ordered.

✦ Order placed for 12 A400M transport aircraft.

Defence Expenditure

Total defence spending: $6.04 billion (2015)

Defence spending in terms of GDP: 1.18% (2015)

Domestic Suppliers: Airbus Group, Santa Barbara, Navantia, Indra, Gamesa, Abengoa

Foreign Suppliers: USA, France, Israel, Germany, Italy

A Spanish Air Force EF-18A Hornet moves into position behind an Air Combat Command KC-135R Stratotanker (not shown) of the 22nd Air Refueling Squadron during a NATO Operation Allied Force Combat Mission. *Image Credit: US Air Force*

Defence Production and R&D

The Spanish defence industry has been producing a variety of military equipment for its armed forces that includes light arms, vehicles, ships, and light transport aircraft. Nearly 150 firms in the country are engaged principally in defence production that primarily involves aerospace, shipbuilding and land armaments. As a member of NATO, Spain is also collaborating with other European countries in a number of defence projects like Eurofighter Typhoon and Airbus A400M programmes. Navantia, Bazan and CASA are some of the leading defence manufacturing companies in the country.

Defence Procurement

Major weapons procurements for all three branches of the military have been going on. While the Army is in the process of acquiring a number of infantry vehicles and other ammunitions, a new class of submarines, frigates and patrol vessels along with new aircraft and helicopters for the navy are scheduled to be inducted in the coming years. The air force is also looking forward to induct the Eurofighter combat jets and A400M strategic transport aircraft to strengthen its capabilities.

CONTACT DETAILS

Ministry of Defence

Paseo de la Castellana,
109 28071, Madrid, Spain
Tel: + 34 91 395 50 00
E-mail: infodefensa@mde.es
Website: www.mde.es

Armed Forces HQ

Vitruvio 1, 28006 Madrid, Spain
Tel. +34-91-7455200

SRI LANKA
(Capital: Colombo)

INTRODUCTION

Area: 65,610 sq km

Population: 22.235 million (July 2016 est.)

Coastline: 1,340 km

Maritime claims: Territorial sea: 12 nm

Contiguous zone: 24 nm

Exclusive Economic Zone: 200 nm

Continental shelf: 200 nm or to the edge of the continental margin

KEY POLITICAL PERSONS

PRESIDENT, MINISTER OF DEFENCE & COMMANDER-IN-CHIEF: Maithripala Sirisena

PRIME MINISTER: Ranil Wickremesinghe

KEY DEFENCE PERSONS

CHIEF OF DEFENCE STAFF: Air Chief Marshal K A Gunatilleke

COMMANDER OF THE ARMY: Lt. Gen. Crishantha de Silva

COMMANDER OF THE NAVY: Vice Adm. Ravindra Wijegunaratne

COMMANDER OF THE AIR FORCE: Air Marshal Kapila Jayampathy

ECONOMY

Sri Lanka is now focusing on long-term strategic and structural development challenges after the end of the three-decade civil conflict. In recent years, the country has experienced strong growth rates in terms of GDP per capita ahead of other countries in the South Asian region. According to the World Bank the Sri Lankan economy has seen robust annual growth at 6.4 percent over the course of 2003 to 2012. The country has also met the Millennium Development Goal (MDG) target of halving extreme poverty and is on track to meet most of the other goals. The main economic sectors of the country are tourism, tea export, apparel, textile, rice production and other agricultural products. In addition to these economic sectors, overseas employment contributes highly in foreign exchange, 90% of expatriate Sri Lankans reside in the Middle East.

GDP: $223 billion (2015 est.), $212 billion (2014 est.)

Real Growth Rate: 5.2% (2015 est.), 4.5% (2014 est.)

Industries: Processing of rubber, tea, coconuts, tobacco and other agricultural commodities; telecommunications, insurance, banking; tourism, shipping; clothing, textiles; cement, petroleum refining, information technology services, construction

Total Exports: $10.5 billion (2015 est.), $11.13 billion (2014 est.)

Export Commodities: Textiles and apparel, tea and spices; rubber manufactures; precious stones; coconut products, fish

Major Markets: US 26%, UK 9%, India 7.2%, Germany 4.3% (2015)

Total Imports: $18.93 billion (2015 est.), $19.42 billion (2014 est.)

Import Commodities: Petroleum, textiles, machinery and transportation equipment, building materials, mineral products, foodstuff

Major Suppliers: India 24.6%, China 20.6%, UAE 7.2%, Singapore 5.9%, Japan 5.7% (2015)

INTERNATIONAL DISPUTES

Sri Lanka has no known international disputes.

DEFENCE

The Ministry of Defence is responsible for the management of the forces. The Joint Operations Command (JOC) is headed by the Chief of the Defence Staff who is the most senior officer in the Armed Forces. The three services have their own respective professional chiefs: the Commander of the Army, the Commander of the Navy and the Commander of the Air Force.

ARMY

Personnel: 190,000

Equipment

Category	Name	In Service
MBT	T-55/AM2	68
	Type 59	50+
	Type 69	
AIFV	BMP-1/II	40+
	Type 85 (Saladin turrets)	10
APC	BTR 80	426
	Type 63	
	WMZ 551 A/B	
	Type 89 VP	
	Type 86	
	South African Buffel	
Artillery	Type 66 152mm	100
	Type 59 130mm	50
	122mm	74
	120mm	50
	85mm	50
	76mm	50
Mortar	107 mm	
	81mm	
AT Gun	HJ-8	
MRL	122mm	20
AA gun	40mm	60
	90mm	24

New Procurements/Upgrades

No major procurement plans are under consideration at present.

SLNS Samudura. *Image Credit: Sri Lanka Navy*

NAVY

Personnel: 54,000

Equipment

Category	Name	In Service
Patrol vessel	SLNS Samudura	1
	SLNS Sayura	1
	SLNS Sagara	1
	Jayesegra class large patrol Boat	1
	Sa'ar 4 class	2
	Rana class patrol craft	3
	Sooraya (Shanghai II) class FAC	8
	Parakramabahu class gun patrol Boat	1
	South Korean Killer type P430	3
	P463 class Israeli Dvora type	6
	Colombo class FAC	10
	Patrol craft-coastal	25
	Harbour patrol craft	40+
	Speed boat-Arrow type	100+
Amphibious Force	Wuhu class LSM	2
	LCM	2
	Yunnan class LCU	1
	M10 Hovercraft	1

New Procurements/Upgrades

✦ India will supply two Offshore Patrol Vessels to Sri Lanka with delivery scheduled for 2017 to 2018.

Marine Infantry: Personnel 5,304 and seven battalions to protect harbor facilities.

Coast Guard: Personnel 957. The Coast Guard is under the Ministry of Defence and Urban Development.

AIR FORCE

Personnel: 38,000

Equipment

Category	Name	In Service
Fighter/attack / AWS role	MiG-23/ 27	8
	F-7/ FT-7	5
	KFIR C-2	5
	KFIR TC-2	2
Transport/ Surveillance	An-32	4
	C-130B	2
	Y-12	5
	Beech 200	3
Helicopter	Mi-24/Mi-35 Hind	
	Bell 212	8
	Bell 206	7
	Bell 412	6
	Mi-17	11

Air Force continued...

Category	Name	In Service
Trainer	K-8	5+
	PT-6	
	Cessna 150	
UAV	Searcher	

New Procurements/Upgrades

✦ An order has been placed for initial batch of six JF-17 Thunder fighters to replace the KFIRs.

Exercise FLASH STYLE is the Annual Joint Combined Exchange Training (JCET) exercise conducted in collaboration with the US Naval Special Warfare Community and the Sri Lankan Armed Forces. *Image Credit Sri Lanka Air Force*

Defence Expenditure

Total defence spending: $2.4 billion (2016); $1.95 billion (2014)

Defence spending in terms of GDP: 2.21% (2015)

Domestic Suppliers: Colombo Dockyard, SLEME

Foreign Suppliers: China, India, Israel, Pakistan

Sri Lanka has allocated high figures for its defence budget for F.Y. 2017. The allocation for the Defence Ministry for 2017 is USD 1.9 billion which is lower than the allocation in 2016, an amount of USD 2.4 billion. The Appropriation Bill was presented to Sri Lanka parliament by Finance Minister Ravi Karunanayake.

Ministry of Defence	Recurrent Expenditure (Rs.)	Capital Expenditure (Rs.)
Operational Activities	4,681,288,000	8,003,690,000
Development Activities	1,171,074,000	3,213,355,000
Sri Lanka Army	151,893,685,000	5,837,520,000
Sri Lanka Navy	48,655,198,000	12,357,700,000
Sri Lanka Air Force	34,047,252,000	19,237,000,000

Source: Budget 2016 - Presentation of the Appropriation Bill (2016), Sri Lanka Parliament

Defence Production and R&D

Sri Lanka does not undertake any defence production activities.

Defence Procurement

Sri Lankan Army/Navy/Air Force have not made any major defence procurements in recent years.

CONTACT DETAILS

Ministry of Defence
15/5 Baladaksha Mawatha
Colombo 03, Sri Lanka
Tel: (+94 1) 243 08 60
Fax: (+94 1) 254 15 29
E-mail: modsec@sltnet.lk
webinfo@defence.lk
www.defence.lk

SUDAN
(Capital: Khartoum)

INTRODUCTION

Area: 861,484 sq km

Population: 36,729,501 (July 2016 est.)

Coastline: 853 km

Maritime claims: Territorial sea: 12 nm

Contiguous zone: 18 nm

Continental shelf: 200 m depth or to the depth of exploitation

KEY POLITICAL PERSONS

PRESIDENT & SUPREME COMMANDER OF ARMED FORCES: Omar Hassan al-Bashir

FIRST VICE-PRESIDENT: Bakri Hassan Salih

KEY DEFENCE PERSONS

DEFENCE MINISTER: Lt. Gen. Awad Mohammed Ahmed bin Auf

CHIEF OF GENERAL STAFF: Lt Gen. Mustafa Osman Obeid Salim

CHIEF OF STAFF OF THE LAND FORCES: Lt. Gen. Yahia Mohamed Khair Ahmed

CHIEF OF STAFF OF THE NAVY: Gen. Fath al-Rahman Muhi al-Din Salih Mohamed

CHIEF OF STAFF OF THE AIR FORCE: Gen. Isam Mubarak Habib Allah Ibrahim

CHAIRMAN OF THE JOINT OPERATIONS STAFF: General Ibrahim Mohamed Al-Hassan

ECONOMY

Sudan is rich in natural resources such as crude oil and petroleum. Its economy is one of the fastest growing economies in the world. The nation began active crude oil exports towards the end of 1999. Sudan's real GDP expanded by 5.2% during 2010 - an improvement over 2009's 4.2% growth but significantly below the more that 10% per year growth experienced prior to the global financial crisis in 2006 and 2007. While the oil sector continues to drive growth, services and utilities play an increasingly important role in the economy with agriculture production remaining important as it employs 80% of the work force and contributes a third of GDP.

GDP: $167 billion (2015 est.), $161.3 billion (2014 est.)

Real Growth Rate: 3.5% (2015 est.), 3.3% (2014 est.)

Industries: Oil, cotton ginning, textiles, cement, edible oils, sugar, soap distilling, shoes, petroleum refining, pharmaceuticals, armaments, automobile/light truck assembly

Oil - proved reserves: 5 billion bbl (1 January 2016 est.)

Natural gas - proved reserves: 21.24 billion cu m (1 January 2016 est.)

Total Exports: $3.169 billion (2015 est.), $4.454 billion (2014 est.)

Export commodities: Oil and petroleum products; cotton, sesame, livestock, groundnuts, gum arabic, sugar

Major Markets: UAE 32%, China 16.2%, Saudi Arabia 15.5%, Australia 4.7%, India 4.2% (2015)

Total Imports: $8.368 billion (2015 est.), $8.106 billion (2014 est.)

Import commodities: Foodstuff, manufactured goods, refinery and transport equipment, medicines and chemicals, textiles, wheat

Major Suppliers: China 26.4%, UAE 10.1%, India 9.1%, Egypt 5.6%, Turkey 4.7%, Saudi Arabia 4.4% (2015)

INTERNATIONAL DISPUTES

✦ The effects of Sudan's almost constant ethnic and rebel militia fighting since the mid-20th century have penetrated all neighbouring states - Chad, Ethiopia, Kenya, Central African Republic, Democratic Republic of the Congo, and Uganda.

✦ In February 2006, Sudan and DROC signed an agreement to repatriate 13,300 Sudanese and 6,800 Congolese.

✦ Sudan accuses Eritrea of supporting Sudanese rebel groups.

✦ Efforts to demarcate the porous boundary with Ethiopia proceed slowly due to civil and ethnic fighting in eastern Sudan.

✦ The boundary that separates Kenya and Sudan's sovereignty is unclear in the "Ilemi Triangle," which Kenya has administered since colonial times.

✦ Sudan claims but Egypt de facto administers security and economic development of Halaib region north of the 22nd parallel boundary.

✦ Periodic violent skirmishes with Sudanese residents over water and grazing rights persist among related pastoral populations along the border with the Central African Republic.

✦ Bilateral relations between South Sudan and Sudan were officially started on 9 July 2011 following the former's independence from the latter.

DEFENCE

ARMY

Personnel: 60,000

Equipment

Category	Name	In Service
MBT	T-62	24
	Type 88	

Army continued...

Category	Name	In Service
MBT	M-60	50
	T-54/55	54
	T-59	50
Light Tank	Type 62 light	65
Amoured car	SALADIN	14
	V-150 COMMANDOS	30
	FERRET	40
	AML-90	6
	BRDM-2	30
AIFV/APC	BMP-2	6
	BTR-50/-152	40
	OT-64	35
	M113	40
	WALID	90
	BTR-80	30
Artillery	M101	20
	M56 (105mm)	6
	D-74	4
	M-1938	18
	Type 54/D-30 (122mm)	80
	Type 59-1/M-46 (130mm)	27
	D-20 (155mm)	60
	M114A1 (155mm) towed	12
	AMX F3 155mm SP	6
Mortar	81mm	120
	M-43	12
	AM-49 (120mm)	
ATGW	SWINGFIRE	
	SAGGER	
AA Gun	M-3 20mm SPAAG	12
	Bofors 40mm	60
	37mm	80
	VULCAN	24
SAM	SA-2	20
	SA-7	
	REDEYE	

New Procurements/Upgrades

No major procurement plans are under consideration at present.

NAVY

Personnel: 500

Equipment

Category	Name	In Service
Light Force	Swiftship type patrol boat	1
	Ex-Yugoslav patrol boat (GIHAD class)	4
	SEWART type patrol craft	3
	Ex-Iranian coastal paatrol craft (KADIR class)	2

New Procurements/ Upgrades

No major procurement plans are under consideration at present.

AIR FORCE

Major air bases: Atbara, Dongola, El Fashir, El Obeid, Geneima, Juba, Khartoum, Malakal, Merowe, Port Sudan, Wad Medani, Waw.

Personnel: 7,000

Equipment

Category	Name	In Service
Fighter	MiG-29SE	20
	MiG-29UB	4
	MiG-23	2
	MiG-21PF	16
	F-5E/F	8
	F-6	12
	F-7	15
	Su-25	11
	COIN	1
	K-8	12
Maritime reconnaissance	C-212 AVIOCAR	6
Transport	F-27 FRIENDSHIP	1
	C-130H	4
	DHC-5D	3
	An-32	6
	An-24	3
	C-212 AVIOCAR	
	EMB-110P	6
	FALCON 20/50	2
Utility/Liaison	TURBO PORTER	8
	TWIN OTTER	1
	KING AIR 90	1
Helicopter	Mi-4	4
	Mi-8	5
	SA.330 PUMA	12
	BO-105	15
	AB-412	8
	Mi-24	7
	Mi-35P	3
Trainer	F-5F	1
	MiG-29UB	2

New Procurements/ Upgrades

No major procurement plans are under consideration at present.

Defence Expenditure

Total defence spending: $4 billion (est.)

Defence spending in terms of GDP: 4.29% (est.)

Domestic suppliers: Military Industry Corporation

Foreign suppliers: China, Russia, Ukraine, UAE, Jordan Saudi Arabia, Iran

Military spending stood at $3.308 billion in 2010 and increased to $3.923 billion in 2011, amounting to 22.2% of GDP. More than three quarter of South Sudan's 2014/2015 budget was spent on government employees and the military. The minister apportions 28% (i.e.; 4.4 billion SSP) to the national army and approximately 13% (i.e.; 1.573 billion SSP) to police, prison and fire services for 2014-2015 fiscal year.

Defence Production and R&D

The Sudanese military research is believed to be limited to conventional and unsophisticated weapons programmes. It is looking to expand its indigenous defence industry with Russian and Chinese assistance. Sudan has signed an agreement with Pakistan on joint military co-operation at all levels. In January 2002, the International Atomic Energy Agency (IAEA) confirmed that the country was not pursuing nuclear weapons technology.

Defence Procurement

Sudan Army/Navy and Air Force have not made any major defence procurements in recent years.

CONTACT DETAILS

Ministry of Defence
P.O. Box 291, Khartoum, Sudan
Tel: +249-11-74910, -72771

Ministry of Foreign Affairs
Intersection of University Avenue
Street Pain Tiger
PO Box 873
Khartoum, Sudan
Tel: (+249 183) 77 31 01
Fax: (+249 183) 77 29 41
E-mail: ministry@mfa.gov.sd
Website: www.mfa.gov.sd

STRATEGIC INFORMATION

SWEDEN
(Capital: Stockholm)

INTRODUCTION

Area: 450,295 sq km

Population: 9,880,604 (July 2016 est.)

Coastline: 3,218 km

Maritime claims:

Territorial sea: 12 nm (adjustments made to return a portion of straits to high seas)

Exclusive Economic Zone: agreed boundaries or midlines

Continental shelf: 200 m depth or to the depth of exploitation

GEOPOLITICAL IMPORTANCE

Sweden is a Scandinavian country in Northern Europe. At 450,295 square kilometres, it is the third-largest country in the European Union by area. Sweden lies west of the Baltic Sea and Gulf of Bothnia, providing a long coastline, and forms the eastern part of the Scandinavian Peninsula. To the west is the Scandinavian mountain chain (*Skanderna*), that separates Sweden from Norway. Finland is located to its north-east. It has maritime borders with Denmark, Germany, Poland, Russia, Lithuania, Latvia and Estonia, and it is also linked to Denmark (south-west) by the Öresund Bridge. Its border with Norway (1,619 km long) is the longest uninterrupted border within Europe. About 15% of Sweden lies north of the Arctic Circle. Southern Sweden is predominantly agricultural, with increasing forest coverage northward. Around

65% of Sweden's total land area is covered with forests.

POLITICAL OVERVIEW

Politics of Sweden takes place in a framework of a parliamentary representative democratic constitutional monarchy. The Swedish institutions are represented by the King Carl XVI Gustav whose functions stay honorary and led by a government called *Regering* and by a parliament "*Riskdag*". The Swedish government is formed by ministers and a Prime Minister called *Statminister*. The Prime Minister is named by the head of Parliament and chooses the ministers of his government. The Judiciary is independent, appointed by the government and employed until retirement.

KEY POLITICAL PERSONS

HEAD OF STATE: King Carl XVI Gustaf

PRIME MINISTER: Stefan Lofven

ECONOMY

Sweden is the seventh-richest country in the world in terms of GDP per capita and a high standard of living. Sweden remains outside the Eurozone because of concerns over its impact on the country's economy, welfare system, and sovereignty. Sweden is an export-oriented mixed economy. Timber, hydropower, and iron ore constitute the resource base of an economy heavily oriented toward foreign trade. The Swedish economy is a developed diverse economy, aided by timber, hydropower, and iron ore. These constitute the resource base of an economy oriented toward foreign trade. The main industries include motor vehicles, telecommunications, pharmaceuticals, industrial machines, precision equipments, chemical goods, forestry, Iron and steel. A high-tech local economy and a comprehensive system of welfare benefits allow Sweden to enjoy one of the highest standards of living in the world. Sweden has one of the most globalised and competitive economies today. According to a forecast for Sweden's economy, the economic recovery that began internationally in 2013 is expected to continue. Sweden's economy experienced modest growth in 2014-15, with real GDP growth above 2%. Sweden is highly dependent on exports, is strongly pro-free trade, and has one of the most internationally integrated economies in the world.

GDP: $473.4 billion (2015 est.), $454.8 billion (2014 est.)

Real Growth Rate: 4.1% (2015 est.), 2.3% (2014 est.)

Industries: Iron and steel, precision equipment (bearings, radio and telephone parts, armaments), wood pulp and paper products, processed foods, motor vehicles

Total Exports: $151.1 billion (2015 est.), $179.6 billion (2014 est.)

Export Commodities: Machinery 35%, motor vehicles, paper products, pulp and wood, iron and steel products, chemicals

Major Markets: Norway 10.3%, Germany 10.3%, US 7.7%, UK 7.2%, Denmark 6.8%, Finland 6.7%, Netherlands 5.2%, Belgium 4.4%, France 4.2% (2015)

Total Imports: $137.3 billion (2015 est.), $160.6 billion (2014 est.)

Import Commodities: Machinery, petroleum and petroleum products, chemicals, motor vehicles, iron and steel; foodstuffs, clothing

Major Suppliers: Germany 17.9%, Netherlands 8.1%, Norway 7.8%, Denmark 7.7%, China 6%, UK 5.5%, Finland 4.6%, France 4.3%, Belgium 4.3% (2015)

DEFENCE & SECURITY

The primary task of the Swedish Armed Forces is to train, organise and deploy military forces, domestically and abroad, while maintaining the long-term ability to defend the country in the event of war. The Ministry of Defence is responsible for Sweden's military defence and civil emergency preparedness. The head of the armed forces is the Supreme Commander, the most senior officer in the country. Swedish units have taken part in peacekeeping operations in the Democratic Republic of Congo, Cyprus, Bosnia and Herzegovina, Kosovo, Liberia, Lebanon, Afghanistan and Chad.

KEY DEFENCE PERSONS

MINISTER FOR DEFENCE: Peter Hultqvist

SUPREME COMMANDER OF ARMED FORCES: Gen. Micael Byden

CHIEF OF ARMY: Maj. Gen. Karl Engelbrektson

CHIEF OF NAVY: Rear Adm. Jens Nykvist

CHIEF OF AIR FORCE: Maj. Gen. Mats Helgesson

Internal Conflict

Sweden is not engaged in any internal conflict in the region.

External Conflict

The Swedish Armed Forces are increasingly active on the international stage. The Swedish Armed Forces are active in over twenty countries..

Swedish Neutrality: Swedish neutrality refers to it's policy of neutrality in armed conflicts, which has been in effect since the early 19th century. After the end of the Cold War and the fall of the Soviet Union, Sweden has officially dropped the principle of neutrality. The present policy of Swedish neutrality is not laid down in the Constitution or required by any international agreement.

Sweden is engaged in armed conflict in Afghanistan as part of an UN-authorised, NATO-led mission, the International Security Assistance Force – ISAF. Sweden does contribute to various NATO and EU battlegroups and is involved in international organisations. Today, Sweden is one of the Allies' most visible supporters among NATO's many partnering countries.

THREAT PERSPECTIVE

The National Centre for Terrorist Threat Assessment (NCT) is a permanent working group within the Swedish Counter-Terrorism Cooperation Council. The NCT is staffed by personnel from the National Defence Radio Establishment (FRA), the Military Intelligence and Security Directorate (MUST) and the Swedish Security Service.

The NCT produces long and short-term strategic assessments of the terrorist threat against Sweden and Swedish interests. The NCT is also tasked with producing strategic analyses of incidents, trends and international developments with a bearing on terrorism that may affect Sweden and Swedish interests, both today and in the future.

STRATEGIC RELATIONS

United States: Sweden was one of the first countries to recognise US independence in 1783 and the two countries have maintained a strong bilateral friendship since then, based on shared values and mutual interests. Sweden is a core member of the North Atlantic Treaty Organization's (NATO) Partnership for Peace programme, and plays an active leadership role on the international stage, from its long-term investment in Afghanistan to its role as a global peacemaker. The United States provides no development assistance to Sweden. Although Sweden has a longstanding policy of political neutrality in international affairs, it is a participant in the Euro-Atlantic Partnership Council. Sweden currently participates with around 500 troops in the International Security Assistance Force (ISAF), under the command of NATO, in Afghanistan. During the Libyan Civil War of 2011, the Swedish Air Force worked close with NATO and USA. Sweden is also a member of the Counter-ISIL Coalition and participates in the United Nations Multidimensional Integrated Stabilisation Mission in Mali (MINUSMA).

Nordic defence cooperation: NORDEFCO (Nordic Defence Cooperation) is a collaboration amongst the Nordic countries in the area of defence. Its five members are Denmark, Finland, Iceland, Norway, and Sweden. The purpose of NORDEFCO is to strengthen the participants' national defence, explore common synergies and facilitate efficient common solutions. NORDEFCO is the organisation responsible for implementing this in areas such as logistics and maintenance, training and exercises, operational activities, such as air and maritime surveillance and international operations. Defence partnerships with other countries are becoming an increasingly important way of enabling Sweden to develop and maintain its military capabilities in Sweden, and on international missions. The cooperation includes troop contributions to international peace missions, development, procurement, maintenance and further development of materiel, officer training and exercises. On 9 November 2016, the Nordic Ministers of Defence made a Joint Statement, after having signed a Memorandum of Understanding on enhanced and easier access to each other's territories in peacetime. The agreement will improve the operational effect and quality of air, land and maritime operations. Also the Northern Group consisting of Denmark, Estonia, Finland, Germany, Iceland, Latvia, Lithuania, the Netherlands, Norway, Poland, the United Kingdom and Sweden, came together informally for discussions on defence and security issues.

India: The ties between India and Sweden were established in 1949 and are founded on shared democratic values. India and Sweden in 2015 signed several agreements to boost ties and decided to restart the bilateral strategic dialogue besides finding ways for investment by the Nordic country in India's defence sector under the ambitious 'Make in India' initiative. A MoU on defence between Sweden and India was signed in November, 2009. The MoU set out several areas for cooperation between the defence authorities of the two countries for the purpose of mutual benefit. Sweden is the third-largest contributor of Foreign Direct Investment (FDI) to India besides being the country's third-largest trade partner after China and Japan.

Multilateral Alliances

Sweden belongs to a number of international organisations, including the United Nations, European Union (EU), Euro Atlantic Partnership Council, the Arctic Council, Organisation for Security and Cooperation in Europe, Organisation for Economic Cooperation and Development, International Monetary Fund, World Bank, and World Trade Organisation. Sweden also is an observer to the Organisation of American States and a participant in the North Atlantic Treaty Organisation's (NATO) Partnership for Peace programme. Sweden is also a member of Partnership for Peace (PfP) a practical partnership programme between NATO and various OSCE states (Organisation for Security and Cooperation in Europe).

DEFENCE CAPABILITIES

ARMY: The Swedish Army has the task of organising and training ground forces and air defence units. Army units operate both on the ground and in the air. They are trained and equipped for combat with all types of enemy in different types of terrain and urban environments. The Army units are categorised into Mechanised Units, Cavalry Units, Artillery Units, Air Defence Units, Command and Control Units, Engineer Units, Support Units and CBRN Units.

Personnel: 19,200

Equipment

Category	Name	In Service
MBT	Stridsvagn 122	100
APC/MICV/IFV	CV-90	500
	Pbv-302	50
	Patria XA-200 6x6	170
	Patria AMV	113
	Bv-206S	93
	RG-32M SCOUT	200+
Artillery	155 mm. towed FH-77B	48
Mortar	120 mm.	50

Army continued...

Category	Name	In Service
Anti-Tank GW	TOW	
	BILL	
	Robot-57	
Recoilless rifle	Carl Gustav	
	AT-4	
AA Gun	Cv90 AA	25
SAM	HAWK	
	I-HAWK	
	RBS-70	
	RBS-23 BAMSE	

New Procurements / Upgrades

- Bohemia Interactive Simulations (BISim) recently was awarded a multi-year, multi-million dollar contract to create new 3D model content and simulation training capabilities in VBS3 for the Swedish Armed Forces. As part of the contract, BISim will create a training application for Sweden's RBS17 Hellfire Shore Defence System for usage with VBS3.
- Sweden intend to buy military logistics vehicles with a total value of over €1 billion along with Norway from Rheinmetall Defence.
- Diehl Defence will deliver IRIS-T SLS surface-to-air missile (SAM) systems to the Swedish Defence Material Administration under a new contract announced on 11 March.
- The Swedish Defence Materiel Administration, FMV, has awarded a USD 100 million contract to BAE Systems for the procurement of 48 of its go-anywhere BvS10 armoured all-terrain vehicles and an extensive initial support package.
- Swedish Defence Procurement Agency (FMV) ordered for six Kodiak armoured engineer vehicles from Rheinmetall Defence.
- The Swedish Defence Materiel Administration, FMV, has awarded a USD 100 million contract to BAE Systems for the procurement of 48 of its go-anywhere BvS10 armoured all-terrain vehicles and an extensive initial support package.
- The Swedish Defence Material Administration signed a contract with Diehl Defence to deliver IRIS-T SLS surface-to-air missile (SAM) systems for the country's armed forces. The new units comprising the IRIS-T missile, missile launching station and fire control system are to improve the national air defence by protection against air attacks from a large variety of threats including missiles, helicopters and aircraft.
- Saab will upgrade Sweden's ground based air defence capabilities following the award of two new orders by the Swedish Defence Materiel Administration (FMV), with a total value of USD 94.7 million. Saab will deliver the system integration and the integration of the command and control systems into the ground based anti-aircraft defence systems, based on Giraffe radar systems.
- The Swedish government has awarded BAE Systems a contract to refurbish 262

Combat Vehicles 90 (CV90) for the Swedish Army. The company's work will include refurbishing the chassis and upgrading the vehicle's survivability and turrets, as well as enhancing combat system performance.

Swedish Pbv 302 (APC).

NAVY: The Swedish Navy has the task of organising and training naval and amphibious units and operates in navigable coastal waters, in the archipelago and on the open sea. Naval units are categorised into Naval flotillas, Submarine flotilla, Amphibious battalion and Naval base. Karlskrona naval base (MarinB) is located at Karlskrona with detachments at Berga, Goteborg and Skredsvik.

Personnel: 9,500

Equipment

Category	Name	In Service
Submarine	Gotland Class	3
	Sodermanland Class	2
Corvette	Visby Class	3
	Goteborg Class	2
	Stockholm Class	2
Mine Warfare Force	Landsort Class	2
	Styrso Class	4
	Koster class	5
Patrol Vessel	Tapper class	10
Auxiliary Force	Submarine rescue ship	1
	Supply ship	1
	Survey Vessel	1
Amphibious Force	Landing Craft Mechanised (LCM)	17
	Landing Craft Utility (LCU)	23
	LCPFM	145
	Light Combat Aircraft (LCA)	80
	Griffon 8100TD Hovercraft	3

New Procurements / Upgrades

✦ Swedish Defence Material Administration (FMV) has awarded defence contract worth SEK147 million to Saab to modify and upgrade two Swedish Navy Koster-class mine countermeasures vessels during 2016-2017 period.

✦ Saab has received an order from the Swedish Defence Material Administration (FMV) for delivery of advanced anti-submarine warfare training, including the autonomous underwater vehicle AUV62 in training configuration. Deliveries will take place during the period 2016-2019.

✦ CAE has been awarded a contract from *Forsvarets Materielverk* (FMV), the Swedish Defence Materiel Administration, to provide a comprehensive Naval Warfare Training System for the Swedish Navy.

✦ The Swedish government has awarded approx. USD 42 million contract to Computing Devices Canada, a General Dynamics company, for three undersea warfare systems to be installed aboard the Swedish Navy's Visby Class corvettes, and land-based training to support the systems.

✦ OSI Maritime Systems (OSI) is awarded a contract for the Company's Tactical Dived Navigation System (TDNS). The Systems will be delivered to ThyssenKrupp Marine Systems AB as part of a Royal Swedish Navy submarine upgrade programme, and includes OSI's subsurface WECDIS software.

✦ Northrop Grumman Corporation's Sperry Marine business unit has won a contract to upgrade the navigation radars on five Swedish navy patrol boats.

✦ Kockums AB signed a contract with the Swedish Defence Materiel Administration (FMV) concerning overall of the design phase of the next-generation submarine. Sweden is planning to purchase two new submarines that will be commissioned by 2018-19 replacing the two submarines of the Södermanland class. The plans also included upgrades to two submarines of the Gotland class. Additional submarines could later be orderd to replace the Gotland class, however this will not be decided before 2020.

✦ The Damen Shipyards Group has signed a contract with the Dutch and Swedish Ministries of Defence for a seven-year, in-service maintenance contract covering a total of five vessels. All of these are tugs purchased from Damen.

AIR FORCE: The Swedish Air Force is tasked with organising and training aircraft units, base and command units and completes its tasks by means of fixed-wing aircraft and helicopters. These tasks include protecting Swedish airspace, conducting rescue operations, performing air transport duties and gathering intelligence. Air Force units are categorised into Fighter Aircraft Units, Transport Aircraft Units, Signal Reconnaissance Units, Radar Surveillance Units, Helicopter Units, Base and Command Units.

Personnel: 6,600

Equipment

Category	Name	In Service
Fighter	JAS 39C/D	100
Transport/ Tanker	C130H	6
	C-17 Globemaster III	(shared with NATO)
	Saab 340	6
Trainer	Saab-105	30
SIGINT	Gulfstream IV	2
UAV	Shadow 200, Elbit Skylark, Skeldar	3+

Swedish Helicopter Flotila

Personnel: 700

Equipment

Category	Name	In Service
Helicopter	NH90	18
	A109	15
	UH-60M Black Hawk	12+
Missile	Sidewinder, Maverick, Hawk, Robot 05, RB 04	

New Procurements / Upgrades

✦ All Swedish Gripen C/D aircrafts will be upgraded to Gripen version 20. The new version brings major changes, both in hardware and software. The biggest change is the Meteor missile. It will give Gripen significantly increased long-range air-to-air capability and give the Gripen fighter conditions to change the balance of air superiority.

✦ *Turbomeca* (Safran group) has signed a contract with defence and security company Saab AB for the repair and overhaul of Arrius 2K2 modules and engines installed on A109 LUH helicopters of the Swedish Air Force.

✦ Saab signed an agreement with the Swedish Defence Materiel Administration (FMV) regarding development and modification of 60 Gripen E for Sweden during the period 2013-2026 as well as a possible order for new production of Gripen E for Switzerland. The Swedish Air Force wants to upgrade 60–80 Gripens to this standard by 2020. The Swedish Riksdag has also approved the purchase of 40 to 60 E/F aircraft, but with an option to cancel if at least 20 aircraft are not ordered by other customers. The Government of Sweden approved the deal for 60 Gripen Es on 17 January 2013. Deliveries will begin in 2018 and be completed in 2027.

✦ Saab has received a serial production order from the Swedish Defence Materiel Administration (FMV) amounting to SEK16, 4 billion for operations during 2013-2026. The order includes modification of 60 Gripen C to Gripen E for Sweden with initial deliveries in 2018.

✦ FMV, the Swedish Defence Materiel Administration, signed a production order contract with United Kingdom MOD regarding the Meteor missile. With the Meteor missile the JAS39 Gripen system will get a significantly increased capability to operate against air targets at long distances with very high performance.

✦ Sweden to receive 15 Sikorsky UH-60M Black Hawk utility helicopters from the US Government. On 18 May 2011, Sikorsky announced the sale of 15 UH-60Ms to Sweden. The UH-60Ms are to be fully operational by 2017.

✦ Sweden and Switzerland have signed a framework agreement for procurement of the next generation Gripen. The Swedish government plans to buy 40-60 new Gripens over the next decade at a cost of 242 million euros.

Major Armaments Producers

Sweden has a well-developed defence industry. The Swedish Security and Defence Association is open to companies whose production of security and defence equipment and services constitutes an essential part of their output, i.e. companies engaged in research, development, production, trade and marketing of military and security equipment, material and services. The Association currently has 65 member companies including 50 SMEs.

3M Svenska AB, 4C Group AB (4C Strategies), Actea Consulting AB, Aimpoint AB, Aqeri AB, BAE Systems Bofors AB, BAE Systems Hägglunds AB, Carmenta AB, CGI Sverige AB, Cobham, Combitech AB, Comex Electronics AB, Condesign AB, CRD Protection AB, CybAero AB, Datapath (f.d Rockwell Collins Sweden AB), Dockstavarvet AB, Eltel Networks Infranet AB, Eurenco Bofors AB, FLIR Systems AB, GSS, GKN Aerospace AB, Habia Cable CS Technology, Hewlett-Packard Sweden AB, IBM Svenska AB, Imdar Systems AB, Kaller, Kitron AB, Life Time Engineering AB, Marinediesel Sweden AB, Micro Systemation AB, Mildef AB, MSE Engineering AB, Nammo Sweden AB, PartnerTech Karlskoga AB, Patria Helicopters AB, Polyamp AB, PrimeKey Solutions AB, Qinetiq Sweden AB, Rolls-Royce AB, Rote Consulting AB, RSG connexion AB, Saab AB, Saab Bofors Test Center AB, SCAMA AB, Scania CV AB, Secana AB, Sepson AB, Sjöland & Thyselius AB, Skyddsprodukter i Sverige AB, SRS Group AB, Swede Ship Marine AB, Svensk Konstruktionstjänst AB, Syntell AB, Systecon AB, Teleanalys AB, T-kartor, Sweden AB, Venatio AB, Volvo Defense AB, Woolpower AB, ÅAC Microtec AB, ÅF Technology AB, Åkers Krutbruk Protection AB, etc. are the member companies.

DEFENCE DEALS

Arms Export

Swedish defence major Saab has unveiled its next generation fighter aircraft, Gripen E, which the company in 2016 said is being offered to India under the 'Make In India' initiative with transfer of technology. Gripen E has significantly improved avionics system when compared to previous versions of the Gripen. The Gripen E is a specific configuration of Gripen NG that has been chosen by the Swedish customer. The exact configuration for another customer such as India will depend on discussions with that customer. Saab is offering the next generation Gripen to India, under 'Make In India' with transfer of technology. The company has not only offered to set up a base here but also help in the development of aerospace capability for the next 100 years. It has also offered to partner in developing the next version of indigenous Light Combat Aircraft Tejas and the Advanced Medium Combat Aircraft (AMCA), being developed and designed by Aeronautical Development Agency. Gripen E is equipped with a highly integrated and sophisticated sensor suite including an Active Electronically Scanned Array (AESA) radar, Infra-Red Search and Track (IRST), Electronic Warfare (EW) suite and datalink technology, which, when combined gives the pilot, and co-operating forces exactly the information needed at all times.

The Swedish Defence and Security Export Agency (FXM) promotes defence export which benefits Swedish defence and security policies. The agency represents the Swedish Government in contracts between governments (G2G contracts). The Government intends to close FXM during 2015. The planned closure of the FXM is expected to trigger a major redevelopment of the country's arms export rules and infrastructure.

DEFENCE BUDGET

Total defence spending: $5.7 billion (2015)

Estimated defence spending in terms of GDP: 1.24% (2015)

Sweden is planning to increase defence spending by 10.2 billion kronor (USD1.18 billion) for the period 2016 to 2020. Large scale of money will be spent by the military on modernising ships that can detect and intercept submarines and also to bring troops to the Baltic Island of Gotland. The Swedish government in 2014 announced that an additional 1.3 billion kronor will be invested in the military in response to increased Russian military activity in the Baltic Sea. The promise comes amid intense infighting in Sweden over military funding, in which all parties, both ruling and opposition, have pledged to increase the military budget in light of Russian military activity in Ukraine. So far, there is no specific threat to Sweden from Russia.

The defence budget for the years 2016 to 2020, will be based on the June 2014 announcement of budgetary increases, amounting to approximately 7 billion SEK (Government Resolution 12 and 15). The announced increase was based on the Defence Commissions analysis. The Swedish Armed Forces have suggested some reallocations of the announced funding for the period of 2016-2020. These suggestions are included in the funding for the period. This includes a reallocation of 1.3 billion SEK from the funding of international missions to the funding of training and readiness.

Sweden's own R & D and defence industry is very much successful. The country's arms manufacturers are among the world's best. The defence industry in Sweden has a long history, dating back to the time when Sweden was a more active military power. Karlskronavarvet, specialising in naval surface vessels and submarines, and Bofors, a producer of artillery systems once owned by Alfred Nobel, have been in business for more than 300 years. Kockums AB is the result of a merger between Kockums in Malmö, Karlskronavarvet in Karlskrona and Marinverkstäderna at Muskö. Bofors today is split in two parts: BAE Systems Bofors (BAE Systems acquired United Defence and its Bofors subsidiary in 2005) and, Saab Bofors Dynamics. Saab AB is an aerospace and defence company based in Sweden. BAE Systems AB is a Swedish defence company and a subsidiary of BAE Systems Land and Armaments, whose ultimate parent is the British defence contractor BAE Systems.

Defence funding 2017-2020 (million SEK)

Armed Forces including increases for Defence Intelligence

	2017	2018	2019	2020
Basic Planning	43087	43881	46026	47576
Additional Funding	1900	2200	2320	2500
Total Sum	44987	46081	48346	50076

The additional funding will mitigate the financial costs of tax changes regarding the employment of young persons as well as increased costs for infrastructure. These costs are expected to be approximately 1.7 billion SEK for the period.

Increase in defence budget will contribute towards:

- Investment in soldier's personal equipment, communication/CIS equipment, radars, logistics, etc
- Upgradation of Main Battle Tanks, Infantry Combat Vehicles and Self-propelled motors
- New anti-tank weapons
- Improving air defence capabilities
- Additional reinforcements for anti-submarine warfare capability
- A renewed civil defence
- Active cyber capabilities. Cyber defence capabilities are an important part of the Swedish Defence.

Material Planning

The task of material planning is carried out jointly by the Armed Forces, the Swedish Defence Materiel Administration (FMV) and the Swedish Defence Research Agency (*Totalförsvarets forskningsinstitut).* These agencies decide how to direct resources, guided by the guidelines set by parliament and the government.

The Armed Forces are engaged in both developing new material and decommissioning that which is outdated. New equipment is often developed in cooperation with other countries in order to spread the costs. The country also decommissions old material, which is often too costly to store when it is no longer in use. Also, materials are donated to certain kinds of organisations that might need it. Otherwise it is sold or destroyed.

Swedish defence and aerospace company Saab has started the assembly of the next generation Gripen, the Gripen E. Following a short and intense period of design using the latest tools and methods, through so called 'Model Based Design', the construction of the Gripen E begins with the manufacturing and assembly of all parts of the fuselage; the largest and most time consuming part of the airframe. These parts will then be joined together and assembled into a complete airframe. This is followed by an intensive construction period to install cables, mount systems, the outer shell and other equipment. Based on the design of previous versions of the Gripen system, the Gripen E offers a next generation sensor suite, new communication links, revolutionary avionics architecture, more thrust, increased flight time, more weapon stations and load capability, a fully digital cockpit including HUD (Head-Up Display) and a brand new electronic warfare system. The test aircraft 39-8 will be the first complete pre-production version of the Gripen E and will be used to demonstrate new features and capabilities. The technological leaps in the Gripen E have been proven in the Gripen demonstrator programme with the Gripen E/F demo aircraft that has flown over 250 hours in countries such as Sweden, the UK, India and Switzerland (in Emmen in October 2012, January and April 2013) since 2008.

Defence Procurement

The Swedish Defence Materiel Administration (*Försvarets materielverk*, FMV) is a Swedish government agency that reports to the Ministry of Defence. The agency is responsible for the supply of materiel to the Swedish defence organisation. Requests for new materials are drafted within the Swedish Armed Forces and then submitted to the FMV for action. FMV is charged with technical preparation work, inspection of materials and equipment, and modifications or upgrades deemed necessary for a given system or piece of equipment. Most military procurement contracts are awarded to Swedish defence companies. However, companies often partner with foreign suppliers. FMV is located in Stockholm.

CONCLUSION

The task of the Swedish Armed Forces is to defend Sweden and safeguard its sovereignty and securing its borders. Sweden gives support to the United Nations (UN) in order to contribute to global peace, security and development. Through its membership in the European Union (EU) Sweden assumes joint responsibility for the security of Europe and the ability of the EU to promote peaceful and democratic development. Sweden's cooperation with NATO allows the Swedish Armed Forces to develop military capabilities and, by contributing to qualified international crisis management operations, Sweden contributes to building security together with partners. The Swedish Armed Forces receive approximately SEK 40 billion to perform important military tasks. Around half of the money goes to the activities conducted by the units. The other half goes to research, development and materiel systems. Sweden takes part in EU rationalisation and cooperation on equipment, and has contributed to every EU crisis management operation. Over a twenty-year period, Sweden has developed close cooperation with the NATO defence alliance within the Partnership for Peace. By taking part in all major NATO operations under a UN mandate and in major exercises with NATO member countries in Europe, Sweden has developed cooperation capabilities with the Alliance and its member countries on a par with that of many NATO members. Sweden has long enjoyed well-developed defence cooperation, including cooperation on equipment, with the Nordic countries, certain large European countries and the US.

CONTACT DETAILS

Ministry of Defence

Jakobsgatan 9, SE-103 33 Stockholm
SwedenTel: (+46 8) 405 10 00
E-mail: registrator@defence.ministry.se
Website: www.sweden.gov.se

Swedish Defence Materiel Administration

Försvarets materielverk (FMV)
Banérgatan 62
S-115 88 STOCKHOLM
Banérgatan 62, Sweden
Telephone: +46 (0)8 - 782 4000
Fax: +46 (0)8 - 661 8220
E-mail: annons@fmv.se
registrator@fmv.se
Website: www.fmv.se

SWITZERLAND
(Capital: Bern)

INTRODUCTION

Area: 41,277 sq km
Population: 8,179,294 (July 2016 est.)
Coastline: 0 km (Landlocked)
Maritime claims: None

KEY POLITICAL PERSONS

PRESIDENT: Doris Leuthard

KEY DEFENCE PERSONS

HEAD OF THE FEDERAL DEPARTMENT OF DEFENCE: Guy Parmelin

CHIEF OF THE ARMED FORCES: Lt. Gen. Andre Blattmann

COMMANDER OF LAND FORCES: Maj. Gen. Daniel Baumgartner

AIR FORCE COMMANDER: Lt. Gen. Aldo C. Schellenberg

ECONOMY

Switzerland is a peaceful, prosperous, and modern market economy with low unemployment, a highly skilled labour force, and a per capita GDP among the highest in the world. Switzerland's economy benefits from a highly-developed service sector, led by financial services, and a manufacturing industry that specializes in high-technology, knowledge-based production. Its economic and political stability, transparent legal system, exceptional infrastructure, efficient capital markets, and low corporate tax rates also make Switzerland one of the world's most competitive economies. The Swiss have brought their economic practices largely into conformity with the EU's to enhance their international competitiveness, but some trade protectionism remains, particularly for its small agricultural sector. The fate of the Swiss economy is tightly linked to that of its neighbours in the euro zone, which purchases half of all Swiss exports. The global financial crisis of 2008 and resulting economic downturn in 2009 stalled export demand and put Switzerland in a recession. The Swiss National Bank (SNB) during this period effectively implemented a zero-interest rate policy to boost the economy as well as prevent appreciation of the franc, and Switzerland's economy recovered in 2010 with 3.0% growth. The sovereign debt crises unfolding in neighboring euro-zone countries, however, coupled with ongoing economic instability in Russia and other eastern European economies continue to pose a significant risk to the Swiss economy, driving up demand for the Swiss franc by investors seeking a safe-haven currency. In January 2015, the SNB abandoned the Swiss franc's peg to the euro, roiling global currency markets and making active SNB intervention a necessary hallmark of present-day Swiss monetary policy. The independent SNB has upheld its zero interest rate policy and conducted major market interventions to prevent further appreciation of the Swiss franc, but parliamentarians have urged it to do more to weaken the currency. The franc's strength has made Swiss exports less competitive and weakened the country's growth outlook; GDP growth fell below 2% per year from 2011-15.

GDP: $482.3 billion (2015 est.); $478.3 billion (2014 est.)

Real Growth Rate: 0.9% (2015 est.); 1.9% (2014 est.)

Industries: Machinery, chemicals, watches, textiles, precision instruments, tourism, banking, and insurance

Total Exports: $303.5 billion (2015 est.); $327.7 billion (2014 est.)

Exports Commodities: Machinery, chemicals, metals, watches, agricultural products

Major Markets: Germany 14.2%, US 10.6%, Hong Kong 8.7%, India 7.3%, China 6.9%, France 6.1%, Italy 5.4%, UK 4.8% (2015)

Total Imports: $247.7 billion (2015 est.); $273.8 billion (2014 est.)

Import Commodities: Machinery, chemicals, vehicles, metals; agricultural products, textiles

Major Suppliers: Germany 20.7%, UK 12.8%, US 8.1%, Italy 7.8%, France 6.7%, China 5.1% (2015)

INTERNATIONAL DISPUTES

Switzerland has no international dispute.

DEFENCE

ARMY

Personnel: 147,075 (including Air Force as of 2013). The total force strength will be reduced to 100,000 under reforms plan which will start from 2018 onwards

Equipment

Category	Name	In Service
MBT	Leopard 2 (Pz 87 Leo)	134
APC, IFV	Duro IIIP	290
	M113	371
	CV90 (Spz 2000)	186
	APC 93 (Piranha)	521
	Duro	290+130 on order
Reconnaissance vehicle	93, 93/97, 97/06	326
Howitzer	M109 calibre	150+
Anti-tank vehicle	Piranha TOW	110
SAM	Stinger	96

New Procurements / Upgrades

✦ 130 Duro IIIP APCs are on order from General Dynamics Land Systems with deliveries going on.

✦ Norway's Kongsberg has received an order to supply the Protector Weapon Control System to the Swiss Army.

AIR FORCE

Equipment

Category	Name	In Service
Fighter	F/A-18 Hornet C/D	31
	F-5 E/F Tiger II	36 (26 operational)
Transport	Beechcraft 1900	1
	Pilatus PC-6 Turbo-Porter	15
	Dassault Falcon 50	1
	Pilatus PC-12	1
	Cessna Citation Excel	1
Trainer	Pilatus PC-7	28
	Pilatus PC-9	8
	Pilatus PC-21	8
UAV	ADS-95 Ranger	15
Helicopter	Super Puma	15
	Cougar	11
	EC 635	20
Air Defence System	Stinger	96
	Rapier	40
	Skyguard	

Cougar helicopter

New Procurements / Upgrades

✦ The search continues for a new fighter aircraft type with no final decision taken at the highest level yet. The new, yet to be procured fighters, will replace the aging fleet of F-5 planes and subsequently the F/A-18 Hornets.

✦ The entire fleet of Boeing F/A-18 Hornet fighters is being modernised under a $535 million programme to keep them operational beyond 2025.

✦ Six Hermes 900 unmanned aerial vehicles (UAVs) built by Israel's Elbit Systems will be procured for the Air Force's ADS-15 project.

✦ Rheinmetall has been awarded a contract in 2016 to upgrade the fire control units and guns of the 35mm medium-calibre Oerlikon Skyguard air defence guns.

Defence Expenditure

Total defence spending: CHF20 billion (2016); CHF17.6 billion (2015)

Estimated defence spending in terms of GDP: 0.6%

Defence Production and R&D

The well-developed and self-sufficient defence and aerospace industry of Switzerland has been catering to domestic needs and also exports military products to international customers. Some of the leading defence and aerospace companies include RUAG Holding, Pilatus Flugzeugwerke AG and Oerlikon-Contraves AG which has been renamed as Rheinmetall Air Defence AG.

A decline in the annual defence budget over the years has adversely affected the domestic defence industry. Reduced military spending has resulted in cutback in armaments production. Besides, a strict export regime has led to the Swiss defence industry facing strong competition from international arms majors.

Defence Procurement

With limited annual defence budget, Switzerland has not yet finalised on the acquisition of a new fighter jet type as a replacement for its aging fleet of frontline fighters even as it has decided to retain and refurbish the existing planes to keep them operational for few more years. The Government has also announced plans to reduce the size of its armed forces from the present strength of over 140,000 to 100,000 starting from 2018 onwards. The major acquisition programmes announced in 2015 and 2016 include procurement of reconnaissance drones and upgrade of medium-calibre Oerlikon Skyguard air defence guns.

CONTACT DETAILS

Federal Department of Defence

Bundeshaus Ost
CH3003 Bern,
Switzerland

Armasuisse (National procurement agency)
Kasernenstrasse 19
CH3003 Bern
Switzerland

SYRIA
(Capital: Damascus)

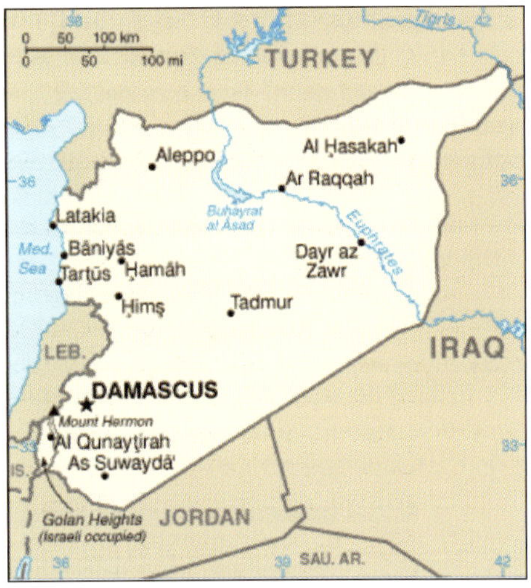

INTRODUCTION

Area: 185,180 sq km

Population: 17,185,170 (July 2016 est.)

Coastline: 193 km

Maritime claims: 12 nm

KEY POLITICAL PERSONS

PRESIDENT& COMMANDER-IN-CHIEF OF THE ARMED FORCES: Bashar al-Asad

PRIME MINISTER: Imad Mohammad Deeb Khamis

KEY DEFENCE PERSONS

DEFENCE MINISTER: Lt. Gen. Fahd Jasem al-Freij

CHIEF OF GENERAL STAFF & ARMY: Gen. Ali Abdullah Ayyoub

CHIEF OF NAVY: Vice Adm. Muhamad Al-Ahmad

CHIEF OF AIR STAFF: Gen. Issam Hallaq

ECONOMY

Since the outbreak of the Syrian civil war in 2011, the Syrian economy has been hit by massive economic sanctions restricting external trade. The GDP is estimated to have contracted by an average of 15.4 percent for the period 2011-14. The sanctions have reversed the previous growth in the economy to a state of decline for the years 2011 and 2012. According to the UN, the total economic damages of the Syrian civil war are estimated at $143 billion as of late 2013. Even if the conflict ceased now and GDP grew at an average rate of five per cent each year, it is estimated that it would take the Syrian economy 30 years to return to the economic level of 2010. Assets and infrastructure have been destroyed, half of the population lives below the poverty line. Before the bloodshed of the civil war in Syria, the country's economy was diverse. The most important sectors included agriculture (22 percent of the economy), industry and excavation (25 percent), retail (23 percent), and tourism (12 percent), according to 2009 figures quoted by the Syrian Central Bank.

GDP: $55.8 billion (2014 est.), $61.9 billion (2013 est.)

Real Growth Rate: -9.9% (2015 est.), -36.5% (2014 est.)

Industries: petroleum, textiles, food processing, beverages, tobacco, phosphate rock mining, cement, oil seeds crushing, automobile assembly

Oil - proved reserves: 2.5 billion bbl (1 January 2016 est.)

Natural gas - proved reserves: 240.7 billion cu m (1 January 2016 est.)

Total Exports: $2.14 billion (2015 est.), $3.015 billion (2014 est.)

Export Commodities: crude oil, minerals, petroleum products, fruits and vegetables, cotton fiber, textiles, clothing, meat and live animals, wheat

Major Markets: Iraq 64.7%, Saudi Arabia 11.2%, Kuwait 7.1%, UAE 6.1%, Libya 4.6% (2015)

Total Imports: $6.663 billion (2015 est.), $8.028 billion (2014 est.)

Import Commodities: machinery and transport equipment, electric power machinery, food and livestock, metal and metal products, chemicals and chemical products, plastics, yarn, paper

Major Suppliers: Saudi Arabia 28%, UAE 13.7%, Iran 10.1%, Turkey 9%, Iraq 8.3%, China 6.1% (2015)

INTERNATIONAL DISPUTES

The province of Hatay in Turkey (containing Antakya) and the Golan Heights were all originally part of French Mandate Syria, but are now outside the boundaries of the modern Syrian state. Syria has never formally recognized Turkish annexation of Hatay Province. Golan Heights is Israeli-occupied with the almost 1,000-strong UN Disengagement Observer Force patrolling a buffer zone since 1964. Since 2000, Lebanon has claimed Shab'a Farms in the Golan Heights. Lacking a treaty or other documentation describing the boundary, portions of the Lebanon-Syria boundary are unclear with several sections in dispute.

DEFENCE

The Syrian military consists of the Air, Ground, Navy and the Air Defence Forces. The armed forces have been fighting insurgents since the beginning of the Syrian civil war in 2011. As the uprising progressed many soldiers have defected from the Syrian Armed Forces and came together under the banner of the Free Syrian Army (FSA). The enlisted members of the Syrian military have dropped by over half from a pre-civil war figure.

The Syrian government has received support from Russia, Iran and Iraq in the ongoing armed conflict. In 2013, Iran-backed Hezbollah entered the war in support of the Syrian Army. On 30 September 2015, Russia started its own air campaign at the request of the government of Syria. According to the United Nations, by June 2013, 90,000 people had been killed in the conflict and by August 2014 that figure had more than doubled to 191,000 - and continued to climb to 250,000 by August 2015.

ARMY

Personnel: 110,000

Equipment

Category	Name	In Service
MBT	T-72	1,400
	T-62	900
	T-55MV	200
	T54/55 (mostly stored)	2,000

Army continued...

Category	Name	In Service
Armoured Car	BRDM-1	400
	BRDM-2	500
AIFV & APC	BMP-1	2,300
	BTR 50/60/152	1,500
	OT-64	
Artillery	122mm, 130mm, 155mm, 180 mm towed	2,600
	2-S1 122mm SP	400
	2-S3 152mm SP	50
Anti-tank gun	100mm	
Mortar	81 mm	
	120 mm	
	160 mm	
MRL	122mm, 140mm, 240mm	
ATGW	Snapper	
	Sagger	
	Swatter	
	Spigot	
	Spandrel	
	Hot	
	Milan	
	Kornet	
AA gun	23mm, 37mm	
	57mm, 85mm	400
	100mm towed	
	ZSU-23-4	300
	ZSU-57-2 SP	10
SAM	SA6/-7/-8/-9/-11/-13/-18	
SSM	Frog-7	24
	Scud B launchers	18
	SS-12	
	SS-21	18

New Procurements/Upgrades

No major procurement plans are under consideration at present.

NAVY

Personnel: 4,000 plus 2,500 reserves

Equipment

Category	Name	In Service
Corvette	ex-USSR Petya II class	2
Patrol craft	ex-USSR Osal/II class missile FAC	8
	ex-USSR Zhuk class	8
Amphibious force	ex-USSR Polnochny class	3
Mine warfare force	ex-USSR Ocean minesweeper (Natya class)	1
	ex-USSR coastal minesweeper (I Sonya class, 2 Vanya class)	3

Naval Aviation: 12 Mi-4, 4 Ka-25, 20 Mi-14

New Procurements/Upgrades

No major procurement plans are under consideration at present.

AIR FORCE

Personnel: 60,000

Equipment

Category	Name	In Service
Fighter	Su-24	20
	MiG-21F	150
	MiG-23	70
	MiG-29	40
	MiG-25	30
	MiG-23BN	40
	Su-20/22	40
Transport	Il-76	4
	An-26	6
	An-12	6
	An-24	2
	Il-14	8
	TU-134	4
Trainer	MiG-21U	20
	L-39 Albatross	83
	L-29 Delphin	60
	MBB-223 Flamingo	30
	MiG-17	30
	PAC Shahbaz	6
Helicopter	Mi-24	36
	Gazelle/Hot	55
	Mi-8	100
	Mi17	100
	Mi-2	10
	Mi-25	50
	Mi-14 Haze	16
	Ka-25 Hormone	5

New Procurements/Upgrades

✦ Syria and Russia has signed a contract for the delivery of Yak-130 (Mitten) jet trainer and light attack aircraft and S-300 missile defence systems.

AIR DEFENCE FORCE

Equipment

Category	Name	In Service
Air Defence Systems	9K37 Buk-M2E	
	SA-2/3	400
	SA-9	200
	SA-5	48
	SA-8	60
	Pantsyr-SIE	36

New Procurements/Upgrades

No major procurement plans are under consideration at present.

Defence Expenditure

Total defence spending: $2,490 (2011); $ 2,346 million (2010)

Defence spending in terms of GDP: 4.1% (est.)

Defence Production and R&D

Limited resources dedicated to research and development has impeded Syria's defence industry making the country dependent on foreign arms suppliers. Its relative backwardness in technological and industrial development, compounded by limited funds, has meant that most of its capability has been developed using outside help.

Defence Procurement

Syria import arms mainly from Russia, Iran and North Korea. It has an arsenal of surface-to-surface missiles. Arms suppliers are extremely limited, given the extensive embargoes placed on the state. However, the UN sanctions do not affect the trade of conventional weapons from Russia to Syria. Russia offers assistance to Syria with the supply and modernization of helicopters, heavy armoured vehicles, tactical missiles, various types of surface warships and other weapons.

Russia has reportedly signed a contract for the delivery of Yak-130 (Mitten) jet trainer and light attack aircraft and S-300 missile defence systems. Syria was reported to have made an advance payment of nearly $100 million for the first six aircraft. The light fighters were to be delivered in three instalments of nine units last year, 12 in 2015 and 15 in 2016.

According to a report published by Index of Security in 2010, Russia has completed a contract to supply thirty-six Pantsir-S1 air defence missile systems to Syria. Negotiations continue on a number of potential projects, in particular for supply of two diesel-electric submarines, MiG-29 SMT fighter jets and upgraded S-125 Neva surface-to-air missiles. Supply contracts also include supplies of SSI 24 MiG-29M/M2 fighter planes.

CONTACT DETAILS

Ministry of Defence

Ommayad Sq.

Damasus, Syria

Tel: +936-11-112101 through -05

112100 through -04

STRATEGIC INFORMATION

TAIWAN
(Capital: Taipei)

INTRODUCTION

Area: 35,980 sq km

Population: 23,464,787 (July 2016 est.)

Coastline: 1,566.3 km

Maritime claims: Territorial Sea: 12 nm

Exclusive Economic Zone: 200 nm

GEOPOLITICAL IMPORTANCE

The island state of Taiwan, also called the Republic of China (ROC), is a sovereign state in East Asia. The ROC constitutes a chain of islands among which Taiwan occupies 99 per cent of its territory. The other islands administered by ROC include Penghu, Kinmen, Matzu and few other small islands. Taiwan is neighboured by the People's Republic of China (PRC) in the west, the Philippines in the south and Japan in the northeast. It is bordered by the Pacific Ocean, East China Sea, South China Sea, Luzon Strait and the Taiwan Strait. The Taiwan Strait separates ROC from mainland China. It is one of the busiest transportation routes in the world with hundreds of commercial ships plying its waters every day. The Strait is also believed to be rich in gas and oil reserves.

The independently positioned Taiwan has direct access to the Pacific Ocean, thereby making it a strategically important country to build up and project naval power in the Pacific waters. The island has often been associated with the term "unsinkable aircraft carrier" since it can be used as an airbase to launch air strikes as done from on-board an aircraft carrier without physically destroying the landmass. Hence, the archipelago holds the possibility of being used to extend power projection by any military force.

POLITICAL OVERVIEW

The political status of Taiwan remains a contentious issue as the autonomously-ruled island has not been recognised as an independent country by the United Nations. Taiwan was part of mainland China or erstwhile Republic of China until 1949. The Chinese Civil War split the entire territory into two parts i.e., Republic of China or Taiwan and People's Republic of China or mainland China. Taiwan claimed independence from mainland China following the end of Chinese Civil War between the ruling party of Republic of China and the Communist Party of China. While the mainland China became People's Republic of China (PRC) and came to be recognised as an independent country after receiving United Nations Permanent Member status in 1971, Taiwan or ROC has struggled to find an independent identity for itself since then.

Taiwan identifies itself as an independent sovereign state. It functions as a multi-party democracy having its own Constitution. The President of Taiwan, who is also the commander-in-chief of the armed forces, is elected by universal suffrage. The island state maintains its own military.

KEY POLITICAL PERSONS

PRESIDENT & SUPREME COMMANDER OF ARMED FORCES:
Tsai Ing-wen

VICE PRESIDENT:
Chen Chien-Jen

ECONOMY

Following its democratic transition from the previous authoritarian regime in the 1980s and 1990s, Taiwan has transformed itself into a modern dynamic market economy. It has established itself as one of the leading economic powerhouses in Asia and is counted as one of the "Four Asian Tigers" along with South Korea, Singapore and Hong Kong. The industrially-advanced Taiwan is ranked 19th largest economy in the world. The country's economy mostly relies on exports carried through marine lanes. Despite confrontations with China over reunification issue, Taiwan has maintained healthy economic and trade relations with it. It has also joined several international organisations, including the World Trade Organisation, under a politically neutral name. However, an ambiguous political and diplomatic status has prevented Taiwan's wider economic participation and integration with the world market. The country had registered an annual GDP growth of 3.8% in 2014-15. However, following a global economic slump and falling domestic exports, the country's economy contracted even as the Government has announced measures to boost infrastructure investment and corporate credit to bolster the economy.

GDP: $1.1 trillion (2015 est.); $1.093 trillion (2014 est.)

Real Growth Rate: 0.6% (2015 est.); 3.9% (2014 est.)

Industries: Electronics, communications and information technology products, petroleum refining, chemicals, textiles, iron and steel, machinery, cement, food processing, vehicles, consumer products, pharmaceuticals

Natural Gas - Proved Reserves: 6.229 billion cu m (31 December 2015 est.)

Crude Oil - Proved Reserves: 10.06 million bbl (1 January 2015 est.)

Total Exports: $335.5 billion (2015 est.); $379.7 billion (2014 est.)

Export Commodities: Semiconductors, petrochemicals, automobile/auto parts, ships, wireless communication equipment, flat display displays, steel, electronics, plastics, computers

Major Markets: China 27.1%, Hong Kong 13.2%, US 10.3%, Japan 6.4%, Singapore 4.4%

Total Imports: $262.9 billion (2015 est.); $318.8 billion (2014 est.)

Import Commodities: Oil/ petroleum, semiconductors, natural gas, coal, steel, computers, wireless communication equipment, automobiles, fine chemicals, textiles

Major Suppliers: Japan 17.6%, China 16.1%, US 9.5%

DEFENCE & SECURITY

Taiwan, until 1945, was ruled by Japan that transferred its administration to mainland China or erstwhile Republic of China (ROC) after facing defeat in World War II. Following the Chinese Civil War of 1949, the mainland China came to be known as People's Republic of China (PRC) but Taiwan retained the original name Republic of China (ROC) which was established in 1912. Both regions claimed to represent the state China and wanted to govern each other's territory. But after the People's Republic of China (PRC) secured the permanent member status in the United Nations and came to be recognised as an independent country, Taiwan or ROC's political status became vague. From time to time, the official status of Taiwan as part of

Chinese territory has remained unclear with some countries, including the United States, maintaining silence over the issue.

The self-proclaimed sovereign state of Taiwan faces the biggest security threat from China which considers the archipelago as part of its territory and intends to unify it with mainland China even by military force. Taiwan's growing military relations with the US has also resulted in tension between China and the US. China, with its rapidly growing economy and expanding military might, has maintained a belligerent stand over the issue of Taiwan which it regards as its "core area of interest."

The ROC's independent military, constituting of the Army, Navy and Air Force, has been groomed with the primary objective of defending the homeland from any invasion by China. Guided by the military strategy of "resolute defence and credible deterrence", the ROC armed forces have focused on enhancing joint operation capabilities with the aim to prevent war and maintain deterrence. Keeping pace with the evolving military technology and new warfare methodology in the information era, the ROC armed forces have strived to develop information and electronic warfare tactics, missile defence systems, joint counter- air, joint sea control and joint ground defence, and acquiring asymmetrical military capabilities even with the country's limited defence personnel and resources. The technologically advanced Taiwanese armed forces have gone on to acquire modern military hardware and weapons, mostly from the US, to counter and prevent any attack by the People's Republic of China.

As part of a comprehensive military restructuring programme, the Taiwanese Defence Ministry announced to implement the "Jingtsui Programme" from 2011 to 2014. The programme gave emphasis on future warfare, enemy situation and threats and the nation's overall resource distribution, as well as "military strategy," "national financial resources and policy guidance" and "voluntarism," finding a balance between "combat requirements" and "financial capabilities." The programme aimed at gathering defence resources together and utilise them to "prevent the enemy from landing and establishing lodgement"; streamlining of high level command organisations to establish a "small but superb, strong and smart" defence force. Enhancement of early warning capabilities and integrating them with the existing intelligence, surveillance and reconnaissance capacity have also become a thrust area for the ROC military so as to counter any sudden and highly intense attack from the PRC.

In the 2013 National Defence Report (defence whitepaper), Taiwan emphasised on building a "Hard ROC" defence force as the primary objective of national defence policy. The Report also put thrust on indigenous development of key technologies and weapon systems besides foreign procurement. While acknowledging a rising cyber security threat from the PRC, the report highlighted on the building up of integrated national information security system and strengthening its network security readiness while reinforcing the security of its command, control, communication and information platforms for unhindered joint operations by its armed forces.

The latest biennial Defence Whitepaper released in October 2015 states that Taiwan's armed forces would continue to bolster their defence capabilities and naval patrol missions to assert the nation's sovereignty over the Pratas Islands and Itu Aba in the South China Sea, both cited as potential conflict areas. The country will issue its next whitepaper on defence in 2017.

KEY DEFENCE PERSONS

MINISTER OF NATIONAL DEFENCE: Feng Shih-kuan

DEPUTY DEFENCE MINISTER: Li Hsi-ming

CHIEF OF GENERAL STAFF: Gen. Yen De-fa

COMMANDER OF ARMY STAFF: Chiu Kuo-cheng

COMMANDER-IN-CHIEF OF NAVY: Adm. Huang Shu-Kuang

COMMANDER-IN-CHIEF OF AIR FORCE: Gen. Shen Yi-ming

Internal Conflict

Terrorism: Though not a potential threat to the state, terrorism has often reared its head in Taiwan. Terrorist incidents like bombing and aircraft hijacking have struck Taiwan in the past. The military often conducts anti-terrorist military drills to counter any terror-related eventualities.

Organised crime: The island state faces the problem of illegal smuggling of firearms and drugs across the Taiwan Strait. Human trafficking, gang wars, extortion and money laundering have also affected its internal security apparatus.

External Conflict

The self-governed Taiwan has not been involved in military conflict with any of its neighbours barring China which has, from time to time, threatened the sovereignty of the former. The cross-Strait relations between Taiwan and China have remained tense, especially after vociferous and rapid Chinese military expansion. To counter the militarily rising China, Taiwan has also intensified its efforts to acquire new weapons, mostly from the US. It is also developing asymmetric warfare capabilities and strengthening its conventional forces to meet internal and external security challenges.

Territorial disputes: Taiwan is engaged in territorial water disputes with Brunei, China, Malaysia, the Philippines, and Vietnam over the Spratly Islands in the South China Sea. The archipelago is also staking claim over the Paracel Islands occupied by China.

Taiwan has been opposing Japan's claims over the uninhabited islands of the Senkaku-shoto (also called Diaoyu Tai) in the East China Sea. The islands, which are around 160 kilometres from Japan's Okinawan chain and about 200 kilometres from Taiwan, lie on vital shipping lanes, and are believed to be near potentially rich gas fields. China also claims sovereignty over the islands.

THREAT PERSPECTIVE

China: The sole and foremost military threat that Taiwan faces is from China. Taiwan's National Defence Report 2013 for the first time made a precise prediction that China, by 2020, will acquire the full capability to launch an attack on the island nation to forcibly reunite it with the mainland. "China has been upgrading its major weapons systems and building up the People's Liberation Army (PLA) as part of its goal to have a strong enough fighting force to attack Taiwan by 2020," the 2015 National Defence Report states.

With its massive all-round military modernisation programme continuing unabated, China appears to have shifted the balance of power in the Taiwan Strait in Beijing's favour. According to the Taiwanese Defence Ministry, China has renewed its earlier military strategy of "using military force to oppose Taiwan independence" to "opposing Taiwan independence and advancing the unity of China" and is rapidly building up military capabilities so as to launch large-scale operations against Taiwan while deterring, delaying or denying any kind of intervention by a foreign military power in the wake of a conflict. China has deployed large number of missiles along its south-eastern coast targeting Taiwan. The Chinese arsenal deployed along its east coast includes some 1,400 tactical ballistic missiles, short-range missiles, cruise missiles, and laser or electronically guided missiles capable of delivering warheads over long distances. The PLA has also continued to deploy amphibious assault vehicles, long-range MLRS, warships, long-range anti-ship missiles, ballistic missiles and advanced fighter jets in its southern provinces of Guangdong and Fujian, according the Taiwanese Defence Ministry. The Defence Ministry's Five-Year Military Reform and Policy Plan Report states that China's Second Artillery Corps (now renamed as PLA Rocket Force) has increased its deployment of Dong Feng 16 (DF-16), DF-21D and DF-31 ballistic missiles, which would enable the People's Liberation Army to carry out multiple-wave precision strikes against Taiwan.

China is also intensely building up its air defence and anti-ship bases along the south-eastern coast. Ships and missile boats armed with long-range anti-ship missiles have also been deployed in the coast. Chinese unmanned aerial vehicles and high-speed anti-radiation missiles are capable of attacking Taiwan's command-and-control bases, according to Taiwan's National Defence Report. Moreover, Taiwan's military strength, which has gone down from 270,000 to around 215,000 active soldiers (by 2014-end) with further cuts looming large, along with its meagre defence budget of just over $10 billion has been dwarfed by China's massive military force consisting of 3 million personnel and a huge defence spending exceeding $100 billion.

STRATEGIC RELATIONS

US: The tiny island state has solely relied on weapons and technology supplied by the US to face any attack from China. The US has continued its efforts to supply arms to Taiwan which has irked Beijing. According to a US report, the United States has announced over $12 billion in arms sales to Taiwan since 2010. In 2008, Pentagon announced an arms package worth $6.6 billion for Taiwan that included the Patriot Advanced Capability-3 guided missile systems, Apache attack helicopters, Harpoon missiles and Javelin anti-tank missiles among other weapons and hardware. The announcement resulted in China suspending bilateral military exchanges with the US. In January 2010, the US government further announced an arms deal worth $6.4 billion for the island state. The arms sales included Patriot missiles, Black Hawk helicopters, and communications equipment for Taiwan's F-16 fighters' fleet. In 2007, Taiwan expressed interest to buy 66 F-16 C/D fighters from the US. China's vehement opposition to the military deals has held back US from supplying some of the weapons and platforms to Taiwan.

DEFENCE CAPABILITIES

ARMY: The primary task of the Taiwanese Army or ROC Army is to defend the integrity of the national territory and sovereignty; guarding the outer islands region; building up its combat strength through military training and acquiring technological specialization; carrying out joint operations with the Navy and Air Force in the event of war.

With China's steadily expanding its naval and air power capabilities in recent times, the Taiwanese Army has faced cut backs in personnel and overall strength as the island state has also shifted its focus more on strengthening the Navy and Air Force. Nevertheless, equipping the ground forces with new combat vehicles, attack helicopters, UAVs, advanced air defence systems and C4I2SR (command, control, communications, computer, information, intelligence, surveillance and reconnaissance) capabilities have been undertaken to enhance the Army's overall strength and prepare it to jointly fight modern-day hi-tech war in consonance with the Navy and Air Force.

Personnel: 150,000

Equipment

Category	Name	In Service	On Order
MBT	M60 A3	250	
	Modified M-48 (CM11/12)	300	
Light Tank	Type 64 (M41)	400	
APC	M113	1090	
	V-150 Commando	300	
	CM-32	200	400
Artillery	M109A5 155mm	50	
	M109A2	100	Being upgraded to M109A5 155mm variant
	M108	150	
	155mm field artillery	6	
	105mm field artillery	500	
	M110A2 203mm SP	25	
	M107 175mm SP	50	
Mortar	81mm towed		
	120mm towed		

Army continued...

Category	Name	In Service	On Order
Mortar	107mm		
	60mm		
MLRS	117mm Kung Feng VI	60	50+ (RT 2000 117mm, 180mm, 230mm)
ATGW	TOW	1000+	
	Javelin	60 launchers 360 missiles	20 launchers, 182 missiles
	Kun Wu	5 launchers	
SAM	Patriot air defence system (PAC-2)	1battalion with 3 batteries	4 batteries on order. The air-defence missile systems are being upgraded to Configurati-on-3
	Tien Kung air defence system	1 battalion with 6 batteries	
	Hawk air defence system (To retire by 2017)	6 battalion each with 6 batteries i.e. each battery equipped with 54 missiles	
	Stringer	1000+	
AA gun	40mm	300	
	20mm SP	20	
SSM	Hsiung Feng IIE (600km SSCM)		
	Ching Feng		

Army Aviation

Equipment

Category	Name	In Service
Combat helicopter	AH-1W Super Cobra	64
	OH-58D Kiowa	13
	AH-64E Apache Longbow	29
Transport/ Utility helicopter	CH-34	7
	UH-1H	100
	Kh-4	8
	Boeing 234 Chinook	3
	CH-47SD	3
	OV-1 Bird Dog	20
	OH-6A	5
	UH-60M Black Hawk	10+ (60 total order)

New Procurements/Upgrades

✦ The Army is evaluating the purchase of around 200 M1A1 Abram main battle tanks from the US to bolster its existing fleet of MBTs.

✦ All 30 AH-64E Apache Longbow combat helicopters have been delivered. One of the choppers crashed during a training exercise in April 2014. The US-made gunships are equipped with Stinger Block I air-to-air missiles and AGM-114L Longbow Hellfire air-to-ground missiles.

✦ Delivery of the advanced Ray Ting-2000 (RT2000) or Thunderbolt 2000 multiple-barrel rocket launch system has started with the first batch handed over to the Army in August 2012. The new MLRS are replacing the existing fleet of 117mm rocket systems.

✦ US defence firm Raytheon has received a $685.7 million Foreign Military Sales (FMS) contract in December 2011 for supplying additional units of the Patriot Air and Missile Defense System to Taiwan. The company is upgrading the existing Patriot systems from configuration-2 to configuration-3 (PAC-3) under a 2009 contract worth $154 million. It is also supplying four batteries of PAC-3 systems to Taiwan under a $1.1 billion contract awarded in 2009.

✦ 182 Javelin anti-tank guided missile systems and 20 launchers are on order from the US.

✦ A total of 60 Black Hawk utility helicopters have been ordered from the US with deliveries going on since December 2014 and will be completed by 2018.

NAVY: The Taiwanese Navy or the Republic of China (ROC) Navy outlines its primary role as safeguarding maritime security in the Taiwan Strait and actively engage in disaster relief operations. It operates with the mission to maintain and defend the safety of the country's maritime regions and secure its international shipping lanes in the Pacific Ocean. The Navy, also consisting of ROC Marine Corps, is deployed in the Taiwan Strait and its surrounding waters. With a stated vision to build "a modernised force" that will be able to control the sea, the Taiwanese Navy, however, has lagged way behind the ferociously growing China. The "Force Build-up Plans and Objectives" outlined in the 2011 National Defence Report of the Taiwanese Defence Ministry emphasised on procurement of diesel-electric submarines, long-range anti-submarine aircraft, new generation missile motorboats and mine-hunters to enhance the armed forces' 3 dimensional joint sea control capabilities, and strengthen counter blockade and joint interdiction capabilities.

In addition, fuel and ammunition supply ships will be constructed to maintain the capabilities of surface operation groups, thus building a sea control force equipped with anti-submarine, air defence and regional ocean surveillance capabilities. In a recent move, the Navy, under a 20-year force modernisation programme, plans to acquire new destroyers, frigates, corvettes and submarines to replace all the US and French-built warships in its fleet. All the new platforms and systems are planned to be built indigenously by the domestic defence industries with Western assistance, if necessary.

Major naval bases: Suao, Keelung, Kenting and Makung

Personnel: 35,000+ (Including Marine Corps)

Equipment

Category	Name	In Service
Destroyer	Keelung-class DDG (ex-US Kidd class)	4
Submarine	Hai Lung (Dragon) class	2
	Hai Shih (Guppy II class for training)	1
Frigate	Cheng Kung class (modified Perry design)	8
	Chi Yang class (ex-US Knox class)	8
	Kang Ding class (French La Fayette type)	6
Patrol vessel	Jin Jiang class OPV	11
	Lung Chiang class missile FAC	2
	Sui Kiang class missile FAC	2
	PLC coastal patrol boat	22
	Kuang-Hwa class (New Hai Ou class) FAC	31
LST	Ex-US	22
	Chung He class (ex-US Newport class)	2
LCU		26
LSM	Ex-US	4
LSD	Xu Hai-Class (ex-US Anchorage class)	1
LCVP		150
Mine sweepers & Mine hunters	Ex-US Osprey class	2
	Yeong Feng-class	4
	Yeongyang class	4
	Adjutant type	4
Auxiliary ship	Wu Yi class supply ship	1
	Combat support ship "Panshih"	1
Weapon	(Naval guns) Mk42 127mm Mk45 127mm Oto Melara 76mm Bofors 40mm	
	Phalanx close-in weapon system	
	(ASM) Hsiung Feng II Hsiung Feng III Syongfong II Harpoon RIM-66 Standard Missile	
	(Torpedo) Mk46 Wire-guided surface and underwater target torpedo	

New Procurements/Upgrades

✦ With an intent to build its own submarines and maritime platforms after failing to acquire the same from the US over the years, the island state under the new Government is planning to spend around $14.8 billion from 2018 through 2040 for indigenous construction of new battleships. The new platforms would include an amphibious transport dock (LPD), high-speed mine-laying vessels and Tuo Jiang-class of corvettes besides submarines.

✦ The country plans to acquire up to 8 conventional submarines. The design/acquisition contract for the programme is expected to be issued in 2016-17 with domestic shipbuilder's keen to take part in the construction of the underwater platforms.

✦ The first Tuo Jiang-class corvette has been commissioned in March 2015. The Navy plans to induct around 12 such 500-ton platforms into service.

✦ The Navy's two existing Hai Lung (Dragon) class submarines will be armed with US-made Harpoon Block II submarine launched anti-ship missiles. The subs will be equipped with over 30 Harpoon missiles having a range of 120 kilometres which are being procured from the US under a 2008 deal amounting to $195.46 million. The US has started delivering the advanced missile systems in 2013. All of the missiles and associated equipment are scheduled to be delivered by 2016-17.

✦ The US House of Representatives in April 2014 has authorised the sale of four Oliver Hazard Perry-class guided missile frigates to the ROC Navy.

✦ The Navy has retired its entire fleet of 50 Hai Ou class (Seagull-class) missile boats. The boats have been replaced by 30 new Kuang Hwa class boats.

✦ A contract has been awarded to an international team, led by Taiwan's Ching Fu Shipbuilding, to deliver six mine countermeasures vessels (MCMVs) by 2024. International partners in the programme include US's Lockheed Martin and Italian shipbuilder Intermarine.

✦ A fast combat support ship "Panshih" has been inducted in service in March 2015. The multi-role vessel will be used as a transport, maritime rescue, and humanitarian assistance platform.

✦ The Defence Ministry has decided to step up production of the indigenous Hsiung Feng IIE subsonic and Hsiung Feng III supersonic anti-ship cruise missiles.

Kang Ding class (French La Fayette type) frigate of Taiwan Navy

Naval Aviation
Equipment

Category	Name	In Service	On Order
ASW Aircraft	P-3C Orion	8	4
	S-2T Turbo Tracker	11	
ASW/ Mine sweeper helicopter	S-70B Sea Hawk	6	
	MH-53E		12
	S-70 CM1/CM2	18	
	MD500 Defender	8	

New Procurements/Upgrades

✦ Eight of the 12 P-3C Orion marine patrol aircraft on order from the US since 2007 have been delivered. The remaining four planes are scheduled for delivery by 2016-17. The new aircraft are replacing the Grumman S-2T anti-submarine warfare planes of the Navy.

✦ The Navy is evaluating options of acquiring new anti-submarine warfare (ASW) helicopters to replace its aging fleet of MD500 Defender choppers. Up to eight to 10 MH-60R Seahawk ASW helicopters could be purchased via the US Foreign Military Sales programme.

✦ Plans are on to upgrade the entire fleet of Sikorsky S-70 CM1/CM2 helicopters.

P3C Orion aircraft of Taiwan *An internet photo*

Marine Corps
Equipment

Category	Name
MBT	M-48
Light tank	M41
APC/AIFV	CM-25
	SIBMAS 6x6
Artillery	155mm towed
	105mm towed
	M109A2 155mm
	M110A2 203mm SP
Mortar	60mm
	81mm
	120mm
RCL	106mm
ATGW	Milan-3
SAM	Dual Mount Stringer
AA gun	M42 DUSTER twin 40mm SP
	M48 SP

AIR FORCE: The Republic of China Air Force (ROCAF) operates with the primary missions of keeping vigil over the Taiwan Strait, maintaining airspace security over the region with effective combat readiness, defending the island's territory against any "surprise enemy attack" and ensuring homeland security.

The Taiwanese Air Force has focused on enhancing its long-range reconnaissance capabilities, integration of C4I2SR systems and battlefield management capabilities effectively. It also aims to procure next generation fighter jets and counter-strike weapons to strengthen its overall combat capability for offensive manoeuvre against the People's Liberation Army of China. Procuring the US-made F-16C/D fighter jets and upgrading the Chingkuo indigenous defence fighters by deploying advanced electronic warfare systems, air-to-ground and air-to-sea missiles are being planned to strengthen joint counter-air operations by the Air Force.

Major air bases: Sungshan, Makung, Tainan, Ching Chuan Kang and Chiayi

Personnel: 35,000+ (Including Air Defence Missile Command)

Equipment

Category	Name	In Service
Fighter/ Attack/ AWS role	F16A/B	144 (to be upgraded)
	F-5E/F / RF-E	65 + 5
	Chingkou IDF	127
	Mirage 2000-5E/5D	55
Trainer	AIDC AT-3	50+
	Beechcraft T-34C	30+
Transport/ Surveillance	C130H/HE	19
	E-2K/E-2T Hawkeye	4+2
	Beech B-1900C	10
Helicopter (SAR)	EC225 Super Puma Mk2	3
	S-70C Blue Hawk	16
Weapon	(AAM) R.550 Magic	
	MICA	
	AIM-7 Sparrow	
	AIM-120 AMRAAM	
	Tien-Chien I	
	Tien-Chien II	
	Wan Chien AGM (under development)	

New Procurements/Upgrades

✦ The entire fleet of F-16A/B fighters are to be upgraded under a contract signed in 2012. The upgrade programme would include the fitment of active electronically scanned array (AESA) radars, embedded global positioning, as well as upgrades to the electronic warfare and other avionics systems of the F-16s. The programme is expected to start in 2017 and continue till 2022, with Lockheed Martin refitting some of the fighters in the US and Taiwan's Aerospace Industrial Development Corporation (ADIC) upgrading the rest of them locally.

✦ The last two upgraded E-2K airborne early warning (EWC) aircraft have been delivered by the US in 2013. The aircraft have been delivered under a deal inked in 2008 to convert a total of four E-2T Hawkeyes to the advanced E-2K configuration. The first two upgraded aircraft were delivered in 2011.

✦ The Defence Ministry announced plans in 2012 to increase the production of the Chingkou indigenous fighters. It submitted a report before the Parliament to upgrade more than half of the fighters and make them ready for deployment within 12 months to maintain credible deterrence against China into the 2030s. Upgrades on the aircraft would include enhanced radar, avionics and electronic warfare capabilities. The remaining fighters will be upgraded by 2017. The first phase of upgradation work involving 71 fighters was completed in 2013 and the Air Force unveiled the refurbished fighters in January 2014. The upgraded IDFs feature new flight control, avionics, radar systems and software updates, and are armed with the Wan Chien stand-off air-to-surface missiles.

✦ Taiwan plans to slash the number of F-16 C/D advanced fighter jets it has been seeking from the US from 66 to 24 due to budgetary constraints even as Washington has held back its decision to supply any new fighter jets to the island state. Taiwan has also expressed interest in buying the F-35 stealth jets in future.

✦ The Defence Ministry announced plans to invest 15 billion Taiwan dollars in 2011-2012 to equip the Chingkou IDFs with a new air-to-ground missile system, 'Wan Chien'. The country has completed testing of the domestically developed long-range ASM and mass production and deployment of the new weapon has started in 2014.

✦ The Government is negotiating to buy six Alenia Aermacchi C-27J Spartan tactical transport planes.

✦ The Defence Ministry has announced plans to purchase advanced trainer jets to replace its aging fleet of F-5E/F jet fighters and AT-3 jet trainers. Provision for the new jet trainers will be made in the 2017 national budget. The Air Force plans to acquire up to 66 new jet trainers. Domestic aerospace firm AIDC has presented the designs of two new jet fighter trainers – AT-3 MAX advanced trainer based on the existing AT-3 jet trainer and the XAT-5 advanced trainer based on the Chingkou indigenous fighter – for the new acquisition project.

✦ The Defence Ministry has decided to scrap the plan to buy second advanced early-warning radar from the US for strengthening its missile defence system. The construction of the first such long-range radar system, ordered from the US in 2003, went live in 2013.

AIR MISSILE DEFENCE COMMAND

Personnel: 5,000+

Nuclear Weapons Programme

To achieve asymmetric warfare capabilities, Taiwan pursued coveted nuclear weapons programme twice in the past. The programme, however, was eventually abandoned following intense US pressure. After China conducted its first nuclear weapons test on 16 October 1964, Taiwan raised concern over its powerful neighbour's military intensions, fearing that use of such weapons could completely annihilate the tiny island state off the map. To counter the threat, the Taiwanese Defence Ministry in the following years decided to take up nuclear weapons development programme which was code-named "Hsin Chu". The Institute of Nuclear Energy Research (INER), established in 1965, was authorised to carry out nuclear research programme, which the Taiwanese government tagged with civilian energy purposes. The INER, however, was set up near the Chung-Shan Institute of Science and Technology (CSIST) – the country's leading military research and development centre and received support from it. Taiwan bought heavy water reactors and other nuclear technology from a number of countries, including the US, Canada, Norway, Germany and France. In 1968, Taiwan signed the Nuclear Non-Proliferation Treaty (NPT) even while it was secretly pursuing nuclear weapons programme, which was later scrapped after strong US opposition. The island state again restarted the stalled nuclear weapons programme for a brief period in the late 1980s, but eventually abandoned it. Some defence experts believe that Taiwan is still capable of successfully developing nuclear weapons as it has already acquired the basic technologies needed to make such weapons. The island state currently has six nuclear reactors and three active nuclear power plants producing around 5,000 MW of nuclear energy.

Major Armaments Producers

Chung-Shan Institute of Science and Technology (CSIST) is Taiwan's leading institution for defence R&D and is engaged in the development, management and integration of advanced weapon systems.

CSBC Corporation, formerly known as China Shipbuilding Corporation is a leading state-owned naval conglomerate that designs and builds warships and other vessels for the ROC Navy.

Aerospace Industrial Development Corporation (AIDC) is a state-owned aviation firm and manufactures fighter jets, trainers and transport aircraft. Taiwan decided to privatise the aircraft manufacturer in 2014.

Indigenous Weapon Systems: Some indigenous weapons produced by Taiwan's defence industry include the **Tien Kung I** and **Tien Kung II** multi-level advanced air-defence missile systems. An advanced variant of the weapon, **Tien Kung III**, is presently under development by the Chungshan Institute of Science and Technology (CSIST). Taiwan plans to spend US$2.5 billion in the next nine years to acquire the new anti-missile system to boost its aerial defences against China. The **Tien Kung III SAMs** will be acquired between 2015 and 2024 to replace the ageing Hawk missiles of the ROC military.

The **Ray Ting-2000** or **Thunder 2000** multiple launch rocket system (MLRS) has been developed with long range and wide impact area for anti-landing operation.

The **Syongfong II** anti-ship missiles have been deployed on Cheng Kung-class and Kang Ding-class frigates and Keelung-series DDGs.

The **Hsiung Feng II** subsonic and **Hsiung Feng III** supersonic anti-ship cruise missiles have been developed to counter Chinese attack.

The **Tien-Chien I** and **Tien-Chien II** air-to-air missiles are deployed on the IDF fighters.

The **Wan Chien** (Ten Thousand Swords) is a new air-to-ground missile system being developed by the CSIST to equip the Chingkou combat jets.

DEFENCE DEALS

Joint Venture/ Collaboration Programmes

USA: Taiwan has achieved self-reliance to a certain extent in military equipment production by tying up with leading US defence firms. The Aerospace Industrial Development Corporation (AIDC) has produced aircraft, helicopters and other aerospace components in partnership with US companies Lockheed Martin, Northrop Grumman, Sikorsky and Bell Helicopters among others.

F-5A/E/F Fighter: The AIDC, under license from the US Government, has co-produced over 300 F-5A/E/F Tiger fighters with Northrop Grumman since the 1970s. A few of the fighters are presently operational with the ROCAF.

Indigenous Defence Fighter: Taiwan began development of the Chingkou indigenous defence fighters in 1980 following US's refusal to sell its F-20 and F-16 fighters to the ROCAF. The AIDC designed, developed and manufactured the new fighters with assistance from US defence firms, including General Dynamics, Honeywell and others.

AT-3 Trainer Jet: The advanced trainer jets, currently operated by the Taiwanese Air Force, were developed by the AIDC in partnership with US' Northrop Grumman between 1984 and 1990.

UH-1 H Helicopter: The multi-utility transport helicopters were assembled by AIDC under license from Bell Helicopter in the 1970s.

Arms Import

Due to the heightened threat posed by its neighbour, Taiwan remains one of the major arms buyers in the Asia Pacific region. In the 1990s, the country was a leading arms importer in the world. The island state has received majority of its military weapons, equipment and systems from the US. Few European countries, including Germany, Italy and France, have also supplied defence equipment to Taiwan. However, many import deals under which Taipei seeks to acquire advanced weapons and platforms for its military, have been stalled due to stiff Chinese opposition.

Meanwhile, defence indigenisation efforts by Taiwan has gathered momentum in recent times with the country planning to set up a military technology department under the MoD in 2017. The new department would work on developing "ground-breaking" and "innovative" defence technologies as well as supporting the advancement of the local defence industrial base, according to the Defence Ministry.

DEFENCE BUDGET

The defence budget of Taiwan has witnessed a declining trend over the last few years. Though in proportion to its small population, Taiwan maintains a large military force with the primary mission to defend its territory against possible Chinese attack, the island's annual military expenditure has been steadily falling in sharp contrast to China's ever-rising defence allocation and military expansion. The Taiwanese military assesses that mainland China's military budget is already 21 times that of Taiwan's. According to experts, Taiwan's closer politico-economic relations with China in recent times have led the island state to reduce its defence expenditure. Moreover, the ROC military was set to be downsized from its present strength of around 240,000 personnel to 215,000 by 2014-end even as the country plans a transition from conscription to an all-volunteer armed forces. The Defence Ministry, in January 2014, further announced to slash the armed forces by up to 20%, from 215,000 to between 170,000 and 190,000 by 2019. The Government, however, reiterated that the country's defence capabilities would not be compromised in the project to build "smaller but leaner and more professional armed forces." The Defence Ministry's deadline to

Budgetary allocation

Year	Allocation	% of GDP
2016-17	TW$323 billion (US$ 10.2 billion)	2.9%
2015-16	TW$321.7 billion (US$ 9.7 billion)	2.8
2014-15	TW$319.3 billion(US$10.7 billion)	2.6
2013-14	Tw$313 billion (US$10.52 billion)	2.1
2012-13	Tw$317.2 billion (US$10.6 billion)	2.2
2011-12	Tw $297.2 billion (US$ 9.2 billion)	2
2010-11	Tw$ 297.4 billion (US$ 9.3) billion	2.1

phase out the decades-old military conscription policy by the end of 2014 has been pushed to 2017-end due to insufficient recruitment for the all-voluntary force.

CONCLUSION

At a time when China has thumped its military feet with a loud bang in the global stage, Taiwan's perennial concern to protect itself from its mighty neighbour's attack has heightened. The large-scale military modernisation programme of China has shifted the balance of power in the Taiwan Strait in Beijing's favour. Though increasing regional tensions in the Asia Pacific region in recent times have involved several littoral states, with China being at the centre, yet Beijing has not entirely shifted its focus away from Taiwan – the region it considers its "core area of interest". Acknowledging the all-round rapid military rise of mainland China, Taiwan has intensified its efforts to acquire advanced defence systems, weapons and platforms to counter the imminent threat. The ROC military has focused on building up asymmetric warfare capabilities as also strengthening its conventional forces to meet the ever-rising PLA power. With strong backing from the US, Taiwan has maintained a tough and defiant poster before a hostile China to counter its intent to reunite the archipelago into its territory by force.

CONTACT DETAILS

Ministry of National Defence
No.409, Bei'an Rd.,
Zhongshan Dist.,
Taipei City 10462,
Taiwan (R.O.C.)
Tel: (+886) 2-2311-6117
Website: www.mnd.gov.tw

TAJIKISTAN
(Capital: Dushanbe)

INTRODUCTION

Area: 144,100 sq km
Population: 8,330,946 (July 2016 est.)
Coastline: 0 km (landlocked)
Maritime claims: None

KEY POLITICAL PERSONS

PRESIDENT & COMMANDER-IN-CHIEF: Emomali Rahmon

PRIME MINISTER: Qohir Rasulzoda

KEY DEFENCE PERSONS

DEFENCE MINISTER: Lt. Gen. Sherali Mirzo

ECONOMY

Tajikistan is one of the poorest countries in the Central Asia region. Economic growth in Tajikistan slowed from an average of 7.5 percent a year over the past decade to 6.4 percent in the first six months of 2015 as a result of the economic recession in Russia, according to the World Bank. The 2009 global economic crises adversely affected Tajikistan but the combination of government's tighter fiscal policy, currency depreciation, and increased support from development partners helped mitigate the adverse impacts of the crises. Nevertheless, continued vulnerabilities to external market shocks, susceptibility to natural disasters, underexploited economic diversification potential, limited arable land, and landlocked location has affected economic development. Tajikistan has rich deposits of gold and silver. The current economic situation remains fragile. In 2014 Tajikistan was the world's most remittance dependent economy with remittances accounting for 49% of GDP and expected to fall by 40% in 2015 due to the economic crisis in Russia.

GDP: $23.31 billion (2015 est.), $22.63 billion (2014 est.)

Real Growth Rate: 3% (2015 est.), 6.7% (2014 est.)

Industries: Aluminium, cement, vegetable oil

Oil - proved reserves: 12 million bbl (1 January 2016 est.)

Natural gas - proved reserves: 5.663 billion cu m (1 January 2016 est.)

Total Exports: $572 million (2015 est.), $526.8 million (2014 est.)

Export Commodities: Aluminium, electricity, cotton, fruits, vegetable oil, textiles

Major Markets: Turkey 19.7%, Kazakhstan 17.6%, Switzerland 13.7%, Iran 8.7%, Afghanistan 7.5%, Russia 5.1%, China 4.9%, Italy 4.8% (2015)

Total Imports: $2.825 billion (2015 est.), $3.528 billion (2014 est.)

Import Commodities: Petroleum products, aluminium oxide, machinery and equipment, foodstuffs

Major Suppliers: China 42.3%, Russia 17.9%, Kazakhstan 13.1%, Iran 4.7% (2015)

INTERNATIONAL DISPUTES

Tajikistan and China have pledged to commence demarcation of the revised boundary agreed to in the delimitation of 2002 and talks continue with Uzbekistan to delimit the border and remove minefields while disputes in Isfara Valley has delayed delimitation with Kyrgyzstan.

DEFENCE

Tajikistan's armed forces consist of Tajik National Army, Tajik Air Force, Presidential National Guard, Tajik Border Troops, Tajik Mobile Forces and Tajik Internal Troops. Due to the presence of Russian forces in the country and civil war Tajikistan only formally legalised the existence of its armed forces in April 1994. The 7000-strong 201st military base in Tajikistan is Russia's largest military contingent in Central Asia. A long term military cooperation agreement with Russian signed in 1999 and renewed in September 2011, establishes the legal framework for the presence of Russian troops in Tajikistan to be maintained for the next 49 years while a five-nation agreement to improve military confidence in border areas has been signed between China, Kazakhstan, Kyrgyzstan, Russia and Tajikistan. Russia has recently agreed to supply USD 200 million worth of weapons and hardware and also to allocate about USD 5 million to Tajikistan's national drug enforcement agency instead of paying rent for the 201st Russian Military Base in Tajikistan. Tajikistan and Iran have signed a Memorandum of Understanding on defence cooperation and an agreement with the United

States allows for US troops to be deployed in and operate from base at Dushanbe and Kulyab.

The total active manpower is reportedly around 25,000. Its military equipment includes: T-62 and T- 72 Main Battle Tanks (MBT) totalling around 35+, some 100 BMP-1/2, few BTR-60/70/80 Armoured Personnel Carriers (APC), some SAM (SA-2, SA-3, etc) missile system, few artillery, MRLs and mortars, a dozen Mi-8, 7 Mi-24 helicopters and L-39C for training or ground attack.

Soldiers of the Armed Forces of Tajikistan during a military parade to mark the 70th anniversary of the Victory in the Great Patriotic War of 1941-1945. *Image Credit: Presidential Press and Information Office | kremlin.ru*

Defence Expenditure

Exact defence expenditure figures are unavailable.

Defence spending in terms of GDP: 1.13% (2014)

Defence Production and R&D

Tajikistan does not undertake any defence production activities.

Defence Procurement

Tajikistan Army/Navy/Air Force have not made any major defence procurements in recent years.

Soldiers of the Armed Forces of Tajikistan

CONTACT DETAILS

Ministry of Defence

Ulitsa Bokhtar 59

734002 Dushanbe

Tajikistan

TANZANIA

(Capital: Dar es Salaam)

INTRODUCTION

Area: 947,300 sq km

Population: 52,482,726 (July 2016 est.)

Coastline: 1,424 km

Maritime claims: Territorial sea: 12 nm

Exclusive Economic Zone: 200 nm

KEY POLITICAL PERSONS

PRESIDENT & COMMANDER-IN-CHIEF: John Magufuli

PRIME MINISTER: Kassim Majaliwa

KEY DEFENCE PERSONS

MINISTER OF DEFENCE: Hussein Ali Mwinyi

CHIEF OF DEFENCE FORCES: Gen. Davis Mwamunyange

COMMANDER OF LAND FORCES: Maj. Gen. James Aloisi Mwakibolwa

COMMANDER OF NAVY: Rear Adm. R S Laswai

COMMANDER OF AIR FORCE: Brig. Gen. William Ingram

ECONOMY

Tanzania is one of the world's poorest economies in terms of per capita income. However, it has achieved high overall growth rates based on gold production and tourism. Tanzania has largely completed its transition to a liberalised market economy, though the government retains a presence in sectors such as telecommunications, banking, energy, and mining. The economy depends on agriculture, which accounts for more than one-quarter of GDP, provides 85% of exports, and employs about 80% of the work force. The World Bank, the IMF, and bilateral donors have provided funds to rehabilitate Tanzania's aging economic infrastructure, including rail and port infrastructure that are important trade links for inland countries. Banking reforms have also helped increase private-sector growth and investment, and the government has increased spending on agriculture. The financial sector in Tanzania has expanded in recent years and foreign-owned banks account for about 48% of the banking industry's total assets. Competition among foreign commercial banks has resulted in significant improvements in the efficiency and quality of financial services, though interest rates are still relatively high, reflecting high fraud risk. Dar es Salaam used fiscal stimulus and loosened monetary policy to ease the impact of the global recession.

GDP: $138.5 billion (2015 est.); $129.4 billion (2014 est.)

Real Growth Rate: 7% (2015 est.); 7% (2014 est.)

Industries: Agricultural processing (sugar, beer, cigarettes, sisal twine); mining (diamonds, gold, and iron), salt, soda ash; cement, oil refining, shoes, apparel, wood products, fertilizer

Total Exports: $5.709 billion (2015 est.); $5.322 billion (2014 est.)

Export Commodities: Gold, coffee, cashew nuts, manufactures, cotton

Major Markets: India 21.4%, China 8.1%, Japan 5.1%, Kenya 4.6%, Belgium 4.3% (2015)

Total Imports: $9.843 billion (2015 est.); $10.92 billion (2014 est.)

Import Commodities: Consumer goods, machinery and transportation equipment, industrial raw materials, crude oil

Major Suppliers: China 34.6%, India 13.5%, South Africa 4.7%, UAE 4.4%, Kenya 4.1% (2015)

INTERNATIONAL DISPUTES

Dispute with Tanzania over the boundary in Lake Nyasa (Lake Malawi) and the meandering Songwe River; Malawi contends that the entire lake up to the Tanzanian shoreline is its territory, while Tanzania claims the border is in the centre of the lake. The conflict was reignited in 2012 when Malawi awarded a license to a British company for oil exploration in the lake.

DEFENCE

ARMY

Personnel: 40,000+

Equipment

Category	Name	In Service
MBT	Type-59	26
	T-62/ Type-63	30+
	Scorpion	36
APC	BTR-40/-152	50
	Type 56	20+
Mortar	M43 82mm	300
	M43 120mm	50
Artillery	ZIS-3 76mm gun	24
	Type 56 85mm	80
	D-30	20
	M-30 122mm	85
	M46 130mm	30+
AA gun	ZPU-2 & ZPU-4 14.5mm	250+
	ZU-23 23mm	40
	Type 55 37mm	100+
SAM	SA-3/6/7	

New Procurements/ Upgrades

No major procurement plans are under consideration at present.

NAVY

Personnel: 1000+

Equipment

Category	Name	In Service
Patrol boat	Ex-Chinese Shanghai II class	2
	Ex-Chinese Huchuan class	4
LCM		1

New Procurements/ Upgrades

✦ Two Haiqing-class patrol boats, TNS *Mwitongo* (P77) and TNS *Msoga* (P78), have been commissioned into service in April 2015. The vessels are built by China.

✦ Two new landing craft have reportedly been acquired from China.

AIR FORCE

Personnel: 900+

Equipment

Category	Name	In Service
Fighter	J-7G	14
	J-6	8
	J-4	4
Helicopter	Bell 412	2
	AB206A/B	4
	AB-205	4
	SA.316B ALOUETTE III	4
	BO-105	1
Trainer	PA-28	4
Transport	DHC-5D	5
	Cessna 404	2
	Cessna 310	7
	HS-748	3
	Y-5	1
	F-28 (VIP)	2

New Procurements/ Upgrades

No major procurement plans are under consideration at present.

Defence Expenditure

Defence spending for 2015-16: Sh1.716 trillion ($808 million) – a hike of 30% over previous year

Estimated defence spending in terms of GDP: 1.5%

Defence Production and R&D

Tanzania does not have the financial or technical resources to pursue any significant defence-related production or research & development. Defence spending remains very low and hence, little funding is available for the defence industry. The country is not subject to any international arms embargo but there has been international concern at to its intentions to develop small-arms manufacturing capabilities.

Defence Procurement

Procurement plans for the Armed Forces have not been officially announced, but the country has acquired some naval platforms from China recently.

CONTACT DETAILS

Ministry of Defence

PO Box 9544

Dar Es Salaam

Tanzania

STRATEGIC INFORMATION

THAILAND

(Capital: Bangkok)

INTRODUCTION

Area: 513,120 sq km

Population: 68,200,824 (July 2016 est.)

Coastline: 3,219 km

Maritime claims: Territorial Sea: 12 nm

Exclusive Economic Zone: 200 nm

Continental Shelf: 200 m depth or to the depth of exploitation

GEOPOLITICAL IMPORTANCE

Situated at the centre of Southeast Asian peninsula, Thailand or the Kingdom of Thailand, shares its land border with Cambodia in the southeast, with Laos in the north and northeast, with Myanmar in the north and west, and with Malaysia in the south. The country shares maritime boundary with Vietnam in the Gulf of Thailand to the southeast, and with India and Indonesia in the Andaman Sea to the southwest. With a varied geographical terrain, Thailand is broadly divided into four different regions: the central region (plains), northern region (mountainous), north-eastern region (plateau) and southern region (peninsula). The central region comprising of the Bangkok metropolitan region and the central plains is often called the "rice bowl" of Thailand and is the most fertile part of the country with high population density.

Sitting at the heart of mainland Southeast Asia, which has traditionally remained a key trading and transition point for the world, Thailand has historically yielded great influence in and around its neighbourhood. Situated almost at equidistance from India and China – the two rapidly growing countries of Asia in the 21st century – Thailand has assumed great strategic importance. Pursuing a goal to rise economically and politically, and transform itself from a developing to a developed nation, Thailand, over the past few years, has steadily expanded its reach across Asia, especially in South Asia, by taking various diplomatic initiatives like forming regional and sub-regional forums and participating in leading multilateral blocks in the region. The country has aimed at a balanced development of the Southeast Asian region where it could emerge as an important geostrategic and economic hub.

POLITICAL OVERVIEW

Thailand functions as a constitutional monarchy where the King (Bhumibol Adulyadej who ruled the country since 1946 died in October 2016), is the head of the state and the Prime Minister is the head of government, as per a new written Constitution adopted in 2007. The King, however, does not hold much real power and is more of a symbol of national unity and identity. The Prime Minister and his council of ministers carry out the executive tasks. The country's legislature and judiciary function independently. The bicameral legislature, called the National Assembly, consists of the Senate of Thailand (upper house), and the House of Representatives of Thailand (lower house). The judiciary consists of all the courts in the country. Despite its democratic credentials, political volatility has continued to mar Thailand since long. After being ruled as an absolute monarchy for several centuries, Thailand transitioned into democratic monarchy in 1932 wherein the king became the titular head of state and the prime minister became the head of government. However, the influence of military over the political dispensation did not cease and the country witnessed military coups several times in its history, including as recently as in 2006.

Once again, the Southeast Asian country witnessed political turmoil in the latter half of 2013 wherein violent anti-Government protests against the elected Yingluck Shinawatra-led dispensation resulted in the Thai Army taking control of the situation and imposing martial law all over the nation in May 2014. This resulted in the ouster of the Shinawatra government and appointment of a military junta led by Army Chief General Prayuth Chan-ocha. The Army scrapped the 2007 charter and rolled out an interim constitution, giving the military sweeping powers. The military dispensation had promised to hold democratic elections in the country in early 2016 but it was later deferred. In August 2016, the country approved a new Constitution in a referendum which gives the military sweeping powers. It now plans to go for the general election in 2017.

KEY POLITICAL PERSONS

HEAD OF STATE:
King Rama X

PRIME MINISTER:
Prayuth Chan-ocha

ECONOMY

Thailand is the second largest economy in Southeast Asia and has witnessed rapid economic growth in the past two decades. The free-market economy, with a well-developed infrastructure, has relied heavily on exports which contribute a major share to its GDP. The service sector and agricultural sector also contribute to the economy. Exports by the country include machinery, electronic components, agricultural commodities, textiles and jewellery. The country is a member of leading financial institutions and forums including the World Bank, World Trade Organisation, International Monetary Fund, Asian Development Bank and Asia-Pacific Economic Cooperation. The World Bank, in July 2011, upgraded Thailand's income categorisation from a lower-middle income economy to an upper-middle income economy. This was due to Thailand's progress in social and economic development, despite facing a number of financial/ economic and political challenges. With an investment-friendly economy, Thailand has also attracted foreign direct investments. After registering strong growth in the period between 2002 and 2007, the country was hit by the global financial crisis and its GDP contracted to well below 3% in 2009. The following year, however, recorded a boom with a GDP growth of 7.8%. In 2011, the economy slowed down again, mainly due to devastating floods in the country along with tsunami and earthquake in the neighbouring Japan which took a toll on Thai exports. Months of political unrest followed by the military coup in May 2014 resulted in a weak economic growth which was also hit by a dry weather, with agricultural output going down. Tourism industry was also adversely affected following the coup. The military junta has taken several stimulus measures to revive and stabilise the economy. The Thai baht, however, has depreciated more than 8% during 2015.

GDP: $1.11 trillion (2015 est.); $1.08 trillion (2014 est.)

Real Growth Rate: 2.8% (2015 est.); 0.8% (2014 est.)

Industries: Tourism, textiles and garments, agricultural processing, beverages, tobacco, cement, light manufacturing such as jewellery and electric appliances, computers and parts, integrated circuits, furniture, plastics, automobiles and automotive parts; world's second-largest tungsten producer and third-largest tin producer

Natural Gas - Proved Reserves: 255.9 billion cu m (1 January 2014 est.)

Crude Oil - Proved Reserves: 461 million bbl (1 January 2015 est.)

Total Exports: $212.1 billion (2015 est.); $224.8 billion (2014 est.)

Export Commodities: Automobiles and parts, computer and parts, jewellery and precious stones, polymers of ethylene in primary forms, refine fuels, electronic integrated circuits, chemical products, rice, fish products, rubber products, sugar, cassava, poultry, machinery and parts, iron and steel and their products

Major Markets: US 11.2%, China 11.1%, Japan 9.4%, Hong Kong 5.5%, Malaysia 4.8%, Australia 4.6%, Vietnam 4.2%, Singapore 4.1% (2015)

Total Imports: $177.5 billion (2015 est.); $200.2 billion (2014 est.)

Import Commodities: Machinery and parts, crude oil, electrical machinery and parts, chemicals, iron & steel and product, electronic integrated circuit, automobile's parts, jewellery including silver bars and gold, computers and parts, electrical household appliances, soybean, soybean meal, wheat, cotton, dairy products

Major Suppliers: China 20.3%, Japan 15.4%, US 6.9%, Malaysia 5.9%, UAE 4% (2015)

DEFENCE & SECURITY

Having deep-rooted affiliation with monarchy since centuries, Thailand professes a policy of avowed allegiance to the Nation, Religion, People and the King. The Kingdom is the only country in mainland Southeast Asia which has never been colonized by any Western power. Being at the heart of mainland Southeast Asia, the country, though, has remained strategically vital for leading global powers since centuries.

In the absence of any specific conventional military threat to Southeast Asia barring some border disputes, Thailand has focused on building up its armed forces in consonance with the evolving global as well as regional strategic changes. The country's defence policy has emphasised on enhancing defence capabilities with a "right-sized Armed Forces" and enhancing security cooperation with neighbouring states, and the regional and international community.

Thailand remains a traditional strategic ally of the US in Southeast Asia for over six decades since the two sides inked the Manila Pact in 1954 to form the Southeast Asia Treaty Organization (now dissolved) and the Rusk-Thanat agreement in 1962 that pledged US support for Thailand's defence and security. The country's military still relies on US military doctrine and structure and receives military training from the latter. In 2003, the Kingdom became the first country in the region to be designated as the US's major non-NATO ally. However, in the wake of evolving geostrategic equations and shifting balance of power, Thailand in past few years has realigned its strategic perspective and tilted to some extent towards its north Asian neighbour, China. The two countries have become important strategic allies in recent times. Thailand, indeed, is the only Southeast Asian country sharing a "comfort zone" with the economically and militarily growing China which has been entangled in maritime disputes with several other regional littoral states, including Vietnam, Philippines, Brunei and Malaysia. While toeing a flexible rebalancing policy with both the US and China, Thailand has also beefed up its bilateral engagements with its South Asian neighbour India in recent times.

Focused on becoming an economically developed country and play a leading role in the regional and international arena, Thailand, despite having few border disputes with its neighbouring countries, does not pursue any exclusive and extensive military goals and rather strives to maintain balanced and cordial relations within its surrounding region and beyond.

KEY DEFENCE PERSONS

DEPUTY PRIME MINISTER & DEFENCE MINISTER: Gen. Prawit Wongsuwan

CHIEF OF JOINT STAFF: Gen. Surapong Suwana-adth

COMMANDER-IN-CHIEF OF ARMY: Gen. Chalermchai Sitthisad

COMANDER-IN-CHIEF OF NAVY: Adm. Na Areenich

COMMANDER-IN-CHIEF OF AIR FORCE: Air Chief Marshal Jom Rungsawang

Internal Conflict

Ethnic insurgency: For several decades, Thailand has been fighting insurgency in its predominantly Muslim-populated southern provinces. The insurgency problem in the south, remotely located from the country's capital, is rooted over the demand for an independent Pattani state and inclusion of Malay as an official language. Five provinces in the remote south – Pattani, Yala, Narathiwat, Satun and Songkhla – have been seeking the restoration of Pattani which was an independent Sultanate before being annexed by Thailand (then Kingdom of Siam) in 1909. The conflict has mainly stemmed from ethnic identity, economic inequality, religious issues and political exclusion. Insurgency-related violence where the armed Islamic militants have targeted civilians has resulted in loss of life and property in the past few years. Proximity of the southern region with the Thai-Malaysian border, which has become a haven for drug-trafficking and illegal arms trading, has further deepened the problem. The Thai government has often linked the violence with terrorism and deployed armed forces in the area to deal with the issue.

Organised crime: The country has been fighting human trafficking, illegal drug trade, money laundering, smuggling, financial fraud and other forms of organised crime.

External Conflict

Territorial disputes: Thailand is engaged in land border dispute with two of its neighbours - Cambodia and Myanmar. With Malaysia, it has a land border as well as maritime sea dispute, though neither country has taken an aggressive stand over it. Thailand is also locked in maritime disputes with Laos and Cambodia.

Cambodia: The two neighbours spar over the 11th century Hindu temple of Preah Vihear in the Thai-Cambodian border. Historically recognized as a Cambodian temple, Preah Vihear (called Phra Viharn by the Thais) was occupied by Thai forces in 1954 soon after Cambodia gained independence. However, the ownership of the temple was eventually given to Cambodia under a 1962 ruling of the International Court of Justice. The dispute over the temple remained subdued for some decades as an internal conflict-ridden Cambodia was busy sorting out its domestic issues. The Preah Vihear dispute resurfaced again in 2008 when Cambodia sought to list the temple as a Unesco World Heritage site which mounted tensions between the two neighbours. While Thailand does not claim any official ownership over the temple, the main area of contention is its surrounding region roughly covering some 4.6 square kilometres, having direct transport links to Thai cities and giving a free access to its citizens and tourists. Hence, the area remains a subject of overlapping territorial claims. Efforts are going on by the two sides to resolve the centuries-old conflict.

The maritime dispute between the two countries involves some 26,000 sq kms of area in the Gulf of Thailand. Both sides have overlapping claims in the region, believed to be rich in oil and natural gas.

Myanmar: Thailand shares a land boundary of 2,401-km with its north-western neighbour Myanmar of which only 60-km is officially demarcated. The border conflict between the two countries dates back to the British colonization of Myanmar (erstwhile Burma) in 1824 and the inking of the Burney Treaty between Thailand and Britain in 1826, which redefined the Thai-Burmese border. Myanmar, since its independence in 1948, has opposed the current border, most of which has not been officially demarcated. The area of dispute, in particular, is the region along the Moei River which forms a natural border between Thailand and Myanmar. The border issue has also given rise to illegal arms and drugs trade, human trafficking, illegal immigration and ethnic insurgency originating in Myanmar and spilling into Thailand.

Laos: Thailand and Laos share a 1,810-km long common border - 702-km of land and 1,108km of the Mekong River. Dispute over the land border resulted in a brief war between the two sides in 1987-88. The two neighbours set up the Thai-Lao Joint Boundary Commission (JBC) in August 1996 to demarcate the 1,810-kilometre common border. Talks are going on to resolve the issue, even as disputes remain over several islands in the Mekong River.

Malaysia: While the maritime boundary dispute involving delimitation of continental shelf in the Gulf of Thailand has been resolved between the two countries, a dispute still remains over an area adjacent to the continental shelves with overlapping claims by Thailand, Malaysia and Vietnam. The land border dispute involving the Bukit Jeli (Jeli Hill) is yet to be resolved between the two countries.

THREAT PERSPECTIVE

Barring the border disputes in its immediate neighbourhood, Thailand has, over the years, successfully maintained a friendly posture and endeavoured to elevate its regional as well as international profile by promoting peace, stability and security. There is no direct military threat to Thailand's national security and sovereignty from any country at present. The only primary threat the country faces today is non-traditional in nature and stems from internal unrest and related violence. At the regional level, Thailand has been making efforts to resolve all boundary disputes with its neighbours. At the global stage, while retaining strong ties with its traditional allies, Thailand is also extending its hand of friendship to newer countries with the larger goal of securing its national interests and rise as an economic power.

STRATEGIC RELATIONS

US: As the US's only "treaty ally" in the entire Southeast Asia, Thailand has immensely benefited from its ties with the world's superpower. While traditional relations between the two countries dates back to 1833 when they inked the bilateral Treaty of Amity and Commerce, strategic alliance between them was established in the post Word War-II phase, when both sides became signatories to the 1954 Manila Pact. The pact led to the formation of Southeast Asia Treaty Organization (SEATO) in 1955 as an "Asian equivalent of Western military alliance NATO". The organisation, with its headquarters at the Thai capital Bangkok, was established as an international forum for collective defence of Southeast Asia, though, barring from Thailand and Philippines, the rest of the member countries in the block were from outside the region. SEATO was eventually dissolved in 1977 after many member nations withdrew from it.

In 1962, Thailand and the US signed the landmark Rusk-Thanat agreement, under which the latter pledged support to Thailand's defence and security that subsequently paved the way for the US military gaining access into the Southeast Asian country. During the US-Vietnam war in the 1950s, Thailand became an indispensable ally of Washington and provided its crucial airfields, ports and other infrastructure for military use against Vietnam. The Thai armed forces, in return, received a wide range of military equipment and training from the US. In subsequent years, the Thai-US military relations were subdued, particularly after the end of the Vietnam War. The relationship revived again in the late 1970s when Vietnam occupied Cambodia in 1978. The US undertook the task of modernising and professionalising the Thai armed forces and in 1982, the two countries staged the first "Cobra Gold" annual military exercise, which today is pegged as one of the largest multi-national field exercises in the Asia Pacific with nearly 30 countries taking part in it.

Over the past few years, US-Thai relations have witnessed changes in the backdrop of evolving regional and global strategic equations. Nevertheless, bilateral ties between the two sides remain firmly in place, with the US providing military as well as economic assistance to Thailand. In 2003, Washington designated Thailand as its major non-NATO ally (MNNA), after the Kingdom contributed troops and support for US military operations in Afghanistan and Iraq. The MNNA tag allowed Bangkok to buy advanced medium-range air-to-air missiles for its US-made F-16 fighters - a first for any Southeast Asian nation.

As Washington has announced plans to shift its strategic focus to the Asia-Pacific by bolstering its military presence in the region, it has begun exploring ways to further reinvigorate its strategic alliance with Thailand by reviving some of its old defence bases there and also increasing joint naval cooperation with the Southeast Asian nation.

However, recent developments at the political level that have jeopardized democratic contours in Thailand following the military-led coup have adversely affected Thai-US relations, with Washington suspending military aid worth US$ 4.7 million to the South East Asian country.

China: Though it initially harboured a policy of suspicion and hostility towards the Communist China, the Thai Kingdom later established diplomatic ties with Beijing in 1975 and since then, the two Asian neighbours have cooperated in several areas of mutual interest. The two sides started the Sino-Thai Defence and Security Consultation in 2001 which later gave way to defence cooperation, including joint development and technological transfers. China began supplying a range of military equipment, including tanks, APCs, small arms, missiles and frigates to the Thai armed forces in the late 1980s. In the past few decades, Beijing has sought to further strengthen its military ties with Thailand, given the strategic importance of the Kingdom in the Asian neighbourhood. China has come to be reckoned as an important strategic partner of Thailand. The two countries hold joint military training exercises and are also partnering in developing equipment for the armed forces. Bilateral ties between the two sides also include cooperation in the areas of trade, infrastructure and energy. The two countries declared their "comprehensive strategic partnership" in 2012.

India: Toeing the policy of balancing its regional and foreign relations, Thailand has made efforts to deepen its ties with India in recent times. The two countries share maritime boundary in the Andaman Sea and hence, reckon each other's strategic importance in the region. While bilateral trade, economic and cultural ties between the two neighbours have flourished in the past, cooperation in the field of defence and security has also assumed significance. India has held regular bilateral naval exercises with Thailand since 1995. The navies are also regularly conducting coordinated maritime patrols in the Indian Ocean Region since 2005. The two countries signed a bilateral Memorandum of Understanding on Defence Cooperation in January 2012. India has also evinced interest in taking up infrastructure development projects with Thailand.

Vietnam: The two neighbours upgraded their partnership into a strategic one as recently as in June 2013. Vietnam thus became Thailand's first strategic partner among the ASEAN members. The two sides have also agreed to effectively carry out bilateral defence cooperation agreement while establishing a defence policy dialogue mechanism.

Australia: Thailand has good defence relations with Australia since World War II. The two countries continue to cooperate in the field by regularly holding bilateral military exchanges and high level defence visits. Thai armed forces also receive training in Australia.

Ukraine: Thailand and Ukraine share defence relations, with the latter extending military and technical cooperation to the Kingdom. The Thai Army has ordered the Ukrainian BTR-3E1 armoured vehicles. Thailand has also started assembling the vehicles locally. The two sides are also planning joint production of other Ukraine-designed vehicles.

Multilateral Alliances

ASEAN: The Association of Southeast Asian Nations (ASEAN), with its headquarters in the Thai capital Bangkok, has emerged as one of the most prominent regional multilateral blocks in Asia. The 10-member ASEAN, established on 8 August 1967 with the primary objective to promote regional trade, peace and stability along with social and cultural development of the region, has successfully pursued its goals and made great strides in achieving regional integration. The absence of any conventional military threat to Southeast Asia aided by the member countries' pledge to resolve all their bilateral issues through political and diplomatic means has further strengthened the ASEAN. The foreign policy of Thailand, which is one of the founding members of ASEAN, has been largely guided by the regional forum. Bangkok has made foremost efforts to transform the association into a regional economic stronghold and a vibrant marketplace. Moreover, having maintained an equation of bonhomie with its immediate and extended neighbourhood, in particular with China which is entangled in maritime dispute with several Southeast Asian littoral states, Thailand's importance in the forum as an arbitrator to resolve conflicts has amplified in recent times.

APEC: Thailand is one of the 12 founding members of the Asia-Pacific Economic Cooperation (APEC). The 21-member organisation, established in 1989 as a regional economic forum among the Pacific Rim nations, has solely focused on free and open trade and investment, promoting and accelerating regional economic integration and encouraging economic and technical cooperation among the member countries to build a dynamic and harmonious Asia-Pacific region.

East Asia Summit: Initially proposed as part of the ASEAN forum, the East Asian Summit, over the years, has evolved as a platform for strategic dialogue and cooperation on key issues in the East Asian region. The first meeting of the forum was held at Kuala Lumpur, Malaysia in 2005. Besides the 10 ASEAN members, the EAS also comprises of Australia, China, India, Japan, New Zealand, South Korea, the US and Russia.

BIMSTEC: The Bay of Bengal Initiative for Multi-Sectoral Technical and Economic Cooperation (BIMSTEC) is a regional block comprising of India, Bangladesh, Sri Lanka, Thailand, Myanmar, Nepal and Bhutan. The multilateral forum with the South Asian and Southeast Asian nations has 13 priority areas, including trade and investment, technology, energy, transport and communication among others. The block, originally named Bangladesh, India, Sri Lanka, and Thailand Economic Cooperation (BISTEC), was formed in 1997 to primarily promote economic and trade cooperation among the member nations. Myanmar, Nepal and Bhutan subsequently joined the forum which was then renamed as BIMSTEC.

DEFENCE CAPABILITIES

ARMY: The Royal Thai Army (RTA) in terms of personnel strength is the largest of all three wings of Thai armed forces, constituting over 60 percent of the total active force. The Army operates with the primary mission to protect the Kingdom's sovereignty and integrity from any internal and external threat. The land forces have played a crucial role in dealing with insurgency-related problems in the country's south. Thailand is also a regular contributor to the UN peacekeeping missions by sending troops for overseas operations. While focusing on its modernisation, the Thai Defence Ministry has emphasised on transforming the Army into "a compact and integrated force." Having an armoured inventory consisting of Cold War-era battle tanks, combat vehicles and other weapons, most of which are now obsolete, Thailand has inked deals with a number of countries including Ukraine and South Africa in last few years to acquire new tanks and armoured vehicles for its land forces. Alongside, it is also planning to acquire new artillery systems, multiple rocket launchers, missiles and helicopters as part of the Army Modernisation Plan: Vision 2020.

Personnel: 190,000

Equipment

Category	Name	In Service	On Order
MBT	M60 A1	53	
	M48A5	70	

Army continued...

Category	Name	In Service	On Order
MBT	M60 A3	125	
	T-69	110	
	T-59	24	
	BM Oplot	10	39
Light Tank	M41(light)	150 (reserve)	
	Stingray	100	
	Scorpion	144	
APC	V150 Commando	150	
	Type 85	600+	
	M113	340	
	WMZ-55B1	96	
	Reva protected Carrier	97	
	BTR-3EI 8X8		200+
Artillery	M109A5 155mm SP	20+	
	Caesar 155mm wheeled SP	6	
	M101 105mm	286	
	LG1Mk2 105mm	24	
	L119 105mm	22	
	M114	60	
	M198	50	
	Type 59 130mm	30+	
	75mm	300+	
Mortar	81mm towed	217	
	120mm towed	86	
	M106A3 120mm mortar carrier	12	
	M125A3 81mm	21	
MLRS	130mm	60	
	DTI-1 300mm	10	
RCL	57mm, 75mm & 106mm		
ATGW	Tow		
	Dragon		
	M901A3 ITOW	18	
SAM	I-Hawk		
	Redeye		
	Spada	1 battery	
	SA-18 Igla		
	Short Blowpipe		
AA gun	Vulcan SP	48	
	40mm	48	
	57mm	24	

Army Aviation

Category	Name	In Service	On Order
Light Transport Aircraft	C-212-300	2	
	Short 330	2	
	OV-10	75	
	Beech 99	1	

Army Aviation continued...

Category	Name	In Service	On Order
Light Transport Aircraft	T-41	23	
	King Air	1	
	JETSTREAM 41	2	
	ERJ-135	2	
Helicopter	Bell 206	12	
	UH-1H	50+	
	UH-60L	9	
	UH-60M	3	4
	UH-72A Lakota		6 (+9 more option)
	AS550 C3 Fennec	8	
	Bell 212/412	20	
	OH13	18	
	Bell 214 B	6	
	AH1F	3	
	Boeing 234 Chinook	3	
	TH300	45	
	Bell 206B	5	
	Mi-17V5	5	
	AW139	2	8

New Procurements/Upgrades

✦ An agreement has been signed with China in 2016 to buy MBT-3000 (export designation "VT4") main battle tanks from Chinese firm NORINCO. The initial order is for 28 platforms with more platforms likely to be procured in future.

MBT-3000 *A NORINCO photo*

✦ Delivery of Oplot MBTs, ordered from Ukraine's state arms exporter Ukrspecexport, is going on. Thailand has signed a deal for 49 Oplots.

✦ Over 200 BTR-3EI 8x8 armoured personnel carriers have been ordered from Ukraine with deliveries going on. In 2013, Ukroboronprom received another contract to deliver additional 21 BTR-3EIs to the Army. Local assembly of the APCs has also started in Thailand.

✦ 21 indigenously developed First Win (4x4) light armoured vehicles (LAVs) have been delivered with more orders expected to be placed.

✦ The Defence Technology Institute of Thailand has tied up with UK company Ricardo for the next phase of development of the indigenous Black Widow Spider 8×8 armoured vehicle programme.

✦ Five Mi-17V5 helicopters have been acquired from Russia. Acquisition of more such choppers is under consideration to replace the Boeing-made Chinook transport helicopters.

✦ Additional four UH-60M Black Hawk tactical transport helicopters are on order under the US's FMS programme.

✦ Delivery of six UH-72A Lakota multi-role helicopters from the US under FMS is going on. Thailand has made a further request for nine more Lakotas from the US to replace its aging fleet of Bell UH-1H helicopters.

✦ Eight AgustaWestland-built AW139 helicopters have been ordered in 2015 to perform transport and utility missions across Thailand.

Mi-17V5 helicopter of Thai Army

NAVY: To defend and protect over 3,000-km long national coastline, Thailand has built up a modern naval force capable of conducting combat missions. The Royal Thai Navy (RTN), indeed, is the only naval power in entire Southeast Asia having an aircraft carrier. Besides, it possesses a number of frigates, corvettes and other naval craft to carry out its missions. However, there are no submarines in its naval flotilla at present, but the country plans to purchase the underwater vessels in near future. The Government in 2015 announced a deal to acquire three Chinese submarines but later put it on hold. In 2016, the country again reiterated its decision to acquire the submarines by 2017.

The RTN mainly operates in the Gulf of Thailand, Malacca Strait and the Indian Ocean region. Though Thailand does not face any conventional external maritime threat and its only areas of focus are fighting piracy, illegal drug trade and related incidents, nevertheless, a naval modernisation programme has been undertaken in recent years to maintain the Kingdom's strategic parity with neighbouring countries and enhance its regional force projection capabilities. Acquisition of new frigates, offshore patrol craft, advanced radars and weapons, along with upgradation of existing fleet of naval vessels are being planned as part of the RTN's modernisation drive.

Major naval bases: Sattahip, Bangkok, Nakhon Phanom, Phang Nga, U-Tapao (naval air base)

Personnel: 56,000

Equipment

Category	Name	In Service
LHD	Chakri Nareubet	1
LPD	HTMS Ang Thong (Endurance class)	1
Frigate	Phutthayotfa class (ex US Knox class)	2
	Naresuan class (Type F52T)	2
	Chao Praya class (Chinese 053HT & 053HTH design)	4
Corvette	Rantanamkosin class	2
	Tapi class	2
Training vessel	Makut Rajakumarn	1
Patrol vessel	Pattani class	2
	Hua Hin class	3
	Khamrosin class	3
	Chonburi class gun FAC	3
	Ratcharit class missile FAC	2
	Prabparapak class missile FAC	3
	Sattahip class	6
	T991 class	3
	T91 class	9
	T81 class	3
	PC30 class	10
	T213	18
LST	Sichang class & ex US	6
LCU	Tongkaew + ex US	12
LSM	Ex-US	2
LCM+LCI	Ex-US	24+1
LCVP	Ex-US	12
Hovercraft	Grifon 100TD	3
Mine hunter	M48 class (to be upgraded)	2
Auxiliary ship		29
Weapon	RIM-7 Sea Sparrow SAM	
	RIM-162 Evolved Sea Sparrow SAM	
	Harpoon ASM	
	Exocet ASM	
	(Torpedo) Mk.32, Mk.46	

Marines

Personnel: 14,000

Equipment

Category	Name	In Service
MBT	M60 A3	5
APC	AAV71A	30+
	BTR-3EI 8X8	12 (on order)
	V-150 Commando	50+
Artillery	GC 45 howitzer	20+
	M101 105mm	30
	LG1Mk2 105mm	6

Naval Aviation

Equipment

Category	Name	In Service
Fighter/ Attack/ AWS role	A-7E Corsair II	14
	TA-7C	4
	AV-8S	7
	TAV-8V	2
	P-3T Orion	2
	UP 3T	1
	F27 Mk200/Mk400	4+2
Transport/ Surveillance	ERJ-135Lr	1
	CL215	2
	Nomad Searchmaster	5
	Do228	6
	Cessna 337	6
Helicopter	S-70B Sea Hawk	6
	MH-60S Sea Hawk	2
	UH-1H	3
	S-76B	4
	Super Lynx 300	2

New Procurements/ Upgrades

✦ The Navy in 2016 has put forth the funding plans to procure three S26T conventional submarines ((Yuan-class / Type-041) from China. The submarines will cost 36 billion baht (about US$1 billion) and the payment will be made under a 10-year instalment plan. The proposal to buy the Chinese subs was put on hold by the Thai Government in 2015, but it has now decided to make the acquisition by 2017.

✦ Government has approved 30bn Baht ($976 million) to buy two medium-sized frigates weighing 2,000 to 3,500 tons. The frigates will replace the two ex-US Knox-class ships. Following the approval, an order was placed with South Korea's Daewoo Shipbuilding & Marine Engineering Co. in August 2013 to acquire a 3,700-ton frigate. The warship will be procured at a cost of $470 million and will be delivered in August 2018. Construction work on the first frigate has started in 2016.

✦ Thai shipbuilder Marsun has delivered three M36 multi-purpose patrol vessels to the Navy in March 2014.

✦ Plans are on to acquire a second Endurance Class LPD from Singapore Technologies Marine Ltd. The first craft of the class is operational in the Navy since 2013.

✦ A new contract has been signed between Bangkok Dock and BAE Systems in 2016 to build a second offshore patrol vessel of the HTMS Krabi class which was commissioned into service in 2013. French company Thales has been awarded a contract in April 2016 to provide an integrated combat suite for the OPV.

✦ Two Naresuan-class guided-missile frigates have been upgraded by SAAB and Raytheon.

✦ Sweden's SAAB has upgraded the command and control system of aircraft carrier Chakri Naruebet which would allow communication between the platform and the Royal Thai Air Force's Gripen fighter jets and the airborne radar system Erieye, carried by the Saab 340 AEW.

✦ The US Government has approved the sale of 16 RIM-162 Evolved SeaSparrow Missile (ESSM) units and associated equipment to Thailand under foreign military sales. The missiles will be deployed on Chinese-built frigates being acquired by the RTN.

✦ Italy's Sistemi Integrati has been awarded a contract to supply the KRONOS radar system to the RTN. The system will equip the Royal Navy Air and Coastal Defence Command and will be deployed in an area in the South of Bangkok. It will be integrated within the air surveillance network of the RTN.

✦ First two of the five light-utility H145M (previously known as EC645 T2) helicopters have been delivered by Airbus Helicopters in April 2016.

H145M light utility helicopter *An Airbus Helicopters Photo*

AIR FORCE: The Royal Thai Air Force (RTAF) is structured and equipped to carry out conventional warfare as well as internal security missions. It is the smallest wing of Thai armed forces in terms of personnel strength and yet the most modern among them. The aerial inventory, which mostly consisted of the US and European fighter jets and other aircraft acquired in the 1980s and 1990s, is being revamped to accommodate new platforms, weapon systems and equipment in recent times. Procurement of European Gripen fighters, reconnaissance aircraft, command and control system and new military radar is going on. Alongside, refurbishing of existing platforms and systems are also being carried out to enhance the combat capabilities of the RTAF which is considered the third largest air force in Southeast Asia after Malaysia and Singapore.

Major air bases: Khorat, Takhli, Udorn, Surat Thani

Personnel: 45,000

Equipment

Air Force continued...

Category	Name	In Service
Fighter/ Attack/ AWS role	F16A/ B	66
	F-5E/ F	23 (10 in store)
	Gripen C/ D	12
	Alpha Jet	25
Trainer	Pilatus PC-9	23
	L-39ZA/ MP Albatros	30
Transport/ Surveilla-nce	C130H-30/J	12 + 6
	A-310-300	1
	Boeing 737-300ER	1
	ATR72-500 (VIP)	4
	N-22B Nomad	16
	BT-67 Basler	6
	CN-235	6
	Merlin IV	3
	Learjet	4
	Saab 340	2
	Sukhoi 100	3 (on order)
Helicopter	S-92 Hawk	3
	Bell 412	15
	H225M	4
	EC725	4 (on order)
	Bell UH-1H	15
Weapon	IRIS-T AAM	
	AIM-9 Sidewinder AAM	
	AIM-120 AMRAAM	
	Python 3 AAM	
	RBS 15-F ASM	

New Procurements/ Upgrades

✦ The Integrated Air Defence System comprising 12 Gripen fighters, two Erieye AEW aircraft and a Saab 340 transport plane, has been commissioned in Sept 2013.

✦ The Air Force has projected the requirement for another six Gripen jets by next decade to supplement the existing fleet of 12 fighters.

✦ A contract worth US$110 million has been signed with Korean Aerospace Industries in 2015 to acquire four T-50 trainer jets having light attack capability. The new jets will be delivered by 2018 and replace the aging fleet of L-39ZA jet trainers.

✦ Lockheed Martin has started delivery of the upgraded F-16A/ B jet fighters to the RTAF under a contract awarded to it in 2012. The firm is carrying out mid-life upgrade (MLU) on the early models of F-16A/ B jet fighters of the RTAF, totalling 18.

✦ Plans are on to replace the aging fleet of Lockheed Martin C-130H Hercules transport aircraft with new planes.

✦ Two more EC725 tactical transport helicopters have been ordered in Oct 2016 with delivery scheduled for 2019. The Air Force has already ordered two such rotorcraft in 2014 with delivery expected in 2016-17. The new choppers will be used for search-and-rescue/combat search-and-rescue (SAR/CSAR) missions as well as troops transport and other roles.

✦ The Lockheed Martin-built TPS-77 ground-based phased array radar has been activated to enhance and expand the RTAF's long-range air surveillance network capability. The radar was ordered in 2007.

✦ Additional AN/TPS-78 air defence and surveillance radar systems are on order from US's Northrop Grumman.

DEFENCE DEALS

Joint Venture/ Collaboration Programmes

China: Thailand and China have agreed to jointly develop multiple rocket launchers with a guidance system. Under the new agreement, the Thai Defence Technology Institute (DTI) would work with China to develop new multiple rocket launchers called "DTI-1G (Guided)" having a higher range than existing systems. Thailand has successfully tested the new rocket launcher in 2015. In an earlier joint deal, the two countries developed the DTI-1 system.

UK: The Bangkok Dock has designed and developed the offshore patrol vessel, HTMS Krabi, for the Royal Thai Navy under a transfer of technology agreement with UK's BAE Systems inked in 2009. The state-run DTI has also joined hands with UK's Ricardo for the next phase of development of the indigenous Black Widow Spider 8×8 infantry fighting vehicle programme.

Defence Industry: The Thai armed forces have mostly relied on foreign military exports as the country lacks a robust defence industry. According to official figures, there are 48 defence-related industries operating under the Ministry of Defence. The defence firms are capable of producing small arms, artillery, rockets and other weapons primarily for the Royal Thai Army and small coastal patrol craft for the Navy. Dearth of defence related research and development programmes, which receive less than 0.1% of the overall defence budget, have led to an underdeveloped domestic defence sector. To augment national defence production, the Government has incorporated technology transfer clause in some of the recently signed military deals with foreign firms, including with Sweden's Saab and UK's BAE Systems. The present military dispensation has particularly focused on defence industrial collaboration with major arms importing nations to galvanize the indigenous defence production capability of Thailand. The Government-owned Defence Technology Institute, set up in January 2009, is also working on few areas of defence research & development programmes including rockets and missile systems, unmanned aerial vehicle (for Army), and vertical take-off and landing unmanned aerial vehicle (for Navy).

DEFENCE BUDGET

Thailand, since the past few years, has allocated a visible amount of its GDP for military expenditure, though when compared with few other Southeast Asian countries, it ranks low.

The annual defence budget of the Kingdom witnessed a northward trend between 2006 and 2011. While in 2010, it allocated 1.5% of its GDP towards defence expenses, in 2011, the defence budget was 1.6% of GDP, at $5.52 billion. In 2013, the government announced a defence spending worth 178.4 billion baht. Following the taking over of power by the country's military in 2014, defence spending has received a major boost with an increase of around 5%. Defence budget for the year 2015 was 193.07 billion baht, while in 2016, the expenditure rose to 206.4 billion baht, an increase of nearly 7% over previous year's spending. For 2017, a budget of 210.7 billion baht ($6 billion) has been approved, which is an increase of 2% over the previous year's expenditure. The country's budget amounts to around 1.5% of its GDP.

A comprehensive modernisation programme undertaken in recent years has led to the rise in defence budget by Thailand as the country is acquiring new fighter jets, military helicopters, armoured vehicles, naval craft and other weapons for its armed forces. The defence expenditure is expected to increase in the coming years as the country fights internal security problems and also intends to maintain strategic parity with other regional players.

CONCLUSION

In the absence of any conventional military threat, Thailand has mainly focused on its socio-economic development, though maintaining military edge in the region has also been prioritised by the Government. Driven by the larger goal of achieving self-reliance and securing its national interests, the Kingdom has adopted a flexible and balancing policy towards its larger Asian neighbours, India and China, even while continuing the deep-rooted relations with its traditional strategic ally, the US. Military acquisitions by the country, however, have been diversified in the past few years and do not exclusively rely on US-supplied weapons and systems. The evolving regional and global geostrategic equations have re-emphasised the strategic significance of Thailand once again in recent times with all the leading powers endeavouring to "win over" the Kingdom to their side.

CONTACT DETAILS

Ministry of Defence
The Department of Defence
Defence Town Hall Road,
Phra Nakhon, Bangkok 10200
Sanamchai Road
Tel: 0-2622-1972

TUNISIA
(Capital: Tunis)

INTRODUCTION

Area: 163,610 sq km
Population: 11,134,588 (July 2016 est.)
Coastline: 1,148 km
Maritime claims: Territorial sea: 12 nm
Contiguous zone: 24 nm
Exclusive Economic Zone: 12 nm

KEY POLITICAL PERSONS

PRESIDENT & COMMANDER-IN-CHIEF: Beji Caid Essebsi

PRIME MINISTER: Youssef Chahed

KEY DEFENCE PERSONS

MINISTER OF DEFENCE: Farhat Horchani

CHIEF OF STAFF OF ARMY: Gen. Ismail Fathalli

CHIEF OF STAFF OF NAVY: Rear Adm. Abderraouf Atallah

CHIEF OF STAFF OF AIR FORCE: Brig. Gen. Bechir Bedoui

ECONOMY

Tunisia's diverse, market-oriented economy has long been cited as a success story in Africa and the Middle East, but it faces an array of challenges following the 2011 revolution. Following an ill-fated experiment with socialist economic policies in the 1960s, Tunisia embarked on a successful strategy focused on bolstering exports, foreign investment, and tourism, all of which have become central to the country's economy. Key exports now include textiles and apparel, food products, petroleum products, chemicals, and phosphates, with about 80% of exports bound for Tunisia's main economic partner, the European Union. Tunisia's liberal strategy, coupled with investments in education and infrastructure, fuelled decades of 4-5% annual GDP growth and improving living standards. Since its establishment in late 2014, Tunisia's new government has faced challenges reassuring businesses and investors, bringing budget and current account deficits under control, shoring up the country's financial system, lowering high unemployment, and reducing economic disparities between the more developed coastal region and the impoverished interior. In 2015, successive terrorist attacks against the tourism sector and worker strikes in the phosphate sector, which combined account for nearly 15% of GDP, slowed growth to less than 1% of GDP.

GDP: $127 billion (2015 est.); $126 billion (2014 est.)

Real Growth Rate: 0.8% (2015 est.); 2.3% (2014 est.)

Industries: Petroleum, mining (particularly phosphate and iron ore), tourism, textiles, footwear, agribusiness, beverages

Total Exports: $14.07 billion (2015 est.); $16.84 billion (2014 est.)

Export Commodities: Clothing, semi-finished goods and textiles, agricultural products, mechanical goods, phosphates and chemicals, hydrocarbons, electrical equipment

Major Markets: France 28.5%, Italy 17.2%, Germany 10.9%, Libya 6.1%, Spain 4.2% (2015)

Total Imports: $19.1 billion (2015 est.); $23.4 billion (2014 est.)

Import Commodities: Textiles, machinery and equipment, hydrocarbons, chemicals, foodstuff

Major Suppliers: France 19.4%, Italy 16.4%, Algeria 8.2%, Germany 7.4%, China 6% (2015)

INTERNATIONAL DISPUTES

Tunisia has no known international disputes.

DEFENCE

ARMY

Personnel: 25,000+

Equipment

Category	Name	In Service
MBT	M60A1	24
	M60A3	50
	Kurassier (tank destroyer)	54
Armoured car	Saladin	22
	AML-90	20
	AML-60	10
	EE-9 Cascavel	15
APC	M113A2	100+
	EE11 URUTU	18
Artillery	M-101A1/A2 (105mm)	45
	M-114 (155mm) towed	12
	M-198 (155mm) towed	50
	M108 105mm SP	10
	M109 155mm SP	18
Mortar	81mm	100
	107mm	50+
	120mm	
SAM	Chaparral	
	RBS-70	
ATGW	SS-11	
	MILAN	
	TOW	
ATK weapon	STRIM-89	
AA gun	M55 20mm	100
	M-1939/Type 55 37mm	

New Procurements/ Upgrades

✦ Delivery of Kirpi (4x4) mine-resistant ambush-protected (MRAP) vehicles from Turkey is going on.

NAVY

Personnel: 4,500

Equipment

Category	Name	In Service
Fleet	Ex-German ALBATROS class FAC (Type 143B)	6
	La Galite class missile FAC (Combattante III type)	3
	BIZERTE class large patrol craft (P48 Type)	3
	UTIQUE class large patrol craft (Chinese HAIZHUI Type)	3
	ISTIQLAL class patrol craft	4
	TAZARKA class attack craft	2

Coast Guard

Category	Name	In Service
Fleet	RAS EL BLAIS class patrol craft	5
	SBEITLA class inshore patrol craft (ex-GDR BREMSE class)	5
	Light patrol boats	10

New Procurements/ Upgrades

✦ Two patrol boats have been received from the US.

✦ ScanEagle UAVs are being procured from the US.

AIR FORCE

Personnel: 3,000

Equipment

Category	Name	In Service
Fighter	F-5E/F	16
	MB326K/L	10
	L-59T	9
	SF-260WT	12
	SF-260C	6
Transport/ Surveillance	C130 B/H Hercules	2
	C-130J Super Hercules	2
	King Air	1
	L410	3
Helicopter	Alouette II, III	11
	UH-1H/ 1N	8
	Puma	1
	Ecureuil	6
	AB-205	15
	Gazelle	6

New Procurements/ Upgrades

✦ The US Government in May 2016 has approved the sale of surplus Bell OH-58D Kiowa Warrior scout and light attack helicopters along with weapon systems and other equipment for a price of US$ 100.8 million.

OH-58D Kiowa Warrior light attack helicopter

✦ The US State Department has approved the sale of 12 Sikorsky UH-60M Black Hawk helicopters to the Tunisian Air Force. The deal, worth $700 million, also includes equipment, parts, training and logistical support.

✦ Under a US FMS deal, Northrop Grumman has been awarded a contract to carry out avionics upgrade for the Tunisian Air Force's 12 F-5 supersonic fighter aircraft fleet. The work is expected to be completed by 2018.

Defence Expenditure

Estimated defence spending in terms of GDP: 1.4%

Leading arms suppliers are the US and Europe.

Defence Production and R&D

The domestic defence industry of Tunisia has very few manufacturing facilities. The country does not undertake any major defence research and development activities.

Defence Procurement

Tunisia plans to modernise its armed forces as most of its military equipment is obsolete. Following its transition to democracy in 2014, the country's military has received military aid and weapons, mostly from the US.

CONTACT DETAILS

Ministry of National Defence
Boulevard Bab Mnara
Tunis 1008, Tunisia
Website: www.defense.tn

STRATEGIC INFORMATION

TURKEY

(Capital: Ankara)

INTRODUCTION

Area: 783,562 sq km

Population: 80,274,604 (July 2016 est.)

Coastline: 7,200 km

Maritime claims:

Territorial sea: 6 nm in the Aegean Sea; 12 nm in Black Sea and in Mediterranean Sea

Exclusive Economic Zone: in Black Sea only: to the maritime boundary agreed upon with the former USSR

GEOPOLITICAL IMPORTANCE

Turkey, known officially as the Republic of Turkey, is a Eurasian country located in Western Asia and in East Thrace in South-eastern Europe. Turkey is bordered by eight countries: Bulgaria to the northwest; Greece to the west; Georgia to the northeast; Armenia, Azerbaijan and Iran to the east; and Iraq and Syria to the southeast. The Mediterranean Sea and Cyprus are to the south; the Aegean Sea to the west; and the Black Sea is to the north. The Sea of Marmara, the Bosphorus and the Dardanelles (which together form the Turkish Straits) demarcate the boundary between East Thrace and Anatolia; they also separate Europe and Asia.

POLITICAL OVERVIEW

Turkey is a democratic, secular, unitary, constitutional republic with an ancient cultural heritage. The President of the Republic is the head of state and has a largely ceremonial role. Executive power is exercised by the Prime Minister and the Council of Ministers which make up the government, while the legislative power is vested in the unicameral parliament, the Grand National Assembly of Turkey. The judiciary is independent of the executive and the legislature, and the Constitutional Court is charged with ruling on the conformity of laws and decrees with the constitution. The Council of State is the tribunal of last resort for administrative cases, and the High Court of Appeals for all others.

KEY POLITICAL PERSONS

PRESIDENT: Recep Tayyip Erdogan

PRIME MINISTER: Binali Yildirim

ECONOMY

Turkey has the world's 17th largest GDP by PPP and 18th largest nominal GDP. The country is among the founding members of the OECD and the G-20. Turkey's largely free-market economy is increasingly driven by its industry and service sectors, although its traditional agriculture sector still accounts for about 25% of employment. An aggressive privatisation programme has reduced state involvement in basic industry, banking, transport, and communication, and an emerging cadre of middle-class entrepreneurs is adding dynamism to the economy and expanding production beyond the traditional textiles and clothing sectors. The automotive, construction, and electronics industries are rising in importance and have surpassed textiles within Turkey's export mix. The economy of Turkey is largely developed. The country is among the world's leading producers of agricultural products; textiles; motor vehicles, ships and other transportation equipment; construction materials; consumer electronics and home appliances. In recent years, Turkey had a rapidly growing private sector, yet the state still plays a major role in industry, banking, transport, and communications. Global economic conditions and tighter fiscal policy caused GDP to contract in 2009, but Turkey's well-regulated financial markets and banking system helped the country weather the global financial crisis and GDP rebounded strongly to 8.2% in 2010, as exports returned to normal levels following the recession. Turkey's public sector debt to GDP ratio has fallen to roughly 40%. Further economic and judicial reforms and prospective EU membership are expected to boost Turkey's attractiveness to foreign investors. However, Turkey's relatively high current account deficit, uncertainty related to monetary policy-making, and political turmoil within Turkey's neighbourhood leave the economy vulnerable to destabilising shifts in investor confidence. Turkey has many modern defence armament manufacturers.

GDP: $1.589 trillion (2015 est.), $1.53 trillion (2014 est.)

Real Growth Rate: 3.8% (2015 est.), 2.9% (2014 est.)

Industries: Textiles, food processing, autos, electronics, mining (coal, chromate, copper, boron), steel, petroleum, construction, lumber, paper

Natural gas - proved reserves: 5.012 billion cu m (1 January 2016 est.)

Crude oil - proved reserves: 300 million bbl (1 January 2016 est.)

Total Exports: $152 billion (2015 est.), $168.9 billion (2014 est.)

Export Commodities: Apparel, foodstuffs, textiles, metal manufactures, transport equipment

Major Markets: Germany 9.3%, UK 7.3%, Iraq 5.9%, Italy 4.8%, US 4.5%, France 4.1% (2015)

Total Imports: $200.1 billion (2015 est.), $232.5 billion (2014 est.)

Import Commodities: Machinery, chemicals, semi-finished goods, fuels, transport equipment

Major Suppliers: China 12%, Germany 10.3%, Russia 9.9%, US 5.4%, Italy 5.1% (2015)

DEFENCE & SECURITY

The Turkish Armed Forces is the second largest standing armed force in NATO, after the US Armed Forces, with a combined strength of just over a million uniformed personnel serving in its five branches. Turkey is considered to be the strongest military power of the Middle East region besides Israel. The Turkish Armed Forces consists of the Army, the Navy and the Air Force. The Gendarmerie and the Coast Guard operate as parts of the Ministry of Internal Affairs in peacetime, although they are subordinated to the Army and Navy Commands respectively in wartime, during which they have both internal law enforcement and military functions. The Chief of the General Staff is the Commander of the Armed Forces. In wartime, he acts as the Commander in Chief on behalf of the President, who represents the Supreme Military Command of the TAF on behalf of the Grand National Assembly of Turkey.

Forces deployed abroad: Afghanistan (1,700 troops in ISAF) , Cyprus (35,000 troops), East Timor (UNMIT), Haiti (MINUSTAH), northern Iraq (1,000 troops), Israel (TIPH), Ivory Coast (UNOCI), , Lebanon (UNIFIL), Liberia (UNMIL), Sudan (UNMIS), Somalia (CTF-151).

KEY DEFENCE PERSONS

MINISTER OF DEFENCE: Fikri Isik

COMMANDER OF ARMED FORCES: Gen. Hulusi Akar

COMMANDER OF LAND FORCES: Gen. Salih Zeki Colak

COMMANDER OF NAVY: Adm. Bulent Bostanoglu

COMMANDER OF AIR FORCE: Gen. Abidin Unal

COMMANDER OF GENDARMERIE: Gen. Galip Mendi

Internal Conflict

2016 failed coup attempt in Turkey: In July 2016, a coup was attempted in Turkey with military tanks in the city of Ankara, and Istanbul. Fethullah Gulen Terrorist Organisation (FETO) was blamed for last months failed coup to topple President Tayyip Erdogan in Turkey. Later, members of Turkey's armed forces took control of the country as explosions, gunfire and a reported air battle between loyalist forces and coup supporters erupted in the capital. Turks also took to the streets, marching and waving Turkish flags to show support to Turkey's President and his government. Erdogan, after the situation became completely under control, dismissed the military action as an attempt at an uprising by a minority within the armed forces, and a probe has been initiated in this regard.

Turkey – Kurdistan Workers' Party conflict: It is an armed conflict between the Republic of Turkey and various Kurdish insurgent groups, which have demanded separation from Turkey to create an independent Kurdistan, or to have autonomy and greater political and cultural rights for Kurds inside the Republic of Turkey. The main rebel group is the Kurdistan Workers' Party or PKK, which is considered a terrorist organisation by Turkey and the United States and was on the European Union terrorist list, but was removed due to court order. Year 2011 showed a sharp increase in violence in recent history of the Kurdish–Turkish conflict. In 2012, the conflict with the PKK took a violent curve. The PKK's recent campaign against ISIS has caused many politicians to reconsider the PKK's position as a terrorist organisation. The conflict between Turkey and PKK escalated following the 2015 Suruc bombing attack. Later, the PKK allegedly carried out a revenge attack. During the Operation Martyr Yalçın, Turkey bombed alleged PKK bases in Iraq and PYD bases in Syria. The conflict then escalated, with pro-PKK Kurdish organisations staging attacks across the country, and Turkish forces attacks in the form of aerial bombardments and military operations.

External Conflict

- ✦ Complex maritime, air, and territorial disputes with Greece in the Aegean Sea. The Aegean dispute is a set of interrelated controversial issues between Greece and Turkey over sovereignty and related rights in the area of the Aegean Sea.
- ✦ The Cyprus dispute is the result of the ongoing conflict between the Republic of Cyprus and Turkey, over the Turkish occupied northern part of Cyprus. Also, the status of Northern Cyprus has become a recurrent issue especially during the recent talks for Turkey's membership of the EU where the division of the island is seen as a major stumbling block in Turkey's long road to membership.
- ✦ Syria and Iraq protest Turkish hydrological projects to control upper Euphrates waters.
- ✦ Turkey has expressed concern over the status of Kurds in Iraq. Recently, in response to a series of attacks by Kurdish rebels, Turkey has deployed active military operations against them. Kurdistan Workers Party (PKK) has threatened Ankara with an open war in case of continuation of previous policies aimed at infringement of the national interests of the Kurds.
- ✦ Border with Armenia remains closed over Nagorno-Karabakh.
- ✦ On 2 September 2011, Turkey downgraded diplomatic ties with Israel to second secretary level and suspended military co-operation between the countries as the UN released its report into Israel's controversial raid of a Gaza-bound ship. As of 2 September 2011, all military agreements between Turkey and Israel have been suspended. Turkey is currently planning to appeal the International Court of Justice in The Hague as soon as next week in order to probe the legality of Israel's naval blockade on the Gaza Strip.
- ✦ Turkey's government has recently requested the deployment of NATO's Patriot surface-to-air missiles on to bolster its defences along its border with Syria and prevent a spill over of the civil war in that nation. NATO allies installed the long-range Patriot batteries on Turkish territory twice before, during the 1991 and 2003 Iraq wars. They were never used and were quietly withdrawn a few months later. Around 600 foreign troops are expected to accompany six Patriot missile systems to be deployed in Turkey to reinforce the NATO member country's air defence.

THREAT PERSPECTIVE

The Kurdistan Workers' Party (*Parti Karkerani Kurdistan*), commonly known as PKK, is a major threat looming in the heart of Turkey. Turkey labelled the organisation as an ethnic secessionist organisation that uses terrorism and the threat of force against both civilian and military targets for the purpose of achieving its political goal. The main aim of the Kurdish terrorist groups in Turkey has been to establish an independent Kurdish state on the south-eastern part of Turkey which is heavily populated by people with Kurdish ethnicity.

Turkey has been fighting against terrorism since its establishment. During its establishment, it faced three threats. Each of the three threats, (1) religious fundamentalism, (2) ethnic (Kurdish) separatism and (3) Soviet (communist) domination somewhat manifested itself in terms of terrorist attacks after the establishment of the Republic. The terrorist organisations in Turkey were strengthened in the 1960s and 1970s especially during the student movements.

Radical Islamist terrorist groups in Turkey are also a dominating threat to the country. The most important Islamist terrorist group in Turkey is the Turkish Hizbullah which has no apparent connection with the Lebanese Hizbullah. After the 2003 Istanbul Bombings were linked to Al-Qaeda, Turkey deployed troops to Afghanistan to fight Taliban forces and Al-Qaeda operatives, with the hope of dismantling both groups. Turkey's responsibilities include providing security in Kabul, as well as in Wardak Province, where it leads PRT Maidan Shahr. Recently both sides have witnessed immense violence with the insurgency escalating in 2012.

Turkey's Kurdish conflict is becoming more violent, with more than 700 dead in fourteen months, the highest casualties in thirteen years including a wide-spread campaign of kidnapping, suicide bombings and terrorist attacks by the PKK. The Turkish government, meanwhile, claims to have killed hundreds of PKK fighters in recent months, both in operations in south-eastern, predominantly Kurdish-populated Turkey and during air raids against suspected PKK camps in the mountains of northern Iraq. 2015-2016 saw eruption in Kurdish–Turkish conflict between various Kurdish insurgent groups and the Turkish government. The 2015 conflict between Turkey and the PKK broke out following two year-long peace negotiations, which began in late 2012. According to the Turkish government, between July 2015 and May 2016, huge number of Kurdish rebels were killed in Turkey and in northern Iraq, while many were killed among Turkish security forces, as reported in local media.

STRATEGIC RELATIONS

Turkey has strengthened its relations both with

the US and the European countries; while further developing its relations with the countries in the Balkans, Middle East and North Africa, Southern Caucasus, South Asia and Central Asia by making use of its close ties. Turkey is also deepening its policies of reaching out to Sub-Saharan Africa, Latin America and Asia-Pacific regions that is home to many emerging powers.

US: US-Turkish relations focus on areas such as strategic energy cooperation, trade and investment, security ties, regional stability, counterterrorism, and human rights progress. Relations were strained when Turkey refused in March 2003 to allow US troops to deploy through its territory to Iraq in Operation Iraqi Freedom, but regained momentum steadily thereafter and mutual interests remain strong across a wide spectrum of issues. Turkey also seek advanced military equipment from US sources, particularly with respect to fighter and drone aircraft, helicopters, and missile defence systems. US also continue to provide cooperation on counterterrorism, law enforcement, and military training and education. U.S. exports to Turkey include aircraft, iron and steel, agricultural products, oil, cotton yarn and fabric, and machinery. US imports from Turkey include vehicles, machinery, iron and steel and their products, agricultural products, travertine, and marble. Syrian Civil War has deteriorated to some extent the relationship shared between Turkey and US especially in regard to Kurdish fighters. Ankara aims to push both IS and Kurdish fighters away from its border. US termed the clashes in places where so-called Islamic State (IS) was not present were a source of deep concern.

Russia: The warm bilateral relations of the past two decades have been severely strained after the November 2015 jet shoot down incident, when a Turkish F-16 combat aircraft shot down a Russian Su-24 fighter aircraft during an airspace dispute close to the Turkish-Syrian border. In response, Russia imposed a number of economic sanctions on Turkey. After the dissolution of Soviet Union, relations between Turkey and Russia dramatically and strongly improved. Russia is Turkey's second largest trade partner with the 2012 bilateral trade figure expected to reach USD 35 billion. The countries aim to bring this figure to USD 100 billion. Russia is the biggest and the most reliable supplier of natural gas to Turkey. The leaders of Russia and Turkey have signed a series of major trade and energy deals in the Turkish capital on December 2014 amid differences over the crises in Syria and Ukraine. In 2016, the normalisation of relations between both the countries began with Erdogan expressing regret to Putin for the downing of the Russian aircraft. The trade ties has since been normalised between both the countries.

Egypt: Egypt and Turkey are bound by strong religious and historical ties, and diplomatic ties between the two have remained extremely friendly. Turkey is planning to sign strategic cooperation agreements with Egypt in the military, diplomatic and commercial fields, after the relationship between Turkey and Israel strained due the 'Palmer Report' on Gaza Strip. Both countries are full members of the Union for the Mediterranean. Turkey and Egypt are among the leading countries of both the Middle East and Mediterranean regions, and are often said to be key in Middle East peace.

Pakistan: Turkey is a close ally of Pakistan. Both the countries have maintained long-standing military ties. Both Turkey and Pakistan are Muslim-majority states and share extensive cultural and geopolitical links. Turkey has been supplying a variety of arms, military equipment to Pakistan. In 2016, Pakistan and Turkey have reaffirmed commitment to transform their special relationship into a strong strategic partnership by signing a joint communique.

India: Turkey and India have undergone an immense social, political and economic transformation since the end of the Cold War. Bilateral relations between both the countries are characterised by warmth and cordiality. They share common values including commitment to secularism and democratic principles. Bilateral relations have been strengthened by the exchange of visits of leaders of both countries in recent times. A Protocol on setting up a Joint Working Group on Counter Terrorism was signed in September, 2003.

Multilateral Alliances

Turkey has become increasingly integrated with the West through membership in organisations such as the Council of Europe, NATO, OECD, OSCE and the G-20 major economies. Turkey began full membership negotiations with the European Union in 2005. The country is also Turkic Council, Joint Administration of Turkic Arts and Culture, Organisation of Islamic Cooperation and the Economic Cooperation Organisation. Turkey, a member of NATO, was granted dialogue partner status in the Shanghai Cooperation Organisation (SCO) at the group's 2012 summit in Beijing. Turkey is also granted chairman of the energy club of SCO for 2017.

NATO: Turkey has maintained forces in international missions under the United Nations and NATO since 1950, including peacekeeping missions in Somalia and former Yugoslavia, and support to coalition forces in the First Gulf War. Turkey also pledged five ships and a submarine as part of its contribution to the March 2011 military intervention in Libya following UN Security Council Resolution 1973 of 2011. Turkey is a non-permanent member of the United Nations Security Council since 1 January 2009. Turkey has attached the utmost importance to NATO's role in maintaining security and stability in the Euro-Atlantic area and in providing a forum for political-military consultations on topics of interest to its members.

DEFENCE CAPABILITIES

ARMY: The Turkish Army or Turkish Land Forces is the main branch of the Turkish Armed Forces responsible for land-based military operations. The land force has a total active manpower of 402,000, including 325,000 conscripts with approximately 258,700 troops in reserve. The army is by far the largest of the three service components. The structure of the Turkish Army has historically had two facets: operational and administrative. The operational chain consists of the field fighting formations, and the administrative the arms and service branches - infantry, armour, artillery etc.

Personnel: 314,000 (800,000 reserves)

Equipment

Category	Name	In Service
MBT	Leopard 1	400
	M60 Patton	923
	Leopard 2	350
	Altay	(250 on order)
	M-48 Patton	750
APC/ MICV/IFV	ACV-15	2000+
	Mortar carrier	178
	TOW carrier	48
	M113	3162
	BTR-80	520
	Cobra MRAP	1000+
	ZPT	1200
	Kirpi MRAP	400
Artillery	105 mm towed M52	250
	105 mm towed M101A1	70
	J-600T Yildirim	20
	203 mm towed M115	159
	155 mm towed Panter	200+
	155 mm M44T	150
	175 mm M107 SP	36
	155 mm M114	500
	155 mm Firtina	200+
	M110 A2	219
Mortar	60 mm	5,342
	81 mm	
	107 mm	
	120 mm	
MLRS	T-300 Kasirga	60
	122 mm T-122 Sakarya	100+
	T-107	80+
	ATACMS rockets	72
Anti-Tank GW	MILAN	400+
	BGM-71 TOW	390+
	ERYX	500

Army continued...

Category	Name	In Service
Anti-Tank GW	Mizrak-U	
	Roketsan Cirit	
Recoilless rifle	57 mm	
	75 mm	3,295
	105 mm	
	Carl Gustaf 84 mm	
SAM	FIM-92 Stinger	3000
	9K38 Igla	30
AA Gun	40 mm Bofors	700+
	Atilgan	70
	Zipkin	78
	35 mm towed	90
	40 mm M42A1 Duster	200

Army Aviation

Category	Name	In Service
Helicopter	CH-47 Chinook	5 (5 on order)
	T/S-70 Black Hawk	45
	AS532 Cougar	20
	AH-1 P/S/W Cobra	40+
	T-129 ATAK	10 (59 on order)
	UH-1H	50
	205 JetRanger	55
Aircraft	T-41	24
	U-17A	71
	T-42	3
	Super King Air 200	5
UAV	Heron, Anka, Firebee, Harpy, CL-89, Bayraktar Mini	

New Procurements/ Upgrades

✦ Ankara is in talks with Moscow regarding the purchase of S-400 long-range air defence systems.

✦ The works on the Hisar national air defence system are continuing at full speed. The Hisar project, expected to be finished by 2020. Aselsan and Roketsan are the prime contractor and subcontractor of the Hisar project.

✦ Menatek has signed the contract for the first 23 ammunition vehicle's Brake System of the Turkish Army.

✦ For Turkey's Next Generation APC, the FNSS Defence Systems is offering the Pars while the Otokar Corporation is offering the Arma.

✦ Turkish Armed Forces have recently received their first T129 ATAK advanced attack and tactical reconnaissance helicopter. This delivery is the first of 59 helicopters, 9 in T129A Early Delivery Helicopter (EDH) configuration and 50 in T129B configuration.

✦ The Undersecretariat for the Defence Industries of Turkey (SSM) has contracted Otokar with the designing of first National Main Battle Tank of Turkey- ALTAY Project. The Turks' official goal is to design, test, and built the first Altay tank in 6.5 years. The deal includes technology transfer worth USD 330 million dollars and the production of 4 prototypes worth USD 70 million dollars. Turkey reportedly plans to produce at 200-250 of the tanks in Turkey.

✦ The Turkish Army has purchased 170 upgraded M60A1's from the Israel Military Industries (IMI). The tanks were upgraded to M60T configuration. IMI was the prime contractor for the USD 687.5 million project, considered to be one of the world's largest tank upgrade programmes, bringing the M60A1 tank to level of the world's leading main battle tanks.

✦ Turkish Army will soon be receiving 336 Firtina and 138 Panter Self Propelled Artillery systems. A requirement has been formulated.

✦ Turkey has selected US's Sikorsky Aircraft Corp. for a contract to co-produce 109 derivative Black Hawk multi-mission helicopters. Under the contract, valued at worth USD 3.5 billion, Sikorsky along with other American and Turkish companies will provide components for the new fleet of T-70 choppers which will be based on the S-70i Black Hawk.

✦ Sikorsky, a Lockheed Martin Company, along with the Government of the Republic of Turkey and members of the Turkish aerospace industry have activated contractual agreements under the Turkish Utility Helicopter Programme (TUHP) to produce the T70 utility helicopter based on Sikorsky's successful multi-role Black Hawk utility aircraft. Deliveries of T70 aircraft could begin as early as 2021 to six Turkish agencies: the Land Forces, Air Force, Gendarmerie, Special Forces, National Police, and the Directorate General of Forestry.

NAVY: The Turkish Navy is one of the three branches of the Turkish Armed Forces. The mission of the Turkish Navy is national defence, although it also participates in coordinated security and humanitarian efforts as part of Black Sea Naval Cooperation Task Group (BLACKSEAFOR). The Turkish Navy maintains several Marines and Special Operations units. These include the Amphibious Marines Brigade (Amfibi Deniz Piyade Tugayi), several commando detachments and two special operations forces. The Turkish Navy is the eighth largest navy in the world. It currently has approximately 55,000 active personnel. The Turkish Navy is also the third largest navy in Europe, in terms of the total displacement of its fleet, which is 258,948 tonnes.

Personnel: 50,000 (55,000 reserve)

Equipment

Category	Name	In Service
Submarine	Gur Class Type 209T2/1400	4
	Preveze Class Type 209T1/1400	4
	Atilay class Type 209/1200	5

Navy continued...

Category	Name	In Service
Frigate	Barbaros Class	4
	Gabya Class	8
	Yavuz Class	4
Corvette	Ada Class	3
	Burak Class	6
Fast Attack Craft	Ruzgar Class	3
	Kilic Class	9
	Yildiz Class	2
	Kartal Class	2
	Dogan Class	4
Patrol Boat	Tuzla-class	8
Mine Warfare Force	Seydi Class	4
	Aydin Class	5
	Engin Class	3
Auxiliary ship	Training	2
	Submarine support ships	2
	Fleet replenishment tankers	2
	Tankers (fleet oil tanker)	3
	Transports	10
	Hydrographic ships	2
Amphibious Force	Osman Gazi Class LST	1
	Ertugrul Class LST	1
	Yeni Class	1
	Landing craft tank (LCT)	27
	Landing Craft Mechanized (LCM)	17

Naval Aviation

Category	Name	In Service
Fixed wing aircraft	CN-235 (Maritime/Transport)	5
	TB-20	7
	ATR 72–600	3+
Helicopter	S-70B SeaHawks	24
	AB-204	3
	AB-212 ASW	10

Turkish corvette TCG Heybeliada from aboard the guided missile destroyer USS Ross (DDG 71) in the Mediterranean Sea. *Image Courtesy: defenseimagery.mil, US*

New Procurements/ Upgrades

✦ Turkish Navy is planning for the acquisition of Multi-Functional Helicopter

Platform Dock (LHD). LHD Project includes LCMs, AAVs, LCVPs, guidance purposed Commander Boat and RHIBs. LHD shall be capable of carrying helicopters, amphibious vehicles including tanks and AAV, and an Amphibious Marine Battalion. The LHD contract has been signed on 1st of June 2015 and the project work has been started in September 2015. It is planned that LHD will be taken into service by April 2020.

✦ The MiLGEM project is the name of Turkish national warship programme. A total of twelve MiLGEM class warships (eight corvettes and four frigates) will be built for the Turkish Navy. The objectives of ADA class corvette project are to enhance the littoral warfare capabilities.

✦ A credit agreement of USD 2.18 billion euro was signed for six submarines to be manufactured by naval shipyards in Turkey with assistance from Germany. The credit will be used to finance manufacturing of six AIP (air-independent propulsion) technology U-214TN submarines in Turkey. According Undersecretariat for Defence Industries of the Turkish Government these submarines will be co-produced with maximum local content at Golcuk Naval Shipyard in Kocaeli, Turkey.

✦ Turkey signed an agreement with Alenia Aeronautica, a Finmeccanica company, for ten ATR 72 ASW aircraft valued at USD 219 million.

✦ The construction of two special operations crafts for the Turkish Naval Forces' Underwater Attack Units (SAT) to take place under New Type SAT Boat Project. Undersecretariat For Defence Industries has issued the tender in September 2015.

✦ Mid-life modernisation of AY Class submarines in accordance with technology.

✦ Mid-life modernisation program is planned for Preveze class submarines.

✦ TF-2000 project will carry Turkish Naval Forces to the new generation Anti-Air Warfare (AAW) frigate. Currently, preparations of Design Phase of the regional air defence frigate is underway.

✦ All Gabya (Ex-Perry) Class frigates furnished with GENESIS have been integrated with combat management system. 4 Barbaros and 4 Gabya Class frigates are gaining ESSM (Evolved SeaSparrow Missile) launching capability as well as 3D radar integration.

Turkish Navy corvette TCG Buyukada. *Image Courtesy: Turkish Navy*

✦ Turkish Naval Forces started the procurement process for a Submarine Rescue Ship (MOSHIP). The first ship of Rescue and Towing ships, ISIN, is planned to be in service on July 2016 and MOSHIP is planned to be in service on July 2017.

AIR FORCE: The Turkish Air Force is one of the oldest air forces in the world and operates one of the largest combat aircraft fleets of NATO. The Turkish Air Force was one of the first air forces in the world to be established. Supported by the TuAF's in-flight refuelling capability, the fighter aircraft of the Turkish Air Force can participate in international operations and exercises on every major continent and return back to their home bases. The Turkish Air Force currently has over 930 different aircraft and it is the third largest air force in NATO, in terms of its fleet size.

Personnel: 65,000 (35,000 reserve)

Equipment

Category	Name	In Service
Fighters/ attack/ Reconna-issance/ bomber	F-4E/ 2020	49
	F-16C	196
	F-16D	44
	NF5/2000	16
	NF-5B	16
	Boeing 737-700 AEW&C	4
Transport/ Tanker	C-160D	12
	C130 Hercules	8
	C130B	13
	CN-235	50
	A400M	2 (8 on order)
	C.550 Citation II (VIP)	2
	Gulfstream IV	1
	CN-235	39
	KC-135R	7
Helicopter	UH-1H Iroquois	15
	Cougar	20
Trainer	T-38M	33
	T-41 Mescalero	25
	T37	58
	SF-260D	35
SAM	Rapier 2000	86
	Hawk XXI	16
	FIM-92 Stinger	108
	Atilgan PMADS	32
AAM	AIM-120 (A/B/C-7) AMRAAM, AIM-7(E/F) Sparrow, AIM-9 Sidewinder	
AGM	SOM, AGM-84, AGM-88 HARM, AGM-65 Maverick, AGM-154 JSOW, SLAM	
UAV	RQ/MQ-1, GNAT, TAI Anka	

At present, the Turkish Air Force operates MALE UAVs such as the TAI Anka, the IAI Heron and the I-GNAT ER. The Turkish Air Force is procuring 10 MQ-1 Predator and 4 MQ-9 Reaper MALE UAV/UCAVs in the short term.

Boeing KC-135R Stratotanker of the Turkish Air Force

An F-4E Phantom II aircraft with the Turkish Air Force takes off from Third Air Force Base Konya, Turkey, during Exercise Anatolian Eagle. *Image Courtesy: SAC Helen Farrer RAF Mobile News Team/MOD, defenceimagery.mod.uk*

New Procurements/ Upgrades

✦ Turkey's F-16C/D fleet will receive a bolstered electronic warfare capability, after Harris was awarded a contract to provide more defensive aids suites for the air force. Awarded by the US Air Force Air Logistics Center at Robins AFB in Georgia, the deal will lead to Harris providing 21 ALQ-211 advanced integrated defensive electronic warfare suite (AIDEWS) pods for integration on the Turkish air force's F-16 fleet. The contract also covers maintenance, spares and engineering support.

✦ The US Government has approved a possible sale of AIM-120C-7 Advanced Medium Range Air-to-Air Missiles (AMRAAM) to Turkey under the Foreign Military Sales programme. The Defence Security Cooperation Agency (DSCA) had notified the Congress of the possible sale and the US State Department has approved the deal, estimated to be worth USD 320 million. Turkey had sought to buy 145 AIM-120C-7 AMRAAM systems, 10 missile guidance sections, and 40 LAU-129 launchers, containers, support equipment, spare and repair parts along with other related items.

✦ Turkish Aerospace Industries (TAI) has rolled out the first locally built primary and basic trainer aircraft Hurkus designed for the Turkish Air Force.

✦ Turkey is a level 3 partner of the F-35 Joint Strike Fighter (JSF) development programme and has agreed to initially purchase 116 F-35A Lightning II aircraft. Turkey is expecting to purchase a second pair of F-35 Joint Strike Fighter Lightening jets. The country will be ordering a second pair of the jointly-made, next-generation, stealth fighter F 35 Joint Strike Fighter Lightning II aircraft, as it did earlier this year following the production of the first two.

✦ Turkey is also a partner nation in the Airbus A400M programme. Although the A400M is essentially a heavy tactical lift aircraft, it can also be transformed into a tanker aircraft for aerial refuelling at short notice. The TuAF (Turkish Air Force) has ordered a total of ten A400M aircraft.

✦ A total of four Boeing 737 AEW&C MESA Peace Eagle aircraft (together with ground support systems) were ordered by the Turkish Air Force, with an option for two more aircraft. Turkish Aerospace Industries (TAI) is the primary subcontractor for the Peace Eagle parts production, aircraft modification, assembly and tests. Another subcontractor, Havelsan, is responsible for system analysis and software support.

✦ Turkey is planning acquire new military/intelligence satellites in the short term. These include a 0.8m resolution reconnaissance satellite (Project Gokturk-1) for use by the Turkish Armed Forces and a 2m resolution reconnaissance satellite (Project Gokturk-2) for use by the Turkish National Intelligence Organisation. Göktürk-2 is currently being produced by the Turkish Aerospace Industries, while Gokturk-1 is in the tender stage. Some electro-optical parts that are required for the Gokturk-1 (0.8m resolution) satellite are beyond TAI's technological know-how, thus a foreign partner is being sought. The official bidders for the project are EADS Astrium (UK), OHB-System (Germany) and Telespazio (Italy).

✦ Havelsan of Turkey and Boeing of the United States are in the process of developing a next generation, high altitude ballistic missile defence shield. It is envisaged that the system will be used by the Turkey, US and other NATO member states.

✦ Turkish Aerospace Industries, Inc. (TAI) and Lockheed Martin unveiled the last Turkish-built F-16 at a ceremony held at TAI's facilities in Ankara. The Turkish Air Force has received 30 new, advanced Block 50 models.

TURKISH GENDARMERIE: The Gendarmerie of the Republic of Turkey, which is responsible for the maintenance of safety and public order as well as carrying out other duties assigned by laws and regulations, is an armed security and law enforcement force, having military nature. As a part of Turkish Armed Forces, the General Command of the Gendarmerie is subordinated to the General Staff in matters related to training and education in connection with the Armed Forces and to the Ministry of Interior in matters related to the performance of the safety and public order duties.

Equipment

Category	Name	In Service
APC	Kirpi	150
	Cobra	100+
	V-150S	75
	BTR-60	
	Condor	
Helicopter	UH-60	20
	UH-1	10
	Mil Mi-17	10+

Turkish Gendarmerie soldiers from Aviation Unit during a military exercise. *Image Courtesy: General Command of Gendarmerie*

Major Armaments Producers

Domestic suppliers: MKEK, ASELSAN, BMC, FNSS, GIRSAN, TISAS, Havelsan, Transvaro, TAI, Otokar, Roketsan, TUBiTAK, Golcuk Naval Shipyard.

Foreign suppliers: United States, Germany, China.

Other defence and aerospace companies: Howaldtswerke-Deutsche Werft (HDW), Lockheed Martin, BAE Systems, Makina ve Kimya Endüstrisi Kurumu (MKEK), Tusas Engine Industries, Inc (TEI), FNSS Defence Systems Co, Istanbul Shipyard.

The Undersecretariat for Defence Industries has made significant achievements in building the blocks for a modern national defence industry in Turkey, with notable results in certain vital areas. Turkey's Defence Industry Executive Committee is the main decision making body of the system. It is chaired by the Prime Minister and consists of the Chief of General Staff and the Minister of National Defence.

Turkish Aerospace Industries, Inc. (TAI) is the center of technology in design, development, manufacturing, integration of aerospace systems, modernisation and after sales support in Turkey. ASELSAN is a Turkish corporation which produces tactical military radios and defence electronic systems for the Turkish Army. The Mechanical and Chemical Industry Corporation (MKEK) established in 1950, is a reorganisation of government-controlled group of factories in Turkey that supplied the Turkish Army with supplies, munitions and equipment. The corporation mainly produces equipment for the Turkish Armed Forces, such as the ammunition for small arms and heavy weapons, heavy weapons and artillery systems, aerial bombs, mines, explosives, and rockets.

Defence Industry Manufacturers Association, SaSaD was established by 12 companies in 1990 with the support of Ministry of National Defence, Turkey. SaSaD is the association for producers of defence systems and equipment for domestic and international markets. As of 2010, SaSaD encompasses more than 120 members and associate members. SaSaD's members include government-owned companies, such as the Machinery and Chemicals Industry, or MKEK, and the military's naval shipyards; firms owned by the Turkish Armed Forces Foundation, including Aselsan, Turkish Aerospace Industries (TAI), Havelsan and Roketsan; and private companies.

DEFENCE DEALS

Arms Export

Turkish defence industry exports rose 35 percent in the first two months of 2016. The country had made significant progress in developing weaponised drones. The country's defence industry is making rapid strides with the mass production of locally-developed infantry rifles and joint production of the aircraft being started in 2016. Turkey will spend up to USD 8 billion in defence purchases as its exports will reach USD 2 billion in 2016, according to a major estimation by the procurement agency, the Under secretariat for Defence Industries (SSM). The expectations in the SSM's updated 2012-2016 strategic programme are realistic given the money Turkey would pay for expensive systems – such as the F-35s or the U-214 submarines from Germany – over the next few years, as well as the rapid increase in its exports mainly to Islamic countries, according to one defence analyst.

Turkey is in talks with four key foreign suppliers on a USD 4 billion Long Range Air and Missile Defence Systems project. The country's mainly exports armoured vehicles of many sorts, rockets and other ammunition, as well as military electronics like radios, to more than 10 Islamic countries. It also sells aviation equipment as part of offset deals. The Turkish government is making determined efforts to promote its indigenously produced armaments. The Turkish defence industry plans to promote its military products at various international defence and security exhibitions in future.

Lockheed Martin, which is competing to supply the Turkish navy with six Salih Reis class frigates, is gridlocked over Turkish demands on some of the technical aspects of the vessels' missile integration. Some Turkish navy officials also oppose mounting AN/ SPY-1 naval radar systems on the frigates, preferring instead to use the indigenously developed Cok Amaclı Faz Dizinli Radar system rather than the US system. The Turkish navy already operates two Salih Reis class frigates, which offer protection from aerial threats and improve the reconnaissance and early warning capabilities of the naval forces. Turkey is co-producing with AgustaWestland a total of 60 T-129 attack helicopters, Turkish versions of the A-129 Mangusta International, for billions of dollars. Turkey earlier also signed a USD 400 million contract with US Boeing to buy six CH-47D heavy-lift helicopters.

Turkish firm under the Ministry of Defence Production collaborated with Pakistan Navy and Karachi Shipyard and Engineering Works (KS&EW) in the design and construction of the 17,000 tonne naval fleet tanker. Pakistan will also receive 34 trainer-cum- fighter aircraft from Turkey free of cost under a major defence deal inked between the two countries. The agreement was reached during the meeting of Pakistan-Turkey High Level Military Dialogue Group which concluded in Ankara.

Arms Import

Turkey's military procurement consists almost exclusively of imports. The majority of weapons, including advanced systems, are imported from the major arms producers. Turkey had the second largest defence budget (after Saudi Arabia) in the Middle East. Turkey is rapidly expanding its military forces by domestically producing many weapons and importing many more. Turkey is using its imports to build up indigenous arms industries by requiring as part of tenders that foreign companies transfer technology, capital and production capabilities to Turkey. Turkey is still reliant on foreign imports for about 35 per cent of critical military technologies.

Russia is planning to develop a long-range air defence complex based on S-300 SAM jointly with Turkey. Russia is ready to offer as part of the tender a joint Russian-Turkish product based on the Antey-2500 system (the export version of the S-300 system). The advanced version of the S-300 missile system, called S-300PMU1, has a range of over 150 kilometres and can intercept ballistic missiles and aircraft at low and high altitudes, making it effective in warding off air strikes, the news report added. The S-300V/Antey 2500 (SA-12 Gladiator/Giant) consists of a new command vehicle, an array of advanced radars and up to six loader vehicles assigned to each launcher. Turkey in 2013 was entering negotiations with the China Precision Machinery Export-Import Corporation (CPMIEC) to buy its first long-range anti-missile system, estimated at USD 4 billion. In 2015 November Turkey scrapped that competition, saying two local companies would instead build the air defence system.

DEFENCE BUDGET

Total defence spending: $18.3 billion (2016)

Estimated defence spending in terms of GDP: 2.2% (2016)

Turkish government has approved nearly USD 6 billion in defence programme. The Turkey's Defence Industry Executive Committee is planning to increase Ankara's own military production capabilities. USD 4.5 billion of the total budget will consist of domestic production.

The SSM Undersecretariat for Turkish Defence Industries has undergone major modernisation projects to achieve their 2023 defence export target of USD 25bn by 2023. Turkey's defence budget saw USD 1.1 billion rise for F.Y. 2015, reaching around TL52.5 billion. 57 percent of the ministry's budget in 2015 will go to personnel spending and 41 percent to acquisitions of goods and services. Turkish government has earlier raised the share of defence budget in its 2014 economic programme. According to Ankara, the budget allocated for national defence and security rose from 45.3 billion liras for this year's budget to 49.6 billion liras for next year, with a 9.5 percent increase. The mid-term economic outlook announced this week spared 21.8 billion liras for the National Defence Ministry, 16.6 billion liras for the Turkish National Police, 452 million liras for the Turkish Coast Guard Command, 6.2 billion liras for the Gendarmerie General Command, 1 billion liras for the National Intelligence Service and 3.5 billion liras for the Ministry of Internal Affairs. After 2020, Turkey is planning to design, develop and produce another fighter plane to close the JSF' s deficiencies either by itself, or through its other, most likely with South Korea. Few years back, Turkish defence companies spent USD 228 million on research and development from their own assets which is a major increase from USD 64 million in 2004 and USD 120 million in 2007.

Turkey's official defence policy is based on Ataturk's principle of 'peace at home, peace in the world'. Turkey's newly published defence industrial strategic plan 2012-2016 will modernise the country's critical technologies, ensure the indigenous development of defence capabilities and progressively reduce dependency on foreign imports. The 2010 budget was 26 billion liras. The total value of the contracts signed in 2011 was some USD 27.3 billion. According to the new plan the total turnover target for the defence industry exports for 2016 is USD 2 billion. The Law on the Court of Accounts provides for external ex-post audits of armed forces' expenditure. It also paves the way for audits of extra budgetary resources earmarked for the defence sector, including the Defence Industry Support Fund.

Turkey's Under secretariat for Defence Industries projects that Turkey will spend up to USD 8 billion in defence purchases as its exports will reach USD 2 billion in 2016. Turkey, besides F-35 aircraft, is planning to purchase U-214 submarines from Germany. The Ministry of National Defence (MND) budget is supplemented by the budgets for the Gendarmerie, Coast Guard and Turkish Defence Industries Under-Secretariat (SSM). Turkish MOD Budget increased 12.2 Billion Dollars in 2011. MND spending accounts for around 70 per cent of the total budget, the Gendarmerie around 17 per cent and SSM procurement 10-12 per cent. The defence ministry is also planning to increase production of indigenous weapons systems.

Turkey's Defence Industry Under secretariat (SSM) is responsible for the modernisation of Turkish Armed Forces and development of defence industry and is currently pursuing 250 projects worth of USD 26 billion. 24 percent of these projects are domestic development, 57 percent is joint production, ten percent is direct procurement and nine percent is consortium. Turkey's Defence Industry Undersecretary is Murad Bayar. The country has bought 50 Husky Visor 2500 US-built mine detection vehicles in a deal worth USD 115 million to use operations to clear the south-eastern Turkish border areas to clear mines planted by Kurdistan Workers Party guerrillas.

CONCLUSION

Turkey's Armed Forces, whose geopolitical and geo-strategic importance is ever increasing, is composed of the Land Forces, Air Forces and Navy that are directly affiliated to the Turkish General Staff. The General Command of the Gendarmerie and the Coast Guard Command serving as elements of inland security during peacetime take their due positions under Land Forces and Navy in wartime. The main elements of Turkish Defence Politics are the commitment towards national defence, NATO solidarity and the Turkish Armed Forces. The Turkish Armed Forces collectively rank as the second largest standing military force in NATO, after the US Armed Forces, with an estimated strength in 2015 of 639,551 military, civilian and paramilitary personnel. In recent years Turkey is standing out as one of the growth defence markets in Asia. The Turkish military perceived itself as the guardian of Kemalist ideology, the official state ideology, especially of the secular aspects of Kemalism. The TAF still maintains an important degree of influence over the decision-making process regarding issues related to Turkish national security, albeit decreased in the past decades, via the National Security Council. The country's defence expenditure is expected to continue growing, largely be driven by a strained relationship with neighbouring countries and Ankara's aspirations in becoming a leading power in the region, along with the need to meet the threat of persistent terror threats and instability within the region. Peacekeeping initiatives will also continue to put pressure on the government to increase defence expenditure. Turkish Armed Forces, till date, has undertaken and is still undertaking peacekeeping and other operations under the aegis of the UN and NATO in order to settle the crises in the Balkans, Caucasus, Middle East, Africa and Asia.

CONTACT DETAILS

Ministry of National Defence
Askerî Yargtay Bakanligi
Ankara Turkey
Tel: (+90 312) 410 65 10
Fax: (+90 312) 418 22 59
E-mail: asyargitay@msb.gov.tr
Website: www.msb.gov.tr

Undersecretariat of Defence Industries (SSM)
Nasuh Akar Mah.
Ziyabey Caddesi 1407.
Sokak No:4 (06520) Balgat
Ankara, Turkey
Tel: +90 (312) 411 90 00 - 400
Fax: +90 (312) 411 93 86
E-Mail: posta@ssm.gov.tr
Website: www.ssm.gov.tr

Defence Industry Manufacturers Association (SASAD)
Paris Caddesi Yazanlar Sok. No: 4/106
06680 Kavaklidere, ANKARA
TURKIYE
Tel: +90 312 426 22 55
Fax: +90 312 426 22 56
Email: sasad@tr.net
Website: www.sasad.org.tr

Turkish Aerospace Industries, Inc.
Fethiye Mahallesi,
Havacılık Bulvarı No:17 06980 Kazan-ANKARA / TÜRKİYE
Tel : +90 312 811 1800
Fax: +90 312 811 1425
Marketing
E-mail: marketing@tai.com.tr
Procurement
E-mail: procurement@tai.com.tr
Website: www.tai.com.tr

ASELSAN
PK 1, 06172, Yenimahalle / Ankara, Turkey
E-mail: marketing@aselsan.com.tr
Website: www.aselsan.com.tr

MKEK
Tandogan 06330 ANKARA
TURKEY
Tel: +90 (312) 296 10 00-09
Website: www.mkek.gov.tr

TURKMENISTAN
(Capital: Ashgabat)

INTRODUCTION

Area: 488,100 sq km
Population: 5,291,317 (July 2016 est.)
Coastline: landlocked
Maritime claims: None

KEY POLITICAL PERSONS

PRESIDENT: Gurbanguly Berdymukhammedov

KEY DEFENCE PERSONS

DEFENCE MINISTER: Maj. Gen. Yaylym Berdiev

ECONOMY

Turkmenistan is largely a desert country with intensive agriculture in irrigated areas, and significant gas and oil resources. The country possesses the world's 4th largest reserves of natural gas and substantial oil resources. Cotton and wheat are the two largest agricultural crops. Turkmenistan is among the top ten producers of cotton in the world. In terms of natural gas reserves, it is ranked 4th in the world. High growth performance sustained over an extended period of time led to a steady increase in income levels and moved the country to an upper middle-income status (GDP per capita exceeded $6,000 by the end of 2012). The Government's National Socio-Economic Development Programme for 2011–2030 and the National Rural Development Programme focus on inclusive economic growth while preserving economic independence, modernizing the country's infrastructure, and promoting foreign direct investment.

GDP: $88.6 billion (2015 est.), $83.19 billion (2014 est.)

Real Growth Rate: 6.5% (2015 est.), 10.3% (2014 est.)

Industries: Natural gas, oil, petroleum products, textiles, food processing

Crude oil - proved reserves: 600 million bbl (1 January 2016 est.)

Natural gas - proved reserves: 7.504 trillion cu m (1 January 2016 est.)

Total Exports: $10.38 billion (2015 est.), $12.39 billion (2014 est.)

Export Commodities: Gas, crude oil, petrochemicals, textiles, cotton fiber

Major Markets: China 68.6%, Turkey 4.9% (2015)

Total Imports: $8.198 billion (2015 est.), $10.17 billion (2014 est.)

Import Commodities: Machinery and equipment, chemicals, foodstuff

Major Suppliers: Turkey 24.9%, Russia 12.3%, China 10.9%, UAE 9.1%, Kazakhstan 5.1%, Germany 4.6%, Iran 4.4% (2015)

INTERNATIONAL DISPUTES

- Field demarcation of the boundaries with Kazakhstan commenced in 2005, but Caspian seabed delimitation remains stalled with Azerbaijan, Iran, and Kazakhstan due to Turkmenistan's indecision over how to allocate the sea's waters and seabed. Bilateral talks continue with Azerbaijan on dividing the seabed and contested oilfields in the middle of the Caspian.

DEFENCE

The Armed Forces of Turkmenistan consists of an Army, Navy, Air and Air Defense Forces, Border Troops, and Internal Troops, and a National Guard. An agreement with Russia formalizes the presence of Russian troops in Turkmenistan.

ARMY

Personnel: 17,000

Equipment

Category	Name	In Service
MBT	T-72	700 (most in storage)
	T-90	10+
AIFV/APC	BMP-1/2	600
	BTR-60/70/80	700 (many in storage)
Artillery	D-20 152mm	125
	D-30 122mm towed	125
	2S3 152mm SP	12
MRL	BM-21 122mm	
	SMERCH 300mm	8

New Procurements/Upgrades

✦ Turkmenistan is expecting delivery of ten more T-90 MBTs to complete the re-equipment of a tank battalion.

NAVY

A naval force has been established for patrol duties on the Caspian Sea which includes two Russian built Type 21418 missile boats (Eldermen class) and Two Sobol class patrol boats. Other vessels in the inventory include: Kalkan-type patrol boats, ten GRIF-T multi-purpose launches and four Dearsan YTKB-type patrol boats (Arkadag class).

AIR FORCE

Personnel: 3,000

Equipment

Category	Name	In Service
Helicopter	MiG-29A/UB	24
	MiG-23	60+
	MiG-25	12
	Su-25	
	Mi-24	10
	Mi-8	10
	S-92	2
Transport	An-26	
Helicopter	AW101	2

New Procurements/Upgrades

No major procurement plans are under consideration at present.

Defence Expenditure

Exact defence expenditure figures are unavailable

Defence spending in terms of GDP: 1.6% (est.)

Defence Production and R&D

Turkmenistan does not undertake any defence production activities.

Defence Procurement

Turkmenistan Army/Navy/Air Force have not made any major defence procurements in recent years.

CONTACT DETAILS

Ministry of Defence
Galkynysh str. 4
744000 Ashgabat
Turkmenistan
Tel: +99-312-3516666, -357948
Fax: +99-312-511430, -397745

UGANDA

(Capital: Kampala)

INTRODUCTION

Area: 241,038 sq km
Population: 38,319,241
Coastline: 0 km (landlocked)
Maritime claims: none (landlocked)

KEY POLITICAL PERSONS

PRESIDENT: Yoweri Kaguta Museveni

PRIME MINISTER: Ruhakana Rugunda

KEY DEFENCE PERSONS

DEFENCE MINISTER: Mwesige Adolf

CHIEF OF DEFENCE FORCES: Gen. Katumba Wamala

COMMANDER OF THE LAND FORCES: Maj. Gen. David Muhoozi

COMMANDER OF THE AIR FORCE: Maj. Gen. Sam Turyagyenda

ECONOMY

Uganda's economy generates export income from coffee, tea, fish, and other products. Uganda has substantial natural resources, including fertile soils, regular rainfall, small deposits of copper, gold, and other minerals, and recently discovered oil. Uganda has never conducted a national minerals survey. Agriculture is the most important sector of the economy, employing over 80% of the work force. Coffee accounts for the bulk of export revenues. The policy changes are especially aimed at dampening inflation and boosting production and export earnings. Uganda's GDP growth is still relatively strong due to past reforms and sound management of the downturn. Oil revenues and taxes will become a larger source of government funding as oil comes on line in the next few years. Instability in southern Sudan is the biggest risk for the Ugandan economy in 2011 because Uganda's main export partner is Sudan and Uganda is a key destination for Sudanese refugees. The global economic downturn has hurt Uganda's exports.

GDP: $79.88 billion (2015 est.), $76.05 billion (2014 est.)

Real Growth Rate: 5% (2015 est.), 4.9% (2014 est.)

Industries: Sugar, brewing, tobacco, cotton textiles; cement, steel production

Oil - proved reserves: 2.5 billion bbl (2014)

Natural gas - proved reserves: 14.16 billion cu m (1 January 2014 est.)

Total Exports: $2.667 billion (2015 est.), $2.725 billion (2014 est.)

Export Commodities: Coffee, fish and fish products, tea, cotton, flowers, horticultural products; gold

Major Markets: Rwanda 10.7%, UAE 9.9%, Democratic Republic of the Congo 9.8%, Kenya 9.7%, Italy 5.8%, Netherlands 4.8%, Germany 4.7%, China 4.1% (2015)

Total Imports: $4.911 billion (2015 est.), $5.1 billion (2014 est.)

Import Commodities: Capital equipment, vehicles, petroleum, medical supplies; cereals

Major Suppliers: Kenya 16.4%, UAE 15.5%, India 13.4%, China 13.1% (2015)

INTERNATIONAL DISPUTES

✦ Uganda is subject to armed fighting among hostile ethnic groups, rebels, armed gangs, militias, and various government forces that extend across its borders.

✦ Uganda hosts 209,860 Sudanese, 27,560 Congolese, and 19,710 Rwandan refugees, while Ugandan refugees as well as members of the Lord's Resistance Army (LRA) seek shelter in southern Sudan and the Democratic Republic of the Congo's Garamba National Park.

DEFENCE

ARMY

Personnel: 38,000

Equipment

Category	Name	In Service
MBT	T-55	90
	T-90	40+
	T-34	7
Light Tank	PT76s	10
APC	BMP-2s	20
	BTR 40/152	20
Artillery	M1942 76mm	55
	M1938 122mm	18
Mortar	82mm	
	Soltam 120 mm	
ATGW	SAGGER	
AA Gun	ZSU23 23 mm	
	M1939 37 mm	
SAM	SA7	

New Procurements/ Upgrades

No major procurement plans are under consideration at present.

NAVY

Personnel: 200

Equipment

Category	Name	In Service
Patrol vessel	Yugoslav AL8K Type	3

New Procurements/ Upgrades

No major procurement plans are under consideration at present.

AIR FORCE

Personnel: 1300

Equipment

Category	Name	In Service
Fighter	MiG-21 MF	5
	MiG-21 Fishbed	6
	Su-30MK2	6
	L-39 ALBATROS	3
	AS-202 BRAVO	1
Transport	Y121V	2
	L-100 HERCULES	1
	GULSTREAM II	1
Helicopter	AB306	2
	AB205	3
	AB212	2
	Mi24PN	6

New Procurements/ Upgrades

✦ Uganda received another two Su-30MK fighters from Russia, after taking delivery of the first aircraft in July, 2011. The country has six on order. Uganda's ministry of defence paid USD 446m as a first part payment to an 'unidentified supplier' from Russia for fighters as well as tanks. In total, Uganda bought between six and eight fighters and other military equipment from Russia, worth USD 744 million.

Defence Expenditure

Total defence spending: $985 million (est.)

Estimated defence spending in terms of GDP: 1.25% (2015)

Foreign suppliers: Russia, China, United States, Italy and Poland

Defence Production and R&D

Uganda runs Luwero Industries Limited, a subsidiary of the National Enterprise Corporation (NEC) established in 1989 by an Act of Parliament, which acts as a commercial arm for the defence forces under the Ministry of Defence. NEC redeploys soldiers who were involved in the war to produce goods and services for the army and use the profit for soldier's welfare. Uganda produces its arms at the country's only production centre located at Nakasongola. The government is rethinking to modernise and professionalise its armed forces, including the procurement of modern equipment.

Defence Procurement

Future procurement possibilities were hindered by the government's obligation to its donors to not spend more than 2 per cent of GDP on defence. Uganda manufactures some ammunition and mainly depends on Eastern Europe for heavy weaponry. Second-hand equipment from the former Warsaw Pact states is likely to continue to dominate procurement through unorthodox channels. Indonesia has invited Uganda to purchase weapon systems such as bulletproof vests, helmets, boots and uniforms. Uganda has shown interest in procuring Indonesian-made weapons.

CONTACT DETAILS

Ministry of Defence

Republic House

PO Box 3798, Kampala

Uganda

Tel: +25641270331

UKRAINE

(Capital: Kyiv)

Crimea Republic is now a part of Russian Federation according to the Federal Constitutional Law dated 21.03.2014 No. 6-FCI/FKZ (Version 31.12.2014) "On accession of Crimean Republic to Russian Federation and on formation of new constituent entities of Crimean Republic and Sevastopol-the city of Federal status." (valid with amendments and addenda from 01.01.2015).

INTRODUCTION

Area: 603,550 sq km

Population: 44,209,733 (July 2016 est.)

Coastline: 2,782 km

Maritime claims: Territorial sea: 12 nm

Exclusive Economic Zone: 200 nm

Continental shelf: 200 m or to the depth of exploitation

KEY POLITICAL PERSONS

PRESIDENT & SUPREME COMMANDER-IN-CHIEF OF THE ARMED FORCES: Petro Poroshenko

PRIME MINISTER: Volodymyr Groysman

KEY DEFENCE PERSONS

DEFENCE MINISTER: Gen. Stepan Poltorak

COMMANDER-IN-CHIEF OF THE ARMED FORCES: Col. Gen. Viktor Muzhenko

COMMANDER OF THE ARMY: Lt. Gen. Serhii Popko

COMMANDER OF THE NAVY: Vice Adm. Ihor Voronchenko

COMMANDER OF THE AIR FORCE: Col. Gen. Serhii Drozdov

ECONOMY

The economy of Ukraine is an emerging free market. The country was greatly affected by the economic crisis of 2008 and as a result a 15.1% decrease in Ukraine's GDP took place over 2008 and 2009. It depends on imports to meet about three-fourths of its annual oil and natural gas requirements and 100% of its nuclear fuel needs. After a dispute that saw gas supplies cut to Europe, Ukraine agreed to 10-year gas supply and transit contracts with Russia in January 2009 that brought gas prices to "world" levels. The strict terms of the contracts have further hobbled Ukraine's cash-strapped state gas company, Naftohaz. Outside institutions - particularly the IMF - have encouraged Ukraine to quicken the pace and scope of reforms to foster economic growth. Due to the War in Donbass, Ukraine's economy shrank by 6.8% in 2014; and further declined in 2015. The World Bank forecast a growth of 1% in 2016. Ukraine's dependence on Russia for energy supplies and the lack of significant structural reform have made the Ukrainian economy vulnerable to external shocks.

GDP: $339.5 billion (2015 est.), $376.7 billion (2014 est.)

Real Growth Rate: -9.9% (2015 est.), -6.6% (2014 est.)

Industries: Coal, electric power, ferrous and nonferrous metals, machinery and transport equipment, chemicals, food processing

Total Exports: $35.5 billion (2015 est.), $50.76 billion (2014 est.)

Export Commodities: Ferrous and nonferrous metals, fuel and petroleum products, chemicals, machinery and transport equipment, food products

Major Markets: Russia 12.7%, Turkey 7.3%, China 6.3%, Egypt 5.5%, Italy 5.2%, Poland 5.2% (2015)

Total Imports: $38.94 billion (2015 est.), $58.24 billion (2014 est.)

Import Commodities: Energy, machinery and equipment, chemicals

Major Suppliers: Russia 20%, Germany 10.4%, China 10.1%, Belarus 6.5%, Poland 6.2%, Hungary 4.2% (2015)

INTERNATIONAL DISPUTES

Crimean Peninsula: Former President Viktor Yanukovych's backtracking on a trade and cooperation agreement with the EU in November 2013 in favour of closer economic ties with Russia led to a three-month protest occupation of Kyiv's central square. Later a military intervention by Russia in the Ukrainian autonomous region of Crimea resulted in the control of Crimea by Russia on 18 March 2014. Crimea is now a part of Russia and Russian authorities claim it as Russian territory while the Ukrainian Government asserts that Crimea remains part of Ukraine.

Dispute over the boundary between Russia and Ukraine through the Kerch Strait and Sea of Azov remains unresolved despite a December 2003 framework agreement. The boundary delimitation treaty (1997) with Belarus is yet to be ratified due to unresolved financial claims.

DEFENCE

The overall strategic situation around Ukraine and the events that took place in 2014 have required the government to revise its approaches towards protecting the sovereignty, territorial integrity and independence of the state. A new military doctrine has been developed. Ukraine has scrapped compulsory military service for young men and brought back military conscription by Presidential decree.

ARMY

Personnel: 2,50,000

Equipment

Category	Name	In Service
MBT	T-80	8
	T-64B	
	T-72	486
	T-84 Oplot	
AIFV/APC	BMP-2	
	BMP-1	
	BTR-60/70/80	
	BRDM-2	1,903
	BTR-3/4	
	D-20 152mm	
	2A36 152mm towed	
	152mm SP	
MRL	BM-21 122mm	
	130mm	342
	220mm	
	300mm	
Mortar	120mm	180
SSM	SS-21 Tockha	
	Tockha-U	12
ATGW	AT-4, 5, 6	
SAM	SA-4, 6, 8, 11, 12, 15	

New Procurements/Upgrades

✦ 200 Dozor-B protected carriers are on order

✦ 75 Saxon wheeled APC are being transferred from the UK surplus.

✦ The Ukrainian military has received a first batch of eight modernised T-80 main battle tanks (MBTs) which would be delivered to a mobile armoured unit of the country's airborne forces.

The Ukrainian military T-80 modernised main battle tanks. *Image credit: Ukraine MoD*

NAVY

Personnel: 10,000

Equipment

Category	Name	In Service
Frigate	Hetman Sahaydachniy class (Krivak III types)	1
	Vinnytsia (Grisha class)	1
Patrol craft	Khmelnits KII corvette (Pauk type)	1
	Zhuk class	1
	Pridniproviya class missile corvette (Tarantul type)	2
	Matka class	1
Mine warfare force	Yevgenya class	1
	Melitopol minehunter (Sonya class)	1
Amphibio-us force	Konstantin class LST (Ropucha type)	2
	Kirovograd LST (Polnochny type)	1
	Pomornik type LCAC	2
Auxiliaries	Slavutych command ship (Bambuk type)	2
	Donbass (Amur type)	1

New Procurements/Upgrades

✦ A programme has been launched for a new class of frigates (Project 58250) to replace the Lutsk and Vinnicia class corvettes.

✦ The lead ship Volodymyr Velykyj is under construction.

✦ Nine Hyurza-M patrol cutters are under construction.

Ukrainian Navy frigate Hetman Sahaydachniy (U 130) participates in a mine counter measure exercise during Exercise Sea Breeze 2012 (SB12). *Image Credit: US Navy*

AIR FORCE

Personnel: 36,100

Equipment

Category	Name	In Service
Fighter	Su-24M	10
	Su-27	18
	MiG-29	20
	Su-25	10
Recce	Su-24MR	5+
Transport	IL-76MD	5
	An-70	1
	Tu-134	18
	An-2	
	An-26	

Air Force continued...

Category	Name	In Service
Helicopter	Mi-24	24
	Mi-8	40+
	Mi-2	8
Trainer	Yak-52	5
	Aero L-39	20+
Air Defence	S-300V	
	2K12 Kub	
	S-300PS	70+
SAM	SA-11	
	SA-5	
	SA-10	
AAM	R-73, R-27, R-60	
ASM	Kh-25, Kh-29	
UAV	RQ-11, Tu-143, Tu-141	

New Procurements/Upgrades

✦ Plans have been formulated for the procurement of a first batch of An-7-transports.

✦ The MiG-29s and Su-25s are being progressively upgraded to the MiG-29MU1 and Su-25M1 configuration.

Defence Expenditure

Total defence spending: $4.3 billion (2016)

Estimated defence spending in terms of GDP: 4% (2015)

Domestic Suppliers: Ukroboronprom

Foreign Suppliers: US, UK, European Union, Canada

According to Ministry of Defence of Ukraine spokesman Oleksii Chernobai, the Defence Ministry budget for 2016 makes up UAH 55.5 billion (about USD 2.22 billion), or 2.45 percent of GDP. Thus, compared to last year the budget has increased by UAH 7.6 billion (about USD 304 million), or 16 percent. 20 percent of the allocated funding (or UAH 11.4 billion) are earmarked for the upgrade of weapons and materiel, and 6 percent or UAH 3.2 billion — for the training of troops. The administrative support expenditures will amount to UAH 6.5 billion, UAH 1 billion is envisaged for the payment of one-time allowances to servicemen and members of their families, UAH 700 million — for acquiring and construction of military housing. An enlisted soldier serving in the Armed Forces of Ukraine under the first contract will get a monthly allowance no less than UAH 7,000 (about USD 280), while a brigade commander will make more than UAH 14,500. Military budget for year 2016 is four times greater than in 2014 according to a statement by Defence Minister of Ukraine Gen. Stepan Poltorak. He stressed that the Defence Ministry worked out a road map, which contains the plan of reforms scheduled on a monthly basis. The reforms include the refinement of Command & Control, military training, logistical support; amend the legislative framework and more.

Ukrainian Defence Industry Companies will be receive USD 600 Million worth contract as per the Memorandum on Cooperation signed between Ukraine's Ukrspecexport and Pakistan's Heavy Industries Taxila (HIT). These agreements are related to tanks technical service and maintenance and modernization. Ukrainian defence industry companies will receive USD 600 million order, according to the Ukraine Ministry of Defence.

Defence Production and R&D

Ukrainian national military exports have been suspended while it focuses on the production of materiel for domestic armed forces during the ongoing unrest in the country.

In 2013, Ukraine was ranked the world's eighth-largest exporter of major conventional arms, according to a report by SIPRI. From 2009 to 2013, Ukrainian defence companies held a 3 percent share of international arms exports with purchases mainly from China, Pakistan and Russia. One of the products that the Ukrainian industry has been successful with in Asia is the Oplot (Fortress) main battle tank (MBT) that is produced by the Malyshev Plant (ZIM) in Kharkiv. The country's intercontinental ballistic missiles, tactical ballistic missiles, and radar and avionics systems are mostly designed in Donetsk and Dnipropetrovsk, while the cities of Kharkiv and Luhansk are production hubs for main battle tanks such as the T-34, T-64, T-80UD, and other types of armoured vehicles. Military aircraft engines are mostly designed and produced in the south-eastern city of Zaporizhia.

Defence Procurement

Ukraine is in great need of new weapons and other military equipment to fight the Russian invasion. The government has increased its defence spending expenditure. In January 2015 Poland announced that it was ready to sell weapons to Ukraine on commercial terms. More countries are also considering selling, or donating, weapons to Ukraine. Meanwhile, the Ukrainian government has approved a change to the procurement law in which the country's State-owned military industries are to be exempted from the need to competitively tender for government work.

CONTACT DETAILS

Ministry of Defence

6, Povitroflots'ky Avenue,
Kyiv, Ukraine 03168
Ph: (044) 253-04-71
E-mail: priymalna_mou@mil.gov.ua
Website: http://www.mil.gov.ua

UNITED ARAB EMIRATES
(Capital: Abu Dhabi)

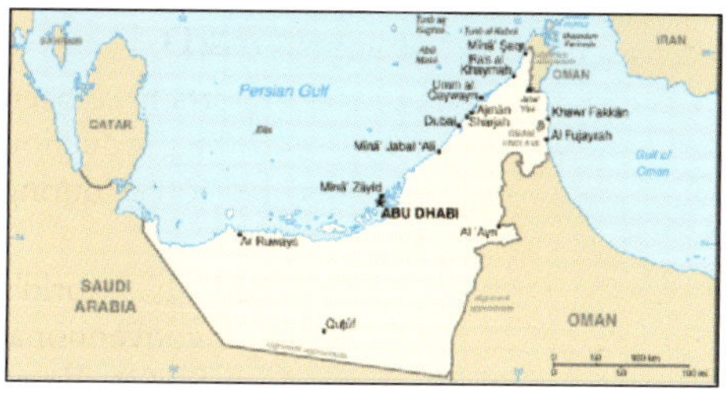

INTRODUCTION

Area: 83,600 sq km

Population: 5,927,482 (July 2016 est.)

Coastline: 1,318 km

Maritime claims: Territorial sea: 12 nm

Contiguous zone: 24 nm

Exclusive Economic Zone: 200 nm

Continental shelf: 200 nm or to the edge of the continental margin

KEY POLITICAL PERSONS

PRESIDENT & COMMANDER OF UNION DEFENCE FORCE: Sheikh Khalifa bin Zayed Al Nahyan

KEY DEFENCE PERSONS

VICE PRESIDENT, PRIME MINISTER & DEFENCE MINISTER: Sheikh Mohammed bin Rashid Al Maktoum

DEPUTY SUPREME COMMANDER OF THE ARMED FORCES: Sheikh Mohamed bin Zayed Al Nahyan

CHIEF OF STAFF: Lt Gen. Hamad Mohammed Thani Al Rumaithy

COMMANDER OF THE ARMY: Maj. Gen. Saleh Mohamed Saleh Al Ameri

COMMANDER OF THE NAVY: RAdm Ibrahim Salim Mohamed Al Musharrakh

COMMANDER OF THE AIR FORCE & AIR DEFENCE: Maj. Gen. Staff Pilot Ibrahim Nasser Al Alawi

ECONOMY

UAE has the second-largest economy in the Arab world which it has been successfully diversifying, helped by a massive construction boom, an expanding manufacturing base, and a thriving services sector. While oil continues to be the mainstay of the economy, non-oil sectors now account for 69 per cent of GDP. In 2011, oil exports accounted for 77% of the UAE's state budget. Tourism is also one of the main sources of revenue. Petroleum and natural gas exports play an important role in the economy. The UAE 2021 Vision aims to place innovation, research, science and technology at the centre of a knowledge-based, highly productive and competitive economy by 2021.

GDP: $647.8 billion (2015 est.), $617.1 billion (2014 est.)

Real Growth Rate: 3.9% (2015 est.), 4.6% (2014 est.)

Industries: Petroleum and petrochemicals; fishing, aluminium, cement, fertilizers, commercial ship repair, construction materials, some boat building, handicrafts and textiles

Crude oil - proved reserves: 98 billion bbl (1 January 2016 est.)

Natural gas - proved reserves: 6.091 trillion cu m (1 January 2016 est.)

Total Exports: $333.3 billion (2015 est.), $367.5 billion (2014 est.)

Export Commodities: crude oil 45%, natural gas, reexports, dried fish, dates (2012 est.)

Major Markets: Iran 14.5%, Japan 9.8%, India 9.2%, China 4.7%, Oman 4.3% (2015)

Total Imports: $243.9 billion (2015 est.), $239.8 billion (2014 est.)

Import Commodities: machinery and transport equipment, chemicals, food

Major Suppliers: China 15.7%, India 12.8%, US 9.7%, Germany 6.8%, UK 4.4% (2015)

INTERNATIONAL DISPUTES

Iran and UAE dispute Tunb Islands and Abu Musa Island, which Iran occupies.

DEFENCE

The United Arab Emirates introduced a law in June 2014 requiring compulsory military service for adult male high school graduates, aged between 18 and 30, to serve nine months and those without a high school diploma to serve two years while military service for females is voluntary. The UAE armed forces are considered highly capable among other militaries in the Arabian Gulf. It is equipped with some of the most modern weapon systems purchased from a variety of countries.

ARMY

Personnel: 60,000

Equipment

Category	Name	In Service
MBT	Leclerc	396
	Scorpion light	80
Amoured car	AML-90	60
AIFV/APC	BMP-3	475
	ACV (all versions)	133
	RG-31	28+
	M-ATV	751
	Fuchs NBC	32
Artillery	M198 155mm towed	13
	105mm towed	6
	AMX MkF3 155mm SP	21
	G6 155mm SP	79
	M-109L47 155mm SP	89
MRL	Astros II	
	Firos-25	
	Smerch	6
Mortar	81mm	
ATGW	Vigilant	
	Tow	
	Milan	50
	Javelin	
SAM	1-Hawk (5 batteries)	
	Rapier	
	Crotale	
	RBS-70	
	SA-14	
	Javelin	
	9K69 Pantsir S1 gun/missile	50
	Avenger fire units with Stinger RMP Block 1 missile	79
	THAAD	(On Order)

New Procurements/Upgrades

✦ A nation-wide ATBM system is to be organised. The programme includes the purchase of both the THAAD system as well as the Patriot PAC-3 system with a total of 10 radar sets, 10 engagement control stations and 37 launchers.

✦ 48 AGRAB 120mm mortar carriers (RG-31 chassis) are on order.

✦ 135 BMP-3 are being upgraded.

✦ The MBDA MPCS (Multi-Purpose Combat System including a turret, sensors, IFF, Mistral missiles, command & firing posts and shelter equipment) would be installed on NIMR's latest HAFEET 640A all-terrain light armoured vehicle platform.

✦ Orders have been placed to buy used Mine Resistant Ambush Protected (MRAP) Vehicles from the United States.

✦ An order has been placed for 750 Oshkosh Mine Resistant Ambush Protected All-Terrain Vehicle (M-ATV).

✦ An order has been placed for an unspecified number of Patria AMV 8x8 AFVs that will receive turrets taken from BMP-3.

✦ UAE plans to buy from US 12 High Mobility Artillery Rocket Systems (HIMARS) Launchers,100 M57 Army Tactical Missile System (ATACMS) T2K (Block IA Unitary) Rockets and 65 M31A1 Guided Multiple Launch Rocket (GMLRS) Unitary Pods.

NAVY

Personnel: 2,500

Equipment

Category	Name	In Service
Corvette	Baynunah class	6
	Muray Jib class	3
	Abu Dhabi class	3
Patrol Craft	Mubarraz class FAC	2
	Ban Yas class FAC	6
	Ardhana class	6
Mine warfare Force	ex-German Type 332 minehunters	2
Amphibious Force	Landing craft tank	7
	Landing craft utility	5
	LCP	4
	LCM	1

The UAE maintains a small battalion sized marine force called the UAE Marines. It is equipped with BMP-3 infantry fighting vehicle. The UAE Coast Guard is primarily responsible for the protection of the UAE's coastline.

New Procurements/Upgrades

✦ CAE has been awarded contracts by the General Headquarters (GHQ) of the United Arab Emirates (UAE) Armed Forces valued at approximately C$145 million. The contracts are for CAE to design and develop a comprehensive Naval Training Centre (NTC) for the UAE Navy and to provide the UAE Joint Aviation Command (JAC) with a suite of helicopter simulators and training devices for the NorthStar Aviation 407 Multi-Role Helicopter (407MRH) as well as the Sikorsky UH-60M Armed Black Hawk (ABH).

✦ A new series of 12 improved 'Ghannatha' class fast combat boats are under construction at ADSB.

✦ 34 16m high-speed launches are on order for the Critical National Infrastructure Authority.

✦ An order has been announced for eight patrol boats to be built in national yards.

The United Arab Emirates navy fast attack craft Mubarraz (P141) participates in International Mine Countermeasures Exercise (IMCMEX) 13 with Combined Task Force (CTF) 523. *Image Credit: US Navy*

AIR FORCE

Personnel: 4,500

Equipment

Category	Name	In Service
Fighter	Mirage 2000-9/9D	68
	F-16E/F Block 60 Desert Falcon	80
Transport	Airbus A330 MRTT	3
	C-17ER	6
	C-130H	3
	CN-235	11
	Cessna 208B	8
Trainer	Hawk Mk102	15
	Hawk Mk63	30
	MB339A	10
	PC-7	30
Helicopter	CH-47C/D	12+
	AH-64A/ 64D	50+
	IAR-99 Puma	6
	AS.550 Fennec	12
	AB-412	4
	UH-60L/60M Black Hawk	25+
Missile	AGM-65, AGM-84, AGM-88, AIM-9M, AIM-120B, Black Shahine	
UAV	Predator XP, Seeker II	

New Procurements/Upgrades

✦ Two additional C-17 transports are on order.

✦ Six CH-47 Chinook helicopters are on order.

✦ Nine AW139 helicopters are on order.

✦ UAE has ordered two high-resolution Pleiades-type Falcon Eye military observation satellites along with two ground facilities and operator training for 20 engineers from France. The satellites are scheduled to be delivered by prime contractor Airbus Defence and Space and payload-maker Thales Alenia Space by 2018.

✦ The Air Force has ordered two Piaggio P.180 Avanti multi-role patrol aircraft (MPA).

✦ Predator UAS are on order.

✦ 17 Thales GM200 air surveillance radars on order.

✦ Two Helios-type military surveillance satellites are on order.

Defence Expenditure

Exact defence expenditure figures are unavailable

Defence spending in terms of GDP: 5.7% (2014)

Defence Production and R&D

The UAE indigenous defence industry is aggressively pushing to enter the global market. Showcasing its defence technology capabilities overseas in markets such as the US, Brazil, France and the Far East is one such step in this direction. Recently the country has secured deals to supply weapons and hardware to Algeria, Turkey, Russia and Italy. Tawazun, the strategic defence investment company, inked agreements with the Algerian and Turkish governments to set up joint ventures for its Nimr Automotive subsidiary. Aerial vehicle maker Adcom Systems also is in talks for the sale of two of its Yabhon United 40 UAVs to the Russian military.

UAE enterprises such as Mubadala, Tawazun, and Emirates Advanced Investments lead the domestic defence industry. Through joint ventures with internationally leading industrial partners, the UAE is now able to design, manufacture and sell a wide array of armaments and services for the UAE armed forces and global clients. For the first time, UAE firms displayed their technology at the Defence and Security Exhibition International in London. They also appeared at the Moscow Airshow, the Gun show in Las Vegas and the Defence and Industry show in Bangkok in a large marketing drive to establish their brand globally. Another Tawazun subsidiary, Caracal, which produces handguns and rifles for use locally and across the Gulf Cooperation Council, has reportedly made export sales to Italian and US entities. The Abu Dhabi Shipbuilding company - ADSB produces a range of ships and is a prime contractor in the Baynunah Programme. It has also produced and is producing ammunition, military transport vehicles and Unmanned Aerial Vehicles. The UAE want to become an arms exporter and a naval vessel maintenance powerhouse over the next five to 10 years.

Defence Procurement

Over the past few years, UAE has made major

investments in its air, land and sea and critical infrastructure defence capabilities, including significant purchases of F-16 aircraft, Patriot and Terminal High Altitude Air Defence (THAAD) systems. The Armed Forces has awarded Tawazun a contract to procure the Talon Laser Guided Rocket. Under the contract, US Company Raytheon will maintain full integration of the Talon system into the UAE Armed Forces to include logistics, training and warranty support. Raytheon is also exploring options to integrate Talon onto NIMR armoured vehicles which are now serving in different units of the UAE Armed Forces. UAE contracted Airbus Defence and Space to manufacture and deliver two Falcon Eye Earth-observation satellites to the national armed forces. The Pleiades-type Falcon Eye satellites offer high-resolution optical capabilities and a ground system for monitoring, receiving and processing images.

Soldiers of UAE armed forces

CONTACT DETAILS

Ministry of Defence
P O Box 2838
Abu Dhabi, UAE
Tel: +9712532330
Fax: +9712531406
Armed Forces HQ
P O Box 309
Abu Dhabi, UAE
Tel: +971234392

STRATEGIC INFORMATION

UNITED KINGDOM
(Capital: London)

INTRODUCTION

Area: 243,610 sq km
Population: 64,430,428 (July 2016 est.)
Coastline: 12,429 km
Maritime claims: Territorial sea: 12 nm
Continental shelf: As defined in continental shelf orders or in accordance with agreed upon boundaries

GEOPOLITICAL IMPORTANCE

Located off the north-western coast of European continent, the United Kingdom comprises a chain of islands including Great Britain and the north-eastern part of the island of Ireland. The Kingdom mainly consists of four countries – England, Scotland, Wales and Northern Island, with greater devolution of power among the last three. Scotland went for a referendum on 18th Sept 2014 for independence, with the voters rejecting the proposal for separation from the 307-year-old political union with Great Britain. The UK is surrounded by the Atlantic Ocean in North West, by the North Sea in North East and by the English Channel in the South. In its West lies the Irish Sea. The Kingdom occupies a unique geostrategic position in the entire Europe. Though separated from mainland Europe, the UK since centuries has wielded more power over the entire continent than any other country in the region. Dubbed as the modern world's first superpower that went on to establish imperial colonies around the globe and on which the "sun never set", the UK even today projects a strong military strength globally, especially its formidable naval fleet which is deployed in several parts of the world.

POLITICAL OVERVIEW

A unitary state, the UK functions as a Constitutional monarchy and a Parliamentary democracy. While the monarch (currently Queen Elizabeth II) is a ceremonial head of the state, the real political power is exercised by the Executive and Legislature constituting of the Prime Minister and Cabinet Ministers (functioning as government) and the two houses of Parliament – House of Lords and House of Commons, respectively. The country does not have a written Constitution. As one of five permanent members of the United Nations Security Council, a founding member of NATO, and of the Commonwealth, the UK pursues a global approach to foreign policy.

KEY POLITICAL PERSONS

HEAD OF STATE:
Queen Elizabeth II

PRIME MINISTER:
Theresa May

ECONOMY

The United Kingdom is the third largest economy in Europe after France and Germany. Based on its GDP, it is the fifth largest economy in the world. It is a leading trading power and financial centre in the world with London being one of the 'command centres' of the global economy. The service sector, primarily constituting of banking, insurance, and business, contributes the largest share of the country's GDP. Once a leading industrial nation in the world, UK's industrial sector, however, has declined over the years. The highly modernised agricultural sector is capable of meeting more than half of the country's food need. Hit by the 2008 global economic recession followed by the eurozone crisis, the UK Government adopted several austerity measures to stabilise the economy and reduce budget deficit. From 2013 onward, the economy witnessed a steady growth and in 2014 it registered a GDP growth of 3%, marking a recovery among other developed economies in the West.

The country is a member of leading

international financial institutions, including the World Trade Organisation (WTO), International Monetary Fund (IMF) and Organisation for Economic Co-operation and Development (OECD) along with the financially powerful G-7 and G-20 group of nations and the Commonwealth states. An active member of the European Union (EU) until June 2016, Britain, in a historical referendum, has decided to exit the block.

GDP: $2.702 trillion (2015 est.); $2.643 trillion (2014 est.)

Real Growth Rate: 2.2% (2015 est.); 3.1% (2014 est.)

Industries: Machine tools, electric power equipment, automation equipment, railroad equipment, shipbuilding, aircraft, motor vehicles and parts, electronics and communications equipment, metals, chemicals, coal, petroleum, paper and paper products, food processing, textiles, clothing, other consumer goods

Crude Oil - Proved Reserves: 2.982 billion bbl (1 January 2015 est.)

Natural Gas - Proved Reserves: 241 billion cu m (1 January 2014 est.)

Total Exports: $436.2 billion (2015 est.); $483.4 billion (2014 est.)

Export Commodities: Manufactured goods, fuels, chemicals; food, beverages, tobacco

Major Markets: US 14.6%, Germany 10.1%, Switzerland 7%, China 6%, France 5.9%, Netherlands 5.8%, Ireland 5.5% (2015)

Total Imports: $627.7 billion (2015 est.); $685.8 billion (2014 est.)

Import Commodities: Manufactured goods, machinery, fuels; foodstuff

Major Suppliers: Germany 14.8%, China 9.8%, US 9.2%, Netherlands 7.5%, France 5.8%, Belgium 5% (2015)

DEFENCE & SECURITY

Driven by commercial and trade interests, the United Kingdom - which was the world's first industrialised nation in the 18th century - had expanded its imperial rule worldwide, mainly through the projection of a powerful and formidable military force. Even during the two World Wars and the Cold War era, Britain remained a mighty military power. By allying with the world's superpower USA as its "indispensable partner" in international military affairs, the United Kingdom has maintained its global military edge over the years. The country is an active member of NATO.

Traditionally, the country's monarch is the Commander-in-Chief of the armed forces (presently Queen Elizabeth II). Constitutionally however, the actual authority over the armed forces is exercised by the Prime Minister, the Secretary of State for Defence and the Cabinet.

UK's defence and security perspective is not merely confined to defending the homeland from internal and external threat, but pursuing wider global interests, ranging from protecting its overseas territories (It still retains some of the colonial territories under its control) to maintaining energy security and conducting international trade. Being a NATO member, its armed forces also take part in international coalition operations, including humanitarian missions. With a robust domestic defence industry, the country has equipped its armed forces with cutting-edge military systems. Its naval fleet maintains a worldwide presence to secure its interests overseas. UK is also one of the leading exporters of military hardware to other countries, thereby becoming a key player in international arms trade.

Hit by economic constraints in the last few years, the British military power, however, is facing an erosion in its overall capability following years of stringent austerity measures. Continuous budgetary cuts on military spending and reduction in troops as well as new military acquisitions could result in the country withdrawing completely from world affairs in the next few years or so.

KEY DEFENCE PERSONS

SECRETARY OF STATE FOR DEFENCE: Michael Fallon

MINISTER OF STATE FOR DEFENCE: Earl Howe

CHIEF OF THE DEFENCE STAFF: Air Chief Marshal Stuart Peach

CHIEF OF THE NAVAL STAFF: Adm. Philip Jones

CHIEF OF THE GENERAL STAFF, ARMY: Gen. Nicholas Carter

CHIEF OF THE AIR STAFF: Air Chief Marshal Stephen Hillier

COMMANDER OF JOINT FORCES COMMAND: Gen. Chris Deverell

Internal Conflict

Terrorism: Terrorism is one of the major threats the country is facing today. In recent years, there have been many attempts to carry out terrorist attacks on different locations across the UK. According to the British Home Office, the current threat level from international terrorism is "substantial" with a strong possibility of a terrorist attack in the country. Terrorism is not associated with any single area of the UK, and individuals convicted of terrorist offences have lived in different parts of the country. Attacks related to terrorism have occurred in London and Glasgow, and thwarted terrorist plots have been aimed at targets outside the capital in recent past. The rise of the extremist terror organisation Islamic State (ISIS) has further heightened the terror threat to the country.

Racism: Racism in one form or other has been widespread in Britain since ages and race-related crimes have taken ugly form at times. Violence against ethnic groups living in the country has increased over the years. Though race relations have improved immensely over the last thirty years, racial segregation, however, remains an important but largely unaddressed problem in the UK.

Organised crime: The country also deals with the issues of drug trafficking, gang wars, human trafficking, financial fraud and other forms of organised crime.

External Conflict

The United Kingdom had previously been engaged in some external conflict with few countries including Iraq and Afghanistan. Although it did not face any direct military threat from any country, especially after the disintegration of Soviet Union in 1991, the British armed forces, nevertheless, have often been involved in international military operations, mostly led by the US and NATO. At present, the country is involved in air strikes over Syria and Iraq to fight the terror outfit Islamic State (ISIS).

Britain's territorial disputes are with Northern Ireland, Argentina, Spain, Denmark and Mauritius.

Overseas Conflict

Iraq: The UK was a major contributor to the US-led invasion of Iraq in 2003. Despite a public mandate going against the country's involvement in the war, UK sent a 46,000-strong army personnel to the region. The British Army controlled the southern regions of Iraq and maintained a peace-keeping presence in the city of Basra until their withdrawal from the war-torn nation on 30 April 2009. Later on, British troops were fully withdrawn from Iraq after the Iraqi government refused to extend their mandate.

UK military operations in Iraq were being conducted under the name 'Operation TELIC'. The UK operated as part of a coalition called the Multi-National Force – Iraq (MNF-I) alongside troops from 25 other nations. In February 2007, the-then British Prime Minister, Tony Blair, announced the first large-scale withdrawal of British troops from Iraq. As of November 2008, there were 4,100 British troops stationed in Iraq. As on 12 February 2009, a total of 179 British Armed Forces personnel or MOD civilians had died serving in Iraq. Presently there are no soldiers in Iraq.

Afghanistan: UK has played a key role in the ongoing Afghanistan conflict which began as a war in October 2001. Led by the US and NATO coalition members, the war against Afghanistan was waged to eliminate Taliban rule in that country as well as to annihilate Al-Qaida and its global terror network. The UK was the second country after the US to deploy a large number of troops in the war-ravaged country. NATO announced withdrawal of its combat troops from Afghanistan by 2014-end, even as the US said its soldiers will remain deployed in that country until 2016. UK also announced that 450 of its troops would remain there for support mission.

Libya: British armed forces took part in the March, 2011 military intervention in Libya led by NATO coalition forces to stop civil war in the North African country. The Royal Navy and Royal Air Force were actively involved in carrying out combat operations along with the French military during the conflict. The war came to an end with the capture and death of Libyan leader Muammar Gaddafi in October, 2011.

Territorial Conflict

Northern Ireland: The centuries-old conflict between Great Britain and Northern Ireland has both political and religious roots. It has revolved around whether the latter should remain a part of the former. While the Catholics in Northern Ireland have opposed partition of the territory from UK, the Protestants have supported the idea of its inclusion in the Republic of Ireland which gained independence from British administration in 1922. In the south, 26 counties formed the separate state of Republic of Ireland while six counties in the north stayed within the UK. Despite several efforts to resolve the dispute, problems still persist between UK and Northern Ireland over devolution of power.

Argentina: The British-Argentina conflict over a chain of islands in the South Atlantic Ocean has again flared up nearly three decades after the Falklands War in 1982. Argentina claims sovereignty over the Falkland Islands, which it calls Las Malvinas. The archipelago, located near Argentina, is a self-governing British Overseas Territory with the UK looking after the region's defence and foreign affairs. The two countries are also engaged in territorial dispute over the South Georgia and the South Sandwich Islands in South Atlantic Ocean. The uninhabited region is a British Overseas Territory over which Argentina claims sovereignty.

Spain: UK and Spain spat over Gibraltar – a British Overseas Territory near the Mediterranean. Spain does not acknowledge British sovereignty over the region and the two sides have fought a war over it in 1704. The self-governed region, spread over an area of 6.8 square kilometres, has rejected proposals of Spanish sovereignty twice in 1967 and 2002. UK looks over the region's defence and foreign affairs.

Mauritius & Seychelles: The region of contention between UK, Mauritius and Seychelles is the Chagos archipelago – a chain of over 60 islands in the Indian Ocean. The area was part of Mauritius till the 18th century, but later came under British rule and presently falls under the British Indian Ocean Territory (BIOT). Mauritius has contended UK's claims over the region and has taken the issue to the United Nations.

Denmark: UK along with Iceland and Ireland dispute Denmark's claim that the Faroe Islands' continental shelf extends beyond 200 nm. The island, situated between the Norwegian Sea and the North Atlantic Ocean, is administered by Denmark.

THREAT PERSPECTIVE

After the collapse of Soviet Union in 1991 and the end of Cold War, UK did not face any direct military threat from any country. Due to the absence of any conventional threat to its territory, the British armed forces were structured as Expeditionary forces. While maintaining a forward military presence globally, Britain's military alliance with the US and NATO has fortified its security position. However, the military resurgence of Russia post the Ukraine crisis followed by the ongoing Syria conflict, and the increasing threat of terrorism from the Islamic State have significantly altered the power equations for the entire European continent, as well as for Britain which in contrast is reeling under defence budgetary cuts and its ever-shrinking military power.

STRATEGIC RELATIONS

US: Over the centuries, the UK and the US have enjoyed deep-rooted historical, cultural, linguistic, and political ties. Alongside, the two countries' governments have shared an intense politico-military and intelligence cooperation since World War II. Bilateral relation between the two countries is based on a long history of "shared interests and common values." Both countries, as permanent members of the UN Security Council and also the NATO, have taken more or less similar stand over key international issues that have influenced or affected their own national interests. Though in terms of projecting power, the US has always exceeded the UK, the latter has nevertheless supported the former, both militarily and diplomatically, in some of its key decisions, including the last decade's wars against Iraq and Afghanistan, despite of domestic public opposition. While the UK sent its troops to take part in the wars, it also provided its territory to the US to carry out combat strikes against Iraq in 2003. The two countries' multidimensional relationship has enormously benefited both in the past. The UK has acknowledged the fact that its strategic allegiance to the world's sole superpower is vital for its own national interest as well as for its security even as the US is steadily shifting its focus away from Europe to Asia in the changing global geostrategic equations. The new equations have in fact resulted in Britain appearing to be steadily going away from the US for whom it was once an indispensable and credible ally.

The US Air Force maintains few military bases in the UK.

France: The UK and France are active members of the NATO, EU and UN Security Council. Both are nuclear weapon states, and have similar national security interests. Though historic rivals since centuries, the two European nations have made efforts to deepen cooperation with a thrust on defence and strategic ties in recent times. The two sides signed a landmark treaty in November 2010 to strengthen bilateral defence cooperation by agreeing to create a Combined Joint Expeditionary Force (CJEF) by 2016 and share nuclear test facilities. The deal also made provision for shared use of aircraft carriers by the two countries' navies as also jointly working on nuclear capabilities. Britain and France have also joined hands to design and develop an unmanned air combat system and are identifying other areas of cooperation to jointly invest and develop critical defence technologies and weapon systems.

Turkey: Acknowledging the importance of Turkey as a regional power, Britain in recent times has made efforts to build up a strong relationship with Ankara. Britain and Turkey signed a "Strategic Partnership" agreement in 2010 which among others includes defence relations, regional stability and international security issues. Bilateral relations between the two countries have strengthened since then with high-level political and diplomatic visits. In 2011, the two sides inked a "Military Co-operation Treaty". The Treaty is aimed at facilitating greater bilateral co-operation across a range of military activities. Britain has also been strongly supporting Turkey's inclusion as a member of the European Union.

India: Reckoning India as a growing economic power, the UK has laid stress on strengthening its ties with New Delhi. The countries share strong economic and social relations besides historical and cultural ties. Their bilateral relationship got the "strategic partnership" stamp in 2004 when the two sides signed an agreement in this regard. Defence cooperation between the two countries has also received a boost in recent times with high-level official visits, joint military training and military exercises and technological collaboration in science and technology. India is also procuring defence equipment from Britain.

Japan: Bilateral relations between Britain and Japan have primarily involved economic and social ties as well as cooperation in nuclear energy sector. The two sides have also been discussing cooperation in the field of defence, including joint research, development and production of defence equipment, and joint military training and exercises.

Multilateral Alliances

NATO: As a founder member of the North Atlantic Treaty Organisation (NATO), Britain has given highest priority to its allegiance to the multinational military block to ensure its own defence and security as well as to protect its interests overseas. The country regards NATO as the "bedrock" of its defence and is a major financial contributor to the alliance. Its armed forces actively take part in most NATO-led combat operations overseas. The country's nuclear forces also make a substantial contribution in ensuring NATO's nuclear deterrence as part of the block's overall military strategy. The multilateral block, formed primarily as a military alliance to counter the then potential Soviet threat to Western Europe, has evolved over the years to meet the present-day changing security challenges. While clearly acknowledging the importance of collective defence through the military alliance to ensure its own territorial sovereignty as well as the security of the European neighbourhood, Britain has laid emphasis on closer cooperation with all its NATO allies.

EU: Although Britain remains outside the Economic and Monetary Union and has not adopted the Euro as its national currency, nevertheless it plays a leading role in the block, mainly formed for greater economic integration of Europe and promote free market economy in the continent. UK's membership of EU is primarily aimed at securing its own national interests besides promoting greater economic growth and prosperity in the region. It also works closely with its other EU allies on other areas such as climate change, migration and transnational crime and international issues and developments.

DEFENCE CAPABILITIES

ARMY: Bestowed with the primary task of safeguarding and protecting the country's interests, the British Army is organised to undertake operations in two broad scenarios – standing tasks (maintaining its presence in the territories controlled by it), and contingent operations in response to international crisis. The country maintains a Territorial Army to carry out exercises and operations overseas. The British Army's worldwide presence is reflected from the fact that its personnel, as of 2010, were deployed in a total of 27 countries besides the British Indian Ocean Territories. It sent the second highest number of troops to Afghanistan after the US and ended its combat operations in that country by 2014, even though few soldiers remain there as part of support mission. Britain aims to transform its Army into a smaller, more flexible and agile force by 2020 to enable it to meet divergent security scenarios in the future and for the first time, regular and reservist soldiers will be fully integrated within a single force structure. With huge defence cuts looming large, the country also plans to further downsize the Army by one-fifth (19,500) which would bring down its total number to 82,000 by 2018 which could further reduce to 50,000 by 2019. The reserve force strength, however, will be increased up to 30,000 by 2020 to fill the manpower gap. Both the forces would be integrated into one.

Personnel: 83,000+

Equipment

Category	Name	In Service	On Order
MBT	Challenger 2	227	
Light Tank	Scimitar	318	
APC/IFV	Warrior	786	
	FV432	541	
	Stormer	151	
	Jackal	400	
	Coyote (TSV light)	70	
	Mastiff	400	
	Spartan & other CVR(T) vehicles	138	
Armoured Car	Land Rover Snatch 2	(To be replaced by Foxhound LPPV)	
	Vector	166	
	Viking BVS10	158	
	Panther	350	
	Husky	325	
	Wolfhound	125	
	Foxhound	400	
Artillery	L115A3		600
	L118 light gun (105 mm towed howitzer)	126	
	AS90 self-propelled gun	89	
Mortar	M6-895 60 mm	640	

Army continued...

Category	Name	In Service	On Order
Mortar	L16A2 81mm	371	
MRLS	M-270	35	
RCL	M72 LAW		
ATGW	Javelin	144	
SAM	Rapier	24	
	Starstreak	147	200
	Javelin	335	
	Stinger	120	

Army Aviation Equipment

Category	Name	In Service	On Order
Light transport aircraft	Defender	9	
	Islander AL1, CC2	6	
Helicopter	Lynx AH-9A (utility)		
	Wildcat	1	33
	Gazelle (reconnaissance)	20+	
	Merlin	28	
	Chinook	48	
	Puma	38	
	Bell 212	8	
	Apache (attack)	50	
	Sea King	28	
UAV	Hermes 450		
	Watchkeeper	32	21
	Desert Hawk		

New Procurements/Upgrades

✦ Plans are on to upgrade the existing fleet of Challenger 2 MBTs or replace them with a new platform.

✦ The Defence Ministry has awarded a contract worth £1 billion (USD 1.6 billion) to upgrade the Army's Warrior infantry combat vehicles. Lockheed Martin UK as the lead contractor in the upgrade programme has received £642 million (USD 1 billion). The upgraded Warriors are expected to enter service by 2020, with the first deliveries beginning in 2018.

✦ A £3.5 billion contract has been awarded to General Dynamics-UK for the delivery of new Ajax armoured fighting vehicles (AFVs) to the Army. Under the contract, GD will

The Ajax AFV will be the Army's first fully digitised armoured vehicle *A UK MoD photo*

supply a total of 589 new vehicles in six variants to provide essential capability to the Armoured Cavalry within Army 2020. The new AFVs will replace the Scimitar light tanks.

✦ Acquisition of new 8x8 mechanised infantry vehicle (MIVs) is under consideration. Over 300 such platforms are expected to be inducted by 2023.

✦ The Defence Ministry is buying 34 Army Wildcat Mk1 helicopters which will replace the Lynx fleet of choppers operated by the Army Air Corps. The new choppers will primarily be used in reconnaissance role and provide protection to soldiers on the ground. Delivery is going on.

✦ Under a £367 million contract awarded by the MoD, Rolls-Royce Turbomeca will provide engine maintenance for 2 of the UK's military helicopter fleets -- for the Army's Apaches and the Merlin helicopters used by the Royal Navy and the Royal Air Force.

✦ Lockheed Martin has been contracted to provide additional Desert Hawk III UAVs to the Army under a $5.1 million contract.

✦ Upgrading of the Apache attack helicopters is under consideration.

NAVY: Britain has maintained its force projection across the globe through its formidable Royal Navy which, for over three centuries, wielded great power and played a crucial role in establishing the British Empire. As an island nation, the country has reckoned the importance of naval power in securing its sea lanes for trade and transport to gain economic prosperity. Hence, the Royal Navy, since long, has been operating as a "blue water" navy, carrying out operations far beyond the UK territory. Solely driven by the task of protecting and promoting the country's interests worldwide, the Royal Navy not only undertakes combat operations wherever and whenever necessary, but is also used strategically to prevent conflict in the international stage. The British Naval fleet is presently deployed in several parts of the globe, including the South Atlantic, Gulf region, Northern Europe (North Atlantic and Mediterranean) and Far East. It is also building up maritime relations in the Asia-Pacific and Indian Ocean regions in recent times. The Royal Navy operates and exercises with allies, coalition partners and other countries to build trust and understanding that contributes to stability in regions of interest to the UK. The Navy also undertakes humanitarian missions overseas.

To meet the challenges of the future, the Royal Navy aims to be an increasingly versatile, network-enabled expeditionary force, continuing to operate on, under and over the sea, on land (along with the Royal Marines) and ever more frequently in the littoral, fully interoperable and integrated with the UK and allied forces. To support this transformation, the Navy is in the process of acquiring a range of new warships, including aircraft carriers, submarines, destroyers and frigates, naval fighter jets along with advanced weapon systems and equipment. The two large aircraft carriers are being built to operate the Joint Strike Fighter and amphibious shipping and would transform the carrier strike capabilities of the Royal Navy and enhance its ability to project power onto the land environment. Under the MoD's Strategic Defence and Security Review (SDSR) programme, plans are on to reduce the Navy by 5,500 troops to maintain a force level of around 30,000 by 2020.

Major naval bases: Gateshead, Tyne & Wear, England, Ceres Division (Leeds), Greenock, Scotland, Govan Division (Glasgow), Liverpool, Whale Island, Portsmouth, Hampshire, Medway Division, Chatham, HMNB Devonport, Plymouth

Personnel: 31,000+ (Including Marines)

Equipment

Category	Name	In Service	On Order
Aircraft carrier	Invincible class (helicopter carrier)	1	
	Elizabeth class		2
Submarine	Vanguard class	4	
	Trafalgar class	4	
	Astute class	3	4
	Successor class		4
Destroyer	Type 45 Daring class	6	
Frigate	Type 23 Duke class	13	
	Type 26 (To replace Type 23)		
Amphibious ship	LPH, LPD, LSD, LCU, LCVP, LCAC etc.	67	
Patrol vessel	Antarctic patrol ship	1	
	River class vessel	4	3
Fast Patrol boat	Archer class	14	
	Scimitar class	2	
Mine sweeper	Hunt class MCMV	3	
	Sandown class SRMH	8	
Auxiliary Ship	Oilers, Tankers, Repair Ships, Ice Patrol Ships etc.	15	
Weapon	(Naval Gun) Mk8 Phalanx 20mm Goalkeeper 30mm		
	(ASM) Harpoon Tomahawk		
	Sea Viper system with Aster 15 & Aster 30 SAM		
	Seawolf SAM		
	Trident D5 SLBM		
	(Torpedo) Spearfish Sting Ray		

Royal Marines

The Marine forces, also called the Royal Marines, constitute of the Marine Corps and amphibious infantry that work alongside the Royal Navy. With specialisation in amphibious warfare capabilities, including the operation of landing craft, mountain warfare, and Arctic warfare, the Royal Marines operate a number of key military equipment, including landing craft, combat vehicles and other weapons.

Equipment

Category	Name
APC	Viking BvS 10
	Land Rover Wolf
	Jackal
	Pinzgauer
Mortar	L9A1 51 mm
	L16A2 81 mm
ATGW	Javelin
	LAW 80 (Light Anti-tank Weapon)

Naval Aviation

Category	Name	In Service	On Order
Fighter	F-35B		
Helicopter	Merlin HM2 (ASW/ASuW)	30	
	Merlin (troops transport)	25	
	Lynx ASW/ASuW	64	
	Sea King (being retired)	48	
	Wildcat		28

New Procurements/Upgrades

✦ The first of the two Elizabeth-class aircraft carriers has been launched in July 2014. Christened HMS Queen Elizabeth, it is the biggest warship ever to enter service with the Royal Navy in 2020. Construction of the second carrier, HMS Prince of Wales, is advancing.

✦ Next-generation Type 26 Global Combat Ships are planned to be procured to replace the Type 23 class frigates. The Navy plans to acquire 13 of the new platforms eight of which will be the BAE-designed "Type 26" warships and the remaining ones "a new class of cheaper, more flexible vessels," all slated to enter service from 2021 onward.

✦ The Government in October 2016 announced an initial £1.3 billion funding for the Successor-class of nuclear submarine programme. Led by BAE Systems, the project will deliver four new submarines to the Royal Navy and replace the current Vanguard class of platforms, with the first submarine, named 'Dreadnought', entering service in the early 2030s. The new class of submarines will provide the UK nuclear deterrence into the 2060s. The total estimated cost for building all four platforms which will be equipped with Trident nuclear weapon system is £31 billion.

✦ Plans are on to acquire a new class of light frigates under the General Purpose Frigate

(GPFF) programme as part of the 2015 Strategic Defence and Security Review (SDSR 15).

✦ A contract worth £452 million has been awarded to South Korea's Daewoo Shipbuilding and Marine Engineering (DSME) to build four new Military Afloat Reach and Sustainability (MARS) tankers. The new 37,000-tonne vessels, planned to enter service from 2017 onwards, will maintain the Navy's ability to refuel at sea and provide fuel to warships and task groups.

✦ BAE Systems has received a £348 million contract to construct three new River-class offshore patrol vessels. The first craft of the class, named Forth, is slated for delivery in 2017.

✦ Reversing its earlier decision to buy the carrier variant of the F-35 Joint Strike Fighter aircraft, the UK government in 2012 announced that it would go for the F-35B short take-off and vertical landing (STOVL) version of the new-generation warplane. The F-35B jump jets will be operated by the Royal Navy's futuristic Elizabeth class carriers as well by the Royal Air Force. 24 of the new fighters are scheduled to be in place on British aircraft carriers by 2023 as part of a total order of 138.

✦ The Government has announced to acquire nine Boeing P-8 Poseidon submarine-hunting aircraft to help protect its nuclear deterrent.

✦ Delivery of Wildcat maritime combat helicopters is going on.

✦ Lockheed Martin has completed the delivery of all 30 upgraded Merlin HM2 ASW/ ASuW helicopters to the Navy in July 2016.

✦ 25 of the AW101 Merlin troop transport helicopters from the Royal Air Force have been transferred to the Navy as per plans announced in the Strategic Defence & Security Review (SDSR) in 2010. The choppers will replace the Sea King helicopters which are set to retire by 2017.

✦ Thales UK has been contracted to produce and deliver 1,000 lightweight multi-role missiles (LMM) for the UK forces. The LMM will first enter service with the UK Armed Forces aboard the new Wildcat Lynx helicopter. The low-cost, lightweight missile, the development of which began in 2008, has been designed to be launched from a variety of air, sea and land platforms against a wide range of threats.

✦ MBDA has bagged a contract worth £250 million from the British Defence Ministry to produce the Sea Ceptor supersonic missile system for the Royal Navy. The Government had unveiled the plan of Sea Ceptor in January 2012 and awarded a contract worth £483m to MBDA to design and develop the cutting-edge air defence system. The new missile is initially being planned to arm the Navy's Type 23 frigates by replacing the Sea Wolf air defence system. Later, it will be deployed on the futuristic Type 26 warships.

AIR FORCE: The Royal Air Force (RAF) is the aerial wing of the British armed forces. The RAF operates with the stated objectives of – defending the UK and its interests; strengthening international peace and stability; and be a force for good in the world. The RAF has been structured to carry out a broad range of activities across the full spectrum of operations ranging from warfighting to humanitarian missions and expeditionary roles. It is one of the powerful and technologically advanced air forces in the world. It is the largest air force in the European Union and the second largest in NATO after the US Air Force. The RAF has played an important role in World War II and in most modern overseas conflicts. In order to meet the divergent security threats and challenges of the present-day world, the RAF has focused on developing itself as a more agile, adaptable and capable expeditionary air power and contribute to the UK's overall defence capability by working alongside the Army and the Royal Navy as also by harmonizing its capability, concepts and doctrine with those of the US Forces. In pursuance of this goal, the RAF has emphasised on developing and strengthening its command and control, and intelligence, surveillance, target acquisition and reconnaissance (ISTAR) capabilities; integrated air-land operation capabilities, and introduction of unmanned air vehicle systems along with the induction of state-of-the-art weapons and platforms. Integrating its various platforms and sensors with an effective network is also a key aspect of the modernisation drive. Under the MoD's SDSR programme, the Air Force is to be curtailed by 8,000 troops to maintain a force level of 32,000 by 2020.

Major air bases: Alconbury, Barkston Heath, Barkway, Boulmer, Brampton, Brize Norton, Coningsby, Cosford, Cranwell, Digby, Donna Nook, Fylingdales, Halton, Henlow, High Wycombe, Holbeach, Honington, Kirton in Lindsey, Leeming, Linton-on-Ouse, Lossiemouth, Marham, Northolt, Odiham, Pembrey Sands, Scampton, Shawbury, Spadeadam, St. Mawgan, Syerston, Tain, Topcliffe, Trimingham, Valley, Waddington, Weston-on-the-Green, Wittering, Woodvale, Wyton

Personnel: 34,000+

Equipment

Air Force continued...

Category	Name	In Service	On Order
Fighter, Attack, AWS role	Eurofighter Typhoon	100+	60
	Tornado GR4/GR4A	87	
	F-35		
Reconnaissance/ Maritime Patrol	Sentinel R1/ASTOR	5	
	RC-135/Rivet Joint	2	1
Tanker, Transport	C-17A Globemaster	8	
	C-130J Hercules	25	
Tanker, Transport	Voyager	14	
	A400M Atlas	7	15
Helicopter	Chinook CH-47	60	
	Griffin	3	
	Puma	20+	
	Sea King	18	
AEW	E3-D Sentry	7	
Trainer	Hawk	150+	
	King Air	5	
	Tucano	55	
	Tutor T Mark 1	100	
	Vigilant T1	50+	
	Viking T1	60+	
UAV	MQ-9 Reaper	10	
Weapon	(AAM) AIM-120B AMRAAM AIM-132 ASRAAM		
	(ASM) Brimstone Storm Shadow		
	Paveway II, III, IV & Enhanced Paveway laser-guided bomb		

New Procurements/Upgrades

✦ The Defence Ministry has announced to buy a total of 138 F-35B (short takeoff/vertical landing - STOVL variant) Lightning II Joint Strike Fighter (JSF) aircraft for its Navy and Air Force.

✦ Seven A400M heavy-lift transport aircraft have been delivered by Airbus so far. A total of 22 A400Ms are being acquired to replace the C-130J planes. Delivery of all the new freighters is scheduled to be completed by 2022.

✦ Delivery of all 14 Voyager future strategic tanker aircraft (FTSA) has been completed with the delivery of the last and final aircraft in July 2016. The fleet has achieved full operational capability in September 2016.

✦ Two Rivet Joint signals intelligence aircraft (RC-135V/W) have been delivered, with the final one due for delivery in 2017.

✦ Raytheon UK has received a contract worth US $165 million to upgrade and maintain the entire fleet of Sentinel R1 Airborne Stand-Off Radar (ASTOR)

Sentinel R1 surveillance aircraft *A Raytheon UK photo*

surveillance aircraft till the fleet's retirement planned for 2021.

✦ Delivery of all 14 Chinook-47 heavy-lift transport helicopters have been completed by Boeing. The last one was delivered in December 2015. A contract worth $1.64 billion was awarded to Boeing for the additional 14 Chinook choppers, taking their total fleet in the RAF up to 60.

✦ The US State Department in 2016 has approved the sale of 26 Predator-B UAVs for worth US$1 billion following a request for the same by the UK MoD.

✦ The MoD has signed a four-year $62.71 million Future Combat Air System (FCAS) research contract with BAE Systems to sustain and develop the country's critical technology and capabilities in unmanned systems.

✦ Engine maker Rolls-Royce has been awarded £865 million contract to upkeep RAF's fleet of Eurofighter Typhoon aircraft which are powered by EJ200 engines.

✦ Rolls Royce has received a £690 million support service contract for Tornado fighters.

✦ BAE Systems has received £8.6 million contract to equip the Tornado GR4 fighters with Brimstone missiles.

✦ An additional C-17 Globemaster has been ordered from Boeing at a cost of £200m, taking the total number of such planes in RAF's fleet to eight.

✦ Delivery of the upgraded SA 330E Puma medium transport helicopters is completed. The choppers are scheduled to be deployed in operational service by 2016.

Major Armaments Producers

BAE Systems: The largest defence manufacturer in the UK and one of the leading defence and aerospace contractors worldwide, BAE Systems designs, develops and produces a whole range of military platforms and equipment for the Army, Navy and Air Force. The company is presently involved in several multi-national, mega-budget defence projects including the F-35 Joint Strike Fighter, Eurofighter Typhoon fighter jet and the A400M heavy-lift transport plane. It is also building the Elizabeth-class aircraft carriers, Astute class submarines and Type-45 destroyers for the Royal Navy. In September 2012, BAE Systems and European defence major EADS revealed that they were in merger talks to create a global aerospace and defence leader. The negotiations, however, broke off due to a lack of consensus among the leading partnering nations and stiff opposition from Germany.

Thales UK: The second largest defence and aerospace company in the UK, Thales's defence portfolio includes missile systems, electronic warfare and radar systems, avionics, C4ISR systems and naval equipment among others.

Rolls Royce: A leading aircraft engine manufacturer, Rolls Royce provides defence aero-engine products and services to 160 customers in over 100 countries around the world. The company produces engines for transport planes, combat jets, reconnaissance aircraft, trainer jets, helicopters, and unmanned aerial vehicles. It also produces equipment and systems for naval vessels.

Babcock: Babcock is the UK's leading engineering support services organisation. The company is a major industrial support partner to the Royal Navy and also supplies equipment and services to a number of other navies in the world, including the US, Canada, Spain and New Zealand. It also provides flight training, aircraft provision and through-life support to the UK's armed forces and manages armoured and support vehicle training fleets of the UK Army. The company has recently bagged a contract worth £15 million from the UK Defence Ministry to design the next generation of nuclear deterrent submarines for the Royal Navy.

Government Organisations in Defence Sector

UKTI Defence & Security Organisation: UKTI Defence & Security Organisation (UKTI DSO) is part of the UK Trade & Investment (UKTI) - a UK Government department working with businesses based in the country to ensure their success in international markets. The UKTI DSO works with the domestic defence industry and the Ministry of Defence to promote UK equipment, products and services all across the world.

ADS Group Limited: The Aerospace Defence Security (ADS) is the premier trade organisation advancing the UK Aerospace, Defence, Security and Space industries. The organisation explores opportunities and priorities in the fields of defence, aerospace, security and space.

DEFENCE DEALS

Joint Ventures/ Collaboration Programmes

US

F-35 Joint Strike Fighter: Britain's BAE Systems is one of the main industrial partners of US's Lockheed Martin in the F-35 Lightening II Joint Strike Fighter programme. Britain is one of leading financial contributors to the programme and over 100 British firms are involved in the project, including Rolls-Royce. The US is developing the fifth-generation stealth fighter in partnership with eight countries, including the UK.

Europe/NATO

Eurofighter Typhoon: Four NATO member countries – Germany, UK, Italy and Spain – have collaborated in the design and development of the multi-role combat aircraft Eurofighter Typhoon. The Eurofighter Typhoon is Europe's largest military collaborative programme.

A400M: Seven NATO nations - Belgium, Britain, France, Germany, Luxembourg, Spain and Turkey – are partnering in the Airbus A400M military transport plane programme led by European conglomerate EADS. It is pegged as Europe's biggest defence project under which the new airlifter is being developed to replace the ageing military cargo carriers of several European air forces.

France

UAV & UCAV: Following the signing of a landmark defence cooperation treaty between UK and France in 2010, Britain's BAE Systems has joined hands with French giant Dassault Aviation to develop a medium altitude long endurance (MALE) unmanned aerial vehicle (UAV) system. The two firms have also agreed to collaborate on designing and building an unmanned combat air system (UCAS) for Europe. Under this, the two Governments in November 2014 have jointly committed £120 million to carry out feasibility study.

Combat Aircraft Engine: Rolls-Royce and French aircraft engine maker Snecma – a subsidiary of Safran Group - have signed a contract with the UK Ministry of Defence to undertake studies into the next generation of UK and French combat aircraft engines, through their 50:50 Rolls-Royce Snecma Ltd joint venture, established in 2001.

India

BAeHAL: A joint venture between BAE Systems and India's Hindustan Aeronautics Ltd (HAL), BAeHAL provides Information Technology solutions and services to aerospace, defence and engineering industries.

DLSI: The Defence Land Systems India (DLSI) is a partnership between BAE Systems and private sector Indian firm Mahindra & Mahindra. The company manufactures military vehicles and select artillery systems.

Arms Export

Home to a vibrant domestic defence industry, UK is one of the leading suppliers of military equipment worldwide. According to official data, total defence exports by the country in 2013 was worth £9.8 billion and in 2014, the country sold £8.5 billion of defence equipment abroad. For 2015, military exports registered a further decline with export order figure standing at £7.7 billion. While in 2014, most of the defence exports went to the Middle East (66%), followed by North America (13%), Asia Pacific (11%) and Europe (10%), for 2015, the figures were more or less same with 63% of export orders coming from Middle East, North America at 18% and Europe at 8%.

The largely privatised defence industry of UK produces a number of advanced military equipment and technology, including artillery, armoured vehicles, aircraft carriers, nuclear submarines, warships, fighter/ trainer jets, avionics and unmanned systems among others. The defence industry also provides upgrade and support services to military hardware sold to other countries. UK has also partnered with the US and other European countries to produce advanced fighter aircraft like the fifth-generation F-35 Lightening II and Eurofighter Typhoon.

Arms Import

UK significantly relies on its large and diverse domestic defence industry to meet the needs of its armed forces. It imports some defence equipment like missiles, armoured vehicles and aircraft, mostly from the US and often from Europe. With gradual reduction in its defence expenditure, the country now intends to cut down its military imports and focus on increasing exports.

DEFENCE BUDGET

UK is one of the world's largest defence spenders, positioning itself among the top ten in the global military expenditure index. In 2015-16 period, it remained the third largest military spender in the world. As a NATO member, Britain has so far met the alliance's commitment to spend at least 2 per cent of its GDP on defence. However, marred by global economic slowdown and European financial crisis, the country announced defence budget cuts, shaving off several major military programmes for all three wings of its armed forces and reduction in overall military personnel strength. The withdrawal of a number of naval vessels, including aircraft carriers and frigates along with fighter jets of the navy, and withdrawal of surveillance aircraft fleet, fighter jets and UAVs earlier planned to be procured for the air force are part of the Government's plan to curtail annual defence spending in the coming years. The Defence Ministry, however, in 2012 said that it was committed to spend £160 billion over the next decade on new state-of-the-art military equipment which would include new armoured vehicles, two new aircraft carriers, six destroyers, 14 Chinook helicopters, A400M military transport planes, air-to-air refueller aircraft and a range of other projects including missiles and other weapons. For FY 2014-15, the official budgetary allocation was £36.43bn – adhering to the NATO stipulated standard of 2% of GDP. For 2015-16, the Government announced a defence spending of £34.3 billion – a reduction of 1.5 per cent over the previous year's budget. For FY 2016-17, £35.1 billion has been allocated for defence.

CONCLUSION

Being a key strategic ally of the US, Britain has over the years maintained its military force projection capability in the world. It has been engaged in a significant number of combat as well as peacekeeping operations around the globe. The nation has followed a cardinal principle of defence and security operations to seek and maintain an edge over potential adversaries, both to increase its chances of success in hostile situations, and also to protect its national assets and interests. However, in the rapidly evolving global geostrategic equations where Europe no longer remains the cynosure of world affairs and Asia Pacific steadily taking up that position, the United Kingdom's military prowess is steadily eroding. After the country decided to exit from the European Union following a referendum in 2016, future military policies likely to be adopted by Britain also appear uncertain. In a recent development, plans to create a European Union (EU) Army with its headquarters at Brussels, Belgium, has been aborted following strong opposition from London. With huge defence cuts staring at its face in the coming years, the UK is set to lose key military capabilities that could deal a deadly blow to its global power projection capabilities as well as in countering home-grown security threats, including terrorism.

CONTACT DETAILS

Ministry of Defence
Main Building
Whitehall, London
SW1A 2HB
United Kingdom
Tel: 020 7218 9000 (For general queries)/ 020 7218 7907 (For national and international media enquiries)

Websites

Ministry of Defence: www.mod.uk
British Army: http://www.army.mod.uk/
Royal Navy: http://www.royalnavy.mod.uk/
Royal Air Force: http://www.raf.mod.uk/

STRATEGIC INFORMATION

UNITED STATES OF AMERICA (Capital: Washington, DC)

INTRODUCTION

Area: 9,833,517 sq km
Population: 323,995,528 (July 2016 est.)
Coastline: 19,924 km
Maritime claims: Territorial sea: 12 nm
Contiguous zone: 24 nm
Exclusive Economic Zone: 200 nm
Continental shelf: not specified

GEOPOLITICAL IMPORTANCE

The United States of America (USA) is situated in North America, and consists of 48 contiguous states and the non-contiguous states of Alaska and Hawaii. Alaska, separated from the contiguous USA by Canada, is the largest state. Hawaii, occupying an archipelago is in the central Pacific, southwest of North America. The 48 states are between the Atlantic and Pacific oceans, south of Canada and north of Mexico. The state of Alaska is in the northwest of the continent, with Canada to the east and Russia to the west across the Bering Strait. The country also possesses several territories in the Caribbean and Pacific. The country shares land borders with Canada and Mexico and maritime (water) borders with Russia, Cuba, and The Bahamas. The United States also has overseas territories with varying levels of independence and organisation.

POLITICAL OVERVIEW

The United States is a Federal Union of fifty states. It is a constitutional republic and representative democracy. Citizens are usually subject to three levels of government: federal, state, and local. The federal government is composed of three branches; Legislative, Executive & Judicial. The House of Representatives has 435 voting members. The Senate has 100 members with each state having two senators. The President of the United States, Congress, and judiciary share powers reserved to the national government, and the federal government shares sovereignty with the state governments. The executive branch consists of the President and those to whom the President's powers are delegated. The President is both the head of state and government, as well as the military commander-in-chief and chief diplomat.

KEY POLITICAL PERSONS

PRESIDENT: Donald J. Trump

VICE-PRESIDENT: Michael R. Pence

SECRETARY OF STATE: Rex W. Tillerson

ECONOMY

The United States has a capitalist mixed economy which is fuelled by abundant natural resources and high productivity. The economy of the country is the world's largest national economy, representing 22% of nominal global GDP and 17% of global GDP (PPP). Its GDP was estimated to be USD 17.914 trillion in 2015. The US economy also maintains a very high level of output per capita. It is also one of the world's wealthiest nations, with abundant natural resources, a well-developed infrastructure, and high productivity. The US is the world's third-largest producer of oil and second-largest producer of natural gas. It has the world's ninth-highest per capita GDP (nominal) and tenth-highest per capita GDP (PPP) as of 2013. Recently, huge reserves of Shale gas have been found which would make it the country with highest reserve of Shale gas. The US is the world's largest producer of oil and natural gas and has abundant natural resources, a well-developed infrastructure, and high productivity. The country has one of the world's largest and most influential financial markets. As per reports, the unemployment, as of November 2015, had declined from a high of 10% to 5%. The overall financial position of US as of 2014 includes USD 269.6 trillion of assets owned by households, businesses, and governments within its borders, representing more than 15.7 times the annual gross domestic product of the United States. Through 2014, the direct costs of the wars totalled more than USD 1.5 trillion, according to US Government figures. US revenues from taxes and other sources are lower, as a percentage of GDP, than those of most other countries. USA is the world's second largest trading nation. The dollar is also used as the standard unit of currency in international markets for commodities such as gold and petroleum.

GDP: $18.04 trillion (2015 est.), $17.58 trillion (2014 est.)

Real Growth Rate: 2.6% (2015 est.), 2.4% (2014 est.)

Industries: Highly diversified, world leading, high-technology innovator, second-largest industrial output in the world; petroleum, steel, motor vehicles, aerospace, telecommunications, chemicals, electronics, food processing, consumer goods, lumber, mining

Crude oil – proved reserves: 36.52 billion bbl (1 January 2015 est.)

Natural gas - proved reserves: 8.734 trillion cu ft (1 January 2013 est.)

Shale gas - production: 13.64 trillion cubic feet (2015)

Total Exports: $1.51 trillion (2015 est.), $1.633 trillion (2014 est.)

Export Commodities: Agricultural products (soybeans, fruit, corn) 9.2%, industrial supplies (organic chemicals) 26.8%, capital goods (transistors, aircraft, motor vehicle parts, computers, telecommunications equipment) 49.0%, consumer goods (automobiles, medicines) 15.0%

Major Markets: Canada 18.6%, Mexico 15.7%, China 7.7%, Japan 4.2% (2015)

Total Imports: $2.273 trillion (2015 est.), $2.386 trillion (2014 est.)

Import Commodities: Agricultural products 4.9%, industrial supplies 32.9% (crude oil 8.2%), capital goods 30.4% (computers, telecommunications equipment, motor vehicle parts, office machines, electric power machinery), consumer goods 31.8% (automobiles, clothing, medicines, furniture, toys)

Major Suppliers: China 21.5%, Canada 13.2%, Mexico 13.2%, Japan 5.9%, Germany 5.5% (2015)

DEFENCE & SECURITY

The US military is one of the largest militaries in terms of number of personnel. They consist of the Army, Navy, Marine Corps, Air Force, and Coast Guard. The Department of Defence (DoD) is America's oldest and largest government agency. The Army, Navy, Air Force and Marine Corps are placed under the direct control of Department of Defence. The DoD is charged with coordinating and supervising all agencies and functions of the government relating directly to national security and the United States armed forces. Among the many DoD agencies are the Missile Defence Agency, the Defence Advanced Research Projects Agency (DARPA), the Pentagon Force Protection Agency (PFPA), the Defence Intelligence Agency (DIA), the National Geospatial-Intelligence Agency (NGA), and the National Security Agency (NSA). The United States has a strong tradition of civilian control of the military. The President is the overall head of the military, and helps form military policy but the United States Department of Defence (DoD), a federal executive department, is the principal organ by which military policy is carried out. The US military is one of the largest armed forces in terms of number of personnel.

The US armed forces as a whole possess large quantities of advanced and powerful equipment, along with widespread placement of forces around the world, gives them significant capabilities in both defence and power projection. Personnel are deployed in many countries around the world, with more than 369,000 of its 1,580,255 active-duty personnel serving outside the United States and its territories. The US is currently moving towards strategic re-balance in the Asia-Pacific region, which is also known as Washington's 'pivot-to-Asia' policy. The rebalancing to this region is a very important part of the new US Defence strategy. US have also appointed a new military commander in Asia-Pacific. Admiral Harry B. Harris, Jr.'s new role with the US Pacific Command comes as the US places more emphasis on its military presence in the region in response to the area's growing economic importance and China's rise as a military power. On 2 June 2012, the erstwhile US Defence Secretary Leon Panetta announced that it will shift the bulk of its naval fleet including as many as six aircraft carriers to the Pacific by 2020 as part of new strategic focus on Asia. Besides the carriers, the redeployment would comprise the majority of navy's cruisers, destroyers, littoral combat ships and submarines and this new posturing would be fortified by an increase in the number and size of military exercises in the Pacific and the number of port visits right up to Indian Ocean. This new strategy focused on Asia would bring more troops and high-tech weapons to project American power in the region over the next decade. The United States continues to face a rapidly changing security environment, as warfare evolves across all domains. US continue to rebalance the Department's posture and presence to the Asia-Pacific while maintaining a focus on the Middle East. It also means working closely with European partners to strengthen their capabilities, maximising the impact of a relatively small US presence in Africa.

KEY DEFENCE PERSONS

DEFENCE SECRETARY: James Mattis

CHAIRMAN, JOINT CHIEFS OF STAFF: Gen. Joseph F. Dunford, Jr.

CHIEF OF STAFF OF ARMY: Gen. Mark A. Milley

CHIEF OF NAVAL PERSONNEL: Vice Adm. Robert P. Burke

CHIEF OF STAFF OF AIR FORCE: Gen. David L. Goldfein

COMMANDANT OF MARINE CORPS: Gen. Robert B. Neller

COMMANDER OF STRATEGIC COMMAND: Gen. John. E. Hyten

Internal Conflict

Terrorism: The Bureau of Counterterrorism and Countering Violent Extremism's mission is to promote US national security by taking a leading role in developing coordinated strategies and approaches to defeat terrorism abroad and securing the counterterrorism cooperation of international partners. The National Counterterrorism Centre (NCTC) is a United States government organisation responsible for national and international counterterrorism efforts. NCTC advises the United States on terrorism. The Counterterrorism Division (CTD) is a division of the National Security Branch of the Federal Bureau of Investigation. CTD investigates terrorist threats inside the United States, provides information on terrorists outside the country, and tracks known terrorists worldwide. The USA PATRIOT Act is an Act of Congress that was signed into law by President George W. Bush on October 26, 2001. The act arms law enforcement with new tools to detect and prevent terrorism, also improves counter-terrorism efforts in several ways. USA PATRIOT Act stands for "Uniting and Strengthening America by Providing Appropriate Tools Required to Intercept and Obstruct Terrorism Act of 2001". On May 26, 2011, President Barack Obama signed the PATRIOT Sunsets Extension Act of 2011. Parts of the Patriot Act expired on June 1, 2015. The USA Freedom Act is a US law enacted on June 2, 2015 that restored in modified form several provisions of the Patriot Act. USA FREEDOM stands for "Uniting and Strengthening America by Fulfilling Rights and Ending Eavesdropping, Dragnet-collection and Online Monitoring Act".

External Conflict

Afghanistan War (Present): On 28 December 2014, NATO formally ended combat operations in Afghanistan and transferred full security responsibility to the Afghan government, via a ceremony in Kabul. In May 2014, the United States announced that its combat operations would end in 2014, leaving just a small residual force in the country until the end of 2016. Obama also announced that all US forces, with the exception of a normal embassy presence, would be removed from Afghanistan by the end of 2016. These plans were confirmed with the signing of the Bilateral Security Agreement between the United States and Afghanistan on 30 September 2014. The relatively small American presence there has been active in executing airstrikes as well as providing close air support for the Afghan forces. The Pentagon in January 2016 has granted new legal authority allowing US military go on the offensive against ISIS-K in Afghanistan. The Nangarhar Military Offensive was carried out against ISIS-K by the Afghan government, with the assistance of ISAF and US airstrikes in 2016.

Iraq (2014-Present): In early June 2014, the Islamic State in Iraq and the Levant (ISIS) took over the cities of Mosul and Tikrit and said it was ready to march on Baghdad. In the summer of 2014 President Obama announced the return of US Forces to Iraq, but only in the form of aerial support, in an effort to halt the advance of ISIS forces, render humanitarian aid to stranded refugees and stabilise the political situation. Currently, the US estimates to have around 4,500 troops in Iraq. US military plans to receive new 500 troops for campaign to take back Mosul from Islamic State by supporting Iraqi and Kurdish troops.

Southeast Asia: Southeast Asia has a long history of important security and economic ties to the United States and is also of strategic interest. The Special Operations Command Pacific, known as SOCPAC. The unit provided unconventional warfare task force support for operations in Southeast Asia. SOCPAC annually conducts small unit exchanges, joint and combined training events, and operational deployments throughout the Pacific. Units are used for counterdrug and humanitarian demining operations, training forces in countries such as Thailand, Laos, Cambodia, Vietnam, and the Philippines.

Africa: US AFRICOM is responsible for all US Department of Defence operations, exercises, and security cooperation on the African continent, its island nations, and surrounding waters. Operation Enduring Freedom-Horn of Africa (OEF-HOA) is the name of the military operations for combating terrorism and piracy in the Horn of Africa and is run by the Africa Command. In North-eastern Africa, allies of the OEFHOA effort include Sudan, Somalia, Djibouti, Ethiopia, Eritrea, Seychelles and Kenya. US Africa Command (AFRICOM) Operation Enduring Freedom Trans Sahara (OEF-TS) provides military support to the Trans Sahara Counter Terrorism Partnership (TSCTP) programme. OEF-TS partnership comprises the United States and ten African countries: Algeria, Burkina Faso, Morocco, Tunisia, Chad, Mali, Mauritania, Niger, Nigeria, and Senegal.

Syria: Since September 2014, the US has been conducting air strikes on IS and other jihadist groups in Syria as part of an international coalition against the jihadist group. Operation Inherent Resolve officially refers to the US's military actions against ISIL, specifically the campaign in Iraq and the campaign in Syria. US and partner nation military forces continued to attack ISIL terrorists in Syria in Nov 2014 using bomber, fighter and remotely piloted aircraft to conduct twenty-three airstrikes. US Special Operations Forces are playing an important role on the ground in the fight against ISIS in Syria.

THREAT PERSPECTIVE

United States face a spectrum of challenges, including violent transnational extremist networks, hostile states armed with weapons of mass destruction, rising regional powers, emerging space and cyber threats, natural and pandemic disasters, and a growing competition for resources. The range of security challenges the country faces right now are beyond Iraq and Afghanistan. There are issues which pertain to Iran, North Korea and modernisation.

Iran: On the 23 July 2015, Barack Obama announced his intention to restore diplomatic relations between the two nations following the agreement on Iran's nuclear programme. This was welcomed by Hassan Rouhani. US fears about Iran's nuclear intentions, mainly after the country said it plans to triple its output of higher-grade uranium. Iran, on the other hand, says the aim of its enrichment programme is energy security. Iran currently has short- and medium- range missiles with improved lethality and more accurate guidance systems. Iran's launch of a medium range, solid fuel ballistic missile demonstrates its ability to hit targets in Israel and southern Europe.

North Korea: The North Korean regime also poses a serious nuclear and missile proliferation concern for the US. North Korea's nuclear weapons capabilities and development of intercontinental ballistic missiles are advancing faster and is therefore a bigger potential threat. North Korea currently deploys a No Dong ballistic missile capable of hitting Japan and South Korea, and is developing a new intermediate-range ballistic missile (IRBM) which can reach Guam and the Aleutian Islands. They also successfully demonstrated the staging and separation technologies required to launch a Taepo-Dong

2 ICBM which has the capability to reach the United States. On January 6, 2016, North Korea conducted a fourth nuclear test. North Korea also claims that US has put Kim Jong-un on its list of sanctioned individuals.

China: The partnership between US & China is such that both nations regards each other as a potential adversary as well as a strategic partner. China and the US have mutual political, economic, and security interests. China is one ascendant country with the potential for competing with the United States. China's arsenal poses serious threat to the United States. Cyber-threats coming from China are also on the rise. China is also developing capacity for attacking satellites and for cyberwarfare. The two countries remain in dispute over territorial issues in the South China Sea. China and the US maintain effective communication and coordination on major regional and international as well as global issues such as Korean Peninsula, Iranian nuclear issue, South Asia and other security issues. In recent years, China and the US have conducted constructive dialogues and productive cooperation on Asia-Pacific affairs. Both the countries are engaged in a race of advancing military technology.

Russia: Ukraine crisis, Syrian civil war and mutual sanctions imposed has casts shadow on US relationship with Russia. On Sept 2015, US and Russia resume military relations to increase fighting against the Islamic State. Representatives from both the countries also met in UN Headquarters to discuss the situation in Syria and Ukraine. On April 2010, US President Barack Obama and Russian President Dmitry Medvedev signed a new START treaty to reduce the number of active nuclear weapons from 2,200 to 1,550. The grant of temporary asylum to US whistle-blower Mr. Edward Snowden by Russia has created a diplomatic turmoil between US and Russia. Both the countries have conducted joint military manoeuvres, training and counter-terrorist exercises in recent past. Currently, as per reports, US-Russian relations have deteriorated sharply, amid a salvo of accusations and disagreements, to their lowest point since the end of the Cold War.

STRATEGIC RELATIONS

India: India and US relations has recently achieved significant milestones with increase in bilateral trade & investment, cooperation on global security matters, upgraded representation in trade & investment forums and admission into multilateral control regimes like MTCR and joint-manufacturing through technology sharing arrangements. US support for India's bid for permanent membership to UN Security Council as well as India's entry to Nuclear Suppliers Group, Wassenaar Arrangement and Australia Group. In 2016, India and United States signed Logistics Exchange Memorandum of Agreement, a bilateral deal on military logistics exchange including defence technology collaboration, expanding military-to-military exchanges. India-US bilateral cooperation is broad-based and multi-sectoral, covering trade and investment, defence and security, education, science and technology, cyber security, high-technology, civil nuclear energy, space technology and applications, clean energy, environment, agriculture and health. India and the US share an extensive cultural, strategic, military, and economic relationship. Over the years, the defence relationship between India and the United States has steadily improved. US recognise India as a key to its strategic interests as India plays a more active role in the broader Asia-Pacific region. India and USA are making efforts to transform defence ties to pursue opportunities for technological collaboration in the field of joint research and co-production of major defence equipment. India and the US have established a Defence Trade and Technology Initiative (DTTI) aimed at simplifying technology transfer policies and exploring possibilities of co-development and co-production to invest the defence relationship with strategic value. Both countries continue to engage under several bilateral institutional mechanisms.

Japan: The relationship between the United States and Japan is very strong which includes political, economic and military alliances. Japan provides bases as well as financial and material support to US forward-deployed forces, which are essential for maintaining stability in the region. Because of the two countries' combined economic and technological impact on the world, the US-Japan relationship has become global in scope. Japan and the United States have closely cooperated on a vast array of global issues. In 2012, both the countries have agreed to discuss updating 15-year-old guidelines on their security alliance in view of China's growing military presence in the region. The two sides would be involved in strategic discussions reviewing their defence roles in the face of China's military build-up and its naval expansion, formulating joint plans and promoting joint use of defence bases. At present, around 50,000 US troops and more than 10,000 marines are stationed in Japan, many of them on the far southern island chain of Okinawa. Japan contributes irreplaceable political, financial, and moral support to US-Japan diplomatic efforts.

Pakistan: Pakistan is a major non-NATO ally of the United States, which is the second-largest supplier of military equipment to Pakistan after China and largest economic aid contributor. US security assistance to Pakistan is focused on strengthening the counter-terrorism (CT) and counterinsurgency (COIN) capabilities of the Pakistan security forces, and promoting closer security ties and interoperability with the US. After the September 11 attacks, Pakistan became a key ally in the war on terrorism with the United States. The country provides key intelligence and logistical support for the United States. Though both the countries are allies and share strong strategic relationship, there were various setbacks and constantly remains a significant mistrust, which continues to deter successful cooperation in defeating terrorist organisations. Though the relationship between both countries became increasingly strained in recent years, US continue to supply military equipment & assistance to Pakistan and is one of Pakistan's largest donors of foreign aids and counterinsurgency funds.

Israel: The United States and Israel maintain a strategic and supportive alliance which is an important part of US Govt's overall Middle East policy. While Israel receives massive security support and a diplomatic umbrella from the United States, the US knows it can view Israel as a military extension of their policies. Israel is one of the United States' two original major non-NATO allies in the Middle East. Both the countries are engaged in extensive strategic, political and military cooperation. This cooperation is broad and includes American aid, intelligence sharing, and joint military exercises. The US-Israel bilateral relationship is strong, anchored by over USD 3.1 billion in Foreign Military Financing annually. In addition to financial support, the US participates in a high level of exchanges with Israel, to include joint military exercises, military research, and weapons development.

South Korea: South Korea is one of America's closest allies. The relationship have been most extensive since 1948, when the US helped establish capitalism in South Korea and fought on its UN-sponsored side in the Korean War (1950–1953). The United States has maintained Army, Air Force, Navy, and Marine personnel in South Korea in support of its commitment under the US-ROK Mutual Defence Treaty to help South Korea defend itself against external aggression. South Korea has also been designated as a major non-NATO ally of USA. US provides significant military and economic assistance to South Korea.

Vietnam: US relations with Vietnam have become deeper and more diverse in the years since political normalisation. The two countries have broadened their political exchanges through regular and regional security. Cooperation in other areas, such as defence, non-proliferation, counterterrorism, and law enforcement, is also expanding steadily. In 2013, Presidents Obama and Sang launched the US-Vietnam Comprehensive Partnership, an overarching framework for advancing the bilateral relationship. In May 2016, President Obama announced the full lifting of the embargo on Vietnam. Vietnam is a partner in non-proliferation regimes, including the Global Initiative to Combat Nuclear Terrorism.

Multilateral Alliances

United States is a permanent member of the United Nations Security Council. It is also a member of the G8, G20, East Asia Summit (EAS) and Organisation for Economic Cooperation and Development and works closely with fellow NATO members on military and security issues. US have contributed support to various UN peacekeeping missions worldwide. The country has strong relationship with UK and close ties with Canada, Australia, New Zealand, the Philippines, Japan, South Korea, Saudi Arabia and Israel and several European countries such as France and Germany. The US military is building up its presence in Australia's Northern Territory in phases and plans to send a 2,500-strong Marine Air Ground Task Force to Darwin on six-month rotations each year starting in 2016.

North Atlantic Treaty Organisation (NATO): NATO is a security alliance of 28 countries from North America and Europe. NATO's fundamental goal is to safeguard the Allies' freedom and security by political and military means.US supports in many of NATO operations. As of May 2012, the US participation in NATO operation are- 824 troops in Kosovo (KFOR, Kosovo Force) and 90,000 troops in Afghanistan (ISAF, International Security Assistance Force). US Missions' operation and issues in NATO includes Ballistic Missile Defence, Assurance Measure & European Security, Afghanistan, NATO Russia Relations, Kosovo, Monitoring the Mediterranean Sea and Counter-piracy off the Horn of Africa.

DEFENCE CAPABILITIES

ARMY: The United States Army is the main branch of the United States Armed Forces responsible for land-based military operations. It is the largest and oldest established branch of the US military. The United States Army is made up of three components: the active component, the Regular Army; and two reserve components, the Army National Guard and the Army Reserve. The Army is headed by the Secretary of the Army (SECARMY), and by a chief military officer, the Chief of Staff of the Army (CSA) who is also a member of the Joint Chiefs of Staff.

Personnel: 475,000 (540,000 Reserve & National Guard)

Equipment

Category	Name	In Service
MBT	M1 Abrams	8,848 (4,393 M1A1 & 3,500 M1 in storage,1,174 M1A2 and M1A2SEP variants)
APC/IFV	M2/M3 Bradley	
	M1120 Series	4,466
	M113	
	M117 ASV	2,777
	M88 Hercules	748
	M93 Fox	108
MRAPs	M-ATV	5,500

Army continued...

Category	Name	In Service
MRAPs	Caiman	2,868
	Couger H/HE	1,317
	International MaxxPro	5,214
	RG 31 / 33 & 33L	595 / 1,167
	Buffalo	750
Artillery	M109 155 mm self-propelled howitzer	950
	M198 155 mm	327
	M777 155 mm gun-howitzer	403
	M119 105 mm howitzer	392
Mortar	60 mm M224	2,000
	81 mm M252	450
	120 mm M120/M120A1	1000
MLRS	M270	857
	M142 HIMARS	340
Anti-Tank GW	TOW launchers	7000
	Hellfire & Javelin	30,000
Recoilless rifle	84 mm M-136 (AT-4)	2,70,000
SAM	THAAD	2
	PATRIOT	10
	Avenger (Pedestal Mounted Stinger)	674
	Stinger MANPADS	

Army Aviation

Category	Name	In Service
Fixed wing aircraft	C-12	127
	RC-12	50
	C-23	51
Trainer	TH-55	200
	TH-67 Creek	137
	OH-58A	
Helicopter	OH-58D	341
	OH-58A/C	281
	AH-64A/D	710
	CH-47D/F	442
	MH-47E/ MH-47G	61 & 1
	TH-55A	147
	UH-60A	948
	MH/UH-60L/K	694
	MH-UH-60M	64
	UH-72A	100

New Procurements/ Upgrades

✦ Boeing has been contracted to remanufacture 117 AH-64D Apaches to the new, more capable AH-64E model. The agreement includes the acquisition of Longbow Crew Trainers, logistical support and spares, carries a total contract value of about USD 1.5 billion.

✦ CACI International Inc. awarded a USD 32.5 million contract to support the development of specialised technologies for the US Army's Night Vision and Electronic Sensors Directorate (NVESD).

✦ Raytheon Company's Patriot missiles, critical components of the Patriot Air and Missile Defence System, have received US Army approval for a second recertification, extending the operational life of the worldwide inventory of Patriot missiles from 30 to 45 years.

✦ The US Army Tank-Automotive and Armaments Command (TACOM) has awarded USD 16 million contract to BAE Systems for the production of 44 additional High Mobility Artillery Rocket System vehicles.

✦ A total of 2,600 M1117 Guardian Armoured Security Vehicle (ASV) have been ordered, with deliveries underway.

✦ Force Protection, Inc. received a modification to contract W56HZV-08-C-0028 from the United States Army Tank-Automotive and Armaments Command (TACOM) for approximately USD62.4 million for 60 Buffalo Mine Protected Clearance Vehicles (MPCV).

✦ The Joint Light Tactical Vehicle or JLTV Technology Development phase industry teams have begun to build government prototypes. The Army plans to purchase 55,000 JLTVs and the Marines plan to buy 5,500.

✦ US Army and US Navy signed a joint five-year contract for the procurement of 537 H-60 helicopters from Sikorsky plus options for an additional 263 aircraft, spares, and kits. The firm order aircraft were valued at USD 7.4 billion.

✦ US Army has ordered 452 CH-47F Chinook helicopters which will be delivered to them by 2020, including 120 newly built and 332 remanufactured aircraft.

✦ BAE Systems has won an USD 850 million contract to run an ammunition factory for the US Army in Virginia.

✦ BAE Systems awarded a contract modification for USD 108.4 million to provide 45 M88A2 Heavy Equipment Recovery Combat Utility Evacuation System (Hercules) vehicles and associated parts to the US Army and Marine Corps.

✦ US Army has ordered 65 additional M1117 Armoured Security Vehicles (ASVs) from Textron Marine and Land Systems at USD 49.61 million.

✦ US Army has awarded a USD 139.6 million contract to Lockheed Martin for supplying 44 combat-proven High Mobility Artillery Rocket Systems (HIMARS) launcher fleet.

✦ ATK has received a USD 36 million order from the US Army to produce 120mm training ammunition for tanks.

✦ Lockheed Martin has received contracts totalling USD 921 million from the US Army Aviation and Missile Command for hardware and services associated with the combat-proven Patriot Advanced Capability-3 (PAC-3)

Missile Segment programme.

✦ The US Army awarded BAE Systems a contract option worth USD 245.3 million to complete the low-rate initial production (LRIP) of the M109A7 self-propelled howitzer and M992A3 ammunition carrier.

✦ Lockheed Martin has been awarded USD 528 Million US Army Contract to build more interceptors for the Terminal High Altitude Area Defence (THAAD) system.

✦ The US Army has awarded a contract worth USD56 million to develop the 3rd Generation Forward Looking Infrared B-Kit, an upgrade to night vision equipment that will allow ground troops to discriminate between friend or foe at twice the distance of current systems.

✦ Lockheed Martin has received an USD 331.8 million contract from the US Army for Lot 11 production of Guided Multiple Launch Rocket System (GMLRS) rockets.

✦ Raytheon Missile Systems has been awarded a USD 23 million US Army contract modification for counter-rocket artillery mortar weapon system's logistics and support.

✦ The US Army has awarded Lockheed Martin a USD 49.3 million contract to upgrade the Modernised Target Acquisition Designation Sight/Pilot Night Vision Sensor (M-TADS/PNVS) on the AH-64E Apache helicopter.

✦ The US Army Intelligence and Security Command (INSCOM) has awarded BAE Systems a five-year contract worth as much as USD 75 million to provide enhanced geospatial intelligence technical and analytical support.

A US Army soldier on patrol with the support of a Humvee vehicle

Lockheed Martin Terminal High Altitude Area Defence (THAAD) system used by the army for ballistic missile protection

NAVY: The United States Navy (USN) is the naval warfare service branch of the United States Armed Forces and one of the seven uniformed services of the United States. There are nine components in the operating forces of the US Navy: the United States Fleet Forces Command, United States Pacific Fleet, United States Naval Forces Central Command, United States Naval Forces Europe, Naval Network Warfare Command, Navy Reserve, United States Naval Special Warfare Command, Operational Test and Evaluation Force, and Military Sealift Command. The Navy is administratively managed by the Department of the Navy, which is headed by the civilian Secretary of the Navy. The Department of the Navy is itself a division of the Department of Defence, which is headed by the Secretary of Defence. The highest ranking naval officer is the Chairman of the Joint Chiefs of Staff.

Major naval bases: China Lake, CA; Lemoore, CA; Monterey, CA; Point Mugu, CA; San Diego, CA; Seal Beach, CA; Groton, CT; Washington, DC; Jacksonville, FL; Key West, FL; Pensacola, FL; Atlanta, GA; Kings Bay, GA; Pearl Harbor, HI; Great Lakes, IL; Crane, IN; New Orleans, LA; Annapolis, MD; Patuxent River, MD; Brunswick, ME; Gulfport, MS; Meridian, MS; Fallon, NV; Earle, NJ; Willow Grove, PA; Newport, RI; Charleston, SC; Millington, TN; Corpus Christi, TX; Fort Worth, TX; Ingleside, TX; Kingsville, TX; Norfolk, VA; Silverdale, WA; Bremerton, WA; Everett, WA; Whidbey Island, WA

Personnel: 329,302 (108,820 reserve)

Equipment

Category	Name	In Service
Aircraft Carrier	CVN Nimitz Class	10
	Gerald R. Ford class	3+ Planned
Destroyer	DDG Arleigh Burke Class	76 (8 on order)
	Zumwalt class	1 (2 Planned)
Submarine	SSBN/SSGN Ohio Class (Ballistic missile sub)	14
	SSBN/SSGN Ohio Class (Cruise missile sub)	4
	SSN Virginia Class	11 (48 planned)
	SSN Seawolf Class	3
	SSN Los Angeles Class	45
Frigate	FFG Oliver Hazard Perry Class	10
	USS Constitution	1
Cruiser	CG Ticonderoga Class	22
Mine Warfare Force	MCM Avenger Class	14
	Austin	1
Command ship	LLC Blue Ridge Class	2
	Coronado	1
	La Salle	1

Navy continued...

Category	Name	In Service
Auxiliary ship	Wichita class	2
	Supply class	4
	Simon Lake class	1
	Kiluauea	4
Amphibious Force	Wasp Class	8
	Tarawa Class	2
	San Antonio Class	9
	Austin Class	4
	Harpers Ferry Class	4
	Whidbey Island Class	8
	Anchorage Class	3
	America Class	11 planned
Missiles	AIM-7 Sparrow, AIM-9 Sidewinder, AIM-120 AMRAAM, AGM-65 Maverick, AGM-84 Harpoon, AGM-88 HARM, AGM-154 Joint Standoff Weapon, BGM-109 Tomahawk (ASM)*, SLAM-ER, SM-3, etc	

*(The United States Navy has a stockpile of around 3,500 Tomahawk cruise missiles of all variants, with a combined worth of approximately USD 2.6 billion.)

Naval Aviation

Category	Name	In Service
Fighter/ attack/ Reconna-issance aircraft	C-2A Transport	34
	E-2C Electronic warfare / E-2D Advanced Hawkeye	67 / 25
	EA-6B Electronic warfare	81
	EA-18 Growler	100
	F/A-18 series (A, B, C, D, E, F)	922
	P-3C Maritime	170
	P-8A Poseidon	18
Helicopter	HH-60H	49
	MH-53E	36
	MH-60R	46 (252 on order)
	MH-60H	159 (272 on order)
	SH-60B	129
	SH-60F	60
Trainer	F-5 series (E, F, N)	36
	T-6A	49
	T-34C	229
	T-44A	52
	T-45C	218
	TH-57B	44
	TH-57C	85

New Acquisition/Procurements

✦ The US Navy has awarded BAE Systems a contract worth USD 600 million for Advanced Precision Kill Weapon System (APKWS) laser-guided rockets. The APKWS rocket is the only US DoD fully qualified guided 2.75-inch rocket.

✦ US Navy has awarded USD 92 million contract modification to Lockheed Martin for work on Lockheed Martin Corp. (LMT) has been awarded a USD 92 million contract modification from the US Navy.

✦ Raytheon Missile Systems is being awarded USD 12,115,859 to a previously awarded cost-plus-fixed-fee, indefinite-delivery/indefinite-quantity contract for up to 52,473 hours of advanced technology insertion and integration support of weapons systems.

✦ Lockheed Martin has been awarded an USD 80,043,707 modification to previously awarded contract to exercise options for Aegis Combat System Engineering Agent (CSEA) efforts for the design, development, integration, test, and delivery of computer programme baselines and associated technology insertion hardware design support for the next/future Advanced Capability Build (ACB).

✦ The US Navy has awarded a USD 2.5 Billion Cyber contract to upgrade the Navy's ship-based computer networks against cybersecurity threats.

✦ The US Navy announced a record USD 17.645 billion contract to build 10 new SSN 774 Virginia-class nuclear-powered attack submarines. The Block IV award covers hull numbers SSN 792 through SSN 801. The 10th ship to be procured under this contract is scheduled for delivery in 2023.

✦ The US Navy has awarded Lockheed Martin's suburban Syracuse plant a contract worth USD 147 million to upgrade its electronic warfare defences against threats such as anti-ship missiles.

✦ The Navy has awarded Sea System Command contract worth up to USD 16.3 million to Accenture for providing mobile maintenance support to the Naval Sea Systems Division.

✦ The US Navy has awarded Northrop Grumman Corporation a USD 3.6 billion contract to deliver 25 new E-2D Advanced Hawkeye aircraft, bringing the total number of aircraft on contract to 50.

✦ The US Navy's Office of Naval Research (ONR) has awarded an USD 11 million contract to Raytheon to develop a vehicle-based laser device capable of defeating low-flying threats such as enemy drones.

✦ The US Navy has awarded Lockheed Martin a USD 10 million cost-plus-fixed-fee contract to provide engineering design services for the MK 41 Vertical Launching System (VLS) that helps protect the fleet from numerous naval warfare threats.

✦ Defence Advanced Research Projects Agency (DARPA) has awarded a USD 150 million worth development contract to Lockheed Martin Corporation for new anti-ship missile.

✦ The United States plans to buy 55 Littoral combat ships (LCS) over times, which are designed to counter submarines, mines, and small surface craft in the world's coastal regions. Two of those ships have already been built. The Navy also has two more ships (one of each type) under construction. The remaining 51 ships would be purchased from 2010 through 2031.

✦ The US Navy is planning to buy 480 F-35C Joint Strike Fighter (JSF), which will replace the F/A-18A, B, C, and D Hornets.

✦ US Navy is planning to purchase 117 of the Boeing Next-Generation 737-based P-8A anti-submarine warfare, anti-surface warfare, intelligence, surveillance and reconnaissance aircraft to replace its P-3 fleet.

✦ Under the USD 1.5 billion contract awarded recently to Huntington Ingalls Industries' (HII) shipbuilding division for detail design and construction of the US Navy's 11th and final San Antonio-class amphibious transport dock ship LPD 27. The navy has received six of 11 LPDs to date, while the remaining vessels are currently under different stages of construction. Delivery of the new vessel is scheduled for mid-2017.

✦ BAE Systems has received a contract worth more than USD 20 million to supply two 57mm Mk 110 MOD 0 guns and engineering support to the US Navy and Coast Guard. The two Mk 110 gun systems include additional engineering support options over the next few years.

✦ The US Missile Defence Agency (MDA) has awarded a contract worth USD 350 million to Raytheon to increase its procurement of the Standard Missile-3 Block IB variants from 8 to 44. The Navy has also awarded Raytheon a contract worth USD 243 million to deliver Standard Missile-6 (SM-6) Block I variant.

✦ A contract worth USD 1.4 billion was awarded to General Dynamics for the construction of USS Zumwalt at Bath Iron Works in Bath, Maine. The future USS Zumwalt (DDG 1000) will be the lead ship of the Navy's newest destroyer class, designed for littoral operations and land attack.

✦ Huntington Ingalls Industries' Ingalls Shipbuilding division has received a USD 602 million contract modification to fund construction of the Arleigh Burke-class (DDG 51) guided missile destroyer DDG 119 for the U S Navy. The modification also includes an additional USD 79.4 million in advance procurement for future destroyers.

✦ The US Navy received the first Lockheed Martin F-35C Lightning II Navy carrier variant (CV) aircraft. The aircraft is the US Navy's first 5th generation F-35 production aircraft. The F-35C, the US Navy and Marine Corps' CV, has larger wings and more robust landing gear than the other F-35 variants. The US Navy plans to declare Initial Operational Capability with the CV in 2019.

✦ The Navy accepted delivery of the USNS Brunswick. The ship will be used for high-speed mobility and transportation of troops, military vehicles, and equipment. Austal is also under contract for the construction of EPFs 8-10 and materials to support the construction of EPF 11.

✦ The US Navy has awarded a production contract for its Surface Electronic Warfare Improvement Program (SEWIP) Block 1B3 to General Dynamics Mission Systems. General Dynamics will deliver Block 1B3 system upgrades over the next five years. The upgrades will be integrated into new and existing Navy ships.

✦ Raytheon has won a USD 102.9 million contract from the US Navy to manufacture, test and deliver fully integrated Navy Multiband Terminals (NMT). Work is scheduled to be completed by Sep 30, 2018.

✦ Harris Corporation has received a three-year, USD 54 million ceiling, single-award IDIQ contract from the US Naval Research Laboratory to provide electronic warfare (EW) technology and engineering services for the Advanced Decoy Architecture Project (ADAP).

✦ Huntington Ingalls Industries Inc.'s HII shipbuilding division won a contract worth USD 200 million from the US Naval Sea Systems Command of the District of Columbia to build LPD 28, the 12th amphibious transport dock of the San Antonio (LPD 17) class.

✦ Austal USA was awarded a USD 326 million contract by the US Navy Friday for two more Expeditionary Fast Transport ships (EFT). The new contract adds ships 11 and 12 to the pre-existing 10-ship contract and takes the value of the program to USD 1.9 billion in total.

✦ The US Navy has awarded Lockheed Martin an initial USD 148.9M contract for full rate production of Surface Electronic Warfare Improvement Program (SEWIP) Block 2 systems with four additional option years to upgrade the fleet's electronic warfare capabilities so warfighters can respond to evolving threats.

✦ Lockheed Martin Rotary and Mission Systems has been awarded a USD 157 million US Navy contract for the procurement of Aegis combat system follow-on support services.

The US Navy amphibious assault ship USS America (LHA-6) returns to Huntington Ingalls Shipyard, Pascagoula, Mississippi (USA), after completing sea trials. *Image Courtesy: US Navy*

MARINE CORPS: The United States Marine Corps (USMC) is a branch of the United States Armed Forces responsible for providing power projection from the sea, using the mobility of the United States Navy to deliver combined-arms task forces rapidly. The Department of the Navy, led by the Secretary of the Navy, oversees both the Marine Corps and the Navy. The most senior Marine officer is the Commandant of the Marine Corps, responsible for organising, recruiting, training, and equipping the Marine Corps so that it is ready for operation under the command of the Unified Combatant Commanders. The Marine Corps is organised into four principal subdivisions: Headquarters Marine Corps (HQMC), the Operating Forces, the Supporting Establishment, and the Marine Forces Reserve (MARFORRES or USMCR).

Personnel: 182,000 (38,500 reserve)

Equipment

Category	Name	In Service
MBT	M-1A1 Abrams	483
APC/IFV	LAV	735
	LVTP-7A1	400
	MRAP	2,200
Artillery	155 mm M777A1 Ultra Light Howitzer	380
	155 mm M-109	140
Mortar	81 mm M252	500
MLRS	MLRS	42
Anti Tank GW	TOW	
	Javelin	
SAM	Stinger	30000
	Avengers	79
	LAV-AD	16

Marine Aviation

Category	Name	In Service
Fighter/ attack/ Reconna-issance aircraft	F/A-18C/CN	168
	F/A-18B/D	72
	AV-8B	160
	TAV-8 Harrier trainer	15
	EA-6B Prowler	20
	KC-130-J/T Hercules	70
Helicopter	AH-1-W/Z Cobra	147
	UH-1 Iroquois	11
	CH-53E Super Stallion	160
	MV-22 Osprey	348 (Planned)
UAV	Wasp, AeroVironment Switchblade, RQ-7 Shadow, RQ-11 Raven, RQ-14 Dragon Eye, RQ-20 Puma, T-20, ScanEagle, RQ-21A Blackjack	

New Acquisition/Procurements

✦ ST Kinetics has been awarded a USD 121.5 million contract to supply 13 amphibious combat vehicle prototypes, known as Terrex 2, to the United States Marine Corps.

✦ The Marine Corps awarded contracts to two manufacturers to build prototypes for the Engineering, Manufacturing and Development of the Amphibious Combat Vehicle Phase 1, Increment 1, worth a total of USD 225 million. BAE's contract is for USD 103.8 million, while SAIC's is for USD 121.5 million.

✦ US Marine Corps Systems Command (MCSC) has awarded a USD 33.2 million contract to General Dynamics Land Systems-Canada, for the delivery 21 RG-31 Mk5EM and 6 RG-31 Mk5E vehicles, for its Mine Resistant Ambush Protected (MRAP) vehicle programme.

✦ F-35B, a short-takeoff and vertical-landing (STOVL) variant slated to replace the US Marine Corps (USMC) AV-8 Harrier IIs and F/A-18 Hornets.

✦ USMC to procure 360 MV-22 Osprey. Currently, a batch of 141 aircraft is on order with deliveries underway. The Marine Corps variant, the MV-22B, is an assault transport for troops, equipment and supplies, capable of operating from ships or from expeditionary airfields ashore. It is replacing the Marine Corps' CH-46E and CH-53D.

✦ US Marine Corps has awarded a USD 125 million dollar contract to Oshkosh Defence, a division of Oshkosh Corporation, to deliver more than 200 Logistics Vehicle System Replacement (LVSR) tractors and nearly 70 LVSR wreckers.

✦ US Marine Corps has awarded a USD 56 million contract to BAE Systems for five separate delivery orders on the RG-33 Mine Resistant Ambush Protected (MRAP).

✦ The USMC Systems Command selected Tactical and Survival Specialties, Inc. to provide and ship Explosive Ordnance Disposal (EOD) kits to Marine Corps EOD units worldwide. The five-year contract is valued at USD 45 million.

✦ US Marine Corps has awarded Booz Allen Hamilton a USD 48.6 million contract for technical, managerial and quantitative and qualitative analytical support.

✦ US Marine Corps has awarded Intelligent Decisions (ID), Inc. a key Indefinite Delivery Indefinite Quantity (IDIQ) multi-award contract – with a potential total value of USD 775 million over the next three- to five-years – for the delivery of commercial-off-the-shelf computer systems (COTS) as part of the Marine Corps Common Hardware Suite (MCHS).

✦ US Marine Corps Logistics Command has awarded a USD 249 million contract to Raytheon to refurbish USMC's vehicles, as part of its Secondary Reparable (SECREP) programme.

Force Recon platoon conduct close quarter combat training while on deployment. *Image Courtesy: US Marine Corps.*

AIR FORCE: The United States Air Force (USAF) is the aerial warfare service branch of the United States Armed Forces. The US Air Force provides air support for surface forces and aids in the recovery of troops in the field. The Department of the Air Force is headed by the civilian Secretary of the Air Force. The Department of the Air Force is a Military Department within the Department of Defence. The highest ranking military officer in the Department of the Air Force is the Chief of Staff of the Air Force who exercises supervision over Air Force units, and serves as a member of the Joint Chiefs of Staff.

Personnel: 308,606 (69,200 reserve)

Equipment

Category	Name	In Service
Fighter/ attack/ Reconna-issance/ bomber	A-10C Thunderbolt II	291 (To be replaced by F-35A)
	AC-130-U/W/J	28
	F-15 Eagle	192 (To be replaced by the F-35A)
	F-15E Strike Eagle	257
	F-16C/D	957 (To be replaced by the F-35A)
	F-22A	184
	F-35A Lightning II	71 (1763 planned, currently 47 for testing)
	U-2 Dragon Lady	31
	WC-130J Hercules	10
	B-1B Lancer	62
	B-2A Spirit	20
	B-52H Stratofortress	77
	E-3 Sentry	30
	E-4B Nightwatch	4
	E-8C Joint STARS	16
	EC-130H Compass Call	14
	EC-130 Commando Solo III	7
	MC-130 Hercules	61
Transport/ Tanker	C-130H Hercules	261
	KC-10A Extender	59
	KC-135 Stratotanker	397
	C-144	2
	VC-25	2
	C-20	7
	C-130J Super Hercules	120
Cargo	C-5-B/C/M Galaxy	55
	C-12 Huron	67
	C-17A Globemaster III	222
	C-21A Learjet	31

Air Force continued...

Category	Name	In Service
Helicopter	UH-1N Twin Huey	59
	TH-1H Iroquois	37
	HH-60 Pave Hawka	100
	CV-22B Osprey	42 (50 on order)
Trainer	T-1 Jayhawk	157
	T-6A Texan II	449
	T-38 Talon	480
UAV	RQ-4 Global Hawk	37
	RQ-11 Raven	Classified
	RQ-170 Sentinel	
	Wasp III	Classified
	MQ-1B Predator	150
	MQ-9B Reaper	93
Missiles	AGM-65, AGM-86, AGM-88, AGM-129, AGM-130, AGM-154, AGM-158, AIM-7, AIM-9,AIM-120, LGM-30G Minuteman III ICBM	

Spacecraft

Category	Name	In Service
Launch vehicle	Atlas V	
	Delta II	
	Delta IV	
Satellite	Defence Satellite Communications System	9
	Global Positioning System	30
	Milstar	5
	Defence Support Programme	
	Defence Meteorological Satellite Programme	
	Boeing X-37B	2

New Acquisition/Procurements

✦ CACI International has been awarded US Air Force contract worth as much as USD 445 million to consolidate work on the Air Force's main satellite control network. The contract is called CAMMO — short for Consolidated Air Force Satellite Control Network Modifications, Maintenance and Operations.

✦ System of Systems Security Inc. has been awarded a maximum USD 99,000,000 other transaction agreement for prototypes in command, control, communications, and cyber, intelligence, surveillance, and reconnaissance (C4ISR) information sharing information systems.

✦ Composite Engineering Inc. has been awarded an USD 18,674,455 modification to previously awarded contract for Air Force Subscale Aerial Target (AFSAT) Lots 11-13 production. The company will provide additional quantity of 21 AFSAT production units being produced under the basic contract (Lot 12) and the associated warranty.

✦ The US Department of Defence (DoD) has signed an agreement with Lockheed Martin for the production and delivery of additional 32 F-35 Lightning II joint strike fighter (JSF) aircraft. The aircraft will be manufactured under the fifth low-rate initial production (LRIP-5) lot, and the unit cost data will be made available following finalisation and contract signing.

✦ US Air Force has awarded a five-year contract to Boeing for engineering sustainment support on the KC-135 Fleet Support Programme. This latest contract, worth up to USD 20 million annually.

✦ F-35A, a conventional takeoff and landing (CTOL) variant slated to replace US Air Force (USAF) F-16 Fighting Falcons.

✦ Boeing's GPS Operational Control Segment (OCS) has gained full operational status with the US Air Force 50th Space Wing, Schriever Air Force Base in Colorado.

✦ The US Air Force has awarded Northrop Grumman Corporation a USD 65 million contract to provide tactical situational awareness to airlift aircrews, fulfilling an urgent operational need for enhanced communications and an operational picture of the battle-space displayed in the cockpit.

✦ Northrop Grumman has received a USD 48.7 million contract from US Air Force for continuing operations and support of the Battlefield Airborne Communications Node (BACN) system.

✦ The Missile Defence Agency (MDA) and US Air Force have awarded Raytheon a USD 125.3 million contract to modernise and add new capability to an Air Force early warning radar (EWR) system. Work is expected to be completed by the end of 2017.

✦ Lockheed Martin Space Systems has been awarded a USD 395 million US Air Force modification contract for two additional GPS III satellites 9 and 10.

✦ Lockheed Martin has received a USD 152 million contract from the US Air Force for Sniper Advanced Targeting Pod (ATP) sustainment support over a five-year period.

✦ Lockheed Martin received an USD 87.6 million contract from the US Air Force for follow-on production of Paveway II Plus Laser Guided Bomb (LGB) kits.

✦ US Missile Defence Agency has awarded Raytheon a USD 230 million contract for 14 Standard Missile-3 Block IA missiles and five SM-3 Block IB missiles.

✦ The US Air Force has awarded a contract worth USD 241.6 million to Lockheed Martin for Lot 10 production of the Joint Air-to-Surface Standoff Missile (JASSM) and its Extended Range (ER) variant. The US Air Force plans to acquire 4,900 JASSMs and its longer range JASSM-ER.

✦ The US Air Force has issued a request for proposals (RFP) on its next-generation long range radar system. The Three-Dimensional Expeditionary Long Range Radar (3DELRR) programme will replace the USAF legacy TPS-75 land-based radar systems.

✦ The United States Air Force Life Cycle Management Center (AFLCMC) has awarded GE Aviation a USD 1 billion contract to continue maturing its three-stream adaptive cycle engine via the Adaptive Engine Transition Programme (AETP). AETP is scheduled to run through 2021 with extensive component, rig and engine testing.

✦ Lockheed Martin in Sept 2016 delivered the 2,000th Joint Air-to-Surface Standoff Missile (JASSM) to the US Air Force.

CV-22 Osprey tiltrotor aircraft. *Image Courtesy: US Air Force*

US STRATEGIC COMMAND

The United States Strategic Command, or USSTRATCOM, is headquartered at Offutt Air Force Base in Nebraska. The command is one of nine US unified commands under the Department of Defence. Strategic Command was established in 1992 as a successor to Strategic Air Command (SAC).

USSTRATCOM is the command and control centre for US strategic forces and controls military space operations, computer network operations, information operations, strategic warning and intelligence assessments as well as global strategic planning. The command is responsible for both early warning of and defence against missile attack and long-range conventional attacks. The command is charged with deterring and defending against the proliferation of weapons of mass destruction. The command has worldwide functional responsibilities not bound by any single area of operations. The command's scope of responsibilities includes the interrelated areas of space operations, information operations, computer network operations, and strategic defence and attack. Tying these areas together is a globally focused command and control, communications and intelligence, surveillance and reconnaissance network (C3ISR).

The US Cyber Command (USCYBERCOM) was activated as a subordinate unified command under USSTRATCOM in September 2009, with a mission to coordinate computer-network defence and direct US cyber-attack operation.

Service Commands

✦ US Army - Army Space and Missile Defence Command/Army Forces Strategic Command

✦ US Navy - Fleet Forces Command

✦ US Marines - Marine Corps Forces Strategic Command

✦ US Air Force - Air Force Global Strike Command & Air Force Space Command

Current Unified Commands

Nine Combatant Commands are responsible for conducting the Department's military operational missions around the world. Six commands have specific military operational mission objectives for geographic areas of responsibility

- US European Command (USEUCOM) is responsible for activities in Europe, Greenland, and Russia.
- US Central Command (USCENTCOM) is responsible for the Middle East, Egypt, and several of the former Soviet republics.
- US Pacific Command (USPACOM) is responsible for China, South and Southeast Asia, Australia, and the Pacific Ocean.
- US Southern Command (USSOUTHCOM) is responsible for Central and South America and the Caribbean.
- US Northern Command (USNORTHCOM) is responsible for North America, including Canada and Mexico.
- US Africa Command (USAFRICOM) is responsible for Africa (except Egypt).

Three Commands have worldwide mission responsibilities focused on a particular function(s):

- US Strategic Command (USSTRATCOM) provides global deterrence capabilities, direction of Global Information Grid operations, and synchronizes Department efforts to combat weapons of mass destruction worldwide.
- US Special Operations Command (USSOCOM) leads, plans, synchronises, and as directed, executes global operations against terrorist networks. RQ-11 Raven & Stalker XE UAVs are used by USSOCOM.
- US Transportation Command (USTRANSCOM) is tasked to coordinate missions worldwide using both military and commercial transportation services.

Currently, the United States nuclear arsenal is deployed in three areas:

- Land-based intercontinental ballistic missiles, or ICBMs;
- Sea-based, nuclear submarine-launched ballistic missiles, or SLBMs; and
- Air-based nuclear weapons of the US Air Force's heavy bomber group

The Pentagon is aiming to invest about 10 percent more over the next five years to upgrade the nation's nuclear deterrent. THE US has begun a USD US10 billion upgrade of its nuclear weapons arsenal after two hard-hitting reviews found decades of neglect have left its most significant line of defence in disrepair. The US had already put itself on the path to modernise ageing weapons at a cost of up to USUSD 1 trillion over the next three decades, including a dozen ballistic missile submarines, up to 100 new bombers and 400 land-based missiles.

US Nuclear Force Structure under the New START Treaty

The New START Treaty establishes central limits on the number of nuclear delivery platforms and warheads associated with them: 700 for deployed ICBMs, deployed SLBMs, and deployed heavy bombers; 800 for deployed and non-deployed ICBM launchers, deployed and non-deployed SLBM launchers, and deployed and non-deployed heavy bombers; and 1,550 warheads on deployed ICBMs, warheads on deployed SLBMs, and nuclear warheads counted for deployed heavy bombers. By February 5, 2018, the Department of Defence (DoD) will transition today's triad to the Treaty-compliant force structure below which fully supports the President's National Security Strategy and Nuclear Weapons Employment Strategy.

New START Treaty Aggregate Numbers of Strategic Offensive Arms

(Data in this Fact Sheet comes from the biannual exchange of data required by the Treaty. It contains data declared current as of 1 September 2016.)

Category of Data	USA
Deployed ICBMs, Deployed SLBMs, and Deployed Heavy Bombers	681
Warheads on Deployed ICBMs, on Deployed SLBMs, and Nuclear Warheads Counted for Deployed Heavy Bombers	1367
Deployed and Non-deployed Launchers of ICBMs, Deployed and Non-deployed Launchers of SLBMs, and Deployed and Non-deployed Heavy Bombers	848

Source: Bureau of Arms Control, Verification, and Compliance; October 1, 2016.

- 700 deployed intercontinental ballistic missiles (ICBMs), deployed submarine-launched ballistic missiles (SLBMs), and deployed heavy bombers equipped for nuclear armaments;
- 1,550 nuclear warheads on deployed ICBMs, deployed SLBMs, and deployed heavy bombers equipped for nuclear armaments (each such heavy bomber is counted as one warhead toward this limit);
- 800 deployed and non-deployed ICBM launchers, SLBM launchers, and heavy bombers equipped for nuclear armaments.

The Treaty also allows the flexibility to modernise the US nuclear deterrent and ensure its safety, security, and effectiveness without constraining US missile defences or long-range conventional strike development. For the duration of New START, the US triad of ICBMs, SLBMs, and nuclear-capable heavy bombers will be maintained, keeping all Ohio-class strategic submarines in the force for the near term and "de-MIRVing" all Minuteman III ICBMs to a single warhead each to increase stability in a crisis. The Treaty provides for 18 on-site inspections per year. There are two basic types of inspections. Type One inspections focus on sites with deployed and non-deployed strategic systems; Type Two inspections focus on sites with only non-deployed strategic systems. Permitted inspection activities include confirming the number of reentry vehicles on deployed ICBMs and deployed SLBMs, confirming numbers related to non-deployed launcher limits, counting nuclear weapons onboard or attached to deployed heavy bombers, confirming weapon system conversions or eliminations, and confirming facility eliminations.

Each side is allowed to conduct ten Type One inspections and eight Type Two inspections annually. According to a study by US Congressional Research Service based on US Strategic Forces, even though the United States plans to reduce the number of warheads deployed on its long-range missiles and bombers, consistent with the terms of the New START Treaty, it also plans to develop new delivery systems for deployment over the next 20-30 years. At the present time, the US land-based ballistic missile force (ICBMs) consists of 450 Minuteman III ICBMs, each deployed with one warhead. The fleet will decline to 400 deployed missiles, while retaining all 450 launchers, to meet the terms of the New START Treaty. The Air Force is also modernising the Minuteman missiles, replacing and upgrading their rocket motors, guidance systems, and other components, so that they can remain in the force through 2030. The US ballistic missile submarine fleet currently consists of 14 Trident submarines; each carries 24 Trident II (D-5) missiles. The Navy converted 4 of the original 18 Trident submarines to carry non-nuclear cruise missiles. The remaining submarines currently carry around 1,000 warheads in total; that number will decline as the United States implements the New START Treaty. The Air Force has also begun to retire the nuclear-armed cruise missiles carried by B-52 bombers, leaving only about half the B-52 fleet equipped to carry nuclear weapons. The Air Force plans to procure both a new long-range bomber during the 2020s and a new cruise missile after 2030.

Data in the below Fact Sheet comes from the biannual exchange of data required by the Treaty. It contains data declared current as of 1 March 2016.

Operational USA Strategic Nuclear Forces, July 1, 2016

ICBMs and ICBM Launchers	Minute man-III	Total
Deployed ICBMs	431	431
Non-deployed ICBMs	255	255
Deployed and Non-deployed Launchers of ICBMs	454	454
Deployed Launchers of ICBMs	431	431
Non-deployed Launchers of ICBMs	23	23
Test Launchers	4	4

SLBMs and SLBM Launchers	Trident II	Total
Deployed SLBMs	230	230
Non-deployed SLBMs	199	199
Deployed and Non-deployed Launchers of SLBMs	324	324
Deployed Launchers of SLBMs	230	230
Non-deployed Launchers of SLBMs	94	94
Test Launchers	0	0

Heavy Bombers	B-2A	B-52H	Total
Deployed Heavy Bombers	12	68	80
Non-deployed Heavy Bombers	8	12	20
Test Heavy Bombers	1	2	3
Heavy Bombers Equipped for Non-nuclear Armaments	0	8	8

Source: Bureau of Arms Control, Verification, and Compliance; July 1, 2016.

Missile Defence Agency

The Missile Defence Agency (MDA) is a research, development, and acquisition agency within the Department of Defence. The MDA's mission is to develop, test, and field an integrated, layered, ballistic missile defence system (BMDS) to defend the United States, its deployed forces, allies, and friends against all ranges of enemy ballistic missiles in all phases of flight. There are several other agencies and military commands which play a role, such as the United States Army Space and Missile Defence Command. Missile defence elements are operated by United States military personnel from US Strategic Command, US Northern Command, US Pacific Command, US Forces Japan, US European Command and others. The United States has missile defence cooperative programmes with a number of allies, including United Kingdom, Japan, Australia, Israel, Denmark, Germany, Netherlands, Czech Republic, Poland, Italy, France, Poland, Romania, Spain, Turkey, Kuwait, Qatar, Saudi Arabia, UAE, and South Korea. The Missile Defence Agency also actively participates in NATO activities to maximise opportunities to develop an integrated NATO ballistic missile defence capability.

Major Armaments Producers

Top 20 US government defence contractors:

Lockheed Martin Corp., Northrop Grumman Corp., Boeing Co., Raytheon Co., General Dynamics Corp., KBR Inc., Science Applications International Corp., L-3 Communications Corp., Computer Sciences Corp., ITT Corp., CACI International Inc., BAE Systems Inc., Hewlett-Packard Co., Harris Corp., United Technologies Corp., URS Corp., Jacobs Engineering Group Inc., DRS Technologies Inc., Dell Computer Corp., and Rockwell Collins Inc.

The *Defence Advanced Research Projects Agency (DARPA)* is the research and development office for the US Department of Defence. DARPA funds unique and innovative research through the private sector, academic and other non-profit organizations as well as government labs. DARPA research runs the gamut from conducting scientific investigations in a laboratory, to building full-scale prototypes of military systems. They fund research in biology, medicine, computer science, chemistry, physics, engineering, mathematics, material sciences, social sciences, neuroscience, and more. DARPA is independent from other more conventional military R&D and reports directly to senior Department of Defence management.

The *Missile Defence Agency (MDA)* is a research, development, and acquisition agency within the Department of Defence. The Missile Defence Agency's mission is to develop, test and prepare for deployment of a missile defence system.

DEFENCE DEALS

Arms Export

The USA alone accounts for one-third of arms exports worldwide. USA remains the leading arms exporter supplying almost USD 23 billion worth of weapons to other countries in 2015. 94 countries imported US made weaponry, with the Middle East accounting for 32 percent of the purchases. USA has been continually restocking supplies in the Middle Eastern crisis region. US has taken a firm lead as the major arms exporter globally, according to new data on international arms transfers published in 2015 SIPRI. Dr. Aude Fleurant, Director of the SIPRI Arms and Military Expenditure Programme noted that, US has long seen arms exports as a major foreign policy and security tool, but in recent years exports are increasingly needed to help the US arms industry maintain production levels at a time of decreasing US military expenditure. The main customers include Saudi Arabia, UAE & Turkey in the Middle East.

The USA's share of the volume of international arms exports was 31 per cent in 2010-14. US has done overseas arms sales agreements with countries like Afghanistan, Algeria, Africa, Bahrain, Bangladesh, Bhutan, British Indian Ocean Territories, Egypt, India, Iran, Iraq, Israel, Jordan, Kuwait, Lebanon, Libya, Maldives, Morocco, Nepal, Oman, Pakistan, Qatar, Saudi Arabia, Sri Lanka, Syria, South Korea, Tunisia, United Arab Emirates, Yemen and many more. Over the past decade, US arms sales have grown from USD 8 billion, and they are likely to continue growing in coming years. Sales to Afghanistan alone totaled nearly USD 20 billion for fiscal years 2009 through 2011.Among the biggest potential arms deals on the table now are armaments sale in India and Brazil, various modernisation programmes for Saudi Arabia, and continuing support for arms sales to Iraq, Afghanistan, Pakistan and Lebanon. The USA delivered weapons to 75 recipients in 2006–10, more than any other supplier. Asia and Oceania was the biggest recipient region of US weapons—accounting for 44 per cent of US deliveries, including 22 per cent in total for South Korea, Japan and Taiwan. The Middle East accounted for 28 per cent and Europe for 19 per cent. Among the USA's deliveries in 2010 were 52 combat aircraft, 2439 armoured vehicles and 108 tanks. Nearly all armoured vehicles went to US allies for use in Afghanistan and Iraq. The US defence market is the largest and most advanced in the world with ample opportunities given the vast size of the defence market.

✦ The State Department has made a determination approving a possible Foreign Military Sale to the United Arab Emirates for exercise participation support. The estimated cost is USD 75 million.

✦ The State Department has made a determination approving a possible Foreign Military Sale to Kuwait for radar field systems and related equipment, training, and support. The estimated cost is USD 194 million.

✦ The State Department has made a determination approving a possible Foreign Military Sale to Egypt for Common Missile Warning System (CMWS) for AH-64E Apache, UH-60 Blackhawks, and CH-47 Chinook Helicopters and related equipment, training, and support. The estimated cost is USD 81.4 million.

✦ The State Department has made a determination approving a possible Foreign Military Sale to Iraq for AC-208 aircraft and related equipment, training, and support. The estimated cost is USD 65.3 million.

✦ The State Department has made a determination approving a possible Foreign Military Sale to Japan for KC-46A aerial refueling aircraft and related equipment, training, and support. The estimated cost is USD 1.9 billion.

✦ The State Department has made a determination approving a possible Foreign Military Sale to Egypt for Sentinel AN/MPQ-64F1 Radars and related equipment, training, and support. The estimated cost is USD 70 million.

✦ The State Department has made a determination approving a possible Foreign Military Sale to Qatar for Mk-V Fast Patrol Boats, equipment, training, and support. The estimated cost is USD 124.02 million.

✦ The State Department has made a determination approving a possible Foreign Military Sale to Afghanistan for individual and crew served weapons, equipment, training, and support. The estimated cost is USD 60 million.

✦ The State Department has made a determination approving a possible Foreign Military Sale to the NATO Support and Procurement Agency for precision guided munitions, equipment, and support. The estimated cost is USD 231 million.

✦ The State Department has made a

determination approving a possible Foreign Military Sale to the Kingdom of Saudi Arabia for M1A2S Saudi Abrams Main Battle Tanks and M88Al/A2 Heavy Equipment Recovery Combat Utility Lift Evacuation System (HERCULES) Armoured Recovery Vehicles (ARV), equipment, training, and support. The estimated cost is USD 1.15 billion.

✦ The State Department has made a determination approving a possible Foreign Military Sale to Argentina for T-6C+ Texan aircraft, training, and support. The estimated cost is USD 300 million.

✦ The State Department has made a determination approving a possible Foreign Military Sale to Israel for Excess SH-60F Sea-Hawk Helicopter equipment and support. The estimated cost is USD 300 million.

✦ The State Department has made a determination approving a possible Foreign Military Sale to the Republic of Korea for SM-2 Block IIIB Standard missiles, containers and support. The estimated cost is USD 65 million.

✦ The State Department has made a determination approving a possible Foreign Military Sale to Australia for AIM-120D Advanced Medium-Range Air-to-Air Missiles and associated equipment, training, and support. The estimated cost is USD 1.22 billion.

✦ The State Department has made a determination approving a possible Foreign Military Sale to Chile for Evolved Seasparrow Missiles (ESSMs), equipment, training, and support. The estimated cost is USD 140.1 million.

✦ The State Department has made a determination approving a possible Foreign Military Sale to Kuwait for F/A-18 C/D services and support. The estimated cost is USD 420 million.

Defence Procurement

Defence Procurement and Acquisition Policy (DPAP) is responsible for all acquisition and procurement policy matters in the Department of Defence (DoD). The DPAP office serves as the principal advisor to the Under Secretary of Defence for Acquisition, Technology and Logistics (AT&L) and the Defence Acquisition Board on acquisition/procurement strategies for all major weapon systems programs, major automated information systems programs, and services acquisitions.

DEFENCE BUDGET

Total defence spending: USD 534.3 billion (base FY-2016); USD 496.1 billion (2015)

Defence spending in terms of GDP: 3.3% (2015)

The FY 2017 budget of USD 582.7 billion complies with the Bipartisan Budget Act of 2015, giving the department both funding stability and protection from the damage of sequestration in FY 2016 and FY 2017. According to US DoD FY 2017 President's Budget Proposal, The base budget of USD 523.9 billion includes an increase of USD 2.2 billion over the FY 2016 enacted budget of USD 521.7 billion. As specified in the budget agreement, DoD's FY 2017 overseas contingency operations budget is USD 58.8 billion, nearly the same as the FY 2016 enacted level of USD 58.6 billion. The combined request represents a total increase of USD 2.4 billion, or less than one percent over FY 2016 enacted levels. The FY-2017 budget request is consistent with the FY 2016 budget request in planning to adjust the size of the force over the next several years to a level of 980,000 soldiers, 308 ships, 182,000 active-duty Marines, and 55 Air Force tactical fighter squadrons. The president's budget funds USD 6.7 billion in FY 2017 for defensive and offensive cyberspace operations, capabilities, and cyber strategy.

Department of Defence Budget (Dollars in Billions):

Budget by Military Dept.	FY 2016 Enacted	FY 2017 Request	USD Change
Army	146.9	148.0	+1.1
Navy	168.8	164.9	-3.9
Air Force	161.8	166.9	+5.1
Defence wide	102.8	102.9	+0.1
Total	**580.3**	**582.7**	**+2.4**

Budget Request: USD 582.7 Billion.

Budget by Appropriation	FY 2016 Enacted	FY 2017 Request	USD Change
Aircraft and Related Systems	50.6	45.3	-5.3
C4I Systems	7.1	7.4	0.3
Ground Systems	9.9	9.8	-0.1
Missile Defence Programmes	9.1	8.5	-0.6
Missiles and Munitions	12.7	13.9	1.2
Mission Support	52.9	52.4	-0.5
Science & Technology (S&T)	13.0	12.5	-0.5
Shipbuilding and Maritime Systems	27.5	27.0	-0.5
Space-Based Systems	7.0	7.1	0.1
Rescissions	-1.8	-	+1.8
Total	**188.0**	**183.9**	**-4.1**

Source: US DoD FY 2017 Budget Request | Defence Budget Overview

According to a press release by US DoD on Fiscal Year 2017 President's Budget Proposal the following summarises the future priorities:

Maritime

✦ The PB 2017 grows the naval fleet by procuring seven major ships in FY 2017 and will increase the battle force by 28 ships, from 280 to 308 ships, in the next five years.

✦ The PB 2017 budget includes USD 5.2 billion in FY 2017 to buy two Virginia-class attack submarines.

✦ The PB 2017 increases funding for advanced undersea capabilities, including USD 37 million for an improved MK-48 torpedo, and USD 106 million for unmanned underwater vehicles in FY 2017.

✦ The budget invests USD 3.2 billion in FY 2017 to buy two DDG-51 Arleigh Burke-class guided missile destroyers.

✦ The budget reduces planned littoral combat ship/fast frigate (LCS/FF) procurement from 52 ships to 40.

Aircraft

✦ The budget includes USD 10.1 billion in FY 2017 for F-35s across the force: 43 F-35As for the Air Force, 16 F-35Bs for the Marine Corps and 4 F-35Cs for the Navy.

✦ For continued development of the Long-Range Strike Bomber, the budget invests USD 1.4 billion in FY 2017.

✦ The budget invests USD 3.1 billion to buy 15 KC-46A Pegasus refuelling tankers in FY 2017.

✦ To ensure enough fighter squadrons are ready to deploy to meet high overseas demand, the Air Force will transition the A-10 fleet two years later than previously planned, enabling a larger near-term force and investment in legacy capabilities.

✦ The Unmanned Carrier-Launched Air Surveillance and Strike (UCLASS) program has been restructured and renamed the Carrier Based Aerial Refuelling System (CBARS). The PB 2017 budget invests USD 89 million in FY 2017 for CBARS.

✦ To maximize capabilities and extend the reach of the airborne systems, PB 2017 invests USD 759 million for two Navy MQ-4C Triton unmanned maritime surveillance and patrol aircraft in FY 2017.

✦ The budget funds USD 2.2 billion in FY 2017 to continue procurement of the P-8A Poseidon maritime patrol and surveillance aircraft.

✦ PB17 supports the Army's Aviation Restructure Initiative by providing USD 1.1 billion for 52 Apache helicopters, and USD 1.0 billion for 36 Black Hawk helicopters in FY 2017.

Land

✦ The PB 2017 will help provide the Army, Marine Corps, and special operations forces with greater lethality to ensure ground forces can deter and, if necessary, fight and win a full-spectrum conflict.

✦ The budget includes a next-generation shoulder-launched weapon, a replacement for the Army Tactical Missile System (ATACMS) that can be used for improved

✦ Counter-battery and long-range strike, and increased firepower for Stryker armoured fighting vehicles, totalling USD 0.8 billion in

FY 2017. The budget adds significant survivability improvements to the Stryker vehicle to complement the lethality upgrades added in FY 2016.

✦ Additionally, the budget invests USD 735 million for 2,020 Joint Light Tactical Vehicles (JLTVs) in FY 2017. The JLTVs are intended to replace Humvees.

✦ For the amphibious combat vehicle, which replaces the Marine Corps' Amphibious Assault Vehicle, the budget invests USD 159 million in FY 2017.

Nuclear

✦ The PB provides USD 19 billion in FY 2017 which reflects the continuing commitment to modernising the nuclear enterprise.

✦ The Ohio Replacement Programme is allocated USD 1.9 billion in FY 2017 and USD 13.2 billion for development and initial construction over the FYDP.

Space

✦ The budget invests USD 1.8 billion for space launch in FY 2017.

✦ The budget allocates USD 108 million over the FYDP to implement the Joint Interagency Combined Space Operations Centre, which will better align joint operations in space across the US government.

NAVAL WARFARE SYSTEMS

The US Navy is the largest, most capable navy in the world, having the world's largest aircraft carrier fleet, with ten in service, two in the reserve fleet, and three new carriers under construction. Today, the increasingly globalised strategic environment provides an increasingly competitive platform for the US Navy in order to operate freely in the maritime environment. The maritime systems have seen an unprecedented growth with global information systems supported by technological boom has changed the way we all do business, including at sea. To stay ahead in the game of protecting the strategic interests, shipbuilding and industrial base have been strengthened which also include the naval aviation enterprise. The US Navy is the cornerstone of American security and prosperity, and are currently adapting to the emerging security environment. The 21st century blue-water US Navy maintains a sizable global presence, deploying in such areas as East Asia, the Mediterranean, and the Middle East, making it an active player in US foreign and defence policy.

Surface Ships

Aircraft Carrier: Aircraft carriers are the flagship of USA's naval forces. Aircraft Carriers support and operate aircraft that engage in attacks, they also engage in maritime security operations, and also provide unique capabilities for disaster response and humanitarian assistance.

Nimitz-class aircraft carrier - The Nimitz-class supercarriers are a class of ten nuclear-powered aircraft carriers in service with the United States Navy. The ships use nuclear power and hence are capable of operating for over 20 years without refuelling and are predicted to have a service life of over 50 years. The 10 Nimitz class aircraft carriers are Nimitz (CVN-68), Dwight D. Eisenhower (CVN-69), Carl Vinson (CVN-70), Theodore Roosevelt (CVN-71), Abraham Lincoln (CVN-72), George Washington (CVN-73), John C. Stennis (CVN-74), Harry S. Truman (CVN-75), Ronald Reagan (CVN-76), George H.W. Bush (CVN-77).

Displacement: 100,000 to 104,600 long tonnes

Length: 332.8m (waterline 317.0m)

Beam: 76.8m (waterline 40.8m)

Draft: 11.3m

Propulsion: 2 × Westinghouse A4W nuclear reactors; 4 × steam turbines; 4 × shafts

Speed: 56+ km/h

Crew: Up to 5,000+

Armament: Multiple NATO Sea Sparrow, Phalanx CIWS, and Rolling Airframe Missile (RAM) mounts.

Aircraft: Approximately 60+ (fixed wings & helicopters)

USS Nimitz (CVN-68) *(Image Credit: United States Navy)*

Gerald R. Ford class aircraft carrier- The Gerald R. Ford class is the future aircraft carrier replacement class for Enterprise and Nimitz class aircraft carriers. Gerald R. Ford (CVN 78) is the first new design for an aircraft carrier since USS Nimitz (CVN 68). The ship is equipped with two newly-designed reactors and has 250 percent more electrical capacity than previous carriers. The improvements will allow the ship to load weapons and launch aircraft faster than ever before. Construction on CVN 78 is nearing completion, as shipbuilders complete final outfitting and testing of the ship. The shipbuilders are on track to deliver Gerald R. Ford to the US Navy in 2016. Ships to be included are PCU Gerald R. Ford (CVN 78) and PCU John F. Kennedy (CVN 79).

Displacement: approximately 100,000 long tonnes full load

Length: 337m

Beam: 41m (waterline); 78m (flight deck)

Height: 76m

Draft: 12m

Speed: 56 km/h

Installed power: Two A1B nuclear reactors

Propulsion: 4 shafts

Crew: 4539

Armament: Evolved Sea Sparrow Missile, Rolling Airframe Missile, CIWS

Aircraft: 75+

USS Gerald R. Ford (CVN-78) *(Image Credit: United States Navy)*

Destroyers: Destroyers are warships that provide multi-mission offensive and defensive capabilities. Destroyers can operate independently or as part of carrier strike groups, surface action groups, amphibious ready groups, and underway replenishment groups. Guided-missile destroyers are multi-mission surface combatants capable of conducting Anti-Air Warfare, Anti-Submarine Warfare, and Anti-Surface Warfare.

Arleigh Burke class Destroyer - The Arleigh Burke class of guided missile destroyers (DDG-51) is the United States Navy's first class of destroyer built around the Aegis Combat System and the SPY-1D multi-function passive electronically scanned array radar. The Arleigh Burke class is comprised of four separate variants or "Flights". DDG 51-71 represent the original design and are designated as Flight I ships; DDG 72-78 are Flight II ships. The Flight IIA design began with DDG 79 and Flight III will continue the evolution of the DDG 51 class with the addition of the Air and Missile Defense Radar (SPY-6). 62 ships are currently operating in the fleet (DDG 51-112). An additional 14 ships are under construction. The US Navy has recently begun a modernisation programme for the Arleigh Burke class.

Displacement: Flight I- 8,315 tonnes; Flight II- 8,400 tonnes; Flight IIA- 9,200 tonnes; Flight III- 9,800 tonnes

Length: Flight I & II- 154m; Flight IIA- 155m

Beam: 18m

Draft: 9.3m

Propulsion: 4 × GE LM2500 gas turbines

Speed: 56 km/h

Range: 37 km/h

Sensors and processing systems: AN/SPY-1D 3D Radar, AN/SPS-67(V)2 Surface Search Radar, AN/SPS-73(V)12 Surface Search Radar, AN/SPG-62 Fire Control Radar,

AN/SQS-53C Sonar Array, AN/SQR-19 Tactical Towed Array Sonar, AN/SQQ-28 LAMPS III Shipboard System

Electronic warfare & decoys: AN/SLQ-32(V)2 Electronic Warfare System, AN/SLQ-25 Nixie Torpedo Countermeasures, MK 36 MOD 12 Decoy Launching System, AN/SLQ-39 CHAFF Buoys

Armament: Standard Missile (SM-2MR); Vertical Launch ASROC (VLA) missiles; Tomahawk; six MK-46 torpedoes (from two triple tube mounts); Close In Weapon System (CIWS), 5-in. MK 45 Gun, Evolved Sea Sparrow Missile (ESSM)

Aircraft: Two LAMPS MK III MH-60 B/R helicopters with Penguin/Hellfire missiles and MK 46/MK 50 torpedoes

Crew: FLIGHT I- 23 officers, 300 enlisted; FLIGHT IIA- 32 Officers, 348 enlisted

USS Arleigh Burke (DDG-51) *(Image Credit: United States Navy)*

Zumwalt-class Destroyer - The Zumwalt-class destroyer, developed under the DD(X) programme, is a class of United States Navy guided missile destroyers designed as multi-mission stealth ships with a focus on land attack. The Zumwalt-class destroyer (DDG 1000) is the lead ship of a class of next-generation multi-mission surface combatants has a 50-fold radar cross section reduction compared to current destroyers, improves strike group defence 10-fold and has 10 times the operating area in shallow water regions against mines. Cutting-edge technologies enable the DDG 1000's advanced capabilities, provide the ship's unmatched versatility and dramatically reduce manning requirements. The US Navy intends to procure three Zumwalt-class destroyers. DDG 1000 was named Zumwalt, DDG 1001 was named Michael Monsoor and DDG 1002 was named Lyndon B. Johnson.

Displacement: 14,798 tonnes

Length: 180m

Beam: 24.6m

Draft: 8.4m

Propulsion: (2) Main Turbine Generators (MTG); (2) Auxiliary Turbine Generators (ATG); (2) 34.6 MW Advanced Induction Motors (AIM)

Speed: 56km/h

Sensors & processing systems: AN/SPY-3 Multi-Function Radar

Armament: (80) Advanced Vertical Launch (AVLS) cells for Tomahawk, ESSM, Standard Missile; (2) Advanced Gun System (AGS) 155 mm guns; Long-Range Land Attack Projectiles (LRLAP) 155 mm rounds; (2) MK 46 Close In Guns (CIGS)

Aircraft: (2) MH60R or (1) MH60R and (3) VTUAVs

Crew: 158

USS Zumwalt undergoing sea trials

(Image Credit: US Navy/General Dynamics Bath Iron Works)

Cruiser: A Cruiser is a large multi-mission surface combatant. The cruisers are the largest and most powerful surface combatant after the aircraft carrier. The role of the cruiser varied according to ship and navy, often including air defence and shore bombardment.

Ticonderoga-class Cruiser - The Ticonderoga class of guided missile cruisers is a class of warships in the United States Navy. These ships are multi-mission which includes Air Warfare, Undersea Warfare, Naval Surface Fire Support and Surface Warfare. These surface combatants are capable of supporting carrier battle groups, amphibious forces, or of operating independently and as flagships of surface action groups. Their Mk 41 Vertical Launch System can launch Tomahawk cruise or fire long-range Standard Missiles for defence against aircraft or anti-ship missiles. Their LAMPS III helicopters and sonar systems empower them to perform anti-submarine missions. Ships under this class include USS Bunker Hill, USS Mobile Bay, USS Antietam, USS Leyte Gulf, USS San Jacinto, USS Lake Champlain, USS Philippine, USS Princeton, USS Normandy, USS Monterey, USS Chancellorsville, USS Cowpens, USS Gettysburg, USS Chosin, USS Hue City, USS Shiloh, USS Anzio, USS Vicksburg, USS Lake Erie, USS Cape St. George, USS Vella Gulf, USS Port Royal. The lead ship of the class, USS Ticonderoga (CG 47) through CG-51 have been decommissioned.

Displacement: 9,600 long tonnes

Length: 173m

Beam: 16.8m

Draft: 10.2m

Propulsion: 4 General Electric LM 2500 gas turbine engines; 2 shafts, 80,000 shaft horsepower total

Speed: 60 km/h

Armament: MK41 vertical launching system Standard Missile (MR); Vertical Launch ASROC (VLA) Missile; Tomahawk Cruise Missile; Six MK-46 torpedoes (from two triple mounts); Two MK 45 5-inch/54 calibre lightweight guns; Two Phalanx close-in-weapons systems

Aircraft: Two SH-60 Sea Hawk (LAMPS III)

Crew: 30 officers, 300 enlisted

A bow view of the destroyer USS SPRUANCE (DD 963), left, and the guided missile cruiser USS TICONDEROGA (CG 47). Although the superstructures are different, these ships use the same basic hull and the same propulsion plant *(Image Credit: United States Navy)*

Littoral Combat Ships: The littoral combat ship (LCS), fast, agile, and networked surface combatant, is a class of comparatively small surface vessels used for operations in the littoral zone by the US Navy. The Freedom class and the Independence class are the first two LCS variants. They have the capabilities of a small assault transport and are also equipped with autonomous air, surface, and underwater vehicles. The LCS is reconfigured for various roles by changing mission packages, each of which includes weapon systems, sensors, etc., carried craft and mission crews. The primary missions for the LCS include countering diesel submarine threats, littoral mine threats, and surface threats, such as small surface craft attacks, to assure maritime access for joint forces.

Freedom-class LCS - The Freedom-class is a relatively new breed of surface warship led by the USS Freedom LCS-1, launched in 2006. Three Freedom-class LCS ships have been commissioned. 8 more are under construction by Lockheed Martin, and 3 more are on order. Ships include USS Fort Worth (LCS-3), USS Freedom (LCS-1), and USS Milwaukee (LCS-5). Other ships are under construction.

Displacement: 3,450 MT full load

Length: 118.1m

Beam: 17.6m

Draft: 4.3m

Speed: 87km/h

Range: 6500km

Crew: 50 core; 65 mission

Sensors & processing systems: TRS-3D radar, COMBATSS-21, AN/SQR-20

EW & Decoys: Argon ST WBR-2000 ESM system, Terma A/S SKWS decoy system

Armament: Mk 110 57 mm gun, Mk 49 launcher with 21 × RIM-116 RAM, 12.7mm machine guns, 30mm Mk44 Bushmaster II guns, AGM-114L Hellfire

Aircraft: MH-60R/S Seahawk, MQ-8B Fire

Scouts or 1 × MQ-8C Fire Scout

The littoral combat ship USS Freedom (LCS 1) *(Image Credit: United States Navy)*

Independence-class LCS – The Independence-class ships are small, agile, stealthy, and speedy surface vessels optimised for littoral operations. Independence ships are specifically designed to defeat anti-access threats in shallow, coastal water regions, including surface craft, diesel submarines and mines. The first ship, USS Independence was laid down at in 2006. The Independence class carries a default armament for self-defence, command and control. Ships in this class includes USS Coronado (LCS-4), USS Independence (LCS-2), and USS Jackson (LCS-6). Other 10 ships are under construction.

Displacement: 3,200 MT full load

Length: 128.5m

Height: 31.6m

Beam: 31.6m

Draft: 4.6m

Sensors & processing systems: AN/SPS-77(V)1 Sea GIRAFFE 3D, BridgeMaster E navigational radar, AN/KAX-2 electro-optical sensor with TV and FLIR, ICMS

EW & decoys: ES-3601 ESM, SRBOC decoy launchers, NULKA active radar decoy system

Armament: Mk 110 57 mm gun, SeaRAM CIWS, .50-cal guns, 30 mm Mk44 Bushmaster II guns, AGM-114L Hellfire missiles

Aircraft carried: MH-60R/S Seahawk, MQ-8B Fire Scouts or 1 × MQ-8C Fire Scout

Crew: 40 core, 35 mission

The littoral combat ship USS Independence (LCS 2) *(Image Credit: United States Navy)*

Amphibious Force

Amphibious Assault Ships: The United States maintains the largest and most capable amphibious force in the world. An amphibious assault ship is a type of amphibious warfare ship employed to land and support ground forces on enemy territory by an amphibious assault. All amphibious warships are capable of Vertical/Short Take-Off and Landing, Short Take-Off Vertical Landing, Vertical Take-Off and Landing tilt-rotor and Rotary Wing aircraft operations. The America-class LHAs and Wasp-class LHDs provide the Marine Corps with a means of ship-to-shore movement by helicopter in addition to movement by landing craft.

America-class amphibious assault ship - The design of the America class is based on that of USS Makin Island (LHD-8). USS America (LHA-6) are designed to put ashore a Marine Expeditionary Unit using choppers and transport aircraft supported by aircraft and attack helicopters. The first of these ships, replacing the Tarawa-class of amphibious assault ships, was commissioned into US Navy in 2014 and a total of 11 are on order. USS America (LHA 6) is in active service and PCU Tripoli (LHA 7) is under construction.

Displacement: 43,745 tonnes (approx.)

Length: 257.3m

Beam: 32.3m

Speed: 37km/h

Propulsion: Two marine gas turbines, two shafts, 70,000 total brake horsepower, two 5,000 horsepower auxiliary propulsion motors

Sensors & processing systems: AN/SPQ-9B fire control radar; AN/SPS-48E air search radar

Electronic warfare & decoys: AN/SLQ-32B(V)2; two Mk53 Nulka decoy launchers

Armament: Two RAM launchers; two NATO Sea Sparrow launchers (with Evolved Sea Sparrow Missile (ESSM)); two 20mm Phalanx CIWS mounts; seven twin .50 cal. machine guns

Aircraft: A mix of: F-35B Joint Strike Fighters (JSF) STOVL aircraft; MV-22 Osprey VTOL tiltrotors; CH-53E Sea Stallion helicopters; UH-1Y Huey helicopters; AH-1Z Super Cobra helicopters; MH-60S Sea Hawk helicopters

Load: 1,687 troops (plus 184 surge)

Crew: 1,059; 65 officers

Amphibious assault ship USS America (LHA-6) *(Image Credit: United States Navy)*

Wasp-class amphibious assault ship - The Wasp-class Landing Helicopter Dock (LHD)s are currently the largest amphibious ships in the world. The Wasp class is a class of LHD amphibious assault ships operated by the United States Navy. The lead ship, USS Wasp (LHD-1) was commissioned in July 1989. The Wasp class can transport United States Marine Corps Marine Expeditionary Unit (MEU), and land them in hostile territory via landing craft or helicopters. The Wasp class is based on the Tarawa-class design. Ships in Wasp Class includes USS Wasp (LHD 1), USS Essex (LHD 2), USS Kearsarge (LHD 3), USS Boxer (LHD 4), USS Bataan (LHD 5), USS Bonhomme Richard (LHD 6), USS Iwo Jima (LHD 7), USS Makin Island (LHD 8).

Displacement: LHDs 1-4: 40,650 tonnes; LHDs 5-7: 40,358 tonnes; LHD 8: 41,772 tonnes

Length: 253.2m

Beam: 31.8m

Draft: 8.1m

Propulsion: (LHDs 1-7) two boilers, two geared steam turbines, two shafts, 70,000 total brake horsepower; (LHD 8) two gas turbines, two shafts; 70,000 total shaft horsepower, two 5,000 horsepower auxiliary propulsion motors

Sensors: AN/SPS-48 or AN/SPS-52; AN/SPS-49; SPS-67; AN/URN-25 TACAN system

Speed: 41km/h

Range: 17,600 km

Armament: Two RAM launchers; two NATO Sea Sparrow launchers; three 20mm Phalanx CIWS mounts (two on LHD 5-8); four .50 cal. machine guns; four 25 mm Mk 38 machine guns (LHD 5-8 have three 25 mm Mk 38 machine guns)

Aircraft: 12 CH-46 Sea Knight helicopters; 4 CH-53E Sea Stallion helicopters; 6 AV-8B Harrier attack aircraft; 3 UH-1N Huey helicopters; 4 AH-1W Super Cobra helicopters. (planned capability to embark MV-22 Osprey VTOL tilt-rotors) and F-35B Joint Strike Fighters (JSF) STOVL aircraft); 3 LCACs or 2 LCUs (Landing/Attack Craft)

Crew: Ships Company: 66 officers, 1,004 enlisted; LHD 8: 65 officers, 994 enlisted; Marine Detachment 1,687 troops (plus 184 surge)

Amphibious assault ships USS Wasp (LHD-1), left, and USS Saipan (LHA-2)

(Image Credit: US Department of Defence)

Amphibious Command Ships: An Amphibious command ship (LCC) of the United States Navy is a large, special purpose ship, originally designed to command large amphibious invasions.

Blue Ridge Class – Both ships in this class, USS Blue Ridge (LCC-19) and USS Mount Whitney (LCC-20), are fleet flagships. Currently, they are assigned to the 6th and 7th fleets as flagships. USS Blue Ridge and USS Mount Whitney's service lives were extended in 2011 by the Chief of Naval Operation to 2039. The primary role of USS Blue Ridge is to provide command, control, communications, computers, and intelligence (C4I) support to the commander and staff of the United States 7th Fleet. USS Mount Whitney is also the command and control ship for the Commander Joint Command Lisbon and the Commander Striking Force NATO.

Displacement: 18,874 tonnes

Propulsion: Two boilers, one geared turbine, one shaft; 22,000 horsepower

Length: 190m

Beam: 32m

Draft: 8.15m

Speed: 42.4km/h

Armament: 2 MK 15 Block 1B Phalanx Close-in-Weapons Systems (CIWS)

Aircraft: All helicopters except the CH-53 Sea Stallion can be carried

Crew: 34 officers, 564 enlisted

Amphibious command ship USS Mount Whitney (LCC-20)
(Image Credit: United States Navy)

Amphibious Transport Docks: Amphibious transport dock ships are warships that embark, transport and land elements of a landing force for a variety of expeditionary warfare missions. They are used to transport and land Marines, their equipment and supplies by embarked Landing Craft, or conventional landing craft and Amphibious Assault Vehicles amplified by helicopters or vertical take-off and landing aircraft.

San Antonio-class - The San Antonio class is a class of amphibious transport docks are used by the United States Navy. The San Antonio class provides the Navy and Marine Corps with modern, sea-based platforms that are networked, survivable, and built to operate with 21st century transformational platforms. Ships under this class includes USS San Antonio (LPD 17), USS New Orleans (LPD 18), USS Mesa Verde (LPD 19), USS Green Bay (LPD 20), USS New York (LPD 21), USS San Diego (LPD 22), USS Anchorage (LPD 23), USS Arlington (LPD 24), USS Somerset (LPD 25), PCU John P. Murtha (LPD 26), PCU Portland (LPD 27), PCU Fort Lauderdale (LPD 28).

Displacement: 24,900 long tonnes

Length: 208m

Beam: 32m

Draft: 7.0m

Propulsion: Four sequentially turbocharged marine Colt-Pielstick Diesels, two shafts, 41,600 shaft horsepower

Speed: 41km/h

Armament: Two Bushmaster II 30 mm Close in Guns, fore and aft; two Rolling Airframe Missile launchers, fore and aft: ten .50 calibre machine guns

Aircraft: Launch or land two CH-53E Super Stallion helicopters or two MV-22 Osprey tilt rotor aircraft or up to four CH-46 Sea Knight helicopters, AH-1 or UH-1 helicopters; Two LCACs or one LCU; and 14 Expeditionary Fighting Vehicles/Amphibious Assault Vehicles

Crew: 28 officers, and 333 enlisted men; Landing force: 66 officers, and 633 enlisted men

USS New Orleans (LPD 18), San Antonio-class amphibious transport dock
(Image Credit: United States Navy)

Landing Craft, Mechanized and Utility - LCM/LCU: The Landing Craft Mechanized (LCM) is a landing craft designed for carrying vehicles. The Landing Craft Utility (LCU) is a type of boat used by amphibious forces to transport equipment and troops to the shore. Landing craft are also used to support civilian humanitarian/maritime operations. Landing craft are capable of transporting cargo, tracked and/or wheeled vehicles and troops from amphibious assault ships to beachheads or piers.

LCU 1610, 1627 and 1646 Class belong to the Landing Craft Utility category. Their characteristics are as follows:

Displacement: 200 tonnes

Length: 41.1m

Beam: 9.1m

Propulsion: 2- Detroit 12V-71 Diesel engines, twin shaft, 680 hp sustained, Kort nozzles

Speed: 20.3km/h

Range: 2,200 km

Military Lift: M1A1 tanks (2), LAVs (10) or 400+ troops; 125 tons of cargo

Armament: 2 - 12.7mm MGs, 2 - 7.62mm MGs

Electronics: Furuno Radar Navigation; C4I up-grade under development

Crew: 13

LCM-8 Type and LCM-6 Type Class belongs to Landing Craft Mechanized category. Their characteristics are as follows:

Displacement: (LCM-8) 105 tonnes; (LCM-6) 64 tonnes

Length: (LCM-8) 22.5m; (LCM-6) 17.1m

Beam: (LCM-8) 6.4m; (LCM-6) 4.3m

Propulsion: (LCM-8) 2- Detroit 12V-71 Diesel engines, 680hp sustained, twin shafts; (LCM-6) 2- Detroit 6-71 Diesel engines, 348 hp sustained, twin shaft, or 2- Detroit 8V-71 Diesel engines, 460 hp sustained, twin shaft

Speed: (LCM-8) 22.2km/h; (LCM-6) 16.6km/h

Military lift: (LCM-8) 1- M48 or 1- M60 tank or 200 troops

Load: (LCM-8) 180 tonnes; 34 tonnes or 80 troops

Crew: 5

A Landing Craft Utility (LCU) assigned to the Surf Riders of Assault Craft Unit One (ACU-1) heads to the beach with embarked US Marines during amphibious assault training in support of exercise Rim of the Pacific (RIMPAC) 2004 *(Image Credit: United States Navy)*

Landing Craft Air Cushion: Landing Craft Air Cushion (LCAC) are hovercrafts for transporting, ship-to-shore and across the beach, personnel, weapons, equipment, and cargo of the assault elements of the Marine Air-Ground Task Force. It is a high-speed, over-the-beach fully amphibious landing craft used by Marine Air-Ground Task Force. LCAC also provides personnel transport, evacuation support, lane breaching, mine countermeasure operations, and Marine and Special Warfare equipment delivery. A total of ninety-one LCAC have now been built and delivered. The final craft, LCAC 91, was delivered to the US Navy in 2001.

Displacement: 170-182 tonnes

Length: 28.0m

Beam: 14.7m

Propulsion: 4-Allied-Signal TF-40B gas turbines (2 propulsion / 2 lift); 16,000 hp sustained; 2-shrouded reversible pitch airscrews; 4-dbl-entry fans, centrifugal or mixed flow (lift)

Speed: 74km/h

Range: 370 km at 75 km/h

Load: 54.43/64.29 tonnes

Armament: 2 - Gun mounts will support: M2HB .50 Cal machine gun;, M240 7.62 mm machine gun; and MK-19 MOD3 40 mm grenade Launcher

Electronics: Radars, Navigation: Marconi LN 66; I band / Sperry Marine Bridge Master E

Crew: 5

Landing Craft, Air Cushion (LCAC) 50. *(Image Credit: United States Navy)*

Sub-surface Ships

Attack Submarine (SSN): Attack submarines (SSN) are the class of nuclear-powered fast attack submarines which seek and destroy enemy submarines and surface ships; utilising Tomahawk cruise missiles and Special Operation Forces; including carrying out of Intelligence, Surveillance, and Reconnaissance missions; support battle group operations; and engage in mine warfare. The US Navy has three classes of SSNs in service.

Virginia-class submarine - The US Navy have built the next-generation attack submarine, the Virginia (SSN 774) class. 12 Virginia's have been commissioned to date and they will replace Los Angeles Class submarines as they retire. Virginia class SSNs have a fly-by-wire ship control system plus special features to support SOF, including a reconfigurable torpedo room. This class has two photonics masts that host visible and infrared digital cameras atop telescoping arms. The new water-backed Large Aperture Bow (LAB) array with two Virginia Payload Tubes (VPTs), each can launch six Tomahawk cruise missiles. Virginia-class submarines will be acquired through 2043, and are expected to remain in service past 2060. Ships under this class includes USS Virginia (SSN 774), USS Texas (SSN 775), USS Hawaii (SSN 776), USS North Carolina (SSN 777), USS New Hampshire (SSN 778), USS New Mexico (SSN 779), USS Missouri (SSN 780), USS California (SSN 781), USS Mississippi (SSN 782), USS Minnesota (SSN 783), USS North Dakota (SSN 784), USS John Warner (SSN 785), plus many others which are still under construction.

Displacement: 7,800 tonnes

Length: 114.8m

US Navy submarine USS North Dakota (SSN-784). *(Image Credit: United States Navy)*

Beam: 10.36m

Propulsion: One nuclear reactor, one shaft

Speed: 46.3km/h

Armament: Tomahawk missiles, twelve VLS tubes (SSNs 774-783) or two VPTs (SSNs 784 and beyond), MK48 ADCAP torpedoes, four torpedo tubes

Crew: 132: 15 officers; 117 enlisted

Seawolf-class submarine - The US Navy also has three Seawolf-class submarines. USS Seawolf (SSN 21) is exceptionally quiet, fast, well-armed, and equipped with advanced sensors. This class has eight torpedo tubes and can hold up to 50 weapons in its torpedo room. Ships under this class includes USS Seawolf (SSN 21), USS Connecticut (SSN 22), and USS Jimmy Carter (SSN 23).

Displacement: (SSNs 21 and 22) 9,138 tonnes; (SSN 23) 12,158 tonnes

Length: (SSNs 21 and 22) 107.6m; (SSN 23) 138.07m

Beam: 12.2m

Propulsion: One nuclear reactor, one shaft

Speed: 46.3km/h

Armament: Tomahawk missiles, MK48 torpedoes, eight torpedo tubes

Crew: 140: 14 Officers; 126 Enlisted

Seawolf (SSN-21). *(Image Credit: Electric Boat Corporation by Jim Brennan)*

Los Angeles-class submarine - 40 Los Angeles (SSN 688)-class submarines have been commissioned till date. 30 Los Angeles-class SSNs are equipped with 12 Vertical Launch System tubes for firing Tomahawk cruise missiles. The class has more operating nuclear submarines than any other in the world. Ships in this class includes USS Bremerton (SSN 698), USS Jacksonville (SSN 699), USS Dallas (SSN 700), USS La Jolla (SSN 701), USS City of Corpus Christi (SSN 705), USS San Francisco (SSN 711), USS Houston (SSN 713), USS Buffalo (SSN 715), USS Olympia (SSN 717), USS Providence (SSN 719), USS Pittsburgh (SSN 720), USS Chicago (SSN 721), USS Key West (SSN 722), plus many others.

Displacement: 6,900 tonnes

Length: 109.73m

Beam: 10.06m

Propulsion: One nuclear reactor, one shaft

Speed: 46.3km/h

Armament: Tomahawk missiles, VLS tubes (SSN 719 and later), MK48 torpedoes, four torpedo tubes

Crew: 16 officers; 127 enlisted

Port bow view of the fore section of the nuclear-powered attack submarine USS SANTA FE (SSN-763) tied up at the pier. *(Image Credit: United States Navy)*

Ballistic Missile Submarines (SSBN): A ballistic missile submarine is a submarine deploying submarine-launched ballistic missiles (SLBMs) with nuclear warheads. The US Navy's ballistic missile submarines serve as an undetectable launch platform for intercontinental missiles.

Ohio-class submarine - Each of the 14 Ohio-class SSBNs can carry up to 24 submarine-launched ballistic missiles (SLBMs) with multiple, independently-targeted warheads. 4 other submarines were later converted to guided missile submarines (SSGN). The SSBN's strategic weapon is the Trident II D5 missile. The Ohio-class submarines are the largest submarines ever built for the US Navy. Ships under this includes USS Henry M. Jackson (SSBN 730), USS Alabama (SSBN 731), USS Alaska (SSBN 732), USS Nevada (SSBN 733), USS Tennessee (SSBN 734), USS Pennsylvania (SSBN 735), USS West Virginia (SSBN 736), USS Kentucky (SSBN 737), USS Maryland (SSBN 738), USS Nebraska (SSBN 739), USS Rhode Island

Trident nuclear powered submarine USS Alaska (SSBN 732) is guided into an explosives handling wharf. *(Image Credit: US Navy photo by Gene Royer).*

(SSBN 740), USS Maine (SSBN 741), USS Wyoming (SSBN 742), USS Louisiana (SSBN 743).

Displacement: 16,674 tonnes

Length: 170m

Beam: 13m

Draft: 10.8m

Propulsion: 1× S8G PWR nuclear reactor; 2× geared turbines; Fairbanks Morse auxiliary diesel

Speed: 37km/h

Armament: (SSBN-734) 24 × Trident II D5 SLBM with up to 12 MIRVed W76 or W88 nuclear warheads each, range 6,100 nmi

Crew: 15 officers, 140 enlisted

Guided Missile Submarines (SSGN): Guided-missile submarines (SSGN) are armed with tactical missiles and equipped with superior communications capabilities. The SSGN Programme Office refuelled and converted four SSBNs into SSGNs. USS Ohio (SSGN 726), USS Michigan (SSGN 727), USS Florida (SSGN 728), and USS Georgia (SSGN 729) were slated for modification, to remain in service carrying conventionally armed guided missiles, and were re-designated as SSGNs. Together, the four SSGNs represent more than half of the Submarine Force's vertical launch payload capacity with each SSGN capable of carrying up to 154 Tomahawk land-attack cruise missiles.

Displacement: 18,750 tonnes

Length: 170.69m

Beam: 12.8m

Propulsion: One nuclear reactor, one shaft

Speed: 36.8 km/h

Armament: Up to 154 Tomahawk missiles, Mk48 torpedoes; 4 torpedo tubes

USS Ohio, one of four Trident Submarines undergoing conversion to a new class of guided missile submarines that will improve their capability to support and launch up to 154 Tomahawk missiles *(Image Credit: United States Navy)*

Unmanned underwater vehicle (UUV): The Unmanned underwater vehicle is used for ISR, Mine Countermeasures, Anti-Submarine Warfare, Inspection/Identification, Oceanography, Communication/Navigation Network Node, Payload Delivery, Information Operations, Time Critical Strike. The growing use of unmanned systems - air, surface, ground, and underwater is continually demonstrating new possibilities that can assist US naval forces maintain maritime superiority around the world. The REMUS (Remote Environmental Monitoring Units) is a three-foot long robot used to clear mines in one square mile with 16 hours. The man-portable class REMUS 100 AUV is used by the US Navy for shallow water mine countermeasures (MCM) and hydrographic reconnaissance. General Motors, the Office of Naval Research and the US Naval Research Laboratory are cooperating to incorporate automotive hydrogen fuel cell systems into the next generation of Navy unmanned undersea vehicles, or UUVs. The Large Displacement Unmanned Undersea Vehicle (LDUUV) programme by the US Office of Naval Research are developing a fully autonomous, long-endurance, land-launched unmanned undersea vehicles capable of operating near shore.

The REMUS 100 is a compact light-weight Autonomous Underwater Vehicle.

(Image Credit: Kongsberg Maritime)

Underwater Search and Recovery Equipment:

1. CURV 21 is a 6,400-pound Remotely Operated Vehicle (ROV) that is designed to meet the US Navy's deep ocean salvage requirements down to a maximum depth of 20,000 feet of seawater.
2. Deep Drone is a 4,100 pound Remotely Operated Vehicle (ROV) that is designed to meet the Navy's mid-water salvage requirements down to a maximum depth of 8,000 feet seawater.
3. MAGNUM is a 3,500 pound Remotely Operated Vehicle (ROV) that is designed to meet the Navy's mid-water requirements down to a maximum depth of 8,000 feet of seawater.
4. The three MINI-ROVs are standard commercial Remotely Operated Vehicles (ROVs) used to meet the Navy's requirements down to a maximum depth of 1,000 feet.
5. Orion is a towed side scan sonar that fulfils the US Navy's deep-water search requirement down to a maximum depth of 20,000 feet of seawater.
6. SWISS is a towed side scan sonar that meets the Navy's mid-water search requirement down to a maximum depth of 7,000 feet of seawater.
7. The TPL-25 System meets the Navy's requirement for locating emergency relocation pingers on downed Navy and commercial aircraft down to a maximum depth of 20,000 feet anywhere in the world.

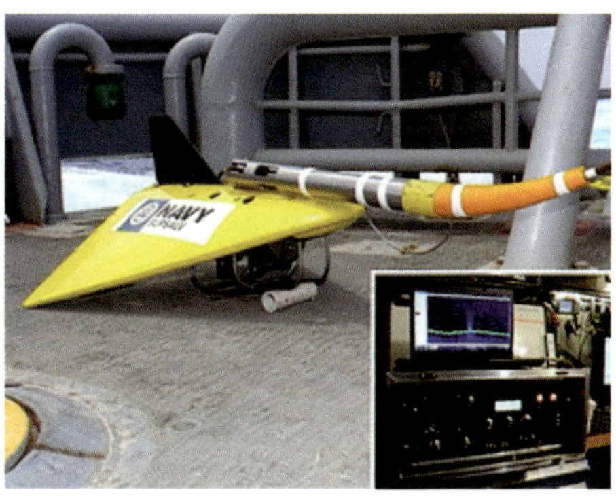

Towed Pinger Locator (TPL) 25 System used for locating emergency relocation pingers on downed Navy and commercial aircraft at a maximum depth of 20,000 feet anywhere in the world *(Image Credit: United States Navy)*

US Navy Supervisor of Salvage and Diving (SUPSALV) is deploying a Deep Drone 8000 Remotely Operated Vehicle (ROV), capable of operating at a depth up to 8000 feet.

(Image Credit: US Air Force)

Auxiliary Ship

Fleet Replenishment Oilers: 15 fleet replenishment oilers of Henry J. Kaiser-class are operated by Military Sealift Command and provide underway replenishment of fuel to US Navy ships at sea and jet fuel for aircraft assigned to aircraft carriers. The class will be replaced under the TAO(X) project which will build 17 double-hulled ships, starting with USNS John Lewis and USNS Harvey Milk.

Military Sealift Command oiler USNS Yukon (T-AO 202)

(Image Credit: United States Navy)

Naval Aviation

Aircraft, Fixed Wing: Aircraft Carrier-based jets are able to strike air, sea, and land targets far from a carrier strike group while protecting friendly forces from enemy aircraft, ships, and submarines. These aircraft also provide logistics support.

C-2 Greyhound - C-2 Greyhound is a high wing, twin-engine monoplane cargo aircraft, designed to land on aircraft carriers. The C-2A Greyhound provides critical logistics support to Carrier Strike Groups. Its primary mission is the transport of high-priority cargo, mail and passengers between carriers and shore bases for US Navy. The C-2A's in-flight ramp open

capability allows airdrop of supplies and personnel. Its on-board Auxiliary Power Unit provides engine-starting capability and ground power self-sufficiency in remote areas provides an operational versatility found in no other cargo aircraft.
Length: 17.30m
Height: 4.85m
Wingspan: 24.60m
Propulsion: Two Allison T56-A-425 turboprop engines; 4,600 shaft horsepower each
Weight: 26,082 Kg
Cruise speed: 465 km/h at 8,750m
Range: 2,400 km
Crew: 4

C-2A Greyhound lands on the flight deck of USS Kitty Hawk (CV 63).
(Image Credit: United States Navy)

C-130 Hercules - The Lockheed C-130 Hercules is a four-engine turboprop military transport aircraft, capable of landing and taking off from short, rough dirt runways. The C-130R and C-130T are US Navy and US Marine Corps models, both equipped with underwing external fuel tanks. The US Navy C-130T is similar, but has additional avionics improvements.
Length: 29.3m
Height: 11.4m
Wingspan: 39.7m
Propulsion: Four Allison T56-A-15 turboprops, each 4,300 horsepower. Horsepower each engine: AC-130A, 3,750 hp; AC-130H, 4,910 hp
Weight: 69,750 kg
Speed: 604.4 km/h at 20,000 feet
Range: 3,770 km with max payload
Crew: 5

Lockheed C-130T Hercules operated by United States Navy Flight Demonstration Squadron ("Blue Angels"), crewed by the United States Marine Corps.
(Image Credit: United States Navy)

C-40 Clipper - The C-40A Clipper provides critical logistics support to the US Navy. The US Naval Reserve operates and maintains the aircraft. It is more fuel-efficient and offers increased range and payload capabilities.
Length: 33.63m
Height: 12.55m
Wingspan: 35.8m
Propulsion: Two CFM56-7 SLST engines
Weight: 77,564 kg
Speed: 940 to 990 km/h
Range: 5,600 km
Crew: 4-6

US Navy C-40A is loaded with cargo *(Image Credit: United States Navy)*

F-35 Lightning II - The F-35 Lightning II is a 5th generation fighter, combining advanced stealth with fighter speed and agility, fully fused sensor information, network-enabled operations and advanced sustainment. The US Navy will receive in 2018 the F-35C carrier-based Catapult Assisted Take-Off But Arrested Recovery (CATOBAR) variant. The Navy intends to buy 480 F-35Cs to replace the F/A-18A, B, C, and D Hornets. The F-35C variant has larger wings and more robust landing gear than the other variants, making it suitable for catapult launches and fly-in arrestments aboard naval aircraft carriers. Its wingtips also fold to allow for more room on the carrier's deck while deployed. The F-35C carries nearly 20,000 pounds of internal fuel for longer range and better persistence.
Length: 15.5m
Wingspan: 13.1m
Empty weight: 15,800 kg
Max take-off weight: 31,800 kg
Range: 2,220 km
Propulsion: 1 × Pratt & Whitney F135 afterburning turbofan
Armament: 25 mm (0.984 in) GAU-22/A 4-barrel rotary cannon; AIM-120 AMRAAM, AIM-9X Sidewinder, IRIS-T, AGM-88 AARGM, AGM-158 JASSM, JAGM, JSM, LRASM, etc
Crew: 1

US Navy Lockheed Martin F-35C Lightning II conducts its first arrested landing aboard the aircraft carrier USS Nimitz (CVN-68).
(Image Credit: United States Navy)

F/A-18E/F Super Hornet - The F/A-18 Super Hornet is a twin-engine, supersonic, all weather multirole fighter jet that is capable of landing and taking off from an aircraft carrier. Super Hornet has longer range, an aerial refuelling capability, increased survivability/lethality and improved carrier suitability. The F/A-18E single-seat and F/A-18F tandem-seat variants are larger and more advanced derivatives of the F/A-18C and D Hornet, used by the US Navy.
Length: 18.31m
Height: 4.88m
Wingspan: 13.62m
Propulsion: Two F414-GE-400 turbofan engines. 22,000 pounds (9,977 kg) static thrust per engine
Weight: 29,932 kg
Speed: 1,915 km/h at 12,190m
Range: 2,346 km clean plus two AIM-9s
Armament: One M61A1/A2 Vulcan 20mm cannon; AIM 9 Sidewinder, AIM-9X (projected), AIM 7 Sparrow, AIM-120 AMRAAM, Harpoon, Harm, SLAM, SLAM-ER (projected), Maverick missiles; Joint Stand-Off Weapon (JSOW); Joint Direct Attack Munition (JDAM); Data Link Pod; Paveway Laser Guided Bomb; various general purpose bombs, mines and rockets.
Crew: (F/A-18E) 1; (F/A-18F) 2

F/A-18F Super Hornet Strike Fighter Squadron 103 on the flight deck of the aircraft carrier USS Dwight D. Eisenhower (CVN 69).

F/A-18 Hornet - F/A-18 Hornet strike fighter is an all-weather fighter and attack aircraft. It was designed for traditional strike applications such as interdiction and close air support without compromising its fighter capabilities. F/A-18 Hornets are currently operating in 37 tactical squadrons from air stations world-wide, and from 10 aircraft carriers. The US Navy's Blue Angels Flight Demonstration Squadron flies F/A-18A, B, C, and D variants. The F/A-18A & F/A-18C are the single-seat variants and the F/A-18B & F/A-18D are the two-seat variants.
Length: 16.8m
Height: 4.6m
Wingspan: 13.5m
Propulsion: Two F404-GE-402 enhanced performance turbofan engines. 17,700 pounds static thrust per engine
Weight: 23,537 kg

Speed: 1,915 km/h at 12,190m

Range: 2,000 km with only two AIM-9s

Armament: One M61A1/A2 Vulcan 20mm cannon; AIM 9 Sidewinder, AIM 7 Sparrow, AIM-120 AMRAAM, Harpoon, Harm, SLAM, SLAM-ER, Maverick missiles; Joint Stand-Off Weapon (JSOW); Joint Direct Attack Munition (JDAM); various general purpose bombs, mines and rockets.

Crew: (F/A-18C) 1; (F/A-18D) 2

F/A-18C Hornet taking off from USS Kitty Hawk *(Image Credit: United States Navy)*

E-2 Hawkeye - The E-2 Hawkeye is the US Navy's all-weather, carrier-based tactical battle management airborne early warning, command and control aircraft. The E-2 Hawkeye is a crucial component of all US Navy carrier air wings. The current version of the Hawkeye, the E-2C, became operational in 1973. The newest variant of the E-2C (Hawkeye 2000) has new mission computer, improved radar displays and Cooperative Engagement Capability (CEC), combined with the shipboard Aegis weapon system. The latest E-2 version is the E-2D Advanced Hawkeye, which features an entirely new avionics suite including the new AN/APY-9 radar, radio suite, mission computer, integrated satellite communications, flight management system, improved T56-A-427A engines, etc.

Length: 17.5m

Height: 5.6m

Wingspan: 28m

Propulsion: Two Allison T-56-A427 turboprop engines; (5,100 shaft horsepower each)

Avionics: AN/APS-145 Radar, OL-483/AP IFF interrogator system, APX-100 IFF Transponder, OL-698/ASQ Tactical Computer Group, AN/ARC-182 UHF/VHF radio, AN/ARC-158 UHF radio, AN/ARQ-34 HF radio, AN/USC-42 Mini-DAMA SATCOM system

Weight: 23,850 kg

Speed: 552 km/h

Crew: 5

US Navy (USN) E-2C Hawkeye 2000 aircraft *(Image Credit: US DoD Media)*

E-6B Mercury - E-6B Mercury provides survivable, reliable, and endurable airborne command, control, and communications between the National Command Authority (NCA) and US strategic and non-strategic forces. It is a communications relay and strategic airborne command post aircraft. The E-6B model, other than basic roles, added further command post capabilities and control of land-based missiles and nuclear-armed strategic bombers, as it is equipped with an airborne launch control system (ALCS). The E-6 fleet was completely modified to the E-6B configuration in 2003.

Length: 45.8m

Height: 12.9m

Wingspan: 45.2m

Weight: 154,400 kg

Propulsion: Four CFM-56-2A-2 High bypass turbofans

Speed: 960 km/h

Range: 12,144 km

Crew: 22

A US Navy Boeing E-6B Mercury *(Image Credit: United States Navy)*

EA-18G Growler - The EA-18G Growler is the newly-designed electronic warfare aircraft which integrates the latest electronic attack technology, including the ALQ-218 receiver, ALQ-99 jamming pods, communication countermeasures, APG-79 AESA radar and satellite communications.

Length: 18.5m

Height: 4.87m

Wingspan: 21,772 kg

Propulsion: Two F414-GE-400 turbofan engines. 22,000 pounds (9,977 kg) static thrust per engine

Range: 2,346 km clean plus two AIM-9s

EA-18G Growler *(Image Credit: United States Navy)*

Armament: Two AIM-120, two AGM-88 HARM, three ALQ-99

Crew: 2

EP-3 ARIES II - The EP-3E ARIES II (Airborne Reconnaissance Integrated Electronic System II) is the US Navy's only land-based signals intelligence (SIGINT) reconnaissance aircraft. The aircraft has sensitive receivers and high-gain dish antennas.

Length: 35.57m

Height: 10.27m

Wingspan: 30.36m

Propulsion: Four Allison T-56-A-14 turboprop engines (4,900 shaft horsepower each)

Weight: 63,394.1 kg

Speed: 745 km/h

Range: 5,556 km

Crew: 22 and more

US Navy Lockheed EP-3E Aries II conversion aircraft *(Image Credit: US Defence Visual Information Centre)*

P-8A Poseidon - The P-8A Poseidon military aircraft is a modified Boeing 737-800ERX. The P-8A Poseidon is the US Navy's replacement platform for the P-3C. The aircraft is designed for long-range anti-submarine warfare; anti-surface warfare; and intelligence, surveillance and reconnaissance missions. It is capable of broad-area maritime and littoral operations and can carry torpedoes, depth charges, Harpoon anti-ship missiles, and other weapons.

Length: 39.47m

Height: 12.83m

Wingspan: 85,820 kg

Propulsion: Two high-bypass CFM56 turbofan engines

Speed: 907 km/h

Range: 8,300 km

Armament: Torpedoes, cruise missiles, bombs, mines

Crew: 9

A US Navy Boeing P-8A Poseidon dropping an anti-submarine torpedo Mark 46. *(Image Credit: United States Navy)*

Aircraft, Rotary Wing: Helicopters, supply platforms with which to conduct search and rescue, special operations, anti-submarine warfare, and anti-surface warfare.

SH-60 / HH-60H / MH-60 Seahawk - The Sikorsky SH-60 Seahawk is a twin turboshaft engine, multi-mission US Navy helicopter used for primary anti-submarine, search & rescue, drug interdiction cargo-lift and anti-surface warfare roles. They are designed to operate from frigates, destroyers, cruisers and aircraft carriers. The SH-60B can deploy sonobuoys (sonic detectors) and torpedoes in an anti-submarine role. The SH-60F is carrier-based and primarily serves as the carrier battle group's primary antisubmarine warfare (ASW) aircraft. The MH-60R is designed to combine the features of the SH-60B and SH-60F. The MH-60S is based on the UH-60L and has many naval SH-60 features. Based on the SH-60F, the HH-60H is the primary combat search and rescue, naval special warfare and anti-surface warfare helicopter. By 2015, the only models of Seahawk in the Navy are the MH-60S and the MH-60R.

Length: 19.6m

Height: 3.9 to 5.1m

Rotor Diameter: 16.4m

Weight: 9,450 to 10,350 kg

Propulsion: Two General Electric T700-GE-700 or T700-GE-701C engines; thrust: up to 1,940 shaft horsepower

Speed: 270 km/h

Range: 600 km

Armament: Mark 46/Mk-54s torpedoes, AGM-114 Hellfire, M60 or M240 or GAU-16/A machine guns, RAMICS

Crew: 3 to 4

An MH-60S Seahawk on the flight deck of the USS Wasp (LHD 1).
(Image Credit: United States Marine Corps)

MH-53E Sea Dragon - The MH-53E is used primarily for Airborne Mine Countermeasures (AMCM), with secondary missions of vertical shipboard delivery and assault support. The MH-53E can tow a variety of mine hunting/sweeping countermeasures systems, including the Mk 105 magnetic minesweeping sled, the AQS-14A side-scan sonar, and the Mk 103 mechanical minesweeping system.

Length: 30.2m

Height: 8.6m

Rotor Diameter: 24.1m

Propulsion: Three General Electric T64-GE-419 turboshaft engines (4,750 shaft horsepower each)

Weight: 31,693 kg

Speed: 278 km/h

Range: 1,000 km

Armament: .50 BMG GAU-15/A machine guns; .50 BMG weapons system; GAU-21 (M3M mounted machine gun)

Crew: 2 to 3

MH-53E Sea Dragon helicopter, on the flight deck of the amphibious assault ship USS Essex (LHD 2) *(Image Credit: United States Navy)*

Weapons: The US Navy has in its arsenal some of the most sophisticated missiles and advanced weapon systems which also includes the Tomahawk Missile, the backbone of US Navy.

Tomahawk Missile - The BGM-109 tomahawk is an all-weather, long range, subsonic cruise missile used for land attack warfare, launched from US Navy surface ships and submarines. Tomahawk can be launched from a ship or submarine and can fly into heavily defended airspace more than 1,000 miles away to conduct precise strikes on high-value targets with minimal collateral damage. The Tomahawk is a highly accurate, GPS enabled precision weapon that has been used over 2,000 times in combat, and flight tested more than 500 times. A major improvement to the Tomahawk is network-centric warfare-capabilities. The "Tactical Tomahawk" cruise missile adds the capability to reprogramme the missile while in-flight to strike any of 15 pre-programmed alternate targets, or redirect the missile to any Global Positioning System (GPS) target coordinates. Tomahawk Block IV (TLAM-E) is the latest improvement to the Tomahawk missile family.

Length: 5.56m; (with booster) 6.25m

Diameter: 51.81m

Wingspan: 2.67m

Propulsion: Block II/III TLAM-A, C & D - Williams International F107 cruise turbo-fan engine; ARC/CSD solid-fuel booster

Weight: 1,315.44 kg

Speed: (Subsonic) 880 km/h

Range: (Block II TLAM-A) 2500 km; (Block III TLAM-C) 1600 km; (Block III TLAM-D) 1250 km; (Block IV TLAM-E) 1600 km

Guidance System: Block II TLAM-A INS, TERCOM, Block III TLAM-C, D & Block IV TLAM-E INS, TERCOM, DSMAC, and GPS

Warhead: Block II TLAM-N W80 nuclear warhead; Block III TLAM-C and Block IV TLAM-E - 1,000-pound class unitary warhead; Block III TLAM-D - conventional sub munitions dispenser with combined effect bomblets

Tactical Tomahawk Cruise Missile launches from the guided missile destroyer USS Stethem (DDG 63). *(Image Credit: United States Navy)*

Harpoon Missile - The AGM/RGM/UGM-84 Harpoon missile is an all-weather, over-the-horizon, anti-ship missile system. It has a low-level, sea-skimming cruise trajectory, active radar guidance. It has been upgraded over the years and is now available as the Harpoon Block II. The Harpoon is capable of executing both land-strike and anti-ship missions. To strike targets on land and ships in port, the missile uses GPS-aided inertial navigation to hit a designated target aim point. The missile's launch platforms include Aircraft (AGM-84), Surface ships (RGM-84) and Submarines (UGM-84).

Length: (Air launched) 3.8m; (Surface & submarine launched) 4.6m

Weight: 690.8 kg (with booster)

Propulsion: Teledyne Turbojet / solid propellant booster for surface and submarine launch. Thrust- greater than 600 pounds (greater than 272.2 kg)

Diameter: 34.3cm

Wingspan: 91.4cm with booster fins and wings

Speed: (High Subsonic) 850 km/h

Range: Over-the-horizon, in excess of 124 km.

Guidance System: Sea-skimming cruise monitored by radar altimeter / active radar terminal homing

Warhead: Penetration / high-explosive blast (224 kg)

UGM-84 surface-to-surface Harpoon missile clears the surface of the water, after being launched from a submarine *(Image Credit: United States Navy)*

AIM-120 AMRAAM - The AIM-120 AMRAAM (Advanced Medium-Range, Air-to-Air missile), a new generation air-to-air missile, has an all-weather, beyond-visual-range (BVR) capability. The missile also provides air defence support. AMRAAM, a follow-on to the AIM-7 Sparrow missiles, is faster, smaller, and lighter, and has improved capabilities against low-altitude targets. The missile has an active radar along with an inertial reference unit and microcomputer system, which makes the missile less dependent on the fire-control system of the launching aircraft. The variants include AIM-120A, AIM-120B, AIM-120C, AIM-120C-4/5/6/7, AIM-120D. AMRAAM-equipped fighters can attack several targets simultaneously.

Length: 3.7m

Diameter: 180mm

Weight: 152 kg

Propulsion: Solid Propellant Rocket

Warhead: Blast Fragmentation; high explosive

Wingspan: AIM-120A/B 21 inches; AIM-120C/D 19 inches

Speed: 4,900 km/h

Platforms: (Navy) F/A-18C/D/E/F Hornet and Super Hornet

Guidance system: inertial guidance, terminal active radar homing

Range: (AIM-120A/B) 55–75 km; (AIM-120C-5) >105 km; (AIM-120D) 160 km

AMRAAM *(Image Credit: Raytheon)*

UGM-133 Trident SLBM - The UGM-133A Trident II, or Trident D5 is the latest generation of the US Navy's submarine-launched fleet ballistic missiles (SLBM). Trident II (D5) missiles are deployed in Ohio-class (Trident) submarines, each carrying 24 missiles. The three-stage, solid-propellant, strategic weapon system is an improved SLBM with greater accuracy, payload, and range than the Trident C-4. Trident II is launched by the pressure of expanding gas within the launch tube. The Trident D5LE (life-extension) version will remain in service until 2042.

Length: 31.41m

Diameter: 2.11m

Weight: 58,500 kg

Propulsion: Three-stage solid-propellant rocket

Range: Greater than 7,360 km

Guidance System: MK 6 Astro Inertial

Warhead: Nuclear MIRV (Multiple Independently Targetable Re-entry Vehicles)

Speed: 29,020 km/h

Launch platform: Ballistic missile submarine

Trident II D5 Fleet Ballistic Missile built by Lockheed Martin

(Image Credit: United States Navy)

AGM-88 HARM - The AGM-88 High-Speed Anti-radiation Missile (HARM) is an air-to-surface tactical missile designed to seek out and destroy enemy radar-equipped air defence systems. The AGM-88 can detect, attack and destroy a target with minimum aircrew input. The proportional guidance system, with a fixed antenna and seeker head in the missile's nose, homes in on enemy radar emissions. The AGM-88E Advanced Anti-Radiation Guided Missile (AARGM), features the latest software, enhanced capabilities intended to counter radar shutdown and passive radar using an additional active millimetre wave seeker. AGM-88E will be initially integrated onto the F/A-18C/D, F/A-18E/F, EA-18G, and Tornado ECR aircraft and later on the F-35.

Length: 4.1m

Diameter: 254mm

Weight: 360 kg

Wingspan: 1.1m

Speed: 1,216 km/h

Propulsion: Thiokol dual-thrust, solid propellant, rocket motor

Range: 91+ km

Guidance System: Radar homing

Warhead: Blast fragmentation; warhead weight 150 pounds (68 kg)

Launch platform: F/A-18, F-4G, F-16, Tornado IDS, F-35 and others

AGM-88 HARM air-to-ground high speed missile, loaded on an F/A-18C Hornet.

(Image Credit: United States Navy)

AGM-114 Hellfire - The AGM-114 Hellfire missile is an air-to-ground, laser guided, subsonic missile with significant antitank capacity. It can also be used as an air-to-air weapon against helicopters or slow-moving fixed-wing aircraft. The Hellfire Air-to-Ground Missile System provides heavy anti-armour capability for attack helicopters. Variants include AGM-114B, AGM-114K, AGM-114M range of missiles.

Length: 1.6246m

Diameter: 17.8 cm

Wingspan: 0.71m

Weight: 44.45 to 48.54 kg

Warhead: Shaped charge and blast fragmentation

Speed: Subsonic

Platforms: (Navy) SH-60B/HH-60H Seahawk

Aviation Ordnance men load an AGM-114 Hellfire missile onto an HH-60H Seahawk aboard the Nimitz-class aircraft carrier USS Abraham Lincoln *(Image Credit: United States Navy)*

AGM-65 Maverick - The AGM-65 Maverick is an air-to-surface tactical missile designed for close air support, interdiction, and defence suppression. It is effective against a wide range of tactical targets, including armour, air defences, ships, ground transportation, and fuel storage facilities. It can carry two types of warheads: a heavy-weight warhead and a light-weight warhead. The heavy-weight version can penetrate the target before detonating and has delayed fuse setting options. Both of the warheads have contact sensors in the nose of the missile. Variants include AGM-65A/B, AGM-65D, AGM-65E, AGM-65F/G, AGM-65H, AGM-65J, and AGM-65K.

Length: 2.49m

Diameter: 30cm

Wingspan: 72cm

Weight: (AGM-65A/B) 210 kg; (AGM-65D) 220 kg; (AGM-65E) 293 kg; (AGM-65F/G) 306 kg; (AGM-65H) 210 kg; (AGM-65J) 297 kg; (AGM-65K) 306 kg

Propulsion: Thiokol SR 109-TC-1 solid-propellant rocket motor; Thiokol SR 114-TC-1 (or Aerojet SR 115-AJ-1) solid-propellant rocket motor

Speed: 1,150 km/h

Range: More than 22 km

Guidance System: Electro-optical (TV) in A and B models; imaging infrared (IIR) in D, F and G models; laser guided in the E model

Warhead: 300-pound penetrating blast-fragmentation warhead; 125-pound shaped-charge

Maverick, a versatile air launched, precision-guided missile that is effective against a wide variety of fixed, stationary, moving and manoeuvring targets both on land and sea.

(Image Credit: Raytheon)

AGM-119B Penguin - The AGM-119 Penguin is a passive IR seeker-based anti-ship missile. The missile flies an indirect flight path to target, and is operated in the "fire-and-forget" mode to allow multiple target prosecutions. The Penguin can be fired singly or in coordinated-arrival salvoes. The Penguin can be launched from surface vessels, fighter aircraft and helicopters. The missile is a highly effective weapon with an accurate navigation system and sophisticated passive infra-red homing array.

Length: 3.06m

Diameter: 28.5cm

Wingspan: 71.1cm

Weight: 387 kg

Propulsion: Solid propellant rocket motor and solid propellant booster

Speed: 1.2 Mach

Range: 33.3 km

Helicopter Antisubmarine Light Five One (HSL-51) fires an AGM-119 Penguin anti-ship missile from an SH-60B Sea Hawk helicopter *(Image Credit: United States Navy)*

Guidance System: Inertial and infrared terminal

Warhead: 110 pounds (50 kg), semi armour piercing

RIM-116 Rolling Airframe Missile - The RIM-116 Rolling Airframe Missile (RAM) is a lightweight, quick-reaction, fire-and-forget missile designed to destroy anti-ship cruise missiles and asymmetric air and surface threats. Its autonomous dual-mode passive radio frequency (RF) and infrared guidance design provide high-firepower capability for engaging multiple threats simultaneously. RAM Block II, the latest step in the development of the Rolling Airframe Missile, is a kinematic and RF receiver upgrade of Block 1/1A. The RIM-116 RAM is operational with US Navy and is being installed on a majority of new ship classes (LHA6, CVN 78 and LCS).

Length: (RAM Block 1A) 2.83m; (RAM Block 2) 2.88m

Diameter: (RAM Block 1A) 12.70 cm; (RAM Block 2) 15.87 cm

Wingspan: (RAM Block 1A) 44.5 cm; (RAM Block 2) 32.17 cm

Weight: (RAM Block 1A) 74.4 kg; (RAM Block 2) 88.2 kg

Propulsion: Solid-propellant rocket

Speed: 2.0+ Mach

Warhead: blast fragmentation

Platforms: Amphibious Assault Ships (LHA/LHD); Landing Platform Dock Ships (LPD); Carriers (CVN); Dock Landing Ships (LSD); Littoral Combat Ship (LCS)

USS Theodore Roosevelt (CVN 71) launches a Rolling Airframe Missile (RAM).

(Image Credit: United States Navy)

RIM-7 Sea Sparrow – The RIM-7 Sea Sparrow is a radar-guided, surface-to-air missile based on the Navy and Marine Corps AIM-7 Sparrow air-to-air missile. The short-range missile is a semi-active homing missile that makes flight corrections via radar uplinks. Evolved Seasparrow Missile (ESSM) (RIM 162D) is an international cooperative upgrade of the RIM-7 Sea Sparrow Missile. ESSM provides self-defence battlespace and firepower against high-speed, highly manoeuvrable anti-ship missiles in the naval environment. RIM-7 Sea Sparrow is launched from the Mk 29 Guided Missile Launching System (GMLS) and the Mk 48 Guided Missile Vertical Launching System (GMVLS). ESSM is fired from the following U.S., NATO and other launcher systems: MK 29 trainable launcher, MK 41 VLS, MK 57 VLS, MK 48 Guided Missile VLS, and the MK 56 Dual Pack ESSM Launching System. ESSM is a medium-range, semi-active homing missile that makes flight corrections via radar and midcourse data uplinks.

Length: 3.64m

Diameter: 20.3 cm

Wingspan: 1m

Propulsion: Alliant TechSystems (Hercules) MK-58 solid-propellant rocket motor; or NAMMO-Raufoss, Alliant

Weight: 225 kg

Speed: 4,256 km/h

Range: 19 km

Warhead: Annular blast fragmentation warhead

Guidance System: Semi-active on continuous wave or pulsed Doppler radar energy or interrupted continuous wave illumination

Platforms: CVN, LHA, and LHD, DDG 51, CG47, DDG 1000 classes (Navy)

Evolved Seasparrow Missile launched from the aircraft carrier USS Carl Vinson.

(Image Credit: United States Navy)

AIM-9 Sidewinder - The AIM-9 Sidewinder is a short-range air-to-air missile. The AIM-9M8/9 a low-cost quick-reaction modification of the missile's infrared counter-countermeasures features. The AIM-9X has a high off-boresight focal-plane array seeker mounted on a highly manoeuvrable airframe with a greatly improved infrared counter-countermeasures capability. Variants include AIM-9D, AIM-9G, AIM-9H, AIM-9L, AIM-9M, AIM-9R, and AIM-9X. The AIM-9X Sidewinder missile is the most advanced infrared-tracking, short-range air-to-air and surface-to-air missile and is configured for easy installation on a wide range of modern aircraft, including the F-15, F-16, F/A-18 and F-4 fighters; A-4, A-6 and AV-8B attack aircraft; and the AH-1 helicopter. The AIM-9X Block II missile is the current version of the AIM-9X Sidewinder short-range missile family. AIM-9X provides full day/night employment, resistance to countermeasures, extremely high off-boresight acquisition and launch envelopes, greatly enhanced manoeuvrability and improved target acquisition ranges.

Length: AIM-9M, 9.3 feet; AIM-9X, 9.9 feet

Diameter: 0.13m

Wingspan: 0.63m

Weight: 85.2754 kg

Propulsion: Solid-propellant rocket
Range: 1.0 to 35.4 km
Guidance System: Passive infrared
Warhead: Annular blast fragmentation (conventional) weighing 9.36 kg
Launch platform: Aircraft

Close-up view showing an AIM-9X Sidewinder short-range, heat-seeking air intercept missile attached to the port side inboard wing pylon of a US Air Force (USAF) McDonnell Douglas F-15C Eagle aircraft *(Image Credit: US Defence Visual Information Center)*

AGM-84K - The AGM-84K Standoff Land Attack Missile-Expanded Response (SLAM-ER) is an air-launched, day/night, adverse weather, over-the-horizon, precision strike missile. SLAM ER is a low-cost, low-risk upgrade to the SLAM. SLAM ER provides the US Navy with surgical strike capability against high-value, fixed land targets, ships in port, or ships at sea. SLAM ER can be launched from safe standoff ranges of more than 150 nautical miles. The SLAM-ER relies on the GPS and infrared imaging for its navigation and control.
Length: 4.4m
Diameter: 34.3m
Wingspan: 2.2m
Weight: 674.5 kg
Range: 250 km
Speed: 855 km/h
Guidance System: Ring Laser Gyro Inertial Navigation System (INS) with multi-channel GPS; infrared seeker for terminal guidance with Man-in-the-Loop control data link from the controlling aircraft. Upgraded missiles incorporate Automatic Target Acquisition (ATA)

US Navy McDonnell Douglas F/A-18C Hornet equipped with an AGM-84 Standoff Land Attack Missile-Expanded Response (SLAM-ER) under the right wing and two AN/AWW-13 Advanced Data Link pods under the left wing *(Image Credit: United States Navy)*

Launch platform: F/A-18C/D Hornet; F/A-18E/F Super Hornet; F-15E Strike Eagle; P-3C Orion; P-8 Poseidon

Aegis Weapon System – Aegis Combat System can simultaneously attack land targets, submarines, and surface ships while automatically protecting the fleet against aircraft, cruise missiles and ballistic missiles. Aegis comprises of AN/SPY-1, an advanced, automatic detect and track, multi-function phased-array radar. It also has the fast-reaction component of the Aegis Anti-Aircraft Warfare (AAW) capability, along with the Phalanx Close In Weapon System (CIWS), and the Mark 41 Vertical Launch System. The Aegis Ballistic Missile Defence (BMD) system is the primary sea-based component of the US missile defence system. Aegis combines robust anti-air warfare (AAW) capabilities with an unprecedented ballistic missile defence (BMD 5.0) functionality into a single integrated air and missile defence (IAMD) system. The Aegis Common Source Library (CSL) enables customers to rapidly integrate new capabilities and processes across the fleet in a "build once, use many times" framework. Aegis CSL can be used across a variety of platforms, ranging from multiple classes and types of ships to land-based systems. The new Baseline 9.C1 in the Aegis Combat System offers advanced defence capabilities and enhanced integration with other systems external to the ship. This Aegis baseline also improves Aegis networking capabilities, allowing Aegis vessels to automatically coordinate defence with input from satellite and ground-based radar assets—forming a true shield of defence over a wide area. There are currently 84 US Navy ships in service with the AEGIS Weapons System installed: 22 Cruisers and 62 Destroyers.

Phalanx Close-In Weapons System (CIWS) - The Phalanx Close-In Weapon System is a rapid-fire, computer-controlled, radar-guided gun system—is designed to defeat anti-ship missiles and other close-in air and surface threats in the sea. A self-contained package, Phalanx automatically carries out functions usually performed by multiple systems: search, detection, threat evaluation, tracking, engagement and kill assessment. The Block 1B version of Phalanx adds control stations that allow operators to visually track and identify targets before engagement. Phalanx is installed on all US Navy surface combatant ship classes.
Height: 4.7m
Weight: (Block 1B) 6,120 kg
Type Fire: (ASM & Aircraft) 4,500 rounds/min, (Asymmetric Threats) 3,000 rounds/min
Magazine Capacity: 1,550 rounds
Calibre: 20mm
Ammunition: Armour Piercing Discarding Sabot
Type: M-61A1 Gatling Gun
Guidance system: Ku-band radar and FLIR
Crew: Automated, with human oversight

Phalanx Block 1B is the only tested and proven system capable of defeating the surface, air and changing littoral threats with which today's fleets must contend
(Image Credit: Raytheon)

CONCLUSION

Today's security environment is dramatically different from the one the United States has been engaged with for the last 25 years, and it requires new ways of thinking and new ways of acting. This security environment is driving the focus of the Defence Department's planning, budgeting for Armed Forces.0020The US Armed Forces has significant capabilities in both defence and power projection due to its large budget, resulting in advanced and powerful equipment, and its widespread deployment of force around the world, including about 800 military bases in foreign locations. The United States is focused on fighting terrorism and countering violent insurgencies, and the country will continue to do so as long as these threats exist. But the security environment is evolving. The US is currently focusing on rebalancing to the Asia-Pacific, maintaining a strong commitment to security and stability in Europe and the Middle East, sustaining a global approach to countering violent extremists, prioritising and protect key investments in technology and to reinvigorate efforts to build innovative partnerships.

The United States has engaged in every type of ground warfare in the last 20 years. From mountains to jungles, and from desert to urban, US have the some of the most experienced warriors in the world. No other country comes close to the amount of combat veterans that US have. Defence spending represents about half of the federal government's discretionary spending, and the military's budget has increased by more than 70 percent since 2001. Although the conflicts in Iraq and Afghanistan have cost the Pentagon upward of USD 1 trillion, nearly half of the growth in defence spending in the past decade has been unrelated to the wars. The steady rise of China and her assertiveness in South China Sea, has compelled US declare that she would favour peaceful adjustments in the region. In the Asia-Pacific leaders summit, President Obama stated his intention to take relations with China to a new level and highlighted the importance of forging a durable strategic partnership with China to ensure its rising power does not destabilise the international system. US is focusing on strategic partnerships with India, Japan, Australia and Vietnam to counter the growing military rise of China. The United States has also been seeking to expand its military cooperation with India as part of the US strategic shift to the Asia-Pacific region. The US Military remains deeply engaged around the world. US is working on battles against cyber threats, terrorism around the globe including Iraq, Syria, Afghanistan and to defeat Al Qaeda and Islamic State of Iraq and the Levant (ISIL). US is also taking steps to counter Russia's stance or policies in Eastern Europe and also involving the country's long-term economic and security interests in Asia-Pacific region.

CONTACT DETAILS

The Department of Defence
Address: 1400 Defence Pentagon
Washington DC 20301-1400
Tel: +1-703-56955261

DARPA's Address
Defence Advanced Research Projects Agency
3701 North Fairfax Drive
Arlington, VA 22203
DARPA General Information (703) 526-6630

Contact DPAP
Defence Procurement and Acquisition Policy
3060 Defence Pentagon
Room 3B855
Washington, DC 20301-3060

URUGUAY
(Capital: Montevideo)

INTRODUCTION

Area: 176,215 sq km
Population: 3,351,016 (July 2016 est.)
Coastline: 660 km
Maritime claims: Territorial sea: 12 nm
Contiguous zone: 24 nm
Exclusive Economic Zone: 200 nm
Continental shelf: 200 nm or edge of continental margin

KEY POLITICAL PERSONS

PRESIDENT: Tabare Vazquez

VICE PRESIDENT: Raul Sendic

KEY DEFENCE PERSONS

MINISTER OF DEFENCE: Jorge Menendez

CHIEF OF DEFENCE: Army General Nelson Pinto

COMMANDER-IN-CHIEF OF ARMY: Gen. Guido Manini Rios

COMMANDER-IN-CHIEF OF NAVY: Rear Adm. Leonardo Alonso

COMMANDER-IN-CHIEF OF AIR FORCE: Air Gen. Alberto Miguel Zanelli Lacoste

ECONOMY

Uruguay's economy is characterised by an export-oriented agricultural sector, a well-educated work force, and high levels of social spending. The 2008-09 global financial crisis put a brake on Uruguay's vigorous growth, which decelerated to 2.9% in 2009. Nevertheless, the country managed to avoid recession and keep positive growth rates, mainly through higher public expenditure and investment. GDP growth reached 8.5% in 2010 and 6% in 2011.

GDP: $71.43 billion (2015 est.), $70.39 billion (2014 est.)

Real Growth Rate: 1.5% (2015 est.), 3.5% (2014 est.)

Industries: Food processing, electrical machinery, transportation equipment, petroleum products, textiles, chemicals, beverages

Total Exports: $9.067 billion (2015 est.), $10.34 billion (2014 est.)

Export Commodities: Beef, soybeans, cellulose, rice, wheat, wood, dairy products; wool

Major Markets: China 15%, Brazil 14.4%, US 6.5%, Argentina 4.9% (2015)

Total Imports: $9.345 billion (2015 est.), $11.25 billion (2014 est.)

Import Commodities: Refined oil, crude oil, passenger and other transportation vehicles, vehicle parts, cellular phones

Major Suppliers: Brazil 18.2%, China 17.4%, Argentina 12.6%, US 9.1%, Germany 4.5%, Nigeria 4.1% (2015)

INTERNATIONAL DISPUTES

Two uncontested boundary disputes with Brazil over Isla Brasilera at the tripoint with Argentina at the confluence of the Quarai/Cuareim and Uruguay rivers, and, in the 235-square kilometre Invernada River region, over which tributary represents the legitimate source of the Quarai/Cuareim River.

DEFENCE

ARMY

Personnel: 16,234

Equipment

Category	Name	In Service
MBT	T-54/55	15
	M-24	17
	M41-A1	47
APC/MICV /IFV	BMP-1	15
	M113	24
	Condor	55
	Gaz 3971 Vodnik	48
	OT M-64 SKOT	53
	OT M-93	47
	MT-LB	3
	Mowag	40
	EE9 Cascavel	15
	EE3 Jararaca	16
Artillery	105 mm. M101/M102	48
	155 mm. M114	8
	122 mm. S1	6
	122 mm. RM-70	4
Mortar	60 mm.	98
	81 mm.	96
	120 mm. SL	44
MRL	122 mm. RM-70	4
Anti-Tank GW	Milan	15
Recoilless rifle	106 mm. M40A1	69
	75 mm. M-20	3
AA gun	40 mm. L/60	8
	20 mm. M167 Vulcan	6
	20 mm. TCM-20	9

New Procurements/ Upgrades

No major procurement plans are under consideration at present.

NAVY

Personnel: 5,403

Equipment

Category	Name	In Service
Frigate	Uruguay Class	2
	Montevideo Class	1
Logistics Support	General Artigas	1
Mine Warfare	Audaz Class	3
Patrol Vessel	15 De Noviembre Class	3
	Rio Negro Class	2
Amphibious Force	LD-41 Class LCM	2
	LD-45 Class LCVP	2

Naval Aviation

Category	Name	In Service
Aircraft	Beechcraft B-200T	1
	2 Jetstream TM Mk2	1

Naval Aviation continued...

Category	Name	In Service
Aircraft	T-34C	2
Helicopter	Bo-105	6
	Wessex HC Mk2	3
	IH-13 Esquilo	1

New Procurements/ Upgrades

No major procurement plans are under consideration at present.

AIR FORCE

Personnel: 2,948

Equipment

Category	Name	In Service
Fighters/ attack/ Reconnai-ssance	Cessna A-37 Dragonfly	8
	IA 58 Pucara	5
Transport/ Tanker	C-130 Hercules	2
	EMB 110 Bandeirante	3
	Beechcraft Twin Bonanza D50	1
	CASA C-212 Aviocar	2
	EMB 120 Brasilia	1
Helicopter	Aerospatiale AS 365 Dauphin	1
	Bell 212 Twin Huey	4
	Bell UH-1 Iroquois	13
Trainer	Aermacchi SF.260	12
	Pilatus PC-7 Turbo Trainer	6
	Cessna T-41 Mescalero	7
Utility	Cessna 206 Stationair	10
	Beechcraft B58 Baron	2

New Procurements/ Upgrades

No major procurement plans are under consideration at present.

Defence Expenditure

Total defence spending: $491 million

Estimated defence spending in terms of GDP: 1.8% (2015)

In 2010, Uruguay increased its defence budget by nearly 25%.

Major defence suppliers: United States, Brazil, Argentina, Israel, Canada, Russia.

Defence Production and R&D

Uruguay has no defence industry, although the navy owns a small shipyard that manufacturers small vessel. There is also a small, army-run explosives production facility known as PESMA.

Defence Procurement

The Uruguayan MoD has announced the creation of a new procurement arm as part of its 2010-2014 Defence Budget in a bid to streamline procurement. The Centralised Defence Procurement Unit (UCAD) will be responsible for all future acquisition programmes. It will ultimately have a chief, deputy chief, three department heads, six advisors and five technical staff.

CONTACT DETAILS

Ministry of National Defence
Avenida 8 de octubre, 2628de
Montevideo, Uruguay
Tel: +5982472828
Fax: +5982474425

Army HQ
Avda. General Garibaldi, 2313
Montevideo, Uruguay
Tel: +5982081542, 2081554
Fax: +598-2036600

Navy HQ
Ranbla 25 de Agost s7n esq. Maciel 4.o Piso
Montevideo, Uruguay
Tel: +59829155555
Fax: +598-2-9169048

Air Force HQ
Avenida Don Pedro de Mendoza, 5553
Montevideo, Uruguay
Tel: +5982224401
Fax: +598-2-223830

UZBEKISTAN
(Capital: Tashkent)

INTRODUCTION

Area: 447,400 sq km

Population: 29,473,614 (July 2016 est.)

Coastline: 0 km (doubly landlocked)

Maritime claims: None

KEY POLITICAL PERSONS

PRESIDENT & PRIME MINISTER: Shavkat Mirziyoyev

KEY DEFENCE PERSONS

DEFENCE MINISTER: Col. Gen. Kabul Berdiev

ECONOMY

Uzbekistan is a dry, landlocked country. The agriculture and manufacturing industries contribute equally to the economy, each accounting for about one-quarter of the GDP. It is a major producer and exporter of cotton. The country is also a big producer of gold, with the largest open-pit gold mine in the world. The country has substantial deposits of copper, strategic minerals, gas, and oil. Uzbekistan is continuing to transition to a more market-oriented economy, according to the World Bank. The country's policy goals and priorities are: to increase the efficiency of infrastructure, especially of energy, transport, and irrigation; to enhance the competitiveness of specific industries, such as agro-processing, petrochemicals, and textiles; to diversify the economy and thereby reduce its reliance on commodity exports; and to improve access to and the quality and outcomes of education, health and other social services in order to become an industrialized, high middle-income country by around 2050.

GDP: $187.9 billion (2015 est.) $174 billion (2014 est

Real Growth Rate: 8 % (2015 est.), 8.1% (2014 est.)

Industries: Textiles, food processing, machine building, metallurgy, mining, hydrocarbon extraction, chemicals

Total Exports: $11.35 billion (2015 est.), $13.31 billion (2014 est.)

Export Commodities: Energy products, cotton, gold, mineral fertilizers, ferrous and nonferrous metals, textiles, food products, machinery, automobiles

Major Markets: China 13.7%, Hong Kong 11.5%, Malaysia 10.8%, Indonesia 8.2%, US 6.9%, Japan 4.4%, South Korea 4.1% (2015)

Total Imports: $11.81 billion (2015 est.), $12.92 billion (2014 est.)

Import Commodities: Machinery and equipment, foodstuff, chemicals, ferrous and nonferrous metals

Major Suppliers: China 14.2%, US 11.2%, Malaysia 11.2%, Japan 6.3%, South Korea 6.1%, Indonesia 4.8% (2015)

INTERNATIONAL DISPUTES

✦ Field demarcation of the boundary with Kazakhstan commenced in 2004. Border delimitation of 130 km of border with Kyrgyzstan is hampered by serious disputes around enclaves and other areas. Prolonged drought and cotton monoculture in Uzbekistan and Turkmenistan created water-sharing difficulties for Amu Darya river states.

DEFENCE

Tashkent declared national sovereignty and the right to defend its own borders on 31 August 1991 which later saw the creation of a Ministry for Defence Affairs and appointment of the first Defence Minister. On 14 January1992, Uzbekistan assumed jurisdiction over all former Soviet ground, air, and air defence units, formations and installations deployed on its soil, with the exception of those strategic forces retained under the Commonwealth of Independent States (CIS).

ARMY

Personnel: 70,000

Equipment

Category	Name	In Service
MBT	T-72	110
	T-62	109
	T-64	
	T-80	100
APC	BTR-60	100
	BTR-70	150
	BTR-80	150
	BMP-2	150

New Procurements/Upgrades

The US State Department has announced that it was giving more than 300 used Mine-Resistant Amour-Protected (MRAP) vehicles to Uzbekistan.

NAVY riverine service operates two Dzaychum class patrol craft (Ukraine-built Gyurza type).

AIR & AIR DEFENCE FORCE

Personnel: 15,000

Equipment

Category	Name	In Service
Fighter	Sukhoi Su-27	6
	Sukhoi Su-25	20
	Sukhoi Su-24	30 (stored)
	Sukhoi Su-17	Unspecified (stored)
	MiG-29	30 (most stored)
Trainer	L-39	20
Helicopter	Mi-8	25
	Mi-17	25
	Mi-24	40
Transport	IL-76	6
	An-12	5
	An-24	1
	An-26	15
SAM	SA-2/-3/-5	1 brigade

New Procurements/Upgrades

No major procurement plans are under consideration at present.

Defence Expenditure

Exact defence expenditure figures are unavailable.

Defence spending in terms of GDP: 3.5% (2010)

Defence Production and R&D

Uzbekistan has seen moderate success in establishing a niche industry focused on aerospace. The TAPO plant manufactured Il-76 military transport aircraft and Il-78 in-flight refuelling aircraft in cooperation with several Russian firms, including the Voronezh aircraft plant. Both aircraft remain the platforms of choice for the Russian Air Force. The Il-76MF version, developed jointly by Uzbek and Russian aircraft design bureaus, has been designated by Russian Ministry of Defence (MoD) as its preferred strategic airlift aircraft.

Defence Procurement

Tashkent has not published any official details of procurement programmes nor discloses data on defence spending. Moscow remains the principal supplier of arms and military equipment to the Uzbek defence and security forces.

CONTACT DETAILS

Ministry of Defence

100, M.Ulugbek str., Tashkent,
Republic of Uzbekistan, 100000
Phones: +998 (71) 269-8200, +998 (71) 269-8228

Key Persons Image(s) Credit:
Kremlin.ru & Mil.ru

STRATEGIC INFORMATION

VENEZUELA

(Capital: Caracas)

INTRODUCTION

Area: 912,050 sq km

Population: 30,912,302 (July 2016 est.)

Coastline: 2,800 km

Maritime claims: Territorial sea: 12 nm

Contiguous zone: 15 nm

Exclusive Economic Zone: 200 nm

Continental shelf: 200 m depth or to the depth of exploitation

GEOPOLITICAL IMPORTANCE

Venezuela, officially called the Bolivarian Republic of Venezuela, is located on the northern coast of South America. It became the first Spanish American colony to declare independence gaining full independence in 1830. As a founding member of the Organization of the Petroleum Exporting Countries (OPEC), Venezuela is an important player in the global oil market. While production has been declining, Venezuela exports of crude oil to the United States have been among the top. It also has the second-largest natural gas reserves in the Americas, behind the United States.

POLITICAL OVERVIEW

Venezuela is a federal republic. Its political system is divided into three groups - Executive, Legislative and Judicial branches. The President is the head of the state. He presides over the executive branch of the government and is responsible for the appointment of Vice-President, members of the Cabinet and other important members of the National Assembly. The legislature of Venezuela is a bicameral legislative assembly, named as the National Assembly or Asamblea Nacional having 167 seats. The members of the house are elected by a popular referendum for a period of five years which can be extended to a two-year term. The Supreme Tribunal of Justice is the highest law court in the judicial system of Venezuela. The magistrates of this court are appointed for a period of 12 years' term. There are lower courts, district courts and municipal courts located at every district.

The country has suffered some coups and military dictatorships. The fall of Marcos Perez Jimenez government in 1958 ended the last dictatorship of the 20th century in Venezuela ushering in a democratic period. The country adopted a new constitution in 1961. Two political parties prevailed during the following decades - the social-democratic Democratic Action (AD) and the Christian-democratic COPEI during the period known as the fourth republic. This system came to an end during the 1998 election when Hugo Chavez won thus beginning the fifth Republic and the left-wing Bolivarian Revolution. The present political system of Venezuela has undergone significant changes as a result of political turmoil and economic backlashes. The current constitution was drafted in 1999 replacing the earlier constitution made in 1961. Hugo Chavez, the central figure of the Venezuelan political landscape since his election to the Presidency in 1998 died in office in early 2013, but was succeeded by his vice-president, Nicolas Maduro, who pledged to continue his policies.

KEY POLITICAL PERSONS

PRESIDENT & COMMANDER-IN-CHIEF: Nicolas Maduro Moros

EXECUTIVE VICE PRESIDENT: Aristobulo Isturiz

ECONOMY

Venezuela's economy is predominantly based on petroleum sector and manufacturing. In 2013, the GDP was USD 373.98 billion making it the world's 30th largest economy. The country is among the largest member of OPEC by oil production and one of the top suppliers of foreign oil to the United States. It has benefited from the historically high international oil prices of the past decade. The World Bank said economic growth and the redistribution of resources associated have led to an important decline in moderate poverty, from 50% in 1998 to approximately 30% in 2012. However, GDP growth slowed to 1.4% in 2013. Its economy is highly vulnerable to fluctuations in oil prices since it represents over 96% of the country's exports and generates nearly half of fiscal income. The United States is Venezuela's largest trading partner. Bilateral trade in goods between both countries reached USD 23.9 billion in 2015. US goods exports to Venezuela totalled USD 8.3 billion in 2015, down 25 percent from 2014. Falling oil prices since 2014 have aggravated Venezuela's economic crisis. Insufficient access to dollars, price controls, and rigid labour regulations have led some US and multinational firms to reduce or shut down their Venezuelan operations. In 2016, Venezuela - which has the world's largest oil reserves - has seen oil production crash to a 13-year low highlighting the economic crisis situation the country is going through.

GDP: $515.7 billion (2015 est.), $546.9 billion (2014 est.)

Real Growth Rate: -5.7% (2015 est.), -3.9% (2014 est.)

Industries: Petroleum, construction materials, food processing, textiles; iron ore mining, steel, aluminium; motor vehicle assembly, chemical products, paper products

Oil - proved reserves: 300 billion bbl (1 January 2016 est.)

Natural gas - proved reserves: 5.617 trillion cu m (1 January 2016 est.)

Total Exports: $38.45 billion (2015 est.), $74.71 billion (2014 est.)

Export Commodities: Petroleum, bauxite and aluminium, minerals, chemicals, agricultural products, basic manufactures

Major Markets: US 26.6%, India 13.7%, China 11.7%, Cuba 6.4% (2015)

Total Imports: $36.46 billion (2015 est.), $47.48 billion (2014 est.)

Import Commodities: Agricultural products, livestock, raw materials, machinery and equipment, transport equipment, construction materials, medical equipment, pharmaceuticals, chemicals, iron and steel products

Major Suppliers: US 18.4%, China 15.3%, Brazil 9.7%, Colombia 5.9%, Mexico 4.2% (2015)

DEFENCE & SECURITY

The President is the commander-in-chief of the Venezuelan National Bolivarian Armed Forces (FANB) and has overall supervision and control over it. The armed forces are trained to fight both a conventional war, and irregular war. The Venezuelan Ministry of the Popular Power for Defence is the federal level organ responsible for maintaining the armed forces. The ministry coordinates numerous counter narcotics operations and oversees the conventional military capabilities of Venezuela.

A new Organic Law of the Armed Forces (LOFAN) approved in September 2005 under the presidency of Hugo Chavez created a civilian military reserve, elite paramilitary groups that will operate independently. The new LOFAN essentially adopts the Cuban model of military organisation, command and control. It legally enshrines three core missions for the Bolivarian armed forces. First, protect the president, his family and his closest associates at all times. Second, maintain internal order against threats to the president. Third, defend Venezuela against external threats. The new doctrine explicitly defines the military's primary mission as defending the stability of the country against internal threats and disruptions and also empowers the Bolivarian military to engage in joint actions with the armed forces of other countries.

KEY DEFENCE PERSONS

MINISTER OF DEFENCE & STRATEGIC OPERATIONAL COMMANDER: Gen. Vladimir Padrino Lopez

COMMANDER GENERAL OF THE ARMY: Gen. Juan de Jesus Garcia Toussaintt

COMMANDER GENERAL OF THE NAVY: Adm. Orlando Maneiro Gaspar

COMMANDER GENERAL OF THE AIR FORCE: Maj. Gen. Edgar Valentin Cruz Arteaga

NATIONAL GUARD: Maj. Gen. Antonio Jose Benavides Torres

NATIONAL MILITIA: Maj. Gen. Cesar Ramon Vega Gonzalez

Internal conflict

Venezuela's core internal threat is insurgency or civil war. Violent crime in Venezuela is pervasive, both in the capital, Caracas, and in the interior. The country's overall per capita murder rate is cited as one of the top five in the world. According to the non-governmental organisation, Venezuelan Violence Observatory (VVO), 2011 was the most "violent year in Venezuela's history." The data said there were 19,336 homicides in 2011, up from 13,080 murders in 2010, an increase of almost 30%. Cross-border violence, kidnapping, drug trafficking, smuggling, and cattle-rustling occur frequently in areas along the 1,000-mile long border between Venezuela and Colombia. Caracas suffers from areas of extreme poverty which provide gangs and criminal elements with an environment conducive to crime and is subsequently difficult to police. Venezuela is witnessing political tensions, economy issues, including expropriations, macroeconomic distortions, physical insecurity, corruption, and a volatile regulatory framework. From January to October 2016, 5,772 protests occurred throughout Venezuela with protests for political rights increasing in late-2016. Nicolas Maduro's government characterised the protests as an undemocratic coup attempt.

External conflict

Venezuela claims most of Guyana west of the Essequibo River. The dispute dates back to the early nineteenth century. Venezuela also has a maritime boundary dispute with Colombia in the Gulf of Venezuela and territorial dispute with Dominica over the Isla Aves archipelago located in the Caribbean.

THREAT PERSPECTIVE

Venezuela faces minimal external threat; however, armed guerrilla groups pose the real peril to regional stability. Many of these home-grown guerrilla groups are no longer contained within the confines of their own nation and continue to increase in militancy. Cross-border violence, kidnapping, drug trafficking, smuggling, and cattle-rustling occur frequently in areas along the 1,000-mile long border between Venezuela and Colombia. Increasing US-led militarisation of Latin America is seen as another threat that appears to be particularly aimed at the country, especially through the US-Colombia agreement for new military bases encircling Venezuela.

US: Venezuela has been vocal about their opposition to the United States, regularly criticising the US government, its policies, and its relations with Latin America. President Maduro has also continued his predecessor's policies, notably what the Venezuelan government refers to as 21st Century Socialism with efforts to expand Venezuelan economic and political influence among nations in Latin America and the Caribbean. The United States is Venezuela's largest trading partner. Bilateral trade in goods between both countries reached USD 23.9 billion in 2015. In 2013, the Department of State announced the re-imposition of non-proliferation sanctions on the Venezuelan Military Industry Company (CAVIM) and other foreign entities and individuals under the Iran, North Korea, and Syria Non-proliferation Act (INKNSA).

STRATEGIC RELATIONS

China: Cooperation between Venezuela and China has blossomed in recent years following the signing of several multi-billion-dollar agreements on investments in oil, energy, construction and high-tech industries. In return for Venezuelan oil exports, China has become a major investor of development projects in Venezuela. The two countries have established military-technological ties with the acquisition of two squadrons of Chinese-built Karakorum-8 trainer jets and ground radars. They have also launched their first joint space satellite, named Venesat-1. China is Venezuela's second-largest trade partner, and Venezuela is China's biggest investment destination in Latin America.

Cuba: Cuba and Venezuela enjoy warm relations during the Presidency of Hugo Chavez. Venezuelan oil and money help to keep the communist-ruled island's troubled economy afloat and the governments have about 30 joint ventures. Venezuela currently ships 100,000 barrels of subsidized oil to Cuba daily. In exchange, Venezuela receives thousands of doctors, teachers and some military advisers. Close ties with Cuba also helped to transform the Military of Venezuela. Recent falling oil prices of Venezuela is giving problems to Cuba in terms of decline of oil shipment which is making Cuba to do energy rationing.

Russia: Russia and Venezuela continue to enjoy warm relations. The leaders of Russia and Venezuela have reaffirmed their wish for continuing their course towards strategic cooperation in all sectors. Ties between the two countries flourished under Chavez, whose 14-year rule earned Moscow a number of lucrative arms and energy deals and a close ally in South America. They have been building closer military relations and have spoken of their opposition to US global dominance. Ministry of Defence of both Russia and Venezuela has signed agreements on military cooperation.

Bolivia: Venezuela has been a major player in Bolivia since the election of President Evo Morales in 2006, helping out its fellow Andean friend with everything from military aid to funding a health programme Venezuelan aid to Bolivia has increased substantially, with Caracas offering important technical support to the re-nationalised Bolivian oil company, Yacimientos Petroliferos Fiscales Bolivianos (YPFB).

Iran: Iran and Venezuela continue to have strong ties. The two nations have signed a Memorandum of Understanding pledging full military support and cooperation. Venezuela has openly supported Iran's nuclear programme. Annual bilateral trade between the two countries is estimated to be in the hundreds of millions of US dollars. In June 2015, the Venezuelan government and a high-level delegation from Iran inked six deals on scientific, technological, economic and health cooperation.

Argentina: Relations between Venezuela and Argentina had been historically friendly until a diplomatic crisis in 2005 forced the two nations apart. Relations began to normalise in 2007 and since then the countries have been moving towards re-establishing an official relationship. The two nations have signed a military cooperation under the framework of Unasur Defence Council. The bilateral agreement also includes areas such as science, technology development and joint military exercises between the armed forces of the two countries.

DEFENCE CAPABILITIES

The national armed force of Venezuela is a conscript force. It does not have nuclear, chemical or biological warfare capability. The Operational Strategic Command is the highest organ of command of the National Armed Forces. The Army, Navy, Air Force and National Guard serves under the Operational Strategic Command and the National Reserve and the Territorial Guard under the National Militia General Command.

The Venezuelan government has been carrying out a massive military purchase programme. This has included negotiations for German submarines and transport aircraft. Agreements have been signed with Russia for aircraft and weapon systems. From Spain, transport aircraft and naval vessels, radars from China, home-made and designed armoured light vehicles and rocket launchers. The Venezuelan Navy's patrol and defence capabilities are being reinforced by an ongoing equipment procurement programme.

ARMY: The National Army of Venezuela is the largest military branch. It has the responsibility for land-based operations against external, or internal threats.

Personnel: 70,000

Equipment

Category	Name	In Service
MBT	AMX-30	82
	Scorpion 90 (light)	66
	AMX 13C90 (light)	31

Army continued...

Category	Name	In Service
MBT	T-72SBIV	200+
IFV/APC	Dragon ASV (90mm gun)	42
	BMP-3	300
	AMX 13VTT	70
	TPz-1 FUCH	10
	V-100/V150- Dragon	115
	BTR-80A	114
Artillery	155 mm. M109 SP (reserve)	5
	155 mm. FMk3 SP (reserve)	20
	155 mm. M114 howitzer (reserve)	24
	105mm M56	80
	105mm M101	40
	2S19 MSTA-S 152mm SP	
Mortar	2S23 Nona-SVK	
	81mm	
	2S12 Sani 120mm towed	
	60mm	
RCL	M40A1 106mm	
MRL	AMX-13/LAR 160mm	12
	BM-30 Smerch	12
	BM-21 Grad 122mm	24
ATK Weapon	Carl Gustaf	
	AT4	
	RPG 7	
ATGW	Mapats 2	
AA Gun	ZSU-23 Zomi twin 23mm towed under CODAI	
SAM	RBS 70 (Giraffe FCS)	
	Mistral	
	Buk-M2	
	Tor-M1	
	IGLA-S (all under CODAI)	1,800

New Procurements/Upgrades

No major procurement plans are under consideration at present.

Army Aviation

Category	Name	In Service
Transport	IAI Arava	5
	M-28 Skytruck	12
	Super King Air 200	2
Helicopter	Mi-17V	18
	Mi-35 (attack)	12
	Mi-26	10
	Bell-205	12 (to be replaced)
	Bell-206	7
	AS-61	3

NAVY: The chain of command of the Venezuelan Navy is: Commanding General in Chief, Inspector General of the Navy and Chief of General Staff of the Navy. There are five major commands: Naval Logistics Command, Naval Personnel Command, Naval Education and Training Command and the Naval Operations Command. The Marine Infantry was reorganised in 2001 into a single division. The country is divided into two naval zone - Western Naval Zone (Hq: Punto Fijo) and Eastern Naval Area (HQ: Carupano) that currently covers the Atlantic coast.

Major naval bases: Puerto Cabello, Puerto Hierro, Punto Fijo

Main naval air stations: Puerto Cabello, La Orchila Island, Caracas, Maiquetia, San Fernando de Apure

Personnel: 18,500 (including 10,000 marines & 500 personnel from Naval Aviation)

Equipment

Category	Name	In Service
Submarine	Sabalo class (Type 209)	2
Frigate	Mariscal Sucre class (Lupo type)	6
Corvette	Guaiquerí class	4
Amphibious Force	Capana class ex-US LST	2
	Los Frajles class LCU	4
Patrol craft	Guaicamacuto class	4
	Point class	3
	Constitucion Class	6
Auxiliaries	Ciudad Bolivar	1
	Oceanographic ship	1
	Sailing training ship	1
	Fleet tug	1
	Survey vessels	3

New Procurements/Upgrades

- ✦ A fourth and last Guaicamacuto class patrol ship is under construction in local yards.
- ✦ A second Damen Stan patrol 4207 patrol boats are under construction. A total of six vessels are on order.
- ✦ The Navy also plans to buy 8 Damen Stan Lander 5612 amphibious warfare craft and six Stan Patrol 5009 and 18 type 1102 fast interceptor boats.

Naval Aviation

Personnel: 2,000

Equipment

Category	Name	In Service
Transport	C-212/200	5
	C-21/400	3
	Super King Air C-200	1
	C-208	1
Helicopter	AB-212 ASW	8
	Bell 412EP	4

Naval Aviation continued...

Category	Name	In Service
Helicopter	Bell 206L	1
	Mi-17V-5	6
	Bell 206	1
	Harbin Z-9	(8 on order)
	Cessna C-210	1

New Procurements/Upgrades

✦ An order for 8 Harbi Z-9 helicopters is expected to equip the new OPVs and patrol ships. The helicopters will be equipped for ASW.

Marine Infantry Division

Personnel: Approximate 15, 000

Equipment

Category	Name	In Service
APC/IFV	E-11 Urutus	38
	AAVT	11
Artillery	M-56 105mm pack howitzer	40
Mortar	120mm mortar	18
SAM	RBS-70	12
AA gun	GAM-BOI 20mm guns	
Patrol vessel	Guardian 25' type	66
	Guardian 22' type	12

New Procurements/Upgrades

✦ Deals have been signed with China to supply unspecified numbers of VN-1 8x8 APCs, VN-16 amphibious light tanks and VN-18 amphibious tracked AFV, SR5 122/220mm MRL and CS/SM4 SP mortar systems (81mm and 120mm weapons).

AIR FORCE: The Bolivarian National Air Force is organised into three major air commands, including a Combat Command, Logistics Command and Air Training Command. There is also a small air defence group. Nationally the air force is organised into five air zones. Within these air zones the air force deploys three fighter groups, three transport groups, two special operations air groups, one tactical air training group, and one air training group. The aviation component of the National Guard functions as a communications and combat reserve force for the Air Force.

Major air bases: Barcelona, Barquisimeto, Caracas (La Carlota), Carrizales, Maracaibo, Maracay (Boca de Rio and Palo Negro).

Personnel: 9,580

Equipment

Category	Name	In Service
Fighter/ Light Attack/ Electronic weapon	Su-30 MK2	24
	F-16A/B	21
	CF-5 Freedom Fighter	16
	North American Rockwell OV-10 Bronco	20

Air Force continued...

Category	Name	In Service
Fighter/ Light Attack/ Electronic weapon	K-8W Karakorum	9
	SF-260U	11
	EMB-312 Tucano	24
	Aermacchi SF.260	12
	Tucano (Embraer EMB 312)	19
	K- 8 w karakorum	18
	Hongdu jl-8	17
Transport/ Tanker/ VIP	C-130 Hercules	3
	Boeing 707-320C	2
	Shorts SD3-60-300	2
	A319CJ	1
	Boeing 737-200	1
	Falcon 50/900XL	5
	AS-532UL Cougar	2
	Falcon 20C/50/900	1
	Cessna Citation I/II/X	6
	BE-200 Super King Air	6
Trainer	Cessna 182R	3
Helicopter	Mi-17	7
	Mi-35	9
Weapon	MistraL (SAM) from France	
	Barak 3 (SAM) from Israel	
	Breda/Bofors 40/L70 mm (SAM)	
	TCM-20 Mk.2 (SAM)	
	Thales Reporter (SAM)	
	Thales Mirador FCS (SAM)	
	Elta EL/M 2106 (SAM)	
	Westinghouse TPS-63 (SAM)	
	Cetec JLY-3D radar (SAM)	10
	JY-11B 3D long range radar (SAM)	3
	Kh-29 (AAM)	
	Kh-59	
	Vympel R-27 (AAM)	
	Vympel R-73 (AAM)	
UAV	Saint Arpia	12+

New Procurements/Upgrades

✦ Venezuela has announced the purchase of 12 more Su-30MK2 from Russia.

✦ A third batch of 8 K-8W armed trainers is on order.

✦ Negotiations are on to buy 24 Hongdu L-15 advanced trainer/LIFT aircraft.

✦ Orders placed for 16 training helicopters from Enstrom Helicopter Corporation.

✦ Orders placed for 36 Diamond DA40 primary training aircraft to replace the SF-260s.

NATIONAL GUARD

The National Guard was renamed as the Bolivarian National Guard of Venezuela, and was expanded even further to include the People's Guards Command in 2011 and the Anti-extortion and Sequestration Command in 2012. The National Guard is a ready reserve force that serves as a border patrol, internal and rear-area security function.

Personnel: 27,720

Equipment

Category	Name	In Service
APC	AUR-416	40+
	Fiat 6646	
	VN-4 (protected carriers)	111
	WTC-1 (water cannons)	10
	ABV-1 (mobile barriers)	10
Aircraft	M-28 Skystruck	11
	201 Arava	4
	BE-200 Super King Air	2
	BE-90 King Air	1
	BE-95-B.58 Baron	1
	Cessna CE-402	1
	CE-206/U206/U206G	5
	CE-182 Skyline II	4
	CE-152 Aerobat	3
	M-26 Isquierka	2
	DA-42MMP Twin Star	6
Helicopter	Mi-17V-5	6
	Bell 412HP/SP/EP	10
	Bell 222	1
	Bell206L	1
	Bell 206B	12
	A-109A Hirundo	3
	AS-350/AS-355F-2	10
	Ecureuil	
	Enstron F28	2
Patrol craft	23m (Sea Ark)	12
	16.5m (Halter Marine)	10
	US-Protector Class	12
	6.4m Cobra design	10
	5.5m	15
	10.6m	8
	9.1m	25
	7.9 raider	25

NATIONAL MILITIA

Personnel: 75,000

The Bolivarian Militia is a derivative of the military reserve and has been established as an organ of the Venezuelan armed forces. They are under the immediate command of the President. Unlike the armed forces, which would battle an external enemy, the National Militia's role is to battle with an internal enemy or the regime's opposition. The General

Command of the National Militia is divided in two parts:

- ✦ The National Reserve, consisting of all Venezuelan citizens who are not in active military service, or have completed military service, or who voluntarily join the reserve units.
- ✦ The Territorial Guard, consisting of all Venezuelan citizens who voluntarily serve to organize local resistance to any external aggression.

DEFENCE PRODUCTION AND R&D

The Venezuelan Defence industry was virtually non-existent until 1975 when the state owned Venezuelan Military Industries Company (Compania Anonima Venezolana de Industrias Militares - CAVIM) was established to encourage growth of the domestic defence industry and to produce the materiel required by the Venezuelan military. The DIANCA and UCOCAR are being involved increasingly in Venezuelan navy's future needs. Today domestic arms production consisted of all-terrain vehicles, trucks, ammunition, rifles, unmanned aircraft, grenades, assembled ships of small and medium-sized ports among other products.

Major Armaments Producers

CAVIM: C.A. Venezolana de Industrias Militares is a company dedicated to the production of weapons, explosive munitions as well as various other products.

CENARECA: Centro Nacional de Repotenciacion C.A is the manufacturer of the all-terrain and high mobility (HMMWV) Tiuna vehicles for the Venezuelan armed forces.

MAZVEN: MAZVEN C.A is a joint venture with the Belarusian Company MAZ which manufactures trucks for military use.

DIANCA: Diques y Astilleros Nacionales C.A is the state shipyard created in 1905 in the city of Puerto Cabello, Carabobo state.

UCOCAR: Unidad Naval Coordinadora de los Servicios de Carenado de la Armada is responsible for the repair, maintenance and construction of ships.

G&F Tecnologia: G&F Tecnologia manufactures unmanned aircraft, communication equipment and other electronic products.

DEFENCE DEALS

Joint Venture Programmes

Venezuela has developed an unarmed aerial drone with help from Iran, China and Russia. The drone has a range of 100 kilometres, can reach an altitude of 3000 meters and stay aloft for up to 90 minutes. It transmits real-time video and images. The 3x4 metre drone is part of Venezuela's defence system, aimed at the monitoring of dams, pipelines and other infrastructure. The drones are the Mohajer-2 class Iranian design, now renamed Sant Arpia. With Russian aid Venezuela plans to build a factory to make AK103 assault rifles. When completed the factory is expected to produce 25,000 rifles and 70 million rounds of ammunition a year.

Arms Export

Venezuela has begun to sell weapons and military vehicles to other Latin American countries within the alliances of ALBA and UNASUR. According to Cavim (Compañía Anónima Venezolana de Industrias Militares), other Latin American states have expressed their interest to buy the Venezuelan drone. Ecuador was the largest importer of Venezuelan aircraft parts and missiles.

Arms Import

Venezuela's most imported military hardware comprised of aircraft, missiles and sensors. Due to military embargo imposed by the US, Venezuela buys arms mainly from Russia and China. Venezuelan leader Nicolas Maduro plans to buy more weapons to strengthen the country's defence against foreign threats.

According to the Stockholm International Peace Research Institute (SIPRI), Venezuela is the largest weapons importer in Latin America. Since 2011, the country spent some $2.6 on weapons, buying the majority of them in Russia. In the period from 2012 to 2015 Venezuela is expected to become the second-largest importer of Russian weapons.

Venezuela and Russia's state-run arms trader Rosobornexport signed more than 30 contracts on arms supplies in 2005-2009. Venezuela received Russian air defense systems (Antey-2500, Pechora-2M, Buk and Igla) as well as T-72M tanks, Grad and Smerch multiple rocket launcher systems, armored vehicles and artillery. Arms supplies from China include long-range JYL-1 radars. Venezuela has reportedly purchased 24 K-8 Karakorum trainer/light fighter planes from China.

DEFENCE BUDGET

Total defence spending: $4,508 million (est.)

Defence spending in terms of GDP: 1.43% (2013)

According to a report by Embassy of the Bolivarian Republic of Venezuela to the US, on Venezuela's defence spending, it noted that Venezuela's military spending is lower than average for Latin America and far less than that of some neighbouring countries in South America. According to the World Bank, defence spending in Venezuela was equivalent to just 0.75% of the country's GDP in 2011 and it has been falling steadily since the 1990s. By this measure, it is two thirds lower than the global average. Venezuela spend to building a sturdy defence, as well as disaster relief, search and rescue operations, and anti-drug trafficking efforts through the Bolivarian National Guard. The report also added that, Venezuela dedicates an enormous amount to social programmes to guarantee access to essential goods like health care, housing and education. Education spending alone will account for 2.5% of GDP in 2013, and in 2012, Venezuela spent nearly 40 times as much on social investments (USD 115 billion) as on defence.

CONCLUSION

The goal of the National Bolivarian Armed Forces of Venezuela is to defend Venezuelan territory from attack, combat drug trafficking, provide search and rescue capabilities, aid the civilian population in case of natural disasters protection, as well as numerous internal security matters. Venezuela with its abundant energy resources continues to be a country of strategic importance. The country is first in petroleum reserves in the world, third in bauxite, fourth in gold, sixth in natural gas and tenth in iron reserves. Late president Chavez was revered by millions of Venezuelans and headed the nation for 14 years. His hand-picked successor, President Nicolas Maduro continues Chavez's socialist programmes. Current concerns include: a weakening of democratic institutions, political polarization, a politicized military, rampant violent crime, overdependence on the petroleum industry with its price fluctuations, foreign exchange controls that discourage private-sector investment, high inflation, a decline in the quality of fundamental human rights, and widespread scarcity of consumer goods.

BTR-80A in Caracas, Venezuela on 5 March 2014 during the commemoration of Hugo Chavez's death. *Image Credit: Cancillería del Ecuador*

CONTACT DETAILS

Ministry of National Defence
Fuerte Tiuna-El Valle
Caracas, Venezuela
Tel: +58212-80131104/0296
Website: www.ejercito.mil.ve

Armed Forces HQ
Fuerte Tiuna-El Valle
Caracas, Venezuela
Tel: +58212-692685

CAVIM HQ
Urb Las Mercedes
Calle Jalisco Cavim building
Caracas, Dtto Capital. Venezuela
Tel: +58(212)9934378 - 9935311- 9932006;
Fax: 9935621

STRATEGIC INFORMATION

VIETNAM

(Capital: Hanoi)

INTRODUCTION

Area: 331,210 sq km

Population: 95,261,021 (July 2016 est.)

Coastline: 3,444 km (excludes islands)

Maritime claims: Territorial sea: 12 nm

Contiguous zone: 24 nm

Exclusive Economic Zone: 200 nm

Continental shelf: 200 nm or to the edge of the continental margin

GEOPOLITICAL IMPORTANCE

Vietnam is bordered by China to the North, Laos to the Northwest, Cambodia to the Southwest, and the South China Sea to the East. The South China Sea is important to international shipping because one-third of the world's shipping transits the sea on the route between Europe and West Asia to East Asia. Its importance also stems from significant oil and gas reserves there; and its considerable fish supplies. Vietnam's Cam Ranh Bay, located about 220 miles (354 kilometres) north of Ho Chi Minh City, provide convenient access to the commercially and strategically vital sea lanes which pass through the South China Sea. Due to its strategic location as well as its particular historical relations with China, Vietnam has emerged as a key player in the Asia -Pacific region. Vietnam also has a high degree of political and economic influence in Laos and it has also emerged as an attractive partner to countries seeking influence in South East Asia.

POLITICAL OVERVIEW

Vietnam is a one-party communist state. Political power lies with the Communist Party of Vietnam (CPV). The party's politburo holds authority over the implementation of all major areas of policy. The politburo, headed by a general secretary, is elected by the party's Central Committee. Only political organisations affiliated with or endorsed by the Communist Party are permitted to contest elections in Vietnam.

The National Assembly is a unicameral legislative body. It exercises three main functions: to legislate, to decide on important national issues, to exercise supreme supervision over all activities of the State. The Assembly has 500 members, elected by popular vote to serve four-year terms. The Communist Party provides policy direction and all members of the assembly must be approved by the CPV before being permitted to stand for election.

The President is elected by the National Assembly for a five-year term and acts as the commander-in-chief of the Vietnam People's Armed Forces and Chairman of the Council for Defence and Security. The president declares the promulgation of the Constitution, laws and ordinances. He also recommend to the National Assembly the election, removal or dismissal of the Vice President, the Prime Minister, Chief Justice of the Supreme People's Court, and Head of the People's Procuracy. The government, the main executive state power of Vietnam, is headed by the prime minister, who has several deputy prime ministers and several ministers in charge of particular activities. The executive branch is responsible for the implementation of political, economic, cultural, social, national defence, security and external activities of the state. The Supreme People's Court is the highest judicial organ of the Socialist Republic of Vietnam. It supervises and directs the judicial work of local People's Courts, Military Tribunals, Special Tribunals and other tribunals, unless otherwise prescribed by the National Assembly at the establishment of such Tribunals.

KEY POLITICAL PERSONS

PRESIDENT & COMMANDER-IN-CHIEF: Tran Dai Quang

PRIME MINISTER: Nguyen Xuan Phuc

ECONOMY

Vietnam has made a shift from a highly centralised planned economy to a socialist-oriented market economy. Manufacturing, information technology and high-tech industries form a large and fast-growing part of the national economy. The country has made remarkable progress in reducing poverty with the percentage of people living in poverty dropped from almost 60% in the 1990s to less than 3% today. In 2014, GDP growth was 6 %. According to World Bank, export value in US dollar terms is estimated to have grown by 11.6% in 2014, outperforming other countries in the region. Vietnam's traditional manufacturing exports such as garments, footwear and furniture, continue to grow while export items such as hi-tech and high-value products (cell phones, computers, electronics, and automobile parts) have also maintained rapid growth, and have now become the largest share of export. Though Vietnam is a relative newcomer to the oil industry, it is currently the third-largest oil producer in Southeast Asia, with an output of 400,000 barrels per day. In 2012 Vietnam unveiled a broad "three pillar" economic reform programme, proposing the restructuring of public investment, state-owned enterprises and the banking sector. Vietnam's economy continues to face challenges from low foreign exchange reserves, an undercapitalized banking sector, and high borrowing costs. Vietnam joined the WTO in January 2007 following more than a decade-long negotiation process. Vietnam is trying to reform its economy by restructuring public investment, state-owned enterprises, and the banking sector.

GDP: $552.3 billion (2015 est.), $517.7 billion (2014 est.)

Real Growth Rate: 6.7% (2015 est.), 6% (2014 est.)

Industries: Food processing, garments, shoes, machine-building; mining, coal, steel; cement, chemical fertilizer, glass, tires, oil, mobile phones

Oil - proved reserves: 4.4 billion bbl (1 January 2016 est.)

Natural gas - proved reserves: 699.4 billion cu m (1 January 2016 est.)

Total Exports: $162.1 billion (2015 est.), $150.2 billion (2014 est.)

Export Commodities: Clothes, shoes, marine products, crude oil, electronics, wooden products, rice, and machinery

Major Markets: US 21.2%, China 13.3%, Japan 8.4%, South Korea 5.5%, Germany 4.1% (2015)

Total Imports: $154.7 billion (2015 est.), $138.1 billion (2014 est.)

Import Commodities: Machinery and equipment, petroleum products, steel products, raw materials for the clothing and shoe industries, electronics, plastics and automobiles

Major Suppliers: China 34.1%, South Korea 14.3%, Singapore 6.5%, Japan 6.4%, Hong Kong 5.1%, Thailand 4.5% (2015)

DEFENCE & SECURITY

Vietnam has established a policy to concurrently develop both its socio-economic and defence capabilities and closely coordinate defence-security and diplomatic activities to support the cause of industrialisation and modernisation, and rapid and sustainable economic development. The country faces diversified and complicated security challenges. According to the Ministry of Defence, although the national security has been maintained, there is socio-political instability caused by both internal and external factors. Territorial disputes in the South China Sea, especially with China, constitute Vietnam's main security challenge. It has stationed small military garrisons on the major islands and rocks that it occupies. Other security issues such as illegal trafficking of weapons and drugs; piracy, organised trans-national crimes, terrorism, illegal migration and immigration; environmental degradation, climate change, and epidemics continue to concern the country. In recent years, Vietnam has stepped up 'defence diplomacy' as a measure to promote military cooperation with partners of interest. It now maintains bilateral defence relations with 65 countries, and Vietnam participates actively in major regional security and defence forums, including the ASEAN Defence Ministers Meeting Plus, the ASEAN Regional Forum, APEC, ASEM, East-Asia summit and the Shangri-La Dialogue.

KEY DEFENCE PERSONS

DEFENCE MINISTER: Gen. Ngo Xuan Lich

DY DEFENCE MINISTER & CHIEF OF GENERAL STAFF OF ARMY: Lt. Gen. Phan Van Giang

COMMANDER OF THE NAVY: Rear Adm. Pham Hoai Nam

COMMANDER OF THE AIR FORCE: Lt. Gen. Le Huy Vinh

External Conflict

Vietnam is involved in complex dispute with China, Brunei, Malaysia, the Philippines, and Taiwan over the Paracel and Spratly Islands. The issues in the South China Sea are not only a conflict over the islands, but also include resources of oil, natural gas, underwater archaeology, hydro geothermal energy and fishery. Vietnam and China have demarcated their land border and are currently putting in place mechanisms to resolve local disputes. There is ongoing cooperation between Military Zones adjacent to the border and between their respective border guards. Its borders with Laos and Cambodia are porous and the main security challenges are transnational. These include trafficking in drugs, weapons and persons; illegal migration; criminal activity and the activities of anti regime political activists.

THREAT PERSPECTIVE

The bilateral relations between Vietnam and China have been turbulent. China has long been the most serious threat to Vietnam's security. In Hanoi's view, the South China Sea relates to almost all aspects of national security and development: protecting territorial integrity and national sovereignty, promoting maritime economic development, maintaining external peaceful environment, especially peaceful relationships with China and other claimants, and safeguarding the regime's legitimacy and internal stability. China claims sovereignty over all islands, rocks, and reefs in the Paracel and Spratly archipelagos. This threat posed by China towards Vietnam is not only its geographical proximity but also in terms of the size and power between the two countries. China is several times larger than Vietnam, while Vietnam's population, despite being the world's 14th largest, stands nowhere near to China in comparison. Economically too China's economic modernisation has caused the power gap between the two countries to become ever wider despite Vietnam's impressive economic performance since the late 1980s. With its economic development, China's military might has grown significantly, posing a formidable threat to Vietnam's security. Its expanding military budget is reportedly concentrated on its air force and navy which makes it particularly worrying for Vietnam as the two nations have competing claims in the South China Sea. Vietnam is also dependent on China for some of its major export industries, while Vietnam's exports to China are just a minuscule portion of China's total imports. So damage to Vietnam's economy would be immense if for some reason China decides to discontinue trade with Vietnam.

STRATEGIC RELATIONS

India: Defence Cooperation has emerged as a significant pillar of India's strategic partnership with Vietnam. India and Vietnam have enhanced their bilateral defence cooperation by signing a joint vision statement in 2015. India-Vietnam defence cooperation currently entails high-level exchange visits, an annual security dialogue, service-to-service interaction, naval port visits, ship construction, training and capacity building, assistance in maintaining military equipment, multilateral exercises, and cooperation at regional forums such as the ASEAN Defence Ministers' Meeting. Vietnam is also looking to purchase the BRAHMOS cruise missiles from India, as per reports. India and Vietnam closely cooperate in various regional forums such as ASEAN, East Asia Summit, Mekong Ganga Cooperation, Asia Europe Meeting (ASEM) besides UN and WTO. Indian investments in Vietnam have remained constant at around USD 1 billion but are expected to rise in the coming years.

Russia: Russia is a principal ally of Vietnam. Cooperation in the military-technical spheres between Russian and Vietnam has contributed towards strengthening the traditional friendship and facilitating the further development of the strategic partnership. Vietnam has become a major customer for Russian weapons, primarily buying submarines and planes. The two nations have been "strategic partners" since 2001 and upgraded to a comprehensive strategic partnership in 2012. Bilateral trade and scientific-cultural exchanges are growing. Russia is among major foreign investors in Vietnam, focusing on mining, processing and manufacturing industries.

China: China and Vietnam considered each other's development as its own development opportunity and hence agreed to deepen their comprehensive strategic co-operation, focusing on maintaining high-ranking contacts through diverse and flexible forms such as bilateral visits, telephone calls via hotline and meetings on the sidelines of multilateral forums to increase strategic exchanges and have a thorough grasp of sound orientations for developing bilateral ties. They also reached consensus on deepening land and sea border defence co-operation with two joint naval patrols to be carried out in the Tonkin Gulf and comprehensive and effective implementation of the Declaration on Conduct of Parties in the East Sea (DOC) to maintain peace and stability together in the

East Sea.

Japan: Vietnam & Japan's defence cooperation took shape in 2011 as tensions in the South China Sea increased. Both countries share a convergence of strategic interests in opposing China's use of civilian maritime law enforcement vessels to assert sovereignty claims. In March 2014, Prime Minister Shinzo Abe and President Truong Tan Sang elevated their strategic partnership to an Extensive Strategic Partnership during the latter's official state visit to Tokyo. The two countries have agreed to a wide-ranging MoU on defence cooperation which included defence exchanges at ministerial, chief of staff and service chief level; naval goodwill visits; annual defence policy dialogue at the deputy defence minister level; cooperation in military aviation and air defence; and personnel training including scholarships for defence personnel to study and train in Japan. Japan has remained committed to promoting cooperation with Vietnam in a wide range of economic areas including investment.

Netherlands: Bilateral relations between the two countries are developing fruitfully. Both sides have regularly held high-ranking meetings and have worked positively together at international and regional forums, particularly within the framework of the United Nations, ASEAN-EU, and ASEM. In recent years, the Netherlands has been one of the largest European investors in Vietnam, ranking11th out of 96 countries. The two countries are also stepping up co-operation in the defence sector.

South Korea: Giving a boost to their strategic partnership, the former Cold War foes have signed a bilateral free trade agreement. The agreement is expected to significantly strengthen trade and investment ties between the two countries. South Korea is already Vietnam's biggest investor, second official development assistance and tourism contributor, and third-largest trading partner after China and the United States.

Germany: Diplomatic relations between Germany and Vietnam have developed consistently well over the past four decades. There have been high-level visits and various other ministerial visits to Vietnam. Germany is Vietnam's most important European Union trade partner by trade value, making up 28 percent of the trade between Vietnam and the EU. Bilateral trade reached US$1.28 billion in the first two months of 2015, up 15.9 percent from the same period in 2014.

UK: Vietnam and the UK have a long history of cooperation. In 2013 the 40th anniversary celebrations of bilateral diplomatic relations was held in Vietnam. Defence cooperation also formed a part of the fourth Vietnam-UK Strategic Dialogue. The two nations have agreed to advance defence cooperation which included increased training cooperation, Defence industrial cooperation, and Maritime cooperation. Both the nations have developed close cooperation at international forums and organisations, especially the United Nations (UN), the Asia-Europe Meeting, and the ASEAN-EU framework. They have also renewed their commitment to expeditiously conclude negotiations on a free trade agreement (FTA) between Vietnam and the EU, which is expected to open new opportunities for both business communities.

Spain: Relations between Spain and Vietnam have developed considerably since they set up a strategic partnership in 2009. The two nations maintain close co-ordination at international and regional forums, particularly those in the framework of the United Nations, ASEAN-EU and Asia-Europe Meeting (ASEM).

Vietnam pledged to work as a bridge to help Spain expand its relations with the Association of Southeast Asian Nations (ASEAN) and member countries while Spain affirmed its support for stronger relations between Vietnam and the EU, including swift recognition of Vietnam's full market-economy status. Spain was the 59th largest investor in 2014 in Vietnam with a registered investment of US$35.5 million. In 2013, Viet Nam's exports to Spain reached US$2.11 billion, up 17.81 per cent over 2012.

Israel: Israel and Vietnam have maintained regular exchange visits and strengthened cooperation in all realms, including economics, trade, education, science-technology, agriculture, health care, as well as in national defence. In recent times defence ties between the two nations have seen increasing. On 2 March 2015, Deputy Defence Minister Nguyen Chi Vinh and Director-General of the Israeli Ministry of Defence Dan Harel signed a memorandum of understanding on bilateral defence cooperation.

United States: US relations with Vietnam have become increasingly cooperative and comprehensive, guided by the 2013 US-Vietnam Comprehensive Partnership, an overarching framework for advancing the bilateral relationship, the 2015 bilateral Joint Vision Statement, and the Joint Statement issued during President Obama's visit to Vietnam in May 2016. In recent years ties between the United States and Vietnam have experienced a significant breakthrough which started with Hillary Clinton's visit to Hanoi in July 2012. The visit paved the way for establishment of the U S-Vietnam Comprehensive Partnership. The US has announced its decision to partially lift its decades-old embargo on providing lethal military support to Vietnam, to help improve its maritime security while Hanoi decided to allow the US Peace Corps to operate in Vietnam. Also facing an enormous challenge - Beijing's maritime expansion in the South China Sea - the two former foes are now prepared to downplay their ideological disagreements to focus on common strategic interests. In May 2016, the United States fully lifted its ban on the sale of lethal weapons to Vietnam and continued to provide Vietnam with maritime security assistance - including through the Maritime Security Initiative, the Cooperative Threat Reduction program, and Foreign Military Financing.

Australia: Australia & Vietnam have decided to work on greater cooperation in the defence relationship by continued personnel exchanges and human resources training, ships visits and by strengthening the exchange of views on regional and security issues of mutual concern through an annual dialogue between officials from the foreign and defence ministries. The two nations have also signed the Declaration on Enhancing the Australia–Vietnam Comprehensive Partnership.

DEFENCE CAPABILITIES

The Vietnam People's Armed Forces - Luc Luong Vu Trang Nhan Dan Viet Nam - comprises four main elements: Vietnam People's Army, Militia and Self Defence Force, Maritime Police, and People's Public Security Force. The Vietnam People's Army comprises the Army, Navy and Air Force. The Commander-in-Chief of the Armed Forces - the President - is a nominal head. The secretary of Central Military Commission, usually the General Secretary of the Communist Party, is the de facto Commander. The Defence Minister oversees operations of the Ministry of Defence, and the VPA. He also oversees such agencies as the General Staff and the General Logistics Department. However, military policy is ultimately directed by the Central Military Commission of the ruling Communist Party of Vietnam.

ARMY: The Vietnamese army is not organised as a separate command, but by military region, combined arms army corps, and arms. The army main forces grouped into fourteen corps headquarters comes under the overall command of the General Staff, while the local and reserve forces come under the control of military commands at regional, province and municipal levels and district/provincial town/city levels. The role of the Army includes the protection and surveillance of its coastline, which stretches the entire length of the country (3,444 km), and search and rescue operations. Most of the arms and numerous units have been awarded the title of Hero of the People's Armed Forces having undergone a lot of trials during the wars for national liberation and defence.

Personnel: 450,000

Equipment

Category	Name	In Service
MBT	Z-62	100
	T-62	220
	T-54/55	600
	T-34/85	200
	M-48	20
	Type-59	100
	PT-76	600
Armoured Car	BRDM	80

Army continued...

Category	Name	In Service
APC	BTR-40/50/60/152	
	Type 56	
	K63	1,500 (total)
	M113	
	V100 Commando	
Artillery	152mm	100
	130mm	200
	76mm	
	100mm	
	105mm	300
	122mm	
	85mm	
Mortar	120mm	
	81mm	
	107mm	
	160mm	
MRL	Type 63 107mm	
	122mm	
	140mm	
SSM	Scud B/C	
ATGW	Sagger	
AA gun	37mm towed	
	23mm towed	
	ZSU-23-4	
	ZSU-57-2 SP	
SAM	SA7	
	SA9	
	IGLA	

New Procurements/Upgrades

No major procurement plans are under consideration at present.

Armoured Personnel Carrier. *Image credit: Vietnam MoD*

NAVY: Vietnam's Navy has transformed into a seagoing, fairly competent, combat-worthy navy from a small coastal patrol force with limited capacity in the 1980s. Equipped with old Soviet-era hardware and an assortment of small seagoing vessels until a decade ago, it has now upgraded itself into a modern fighting force. The force today consists of frigates, corvettes, patrol craft, missile boats, maritime patrol aircraft (MPA) and even submarines.

The Navy Headquarters has the Commander and Vice Commanders, the Commissar and the Deputy Commissar. The Service is divided into five Naval Regions and affiliated units. The main forces of the Navy are the units of surface ships, coastal artillery and land-to-sea missiles, marine, naval command and island defence forces. The Service has been awarded the title of Hero of the People's Armed Forces.

The Navy is equipped with two Russian Gepard class ASW frigates and three submarines. Russia will supply 12 Su-30 Mk-2 multirole fighter aircraft to provide air cover to its naval fleet and India has offered assistance in procuring patrol boats. Vietnam plans to buy four high-end Sigma-class corvettes from the Netherlands. It has acquired three twin Otter MPAs from Canada while Japan plans to give Vietnam six used naval ships for patrols in the South China Sea.

Major naval bases: Cam Ranh Bay, Can Tho, Chu Lai, Haiphong, Hue, Quang Khe, Qui Nhon, Vinh

Personnel: 40,000 (including 25000 Naval Infantry)

Equipment

Category	Name	In Service
Submarine	Hanoi class (Kilo type)	5
	Midget	2
Frigate	Petya-II class	5
	Gepard 3.9 type	2 (more on order)
Corvette	Tarantul-I class	4
	Molniya class	6 (more on order)
	Pauk class	1
Missile Boat	Svetlak class	5
	TT400TP gun boat	4
	OSA -II class missile FAC	8
	Turya class	5
	SO-1 class large patrol craft	4
	Zhuk class	11
	Poluchat class	2
Minesweeper	Yurka class	2
	Yevgenya class	2
	Sonya class	4
Amphibious Force	US LST	3
	US LSM	3
	Polnochny class	3
	LCM & LCU	24
	Truong SA	1
Training Vessel	Shanghai class	6
Weapon	P-5 Shaddock (ASM)	5 batteries
	SSC-5 Bastion (ASM)	2
	Kh-35E	
	Kalibr 3M54	
	3M-14/54 Klub	5+

Naval Aviation

Category	Name	In Service
Helicopter	Ka 27/29 ASW	5
	Mi-4 (SAR)	10
	Beriev Be12	4
Aircraft	DHC-6 Twin Otter	6

New Procurements/Upgrades

✦ Larsen & Toubro (L&T) has signed the principle contract with Vietnam Border Guard valued at USD 99.7 Million for design and construction of high speed patrol vessels in India as well as for transfer of design and technology along with supply of equipment and material kits for construction of follow-on vessels at a Vietnam shipyard.

✦ The Vietnamese Navy has taken delivery of the Project 1241 Tarantul-class or Molniya corvettes. The ships, coded HQ 379 and HQ 380, are among the six ordered by the navy, with the first two – HQ 377 and HQ 378 – delivered in 2014. The vessels are equipped with 16 Uran-E missiles of 130 km in range, two AK-630 ship-borne artillery systems, and an AK-176M automatic gun.

✦ The Navy has taken delivery of the fourth Russian-built Kilo-class submarine HQ-185 Da Nang. While the first two subs - HQ-182 Hanoi and HQ-183 Ho Chi Minh City - were commissioned into service in April 2014, the third vessel arrived in the country in February 2015. Two more submarines Hai Phong and Khanh Hoa were delivered in August 2015.

✦ Two new Gepard 3.9 corvettes designed for anti-submarine warfare will be delivered to the Vietnam Navy in 2016 and 2017, equipped with the Palma close-in weapon system (CIWS).

✦ The PAVN Navy has taken delivery of its fourth TT400TP-class patrol boat HQ 275 following on from the induction of the third HQ 274 in May. The first two ships in the class, HQ 272 and HQ 273, were commissioned in January 2012 and August 2012 respectively. The TT400TP class is tasked with conducting maritime border patrols, reconnaissance, and maritime surveillance operations.

✦ Dutch shipyard Damen unveiled for the first time the design of the Sigma 9814 Corvette ordered by Vietnam during Vietship 2014 Nava Exhibition. Netherlands will build four Sigma corvettes for the Vietnamese Navy. The first two ships will be built in Vlissingen (Netherlands), and the last two (options) will be built in Vietnam, under Dutch supervision.

✦ Vietnam has taken delivery of three DHC-6 Twin Otter Series 400 aircraft from Viking Air of Victoria, BC, Canada of the total six ordered in 2010.

AIR FORCE: The current Vietnam People's Air Force (Không quân Nhân dân Việt Nam) is the successor of the former North Vietnamese Air Force (VPAF) and the Republic of Vietnam Air Force (VNAF), absorbed following the re-unification of North- and South Vietnam in 1975. Vietnam

People's Air Force (VPAF) is one of three main branches in Vietnam People's Army, a part of the Ministry of Defence. The Air Force is organised into the Service Command, combat units, support units, educational institutions and production units. The Service Command has the Commander and Vice Commanders, the Commissar and the Deputy Commissar. The main combat units of the Service are air divisions, air defence divisions, and affiliated air units. By virtue of their achievements in combat operations and other tasks, both the Air Defence and Air Force have been awarded the title of Hero of the People's Armed Forces.

Major Air Bases: Bienhoa, Binh Thuy, Cana, Dalat, Da Nang, Dien Bien Phu, Dong Hoi, Gia Lam. Hoa Lae, Phan Rang, Pleiku, Ho Chi Minh City, Vinh

Personnel: 30,000

Equipment

Category	Name	In Service
Fighter/Attack	Su-27SK/UBK	10
	Su-30 MK2	30+
	Su-22 M4	38
Trainer	L-39	38
Transport	An-26	10
	An-2	15
	An-24	12
	Li-2	30
	Il-18 (VIP)	2
Helicopter/ Attack/ Transport	Mi-24	35
	Mi-8/Mi-17/Mi17I	60
	Ka-25	15
	Mi-6	10
	Mi-4	4
	Mi-2	28
Maritime patrol	PZL M-28 MP	2
AA Gun	37mm	
	57mm	
	85mm	
	100mm	
	ZSU-23-4	75+
	ZSU-57-2	75+
	130mm	
SAM	SA-2	
	SA-7	
	SA-3	
	S-300 PMU-1	10
	2K12 Kub	10
	9K35 Strela-10	
AAM	AA-11, AA-2, AA-8, RVV-AE	
Anti-Ship	AS-13, AS-14, AS-17	

New Procurements/Upgrades

✦ Vietnam's three C295 transport aircraft are ordered from Airbus Defence and Space. The contract includes the supply of spares, maintenance, and training to the air force.

✦ 18 Su-30K fighters have been purchased from Russia and will be upgraded the Su-30KN configuration.

✦ A contract has been signed with Russia to deliver another batch of 12 Su-30MK2 fighter aircraft. The previous contracts for delivery of eight and twelve fighters respectively have been successfully implemented by the enterprise.

Missiles

✦ Vietnam is reportedly in the process of acquiring 50 anti-ship and land attack 3M-14E Klub supersonic cruise missiles for its fleet of SSK Kilo-class diesel-electric submarines. Russia has already delivered 28 missiles over the last two years to Hanoi although the precise number remains unknown.

✦ Vietnam's Varshavianka-class submarine is armed with unique Club-S cruise missiles. The missiles have a range of 300 kilometres and initially fly at a subsonic speed, but the 400-kilogram warhead separates when it approaches the target and accelerates to three times the speed of sound.

✦ The Air Defence-Air Force of Vietnam has reportedly upgraded the Russian-made air defence missile system S-125-2TM Pechora after a test launch in Hanoi. The S-125-2TM Pechora, also known as the Pechora-2TM, is an upgraded version of the S-125 / Pechora air defence missile system that can destroy advanced, small aerial attack platforms flying at low altitudes.

✦ The Army possesses 2,000 Scud missiles with approximately 60 to 96 launchers. The Scud missiles are reportedly capable of striking targets on Hainan Island including the Yalong Naval Base near Sanya, and cities within its 500 km range in southern China such as Nan Ning.

✦ The Navy has also acquired Russian Yakhont anti-ship missiles enough to equip a battalion.

✦ Hanoi and New Delhi are in talks for supply of BRAHMOS missiles to the South East Asian country which is locked in a territorial dispute with China.

Coast Missile. *Image credit: Vietnam MoD*

Vietnam Air Force officials at parade to mark the 70th anniversary of the August Revolution and the National Day. *Image credit: Vietnam MoD*

DEFENCE PRODUCTION AND R&D

Vietnam does not have a modern national defence industry and its capacity is generally limited to small arms production. According to the 2004 Defence White Paper, the Vietnamese defence industry is being developed in the direction of mainly self-reliance but at the same time, various forms of associate and co-operative relations with friendly countries are being broadened so as to acquire advanced technologies with a view to bringing the Vietnamese defence industry to a higher level of development.

The VPA's Technology General Department is undertaking efforts to improve its capacities in the research, development and application of new technologies. In recent years, Vietnam has advanced its defence cooperation relations with foreign countries by conducting joint research programmes. The country has progressed in its rocket propellant research with its Institute of Technology (Air Defence - Air Force), having successfully studied the production of propellant for R-17E (scud) missiles. The research embarked by scientist from the institute was aimed to locally produce the highly prized rocket propellant in which its distribution has been strictly controlled by other foreign countries.

Major Armaments Producers

✦ Hong Ha Shipbuilding Company, Vietnam Shipbuilding Company, Song Thu Shipbuilding Company,

✦ Service Flight Corporation of Vietnam, Group 559, Ba Son trust-company.

✦ Research Institutes directly attached to the Ministry of National Defence:

✦ The Institute for Military Strategy - a consultative organ for the Ministry of National Defence

✦ The Institute for Military Science and Technology - is a research institution on military science-technologies

✦ The Institute for Military History - is the leading historical research institution of the VPA

✦ The Institute for Defence International Relations - is responsible for research on the world defence relations.

DEFENCE DEALS

India in Sept 2016 has extended a USD 500 million line of credit to Vietnam to deepen their defence cooperation and signed 12 agreements including a deal to construct offshore patrol boats, amid China's muscle flexing in the disputed South China Sea and 'emerging regional challenges'. According to media reports, Prime Minister Narendra Modi, who held wide-ranging talks with his Vietnamese counterpart Nguyen Xuan Phuc here, said the two countries have decided to elevate their strategic ties to a Comprehensive Strategic Partnership to provide it a new momentum. The 12 agreements were signed in a wide range of areas covering defence, IT, space, cyber security and sharing white shipping information. Vietnam has shown a keen interest cooperating with India in air and defence production. India's L&T will build offshore high speed patrol boats for Vietnamese Coast Guards, while a pact was signed on cooperation in UN peacekeeping matters. Indian Navy and Vietnam Navy will cooperate in sharing of white shipping information.

Joint Venture Programmes

Sweden: Vietnam and Sweden have signed an agreement to jointly develop an unmanned aerial vehicle (UAVs) according to the Vietnam Aerospace Association (VASA). In Phase One, Sweden will provide funds and equipment for Vietnam to manufacture two Magic Eye 1 UAVs. It will send experts from Sweden's Unmanned Group; provide intellectual property and the model design to Vietnam. During Phase Two, the two sides will cooperate in aeronautical electronic science development, and in the final phase, they will manufacture UAVs at each other's demand and capacity.

Russia: Vietnam will jointly develop a modified SS-N-25 Switchblade/ Kh-35 Uran subsonic anti-ship missile with Russia. The Uran subsonic anti-ship missile can be launched from helicopters, surface ships and coastal defence batteries. It has a range of up to 130 kilometres (70 nautical miles) and carries a 145-kilogram high explosive warhead.

DEFENCE BUDGET

Total defence spending: $4 billion (2015)

Defence spending in terms of GDP: 2.36% (2015)

Domestic Suppliers: Viettel Mobile, Precision Engineering 17 Company, Service Flight Corporation, 189 Shipbuilding Company, Hong Ha shipbuilding company

Foreign Suppliers: Russia, India, Germany, Israel, Netherlands, Sweden, USA, Turkey, South Korea, UK, Spain, Bulgaria, Portugal

Vietnam has increased its defence spending with military expenditure in 2015 amounted to USD 4.3 billion (approx.) of total government spending. Vietnam will increase its defence spending in future after US President Obama announced an end to decades-old military sanctions banning US arms exports to Vietnam, though currently more than 75% military imports come from Russia.

Vietnam only rarely provides information on arms procurements, servicing agreements and defence industry cooperation. In recent years, Vietnam has invested heavily in its navy partly due to tensions with China in the South China Sea. The December 2009 White Paper said Vietnam's defence budget in 2008 was USD 1.46 billion, or 1.8 percent of gross domestic product (GDP). Hanoi claimed to have increased the 2011 budget by 70% to around USD 2.5 billion, but some analysts estimate that the true budget is already around US $4 billion (roughly 3.5% of GDP). The country's economy as of 2012 is growing rapidly, defence investment is increasing and there have been efforts by the government to open up its defence markets and set up agreements with international suppliers beyond its historic partner, Russia.

Defence Procurement

A limited domestic defence capability drives the country's defence imports. The country relies on foreign suppliers to meet its demand for advanced military hardware. Vietnam has slowly built up a modest naval and maritime air capacity to monitor its territorial waters, continental shelf and Exclusive Economic Zones. Numerous government level defence co-operation agreements between Vietnam and Western countries have been signed including with Germany, the Netherlands and the UK. It has also signed a major agreement with Russia which provided a framework for continuing and future Russian defence sales and support to Vietnam.

Government has also taken steps to develop a wider and more diversified procurement network for Soviet-compatible equipment from India, the Ukraine, the Czech Republic and Poland. It is also working out the possibilities of arms procurements and technology transfers through co-production. Senior defence officials have led high-level delegations to Europe, Israel, East Asia, South Africa and Brazil to explore the arms market and to upgrade their knowledge of current developments in military technology. Ukraine and Vietnam have agreed to explore the possibility of arms co-production, while Belarus raised the possibility of cooperation in the 'military-technical area' in discussions between the two defence ministers. Vietnam and China have discussed stepping up cooperation in the research and transfer of military technologies. With Hungary it has held discussions on the transfer of technology to enable Vietnam to produce a variant of the Russian SA-18 man-portable surface-to-air missile. The US has partly lifted a 40-year ban on arms sales to former foe Vietnam.

In recent years, Vietnam has signed a number of large arms procurement contracts with an aim to modernise its navy. Russia is a significant defence trade partner of the country. Ships and aircraft account for majority of the defence imports. The biggest programme is delivery of six Project 636 (Kilo-class) diesel-electric submarines to Vietnamese Navy and the second is the Molniya-class boat delivery and license production programme. Another large project being implemented with Vietnam is delivery of frigates. Russia's Rosoboronexport has signed a $350 million contract for delivery of two Project 11661 Gepard-3.9 frigates to Vietnamese Navy.

CONCLUSION

Vietnam's strategic significance has increased dramatically in recent years due to transformations in its economic performance and foreign-policy orientation. Deepening and upgrading relations with leading partners have been prioritized. Vietnam played a pivotal role in helping to establish Asia's emerging security order. Relations with traditional friends and international organizations are also being strengthened. Vietnam now has bilateral defence relations with 65 countries and participates actively in regional security and defence forums such as the Shangri-La Dialogue, the ASEAN Defence Ministers Meeting Plus (ADDM+), and the ASEAN Regional Forum (ARF). Vietnam has become one of the fastest growing defence spenders in the world in 2016.

CONTACT DETAILS

Ministry of National Defence
No 7, Nguyen Tri Phuong St.
Ba Dinh Dist.
Hanoi, Vietnam
Tel: 069 534223 - 069 532090
Fax: +84-4-069532090

YEMEN
(Capital: Sanaa)

INTRODUCTION

Area: 527,968 sq km

Population: 27,392,779 (July 2016 est.)

Coastline: 1,906 km

Maritime claims: Territorial sea: 12 nm

Contiguous zone: 24 nm

Exclusive Economic Zone: 200 nm

Continental shelf: 200 nm or to the edge of the continental margin

KEY POLITICAL PERSONS

PRESIDENT & COMMANDER-IN-CHIEF OF THE ARMED FORCES: Abdrabuh Mansour Hadi

PRIME MINISTER: Ahmed Obeid bin Daghr

Note: Yemen has been in a state of political crisis since 2011 when a fresh wave of protests started, inspired by the Arab Spring. The country has become a base for militant groups, like Al-Qaeda and Islamic State and spiralled into civil war in 2014. Fighting continues despite peace initiatives. Currently the country has no functioning central government. However, on 21 February 2015, Abdrabbuh Mansur Al-Hadi rescinded his resignation and declared he was still the legitimate president. He called on government institutions to gather in Aden which he proclaimed on 21 March 2015 was Yemen's "economic and temporary capital" while Sana'a remains under Houthi Shiite control. The Yemeni Civil War is an ongoing conflict that began in 2015 between two factions claiming to constitute the Yemeni government, along with their supporters and allies. A Saudi-led intervention, with a coalition of nine Middle Eastern countries, has taken place since 2015. The coalition received logistical and intelligence support from the US, UK and France. In September 2015, Abdrabbuh Mansur Hadi returned to Aden as Saudi-backed government forces recaptured the city. After more than a year-and-a-half of fighting, no side appears close to a decisive military victory.

KEY DEFENCE PERSONS

DEFENCE MINISTER: Maj. Gen. Mahmoud al-Subaihi

CHIEF OF STAFF: Maj. Gen. Mohamed Ali al-Makdashi

ECONOMY

Oil and gas revenues account for roughly 25% of GDP and 65% of government revenue. As part of the diversification effort to bolster its economy, Yemen exported its first liquefied natural gas in October 2009. However, the unrest that began in early 2011 caused the GDP to plunge almost 11% in that year. Yemen continues to face difficult long-term challenges including declining water resources, high unemployment, severe food scarcity, and a high population growth rate. In August 2014, the IMF approved a three-year, $570 million Extended Credit Facility for Yemen. Progress toward achieving stability has been slow and uneven.

GDP: $75.54 billion (2015 est.), $105.1 billion (2014 est.)

Real Growth Rate: -28.1% (2015 est.), -0.2% (2014 est.)

Industries: crude oil production and petroleum refining; small-scale production of cotton textiles, leather goods; food processing; handicrafts; aluminium products; cement; commercial ship repair; natural gas production

Oil - proved reserves: 3 billion bbl (1 January 2016 est.)

Natural gas - proved reserves: 478.5 billion cu m (1 January 2016 est.)

Total Exports: $1.364 billion (2015 est.), $7.601 billion (2014 est.)

Export Commodities: crude oil, coffee, dried and salted fish, liquefied natural gas

Major Markets: China 24.5%, UAE 16.5%, South Korea 10%, Saudi Arabia 10%, Kuwait 9.1%, India 8.5% (2015)

Total Imports: $4.793 billion (2015 est.), $10.86 billion (2014 est.)

Import Commodities: food and live animals, machinery and equipment, chemicals

Major Suppliers: UAE 20.9%, China 14.3%, Saudi Arabia 9.9%, Kuwait 7.4%, India 4.6% (2015)

INTERNATIONAL DISPUTES

✦ An international coalition, led by Saudi Arabia, launched air strikes against the Huthi armed group in Yemen on 25 March 2015 sparking a full-blown armed conflict. While the country is being bombarded by coalition forces from the air, rival factions are fighting against each other on the ground. On one side are the Huthis, an armed group whose members belong to a branch of Shia Islam known as Zayidism. The Huthis are allied with supporters of Yemen's former President Ali Abdullah Saleh. On the other side are anti-Huthi forces that are allied with the Saudi Arabian-led coalition. North and South Yemen were merged to become the Republic of Yemen. Tensions between North and South continue with sporadic fighting, according to Amnesty International.

✦ Saudi Arabia has reinforced its concrete-filled security barrier along sections of the fully demarcated border with Yemen to stem illegal cross-border activities.

DEFENCE

Yemen being in a state of civil war, proper structure of the armed forces no longer exists. Past data of the country's military equipment has been compiled below:

ARMY

Personnel: 55,000 (est.)

Equipment

Category	Name	In Service
MBT	T-72	39
	T-62	200
	T-54	300
	T-55	300
	M-60A1	140
Armoured car	Saladin	60
	BRDM-2	100
	AML-90	125
	Ferret	10+
AIFV/APC	BMP-1/2	150
	BMD-1	
	M-113A1	76
	BTR- 40/60/152	650
	M-577A1	6
	M-578	6
	Walid	
Artillery	76mm	100
	85mm	100
	100mm	100
	105mm	100
	122mm	150
	130mm	75
	M114 155mm towed	12

Army continued...

Category	Name	In Service
Mortar	81mm	
	120mm	
	160mm	
MRL	BM-21 122mm	280
	BM-14	11
ATK weapon	LAW	
ATGW	Vigilant	
	Dragon	24 launchers, 1500 missiles
	Tow	
AA Gun	ZU-23-2 2mm	50+
	37mm	50+
	57mm	150+
	85mm towed	50+
	ZU-23-4	30+
	Vulcan 20mm	72
SAM	SA-2	21
	SA-6	21
	SA-7	20
SSM	Scud	15+
	Frog-7	12
	Scud-B	6
	SS-21	4

New Procurements/Upgrades

No major procurement plans are under consideration at present.

NAVY

Personnel: 2,500

Equipment

Category	Name	In Service
Patrol craft	P-1022	10
	ex-USSR Tarantul class	2
	Chinese-built Hounan class missile FAC	3
	ex-USSR OSA II class missile FAC	3
	ex-USSR Zhuk class	2
	ex-US patrol boat	7
	Broadsword class	1
	CMN 15-60 fast patrol launches	6
Mine warfare Force	ex-USSR Natya class	2
	ex-USSR Yevgenya class	5
Amphibious Force	Bilquis (modified Polnocny type)	1
	Saba class LCU (modified Deba type)	3
Marine police	Fairey TRACKER Mk2 type coastal patrol craft	5
	Spear type coastal patrol craft	4
Coastal defence	SS-C-3 missile batteries	
	130mm guns & tank turrets	

New Procurements/Upgrades

No major procurement plans are under consideration at present.

AIR FORCE

Personnel: 3,500

Equipment

Category	Name	In Service
Fighter	SU-22	48
	MiG-29 SMT	13
	MiG-29 UBT	5
Transport	An-24	6
	IL-14	8
	An-12	3
	An-26	9
	Fokker F.27	3
	C130H HERCULES	2
	SKYVANS	2
Helicopter	Mi-8/Mi-17	40
	AB-204	2
	AB-205	2
	Mi-24	15
	AB-212	6
	ALOUETTE	2
	AB-206	6
	Mi-4	4
	Ka-27	2
	HUEY II	4
Trainer	F-5-B	4
	Yak-11	18
	L-39C	12

New Procurements/Upgrades

No major procurement plans are under consideration at present.

Defence Expenditure

Total defence spending: $1,222 million (2008); $1,271 million (2007)

Estimated defence spending in terms of GDP: 4.59% (2013)

Defence Production and R&D

Yemen currently being affected by war has not undertaken any defence production activities.

Defence Procurement

The Yemen Army/Navy/Air Force has not made any major defence procurements in recent years.

CONTACT DETAILS

Ministry of Defence

P.O. Box 1399, Sana'a

Tel: 967-1-252-374

Fax: 967-1-252-378

ZAMBIA
(Capital: Lusaka)

INTRODUCTION

Area: 752,618 sq km

Population: 15,510,711 (July 2016 est.)

Coastline: 0 km (landlocked)

Maritime claims: None (landlocked)

KEY POLITICAL PERSONS

PRESIDENT: Edgar Chagwa Lungu

VICE PRESIDENT: Inonge Mutukwa Wina

KEY DEFENCE PERSONS

MINISTER OF DEFENCE: Chama Davies

COMMANDER OF ARMY: Lt. Gen. Paul Mihova

COMMANDER OF AIR FORCE: Lt. Gen. Eric Chimese

ECONOMY

Zambia has had a decade of rapid economic growth. A combination of prudent macroeconomic management, market liberalisation policies, and steep increase in copper prices helped drive investments in the copper industry and related infrastructure to achieve an average annual growth of about 6.7% during the last decade, according to the World Bank. Privatisation of government-owned copper mines in the 1990s relieved the government from covering mammoth losses generated by the industry and greatly increased copper mining output and profitability to spur economic growth. However, the high population growth rate of 2.3% per annum makes it difficult for per capita income to increase. Poverty remains a significant problem in Zambia. The low GDP per capita, which stands at $1400, places the country among the world's poorest nations.

GDP: $62.45 billion (2015 est.), $61.39 billion (2014 est.)

Real Growth Rate: 3% (2015 est.), 5% (2014 est.)

Industries: copper mining and processing, emerald mining, construction, foodstuffs, beverages, chemicals, textiles, fertilizer, horticulture

Total Exports: $6.998 billion (2015 est.), $10.22 billion (2014 est.)

Export Commodities: copper/cobalt, cobalt, electricity; tobacco, flowers, cotton

Major Markets: China 25.5%, Democratic Republic of the Congo 13%, South Africa 6.4%, South Korea 4.9%, India 4.3% (2015)

Total Imports: $7.711 billion (2015 est.), $8.595 billion (2014 est.)

Import Commodities: machinery, transportation equipment, petroleum products, electricity, fertilizer, foodstuffs, clothing

Major Suppliers: South Africa 34.5%, Democratic Republic of the Congo 18.2%, Kenya 9.7%, China 7.3%, India 4.4% (2015)

INTERNATIONAL DISPUTES

✦ In 2004, Zimbabwe dropped objections to plans between Botswana and Zambia to build a bridge over the Zambezi River, thereby de facto recognising a short, but not clearly delimited, Botswana-Zambia boundary in the river

DEFENCE

The Zambian Defence Force (ZDF) consists of the Army, the Air Force, and Zambian National Service (ZNS). The ZDF is designed primarily for internal defence in Zambia. Being a landlocked country, Zambia has no navy.

ARMY

Personnel: 21,500

Equipment

Category	Name	In Service
MBT	Type 59	16
	T54/55	8
	PT-76 light	20
Armoured Car	Ferret	24
	BRDM-1/2	76
	BTR-60	13
	BTR-70	20
Artillery	M1942 76mm	35
	Mod.56 105mm	18
	D-30 122mm	25
	M46 130mm	18
Mortar	81 mm	55
	82mm	24
	120mm	12
MRL	BM-21 122mm	50
RCL	M18 57mm	10
	M20 75 mm	10
	Carl Gustav 84 mm	10
ATGW	SAGGER	9
AA gun	M55 20mm	50
	M1939 37mm	40
	S60 57mm	30
	KS-12 85mm	16
SAM	SA-7	9

New Procurements/Upgrades

No major procurement plans are under consideration at present.

AIR FORCE

Personnel: 1,500

Equipment

Category	Name	In Service
Fighter	MiG-21	12
	J-6	12
	K-8	15
Transport	Do-28 Skyservant	9
	DHC-5D Buffalo	5
	Yak-40	3
	An-26	4
	Y-12	9
	MA 60	2
Liasion	MFI-17 Safari	20
Trainer	SF-260MZ	8
Helicopter	AB205As	15
	Z-9	4

New Procurements/Upgrades

The air force plans to procure a single C-27J transport, L-15 and additional SF-260 trainers and Mi-17 helicopters and air defence equipment P-18 radar.

Defence Expenditure

Total defence spending: $486 million (2015), $443 million (2014)

Defence spending in terms of GDP: 1.8% (2015)

Defence Production and R&D

Zambia does not undertake any defence production activities.

Defence Procurement

Zambia receives military support mainly from China and United Kingdom. The nation has setup a Defence Procurement Bureau under the Ministry of Defence as part of Zambia's Public Service Reform Programme. It lacks funds for its defence procurement requirements. China provides cheap new equipment and assists the rehabilitating existing equipment of the nation since 1990. Zambia has received $500,000 for Foreign Military Financing (FMF), congressionally appropriated grants from foreign governments to finance the purchase of American-made weapons, services and training. India has also provided some light arms and medium guns to Zambia in 2006 and had offered to sell further items of this type, as well as ammunition.

CONTACT DETAILS

Ministry of Defence

P.O Bag RW 17X
Lusaka, Zambia
Tel: +2601-251-211., 254-667
Fax: +2601-254-670, 221-339, 253-875

ZIMBABWE

(Capital: Harare)

INTRODUCTION

Area: 390,757 sq km
Population: 14,546,961 (July 2016 est.)
Coastline: 0 Km (Landlocked)
Maritime claims: None

KEY POLITICAL PERSONS

PRESIDENT & COMMANDER-IN-CHIEF: Robert Gabriel Mugabe

VICE PRESIDENT: Emmerson Mnangagwa

KEY DEFENCE PERSONS

DEFENCE MINISTER: Sydney Tigere Sekeramayi

CHIEF OF DEFENCE FORCES: Gen. Constantine Guveya Chiwenga

COMMANDER, ARMY: Lt. Gen. Phillip Valerio Sibanda

COMMANDER, AIR FORCE: Air Mar. Perrance Shiri

ECONOMY

Zimbabwe's economy depends heavily on its mining and agriculture sectors. Following a decade of contraction from 1998 to 2008, the economy recorded real growth of more than 10% per year from 2010-13, before slowing to roughly 3% in 2014 due to poor harvests, low diamond revenues, and decreased investment. Infrastructure and regulatory deficiencies, a poor investment climate, a large public and external debt burden, and extremely high government wage expenses impede the country's economic performance. Until early 2009, the Reserve Bank of Zimbabwe (RBZ) routinely printed money to fund the budget deficit, causing hyperinflation. Dollarization in early 2009 - which allowed currencies such as the Botswana pula, the South Africa rand, and the US dollar to be used locally - ended hyperinflation and reduced inflation below 10% per year, but exposed structural weaknesses that inhibit broad-based growth. The RBZ introduced bond coins denominated in 1, 5, 10, and 25 cent increments at par with the US dollar in December 2014, more than five years after the Zimbabwe dollar was taken out of circulation. In January 2015, as part of the government's effort to boost trade and attract foreign investment, the RBZ announced that the Chinese renminbi, Indian rupee, Australian dollar, and Japanese yen would be accepted as legal tender in Zimbabwe. The country's government entered a second Staff Monitored Programme with the International Monetary Fund in 2014 and undertook other measures to reengage with international financial institutions.

GDP: $28.1 billion (2015 est.); $27.69 billion (2014 est.)

Real Growth Rate: 1.5% (2015 est.); 3.9% (2014 est.)

Industries: Mining (coal, gold, platinum, copper, nickel, tin, diamonds, clay, numerous metallic and non-metallic ores), steel; wood products, cement, chemicals, fertilizer, clothing and footwear, foodstuffs, beverages

Total Exports: $3.551 billion (2015 est.); $3.703 billion (2014 est.)

Export Commodities: Platinum, cotton, tobacco, gold, ferroalloys, textiles/clothing

Major Markets: China 27.8%, Democratic Republic of the Congo 14%, Botswana 12.5%, South Africa 7.6% (2015)

Total Imports: $6.016 billion (2015 est.); $6.307 billion (2014 est.)

Import Commodities: Machinery and transport equipment, other manufactures, chemicals, fuel, food products

Major Suppliers: South Africa 48.1%, China 12.1%, India 5.2%, Zambia 4.6% (2015)

INTERNATIONAL DISPUTES

- Namibia has supported, and in 2004 Zimbabwe dropped objections to, plans between Botswana and Zambia to build a bridge over the Zambezi River, thereby de facto recognizing a short, but not clearly delimited, Botswana-Zambia boundary in the river.
- South Africa has placed military units to assist police operations along the border of Lesotho, Zimbabwe, and Mozambique to control smuggling, poaching, and illegal migration.

DEFENCE

ARMY

Personnel: 25,000

Equipment

Category	Name	In Service
MBT	Type 59	35
	Type 69	10
Armoured Car	AML-90	20
	Elands Mk7	20
	EE9 CASCAVEL	90
	Panhard AML	30
APC	UR 416	30+
	BTR 50	40
	MPCV 4x4	100+
Artillery	25 pdr 88mm	24
	Type 60 122 mm	18
	Type 54 122mm	12
Mortars	L16 81mm/ 82mm	600
	2B11 120 mm	140
AA Gun	ZPU1/2/4 14.5mm	
	ZU23 23mm	200+
	M1939 37mm	
SAM	SA-7/SA-18 Igla	30

New Procurements/ Upgrades

No procurement plans are under consideration at present.

AIR FORCE

Personnel: 1,000

Equipment

Category	Name	In Service
Fighter/ Trainer	F-7	6
	MiG-23	3
	SF-260	10+
	K-8 Karakoram	11
Transport, Reconnaissance	Islander	6
	CASA 212-200	10
	An-12	1
Helicopter	Alouette III	
	Mi-17	1
	Mi-35	6

New Procurements/ Upgrades

✦ 12 JF-17 Thunder fighter jets are reportedly on order from China.

Mi-35 helicopter of the Zimbabwe Air Force

Alouette III chopper of Zimbabwe Air Force

Defence Expenditure

Total defence spending: $380 million (2015)

Estimated defence spending in terms of GDP: 2%

Defence Production and R&D

Zimbabwe Government has set up the Zimbabwe Defence Industries (ZDI) which is the country's only primary defence contractor to meet the requirements of the Government.

Defence Procurement

No confirmed data or information is available on the current acquisition programmes by the Zimbabwean military. Defence procurement by the country was affected after the US and EU imposed an arms embargo on it in 2002. The EU in February 2012 further extended the embargo which still continues. The African nation has eyed military imports from non-EU countries, especially from China in recent times, to modernise its armed forces.

CONTACT DETAILS

Ministry of Defence

Public Relations Office

Ministry of Defence - H/Q

Defence House

cnr Kwame Nkuruma / 3rd Street

Harare

Tel.: +263-4 250042, 252050 - 7, 700155 - 8, 700077 – 8

Website: http://www.mod.gov.zw/

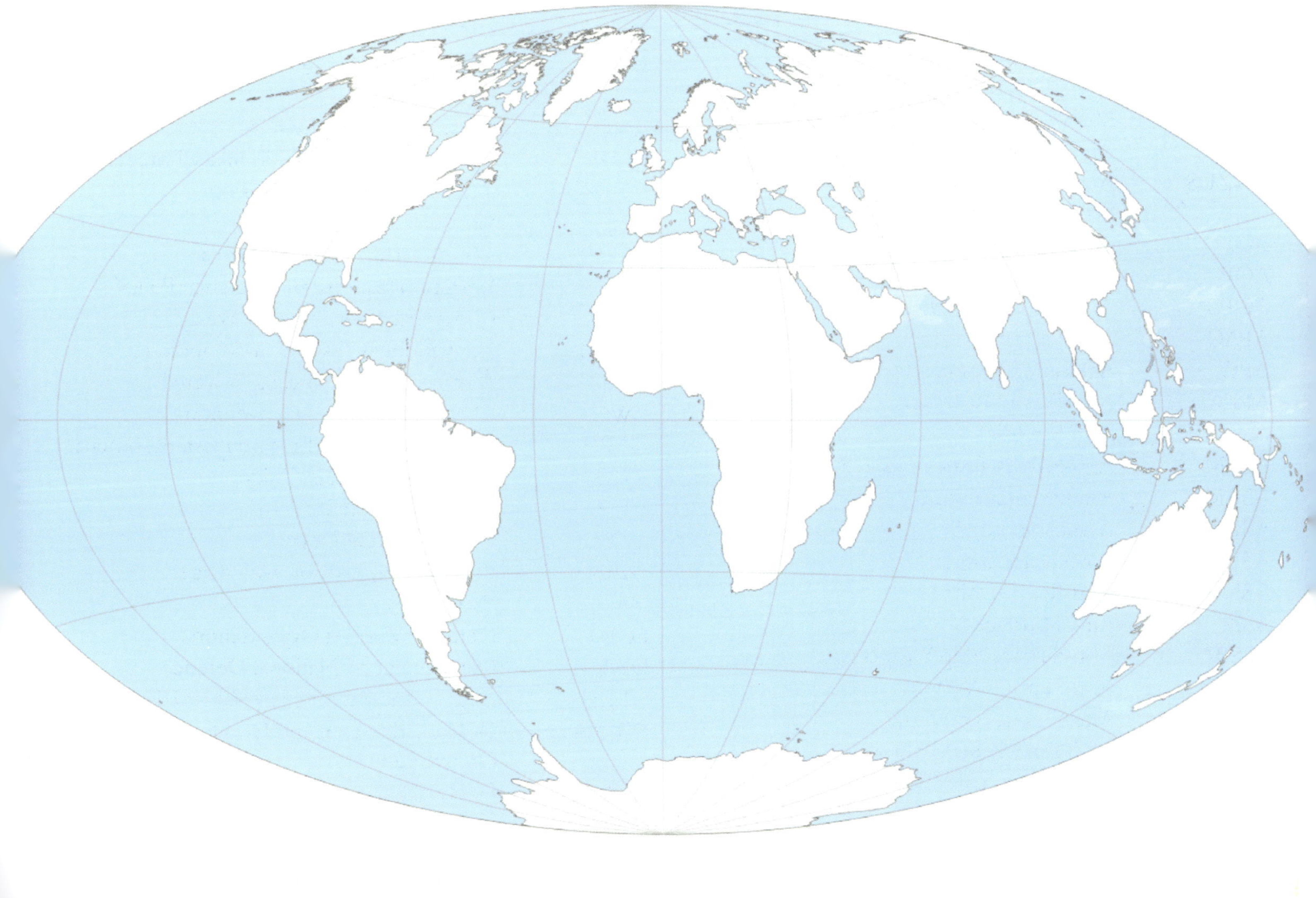

ABBREVIATIONS

AA Gun	Anti-Aircraft Gun
AAM	Air-to-Air Missile
AAW	Anti-Air Warfare
AB	Airborne
ABM	Anti-Ballistic Missile
ACD	Arms Control & Disarmamanent
ACS	Association of Caribbean States
ACV	Air Cushion Vehicle
ADF	Air Defence Force
AESA	Active Electronically Scanned Array
AEV	Armoured Engineering Vehicle
AEW	Airborne Early Warning
AEW&C	Airborne Early Warning and Control
AFV	Armoured Fighting Vehicle
AGM	Air-to-Ground Missile
AGSS	Auxilliary Research Submarine
AIFV	Armoured Infantry Fighting Vehicle
AIP	Air Independent Propulsion
ALCM	Air-Launched Cruise Missile
ALH	Advanced Light Helicopter
ALMDS	Airborne Laser Mine Detection System
ALOC	Air Line of Communication
AMRAAM	Advanced Medium-Range Air-to-Air Missile
ANG	Air National Guard
ANZUS	Australia, New Zealand, United States Security
AOI	Arab Organisation for Industrialisation
AOP	Airborne Observation Post
AOPS	Arctic/ Offshore Patrol Ship
APAC	Asia-Pacific
APC	Armoured Personnel Carrier
APEC	Asia-Pacific Economic Cooperation
ASAT	Anti-Satellite
ASBM	Anti-Ship Ballistic Missile System
ASCM	Anti-Ship Cruise Missile
ASEAN	Association of Southeast Asian Nations
ASI	Air-Surface Integration
AShM	Anti-Ship Missile
ASM	Air-to-Surface Missile
ASSW	Anti-Surface Ship Weapon
ASuW	Anti-Surface Warfare
ASW	Anti-Submarine Warfare
ATGM	Anti-Tank Guided Missile
ATGW	Anti-Tank Guided Weapon
Atk	Anti-Tank
ATV	All-Terrain Vehicle
AV	Armoured Vehicle
AWACS	Airborne Warning and Control System
AWD	Air-Warfare Destroyer
AAW	Anti-Air Warfare
BDL	Bharat Dynamics Ltd.
BIMSTEC	Bay of Bengal Initiative for Multi-Sectoral Technical and Economic Cooperation
BISTEC	Bangladesh, India, Sri Lanka, and Thailand Economic Cooperation
BMD	Ballistic Missile Defence
BWC	Biological and Toxin Weapons Convention
C4I	Command, Control, Communications, Computers and Intelligence
C4I2SR	Command, Control, Communications, Computer, Information, Intelligence, Surveillance and Reconnaissance
CATOBAR	Catapult Assisted Take Off but Arrested Recovery
CAEW	Conformal Airborne Early Warning
CAGR	Compound Annual Growth Rate
CDCM	Coastal Defence Cruise Missile
CDF	Chief of Defence Staff
CELAC	Community of Latin American and Caribbean States
CIWS	Close-In Weapon System
CJEF	Combined Joint Expeditionary Force
CMC	Central Military Commission
CN	Chief of Navy
CORPAT	Coordinated Maritime Patrols
CSDP	Common Security and Defence Policy
CTBT	Comprehensive Test Ban Treaty
CTOL	Conventional Takeoff and Landing
CVN	Aircraft Carrier, Nuclear Powered
CWC	Chemical Weapons Convention
DARE	Defence Avionics Research Establishment
DCG	Defence Capability Guide
DCP	Defence Capability Plan
DDG	Guided Missile Destroyer
DECO	Defence Export Control Office
DF	Dong Feng
DMO	Defence Materiel Organisation
DND	Department of National Defence
DPRK	Democratic People's Republic of Korea
DRDO	Defence Research and Development Organisation
DSTO	Defence Science and Technology Organisation
EAC	Eastern Area Command
ECM	Electronic Countermeasures

ELINT	Electronic Intelligence
EMP	Euro-Mediterranean Partnership
ENP	European Neighbourhood Policy
ENPI	European Neighbourhood and Partnership Instrument
EPCO	Equipment Procurement and Construction Office
ESSM	Evolved Sea Sparrow Missile
EU	European Union
EW	Electronic Warfare
FAC	Fast Attack Craft
FASGW-ANL	Future Anti-Surface Guided Weapon - Anti-Navire Leger
FF	Frigate
FFG	Guide Missile Frigate
FGFA	Fifth Generation Fighter Aircraft
FLN	National Liberation Front
FMC	Fast Missile Craft
FMF	Foreign Military Financing
FMS	Foreign Military Sales
FPDA	Five Power Defence Arrangements
G20	Group of Twenty Finance Ministers and Central Bank Governors
GDP	Gross Domestic Product
GPS	Global Positioning System
GRSE	Garden Reach Ship Builders and Engineers
GW	Guided Weapon
HAIC	Hongdu Aviation Industry Corporation
HAL	Hindustan Aeronautics Limited
HALE	High Altitude Long Endurance
IAC	Indigenous Aircraft Carrier
IAEA	International Atomic Energy Agency
IAF	Indian Air Force
IAI	Israel Aerospace Industries
IBSAMAR	India-Brazil-South Africa Maritime
ICBM	Intercontinental Ballistic Missile
ICJ	International Court of Justice
IFV	Infantry Fighting Vehicles
IGMDP	Integrated Guided Missile Development Programme
IISS	International Institute for Strategic Studies
IMB	International Maritime Bureau
IMF	International Monetary Fund
IOR	Indian Ocean Region
IRB	Industrial and Regional Benefits
IRBM	Intermediate-Range Ballistic Missile
IRGC	Islamic Revolutionary Guard Corps
IRIAF	Islamic Republic of Iranian Air Force
ISAF	International Security Assistance Force
ISC	Integrated Space Cell
ISRO	Indian Space Research Organisation
JASSM	Joint Air-to-Surface Standoff Missiles
JASSM-ER	Joint Air-to-Surface Standoff Missiles - Extended Range
JSS	Joint Support Ship
KMW	Krauss-Maffei Wegmann
KSEW	Karachi Shipyard and Engineering Works
LACM	Land Attack Cruise Missile
LCA	Light Combat Aircraft
LCAC	Landing Craft, Air Cushioned
LCM	Landing Craft, Mechanised
LCU	Landing Craft, Utility
LCVP	Landing Craft, Vehicle, Personnel
LHA	Landing Helicopter Assault
LHD	Landing Helicopter Dock
LPD	Landing Platform Dock
LRST	Long Range Surveillance and Tracking
LSM	Landing Ship, Medium
LST	Landing Ship, Tank
LUH	Light Utility Helicopter
MALE	Medium Altitude Longue Endurance
MARLANT	Maritime Forces Atlantic
MBT	Main Battle Tank
MDIC	Malaysian Defence Industry Council
MDL	Mazagon Dockyards Limited
MEADS	Medium Extended Air Defence System
MECC	Maritime-Enforcement Co-ordination Centre
MFN	Most Favoured Nation
MHI	Mitsubishi Heavy Industries
MICV	Mechanised Infantry Combat Vehicle
MIDES	Malaysian Defence, Security and Enhancement Industry Council
MINFAR	Ministry of the Revolutionary Armed Forces
MININT	Ministry of Interior
MLRS	Multiple Launch Rocket System
MLU	Mid-Life Update
MMA	Multi-Mission Maritime Aircraft
MMRCA	Medium Multi-Role Combat Aircraft
MoU	Memorandum of Understanding
MRBM	Medium Range Ballistic Missile
MRL	Mount Rocket Launcher
MRTT	Multi-Role Tanker Transport
MTA	Multirole Transport Aircraft
MTC	Military-Technical Cooperation

MTCR	Missile Technology Control Regime
MTTF	Mission Transition Task Force
NAM	Non-Aligned Movement
NATO	North Atlantic Treaty Organisation
NCA	Nuclear Command Authority
NDHQ	National Defence Headquarters
NEFA	North East Frontier Agency
NHQ	Naval Headquarters
NMRH	Naval Multi Role Helicopter
NORAD	North American Aerospace Defence Command
NPC	National People's Congress
NPT	Non-Proliferation Treaty
NSFC	Naval Strategic Force Command
NSPO	National Service Products Organisation
NTM-A	NATO Training Mission–Afghanistan
OAS	Organisation of American States
OECD	Organisation for Economic Cooperation and Development
OEM	Original Equipment Manufacturer
OIC	Organisation of Islamic Conference
OPEC	Organisation of the Petroleum Exporting Countries
OPV	Off Shore Patrol Vessel
PAC-3	Patriot Advanced Capability 3
PAK-FA	Perspective Aviation Complex of Frontline Aviation
PCU	Pre-Commissioning Unit
PLA	People's Liberations Army
PLO	Palestine Liberation Organisation
PoK	Pak-occupied-Kashmir
PRC	People's Republic of China
QIZ	Qualified Industrial Zone
RAM	Rolling Airframe Missile
RCL	Recoilless Rifle
RCN	Royal Canadian Navy
RFI	Request for Information
RFP	Request for Proposal
RL	Rocket Launcher
RMA	Revolution in Military Affairs
RPP	Report on Plans and Priorities
SAARC	South Asian Association for Regional Cooperation
SAATEG	System for Imagery Reconnaissance Deep in the Area of Operations
SAC	Southern Area Command
SAF	Second Artillery Force
SAM	Surface-to-Air Missile
SAR	Search and Rescue
SATCOM	Satellite Communication
SCAF	Supreme Council of the Armed Forces
SCO	Shanghai Cooperation Organisation
SIATI	Society of Indian Aerospace and Technologies Industries
SIGINT	Signals intelligence
SIMPRO	Simulator Research and Development Centre
SIPRI	Stockholm International Peace Research Institute
SLBM	Submarine-Launched Ballistic Missile
SPD	Strategic Plans Division
SRBM	Short-Range Ballistic Missile
SRSAM	Short-Range Surface-to-Air Missile
SSBN	Submersible Ship Ballistic, Nuclear / Nuclear-Powered Ballistic Missile-Carrying Submarine
SSGN	Cruise Missile Carrying Nuclear Submarine
SSI	Submersible Ship, Conventional with AIP
SSK	Submersible Ship, Conventional
SSM	Surface-to-Surface Missile
SSN	Submersible Ship, Nuclear / Nuclear-Powered Attack Submarine
STOL	Short Takeoff and Landing
STOVL	Short Takeoff and Vertical Landing
TAPV	Tactical Armoured Patrol Vehicles
TAR	Tibetan Autonomous Region
TTH	Tactical Transport Helicopter
TUAV	Tactical Uninhabited Aerial Vehicle
UAC-TA	United Aircraft Corporation -Transport Aircraft
UAS	Unmanned Aircraft System
UAV	Unmanned Aerial Vehicle
UCAV	Unmanned Combat Aerial Vehicle
UNASUR	Union of South American Nations
UNSC	United Nations Security Council
USS	United States Ship
VADS	Vulcan Air Defence System
VLA	Vertical Launch Anti-Submarine Missile
VTOL	Vertical Take-off and Landing
WAC	Western Area Command
WMD	Weapons of Mass Destruction
WPK	Workers' Party of Korea
WTO	World Trade Organisation